CATIA V5-6R2022 for Designers

(20th Edition)

CADCIM Technologies
525 St. Andrews Drive
Schererville, IN 46375, USA
(www.cadcim.com)

Contributing Author
Sham Tickoo
Professor
Department of Mechanical Engineering Technology
Purdue University Northwest
Hammond, Indiana, USA

CADCIM Technologies

CATIA V5-6R2022 for Designers
Sham Tickoo

CADCIM Technologies
525 St Andrews Drive
Schererville, Indiana 46375, USA
www.cadcim.com

ISBN 978-1-64057-148-8

www.cadcim.com

DEDICATION

To teachers, who make it possible to disseminate knowledge
to enlighten the young and curious minds
of our future generations

To students, who are dedicated to learning new technologies
and making the world a better place to live in

THANKS

To the faculty and students of the MET department of
Purdue University Northwest for their cooperation

To employees of CADCIM Technologies for their valuable help

Table of Contents

Chapter 3: Drawing Sketches in the Sketcher Workbench-II

Chapter 4: Constraining Sketches and Creating Features

Chapter 5: Reference Elements and Sketch-Based Features

Chapter 6: Creating Dress-Up and Hole Features

Chapter 7: Editing Features

Chapter 8: Transformation Features and Advanced Modeling Tools-I

Chapter 9: Advanced Modeling Tools-II

Chapter 10: Working with the Wireframe and Surface Design Workbench

Chapter 11: Editing and Modifying Surfaces

Chapter 12: Assembly Modeling

Chapter 13: Working with the Drafting Workbench-I

Chapter 16: DMU Kinematics

CHAPTERS AVAILABLE FOR FREE DOWNLOAD

In this textbook, two chapters have been given for free download. You can download these chapters from our website *www.cadcim.com*. To download these chapters, follow the path: *Textbooks > CAD/CAM > CATIA > CATIA V5-6R2022 for Designers > Chapters for Free Download.* Next, click on the chapter name that you want to download.

Chapter 17: Introduction to Generative Shape Design

Chapter 18: Working with the FreeStyle Workbench

Chapter 19: Introduction to FEA and Generative Structural Analysis

PROJECTS AVAILABLE FOR FREE DOWNLOAD

In this textbook, four projects are available for free download. You can download these projects from our website **www.cadcim.com**.

To download these projects, follow the path: *Textbooks > CAD/CAM > CATIA > CATIA V5-6R 2022 for Designers > Projects for Free Download*. Next, click on the project name that you want to download.

Note

For additional projects, visit **www.cadcim.com** and follow the path: *Textbooks > CAD/CAM > Parametric Solid Modeling Projects > Parametric Solid Modeling Projects*

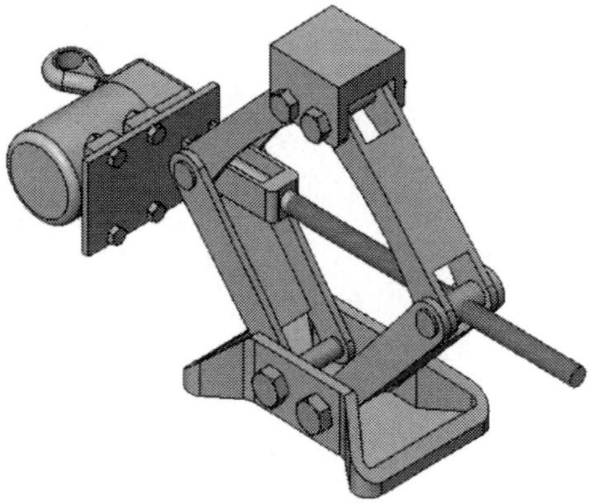

Project 1: *Car Jack Assembly*

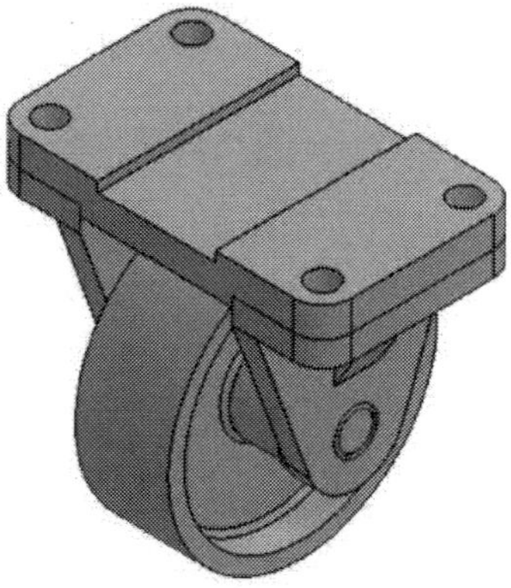

Project 2: *Wheel Assembly*

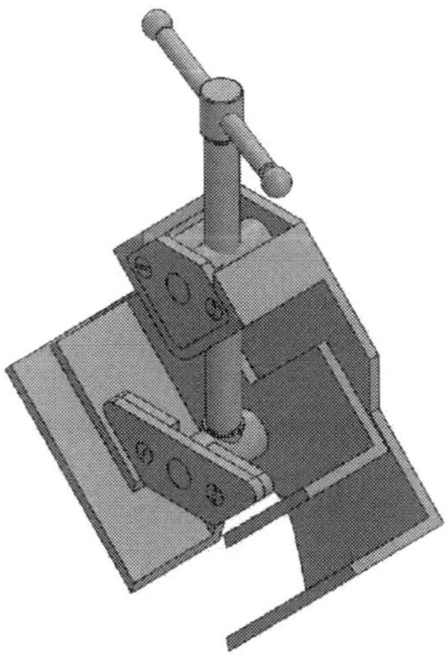

Project 3: *Angle Clamp Assembly*

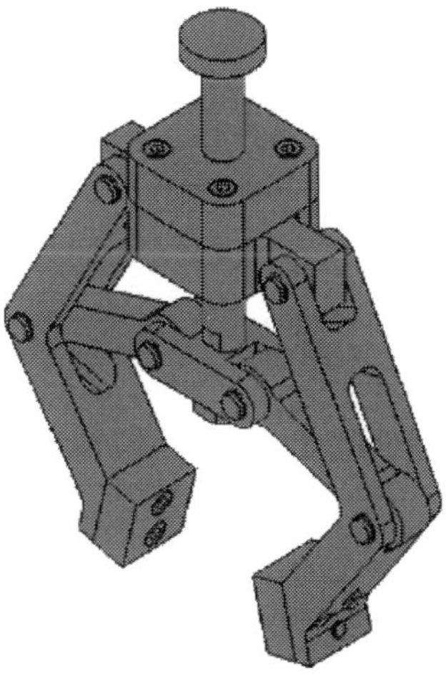

Project 4: *Pneumatic Gripper Assembly*

This page is intentionally left blank

Preface

CATIA V5-6R2022

CATIA, developed by Dassault Systemes, is one of the world's leading CAD/CAM/CAE packages. Being a solid modeling tool, it not only unites the 3D parametric features with 2D tools, but also addresses every design-through-manufacturing process. Besides providing an insight into the design content, the package promotes collaboration between companies and provides them an edge over their competitors.

In addition to creating solid models, sheet metal components, and assemblies, 2D drawing views can also be generated in the **Drafting** workbench of CATIA. The drawing views that can be generated include orthographic, section, auxiliary, isometric, and detail views. You can also generate model dimensions and create reference dimensions in the drawing views. The bi-directional associative nature of this software ensures that the modifications made in the model are reflected in the drawing views and vice versa.

CATIA V5-6R2022 for Designers textbook is written with the intention of helping the readers effectively use the CATIA V5-6R2022 solid modeling tool. This textbook provides a simple and clear explanation of tools used in common workbenches of CATIA V5-6R2022. After reading this textbook, you will be able to create models, and assemble and draft them. The chapter on DMU Kinematics workbench will enable the users to create, edit, simulate, and analyze different mechanisms dynamically. The chapter on FreeStyle workbench will enable the users to dynamically design and manipulate surfaces. The chapter on Generative Shape Design explains mechanical engineering industry examples and tutorials used in this textbook ensure that the users can relate the knowledge of this textbook with the actual mechanical industry designs. In addition, a chapter on FEA and structural analysis has been added to help users to analyze their own designs by calculating stresses and displacements using various tools available in the Advanced Meshing Tools and Generative Structural Analysis workbenches of CATIA V5-6R2022. Furthermore, four projects are also added in this book for the students to practice the skills learned. The main features of this textbook are as follows:

- **Tutorial Approach**
  The author has adopted the tutorial point-of-view and the learn-by-doing approach throughout the textbook. This approach guides the users through the process of creating the models in the tutorials.

- **Real-World Projects as Tutorials**
  The author has used about 43 real-world mechanical engineering projects as tutorials in this textbook. This enables the readers to relate the tutorials to the real-world models in the mechanical engineering industry. In addition, there are about 40 exercises that are also based on the real-world mechanical engineering projects.

- **Additional Students Projects for Practice**
  In this book, four projects are provided for the students to practice the tools learned and enhance their skills.

- **Tips and Notes**
  Additional information related to various topics is provided to the users in the form of tips and notes.

- **Command Section**
  In every chapter, the description of a tool begins with the command section that gives a brief information about various methods of invoking that tool.

- **Heavily Illustrated Text**
  The text in this book is heavily illustrated with about 1100 line diagrams and screen capture images.

- **Learning Objectives**
  The first page of every chapter summarizes the topics that are covered in that chapter.

- **Self-Evaluation Test, Review Questions, and Exercises**
  Every chapter ends with Self-Evaluation Test so that the users can assess their knowledge of the chapter. The answers to Self-Evaluation Test are given at the end of the chapter. Also, Review Questions and Exercises are given at the end of the chapters and they can be used by the instructors as test questions and exercises.

Symbols Used in the Textbook

Note

The author has provided additional information to the users about the topic being discussed in the form of Notes.

Tip

Special information and techniques are provided in the form of Tips that will increase the efficiency of the users.

Enhanced

This icon indicates that the command or tool being discussed has been enhanced in CATIA V5-6R2022.

Formatting Conventions Used in the Textbook

Please refer to the following list for the formatting conventions used in this textbook.

- Names of tools, buttons, options, toolbars, and sub-toolbars are written in boldface.
  Example: The **Extrude** tool, the **OK** button, the **Modify** toolbar, the **Pads** sub-toolbar, and so on.

- Names of dialog boxes, drop-downs, drop-down lists, list boxes, areas, edit boxes, check boxes, and radio buttons are written in boldface.
  Example: The **Revolve** dialog box, the **Type** drop-down list in the **Pad Definition** dialog box, the **Depth** edit box in the **Pocket Definition** dialog box, the **Thick Profile** check box in the **Shaft Definition** dialog box, the **Manual** radio button of the **General** tab in the **Options** dialog box, and so on.

- Values entered in edit boxes are written in boldface.
  Example: Enter **5** in the **Radius** edit box.

- Names and paths of the files are written in italics.
  Example: *C:\CATIA\c03*, *c03tut03.prt*, and so on

- The methods of invoking a tool/option from the **Ribbon**, **Quick Access Toolbar**, **Application Menu**, and so on are enclosed in a shaded box.

  Menubar: Insert > Profile > Line > Line

  Toolbar: Profile > Line sub-toolbar > Line

Naming Conventions Used in the Textbook

Tool

If you click on an item in a toolbar and a command is invoked to create/edit an object or perform some action, then that item is termed as **tool**.

For example:
To Create: **Line** tool, **Dimension** tool, **Extrude** tool
To Modify: **Fillet** tool, **Draft** tool, **Trim Surface** tool
Action: **Zoom All** tool, **Pan** tool, **Copy** tool

If you click on an item in a toolbar and a dialog box is invoked wherein you can set the properties to create/edit an object, then that item is also termed as **tool**, refer to Figure 1.

For example:
To Create: **Pad** tool, **Groove** tool, **Shaft** tool
To Modify: **Concatenate** tool, **Global Deformation** tool

Button

The item in a dialog box that has a 3D shape like a button is termed as **Button**. For example, **OK** button, **Cancel** button, **Apply** button, and so on.

Dialog Box

The naming conventions for the components in a dialog box are mentioned in Figure 1.

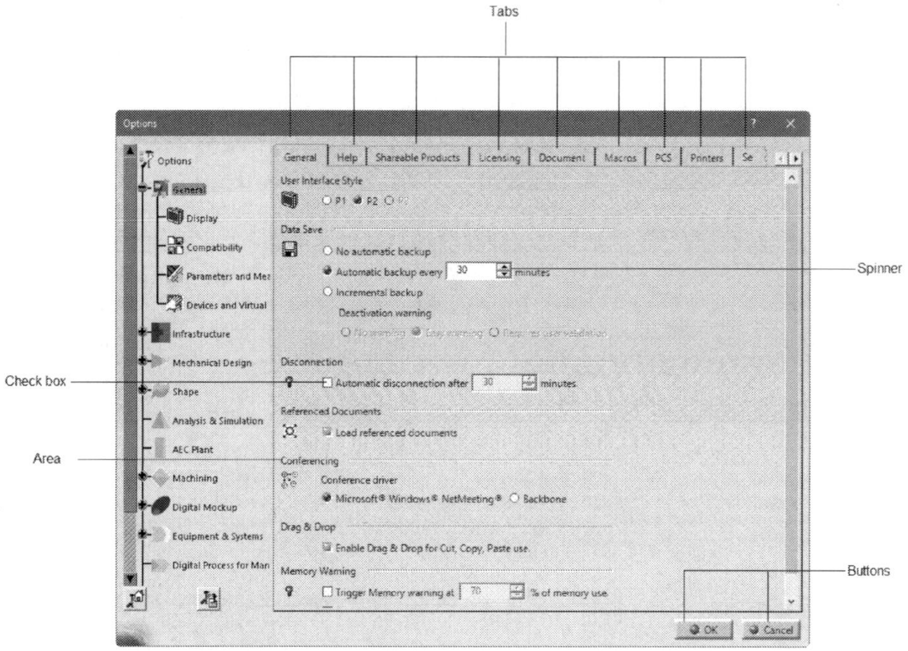

Figure 1 The components in a dialog box

Sub-toolbar

A sub-toolbar is available in the toolbar and is the one in which a set of common tools are grouped together for performing an action. You can identify a sub-toolbar with a down arrow on it. For example, **Walls** sub-toolbar, **Components** sub-toolbar, **Regions** sub-toolbar, and so on; refer to Figure 2.

The sub-toolbar, which appears on clicking the down arrow, can be detached from its main toolbar if you click on the line appearing on its top, refer to Figure 3. This line can be on the top or left of the sub-toolbar, depending on whether the sub-toolbar is horizontal or vertical. In this textbook, the path to invoke a tool is given as:

Menubar:	Insert > Walls > Swept Walls > Flange
Toolbar:	Walls > Swept Walls sub-toolbar > Flange

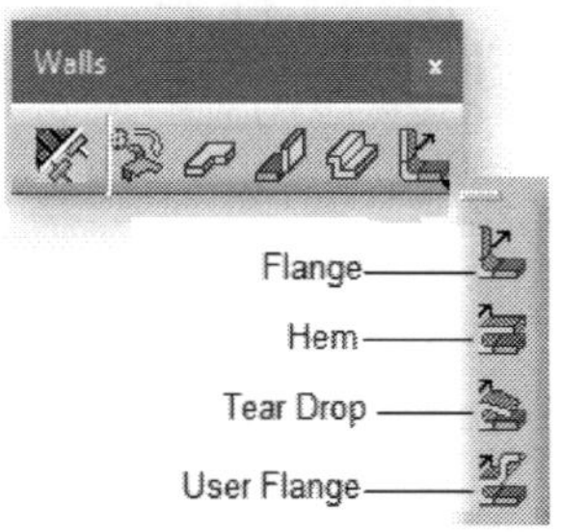

*Figure 2 Tools in the **Walls** sub-toolbar*

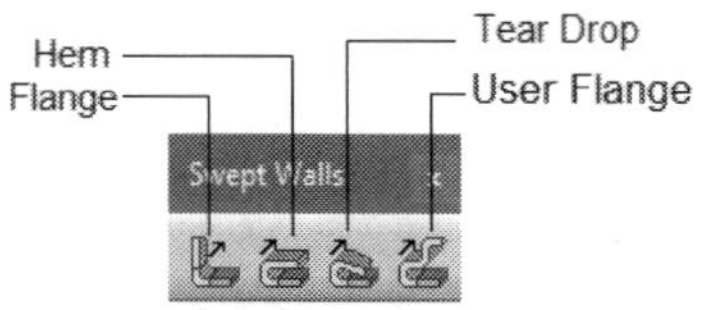

*Figure 3 Tools in the **Swept Walls** sub-toolbar*

Free Companion Website

It has been our constant endeavor to provide you the best textbooks and services at affordable price. In this endeavor, we have come out with a Free Companion website that will facilitate the process of teaching and learning of CATIA V5-6R2022. If you purchase this textbook, you will get access to the files on the Companion website.

The following resources are available for the faculty and students in this website:

Faculty Resources

- **Technical Support**
  You can get online technical support by contacting ***techsupport@cadcim.com***.

- **Instructor Guide**
  Solutions to all review questions and exercises in the textbook are provided in this link to help the faculty members test the skills of the students.

- **Part Files**
  The part files used in illustration, tutorials, and exercises are available for free download.

- **Free Download Chapters**
  Chapters available for free download.

- **Free Download Projects**
  In this book, four projects are available for free download.

Student Resources

- **Technical Support**
  You can get online technical support by contacting ***techsupport@cadcim.com***.

- **Part Files**
  The part files used in illustrations and tutorials are available for free download.

- **Free Download Chapters**
  Chapters available for free download.

- **Free Download Projects**
  In this book, four projects are available for free download.

Note that you can access the faculty resources only if you are registered as faculty at ***www.cadcim.com/Registration.aspx***

If you face any problem in accessing these files, please contact the publisher at ***sales@cadcim.com*** or the author at ***stickoo@pnw.edu*** or ***tickoo525@gmail.com***.

Video Courses

CADCIM offers video courses in CAD, CAE Simulation, BIM, Civil/GIS, and Animation domains on various e-Learning/Video platforms. To enroll for the video courses, please visit the CADCIM website using the link ***https://www.cadcim.com/video-courses***.

Stay Connected

You can now stay connected with us through Facebook and Twitter to get the latest information about our text books, videos, and teaching/learning resources. To stay informed of such updates, follow us on Facebook (***www.facebook.com/cadcim***) and Twitter (***@cadcimtech***). You can also subscribe to our You Tube channel ***(www.youtube.com/cadcimtech***) to get the information about our latest video tutorials.

Chapter 1

Introduction to CATIA V5-6R2022

Learning Objectives

After completing this chapter, you will be able to:

- *Understand the benefits of using CATIA V5*
- *Use various workbenches of CATIA V5*
- *Get familiar with important terms and definitions used in CATIA V5*
- *Understand the system requirements to install CATIA V5*
- *Understand the functions of the mouse buttons*
- *Understand the use of Hot Keys*
- *Modify the color scheme in CATIA V5*

INTRODUCTION TO CATIA V5-6R2022

Welcome to CATIA (Computer Aided Three Dimensional Interactive Application). As a new user of this software package, you will join hands with thousands of users of this high-end CAD/CAM/CAE tool worldwide. If you are already familiar with the previous releases, you can upgrade your designing skills with tremendous improvement in this latest release.

CATIA V5, developed by Dassault Systemes, France, is a completely re-engineered, next-generation family of CAD/CAM/CAE software solutions for Product Lifecycle Management. Through its exceptionally easy-to-use and state-of-the-art user interface, CATIA V5 delivers innovative technologies for maximum productivity and creativity, from the inception concept to the final product. CATIA V5 reduces the learning curve, as it allows the flexibility of using feature-based and parametric designs.

CATIA V5 provides three basic platforms: P1, P2, and P3. P1 is used for small and medium-sized process-oriented companies that wish to grow toward the large scale digitized product definition. P2 is used for the advanced design engineering companies that require product, process, and resource modeling. P3 is for high-end design applications and is basically for Automotive and Aerospace Industry where high quality surfacing or Class-A surfacing is used.

The subject of interpretability offered by CATIA V5 includes receiving legacy data from other CAD systems and from its own product data management modules. The real benefit of CATIA V5 is that the links remain associative. As a result, any change made to this external data gets notified and the model can be updated quickly.

The latest application launched by Dassault Systemes in the family of CATIA is CATIA V6. This application is based entirely on a database PLM structure and is used for creating the business processes to get the work done in a production environment. For the users of CATIA V5, the transition from CATIA V5 to V6 is one of the requirements. CATIA V5-6R2022 is a product from the family that supports the file format of CATIA V5 as well as CATIA V6. Also, all the features in Part Design, Generative Surface Design, and Sketcher workbenches related to 3D parametric geometry creation are preserved so that they can be used in both CATIA V5 and CATIA V6.

CATIA V5 WORKBENCHES

CATIA V5 serves the basic design tasks by providing different workbenches. A workbench is defined as a specified environment consisting of a set of tools that allows the user to perform specific design tasks. The basic workbenches in CATIA V5 are **Part Design**, **Wireframe and Surface Design**, **Assembly Design**, **Drafting**, **Generative Sheetmetal Design**, **DMU Kinematics**, **FreeStyle**, **Generative Shape Design**, **Advanced Meshing Tools**, **Generative Structural Analysis**, and so on. These workbenches are discussed next.

Part Design Workbench

The **Part Design** workbench is a parametric and feature-based environment in which you can create solid models. The basic requirement for creating a solid model in this workbench is sketch. The sketch for the features is drawn in the **Sketcher** workbench that can be invoked a within the **Part Design** workbench. You can draw the sketch using the tools in this workbench. While drawing the sketch, some constraints are automatically applied to it. You can also apply additional constraints and dimensions manually. After drawing the sketch, exit the **Sketcher** workbench and convert it into a feature. The tools in the **Part Design** workbench can be used to convert the sketch into a sketch-based feature. This workbench also provides other tools to apply the placed features, such as fillets, chamfers, and so on. These features are called the dress-up features. You can also assign materials to the model in this workbench.

Wireframe and Surface Design Workbench

The **Wireframe and Surface Design** workbench is also a parametric and feature-based environment and is used to create wireframe or surface models. The tools in this workbench are similar to those in the **Part Design** workbench with the only difference that the tools in this environment are used to create basic and advanced surfaces.

Assembly Design Workbench

The **Assembly Design** workbench is used to assemble the components using the assembly constraints available in this workbench. There are two types of assembly design approaches:

1. Bottom-up
2. Top-down

In the bottom-up approach of the assembly design, the components are assembled together to maintain their design intent. In the top-down approach, components are created inside the assembly in the **Assembly Design** workbench. You can also assemble an existing assembly to the current assembly. The **Space Analysis** toolbar provides the **Clash** analysis tool that helps in detecting clash, clearance, and contact between components and subassemblies.

Drafting Workbench

The **Drafting** workbench is used for the documentation of the parts or assemblies created earlier in the form of drawing views and their detailing. There are two types of drafting techniques:

1. Generative drafting
2. Interactive drafting

The generative drafting technique is used to automatically generate the drawing views of the parts and assemblies. The parametric dimensions added to the component in the **Part Design** workbench during its creation can be generated and displayed automatically in the drawing views. The generative drafting is bidirectionally associative in nature. You can also generate the Bill of Material (BOM) and balloons in the drawing views.

In interactive drafting, you need to create the drawing views by sketching them using the normal sketching tools and then adding the dimensions.

Generative Sheetmetal Design Workbench

The **Generative Sheetmetal Design** workbench is used for the designing of the sheet metal components. Generally, the solid models of the sheet metal components are created to generate the flat pattern of the sheet, study the design of the dies and punches, study the process plan for designing. This workbench also provides the tools needed for manufacturing the sheet metal components.

DMU Kinematics Workbench

The **DMU Kinematics** workbench is used to design mechanisms by adding joints between the components. It also allows you to simulate and analyze the working of the mechanisms dynamically.

FreeStyle Workbench

The **FreeStyle** workbench is used to manipulate and refine an existing surface. In addition, you can create independent surfaces and analyze the surfaces using the tools available in this workbench.

Generative Shape Design Workbench

The **Generative Shape Design** workbench allows you to easily model both simple and complex shapes using the wireframe and surface features. It is used to create the Solid-based hybrid designs. It provides the tools to create the design as well as edit the hybrid designs.

Advanced Meshing Tools workbench

The **Advanced Meshing Tools** workbench provides you tools to quickly generate a finite element model for the complex parts with advanced control on mesh specifications whether they are surface parts or solid parts.

Generative Structural Analysis workbench

The **Generative Structural Analysis** workbench also allows the user to perform an analysis on a volume part, any surface part, and any wireframe geometry.

SYSTEM REQUIREMENTS

The system requirements to ensure the smooth running of CATIA V5-6R2022 on your system are as follows:

- System unit: Microsoft Windows 10, Windows 8 (64 bit only) or Windows 7 (SP1 required).
- CPU quad-core or hexa-core Intel i7/Intel i9/Threadripper/Xeon/.
- GeForce GTX GPU compatible with OpenGL 3.2 and 2 GB RAM.
- Hard disk: SSD.
- Small projects (under 100 images at 14 MP): 8 GB RAM, 15 GB SSD Free Space.
- Medium projects (between 100 and 500 images at 14 MP): 16GB RAM, 30 GB SSD Free Space.
- Large projects (over 500 images at 14 MP): 32 GB RAM, 60 GB SSD Free Space.
- Very Large projects (over 2000 images at 14 MP): 64 GB RAM, 120 GB SSD Free Space.

GETTING STARTED WITH CATIA V5-6R2022

Install CATIA V5-6R2022 on your system and then start it by double-clicking on its shortcut icon displayed on the desktop of your computer. After the system has loaded all the required files to start CATIA V5-6R2022, a new product file with the default name **Product1** will start automatically, as shown in Figure 1-1.

Close this file by choosing **File > Close** from the menu bar. Figure 1-2 shows the interface that appears after closing the initial Product file.

Figure 1-1 *The initial interface that appears after starting CATIA V5-6R2022*

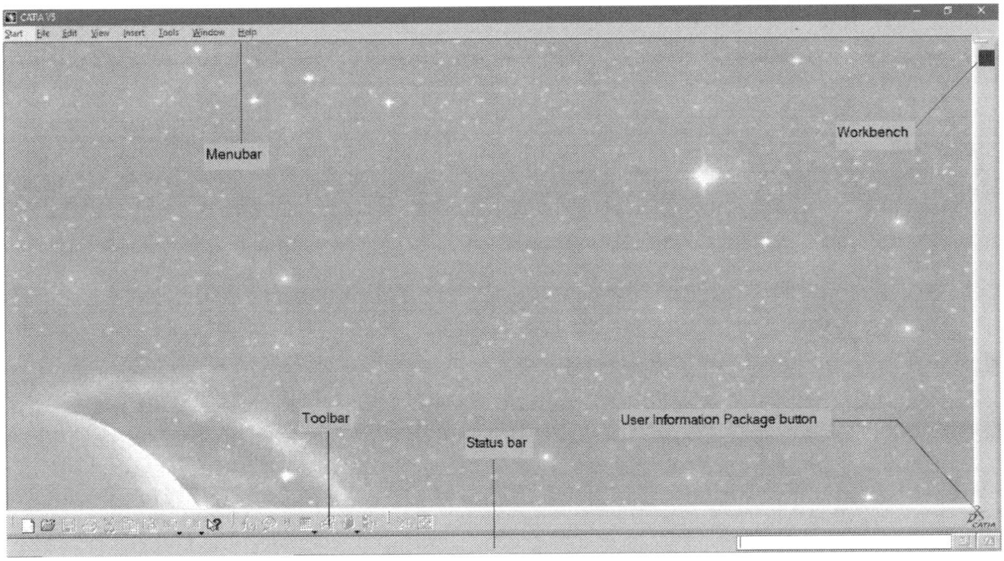

Figure 1-2 *The interface that appears after closing the initial product file*

IMPORTANT TERMS AND DEFINITIONS

Some important terms and definitions used in CATIA V5-6R2022 are discussed next.

Feature-based Modeling

A feature is defined as the smallest building block that can be modified individually. A model created in CATIA V5 is a combination of a number of directly or indirectly related individual features. You can modify these features any time during the design process. If a proper design intent is maintained while creating the model, then these features automatically adjust according to the change occurred in their surroundings. This provides greater flexibility to the design.

Parametric Modeling

The parametric nature of a software package is defined as its ability to use the standard properties or parameters in defining the shape and size of a geometry. The main function of this property is to transform the selected geometry to a new size or shape without considering its original dimensions. You can change or modify the shape and size of any feature at any stage of the design process. This property makes the designing process very easy. For example, consider the design of the body of a pipe housing shown in Figure 1-3.

To change the design by modifying the diameter of the holes and their number on the front, top, and bottom faces, you have to simply select the feature and change the diameter and the number of instances in the pattern. The modified design is shown in Figure 1-4.

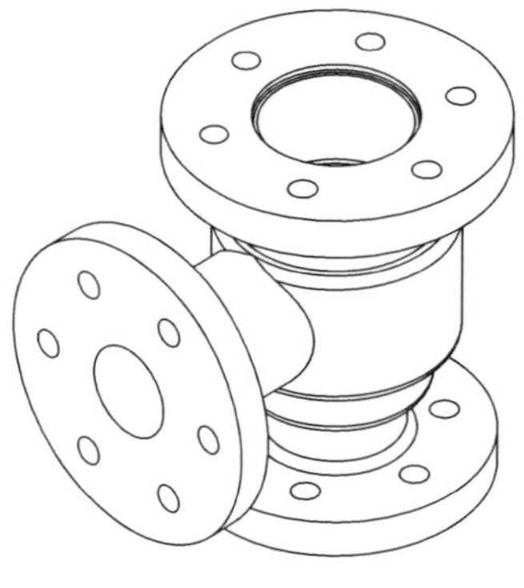

Figure 1-3 *Body of a pipe housing*

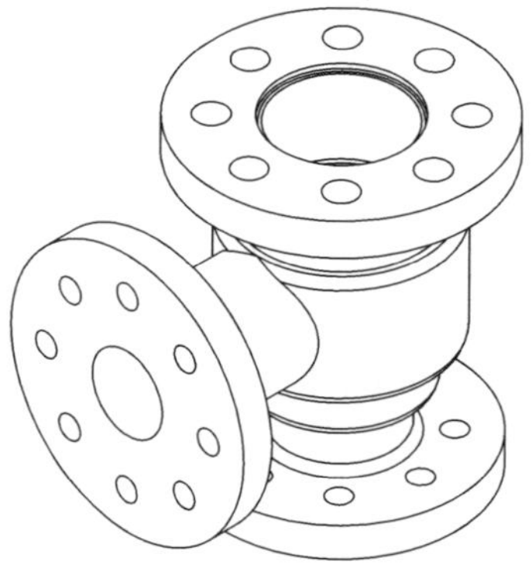

Figure 1-4 *Modified body of a pipe housing*

Bidirectional Associativity

As mentioned earlier, CATIA V5 has different workbenches such as **Part Design**, **Assembly Design**, **Drawing**, **Generative Sheetmetal Design**, and so on. The bidirectional associativity that exists between all these workbenches ensures that any modification made in the model in any of the workbenches of CATIA V5, reflects automatically and immediately in other workbenches also. For example, if you modify the dimension of a part in the **Part Design** workbench, the change will reflect in the **Assembly Design** and **Drawing** workbenches also. Consider the drawing views of the pipe housing shown in Figure 1-5. When you modify the model in the **Part Design** workbench, the changes are reflected in the **Drawing** workbench automatically. Figure 1-6 shows the drawing views of the pipe housing after increasing the diameter and number of holes at the top flange.

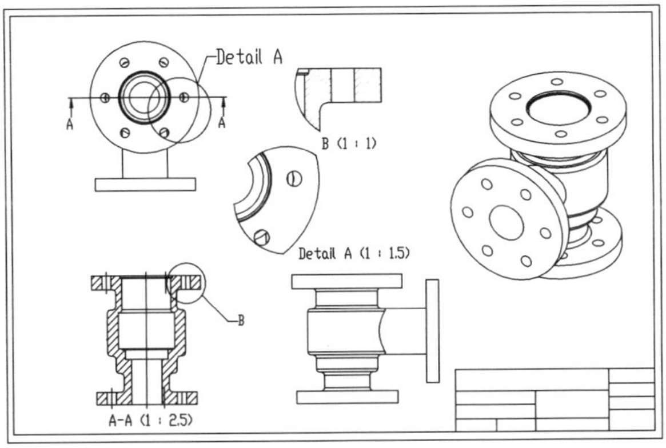

Figure 1-5 The drawing views of the pipe housing before modifications

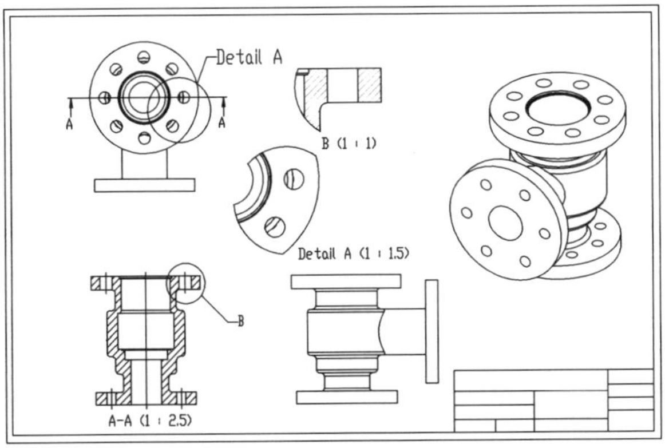

Figure 1-6 The drawing views of the pipe housing after modifications

CATPart

CATPart is a file extension associated with all those files that are created in the **Sketcher**, **Part Design**, **Generative Sheetmetal Design**, **Generative Shape Design**, and **Wireframe and Surface Design** workbenches of CATIA V5.

CATProduct

CATProduct is a file extension associated with all those files that are created in the **Assembly Design** workbench of CATIA V5.

CATDrawing

CATDrawing is a file extension associated with all those files that are created in the **Drafting** workbench of CATIA V5.

Specification Tree

The Specification tree displays all the operations carried out on the part in a sequence. Figure 1-7 shows the Specification tree that appears when you start a new file under the **Part Design** workbench.

Compass

The compass is used to manipulate the orientation of parts, assemblies, or sketches. You can also orient the view of the parts and assemblies. The compass is shown in Figure 1-8. By default, it appears at the top right corner of the geometry area.

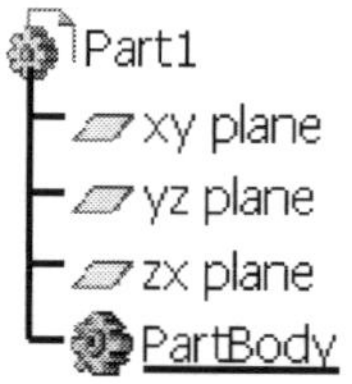

Figure 1-7 *The Specification tree that appears on starting a new CATPart file*

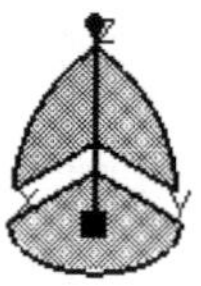

Figure 1-8 *The compass*

Constraints

Constraints are the logical operations that are performed on the selected element to define its size and location with respect to the other elements or reference geometries. There are two types of constraints in CATIA V5. The constraints in the **Sketcher** workbench are called sketcher constraints and are used to precisely define the size and position of the sketched elements with respect to the surroundings. The assembly constraints available in the **Assembly Design** workbench are used to define the precise position of the components in the assembly. These constraints are discussed next.

Sketcher Constraints

Sketcher constraints are the logical operations performed on the sketched elements to define their size and position with respect to the other elements. These are two methods to apply geometric constraints; automatic and manual. While drawing the sketch, some constraints are automatically applied to it. For applying constraints manually, you need to invoke the **Constraints Defined in Dialog Box** tool and select appropriate check box.

The constraints in the **Sketcher** workbench are discussed next.

Distance

This constraint is used to apply a distance dimension between any two selected entities.

Length

This constraint is used to apply a linear dimension to the selected line.

Angle

This constraint is used to apply an angular dimension between any two selected lines.

Radius / Diameter

This constraint is used to apply a radius or diameter dimension to the selected circular entity.

Semimajor axis

This constraint is used to apply a dimension to the major axis of the selected ellipse.

Semiminor axis
This constraint is used to apply a dimension to the minor axis of the selected ellipse.

Symmetry
This constraint is used to force the selected entity to become symmetrical about an axis. A line segment can be used as an axis.

Curvilinear distance
This constraint is used to apply the curvilinear distance to the curve. You can apply curvilinear distance on different type of curves such as spline, arc, circle, and conics.

Midpoint
This constraint forces a selected point to be placed on the midpoint of the selected line.

Equidistant point
This constraint forces a selected point to be placed at an equal distance from any two preselected points.

Fix
This constraint is used to fix a selected entity in terms of its position with respect to the coordinate system of the current sketch.

Coincidence
This constraint is used to make two points, two lines, a point and a line, or a point and a curve coincident.

Concentricity
This constraint is used to make two circles, two arcs, an arc and a circle, a point and a circle, or a point and an arc concentric.

Tangency
This constraint is used to force the selected line segment or curve to become tangent to another curve.

Parallelism
The Parallelism constraint is used to force any two selected line segments to become parallel to each other. The selected line segments can be axes as well.

Perpendicular
The Perpendicular constraint is used to force any two selected line segments to become perpendicular to each other. The selected line segments can be axes as well.

Horizontal
The Horizontal constraint forces the selected line segment to become horizontal.

Vertical
The Vertical constraint forces the selected line segment to become vertical.

Assembly Constraints

The constraints in the **Assembly Design** workbench are the logical operations performed to restrict the degrees of freedom of a component and to define its precise location and position with respect to the other components in the assembly. The constraints in this workbench are discussed next.

Coincidence

This constraint is used to force two selected entities to coincide with each other. The selected entities can be central axes of circular components, two adjacent or opposite faces, or two adjacent planes.

Contact

This constraint is used to force two selected faces to maintain contact with each other.

Offset

This constraint is used to place two different selected faces, planes, or central axes at a distance with respect to each other.

Angle

This constraint is used to place two selected entities at an angle with respect to each other. These entities can be the central axes of circular components, two faces, two planes, a combination of an axis and a face, a plane and a face, or an axis and a plane.

Fix

This constraint fixes the position of the selected part in the 3D space.

Fix Together

This constraint fixes the position of two different selected parts with respect to each other.

Quick

The **Quick Constraint** tool is used to apply the most appropriate constraint to the elements in the current selection set. You can set the priority depending on which CATIA V5 will perform the constraint selection.

PartBody

The PartBody is the default body in the **Part Design** workbench. All the solid features, such as pad, pocket, shaft, and so on are placed inside it. Other bodies that will be inserted under the **Part Design** workbench will be named as **Body.2**, **Body.3**, and so on.

Geometrical Set

The geometrical set is defined as a body that includes the created planes, surfaces, wireframe elements, and reference elements.

Wireframe

The wireframe construction elements aid in creating surfaces. They generally consist of points, lines, and arcs, and are used as substitutes of entities drawn in the **Sketcher** workbench.

Surface

Surfaces are geometric features which have no thickness. They are used to create complex shapes that are difficult to make using solid features. After creating a surface, you can assign a thickness to it to convert it into a solid body.

Feature

A feature is defined as a basic building block of a solid model. The combination of various features results in a complete model. In the **Part Design** workbench of CATIA V5, the features are of the following four types:

1. Sketch-Based Features
2. Dress-Up Features
3. Transformation Features
4. Surface-Based Features

Reframe on

Sometimes, a feature, a body, or a sketch may not be visible in the available space of the geometry area. The **Reframe on** option, available in the contextual (shortcut) menu, is used to view the particular selection in the available display space.

Center Graph

The **Center graph** option, available in the contextual menu, is used to bring the selected feature, body, or sketch in the Specification tree to the middle left portion of the geometry area.

UNDERSTANDING THE FUNCTIONS OF THE MOUSE BUTTONS IN CATIA

To work in CATIA V5 design workbenches, it is necessary that you understand the functions of the mouse buttons. The efficient use of these buttons along with the Ctrl key on the keyboard can reduce the time required to complete the design task. The different combinations of the Ctrl key and mouse buttons are listed next:

1. The left mouse button is used to make a selection by dragging a window or by simply selecting a face, surface, sketch, or object from the geometry area or from the Specification tree. For multiple selections, press and hold the Ctrl key and select the entities using the left mouse button.

2. The right mouse button is used to invoke the contextual (shortcut) menu after selecting an element or invoking a tool.

3. Press and hold the middle mouse button and drag the mouse to pan the view of the model on the screen.

4. Press and hold the middle mouse button and then click the right mouse button once to invoke the **Zoom** mode. Now, drag the mouse up to zoom in the view of the model. Similarly, drag it down to zoom out the view of the model. You can also invoke the zoom tool by first

pressing and holding the Ctrl key and then pressing and holding the middle mouse button. Now, release the Ctrl key and drag the cursor to zoom in and out the view of the model. Figure 1-9 shows how to use a mouse for zooming in and zooming out operations.

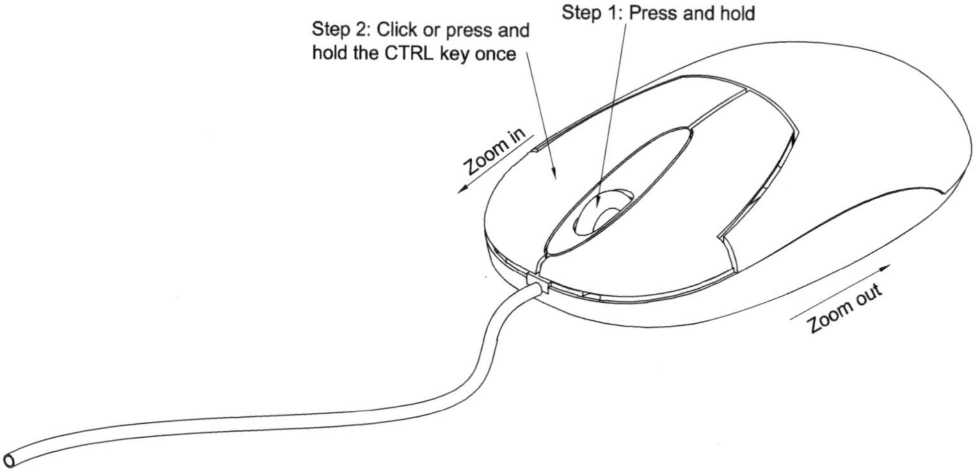

Figure 1-9 *Using mouse to perform the zoom in and zoom out operations*

5. Press and hold the middle mouse button. Then, press and hold the right mouse button or left mouse button to invoke the **Rotate** mode. Next, drag the mouse to dynamically rotate the view of the model in the geometry area and view it from different directions. You can rotate the model in the geometry area by pressing and holding the middle mouse button and then pressing and holding the Ctrl key. Next, drag the cursor to rotate the view of the model. Figure 1-10 shows how to use the three-button mouse to perform the rotate operation.

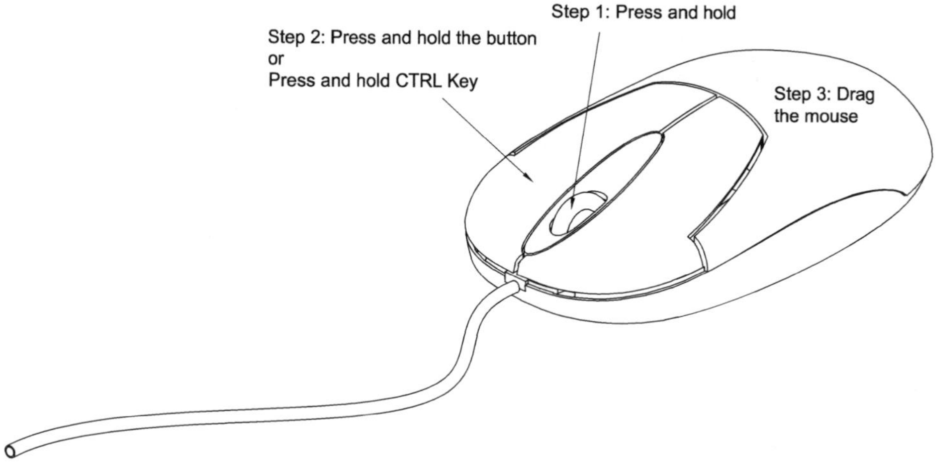

Figure 1-10 *Using the three-button mouse to perform the rotate operation*

Note

It is assumed that a three-button mouse is configured on your system.

TOOLBARS

CATIA V5 offers a user-friendly design environment by providing specific toolbars to each workbench. Therefore, it is important that you get acquainted with various toolbars and buttons that appear in the workbenches of CATIA V5. These toolbars are discussed next.

Standard Toolbar

This toolbar is common to all the workbenches of CATIA V5. Figure 1-11 shows the **Standard** toolbar.

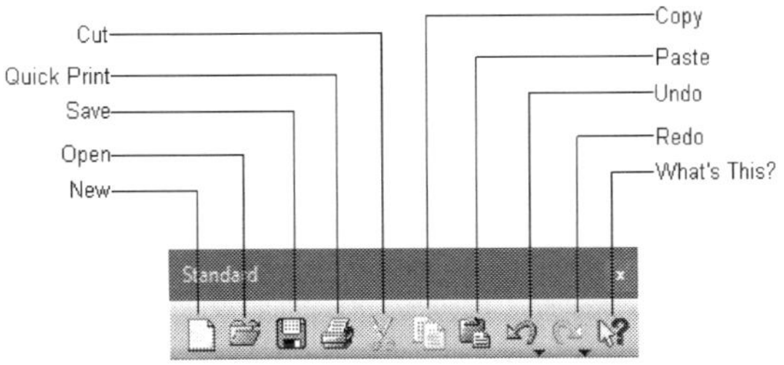

Figure 1-11 *The **Standard** toolbar*

The tools in this toolbar are used to start a new file, open an existing file, save a file, and print the current document. These buttons are also used to cut and place the selection on a temporary clipboard, copy a selection, paste the content from the clipboard to a selected location, undo, redo, and invoke the help topics. The **What's This?** button provides help on the toolbar icons.

Status Bar

The status bar, which is located at the bottom of the CATIA V5 window, comprises of three areas, as shown in Figure 1-12. These areas are discussed next.

Figure 1-12 *The status bar*

Current Information or dialog box

The **Current Information or dialog box** area displays the current information about the selected feature or current tool.

Power Input Field Bar

The **Power Input Field Bar** is used to invoke the commands and enter the data or value that can be directly associated with the feature.

Note

*1. In case an incorrect command is entered in the **Power Input Field Bar**, the **Power input message** dialog box appears informing about unknown command or syntax error. Choose the **OK** button from this dialog box.*

*2. To launch any command using the **Power Input Field Bar**, the general syntax of the command will be C: <name of the command>. For example, to start a new file, enter C: New.*

Tip

Entering commands in the ***Power Input Field Bar*** *is a tedious process. It is recommended that you invoke the tools from the toolbars or menu bar instead of using the* ***Power Input Field*** *bar.*

Dialog Box Display Button

Choosing the **Dialog Box Display** button will turn on or off the display of the current dialog box.

Part Design Workbench Toolbars

You can invoke the **Part Design** workbench by choosing the **New** button from the **Standard** toolbar and selecting **Part** from the **New** dialog box displayed. Alternatively, you can choose **Start > Mechanical Design > Part Design** from the menubar. The toolbars in the **Part Design** workbench are discussed next.

View Toolbar

The buttons in the **View** toolbar, refer to Figure 1-13, are used for manipulating the view of the model using the tools such as pan, zoom, normal viewing about a planar surface, face or plane, defining a render style, and so on. The **View** toolbar is available in all the workbenches.

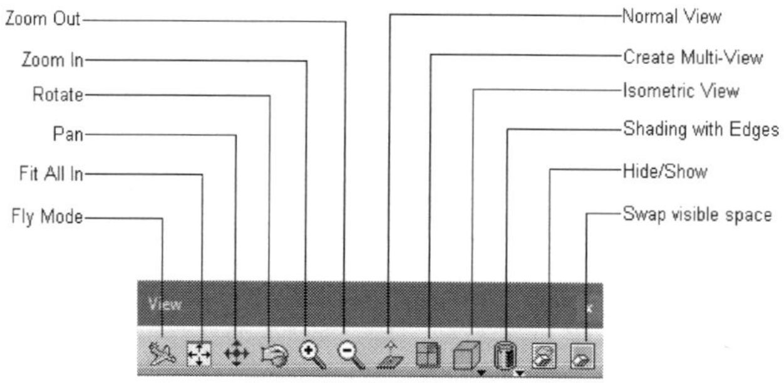

Figure 1-13 *The* ***View*** *toolbar*

Note

Buttons such as ***Fly Mode****,* ***Normal View****,* ***Isometric View****,* ***View Mode****, and* ***Rotate*** *are not available in the* ***Drafting*** *workbench. For all the other workbenches discussed in this book, the function of the* ***View*** *toolbar is the same.*

Select Toolbar

The **Select** tool is invoked from the **Select** toolbar to select a particular object or sketch. When you invoke the **Select** tool, you are prompted to select an object or a tool. By default, the **Select** tool remains active until another tool or object is selected. Figure 1-14 shows the **Select** toolbar.

Sketcher Toolbar

The **Sketcher** button in the **Sketcher** toolbar is used to invoke the **Sketcher** workbench. You can also invoke it from the main menu bar by choosing **Start > Mechanical Design > Sketcher**. Figure 1-15 shows the **Sketcher** toolbar.

After choosing the **Sketcher** button, select a plane or a planar face to invoke the **Sketcher** workbench. The toolbars in the **Sketcher** workbench are discussed next.

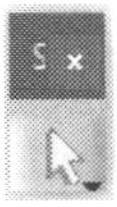

*Figure 1-14 The **Select** toolbar*

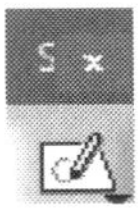

*Figure 1-15 The **Sketcher** toolbar*

Profile Toolbar

The tools in the **Profile** toolbar are used to draw sketches. It is one of the most important toolbars in the **Sketcher** workbench. Figure 1-16 shows the **Profile** toolbar.

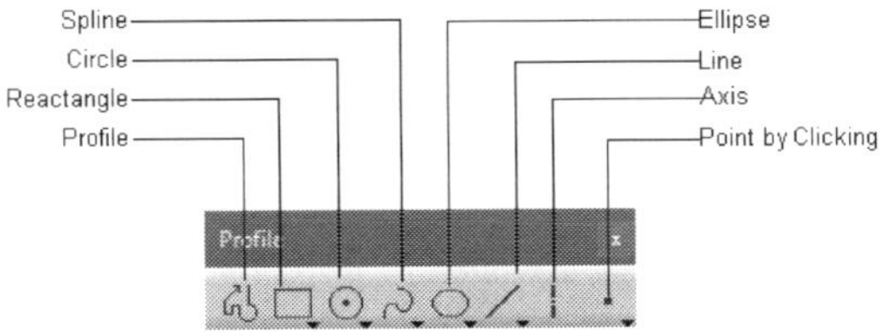

*Figure 1-16 The **Profile** toolbar*

Constraint Toolbar

The tools in the **Constraint** toolbar are used to apply constraints to the geometric entities, and assign dimensions to a drawn sketch. You can make a sketch fully defined by using the tools from this toolbar. A fully defined sketch is known as an Iso-constraint sketch and is discussed in the later chapters. Figure 1-17 shows the **Constraint** toolbar.

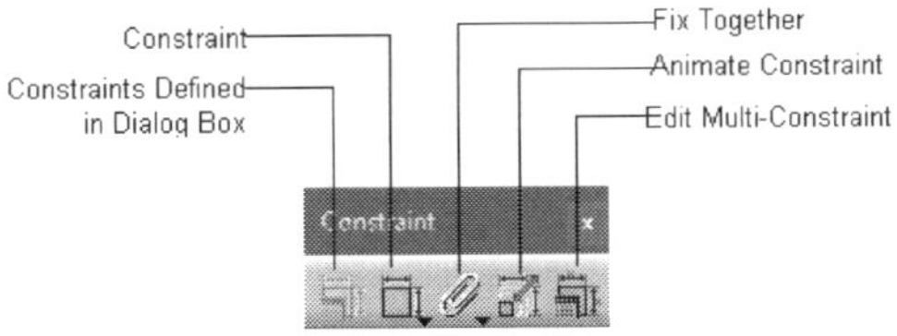

*Figure 1-17 The **Constraint** toolbar*

Operation Toolbar

The tools in the **Operation** toolbar are used to edit the drawn sketches. Figure 1-18 shows the **Operation** toolbar.

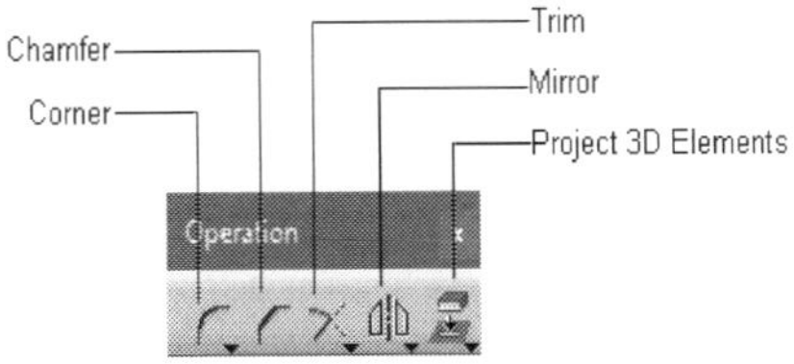

*Figure 1-18 The **Operation** toolbar*

Sketch tools Toolbar

The tools in the **Sketch tools** toolbar are used to set the sketcher settings such as setting the snap, switching between the standard and construction elements, and so on. Figure 1-19 shows the **Sketch tools** toolbar.

Once the basic sketch is complete, you need to convert it into a feature. Choose the **Exit workbench** button from the **Workbench** toolbar and switch back to the **Part Design** workbench.

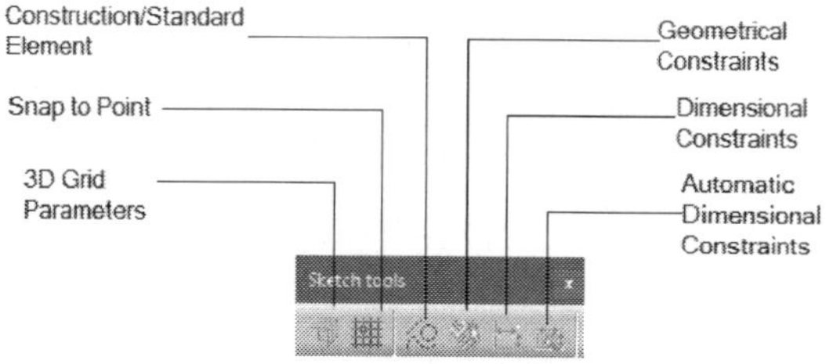

*Figure 1-19 The **Sketch tools** toolbar*

The remaining toolbars of the **Part** workbench are discussed next.

Sketch-Based Features Toolbar

The tools in this toolbar are used to convert a sketch drawn in the **Sketcher** workbench into a feature. Figure 1-20 shows the buttons in the **Sketch-Based Features** toolbar.

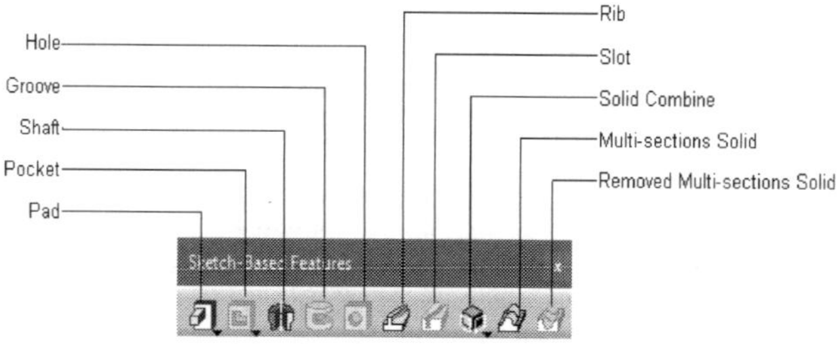

*Figure 1-20 The **Sketch-Based Features** toolbar*

Dress-Up Features Toolbar

The tools in the **Dress-Up Features** toolbar are used to apply the dress-up features such as fillet, chamfer, shell, and so on. Figure 1-21 shows the **Dress-Up Features** toolbar.

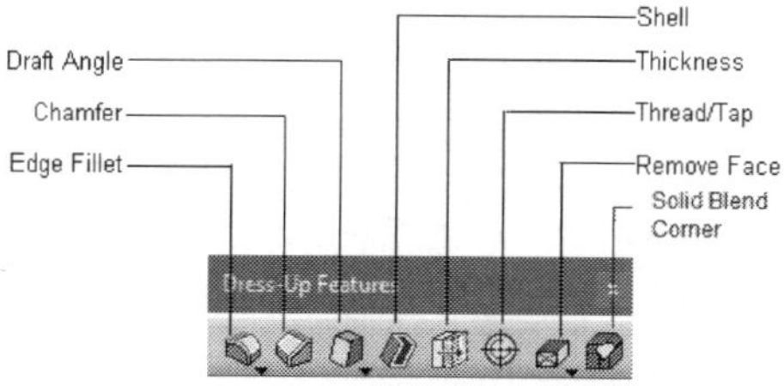

*Figure 1-21 The **Dress-Up Features** toolbar*

Measure Toolbar

The tools in the **Measure** toolbar are used to measure a single item, measure the distance between two geometries, or calculate the mass properties of the object. Figure 1-22 shows the **Measure** toolbar.

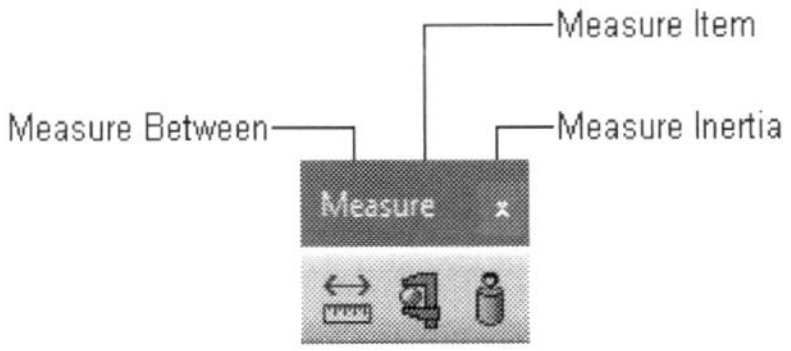

***Figure 1-22** The **Measure** toolbar*

Transformation Features Toolbar

The tools in the **Transformation Features** toolbar are used to apply the transformation features to the parts such as moving, mirroring, patterning, and so on. Figure 1-23 shows the **Transformation Features** toolbar.

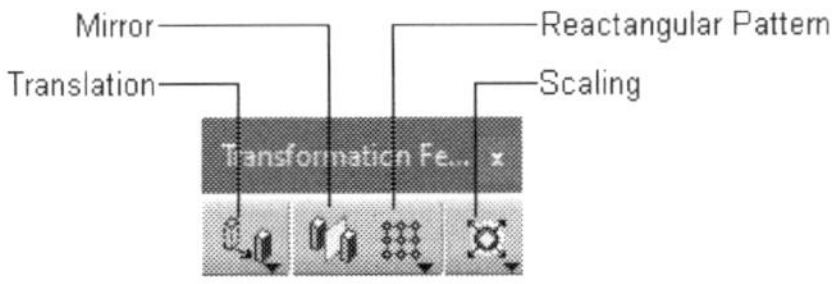

***Figure 1-23** The **Transformation Features** toolbar*

Dynamic Sectioning Toolbar

The tool in the **Dynamic Sectioning** toolbar is used to visualize the 3D section view of a part at a position specified by using a section plane. Figure 1-24 shows the **Dynamic Sectioning** toolbar.

Apply Material Toolbar

The tool in the **Apply Material** toolbar is used to assign a material to the part body. Figure 1-25 shows the **Apply Material** toolbar.

***Figure 1-24** The **Dynamic Sectioning** toolbar*

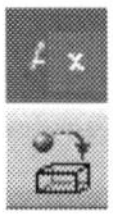

***Figure 1-25** The **Apply Material** toolbar*

Surface-Based Features Toolbar

The tools in the **Surface-Based Features** toolbar are used to perform surface-based operations on part bodies or to convert a surface body into a solid body. Figure 1-26 shows the tools in the **Surface-Based Features** sub-toolbar.

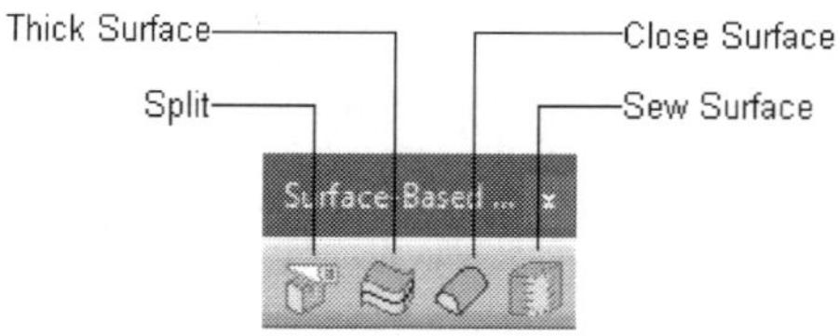

*Figure 1-26 The **Surface-Based Features** sub-toolbar*

Wireframe and Surface Design Workbench Toolbars

You can invoke the **Wireframe and Surface Design** workbench from the main menu bar by choosing **Start > Mechanical Design > Wireframe and Surface Design**. The toolbars in the **Wireframe and Surface Design** workbench are discussed next.

Surfaces Toolbar

The tools in the **Surfaces** toolbar are used to create surfaces. Figure 1-27 shows the tools in the **Surfaces** toolbar.

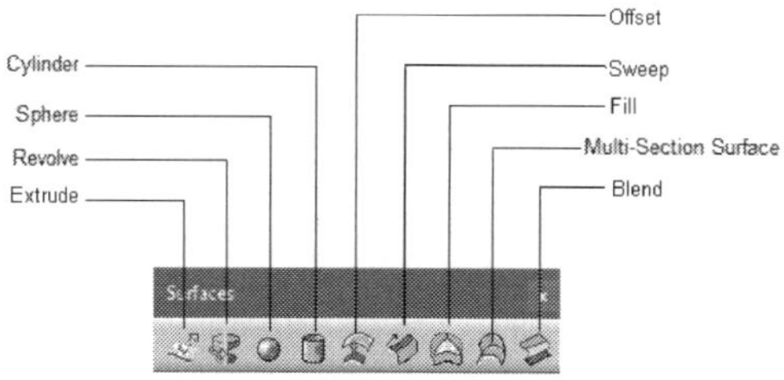

*Figure 1-27 The **Surfaces** toolbar*

Operations Toolbar

The tools in the **Operations** toolbar are used for surface editing operations. Figure 1-28 shows the tools in the **Operations** toolbar.

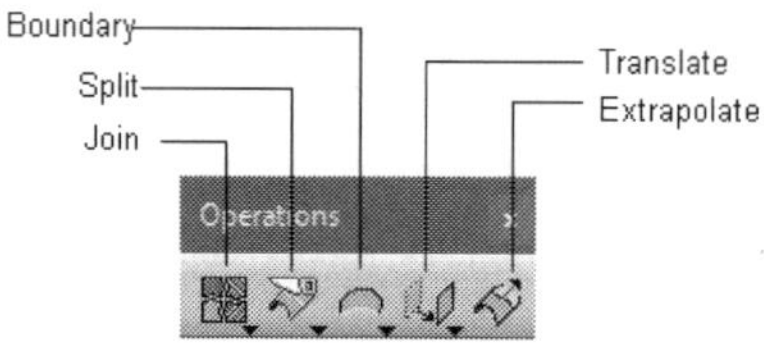

*Figure 1-28 The **Operations** toolbar*

Wireframe Toolbar

The tools in the **Wireframe** toolbar are used to create 2D or 3D curves using points, lines, and splines. User-defined planes can also be created by choosing the **Plane** button from this toolbar. You will learn about creating a user-defined plane in the later chapters. Figure 1-29 shows the tools in the **Wireframe** toolbar.

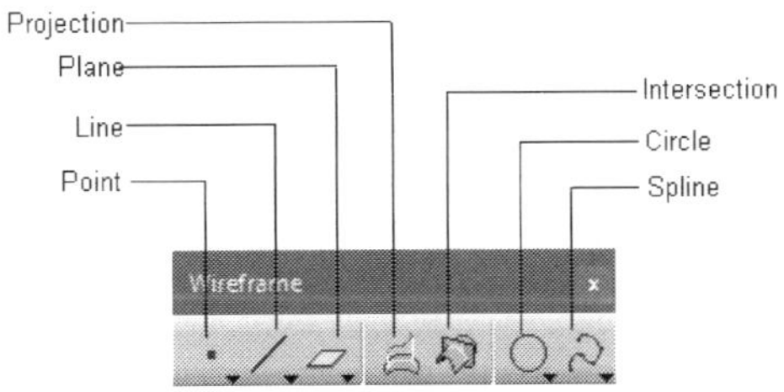

*Figure 1-29 The **Wireframe** toolbar*

Assembly Design Workbench Toolbars

You can invoke the **Assembly Design** workbench by choosing the **New** button from the **Standard** toolbar and then selecting **Product** from the **New** dialog box. Alternatively, you can choose **Start > Mechanical Design > Assembly Design** from the menubar. The toolbars in the **Assembly Design** workbench are discussed next.

Product Structure Tools Toolbar

The tools in the **Product Structure Tools** toolbar are used to insert an existing part or assembly in the current product file. You can also create a new assembly or part inside the Product file using the tools in this toolbar. Figure 1-30 shows the buttons in the **Product Structure Tools** toolbar.

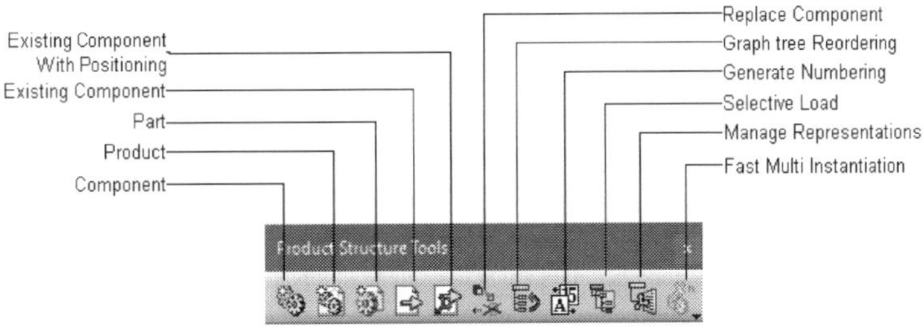

*Figure 1-30 The **Product Structure Tools** toolbar*

Constraints Toolbar

The **Constraints** toolbar is used to apply constraints to the components of the assembly to restrict degrees of freedom (DOFs) of the component with respect to the surroundings. Figure 1-31 shows the tools in the **Constraints** toolbar.

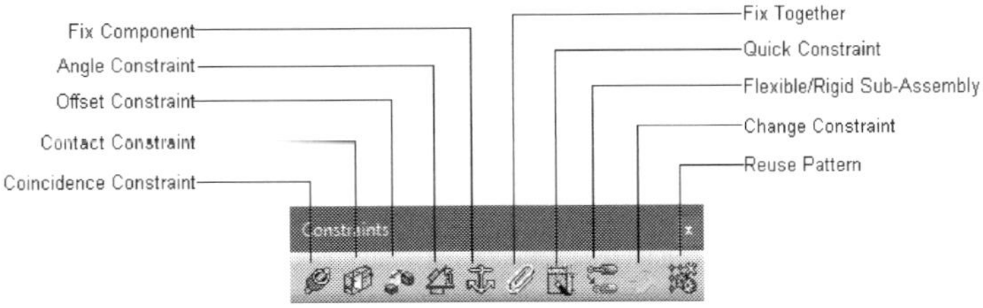

*Figure 1-31 The **Constraints** toolbar*

Move Toolbar

The tools in this toolbar are used to perform tasks such as moving and snapping the parts or exploding an assembly in the **Assembly Design** workbench. Figure 1-32 shows the **Move** toolbar.

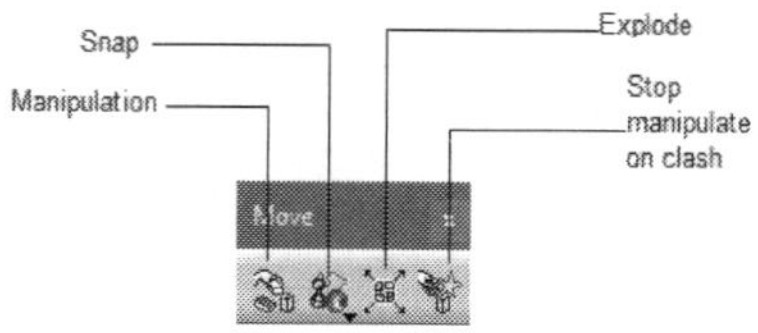

*Figure 1-32 The **Move** toolbar*

Space Analysis Toolbar

The tools in the **Space Analysis** toolbar are used to check any interference and clash in the assembly, create a section of the assembly, and perform the distance analysis. Figure 1-33 shows the **Space Analysis** toolbar.

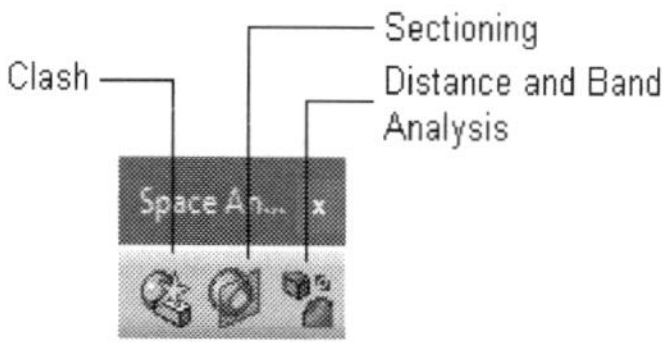

*Figure 1-33 The **Space Analysis** toolbar*

Drafting Workbench Toolbars

To invoke the **Drafting** workbench, choose the **New** button from the **Standard** toolbar and select the **Drawing** option from the **New** dialog box. Choose the **OK** button in the **New Drawing** dialog box. Alternatively, choose **Start > Mechanical Design > Drafting** from the menubar to invoke this workbench. The toolbars in the **Drafting** workbench are discussed next.

Drawing Toolbar

The tools in the **Drawing** toolbar are used to insert a new sheet, create a new view, and so on. Figure 1-34 shows the **Drawing** toolbar.

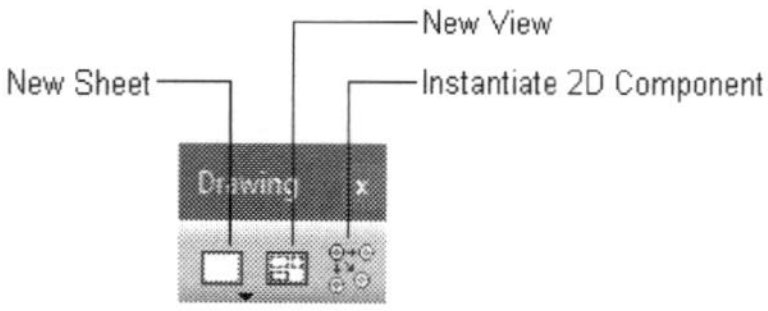

*Figure 1-34 The **Drawing** toolbar*

Views Toolbar

The tools in the **Views** toolbar are used to generate orthographic, section, detail, or clipped view for a solid part or assembly. Figure 1-35 shows the **Views** toolbar.

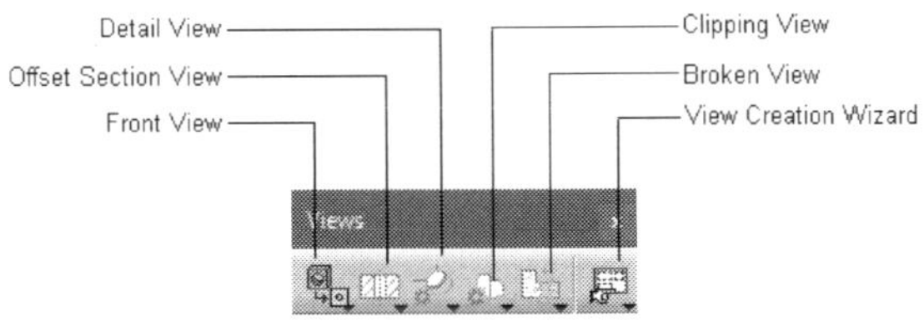

*Figure 1-35 The **Views** toolbar*

Generation Toolbar

The tools in the **Generation** toolbar are used to generate dimensions and assign balloons to the assembly drawings. Figure 1-36 shows the **Generation** toolbar.

*Figure 1-36 The **Generation** toolbar*

Other toolbars in the **Drafting** workbench are similar to those discussed in the **Sketcher** workbench. The tools in these toolbars are discussed in the later chapters.

You will notice a down arrow at the bottom right corner of the buttons in most of the toolbars. When you click on this arrow, a flyout will appear. Figure 1-37 shows the flyout that appears when you click on the down arrow of the **Line** button in from the **Geometry Creation** toolbar.

Note

The sub-toolbar which appears on choosing the down arrow becomes an independent toolbar if you click on the line available on its top, refer to Figure 1-37. This line can be on the top or left of the sub-toolbar, depending on whether the sub-toolbar is horizontal or vertical.

Generative Sheetmetal Workbench Toolbars

You can invoke the **Generative Shape Design** workbench by choosing **Start > Mechanical Design > Generative Sheetmetal Design** from the menu bar. The toolbars in the **Generative Sheetmetal Design** are discussed next.

Walls Toolbar

The tools in the **Walls** toolbar are used to create different types of walls. Figure 1-38 shows the **Walls** toolbar.

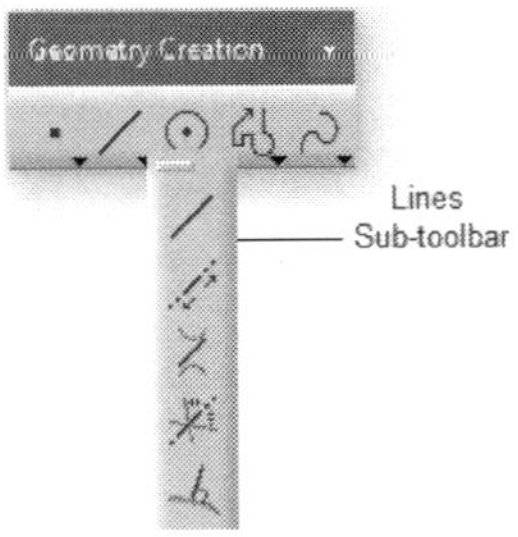

Figure 1-37 The sub-toolbar that appears on clicking a down arrow

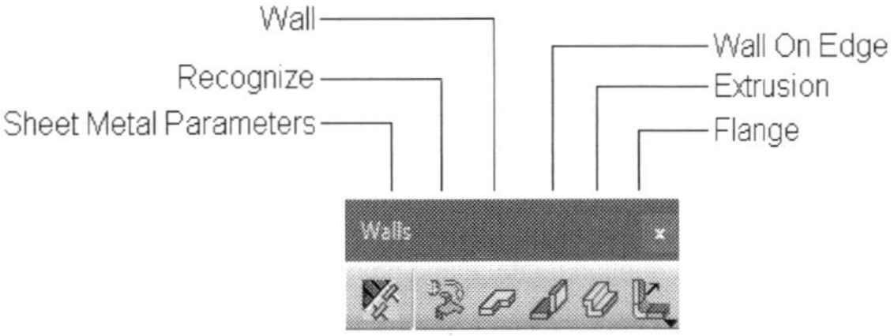

*Figure 1-38 The **Walls** toolbar*

Swept Walls Toolbar

The tools in the **Swept Walls** toolbar are used to create sweep features. These features are created by sweeping a profile along the spline. Figure 1-39 shows the **Swept Walls** toolbar.

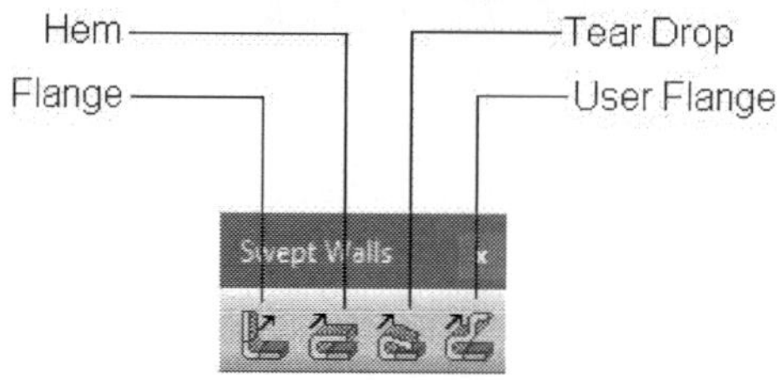

*Figure 1-39 The **Swept Walls** toolbar*

Stamping Toolbar

The tools in the **Stamping** toolbar are used to create different types of stamping features. Figure 1-40 shows the **Stamping** toolbar.

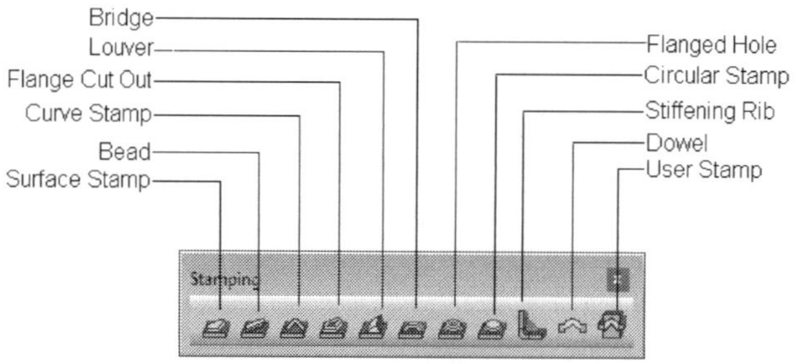

*Figure 1-40 The **Stamping** toolbar*

DMU Kinematics Workbench Toolbars

This **DMU Kinematics** workbench can be invoked by choosing **Start > Digital Mockup > DMU Kinematics** from the menu bar. The toolbars in the **DMU Kinematics** environment are discussed next.

DMU Kinematics Toolbar

The tools in the **DMU Kinematics** toolbar are used to create mechanisms and simulate the mechanism. Figure 1-41 shows the **DMU Kinematics** toolbar.

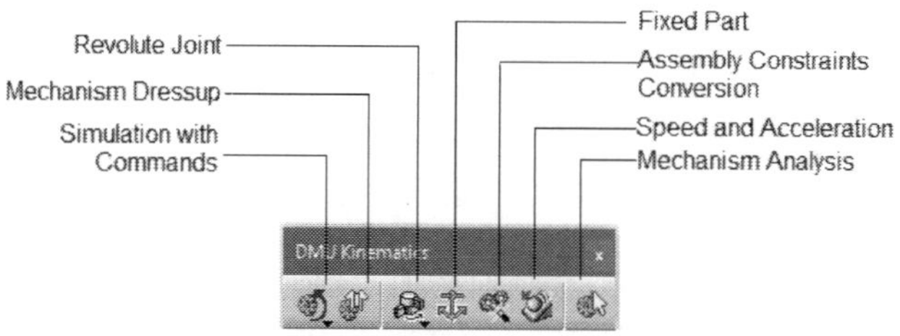

*Figure 1-41 The **DMU Kinematics** toolbar*

Kinematics Joints Toolbar

The tools in the **Kinematics Joints** toolbar are used to create different type of joints in the mechanism. Figure 1-42 shows the **Kinematics Joints** toolbar.

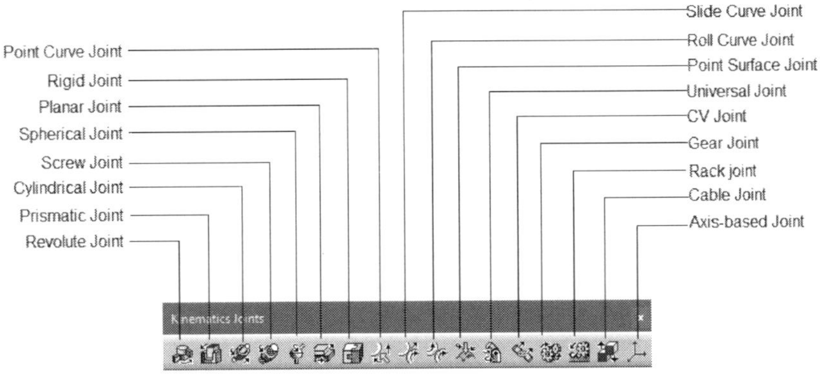

***Figure 1-42** The **Kinematics Joints** toolbar*

Generative Shape Design Workbench Toolbars

The **Generative Shape Design** workbench can be invoked by choosing **Start > Shape > Generative Shape Design** from the menubar. The toolbars in the **Generative Shape Design** workbench are discussed next.

Volumes Toolbar

The tools in this toolbar are used to convert a sketch drawn in the Sketcher workbench into a volume feature. Figure 1-43 shows the buttons in the **Volumes** toolbar.

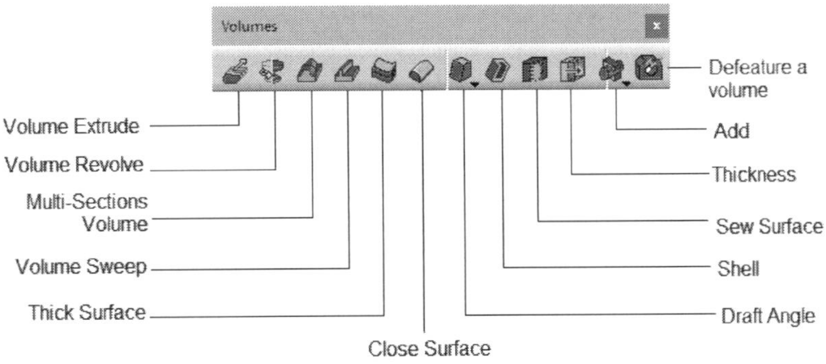

***Figure 1-43** The **Volumes** toolbar*

FreeStyle Toolbars

The **FreeStyle** workbench can be invoked by choosing **Start > Shape > FreeStyle** from the menubar. The toolbars in the **FreeStyle** workbench are discussed next.

Surface Creation Toolbar

The tools in the **Surface Creation** toolbar are used to create surfaces by using different surface creation tools. Figure 1-44 shows the **Surface Creation** toolbar.

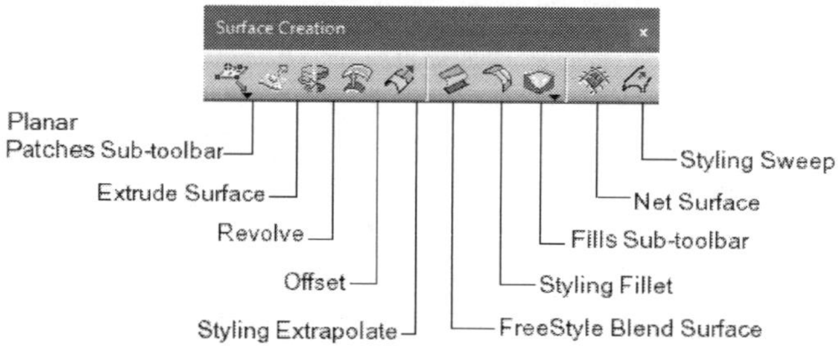

*Figure 1-44 The **Surface Creation** toolbar*

Shape Modification Toolbar

The tools in the **Shape Modification** toolbar are used to modify the shape of surfaces. Figure 1-45 shows the **Shape Modification** toolbar.

Operations Toolbar

The tools in the **Operations** toolbar are used to perform different operations on a surface like trimming, untrimming, fragmentation, and so on. Figure 1-46 shows the **Operations** toolbar.

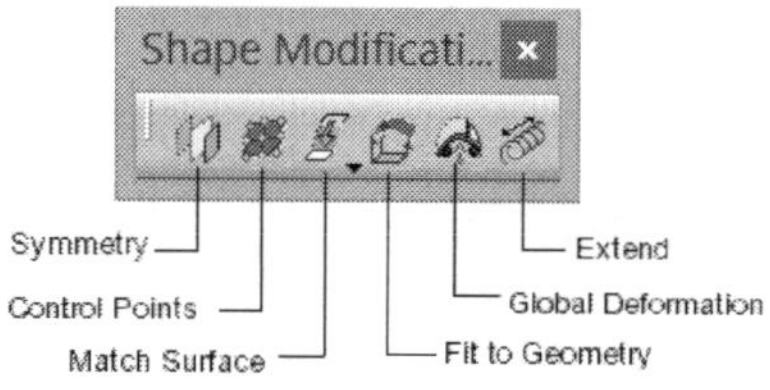

*Figure 1-45 The **Shape Modification** toolbar*

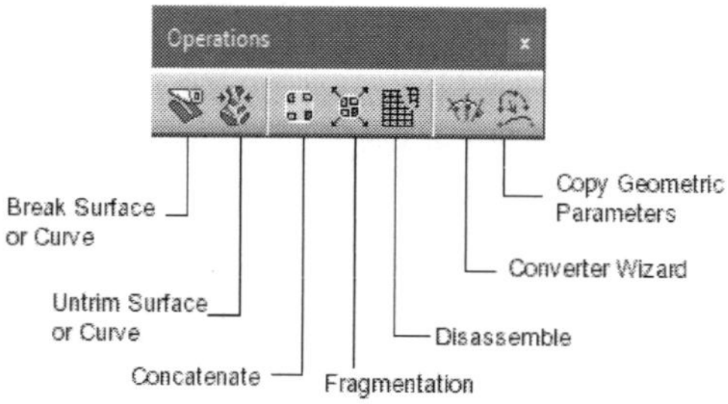

*Figure 1-46 The **Operations** toolbar*

Advanced Meshing Tools Workbench Toolbars

The workbench can be invoked by choosing **Start > Analysis and Simulation > Advanced Meshing Tools** from the menubar. The toolbars in this workbench are discussed next.

Meshing Methods Toolbar

The tools in the **Meshing Methods** toolbar are used to create mesh. Figure 1-47 shows the **Meshing Methods** toolbar.

Mesh Transformation Toolbar

The tools in the **Mesh Transformation** toolbar are used to transform a mesh by using different mesh transformation tools. Figure 1-48 shows the **Mesh Transformation** toolbar.

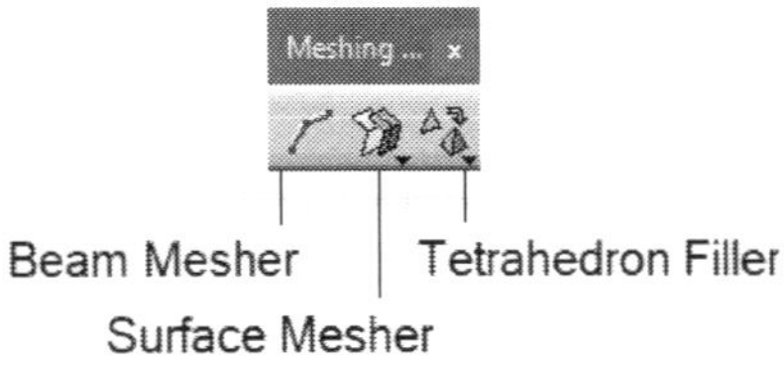

*Figure 1-47 The **Meshing Methods** toolbar*

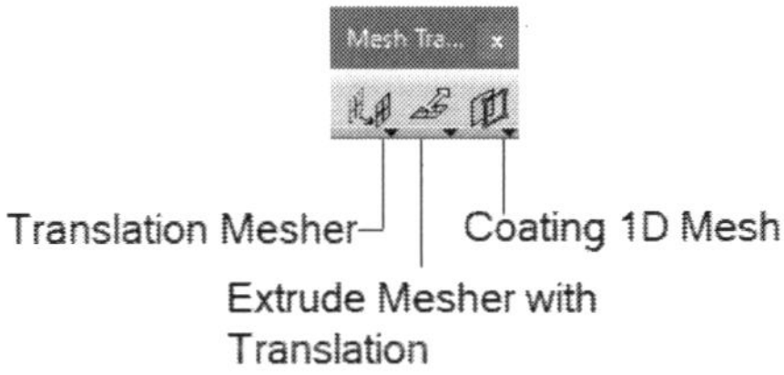

*Figure 1-48 The **Mesh Transformation** toolbar*

Mesh Operators Toolbar

The tools in the **Mesh Operators** toolbar are used to perform various operations such as moving mesh nodes, specifying meshing offset, split quads on the mesh created. Figure 1-49 shows the **Mesh Operators** toolbar.

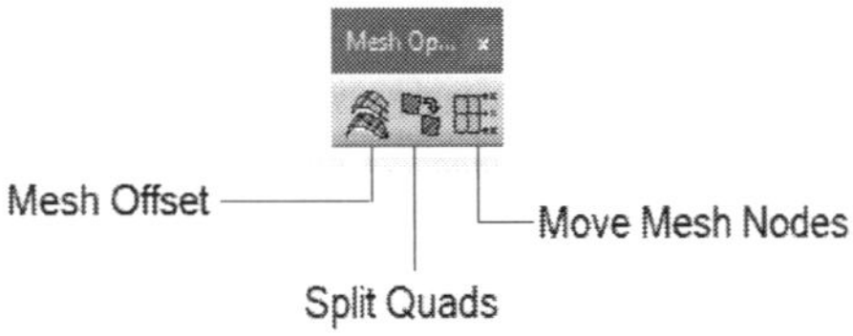

*Figure 1-49 The **Mesh Operators** toolbar*

Generative Structural Analysis Toolbars

The **Generative Structural Analysis** workbench can be invoked by choosing **Start > Analysis and Simulation > Generative Structural Analysis** from the menubar. The toolbars in this workbench are discussed next.

Model Manager Toolbar

The tools in the **Model Manager** toolbar are used to change physical properties of a shape design by selecting a meshed wireframe geometry by using different mesh creation tools. Figure 1-50 shows the **Model Manager** toolbar.

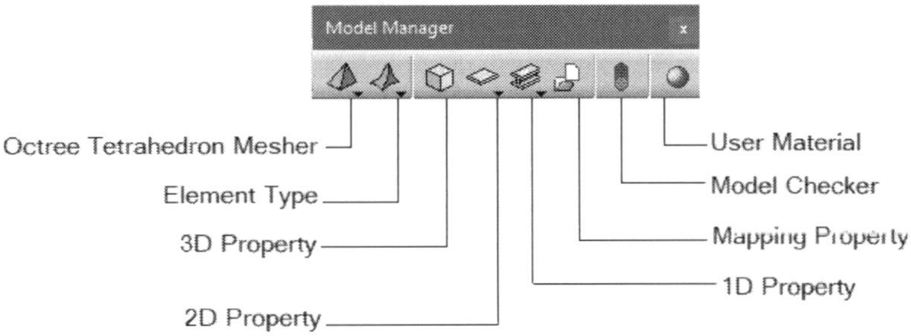

*Figure 1-50 The **Model Manager** toolbar*

Virtual Parts Toolbar

The tools in the **Virtual Parts** toolbar are used to create virtual parts. Figure 1-51 shows the **Virtual Parts** toolbar.

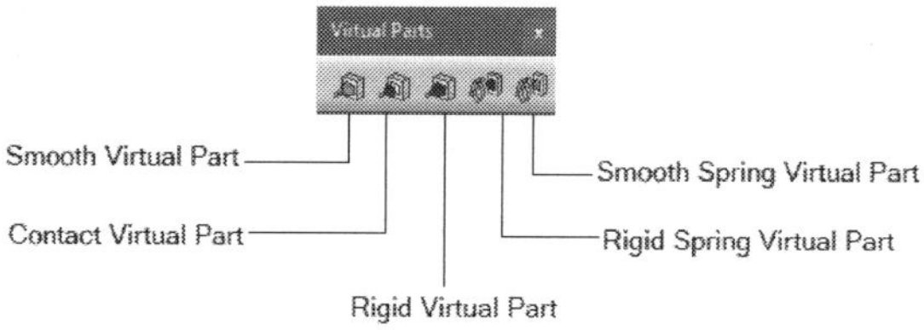

Figure 1-51 The ***Virtual Parts*** *toolbar*

Restraints Toolbar

The tools in the **Restraints** toolbar are used to restrict the degrees of freedom of a part. Figure 1-52 shows the **Restraints** toolbar.

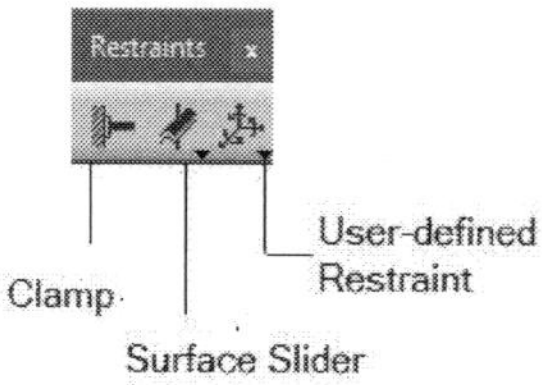

Figure 1-52 The ***Restraints*** *toolbar*

Loads Toolbar

The tools in the **Loads** toolbar are used to apply loads on the model in different ways. Figure 1-53 shows the **Loads** toolbar.

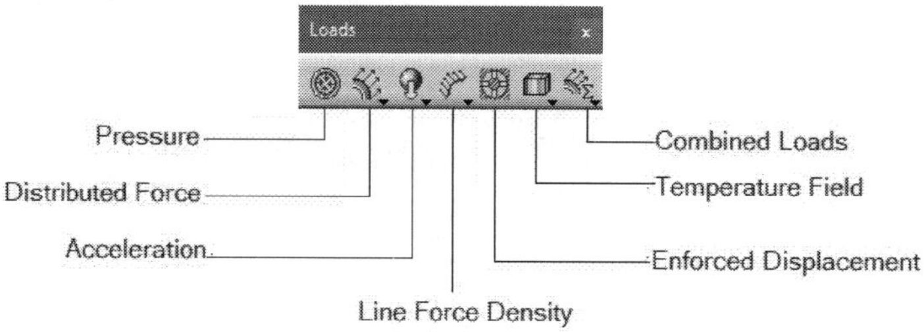

Figure 1-53 The ***Loads*** *toolbar*

Note

To bring back any deleted toolbar, right-click anywhere in the toolbars area and select the name of the corresponding toolbar from the shortcut menu. You can also restore the positions of the toolbars to their default state. To do so, right-click anywhere in the toolbar area and choose ***Customize*** *from the shortcut menu; the* ***Customize*** *dialog box will be displayed. Next, choose* ***Restore all contents*** *and* ***Restore position*** *from the* ***Toolbars*** *tab.*

HOT KEYS

CATIA V5 is more popularly known for its icon-driven structure. However, you can use the keys of the keyboard to invoke some tools. These keys are called as hot keys. Some hot keys along with their functions are listed in the table given next.

Hot Key	Function
Ctrl+Z	Invokes the **Undo** tool
Ctrl+Y	Invokes the **Redo** tool
Ctrl+S	Saves the current document
Alt+Enter	Invokes the **Properties** tool
Ctrl+F	Invokes the **Search** tool
Ctrl+U	Invokes the **Update** tool
Shift+F2	Invokes the **Specification Overview** tool
F3	Toggles the display of the Specification tree
Shift+F1	Invokes the **What's This?** tool
F1	Invokes the CATIA V5 **Help** tool
Ctrl+D	Invokes the **Fast Multi-Instantiation** tool in the **Assembly Design** workbench
Alt + F4	Exits CATIA V5
Alt + F8	Opens the **Macros** dialog box on screen
Alt + F11	Opens the **Launch V.B.A** message box on screen
Ctrl+C	Copies the selected feature
Ctrl+V	Pastes the copied feature
Ctrl+X	Cuts the selected feature

Alt + E	Opens the **Edit** menu
Alt + F	Opens the **File** menu
Alt + H	Opens the **Help** menu
Alt + I	Opens the **Insert** menu
Alt + V	Opens the **View** menu
Alt + S	Opens the **Start** menu
Ctrl + N	Opens the **New** dialog box
Ctrl + O	Opens the **File selection** window
Ctrl + P	Prints document

COLOR SCHEME

CATIA V5 allows you to use various color schemes as the background screen color, as well as to display the entities on the screen. To change the color scheme, choose **Tools > Options** from the menu bar; the **Options** dialog box will be displayed. Choose **General > Display** option from the left pane of the **Options** dialog box; the tabs corresponding to this selection will be displayed on the right in the **Options** dialog box. Choose the **Visualization** tab and select the White color from the **Background** drop-down list. Choose **OK** to apply the scheme to the CATIA V5 environment. Note that all files that you open henceforth will use this color scheme. The default color of the sketched elements is white, therefore, you need to change the colors of the sketch elements to black. To change the default color, invoke the **Options** dialog box and expand the **Mechanical Design** branch from the left of this dialog box. Next, expand the **Sketcher** branch and choose **Black** from the **Default color of the elements** drop-down list. Choose the **OK** button from the **Options** dialog box. All sketches drawn after setting this option will be displayed in black.

To distinguish between a plane and a background, you need to change the default color of the planes. To do so, select the plane and right-click to invoke the shortcut menu. Select the **Properties** option from it; the **Properties** dialog box is displayed. Choose the **Graphic** tab and then select the Black color from the drop-down list in the **Lines and Curves** area. Choose the **OK** button to apply the scheme to the CATIA V5 environment. Similarly, change the color of the remaining two planes. Figure 1-54 shows the options in the **Graphic** tab of the **Properties** dialog box.

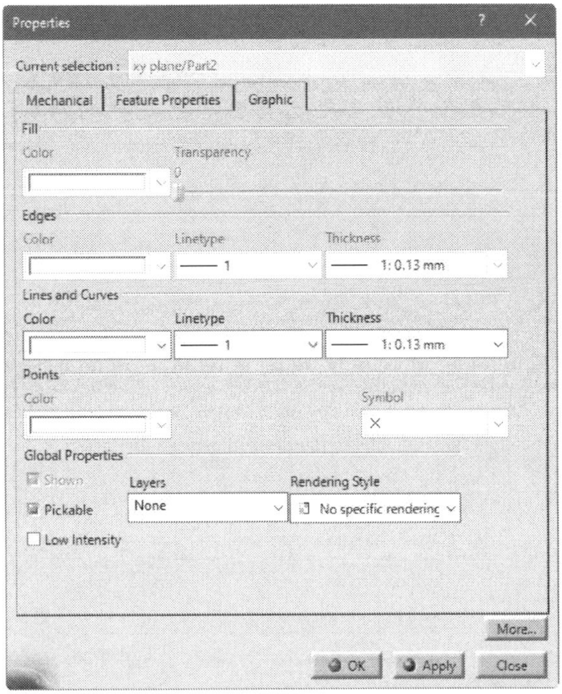

*Figure 1-54 The **Properties** dialog box with **Graphic** tab chosen*

Note

If you change the background color of the screen to white, then you will have to change the color of all the planes. The planes include the default xy, yz, and zx planes and also the planes that will be created while developing the design. As it is very cumbersome to perform this activity every time, it is recommended that you do not change the interface color of the CATIA V5.

For the purpose of printing, this book will follow the white background of the CATIA V5 environment. However, for better understanding and clear visualization at various places, this book will follow other color schemes too.

Self-Evaluation Test

Answer the following questions and then compare them to those given at the end of this chapter:

1. The______________ property ensures that any modification done in a model in any one of the workbenches of CATIA V5 is automatically reflected in the other workbenches immediately.

2. The ______________ geometric constraint forces two selected arcs, circles, a point and an arc, a point and a circle, or an arc and a circle to share the same center point.

3. The ______________ geometric constraint is used to make two points, a point and a line, or a point and an arc coincident.

4. The ________ geometric constraint is used to force the selected sketched entities to become symmetrical about an axis.

5. To __________the view of the model, you need to press and hold the middle mouse button and drag the mouse.

6. ____________ are the logical operations that are performed on the selected element to define its size and location with respect to the other elements or reference geometries.

7. The **Part Design** workbench is a parametric and feature-based environment in which you can create solid models. (T/F)

8. A solid model created in CATIA V5 is an integration of a number of features. (T/F)

9. Generative drafting is a process of generating drawing views of an existing part or assembly created earlier. (T/F)

10. By default, the compass is located on the left side of the CATIA V5 window. (T/F)

Review Questions

Answer the following questions:

1. In the **FreeStyle** workbench, the tools available in the _________ toolbar are used to create surfaces.

2. In the **Assembly** workbench, the tools in the___________ toolbar are used to insert an existing part or assembly in the current product file.

3. The model created in the **Part** workbench can be edited in the **Wireframe and Surface Design** workbench. (T/F)

4. Features created in **FreeStyle** workbench do not have a history. (T/F)

5. The **View** toolbar is available in all the workbenches. (T/F)

Answers to Self-Evaluation Test

1. Bidirectional Associativity, **2. Concentric**, **3. Coincident**, **4. Symmetry**, **5.** Pan, **6.** Constraints, **7.** T, **8.** T, **9.** T, **10.** F

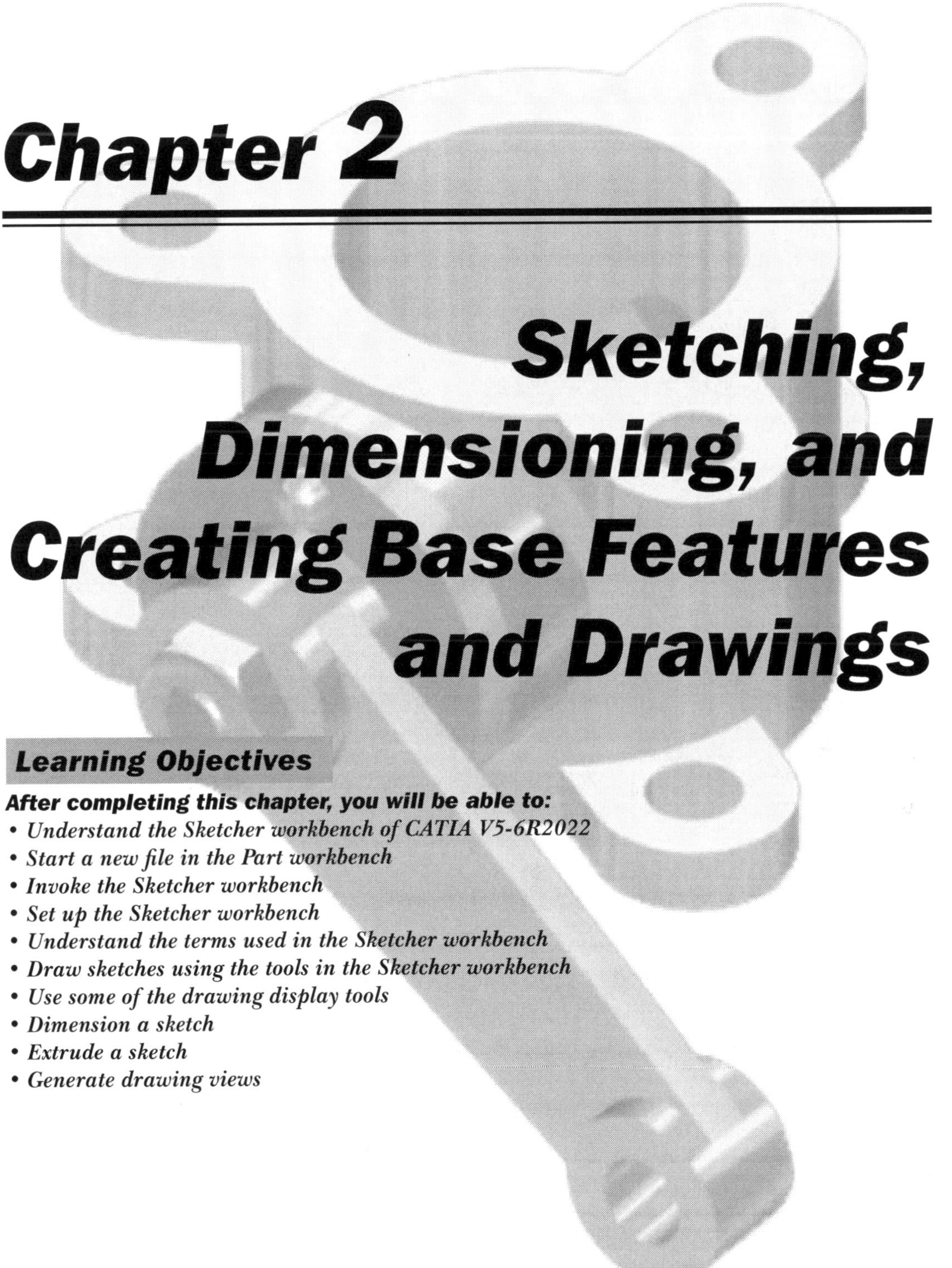

Chapter 2

Sketching, Dimensioning, and Creating Base Features and Drawings

Learning Objectives

After completing this chapter, you will be able to:

- *Understand the Sketcher workbench of CATIA V5-6R2022*
- *Start a new file in the Part workbench*
- *Invoke the Sketcher workbench*
- *Set up the Sketcher workbench*
- *Understand the terms used in the Sketcher workbench*
- *Draw sketches using the tools in the Sketcher workbench*
- *Use some of the drawing display tools*
- *Dimension a sketch*
- *Extrude a sketch*
- *Generate drawing views*

THE SKETCHER WORKBENCH

Most components designed using CATIA V5-6R2022 are a combination of sketched features, placed features, and derived features. Placed features are created without drawing a sketch whereas sketched features require a sketch that defines its shape. Generally, the base feature of any design is a sketched feature. For example, refer to the solid model of a chain link shown in Figure 2-1. The base sketch to create this solid model is shown in Figure 2-2.

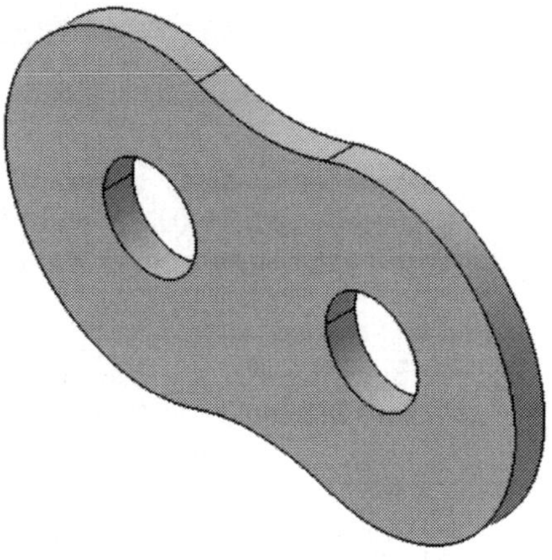

Figure 2-1 Solid model of the chain link

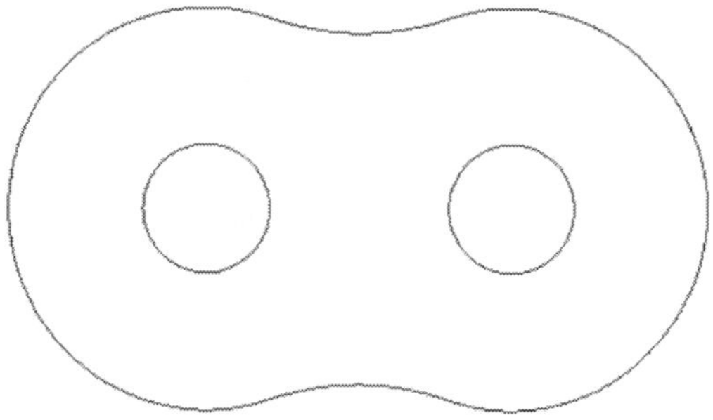

Figure 2-2 Base sketch for the solid model

The **Sketcher** workbench provides the space and tools to draw the sketches of the solid model. Generally, the first sketch drawn to start the design is called the base sketch which is then converted into a base feature. However, once you get familiar with the advanced options of CATIA V5, you will also be able to use a derived feature or a derived part as the base feature. In this chapter, you will learn more about the sketching tools in the **Sketcher** workbench that are used for drawing and displaying the sketches.

To draw a sketch, invoke the **Sketcher** workbench by choosing the **Sketch** tool from the **Sketcher** toolbar. Next, select a plane to draw the sketch. Draw the sketch and proceed to the **Part Design** or **Wireframe and Surface Design** workbench to convert it into a solid model or a surface model.

STARTING A NEW FILE

When you start CATIA V5-6R2022, the initial interface of CATIA V5-6R2022 is displayed, as shown in Figure 2-3. Start a new file in the **Part Design** workbench.

Choose **Part Design** from **Start > Mechanical Design**; the **New Part** dialog box will be displayed, as shown in Figure 2-4. Enter part name in the **Enter part name** edit box and select the **Enable hybrid design** radio button if not selected by default. Choose the **OK** button; the **Part Design** workbench will be displayed. Alternatively, choose **New** from the **File** menu; the **New** dialog box will be displayed, as shown in Figure 2-5. Select **Part** from the **List of Types** list box in the **New** dialog box or write the word **Part** in the **Selection** edit box available in this dialog box.

Figure 2-3 Initial interface of CATIA V5-6R2022

Figure 2-4 The ***New Part*** *dialog box*

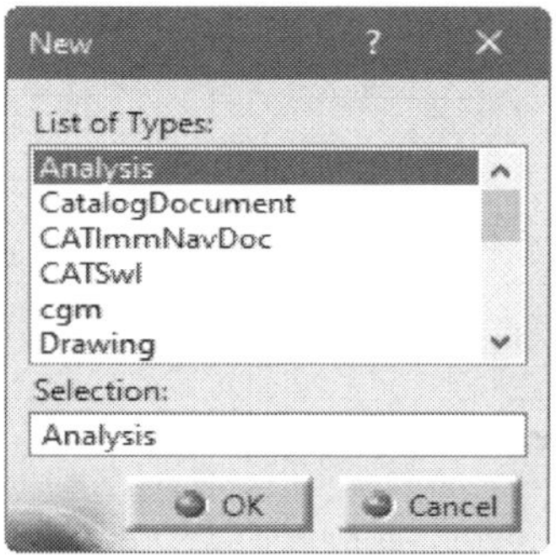

Figure 2-5 The ***New*** *dialog box*

Next, choose **OK** button; the **New Part** dialog box will be displayed. Enter the file name in the **Enter part name** edit box and choose the **OK** button; a new file in the **Part Design** workbench will be displayed on the screen, as shown in Figure 2-6. The standard tools like the Specification tree, Compass, and Geometry Axes will help you complete the design. The Specification tree is displayed on the top left corner of the screen. The **Compass** is displayed on the top right corner while the Geometry Axes is displayed on the bottom right corner of the screen.

If you select the **Enable hybrid design** check box in the **New Part** dialog box, you will be able to work in a hybrid design mode. In this design mode, part body includes solid, wireframe, and surface elements. The color of the **PartBody** node icon will be green in both hybrid and non-hybrid design modes. But, when both design modes are opened together, the color of the **PartBody** node icon will be displayed as green in the hybrid design mode and as gray in the non hybrid design mode.

Note

The color of the ***PartBody*** *node icon will be gray only when the* ***Enable hybrid design inside part bodies and bodies*** *check box is selected in the* ***Hybrid design*** *area of the* ***Part Document*** *tab of the* ***Part Infrastructure*** *sub-node of the* ***Infrastructure*** *node in the* ***Options*** *dialog box.*

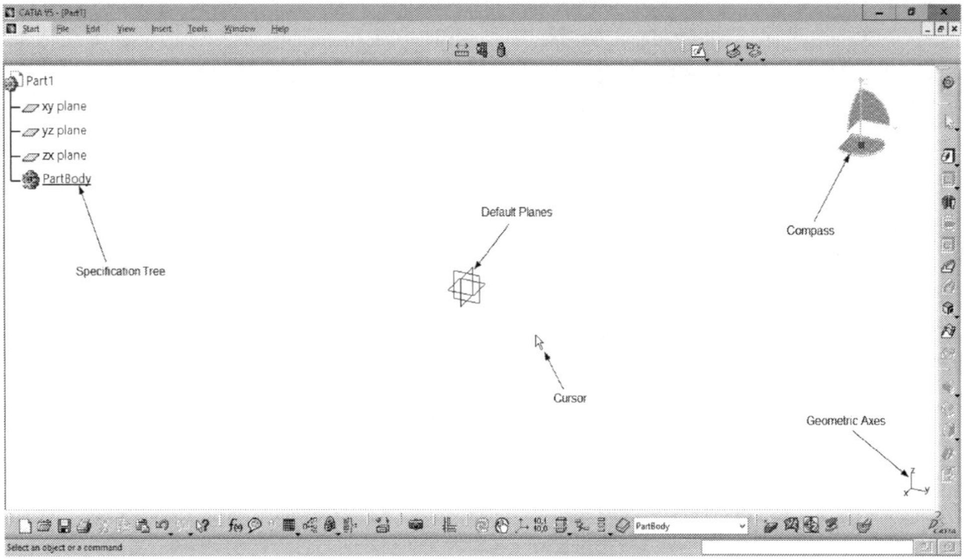

Figure 2-6 *A new file opened in the* ***Part Design*** *workbench*

If you select the **Create a geometrical set** check box, you will be able to create a geometrical set as well as a new part. Geometrical set created with this option is located above the Part Body node in the specification tree.

If you select the **Create an ordered geometrical set** check box, you will able to create an ordered geometrical set as well as a new part. The ordered geometrical set created using this option are located under the PartBody node in the specification tree.

If you clear the **Enable hybrid design** check box from the **New Part** dialog box, the new file will start in the conventional design mode.

Note

1. In this textbook, the hybrid design mode has been used. Therefore, it is recommended that you keep the ***Enable hybrid design*** *check box selected whenever you start a new file.*

2. You can hide the Compass, the Specification tree, or the Geometry Axis by using the ***View*** *menu. By default, check marks are displayed on the left of* ***Geometry****,* ***Specifications****, and* ***Compass*** *in the menu bar indicating their display is turned on. Choose these options again to turn off their display. The display of these tools should be turned off only when the geometry area is too small to view the model, else it is recommended that you do not hide these standard tools. You can also use the F3 key to toggle the display of the Specification tree.*

INVOKING THE SKETCHER WORKBENCH

Sketch is the basic step to create the base feature of a solid model. In CATIA V5, a sketch is drawn in the Sketcher workbench. To invoke the Sketcher workbench, choose the down arrow on the lower right corner of the **Sketcher** toolbar; the **Sketcher** sub-toolbar will appear, as shown in Figure 2-7. The two buttons in the **Sketcher** sub-toolbar are **Sketch** and **Positioned Sketch**. The next section focuses on invoking the **Sketcher** workbench using these two buttons.

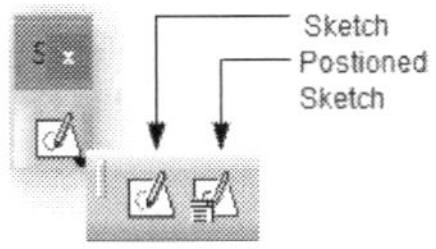

*Figure 2-7 The **Sketcher** sub-toolbar*

Invoking the Sketcher Workbench using the Sketch Tool

To invoke the **Sketcher** workbench using this method, choose the **Sketch** tool from the **Sketcher** sub-toolbar; you will be prompted to select a plane, a planar face, or a sketch. Select a plane from the three default planes in the Specification tree or from the geometry area; the **Sketcher** workbench will be invoked and the selected plane will be oriented parallel to the screen, refer to Figure 2-8. Also, you will be prompted to select an object or a command. The sketching components that are displayed in the geometry area are discussed later in this chapter.

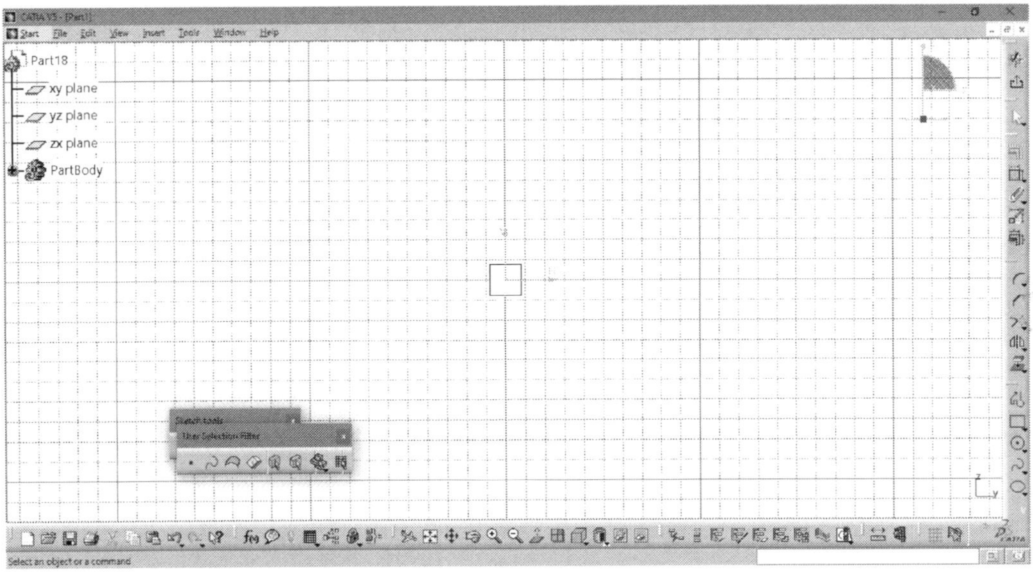

*Figure 2-8 The **Sketcher** workbench invoked on selecting a plane as the sketching plane*

Note

*Remember once the **Sketcher** workbench is invoked, you will stay in the **Select** mode till you exit the **Sketcher** workbench. To exit, choose the **Exit workbench** tool from the **Workbench** toolbar.*

Invoking the Sketcher Workbench using the Positioned Sketch Tool

In CATIA V5, you can also create a user-defined absolute axis while invoking the **Sketcher** workbench by using the **Positioned Sketch** tool. To invoke the **Sketcher** workbench using this option, choose the **Positioned Sketch** tool from the **Sketcher** sub-toolbar; the **Sketch Positioning** dialog box will be displayed, as shown in Figure 2-9. Also, you will be prompted to select a plane, a planar face, a sketch, an axis system, or two lines. You can set the absolute axis by using the options in this dialog box.

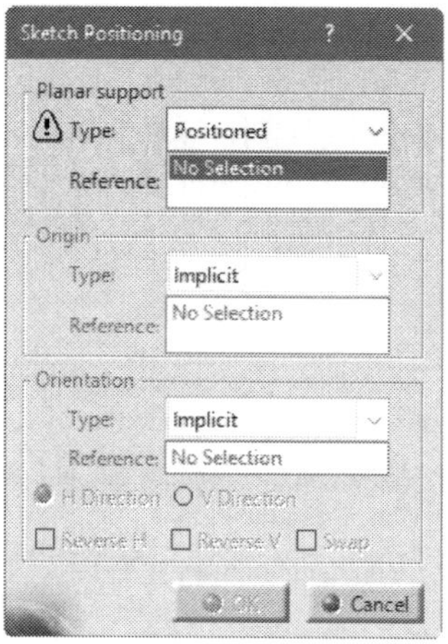

*Figure 2-9 The **Sketch Positioning** dialog box*

SETTING THE SKETCHER WORKBENCH

After invoking the **Sketcher** workbench, you need to set the workbench as per the sketching or drawing requirements. These requirements include modifying units, grid settings, and so on. The next section focuses on setting these parameters.

Modifying Units

To modify units, invoke the **Options** dialog box by choosing **Options** from the **Tools** menu. Next, click on the + sign on the left of the **General** option to expand the tree, if it is not already expanded. Choose the **Parameters and Measure** option; the tabs corresponding to this selection appear on the right in the **Options** dialog box. Next, choose the **Units** tab. The **Options** dialog box after choosing the **Units** tab is shown in Figure 2-10.

You can set the units for length, angle, time, mass, and so on, by using the options in the **Units** area. After specifying the value of the units, choose the **OK** button from the **Options** dialog box.

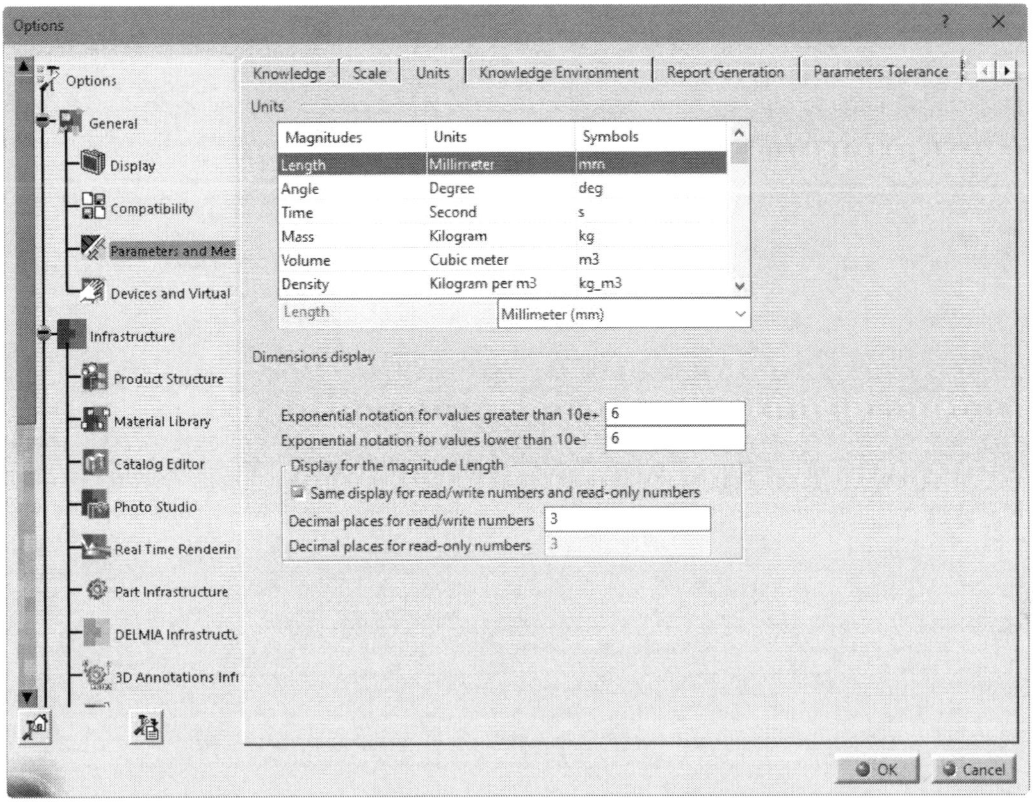

*Figure 2-10 The **Options** dialog box with the **Units** tab chosen*

Modifying the Grid Settings

When you invoke the **Sketcher** workbench, two types of lines are displayed in the geometry area: one in the horizontal direction and the other in the vertical direction. These horizontal and vertical lines are continuous and dotted lines. The spacing between two continuous lines is called primary spacing and the spacing between two dotted lines is called graduation. The mesh that is formed due to the intersection of these lines in the horizontal and vertical directions is called grid. In other words, primary spacing and graduation define the grid.

By default, the value of the **Graduations** parameter is set to 10 in both horizontal and vertical directions. The default value of the **Primary spacing** parameter is 100mm. Though you can change the **Primary spacing** and **Graduations** values in the horizontal and vertical directions individually, yet it is recommended not to change them. If the values of **Primary spacing** or **Graduations** in the horizontal direction are different from those in the vertical direction, then the **Grid** will be distorted. To change the values of **Primary spacing** and **Graduations**, choose **Options** from the **Tools** menu; the **Options** dialog box will be displayed. Choose the **Mechanical Design** node from the tree on the left of the dialog box. Next, choose the **Sketcher** option to display the **Sketcher** tab on the right in the **Options** dialog box, refer to Figure 2-11.

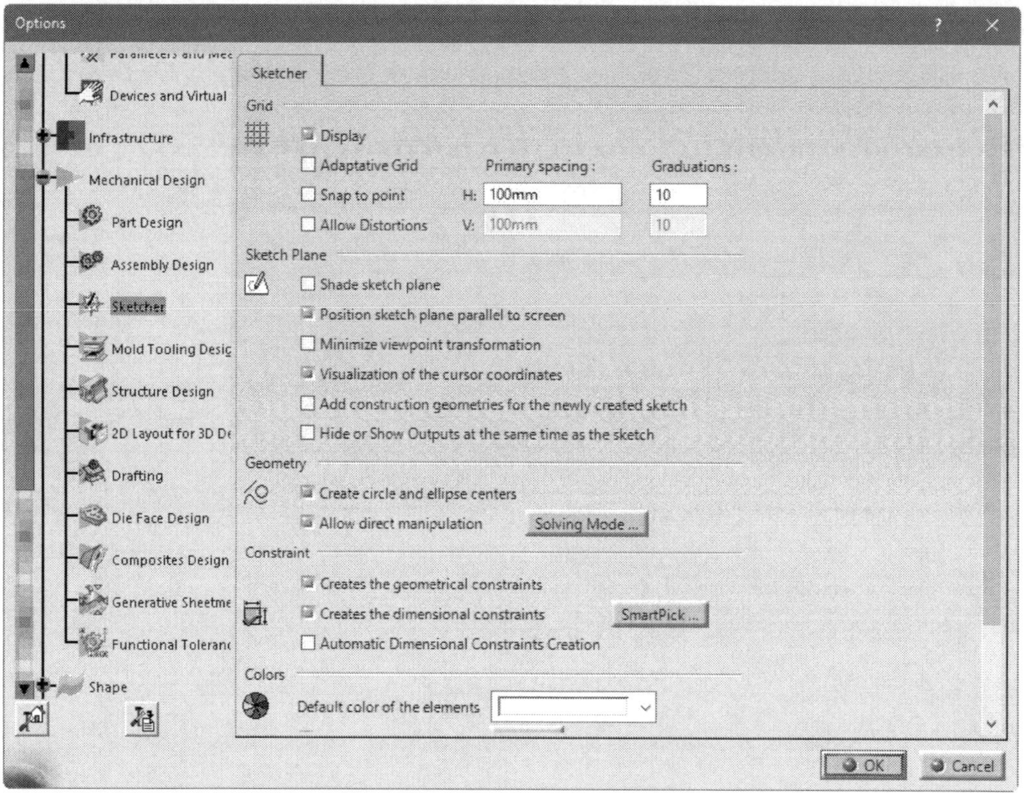

*Figure 2-11 The **Options** dialog box with the **Sketcher** option chosen*

The edit boxes of **Primary spacing** and **Graduations** on the right of the **H** row are already enabled. Here, **H** refers to the horizontal direction. To enable the edit boxes of **Primary spacing** and **Graduations** under the **V** row, select the **Allow Distortions** check box. Here, **V** refers to the vertical direction. Next, enter the values in the edit boxes corresponding to the **H** and **V** directions and then choose the **OK** button; the newly formed **Grid** will be applied to the **Sketcher** workbench. Note, henceforth all the files that you open or start in the **Sketcher** workbench will use these values for **Grid**.

UNDERSTANDING SKETCHER TERMS

Before learning about the sketching tools, it is important for you to understand some of the terms used in the **Sketcher** workbench. These terms are discussed next.

Specification Tree

The Specification tree is a manager that keeps track of all operations performed on the model. When you invoke the **Sketcher** workbench, a new member or branch, **Sketch.1**, is added to the Specification tree under the **PartBody** node. Click on the + sign on the left of the **PartBody** to expand it; you can view the **Sketch.1** member of the Specification tree. A + sign is associated with the **Sketch.1** on the branch. Click on this sign once to further expand the branch. Figure 2-12 shows the expanded Specification tree.

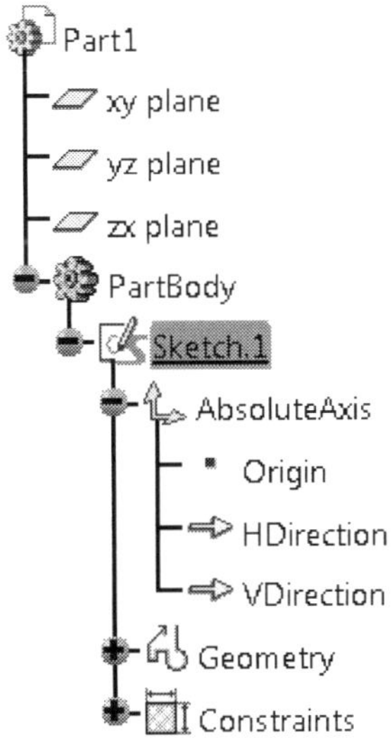

Figure 2-12 *The expanded Specification tree*

Various levels under **Sketch.1** in the Specification tree are discussed next.

AbsoluteAxis

In the **Sketcher** workbench, the default horizontal and vertical axes passing from the origin (0,0) to infinity are referred to as **AbsoluteAxis**. These axes will be highlighted in the geometry area, when **AbsoluteAxis** is selected from the Specification tree. Notice the + sign available on the left of **AbsoluteAxis** in the Specification tree. Click on this + sign once to expand the branch by one level. The levels associated with this branch are discussed next.

Tip

While expanding the branch of the Specification tree, if the branch lines are accidentally clicked then the Specification tree will get activated. Consequently, the geometry area will be frozen. Now, zooming and panning will resize or reposition the Specification tree instead of the geometry view. The geometry area can be activated again by clicking on the branch line again or on the Geometry Axis available on the bottom right corner of the geometry area.

Origin

The **Origin** in the **Sketcher** workbench is the point where the absolute horizontal axis intersects the absolute vertical axis. The coordinates for **Origin** are (0,0). It is widely used while applying dimensional constraints to the sketches. You will learn more about dimensional constraints in later chapters.

HDirection

The direction parallel to horizontal axis is referred as **HDirection** and is mostly used to constrain a sketch. It is represented by **H** icon in the drawing window and displayed as **HDirection** in the Specification tree.

VDirection

The direction parallel to vertical axis is referred to as **VDirection** and is mostly used to constrain a sketch.

The branches of the Specification tree will increase as the design process continues. You will learn more about the branches associated with the Specification tree in the **Sketcher** workbench while drawing and constraining sketches.

Grid

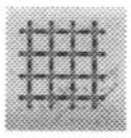

This option is used to display or hide the Graduations and Primary Spacing lines from the graphic area. To activate or deactivate it, choose the **Grid** button from the **Visualization** toolbar which appears only when you are in the **Sketcher** workbench.

Snap to Point

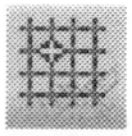

This option is used to snap to the point of intersection of the primary spacing and the graduation lines while sketching. By default, the snap mode is active. To activate or deactivate it, choose the **Snap to Point** button from the **Sketch tools** toolbar which appears only when you invoke the **Sketcher** workbench.

Construction/Standard Element

An element that is not a part of the profile while creating features and is used only as a reference or to constrain the elements of the sketch in the **Sketcher** workbench, is called a **Construction** element. This element can be used only in the **Sketcher** workbench. A **Standard** element is one that takes part in the feature creation. Depending on the requirement of the design, you can convert a standard element into a construction element or vice-versa using the **Construction/Standard Element** button.

Select Toolbar

While drawing a sketch, you often need to select some elements. The tools required to make a selection are available in the **Select** toolbar, as shown in Figure 2-13. Various tools such as **Select**, **Rectangle Selection Trap**, and so on are available in this toolbar. By default, the **Select** tool is activated in the sketcher workbench unless any other tool or object is selected.

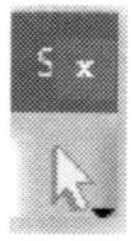

Figure 2-13 *The* ***Select*** *toolbar*

The tools in the **Select** toolbar can be invoked by choosing the down arrow on the lower-right of the **Select** toolbar. When you click on the down arrow, the **Select** sub-toolbar will be displayed, as shown in Figure 2-14. The methods of selecting an entity using the tools in the **Select** toolbar are discussed next.

Note

You can detach a sub-toolbar by dragging the vertical/horizontal line displayed at its top or its extreme left and placing it in the geometry area. On doing so, it will act as a toolbar. Therefore, in this textbook, the sub-toolbar will be referred by the name of the toolbar that will be obtained by detaching from the parent toolbar.

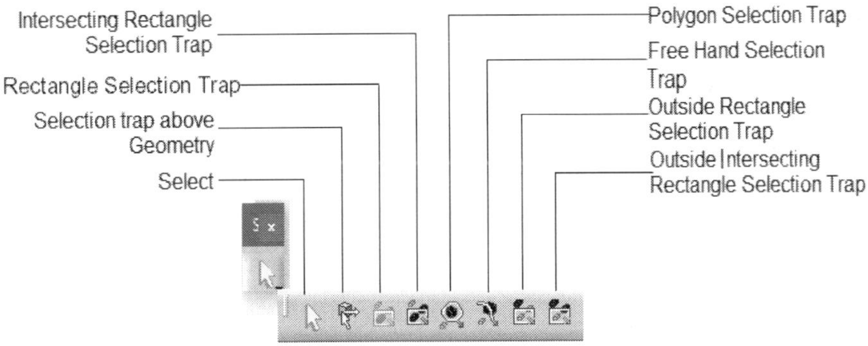

***Figure 2-14** The **Select** sub-toolbar*

Select

This tool allows you to make a selection of the elements. As you move the arrow cursor near the element with the **Select** tool activated, the arrow cursor will be replaced with a hand cursor. Click on the element to select it.

Rectangle Selection Trap

This tool is used to select entities by creating a selection trap. A trap is a rectangular box drawn by dragging the mouse to define the diagonally opposite corners. All the objects that lie completely inside the selection trap are selected. This tool is active by default. If not, choose the **Rectangle Selection Trap** tool from the **Select** toolbar. Note that this method of selection will not allow you to start creating trap from the top of an entity.

Selection trap above Geometry

This tool works same as the **Rectangle Selection Trap** tool with the only difference that this tool allows you to start creating trap from the top of given sketched entity.

Intersecting Rectangle Selection Trap

The **Intersecting Rectangle Selection Trap** tool is similar to the **Selection trap above Geometry** tool. The difference between them is that this tool allows you to select elements of a sketch that are inside or are intersected by the trap. To create an intersecting trap, choose the **Intersecting Rectangle Selection Trap** tool from the **Select** toolbar. Next, specify the first corner and then drag the mouse to specify the second corner.

Polygon Selection Trap

This method includes selection of elements by drawing a closed polygon as the selection trap. You can select the elements of a sketch that are completely inside the polygon by using this method. To use this method, choose the **Polygon Selection Trap** tool from the **Select** toolbar and draw a closed polygon by specifying its adjacent corners. The polygon creation can be terminated by double-clicking in the geometry area.

Free Hand Selection Trap

This method includes selection of elements by dragging the mouse to draw a free sketch across them. The elements intersected by the free sketch are selected.

Outside Rectangle Selection Trap

This method is used to select the elements that are outside the selection trap. The elements that are intersected by the trap are not selected.

Outside Intersecting Rectangle Selection Trap

The elements that are outside the selection trap or are intersected by the selection trap are selected by using this method.

Inferencing Lines

The inferencing lines are temporary lines that are used to track a particular point on the screen. When a sketching tool is selected in the sketcher environment, the inferencing lines are automatically displayed from the endpoints of the sketched elements or from the origin. Consider a case in which you want to draw a line such that its endpoint is tangent to the circle. Specify the start point of the line and then move the cursor in the direction tangent to the circle. You will note that the inference line is shown tangent to the existing circle. Next, specify the endpoint of the line. Figure 2-15 shows the use of the inferencing line to draw a tangent line. The inferencing lines are available only in the sketcher workbench.

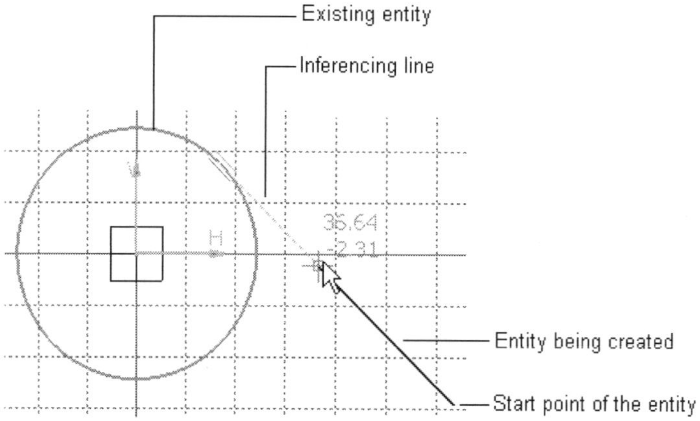

Figure 2-15 *Use of inferencing line to draw a tangent line*

DRAWING SKETCHES USING SKETCHER TOOLS

The tools to draw the sketches in the **Sketcher** workbench are available in the **Profile** toolbar. Most of the tools have a down arrow indicating that they have some more tools. To access these tools, click on the arrow and choose them from the sub-toolbar. As mentioned earlier, the name of the sub-toolbar will be the name of the toolbar that will be obtained by detaching it from the parent toolbar. The procedure to draw sketches using the sketch tools is discussed next.

Drawing Lines

Menubar:	Insert > Profile > Line > Line
Toolbar:	Profile > Line sub-toolbar > Line

The **Line** tool is one of the basic sketching tools in the **Sketcher** workbench. The general definition of a line is the shortest distance between two points. As CATIA V5 is parametric in nature, it allows user to first draw a line of any length at any angle and then change it to the desired length and angle. To draw a line, choose the **Line** tool from the **Profile** toolbar. The methods to draw a line in CATIA V5 are discussed next.

Drawing Lines by Specifying Points in the Geometry Area

To draw a line by specifying points in the geometry area, choose the **Line** tool from the **Profile** toolbar. You will observe that as you move the cursor in the geometry area, the coordinates corresponding to the current location of the cursor are displayed above it.

On invoking the **Line** tool, you will be prompted to select a point or click to locate the start point of the line. The prompt sequence will be displayed in the current information area of the status bar below the geometry area. Click anywhere in the geometry area to specify the start point of the line; you will be prompted to specify the endpoint. Move the cursor away from the start point. On doing so, a rubber band line will be attached to the cursor. Click anywhere in the geometry area to specify the endpoint of the line. Figure 2-16 shows the line drawn by selecting points from the geometry area. The orange color of the line indicates that it is selected. Click anywhere on the screen to end the selection mode.

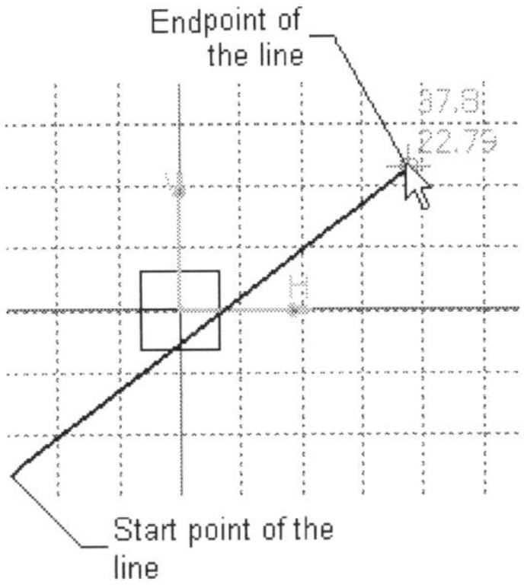

Figure 2-16 *The line drawn by selecting its start and end points from the geometry area*

Note
A line in CATIA V5 consists of three geometric elements: start point, line segment, and endpoint. The start point and endpoint are construction elements while the line segment is a standard element.

Drawing Lines by Using the Sketch tools Toolbar

Lines can also be drawn using the **Sketch tools** toolbar, which expands when you invoke the **Line** tool. Figure 2-17 shows the **Sketch tools** toolbar after invoking the **Line** tool. The two methods to draw a line using the **Sketch tools** toolbar are discussed next.

Figure 2-17 *The expanded **Sketch tools** toolbar displayed after invoking the **Line** tool*

Drawing Lines by Entering the Values of the Start and End Points

To draw a line using the start point and endpoint values, invoke the **Line** tool. On doing so, the **Sketch tools** toolbar will expand. In the **H** and **V** edit boxes, specify the horizontal and vertical coordinate values of the start point, respectively, and then press Enter; you will be prompted to select the endpoint. Specify values in the **H** and **V** edit boxes and press Enter again; a line will be drawn in the geometry area corresponding to the values entered in the start point and endpoint edit boxes. Also, the horizontal and vertical dimensions of the start point and endpoint with respect to the origin are displayed. On completion of the line, you will observe that the **Sketch tools** toolbar is compressed to its original size after the line is drawn. The color of the created line is orange indicating that it is selected. To end the selection mode, click anywhere in the geometry area. The line will appear green in color which means that it is fully constrained. You will learn more about constraints in the later chapters.

Similarly, you can draw a line by specifying the start point and by entering the length and angle values. The positive angular value is measured in counterclockwise direction with respect to the H axis and the negative angular value is measured in clockwise direction with respect to the H axis.

Note

*As the **Dimensional Constraints** button is chosen in the **Sketch tools** toolbar, the specified dimension values for the start point and the endpoint will be displayed. Let these values remain in the geometry area.*

You will also notice that the color of the construction elements such as the start and endpoints of the line is gray. This suggests that the element is fully constrained.

Drawing Lines with a Symmetrical Extension

To draw a line with a symmetrical extension, invoke the **Line** tool and choose the **Symmetrical Extension** tool from the expanded **Sketch tools** toolbar. When you draw the line using this option, its total length is double the distance you have moved while specifying the start point and the endpoint.

In CATIA V5, a few more types of lines such as the infinite line, bisecting line, line normal to curve, and bi-tangent line can be drawn. To draw these lines, choose the down arrow on the right of the **Line** tool from the **Profile** toolbar; the **Line** sub-toolbar will appear, as shown in Figure 2-18. The types of lines that can be drawn using the tools available in the **Line** sub-toolbar are discussed next.

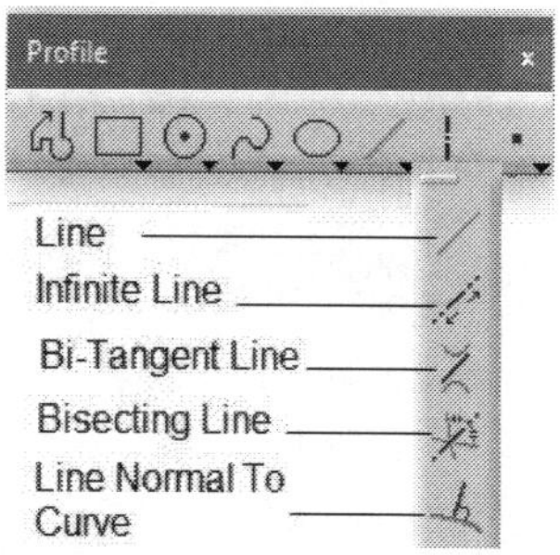

***Figure 2-18** The **Line** sub-toolbar*

> **Tip**
> *The **Grid** button in the **Visualization** toolbar is used to toggle the display of the grid. While sketching, you can choose the **Grid** button any time to turn on or off the display of the grid.*

Drawing Infinite Lines

Menubar:	Insert > Profile > Line > Infinite Line
Toolbar:	Profile > Line sub-toolbar > Infinite Line

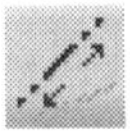

To draw an infinite line, invoke the **Infinite Line** tool from the **Line** sub-toolbar in the **Profile** toolbar; the **Sketch tools** toolbar will expand. You can draw a horizontal infinite line, vertical infinite line, and infinite line passing through any two points by using the options in this toolbar.

Drawing Bi-Tangent Lines

Menubar:	Insert > Profile > Line > Bi-Tangent Line
Toolbar:	Profile > Line sub-toolbar > Bi-Tangent Line

A line that is tangent to any two curved entities is called bi-tangent lines. The curved entities can be circle, ellipse, arc, conic, and spline. You will learn more about these curved geometries later in this chapter. To draw a bi-tangent line, invoke the **Bi-Tangent Line** tool from the **Line** sub-toolbar in the **Profile** toolbar. Next, select the first curved geometry and then select the second curved geometry; a line will be drawn between the two selected curved elements. Also, the coincidence symbol will be displayed at the endpoints of the bi-tangent line. These are the coincidence constraints. You will learn more about these constraints in the later chapters.

Drawing Bisecting Lines

Menubar:	Insert > Profile > Line > Bisecting Line
Toolbar:	Profile > Line sub-toolbar > Bisecting Line

A bisecting line passes through the intersection of two non-parallel lines, such that the angle formed between them is divided equally. The intersection point of the non-parallel lines can be actual or apparent obtained by extending the lines virtually. To draw a bisecting line, invoke the **Bisecting Line** tool from the **Line** sub-toolbar in the **Profile** toolbar. Select the first line and then select the second line; a bisecting line of infinite length will be drawn.

Drawing Lines Normal to a Curve

Menubar:	Insert > Profile > Line > Line Normal To Curve
Toolbar:	Profile > Line sub-toolbar > Line Normal To Curve

To draw a line normal to a curve, invoke the **Line Normal To Curve** tool from the **Line** sub-toolbar in the **Profile** toolbar; you will be prompted to select the curve. Specify the start point of the line anywhere on the periphery of the curve; you will be prompted to specify the other end point of the line. Click to specify the endpoint; a line normal to it will be drawn.

Drawing Center Lines

Menubar: Insert > Profile > Axis
Toolbar: Profile > Axis

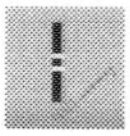

You can draw a center line in CATIA V5 using the **Axis** tool. Generally, this tool is used to create the axis for the revolved feature. You will learn more about the revolved features in the later chapters. To draw an axis, invoke the **Axis** tool from the **Profile** toolbar; the **Sketch tools** toolbar will expand and you will be prompted to specify the start point of the axis. Click in the geometry area to specify the start point; you will be prompted to specify the endpoint of the axis. Move the cursor and click to specify the endpoint; an axis with the specified points will be displayed in the geometry area, as shown in Figure 2-19. You can also draw an axis by entering the parameters in the respective edit boxes of the expanded **Sketch tools** toolbar.

Drawing Rectangles, Oriented Rectangles, and Parallelograms

CATIA V5 provides some set of tools that help you draw predefined profiles faster. These tools are grouped together in the **Predefined Profile** sub-toolbar. To view this sub-toolbar, choose the arrow on the right of the **Rectangle** tool in the **Profile** toolbar; the **Predefined Profile** sub-toolbar will be displayed, as shown in Figure 2-20. The tools in this toolbar are **Rectangle**, **Oriented Rectangle**, **Parallelogram**, and so on. Some of these tools are discussed here and the remaining will be discussed in the next chapter.

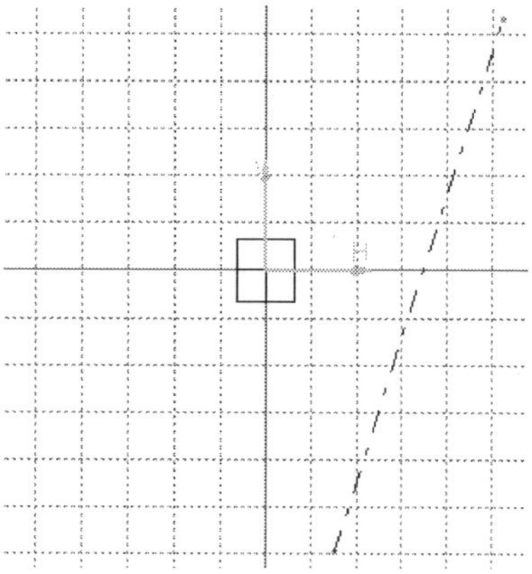

Figure 2-19 *An axis drawn in the geometry area*

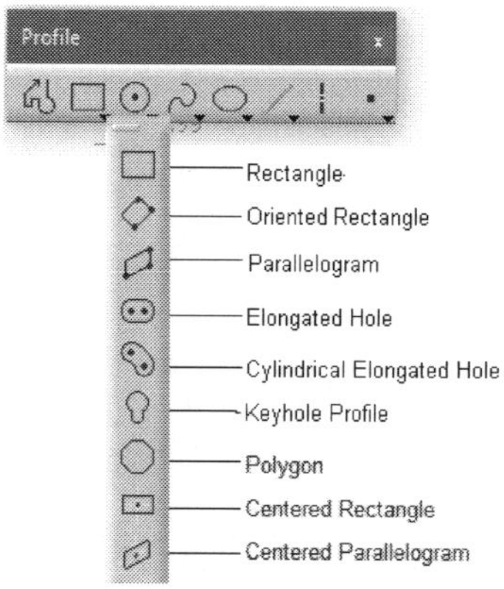

Figure 2-20 *The* ***Predefined Profile*** *sub-toolbar*

Drawing Rectangles

Menubar: Insert > Profile > Predefined Profile > Rectangle
Toolbar: Profile > Predefined Profile sub-toolbar > Rectangle

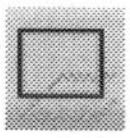

To draw a rectangle, invoke the **Rectangle** tool from the **Predefined Profile** sub-toolbar, refer to Figure 2-20; the **Sketch tools** toolbar will expand and you will be prompted to specify the first point for the rectangle. Click in the geometry area to specify the first point or the first corner of the rectangle; you will be prompted to specify the second point. Move the cursor away from the first point in the geometry area; preview of the rectangle is displayed. Click to specify the diagonally opposite corner of the rectangle. You can also draw a rectangle by entering the values in the **Sketch tools** toolbar. On drawing a rectangle by using this method, you will notice that dimensions and constraints are applied to the resulting rectangle. You will learn more about dimensioning and constraining in the later chapters.

Drawing Oriented Rectangles

Menubar: Insert > Profile > Predefined Profile > Oriented Rectangle
Toolbar: Profile > Predefined Profile sub-toolbar > Oriented Rectangle

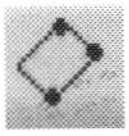

To draw an oriented rectangle, invoke the **Oriented Rectangle** tool from the **Predefined Profile** sub-toolbar in the **Profile** toolbar; you will be prompted to specify the start point. Click in the geometry area to specify it; you will be prompted to specify the end point of the first side. Move the cursor away from the first point in any direction and specify the end point of the first side; you will be prompted to define the second side. The angle formed between the line and horizontal reference is called the orientation angle of the rectangle.

Also, the symbol of the perpendicular constraint will be displayed between the lines. You will learn more about constraints in later chapters. Click in the geometry area to specify the third corner of the rectangle. Figure 2-21 shows the oriented rectangle being drawn.

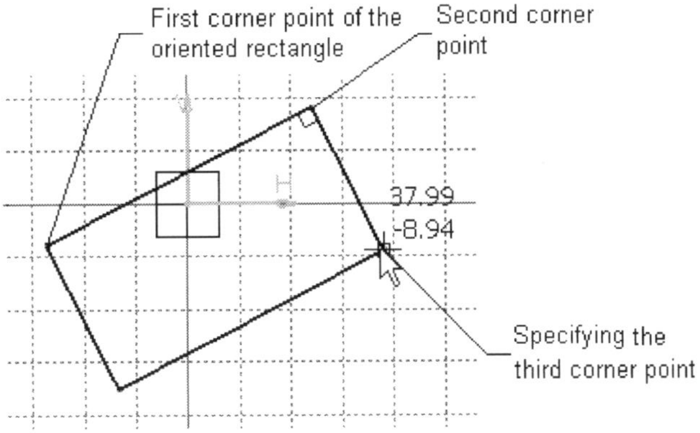

Figure 2-21 *Oriented rectangle being drawn by specifying the corner points*

Note

*You can also use the **Sketch tools** toolbar to enter the coordinate values for the first, second, and third corners in the respective edit boxes. To specify the orientation of the rectangle, enter the value of the orientation angle in the **A** edit box of the **Sketch tools** toolbar. Once you have specified the values, you need to press the Enter key to accept them.*

Drawing Parallelograms

Menubar: Insert > Profile > Predefined Profile > Parallelogram
Toolbar: Profile > Predefined Profile sub-toolbar > Parallelogram

A parallelogram is a quadrilateral whose opposite sides are parallel to each other. To draw it, invoke the **Parallelogram** tool from the **Predefined Profile** sub-toolbar in the **Profile** toolbar; the **Sketch tools** toolbar will expand and you will be prompted to specify the start point of the parallelogram. Click in the geometry area to specify its first corner; you will be prompted to specify the end point of its first side. On moving the cursor away from the first corner, you will notice a line attached to the cursor. The line represents the first side of the parallelogram. Click in the geometry area to specify the endpoint of the line; you will be then prompted to specify the second side. Move the cursor away from the second corner; preview of the parallelogram will be displayed. Click to specify the second side of the parallelogram; the parallelogram will be created, as shown in Figure 2-22.

Note

In CATIA V5, you can use the ***Sketch tools*** *toolbar to enter the coordinate values of the corner points of the parallelogram. You can also enter the values for width, angle, and height of the parallelograms in the respective edit boxes in the expanded* ***Sketch tools*** *toolbar.*

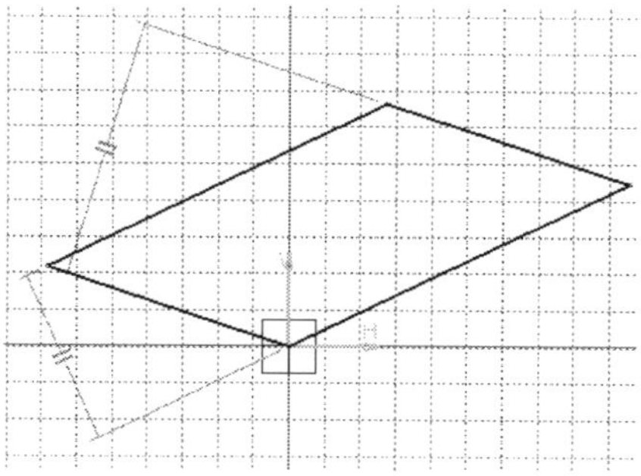

Figure 2-22 *A parallelogram created by specifying the corner points*

Creating Points

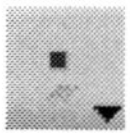

A point is defined as the geometrical element that has no magnitude of length, width, or thickness. It is only specified by its position. In CATIA V5, you can create points by clicking in the geometry area or by specifying the coordinates. You can also locate an intersection point or project a point on an element. To invoke any of the tools for creating a point, choose the down arrow on the right of the **Point** tool in the **Profile** toolbar; the **Point** sub-toolbar will be displayed, as shown in Figure 2-23. The methods for creating points are discussed next.

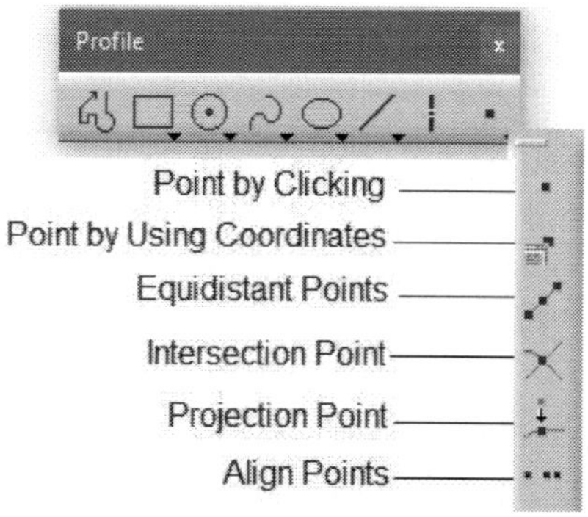

Figure 2-23 *The* ***Point*** *sub-toolbar*

Creating Points by Clicking

Menubar:	Insert > Profile > Point > Point
Toolbar:	Profile > Point sub-toolbar > Point by Clicking

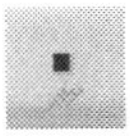

To create points by clicking, invoke the **Point by Clicking** tool from the **Point** sub-toolbar in the **Profile** toolbar; the **Sketch tools** toolbar will expand and you will be prompted to click to create the point. Click anywhere in the geometry area to create the point. You can also enter the horizontal and vertical coordinate values in the **H** and **V** edit boxes of the **Point Coordinates** area displayed in the expanded **Sketch tools** toolbar. You can create points by defining their coordinates using the other options in the **Point** sub-toolbar.

Creating Points by using Coordinates

Menubar:	Insert > Profile > Point > Point by Using Coordinates
Toolbar:	Profile > Point sub-toolbar > Point by Using Coordinates

To create points by using coordinates, invoke the **Point by Using Coordinates** tool from the **Point** sub-toolbar in the **Profile** toolbar; the **Point Definition** dialog box will be displayed, as shown in Figure 2-24. In this dialog box, you can use either cartesian (H and V) or polar coordinates to define a point. You can also select a previously created point as a reference for the point you want to create. On choosing the **Cartesian** tab, you need to specify **H** and **V** parameters or use the corresponding spinners. On choosing the **Polar** tab, you need to specify the values for **Radius** and **Angle** in the corresponding edit boxes.

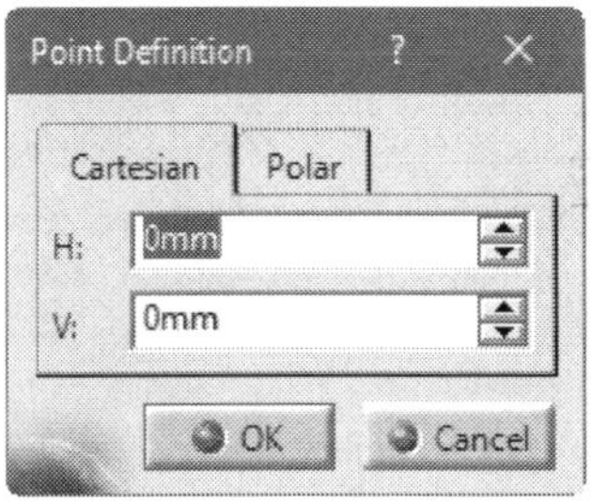

***Figure 2-24** The **Point Definition** dialog box*

Creating Equidistant Points

Menubar:	Insert > Profile > Point > Equidistant Points
Toolbar:	Profile > Point sub-toolbar > Equidistant Points

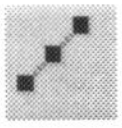

To create equidistant points on a line, curve, or between two points, invoke the **Equidistant Points** tool from the **Point** sub-toolbar in the **Profile** toolbar; you will be prompted to select the origin point or a curve on which the points are to be created. Select the object; the **Equidistant Point Definition** dialog box will appear, as shown in Figure 2-25. Specify the number of equidistant points to be created in the **New Points** spinner. If you select a line or a curve and then click on the extremity of that line or curve, the options in the **Parameters** drop-down will become available. You can select the **Points & Length** option from this drop-down. By using this option, you can specify the number of points and length in the **New Points** and **Length** spinners, respectively. If you select the **Points & Spacing** option, you can specify the number of points and spacing between the points in the **New Points** and **Spacing** spinners, respectively. If you choose the **Spacing & Length** option, you can specify the spacing and length for the

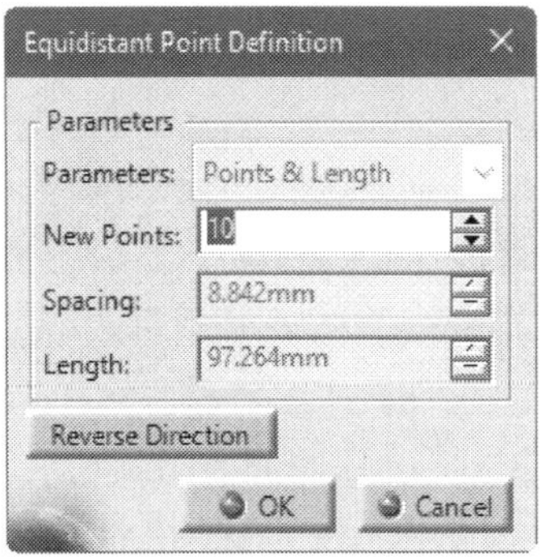

***Figure 2-25** The **Equidistant Point Definition** dialog box*

points. Choose the **Reverse Direction** button to reverse the direction of the points if required. Choose **OK** to create equidistant points, refer to Figure 2-26.

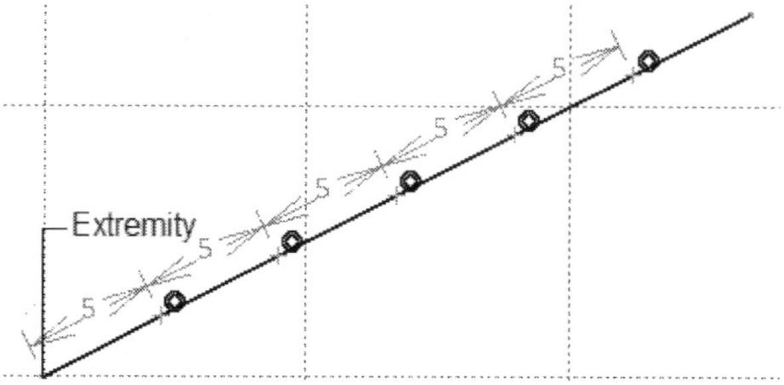

Figure 2-26 *The equidistant points created on the line*

Creating Intersection Points

Menubar:	Insert > Profile > Point > Intersection Point
Toolbar:	Profile > Point sub-toolbar > Intersection Point

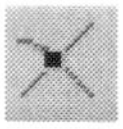

To create points at the intersection of the selected elements, invoke the **Intersection Point** tool from the **Point** sub-toolbar of the **Profile** toolbar. On doing so, you will be prompted to select the set of the elements to be intersected with another element. Select the set of the elements; the intersection points will be created on the intersection of the selected elements, as shown in Figure 2-27.

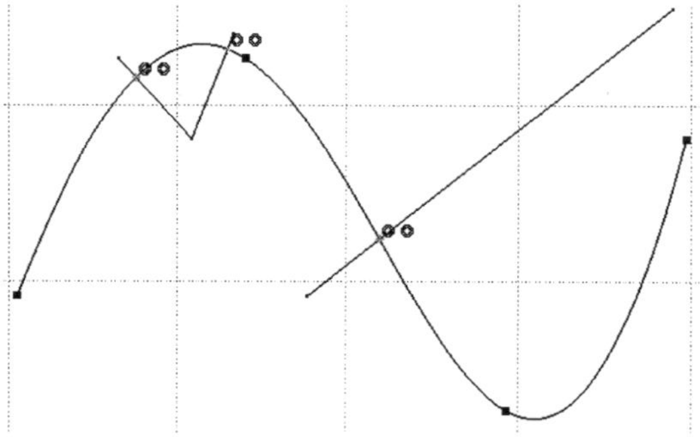

Figure 2-27 *The intersection points created on the curve*

Creating Projection Points

Menubar:	Insert > Profile > Point > Projection Point
Toolbar:	Profile > Point sub-toolbar > Projection Point

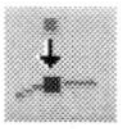

You can create one or more points by projecting points onto a curve element. To create projection points on a curve, select the points that are to be projected and choose the **Projection Point** tool from the **Point** sub-toolbar in the **Profile** toolbar; you will be prompted to select the element on which the selected points will be projected. Select the curve; the points will be projected automatically on it and construction lines representing the direction of the projection will also appear in the graphics area. The projection options available in the **Sketch tools** toolbar are discussed next.

Orthogonal Projection

This option is selected by default. Select the points and then select the curve. All the selected points will be projected on the curve according to a normal direction on this curve, as shown in Figure 2-28.

Along a Direction

Choose the **Along a Direction** option from the **Sketch tools** toolbar. Select a point and then define the position of the point on the curve; the selected points will get projected along the given direction, as shown in Figure 2-29.

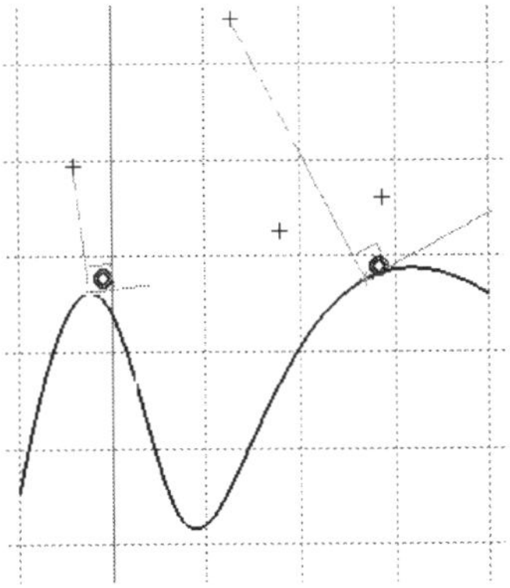

Figure 2-28 *The projection points created using the* ***Orthogonal Projection*** *option*

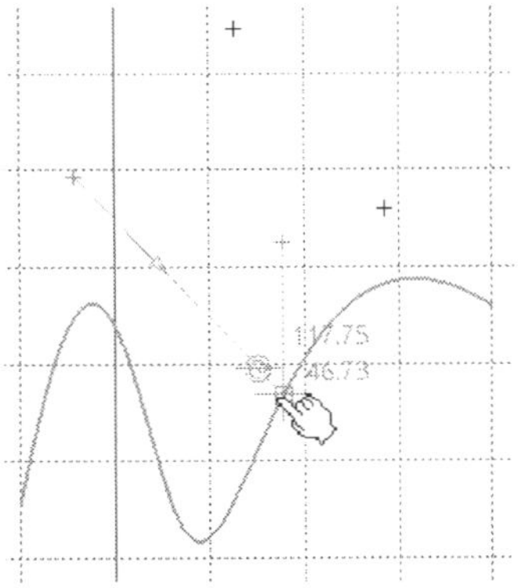

Figure 2-29 *The projection points created using the* ***Along a Direction*** *option*

Aligning Points

Menubar:	Insert > Profile > Point > Align Points
Toolbar:	Profile > Point sub-toolbar > Align Points

You can use this tool to align points in a particular direction. Choose the **Align Points** tool from the **Point** sub-toolbar in the **Profile** toolbar and select all the points that you want to align; a small arrow indicating the direction of alignment will be displayed at the origin point. The direction of this arrow will change with respect to the pointer position in the graphics area, refer to Figure 2-30.

By default, the **Along a Direction** button is chosen in the **Sketch tools** toolbar. If you want to change the origin point, choose the **Change Origin Point** button in the **Sketch tools** toolbar and select a point to make it the origin point. If you choose the **Horizontal Alignment** button then the points will be aligned horizontally. Similarly, choose the **Vertical Alignment** button to align the points vertically. In the **Sketch tools** toolbar, choose the **Align along selected linear element** button and select the line; the points will be aligned along the direction of the selected line, as shown in Figure 2-31.

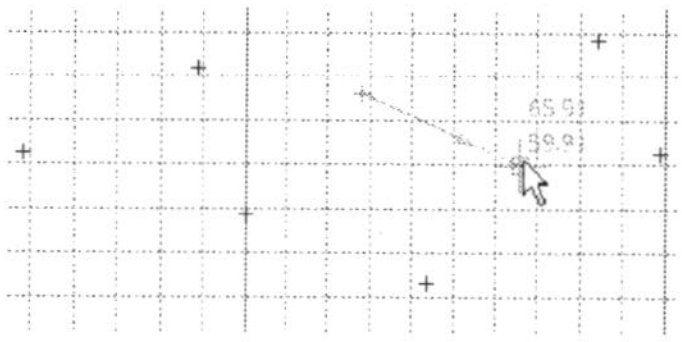

Figure 2-30 The point alignment direction

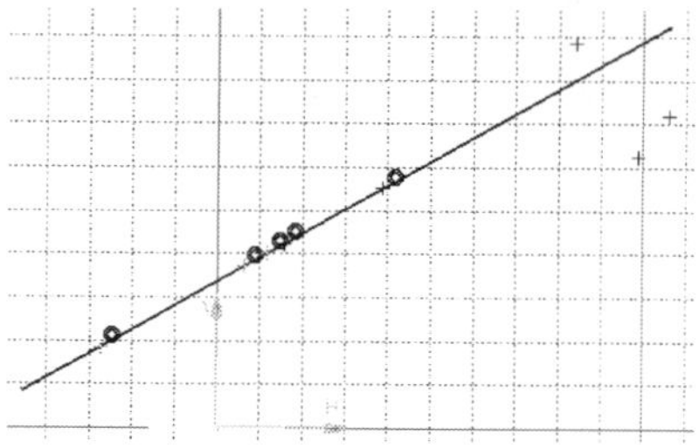

Figure 2-31 The points aligned along the line

Drawing Circles

You can draw a circle in CATIA by defining its center and radius, by specifying three points on its periphery, by using coordinates or so on. All the tools to draw a circle in CATIA are grouped in the **Circle** sub-toolbar. To view the **Circle** sub-toolbar, choose the down arrow on the right of the **Circle** tool in the **Profile** toolbar; the **Circle** sub-toolbar will be displayed, as shown in Figure 2-32. The tools available in this sub-toolbar will help you to draw various types of circles and arcs. Different methods to draw circles and arcs are discussed next.

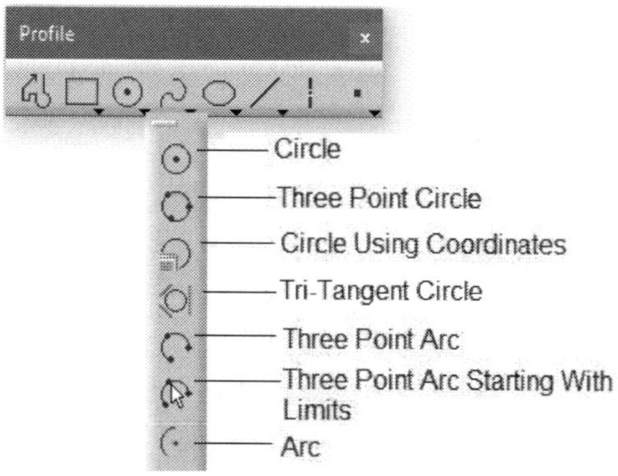

*Figure 2-32 The **Circle** sub-toolbar*

Drawing Circles using the Circle Tool

Menubar:	Insert > Profile > Circle > Circle
Toolbar:	Profile > Circle sub-toolbar > Circle

To draw a circle using this method, invoke the **Circle** tool from the **Circle** sub-toolbar; you will be prompted to specify its center. Click anywhere in the geometry area to specify the center point; you will be prompted to specify a point that determines the radius of the circle. Move the cursor away from the center point to specify the radius; preview of the circle will be displayed. Click in the geometry area to define its radius.

Note

*You can also draw a circle by specifying the coordinates of its center point in the **Circle Center** edit box and the radius value in the **R** edit box of the expanded **Sketch tools** toolbar.*

Drawing a Three Point Circle

Menubar: Insert > Profile > Circle > Three Point Circle
Toolbar: Profile > Circle sub-toolbar > Three Point Circle

In CATIA V5, a circle can also be drawn by specifying any three points that will lie on its circumference. To draw a three point circle, invoke the **Three Point Circle** tool from the **Circle** sub-toolbar in the **Profile** toolbar; the **Sketch tools** toolbar will expand and you will be prompted to specify the start point of the circle. Click anywhere in the geometry area to specify the start point; you will be prompted to specify the second point through which the circle will pass. As you move the cursor away from the first point, a dotted line that originates from the first point and moves along with the cursor will be displayed. This is the chord of the circle. Click in the geometry area to specify the second point on the circle; you will be then prompted to specify the last point. As you move the cursor to specify the third point, preview of the circle will be displayed. Click to specify the third point to create the circle.

Note

*You can also enter radius value in the **R** edit box of the expanded **Sketch tools** toolbar. Remember that when you enter radius value, the other two points that lie on the circle should be specified within the reach of the radius value.*

Drawing Circles using Coordinates

Menubar: Insert > Profile > Circle > Circle Using Coordinates
Toolbar: Profile > Circle sub-toolbar > Circle Using Coordinates

In CATIA V5, a circle can also be drawn by specifying the absolute coordinate values for the center and the radius. To do so, invoke the **Circle Using Coordinates** tool from the **Circle** sub-toolbar in the **Profile** toolbar; the **Circle Definition** dialog box will be displayed, as shown in Figure 2-33. You can specify the coordinate values of the center point and radius using the options in this dialog box.

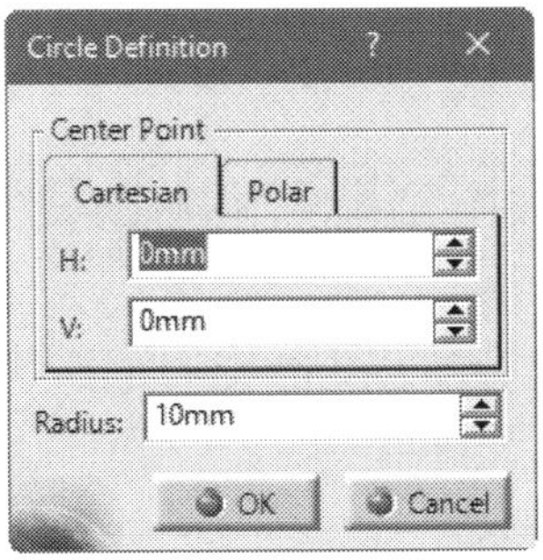

Figure 2-33 *The **Circle Definition** dialog box*

Drawing Tri-Tangent Circles

Menubar: Insert > Profile > Circle > Tri-Tangent Circle
Toolbar: Profile > Circle sub-toolbar > Tri-Tangent Circle

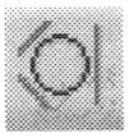

A tri-tangent circle is the one that is tangent to three sketched elements. The circle thus formed has a tangent relation with all the three elements. To draw it, you first need to draw three elements which can be lines, circles, ellipses, arcs, or any geometrical element with which a circle can form a tangent relation. Next, invoke the **Tri-Tangent Circle** tool from the **Circle** sub-toolbar in the **Profile** toolbar and select the three elements one by one; a circle tangent to all these three elements will be displayed in the geometry area. Also, you will notice that some constraints are applied to the circle. You will learn more about them in the later chapters.

Note

The location of the elements to be selected for creating a tri-tangent circle is important because its creation depends on the orientation of these selected elements. Also, the tangents are created as close as possible to the point where you click to select the elements. In case the element has to be extended to fulfill the need of the tangent relation, CATIA V5 will form a circle tangent at an apparent intersection.

Drawing Arcs

An arc is a geometric element that forms a sector of a circle or ellipse. Each arc must include at least two points. The tools to draw arcs are available in the **Circle** sub-toolbar. In CATIA V5, there are three methods to draw arcs. These methods are discussed next.

Drawing Arcs by Defining the Center Point

Menubar:	Insert > Profile > Circle > Arc
Toolbar:	Profile > Circle sub-toolbar > Arc

To draw an arc by defining its center point, invoke the **Arc** tool from the **Circle** sub-toolbar in the **Profile** toolbar; you will be prompted to specify the center point. Click to specify the arc center; you will be prompted to define the radius and start point of the arc. Move the cursor away from the center point; the preview of the circle is displayed in the geometry area. Click to specify the start point of the arc; you will be then prompted to specify the endpoint of the arc. As you move the cursor, preview of the arc is displayed, as shown in Figure 2-34. Click in the geometry area to specify the endpoint.

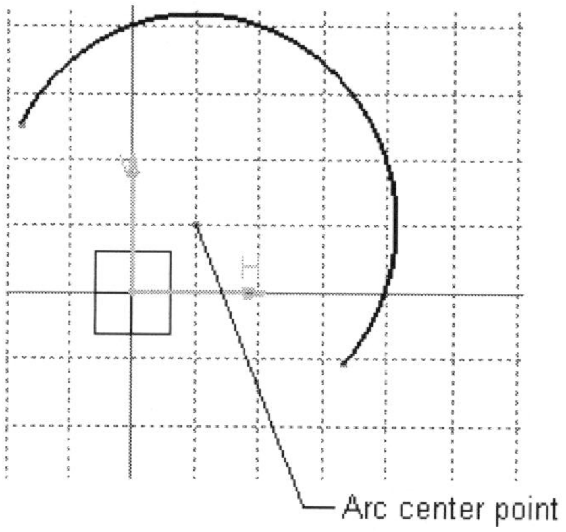

Figure 2-34 *An arc drawn by defining its center point*

Drawing Three Point Arcs

Menubar:	Insert > Profile > Circle > Three Point Arc
Toolbar:	Profile > Circle sub-toolbar > Three Point Arc

To draw a three point arc, choose the **Three Point Arc** tool from the **Circle** sub-toolbar in the **Profile** toolbar; you will be prompted to specify the start point of the arc. Click anywhere in the geometry area to specify the start point; you will be prompted to select the second point through which the arc will pass. As you move the cursor away from the first point, a dotted chord will be displayed. Click to specify the second point; you will be prompted to specify the endpoint of the arc. As you move the cursor away to specify this point, the preview of the arc is displayed. Click in the geometry area to specify the endpoint. Figure 2-35 shows first, second, and third points being selected to draw a three point arc.

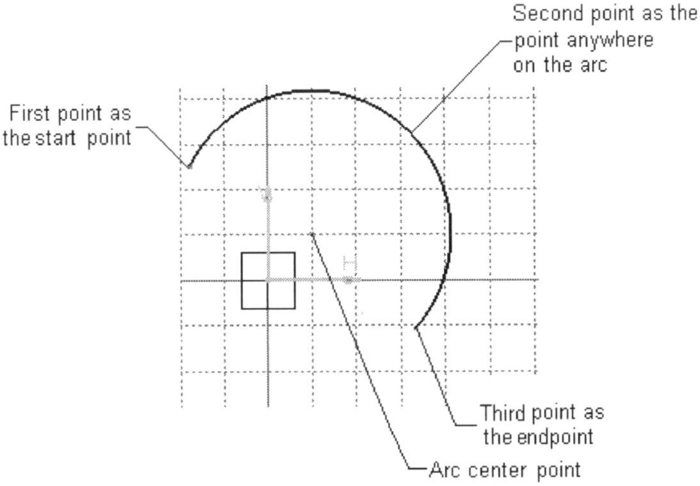

Figure 2-35 *Selecting points to draw a three point arc*

Drawing Three Point Arc Starting with Limits

Menubar:	Insert > Profile > Circle > Three Point Arc Starting With Limits
Toolbar:	Profile > Circle sub-toolbar > Three Point Arc Starting With Limits

While drawing a three point arc starting with limits, you can first specify the start point and the endpoint of the arc and then the third point anywhere on it. To draw this arc, invoke the **Three Point Arc Starting With Limits** tool from the **Circle** sub-toolbar in the **Profile** toolbar; you will be prompted to specify the start point of the arc. Click in the geometry area to specify the start point; you will be prompted to specify the endpoint of the arc. Move the cursor away from the start point and click to specify the endpoint; you will be prompted to specify the second point through which the arc will pass. As you move the cursor to specify this point, the preview of the arc will be displayed. Click to specify the point on the arc. Figure 2-36 shows the selection of first, second, and third points to draw an arc using this option.

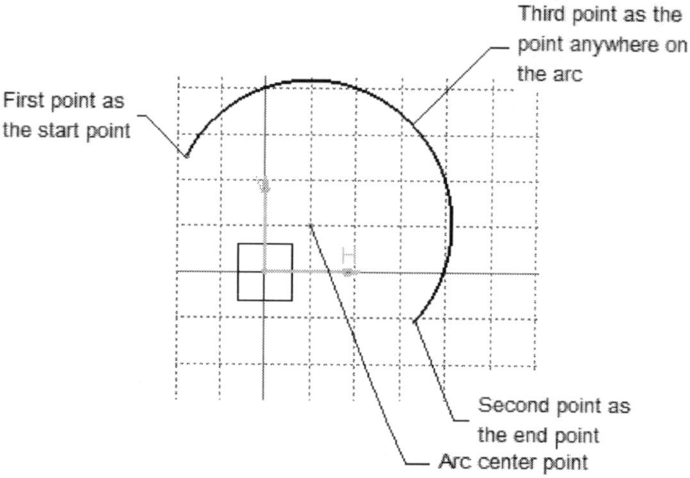

Figure 2-36 *Selecting points to draw an arc using the* ***Three Point Arc Starting With Limits*** *tool*

Drawing Profiles

Menubar:	Insert > Profile > Profile
Toolbar:	Profile > Profile

In CATIA V5, a profile is defined as a combination of continuous lines and arcs. Drawing a continuous line means that the line automatically starts at the endpoint of the previous line. A profile can be an open or a closed contour. To draw the profile, invoke the **Profile** tool from the **Profile** toolbar; the **Sketch tools** toolbar will expand and the **Line** tool will be chosen in it. Also, you will be prompted to select the start point of the profile.

Click anywhere in the geometry area to specify the start point. Next, move the cursor away from the first point; a rubber-band line will get attached to the cursor with its first point fixed to the point you had specified. Click anywhere in the geometry area to specify the endpoint of the line or the second point of the profile. Move the cursor away from the second point to draw the second line that is in continuation with the first line. You will notice that the second line originates from the endpoint of the first line. Click anywhere in the geometry area to specify the endpoint of the second line or the third point of the profile. To exit the **Profile** tool after drawing an open profile, choose the **Profile** tool again. If you draw a closed profile, you do not need to exit the **Profile** tool by choosing the **Profile** tool from the **Profile** toolbar. The tool will be automatically terminated when you specify the point to close the profile. Figure 2-37 shows an open profile.

You will notice that the expanded **Sketch tools** toolbar has three buttons: **Line**, **Tangent Arc**, and **Three Point Arc**, as shown in Figure 2-38. When you invoke the **Profile** tool, the **Line** button will be chosen by default. The profile that you have been drawing so far, using the **Profile** tool, is a combination of continuous lines. The process to draw an arc in continuation with the line using the **Profile** tool is discussed next.

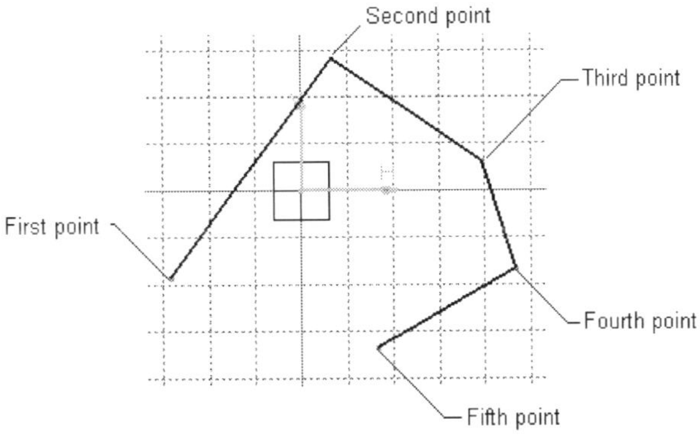

*Figure 2-37 An open profile drawn using the **Profile** tool*

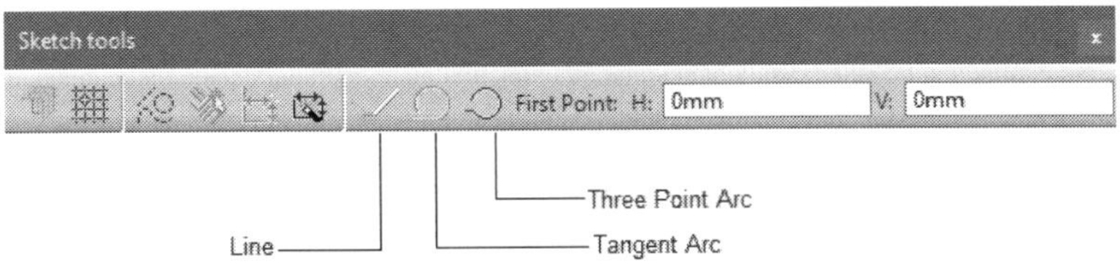

*Figure 2-38 The **Sketch tools** toolbar displayed on choosing the **Profile** tool*

Drawing a Tangent Arc using the Profile Tool

To draw a tangent arc in continuation with the line, invoke the **Profile** tool from the **Profile** toolbar. You will notice that the **Tangent Arc** button is disabled. This is because you first need to draw at least one line. After drawing the line, the **Tangent Arc** button will get enabled. Choose the **Tangent Arc** button from the expanded **Sketch tools** toolbar; the preview of the arc will be displayed in the geometry area and you will be prompted to specify the endpoint of the arc. Click in the geometry area to specify the endpoint, an arc and tangent to the line will be drawn and displayed in the geometry area. Figure 2-39 shows a tangent arc being drawn using the **Profile** tool. After drawing the arc, the **Line** tool will again be chosen in the **Sketch tools** toolbar and you will be prompted to specify the endpoint of the current line.

Alternatively, you can invoke the **Profile** tool and specify the start and end points of the line segment. While specifying the end point, hold and drag the left mouse button; an orange color rectangle gets attached to the cursor. Release the left mouse button; a tangent arc to the line will get drawn.

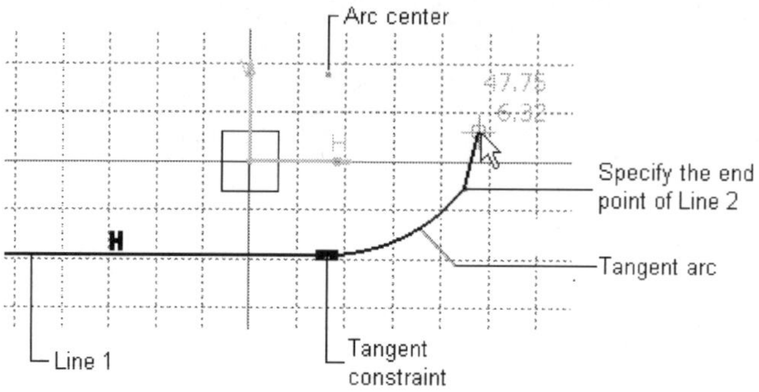

*Figure 2-39 A tangent arc being drawn using the **Profile** tool*

Drawing Three Point Arcs using the Profile Tool

To draw a three point arc using the **Profile** tool, invoke this tool from the **Profile** toolbar. You will notice the **Three Point Arc** button available in the **Sketch tools** toolbar. Draw a line using the **Profile** tool. Now, instead of specifying the third point of the profile, choose the **Three Point Arc** button from the expanded **Sketch tools** toolbar; you will be prompted to specify the second point of the arc. Remember that the first point of the three point arc is the endpoint of the line you have drawn. Click in the geometry area to specify the second point of the arc; you will be prompted to specify the last point. Move the cursor and click to specify it; the three point arc will be displayed in the geometry area. Also, the **Profile** tool will be still active and you will be prompted to specify the endpoint of the current line. You can choose the **Profile** tool again to exit the **Profile** tool or continue with the tool by specifying more points in the geometry area.

DRAWING DISPLAY TOOLS

The drawing display tools for viewing drawing elements or geometries are available in the **View** toolbar as shown in Figure 2-40. The basic tools such as **Fly Mode**, **Fit All In**, **Pan**, **Zoom In**, **Zoom Out**, **Zoom Area, Rotate**, **Normal View**, and **Hide/Show** are discussed next. You will learn about the remaining tools in the later chapters.

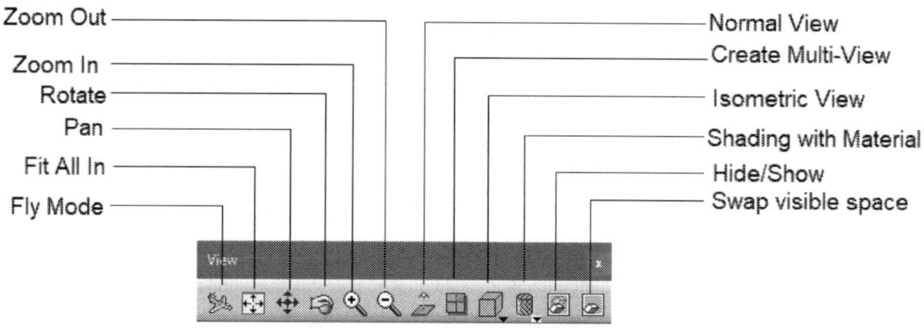

*Figure 2-40 The **View** toolbar*

Fly Mode

Toolbar: View > Fly Mode

The **Fly Mode** tool is used to view sketched elements in flying mode. On choosing this button, you will enter into the mode and notice that five new buttons are added to this toolbar namely, **Examine Mode**, **Turn Head**, **Fly**, **Accelerate**, and **Decelerate**.

On choosing the **Fly** button and by clicking the left mouse button in the graphics window, you will begin to fly forward in the chosen direction. Dragging the mouse in left, right, top, and down direction while holding the left button changes the direction accordingly. The **Accelerate** and **Decelerate** buttons are used to increase and decrease the flying speed and the **Examine Mode** button is used to exit the fly mode.

Note

*When you switch to fly mode, the **Parallel** view changes into the **Perspective** view. To switch back to the **Parallel** view, choose **View > Render Style > Parallel**.*

Fit All In

Menubar: View > Fit All In
Toolbar: View > Fit All In

The **Fit All In** tool is used to display all sketched elements or geometries in the visible space. Note, if a drawing consists of dimensions that are beyond the visible space, invoking this tool will also include them in the visible space. You will learn more about dimensions in the later chapters.

Pan

Menubar: View > Pan
Toolbar: View > Pan

The **Pan** tool is used to drag the current view in the geometry area. This option is generally used to display the elements or part of the elements that are outside the geometry area without actually changing the magnification of the current drawing. This is similar to holding a portion of the sketch and dragging it across the geometry area.

Zoom In

Menubar: View > Modify > Zoom In
Toolbar: View > Zoom In

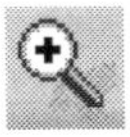

The **Zoom In** tool is used to zoom in the sketches in increments. Choose this button once to zoom in the sketch.

Zoom Out

Menubar:	View > Modify > Zoom Out
Toolbar:	View > Zoom Out

The **Zoom Out** tool is used to zoom out of the sketch in increments. Choose this button once from the **View** toolbar to zoom out of the sketch.

Tip

*You can also dynamically zoom in or zoom out an entity by selecting the **Zoom In Out** option from the **View** menu. To zoom in using this option, press and hold the left mouse button and then drag the mouse upward. To zoom out of the sketches, press and hold the left mouse button and then drag the mouse downward. The tool is automatically terminated once you release the mouse button.*

Zoom Area

Menubar:	View > Zoom Area

The **Zoom Area** tool is used to define an area which is to be magnified and viewed in the available geometry area. The area is defined using the two diagonal points of a rectangular box in the geometry area. Choose the **Zoom Area** option from the **View** menu and press and hold the left mouse button to specify the first corner point. Then, drag the mouse to specify the other corner point of the box. The area that is enclosed inside the window will be magnified and displayed.

Normal View

Menubar:	View > Modify > Normal View
Toolbar:	View > Normal View

The **Normal View** tool is used to orient the view normal to the view direction. If, the current view is already normal to the view direction and you choose the **Normal View** button from the **View** toolbar; the viewing plane will be reversed. In other words, on choosing this button, if the front plane is the current viewing plane; the back plane will become active for viewing.

Note

*1. By default, whenever you invoke the **Sketcher** workbench without defining any particular orientation, the positive horizontal reference direction will point toward the right of the geometry area and also towards its upper area. If you choose the **Normal View** button, the direction of the horizontal reference will be reversed by 180-degree. This means that the positive horizontal reference direction will point toward the left of the geometry area. Note that the vertical reference direction remains unchanged.*

*2. While working in the **Sketcher** workbench, you can choose the **Normal View** button to orient the sketch normal to the sketching plane.*

Create Multi View

Toolbar: View > Create Multi-View

The **Create Multi-View** tool is used to split the geometry area into four viewports. These viewports can be used to display different views of the model. To restore the single viewport configuration, choose this tool again.

Hide/Show Geometric Elements

Menubar: View > Hide/Show > Hide/Show
Toolbar: View > Hide/Show

The **Hide/Show** tool is used to hide sketcher elements or geometric elements from the current display. To do so, invoke this tool by choosing the **Hide/Show** button from the **View** toolbar; you will be prompted to select an element. Click on the element to be hidden from the geometry area. You will notice that the selected element is no longer visible.

Swap Visible Space

The hidden elements are stored in a space different from the current display space. To view the space where the hidden elements are stored, invoke the **Swap visible space** tool from the **View** toolbar; you will notice that the background of the current geometry area changes to green and only the hidden elements are visible. Invoke the **Hide/Show** tool and select the hidden elements to be redisplayed in the visible space. To return to the geometry area, choose the **Swap visible space** button again. Note that when you hide an element, only its display is turned off, but it still participates in the feature creation.

Tip
If you draw a sketch in the space containing the hidden elements, it will not be visible there. It will only be displayed after you return to the visible geometry area.

APPLYING DIMENSIONAL CONSTRAINTS

Menubar: Insert > Constraint > Constraint Creation > Constraint
Toolbar: Constraint > Constraint Creation sub-toolbar > Constraint

After applying the geometric constraints, you need to apply the dimensional constraints to fully define the sketches. It is recommended that you apply the dimensional constraints to the sketches using the **Constraint** tool. To invoke this tool, choose the down arrow on the right of the **Constraint** tool; the **Constraint Creation** sub-toolbar will be displayed, as shown in Figure 2-41. Now, choose the **Constraint** tool from this sub-toolbar.

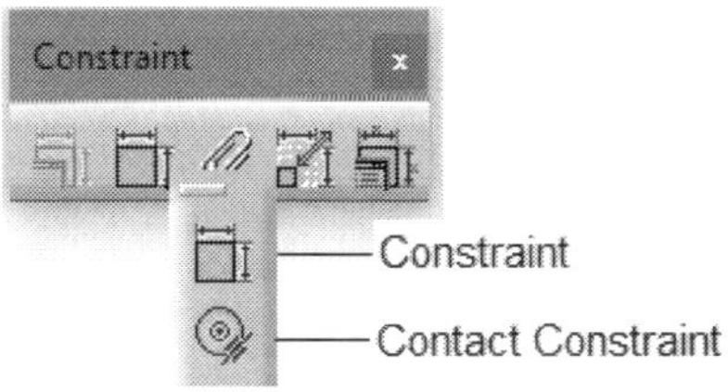

Figure 2-41 *The* ***Constraint*** *toolbar with the* ***Constraint Creation*** *sub-toolbar*

When you choose this tool, you will be prompted to select an element to be constrained. Select the element that you need to dimension; a dimension will be attached to the cursor. Click anywhere in the geometry area to place the dimension. Using this tool, you can apply various types of dimensions such as linear dimensions, radius dimensions, diameter dimensions, and so on. These dimensions are discussed next.

Applying Linear Dimension to a Line or between Two Points

You can apply a linear dimension to a line or between two points using the **Constraint** tool. To do so, choose the **Constraint** tool and select a line; a linear dimension will be attached to the cursor. Move the cursor and click in the geometry area at a location where you need to place the dimension; a linear dimension will be placed in the geometry area.

To apply a linear dimension between two points, choose the **Constraint** tool, and then select two points from the geometry area. Next, right-click to display the contextual menu, as shown in Figure 2-42.

Choose the **Horizontal Measure Direction** option to place the linear dimension along the horizontal axis, or the **Vertical Measure Direction** option to place it along the vertical axis. On doing so, a linear dimension will be attached to the cursor. Click in the geometry area to place the dimension. Figure 2-43 shows the line to be selected for dimensioning and Figure 2-44 shows the resultant linear dimension. Figures 2-45 shows the points to be selected and Figure 2-46 shows the resultant linear dimensions.

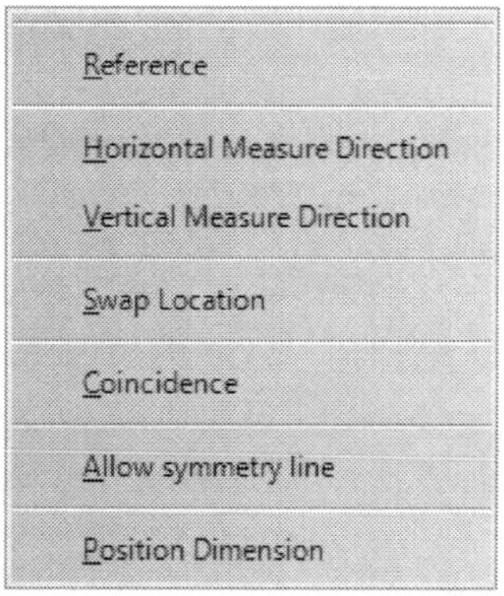

***Figure 2-42** The contextual menu for the **Constraint** tool*

Tip
*If you choose the **Reference** option from the contextual menu, the selected dimension will be displayed as a reference dimension and will be displayed inside parenthesis. A reference dimension does not drive the geometry but it is the geometry which drives the reference dimension.*

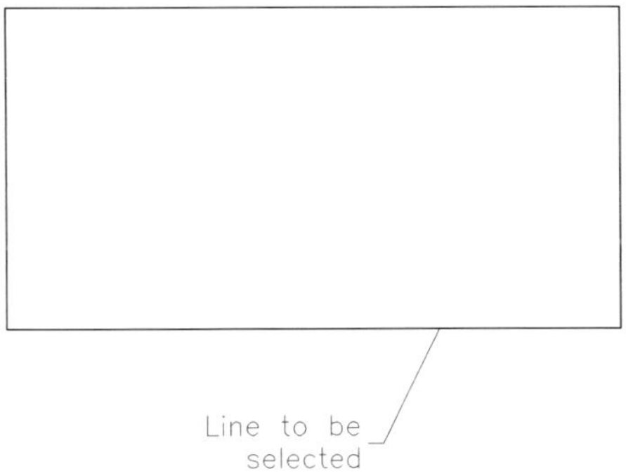

Figure 2-43 Line to be selected for dimensioning

Figure 2-44 Resultant linear dimension

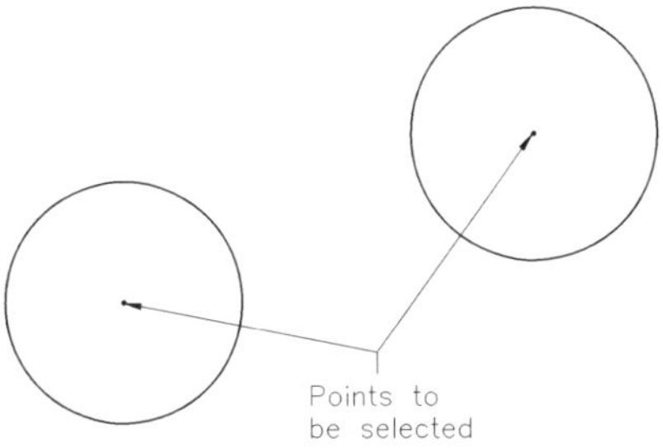

Figure 2-45 Points to be selected for dimensioning

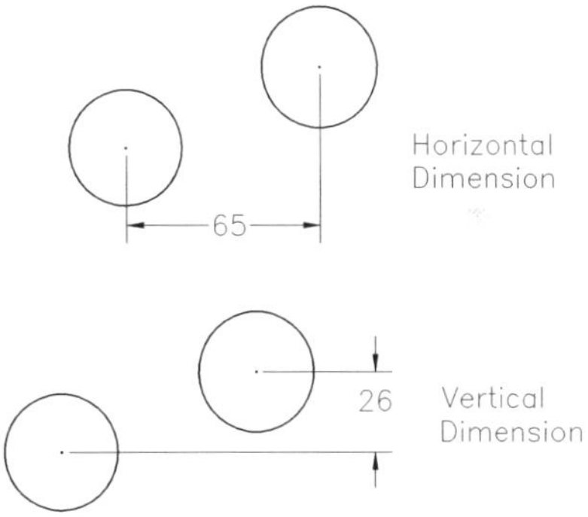

Figure 2-46 Resultant linear dimension

Dimensioning an Inclined Line

By default, whenever you select an inclined line, the aligned dimension is applied to it. You can also apply a horizontal or vertical dimension to it. To apply a horizontal dimension to an inclined line, choose the **Constraint** tool and select the line; an aligned dimension is attached to the cursor. Right-click to invoke the contextual menu. Choose the **Horizontal Measure Direction** or **Vertical Measure Direction** option depending on whether you want to place the horizontal or vertical dimension. To switch back to the aligned dimensioning, right-click in the geometry area again before placing the dimension, and choose the **No Measure Direction** option from the contextual menu.

Dimensioning an Arc or a Circle

To dimension an arc or a circle, choose the **Constraint** tool and select the arc or the circle that you need to dimension. By default, the diameter dimension is applied to circles and the radius dimension is applied to arcs. Move the cursor and click in the geometry area to place the dimension.

Note

1. While applying dimensions, you can convert a radius dimension into a diameter dimension. To do so, before placing the dimension, invoke the contextual menu by right-clicking. Choose the ***Diameter*** *option from the contextual menu. Similarly, you can convert a diameter dimension into a radius dimension.*

*2. To convert a radius dimension already applied to an arc into a diameter dimension, double-click on the radius dimension; the **Constraint Definition** dialog box is displayed. Select the **Diameter** option from the **Dimension** drop-down list. You will notice that the radius dimension will be replaced by the diameter dimension. Choose the **OK** button from the **Constraint Definition** dialog box to apply the change. Similarly, you can change a diameter dimension into a radius dimension. You will learn more about applying dimensions using the **Constraint Definition** dialog box later in this chapter.*

Applying Angular Dimensions

To apply an angular dimension, choose the **Constraint** tool and select the first line; a linear dimension is attached to the cursor. Select the second line; an angular dimension is attached to the cursor. Next, move the cursor and place the angular dimension. Remember that the type of angular dimension depends on its placement point. Figures 2-47 through 2-50 show the angular dimensions placed at different locations.

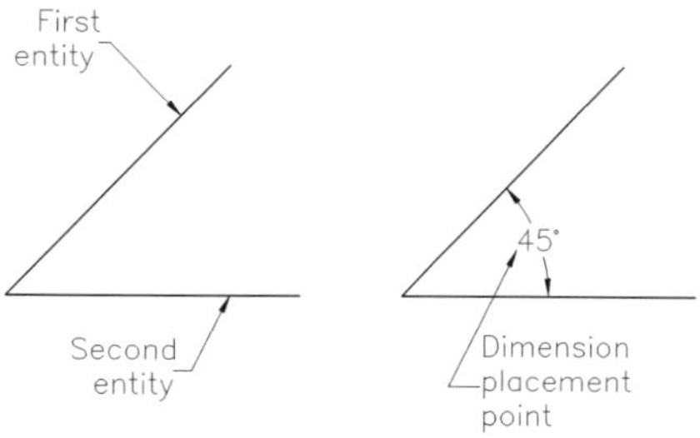

***Figure 2-47** The angular dimension placed according to the placement point*

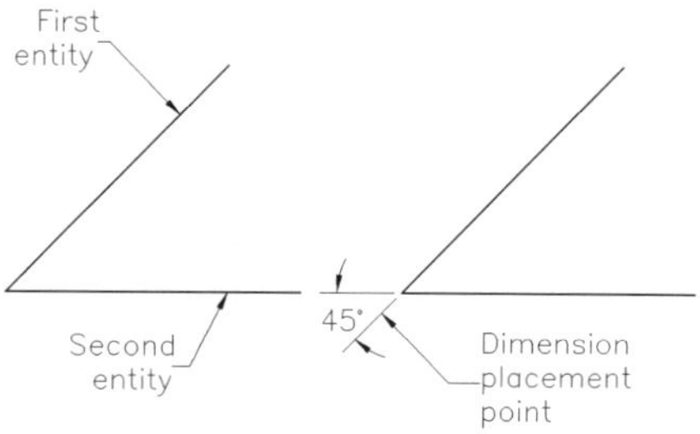

***Figure 2-48** The angular dimension placed according to the placement point*

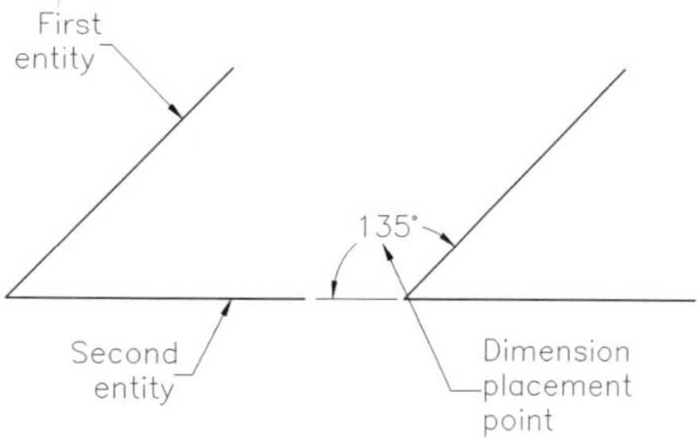

***Figure 2-49** The angular dimension placed according to the placement point*

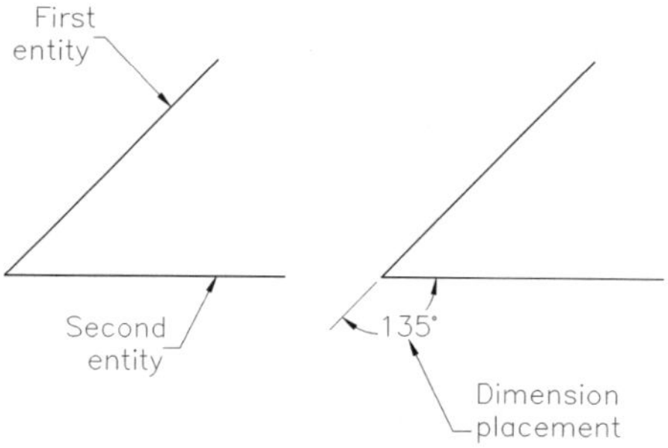

***Figure 2-50** The angular dimension placed according to the placement point*

Applying Linear Diameter Dimensions

Linear diameter dimensions are applied to the sketches of the features that need to be created by revolving the sketch using the **Shaft** or **Groove** tool. Figure 2-51 shows a component created using the revolved feature. Note that the sketch must have a center line drawn using the **Axis** tool, around which the sketch will be revolved. To apply a linear diameter dimension, choose the

Constraint tool and then select the entity that you need to dimension; a linear dimension will be attached to the cursor. Next, select the centerline that was drawn using the **Axis** tool; a linear dimension will be attached to the cursor. Right-click in the geometry area to invoke the contextual menu. Choose the **Radius / Diameter** option from the contextual menu; a linear diameter will be attached to the cursor. Now, move the cursor and place the dimension. Figure 2-52 shows the element and the centerline to be selected and the resultant linear diameter dimension.

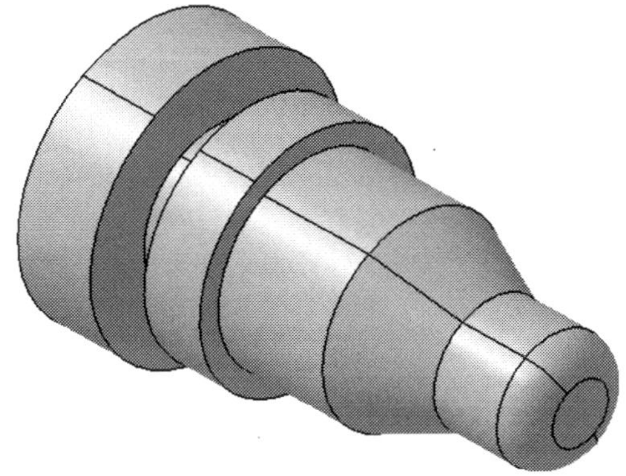

Figure 2-51 Model created by revolving a sketch around the horizontal centerline

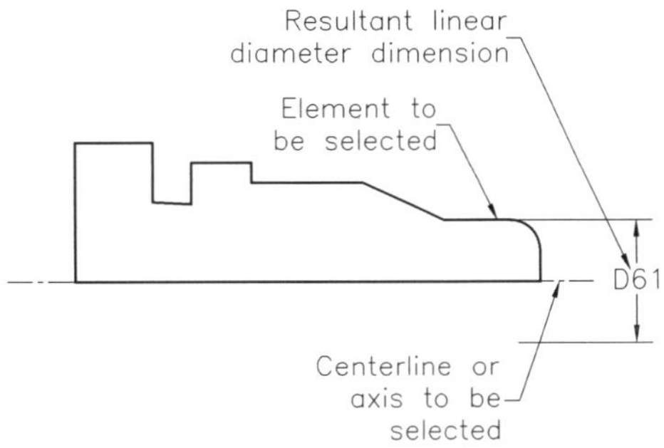

Figure 2-52 Element and centerline selected and the resultant linear diameter dimension

Modifying Dimensions after Placing Them

As discussed earlier, CATIA V5 is a parametric software. Therefore, you can change the dimensions and modify the design at any stage of the design cycle. In the previous section, you learned to place various types of dimensions. Note that the default value placed while dimensioning the entity may not be the required value. To modify the dimension value, double-click on it; the **Constraint Definition** dialog box will be displayed, refer to Figure 2-53. Set the value of the dimension in the **Value** spinner and choose the **OK** button from the **Constraint Definition** dialog box.

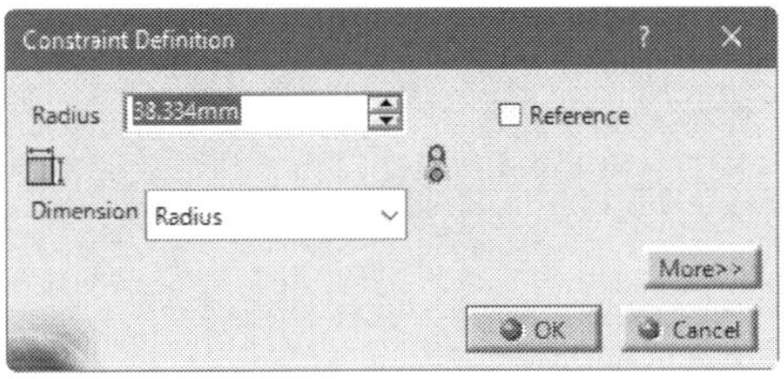

*Figure 2-53 The **Constraint Definition** dialog box*

Tip

*If you choose the **More** button from the **Constraint Definition** dialog box, the dialog box will expand and the **Supporting Elements** area will be displayed. The **Type** column of this area displays the type of element on which the dimensional constraint is applied. The **Component** column displays the name of the element or the elements on which it is applied. The **Status** column displays the status of the constraint. You will learn more about the status of the constraints later in this chapter.*

*The name of the dimension is displayed in the **Name** edit box provided above the **Supporting Elements** area. You can modify the name of the dimension using this edit box.*

CREATING BASE FEATURES BY EXTRUSION

Menubar:	Insert > Sketch-Based Features > Pad
Toolbar:	Sketch-Based Features > Pad sub-toolbar > Pad

The **Pad** tool is one of the most widely used tool to create the base features. To create a base feature using this tool, draw the sketch and exit the **Sketcher** workbench. Then, choose the down arrow on the right of the **Pad** tool in the **Sketch-Based Features** toolbar; the **Pad** sub-toolbar will be displayed, as shown in Figure 2-54. Select the sketch and then choose the **Pad** tool from the **Pad** sub-toolbar; the **Pad Definition** dialog box will be displayed, as shown in Figure 2-55 and you will be prompted to enter the required data to modify the pad. Also, the name of the selected sketch will be displayed in the **Selection** display box and the preview of the pad feature will be displayed in the geometry area.

Set the value of the depth of extrusion in the **Length** spinner. You can also define the extrusion depth dynamically in the geometry area. To do so, move the cursor close to **LIM1** in the default preview of the extrude feature. When the Limit drag handle (hand cursor) is displayed, refer to Figures 2-56 and 2-57, press and hold the left mouse button and drag the cursor; the depth of the extrusion will get dynamically defined. Then, choose the **OK** button from the **Pad Definition** dialog box. Figure 2-58 shows the model after creating the base feature by extruding the sketch.

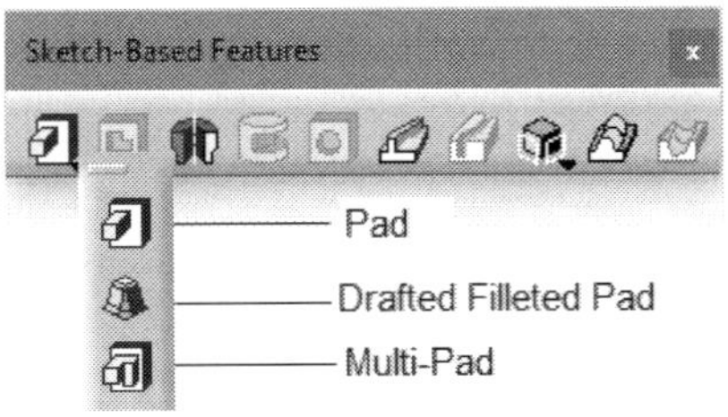

***Figure 2-54** The **Sketch-Based Features** toolbar with the **Pad** sub-toolbar*

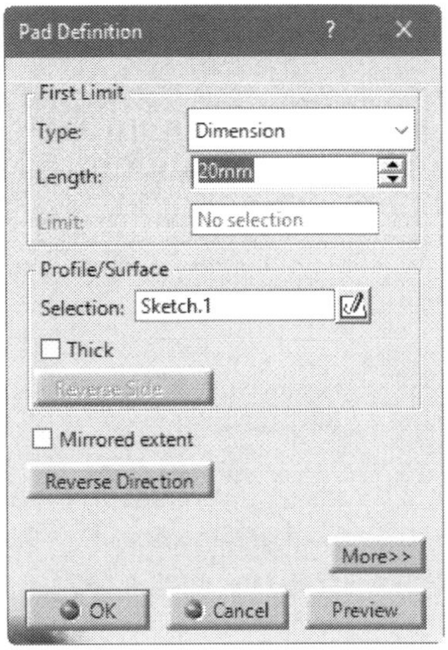

***Figure 2-55** The **Pad Definition** dialog box*

You can extrude the sketch feature about the sketch plane. To do so, choose the **More** button; the **Pad Definition** dialog box will expand. Enter the depth of extrusion for the second direction in the **Length** spinner of the **Second Limit** area; the changes will be displayed in the geometric area.

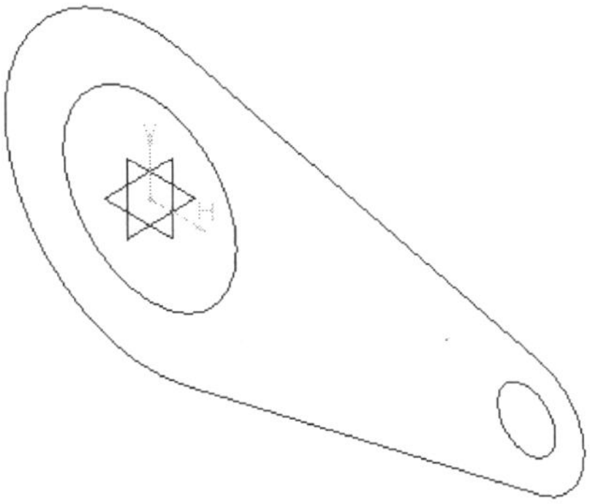

***Figure 2-56** The sketch after exiting the **Sketcher** workbench*

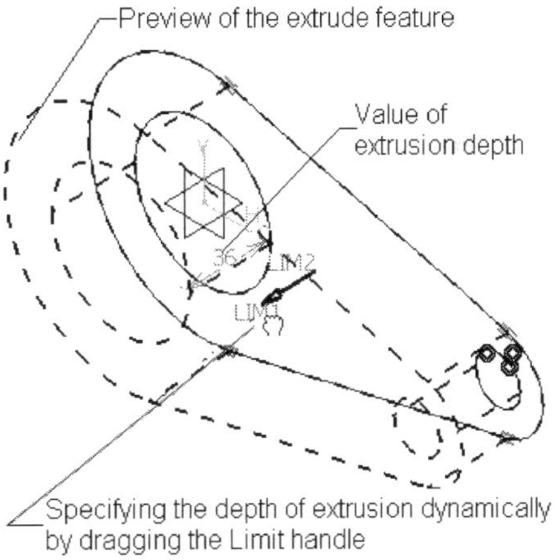

***Figure 2-57** Dynamically dragging the Limit drag handle to specify the depth of extrusion*

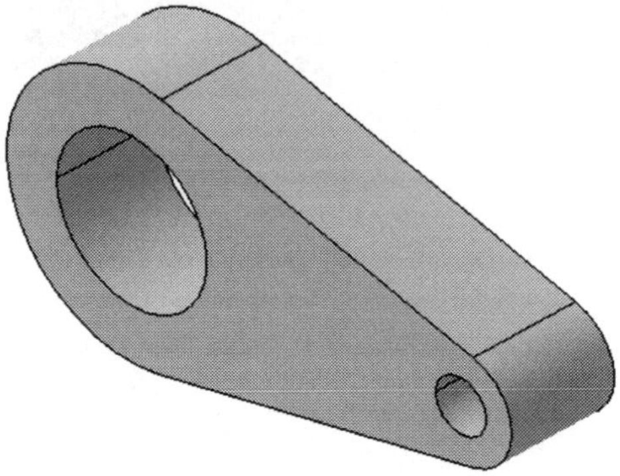

Figure 2-58 *The model after extruding the sketch*

GENERATING DRAWING VIEWS

After creating a solid model or an assembly, you need to generate its drawing views. Drawing views are the two-dimensional (2D) representations of a solid model or an assembly. CATIA V5 provides you with the **Drafting** workbench which is a specialized environment for generating 2D drawing views. This workbench provides all tools required to generate drawing views, modify, and apply dimensions and add annotations. In other words, you can get the final shop floor drawing using this workbench of CATIA V5. There are two types of drafting techniques in CATIA V5: Generative drafting and Interactive drafting. Generative drafting is a technique of generating the drawing views using a solid model or an assembly model. Interactive drafting is a technique, in which the sketcher tools are used to draw the 2D drawing views.

TUTORIALS

Tutorial 1

Draw the sketch shown in Figure 2-59. Extrude the sketch by 60 units and then generate a drawing with the orthographic and isometric views of the model. The isometric view of the model is shown in Figure 2-60. You do not need to dimension the drawing.

(Expected time: 30 min)

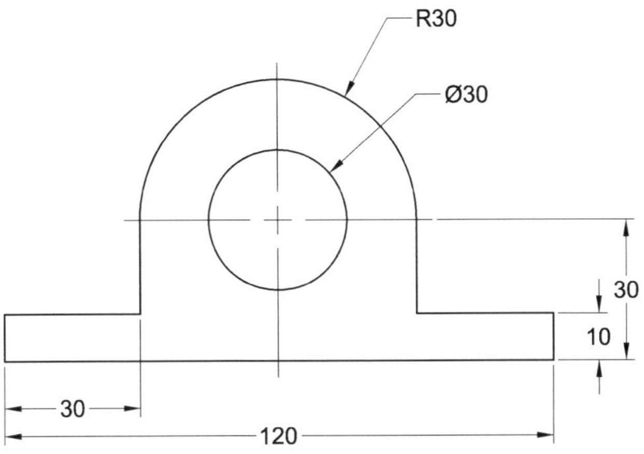

Figure 2-59 The sketch for the solid model

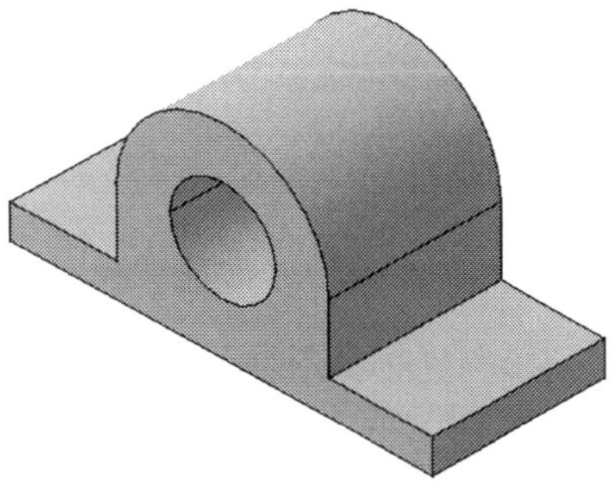

Figure 2-60 The solid model for Tutorial 1

The following steps are required to complete this tutorial:

a. Start CATIA V5 and then start a new Part file.
b. Draw the sketch of the model using the **Profile** and **Circle** tools.
c. Extrude the sketch upto a distance of 60 mm using the **Extrude** tool.
d. Save the model and then generate its drawing views.

Starting a New Part File

1. Start CATIA V5 by double-clicking on the shortcut icon of **CATIA V5-6R2022** on the desktop of your computer; a new file, **Product1** is started.

2. Next, choose **Start > Mechanical Design > Part Design** from the menu bar; the **New Part** dialog box is displayed, as shown in Figure 2-61.

3. Enter **C02_Tut1** as the name of the file in the **Enter part name** edit box.

4. Select the **Enable hybrid design** check box from the **New Part** dialog box if not selected.

*Figure 2-61 The **New Part** dialog box*

5. Choose the **OK** button; a new file in the **Part Design** workbench is started.

6. Choose the **Sketch** tool from the **Sketcher** toolbar and then select the yz plane from the Specification tree as the sketching plane; the **Sketcher** workbench screen is displayed, as shown in Figure 2-62.

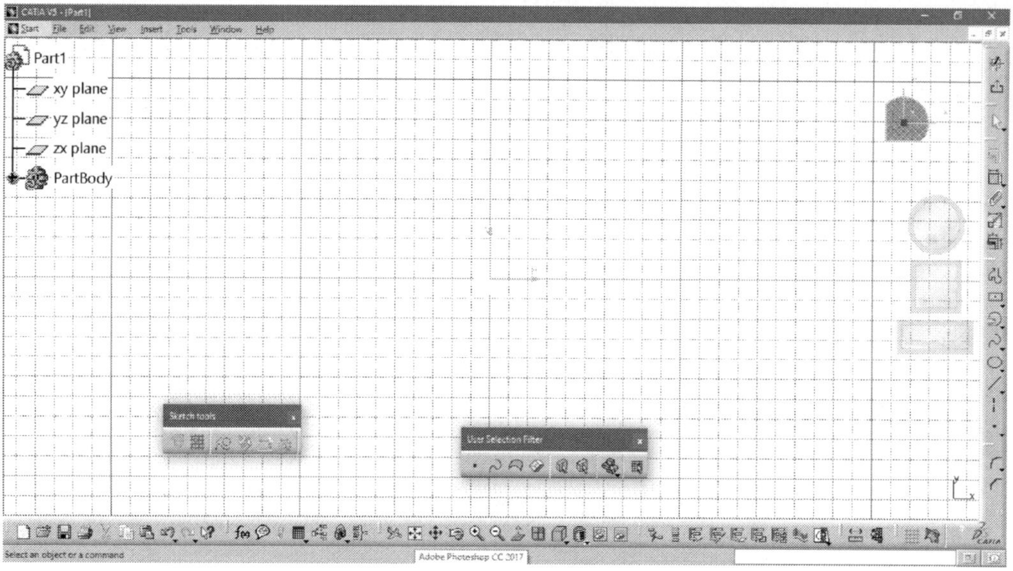

*Figure 2-62 The **Sketcher** workbench screen*

In this tutorial, you need to draw the sketch in two parts: first draw the outer loop and second draw the inner circle in the Sketcher workbench.

Drawing the Outer Loop of the Sketch

In this section, you need to draw the outer loop of the sketch using the **Profile** tool. In the sketch, the lower left corner of the sketch will be coincident with the origin. The resulting sketch will be in the first quadrant.

1. Choose the **Profile** tool from the **Profile** toolbar.

2. Choose the **Snap to Point** button from the **Sketch tools** toolbar if not chosen.

3. Move the cursor to the location whose coordinates are (0,0) (at the origin) and click to specify the start point of the line. Note that the coordinates of the point are displayed on the cursor.

4. Move the cursor horizontally toward right, the line turns blue. Click to specify the endpoint of the line where the coordinates are 120,0.

Note

The constraints that get automatically applied to the sketch drawn will be explained in the later chapters.

5. Move the cursor vertically upward and click to specify the endpoint of the line when the value of the coordinates is 120,10.

6. Move the cursor horizontally toward the left and click to specify the endpoint of the line when the value of the coordinates is 90,10.

7. Move the cursor vertically upward and click to specify the endpoint of the line when the value of the coordinates is 90,30.

   After drawing these four lines, you need to draw a tangent arc using the **Tangent Arc** tool.

8. Choose the **Tangent Arc** tool from the **Sketch tools** toolbar to switch to the **Tangent Arc** mode. 

Note

*If the **Sketch tools** toolbar is not displayed in the graphics area then right-click on any of the available toolbars and select the **Sketch tools** from the shortcut menu displayed, the **Sketch tools** toolbar will become available in the graphics area.*

9. Move the cursor to the location whose coordinates are 30,30 and specify the endpoint of the tangent arc at that location. Note that, after specifying the endpoint of the tangent arc, the **Line** mode is activated and a line is attached to the cursor again.

Note

While drawing an arc, you will notice that the inferencing lines are displayed in the geometry area. These lines indicate the relations that they can have with other entities.

10. Move the cursor vertically downward and click to specify the endpoint of the line when the value of the coordinates is 30,10.

11. Move the cursor horizontally toward the left and click to specify the endpoint of the line when the value of the coordinates is 0,10.

12. Move the cursor vertically downward and click to specify the endpoint of the line such that the endpoint is coincident to the start point of the first line.

    The sketch after drawing the outer loop is shown in Figure 2-63. In this figure, the constraints are hidden for better visualization.

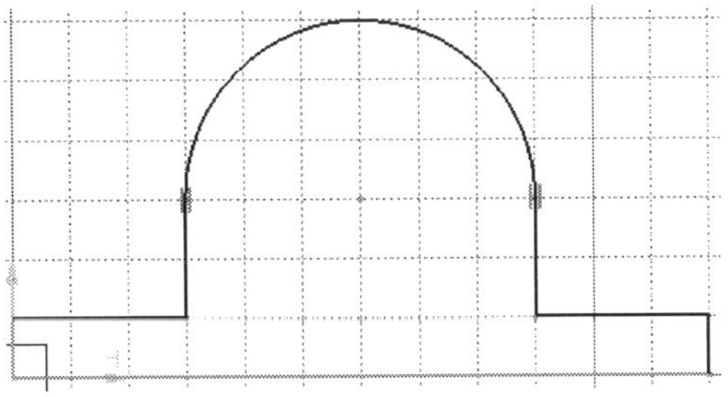

Figure 2-63 *The sketch after drawing the outer loop*

Drawing the Inner Loop of the Sketch

The inner loop of the sketch consists of a circle. You need to draw the circle using the **Circle** tool such that it is concentric to the arc of the outer loop.

1. Choose the **Circle** button from the **Profile** toolbar.

2. Move the cursor to the center point of the circular arc and click to specify the center point of the circle.

3. Enter **15** as the radius value of the circle in the **R** edit box provided in the **Sketch tools** toolbar and press Enter.

   The final sketch after drawing the inner loop is shown in Figure 2-64. Note that in this figure, the display of constraints is turned on.

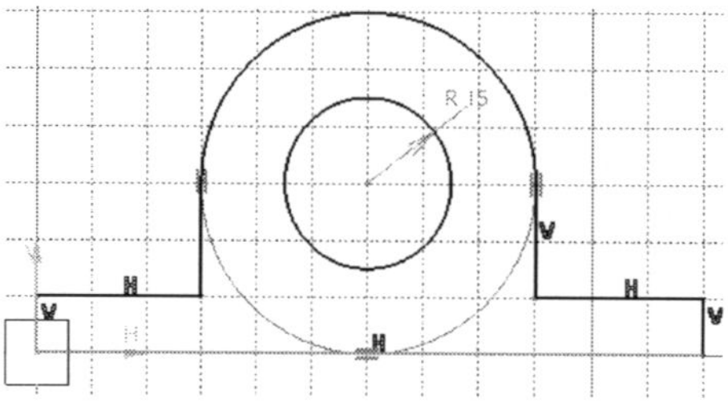

Figure 2-64 *The final sketch for Tutorial 1*

Extruding the Sketch

Next, you need to extrude the sketch.

1. Choose the **Pad** button from the **Sketch-Based Features** toolbar; the **Pad Definition** dialog box is displayed and a preview of the extruded feature is displayed in the geometry area.

2. Set **60** as the value in the **Length** spinner of the **Pad Definition** dialog box, and then choose **OK** from this dialog box.

3. Click once in the geometry area to remove the newly created base feature from the current selection set.

4. Orient the view of the model to the isometric view using the **Isometric View** tool from the **View** toolbar. The model after creating the extruded feature is shown in Figure 2-65.

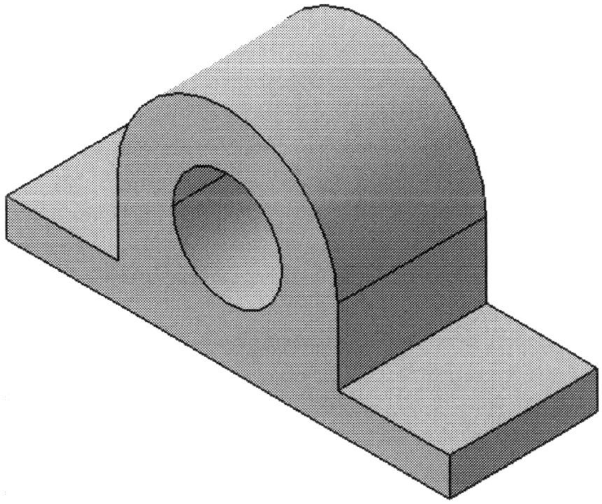

Figure 2-65 *The final model for Tutorial 1*

Saving the Model

After completing the model, you need to save it.

1. Choose the **Save** button from the **Standard** toolbar to invoke the **Save As** dialog box. Using this dialog box, create a folder named ***CATIA*** inside the *C:* drive. Then create the folder ***c02*** inside the CATIA folder.

2. Next, choose the **Save** button; the file is saved at *C:\CATIA\c02.*

Generating Drawing Views of Model

1. Choose **Start > Mechanical Design > Drafting** from the menubar to display the **New Drawing Creation** dialog box.

2. Choose the **OK** button from the **New Drawing Creation** dialog box: a new file is started in the **Drafting** workbench.

3. Choose the **Front View** tool from the **Projections** sub-toolbar in the **Views** toolbar; you are prompted to select a reference plane on a 3D geometry.

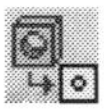

4. Choose **Window > c02tut1.CATPart** from the menu bar; the part file is displayed.

5. Select the **yz** plane (of the front planar face) from the Specification tree; the drawing file is invoked.

   A preview of the front view is displayed and a knob appears on the view to set the orientation of the front view.

6. Move the cursor on the frame that shows the preview of the front view; the cursor is replaced by a hand cursor. Press and hold the left mouse button and drag the cursor close to the lower left corner of the drawing sheet.

7. Click anywhere on the drawing sheet to generate the front view.

8. Next double click on the **Projection View** tool from the **Projections** sub-toolbar in the **Views** toolbar then place the cursor on the front view; a preview of the projected view is attached to the cursor.

9. Similarly, taking the front view as the parent view, generate the drawing of right-side view.

10. Choose the **Isometric View** tool from the **Projections** sub-toolbar in the **Views** toolbar; you are prompted to select a reference plane on a 3D geometry. 

11. Choose **Window > c02tut1.CATPart** from the menu bar; the part file is displayed.

12. Select the front face of the model from the geometry area; the drawing sheet is displayed. Also, a preview of the isometric view of the model is displayed on the drawing sheet along with a knob which can be used to orient the isometric view.

13. Click anywhere on the drawing sheet to generate the isometric view. The drawing sheet after generating the isometric view is shown in Figure 2-66.

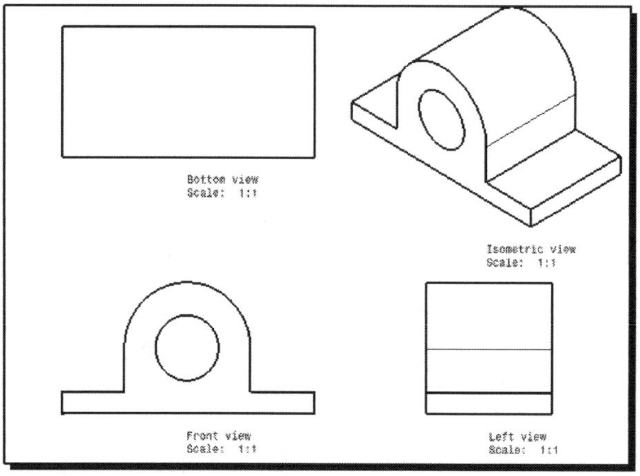

***Figure 2-66** Drawing sheet after generating the drawing views*

Tutorial 2

Draw the sketch shown in Figure 2-67. Extrude the sketch by 20 units and then generate a drawing with the orthographic and isometric views of the model. The isometric view of the model is shown in Figure 2-68. You do not need to dimension the drawing.

(Expected time: 30 min)

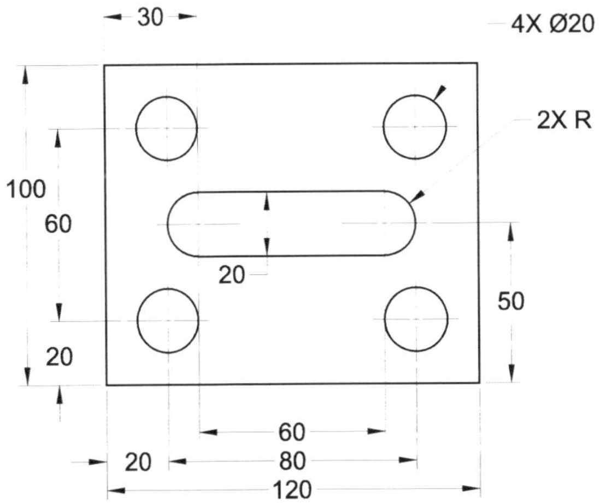

Figure 2-67 The sketch for the solid model

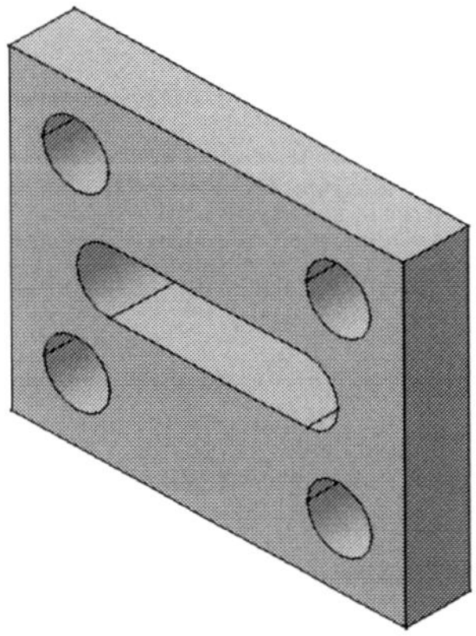

Figure 2-68 The solid model for Tutorial 2

The following steps are required to complete this tutorial:

a. Start a new Part file.
b. Draw the sketch of the model using the **Rectangle**, **Profile**, and **Circle** tools.
c. Extrude the sketch upto a distance of 20 mm using the **Extrude** tool.
d. Save the model and then generate its drawing views.

Starting a New Part File

If you are starting a new session of CATIA, close the default Product file.

1. Choose **File > New** from the menu bar; the **New** dialog box is displayed, as shown in Figure 2-69. Alternatively, choose the **New** tool from the **Standard** toolbar to invoke the **New** dialog box.

2. In this dialog box, select **Part** from the **List of Types** list box and then choose the **OK** button; the **New Part** dialog box is displayed.

3. Enter **c02tut2** as the name of the file in the **Enter part name** edit box and choose the **OK** button from the **New Part** dialog box; the new **Part** file opens in the **Part Design** workbench.

4. Choose the **Sketch** tool from the **Sketcher** toolbar and then select the yz plane from the Specification tree as the sketching plane; the **Sketcher** workbench is invoked.

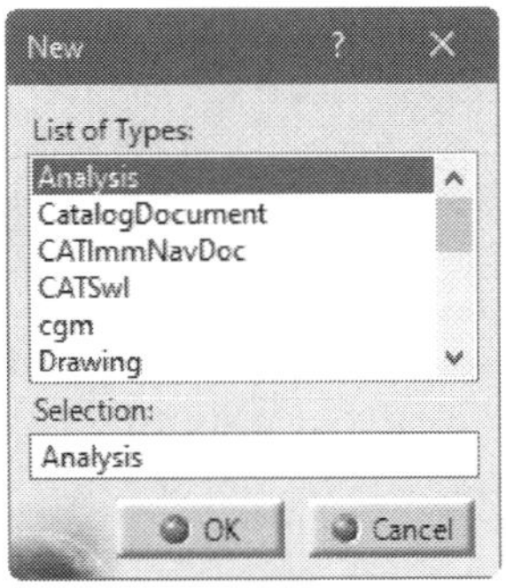

*Figure 2-69 The **New** dialog box*

In this tutorial, you need to draw the sketch in two parts.

Initially, you need to draw the outer loop of the sketch, a rectangle, and then the inner loops of the sketch, which consist of four circular holes and an elongated hole. First, draw an elongated hole using the **Profile** tool and then the four holes using the **Circle** tool.

Drawing the Outer Loop of the Sketch

In this section, you need to draw the outer loop of the sketch using the **Rectangle** tool.

1. Choose the **Rectangle** tool from the **Profile** toolbar.

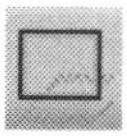

2. Move the cursor to the location whose coordinates are -60,-50 and click to specify the lower left corner of the rectangle.

3. Move the cursor to the location whose coordinates are 60,50 and click to specify the upper right corner of the rectangle. Figure 2-70 shows the outer loop of the sketch drawn using the **Rectangle** tool.

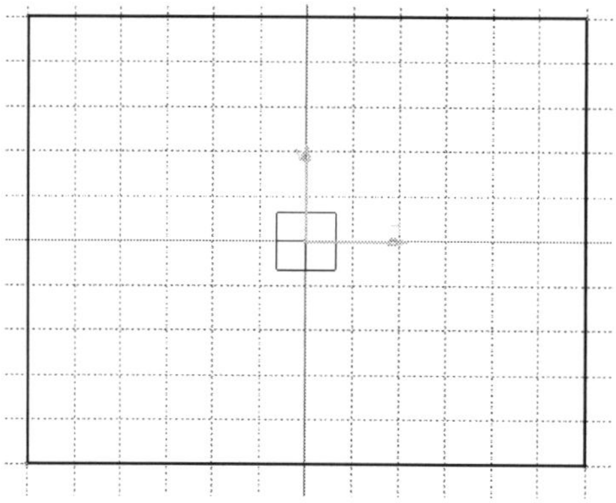

Figure 2-70 *The outer loop of the sketch*

Drawing the Inner Loop of the Sketch

After drawing the outer loop of the sketch, Now, you need to draw its inner loop.

1. Choose the **Profile** tool from the **Profile** toolbar.

2. Move the cursor to the location whose coordinates are -30,10 and click to specify this point as the start point of the line.

3. Move the cursor horizontally toward the right and click to specify the endpoint of the line where the coordinates are 30,10.

   Next, you need to draw a tangent arc by switching over to the **Tangent Arc** option using the **Sketch tools** toolbar.

4. Choose the **Tangent Arc** tool from the **Sketch tools** toolbar to switch over to the **Tangent Arc** mode.

5. Move the cursor to the location whose coordinates are 30,-10 and click to specify this point as the endpoint of the tangent arc.

Note that after specifying the endpoint of the tangent arc, the **Line** mode is activated and the line is attached to the cursor again.

6. Move the cursor to the location whose coordinates are -30,-10 and click to specify the endpoint of the line.

7. Choose the **Tangent Arc** tool from the **Sketch tools** toolbar to switch to the **Tangent Arc** mode.

8. Move the cursor to the start point of the first horizontal line and then click to specify the endpoint of the arc when it snaps to the start point.

The sketch of the elongated hole is shown in Figure 2-71. In this figure, the constraints are hidden for better visualization.

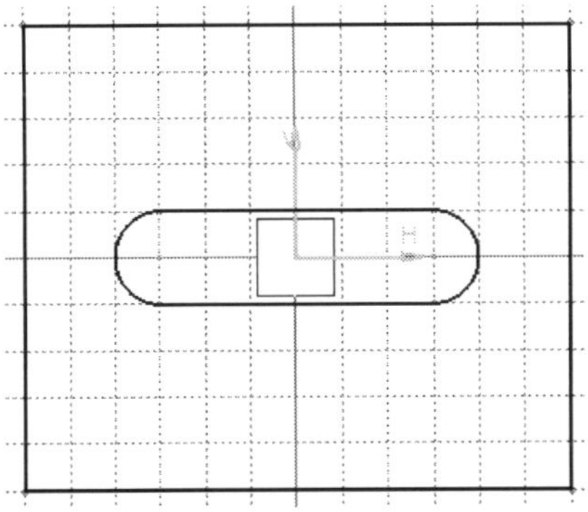

Figure 2-71 *The sketch of the elongated hole*

9. Choose the **Circle** tool from the **Profile** toolbar.

10. Move the cursor to the location whose coordinates are 40,30 and click to specify the center point of the circle.

11. Enter **10** as the radius value for the circle in the **R** edit box of the **Sketch tools** toolbar and press Enter; you will notice that a radius dimension is displayed attached to the circle.

12. Choose the **Circle** tool from the **Profile** toolbar.

13. Move the cursor to the location whose coordinates are 40,-30 and click to specify the center point of the circle.

14. Enter **10** as the radius value for the circle in the **R** edit box of the **Sketch tools** toolbar and press Enter.

15. Similarly, draw the other two circles. The coordinates of the center point of the other two circles are -40,30 and -40,-30, respectively. The final sketch with the display of constraints turned on is shown in Figure 2-72.

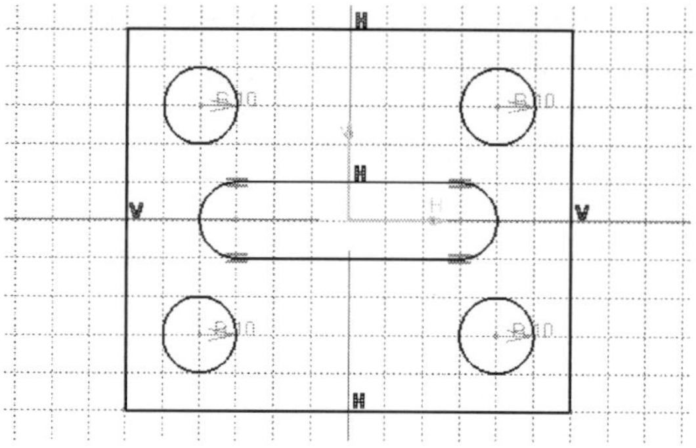

Figure 2-72 *The final sketch*

Extruding the Sketch

Next, you need to extrude the sketch.

1. Choose the **Pad** button from the **Sketch-Based Features** toolbar; the **Pad Definition** dialog box is displayed and a preview of the extruded feature is displayed in the geometry area.

2. Set **20** as the value in the **Length** spinner of the **Pad Definition** dialog box, and then choose **OK** button from this dialog box.

3. Click once in the geometry area to remove the newly created base feature from the current selection set.

4. Orient the view of the model to the isometric view using the **Isometric View** tool from the **View** toolbar. The model after creating the extruded feature is shown in Figure 2-73.

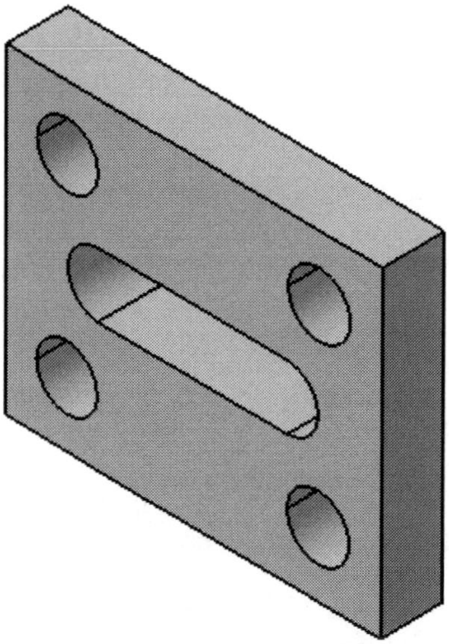

Figure 2-73 *The final model for Tutorial 2*

Saving the Model

After completing the model, you need to save it.

1. Choose the **Save** button from the **Standard** toolbar to invoke the **Save As** dialog box. Using this dialog box, create a folder named ***CATIA*** inside the *C:* drive. Then create the folder ***c02*** inside the CATIA folder.

2. Next, choose the **Save** button; the file is saved at *C:\CATIA\c02*.

Generating Drawing Views of Model

1. Choose **Start > Mechanical Design > Drafting** from the menubar to display the **New Drawing Creation** dialog box.

2. Choose the **OK** button from the **New Drawing Creation** dialog box: a new file is started in the **Drafting** workbench.

3. Choose the **Front View** tool from the **Projections** sub-toolbar in the **Views** toolbar; you are prompted to select a reference plane on a 3D geometry. 

4. Choose **Window > c02tut2.CATPart** from the menu bar; the part file is displayed.

5. Select the **yz** plane (of the front planar face) from the Specification tree; the drawing file is invoked.

   A preview of the front view is displayed and a knob appears on the view to set the orientation of the front view.

6. Move the cursor on the frame that shows the preview of the front view; the cursor is replaced by a hand cursor. Press and hold the left mouse button and drag the cursor close to the lower left corner of the drawing sheet.

7. Click anywhere in the drawing sheet to generate the front view.

8. Next, double-click on the **Projection View** tool from the **Projections** sub-toolbar in the **Views** toolbar then place the cursor on the front view; a preview of the projected view is attached to the cursor.

9. Similarly, taking the front view as the parent view, generate the drawing of right-side view.

10. Choose the **Isometric View** tool from the **Projections** sub-toolbar in the **Views** toolbar; you are prompted to select a reference plane on a 3D geometry. 

11. Choose **Window > c02tut2.CATPart** from the menu bar; the part file is displayed.

12. Select the front face of the model from the geometry area; the drawing sheet is displayed. Also, a preview of the isometric view of the model is displayed on the drawing sheet along with a knob which can be used to orient the isometric view.

13. Click anywhere on the drawing sheet to generate the isometric view. The drawing sheet after generating the isometric view is shown in Figure 2-74.

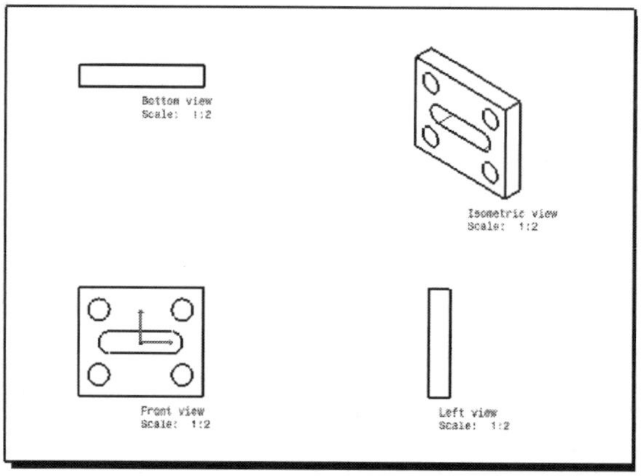

Figure 2-74 *Drawing sheet after generating the drawing views*

Tutorial 3

Draw the sketch shown in Figure 2-75. Extrude the sketch by 40 units and then generate a drawing with the orthographic and isometric views of the model. The isometric view of the model is shown in Figure 2-76. You do not need to dimension the drawing.

(Expected time: 20 min)

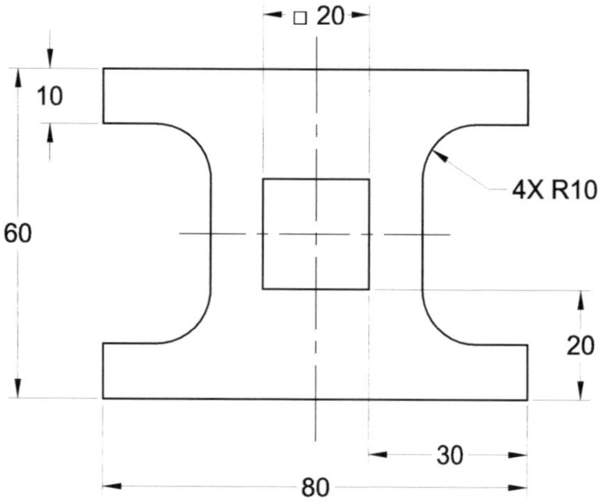

Figure 2-75 *The sketch of the model*

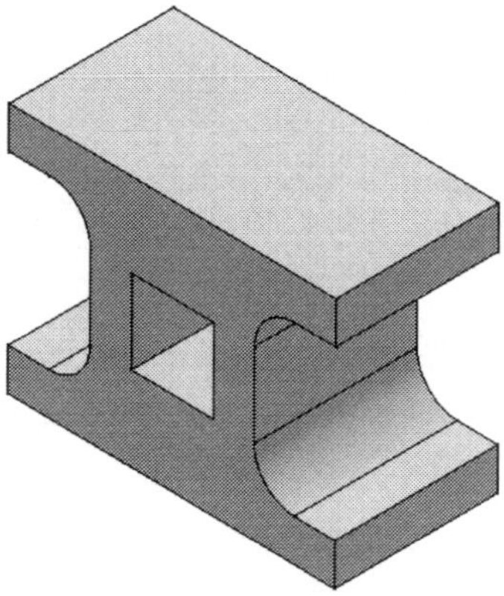

Figure 2-76 *The solid model for Tutorial 3*

The following steps are required to complete this tutorial:

a. Start a new Part file.
b. Draw the sketch of the model using the **Profile** and **Rectangle** tools.
c. Extrude the sketch upto a distance of 40 mm using the **Extrude** tool.
d. Save the model and then generate its drawing views.

Starting a New Part File

1. Choose **File > New** from the menu bar; the **New** dialog box is displayed.

2. In the **New** dialog box, select **Part** from the **List of Types** list box and choose the **OK** button; the **New Part** dialog box is displayed.

3. Enter **c02tut3** as the name of the file in the **Enter part name** edit box. Retaining rest of the default settings in the **New Part** dialog box, choose the **OK** button; a new **Part** file opens in the **Part Design** workbench.

4. Choose the **Sketch** tool from the **Sketcher** toolbar and then select the yz plane as the sketching plane to invoke the **Sketcher** workbench.

   Now, you need to draw the sketch in two parts: first the outer loop and then the inner cavity.

Drawing the Outer Loop of the Sketch

In this section, you need to draw the outer loop of the sketch using the **Profile** tool. Start drawing the outer loop from the lower left corner of the sketch. It is recommended that you keep the origin in the middle of the sketch drawn as it will reduce the time required for constraining and dimensioning the sketches. Also, it will help you capture the design intent easily.

1. Choose the **Profile** tool from the **Profile** toolbar.

2. Click to specify the start point of the line at the location whose coordinates are -40,-30 and then move the cursor horizontally toward the right; you will notice that the line turns blue.

3. Move the cursor to the location whose coordinates are 40,-30. Click to specify the endpoint of the line; a rubber band line is attached to the cursor.

4. Move the cursor vertically upward and click to specify the endpoint of the second line whose coordinates are 40,-20.

5. Move the cursor horizontally toward the left where the coordinates are 30,-20 and click to specify the endpoint of the third line.

   After drawing these three lines, you need to draw a tangent arc using the **Tangent Arc** button from the **Sketch tools** toolbar.

6. Choose the **Tangent Arc** tool from the **Sketch tools** toolbar.

7. Move the cursor to the location whose coordinates are 20, -10 and click to specify the endpoint of the tangent arc; the **Line** mode is activated and the line gets attached to the cursor again. Figure 2-77 shows the sketch after drawing three lines and a tangent arc. In this figure, the constraints are hidden for better visualization.

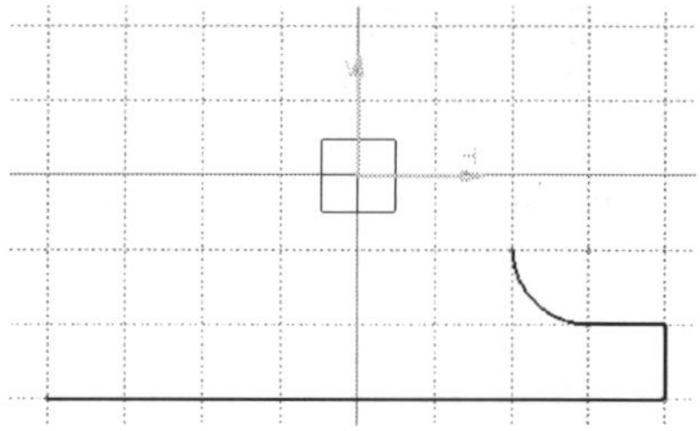

Figure 2-77 *The sketch after drawing three lines and a tangent arc*

8. Move the cursor vertically upward to the location whose coordinates are 20,10 and then click to specify the endpoint of the line at this location.

   Next, you need to draw a tangent arc using the **Tangent Arc** button from the **Sketch tools** toolbar.

9. Choose the **Tangent Arc** tool from the **Sketch tools** toolbar; the **Tangent Arc** mode is activated.
10. Move the cursor to the location whose coordinates are 30, 20 and click to specify the endpoint of the tangent arc. As soon as, you specify the end point of the tangent arc, the **Line** mode is activated again.
11. Move the cursor horizontally toward the right and click to specify the endpoint of the line when the coordinates are 40,20.
12. Move the cursor vertically upward and click to specify the endpoint of the line when the coordinates are 40,30.
13. Move the cursor horizontally toward the left and click to specify the endpoint of the line when the coordinates are -40,30.
14. Move the cursor vertically downward and click to specify the endpoint of the line when the coordinates are -40,20.
15. Move the cursor horizontally toward the right and click to specify the endpoint of the line where the coordinates are -30,20.

    Next, you need to draw a tangent arc by choosing the **Tangent Arc** tool from the **Sketch tools** toolbar.

16. Choose the **Tangent Arc** button from the **Sketch tools** toolbar; the **Tangent Arc** mode is activated.
17. Move the cursor to the location whose coordinates are -20,10 and click to specify the endpoints of the arc at this location; the **Line** mode is activated and line is attached to the cursor.
18. Move the cursor vertically downward and specify the endpoint of the line when the coordinates are -20,-10.

19. Switch to the **Tangent Arc** mode by choosing the **Tangent Arc** tool from the **Sketch tools** toolbar and then move the cursor to the location whose coordinates are -30,-20 and click to specify the endpoint of the tangent arc at this location.

20. Move the cursor horizontally toward the left and click to specify the endpoint of the line where the coordinates are -40,-20.

21. Move the cursor vertically downward and click to specify the endpoint of the line when it snaps to the start point of the outer loop. The sketch after drawing the outer loop and hiding the constraints is shown in Figure 2-78.

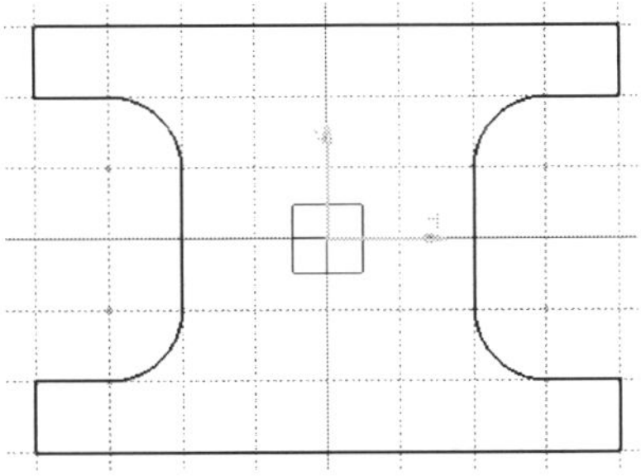

Figure 2-78 *The sketch after drawing the outer loop and hiding the constraints*

Drawing the Inner Cavity of the Sketch

After drawing the outer loop of the sketch, you need to draw its inner rectangular cavity using the **Rectangle** tool.

1. Choose the **Rectangle** tool from the **Profile** toolbar.

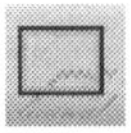

2. Move the cursor to the location whose coordinates are -10,10 and click to specify the upper-left corner of the rectangle at this location.

3. Move the cursor to the location whose coordinates are 10,-10 and click to specify the lower-right corner of the rectangle at this location.

4. Choose the **Fit All In** button from the **View** toolbar to fit the sketch into the geometry area.

   The final sketch after drawing the inner loop is shown in Figure 2-79. Note that in this figure, the display of constraints has been turned on.

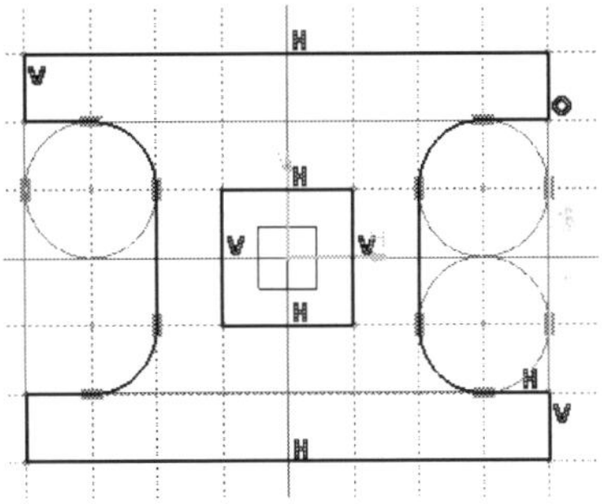

Figure 2-79 *The final sketch after drawing the inner loop*

Extruding the Sketch

Next, you need to extrude the sketch.

1. Choose the **Pad** button from the **Sketch-Based Features** toolbar; the **Pad Definition** dialog box is displayed and a preview of the extruded feature is displayed in the geometry area.

2. Set **40** as the value in the **Length** spinner of the **Pad Definition** dialog box, and then choose **OK** button from this dialog box.

3. Click once in the geometry area to remove the newly created base feature from the current selection set.

4. Orient the view of the model to the isometric view using the **Isometric View** tool from the **View** toolbar. The model after creating the extruded feature is shown in Figure 2-80.

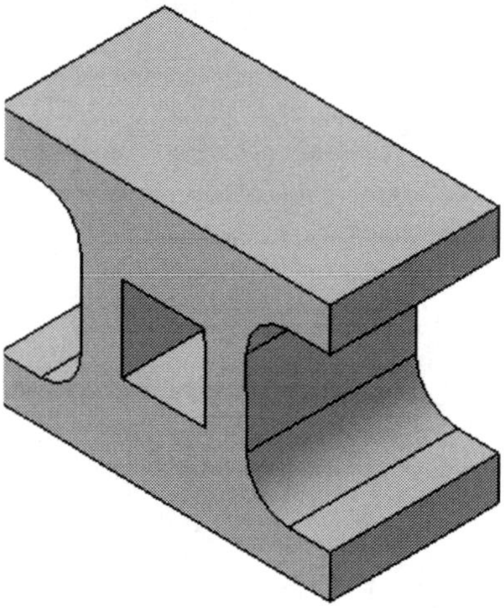

Figure 2-80 *The final model for Tutorial 3*

Saving the Model

After completing the model, you need to save it.

1. Choose the **Save** button from the **Standard** toolbar to invoke the **Save As** dialog box. Using this dialog box, create a folder named ***CATIA*** inside the *C:* drive. Then create the folder ***c02*** inside the CATIA folder.

2. Next, choose the **Save** button; the file is saved at *C:\CATIA\c02*.

Generating Drawing Views of Model

1. Choose **Start > Mechanical Design > Drafting** from the menubar to display the **New Drawing Creation** dialog box.

2. Choose the **OK** button from the **New Drawing Creation** dialog box: a new file is started in the **Drafting** workbench.

3. Choose the **Front View** tool from the **Projections** sub-toolbar in the **Views** toolbar; you are prompted to select a reference plane on a 3D geometry. 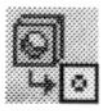

4. Choose **Window > c02tut3.CATPart** from the menu bar; the part file is displayed.

5. Select the **yz** plane (of the front planar face) from the Specification tree; the drawing file is invoked.

   A preview of the front view is displayed and a knob appears on the view to set the orientation of the front view.

6. Move the cursor on the frame that shows the preview of the front view; the cursor is replaced by a hand cursor. Press and hold the left mouse button and drag the cursor close to the lower left corner of the drawing sheet.

7. Click anywhere on the drawing sheet to generate the front view.

8. Next double-click on the **Projection View** tool from the **Projections** sub-toolbar in the **Views** toolbar then place the cursor on the front view; a preview of the projected view is attached to the cursor.

9. Similarly, taking the front view as the parent view, generate the drawing of right-side view.

10. Choose the **Isometric View** tool from the **Projections** sub-toolbar in the **Views** toolbar; you are prompted to select a reference plane on a 3D geometry.

11. Choose **Window > c02tut3.CATPart** from the menu bar; the part file is displayed.

12. Select the front face of the model from the geometry area; the drawing sheet is displayed. Also, a preview of the isometric view of the model is displayed on the drawing sheet along with a knob which can be used to orient the isometric view.

13. Click anywhere on the drawing sheet to generate the isometric view. The drawing sheet after generating the isometric view is shown in Figure 2-81.

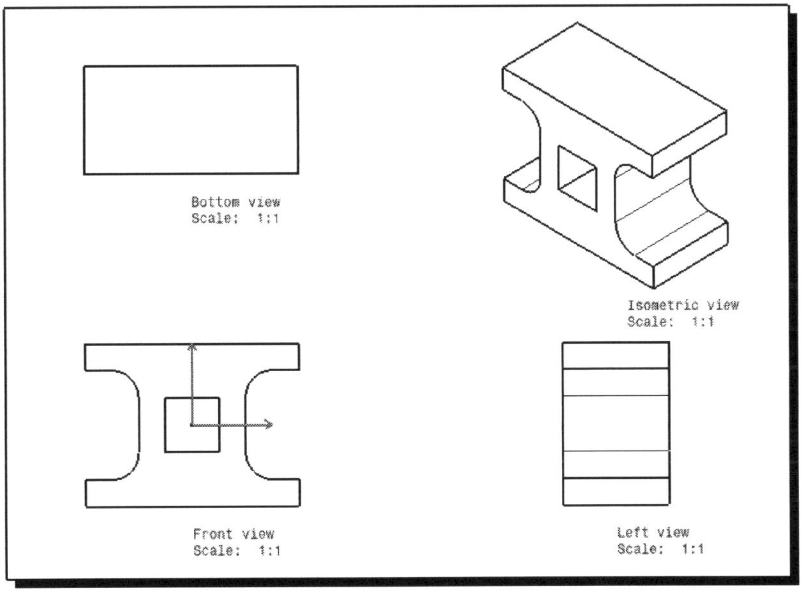

Figure 2-81 *Drawing sheet after generating the drawing views*

Tutorial 4

Draw the sketch shown in Figure 2-82. Extrude the sketch by 60 units and then generate a drawing with the orthographic and isometric views of the model. The isometric view of the model is shown in Figure 2-83. You do not need to dimension the drawing.

(Expected time: 30 min)

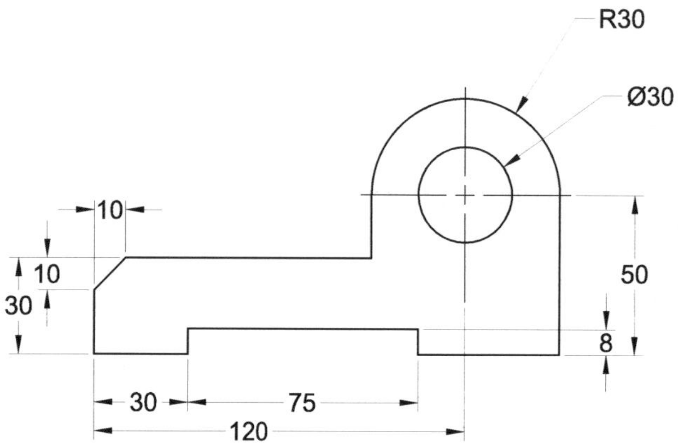

Figure 2-82 *The sketch of the model*

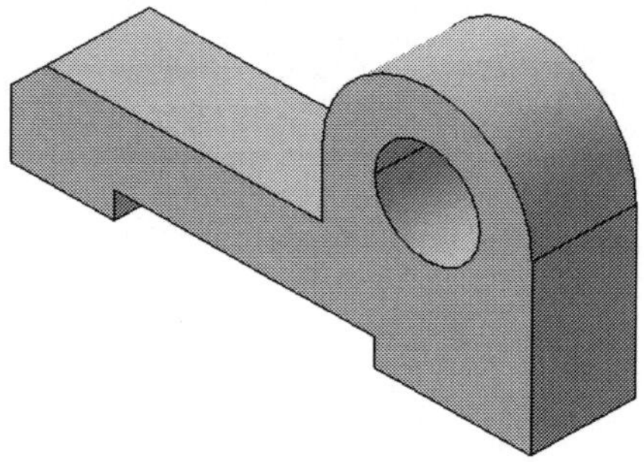

Figure 2-83 *The solid model for Tutorial 4*

The following steps are required to complete this tutorial:

a. Start CATIA V5 and then start a new **Part** file.
b. Draw the sketch of the model using **Line**, **Arc**, and **Circle** tools.
c. Extrude the sketch upto a distance of 60 mm using the **Extrude** tool.
d. Save the model and then generate its drawing views.

Starting CATIA V5 and then a New Part File

1. Choose **File > New** from the menu bar; the **New** dialog box is displayed.

2. In this dialog box, select **Part** from the **List of Types** list box and choose the **OK** button; the **New Part** dialog box is displayed.

3. In the **New Part** dialog box, enter **c02tut4** as the name of the file in the **Enter part name** edit box. Accept the rest of the default setting in the **New Part** dialog box and choose the **OK** button; a new **Part** file opens in the **Part Design** workbench.

4. Choose the **Sketch** tool from the **Sketcher** toolbar and then select the yz plane as the sketching plane to invoke the **Sketcher** workbench.

Now, you need to draw the sketch in two parts: first as the outer loop and second as the inner circle.

Drawing the Outer Loop of the Sketch

In this section, you need to draw the sketch symmetrically around the origin because it will reduce the time required for constraining and dimensioning it. You will draw the outer loop of the sketch using the **Line** and **Arc** tools.

1. Invoke the **Line** tool by choosing the **Line** tool from the **Profile** toolbar.

2. Choose the **Snap to Point** button from the **Sketch tools** toolbar if not chosen.

3. Move the cursor in the third quadrant; the coordinates of the point are displayed above the cursor.

4. Click to specify the point whose coordinates are -50,-30. Next, move the cursor horizontally toward the right.

   It is evident from Figure 2-82 that the length of the first horizontal line at the lower left corner of the sketch is 30mm. Therefore, you need to move the cursor until the length of the line is shown as 30mm in the **L** edit box of the **Sketch tools** toolbar.

5. Click to draw the line when the **L** edit box displays a value of 30mm

   After drawing the first horizontal line, you will notice that a Horizontal constraint is applied to it. Note that the line is still selected and displayed in orange. Click anywhere in the geometry area to remove it from the selection set.

   As soon as you specify the endpoint of the line, the **Line** tool gets terminated. Therefore, you need to choose the **Line** tool again and again to draw multiple lines. You can avoid it by double-clicking on the **Line** tool in the **Profile** toolbar. On doing so, the **Line** tool will not terminate until you press the Esc key twice.

6. Double-click on the **Line** tool to invoke the **Line** tool and select the endpoint of the first horizontal line.

7. Press the TAB key twice to highlight the value displayed in the **L** edit box of the **Sketch tools** toolbar. Enter **8** in this edit box and then press the Enter key.

8. Now, move the cursor vertically upward and click when a vertical line is displayed in blue; a vertical line of length 8mm is drawn. You will notice that this line is no longer in the selection mode and you are prompted to select the start point of the next line. This happens because of double-clicking on the **Line** tool. It keeps the **Line** tool active until another tool is invoked.

9. Select the endpoint of the vertical line as the start point of the second horizontal line. Enter **75** in the **L** edit box of the **Sketch tools** toolbar and press Enter. Now, move the cursor horizontally toward the right and click when a horizontal line is displayed; the second horizontal line of length 75mm is drawn.

10. Select the endpoint of the second horizontal line as the start point of the second vertical line and move the cursor vertically downward. Click when the **L** edit box displays a value of 8mm; the second vertical line of length 8mm is drawn.

11. Select the endpoint of the second vertical line as the start point of the third horizontal line and move the cursor horizontally toward the right. Click to draw the line when the length in the **L** edit box shows a value of 45mm.

12. Select the endpoint of the previous line as the start point of the third vertical line and move the cursor vertically upward. Click to draw the line when the **L** edit box displays a value of 50mm; the third vertical line of length 50mm is drawn.

    Next, you need to draw a three point arc using the **Three Point Arc** tool.

13. Choose the **Three Point Arc** tool from the **Circle** sub-toolbar.

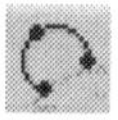

14. Select the start point of the arc as the endpoint of the previous vertical line.

15. Move the cursor to the point whose coordinates are 70,50. These coordinates are displayed in the **Sketch tools** toolbar and also on the top of the cursor. Now, click in the geometry area to specify the second point.

16. Move the cursor to a location 40,20 in the geometry area to specify the third point of the arc and then click; the coordinate values are displayed on the top of the cursor.

    This draws the arc of the outer loop. The arc is in the selection mode. Click anywhere in the geometry area to exit the selection mode.

17. Now, to continue drawing the outer loop, you need to invoke the **Line** tool again. Double-click on the **Line** tool in the **Profile** toolbar to invoke the **Line** tool.

18. Select the endpoint of the arc as the start point of the fourth vertical line. Move the cursor vertically downward and click to draw the line when the length in the **L** edit box shows a value of 20mm in the **Sketch tools** toolbar.

    The fourth vertical line of length 20mm is drawn. You will notice that the line is no longer in the selection mode and you are prompted to specify the start point of the next line.

19. Select the endpoint of the previous line as the start point of the fourth horizontal line. Move the cursor horizontally toward the left. Click to draw the line when the length in the **L** edit box shows a value of 80mm in the **Sketch tools** toolbar.

    The fourth horizontal line of length 80mm is drawn. Note that the line is green in color because it passes through the origin.

20. Select the endpoint of the previous line as the start point of the inclined line. Move the cursor such that the line is drawn at an angle of 225°. The current angle is displayed in the **A** edit box of the **Sketch tools** toolbar. Click when a vertical inferencing line is displayed

between the endpoint of the inclined line and the start point of the first horizontal line; a horizontal inclined line of 10 unit is drawn.

21. Select the endpoint of the inclined line as the start point of the next line. Move the cursor vertically downward and click when the **L** edit box displays a value of 20mm. Next, press the Esc key to exit the active tool.

    This completes the sketch of the outer loop. It is recommended that you modify the geometry area such that the sketch fits inside the screen. This can be done by using the **Fit All In** tool.

22. Choose the **Fit All In** button from the **View** toolbar to fit the current sketch into the screen. The completed outer loop of the sketch is shown in Figure 2-84. Note that in this figure, the display of constraints and dimensions is turned off using the **Hide/Show** tool for clarity.

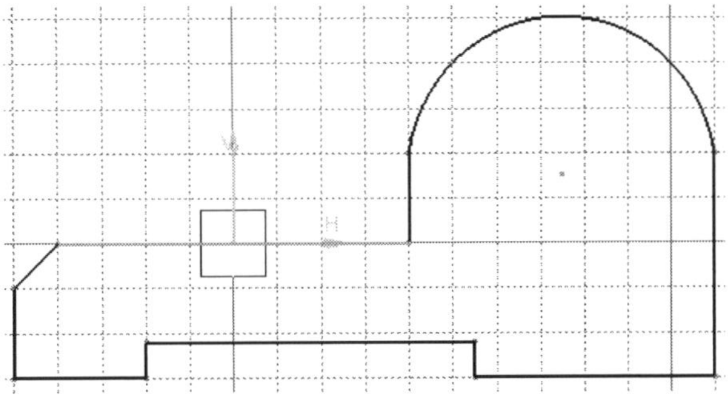

Figure 2-84 *The completed outer loop of the sketch*

Drawing the Inner Circle

Now, you need to draw a circle using the **Circle** tool.

1. Choose the **Circle** tool from the **Circle** sub-toolbar to invoke it; you are prompted to define the center point of the circle.

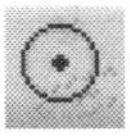

2. Move the cursor to the point whose coordinates are 70,20. Click when the cursor snaps to this point.

3. Move the cursor horizontally toward the right and click when the **R** edit box of the **Sketch tools** toolbar displays a value of 15mm. Click anywhere in the geometry area to remove the circle from the selection.

   The final sketch after drawing the circle is shown in Figure 2-85. Note that in this figure, the display of the constraints has been turned on.

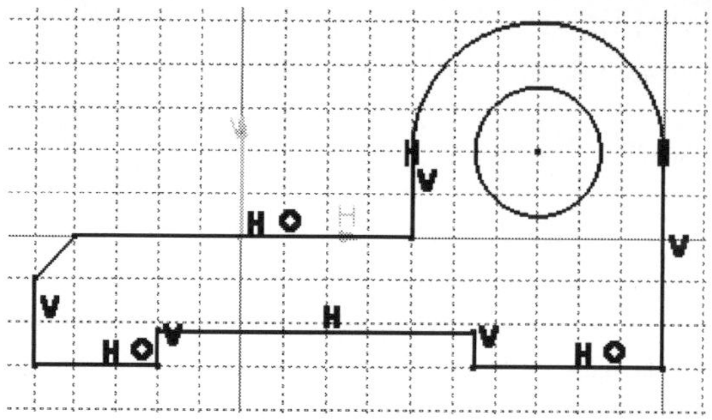

Figure 2-85 The final sketch for Tutorial 4

Extruding the Sketch

Next, you need to extrude the sketch.

1. Choose the **Pad** button from the **Sketch-Based Features** toolbar; the **Pad Definition** dialog box is displayed and a preview of the extruded feature is displayed in the geometry area.

2. Set **60** as the value in the **Length** spinner of the **Pad Definition** dialog box, and then choose **OK** from this dialog box.

3. Click once in the geometry area to remove the newly created base feature from the current selection set.

4. Orient the view of the model to the isometric view using the **Isometric View** tool from the **View** toolbar. The model after creating the extruded feature is shown in Figure 2-86.

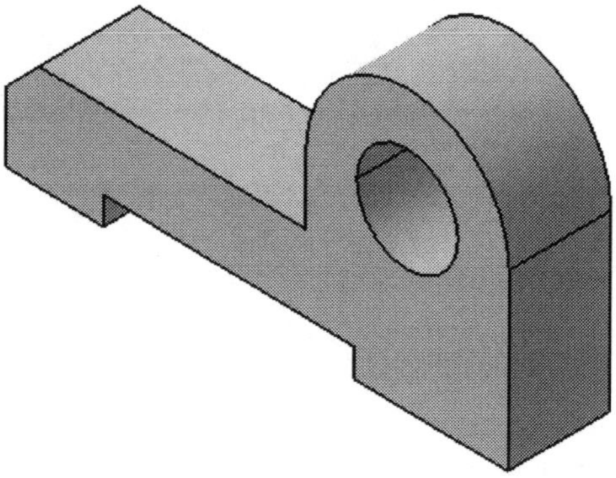

Figure 2-86 The final model for Tutorial 4

Saving the Model

After completing the model, you need to save it.

1. Choose the **Save** button from the **Standard** toolbar to invoke the **Save As** dialog box. Using this dialog box, create a folder named ***CATIA*** inside the *C:* drive. Then create the folder ***c02*** inside the CATIA folder.

2. Next, choose the **Save** button; the file is saved at *C:\CATIA\c02*.

Generating Drawing Views of Model

1. Choose **Start > Mechanical Design > Drafting** from the menubar to display the **New Drawing Creation** dialog box.

2. Choose the **OK** button from the **New Drawing Creation** dialog box: a new file is started in the **Drafting** workbench.

3. Choose the **Front View** tool from the **Projections** sub-toolbar in the **Views** toolbar; you are prompted to select a reference plane on a 3D geometry.

4. Choose **Window > c02tut4.CATPart** from the menu bar; the part file is displayed.

5. Select the **yz** plane (of the front planar face) from the Specification tree; the drawing file is invoked.

   A preview of the front view is displayed and a knob appears on the view to set the orientation of the front view.

6. Move the cursor on the frame that shows the preview of the front view; the cursor is replaced by a hand cursor. Press and hold the left mouse button and drag the cursor close to the lower left corner of the drawing sheet.

7. Click anywhere in the drawing sheet to generate the front view.

8. Next, double-click on the **Projection View** tool from the **Projections** sub-toolbar in the **Views** toolbar then place the cursor on the front view; a preview of the projected view is attached to the cursor.

9. Similarly, taking the front view as the parent view, generate the drawing of right-side view.

10. Choose the **Isometric View** tool from the **Projections** sub-toolbar in the **Views** toolbar; you are prompted to select a reference plane on a 3D geometry.

11. Choose **Window > c02tut4.CATPart** from the menu bar; the part file is displayed.

12. Select the front face of the model from the geometry area; the drawing sheet is displayed. Also, a preview of the isometric view of the model is displayed on the drawing sheet along with a knob which can be used to orient the isometric view.

13. Click anywhere on the drawing sheet to generate the isometric view. The drawing sheet after generating the isometric view is shown in Figure 2-87.

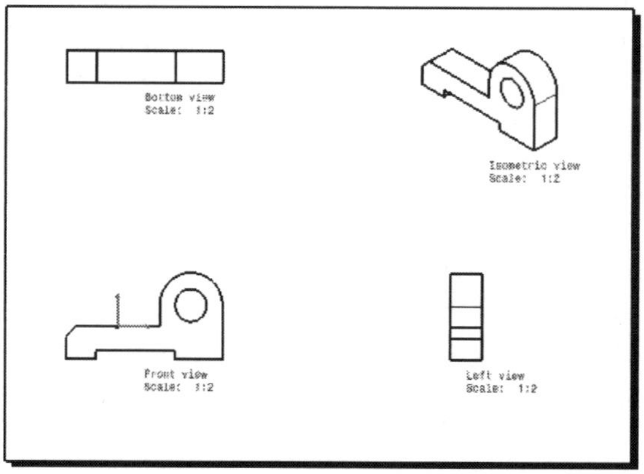

Figure 2-87 *Drawing sheet after generating the drawing views*

Self-Evaluation Test

Answer the following questions and then compare them to those given at the end of this chapter:

1. You can convert a sketched element into a construction element by using the _________ tool.
2. To draw a rectangle at an angle, you need to use the _________ tool.
3. The rectangle is considered as a combination of individual _________.
4. The _________ tool is used to draw continuous lines.
5. Using the _________ tool, you can create a circle by specifying the coordinates of its center point.
6. _________ are temporary lines that are used to track a particular point on the screen.
7. The base feature of any design is a sketched feature which is created by drawing the sketch. (T/F)
8. You can draw an arc while working with the **Profile** tool. (T/F)
9. To enter the **Sketcher** workbench, you need to choose the **Sketch** tool. (T/F)
10. When you open a file that has been saved in the sketching environment, it opens in the part modeling environment. (T/F)

Review Questions

Answer the following questions:

1. In CATIA V5, a combination of which of the following elements is considered as a rectangle?

   (a) **Lines** (b) **Arcs**
   (c) **Splines** (d) None of these

2. Which of the following tools is not available in the **Predefined Profile** sub-toolbar?

   (a) **Rectangle** (b) **Oriented Rectangle**
   (c) **Parallelogram** (d) **Circle**

3. Which one of the following elements will not be considered while converting a sketch into a feature?

   (a) **Sketched circles** (b) **Sketched lines**
   (c) **Construction elements** (d) None of these

4. Which one of the following tools is available in the **Line** toolbar?

   (a) **Line** (b) **Infinite Line**
   (c) **Bisecting Line** (d) All of these

5. In which workbench of CATIA V5, you can draw the sketches that can be used to create features?

   (a) **Part** (b) **Assembly**
   (c) **Shape** (d) None of these

6. The 3 point arc is the arc that is drawn by defining a start point, an endpoint, and a point on the arc. (T/F)

7. The **Parallelogram** tool is available in the **Predefined Profile** sub-toolbar. (T/F)

8. The **Symmetrical Extension** button when chosen from the **Sketch tools** toolbar draws a simple line. (T/F)

9. In CATIA V5, circles are drawn by specifying the center point of the circle and then entering radius value in the dialog box that is displayed. (T/F)

10. When you start CATIA V5, a file in the **Product** workbench is started by default. (T/F)

EXERCISES

Exercise 1

Draw the sketch shown in Figure 2-88. Extrude the sketch by 20 units and then generate a drawing with the orthographic and isometric views of the model. The isometric view of the model is shown in Figure 2-89. You do not need to dimension the drawing.

(Expected time: 20 min)

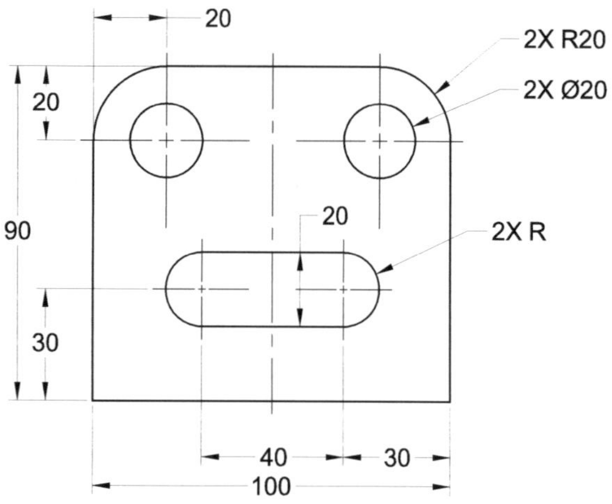

Figure 2-88 *The sketch of the model*

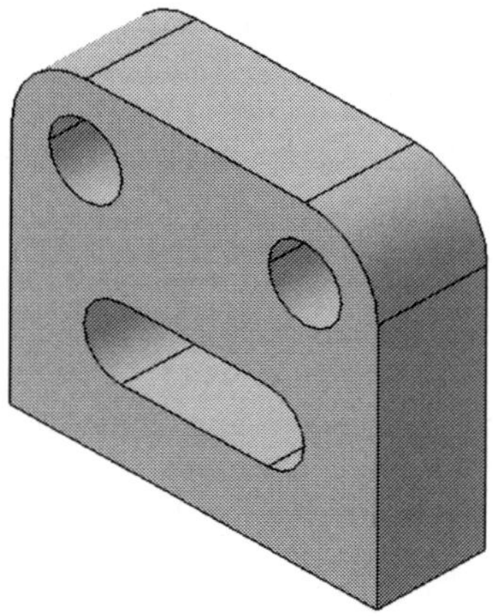

Figure 2-89 *The solid model for Exercise 1*

Exercise 2

Draw the sketch shown in Figure 2-90. Extrude the sketch by 30 units and then generate a drawing with the orthographic and isometric views of the model. The isometric view of the model is shown in Figure 2-91. You do not need to dimension the drawing.

(Expected time: 20 min)

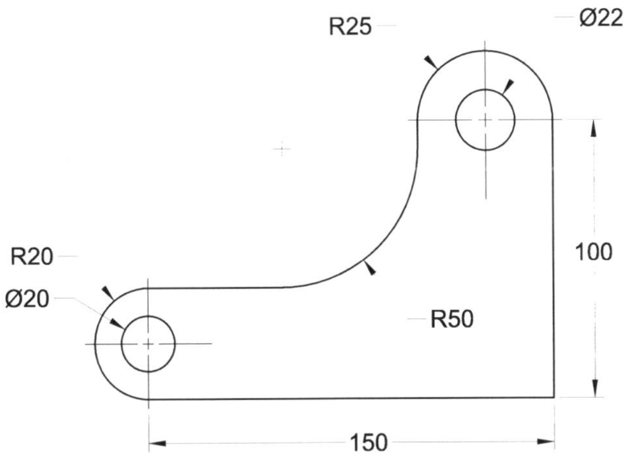

Figure 2-90 The sketch of the model

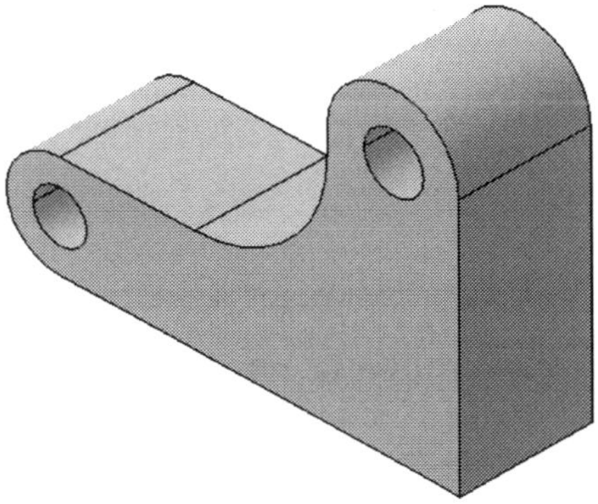

Figure 2-91 The solid model for Exercise 2

Exercise 3

Draw the sketch shown in Figure 2-92. Extrude the sketch by 25 units and then generate a drawing with the orthographic and isometric views of the model. The isometric view of the model is shown in Figure 2-93. You do not need to dimension the drawing.

(Expected time: 25 min)

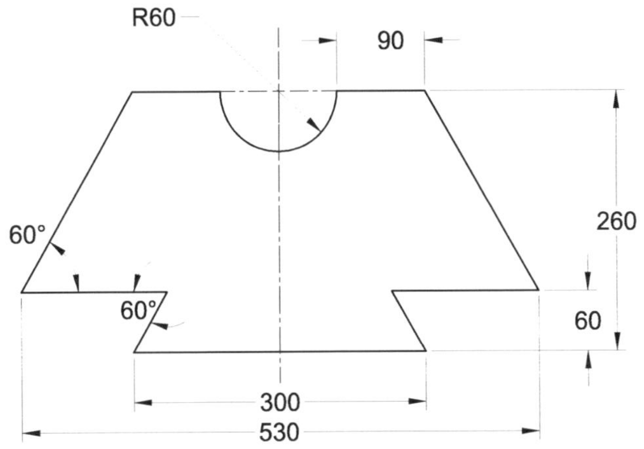

Figure 2-92 The sketch of the model

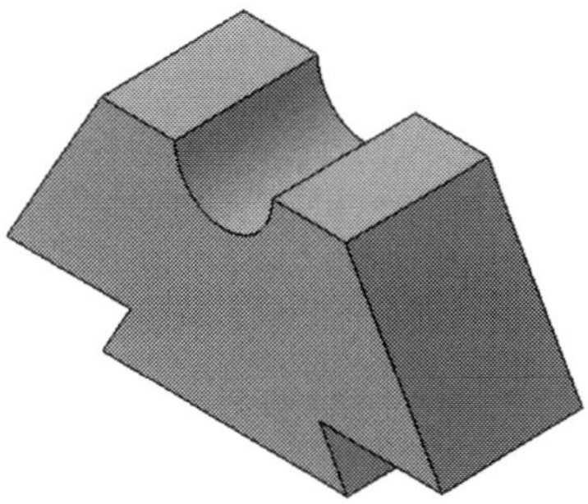

Figure 2-93 The solid model for Exercise 3

Exercise 4

Draw the sketch shown in Figure 2-94. Extrude the sketch by 15 units and then generate a drawing with the orthographic and isometric views of the model. The isometric view of the model is shown in Figure 2-95. You do not need to dimension the drawing.

(Expected time: 30 min)

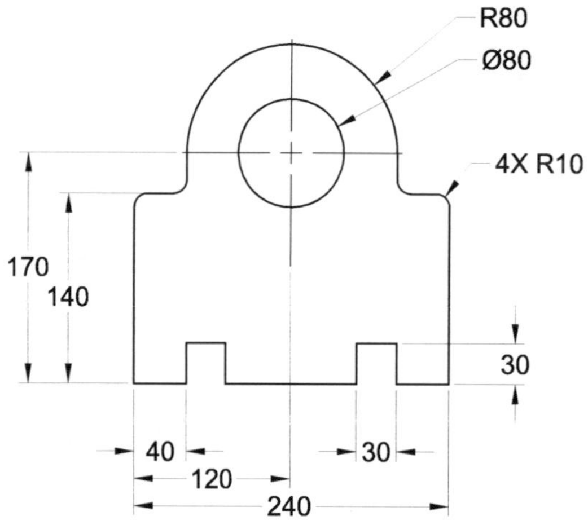

Figure 2-94 *The sketch of the model*

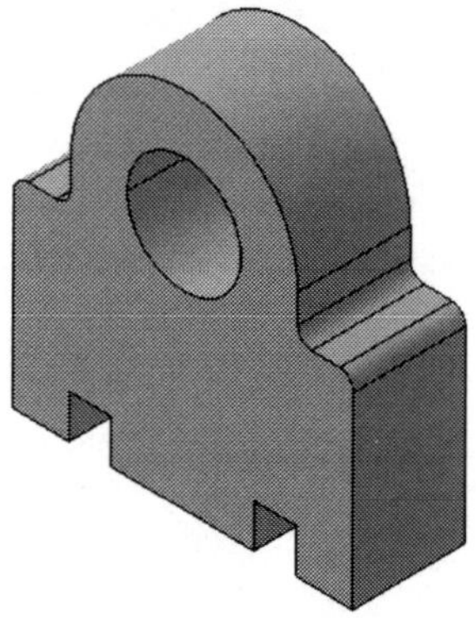

Figure 2-95 *The solid model for Exercise 4*

Answers to Self-Evaluation Test

1. Construction/Standard Element, **2. Oriented Rectangle**, **3.** lines, **4. Profile**, **5. Circle Using Coordinates**, **6.** Inferencing lines, **7.** T, **8.** T, **9.** T, **10.** F

Chapter 3

Advance Sketching Tools

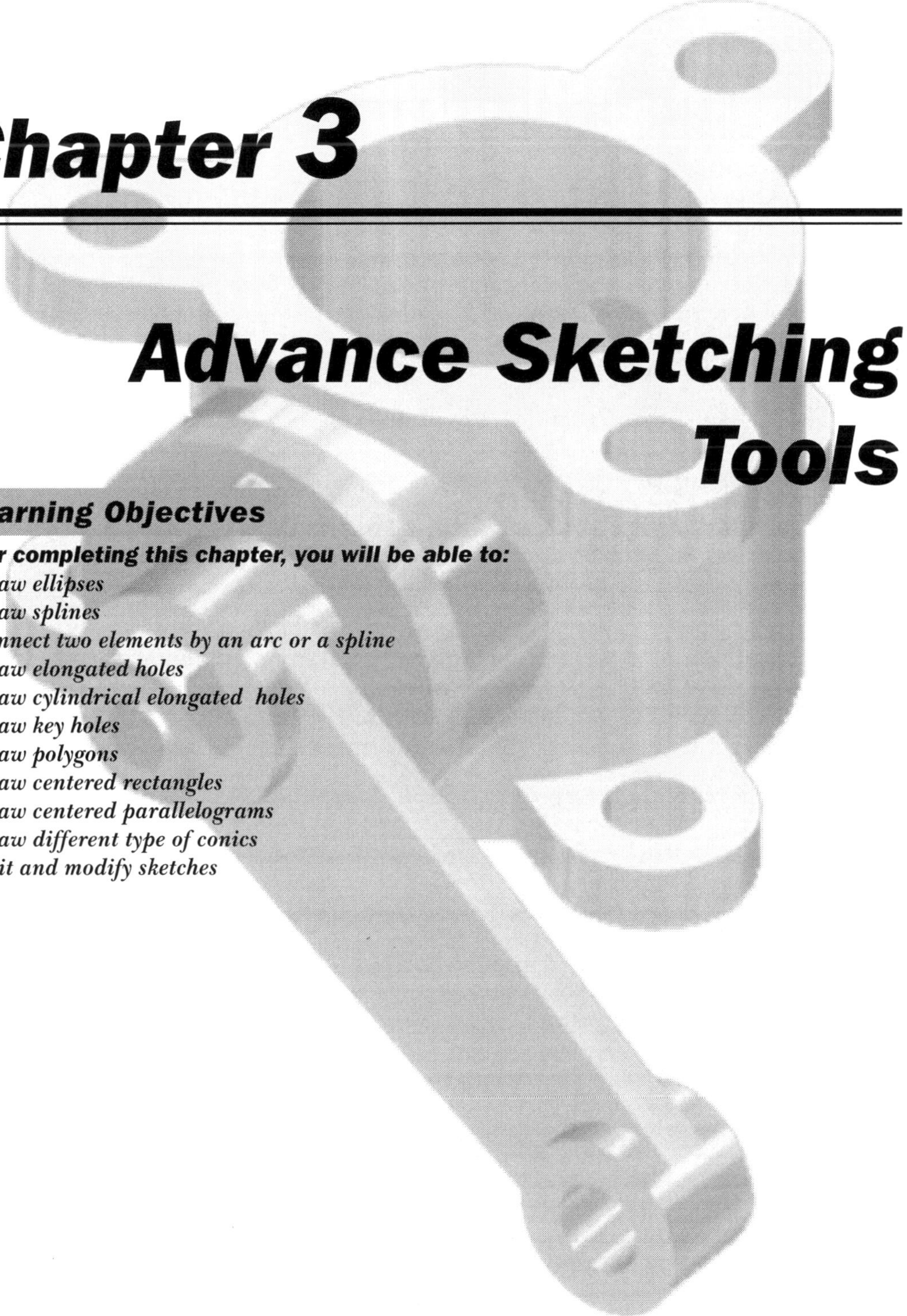

Learning Objectives

After completing this chapter, you will be able to:

- *Draw ellipses*
- *Draw splines*
- *Connect two elements by an arc or a spline*
- *Draw elongated holes*
- *Draw cylindrical elongated holes*
- *Draw key holes*
- *Draw polygons*
- *Draw centered rectangles*
- *Draw centered parallelograms*
- *Draw different type of conics*
- *Edit and modify sketches*

OTHER SKETCHING TOOLS IN THE SKETCHER WORKBENCH

You have learned about some of the sketching tools in the last chapter. In this chapter, you will learn about the remaining sketching tools in the **Sketcher** workbench.

Drawing Conics

Conics are the geometrical elements that are formed by the intersection of a plane and a cone. By changing the angle and location of the intersection, you can produce an ellipse, parabola, or hyperbola. To draw a conic in CATIA V5, click on the down arrow available on the right of the **Ellipse** tool in the **Profile** toolbar; the **Conic** sub-toolbar will be displayed. The tools available in this sub-toolbar are discussed next.

Drawing Ellipses

Menubar:	Insert > Profile > Conic > Ellipse
Toolbar:	Profile > Conic sub-toolbar > Ellipse

To draw an ellipse, choose the **Ellipse** tool from the **Conic** sub-toolbar in the **Profile** toolbar. Figure 3-1 shows the **Profile** toolbar with the **Conic** sub-toolbar.

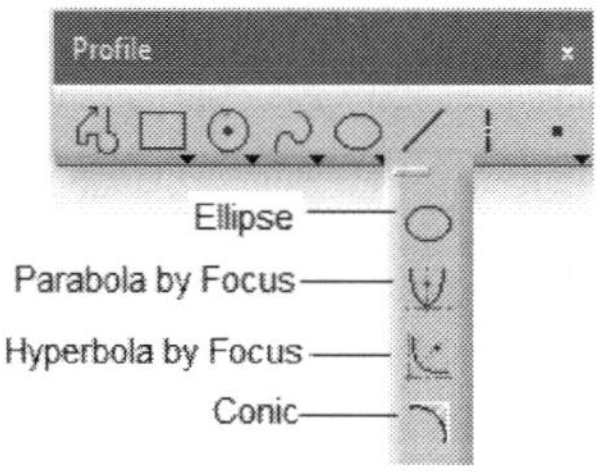

*Figure 3-1 The **Profile** toolbar with the **Conic** sub-toolbar*

On choosing the **Ellipse** tool, the **Sketch tools** toolbar will expand and you will be prompted to specify the ellipse center. Click in the geometry area to specify the center of the ellipse; you will be prompted to define the major axis and the orientation of the ellipse. In CATIA V5, the first axis of an ellipse is the major axis. To define it, you need to specify a point on the ellipse. The orientation of the ellipse depends on the angle formed between the major axis and the **H** direction. Move the cursor away from the center point; the preview of the ellipse is also displayed. Click in the geometry area to define the major axis; you will now be prompted to specify a point on the ellipse, which will determine the other axis. Figure 3-2 shows a point being specified on the ellipse. You will notice a few construction elements displayed on it. These elements define its major axis and orientation. Click in the geometry area to specify the third point on the ellipse; an ellipse, based on the specified parameters, is displayed in the geometry area, as shown in Figure 3-3.

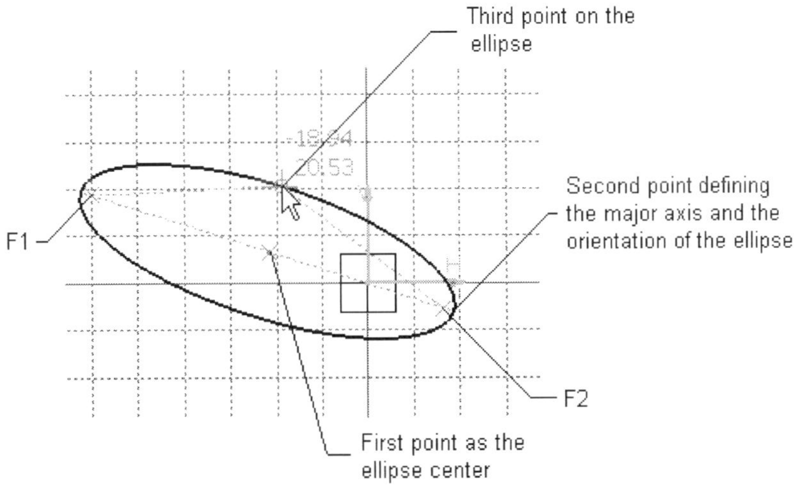

Figure 3-2 *Specifying three points to draw an ellipse*

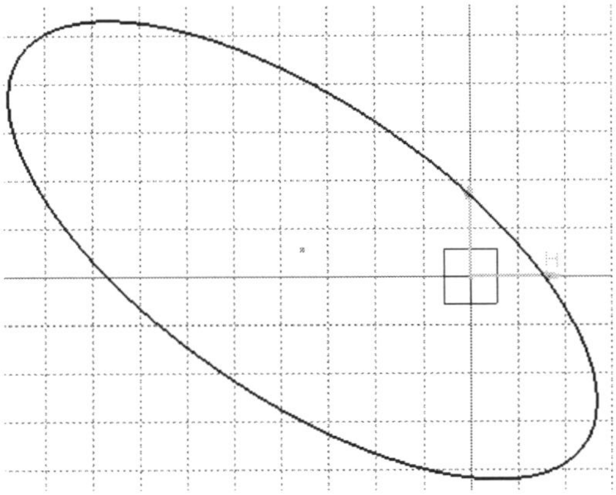

Figure 3-3 *The resulting ellipse*

Note
In CATIA V5, the first axis of an ellipse should be the major axis.

Drawing a Parabola by Focus

Menubar:	Insert > Profile > Conic > Parabola by Focus
Toolbar:	Profile > Conic sub-toolbar > Parabola by Focus

To draw a parabola by focus, choose the **Parabola by Focus** tool from the **Conic** sub-toolbar in the **Profile** toolbar; the **Sketch tools** toolbar will expand and you will be prompted to specify the focus. Click in the geometry area to specify the focus; you will be prompted to specify the apex. Move the cursor away from the focus; the preview of the parabola, attached to the cursor, is displayed. Click to specify the apex; you will be prompted to specify the start point. Move the cursor away from the apex and specify the start point; you will be prompted to specify the endpoint. Move the cursor along the path of the parabola and click to specify its end point. Figure 3-4 shows the points used to draw the parabola and the resulting parabola.

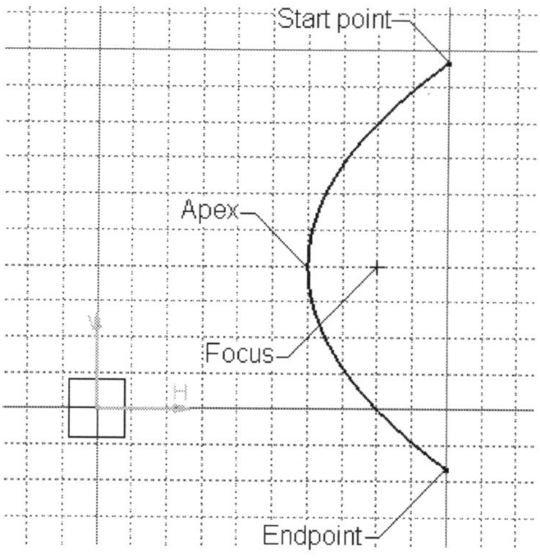

Figure 3-4 Points used to draw a parabola

Drawing a Hyperbola by Focus

Menubar:	Insert > Profile > Conic > Hyperbola by Focus
Toolbar:	Profile > Conic sub-toolbar > Hyperbola by Focus

To draw a hyperbola by focus, choose the **Hyperbola by Focus** tool from the **Conic** sub-toolbar in the **Profile** toolbar; the **Sketch tools** toolbar will expand and you will be prompted to specify the focus. Click to specify the focus which is referred to as F1 in Figure 3-5; you will be prompted to specify the center. Move the cursor away from the focus. As you move the cursor, you will notice that the preview of the hyperbola is attached to the cursor. Click to specify its center which is referred to as F2 in Figure 3-5; you will be prompted to specify the apex of the hyperbola. Move the cursor toward focus F1 to specify the apex. You will notice that the preview of the hyperbola moves along with the cursor. Also, in the **Sketch tools** toolbar, the value of eccentricity in the **e** edit box changes accordingly. Eccentricity, in case of hyperbola, is defined as the ratio of the distance of the apex from the center point to the distance of the center point from the focus point.

Click to specify the apex; you will now be prompted to specify the start point of the hyperbola. Move the cursor away from the apex and specify the start point, as shown in Figure 3-6. You can move the cursor in either direction to specify the start point. On doing so, you will be prompted to specify the endpoint. Move the cursor in the opposite direction of the start point; the preview of the hyperbola will follow the cursor. Click to specify the endpoint.

Note

In case of a parabola/hyperbola, if the focus, center point, or both do not lie on any of the axes or any sketched element, they will not be displayed as construction points after the parabola/hyperbola is drawn.

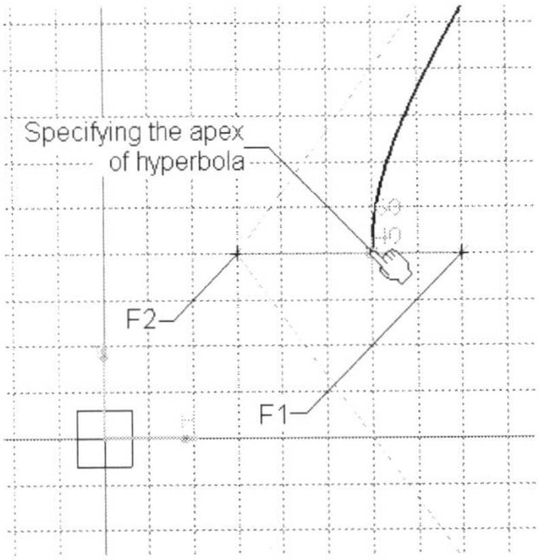

Figure 3-5 Specifying the focus and apex of the hyperbola

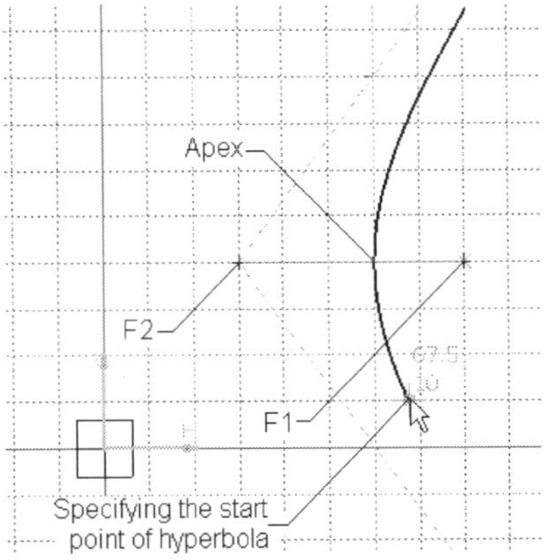

Figure 3-6 Specifying the start point of the hyperbola

Drawing Conics

Menubar:	Insert > Profile > Conic > Conic
Toolbar:	Profile > Conic sub-toolbar > Conic

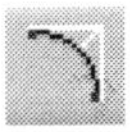

To draw a conic, choose the **Conic** tool from the **Conic** sub-toolbar in the **Profile** toolbar; the **Sketch tools** toolbar will expand. In the expanded toolbar, by default the **Nearest End Point**, **Two Points**, and **Start and End Tangent** tools will be chosen. Also, you will be prompted to specify the first endpoint. Click in the geometry area to specify the first endpoint of the conic; you will be prompted to specify the tangent at the first endpoint. Move the cursor away from the endpoint to define the constructional tangent line and then specify a point, as shown in Figure 3-7. Similarly, specify the second endpoint of the conic and its tangent line. Next, move the cursor between the two specified endpoints; the preview of the conic will be displayed. Finally, define a point on the preview to create the conic. Figure 3-7 shows the preview of conic.

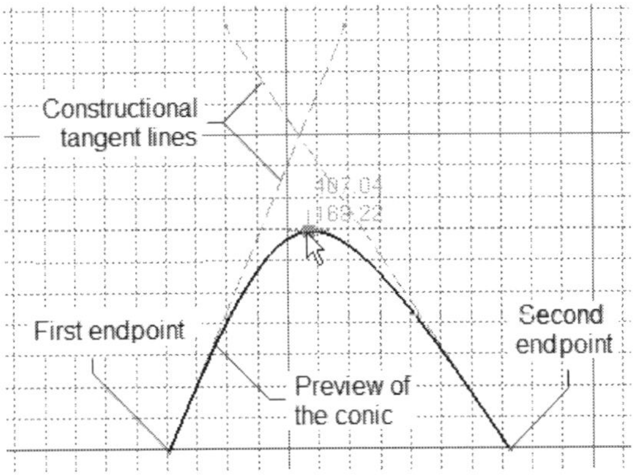

Figure 3-7 The preview of conic

Drawing Splines

Menubar:	Insert > Profile > Spline > Spline
Toolbar:	Profile> Spline sub-toolbar > Spline

Splines are the curves whose behavior is defined by piecewise function of polynomial equations. To draw a spline, choose the down arrow on the right of the **Spline** tool in the **Profile** toolbar; the **Spline** sub-toolbar will be displayed, as shown in Figure 3-8.

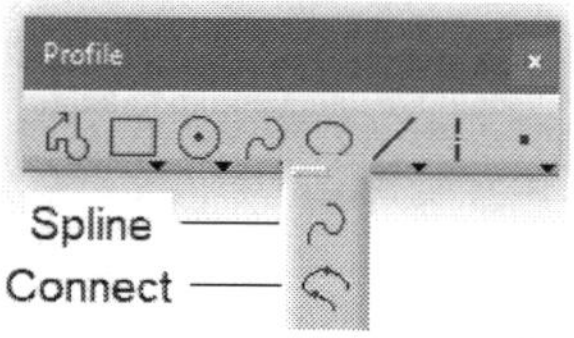

*Figure 3-8 The **Profile** toolbar with the **Spline** sub-toolbar*

Choose the **Spline** tool from the **Spline** sub-toolbar; you will be prompted to specify the first control point of the spline. Click to specify the first point; you will be prompted to specify the next point of the spline or double-click to specify the endpoint and exit the **spline** tool. Move the cursor; the preview of the spline is displayed. Click to specify the second control point. Similarly, you can specify multiple points to draw a spline. Figure 3-9 shows a spline being drawn by specifying multiple points.

Note

In a spline, control points are construction elements, while curve is a standard element.

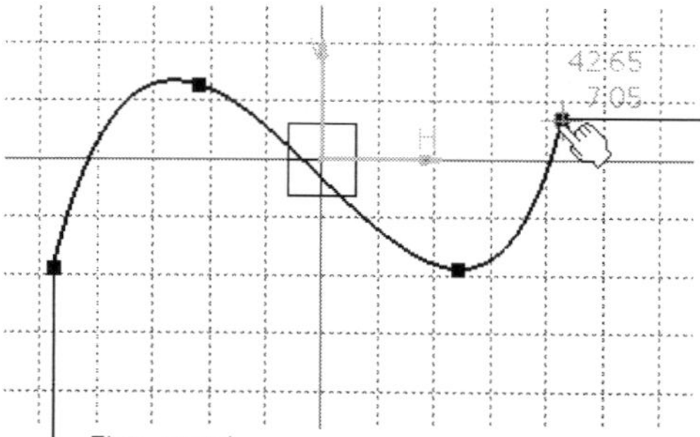

Figure 3-9 Drawing a spline by specifying multiple points

Connecting Two Elements by a Spline or an Arc

Menubar:	Insert > Profile > Spline > Connect
Toolbar:	Profile > Spline sub-toolbar > Connect

Two elements such as lines, arcs, ellipses, circles, or splines can be connected together by using an arc or a spline. To do so, choose the **Connect** tool from the **Profile** toolbar; the **Sketch tools** toolbar will expand, as shown in Figure 3-10.

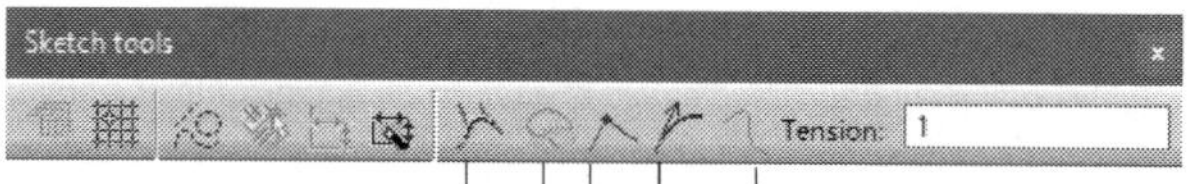

*Figure 3-10 The **Sketch tools** toolbar after choosing the **Connect** tool*

The next section discusses how the two selected elements can be connected by a spline or an arc.

Connecting Two Elements with a Spline

By default, the **Connect with a Spline** tool is chosen in the **Sketch tools** toolbar. Also, you are prompted to select the first element to be connected. Select the first element; you will be prompted to select the last element. Select the last element; both the selected elements will get connected by a spline in the geometry area.

When you connect the elements, you will notice that, by default, the **Continuity in curvature** tool is chosen in the **Sketch tools** toolbar. As a result, the resulting spline will maintain a curvature continuity with the selected elements. You can set the tension value in the **Tension** edit box, if required.

If you choose the **Continuity in tangency** tool from the **Sketch tools** toolbar, the resulting spline will maintain a tangent continuity with the selected elements. You can also set the tension value in the **Tension** edit box.

If you choose the **Continuity in point** tool from the **Sketch tools** toolbar, the resulting spline will maintain a point continuity with the selected elements. In this case, the resulting element will be a straight spline with two control points.

You can also assign different attributes at the two ends of the Connect Curve. To do so, choose the required continuity tool from the **Sketch tools** toolbar and then select the first element. Next, choose the required continuity tool for the second element and then select the second element. You can also set different tension values for curves to be connected in the **Tension** edit box.

Connecting Two Elements with an Arc

To connect two selected elements with an arc, choose the **Connect** tool from the **Profile** toolbar. Next, choose the **Connect with an Arc** tool from the **Sketch tools** toolbar; you will be prompted to select the first element to be connected. Select the first element; you will be prompted to select the last element. Once you specify the last element, the connecting arc generated using this tool will become tangent to the two elements.

Drawing Elongated Holes

Menubar:	Insert > Profile > Predefined Profile > Elongated Hole
Toolbar:	Profile > Predefined Profile sub-toolbar > Elongated Hole

An elongated hole is a geometry that consists of two parallel lines and two tangent arcs, as shown in Figure 3-11. To draw an elongated hole, choose the **Elongated Hole** tool from **Predefined Profile** sub-toolbar in the **Profile** toolbar; you will be prompted to specify the center to center distance. This is the distance between the centers of the two arcs in the elongated hole. Click on the geometry area to specify the first center point; you will be prompted to locate the endpoint of the distance. Move the cursor away from the first center point; a

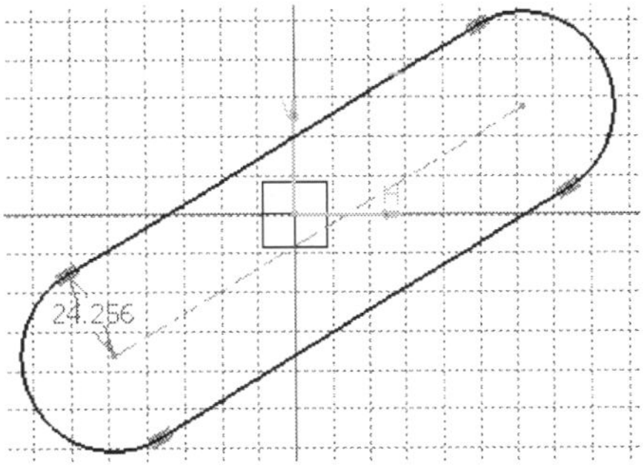

Figure 3-11 *An elongated hole profile*

center line will be attached to the cursor. Click to specify the endpoint; you will be prompted to define a point on the elongated hole. Move the cursor to specify the point. While moving the cursor, the preview of the elongated hole will be displayed in the geometry area. Figure 3-11 shows an elongated hole with the tangent and parallel constraints applied. These constraints will be discussed in later chapters.

Note

You can enter the parameters required to define the elongated hole in the respective edit boxes of the expanded ***Sketch tools*** *toolbar. The parameters include the coordinate values of the start point and endpoint of the line, angle value formed between the line and the horizontal reference, radius of the elongated hole, or the coordinate value of the point on the elongated hole.*

Drawing Cylindrical Elongated Holes

Menubar: Insert > Profile > Predefined Profile > Cylindrical Elongated Hole
Toolbar: Profile > Predefined Profile sub-toolbar > Cylindrical Elongated Hole

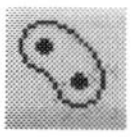

A cylindrical elongated hole is a geometry that comprises of four arcs. Each arc is tangent to its adjacent arcs, as shown in Figure 3-12. To draw a cylindrical elongated hole, choose the **Cylindrical Elongated Hole** tool from the **Predefined Profile** sub-toolbar in the **Profile** toolbar. On doing so, the **Sketch tools** toolbar will expand and you will be prompted to specify the center to center arc. Click in the geometry area to specify the center point; you will be prompted to specify the radius and the start point of the arc. Move the cursor away from the center point; a construction circle will get attached to the cursor. Click to specify the start point; you will now be prompted to move the cursor and specify the end point of the arc. Move the cursor away from the start point; a construction arc will get attached to the cursor. Click in the geometry area to specify its endpoint; you will be prompted to specify a point on the cylindrical elongated hole. Move the cursor away from the third point to specify a point; the preview of the cylindrical elongated hole is displayed. Click on it to specify a point; the cylindrical elongated hole will be created, as shown in Figure 3-12.

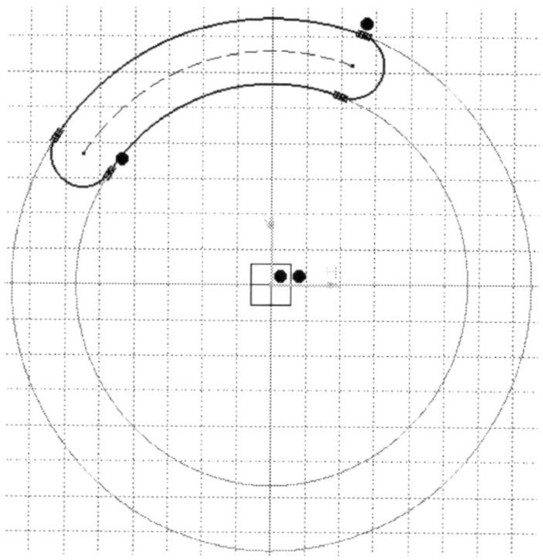

Figure 3-12 *A cylindrical elongated hole profile*

To draw a cylindrical elongated hole, you can also enter its parameters in various edit boxes of the **Sketch tools** toolbar.

Note

You will observe that sometimes while moving the cursor to specify a point on the geometry or define its shape and size, a ⊖ sign is displayed over the cursor. This sign suggests that you cannot specify a point for the element at the current location of the cursor.

Drawing Keyhole Profiles

Menubar: Insert > Profile > Predefined Profile > Keyhole Profile
Toolbar: Profile > Predefined Profile sub-toolbar > Keyhole Profile

A keyhole profile is a keyhole shaped geometry that comprises of two arcs and two lines, as shown in Figure 3-13. To draw a keyhole profile, invoke the **Keyhole Profile** tool from **Predefined Profile** sub-toolbar in the **Profile** toolbar; the **Sketch tools** toolbar will expand and you will be prompted to specify the start point. Click in the geometry area to specify the start point; you will be prompted to define the center point of the smaller radius arc. Click in geometry area to specify the center point of smaller arc; a dashed line will be displayed which defines the length of the keyhole profile. Keyhole attached to the cursor will be displayed in the geometric area and you will be prompted to specify a point on the keyhole profile to define the radius of the small arc. Move the cursor away from the center point of the small arc to preview the keyhole profile. Click on the preview to define the smaller radius; you will be prompted to specify a point on the keyhole profile to define the radius of the larger arc. Click on the preview of the keyhole to specify it. The final keyhole profile with the specified values will be displayed in the geometry area, refer to Figure 3-13. You can also specify the required parameters in the **Sketch tools** toolbar to draw a keyhole profile.

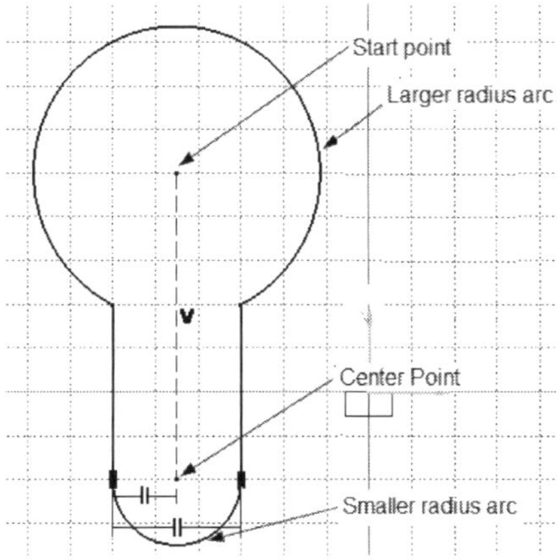

Figure 3-13 *A keyhole profile*

Drawing Polygons

Menubar: Insert > Profile > Predefined Profile > Polygon
Toolbar: Profile > Predefined Profile sub-toolbar > Polygon

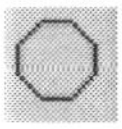

To draw a polygon, choose the **Polygon** tool from the **Predefined Profile** sub-toolbar in the **Profile** toolbar; you will be prompted to select the center of the circum circle or incircle. Also, the **Sketch tools** toolbar will expand displaying the coordinates of the polygon center. You can choose the **Circumcircle** or **Incircle** tool from the **Sketch tools** toolbar to create a circum circle polygon or an incircle polygon. After defining the center of the polygon, specify the radius and angle of the polygon. Specify the number of sides in the **Number of Sides** edit box in the **Sketch tools** toolbar or it can be directly varied by moving the mouse in clockwise and anti-clockwise direction to increase or decrease it respectively; a preview of the polygon with number of sides is displayed. Note that you can create a polygon with minimum and maximum number of sides as 3 to 24 respectively. You can also lock the number of sides of the polygon for next command by selecting the **As default number of sides** tool from the **Sketch tools** toolbar. This tool will be available on specifying the center point of the polygon. On selecting this tool the **Number of Sides** edit box becomes inactive at the pre specified value. Click in the graphics area to create the polygon. A pentagon created using the **Polygon** tool is shown in Figure 3-14.

Note

1. A circle passing through the vertices of the polygon is a circumcircle.
2. A circle touching all the sides of the polygon tangentially is called an incircle.
3. The default polygon that you can create will be a hexagon.

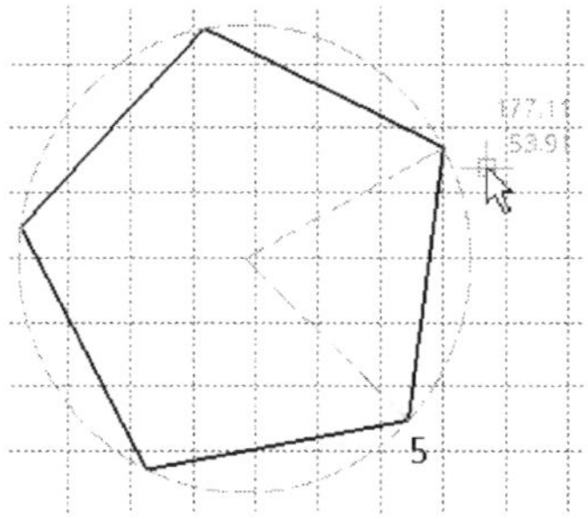

***Figure 3-14** Pentagon drawn using the **Polygon** tool*

Drawing Centered Rectangles

Menubar:	Insert > Profile > Predefined Profile > Centered Rectangle
Toolbar:	Profile > Predefined Profile sub-toolbar > Centered Rectangle

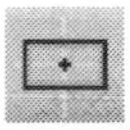

In CATIA V5, you can draw a rectangle that is centered about a point. To draw a centered rectangle, choose the **Centered Rectangle** tool from the **Predefined Profile** sub-toolbar in the **Profile** toolbar; you will be prompted to select a point to create the center of the rectangle. Click in the geometry area to specify the point and move the cursor away from that point; a preview of the rectangle will be displayed. Now, specify a point on any corner of the previewed rectangle. Figure 3-15 shows a centered rectangle, along with its center point and the point at its corner.

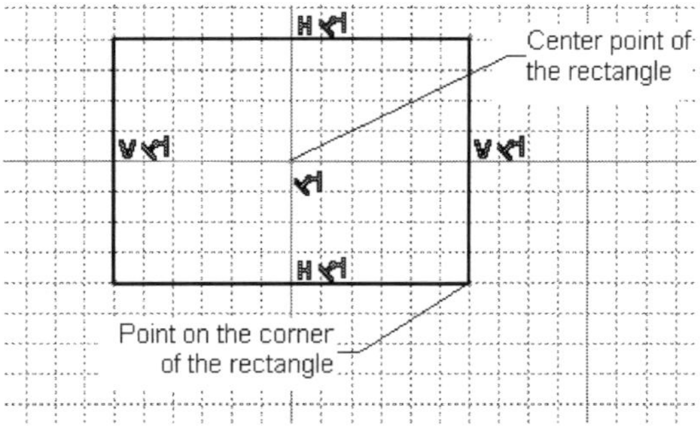

***Figure 3-15** Centered rectangle along with the center point and the point at its corner*

Drawing Centered Parallelograms

Menubar: Insert > Profile > Predefined Profile > Centered Parallelogram
Toolbar: Profile > Predefined Profile sub-toolbar > Centered Parallelogram

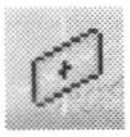

CATIA V5 also allows you to draw a centered parallelogram. Note that to draw such a parallelogram, you need to select two lines or construction lines. The opposite sides of the parallelogram will be parallel to these two lines. To create this type of parallelogram, choose the **Centered Parallelogram** tool from the **Predefined Profile** sub-toolbar in the **Profile** toolbar; you will be prompted to select the first line. Select the first line to which one set of sides of the parallelogram will be parallel. Next, select the second line; the parallelogram will be created with its center at the intersection point of the selected lines, and the second set of the opposite sides parallel to second selected line. Also, you will be prompted to select the endpoint to create a centered parallelogram. Move the cursor and click to specify a point on any one of the corners of the parallelogram. Figure 3-16 shows the centered parallelogram with the first and second reference lines and the point on the parallelogram.

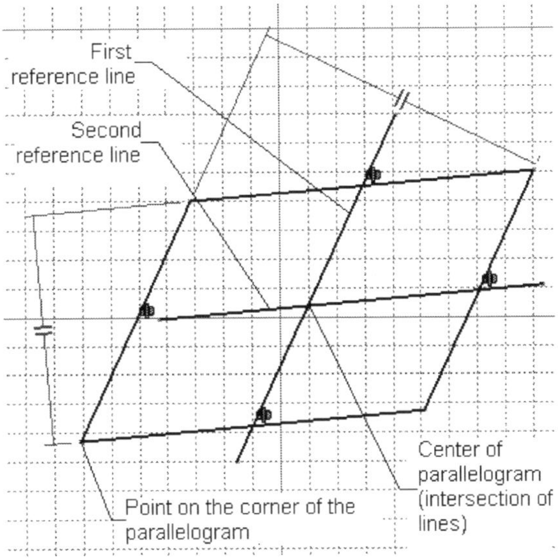

Figure 3-16 *Centered parallelogram with the first and second reference lines and the point on the parallelogram*

Note

After drawing the centered parallelogram, you can convert the reference lines of standard elements into construction elements by selecting them and using the ***Construction/Standard Element*** *tool in the* ***Sketch tools*** *toolbar.*

EDITING AND MODIFYING SKETCHES

In this section of the chapter, you will learn about the editing and modification tools used in the **Sketcher** workbench. These tools are used for trimming the sketches using the quick trim, breaking a sketched element, filleting the sketches, adding chamfer to the sketches, and so on. These tools are discussed next.

Trimming Unwanted Sketched Elements

Menubar:	Insert > Operation > Relimitations > Trim
Toolbar:	Operation > Relimitations sub-toolbar > Trim

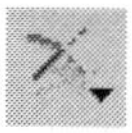

In the **Sketcher** workbench, you are provided with the **Trim** tool to remove the unwanted intersected portion of a sketched element. To do so, invoke the **Relimitations** sub-toolbar by choosing the down arrow provided on the right of the **Trim** tool in the **Operation** toolbar. The **Relimitations** sub-toolbar is shown in Figure 3-17.

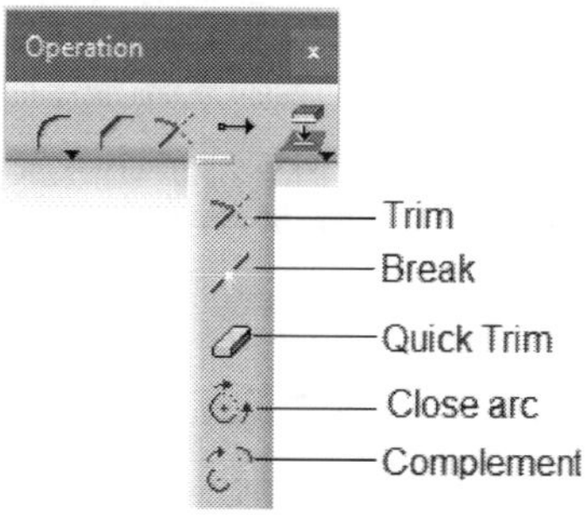

***Figure 3-17** The **Operation** toolbar with the **Relimitations** sub-toolbar*

Choose the **Trim** tool from the **Relimitations** sub-toolbar in the **Operation** toolbar; the **Sketch tools** toolbar will expand and you will be prompted to select a point or a curve type element. By default, the **Trim All Elements** tool is chosen in the expanded **Sketch tools** toolbar. Select the side of the first element that you need to retain. Next, select the second element that will act as the cutting edge to trim the first element. Figure 3-18 shows the elements selected to be trimmed and Figure 3-19 shows the resulting trimmed elements. Note that the sides that you click while selecting the elements will be retained after trimming.

After invoking the **Trim** tool, if you choose the **Trim First Elements** tool from the **Sketch tools** toolbar, only the first element will be trimmed with respect to the second element. Figure 3-20 shows the elements selected to be trimmed and Figure 3-21 shows the resulting trimmed elements.

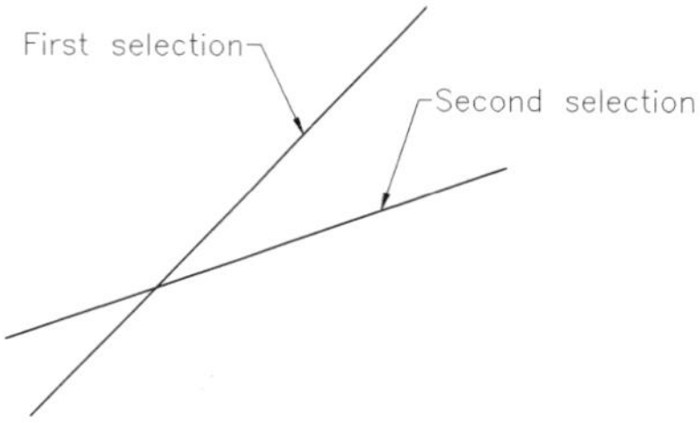

***Figure 3-18** Elements to be selected for trimming*

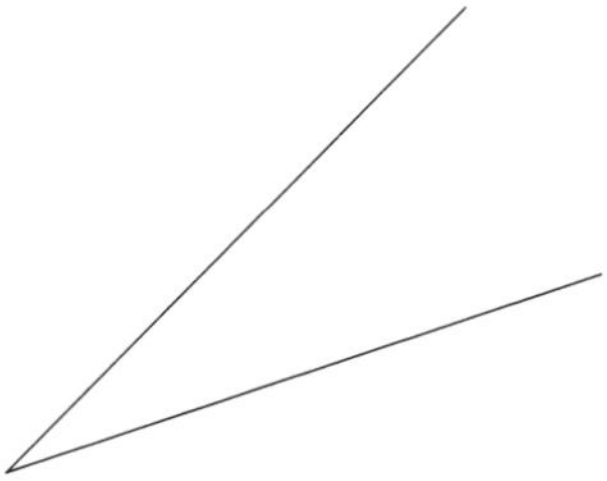

***Figure 3-19** The resulting trimmed elements*

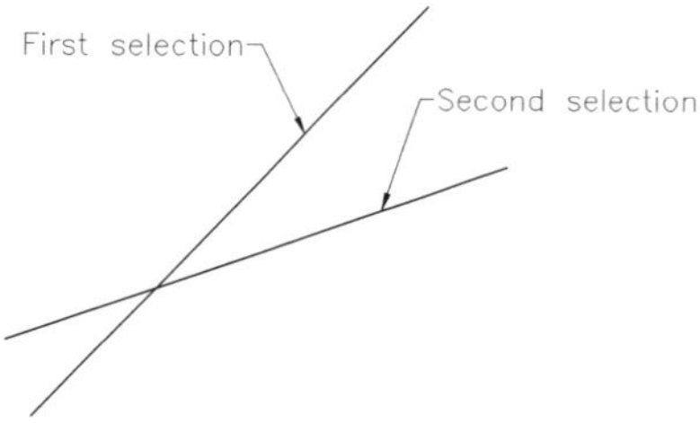

***Figure 3-20** Elements to be selected for trimming*

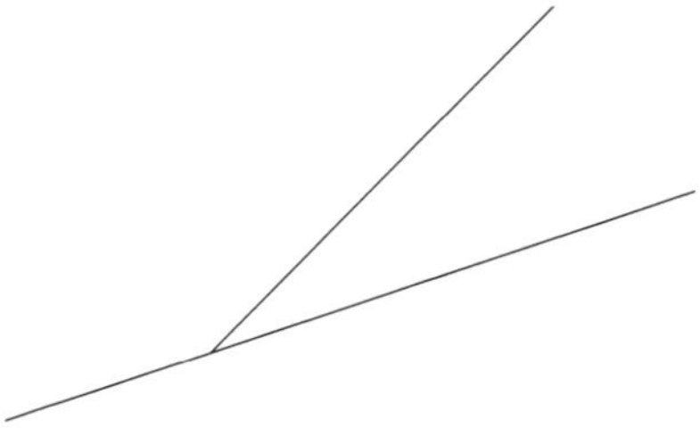

***Figure 3-21** The resulting trimmed element*

Extending Sketched Elements

Menubar:	Insert > Operation > Relimitations > Trim
Toolbar:	Operation > Relimitations sub-toolbar > Trim

In CATIA V5, you can also extend the sketched elements by using the **Trim** tool. Invoke this tool; you will be prompted to select a point or a curve type element. Select the sketched element to be extended and then select the destination up to which you need to extend it. You can also click anywhere in the drawing window to dynamically extend the selected element. If you are using the **Trim** tool to extend the elements, it is recommended to choose the **Trim First Element** tool from the **Sketch tools** toolbar. This is because if the destination to extension is another element, then the other portion of the element will be deleted. Figure 3-22 shows the element selected to be extended and also the destination element. Figure 3-23 shows the resulting extended element.

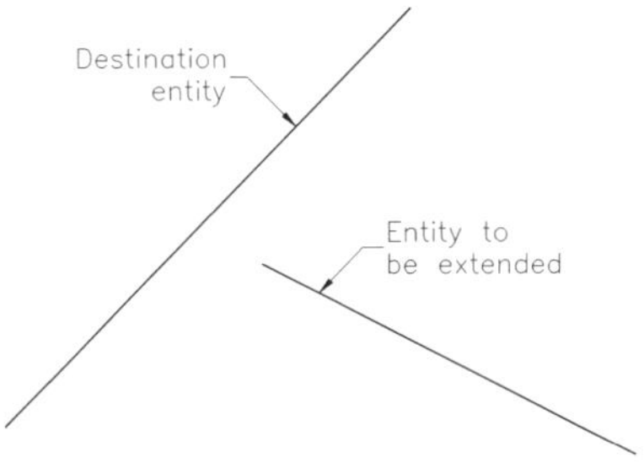

Figure 3-22 *Element selected to be extended*

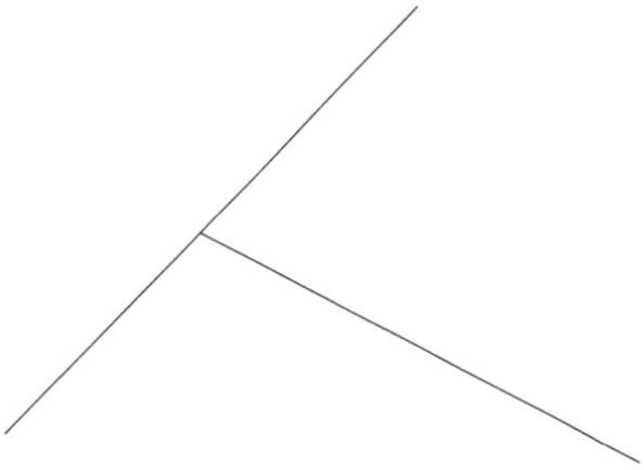

Figure 3-23 *The resulting extended element*

Breaking Elements

Menubar:	Insert > Operation > Relimitations > Break
Toolbar:	Operation > Relimitations sub-toolbar > Break

You can break a line or a curve at a desired position by using the **Break** tool. Different methods for breaking a line are discussed next.

To break a line at a point, choose the **Break** tool from the **Relimitations** sub-toolbar in the **Operation** toolbar; you will be prompted to select the element to be broken. Select the element to be broken; the start point and the probable break point will be displayed immediately. Next, click at the desired position to create the break point; the line will be broken at the desired point. The line is now divided into two segments and coincident constraints are created with both segments. These newly created coincident constraints make segments collinear to each other, refer to Figures 3-24 and 3-25.

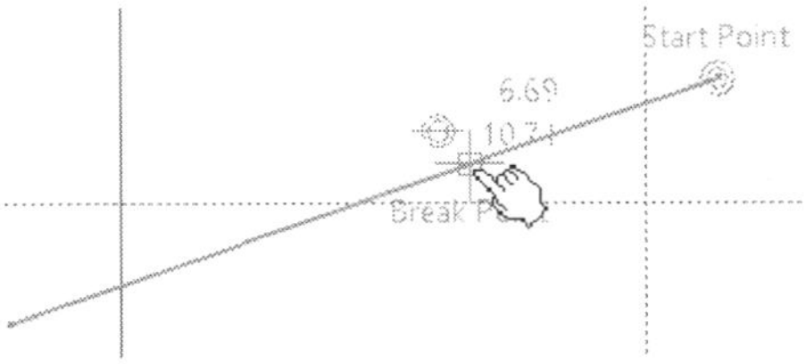

Figure 3-24 *Line selected to be broken at a point*

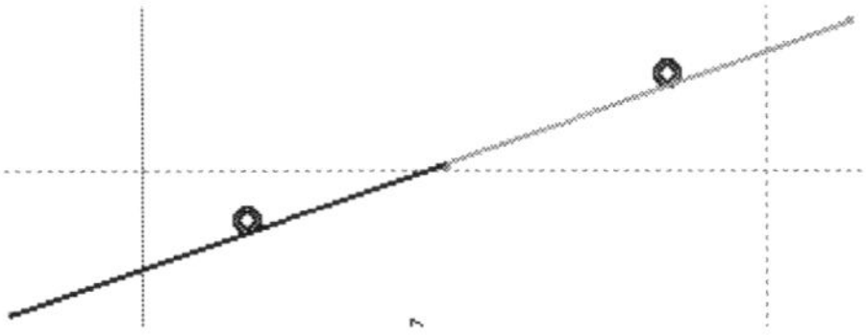

Figure 3-25 *Break point with coincident constraints created*

You can also break a line using a point that belongs to another line. To do so, select the line to be broken; the start point and the probable break point will be displayed immediately. Select the second line; the projection of the selected line will be displayed along with the possible constraint. The first line will be broken from the point where the projection of the second line intersects the first line. The line now comprises two segments with coincident constraint applied to them, refer to Figures 3-26 and 3-27.

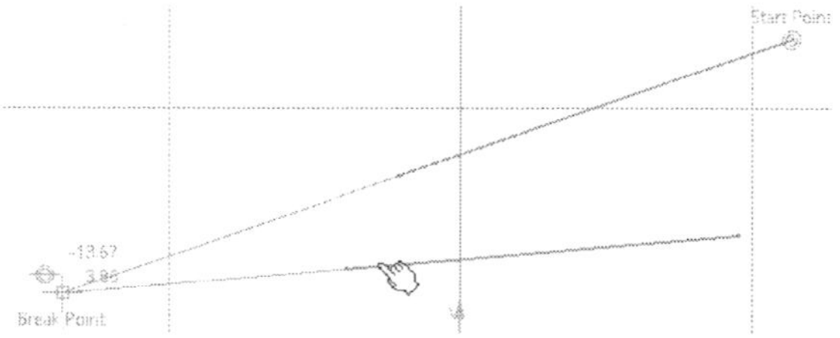

Figure 3-26 *Projection of the break point with possible constraints*

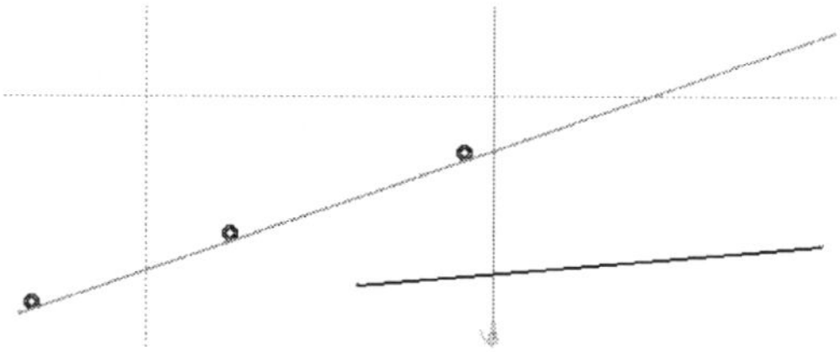

Figure 3-27 *Break point created at the projection of the selected line*

You can also break a line from the projection of any selected point. To do so, choose the **Break** tool and then select a line and a point; the line will be broken at the projection of the selected point. The resulting line will comprise two segments with coincident constraints applied to them, refer to Figures 3-28 and 3-29.

Note

If you do not want to consider the possible constraints while breaking the element, hold down the SHIFT key while selecting the second point.

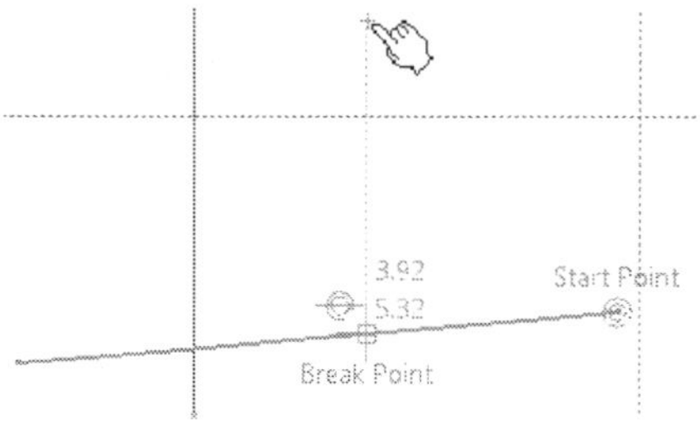

Figure 3-28 *Projection of the point displaying the break point*

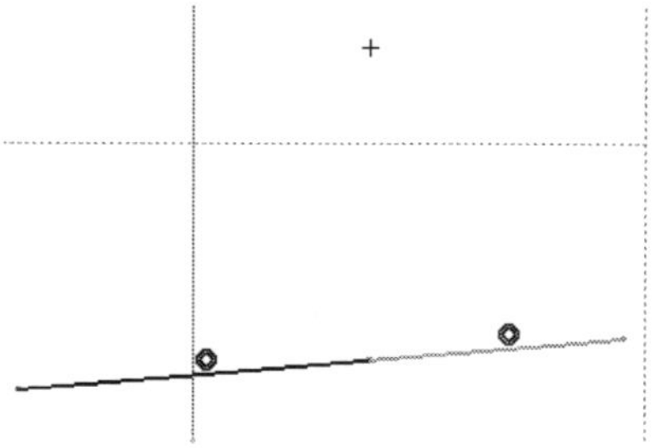

Figure 3-29 *Break point with coincident constraint created*

Closing Elements

Menubar: Insert > Operation > Relimitations > Close arc
Toolbar: Operation > Relimitations sub-toolbar > Close arc

The **Close arc** tool is used to close trimmed circles, ellipses, or splines. To close elements, choose this tool from the **Relimitations** sub-toolbar in the **Operation** toolbar and then select one or more elements to be closed. In case of an arc, the resultant entity will be a circle and in case of a trimmed spline, the resultant entity will be the original spline.

Tip

You can close an arc or a trimmed circle to form a complete circle using the ***Close arc*** *tool in the* ***Relimitations*** *sub-toolbar. Choose the* ***Close arc*** *tool from the* ***Relimitations*** *sub-toolbar and select the arc or trimmed circle to be closed. You can also close an arc or a trimmed circle by using the shortcut menu. To do so, right click to invoke the shortcut menu and then choose* ***"Name of the Element" > Close arc****; the trimmed circle or arc will be closed.*

You can also convert the complementary side of the trimmed circle or an arc into a standard element and remove its existing portion. To do so, choose the ***Complement*** *tool from the* ***Relimitations*** *sub-toolbar and select the element. You can also use the shortcut menu to convert the complementary portion into an element, as discussed above.*

Trimming by Using the Quick Trim Tool

Menubar: Insert > Operation > Relimitations > Quick Trim
Toolbar: Operation > Relimitations sub-toolbar > Quick Trim

The **Quick Trim** tool is used to quickly trim the unwanted sketched elements. To trim an element, choose the **Quick Trim** tool from the **Relimitations** sub-toolbar in the **Operation** toolbar; the **Sketch tools** toolbar will expand and you will be prompted to select a curve type element. By default, the **Break And Rubber In** option is chosen in the **Sketch tools** toolbar. This option results in the breakage of selected element with respect to the intersecting elements and the selected portion of the first element will be removed from the geometry. The **Break And Rubber Out** option also breaks the first selected element. But the selected portion will be retained. The **Break And Keep** option in the **Sketch tools** toolbar is used to break the selected element at the point of intersection. In this case, no portion will be removed from geometry. You can also remove the non-intersecting sketched elements using the **Quick Trim** tool. As a result, this tool also works as the **Delete** tool on the entities that are not intersected by any other entity.

Filleting Sketched Elements

Menubar: Insert > Operation > Corner > Corner
Toolbar: Operation > Corner sub-toolbar > Corner

In the **Sketcher** workbench of CATIA V5, you are provided with the **Corner** tool to fillet the sketched elements. When you choose this tool, the **Sketch tools** toolbar will expand and you will be prompted to select the first curve or a common point. Select the first element to be filleted; you will be prompted to select the second curve. Select it and specify the fillet radius in the **Radius** edit box in the expanded **Sketch tools** toolbar. You can also specify the fillet radius by dynamically moving the cursor and then specifying a point on the arc.

Note

The creation of the fillet depends on the point that is selected to specify the fillet radius in the dynamic fillet creation. You can also fillet two parallel lines using the ***Corner*** *tool.*

The **Sketch tools** toolbar, which expands on invoking the **Corner** tool, displays various options that are used to create a fillet with different types of trimming options. If you choose:

- the **Trim All Elements** tool, both the selected elements will be trimmed beyond the fillet region. This button is chosen by default. 
- the **Trim First Element** tool and then fillet the sketched elements, the resulting fillet will be created by trimming only the first element. The second element will be retained. 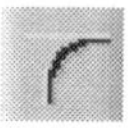
- the **No Trim** tool, the resulting fillet will be created by retaining both the selected elements. 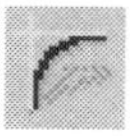
- the **Standard Lines Trim** tool, the resulting fillet will be created by retaining both the selected elements, and the retained elements will remain as standard elements. But if the elements extend beyond the corner selected to be trimmed, the extended portion will be removed. 
- the **Construction Lines Trim** tool, the resulting fillet will be created by retaining the selected elements, but the retained elements will be converted to construction elements. 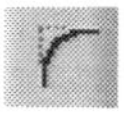
- the **Construction Lines No Trim** tool, the lines that extend beyond the corner will be retained as the construction elements. 

Figure 3-30 shows the elements to be selected and the fillet created using the **Trim All Elements** button. Figure 3-31 shows the fillet created using the **Trim First Element** tool and the **No Trim** tool.

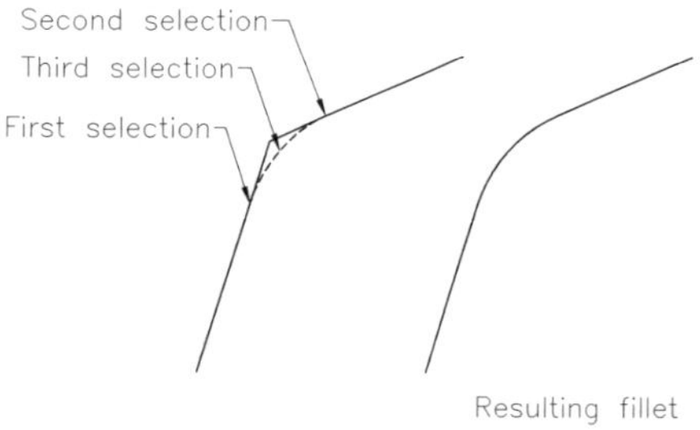

Figure 3-30 *Fillet created using the **Trim All Elements** button*

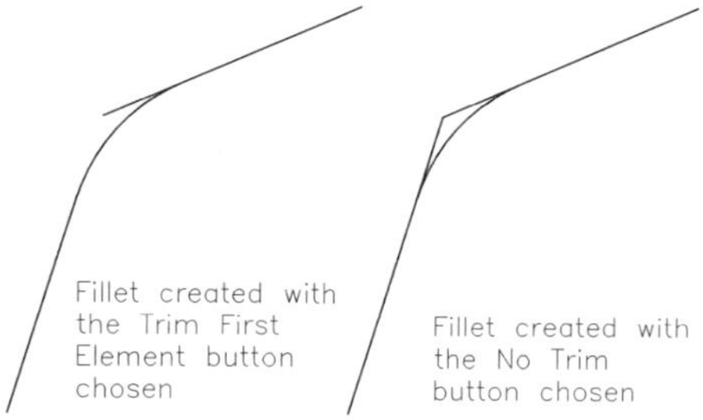

Figure 3-31 *Fillets created using the **Trim First Element** and **No Trim** buttons*

Creating a Tangent Arc

Menubar:	Insert > Operation > Corner > Tangent Arc
Toolbar:	Operation > Corner sub-toolbar > Tangent Arc

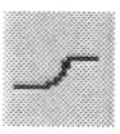

The **Tangent Arc** tool is used to create an arc tangent to a line. To create a tangent arc, choose this tool from the **Corner** sub-toolbar in the **Operation** toolbar; you will be prompted to select a curve. Select a line or curve; you will be prompted to select the end point of the corner. Select the end point of the curve; you will be prompted to locate the corner radius. Move the cursor to the desired location in the geometry area to locate the corner and click to create the tangent arc. As the tangent arc is created, rest of the line will disappear if the **Trim First Element** tool is chosen in the **Sketch tools** toolbar, as shown in Figure 3-32. To retain the rest of the line, choose the **No Trim** tool in the **Sketch tools** toolbar. If you want to keep the same radius value while creating other corners, click **Keep as default radius for next** in the **Sketch tools** toolbar. To create an arc tangent to a line/curve upto the end point of another line/curve, select a line/curve and then select the end point of the other line/curve; the tangent arc will be created between the selected entities, as shown in Figure 3-33. Note that the options in the **Sketch tools** toolbar will be available on selecting an element.

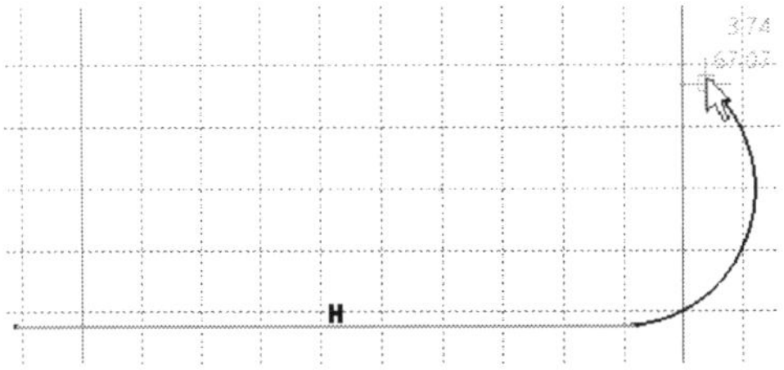

Figure 3-32 *Arc created tangent to a line*

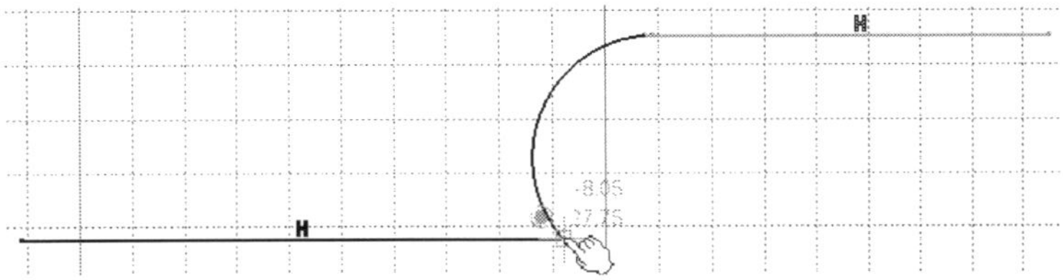

Figure 3-33 *Tangent arc created upto the end point of another curve*

Chamfering Sketched Elements

Menubar:	Insert > Operation > Chamfer
Toolbar:	Operation > Chamfer

The **Sketcher** workbench of CATIA V5 also provides you with the **Chamfer** tool to chamfer the sketched elements. On invoking this tool, the **Sketch tools** toolbar will expand and you will be prompted to select the first curve or a common point. Select the first element; you will be prompted to select the second element. When you select the second element, the **Sketch tools** toolbar will expand and the **Angle** and **Length** edit boxes will be activated. Specify the values in these edit boxes and press the ENTER key; the chamfer will be created and some dimensions will be applied to it. You can also dynamically specify the parameters of a chamfer. Figure 3-34 shows the elements selected and the resultant chamfer.

After selecting the geometries to be chamfered, the **Sketch tools** toolbar expands providing you with some options to specify the parameters of the chamfer. These options are explained next.

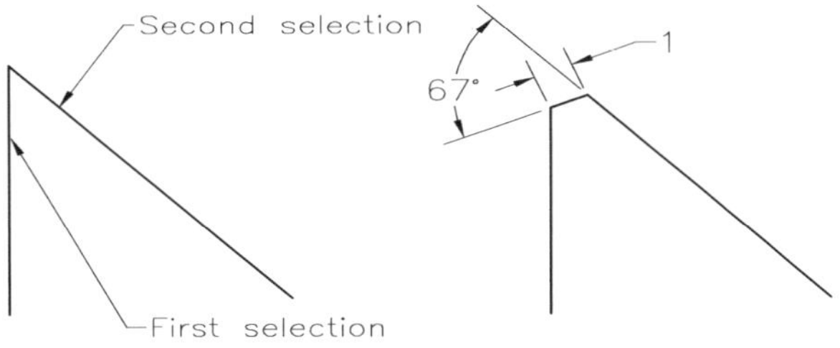

Figure 3-34 *Elements to be selected and the resulting chamfer*

If you choose the **Hypotenuse and Angle** tool, you need to specify the angle and length of the hypotenuse in the edit boxes in the **Sketch tools** toolbar. This button is chosen by default.

If you choose the **First and Second Length** tool, then you need to specify the chamfer distances in the **First length** and **Second length** edit boxes.

If you choose the **First Length and Angle** tool, then you need to specify the length of the chamfer from the first selection and also the angle of the chamfer.

You can also specify whether you want to trim or retain the elements using the other tools in the **Sketch tools** toolbar. These options are the same as those discussed while filleting the elements.

Mirroring Sketched Elements

Menubar:	Insert > Operation > Transformation > Mirror
Toolbar:	Operation > Transformation sub-toolbar > Mirror

You can mirror the sketched elements along the mirror line using the **Mirror** tool in the **Sketcher** workbench of CATIA V5. For mirroring the sketched elements, choose the down arrow on the right of the **Mirror** tool provided in the **Operation** toolbar; the **Transformation** sub-toolbar will be displayed, as shown in Figure 3-35. The tools in this sub-toolbar are known as the transformation tools.

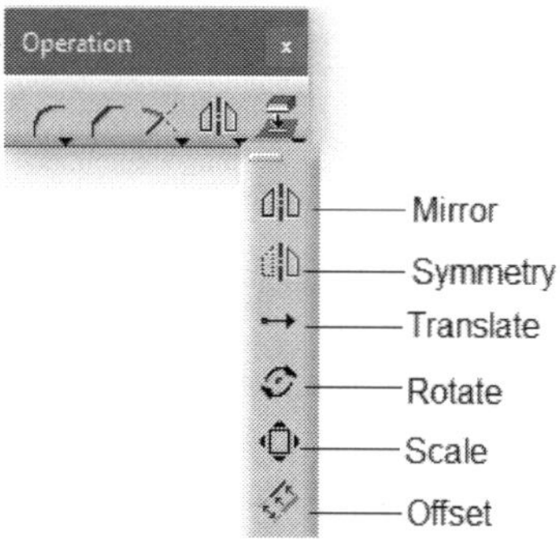

*Figure 3-35 The **Operation** toolbar with the **Transformation** sub-toolbar*

Select the sketched elements that you need to mirror by dragging a window around them. Alternatively, you can press and hold the CTRL key and select the elements for multiple element selection. Next, choose the **Mirror** tool from the **Transformation** sub-toolbar; you will be prompted to select the line or axis from which the elements will remain equidistant. Select a line, center line, or any of the axes as the mirror axis; the selected elements will be mirrored about the mirror axis and the symmetry constraints will be applied to the sketch on both sides of the mirror axis. Figure 3-36 shows the elements selected to be mirrored and the mirror line to be selected. Figure 3-37 shows the resulting mirrored sketch.

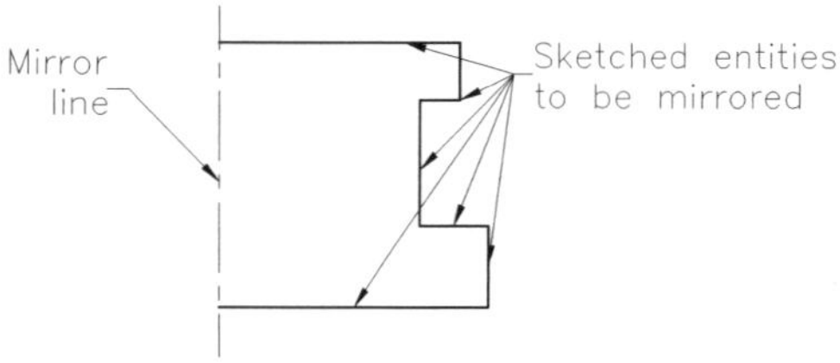

Figure 3-36 Elements selected to be mirrored and the mirror line to be selected

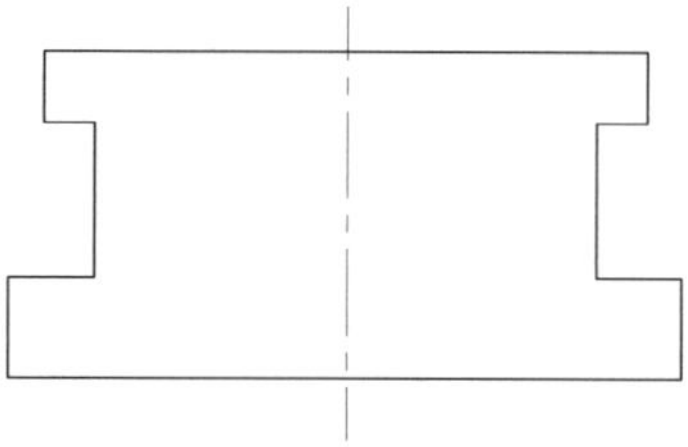

Figure 3-37 The resulting mirrored sketch

Mirroring Elements without Duplication

Menubar: Insert > Operation > Transformation > Symmetry
Toolbar: Operation > Transformation sub-toolbar > Symmetry

The **Symmetry** tool mirrors the sketched elements about a mirror axis but deletes the original elements. To mirror the elements without duplication, select the elements by dragging a window around them. Next, choose the **Symmetry** tool from the **Transformation** sub-toolbar in the **Operation** toolbar; you will be prompted to select the line or axis from which the elements will remain equidistant. Select the symmetry line; the selected elements will be mirrored on the other side of the symmetry line, while the original elements will be removed.

Tip
If you select the elements after invoking any of the transformation tools, you need to drag a window to select multiple elements. In such a case, you are not allowed to hold the CTRL key and select multiple elements.

Translating Sketched Elements

Menubar: Insert > Operation > Transformation > Translate
Toolbar: Operation > Transformation sub-toolbar > Translate

The **Sketcher** workbench provides you with the **Translate** tool to move the selected sketched elements from their initial position to the required place. To move the sketched elements, select them and then choose the **Translate** tool from **Transformation** sub-toolbar in the **Operation** toolbar; the cursor will be replaced by a point cursor and the **Translation Definition** dialog box will be displayed, as shown in Figure 3-38. Also, you will be prompted to select the transition start point. Select a point in the geometry area that will be used as the base point of translation. On selecting the base point, the **Value** spinner becomes active. Set the incremental translation distance in the **Value** spinner in the **Length** area of the **Translation Definition** dialog box and press the ENTER key; the dialog box will not be displayed any more. Specify a point in the geometry area to place the selected sketched element. As the **Duplicate mode** check box is selected by default and the value of the instance is set to 1, a copy of the selected element will be created at the specified distance. You can also increase the value of the increment using the **Instance(s)** spinner.

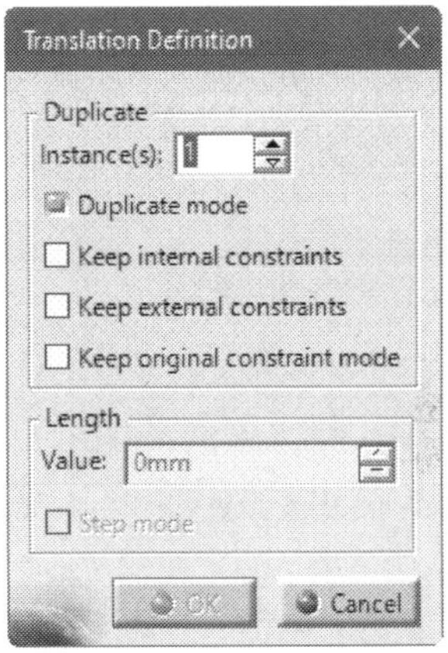

*Figure 3-38 The **Translation Definition** dialog box*

You can select the **Keep internal constraints** and **Keep external constraints** check boxes to retain the internal and external constraints, respectively. Select the **Keep original constraint mode** check box to keep the constraint in original mode. You will learn more about them in the later chapters.

Tip

*You can also specify the translate distance dynamically. To translate an element using this method, select it and invoke the **Translation Definition** dialog box. Next, specify the start point of the translation and then move the cursor to specify a location where you need to place it.*

Note

*If the **Duplicate mode** check box is cleared, then you can only move the selected elements but cannot copy them.*

Rotating Sketched Elements

Menubar:	Insert > Operation > Transformation > Rotate
Toolbar:	Operation > Transformation sub-toolbar > Rotate

The **Rotate** tool is used to rotate the sketched elements around a rotation center point. Select the elements by drawing a window around them and then choose the **Rotate** tool from the **Transformation** sub-toolbar in the **Operation** toolbar; the cursor will be replaced by a point cursor and the **Rotation Definition** dialog box will be displayed, as shown in Figure 3-39. Also, you will be prompted to select the rotation center point. Click to specify a point around which the selected sketched elements will be rotated; you will be prompted to select a point to define a reference line for the angle. Click to specify a point; you will be prompted to select a point to define angle. As you move the cursor to specify the third point, a preview of the rotated selected elements will also be displayed. Select a point to specify the rotation angle.

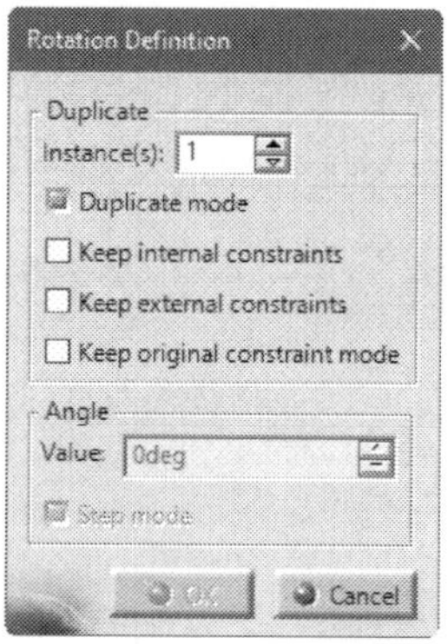

Figure 3-39 *The **Rotation Definition** dialog box*

Since the **Duplicate mode** check box is selected by default, another copy of the rotated element will be created. Figure 3-40 shows the points to be selected and the preview of the rotated instance of the selected elements. You can rotate the sketch elements in either direction (clockwise or counterclockwise) by entering a negative or positive value for the rotational angle in the **Value** spinner.

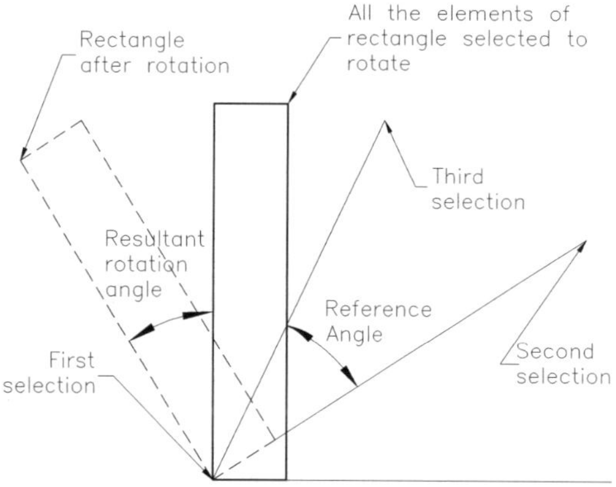

Figure 3-40 *Points to be selected and the preview of the rotated elements*

Scaling Sketched Elements

Menubar: Insert > Operation > Transformation > Scale
Toolbar: Operation > Transformation sub-toolbar > Scale

To scale the sketched elements, select them and then choose the **Scale** tool from **Transformation** sub-toolbar in the **Operation** toolbar; the **Scale Definition** dialog box will be displayed, as shown in Figure 3-41, and you will be prompted to select the scaling center point. Select a point in the drawing window; you will be prompted to select a point to define the scaling value. You can define the scaling factor dynamically in the geometry area or set its value in the **Value** spinner in the **Scale** area of the **Scale Definition** dialog box.

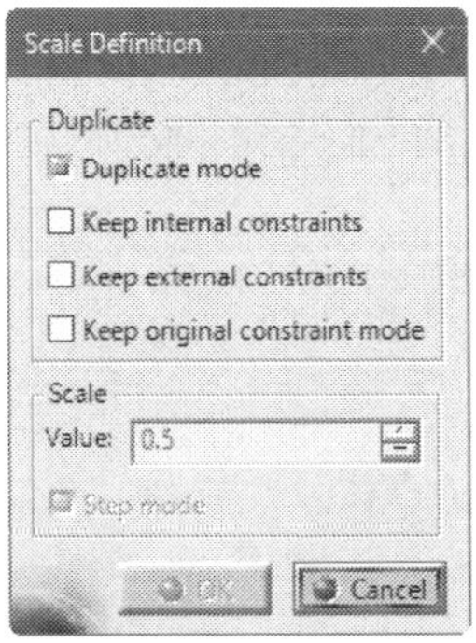

*Figure 3-41 The **Scale Definition** dialog box*

Offsetting Sketched Elements

Menubar: Insert > Operation > Transformation > Offset
Toolbar: Operation > Transformation sub-toolbar > Offset

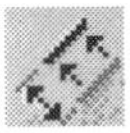

To offset the sketched elements, select them and then choose the **Offset** tool from the **Transformation** sub-toolbar in the **Operation** toolbar; the **Sketch tools** toolbar will expand. Specify the direction to offset the selected sketched elements and also the offset distance. Move the cursor to the side on which you want to specify the direction of the offset and then click in the geometry area; the selected element will be offset. You can also specify the offset distance in the **Offset** edit box in the expanded **Sketch tools** toolbar.

There are four additional tools in the expanded **Sketch tools** toolbar, as shown in Figure 3-42: **No Propagation**, **Tangent Propagation**, **Point Propagation**, and **Both Side Offset** tools. These tools are used to define the elements that will be selected to offset. By default, the **No Propagation** tool is chosen. As a result, only the selected element will be offset. If you choose the **Tangent Propagation** tool, all elements that are tangent to the selected element will be automatically selected. If you choose the **Point Propagation** tool, all elements connected end to end with the selected element and forming a closed loop will be selected automatically.

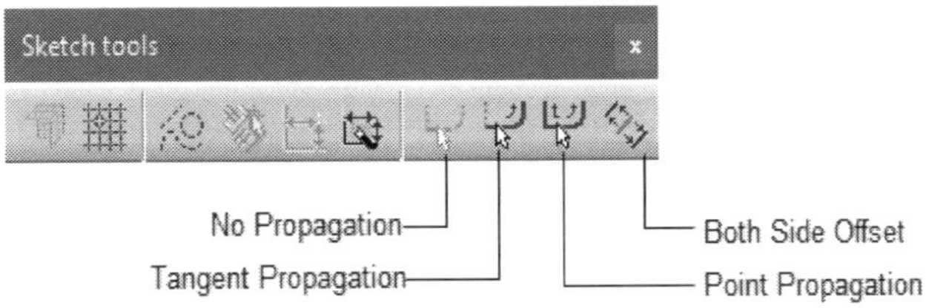

*Figure 3-42 The options in the expanded **Sketch tools** toolbar*

If you choose the **Both Side Offset** tool, the offset elements will be created on both sides of the selected element. Figure 3-43 shows the elements selected to be offset and the elements created after offsetting. In this figure, only the horizontal line is selected and then the **Point Propagation** tool is chosen. As a result, the entire closed loop is selected. Now click to create the offset of the closed loop.

Modifying the Offset Elements

You can modify the value of the offset elements using the **Profile Offset Definition** dialog box. To do so, expand the **Constraints** sub node of the Sketch node from the specification tree. Now double-click on the Profile Offset to be modified; the **Profile Offset Definition** dialog box will be displayed, as shown in Figure 3-44. You can modify the value of global offset distance in the **Value** spinner and the offset value of any particular element in the **Value** spinner from the **Local** area. The modified value of the element can be seen in the **Value** column next to the **Base** column. After modifying the parameters, choose the **OK** button from the **Profile Offset Definition** dialog box.

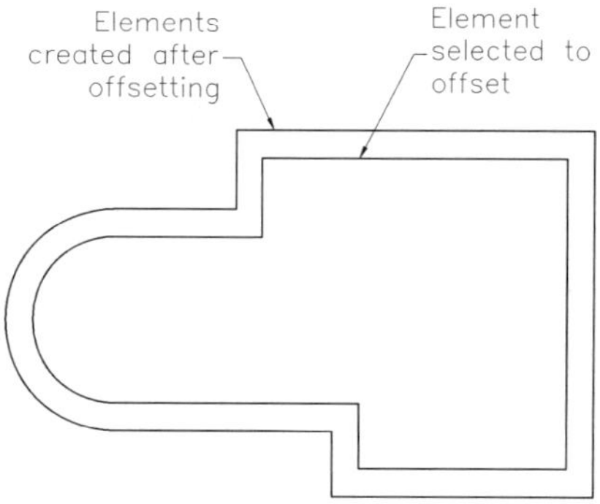

Figure 3-43 Elements created after offsetting the selected element

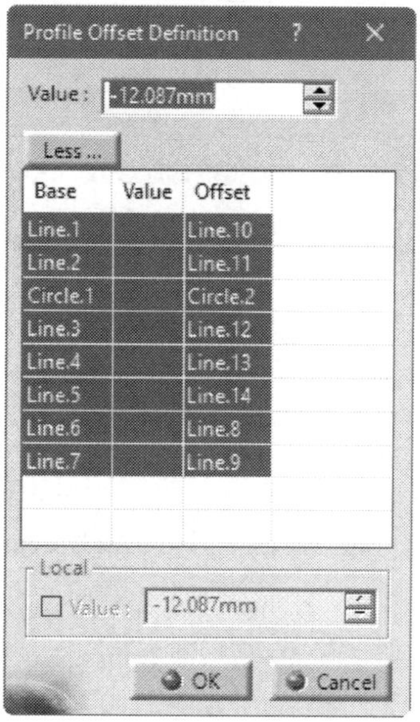

*Figure 3-44 The **Profile Offset Definition** dialog box*

Modifying Sketched Elements

In the **Sketcher** environment of CATIA V5, you can modify sketched elements by double-clicking on them. The process of modification of various sketched elements is discussed next.

Modifying the Sketched Line

You can modify a sketched line using the **Line Definition** dialog box. To modify a sketched line, double-click on it; the **Line Definition** dialog box will be displayed, as shown in Figure 3-45. You can modify the start point, endpoint, length, and angle of the line using the options available in it. After modifying the parameters, choose the **OK** button from the **Line Definition** dialog box. You can also convert the standard element to the construction element by selecting the **Construction element** check box at the bottom.

Modifying the Sketched Circle

You can modify a sketched circle by using the **Circle Definition** dialog box. You can invoke this dialog box by double-clicking on the sketched circle. The **Circle Definition** dialog box is shown in Figure 3-46. Using this dialog box, you can modify the coordinates of the center point and the radius of the circle. You can also change the standard element into a construction element by selecting the **Construction element** check box.

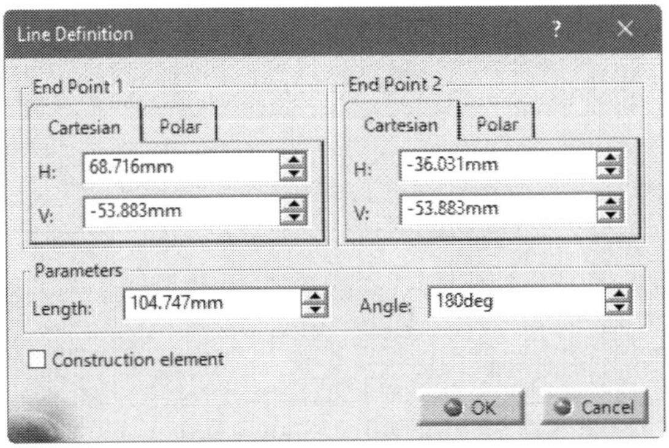

Figure 3-45 The ***Line Definition*** *dialog box*

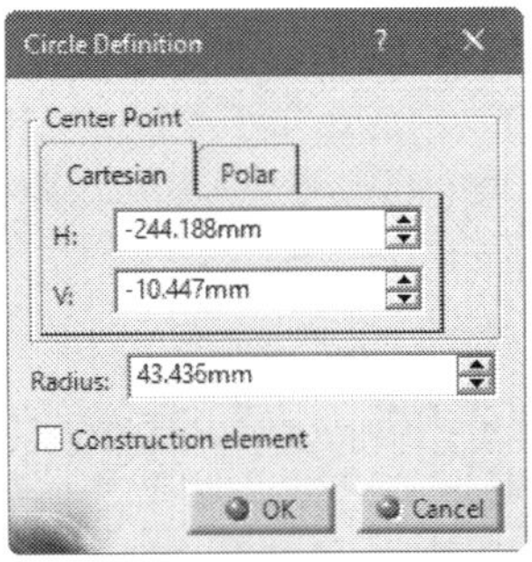

Figure 3-46 The ***Circle Definition*** *dialog box*

Modifying the Sketched Arc

The arcs are also modified using the **Circle Definition** dialog box. To invoke this dialog box, double-click on the arc to be modified. You can modify the coordinates of the center point and radius of the arc using the options in this dialog box.

Modifying the Sketched Spline

You can modify a spline using the **Spline Definition** dialog box which is displayed when you double-click on the spline. The **Spline Definition** dialog box is shown in Figure 3-47. The main objective of modifying a spline is to reshape it by selecting a sketched point that will be added as a control point to it. By default, the **Add Point After** radio button is selected in this dialog box and it is used to add a control point to the spline after the specified control point. As a result, you will be prompted to select the new control point. Click in the geometry area; the new point will be added in the spline as a control point. Alternatively, in the selection area of the **Spline Definition** dialog box, select a control point after which the new control point is to be added.

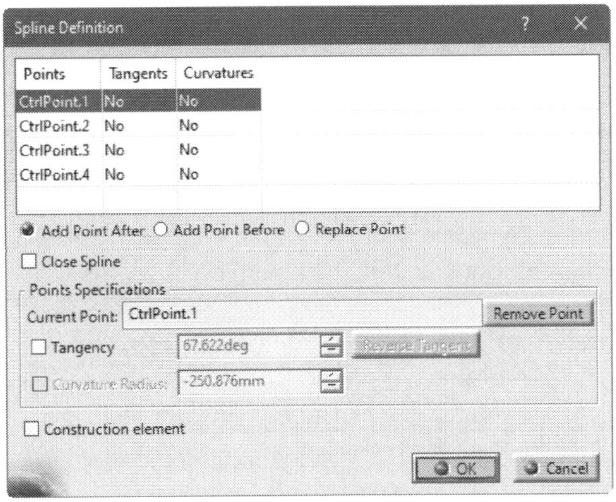

Figure 3-47 The ***Spline Definition*** *dialog box*

The **Add Point Before** radio button is selected to add a new control point before the selected control point. The **Replace Point** radio button is selected to replace the selected control point with the new control point. The **Close Spline** check box is used to close the endpoints of the spline. You can also set the tangency and curvature radius of the selected control point using the other options in this dialog box.

To change the control points of a spline, double-click on the control point that you want to edit; the **Control Point Definition** dialog box will be displayed, as shown in Figure 3-48. Also, the tangency and curvature radius manipulators will be displayed over the spline, as shown in Figure 3-49. Enter new coordinates in the **H** (horizontal) and **V** (vertical) boxes and select the **Tangency** check box to impose tangency on this control point. The tangency manipulator over the spline will be highlighted and the **Reverse Tangent** button and the **Curvature Radius** check box will get activated in the **Control Point Definition** dialog box. Specify the value of angle and radius in the **Tangency** and **Curvature Radius** edit boxes. Alternatively, you can use the manipulator to get the desired effect of angle and radius with respect to the H direction.

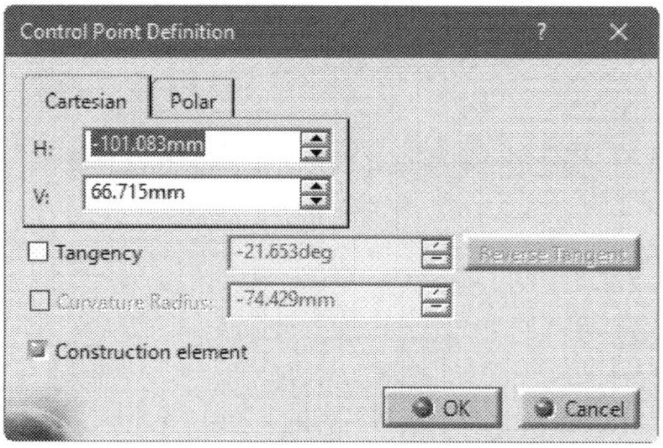

*Figure 3-48 The **Control Point Definition** dialog box*

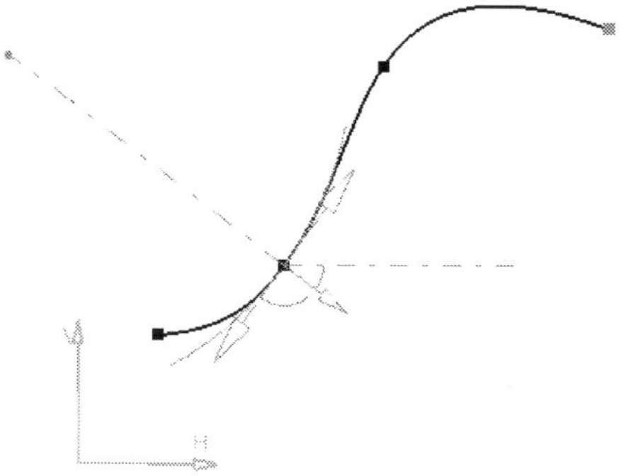

Figure 3-49 The spline with manipulators

Modifying the Sketched Point

To modify a sketched point, double-click on it; the **Point Definition** dialog box will be displayed, as shown in Figure 3-50. You can modify the coordinates of the point using the options available in this dialog box.

Modifying the Sketched Ellipse

To modify a sketched ellipse, double-click on it; the **Ellipse Definition** dialog box will be displayed, as shown in Figure 3-51. You can modify the coordinates of the center point, major radius, minor radius, and angle of the ellipse using the options available in this dialog box.

Similarly, you can modify the other sketched elements such as parabola, hyperbola, and so on.

Modifying the Sketched Elements by Dragging

You can also modify the parameters such as the size, shape, and position of the sketched elements by dragging. The modification of the sketched element can be done by dragging its start point, endpoint, profile, or control points.

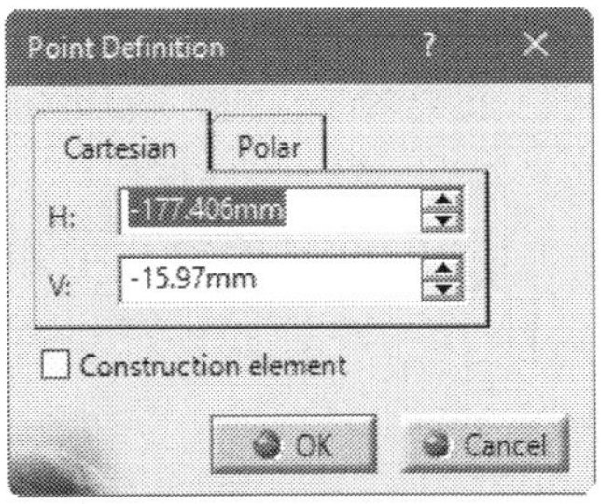

*Figure 3-50 The **Point Definition** dialog box*

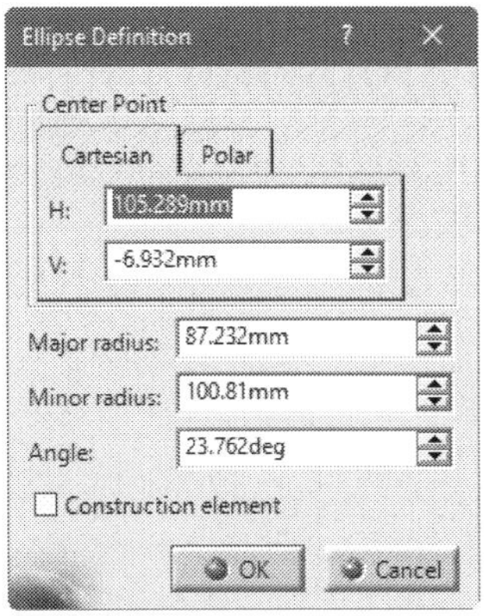

*Figure 3-51 The **Ellipse Definition** dialog box*

Deleting Sketched Elements

To delete a sketched element, select it and press the DELETE key. Alternatively, select the entity to be deleted and then right-click to invoke the contextual menu. Then, choose the **Delete** option from it.

TUTORIALS

Tutorial 1

In this tutorial, you will draw the sketch of the model shown in Figure 3-52. Its sketch is shown in Figure 3-53. Do not dimension the sketch. The solid model and its dimensions are given only for reference. **(Expected time: 30 min)**

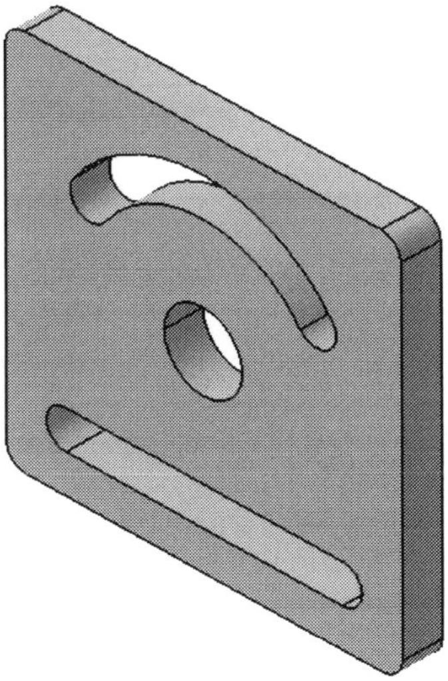

Figure 3-52 The model for Tutorial 1

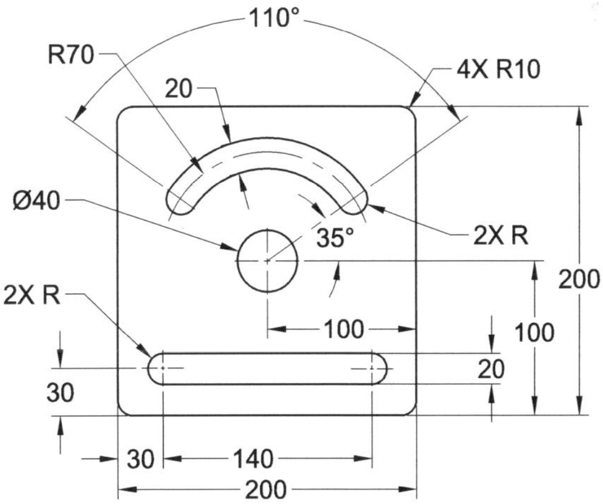

Figure 3-53 The sketch for Tutorial 1

The following steps are required to complete this tutorial:

a. Start a new part file in the **Part** workbench.
b. Draw the outer loop of the sketch using the **Rectangle** tool and then fillet it using the **Corner** tool.
c. Draw the inner loop of the sketch using the **Circle**, **Elongated Hole**, and **Cylindrical Elongated Hole** tools.
d. Save and close the file.

Starting a New File in the Part Workbench and Invoking the Sketcher Workbench

1. Choose the **New** button from the **Standard** toolbar to display the **New** dialog box.

2. Select the **Part** option from the **List of Types** list box and choose the **OK** button; the **New Part** dialog box is displayed.

3. Enter the name of the part as *c03tut1* in the **Enter part name** edit box of the **New Part** dialog box and select the **Enable hybrid design** check box if not already selected. Next, choose the **OK** button; a new file in the **Part** workbench is started.

4. Choose the **Sketch** tool from the **Sketcher** toolbar and select the **yz plane** from the Specification tree to enter in the **Sketcher** environment.

Drawing the Outer Loop of the Sketch

To draw the outer loop of sketches, you need to draw a rectangle using the **Centered Rectangle** tool. Next, you need to edit it by filleting its corners using the **Corner** tool. Before you draw the rectangle, you need to zoom out the geometry area to draw the rectangle conveniently.

1. Choose the **Zoom Out** tool from the **View** toolbar and make sure that the **Snap to Point** tool is chosen in the **Sketch tools** toolbar.

2. Choose the **Centered Rectangle** tool from the **Predefined Profile** sub-toolbar in the **Profile** toolbar; you are prompted to specify a point to create the center of the rectangle.

3. Move the cursor to the origin and click to specify a point to define the center point of the rectangle when the value of the coordinates above the cursor is displayed as 0,0; you are prompted to specify the second point to create a centered rectangle.

4. Move the cursor to a location whose coordinates are close to 100,100 and click when 200 is displayed in the **Height** and **Width** edit boxes in the **Sketch tools** toolbar; a rectangle is drawn, as shown in Figure 3-54. Now, click anywhere in the geometry area to make sure that the rectangle is no more selected.

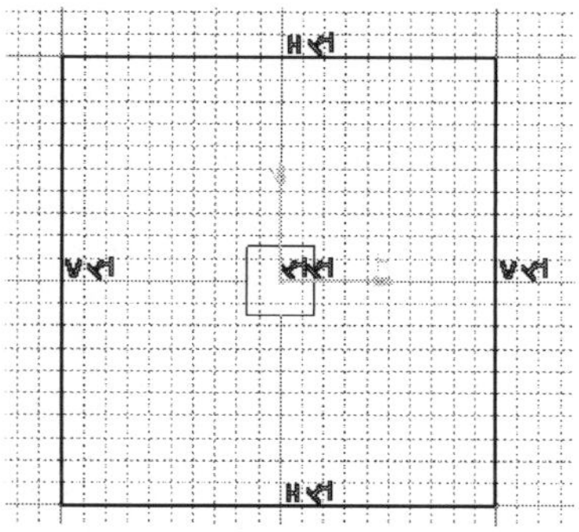

***Figure 3-54** The sketch after drawing the centered rectangle*

Next, you need to edit the rectangle by filleting its corners using the **Corner** tool.

The vertices to be selected are shown in Figure 3-55.

Note

*To invoke the **Predefined Profile** sub-toolbar, choose the down arrow available on the right of the **Rectangle** tool in the **Profile** toolbar; the **Predefined Profile** sub-toolbar is displayed.*

5. Choose the **Corner** tool from the **Operation** toolbar; you are prompted to select the first curve or a common point.

6. Select the upper right corner of the rectangle; the **Sketch tools** toolbar expands.

7. Enter **10** in the **Radius** edit box. Next, press the ENTER key; the selected corner of the rectangle is filleted and the radius value is displayed on the fillet.

8. Similarly, fillet the other corners of the rectangle by following the procedure mentioned in the previous steps. The final outer loop of the sketch after filleting all vertices is shown in Figure 3-56.

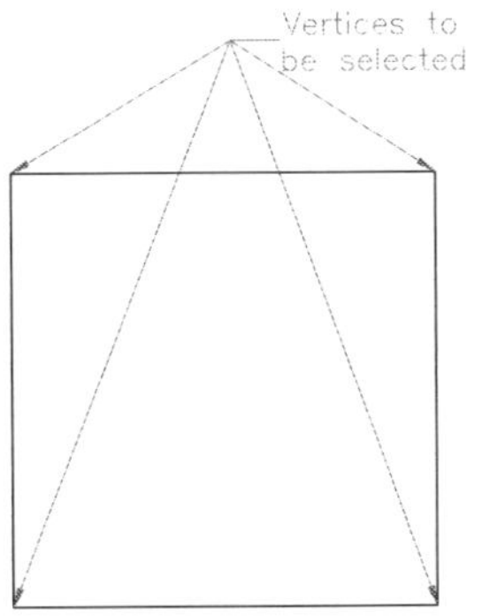

***Figure 3-55** The vertices to be selected*

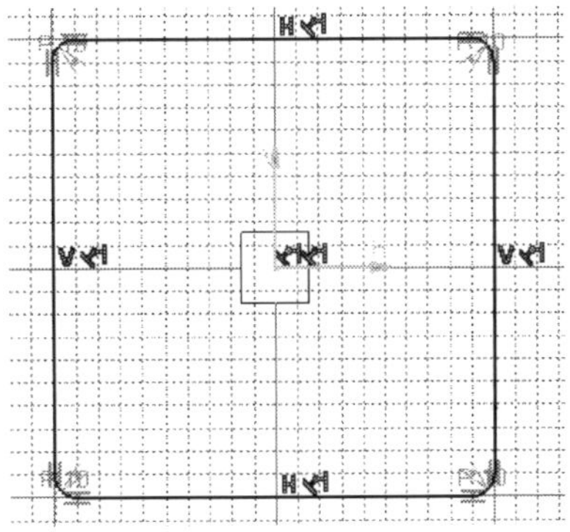

***Figure 3-56** The final outer loop of the sketch*

Drawing the Inner Loop of the Sketch

The inner loop will be drawn using the **Circle**, **Elongated Hole**, and **Cylindrical Elongated Hole** tools.

1. Choose the **Circle** tool from the **Profile** toolbar; you are prompted to select a point to define the center of the circle. Click to specify the center point of the circle at the origin.

2. Move the cursor horizontally toward the right and click to specify a point on the circle when the value of the radius is displayed as **20** in the **R** edit box. The sketch after drawing the circle is shown in Figure 3-57. Next, you need to draw an elongated hole using the **Elongated Hole** tool.

3. Choose the **Elongated Hole** tool from the **Predefined Profile** sub-toolbar in the **Profile** toolbar; you are prompted to define the center to center distance of the elongated hole.

4. Move the cursor to a location whose coordinates are -70,-70 and click to specify the start point of the center to center distance of the elongated hole.

5. Move the cursor horizontally toward the right and click to specify the endpoint at the location whose coordinates are 70,-70; you are prompted to define a point on the elongated hole.

6. Move the cursor vertically upward and click to specify a point on the elongated hole when the radius value is displayed as **10** in the **Radius** edit box. Figure 3-58 shows the sketch after drawing the elongated hole.

   After drawing the elongated hole, you need to draw a cylindrical elongated hole.

7. Choose the **Cylindrical Elongated Hole** tool from the **Predefined Profile** sub-toolbar in the **Profile** toolbar; you are prompted to define the center to center arc.

8. Move the cursor to the origin and click to specify it as the center point of the reference arc.

9. Enter **70** in the **R** edit box to specify the radius at the start point of the elongated hole. Next, press the ENTER key.

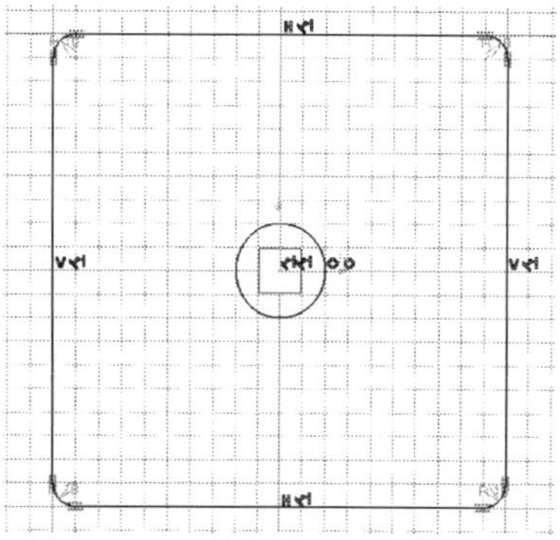

Figure 3-57 *The sketch after drawing the circle*

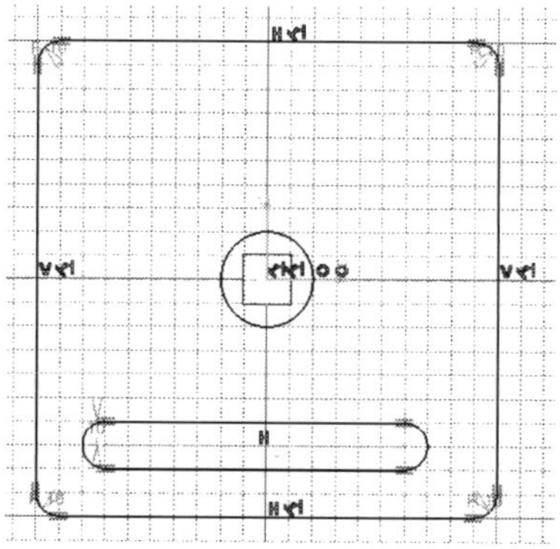

Figure 3-58 *The sketch after drawing the elongated hole*

10. Enter **35** in the **A** edit box and press the ENTER key.

11. Enter **110** in the **S** edit box to specify the angle between the start point and the endpoint of the elongated hole. Next, press the ENTER key.

Note

You will notice that some dimensional and geometrical constraints are applied to the sketch because the ***Geometrical Constraints*** *and* ***Dimensional Constraints*** *buttons are chosen in the* ***Sketch tools*** *toolbar by default.*

12. Enter the value **10** in the **Radius** edit box. Next, press the ENTER key; the final sketch is created, as shown in Figure 3-59. Now, click anywhere in the geometry area to make sure that the cylindrical elongated hole is no more selected.

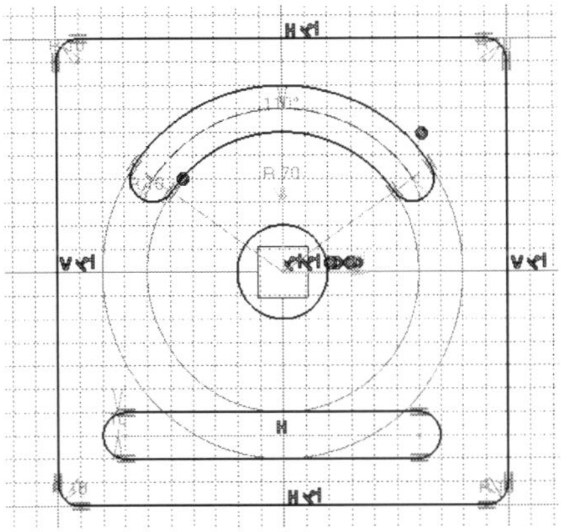

Figure 3-59 *The final sketch*

Saving and Closing the Sketch

1. Choose the **Save** button from the **Standard** toolbar to invoke the **Save As** dialog box. Create the *c03* folder inside the *CATIA* folder.

2. Choose the **Save** button from this dialog box; the file is saved at *C:\CATIA\c03*.

3. Close the part file by choosing **File > Close** from the menu bar.

Tutorial 2

In this tutorial, you will draw the sketch of the model shown in Figure 3-60. The sketch is shown in Figure 3-61. Do not dimension the sketch. The solid model and its dimensions are given only for your reference. **(Expected time: 30 min)**

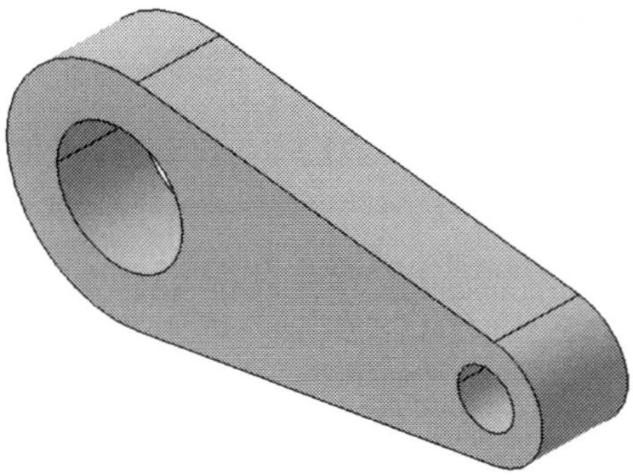

Figure 3-60 *The model for Tutorial 2*

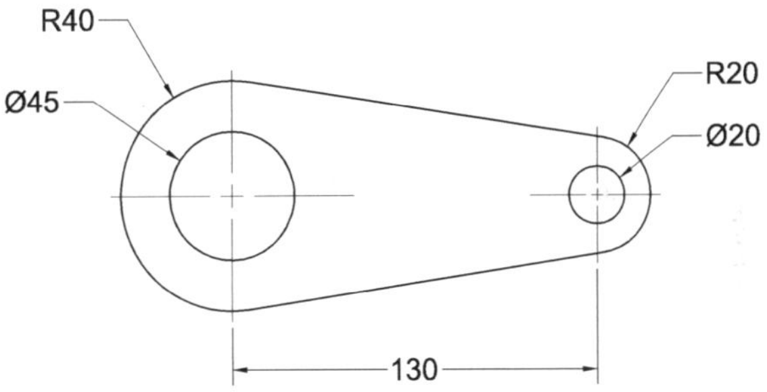

Figure 3-61 *The sketch of the model for Tutorial 2*

The following steps are required to complete this tutorial:

a. Start a new part file in the **Part** workbench and draw the outer loop of the sketch using the **Circle** and **By-Tangent Line** tools.
b. Trim the unwanted portion of the outer loop of the sketch using the **Quick Trim** tool.
c. Draw the inner loops of the sketch using the **Circle** tool.
d. Save and close the file.

Starting a New File in the Part Workbench

Before proceeding further, you need to start a new file in the **Part** workbench.

1. Choose the **New** button from the **Standard** toolbar; the **New** dialog box is displayed.

2. Select the **Part** option from the **List of Types** list box and choose the **OK** button; the **New Part** dialog box is displayed.

3. Specify the name of the part as *c03tut2* in the **Enter part name** edit box of the **New Part** dialog box. Select the **Enable hybrid design** check box from the **New Part** dialog box if not already selected, and then choose the **OK** button; a new **Part** file is started in the **Part** workbench.

4. Choose the **Sketch** tool from the **Sketcher** toolbar.

5. Select the **yz plane** from the Specification tree or from the graphics area to enter the **Sketcher** workbench.

Drawing the Sketch

The outer loop of the sketch is drawn using the **Circle** and **Bi-Tangent Line** tools.

1. Choose the **Circle** tool from the **Profile** toolbar; you are prompted to select a point to define the center of the circle.

2. Move the cursor toward the origin and click to specify the center point of the circle when the coordinates above the cursor display 0,0. Make sure the **Snap to Point** tool is chosen in the **Sketch tools** toolbar.

3. Move the cursor horizontally toward the right and click when the **R** edit box in the **Sketch tools** toolbar displays **40**; the circle is drawn, as shown in Figure 3-62.

4. Again, choose the **Circle** tool from the **Profile** toolbar.

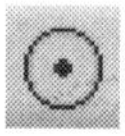

5. Move the cursor to a location whose coordinates are 130,0 and click to specify the center point of the circle.

6. Move the cursor toward the right and specify a point on the circle when **20** is displayed as the radius value in the **R** edit box of the **Sketch tools** toolbar. The sketch, after drawing the second circle, is shown in Figure 3-63.

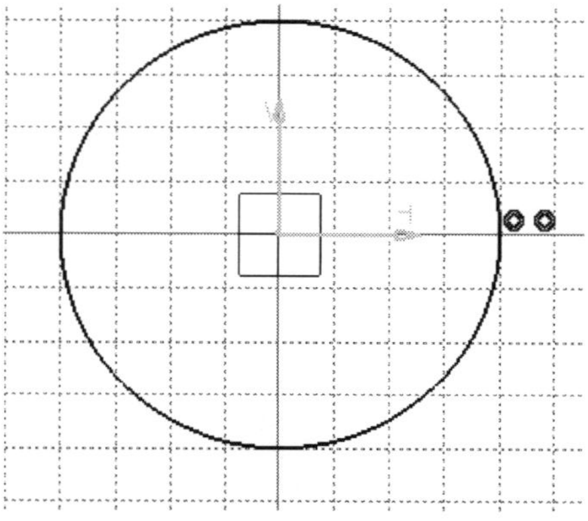

***Figure 3-62** The sketch after drawing the first circle*

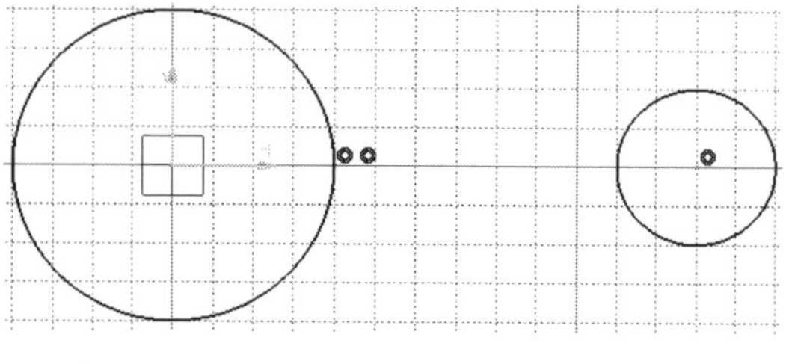

***Figure 3-63** The sketch after drawing the second circle*

After drawing both circles, you need to draw two lines in such a way that they are tangent to both of them. These lines will be drawn using the **Bi-Tangent Line** tool.

7. Choose the **Bi-Tangent Line** tool from the **Line** sub-toolbar; you are prompted to select the geometry to create a tangent line.

Note

*Choose the down arrow available on the right of the **Line** tool in the **Profile** toolbar; the **Line** sub-toolbar is displayed.*

8. Move the cursor to the first quadrant of the first circle and specify the start point of the line on its circumference; you are prompted to select the geometry to create a tangent line.

9. Move the cursor to the first quadrant of the second circle and specify the endpoint of the line on its circumference; a tangent line is drawn, as shown in Figure 3-64.

10. Similarly, draw a tangent line on the lower side of the sketch by selecting the fourth quadrants of the first and second circles. Figure 3-65 shows the sketch after drawing the second tangent line.

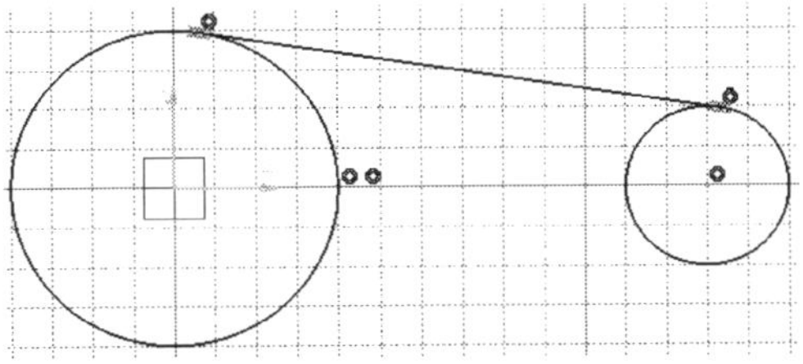

Figure 3-64 *The sketch after drawing the first tangent line*

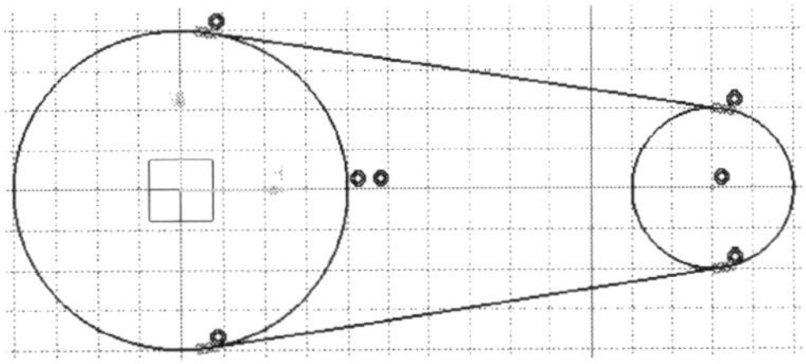

Figure 3-65 *The sketch after drawing the second tangent line*

Trimming the Unwanted Portion of the Outer Loop of the Sketch

After drawing the outer loop of the sketch, you need to trim its unwanted portion using the **Quick Trim** tool.

1. Click on the arrow available on the right of the **Trim** tool in the **Operation** toolbar; the **Relimitations** sub-toolbar is displayed. Double-click on the **Quick Trim** tool from the **Relimitations** sub-toolbar.

2. Click on the unwanted portion of the sketch, refer to Figure 3-66. The final sketch is shown in Figure 3-67.

Tip

If you click on any tool once to invoke a tool, the tool will be active for only one time use. However, if you double-click on the button to invoke a tool, the tool will remain active unless you terminate/exit it.

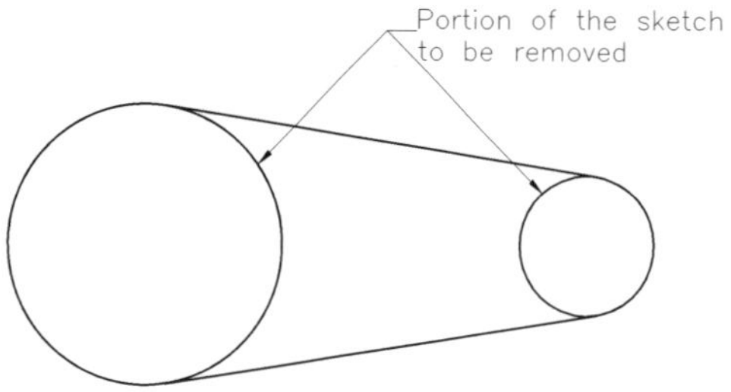

Figure 3-66 The unwanted portion of the sketch to be trimmed

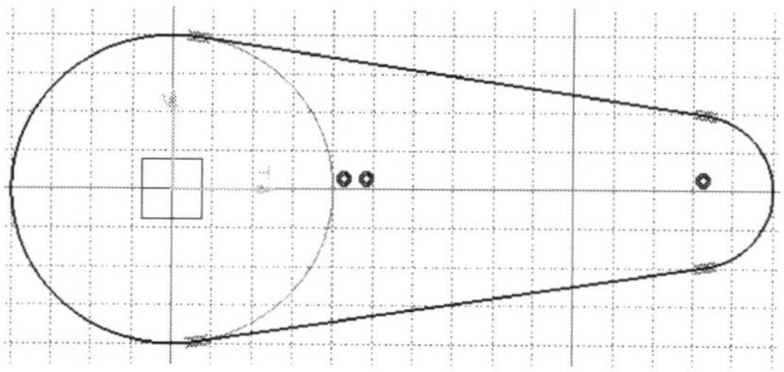

Figure 3-67 The sketch after trimming the unwanted portion

Drawing the Inner Loop of the Sketch

After drawing and trimming the outer loop of the sketch, you need to draw its inner loops consisting of two circles by using the **Circle** tool.

1. Double-click on the **Circle** tool from the **Profile** toolbar; you are prompted to select a point to define the center of the circle.

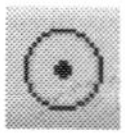

2. Move the cursor to the origin and click to specify the center point of the circle when the value of the coordinates is 0,0.

   As the radius of this circle is not in multiples of 10, you cannot define the radius in the geometry area. So, specify the radius of the circle in the **R** edit box of the expanded **Sketch tools** toolbar.

3. Enter **45/2** (radius of the circle) in the **R** edit box in the **Sketch tools** toolbar. Next, press the ENTER key; the inner circle is created.

4. As you double-clicked on the **Circle** tool, the **Circle** tool will still be active. Specify the center point of the second circle at a location whose coordinates are 130,0.

5. Move the cursor horizontally toward the right and click when the coordinate values above the cursor are 140,0. The final sketch after creating the outer and inner loops is shown in Figure 3-68. Press the ESC key to exit the selection set and the **Circle** tool.

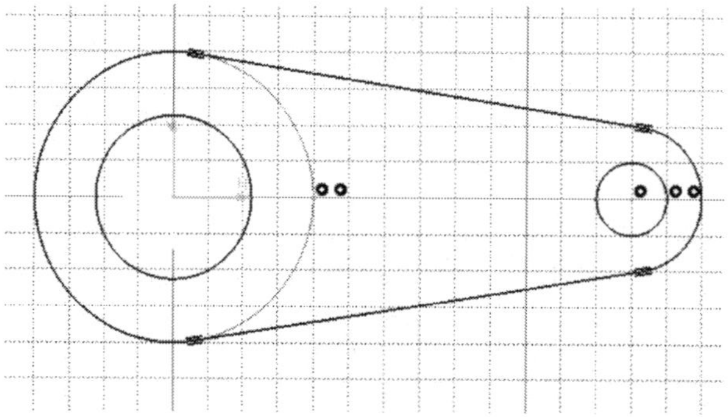

Figure 3-68 Final sketch after creating the outer and inner loops

Saving and Closing the Sketch

1. Choose the **Save** button from the **Standard** toolbar to invoke the **Save As** dialog box.

2. Choose the **Save** button from this dialog box; the file is saved at *C:\CATIA\c03*.

3. Close the part file by choosing **File > Close** from the menu bar.

Tutorial 3

In this tutorial, you will draw the sketch of the model shown in Figure 3-69. The sketch is shown in Figure 3-70. Do not dimension the sketch. The solid model and its dimensions are given only for your reference. **(Expected time: 30 min)**

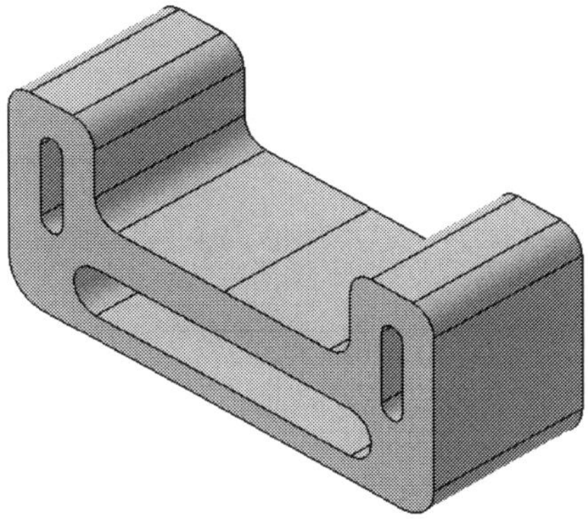

Figure 3-69 *The model for Tutorial 3*

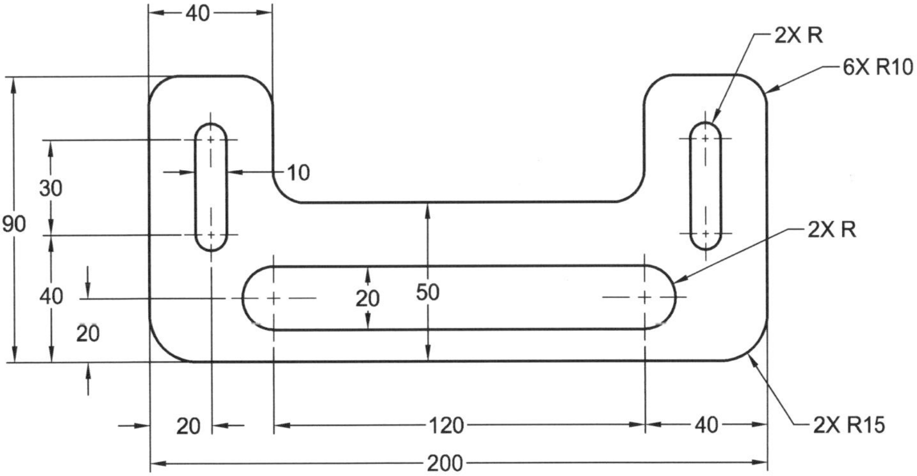

Figure 3-70 *The sketch of the model for Tutorial 3*

The following steps are required to complete this tutorial:

a. Start a new part file in the **Part** workbench and draw the right half of the sketch using the **Profile** and **Elongated Hole** tools.
b. Mirror the sketch along the vertical axis of origin.
c. Draw the elongated hole in the lower portion of the sketch.
d. Save and close the file.

Starting a New File and Invoking the Sketcher Workbench

1. Start a new file with the name *c03tut3* in the **Part** workbench.

2. Choose the **Sketch** button from the **Sketcher** toolbar and select the **yz plane** from the specification tree; the **Sketcher** workbench is invoked.

Drawing the Right Portion of the Sketch

It is evident from the figure that the sketch is symmetrical about the vertical axis. Therefore, you will draw only the right portion of the sketch and then mirror it about the vertical axis of the origin.

1. Choose the **Profile** tool from the **Profile** toolbar; you are prompted to select the start point of the profile.

2. Move the cursor to the origin and click to specify the start point of the line at this location.

3. Move the cursor horizontally toward the right and click to specify the endpoint at a location where the coordinates are 100,0; a rubber-band line is attached to the cursor.

4. Move the cursor vertically upward and click to specify the endpoint at a location where the coordinates are 100,90; another rubber-band line is attached to the cursor.

5. Move the cursor horizontally toward the left and click to specify the endpoint at a location where the coordinates are 60,90.

6. Move the cursor vertically downward and click to specify the endpoint at a location where the coordinates are 60,50.

7. Move the cursor horizontally toward the left and click to specify the endpoint at a location where the coordinates are 0,50.

8. Again, choose the **Profile** tool from the **Profile** toolbar to exit the tool.

   Next, you need to fillet the corners of the sketch using the **Corner** tool.

9. Choose the **Corner** tool from the **Operation** toolbar.

10. Select the lower right vertex of the sketch and set **15** as the value of the radius in the **Radius** edit box in the **Sketch tools** toolbar. Next, press ENTER.

11. Similarly, fillet other corners of the sketch with a radius value 10, refer to Figure 3-70.

12. Draw a vertical elongated hole on the right of the sketch using the **Elongated Hole** tool such that the coordinate value of start and end points are 80,40 and 80,70, respectively. Specify the value of radius as **5** in the **Radius** edit box of the **Sketch tools** toolbar, refer to Figure 3-71.

Mirroring the Sketch

After drawing the right-half of the sketch, you need to mirror it using the **Mirror** tool about the vertical axis of the origin.

1. Drag a window around all the sketched elements to select them. Next, press and hold the CTRL key and select the vertical and horizontal axes displayed at the origin to remove them from the selection set if they are also selected.

2 Choose the **Mirror** tool from the **Operation** toolbar; you are prompted to select the line or axis from which the elements will remain equidistant.

3. Select the vertical axis; the sketch is mirrored to the other side of the selected axis, as shown in Figure 3-72.

4. Draw the horizontal elongated hole such that the coordinate values of the start and end points are 60,20 and -60,20, respectively. Specify the value of radius as **10** in the **Radius** edit box of the **Sketch tools** toolbar. The final sketch is shown in Figure 3-73. Press the ESC key to exit the selection set.

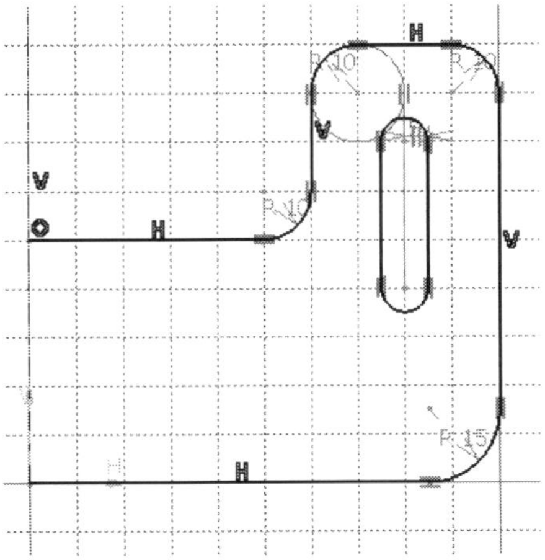

Figure 3-71 *The sketch after drawing the elongated hole*

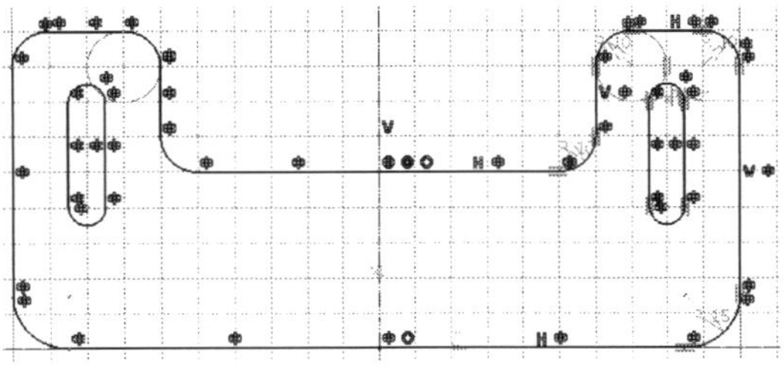

Figure 3-72 *The sketch after mirroring*

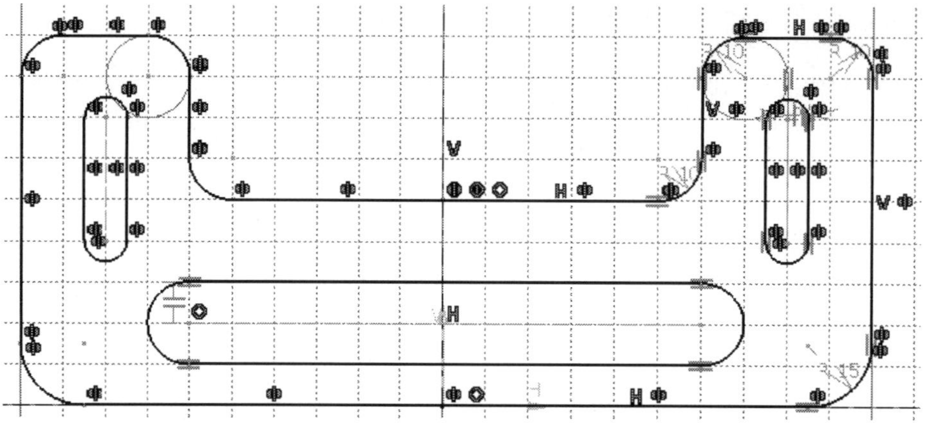

Figure 3-73 The sketch after creating the horizontal elongated hole

Saving and Closing the Sketch

1. Choose the **Save** button from the **Standard** toolbar to invoke the **Save As** dialog box.

2. Choose the **Save** button in this dialog box; the file is saved at *C:\CATIA\c03*.

3. Close the part file by choosing **File > Close** from the menu bar.

Self-Evaluation Test

Answer the following questions and then compare them to those given at the end of this chapter:

1. After invoking the **Quick Trim** tool if you choose the __________ button from the **Sketch tools** toolbar and then select the sketched element, the selected element will break at the intersection.

2. The __________ tool is also used to extend the sketched elements.

3. You can offset the sketched elements by selecting them and then choosing the __________ tool from the **Transformation** sub-toolbar.

4. You can modify a sketched arc using the __________ dialog box.

5. You can modify a sketched ellipse using the __________ dialog box.

6. You can create a parabola by using the **Parabola by Focus** tool from the **Conic** toolbar. (T/F)

7. In CATIA V5, you can draw a hexagon using the **Rectangle** tool. (T/F)

8. You cannot draw an n-sided polygon by using the **Profile** tool. (T/F)

9. In the **Sketcher** workbench of CATIA V5, you cannot trim the sketched elements. (T/F)

10. You can draw a key hole profile in the **Sketcher** workbench of CATIA V5. (T/F)

Review Questions

Answer the following questions:

1. Which of the following dialog boxes is used to modify a sketched point?

   (a) **Sketched Point** (b) **Point Definition**
   (c) **Modify Point** (d) None of these

2. Which of the following properties of a line cannot be modified using the **Line Definition** dialog box?

   (a) **End Point 1** (b) **End Point 2**
   (c) **Length** (d) **Color**

3. Which of the following tools is used to fillet the sketched elements?

   (a) **Fillet** (b) **Corner**
   (c) **Chamfer** (d) None of these

4. Which of the following tools is used to draw a parallelogram by specifying the center point?

   (a) **Parallelogram with mid point** (b) **Centered Rectangle**
   (c) **Centered Parallelogram** (d) **Circle**

5. Which of the following drop-downs is used to invoke the **Keyhole Profile** tool?

   (a) **Transformation** (b) **Relimitations**
   (c) **Operation** (d) **Predefined Profile**

6. You can scale the sketched elements by selecting them and then choosing the **Translate** tool from the **Transformation** toolbar. (T/F)

7. The **Rotate** tool is used to rotate the sketched elements. (T/F)

8. You can create complementary portion of an arc or a trimmed circle by choosing the **Complement** tool from the **Relimitations** sub-toolbar and then by selecting the element. (T/F)

9. In the **Sketcher** environment of CATIA V5, you can modify a sketched element by double-clicking on it. (T/F)

10. You can create a cylindrical elongated hole by using the **Elongated Hole** tool. (T/F)

EXERCISES

Exercise 1

Draw the sketch of the model shown in Figure 3-74. The sketch to be drawn is shown in Figure 3-75. Do not dimension the sketch. The solid model and dimensions are given only for your reference.
(Expected time: 30 min)

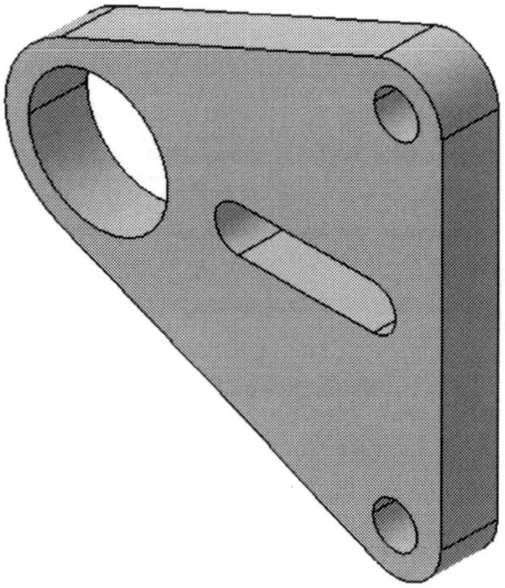

Figure 3-74 *The model for Exercise 1*

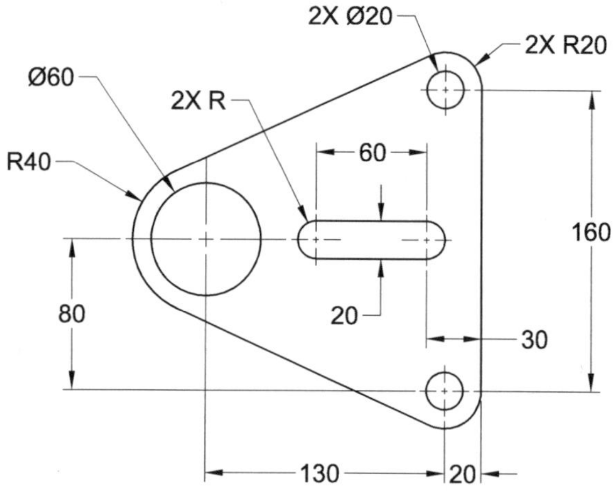

Figure 3-75 *The sketch for Exercise 1*

Exercise 2

Draw the sketch of the model shown in Figure 3-76. The sketch to be drawn is shown in Figure 3-77. Do not dimension the sketch. The solid model and dimensions are given only for your reference.
(Expected time: 30 min)

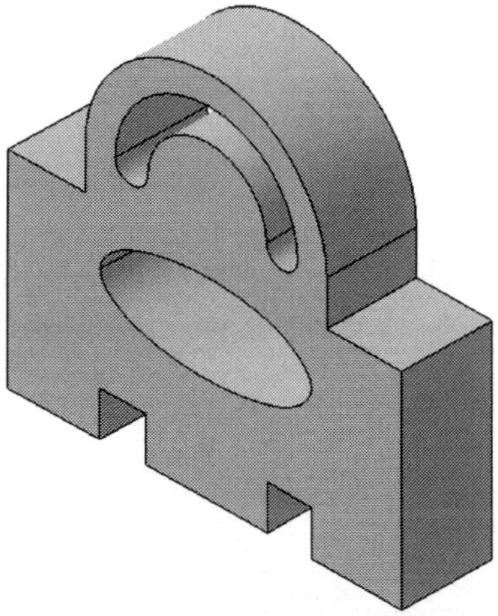

Figure 3-76 *The model for Exercise 2*

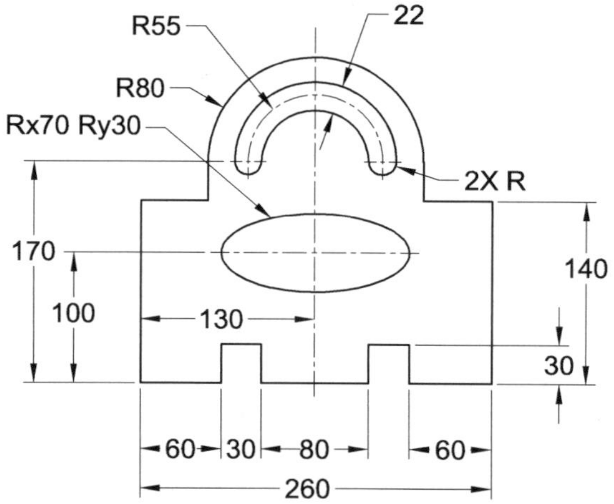

Figure 3-77 *The sketch for Exercise 2*

Exercise 3

Draw the sketch of the model shown in Figure 3-78. The sketch to be drawn is shown in Figure 3-79. Do not dimension the sketch. The solid model and dimensions are given only for your reference. **(Expected time: 30 min)**

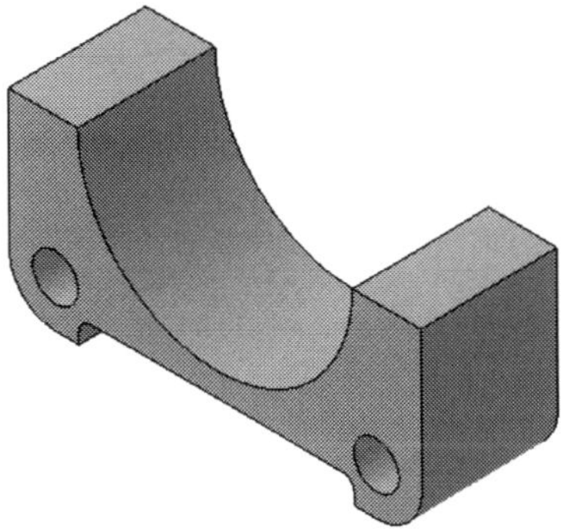

Figure 3-78 The model for Exercise 3

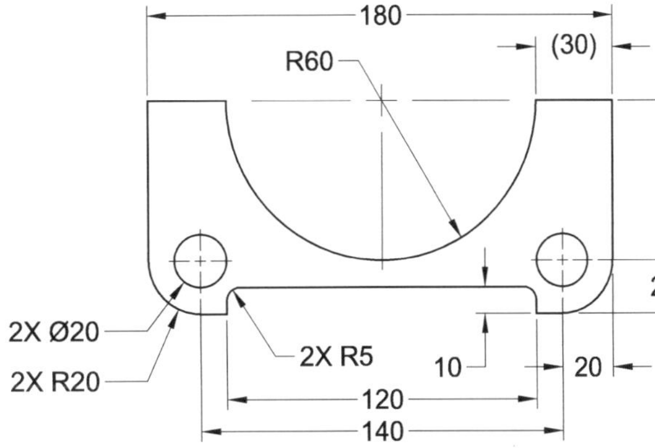

Figure 3-79 The sketch for Exercise 3

Exercise 4

Draw the sketch of the model shown in Figure 3-80. The sketch to be drawn is shown in Figure 3-81. Do not dimension the sketch. The solid model and dimensions are given only for your reference. **(Expected time: 30 min)**

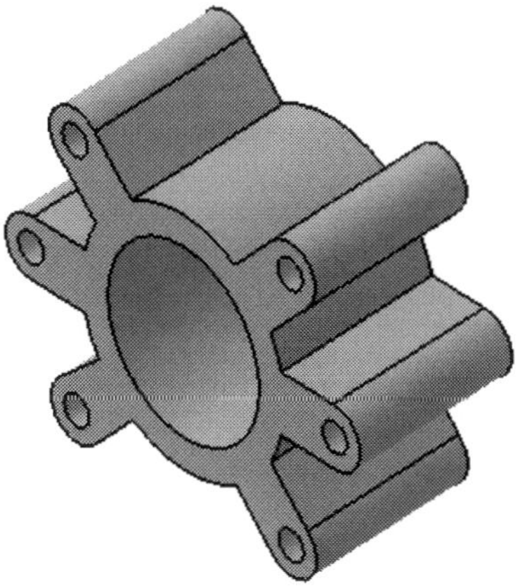

Figure 3-80 The model for Exercise 4

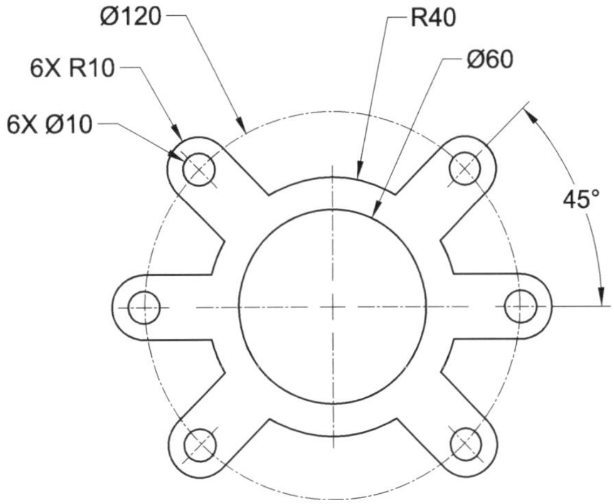

Figure 3-81 The sketch for Exercise 4

Exercise 5

Draw the sketch of the model shown in Figure 3-82. The sketch to be drawn is shown in Figure 3-83. Do not dimension the sketch. The solid model and dimensions are given only for your reference.

(Expected time: 30 min)

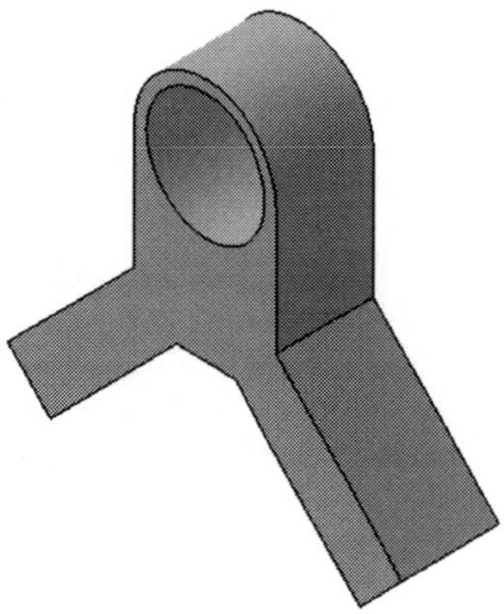

Figure 3-82 The model for Exercise 5

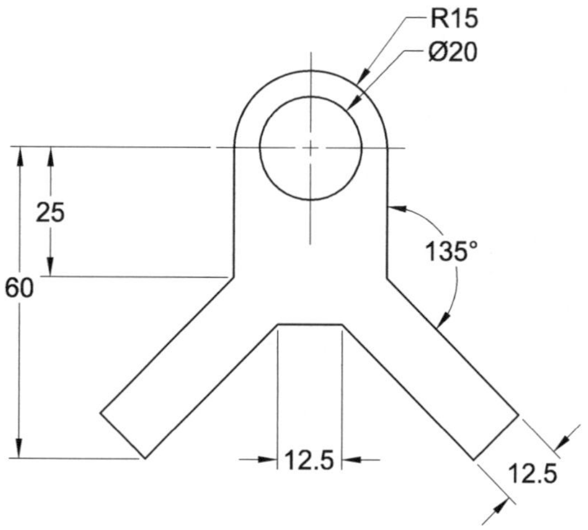

Figure 3-83 The sketch for Exercise 5

Answers to Self-Evaluation Test

1. **Break And Keep**, **2.** **Trim**, **3.** **Offset**, **4.** **Circle Definition**, **5.** **Ellipse Definition**, **6.** T, **7.** F, **8.** F, **9.** F, **10.** T

Chapter 4

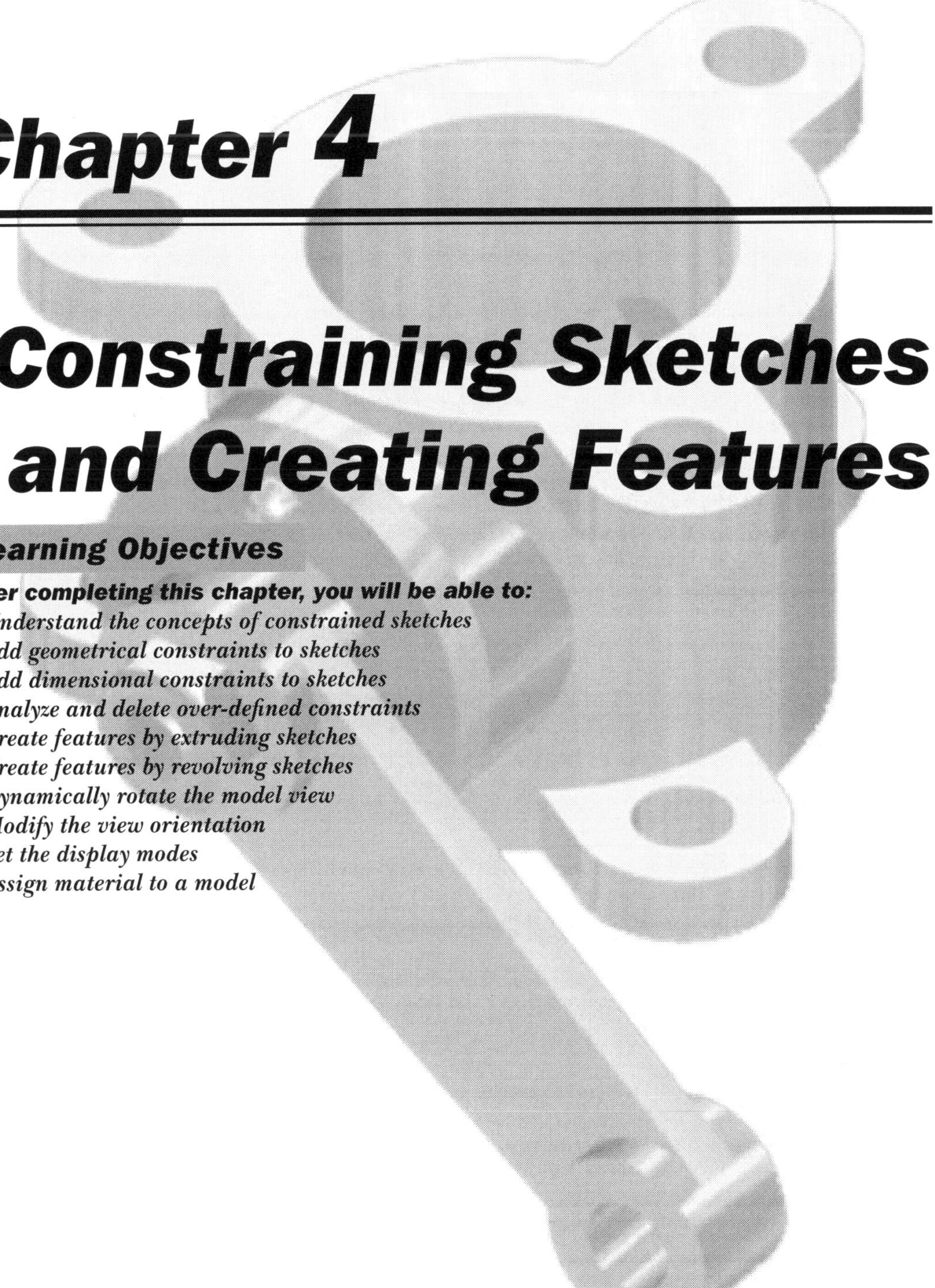

Constraining Sketches and Creating Features

Learning Objectives

After completing this chapter, you will be able to:

- *Understand the concepts of constrained sketches*
- *Add geometrical constraints to sketches*
- *Add dimensional constraints to sketches*
- *Analyze and delete over-defined constraints*
- *Create features by extruding sketches*
- *Create features by revolving sketches*
- *Dynamically rotate the model view*
- *Modify the view orientation*
- *Set the display modes*
- *Assign material to a model*

CONSTRAINING SKETCHES

In earlier chapters, you learned to draw, edit, and modify the sketches in the **Sketcher** workbench of CATIA V5. Now, you will learn to constrain them so that you can restrict their degrees of freedom to make them stable. The stability ensures that the size, shape, and location of the sketches do not change unexpectedly. The geometrical constraint are applied first, some of which are automatically applied while drawing. After applying the remaining geometrical constraints, you need to add dimensional constraints using the tools in the **Constraint** toolbar, as shown in Figure 4-1.

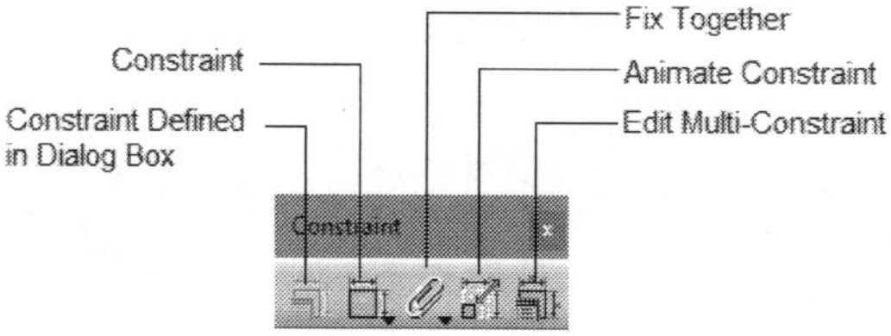

Figure 4-1 *The* ***Constraint*** *toolbar*

CONCEPT OF CONSTRAINED SKETCHES

After drawing and applying constraints, generally the sketch can exist in any one of the following five states:

1. Iso-Constrained
2. Under-Constrained
3. Over-Constrained
4. Inconsistent
5. Not Changed

These states are described next.

Iso-Constrained

An iso-constrained sketch, also known as a fully constrained sketch, is the one in which all the degrees of freedom of each element are defined using the geometric and dimensional constraints. As a result, the sketch cannot change its position, shape, or size unexpectedly. These dimensions can change only if they are modified deliberately by the user. The elements of an iso-constrained sketch are displayed in green by default.

Under-Constrained

An under-constrained sketch is the one in which some degrees of freedom (DOF) of the entity are not completely defined using constraints. The under-constrained elements of the sketch are displayed in white. You need to apply additional constraints to constrain their degrees of freedom. Under-constrained sketches tend to change their position, size, or shape unexpectedly. Therefore, it is necessary to fully-define the sketched elements.

Over-Constrained

An over-constrained sketch is the one in which some extra constraints get applied. The over-constrained entities are displayed in purple. The entities that are affected due to over constraining are also displayed in purple. It is always recommended to delete the extra constraints and make the sketch iso-constrained before exiting the **Sketcher** environment.

Inconsistent

When the geometry cannot accommodate the changes that you applied to it, the geometry (sketch) becomes inconsistent state, and the color of the geometry turns red.

Not Changed

The Not changed state of the geometry occurs when the geometry cannot accommodate the changes because the geometry on which it is dependent is inconsistent or overdefined. In this state, the system is not able to calculate the position of the geometry and the geometry turns brown.

Note

*Sometimes, all the sketched elements of a sketch are displayed as iso-constrained but actually the sketch is under-constrained. To check the status of the sketch, choose the **Sketch Solving Status** tool from the **Tools** toolbar; the **Sketch Solving Status** dialog box with the status of the sketch is displayed. If the sketch is under-constrained, the entities that need to be constrained are highlighted in the geometry area.*

APPLYING CONSTRAINTS

In the **Sketcher** workbench, you can apply constraints to the sketches using the methods discussed next.

Applying Geometrical Constraints Automatically

Toolbar: Sketch tools > Geometrical Constraints

The **Geometrical Constraints** tool is active by default, and therefore, as you draw an element, some constraints are automatically applied to it. Consider a case of drawing a circle. First, you specify its center point in the geometry area and then a point on the circle to define its radius. If you move the cursor and specify the point on the circle when it snaps the endpoint of a line, the coincidence constraint is applied automatically between the circle and the line endpoint. This is because the automatic constraining is active. Figure 4-2 shows a sketched circle, sketched line, and a coincidence constraint applied automatically between the profile of the circle and the endpoint of the line.

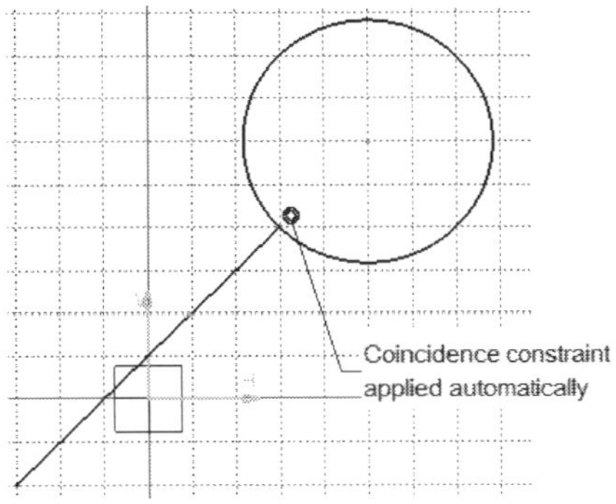

***Figure 4-2** The coincidence constraint applied automatically to the line and the circle*

By default, the automatic constraining is enabled because the **Geometrical Constraints** tool in the **Sketch Tools** toolbar is chosen. If you disable this button by choosing it again, the geometrical constraints will not be applied automatically when you sketch the elements.

Applying Additional Constraints to the Sketch

Menubar:	Insert > Constraint > Constraint
Toolbar:	Constraint > Constraints Defined in Dialog Box

Generally, the constraints applied automatically to the sketch are not enough to make it iso-constrained. You need to manually apply some additional constraints to the sketch by defining them using the **Constraint Definition** dialog box. To apply additional constraints, select the elements. You can use the Ctrl key to select more than one element. Next, choose the **Constraints Defined in Dialog Box** tool from the **Constraint** toolbar; the **Constraint Definition** dialog box will be displayed, as shown in Figure 4-3.

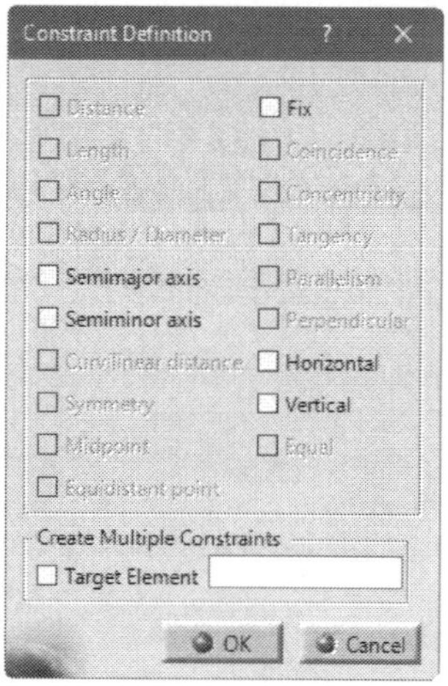

***Figure 4-3** The **Constraint Definition** dialog box*

The most suitable constraints for the current selection set are available in this dialog box. The constraints that cannot be applied to the current selection set are inactive in this dialog box.

The constraints that are provided in the **Constraint Definition** dialog box are discussed next.

Distance

The Distance constraint when applied to two selected entities will add a distance dimension between them. Consider a case in which you have selected the center points of two circles. If you select the **Distance** check box from the **Constraint Definition** dialog box, a distance dimension will be applied between the selected center points.

Length

The Length constraint applies a linear dimension to the selected line. Select the line on which you need to apply this constraint and invoke the **Constraint Definition** dialog box. Select the **Length** check box; a linear dimension will be displayed on the selected lines.

> **Tip**
> *You can apply a constraint to a single element or between two or more than two elements. You can also apply it between a sketched element and an edge or a vertex of the model. You will learn more about the edges and vertices of a model in the later chapters. The projected sketched elements can also be used to apply constraints to other sketched elements.*

Angle

The Angle constraint applies an angular dimension between two selected lines, or between a line and an existing edge of a model. Select the lines to apply this constraint and invoke the **Constraint Definition** dialog box. Select the **Angle** check box; an angular dimension will be displayed on the selected lines.

Radius / Diameter

The Radius / Diameter constraint applies a radius or diameter dimension to the selected arc or circle. Select an arc or circle to apply this constraint and invoke the **Constraint Definition** dialog box.

Select the **Radius / Diameter** check box. If the selected element is an arc, the radius dimension will be applied. If the selected element is a circle, the diameter dimension will be applied.

Semimajor axis

The Semimajor axis constraint applies the diameter dimension of the major axis of an ellipse. Select the ellipse and then invoke the **Constraint Definition** dialog box. On selecting the **Semimajor axis** check box, the diameter dimension of the major axis of ellipse will be displayed.

Semiminor axis

The Semiminor axis constraint is used to apply the diameter dimension of the minor axis of an ellipse. Select the ellipse and invoke the **Constraint Definition** dialog box. On selecting the **Semiminor axis** check box from this dialog box, the diameter dimension of the minor axis of ellipse will be displayed.

Curvilinear distance

The Curvilinear distance constraint is used to apply the curvilinear distance to a curve. To add the Curvilinear distance constraint, select two points on the curve and then the curve by pressing the Ctrl key. Next, invoke the **Constraint Definition** dialog box. In this dialog box, select the **Curvilinear distance** check box; the curvilinear distance will be applied between the selected points. You can apply curvilinear distance on different types of curves such as splines, circles, arcs, and conics. In case of circle, arc, and conic, you need to create points to add curvilinear distance.

Symmetry

The Symmetry constraint forces the selected elements to be symmetric about a given axis. Select the elements that need to be symmetric and then select the symmetry axis which can be a line, axis, or the vertical and horizontal axis of the coordinate system. Note that you need to follow this particular sequence to select the entities, otherwise you will not get the desired result. Next, invoke the **Constraint Definition** dialog box. Then, select the **Symmetry** check box and choose the **OK** button. You will notice that the selected entities are symmetric along the selected symmetry axis and the symmetric symbol is displayed on the entities.

Midpoint

The Midpoint constraint forces a selected point to get placed at the middle of a selected line. Select a point and a line. The point can be a sketched point, center point, endpoint, or control point of a spline. Next, invoke the **Constraint Definition** dialog box and select the **Midpoint** check box; the selected point will be placed at the middle of the selected line.

Equidistant point

The Equidistant point constraint forces two selected points to maintain an equal distance from the third point that lies between these two points. Select two points that are at an equal distance. Next, select a point that will be used as a reference to make the two points equidistant. This point should lie between the two points selected earlier. Make sure that you follow this sequence of selection. Next, invoke the **Constraint Definition** dialog box and select the **Equidistant point** check box. Choose the **OK** button from the **Constraint Definition** dialog box; the two selected points will be placed at an equal distance from the third point.

Fix

The Fix constraint forces a selected entity to be locked at its location. This constraint ensures that the current location of the selected entity cannot be modified. To apply this constraint, select the entity that needs to be fixed and then invoke the **Constraint Definition** dialog box. Next, select the **Fix** check box to fix the selected element.

Coincidence

The Coincidence constraint forces two selected entities to coincide with each other. To apply this constraint, select the entities that need to be coincident. The entities to be coincided could be two points, a point and a line, a curve, or a spline. After selecting the entities, invoke the **Constraint Definition** dialog box and select the **Coincidence** check box to apply the **Coincidence** constraint.

Concentricity

The Concentricity constraint forces the selected arcs or circles to share same center point. You can also select an arc and a point, or a circle and a point. To apply this constraint, select two arcs, or two circles, or an arc and a circle. Next, invoke the **Constraint Definition** dialog box and select the **Concentricity** check box; the selected elements will share the same center point.

Tangency

The Tangency constraint forces the selected elements to become tangent to each other. Select the two elements to be made tangent and invoke the **Constraint Definition** dialog box. Select the **Tangency** check box; the tangency constraint is applied between the selected elements.

Parallelism

The Parallelism constraint forces two selected lines to become parallel to each other. Select the two lines and invoke the **Constraint Definition** dialog box. Select the **Parallelism** check box; the selected lines will become parallel to each other.

Perpendicular

The Perpendicular constraint forces the selected lines to become perpendicular to each other. The element selected first will remain constant, while the second will change its angle, if it is not constrained. To apply this constraint, select two lines and invoke the **Constraint Definition** dialog box. Select the **Perpendicular** check box; the selected entities will become perpendicular and the perpendicular symbol will be displayed on them.

Horizontal

The Horizontal constraint forces the selected element to become parallel to the horizontal axis. Select a line and invoke the **Constraint Definition** dialog box. Next, select the **Horizontal** check box; the selected line will become parallel to the horizontal axis.

Vertical

The Vertical constraint forces the selected element to become parallel to the vertical axis. To apply this constraint, select a line, and invoke the **Constraint Definition** dialog box. Next, select the **Vertical** check box; the selected line will become parallel to the vertical axis.

Equal

The Equal constraint forces two selected lines to become equal to each other. Select the two entities and invoke the **Constraint Definition** dialog box. Select the **Equal** check box; the selected lines will become equal to each other.

Target Element

The **Target Element** check box is used to apply multiple constraints to the selected entities. To apply multiple constraints to an entity, select the entity and then invoke the **Constraint Definition** dialog box. Next, select the **Target Element** check box from the **Create Multiple Constraints** element area in the **Constraint Definition** dialog box. On doing so, you will be prompted to select the element for applying constraint(s) and the constraint type. Select the element for applying constraint; the corresponding constraints check boxes will be activated in the **Constraint Definition** dialog box. Select the appropriate constraint check box to apply the constraint; the selected element will get constrained. Next, choose the **OK** button from the dialog box.

Applying Contact Constraints

Menubar: Insert > Constraint > Constraint Creation > Contact Constraint
Toolbar: Constraint > Constraint Creation sub-toolbar > Contact Constraint

The **Contact Constraint** tool is used to automatically apply geometrical constraints to the selected elements depending on their position and geometry. The sequence of the constraints applied is Concentricity, Coincidence, and Tangency. Only the above mentioned constraints are applied using this tool. To apply constraints using this tool, choose the **Contact Constraint** tool from the **Constraint Creation** sub-toolbar of the **Constraint** toolbar; you will be prompted to select an element to be constrained. Select the element from the geometry area; you will be prompted to select another element. Select the second element from the geometry area. The most suitable constraint from the above mentioned list will be applied to the current selection set.

Applying Fix Together Constraints

Menubar: Insert > Constraint > Constraint Creation > Fix Together
Toolbar: Constraint > Constrained Geometry sub-toolbar > Fix Together

The **Fix Together** tool is used to fix two or more than two entities together. Remember that these entities are fixed only with respect to each other and not with the other elements in the sketch. If you move one of the entities, the other entities fixed to that entity will also move. To apply this constraint, select all the entities that you want to fix together and choose the **Fix Together** tool. This tool is available in the **Constrained Geometry** sub-toolbar. To display the **Constrained Geometry** sub-toolbar, choose the arrow on the right of the **Fix Together** tool. On choosing the **Fix Together** tool, the **Fix Together Definition** dialog box will be displayed, as shown in Figure 4-4. All the selected entities will be displayed in the **Geometry** area of this dialog box and the fix together symbol will be placed among the selected entities in the drawing window.

Figure 4-5 shows the **Fix Together** constraint applied to the sketched entities.

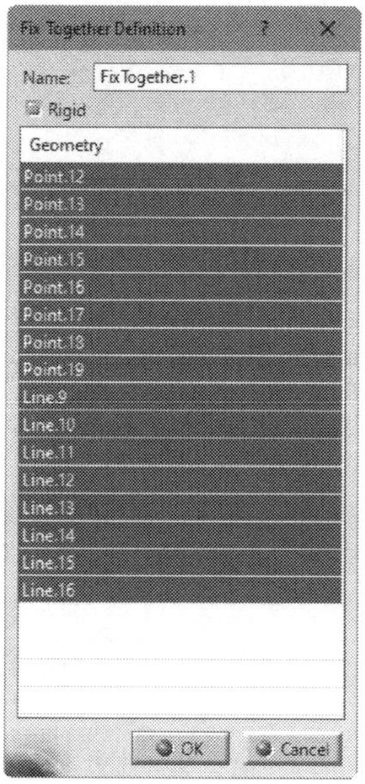

Figure 4-4 The ***Fix Together Definition*** *dialog box*

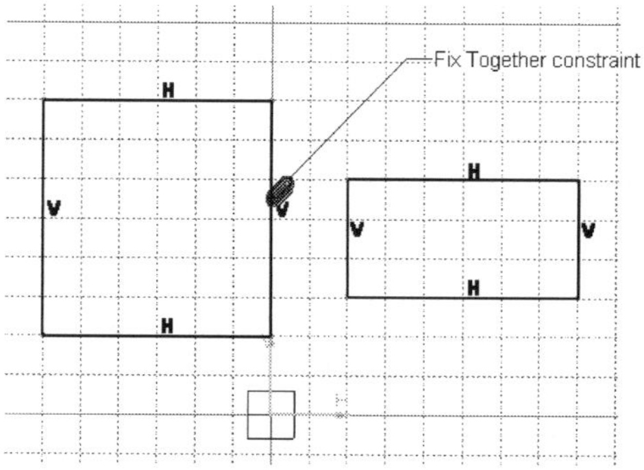

Figure 4-5 The ***Fix Together*** *constraint applied to the sketched entities*

Note

1. Instead of selecting the entity and then choosing the ***Fix Together*** *tool, you can also choose the* ***Fix Together*** *tool and then select the entity.*

2. To delete any entity, right-click on the required entity in the ***Geometry*** *area of the* ***Fix Together Definition*** *dialog box; the contextual menu will be displayed. Choose the* ***Delete*** *option from the contextual menu.*

Applying Auto Constraints

Menubar:	Insert > Constraint > Constraint Creation > Auto Constraint
Toolbar:	Constraint > Constrained Geometry sub-toolbar > Auto Constraint

The **Auto Constraint** tool automatically applies geometric and dimensional constraints to a sketch. To apply constraints, select the elements to be constrained and choose the **Auto Constraint** tool from the **Constraint Geometry** sub-toolbar; the **Auto Constraint** dialog box will be displayed, as shown in Figure 4-6. In this dialog box, the number of elements selected will be displayed in the **Elements to be constrained** display box.

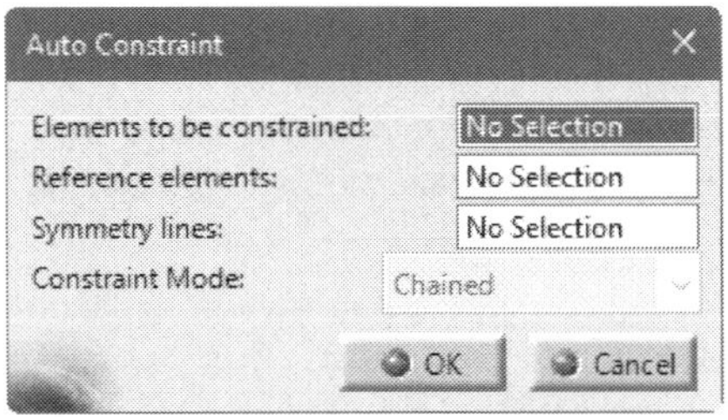

***Figure 4-6** The **Auto Constraint** dialog box*

You can also invoke the **Auto Constraint** dialog box first and then select the elements to be constrained. Next, you need to select the references with respect to which the selected elements will be constrained. To do so, click on the **Reference elements** display box; you will be prompted to select the reference elements. From the geometry area, select two elements. You can also select the horizontal and vertical axis as the reference. Select the type of dimensioning technique from the **Constraint Mode** drop-down list. The options in this drop-down list are used to create dimensional constraints in the chained or the stacked form. Next, choose the **OK** button in the **Auto Constraint** dialog box. All possible constraints applied to the selected elements will be displayed. All the selected entities will turn green indicating that the selected elements are iso-constrained, which means that all their degrees of freedom are completely defined. You will learn more about iso-constrained elements later in this chapter. Note that you may need to move the dimensional constraints to arrange them properly in the geometry area. Figures 4-7 and 4-8 show the constraints applied using the **Auto Constraint** tool by selecting the **Chained** and **Stacked** options, respectively from the **Constraint Mode** drop-down list. Finally, they are moved and properly arranged.

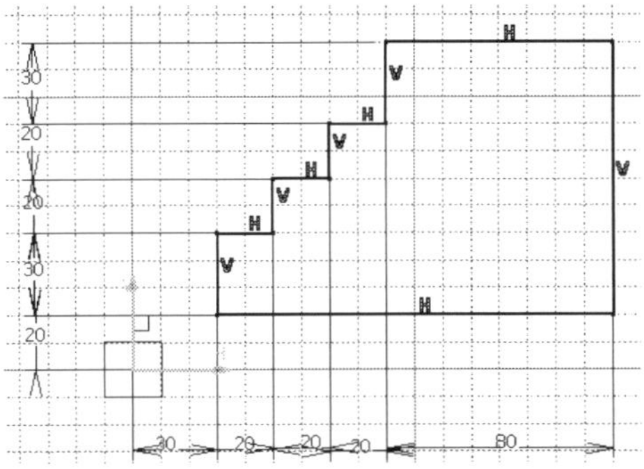

***Figure 4-7** Automatic dimensions applied using the **Chained** option*

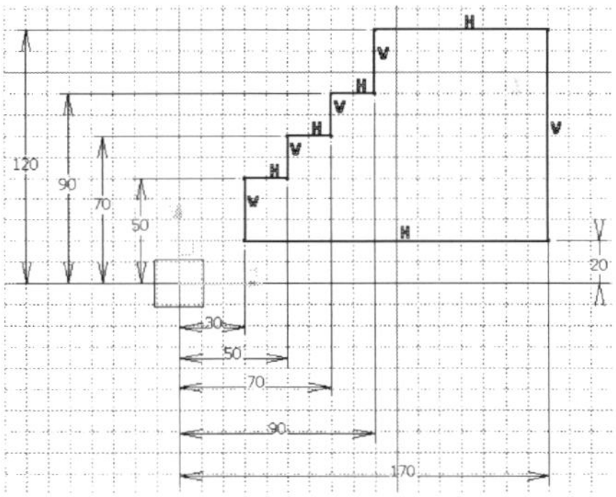

***Figure 4-8** Automatic dimensions applied using the **Stacked** option*

Tip

*Note that the **Symmetry lines** display area in the **Auto Constraint** dialog box is used to select a symmetry line. If you select it, all the symmetric elements along the symmetric line will be detected and the symmetric constraint will be applied to them.*

Animate Constraint

Menubar:	Insert > Constraint > Animate Constraint
Toolbar:	Constraint > Animate Constraints

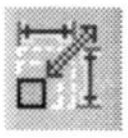

The **Animate Constraint** tool is used to animate a dimensional constraint. In order to animate the constraint, you need to assign dimension to the selected elements and make the sketch constrained. Figure 4-9 shows a fully constrained sketch. Select the angle constraint and choose the **Animate Constraints** tool from the **Constraint** toolbar; the **Animate Constraint** dialog box appears, as shown in Figure 4-10. Specify value for angle/length range in the **First value** and **Last value** edit boxes. Next, specify number of steps between the first value and last value in the **Number of steps** edit box. Select the **Hide Constraints** check box if you want to hide the constraints. To animate the selected constraint, choose the **Loop** button in the **Options** area and then choose the **Run Animation** button from the **Actions** area; the selected constraint will be animated along with the geometry.

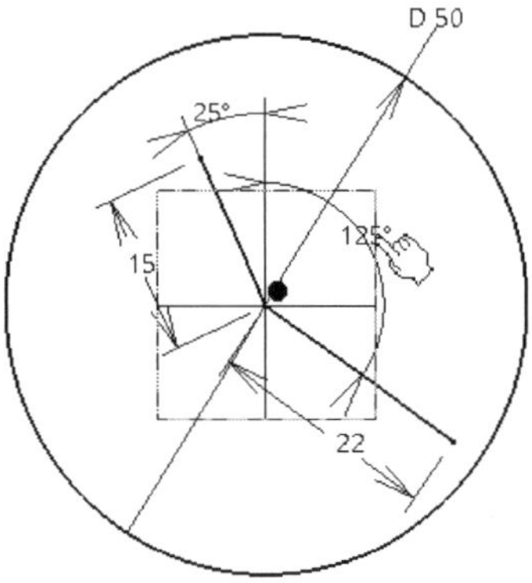

Figure 4-9 Fully constrained sketch

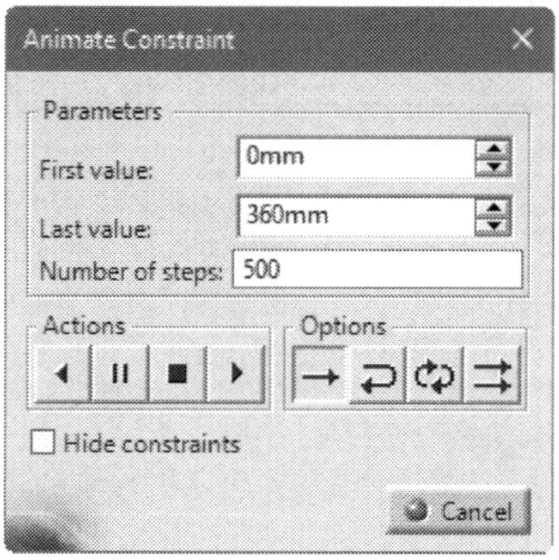

*Figure 4-10 The **Animate Constraint** dialog box*

Editing Multiple Dimensions

Menubar:	Insert > Constraint > Edit Multi-Constraint
Toolbar:	Constraint > Edit Multi-Constraint

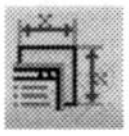

The **Edit Multi-Constraint** tool allows you to edit multiple dimensions using a single dialog box. To use this tool, first apply all the required dimensions to the sketch. Next, choose the **Edit Multi-Constraint** tool from the **Constraint** toolbar; the **Edit Multi-Constraint** dialog box will be displayed, as shown in Figure 4-11. All the dimensions applied to the sketch will be displayed in this dialog box. Select the dimension that you need to edit; the selected dimension will be highlighted in orange in the drawing area. Set the dimension value in the **Current value** spinner. Next, select another dimension to edit. You will notice that the dimension edited earlier is displayed in cyan. Follow this procedure to edit other

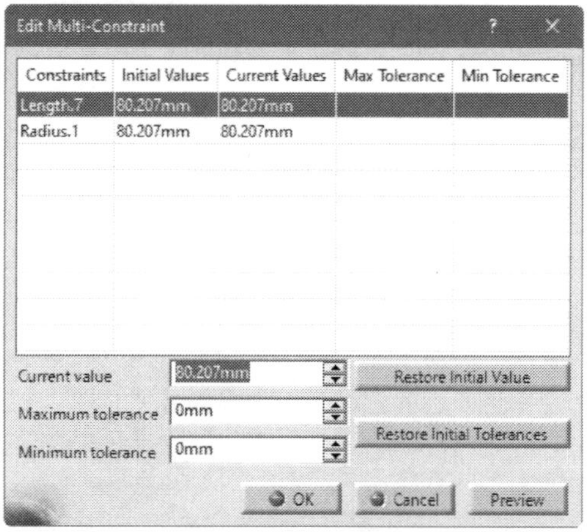

*Figure 4-11 The **Edit Multi-Constraint** dialog box*

dimensions. Next, choose the **Preview** button; the sketch will be regenerated based on the values of the edited dimension. You can also choose this button after modifying any dimension value.

The **Restore Initial Value** button is used to restore the initial dimension values. You can also modify or assign maximum and minimum tolerance values to the dimensions by using the **Maximum tolerance** and **Minimum tolerance** spinners in the dialog box. The assigned or modified tolerance value of the dimensions will be displayed under the **Max Tolerance** and **Min Tolerance** columns of this dialog box. The **Restore Initial Tolerances** button is used to restore the initial tolerance values of the dimension. Choose the **OK** button to confirm the changes made in the dimensions and exit the **Edit Multi-Constraint** dialog box.

ANALYZING AND DELETING OVER-DEFINED CONSTRAINTS

A sketch may be over-constrained after applying dimensions to it. These constraints are displayed in magenta. You need to delete the over-defined constraints from the sketch. To do so, select the over-defined constraint and press Delete. Next, if the sketched elements and the constraints are not displayed in green, you need to define the remaining constraints to make the sketch fully constrained sketch.

Analyzing Sketch Using the Sketch Analysis Tool

Menubar:	Tools > Sketch Analysis
Toolbar:	Tools > 2D Analysis Tools sub-toolbar > Sketch Analysis

The **Sketch Analysis** tool is used to analyse different types of sketch errors. The errors may include open profiles, overlapping profiles, isolated elements, and so on. Also, this tool is used to check the authenticity of the entire sketch and modify it as per your requirements. You can invoke this tool by choosing the **Sketch Analysis** tool from the **2D Analysis Tools** sub-toolbar. On doing so, the **Sketch Analysis** dialog box will be displayed, as shown in Figure 4-12. In this dialog box, the **Geometry** tab is chosen by default. It displays whether the geometry made in the sketch is valid or not. Also, it displays whether the profiles used in the sketch are open or closed. If you choose the **Use-edges** tab, you will get information about all the sketches created by using the **Project 3D Elements** tool in the **Sketcher** workbench. If you choose the **Diagnostic** tab, you will get information about the constraining status of the sketches. The different constraining status are Iso-Constraint, Under-Constraint, and Over-Constrained. You can further edit the sketch by using the tools in the **Actions** area in each of the specified tabs.

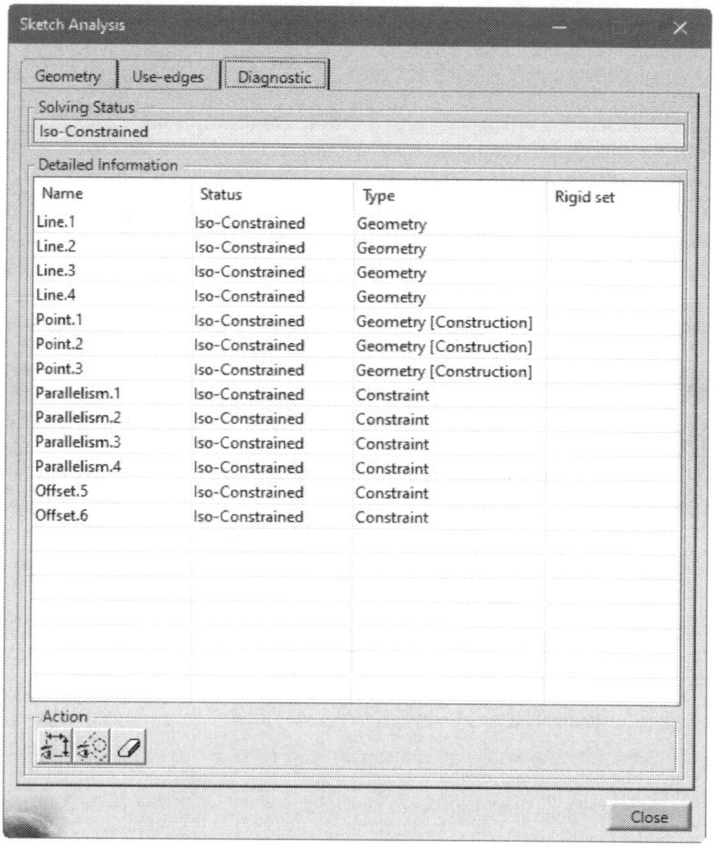

***Figure 4-12** The **Sketch Analysis** dialog box*

Tip

*For some reasons, if you need an over-defined dimensional constraint to be displayed with the sketch, double-click on the sketch to invoke the **Constraint Definition** dialog box. Select the **Reference** radio button and choose the **OK** button from this dialog box; the selected dimension will be displayed as reference dimension.*

*You can also deactivate an over-defined constraint to make the sketch iso-constrained. To do so, select the constraint and right-click to invoke the contextual menu. Choose the **Name of the constraint** option to invoke the cascading menu. Next, choose the **Deactivate** option from it; the deactivated geometrical constraint will be displayed in grey. The deactivated dimension constraint shows the deactivation symbol on the left of the dimensional value. Similarly, you can activate a deactivated constraint by choosing the **Activate** option from the cascading menu.*

*From the **Constraint Definition** dialog box, you can find out whether the selected constraint is resolved or is over-defining the sketch. If it is fully resolved, the symbol will be displayed in the **Constraint Definition** dialog box. If the selected constraint is over-defining the sketch, the symbol will be displayed.*

EXITING THE SKETCHER WORKBENCH

Toolbar: Workbench > Exit workbench

After the required sketch is drawn, you need to exit the **Sketcher** workbench to convert it into a feature. Choose the **Exit workbench** button from the **Workbench** toolbar; you will exit the **Sketcher** workbench and the **Part Design** workbench will be invoked again.

CREATING FEATURES BY EXTRUSION

Menubar: Insert > Sketch-Based Features > Pad
Toolbar: Sketch-Based Features > Pad sub-toolbar > Pad

The **Pad** tool is one of the most widely used tool to create the base features. To create a base feature using this tool, draw the sketch and exit the **Sketcher** workbench. Then, choose the down arrow on the right of the **Pad** tool in the **Sketch-Based Features** toolbar; the **Pad** sub-toolbar will be displayed, as shown in Figure 4-13. Select the sketch and then choose the **Pad** tool from the **Pad** sub-toolbar; the **Pad Definition** dialog box will be displayed, as shown in Figure 4-14 and you will be prompted to enter the required data to modify the pad. Also, the name of the selected sketch will be displayed in the **Selection** display box and the preview of the pad feature will be displayed in the geometry area.

Set the value of the depth of extrusion in the **Length** spinner. You can also define the extrusion depth dynamically in the geometry area. To do so, move the cursor close to **LIM1** in the default preview of the extrude feature. When the Limit drag handle (hand cursor) is displayed, refer to Figures 4-15 and 4-16, press and hold the left mouse button and drag the cursor; the depth of the extrusion will get dynamically defined. Then, choose the **OK** button from the **Pad Definition** dialog box. Figure 4-17 shows the model after creating the base feature by extruding the sketch.

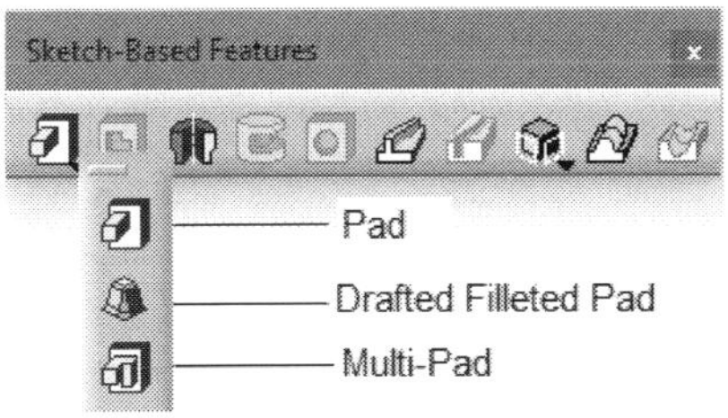

*Figure 4-13 The **Sketch-Based Features** toolbar with the **Pad** sub-toolbar*

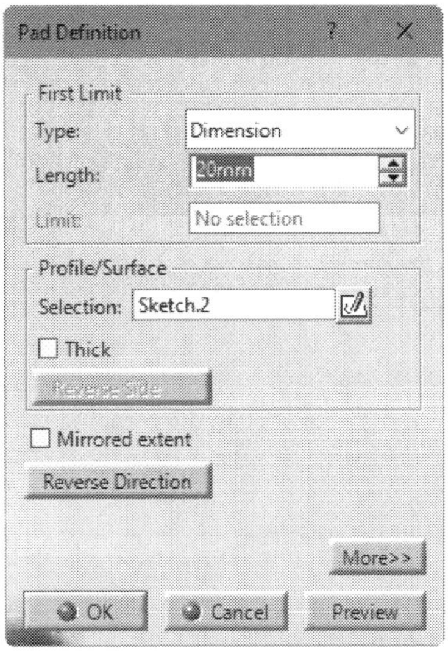

*Figure 4-14 The **Pad Definition** dialog box*

You can extrude the sketch feature about the sketch plane. To do so, choose the **More** button; the **Pad Definition** dialog box will expand. Enter the depth of extrusion for the second direction in the **Length** spinner of the **Second Limit** area; the changes will be displayed in the geometric area.

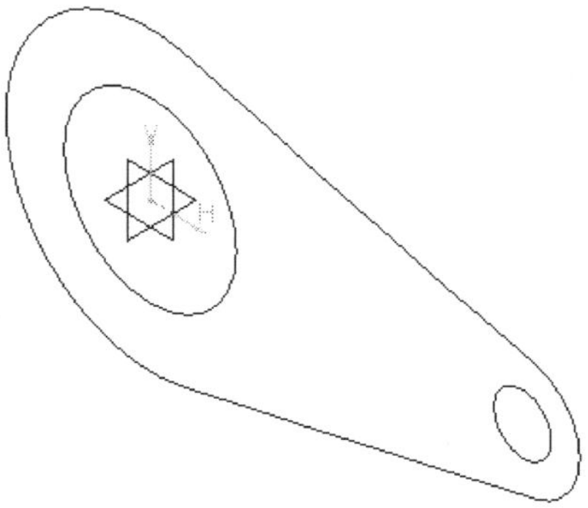

Figure 4-15 *The sketch after exiting the* ***Sketcher*** *workbench*

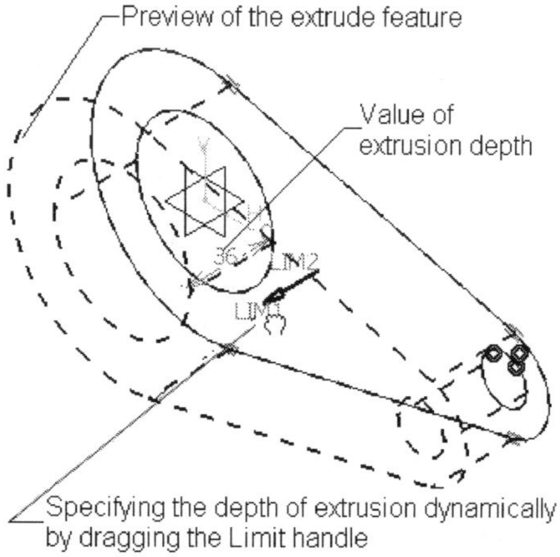

Figure 4-16 *Dynamically dragging the Limit drag handle to specify the depth of extrusion*

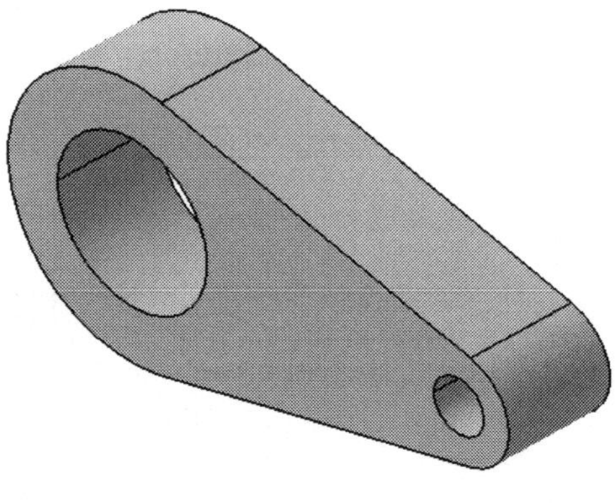

Figure 4-17 *The model after extruding the sketch*

Creating a Thin Extruded Feature

You can also create a thin extruded feature using the **Pad Definition** dialog box. To create a thin extruded feature, select the **Thick** check box from the **Pad Definition** dialog box; the dialog box will expand and the options in the **Thin Pad** area of this dialog box will be displayed, as shown in Figure 4-18.

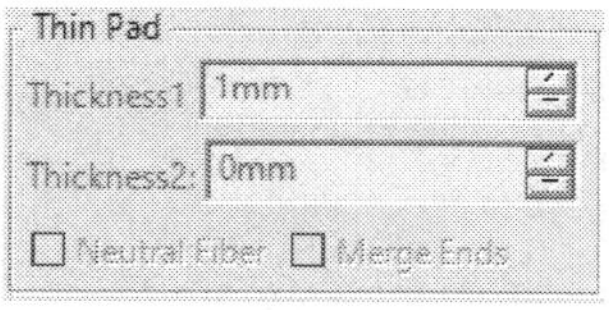

Figure 4-18 *The* ***Thin Pad*** *area of the* ***Pad Definition*** *dialog box*

The preview of the extrude feature is modified according to the default thickness value. You can specify the inside thickness value in the **Thickness1** spinner, and the thickness in the other direction that is outside the sketch profile can be specified in the **Thickness2** spinner. If you select the **Neutral Fiber** check box from the **Thin Pad** area, the thickness provided in the **Thickness1** spinner will be specified equally outside and inside the sketch. Figure 4-19 shows the thickness added to the sketch using various options. Figure 4-20 shows a thin extruded feature. You will learn about the **Merge Ends** option of this dialog box in the later chapters.

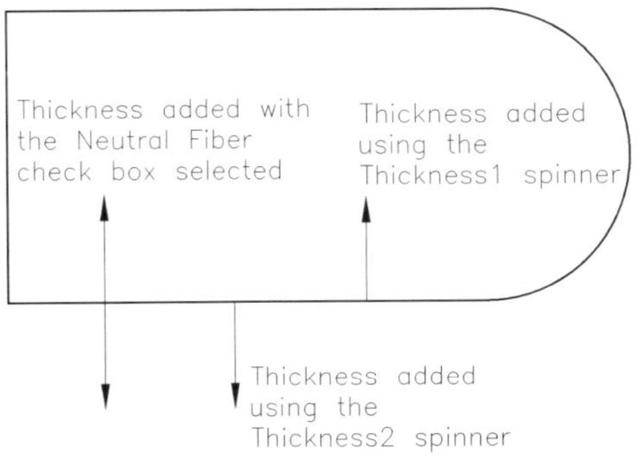

Figure 4-19 *Thickness added to the sketch using different options*

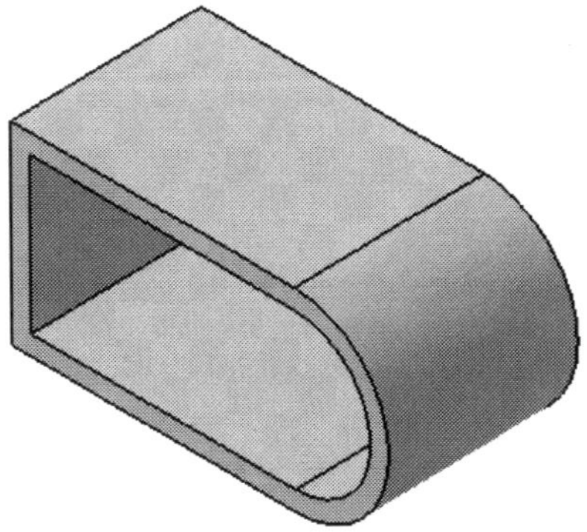

Figure 4-20 *A thin extruded feature created using the **Thin** option*

Tip

*You can also invoke the **Pad** tool first and then draw the sketch. To do so, choose the **Pad** tool from the **Pads** sub-toolbar; the **Pad Definition** dialog box will be displayed. Next, choose the Sketch button next to the **Selection** display box in the **Profile/Surface** area of the **Pad Definition** dialog box; the **Running Commands** dialog box will be displayed. This dialog box will provide information about the running commands and you will be prompted to select a sketch plane. Select the plane or the planar face on which you want to draw the sketch; the **Sketcher** workbench will be invoked. Draw the sketch using the tools in this workbench. Next, exit the workbench and specify the parameters in the **Pad Definition** dialog box.*

Note

*You can also draw an open sketch to extrude it as a thin feature. To do so, draw an open sketch and then choose the **Pad** tool from the **Pad** sub-toolbar; the **Pad Definition** dialog box will be displayed. Also, you will be prompted to select the sketch to be extruded. Select the open sketch; the **Feature Definition Error** message box will be displayed with the message that you must use the thick option in order to use an open profile. Choose the **Yes** button from the **Feature Definition Error** message box. Next, select the **Thick** check box from the **Pad Definition** dialog box; the dialog box will be expanded. Specify the thickness parameters in the **Thickness1** and **Thickness2** spinners and then choose the **OK** button from the **Pad Definition** dialog box.*

Extruding Sketch Using the Profile Definition Dialog Box

You can select a particular contour from a multi-contour sketch using the **Profile Definition** tool. You can also select a particular sketch from a multi-contour sketch that has already been used to create a feature. This implies that the sketch has an extrusion ring option available while creating features in CATIA V5. To create a feature by selecting a particular contour from a multi-contour sketch, draw the sketch and invoke the **Pad Definition** dialog box. Right-click in the **Selection** area and choose the **Go to profile definition** option; the **Profile Definition** dialog box will be displayed, as shown in Figure 4-21.

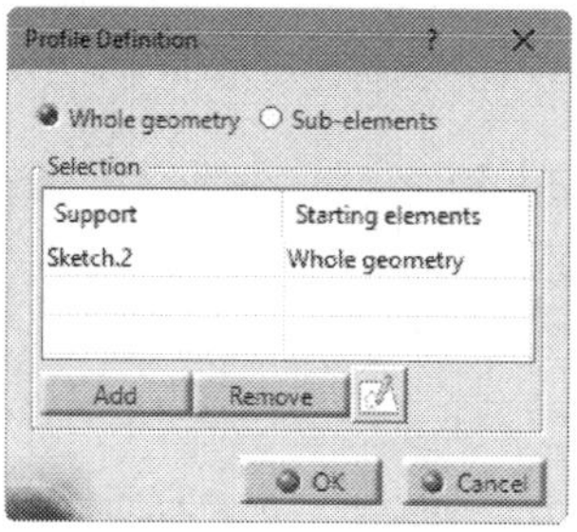

Figure 4-21 *The **Profile Definition** dialog box*

If you select the sketch before invoking the **Profile Definition** dialog box, the name of the selected sketch will be displayed in the **Support** column. Also, the entity used to select it will be displayed in the **Starting elements** column. You will notice that **Whole geometry** is displayed in this column. As a result, the entire sketch is selected for extruding. Select the name of the sketch and choose the **Remove** button from this dialog box. Now, select the **Sub-elements** radio button and then select any element of the contour that you need to extrude from the multi-contour sketch. The names of the sketch and the selected entity will be displayed in the **Support** and **Starting elements** columns, respectively. Also, a preview of the extruded feature will be displayed in the geometry area. Remember that you can select more than one contour. After selecting the contours, choose the **OK** button; the **Pad Definition** dialog box will be displayed. Set the parameters in the **Pad Definition** dialog box and choose the **OK** button to complete the feature creation. Figure 4-22 shows a multi-contoured sketch. Figure 4-23 shows the extruded contours.

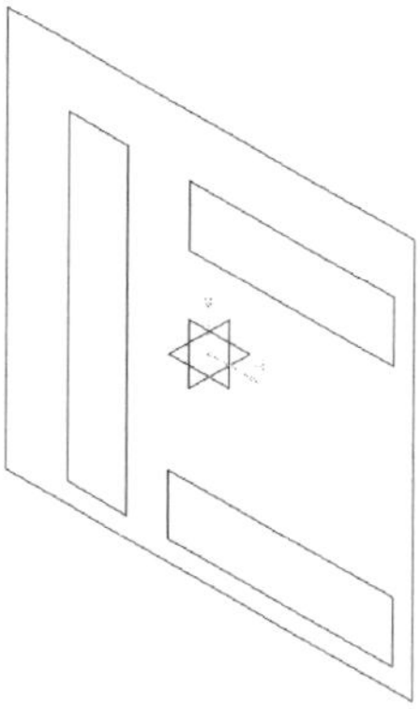

Figure 4-22 *A multi-contoured sketch*

Figure 4-23 *Extruded contours*

Extruding the Sketch along a Directional Reference

By default, a sketch is always extruded normal to the sketching plane. However in CATIA V5, you can also extrude a sketch along the directional reference. You can select a sketched line, axis, or straight edge as directional reference. To extrude a sketch along the directional reference, draw the sketch to be extruded and also the sketch to be used as the directional reference. Note that the directional reference should be drawn in a plane other than that of the sketch to be extruded. Next, select the sketch to be extruded and invoke the **Pad Definition** dialog box. Choose the **More** button from the dialog box; the dialog box will expand. The **Normal to profile** check box is selected by default in this dialog box. Click in the **Reference** display box of the **Direction** area and select the directional reference. Next, set the parameters in the **Pad Definition** dialog box and choose the **OK** button. Figure 4-24 shows the sketch to be extruded and the directional reference to be selected. Figure 4-25 shows the extruded feature created by extruding the sketch along a directional reference.

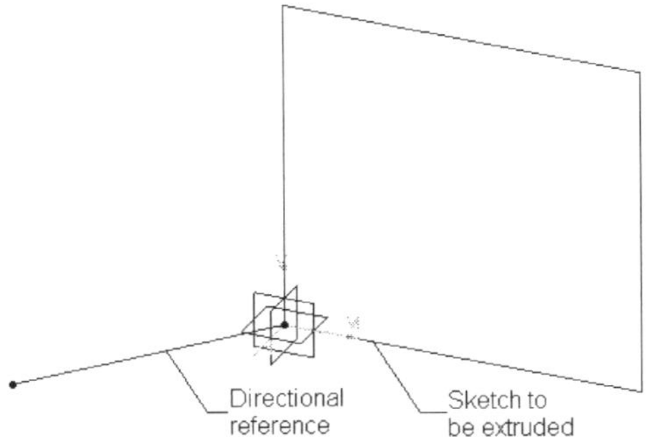

Figure 4-24 *Sketch to be extruded and the directional reference to be selected*

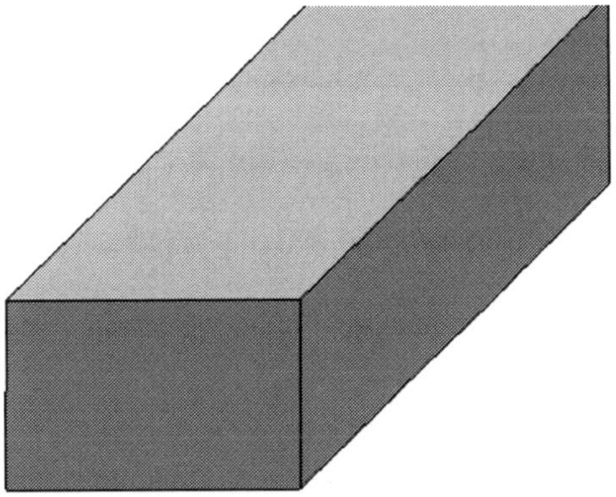

Figure 4-25 *The extruded feature created by extruding a sketch along a directional reference*

Tip

*You can reverse the direction of extrusion by using the **Reverse Direction** button in the **Pad Definition** dialog box or by clicking on the arrow displayed in the preview of the extrusion feature.*

*In CATIA V5, you can define the condition for feature termination in the first direction using the options in the **First Limit** area. Similarly, you can also define the feature termination in the second direction using the options in the **Second Limit** area. To specify the feature termination in the second direction, expand the **Pad Definition** dialog box. Set the value of the depth in the **Length** spinner provided in the **Second Limit** area. You can also dynamically drag the second limit handle, named **LIM2**, from the preview in the geometry area to define the depth of the second limit.*

*You can also extrude the sketch symmetrically on both sides of the sketching plane using the **Mirrored extent** check box in the **Pad Definition** dialog box. On selecting this check box, the extrude feature with the same depth will also be created on the other side of the sketch plane.*

CREATING FEATURES BY REVOLVING SKETCHES

Menubar:	Insert > Sketch-Based Features > Shaft
Toolbar:	Sketch-Based Features > Shaft

To create a revolved feature, draw a sketch that will revolve around the center line also known as axis. Next, exit the **Sketcher** workbench and choose the **Shaft** button from the **Sketch-Based Features** toolbar; the **Shaft Definition** dialog box will be displayed, as shown in Figure 4-26. Also, the preview of the shaft feature using the default parameters will be displayed in the geometry area.

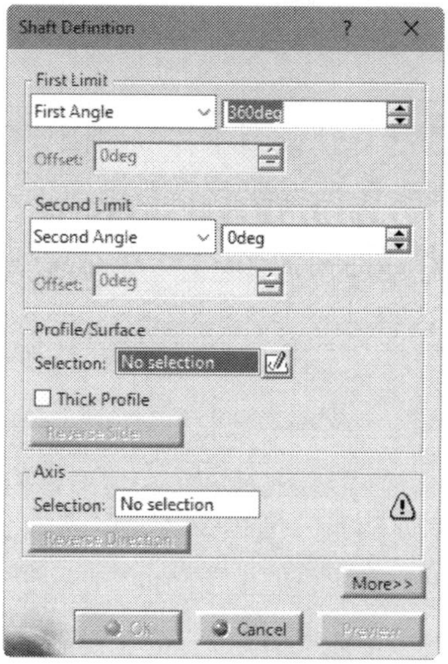

*Figure 4-26 The **Shaft Definition** dialog box*

The name of the sketch will be displayed in the **Selection** display box of the **Profile/Surface** area. By default, the **First angle** option is selected in the drop-down list available in the **First Limit** area and the angle spinner next to this drop-down list displays **360deg**. Therefore, the sketch revolves by 360-degree. You can also set the angular values in the **First Angle** and **Second Angle** spinners to define the angle in the first and second directions. Note that sum of both the angles cannot be greater than 360-degree. On selecting the **Up to next** option from the drop-down list available in the **First Limit** area, the angle spinner will not be available anymore and the feature will be created up to the next intersecting feature. If you select the **Up to last** option from the drop-down list, the feature will be created up to the last intersecting feature. If you select the **Up to plane** option from the drop-down list, a selection box will be displayed next to this drop-down list and you are prompted to select the first limit; select the required plane. The feature will be created up to the defined plane. If you select the **Up to surface** option from the drop-down list and select the required surface, the feature will be created up to the defined surface. The same process of selection is adopted in case of the drop-down list available in the **Second Limit** area of this dialog box.

The name of the axis is displayed in the **Selection** display box in the **Axis** area. You can also use the **Reverse Direction** button from the **Shaft Definition** dialog box to reverse the default direction of rotation. After specifying the angular values, choose the **Preview** button to preview the shaft feature created using the current values. If the preview seems fine, then choose the **OK** button from the **Shaft Definition** dialog box. Figure 4-27 shows the sketch and the axis around which the sketch will be revolved. Figure 4-28 shows the model after creating the shaft feature.

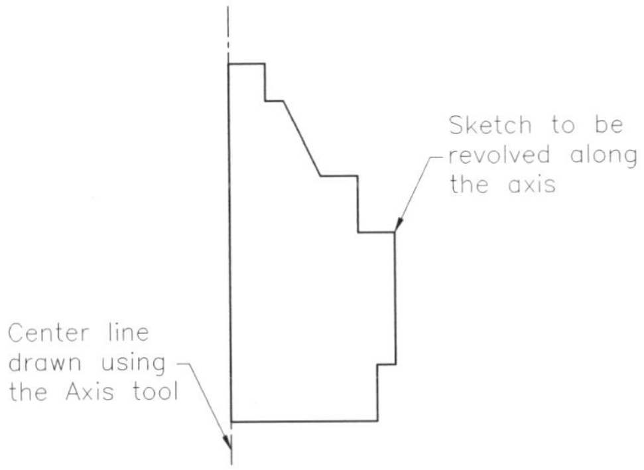

Figure 4-27 *The sketch and the axis around which the sketch will be revolved*

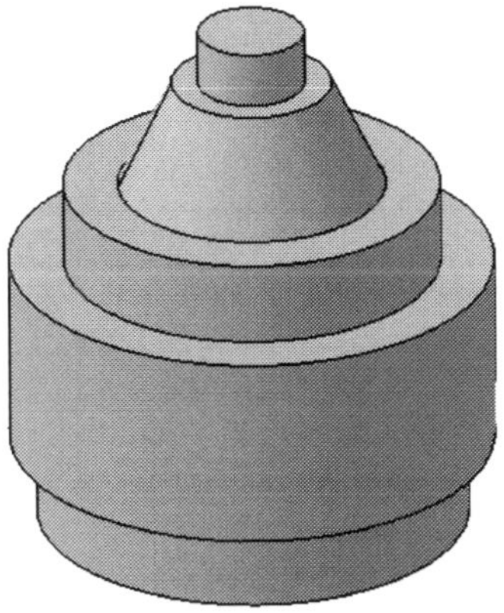

Figure 4-28 *The model after creating the shaft feature*

Tip

*The **Shaft** tool follows the right-hand thumb rule. This rule states if the thumb of the right hand points in the direction of the axis of rotation, the direction of the curled fingers will define the positive direction of rotation. You can also specify the angular values dynamically by dragging the **LIM1** and **LIM2** handles in the geometry area.*

*You can also draw a sketch after invoking the **Shaft Definition** dialog box. To do so, invoke the **Shaft Definition** dialog box and choose the **Sketch** button in the **Profile/Surface** area; the **Running Commands** dialog box will be displayed. On selecting the sketching plane, the **Sketcher** workbench will be invoked. Draw the profile to be revolved and the axis about which the profile will be revolved and then exit the **Sketcher** workbench. The preview of the shaft feature will be displayed in the geometry area. Set the parameters and choose the **OK** button from the **Shaft Definition** dialog box to create the feature.*

*To create a shaft feature using a sketch that does not have an axis drawn along with it, invoke the **Shaft Definition** dialog box after drawing the sketch. The preview of the shaft feature will not be displayed in the geometry area. Therefore, to create a shaft feature, you need to specify the axis of revolution using the **Selection** display box in the **Axis** area. You can select any edge, sketch line, or any existing axis, such as **Horizontal** and **Vertical** axes from the geometry area to specify the axis of revolution.*

Creating Thin Shaft Features

You can also create a thin revolved feature using the **Shaft** tool. Select the **Thick Profile** check box in the **Shaft Definition** dialog box; the dialog box will expand and the **Thin Shaft** area will be displayed in the expanded dialog box. Now, you can create a thin shaft feature by using the options available in this area. Figure 4-29 shows a thin shaft feature created using the **Shaft** tool. Note that for the thin shaft feature, you can also use an open profile as sketch.

DYNAMICALLY ROTATING VIEW OF A MODEL

CATIA V5 allows you to rotate the view of a model dynamically in the 3D space so that the solid model can be viewed from all directions. This characteristic allows you to visually maneuver around the model so that all the features in the model can be clearly viewed. The tool to rotate a model can be invoked even when you are using some other tool. Consider a case in which you need to create the base feature of a model using the **Pad** tool and view its preview using the **Preview** button. While the **Pad Definition** dialog box is invoked, you can freely rotate the view of the model in the 3D space. In CATIA V5, you can rotate the view of the model using the Compass on the top right corner of the geometry area. In the following section, you will learn how to rotate the view of the model using both the options.

Rotating the View Using the Rotate Tool

Menubar:	View > Rotate
Toolbar:	View > Rotate

To rotate a view freely in the 3D space, choose the **Rotate** tool from the **View** toolbar. Next, press and hold the left mouse button at any point close to the model in the graphics window; a reference circle and a cross mark will be displayed, as shown in Figure 4-30. Drag the cursor to rotate the view of the model in the 3D space. After rotating the view of the model, release the left mouse button. Figure 4-30 shows the view of the model being rotated. You can also rotate a view freely in the 3D space when any other tool is activated. In this case, when you invoke the **Rotate** tool, the display of the current dialog box will be turned off until you rotate the view of the model.

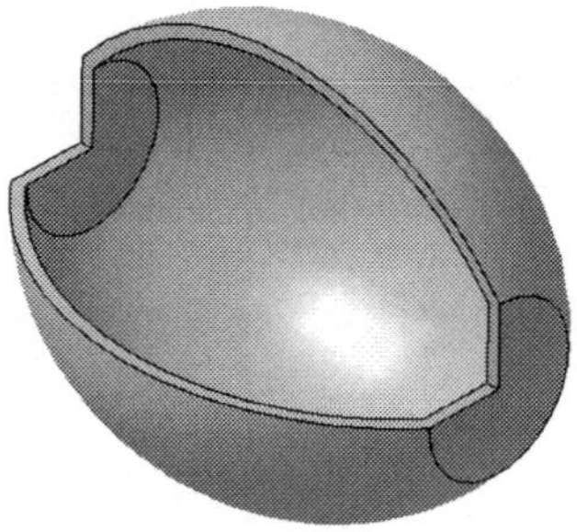

Figure 4-29 *The model after creating the thin shaft feature using the* ***Shaft*** *tool*

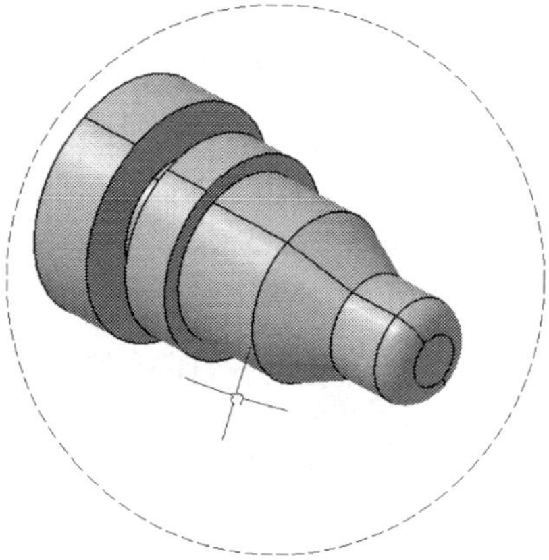

Figure 4-30 *View of the model being rotated using the* ***Rotate*** *tool*

Tip

The alternative method of rotating the view of the model freely in the 3D space is to press and hold the middle mouse button, and then drag the cursor keeping the left or right mouse button pressed.

Rotating the View Using the Compass

You can also rotate the view of the model using the Compass at the top right corner of the geometry area. The Compass is shown in Figure 4-31.

To rotate the view of the model in the XY plane using the Compass, move the cursor to the circular arc on the lower horizontal plane of the Compass. Next, press and hold the left mouse button when the arc is displayed in orange. Next, drag the cursor to rotate the Compass as well as the view of the model.

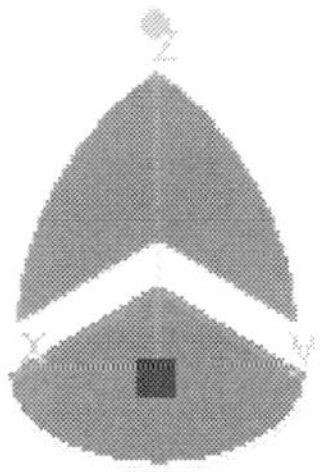

Figure 4-31 The Compass

To rotate the view of the model in the YZ plane, move the cursor to the circular arc on the right vertical plane. Press and hold the left mouse button when the arc is highlighted in orange. Drag the cursor to rotate the view of the model.

Similarly, you can rotate the view of the model along the ZX plane. You can also rotate it in the 3D space by holding the Compass from the dot displayed on top of the Z-axis, and then dragging the cursor.

Tip

You can also pan the view of the model using the Compass. To do so, move the cursor to any plane of the Compass along which you need to pan. When the edges are highlighted in Compass, press and hold the left mouse button, and then drag the cursor to pan the model. To pan the model along a particular axis, place the cursor on the edge corresponding to that axis on the Compass; the corresponding edge will be highlighted on the Compass, press and hold the left mouse button and then drag the cursor to pan the model.

MODIFYING THE VIEW ORIENTATION

When you exit the **Sketcher** workbench, the view of the sketch automatically changes to isometric. CATIA V5 allows you to manually change the view orientation using some predefined standard views. The tools used to modify the view orientation of the model are available in the **Quick view** sub-toolbar, as shown in Figure 4-32. To display this sub-toolbar, choose the down arrow on the right of the **Isometric View** button in the **View** toolbar.

You can set the orientation of the view of the model to isometric, front, back, left, right, top, or bottom by using the tools in the **View** toolbar. The **Named Views** tool is also available in this toolbar. When you choose the **Named Views** button, the **Named Views** dialog box will be displayed, as shown in Figure 4-33.

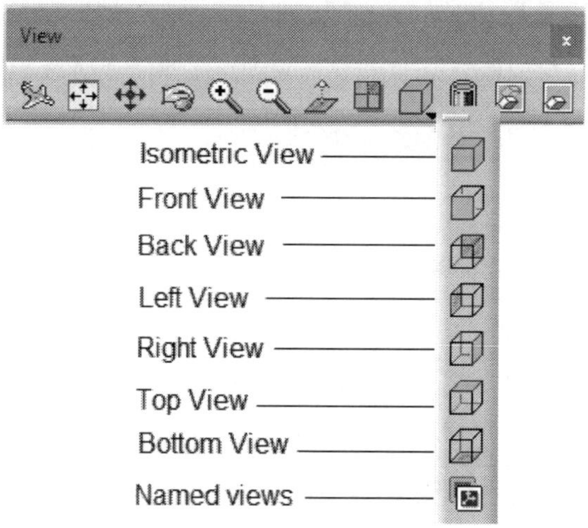

*Figure 4-32 The **View** toolbar with the **Quick** view sub-toolbar*

To create a user-defined view, set the orientation of the view of the model and invoke the **Named Views** dialog box. Next, choose the **Add** button from this dialog box; a new view will be created with a default name. The **Delete** button is used to delete the selected view. Remember that the default views displayed in the dialog box cannot be deleted. The **Reverse** button is used to reverse the currently selected view.

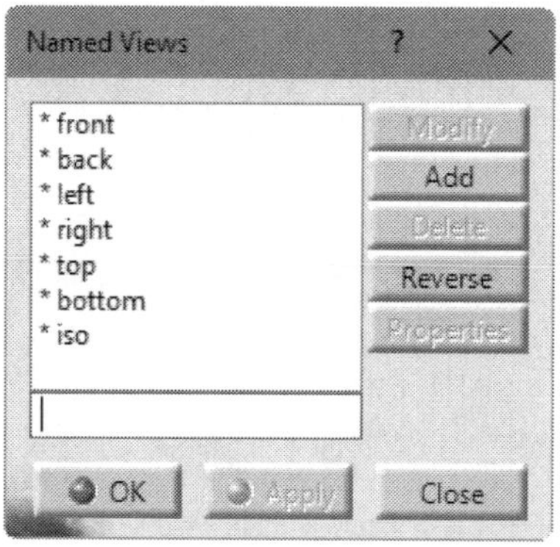

*Figure 4-33 The **Named Views** dialog box*

If you select a user-defined view and choose the **Properties** button, the **Camera Properties** dialog box will be displayed. The options in this dialog box are used to set the position of the camera and also the type of viewing, which can be parallel and perspective.

If you select a default view and choose the **Properties** button, the **Views & Layout** dialog box will be displayed. Using the options in the dialog box, you can change the orientation of the default view. The orientation of all the other default views will also change accordingly. To restore the default orientation, choose the **Reset All** button from the **Standard Views** tab of this dialog box.

The **Modify** button in the **Named Views** dialog box is used to modify the orientation of a user-defined view. To do this, activate the user-defined view by clicking on it and then change the orientation of the view of the model by rotating it. Next, choose the **Modify** button.

DISPLAY MODES OF THE MODEL

In CATIA V5, you can use various predefined modes provided in the **View mode** sub-toolbar to display the models. To invoke the **View mode** sub-toolbar, choose the down arrow on the right of the **Shading with Edges** tool of the **View** toolbar; the **View mode** sub-toolbar will be displayed, as shown in Figure 4-34. You can select the required display mode from this sub-toolbar.

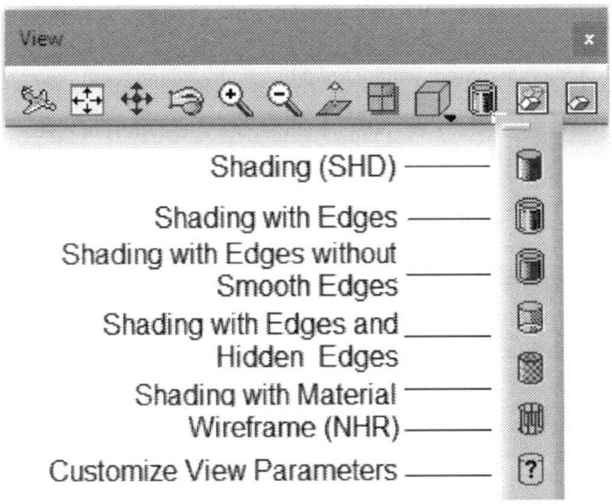

Figure 4-34 *The* ***View*** *toolbar with the* ***View mode*** *sub-toolbar*

The tools available in the **View mode** sub-toolbar are discussed next.

Shading (SHD)

When you choose the **Shading (SHD)** tool from the **View mode** sub-toolbar, the model will be displayed in the shaded mode without highlighting the edges.

Shading with Edges

The **Shading with Edges** tool is the default mode in which the model is displayed. When you start a new file and create the base feature, it is automatically displayed as shaded along with the edges of the model.

Shading with Edges without Smooth Edges

When you choose the **Shading with Edges without Smooth Edges** tool, the model will be displayed in shading with edges, but the display of the smooth edges will be removed.

Shading with Edges and Hidden Edges

When you choose the **Shading with Edges and Hidden Edges** tool, the model will be displayed in shading with edges and the hidden edges will be displayed with dashed lines along with the visible edges.

Shading with Material

When you choose the **Shading with Material** tool, the model will be displayed in the rendered mode. Before choosing this tool, you need to assign a material to the model. Else, it will be displayed in the shaded mode. You will learn more about assigning materials later in this chapter.

Wireframe (NHR)

When you choose the **Wireframe (NHR)** tool, the model will be displayed in the wireframe mode.

Customize View Parameters

When you choose the **Customize View Parameters** tool, the **View Mode Customization** dialog box will be displayed. You can set a customized view mode using the options in this dialog box. After setting the required options, choose the **OK** button to set the current view mode to the customized view mode.

CREATING SECTIONS DYNAMICALLY

Sometimes, during the process of designing, checking, or reviewing of a part body, model, or assembly, the conventional views of the model are not sufficient. In CATIA V5, this issue can be tackled by using the **Dynamic Sectioning** tool available in the **Dynamic Sectioning** toolbar. This tool allows you to visualize the model sections at a position specified by section plane. When you invoke this tool, the section plane will be displayed at the origin. The section plane is displayed as a dashed rectangle along with the U, V, and W directions displayed at the local axis system of the section. Figure 4-35 shows the section plane displayed on choosing the **Dynamic Sectioning** tool. You can translate, rotate, resize, and relocate the section plane to a position you need to check the section of the model. When you move, rotate, or relocate the section plane, the updated section of the model will be displayed dynamically. The procedure to work with this tool is discussed next.

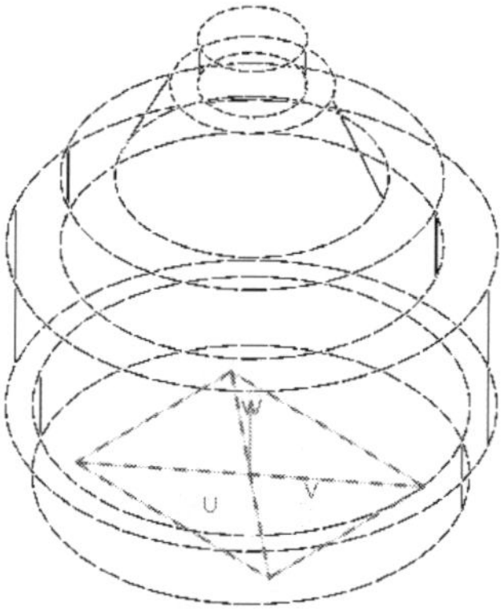

Figure 4-35 *The section plane displayed after using the* ***Dynamic Sectioning*** *tool*

Maneuvering the Section Plane

As discussed earlier, you can move, rotate, resize, and reposition the section plane at a place where you want to cut the part and view the section dynamically. To move the plane, move the cursor over the diagonal of the plane; two arrows will be displayed perpendicular to the plane pointing in opposite direction. Click on the desired diagonal and drag it to the required location; the plane will move to that location. Also, the section view of the model will be displayed automatically where the section plane is moved. Figure 4-36 displays the section view of the model when the section plane is moved vertically upward. Similarly, you can rotate the section plane to view the section at that angle. When you move the cursor over the arc displayed at U-V, V-W or W-U, the possible axis of rotation and its angular movement will be displayed. The angular movement will be displayed as an arc with arrows on both sides. Click on the desired rotation axis and drag the cursor to define the rotation angle. Figure 4-37 displays the section of the model on rotating the section plane.

Figure 4-36 Section view of the model displayed on moving the section plane

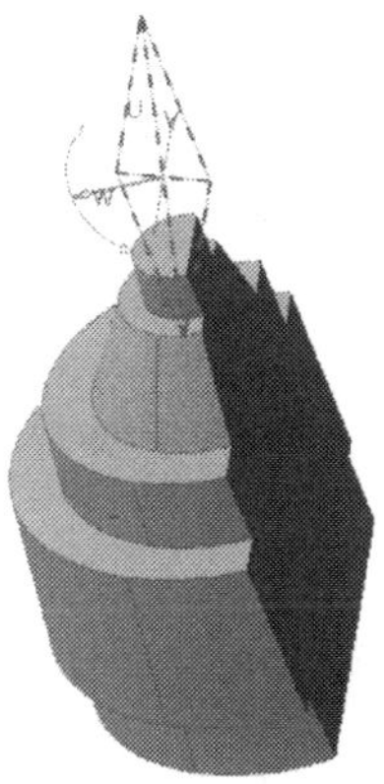

Figure 4-37 Section view of the model displayed on rotating the section plane

To resize the section plane, move the cursor on the rectangular edges of the section plane; the possible directions of movement will be displayed represented by two arrows pointing in opposite directions. Click and drag the cursor to resize the section plane.

Position of Section Planes

By default, the section plane is positioned at the origin and according to the selected geometry. Some of the geometries are discussed next.

Selecting Planes

If you select any plane from the drawing area or from the specification tree, a section plane displayed at the origin will be aligned with the selected plane. Also, the dynamic section view will be displayed.

Selecting Planar Faces

If you select any planar face of a model, then the section plane will be displayed at the selection point of the face.

Selecting Cylindrical Faces

On selecting the cylindrical face of the model, the section plane passes through the axis of the cylindrical face. The origin of the section plane will be at the projection of the selection point on the axis. The normal of the section plane will be on the line defined by the origin of the section plane and the selection point. If you press the Shift key while selecting the cylindrical face, then the normal will be created along the axis. But when you press the Ctrl key while selecting the cylindrical face, the section plane will be placed on the selected face at the selection point.

Selecting Cylindrical Edges

On selecting the cylindrical edge of the model, the normal of the section plane will be tangent to the edge and will be placed at the selection point. But, if you press the Shift key while selecting the edges, then the section plane will be tangent and the normal will point along the radial direction.

ASSIGNING MATERIAL TO A MODEL

Toolbar:	Apply Material > Apply Material

CATIA V5 also allows you to assign a material to the model in the **Part Design** workbench. All physical properties of the material are also assigned to the model. These physical and thermal properties include the type of material, Young's Modulus, Poisson Ratio, Density, Thermal Expansion, and Yield Strength. To assign a material, select the model from the Specification tree and choose the **Apply Material** button from the **Apply Material** toolbar. On doing so, the **Library (ReadOnly)** dialog box will be displayed, as shown in Figure 4-38. Select the material to be assigned using the tabs in the **Library (ReadOnly)** dialog box and then choose the **Apply Material** button to apply the selected material. Next, choose the **Close** button to exit the dialog box. You can also customize the view parameters. To do so, choose the **Customize View Parameters** button from the **View mode** sub-toolbar of the **View mode** toolbar; the **View Mode Customization** dialog box will be displayed, refer to Figure 4-39.

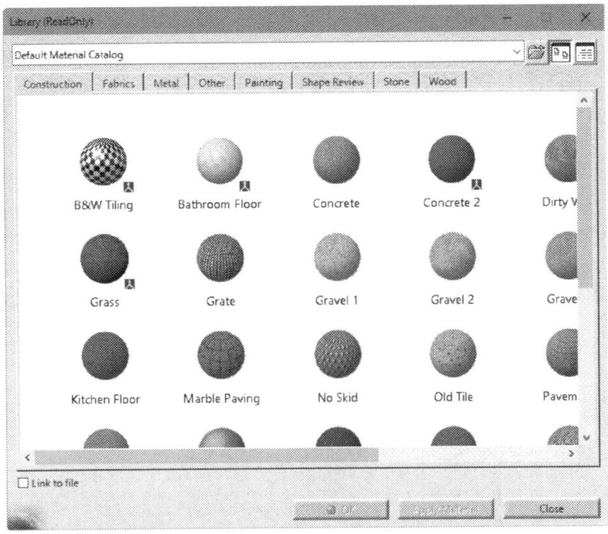

Figure 4-38 *The* ***Library (ReadOnly)*** *dialog box*

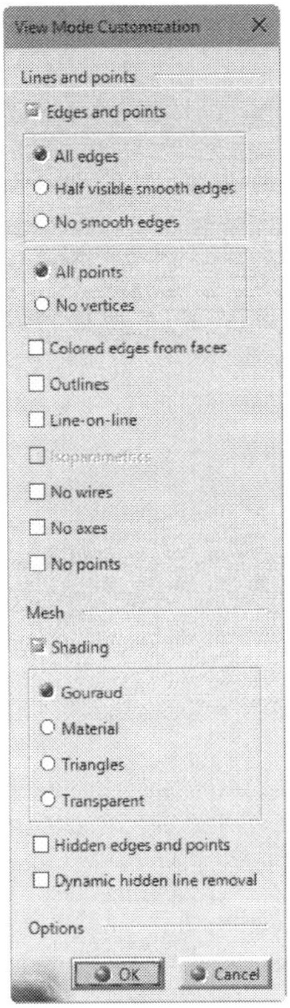

Figure 4-39 *The* ***View Mode Customization*** *dialog box*

Make sure that the **Shading** check box and the **Material** radio buttons are selected in the **Mesh** area of this dialog box. Next, choose the **OK** button to display the assigned material. You can also choose the **Shading with Material** button from the **View mode** toolbar to display the assigned material. Figure 4-40 shows the model with the Steel material applied to it.

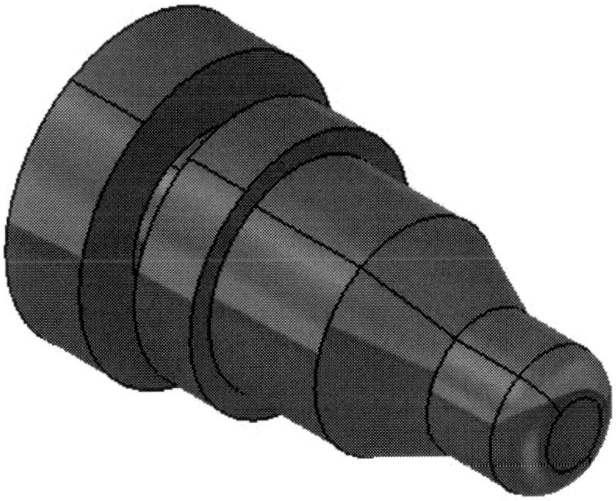

Figure 4-40 *The model after applying the Steel material*

TUTORIALS

Tutorial 1

In this tutorial, you will create the model shown in Figure 4-41. Its views and dimensions are shown in Figure 4-42. After creating this model, you will apply the Copper material to it and then rotate its view in the 3D space. **(Expected time: 30 min)**

Figure 4-41 *Solid model for Tutorial 1*

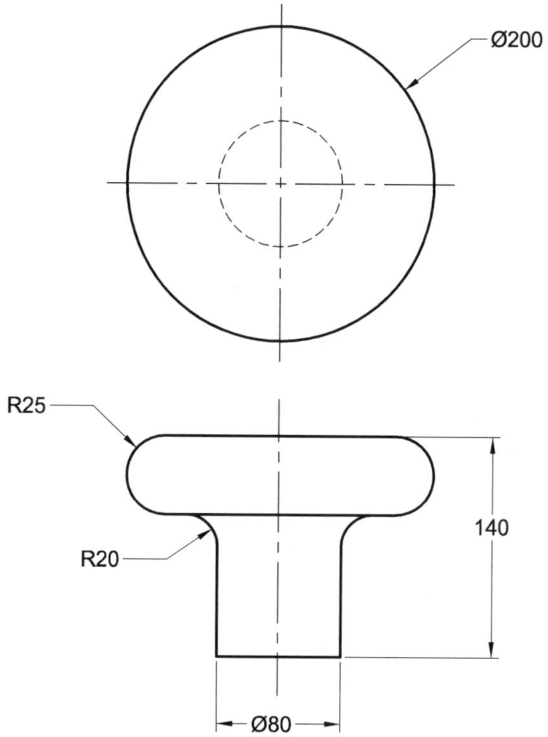

Figure 4-42 *Orthographic views and dimensions for Tutorial 1*

The following steps are required to complete this tutorial:

a. Start a new file in the **Part Design** workbench and draw the sketch on the YZ plane using the tools available in the **Sketcher** workbench of CATIA V5.
b. Apply the required geometrical and dimensional constraints.
c. Use the **Shaft** tool to create the base feature.
d. Assign the material to the model.
e. Rotate the model view in 3D space.
f. Save and close the file.

Starting a New File and Drawing the Sketch of the Base Feature

1. Choose **Start > Mechanical Design > Part Design** from the menu bar; the **New Part** dialog box is invoked.

2. In this dialog box, enter **c04tut01** in the **Enter part name** edit box and then choose the **OK** button.

   In this tutorial, you need to create the base feature of the model by revolving a sketch around the revolving axis. For this, first you need to draw the sketch of the base feature and the revolving axis on the yz plane and then apply the geometrical and dimensional constraints to the sketch to make the sketch iso-constrained.

3. Select the yz plane from the Specification tree or the geometry area and then choose the **Sketch** tool from the **Sketcher** toolbar; the **Sketcher** workbench is invoked.

4. Draw a vertical centerline that passes through the origin using the **Axis** tool of the **Profile** toolbar.

5. Using the **Profile** tool, draw the required sketch, as shown in Figure 4-43. Use the **Tangent Arc** option and the **R** edit box of the **Sketch tools** toolbar to create the radial shape.

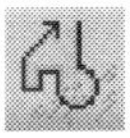

   Note that in this figure, the display of grid lines has been turned off for better visualization. Also, the sketch need not be of exact dimensions.

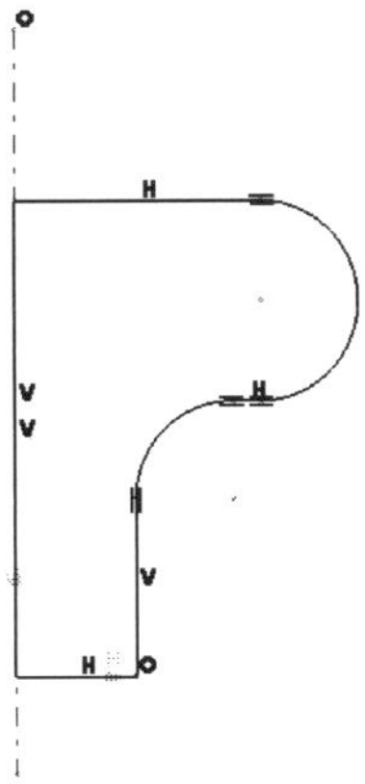

Figure 4-43 *Sketch drawn using the **Axis** and **Profile** tools*

Applying Geometrical and Dimensional Constraints to the Sketch

After drawing the sketch, you need to apply geometrical and dimensional constraints to make it iso-constrained. As discussed earlier, an iso-constrained sketch is the one in which the position, shape, and size of all the elements are completely defined. Most of the geometrical constraints are applied to the sketched entities while drawing the sketch. If required, you can also apply additional geometrical constraints before applying the dimensional constraints. The dimensions required for this sketch are shown in Figure 4-42.

1. Double-click on the **Constraint** tool in the **Constraint** sub-toolbar; you are prompted to select the element to be constrained.
2. Select the vertical line on the right of the sketch from the geometry area. You will notice that a vertical dimension is attached to the cursor.
3. Next, select the vertical centerline from the geometry area; a horizontal dimension between the vertical line on the right and the vertical centerline is attached to the cursor.
4. Right-click in the geometry area; a contextual menu is invoked. Choose the **Radius /Diameter** option from it; a linear diameter dimension is attached to the cursor.
5. Move the cursor vertically downward and click once in the geometry area to place the dimension, refer to Figure 4-44.
6. Select the centerline from the geometry area; a vertical dimension is attached to the cursor.
7. Select the semi-circular arc on the right of the sketch; a linear horizontal dimension is attached to the cursor.
8. Right-click in the geometry area and choose the **Radius/Diameter** option from the contextual menu.
9. Move the cursor vertically downward and place the new linear diameter dimension below the previous dimension, refer to Figure 4-44.
10. Next, select the right semicircular arc; a radial dimension is attached to the cursor. Move the cursor to the desired location and place the dimension. Similarly, dimension the other arc.
11. Select the horizontal lines at the bottom and top portion of the sketch; a vertical dimension is attached to the cursor.
12. Move the cursor horizontally toward the right and place the dimension.

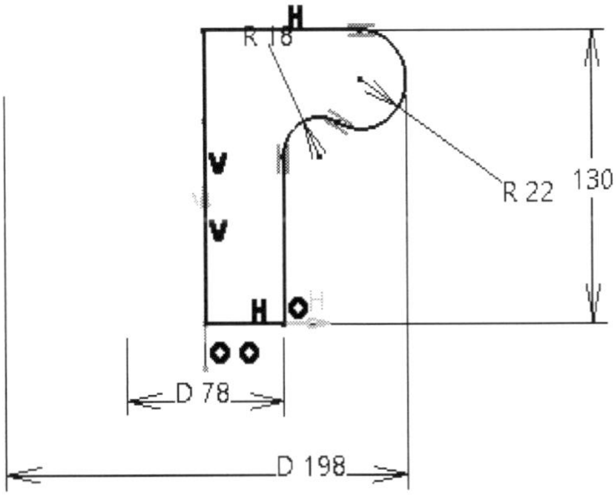

Figure 4-44 The sketch after applying all the constraints

Note

The default values of dimensions displayed in Figure 4-44 may differ from the actual dimensions while dimensioning the sketch. The advantage of using parametric software is that you can draw a sketch of any size and then modify its dimensions.

All the sketched elements are displayed in green suggesting that all the degrees of freedom are restricted and the sketch is in the iso-constrained state. After adding all the dimensional constraints, you need to modify its dimensions, as shown in Figure 4-45.

Before modifying dimensions, you need to exit the **Constraint** tool.

13. Press the Esc key to exit the **Constraint** tool. Double-click on the linear diameter dimension of the right arc; the **Constraint Definition** dialog box is displayed.

14. Set the value in the **Diameter** spinner to **200** and then choose the **OK** button from the **Constraint Definition** dialog box.

15. Similarly, modify other dimensional values, as shown in Figure 4-45. This figure shows the final sketch after modifying all the dimensions.

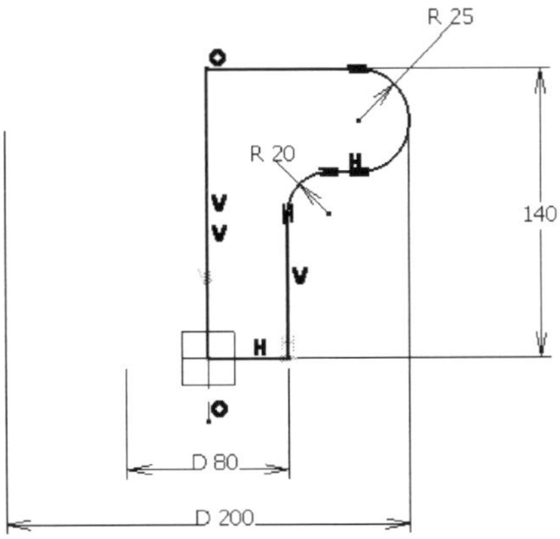

Figure 4-45 Final sketch for the base feature

Next, you need to exit the **Sketcher** workbench to convert the sketch into a feature.

16. Choose the **Exit** workbench button from the **Workbench** toolbar to exit the **Sketcher** workbench.

Creating the Base Feature

After drawing the sketch, you need to convert it into a base feature. You will revolve the sketch about the revolving axis using the **Shaft** tool.

After exiting the **Sketcher** workbench, the sketch is displayed in orange implying that it is selected by default.

1. Choose the **Shaft** tool from the **Sketch-Based Features** toolbar; the **Shaft Definition** dialog box is displayed. Also, a preview of the shaft feature is displayed in the geometry area. 

   By default, the **First Angle** option is selected in the drop-down list available in the **First Limit** area and the default value of the angle in the **First Angle** spinner is **360deg.** You need the same angular value for creating this feature. Therefore, you do not need to change the default values.

2. Choose the **OK** button from the **Shaft Definition** dialog box; a revolved feature is created.

3. Click once in the geometry area to remove the newly created base feature from the selection set.

4. Choose the **Isometric View** button from the **View** toolbar to orient the view of the model from 3D view to isometric view. 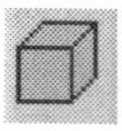

5. Choose the **Fit All In** button from the **View** toolbar to fit the view of the model in the geometry area. The model after creating the base feature is shown in Figure 4-46. 

***Figure 4-46** Final model for Tutorial 1*

Assigning the Material to the Model

Next, you need to assign Copper material to the model and then turn on the display of the material by customizing the display mode.

1. Select **PartBody** from the Specification tree or select any face of the model from the geometry area. Next, choose the **Apply Material** tool from the **Apply Material** toolbar; the **Library (ReadOnly)** dialog box is displayed.

2. Choose the **Metal** tab from this dialog box; all the default metals in this tab are displayed in the display area.

3. Select **Copper** from the display area and then choose the **Apply Material** button from the **Library (ReadOnly)** dialog box.

4. Choose the **OK** button to exit the **Library (ReadOnly)** dialog box.

5. Next, choose the **Shading with Material** button from the **View** toolbar. The model after assigning the Copper material and choosing the **Shading with Material** button is shown in Figure 4-47.

Rotating the View of the Model in 3D Space

After assigning the material to the model, you need to rotate its view in 3D space.

1. Press and hold the middle mouse button and then drag the cursor keeping the left or right mouse button pressed to rotate the view of the model.

2. Next, release both the mouse buttons. Figure 4-48 shows the view of the model being rotated.

Figure 4-47 *The model after assigning the Copper material*

Figure 4-48 *The view of the model being rotated*

3. Choose the **Isometric View** tool from the **Quick view** sub-toolbar to restore the isometric view of the model.

Saving and Closing the File

1. Choose the **Save** button from the **Standard** toolbar to invoke the **Save As** dialog box. Create the *c04* folder inside the *CATIA* folder.

2. Choose the **Save** button; the file is saved at *C:\CATIA\c04*.

3. Close the part file by choosing **Close** from the **File** menu.

Tutorial 2

In this tutorial, you will create the base feature of the model shown in Figure 4-49 by extruding a sketch drawn on the YZ plane. Then, you will apply the Aluminium material to the model and rotate its view in 3D space. The views and dimensions of the model are shown in Figure 4-50.

(Expected time: 30 min)

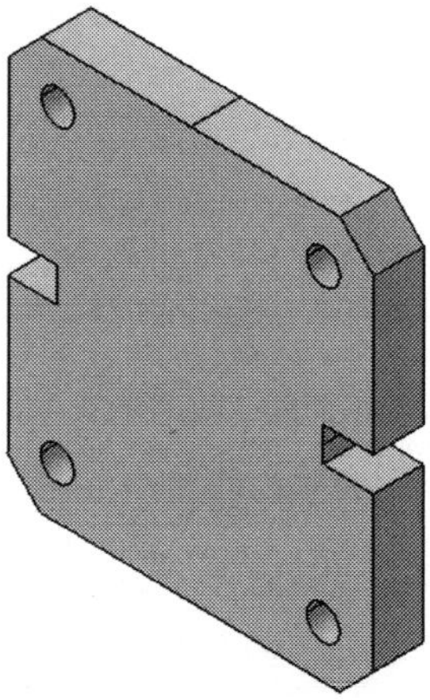

Figure 4-49 *Solid model for Tutorial 2*

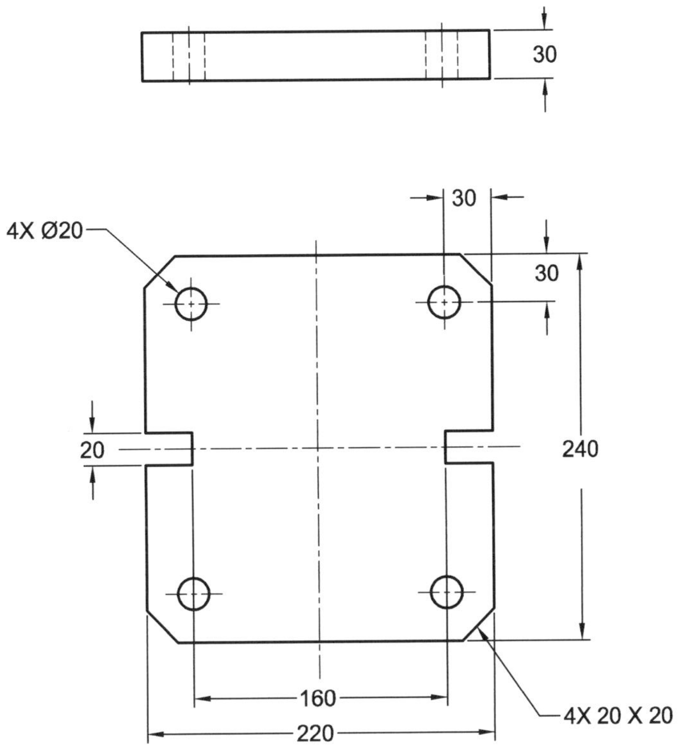

Figure 4-50 *Views and dimensions for Tutorial 2*

The following steps are required to complete this tutorial:

a. Start a new file in the **Part Design** workbench and draw the sketch on the yz plane using the tools in the **Sketcher** workbench of CATIA V5.
b. Apply the required geometrical and dimensional constraints to the sketch.
c. Extrude the sketch to the given depth.
d. Assign material to the model and then rotate its view in 3D space.
e. Save and close the file.

Starting a New File and Drawing Sketches

1. Choose **Start > Mechanical Design > Part Design** from the menu bar; the **New Part** dialog box is invoked.

2. In this dialog box, enter **c04tut02** in the **Enter part name** edit box and then choose the **OK** button.

   The base feature of this model is a pad feature. You need to create the pad feature by extruding the sketch drawn on the yz plane. Therefore, first you need to invoke the **Sketcher** environment using the yz plane as the sketching plane.

3. Select the yz plane from the Specification tree or the geometry area, and then choose the **Sketch** tool from the **Sketcher** toolbar, the **Sketcher** workbench will be invoked.

   As is evident from Figure 4-50, the sketch is symmetric to both the axes. Therefore, you need to draw only quarter of the sketch, and then mirror the quarter about the horizontal centerline and then mirror the entire sketch about the vertical centerline. You do not need to add extra dimension to the sketch because symmetric constraints are applied automatically maintaining the design intent of the sketch.

4. Draw vertical and horizontal centerlines in the first quadrant using the **Axis** tool.

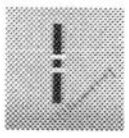

   Note that on drawing the second center line the first one will automatically get converted into a construction line.

5. Draw first quarter of the sketch, as shown in Figure 4-51, using the **Profile** and **Circle** tools.

   After drawing the first quarter of the sketch, you need to mirror it below the horizontal centerline using the **Mirror** tool.

6. Select all the sketched elements except centerlines if required press and hold the Ctrl key to select sketched elements, refer to Figure 4-52.

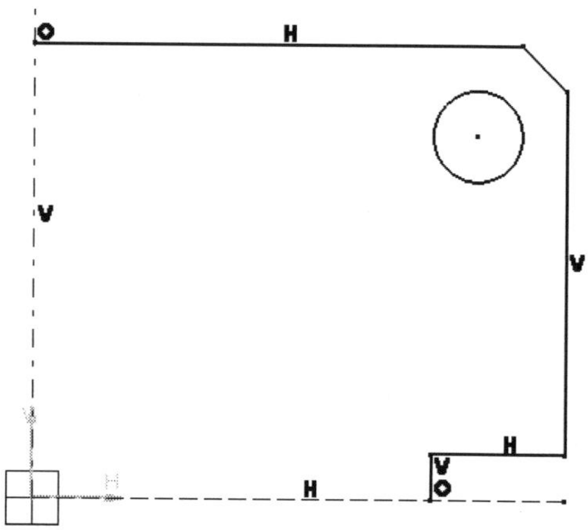

Figure 4-51 *First quarter of the sketch*

7. Choose the **Mirror** tool from the **Operation** toolbar; you are prompted to select the line or axis from which the elements will remain equidistant.

8. Select the horizontal centerline, as shown in Figure 4-52; the selected sketched elements are mirrored on the other side of the horizontal centerline, as shown in Figure 4-53.

   Next, you need to mirror the sketched entities about the vertical centerline.

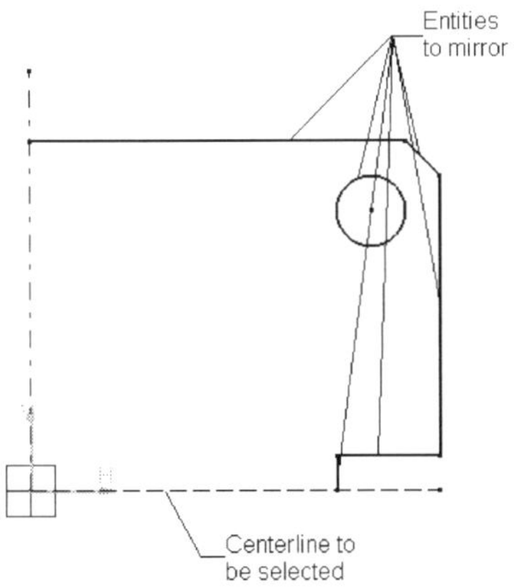

Figure 4-52 Sketched entities and the centerline to be selected

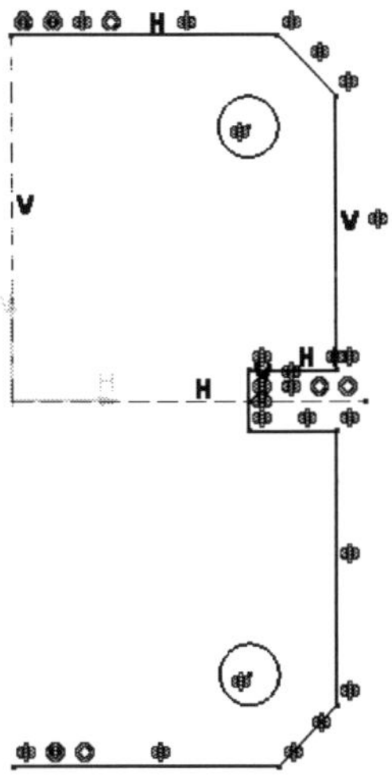

Figure 4-53 Resulting mirrored sketch

9. Press and hold the Ctrl key and select all the sketched elements except the center lines.

10. Choose the **Mirror** tool from the **Operation** toolbar and select the vertical center line; all the selected sketched elements are mirrored about the vertical centerline. The sketch after mirroring all the elements and hiding the symbols of constraint is shown in Figure 4-54.

Note

1. In Figures 4-51 through 4-53, the display of the grid lines is turned off and the constraints are hidden for better visualization. To turn the display of the grid lines on/off, choose the ***Grid*** *button from the* ***Visualization*** *toolbar.*

2. To hide the constraints, expand the ***Sketch.1*** *node from the Specification tree; the* ***Constraints*** *node is displayed. Select all the constraints from this node and right-click; a contextual menu is displayed. Choose the* ***Hide / Show*** *option from the contextual menu. You can use the same option to display the hidden constraints. Alternatively, you can hide the constraints by using the* ***Dimensional Constraints*** *and* ***Geometrical Constraints*** *buttons from the* ***Visualization*** *toolbar.*

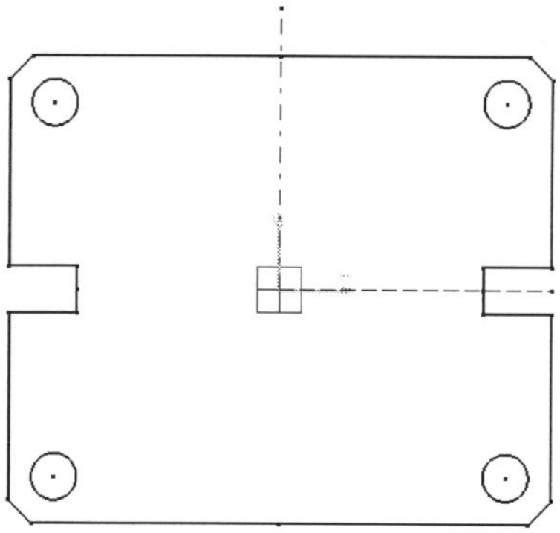

Figure 4-54 Resulting mirrored sketch with Geometric constraints turned off

Applying Dimensional Constraints to the Sketch

After drawing the sketch, you need to apply constraints to the sketch in order to make it stable. In this tutorial, you will only apply dimensional constraints to the sketch. This is because the required geometrical constraints have already been applied to the sketch while drawing and mirroring it.

1. Double-click on the **Constraint** tool from the **Constraint** toolbar.

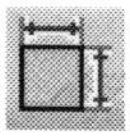

2. Select the lower left vertical line from the geometry area; a vertical dimension is attached to the cursor. Select the lower right vertical line; a horizontal dimension is attached to the cursor. Move the cursor vertically downward and place the dimension.

3. Select the lower horizontal line; a horizontal dimension is attached to the cursor. Select the upper horizontal line of the sketch; a vertical dimension is attached to the cursor. Move the cursor horizontally toward the right and place the dimension.

4. Select the center point of the upper right circle and then select the upper right vertical line; a horizontal dimension is attached to the cursor. Move the cursor vertically upward and place the dimension.

5. Select the center point of the same circle and then select the upper horizontal line; a vertical dimension is attached to the cursor. Move the cursor horizontally toward the right and place the dimension.

6. Select the upper right circle; a diametrical dimension is attached to the cursor. Move the cursor vertically upward and place the dimension.

7. Now, select the upper left inclined line of the sketch; an aligned dimension is attached to the cursor. Right-click in the geometry area to invoke the contextual menu, and then choose the **Horizontal Measure Direction** option from it; a horizontal dimension is attached to the cursor.

8. Move the cursor vertically upward and place the dimension.

9. Select the same inclined line and invoke the contextual menu by right-clicking in the geometry area. Next, choose the **Vertical Measure Dimension** option from the contextual menu and place the dimension on the left of the sketch.

   You will notice that all the sketched elements are displayed in green except the rectangular open slot on the right and left of the sketch. You need to dimension the right slot to make the sketch iso-constrained (fully constrained).

10. Apply the horizontal and vertical dimensions to the slot, refer to Figure 4-55.

    Next, you need to modify the dimensions.

11. Double-click on the horizontal dimension that defines the width of the sketch; the **Constraint Definition** dialog box is displayed. In this dialog box, set the value in the **Value** spinner to **220** and then choose the **OK** button to exit the dialog box.

12. Double-click on the vertical dimension that defines the height of the sketch; the **Constraint Definition** dialog box is displayed. Set the value in the **Value** spinner to **240** and choose **OK** from the dialog box.

13. Double-click on the vertical dimension defining the length of the inclined line; the **Constraint Definition** dialog box is displayed. Set **20** as the value in the **Value** spinner of this dialog box and then choose **OK**.

14. Similarly, modify other dimensions. The sketch after modifying all the dimensions is displayed, as shown in Figure 4-55.

    After completing the sketch, you need to exit the **Sketcher** workbench.

15. Choose the **Exit workbench** tool from the **Workbench** toolbar to exit the **Sketcher** workbench.

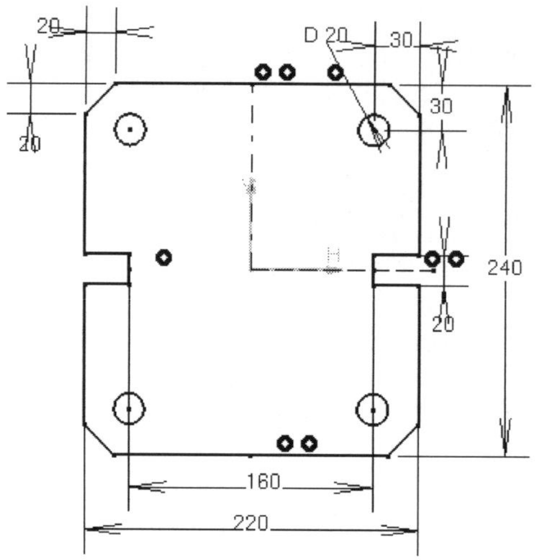

Figure 4-55 The sketch after applying all dimensions

Extruding the Sketch

Next, you need to extrude the sketch.

1. Choose the **Pad** button from the **Sketch-Based Features** toolbar; the **Pad Definition** dialog box is displayed and a preview of the extruded feature is displayed in the geometry area.

2. Set **30** as the value in the **Length** spinner of the **Pad Definition** dialog box, and then choose **OK** from this dialog box.

3. Click once in the geometry area to remove the newly created base feature from the current selection set.

4. Orient the view of the model to the isometric view using the **Isometric View** tool from the **View** toolbar. The model after creating the extruded feature is shown in Figure 4-56.

Assigning the Material to the Model

After creating the model, you need to assign a material to it and turn on the display of the material by setting the display mode to view the customized view parameters.

1. Select **PartBody** from the Specification tree and choose the **Apply Material** button from the **Apply Material** toolbar; the **Library (ReadOnly)** dialog box is displayed.

2. Choose the **Metal** tab from this dialog box; all the default metals in this tab are displayed in the display area of the dialog box.

3. Select **Aluminium** from the display area of the **Metal** tab and choose the **Apply Material** button from the **Library (ReadOnly)** dialog box.

4. Exit the **Library (ReadOnly)** dialog box by choosing the **OK** button from it.

5. Now, choose the **Shading with Material** button from the **View mode** sub-toolbar of the **View** toolbar. The model after applying the Aluminium material is shown in Figure 4-57.

6. Next, rotate the view of the model in the 3D space using the **Rotate** tool.

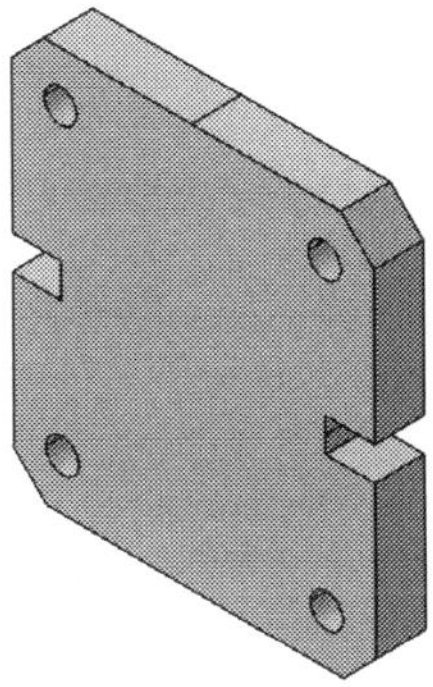

Figure 4-56 *The model after creating the base feature*

Figure 4-57 *The model after applying the Aluminium material*

Saving and Closing the File

1. Choose the **Save** button from the **Standard** toolbar to invoke the **Save As** dialog box.

2. In this dialog box, enter the name of the file as ***c04tut02*** in the **File name** edit box and choose the **Save** button. The file is saved at *C:\CATIA\c04*.

3. Close the part file by choosing **Close** from the **File** menu.

Self-Evaluation Test

Answer the following questions and then compare them to those given at the end of this chapter:

1. When you choose the __________ button from the **View mode** toolbar, a shaded model is displayed with edges and hidden edges.

2. To create a thin extrude feature, select the __________ check box from the **Pad Definition** dialog box.

3. When you choose the __________ button from the **View mode** toolbar, the model is displayed in the shaded mode.

4. When the __________ constraint is applied to two selected entities, the distance dimension is added between them.

5. The __________ constraint forces two selected arcs or circles to share the same center point.

6. In the **Sketcher** workbench of CATIA V5, you can apply geometrical constraints to the sketches. (T/F)

7. You can invoke the **Pad** tool first and then draw a sketch. (T/F)

8. The **Shaft** tool does not follow the right hand thumb rule. (T/F)

9. You can also pan the view of a model using a Compass. (T/F)

10. Using the **Profile Definition** tool, you can select a particular contour from a multi-contour sketch to extrude it. (T/F)

Review Questions

Answer the following questions:

1. Which dialog box is used to specify the user-defined settings to change the view of the model?

   (a) **Shading** (b) **Customize View Parameters**
   (c) **Wireframe** (d) None of these

2. Which tool is used to extrude a sketch?

   (a) **Pocket** (b) **Shaft**
   (c) **Pad** (d) None of these

3. Which state of a sketch is the most stable?

   (a) **Iso-Constrained** (b) **Over-Constrained**
   (c) **Under-Constrained** (d) All of these

4. Which of the following geometrical constraints forces the selected element to remain in its position?

   (a) **Fix** (b) **Parallelism**
   (c) **Perpendicular** (d) **Concentricity**

5. Which of the following tools in the **Constraint** toolbar is used to apply dimensions to a sketch?

   (a) **Constraint** (b) **View**
   (c) **Pad** (d) None of these

6. To rotate a view freely in the 3D space, choose the **Pan** tool from the **View** toolbar. (T/F)

7. You cannot apply a constraint between a sketched element and an edge or a vertex of a model. (T/F)

8. To apply the dimension, if you select an inclined line, the aligned dimension is applied to it. (T/F)

9. You can assign a material to a model using the **Pad Definition** dialog box. (T/F)

10. You can create a thin shaft feature using the options in the **Thin Shaft** area of the **Shaft Definition** dialog box. (T/F)

EXERCISES

Exercise 1

Create the model shown in Figure 4-58. The dimensions of the model are shown in Figure 4-59. **(Expected time: 30 min)**

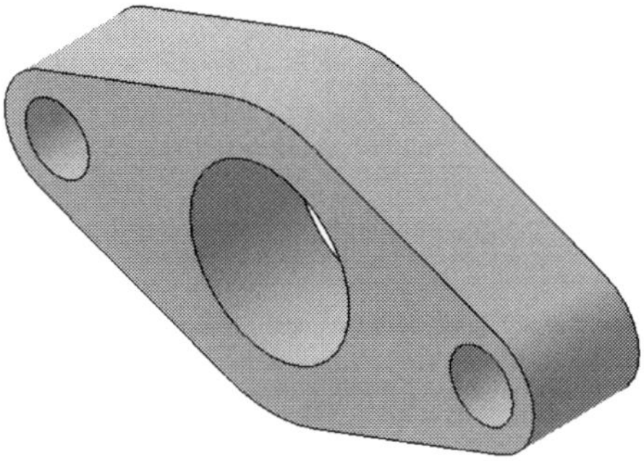

Figure 4-58 *Solid model for Exercise 1*

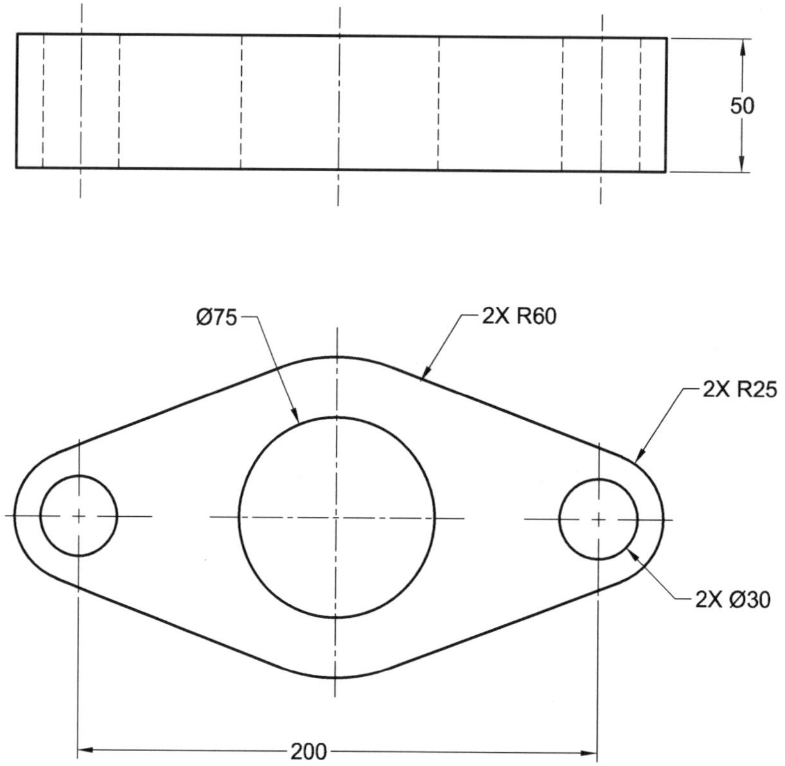

Figure 4-59 Views and dimensions for Exercise 1

Exercise 2

Create the model shown in Figure 4-60. The dimensions of the model are shown in Figure 4-61. **(Expected time: 30 min)**

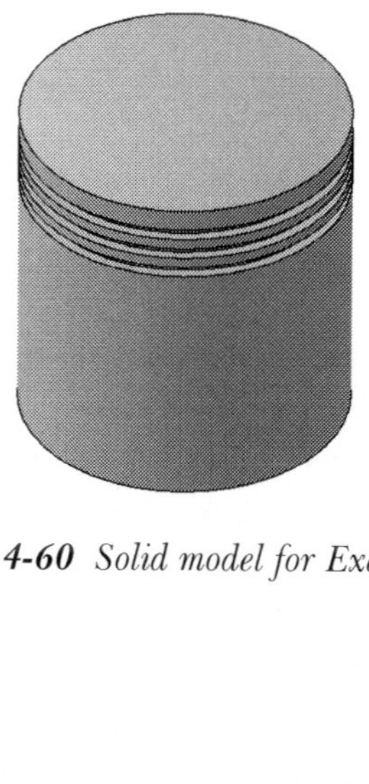

Figure 4-60 Solid model for Exercise 2

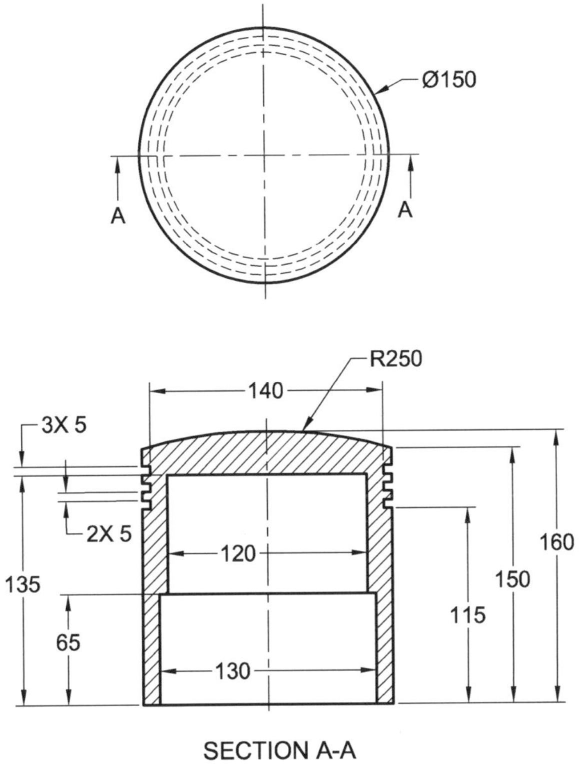

Figure 4-61 Views and dimensions for Exercise 2

Exercise 3

Create the model shown in Figure 4-62. The dimensions of the model are shown in Figure 4-63. **(Expected time: 30 min)**

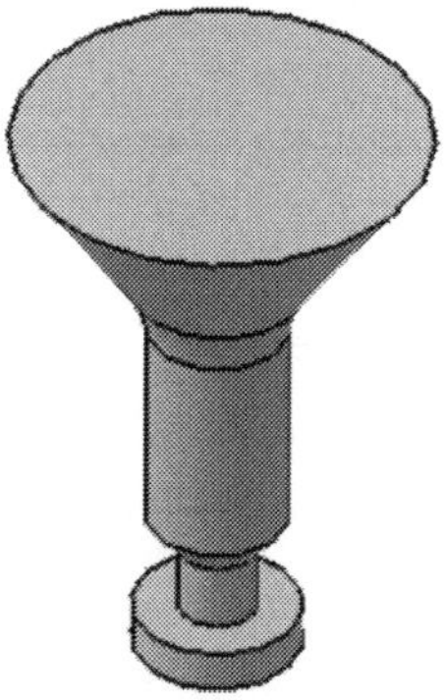

Figure 4-62 Solid model for Exercise 3

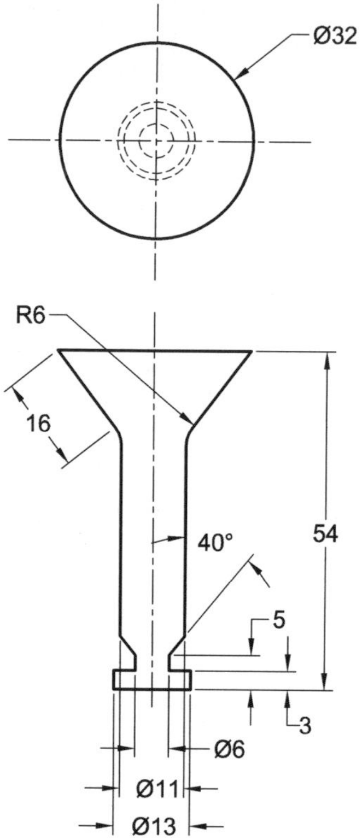

Figure 4-63 Views and dimensions for Exercise 3

Exercise 4

Create the model shown in Figure 4-64. The dimensions of the model are shown in Figure 4-65. **(Expected time: 30 min)**

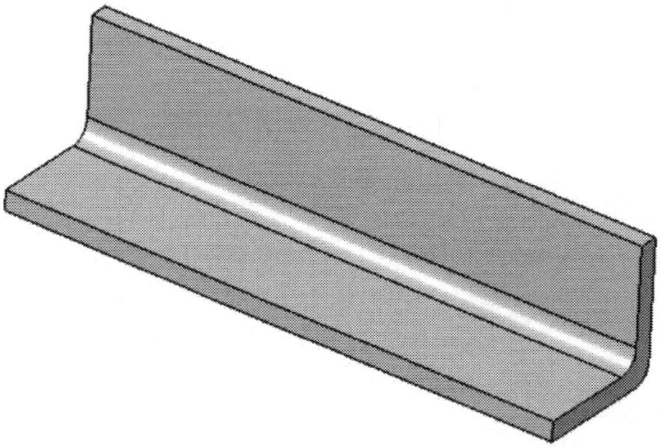

Figure 4-64 Solid model for Exercise 4

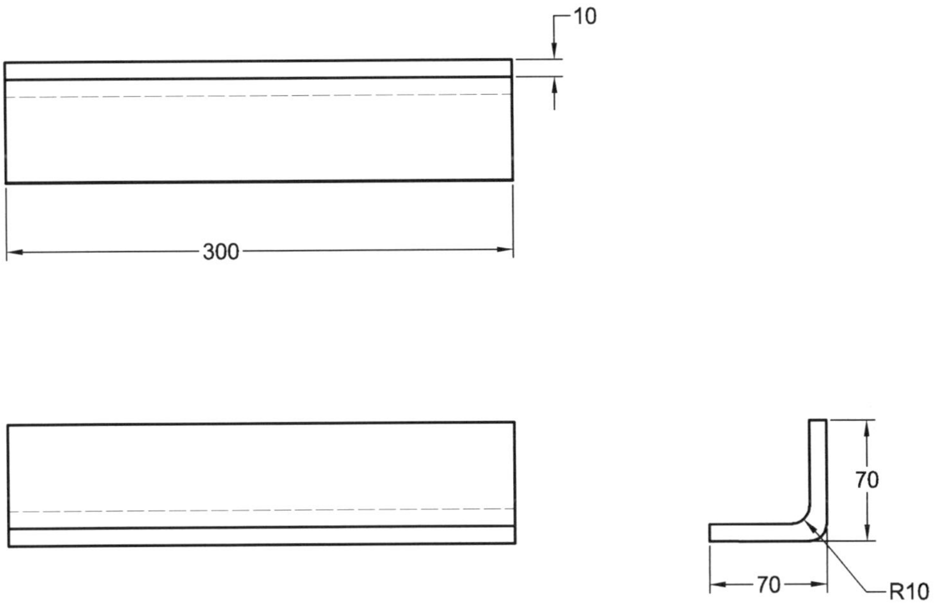

Figure 4-65 *Views and dimensions for Exercise 4*

Answers to Self-Evaluation Test

1. Shading with Edges and Hidden Edges, **2. Thick**, **3. Shading (SHD)**, **4.** Distance, **5.** Concentricity, **6.** T, **7.** T, **8.** F, **9.** T, **10.** T

Chapter 5

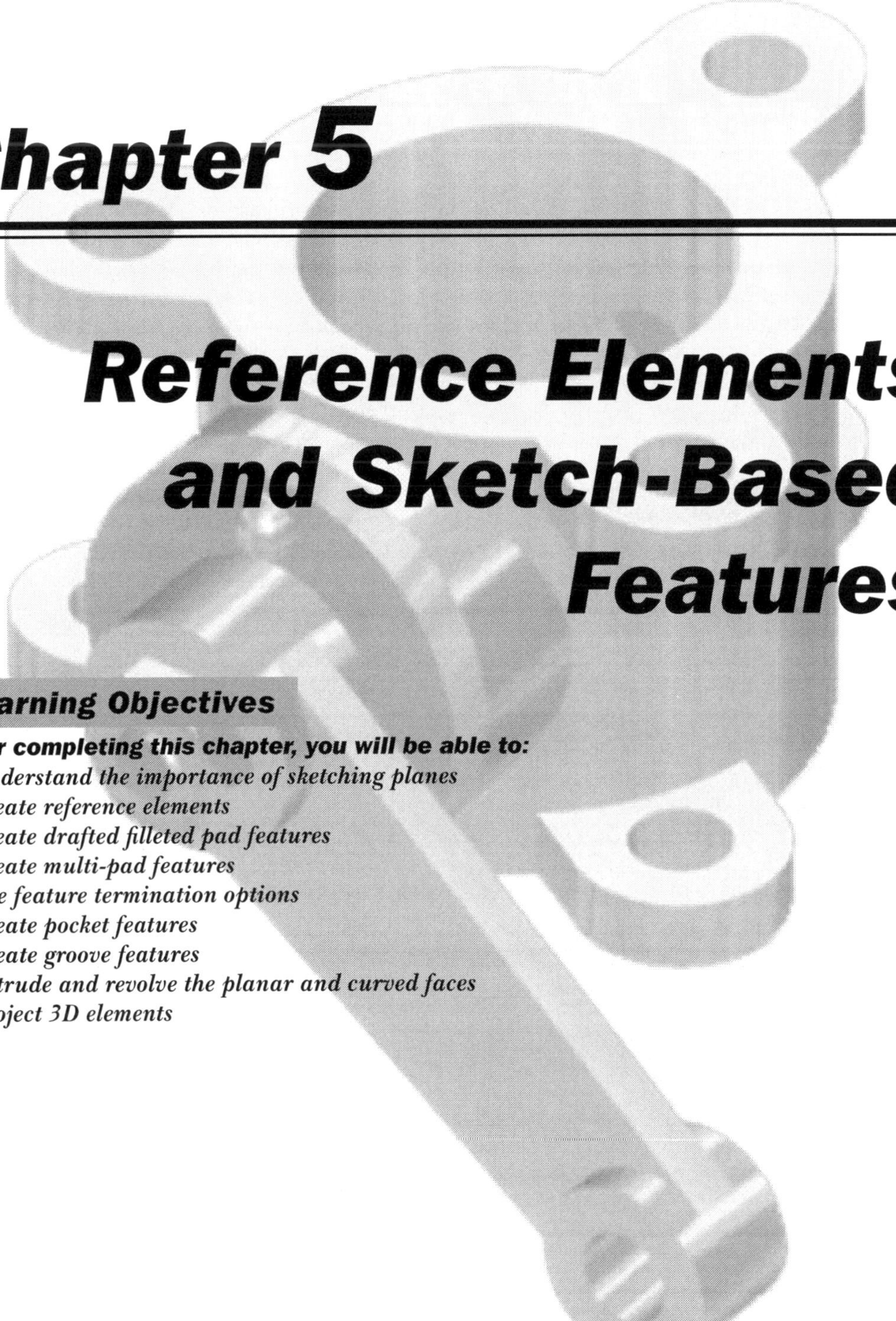

Reference Elements and Sketch-Based Features

Learning Objectives

After completing this chapter, you will be able to:

- *Understand the importance of sketching planes*
- *Create reference elements*
- *Create drafted filleted pad features*
- *Create multi-pad features*
- *Use feature termination options*
- *Create pocket features*
- *Create groove features*
- *Extrude and revolve the planar and curved faces*
- *Project 3D elements*

IMPORTANCE OF SKETCHING PLANES

In previous chapters, you learned to create the base feature of the models by extruding and revolving the sketches. All these features were created on a single sketching plane, the yz plane. Most mechanical designs consist of multiple features such as sketched features, referenced elements, and placed features, which are integrated to complete a model. You can select any one of the default planes as the sketching plane to create the sketch of the base feature. You need to be careful while selecting the sketching plane for creating the base feature because the orientation of the component depends on the orientation of the sketching plane. Next, you need to create other features of the model. If they are sketch-based features, you need to select a sketching plane to create the sketch. You can select the default plane, a planar face of the model, or create reference planes that will be selected as the sketch plane for creating other sketch-based features. For example, consider the model shown in Figure 5-1. Its base feature is shown in Figure 5-2.

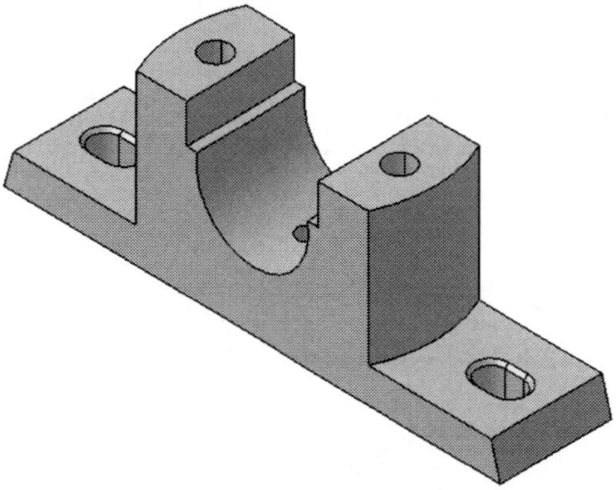

Figure 5-1 *A multi-featured model*

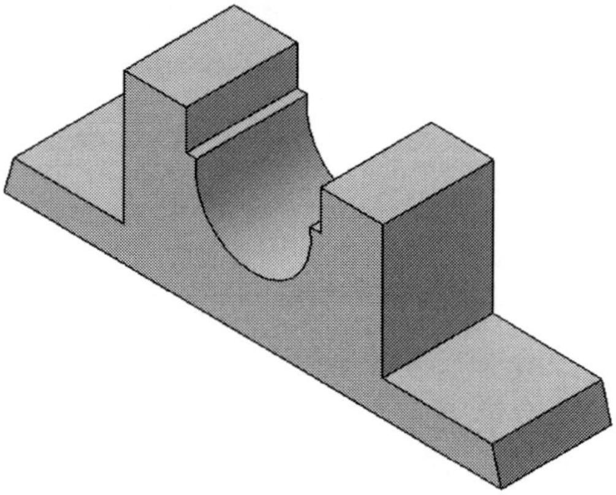

Figure 5-2 *Base feature of the model*

The sketch for the base feature is drawn on the yz plane and then extruded using the **Pad** tool. Next, you need to create other features of the model which include sketch-based features, reference elements, and placed features, refer to Figure 5-3.

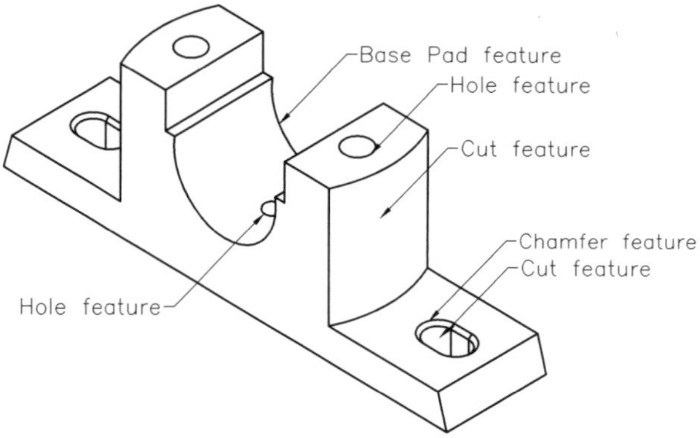

Figure 5-3 *Other features of the final model*

It is evident from Figure 5-3 that the features added to the base features are not created on the same plane on which the sketch of the base feature is drawn. Therefore, to draw the sketches of the other sketch-based features, you need to define other sketching planes.

REFERENCE ELEMENTS

Reference elements are the features that have no mass and volume and are used only to assist you in the creation of the models. They act as a reference for drawing sketches for features, defining the sketch plane, placing placed features, assembling components, creating sketch-based features, and so on. The reference elements are widely used in creating complex models. Therefore, you need to have a good understanding of reference elements before you start creating them. In CATIA V5, reference elements exist as a plane, line, and point. These reference elements are discussed in the following section. You can invoke the tools to create reference elements using the **Reference Elements** toolbar, as shown in Figure 5-4.

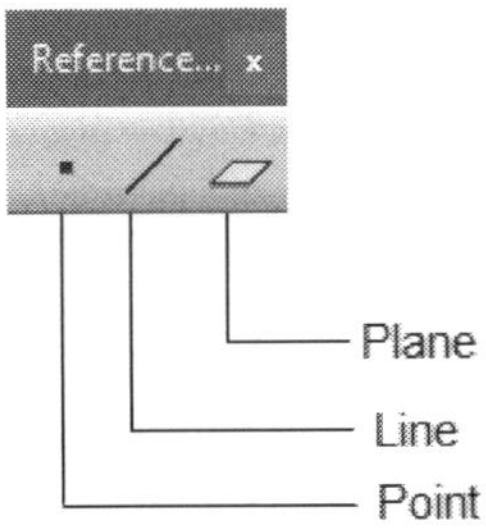

Figure 5-4 *The* ***Reference Elements*** *toolbar*

Note

Sometimes toolbars are not visible on the screen even after they are invoked. To display such toolbars on the screen, double-click on the double arrow displayed on the lower or upper right corner of the toolbar area; the hidden toolbars will be displayed.

Reference Planes

As discussed earlier, most of the mechanical engineering components or designs are multi-featured models, and the features are generally not created on the same plane on which the base feature is created. Therefore, you need to select other default planes or create new planes to be used as the sketching plane for other features. It is clear from the above discussion that you can use the default planes as the sketching planes or create a new sketching plane. In the next section, you will learn about default planes and also about the process of creating new planes.

Default Planes

On starting a new file in the **Part Design** workbench, you are provided with the following three default planes:

1. xy plane
2. yz plane
3. zx plane

As discussed earlier, the orientation of the model depends on the sketch of the base feature. Therefore, it is recommended that you carefully select the sketching plane for drawing the sketch of the base feature, which can be drawn on any one of the three datum planes provided by default.

Creating New Planes

Toolbar: Reference Elements > Plane

As mentioned earlier, planes are used as the sketching planes for drawing sketches for the sketch-based features, applying references to the placed features, and so on. The sketch of the base feature is generally drawn using one of the default planes as the sketching plane. After creating the base feature, you can select one of its faces as the sketching plane to draw the sketch for the other sketch-based features. However, sometimes you may need to draw a sketch on a plane, which is at an offset distance from the planar face of the base feature. In this case, you need to create a plane at an offset distance from the selected planar face or default plane.

In CATIA V5, there are twelve different methods of creating new planes. To create a new plane, choose the **Plane** tool from the **Reference Elements (Extended)** toolbar; the **Plane Definition** dialog box is displayed, as shown in Figure 5-5. The methods of creating new planes are discussed next.

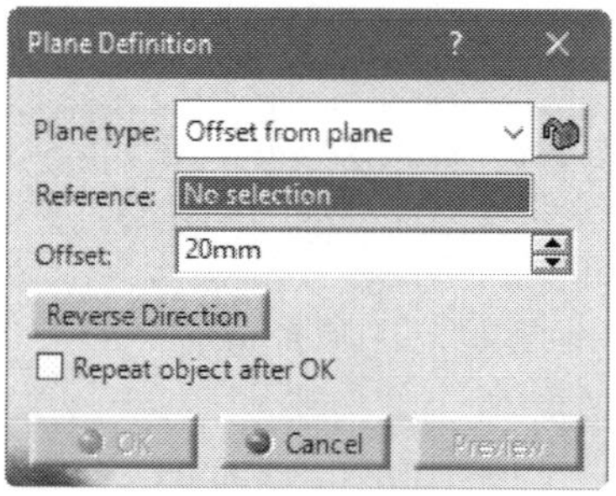

*Figure 5-5 The **Plane Definition** dialog box*

Creating a Plane at an Offset from an Existing Plane/Planar Face

When you invoke the **Plane Definition** dialog box, the **Offset from plane** option is selected by default in the **Plane type** drop-down list and you are prompted to select a plane to offset. Select the plane or the planar face from which the offset plane needs to be created; the preview of the offset plane with the default offset value will be displayed in the geometry area. You can set the value of the offset distance in the **Offset** spinner. Alternatively, you can use the **Offset** drag handle attached to the preview of the plane to dynamically define it. You can use the **Reverse Direction** button from the dialog box or select the reverse direction arrow from the preview to flip the direction of the offset. Next, choose the **OK** button from the **Plane Definition** dialog box to create a new plane at an offset distance from the selected face or plane. Figure 5-6 shows the face selected as the reference and also the resultant plane created at an offset distance.

Select the **Repeat object after OK** check box and then choose the **OK** button; the **Object Repetition** dialog box will be displayed. You can specify the number of copies of the new plane in the **Instance(s)** spinner. Note that the **Create in a new Body** check box is selected by default in the **Object Repetition** dialog box. You will learn more about the new bodies in later chapters. After setting the number of instances, choose the **OK** button from the **Object Repetition** dialog box. In addition to the original offset plane, the number of offset planes specified in the **Instance(s)** spinner will also be created.

Creating a Plane Parallel to an Existing Plane and Passing through a Point

The **Parallel through point** option in the **Plane type** drop-down list is used to create an offset plane that is parallel to a reference plane or planar face and passes through a specified point. To create a plane using this option, select the **Parallel through point** option from the **Plane type** drop-down list; you are prompted to select a plane to offset. Select a plane or planar face from the geometry area; you are prompted to select a point. Select a point or vertex from the geometry area; the preview of the plane is displayed. Choose the **OK** button from the **Plane Definition** dialog box. Figure 5-7 shows the reference planar face, the sketch point, and the resultant plane.

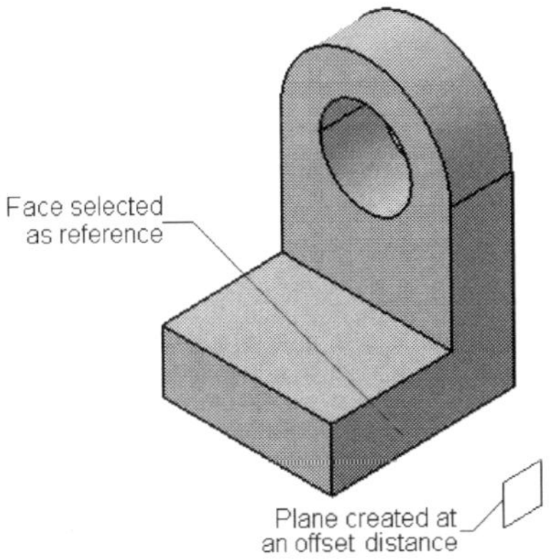

***Figure 5-6** Reference planar face and resultant plane*

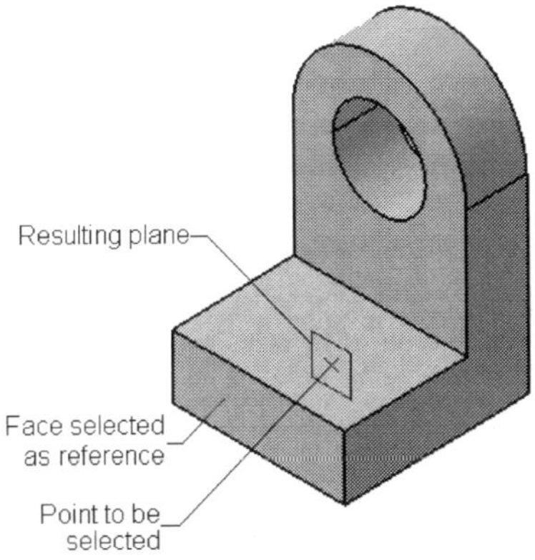

***Figure 5-7** Reference planar face, sketch point, and resultant plane*

Creating a Plane at an Angle/Normal to a Plane

The **Angle/Normal to plane** option in the **Plane type** drop-down list is used to create a plane at an angle to a reference plane or face. You can also create a plane normal to the selected plane or face. To create a plane using this option, select the **Angle/Normal to plane** option from the **Plane type** drop-down list; you will be prompted to select the rotation axis. Select an edge of the model, sketched line, or axis from the geometry area that will be used as the axis of rotation; you will be prompted to select the reference plane. Select the reference plane or planar face from the geometry area such that the rotation axis and the selected reference plane are parallel to each other; the preview of the plane is displayed. Set the value of the rotation angle in the **Angle** spinner. To reverse the direction of the plane creation, specify a negative angular value. Next, choose the **OK** button from the **Plane Definition** dialog box. Figure 5-8 shows the rotation axis, reference face, and resultant plane at an angle.

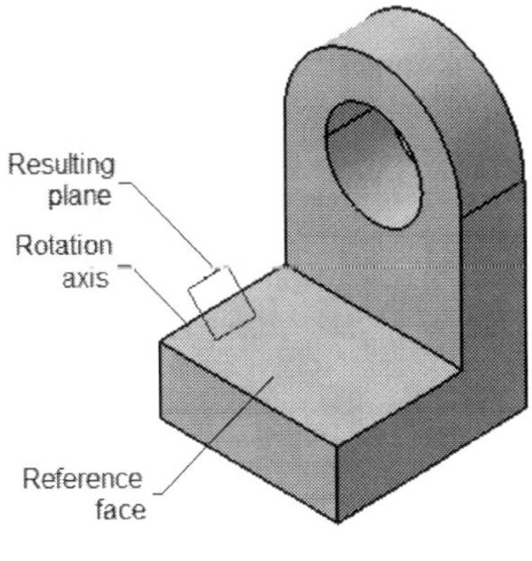

***Figure 5-8** Rotation axis, reference face, and resulting plane at an angle*

If you choose the **Normal to plane** button from the **Plane Definition** dialog box, a plane will be created normal to the reference plane. You can also create multiple copies of the plane using the **Repeat object after OK** check box.

Creating a Plane through Three Points

The **Through three points** option in the **Plane type** drop-down list is used to create a plane that passes through three selected points. On selecting this option, you will be prompted to select the first point. You can select a point, vertex, or an endpoint of a line from the geometry area. On doing so, you will be prompted to select the second point. Select the second point from the geometry area. Similarly, select the third point; the preview of the plane will be displayed. Choose the **OK** button from the **Plane Definition** dialog box. Figure 5-9 shows the three vertices to be selected and also the resulting plane.

Creating a Plane through Two Lines

The **Through two lines** option in the **Plane type** drop-down list is used to create a plane that will pass through two selected lines, edges, or an edge and a line. When you select this option, you are prompted to select the first line. Select an edge, sketched line, or an axis from the geometry area; you will be prompted to select the second line. After you select the second line, the preview of the plane will be displayed. Choose the **OK** button from the **Plane Definition** dialog box. Figure 5-10 shows the reference edges to be selected and the resulting plane.

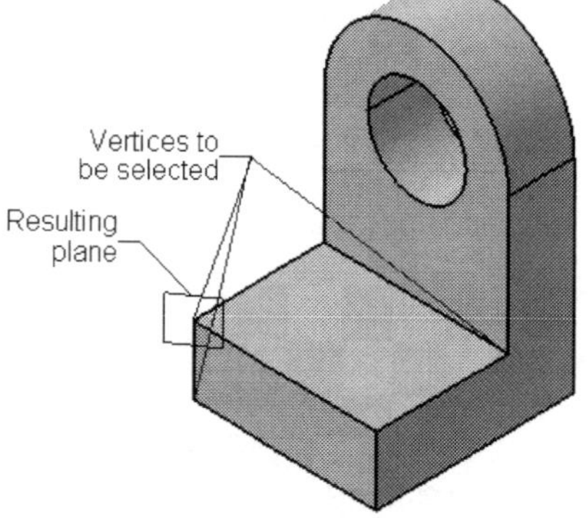

Figure 5-9 *Three vertices to be selected and the resulting plane*

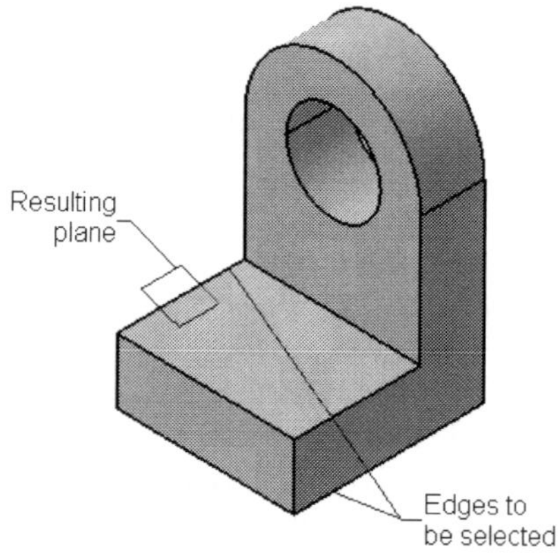

Figure 5-10 *Edges to be selected and the resulting plane*

If you select the **Forbid non coplanar lines** check box, you can only select coplanar lines to create the plane.

Creating a Plane through a Point and a Line

The **Through point and line** option in the **Plane type** drop-down list is used to create a plane that passes through a point and a line. When you select this option, you will be prompted to select a point. Select the point through which you want the plane to pass; you will be prompted to select a line. Select a line, axis, or an edge from the geometry area through which the plane will pass; the preview of the plane will be displayed in the geometry area. Choose the **OK** button from the **Plane Definition** dialog box. Figure 5-11 shows the point and the line to be selected and the resulting plane.

Creating a Plane through a Planar Curve

The **Through planar curve** option is used to create a plane that will be coplanar to the selected arc. To create a plane using this option, select the **Through planar curve** option from the **Plane type** drop-down list; you will be prompted to select the curve. Select the curve from the geometry area. You will notice that the preview of the plane is placed coplanar to the selected curve. Choose the **OK** button from the **Plane Definition** dialog box. Figure 5-12 shows the planar curve to be selected and also the resulting plane.

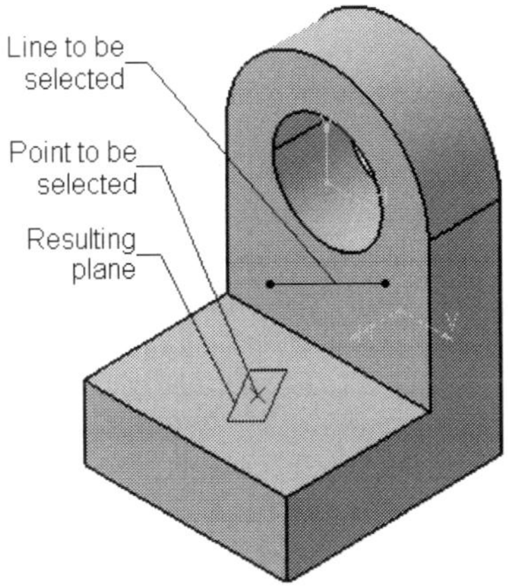

Figure 5-11 The point and the line to be selected and the resulting plane

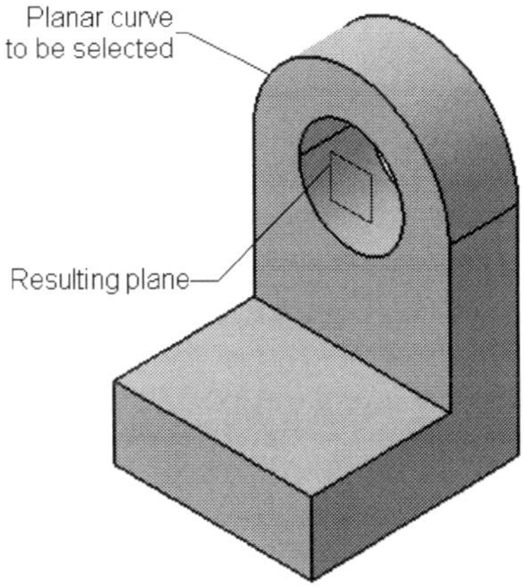

Figure 5-12 The planar curve to be selected and the resulting plane

Creating a Plane Normal to a Curve

The **Normal to curve** option is used to create a plane that is normal to the selected curve. On selecting this option, you will be prompted to select the reference curve. Select the curve from the geometry area; the preview of the normal plane, placed at the midpoint of the selected curve, is displayed. Note that, by default, the midpoint of the selected curve is considered as a point to place the plane normal to curve. If you do not want to create a plane in the middle of the selected curve, select a point on the curve/vertex/edge where the plane will be placed. If you want to create a plane at a specific location on the curve, select the **Ratio of Curve Length** check box and specify the required ratio value in the **Ratio** spinner box. Select the **Reverse Direction** button to reverse the direction of the plane.

Next, choose the **OK** button from the **Plane Definition** dialog box. Figure 5-13 shows the curve and the point to be selected and the resulting plane.

Creating a Plane Tangent to a Surface

The **Tangent to surface** option is used to create a plane tangent to a selected surface and passing through a selected point. On selecting this option, you will be prompted to select the reference surface. Select the surface from the geometry area; you will be prompted to select a tangency point. Select a point. The point should not necessarily be placed on the selected surface. The preview of the plane, tangent to the selected surface, is displayed. Choose the **OK** button from the **Plane Definition** dialog box. Figure 5-14 shows the reference surface to be selected, point to be selected, and the resulting plane.

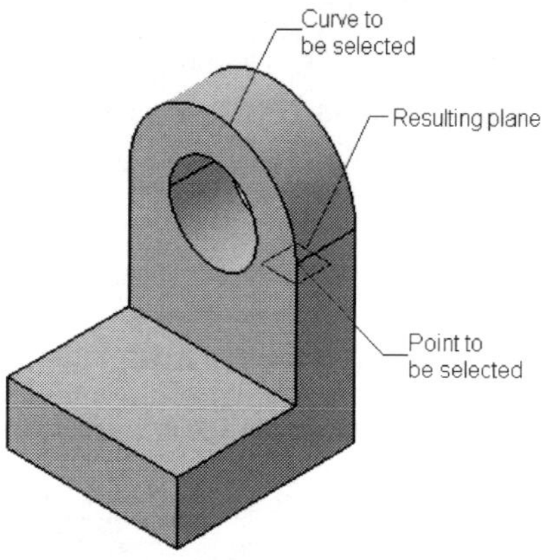

Figure 5-13 The curve and the point to be selected, and the resulting plane

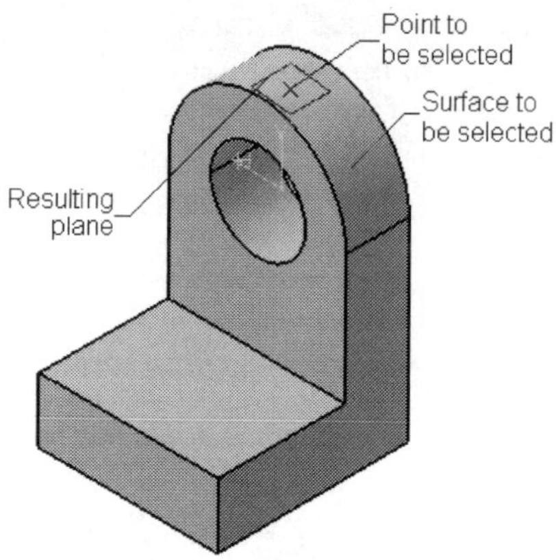

Figure 5-14 The reference surface and the point to be selected, and the resulting plane

Creating a Plane Using an Equation

The **Equation** option is used to create a plane using the equation Ax+By+Cz = D, where the values of A, B, C, and D are variable and can be changed to modify the orientation of the plane. When you select this option, the **Plane Definition** dialog box will expand and the **A**, **B**, **C**, and **D** spinners will be displayed. You can set the values in these spinners to create a plane using the above mentioned equation. The **Normal to compass** button in this dialog box is used to create a plane normal to the compass. The **Parallel to screen** button is used to create a plane normal to the current view of the screen. After specifying the parameters, choose the **OK** button from the **Plane Definition** dialog box.

Creating a Plane Using Mean through Points

The **Mean through points** option is used to create a plane at an orientation defined by the mean of the selected points. On selecting this option, you will be prompted to select points. Select the required points or vertices from the geometry area; the names of the selected points and vertices are displayed in the **Points** display box in the **Plane Definition** dialog box. The preview of the plane, created with its orientation depending on the mean of the selected points, is displayed in the geometry area. Choose the **OK** button from the **Plane Definition** dialog box.

Creating a Plane between two Planes/Faces

The **Between** option is used to create a plane between two planes/faces. On selecting this option, you will be prompted to select the first plane. Select the plane/face as the first plane from the geometry area; you will be prompted to select the second plane. Select the second plane/face; a preview of the plane between two planes/faces is displayed. Choose the **OK** button from the **Plane Definition** dialog box.

Creating Points

Toolbar:	Reference Elements (Extended) > Point

Sometimes, you need to define reference points to create features, surfaces, or a plane. As discussed earlier, you can create these points in the **Sketcher** workbench. You can also create the points in the **Part** workbench using the **Point** tool.

To create points, choose the **Point** tool from the **Reference Elements (Extended)** toolbar; the **Point Definition** dialog box will be displayed, as shown in Figure 5-15. You can create points by using different options in the **Point type** drop-down list of this dialog box. In CATIA V5, you can create points using seven different methods. These methods of creating points are discussed next.

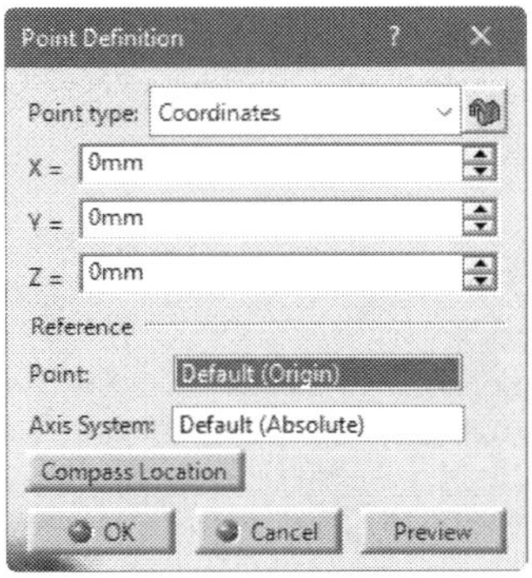

*Figure 5-15 The **Point Definition** dialog box*

Creating a Point Using Coordinates

The **Coordinates** option is used to create a point by specifying the values of its coordinates. To create a point using coordinates, invoke the **Point Definition** dialog box. The **Coordinates** option in the **Point type** drop-down list is selected by default. Specify the coordinates of the new point in the **X**, **Y**, and **Z** edit boxes. By default, the point will be created with reference to the origin refer to the **Point** display box in the **Reference** area. You can also specify a user-defined reference point by clicking in this area and then selecting the point. The **Axis System** display box displays the default axis system, the point's coordinates will work with respect to this axis system. The **Compass Location** button is used to create a point at the location of the compass on the model. After setting the parameters, choose the **OK** button from the **Point Definition** dialog box.

Creating a Point on a Curve

The **On curve** option is used to create a point on the selected curve. When you select this option, the dialog box will expand and you will be prompted to select the curve. Select a curve, on which you need to create the point from the geometry area. Specify the parameters in the **Point Definition** dialog box to define the position of the point. After setting the parameters, choose the **OK** button from the **Point Definition** dialog box. You can select the desired radio button from the **Distance to Reference** area to determine where the new point is to be created. Select the **Distance on curve** radio button to specify the point along the curve at a specified distance from the reference point. Select the **Distance along direction** radio button to create a point along a specified direction at an offset distance from the reference point. Select the **Ratio of curve length** radio button to create a point at a given ratio between the reference point and the curve's extremity. In this case, a ratio value needs to be specified. After setting the parameters, choose the **OK** button from the **Point Definition** dialog box.

Creating a Point on a Plane

The **On plane** option is used to create a point on the selected plane. On selecting this option, you will be prompted to select a plane. Select a plane or planar face on which the point is to be placed. You will notice that a point along with its dimensions is attached to the cursor. These dimensions are with respect to the origin on the selected plane. This is because the origin is selected by default as the reference point. You can also select a user-defined point as the reference point using the **Reference** area. Next, click on a location to place the point in the geometry area or set the values of the **H** and **V** spinners to specify the distances of the point from the origin. You can also project this point on a surface. To do so, click once in the **Surface** display box and then select the surface on which you need to project the point. Next, choose the **OK** button from the **Point Definition** dialog box.

Creating a Point on a Surface

The **On surface** option is used to create a point on the selected surface. When you select this option, you will be prompted to select the surface. Select the surface to create the point. Next, move the cursor on the selected surface and click once to specify the placement of the point. By default, the distance of the point is specified from the reference point placed at the middle of the surface. Next, choose the **OK** button from the **Point Definition** dialog box.

Creating a Point at the Center of Circle/Sphere/Ellipse

The **Circle / Sphere / Ellipse center** option is used to create a point at the center of the selected circle/sphere/ellipse. On selecting this option, you will be prompted to select a circle/sphere/ellipse to get its center. Select a sketched circle/sphere/ellipse or a circular/spherical/elliptical edge; the preview of the point is displayed at the center of the selected sketched circle/sphere/ellipse. Choose the **OK** button from the **Point Definition** dialog box.

Creating a Point Tangent to a Curve

The **Tangent on curve** option is used to create a point tangent to the selected arc. On selecting this option, you will be prompted to select a curve. Select the curve from the geometry area; you will be prompted to select the reference direction. Select a line, planar face, or plane as the directional reference. Choose the **OK** button from the **Point Definition** dialog box.

Creating a Point between Two Points

The **Between** option is used to create a point between any two selected points by defining the ratio of the distance from the two points. On selecting this option, you will be prompted to select the first point. Select a point or vertex from the geometry area; you will be prompted to select the second point; select it. Next, specify the ratio to create the point between the two selected points. The **Middle Point** button is used to create the point at the middle of the selected points. You can also select the supporting surface or plane for creating the point by using the **Support** edit box. Choose the **OK** button from the **Point Definition** dialog box.

Creating Reference Lines

Toolbar:	Reference Elements(Extended) > Line

You can also create reference lines using the **Line** tool in the **Reference Elements** toolbar. To create the reference lines, choose the **Line** tool from the **Reference Elements (Extended)** toolbar; the **Line Definition** dialog box will be displayed. The options in this dialog box are similar to those discussed earlier.

OTHER SKETCH-BASED FEATURES

In the previous chapter, you learned about the **Pad** and **Shaft** features. In this chapter, some of the other sketch-based features are discussed.

Creating Drafted Filleted Pad Features

Menubar: Insert > Sketch-Based Features > Drafted Filleted Pad
Toolbar: Sketch-Based Features > Pads sub-toolbar > Drafted Filleted Pad

You can create an extruded feature with drafted faces and filleted edges using the **Drafted Filleted Pad** tool. To create this feature, draw a sketch and exit the **Sketcher** workbench. Next, choose the **Drafted Filleted Pad** tool from the **Pads** sub-toolbar. If the sketch to be extruded is not selected earlier, you need to select it now. On doing so, the **Drafted Filleted Pad Definition** dialog box will be displayed, as shown in Figure 5-16. Next, you need to specify a reference for the second limit. Select the sketching plane on which the sketch is drawn as the second limit; the name of the selected plane is displayed in the **Limit** display box of the **Second Limit** area and the preview of extruded feature is displayed in geometric area. Set the value of the depth of extrusion in the **Length** spinner. Specify the value of the draft angle in the **Angle** spinner. By default, the **First limit** radio button is selected in the **Draft** area. Therefore, the first limit of the extruded feature will be considered as a neutral plane for adding a draft. If you select the **Second limit** radio button, then the second reference will be considered as the neutral plane. You can modify the default values of the fillet in the spinners in the **Fillets** area.

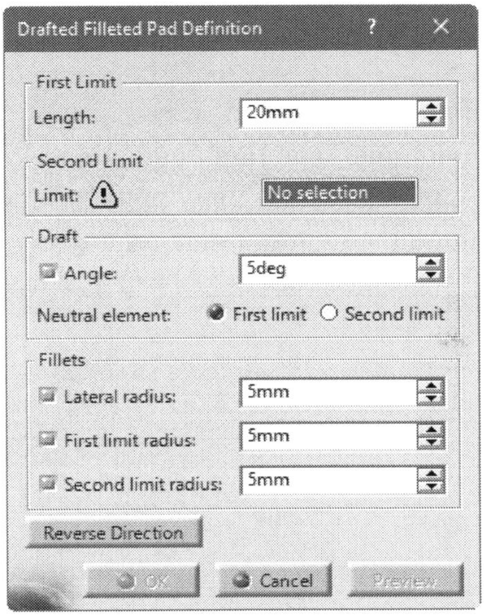

Figure 5-16 The ***Drafted Filleted Pad Definition*** *dialog box*

After setting the parameters, choose the **OK** button from the **Drafted Filleted Pad Definition** dialog box. Figure 5-17 shows the sketch to be extruded and the plane to be selected as the second limit. Figure 5-18 shows the resulting drafted fillet pad.

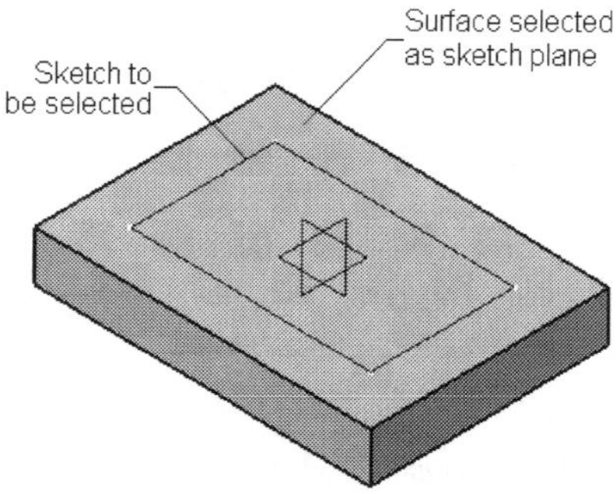

Figure 5-17 The sketch to be extruded and the plane to be selected as the second limit

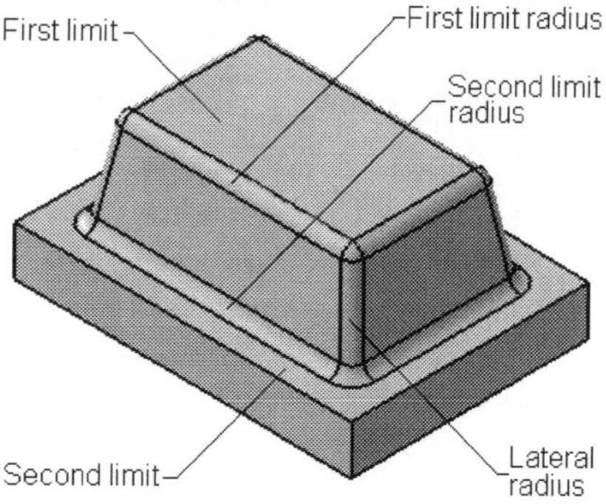

Figure 5-18 The resulting drafted filleted pad

Creating Multi-Pad Features

Menubar:	Insert > Sketch-Based Features > Multi-Pad
Toolbar:	Sketch-Based Features > Pads sub-toolbar > Multi-Pad

The **Multi-Pad** tool is used to create an extruded feature in which you can specify different extrusion depths in each closed loop in the sketch, refer to the sketch shown in Figure 5-19. The resulting multi-pad feature is displayed in Figure 5-20.

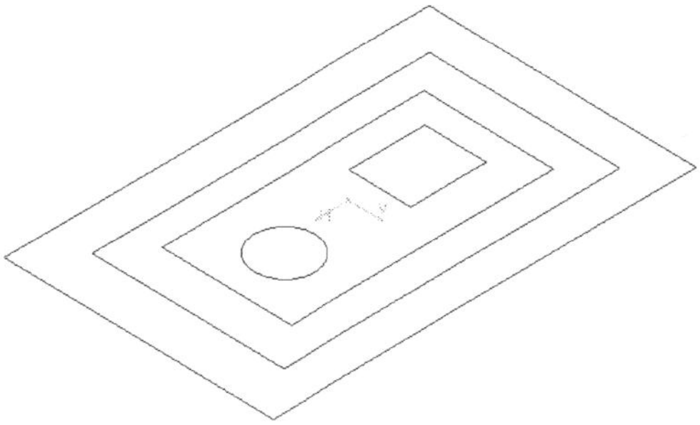

Figure 5-19 The multi-loop sketch to be extruded

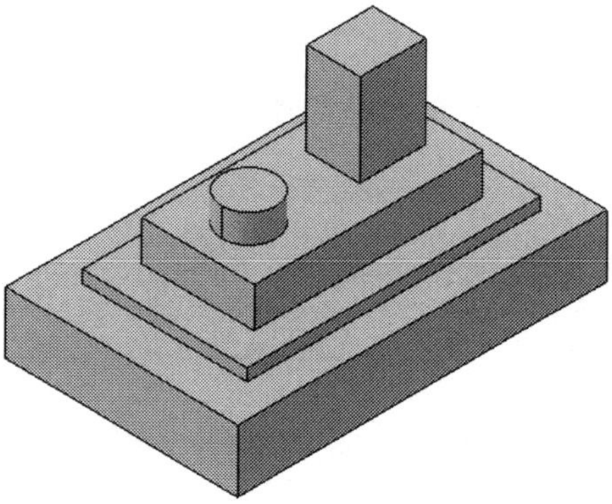

*Figure 5-20 The extruded feature created using the **Multi-Pad** tool*

To create a multi-pad feature, choose the **Multi-Pad** tool from the **Pads** sub-toolbar and then select the sketch to be extruded; the **Multi-Pad Definition** dialog box will be displayed, as shown in Figure 5-21.

The names of all loops are displayed in the list box provided in the **Domains** area of the **Multi-Pad Definition** dialog box. Select the name of the loop from the list box; it is highlighted in the geometry area. Set the value of depth of the selected loop using the **Length** spinner. Similarly, define the extrusion depth of all loops. Next, choose the **Preview** button from the **Multi-Pad Definition** dialog box; the resulting preview is displayed in the geometry area. Choose the **OK** button from the **Multi-Pad Definition** dialog box.

If you choose the **More** button from the **Multi-Pad Definition** dialog box, the dialog box expands. Using the available options, you can specify the extrusion depths in the other direction. You can also define a direction vector to extrude the sketch along that particular vector.

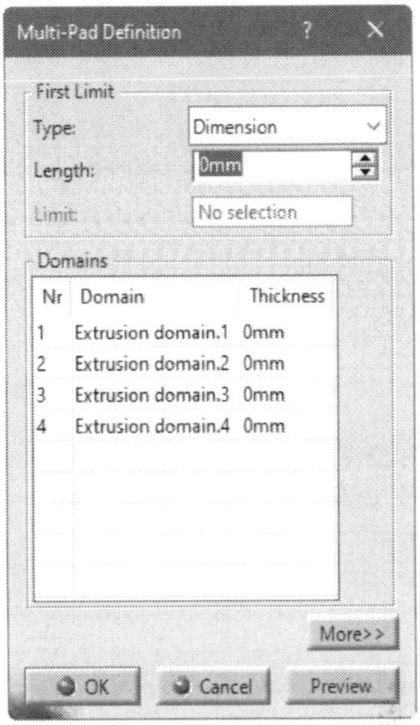

Figure 5-21 *The* ***Multi-Pad Definition*** *dialog box*

Feature Termination Options

In the previous chapter, you learned to create a pad feature by extruding sketch using the **Dimension** (default) option in the **Type** drop-down list. In this section, you will learn about termination options available in the **Type** drop-down list of the **Pad Definition** dialog box. These feature termination options are discussed next.

Up to next

The **Up to next** option is used to extrude the selected sketch from the sketching plane to the next surface that intersects the feature. To create a pad feature using this option, invoke the **Pad Definition** dialog box. In this dialog box, select the **Up to next** option from the **Type** drop-down list in the **First Limit** area. Next, choose the **OK** button from this dialog box after previewing the pad feature. Figure 5-22 shows the **Pad 1** feature created using the **Up to next** option.

Up to last

The **Up to last** option is used to extrude the selected sketch up to the last surface of the model that intersects the feature. To create a pad feature using this option, invoke the **Pad Definition** dialog box, and then select the **Up to last** option from the **Type** drop-down list of this dialog box. Next, choose the **OK** button of this dialog box after previewing the pad feature. Figure 5-22 shows the **Pad 2** feature created using the **Up to last** option.

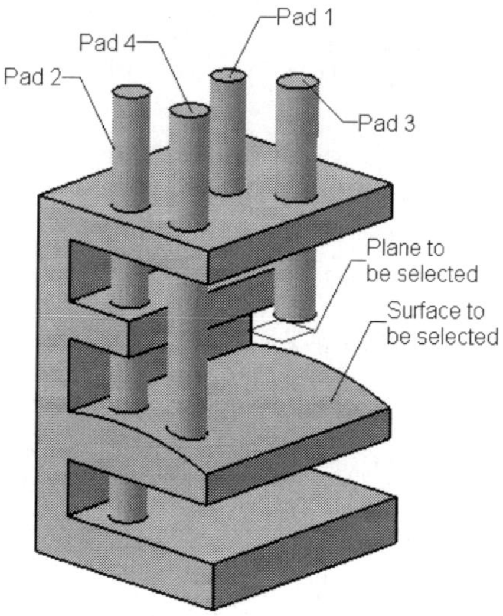

***Figure 5-22** Pad features created using different feature termination options*

Up to plane

The **Up to plane** option is used to extrude the selected sketch from the sketch plane up to the selected plane or planar face. To create a pad feature using this option, invoke the **Pad Definition** dialog box, and then select the **Up to plane** option from the **Type** drop-down list of this dialog box; you will be prompted to select the first limit. Select the plane or planar face from the geometry area as the first limit; the preview of the pad feature is displayed. Figure 5-22 shows the **Pad 3** feature created using the **Up to plane** option.

Up to surface

The **Up to surface** option is used to extrude the selected sketch from the sketch plane to the selected surface or planar face. To create a pad feature using this option, invoke the **Pad Definition** dialog box, and then select the **Up to surface** option from the **Type** drop-down list of this dialog box; you will be prompted to select the first limit. Select a surface from the geometry area as the first limit; the preview of the pad feature will be displayed. Figure 5-22 shows the **Pad 4** feature created using the **Up to surface** option.

Feature Termination at an Offset

You can also terminate a feature to be created at an offset distance from the planes or faces selected for feature termination. This option works in combination with all the feature termination options discussed earlier. To use this option, set the offset distance in the **Offset** spinner of the **Pad Definition** dialog box; the feature will be terminated at an offset distance from the selected plane/face. You can specify a positive or a negative offset distance value.

Creating Pocket Features

Menubar:	Insert > Sketch-Based Features > Pocket
Toolbar:	Sketch-Based Features > Pockets sub-toolbar > Pocket

Pocket is a material removal tool. This tool removes the material from an existing feature by extruding the sketch to the given depth or feature termination condition. To create a pocket feature, draw the sketch and then choose the **Pocket** tool from either the **Sketch-Based Features** toolbar or the **Pockets** sub-toolbar; the **Pocket Definition** dialog box will be displayed, as shown in Figure 5-23.

Select the sketch to be extruded, if it was not selected earlier; the preview of the pocket feature is displayed in the geometry area using the default values. You can set the extrusion depth using the termination options in the **Type** drop-down list in the **First Limit** area of the **Pocket Definition** dialog box. After setting the parameters, choose the **OK** button from the **Pocket Definition** dialog box. Figure 5-24 shows the sketch to be extruded to create the pocket feature and Figure 5-25 shows the resulting pocket feature.

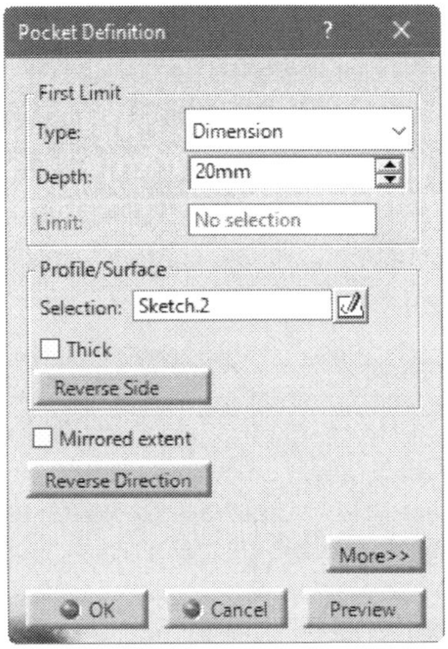

Figure 5-23 The ***Pocket Definition*** *dialog box*

If you select the **Thick** check box in the **Profile/Surface** area, the **Pocket Definition** dialog box will expand and the options in the **Thin Pocket** area will be displayed. You can create a thin pocket feature using these options.

The **Reverse Side** button in the **Profile/Surface** area is available only when the **Thick** check box is cleared. This button is chosen to flip the direction of the material removal. You can also click on the arrow displayed in the sketch in the geometry area to flip the direction. Figure 5-26 shows the cut feature with the default material removal side selected. Figure 5-27 shows the cut feature after flipping the material removal side.

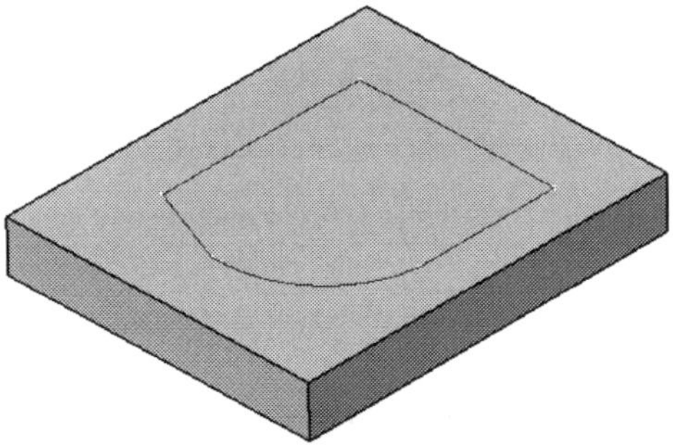

Figure 5-24 *The sketch to be extruded to create the pocket feature*

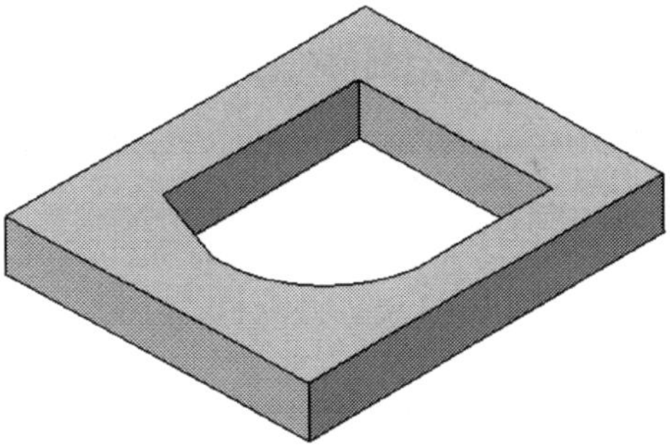

Figure 5-25 *The resulting pocket feature*

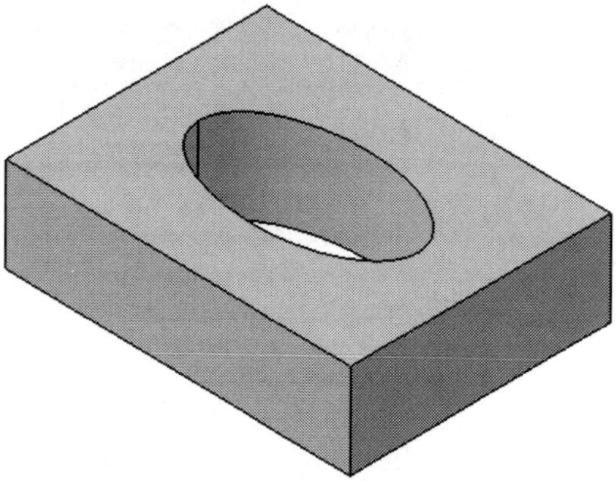

Figure 5-26 Pocket feature with the default material removal side selected

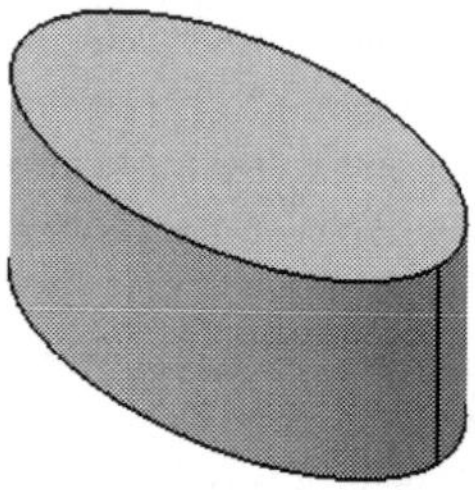

Figure 5-27 Pocket feature after flipping the material removal side

Creating Drafted Filleted Pocket Features

Menubar: Insert > Sketch-Based Features > Drafted Filleted Pocket
Toolbar: Sketch-Based Features > Pockets sub-toolbar > Drafted Filleted Pocket

You can use the **Drafted Filleted Pocket** tool to create a drafted and filleted cut feature. The procedure of creating this feature is the same as for creating the drafted filleted pad. Figure 5-28 shows the drafted filleted pocket feature.

Creating Multi-Pocket Features

Menubar: Insert > Sketch-Based Features > Multi-Pocket
Toolbar: Sketch-Based Features > Pockets sub-toolbar > Multi-Pocket

The **Multi-Pocket** tool is used to create a multi-depth cut feature using multiple closed loops from the sketch. The procedure of creating this feature is the same as that discussed while creating the multi-pad feature. Figure 5-29 shows a multi-pocket feature.

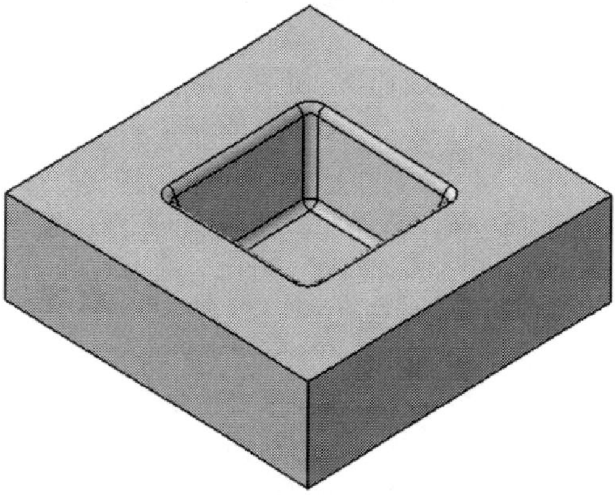

Figure 5-28 The drafted filleted pocket feature

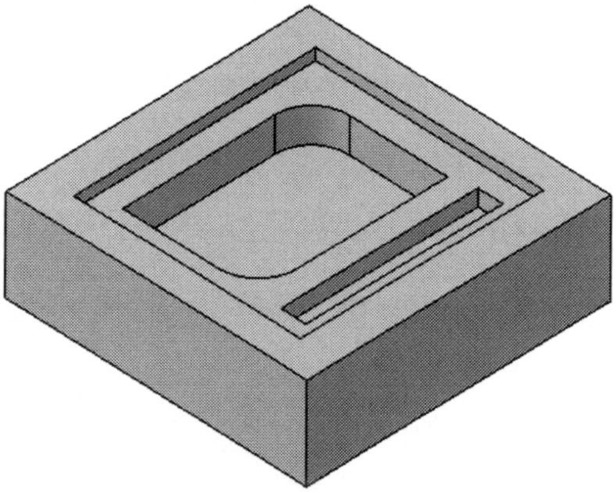

Figure 5-29 The multi-pocket feature

Creating Groove Features

Menubar: Insert > Sketch-Based Features > Groove
Toolbar: Sketch-Based Features > Groove

The **Groove** tool is used to remove the material by revolving the sketch around the axis of revolution. The working of this tool is similar to that of the **Shaft** tool. The only difference is that this is a material removal operation. To create a groove feature, draw the sketch and then choose the **Groove** tool from the **Sketch-Based Features** toolbar; the **Groove Definition** dialog box will be displayed, as shown in Figure 5-30.

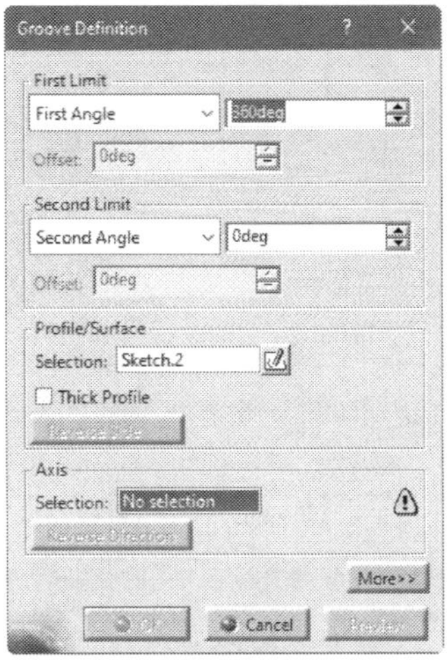

*Figure 5-30 The **Groove Definition** dialog box*

Select the sketch and the axis of rotation from the drawing area, if they are not selected. The preview of the groove feature is displayed in the geometry area. Choose the **OK** button from the **Groove Definition** dialog box. Figure 5-31 shows the sketch for the groove feature and Figure 5-32 shows the resulting groove feature.

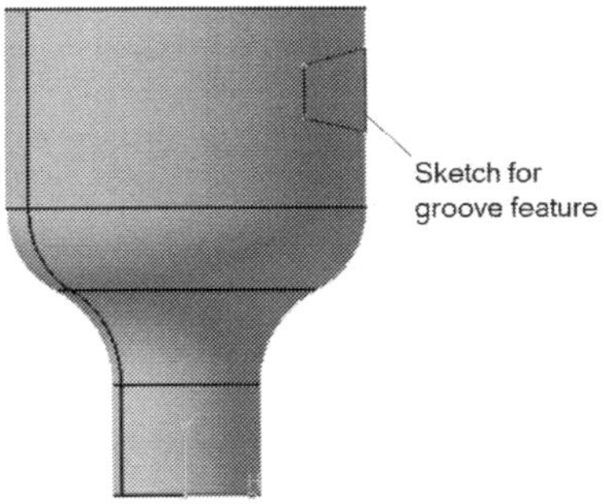

Figure 5-31 Sketch for creating the groove feature

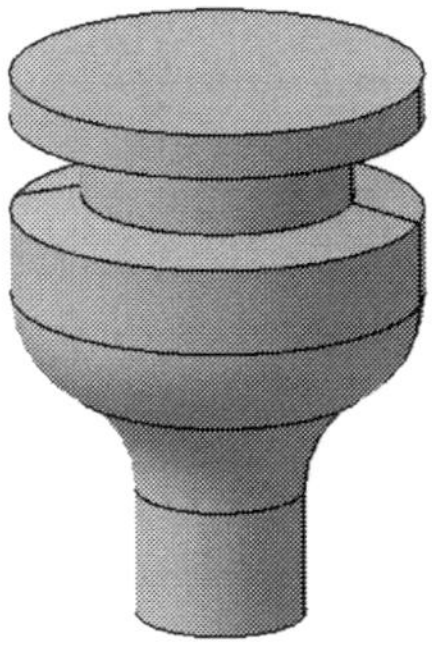

Figure 5-32 The resulting groove feature

Extruding and Revolving Planar and Non-planar Faces

While creating a pad, pocket, shaft, or groove feature, you can also select a face rather than selecting a sketch. The face could be planar or curved. When you select a face to be extruded, the **Warning** message box is displayed, as shown in Figure 5-33. This message box informs that you have selected a surface/face as profile and therefore, you must specify a direction of extrusion. Choose the **Yes** button from the **Warning** message box and then select the direction of extrusion from the geometry area. You can select an edge or a sketched line to specify the direction of extrusion. Next, specify other parameters to create the feature. Figure 5-34 shows the face to be selected and the direction of extrusion and Figure 5-35 shows the resulting pad feature.

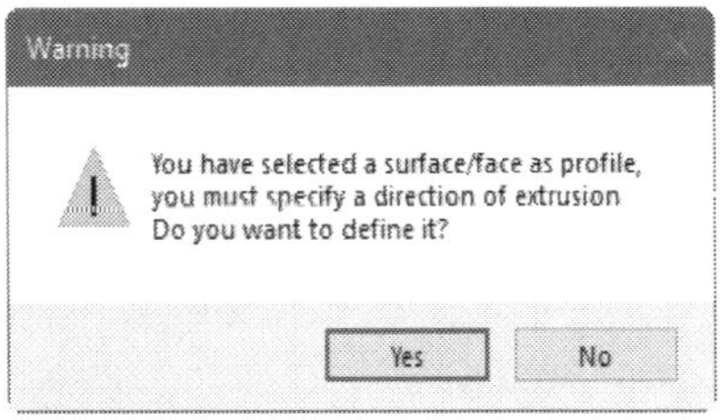

*Figure 5-33 The **Warning** message box*

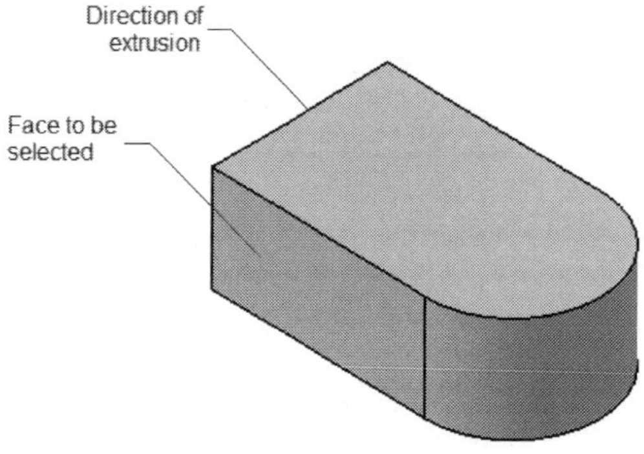

Figure 5-34 Face to be selected and the direction of extrusion

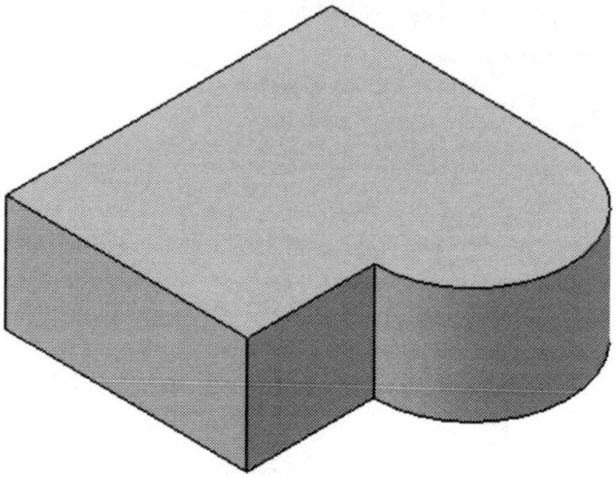

Figure 5-35 The resulting pad feature

Projecting 3D Elements

Menubar: Insert > Operation > 3D Geometry > Project 3D Elements
Toolbar: Operation > 3D Geometry sub-toolbar > Project 3D Elements

The **Project 3D Element** tool is used to project a 2D or 3D element on the current sketch plane. This tool is available only in the sketcher environment and can be invoked from the **Operation** toolbar or the **3D Geometry** sub-toolbar. After invoking this tool, you can create sketched entities by projecting the existing elements on the active sketch plane. The 3D elements that can be projected are edges, faces, or sketched entities. To project elements, select the 3D elements to be projected on to the sketch plane and then choose the **Project 3D Elements** tool from the **Operation** toolbar; the **Projection** dialog box will be displayed. If a face is selected for the projection, all its edges are projected on to the sketch plane. After the entities are projected, they remain selected and appear in orange. Click anywhere in the geometry area to clear the selected elements. As the projected entities have an associative relationship with their parent entities, they appear yellow and cannot be moved. Figure 5-36 shows the face and the edge selected for projection and also the resulting projections.

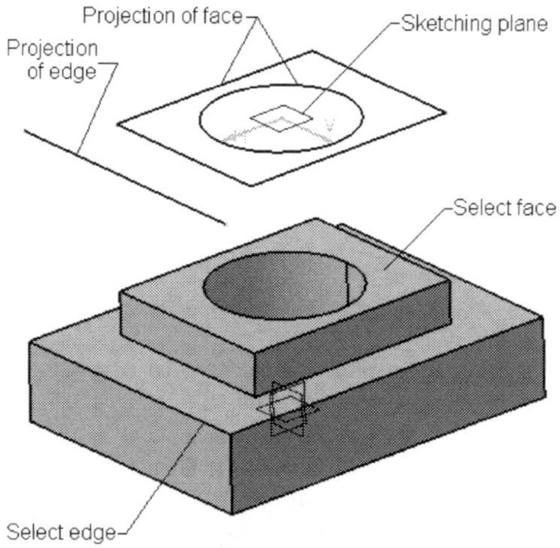

Figure 5-36 The projected elements

The association of the projected element with its parent element can be removed by isolating it. To do so, invoke the contextual menu by right-clicking on the projected element, and then choose **Mark.1 object > Isolate** from it. After isolating the sketch, you need to apply relevant constraints and dimensions to make it iso-constrained.

TUTORIALS

Tutorial 1

In this tutorial, you will create the model shown in Figure 5-37. Its views and dimensions are shown in Figure 5-38. **(Expected time: 30 min)**

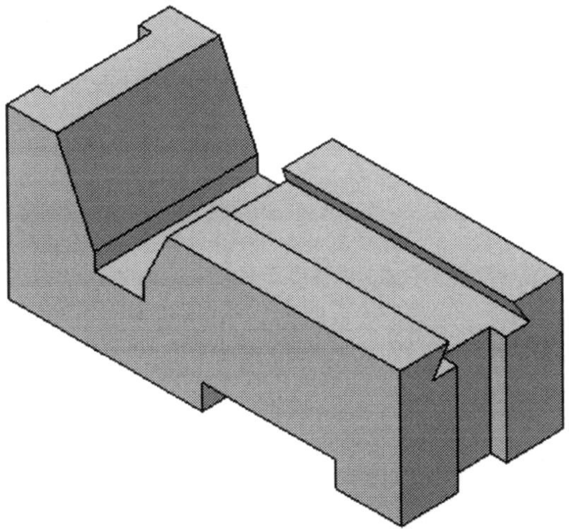

Figure 5-37 *Model for Tutorial 1*

The following steps are required to complete this tutorial:

a. Start a new file in the **Part** workbench, draw the sketch of the base feature on the yz plane, and extrude the sketch to a distance of 40 units using the **Pad** tool.
b. Create the second feature which is a pocket feature.
c. Create the third feature by extruding a sketch drawn on the right face of the base feature using the **Pocket** tool.
d. Create the fourth feature which is also a pocket feature.
e. Save and close the file.

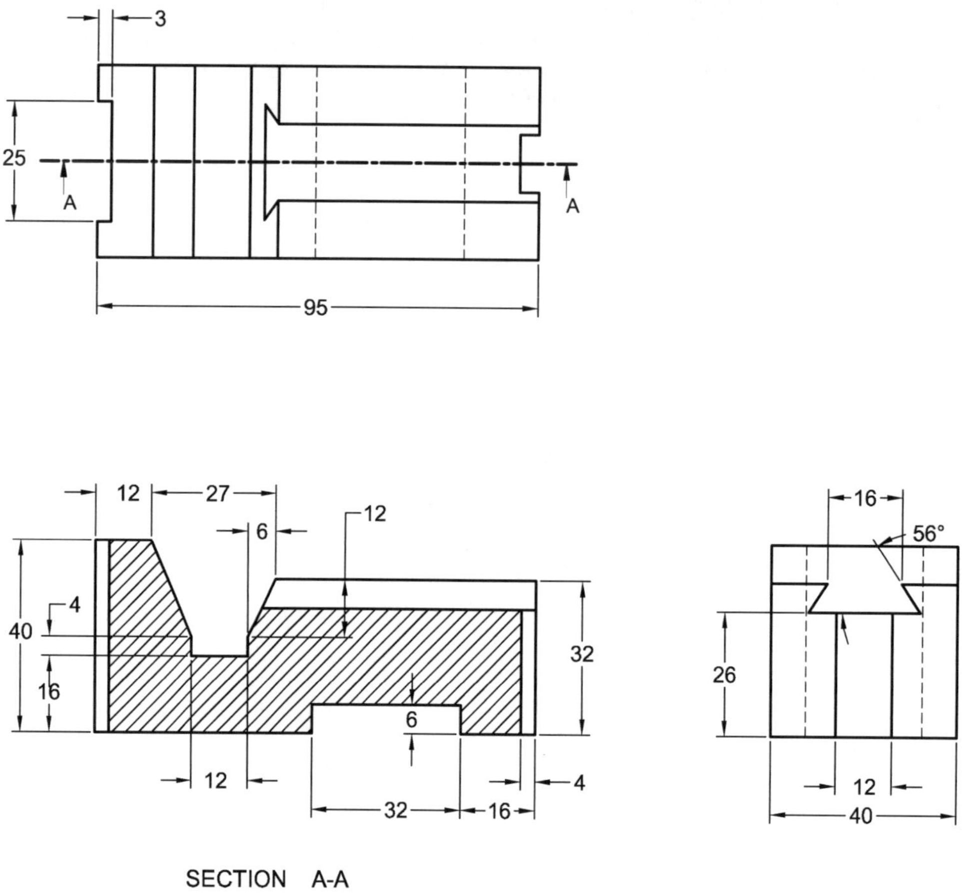

Figure 5-38 Views and dimensions for Tutorial 1

Creating the Base Feature of the Model

1. Start a new file in the **Part** workbench. Select the **yz plane** from the Specification tree and invoke the **Sketcher** workbench.

2. Draw the sketch, as shown in Figure 5-39, and exit the **Sketcher** workbench.

3. Choose the **Pad** tool from the **Sketch-Based Features** toolbar; the **Pad Definition** dialog box is displayed. Next, select the **Mirrored extent** check box from the dialog box. 

4. Set the value of the **Length** spinner to **20** if not already set and then choose the **OK** button from the dialog box. The isometric view of the model after creating the base feature is shown in Figure 5-40.

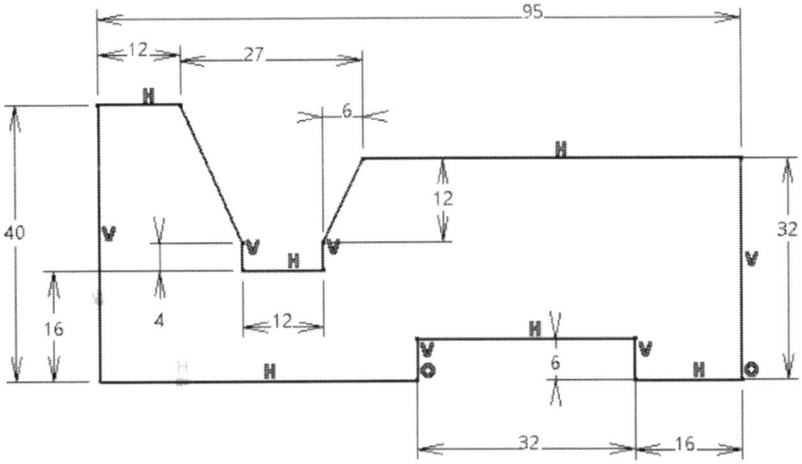

Figure 5-39 Sketch of the base feature

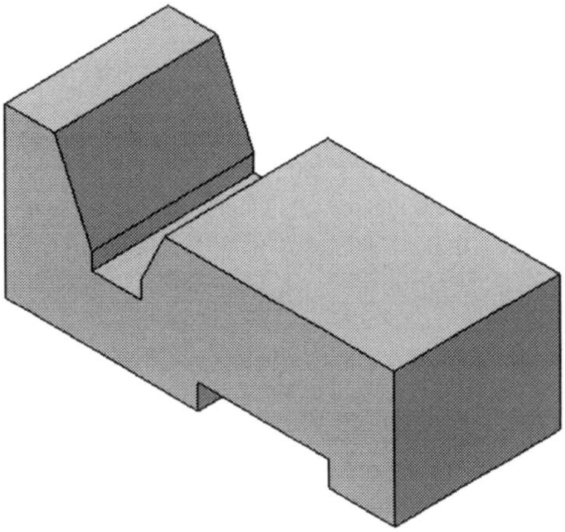

Figure 5-40 Base feature of the model

Creating the Second Feature of the Model

After creating the base feature, you need to create the pocket feature of the model by extruding the sketch drawn on the right face of the base feature.

1. Invoke the **Sketcher** workbench by using the right face of the base feature.

2. Draw the sketch, as shown in Figure 5-41, and exit the **Sketcher** workbench.

3. Invoke the **Pocket** tool from the **Sketch-Based Features** toolbar; the **Pocket Definition** dialog box as well as the preview of the **Pocket** feature is displayed. 

4. Select the **Up to next** option from the **Type** drop-down list and choose the **OK** button from the **Pocket Definition** dialog box to end feature creation. The model after creating the second feature is shown in Figure 5-42.

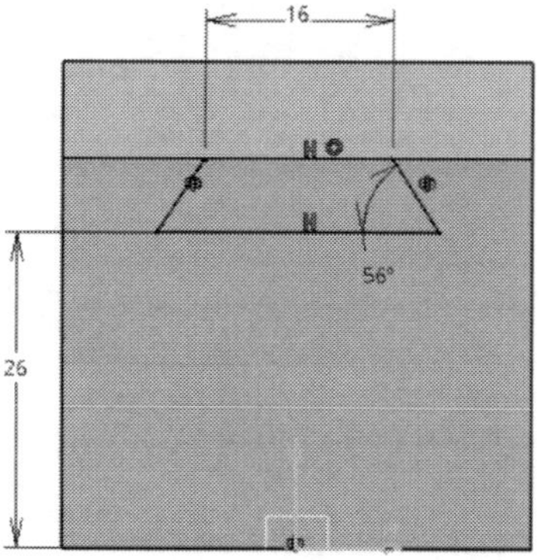

Figure 5-41 Sketch of the second feature

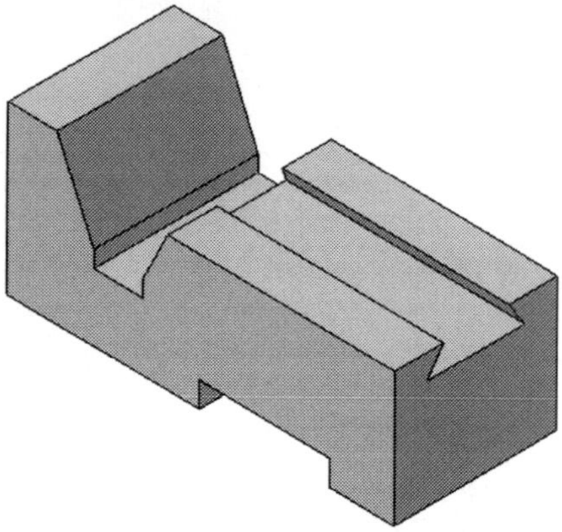

Figure 5-42 The model after creating the second feature

Creating the Third Feature of the Model

The third feature of the model is also a pocket feature. You will create this feature by extruding the sketch drawn on the right face of the base feature.

1. Select the right face of the base feature and invoke the **Sketcher** workbench.
2. Draw the sketch, as shown in Figure 5-43, and then exit the **Sketcher** workbench.
3. Invoke the **Pocket Definition** dialog box and then select the **Dimension** option from the **Type** drop-down list in it.
4. Set the value of the **Depth** spinner to **4** and choose the **OK** button to create the pocket feature. The model after creating the third feature is shown in Figure 5-44.

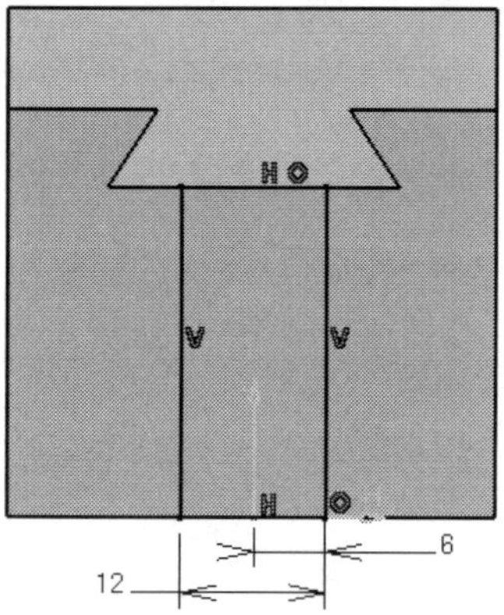

Figure 5-43 Sketch of the third feature

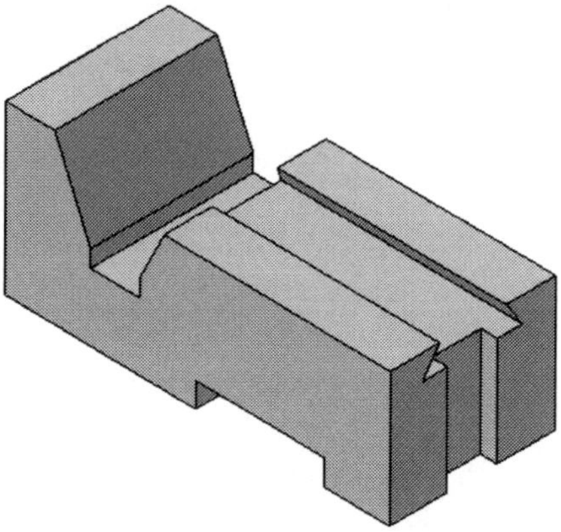

Figure 5-44 The model after creating the third feature

Creating the Fourth Feature of the Model

The fourth feature of the model is a rectangular pocket feature. You will create this feature by extruding the sketch drawn on the top face of the model.

1. Select the topmost face of the model and invoke the **Sketcher** workbench.

2. Draw the sketch for the fourth feature, refer to Figure 5-38 for dimensions, and exit the **Sketcher** workbench.

3. Invoke the **Pocket Definition** dialog box and extrude the sketch using the **Up to last** option in the **Type** drop-down list of the dialog box. Next, choose the **OK** button from the dialog box. The final model after creating all the features is shown in Figure 5-45.

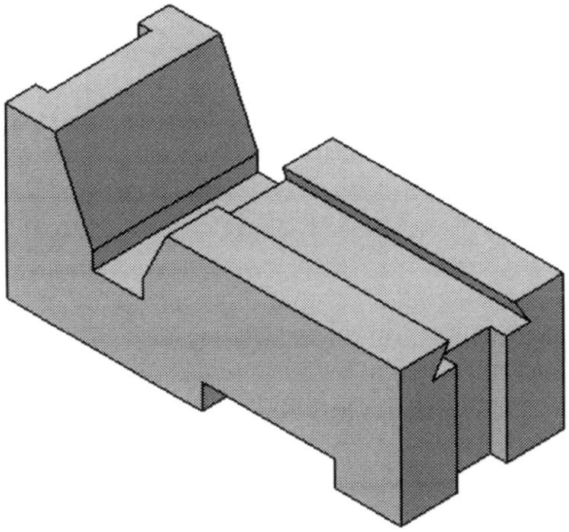

Figure 5-45 *Final model*

Saving and Closing the File

1. Choose the **Save** button from the **Standard** toolbar to invoke the **Save As** dialog box. Create the *c05* folder inside the *CATIA* folder.

2. Enter the name of the file as **c05tut1** in the **File name** edit box and choose the **Save** button from the **Save As** dialog box. The file is saved at *C:\CATIA\c05*.

3. Close the part file by choosing **File > Close** from the menu bar.

Tutorial 2

In this tutorial, you will create the model shown in Figure 5-46. Its views and dimensions are shown in Figure 5-47. **(Expected time: 45 min)**

The following steps are required to complete this tutorial:

a. Start a new file in the **Part** workbench and create the base feature of the model by extruding the sketch drawn on the yz plane.
b. Create the second feature by extruding the sketch drawn on the front face of the base feature by using the **Pocket** tool.
c. Create the third feature by extruding the rectangular sketch drawn on the front face of the base feature.

d. Create the fourth feature by extruding the circular sketch drawn on the front face of the base feature.
e. Create the fifth feature by extruding the sketch drawn on the top face of the third feature using the **Pocket** tool.
f. Save and close the file.

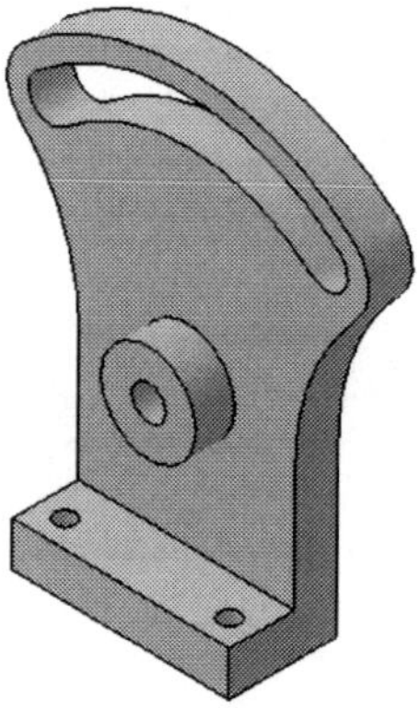

Figure 5-46 *Model for Tutorial 2*

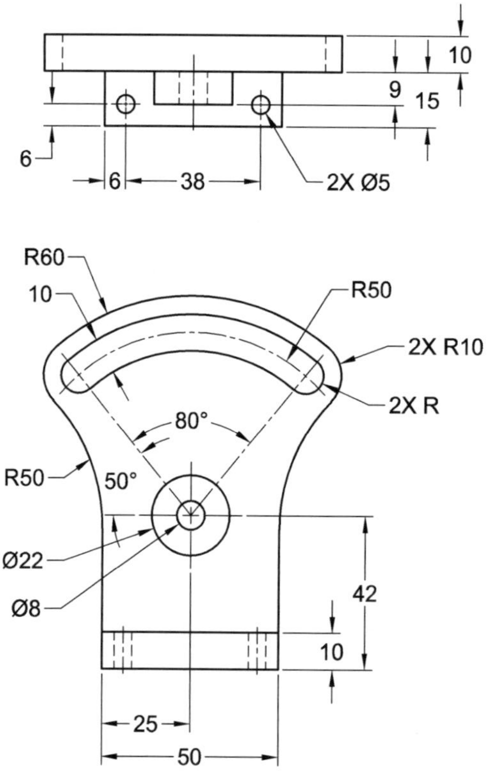

Figure 5-47 *Orthographic views and dimensions for Tutorial 2*

Creating the Base Feature of the Model

1. Start a new part file. Select the yz plane from the geometry area and then invoke the **Sketcher** workbench.

2. Draw the sketch, as shown in Figure 5-48, and exit the **Sketcher** workbench.

3. Invoke the **Pad** tool from the **Sketch-Based Features** toolbar; the **Pad Definition** dialog box is displayed. 

4. Select the **Mirrored extent** check box in the **Pad Definition** dialog box.

5. Set **5** in the **Length** spinner and choose the **OK** button from the dialog box; the base feature is created, as shown in Figure 5-49.

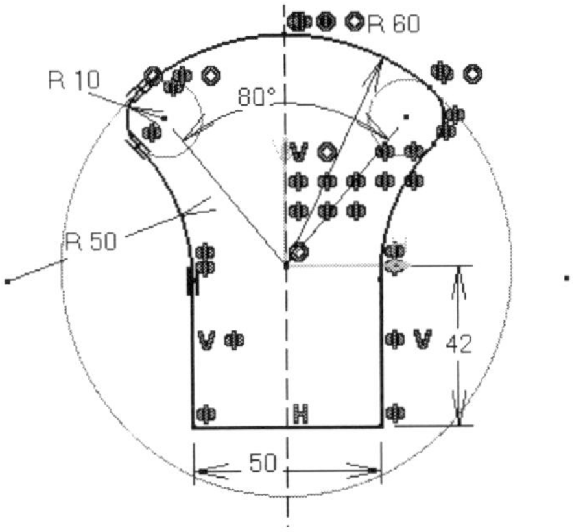

Figure 5-48 *Sketch of the base feature*

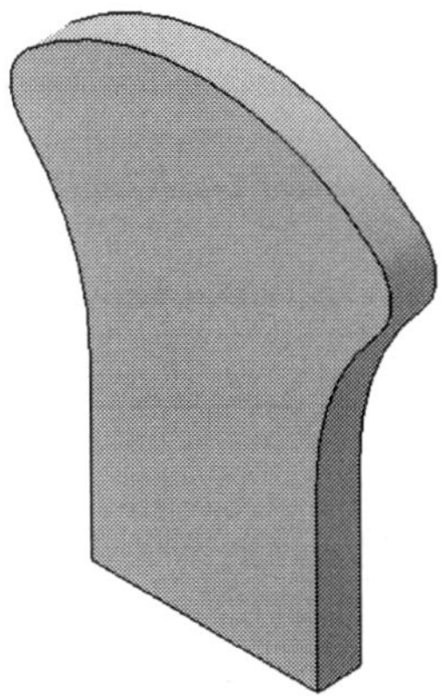

Figure 5-49 *Base feature of the model*

Creating the Second Feature of the Model

The second feature is a pocket feature. You will create this feature by extruding the sketch drawn on the front face of the base feature.

1. Select the front face of the base feature and invoke the **Sketcher** workbench.

2. Draw the sketch, as shown in Figure 5-50, by using the **Cylindrical Elongated Hole** tool and exit the **Sketcher** workbench.

3. Invoke the **Pocket** tool from the **Sketch-Based Features** toolbar; the **Pocket Definition** dialog box is displayed. 

4. In this dialog box, select the **Up to last** option from the **Type** drop-down list in the **First Limit** area and then choose the **OK** button to create the second feature, as shown in Figure 5-51.

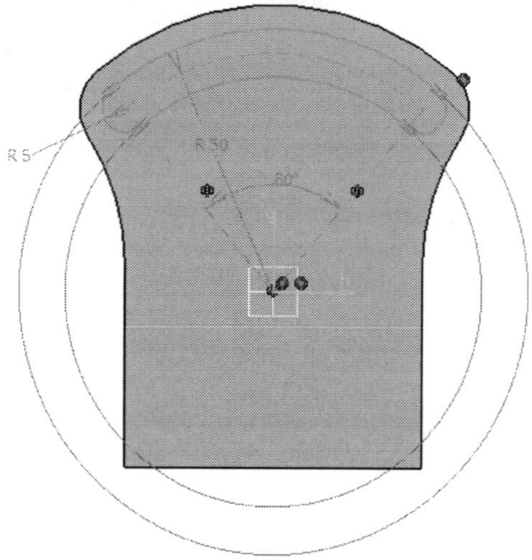

Figure 5-50 Sketch of the second feature

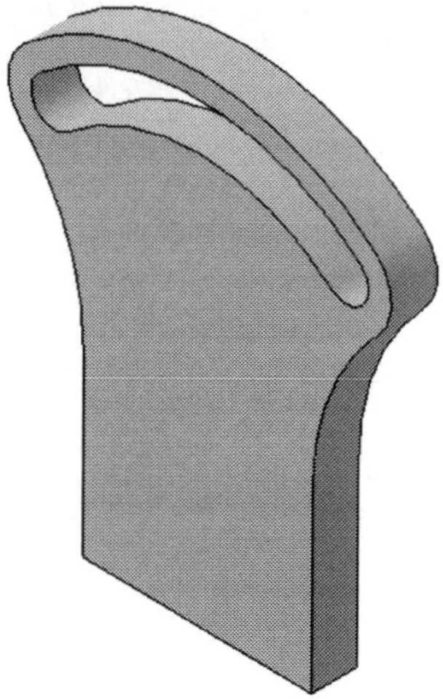

Figure 5-51 Model after creating the second feature

Creating the Third, Fourth, and Fifth Features of the Model

You will create the third feature of this model by extruding the sketch drawn on the front face of the base feature.

1. Select the front face of the base feature and invoke the **Sketcher** workbench.
2. Draw the sketch, as shown in Figure 5-52. The lower horizontal line in this sketch is drawn by projecting the bottom edge of the base feature. Exit the **Sketcher** workbench.
3. Extrude the sketch to a distance of 15 mm. The model after creating the third feature is shown in Figure 5-53.

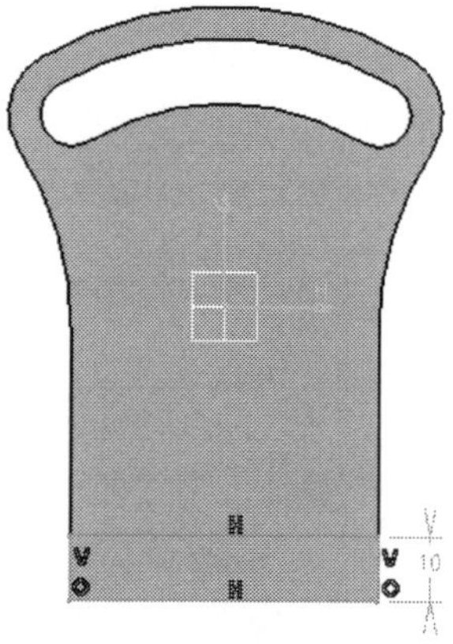

Figure 5-52 Sketch of the third feature

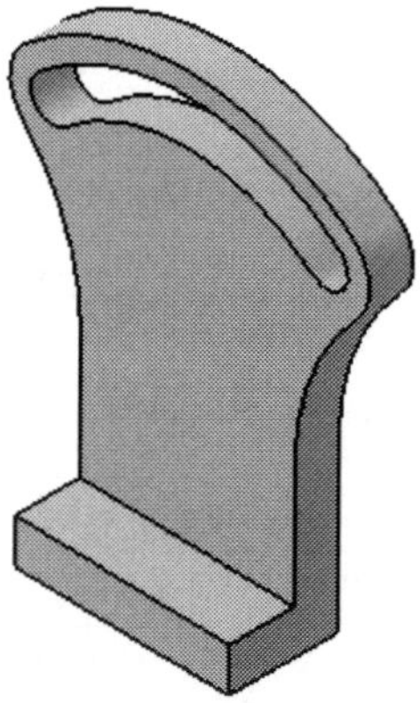

Figure 5-53 Model after creating the third feature

4. Similarly, create fourth and fifth features of the model. The final model after creating the fourth and fifth features is shown in Figure 5-54.

Saving and Closing the File

1. Choose the **Save** button from the **Standard** toolbar to invoke the **Save As** dialog box.

2. Enter **c05tut2** as the name of the file in the **File name** edit box and choose the **Save** button. The file is saved at *C:\CATIA\c05*.

3. Close the part file by choosing **File > Close** from the menu bar.

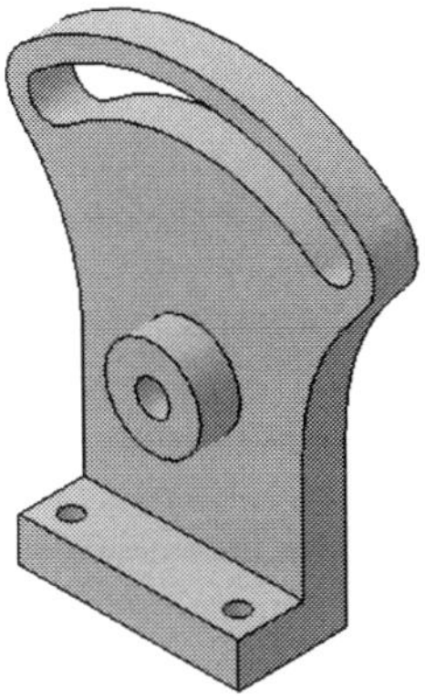

Figure 5-54 *Final model*

Tutorial 3

In this tutorial, you will create the model shown in Figure 5-55. The views and dimensions of the model are shown in Figure 5-56. **(Expected time: 30 min)**

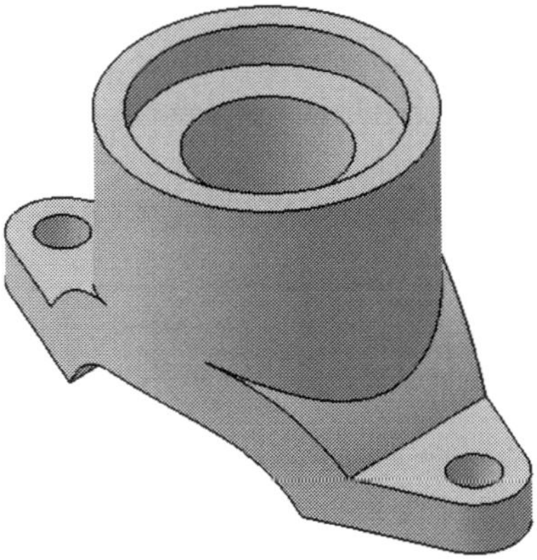

Figure 5-55 *Solid model for Tutorial 3*

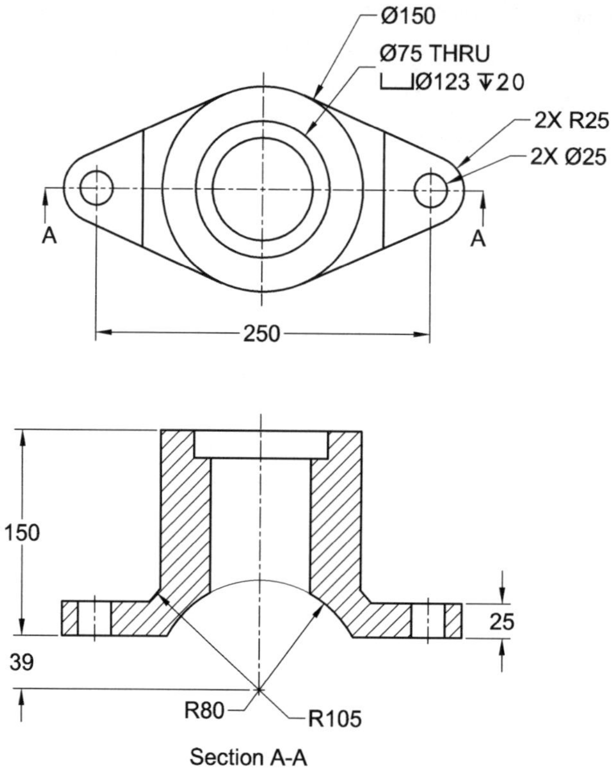

Figure 5-56 Views and dimensions for Tutorial 3

The following steps are required to complete this tutorial:

a. Start a new file in the **Part** workbench, draw the sketch for the base feature on the yz plane, and extrude the sketch to the required distance by using the **Pad** tool.
b. Create the second feature, which is a pocket feature.
c. Create the third feature by extruding the sketch drawn on a plane at an offset distance from the xy plane.
d. Create the fourth feature, which is a groove feature.
e. Create the last feature of the model, which is the pocket feature.
f. Save and close the file.

Creating the Base Feature of the Model

It is evident from the model that its base comprises of a complex geometry. Therefore, you need to create the base first and then create the pocket feature on the base to get the desired shape. You will create the base feature by extruding the sketch drawn on the yz plane using the **Mirrored extent** option.

1. Start a new file in the **Part** workbench.

2. Select the yz plane from the geometry area and invoke the **Sketcher** workbench.

3. Draw the sketch of the base feature, as shown in Figure 5-57, and exit the **Sketcher** workbench.

4. Invoke the **Pad** tool from the **Sketch-Based Features** toolbar; the **Pad Definition** dialog box is displayed.

5. Select the **Mirrored extent** radio button from the **Pad Definition** dialog box.

   The required extrusion depth of the base feature should be 150. As the **Mirrored extent** radio button is selected, you need to specify the extrusion depth as half of the actual depth.

6. Set **75** in the **Length** spinner and choose the **OK** button from the **Pad Definition** dialog box. The isometric view of the model after creating the base feature is shown in Figure 5-58.

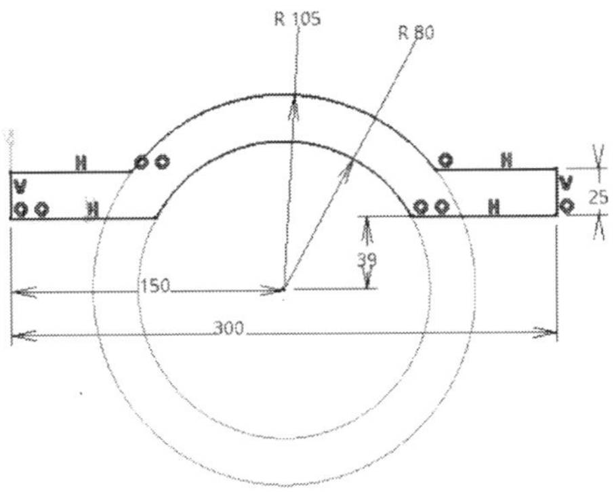

Figure 5-57 *The sketch of the base feature*

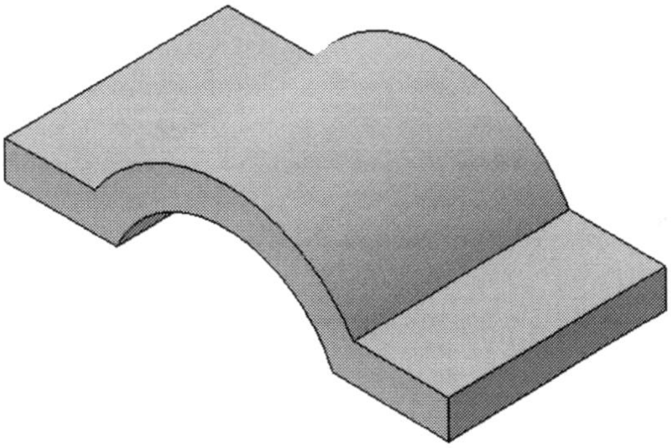

Figure 5-58 *The model after creating the base feature*

Creating the Second Feature of the Model

Next, you need to create the second feature which is a pocket feature. You will create the sketch of the second feature on the xy plane.

1. Invoke the **Sketcher** workbench by selecting the xy plane as the sketching plane and then draw the sketch, as shown in Figure 5-59.

2. Exit the **Sketcher** workbench.

3. Invoke the **Pocket** tool from the **Sketch-Based Features** toolbar; the **Pocket Definition** dialog box is displayed. Also, preview of the pocket feature is displayed in the graphics area with default values. You will notice that the default side of the extrusion depth is in a direction opposite to the required direction. Therefore, you need to reverse the direction of extrusion depth.

4. Choose the **Reverse Direction** button from the **Pocket Definition** dialog box.
   The default side of the material removal is also in a direction opposite to the required direction. Therefore, you need to reverse the direction of material removal.

5. Choose the **Reverse Side** button from the **Pocket Definition** dialog box.

6. Select the **Up to last** option from the **Type** drop-down list and choose the **OK** button from the **Pocket Definition** dialog box. The model after creating the second feature is shown in Figure 5-60.

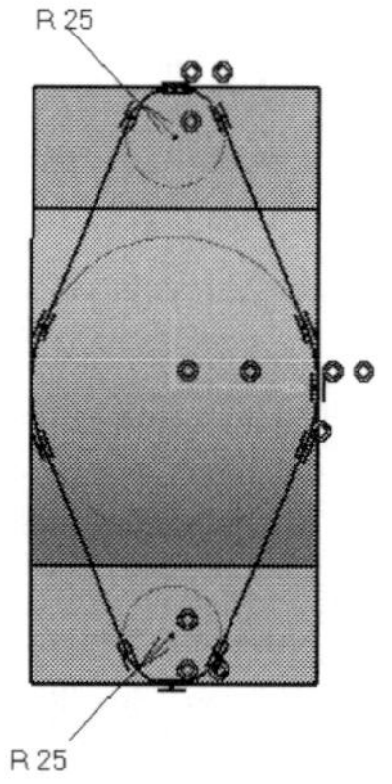

Figure 5-59 *The sketch of the second feature*

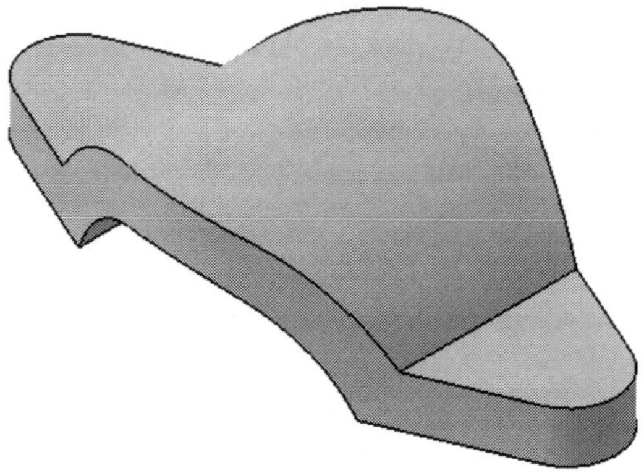

Figure 5-60 *Model after creating the second feature*

Creating the Third Feature of the Model

The third feature of the model is a pad feature which is created by using the **Up to surface** termination option. You will draw the sketch for this feature on a plane that is at some offset distance from the xy plane.

1. Select the xy plane from the geometry area or the Specification tree and choose the **Plane** tool from the **Reference Elements (Expanded)** toolbar; the **Plane Definition** dialog box is displayed. Also, preview of the offset plane with default values is displayed in the geometry area. 

2. Set **189** in the **Offset** spinner and choose the **OK** button from the **Plane Definition** dialog box; a plane is created at an offset distance of 189 mm.

3. Invoke the sketching environment using the newly created plane as the sketching plane.

4. Draw the sketch, as shown in Figure 5-61, and exit the **Sketcher** workbench.

5. Invoke the **Pad Definition** dialog box; preview of the **Pad** feature is displayed in the geometry area.

   Note that the direction of extrusion is opposite to the required direction. Therefore, you need to reverse the direction of the extrusion.

6. Choose the **Reverse Direction** button from the **Pad Definition** dialog box to reverse the direction of extrusion.

7. Select the **Up to surface** option from the **Type** drop-down list and then select the surface shown in Figure 5-62.

8. Choose the **OK** button from the **Pad Definition** dialog box. The model after creating the third feature is shown in Figure 5-63.

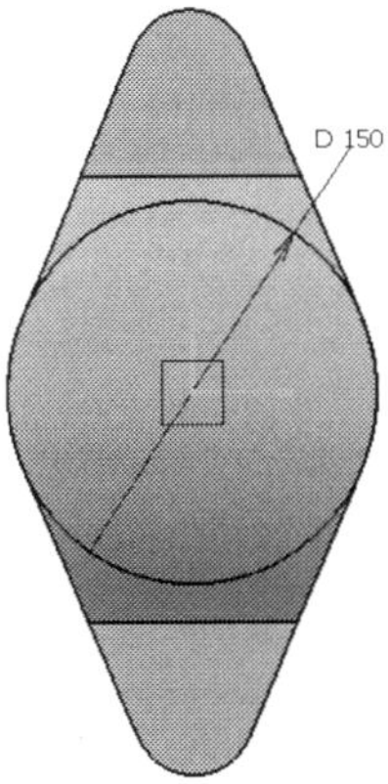

Figure 5-61 *Sketch of the third feature*

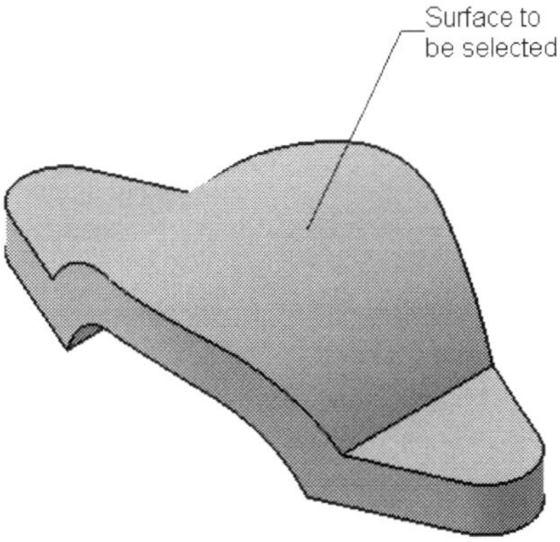

Figure 5-62 *Surface to be selected*

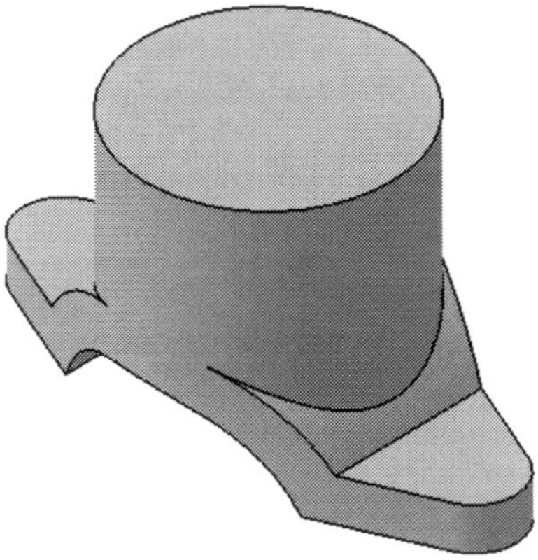

Figure 5-63 *The model after creating the third feature*

As you do not need the offset plane anymore, you can turn off its display. Note that you should not delete this plane; if you do so, the feature created using this plane will also be deleted.

9. Expand the **PartBody** node and right-click on the **Plane.1** from the Specification tree or the geometry area; a shortcut menu is displayed. Choose the **Hide/Show** option from this contextual menu; the plane gets hidden.

Creating the Fourth Feature of the Model

The fourth feature of this model is a groove feature. This feature will be created by revolving the sketch about an axis.

1. Select the yz plane from the geometry area and invoke the **Sketcher** workbench.
2. Draw the sketch, as shown in Figure 5-64, and exit the **Sketcher** workbench.

3. Invoke the **Groove** tool from the **Sketch-Based Features** toolbar; the **Groove Definition** dialog box is displayed. Select the newly created sketch if not selected by default; a preview of the **Groove** feature is displayed in the geometry area.

4. Choose the **OK** button from the **Groove Definition** dialog box. The model after creating the fourth feature is shown in Figure 5-65.

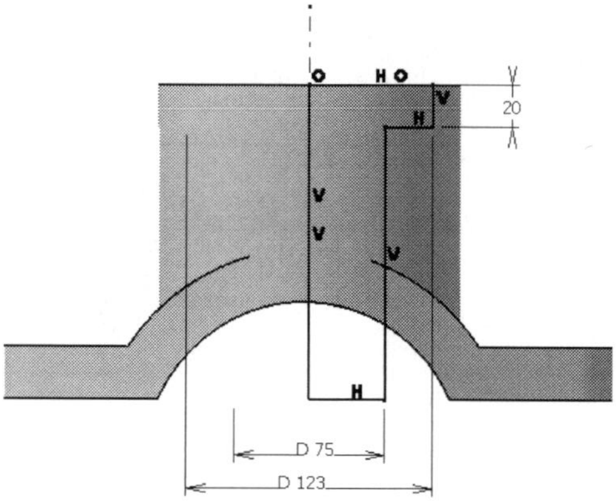

Figure 5-64 *Sketch of the fourth feature*

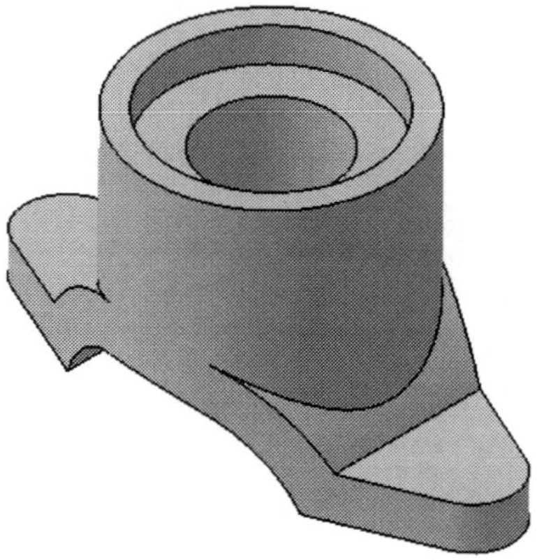

Figure 5-65 *The model after creating the fourth feature*

Creating the Fifth Feature of the Model

Next, you need to create holes by removing the material using the pocket feature. You will create this feature by extruding the sketch drawn on the top planar face of the base feature to remove the material.

1. Select the top planar face of the base feature as the sketching plane and invoke the **Sketcher** workbench.

2. Draw the sketch of the fifth feature, as shown in Figure 5-66, and exit the **Sketcher** workbench.

3. Invoke the **Pocket Definition** dialog box and select the **Up to last** option from the **Type** drop-down list.

4. Next, choose the **OK** button from the **Pocket Definition** dialog box. The final model after creating the fifth feature is shown in Figure 5-67.

Saving and Closing the File

1. Choose the **Save** button from the **Standard** toolbar to invoke the **Save As** dialog box.

2. Enter **c05tut3** in the **File name** edit box and choose the **Save** button. The file is saved at *C:\CATIA\c05*.

3. Close the part file by choosing **File > Close** from the menu bar.

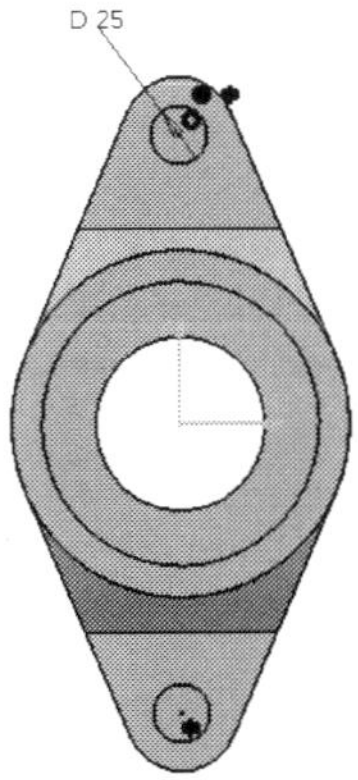

Figure 5-66 Sketch of the fifth feature

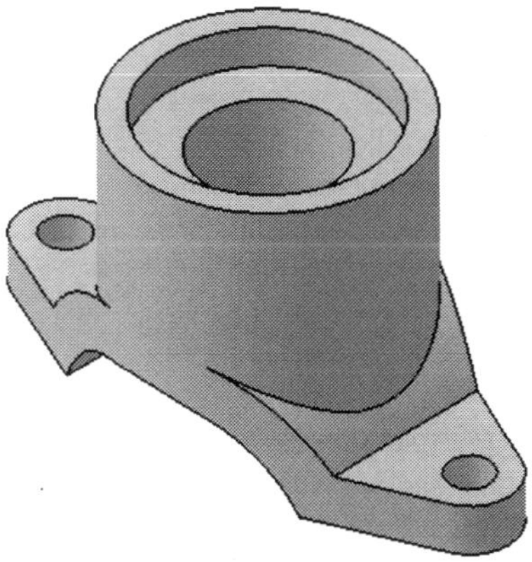

Figure 5-67 The final model

Tutorial 4

In this tutorial, you will create the model shown in Figure 5-68. The views and dimensions of the model are shown in Figure 5-69. (**Expected time: 30 min**)

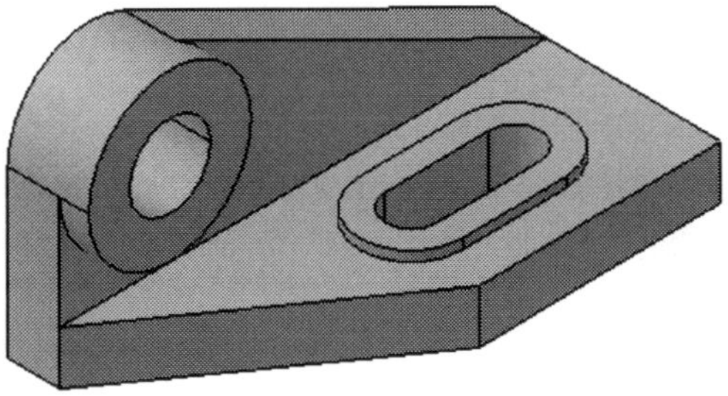

Figure 5-68 Solid model for Tutorial 4

The following steps are required to complete this tutorial:

a. Start a new file in the **Part** workbench, draw the sketch of the base feature on the xy plane, and then extrude the sketch to the required distance using the Pad tool.
b. Create the second feature, which is a Pad feature.
c. Create the third feature, which is a Pad feature, by extruding the sketch drawn on the back face of the second feature.
d. Create the fourth feature by extruding the sketch drawn on the front face of the third feature.
e. Create the fifth feature of the model, which is Pocket feature, by drawing the sketch on the front face of the fourth feature.
f. Create the last feature of the model, which is also a Pocket feature, by drawing the sketch on the top face of the base feature.
g. Save and close the file.

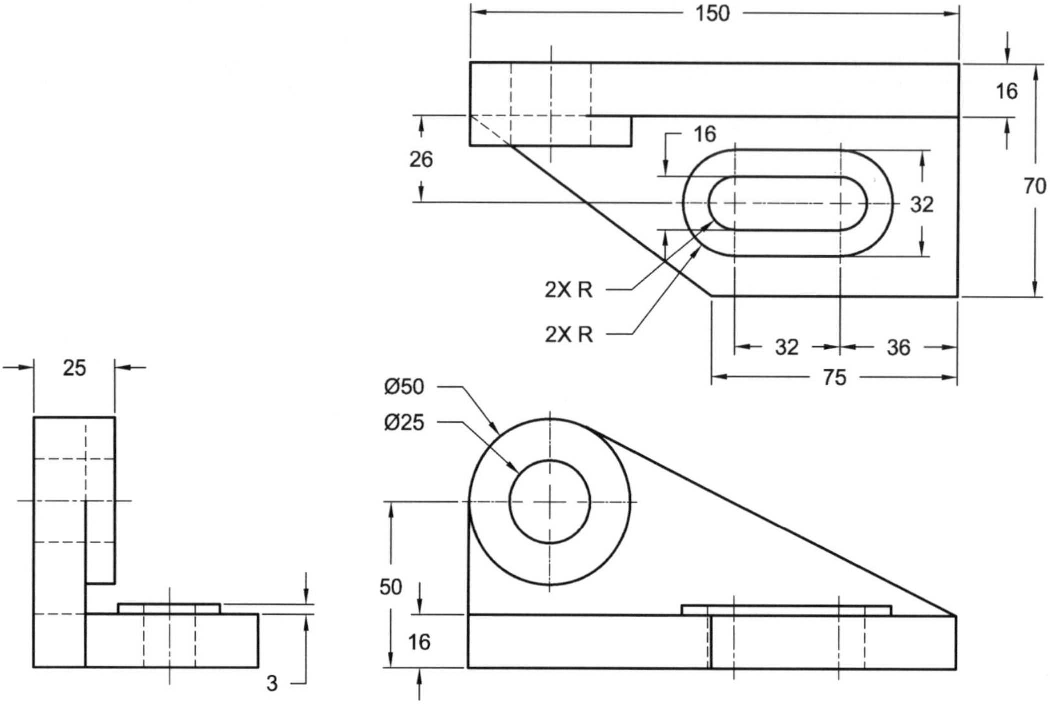

Figure 5-69 Views and dimensions for Tutorial 4

Creating the Base Feature of the Model

1. Start a new file in the **Part** workbench.

2. Select the xy plane as the sketching plane and invoke the **Sketcher** workbench.

3. Draw the sketch, as shown in Figure 5-70, and exit the **Sketcher** workbench.

4. Invoke the **Pad** tool from the **Sketch-based Features** toolbar; the **Pad Definition** dialog box is displayed.

5. Set **16** in the **Length** spinner and choose the **OK** button from the dialog box; the base feature is created, as shown in Figure 5-71.

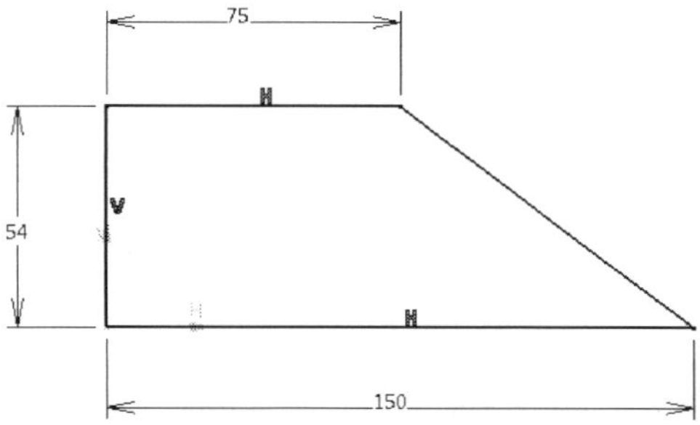

Figure 5-70 The sketch of the base feature

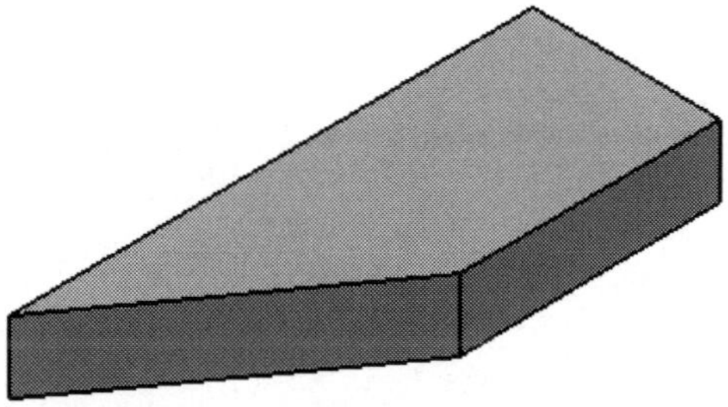

Figure 5-71 The model after creating the base feature

Creating the Second Feature of the Model

The second feature is a **Pad** feature. You will create this feature by extruding the sketch drawn on the top face of the base feature.

1. Select the top planar face of the base feature as the sketching plane and invoke the **Sketcher** workbench.

2. Draw the sketch, as shown in Figure 5-72, using the **Elongated Hole** tool and exit the **Sketcher** workbench.

3. Invoke the **Pad** tool from the **Sketch-Based Features** toolbar; the **Pad Definition** dialog box is displayed.

4. Set **3** in the **Length** spinner and then choose the **OK** button from the dialog box to create the feature, as shown in Figure 5-73.

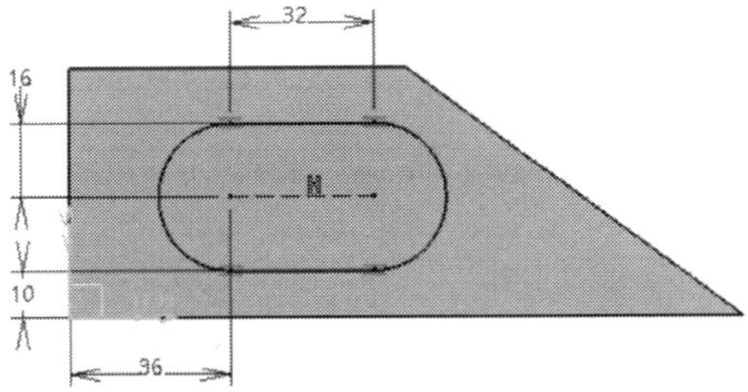

Figure 5-72 *Sketch for the second feature*

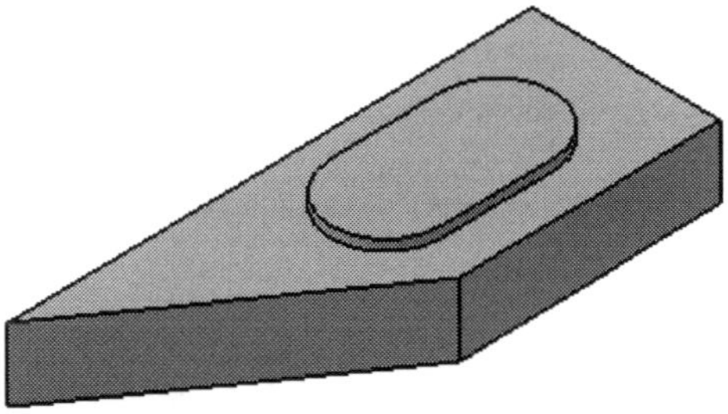

Figure 5-73 *Solid model after creating the second feature*

Creating the Third Feature of the Model

The third feature of the model is a **Pad** Feature. You will create this feature by drawing the sketch on the back face of the base feature and then adding the required distance.

1. Select the back face and invoke the **Sketcher** workbench.

2. Draw the sketch, as shown in Figure 5-74, and then exit the **Sketcher** workbench.

3. Invoke the **Pad** tool from the **Sketch-Based Features** toolbar; the **Pad Definition** dialog box is displayed.

4. Set **16** in the **Length** spinner and choose the **OK** button from the dialog box to create the feature, as shown in Figure 5-75. You can choose the **Reverse Direction** button to change the direction of extrusion, if required.

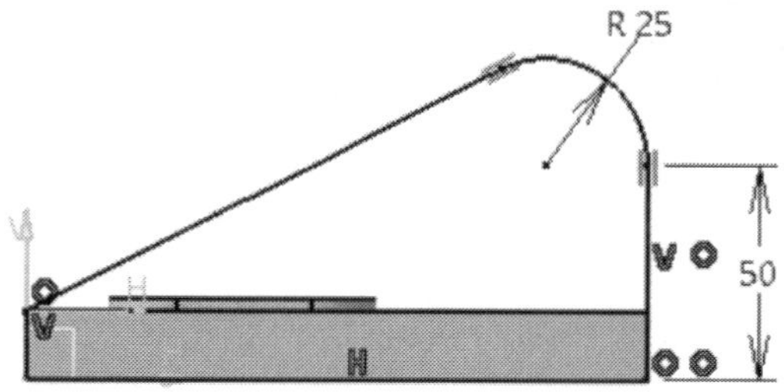

Figure 5-74 Sketch for the third feature

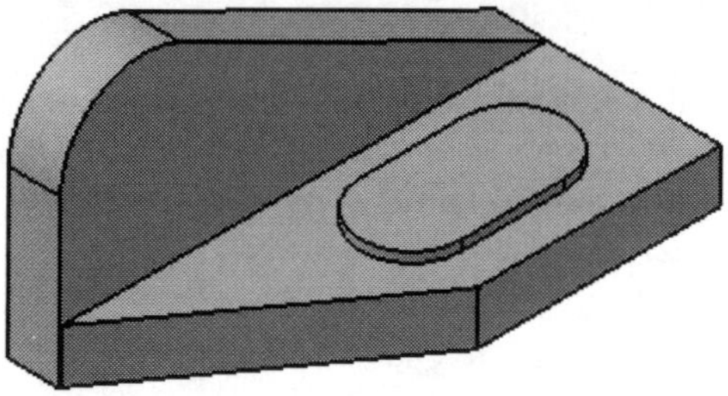

Figure 5-75 Model after creating the third feature

Creating the Fourth Feature of the Model

The fourth feature is also a **Pad** feature. You will create this feature by extruding the sketch drawn on the front face of the third feature.

1. Select the front face of the third feature and invoke the **Sketcher** workbench.
2. Draw the sketch for the fourth feature which is a circle of diameter **50**. Next, apply concentricity constraint between the circle and the curved edge, refer to Figure 5-76, and then exit the **Sketcher** workbench.
3. Invoke the **Pad** tool from the **Sketch-Based Features** toolbar; the **Pad Definition** dialog Box is displayed.
4. Set **9** in the **Length** spinner and choose the **OK** button from the dialog box to create the feature, as shown in Figure 5-77.

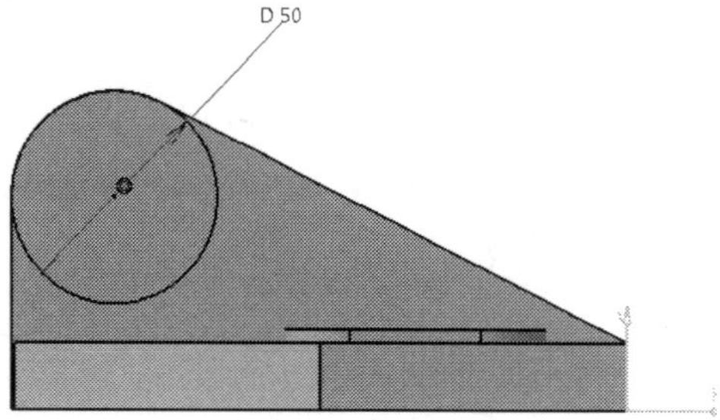

Figure 5-76 Sketch for the fourth feature

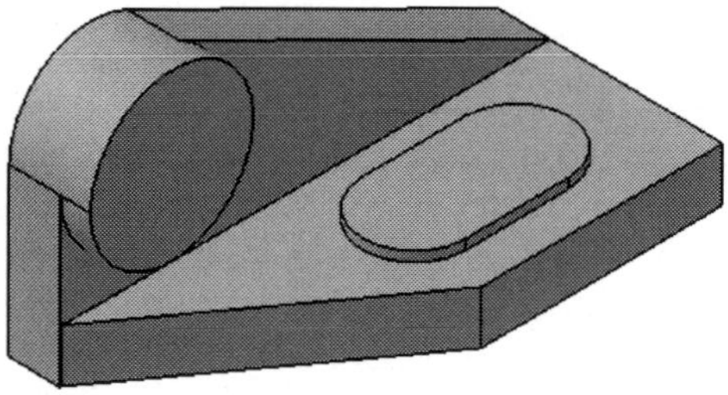

Figure 5-77 Model after creating the fourth feature

Creating the Fifth Feature of the Model

The fifth feature is a **Pocket** feature. You will create this feature by extruding the sketch drawn on the front face of the model.

1. Select the front face of the fourth feature and invoke the **Sketcher** workbench.
2. Draw the sketch for the fifth feature, as shown in Figure 5-78, and then exit the **Sketcher** workbench.

3. Invoke the **Pocket Definition** dialog box and remove the material using the **Up to last** option in the **Type** drop-down list of the dialog box. Next, choose the **OK** button from the dialog box. The model after creating the fifth feature is shown in Figure 5-79.

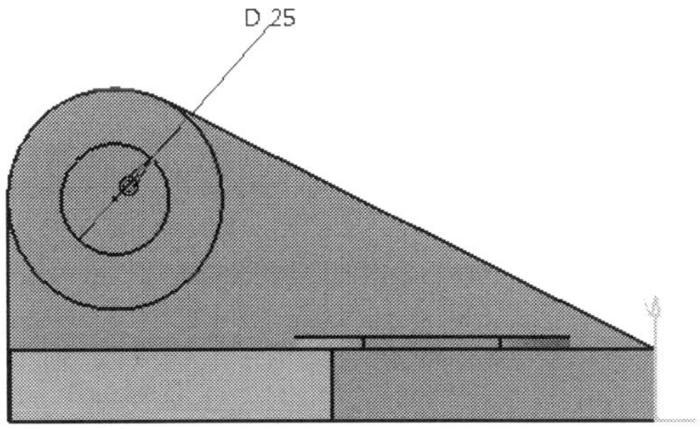

Figure 5-78 *Sketch for fifth feature*

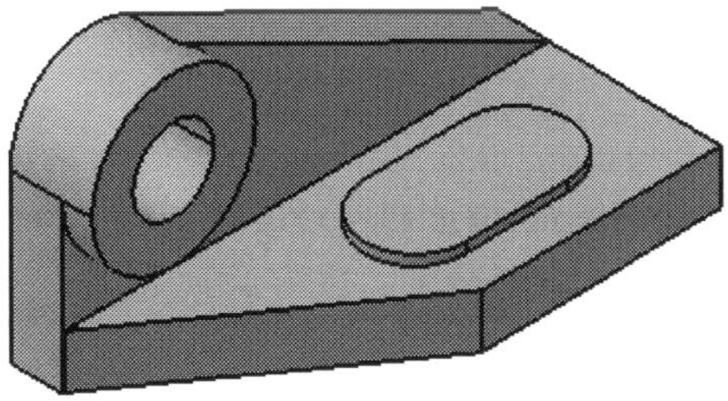

Figure 5-79 *Model after creating the fifth feature*

Creating the Sixth Feature of the Model

The sixth feature of the model is again a **Pocket** feature. You will create this feature by extruding the sketch drawn on the top face of the base feature.

1. Select the top face of the second feature and invoke the **Sketcher** workbench.

2. Draw the sketch for the sixth feature, as shown in Figure 5-80, and exit the **Sketcher** workbench.

3. Invoke the **Pocket Definition** dialog box and remove the material using the **Up to last** option in the **Type** drop-down list of the dialog box. Next, choose the **OK** button from the dialog box. The final model after creating the sixth feature is shown in Figure 5-81.

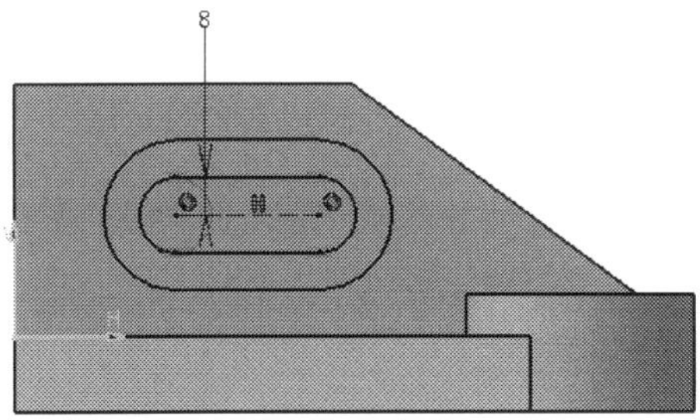

Figure 5-80 *Sketch for the sixth feature*

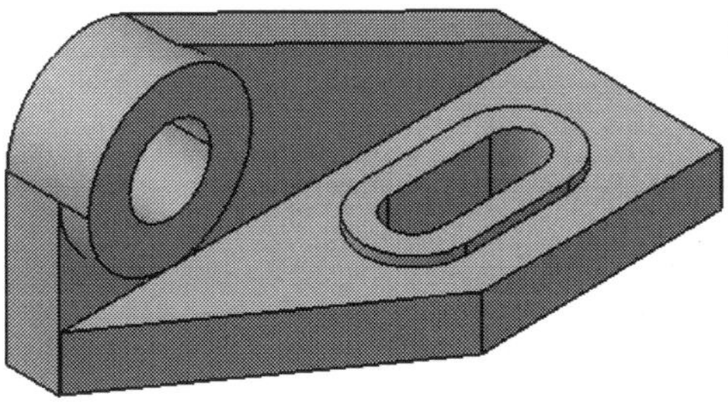

Figure 5-81 *Final model*

Saving and Closing the File

1. Choose the **Save** button from the **Standard** toolbar to invoke the **Save As** dialog box.

2. Enter **c05tut4** as the file name in the **File name** edit box and choose the **Save** button. The file is saved at *c:/CATIA/c05*.

3. Close the part file by choosing **File > Close** from the menu bar.

Tutorial 5

In this tutorial, you will create the model of the shaft shown in Figure 5-82 by using the **Multi-Pad** tool. The dimensions of the model are shown in Figure 5-83.

(Expected time: 30 min)

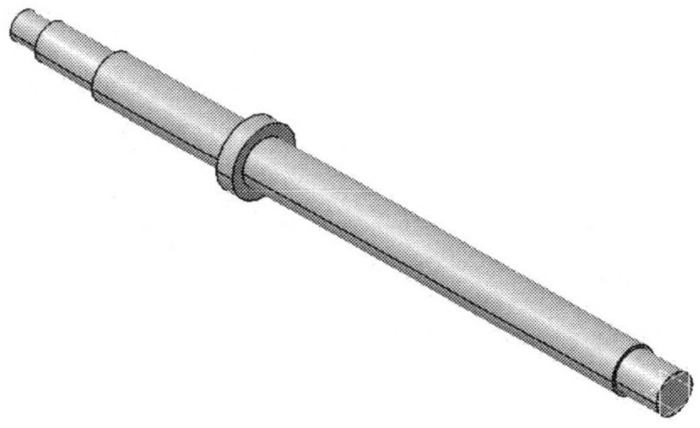

Figure 5-82 *The model of the shaft*

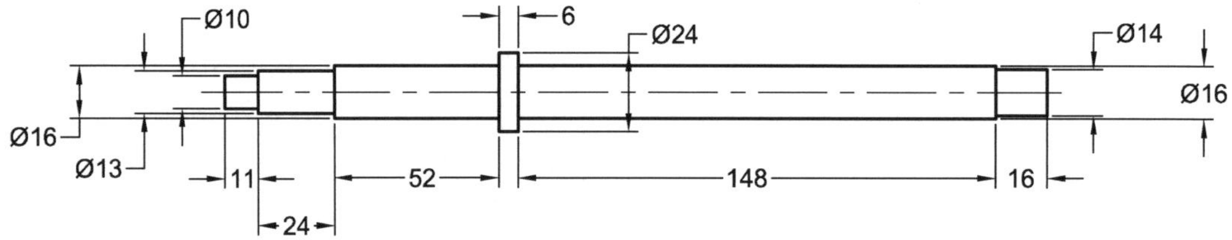

Figure 5-83 *Dimensions for Tutorial 5*

The following steps are required to complete this tutorial:

a. Start a new file in the **Part** workbench and then draw the sketch on the zx plane, refer to Figure 5-84.
b. Create the shaft using the **Multi-pad** tool, create the solid, as shown in Figure 5-85 by specifying different distances.
c. Save and close the file.

Creating the Sketch of the Model

1. Start a new file in the **Part** workbench.

2. Select the zx plane as the sketching plane and invoke the **Sketcher** workbench.

3. Draw the sketch, as shown in Figure 5-84, and then exit the **Sketcher** workbench.

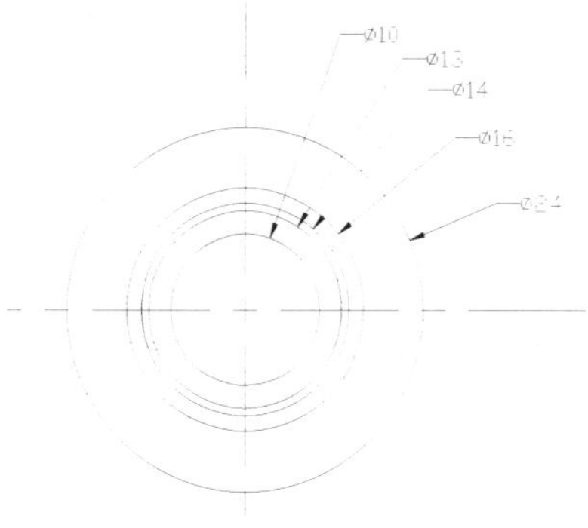

Figure 5-84 *Sketch of the model*

Creating the Model of Shaft

The model of shaft can be created by using the **Multi-Pad** tool.

1. Choose the **Multi-Pad** tool from the **Pad** sub-toolbar of the **Sketch-Based Features** toolbar.

2. Select the sketch; the **Multi-Pad Definition** dialog box is displayed, as shown in Figure 5-85.

3. Click on the arrow displayed on the sketch to flip the direction of extrusion if it is not as shown in Figure 5-86.

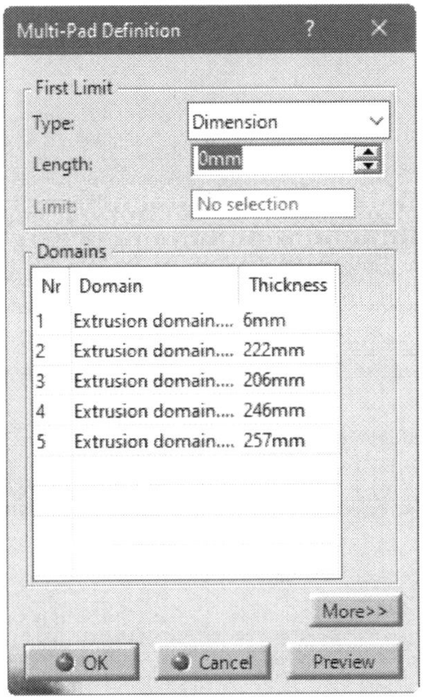

*Figure 5-85 The **Multi-Pad Definition** dialog box*

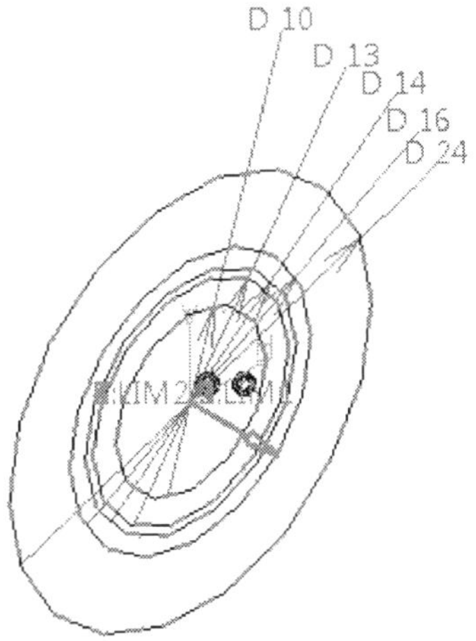

Figure 5-86 Direction of extrusion of sketch

4. In the **Multi-Pad Definition** dialog box, select the closed region created by the circle having 10 as diameter and set **257** in the **Length** spinner of the **First Limit** area; wireframe preview of the extrusion is displayed in the drawing area on clicking once in the drawing area.

5. Select the closed region created between the circles having 10 and 13 as diameter from the dialog box and set **257** in the **Length** spinner of the **First Limit** area; preview of the extrusion is displayed in the drawing area if not then click once in the drawing area.

6. Select the region created between the circles having 14 and 13 as diameter from the dialog box and set **257** in the **Length** spinner of the **First Limit** area.

7. Select the region created between the circles having 16 and 14 as diameter from the dialog box and set **241** in the **Length** spinner of the **First Limit** area.

8. Similarly, select the region created between the circles having diameter 24 and 16 from the dialog box and set **93** in the **Length** spinner of the **First Limit** area.

9. Choose the **More>>** button from the **Multi-Pad Definition** dialog box; the dialog box gets expanded. Now, set **-11** in the **Length** spinner of the **Second Limit** area of the closed region created between the circles having diameter 10 and 13. After that set **-35** in the **Length** spinner in the **Second Limit** area of the closed region created between the circles having diameter 13 and 14 in the dialog box. On selecting the **Preview** button, a preview of the shaft is displayed, as shown in Figure 5-87.

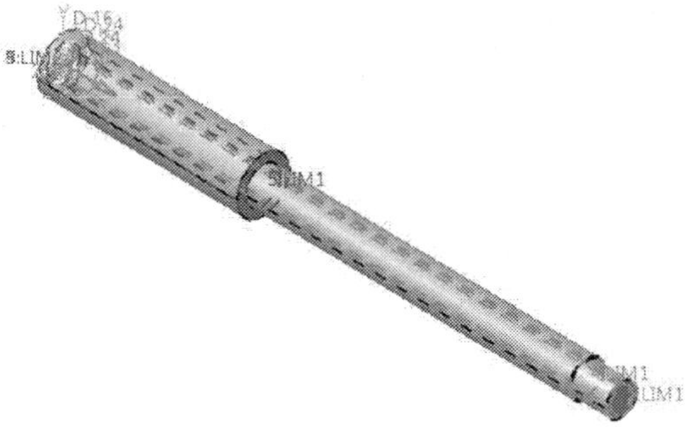

Figure 5-87 *Preview of the shaft*

10. Similarly, set **-35** in the **Length** spinner of the **Second Limit** area of the closed region created between circles having diameter 14 and 16. Next set **-87** in the **Length** spinner in the **Second Limit** area of the closed region created between the circles having diameter 16 and 24 from the dialog box.

11. Choose the **OK** button from the dialog box, refer to Figure 5-88; the shaft is created as shown in Figure 5-89.

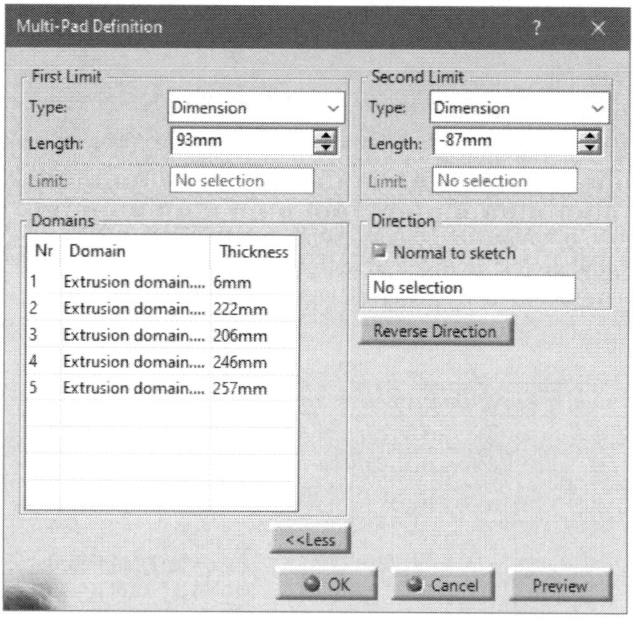

Figure 5-88 *The **Multi-Pad Definition** dialog box*

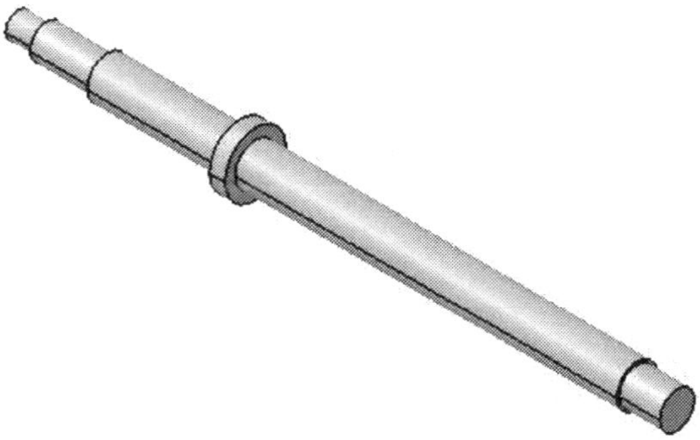

Figure 5-89 The shaft created

Saving and Closing the File

1. Choose the **Save** button from the **Standard** toolbar to invoke the **Save As** dialog box.

2. Enter **c05tut5** as the file name in the **File name** edit box and choose the **Save** button. The file is saved at *c:/CATIA/c05*.

3. Close the part file by choosing **File > Close** from the menu bar.

Self-Evaluation Test

Answer the following questions and then compare them to those given at the end of this chapter:

1. The _________ tool is used to remove material from an existing feature by extruding a sketch upto a given depth.

2. The _________ tool is used to remove material by revolving a sketch about an axis.

3. You can create points in the **Part Design** workbench using the _________ tool.

4. The _________ option is used to create a plane using the equation Ax+By+Cz = D.

5. The **Offset from plane** option is selected by default in the **Plane type** drop-down list of the _________ dialog box.

6. Reference Planes are used as sketching planes to draw sketches for the sketch-based features. (T/F)

7. When you start a new file in the **Part Design** workbench, CATIA V5 does not provide any default plane. (T/F)

8. While creating a pad, pocket, shaft, or groove feature, you can also select a face instead of the sketch. (T/F)

9. The **Multi-Pad** tool is used to create an extrude feature in which you can specify different extrusion depths to each closed loop in the sketch. (T/F)

10. The **Up to plane** option is used to extrude a sketch from the sketch plane to the selected plane or the planar face. (T/F)

Review Questions

Answer the following questions:

1. The **Up to surface** option is used to extrude a sketch from the sketch plane to the selected surface. (T/F)

2. The **Mean through points** option is used to create a plane at an orientation defined by the mean of the selected points. (T/F)

3. The **Through three points** option in the **Plane type** drop-down list is used to create a plane that passes through three selected points. (T/F)

4. The **Up to next** option is used to extrude a sketch from the sketch plane to the selected surface. (T/F)

5. The **Between** option is used to create a point between two selected points. (T/F)

EXERCISES

Exercise 1

Create the model shown in Figure 5-90. The views and dimensions of the model are shown in Figure 5-91. **(Expected time: 30 min)**

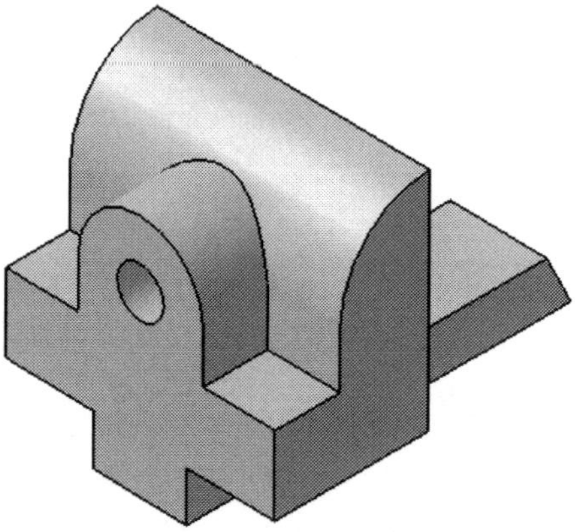

Figure 5-90 *Model for Exercise 1*

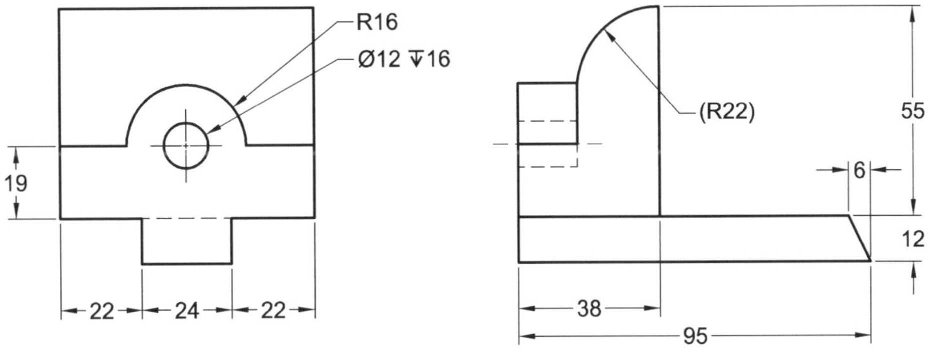

Figure 5-91 Views and dimensions for Exercise 1

Exercise 2

Create the model shown in Figure 5-92. The views and dimensions of the model are shown in Figure 5-93. **(Expected time: 30 min)**

Figure 5-92 Model for Exercise 2

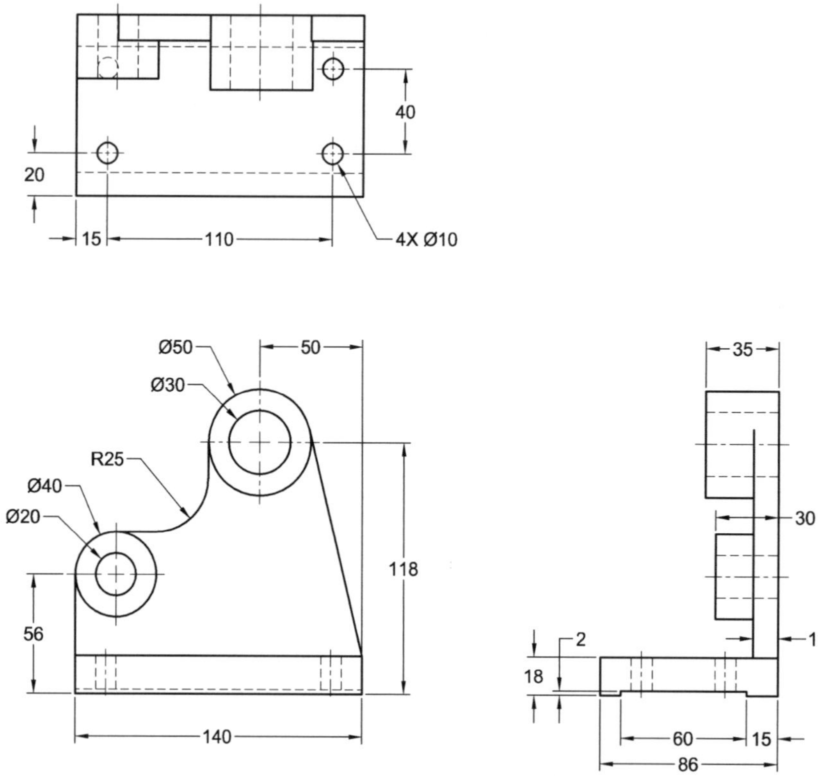

Figure 5-93 *Views and dimensions for Exercise 2*

Exercise 3

Create the model shown in Figure 5-94. The views and dimensions of the model are shown in Figure 5-95. **(Expected time: 30 min)**

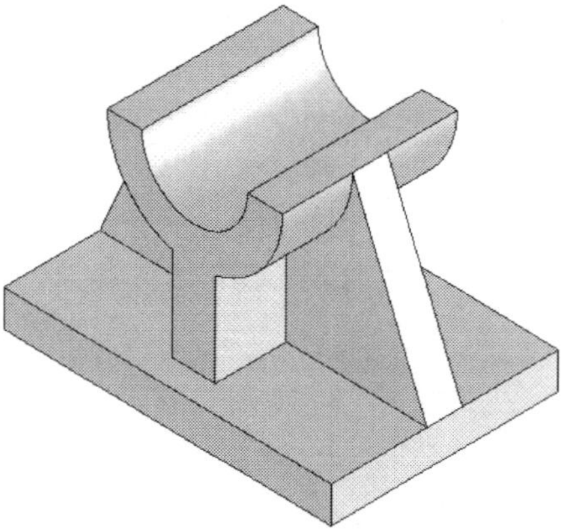

Figure 5-94 *Solid model for Exercise 3*

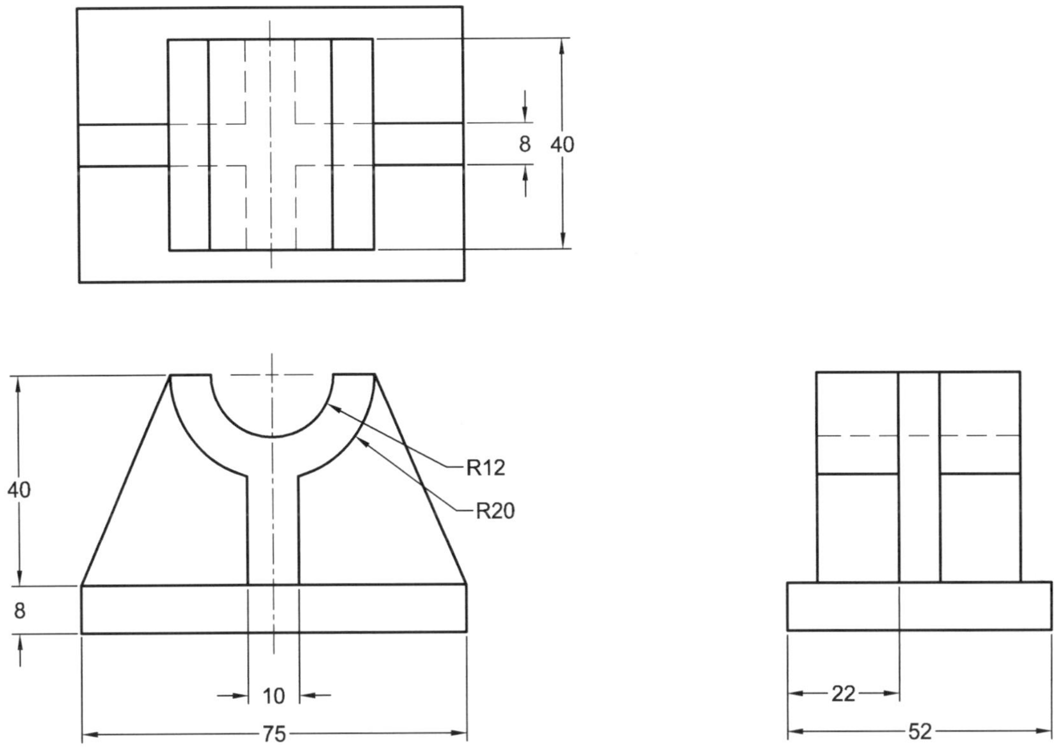

Figure 5-95 Views and dimensions for Exercise 3

Exercise 4

Create the model shown in Figure 5-96. The views and dimensions of the model are shown in Figure 5-97. **(Expected time: 30 min)**

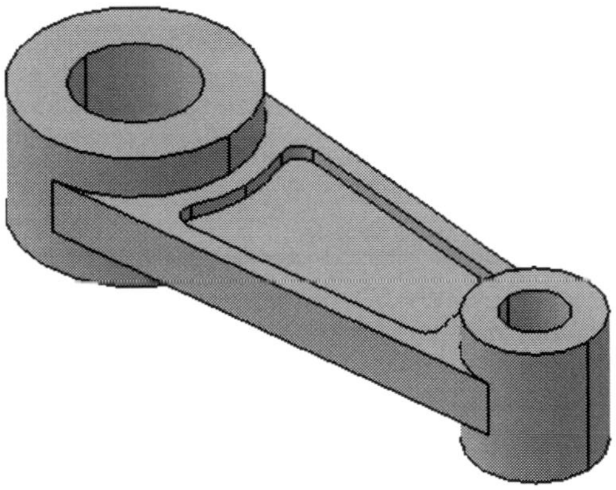

Figure 5-96 Solid model for Exercise 4

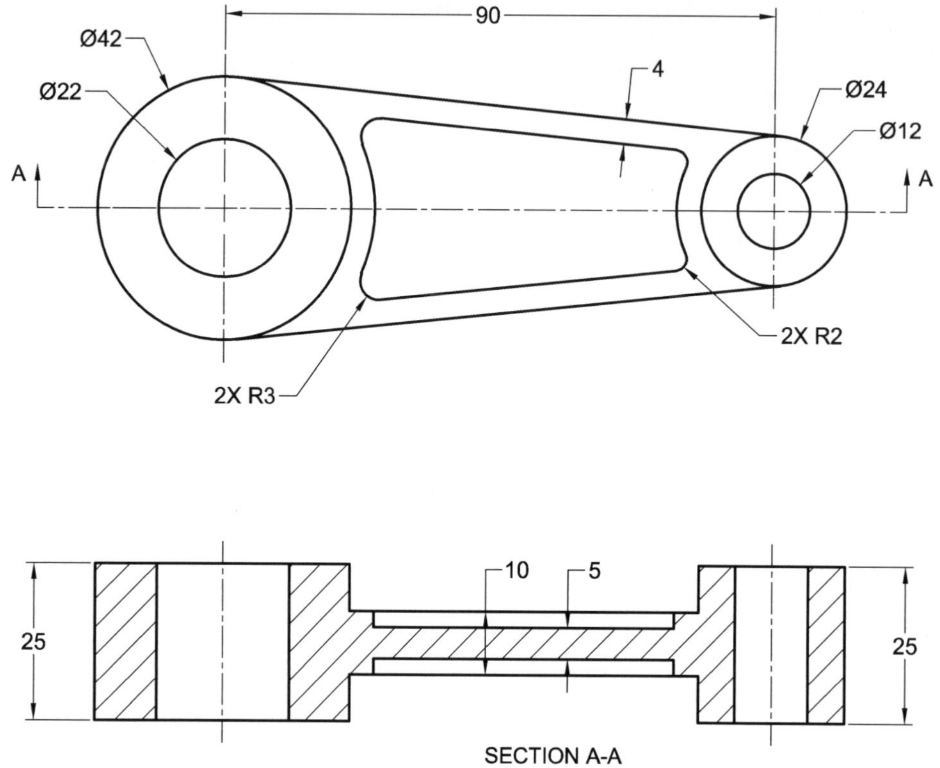

Figure 5-97 *Views and dimensions for Exercise 4*

Answers to Self-Evaluation Test

1. Pocket, **2. Groove**, **3. Point**, **4. Equation**, **5. Plane Definition**, **6.** T, **7.** F, **8.** T, **9.** T, **10.** T

Chapter 6

Creating Dress-Up and Hole Features

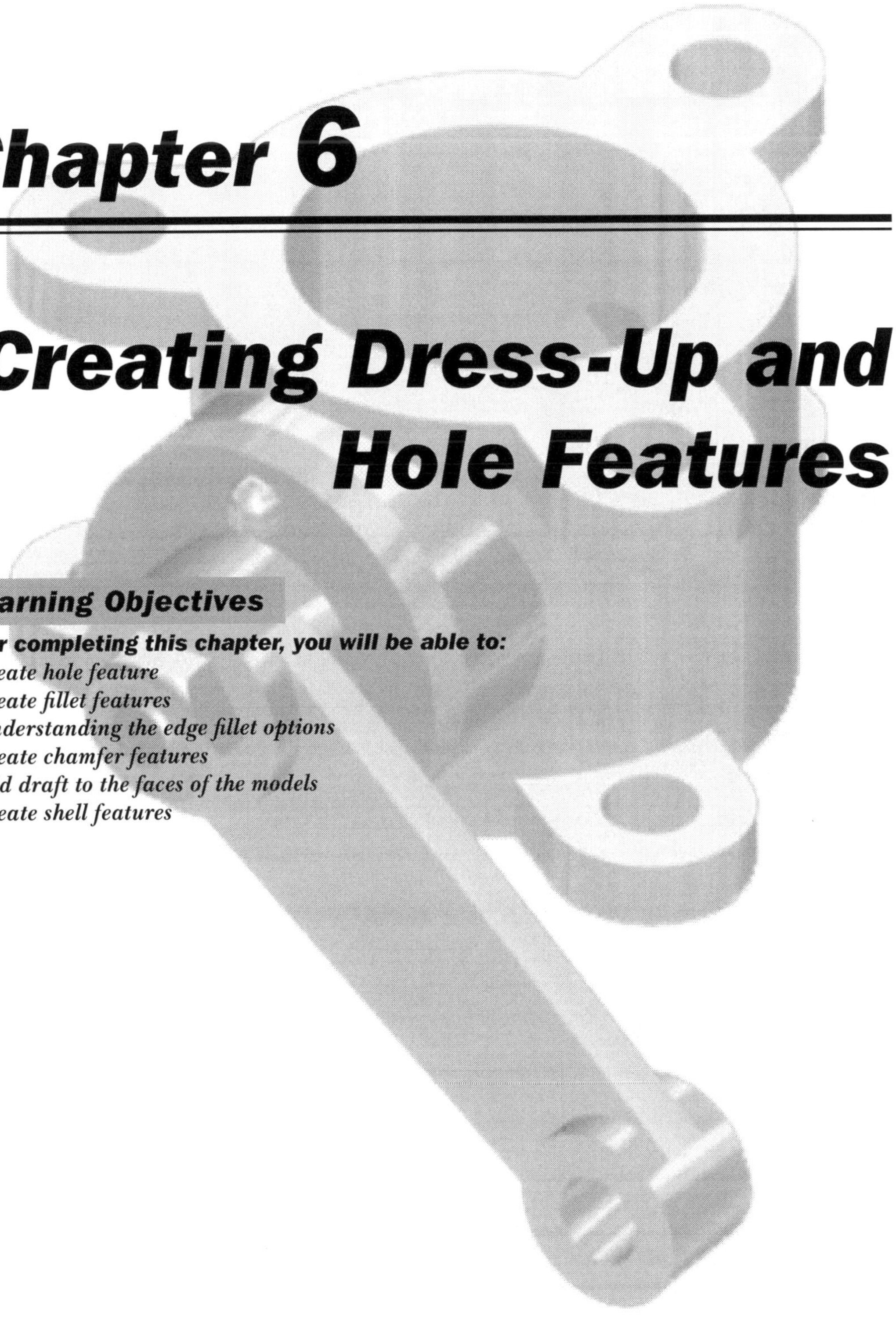

Learning Objectives

After completing this chapter, you will be able to:

- *Create hole feature*
- *Create fillet features*
- *Understanding the edge fillet options*
- *Create chamfer features*
- *Add draft to the faces of the models*
- *Create shell features*

ADVANCED MODELING TOOLS

In this chapter, you will learn to create some of the placed features that aid in constructing a model. For example, in the previous chapter, you learned to create holes by extruding a circular sketch using the **Pocket** tool. In this chapter, you will learn to create holes using the **Hole** tool. You will also learn about some other advanced modeling tools such as fillets, chamfer, draft, shell, and so on.

Creating Hole Features

Menubar:	Insert > Sketch-Based Features > Hole
Toolbar:	Sketch-Based Features > Hole

You can create a simple hole, threaded hole, tapered hole, counterbore hole, countersunk hole, and a counterdrilled hole using the **Hole** tool. You can also create threads in the holes using this tool. However, you can create only one hole feature at a time. To create a hole, invoke the **Hole** tool from the **Sketch-Based Features** toolbar; you will be prompted to select a face or plane. Select the face or plane from the geometry area on which you want to place the hole; the preview of the hole feature and the **Hole Definition** dialog box will be displayed as shown in Figure 6-1.

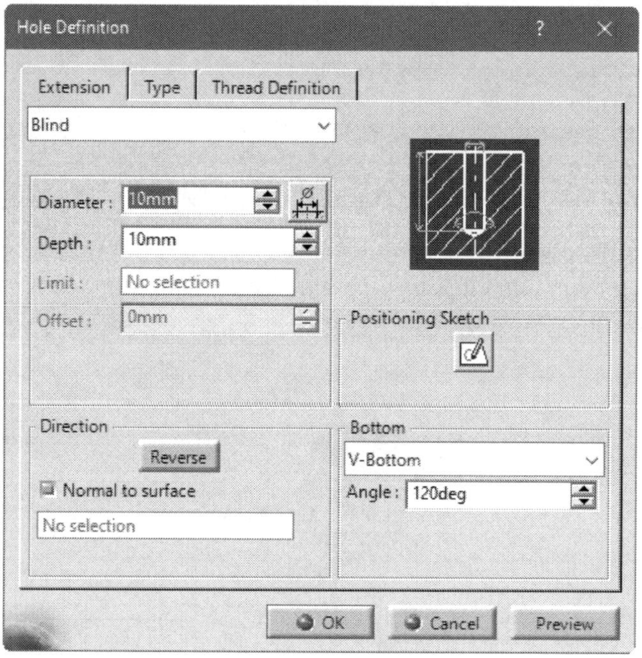

Figure 6-1 *The* ***Hole Definition*** *dialog box*

Creating a Simple Hole

Choose the **Type** tab from the **Hole Definition** dialog box. By default, the **Simple** option is selected in the **Type** drop-down list of this tab. Therefore, a simple hole will be created using the current option. Next, you need to position the center point of the hole. To do so, choose the **Extension** tab. Then, choose the **Sketcher** button from the **Positioning Sketch** area; the Sketcher workbench will be invoked and the **Hole Definition** dialog box will disappear. In the **Sketcher** workbench, the center point of the hole is displayed as a sketched point. Specify the location of the sketched point using the dimensions and constraints and exit the **Sketcher** workbench; the **Hole Definition** dialog box will be displayed again. Next, set the feature termination condition and the diameter of the hole using the options in the **Extension** tab. You can also reverse the direction of the feature creation using the **Reverse** button from the **Direction** area. By default, the **Normal to surface** check box is selected in the **Direction** area. You can also create a hole along a specified direction by clearing the **Normal to surface** check box and selecting the direction along which you need to create it.

The drop-down list in the **Bottom** area is used to specify the shape of the end of the hole. This drop-down list will not be activated if you select the **Up To Next** or **Up To Last** termination option from the drop-down list above the **Diameter** spinner in the dialog box. In such a case, the ends of the holes are automatically created using the **Trimmed** option from the drop-down list in the **Bottom** area. If you select the **Flat** option from this drop-down list, the resulting hole will be flat at the bottom. If you select the **V-Bottom** option from the drop-down list, the **Angle** spinner will be activated. You can set the angle of V-shape in this spinner. The resulting end of the hole thus created will be of V-shape as per the angle value specified in the **Angle** spinner. Figure 6-2 shows all the three types of bottom termination options.

After setting the hole parameters, choose the **OK** button from the **Hole Definition** dialog box to create a simple hole. Figure 6-3 shows a base plate after creating two simple holes using the **Hole** tool.

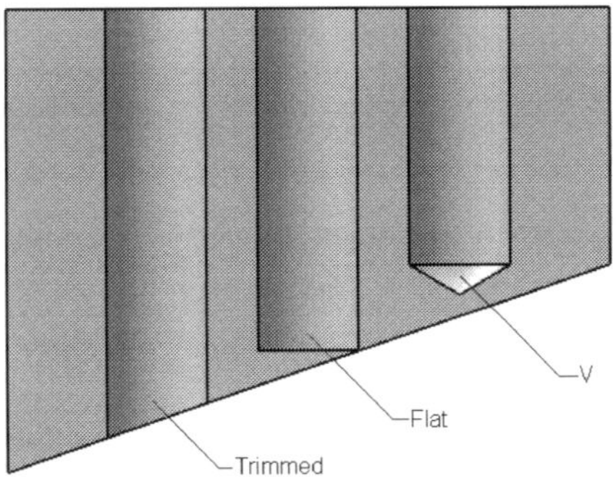

Figure 6-2 *Types of bottom termination options for a hole feature*

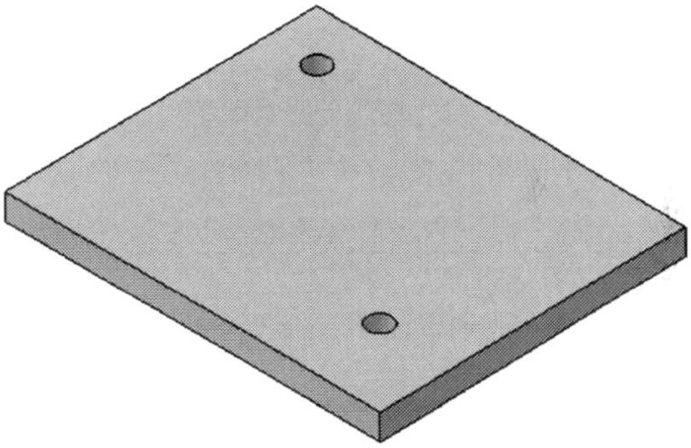

Figure 6-3 *Base plate with holes created using the **Hole** tool*

Tip

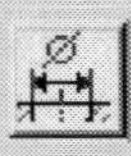

*While creating a hole using the **Hole** tool, you can also use a hole callout to display the hole tolerance. To do so, choose the **Limit of Size Definition** button from the **Extension** tab of the **Hole Definition** dialog box; the **Limit of Size Definition** dialog box will be displayed. The preview of the hole tolerance callout will also be displayed on the hole feature in the geometry area. Set the value of the hole tolerance using the options in the **Limit of Size Definition** dialog box and choose the **OK** button. Now, set the parameters of the hole and exit the **Hole Definition** dialog box to complete the feature creation. The annotation set and the information about the hole tolerance callout are displayed in the Specification tree.*

Creating a Threaded Hole

To create a threaded hole, choose the **Thread Definition** tab. By default, the **Threaded** check box is cleared. Select the **Threaded** check box to activate the options in the **Thread Definition** area, as shown in Figure 6-4.

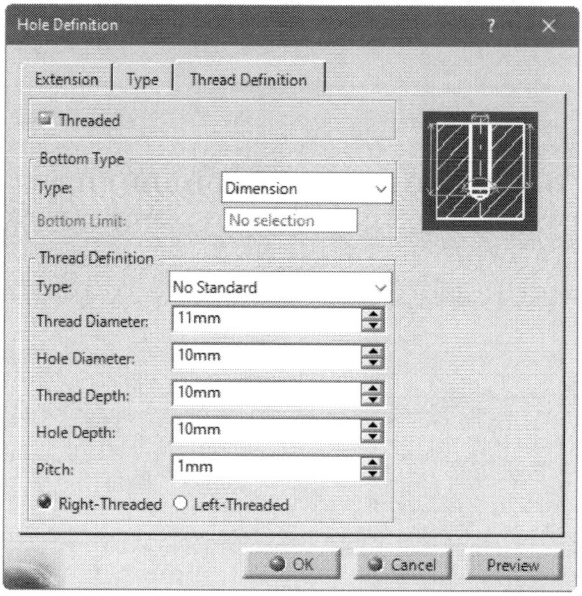

Figure 6-4 *Options displayed on selecting the* ***Threaded*** *check box*

The **Bottom Type** area in this dialog box is used to select the options to specify the thread depth. By default, the **Dimension** option is selected in the **Type** drop-down list in the **Bottom Type** area. This option allows you to specify the desired thread depth in numerical values. To specify the thread depth, use the **Thread Depth** spinner in the **Thread Definition** area. The **Support Depth** option in the **Type** drop-down list in the **Bottom Type** area sets the thread depth equal to the hole depth. The **Up-To-Plane** option in the **Type** drop-down list allows you to specify the thread depth by selecting a termination plane or the surface up to which the thread will be created. As soon as you select the **Up-To-Plane** option, the **Bottom Limit** selection box will become active and you will be prompted to select the termination plane or the surface. Select the plane or the surface from the drawing area; the thread in the hole will terminate at the specified plane.

By default, the **No Standard** option is selected in the **Type** drop-down list in the **Thread Definition** area. Therefore, you need to manually specify the parameters to define the thread. Set the value of the thread diameter in the **Thread Diameter** spinner and the value of the hole diameter in the **Hole Diameter** spinner. By default, these values are based on the diameter value specified in the extension tab. Set the thread depth and the hole depth in the **Thread Depth** and the **Hole Depth** spinners, respectively. Also, set the pitch value in the **Pitch** spinner. By default, the **Right-Threaded** radio button is selected. To create a left hand thread, select the **Left-Threaded** radio button. After setting the parameters, choose the **OK** button from the **Hole Definition** dialog box; a threaded hole will be created. Note that the thread will not be displayed in the hole because only a cosmetic thread can be added to a hole feature. When you generate the drawing view, the thread convention will be displayed in it. You will learn more about generating drawing views in the later chapters.

To create standard threaded holes, select the **Metric Thin Pitch** or **Metric Thick Pitch** option from the **Type** drop-down list in the **Thread Definition** area, refer to Figure 6-4. You will notice that on selecting the **Metric Thick Pitch** or **Metric Thin Pitch** option from the **Type** drop-down list, the **Thread Diameter** spinner will change into the **Thread Description** drop-down list. You can select the standard size of the thread from this drop-down list. In this case, you need to specify the thread depth and the hole depth, the hole diameter, thread diameter, and thread pitch are automatically defined on the basis of the selected standard size of the thread from the **Thread Description** drop-down list.

If the value of hole diameter matches the standard value, then the relevant thread parameters will be applied automatically as per the standards selected from the **Type** drop-down list in the **Thread Definition** area. If the value of the hole diameter does not match any standard value, a warning message box will be displayed. Choose the **OK** button from the **Warning** message box; the hole diameter and the other thread diameter parameters will be set in accordance with the smallest standard value of the diameter available in the **Thread Description** drop-down list.

Creating a Tapered Hole

To create a tapered hole, choose the **Type** tab in the **Hole Definition** dialog box and select the **Tapered** option from the drop-down list, as shown in Figure 6-5. The preview of the tapered hole will be displayed in the geometry area with the default values. Specify the taper angle in the **Angle** spinner of the **Parameters** area, refer to Figure 6-5.

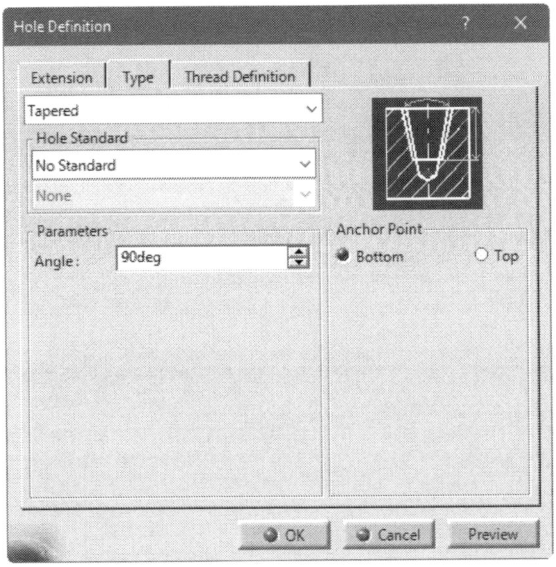

*Figure 6-5 The **Hole Definition** dialog box displayed after selecting the **Tapered** option*

Note that, by default, the **Bottom** radio button is selected in the **Anchor Point** area of the **Hole Definition** dialog box. As a result, the specified angle and diameter will be applied with respect to the bottom face of the tapered hole. If you select the **Top** radio button from the **Anchor Point** area, the specified angle and diameter will be applied with respect to the selected placement plane of the hole. After setting all parameters, choose the **OK** button from the **Hole Definition** dialog box to create the tapered hole.

Creating a Counterbore Hole

A counterbore hole is a stepped hole and has two hole diameters, a bigger diameter and a smaller diameter. The bigger diameter is called the counterbore diameter and the smaller diameter is called the hole diameter. This hole type requires you to specify two depths, counterbore depth and hole depth. The counterbore depth is the depth up to which the bigger diameter will be defined. The hole depth is the total depth of the hole including the counterbore depth. Figure 6-6 shows the sectional view of a counterbore hole. Figure 6-7 shows a base plate with the counterbore holes.

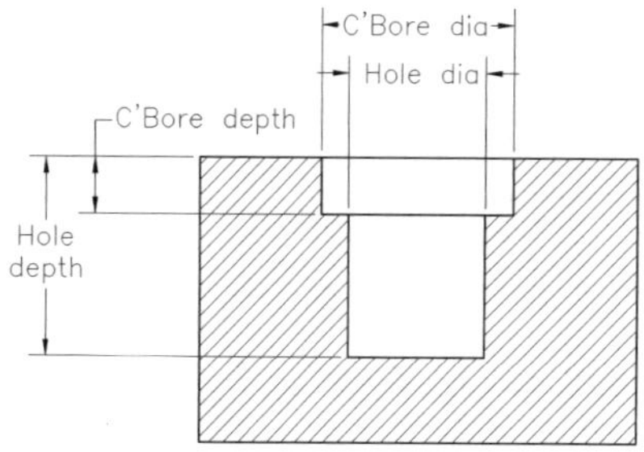

Figure 6-6 *The sectional view of a counterbore hole*

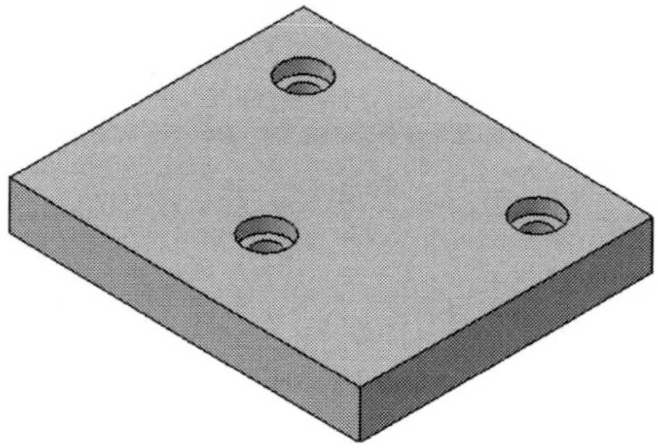

Figure 6-7 *The base plate with counter bore holes*

To create a counterbore hole, select the **Counterbored** option from the drop-down list in the **Type** tab of the **Hole Definition** dialog box, as shown in Figure 6-8; the preview of the counterbore hole will be displayed in the geometry area.

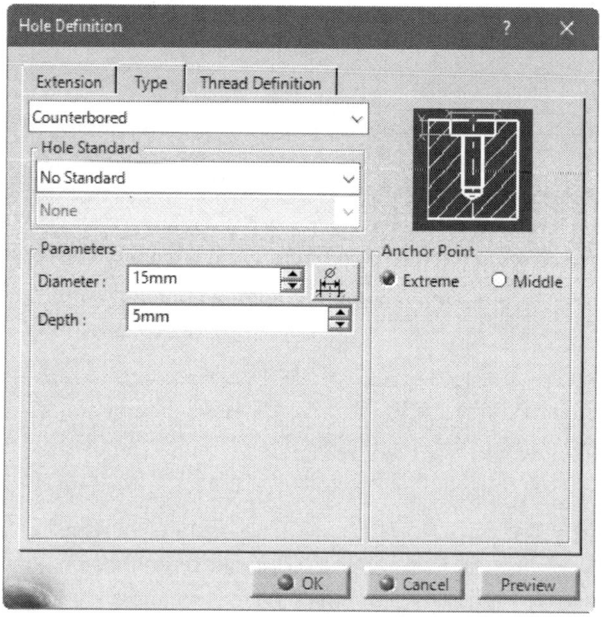

Figure 6-8 *The* ***Hole Definition*** *dialog box displayed after selecting the* ***Counterbored*** *option*

By default, the **No Standard** option is selected in the drop-down list in the **Hole Standard** area. Therefore, you need to manually specify the parameters to define the hole. Set the value of the counterbore diameter using the **Diameter** spinner and the counterbore depth using the **Depth** spinner in the **Parameters** area. You can also set the diameter and depth of the hole using the options in the **Extension** tab.

To create standard holes, select the **Metric_Cap_Screws** or **Socket_Head_Cap_Screws** option from the **Hole Standard** area of the **Hole Definition** dialog box; the drop-down list under this list in the **Hole Standard** area will display various options for standard sizes of the hole. Select any standard size from the drop-down list; the diameter and depth will automatically be defined in the respective edit boxes. Note that the options for creating standard holes are available only when you select the **Counterbored** option from the **Type** tab. Also, note that on selecting the **Socket_Head_Cap_Screws** option from the drop-down list under the **Hole Standard** area of this dialog box, the **Feature Definition Error** message box will be displayed. The message box informs you that the standard units are different from the model units and will be converted. Choose the **OK** button from this error message box.

You will notice that the **Extreme** radio button is selected in the **Anchor Point** area of the **Type** tab. If you select the **Middle** radio button, the bottom face of the counter will be placed on the selected placement plane of the hole. You can also define the thread parameters for a counterbore hole. After setting the parameters, choose the **OK** button from the **Hole Definition** dialog box.

Creating a Countersunk Hole

A countersunk hole also has two hole diameters, but the transition between the bigger diameter and the smaller diameter is in the form of a tapered cone. The parameters of a countersunk hole are: countersunk diameter, hole diameter, depth of the hole, and the countersunk angle. You need to specify these parameters to create a countersunk hole. Figure 6-9 shows the sectional view of a countersunk hole. Figure 6-10 shows the spacer plate after adding the countersunk holes.

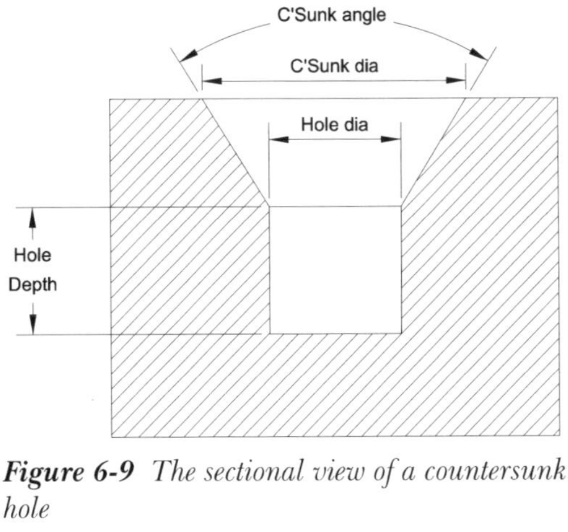

Figure 6-9 The sectional view of a countersunk hole

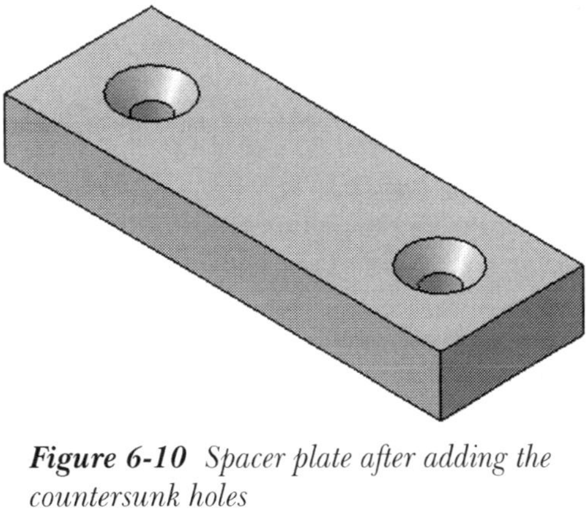

Figure 6-10 Spacer plate after adding the countersunk holes

To create a countersunk hole, select the **Countersunk** option from the drop-down list in the **Type** tab, as shown in Figure 6-11. Its preview will be displayed in the geometry area.

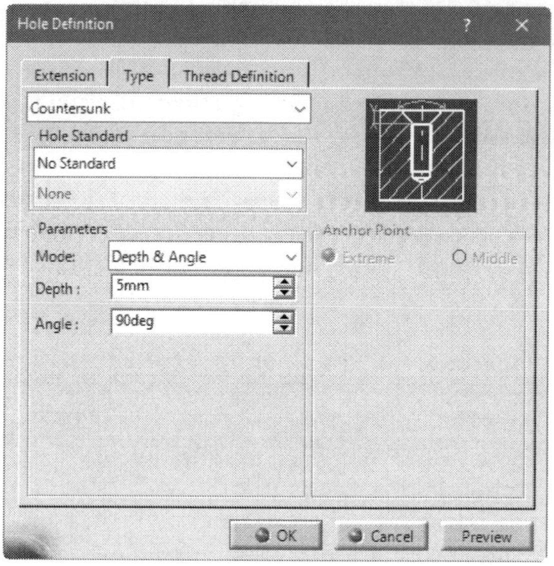

***Figure 6-11** The **Hole Definition** dialog box displayed after selecting the **Countersunk** option*

You can choose the option for specifying the parameters of the countersunk using the **Mode** drop-down list in **Parameters** area. By default, the **Depth & Angle** option is selected. Therefore, you need to define the depth and angle of the countersunk in the **Depth** and **Angle** spinners, respectively. If you select the **Depth & Diameter** option from the **Mode** drop-down list, you need to define the depth and diameter of the countersunk in the **Depth** and **Diameter** spinners, respectively. Similarly, if you select the **Angle & Diameter** option from the **Mode** drop-down list, you need to set the value of the angle and diameter in the respective spinners. Now, set the other parameters of the hole feature in the **Extension** tab. You can also specify the thread parameters of the countersunk hole. After setting the parameters, choose the **OK** button from the **Hole Definition** dialog box.

Creating a Counterdrilled Hole

A counterdrilled hole is a combination of a counterbore and a countersunk hole. This hole type has two hole diameters and the transition between the bigger diameter and the smaller diameter, after applying the counterbore depth is in the form of a tapered cone, refer to Figure 6-12. You need to define the counterbore diameter, hole diameter, depth of counterbore, depth of the hole, and countersink angle to create a counterdrill hole. Figure 6-12 shows the sectional view of a counterdrilled hole. Figure 6-13 shows the spacer plate with the counterdrilled holes.

To create a counterdrilled hole, select the **Counterdrilled** option from the drop-down list in the **Type** tab of the **Hole Definition** dialog box; its preview is displayed in the geometry area. Figure 6-14 shows the **Hole Definition** dialog box displayed on selecting the **Counterdrilled** option from the drop-down list. You need to set the value of the diameter of the counter using the **Diameter** spinner and the value of its depth using the **Depth** spinner. Next, you need to set the value of the drill angle in the **Angle** spinner. You can also specify the thread parameters while creating a counterdrilled hole. After specifying all the parameters, choose the **OK** button from the **Hole Definition** dialog box.

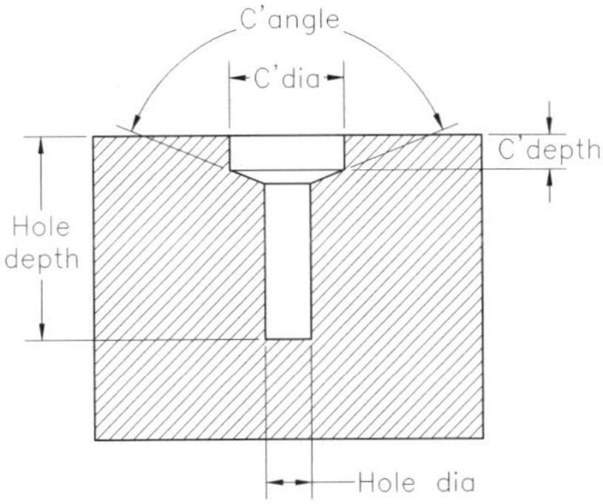

Figure 6-12 *The sectional view of a counterdrilled hole*

Figure 6-13 *The spacer plate after adding the counterdrilled holes*

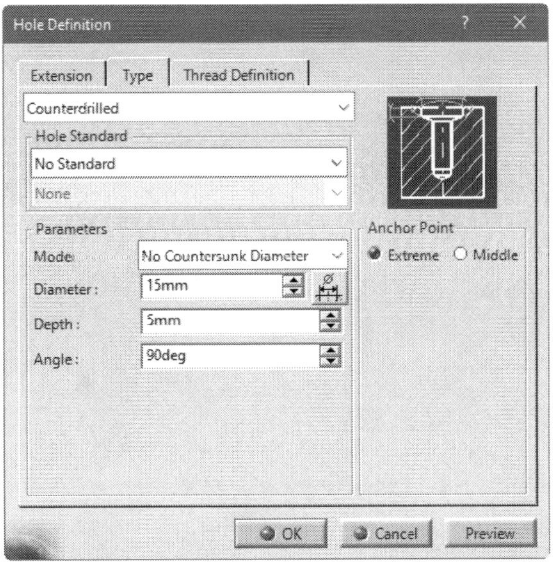

Figure 6-14 *The* ***Hole Definition*** *dialog box displayed after selecting the* ***Counterdrilled*** *option*

Tip

1. To apply cosmetic threads on plane holes or plane cylindrical shafts, choose the ***Thread/Tap*** *button from the* ***Dress-Up Features*** *toolbar; the* ***Thread/Tap Definition*** *dialog box will be displayed. Select the cylindrical surface on which you want to apply the thread and then select the face from which the thread will start. Now, set the thread parameters in the* ***Numerical Definition*** *area and choose the* ***OK*** *button from the* ***Thread/Tap Definition*** *dialog box; the* ***Thread.1*** *feature will be displayed in the Specification tree.*

2. You need to make sure that the ***Thread/Tap*** *tool is not used for applying threads to the cylindrical holes created using the* ***Hole*** *tool. If you do so, an error message window will be displayed which will prompt you to use the* ***Hole*** *command to tap a hole.*

Creating Fillet Feature

A fillet is a curved face of a constant or variable radius that is tangent to two surfaces. In general, a fillet is provided to reduce stress concentration at the corners and edges of the model. The **Part** workbench of CATIA V5 provides you with the tools to fillet the sharp edges of the models. You can create simple edge fillets, variable radius fillets, chordal fillet, face to face fillets and tritangent fillets using tools in the **Part** mode of CATIA V5. Choose the black arrow on the right of the **Edge Fillet** tool in the **Dress-Up Features** toolbar; the **Fillets** sub-toolbar will be displayed, as shown in Figure 6-15. The procedures of creating various types of fillets are discussed next.

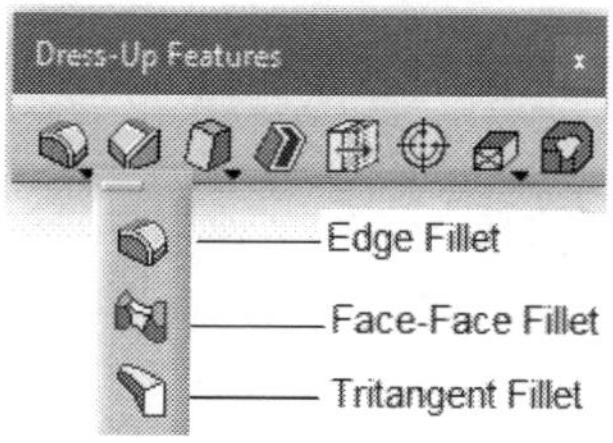

*Figure 6-15 The **Fillets** sub-toolbar in the **Dress-Up Features** toolbar*

Creating the Edge Fillet

Menubar:	Insert > Dress-Up Features > Edge Fillet
Toolbar:	Dress-Up Features > Fillets sub-toolbar > Edge Fillet

The **Edge Fillet** tool is used to create constant as well as variable radius fillet by specifying the radius or chord length of the fillet. To create an edge fillet, choose the **Edge Fillet** tool from the **Fillets** sub-toolbar in the **Dress-Up Feature** toolbar; the **Edge Fillet Definition** dialog box will be displayed, as shown in Figure 6-16. Also, you will be prompted to select the edge or face to be filleted.

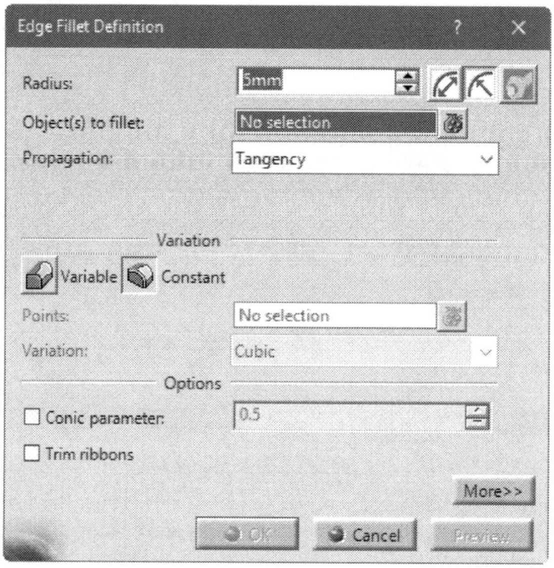

*Figure 6-16 The **Edge Fillet Definition** dialog box*

Select the edge that you need to fillet; the total number of the selected edges and faces will be displayed in the **Object(s) to fillet** display box. Note that the radius option will be displayed only when the Radius button  is chosen. If you choose the Chordal length button  the **Radius** spinner is changed into the **Chordal length** spinner and the value specified in this spinner will be the chord length of the fillet. You can also create variable chord length fillet along the length of the edge. Set the value of the fillet radius using the **Radius** spinner and choose the **Preview** button from the **Edge Fillet Definition** dialog box. Figure 6-17 shows the edge selected to be filleted and Figure 6-18 shows the resulting filleted edge. Figure 6-19 shows the face selected to be filleted and Figure 6-20 shows the resulting filleted face.

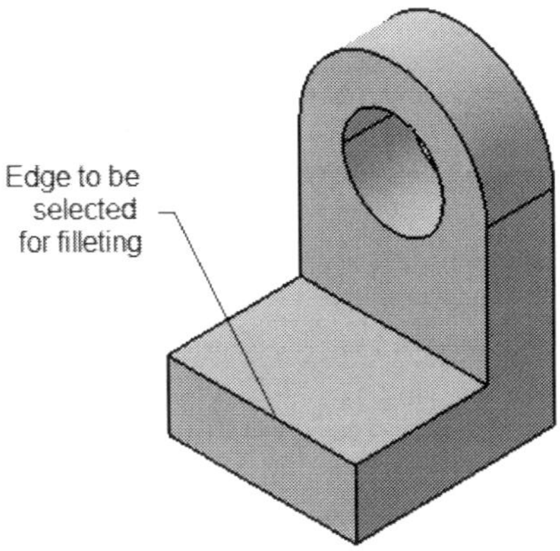

Figure 6-17 *Edge selected to be filleted*

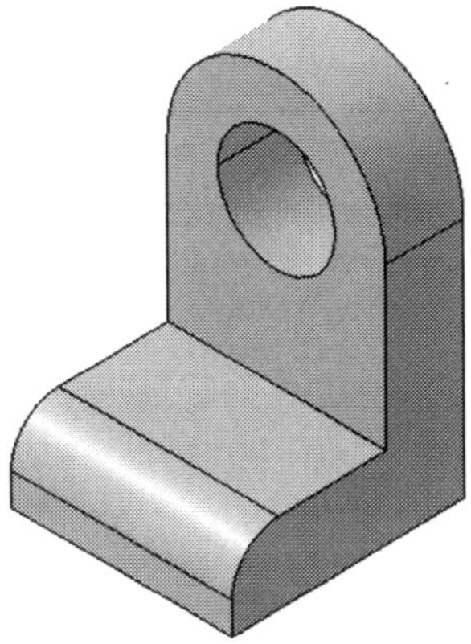

Figure 6-18 *Resulting filleted edge*

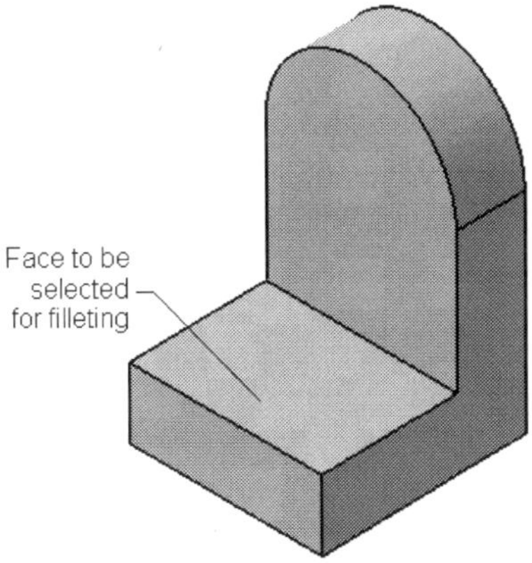

Figure 6-19 *Face selected to be filleted*

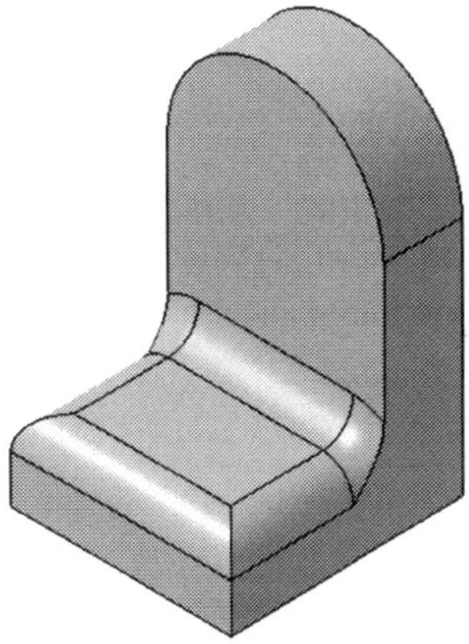

Figure 6-20 *Resulting filleted face*

The other options in the **Edge Fillet Definition** dialog box for creating edge fillets are discussed next.

Selection Manager

In CATIA V5, you can manage the entities in the current selection set. To do so, choose the Selection Manager button on the right of the **Object(s) to fillet** display box; the **Fillet edges** dialog box will be displayed. All the selected entities are listed in this dialog box. The **Remove** button in this dialog box is used to remove an entity from the current selection set. The **Replace** button is used to replace an entity from the current selection set with another entity from the model. To replace any entity, select it from the **Fillet edges** dialog box and choose the **Replace** button followed by the replacement edge.

Propagation

The options available in the **Propagation** drop-down list are used to manage the propagation of the fillet. By default, the **Tangency** option is selected in the **Propagation** drop-down list. Therefore, the edges tangent to the selected edge will also be selected and filleted. If you select the **Minimal** option, only the selected edge will be filleted, as shown in Figure 6-21. Figure 6-22 shows the edge filleted using the **Tangency** option and Figure 6-23 shows the edge filleted using the **Minimal** option.

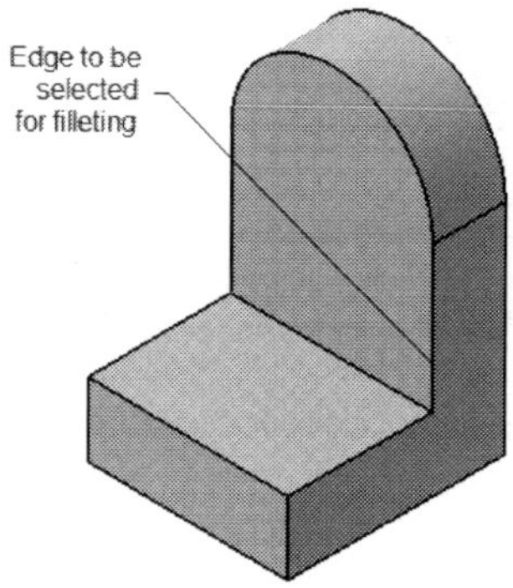

Figure 6-21 Edge selected to be filleted

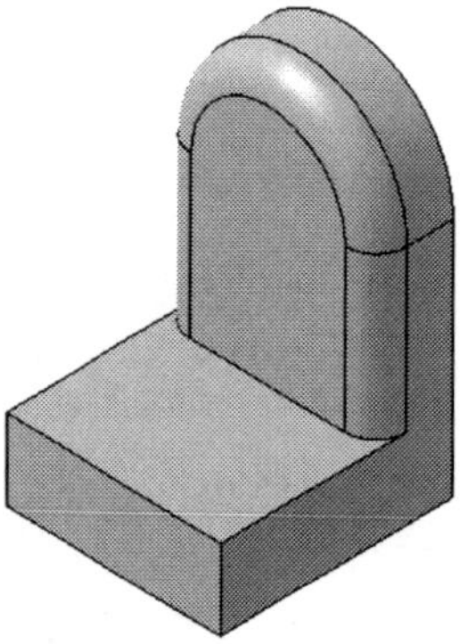

*Figure 6-22 Fillet created using the **Tangency** option*

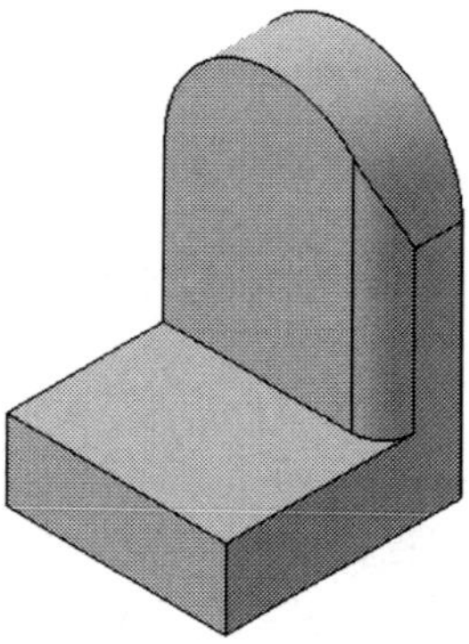

*Figure 6-23 Fillet created using the **Minimal** option*

To create a fillet at the edges created by the intersection of multiple features, select the **Intersection** option from the **Propagation** drop-down list and select the object to fillet. Choose the **OK** button; the fillet will be created at the edges formed by the intersection of the selected object with the other existing features, as shown in Figure 6-24.

If you select the **Intersection with selected features** option in the **Propagation** drop-down list, you will be prompted to select the feature to be filleted. Select the features, next click on the **Selected features** selection box and select the intersecting features. On doing so, only the edges created between the intersection of the selected features will be filleted, as shown in Figure 6-25.

Variation

By using the options available in the **Variation** area, you can create a fillet with constant radius or variable radius along the selected edge.

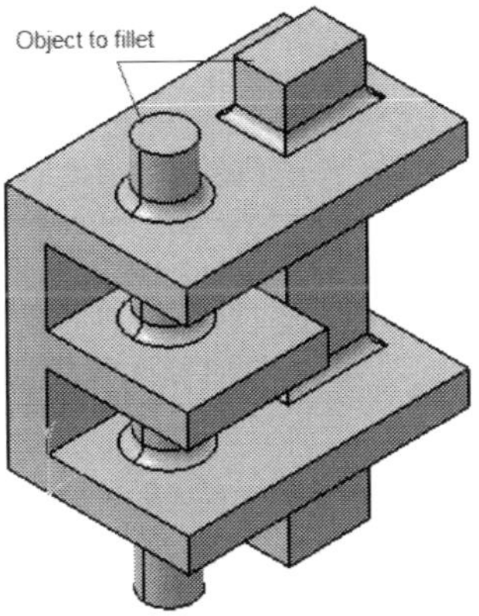

Figure 6-24 *Fillet created using the* ***Intersection*** *option*

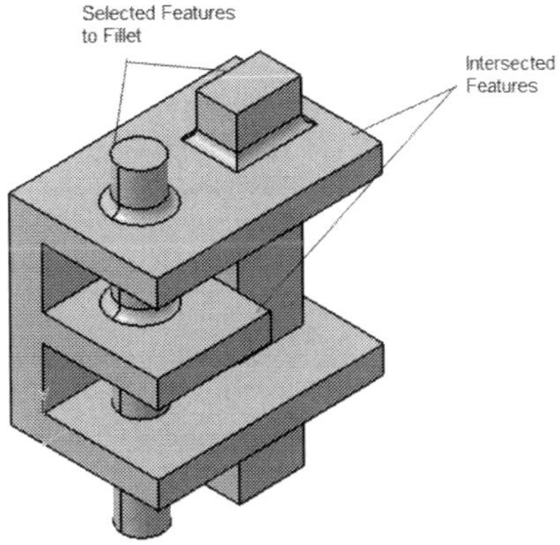

Figure 6-25 *Fillet created using the* ***Intersection with selected features*** *option*

To create a variable radius fillet, choose the **Variable** button from the **Variation** area and select the edge that you need to fillet; two radius callouts will be attached to the endpoints of the selected edge. You can also select multiple edges for applying the fillets by using the Selection Manager button on the right of the **Object(s) to fillet** display box. Next, double-click on one of the radius callouts from the geometry area, the **Parameter Definition** dialog box will be invoked. Now, set the value of the radius in the **Value** spinner of this dialog box and choose the **OK** button. Similarly, double-click on the other callout and set the value of the second radius in the **Value** spinner. Now, choose the **OK** button from the **Edge Fillet Definition** dialog box. The model after creating the variable radius fillet is shown in Figure 6-26.

You can also define additional control points on the selected edge. To do so, click in the **Points** selection box in the **Variation** area of the **Edge Fillet Definition** dialog box and then click anywhere on the edge; a callout will be attached to the specified point. You can double-click on the callout value to modify the fillet radius. You can also create as many control points as you need by repeating this procedure. Figure 6-27 shows a variable radius fillet after specifying the radii at additional control points.

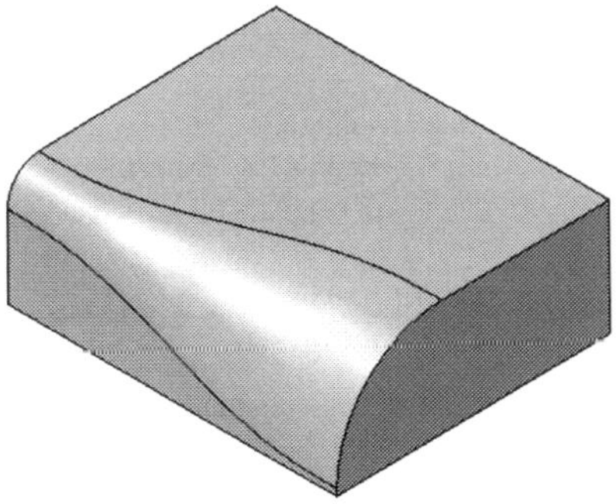

Figure 6-26 *Model after creating the variable radius fillet*

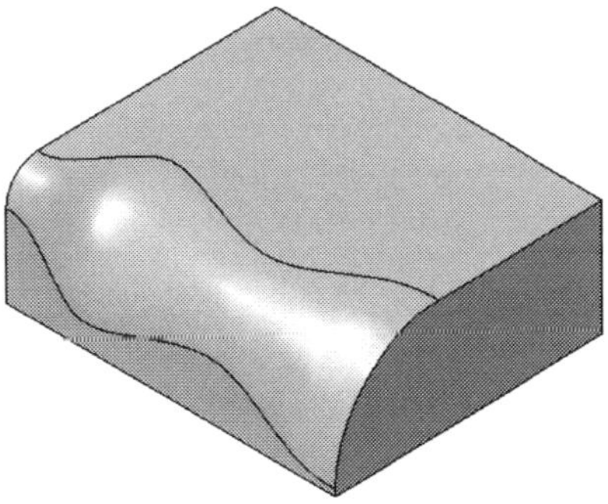

Figure 6-27 *Variable radius fillet after specifying additional control points*

You can also manage the transition of the variable radius fillet. By default, the **Cubic** option is selected in the **Variation** drop-down list of the **Variation** area of the **Edge Fillet Definition** dialog box. This option will result in smooth transition of the fillet surface. If you select the **Linear** option from the **Variation** drop-down list, it will result in straight transition of the fillet surface. Figure 6-28 shows the variable radius fillet with the **Cubic** option selected and Figure 6-29 shows the variable radius fillet with the **Linear** option selected.

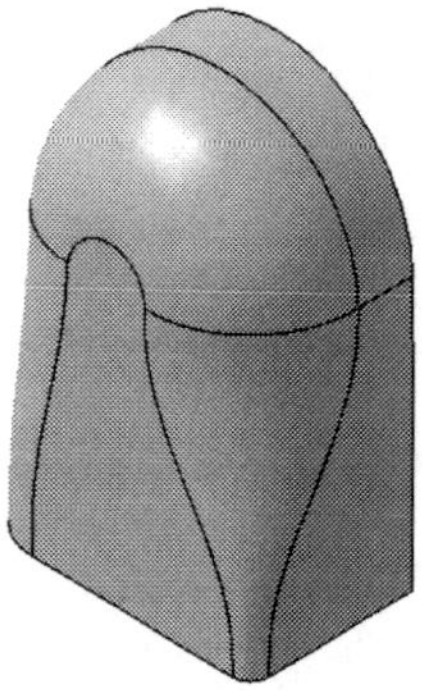

Figure 6-28 *Variable radius fillet created with the **Cubic** option selected*

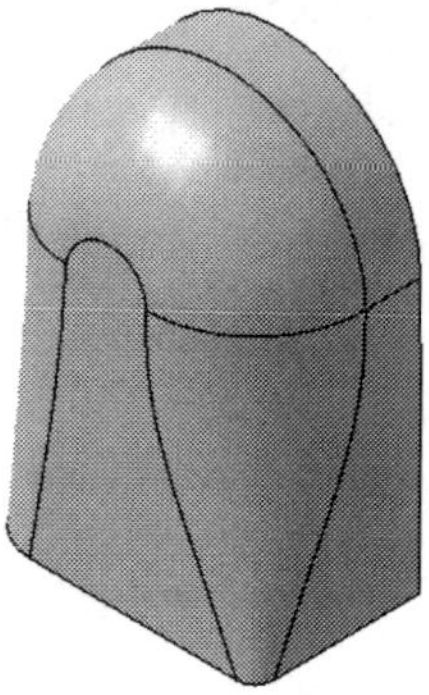

Figure 6-29 *Variable radius fillet created with the **Linear** option selected*

Conic parameter

To create a conical fillet, select the **Conic parameter** check box in the **Options** area of the **Edge Fillet Definition** dialog box; the **Conic parameter** spinner will get activated. The value in the **Conic parameter** spinner should lie between 0 and 1. A value between 0 and 0.5 will result in a fillet with hyperbolic shape. The value 0.5 will result in a fillet with circular shape whereas a value between 0.5 and 1 will result in a fillet with parabolic shape, refer to Figure 6-30.

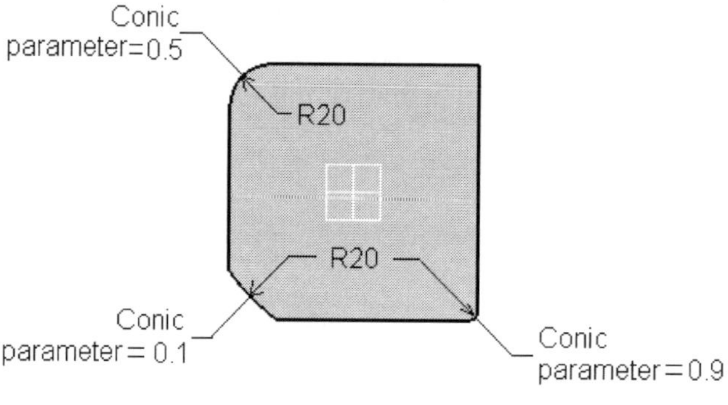

Figure 6-30 *Fillets created with different values of Conic parameter*

Trim ribbons

You can use the options in the **Edge Fillet Definition** dialog box to trim the intersecting surfaces. Assume that you have to fillet the model shown in Figure 6-31 and the fillets thus created are overlapping. If the **Trim ribbons** check box is selected in the **Edge Fillet Definition** dialog box then on choosing the **OK** button, the intersecting surfaces created as a result of filleting will be trimmed, as shown in Figure 6-32. Else, an error message will be displayed on screen. Note that the **Trim ribbons** check box will not be enabled when the **Minimal** option is selected in the **Propagation** drop-down list.

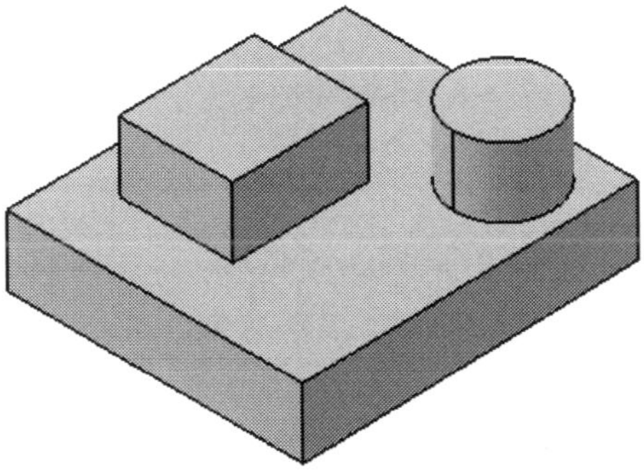

Figure 6-31 The model for creating the fillet

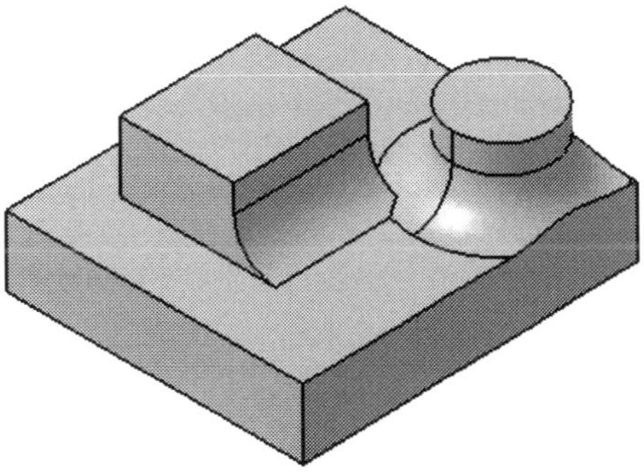

Figure 6-32 Fillet created after trimming of intersecting surfaces

Edge(s) to keep

Sometimes while filleting, some of the edges get distorted in order to accommodate the fillet radius, as shown in Figure 6-33. In this model, the bottom edge of the elliptical extruded feature is filleted. The inclined edges are distorted in order to accommodate the fillet radius. To avoid this distortion, choose the **More** button from the **Edge Fillet Definition** dialog box; the **Edge Fillet Definition** dialog box will expand. Click once in the **Edge(s) to keep** display box and select the distorted edges. Now, choose the **OK** button from the **Edge Fillet Definition** dialog box; the edges will not be distorted, as shown in Figure 6-34. Note that the **Edge(s) to keep** display box will not be enabled when the **Conic parameter** edit box is selected.

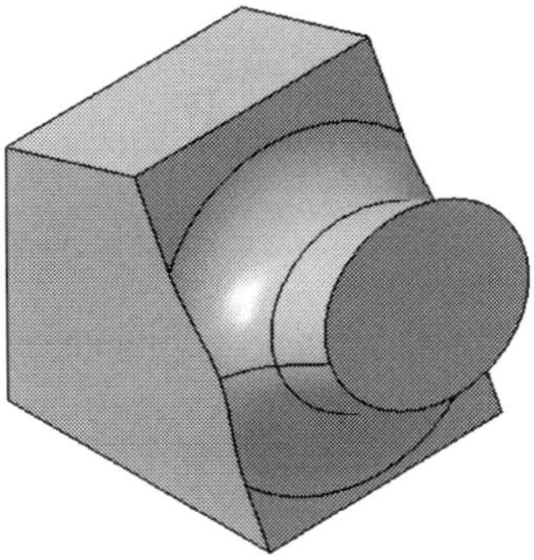

Figure 6-33 Edges distorted to accommodate the fillet radius

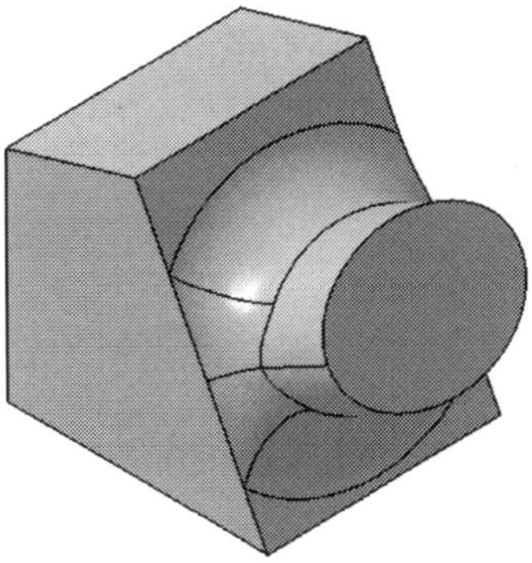

Figure 6-34 The model after retaining the edges

Note

If the fillet radius is too large to retain the selected edges, the ***Update Diagnosis*** *message box will be displayed with the message* ***The operation cannot be initialized for one or more ribbon(s)*** *in the diagnosis area of this message box. To avoid this error, you need to reduce the fillet radius.*

Circle Fillet

This option is used to control the shape of the fillet and will be enabled only when you select the **Variable** button available in the **Variation** area of the **Edge Fillet Definition** dialog box. Select this option and you will be prompted to select the spine. The spine can be a wireframe element or a sketched element. Select the spine and click **OK**. You will notice that the circles of the resultant fillet will become normal to the selected spine. Figure 6-35 shows a standard fillet created on the selected edge. Figure 6-36 shows a fillet created using the **Circle Fillet** option.

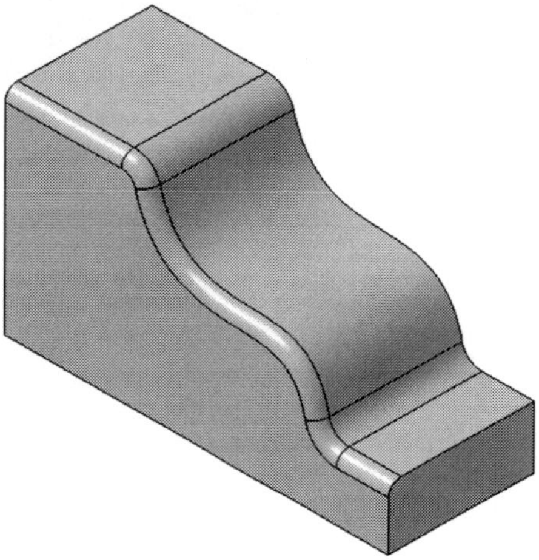

Figure 6-35 *Fillet created without the* ***Circle Fillet*** *option selected*

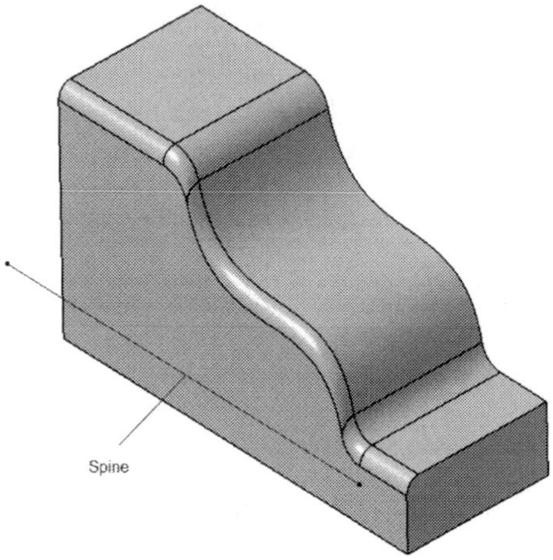

Figure 6-36 *Fillet created with the* ***Circle Fillet*** *option selected*

Limiting element(s)

You can also set the limit of the fillet along the selected edge upto which the fillet will be created. Select the edge or edges to be filleted and set the value of the radius. Now, expand the **Edge Fillet Definition** dialog box using the **More** button if it is not expanded by default. Click once in the **Limiting element(s)** display box and select the plane or planer surface upto which you need to create the fillet. An arrow displayed in the geometry area will define the direction of the fillet creation. You can flip the direction by clicking on the arrow in the preview of the fillet. You can also create a point or plane when the **Edge Fillet** tool is still active to define the limit of the fillet. To do so, right-click in the **Limiting element(s)** selection box; a contextual menu will be displayed. Define the limit using the options in the contextual menu. Figure 6-37 shows the edge to be filleted and the limiting element to be selected and Figure 6-38 shows the resulting fillet.

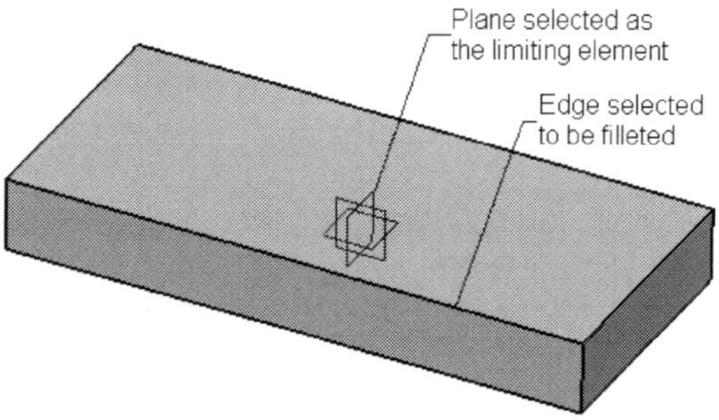

Figure 6-37 *Edge to be filleted and the limiting element to be selected*

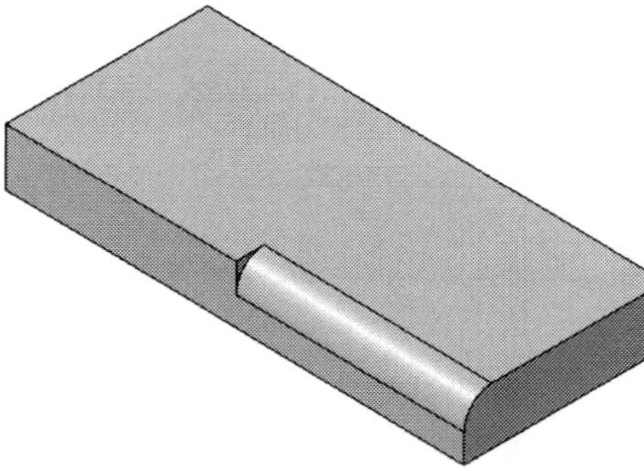

Figure 6-38 *Resulting fillet*

Note

*Instead of selecting or creating a limiting element, you can specify the limit of a fillet by directly selecting points on the edge to be filleted. To define the limits using this method, select the edge to fillet and define the fillet radius. Now, expand the **Edge Fillet Definition** dialog box and click once in the **Limiting element(s)** selection box. Click on the selected edge where you want to define the limit of the fillet; a blue circle will be displayed on the current selection. The arrow defining the direction of the fillet creation will also be displayed. If you have selected two elements to limit the fillet, you need to make sure that the arrows of both the limits point in the opposite directions. You can flip the direction of arrows by clicking on them. Figure 6-39 shows the fillet after specifying two limit elements. In this figure, the arrows of both the limits point toward the midpoint of the edge.*

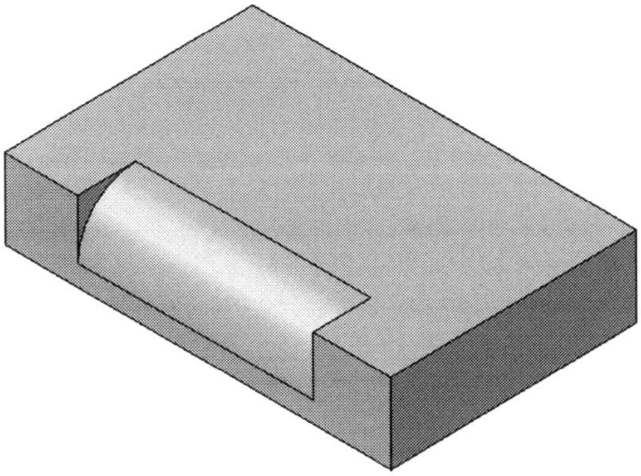

Figure 6-39 *Fillet after specifying the limit elements*

Blend corner(s)

The setback fillet can be created on a vertex where three or more than three edges merge. This fillet is used to smoothly blend the transition surfaces generated from the edges up to the fillet vertex. To create a setback fillet, select the edges on which you want to create the fillet and then specify the value of the fillet radius. Now, expand the **Edge Fillet Definition** dialog box by choosing the **More** button from the dialog box. Next, right-click in the **Blend corner(s)** selection box in the **Edge Fillet Definition** dialog box; a contextual menu will be displayed. Choose the **Create by edges or vertex** option from the contextual menu; all vertices of the selected edges will be selected automatically. Click the button on the right side of the **Blend corner(s)** to display the **Corner Setback Value Panel** dialog box. In this dialog box, you can modify the Setback Value of the selected edges individually. As the vertex is selected, the **Corner.1** callout will be displayed on the vertex. You will notice that individual setback dimensions are also attached to the selected edges. Select any one of the dimensions and set its value in the **Setback distance** spinner. Similarly, set the setback distance for the other edges. Figure 6-40 shows the edges selected to be filleted. Figure 6-41 shows the preview of the setback fillet after setting the setback distance. Figure 6-42 shows the resulting setback fillet.

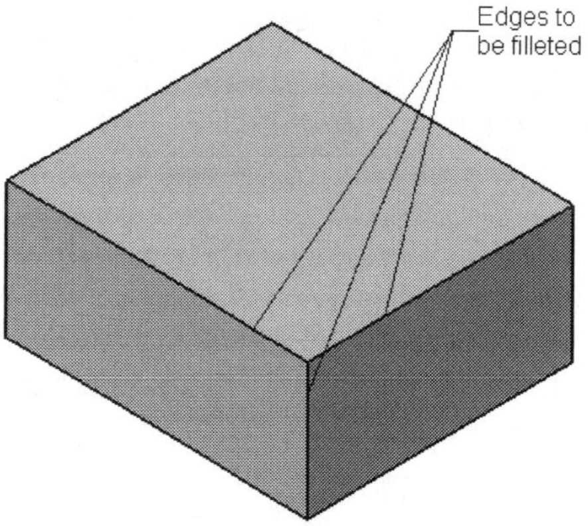

Figure 6-40 Edges selected to be filleted

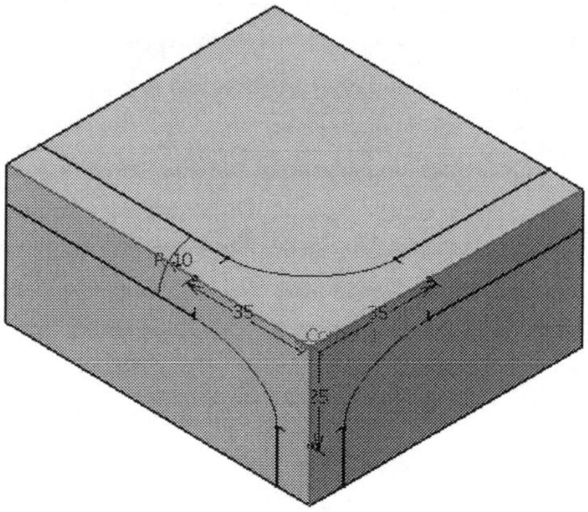

Figure 6-41 Preview of the setback fillet

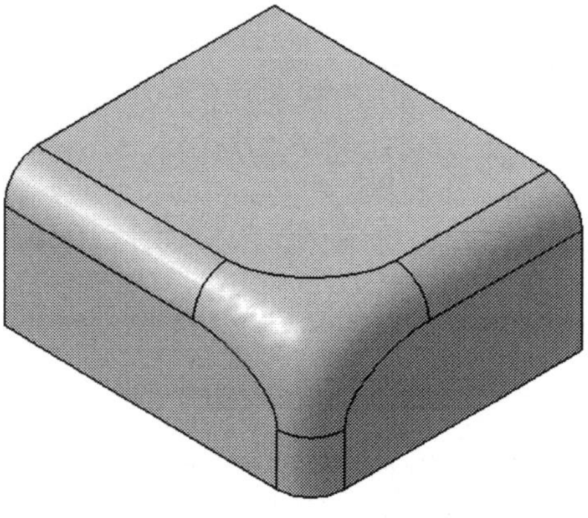

Figure 6-42 Resulting setback fillet

No internal sharp edge

While creating variable radius fillets, the software may generate unexpected sharp edges when the surfaces to be connected are continuous in tangency but not continuous in curvature. In order to improve the design, select the **No internal sharp edge** radio button. This will remove all possibly generated sharp edges.

Note

Make sure the setback distance is equal to or greater than the fillet radius. Else, the fillet will not be created.

Creating Face-Face Fillets

Menubar:	Insert > Dress-Up Features > Face-Face Fillet
Toolbar:	Dress-Up Features > Fillets sub-toolbar > Face-Face Fillet

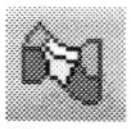

To fillet the selected faces of the model, invoke the **Face-Face Fillet** tool from the **Fillets** sub-toolbar in the **Dress-Up Features** toolbar; the **Face-Face Fillet Definition** dialog box will be displayed, as shown in Figure 6-43.

Select the first and second faces from the geometry area and then set the value of the radius of the fillet using the **Radius** spinner. Choose the **Preview** button from the **Face-Face Fillet Definition** dialog box. If the **Feature Definition Error** window is displayed, you need to modify the value of the fillet radius. Figure 6-44 shows the faces to be selected to create the face-face fillet and Figure 6-45 shows the resulting face-face fillet.

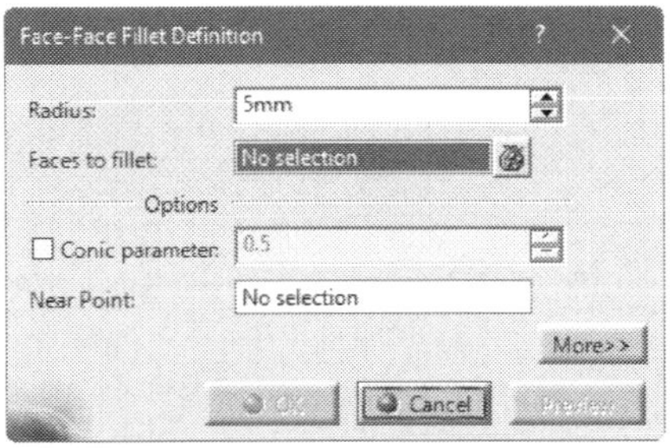

*Figure 6-43 The **Face-Face Fillet Definition** dialog box*

By using the **Near Point** selection box, you can maintain the curvature between the selected faces while creating the face-face fillet.

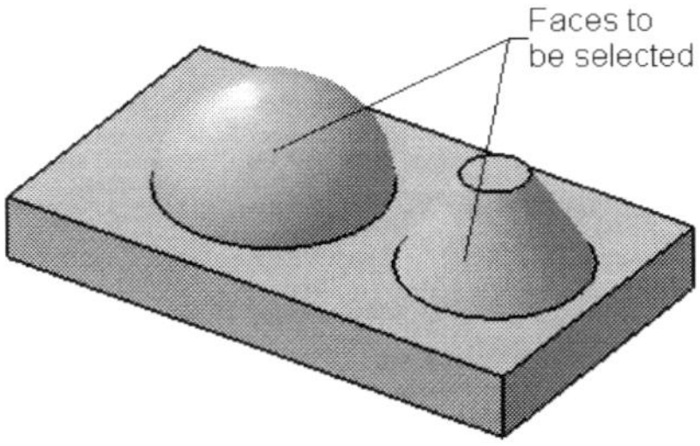

Figure 6-44 Faces to be selected

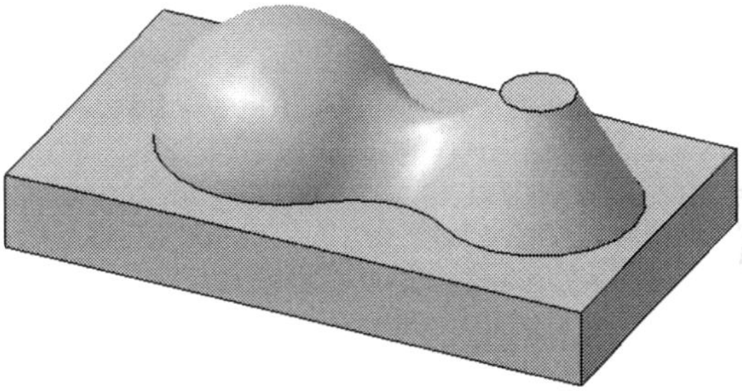

Figure 6-45 Resulting face-face fillet

You can also create a face-face fillet by using the hold curve and spine. Note that in this case you do not need to specify the radius for the fillet, as the fillet is controlled by the hold curve and the spine. To create a face-face fillet by using the hold curve and spine, expand the **Face-Face Fillet Definition** dialog box by choosing the **More** button. In the expanded dialog box, activate the **Hold Curve** selection box and then select the hold curve from the drawing area. As soon as you select the hold curve, the **Spine** display area of the expanded dialog box will be enabled and you will be prompted to select a spine. Select a spine from the drawing area. Next, choose the **OK** button from the dialog box. Figure 6-46 shows the hold curve and spine and Figure 6-47 shows the resulting face-face fillet. Note that the hold curve must be sketched on the selected face.

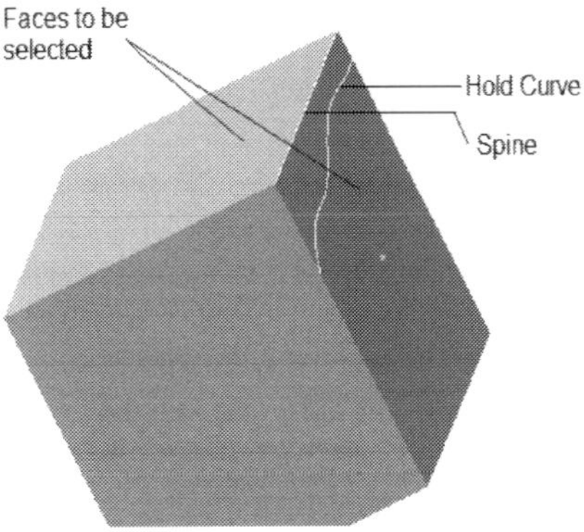

Figure 6-46 Hold curve and spine

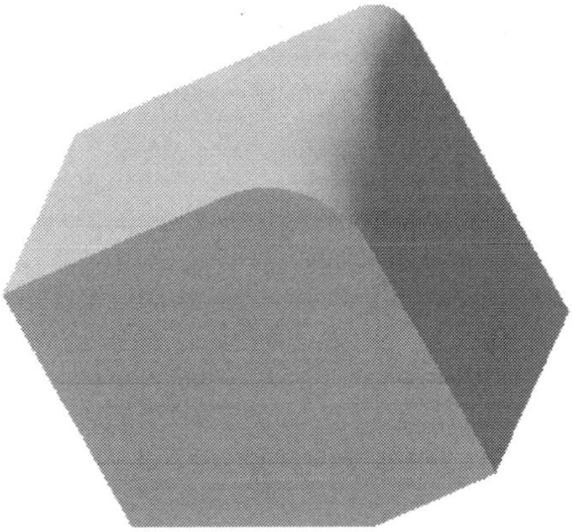

Figure 6-47 Resulting face-face fillet

Creating Tritangent Fillets

Menubar:	Insert > Dress-Up Features > Tritangent Fillet
Toolbar:	Dress-Up Features > Fillets sub-toolbar > Tritangent Fillet

To create a fillet feature that is tangent to three selected faces, invoke the **Tritangent Fillet** tool from the **Fillets** sub-toolbar in the **Dress-Up Features** toolbar; the **Tritangent Fillet Definition** dialog box will be displayed, as shown in Figure 6-48.

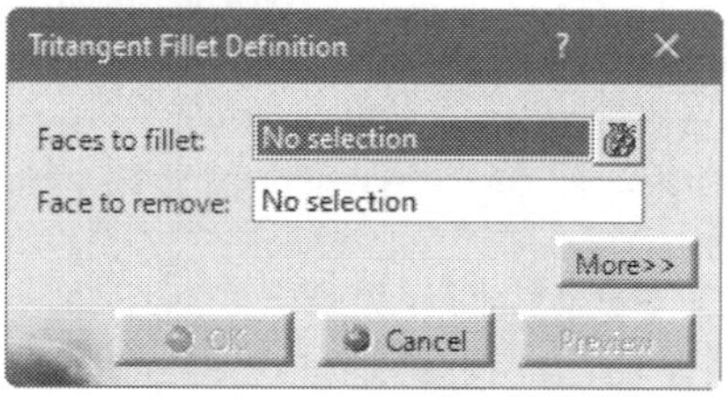

***Figure 6-48** The **Tritangent Fillet Definition** dialog box*

Also, you will be prompted to select the first face. Select the first face; you will be prompted to select the second face. Select the second face; you will be prompted to select the face to be removed. Select the face from the geometry area, refer to Figure 6-49. Choose the **Preview** button from the **Tritangent Fillet Definition** dialog box to preview the tritangent fillet. Figure 6-49 shows the faces to be selected and Figure 6-50 shows the resulting tritangent fillet.

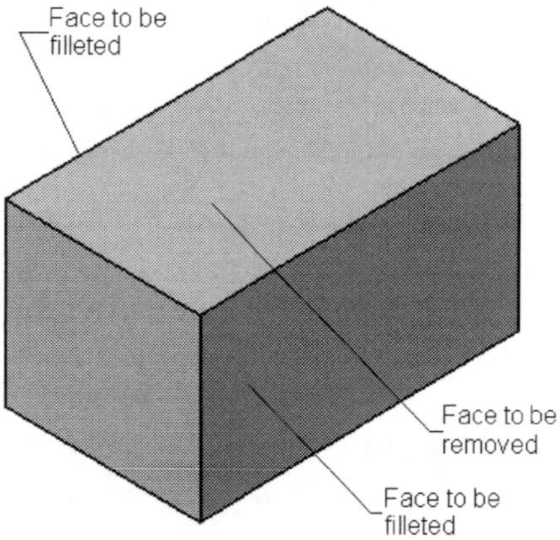

***Figure 6-49** Faces to be selected*

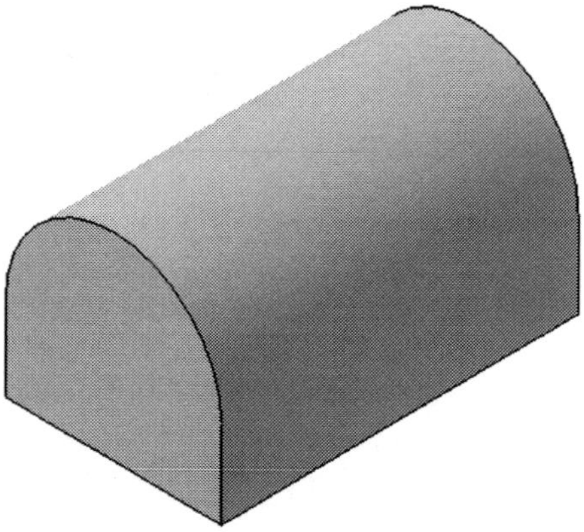

***Figure 6-50** Resulting tritangent fillet*

Creating Chamfers

Menubar:	Insert > Dress-Up Features > Chamfer
Toolbar:	Dress-Up Features> Chamfer

Chamfering is defined as a process by which the sharp edges are beveled in order to reduce the stress concentration in the model. This process also eliminates the sharp edges that are not desirable. To chamfer the edges of a model, choose the **Chamfer** tool from the **Dress-Up Features** toolbar; the **Chamfer Definition** dialog box will be displayed, as shown in Figure 6-51. Also, you will be prompted to specify the required data to define the chamfer.

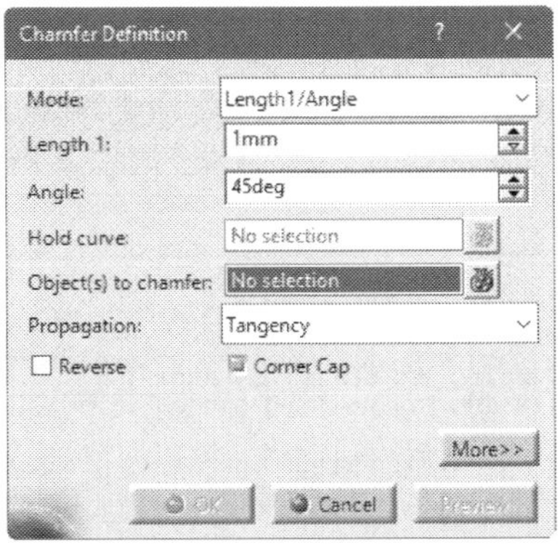

***Figure 6-51** The **Chamfer Definition** dialog box*

First, you need to select the edges or faces that are to be chamfered. If you select a face to chamfer, all edges of that face will be chamfered. The number of the selected elements are displayed in the **Object(s) to chamfer** display box. You can also use the Selection Manager button on the right of the **Object(s) to chamfer** display to filter the selection.

You will notice that the **Length1/Angle** option is selected by default in the **Mode** drop-down list. Therefore, you need to define the values of the length of the chamfer and its angle in the **Length 1** and **Angle** spinners, respectively. On selecting the **Length1/Length2** option from the **Mode** drop-down list, you will be prompted to define the value of the first and second lengths of the chamfer in the **Length 1** and **Length 2** spinners, respectively. If you select the **Symmetric extent** check box, then the **Length 2** spinner will become inactive and the length will be symmetric in both the directions. Note that this option will be available only when the **Length1/Length 2** option is selected in the **Mode** drop-down list. On selecting the **Chordal length/Angle** option from the **Mode** drop-down list, you will be prompted to define the values of chordal length and its angle in the **Chordal length** and **Angle** spinners, respectively. If you select the **Height/Angle** option from the **Mode** drop-down list, you will be prompted to define the values of height and angle in the **Height** and **Angle** spinners, respectively. On selecting the **Hold curve/Angle** option from the **Mode** drop-down list, you will be prompted to define the angle and hold curve in the **Angle** spinner and the **Hold curve** display box, respectively. On selecting the **Hold curve/ Length** option from the **Mode** drop-down list, you will be prompted to define the length and hold curve in the **Length** spinner and the **Hold curve** display box, respectively.

To chamfer all edges tangent to the selected edges, select the **Tangency** option from the **Propagation** drop-down list. To chamfer only the selected edge, select the **Minimal** option from the **Propagation** drop-down list. The **Reverse** check box is selected to flip the direction of the first length. Figure 6-52 shows the edge selected to be chamfered and Figure 6-53 shows the resulting chamfer. If you select the **Corner Cap** check box, you can create corner cap among the three selected edges.

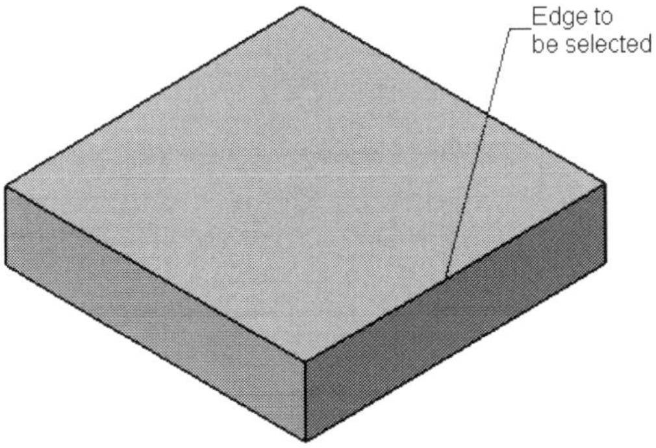

***Figure 6-52** Edge selected to be chamfered*

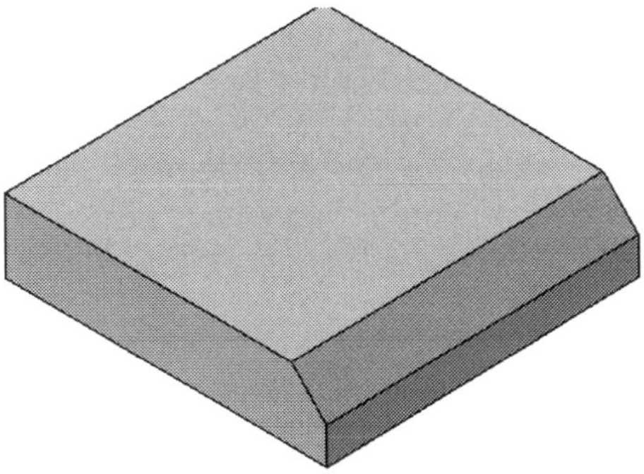

***Figure 6-53** Resulting chamfer*

Adding Draft to the Faces of the Model

Menubar: Insert > Dress-Up Features > Draft
Toolbar: Dress-Up Features > Drafts sub-toolbar > Draft Angle

Drafting is defined as the process of adding a taper angle to the faces of a model. Adding a draft to the faces of a model is one of the most important operations, especially while creating the components that need to be cast, molded, or formed. Draft angles enable components to be easily ejected from the die. The **Part** workbench of CATIA V5 provides you with various tools to draft the faces of the model. These tools are discussed next.

Adding a Simple Draft

The **Draft Angle** tool is the most widely used tool to add a draft to the faces of a model. To add a draft, invoke the **Draft Angle** tool from the **Drafts** sub-toolbar in the **Dress-Up Features** toolbar; the **Draft Definition** dialog box will be displayed, as shown in Figure 6-54. Also, an arrow will be displayed at the origin, pointing in the default pull direction.

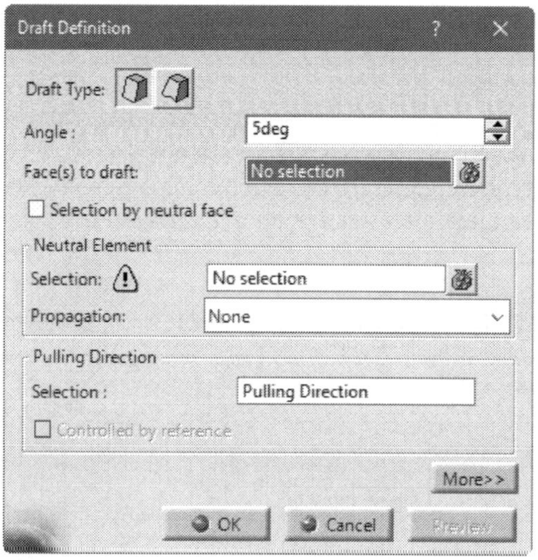

Figure 6-54 *The* ***Draft Definition*** *dialog box*

Select the faces from the geometry area on which you need to add the draft angle; the selected faces will be displayed in brown. The faces tangent to the selected face will be automatically selected. You can also filter the selection using the Selection Manager button on the right of the **Face(s) to draft** display box. Next, you need to define a neutral plane. Click once in the **Selection** display box in the **Neutral Element** area and then select a face or plane that will be defined as the neutral plane. By default, the **None** option is selected in the **Propagation** drop-down list. If you select the **Smooth** option, the faces tangent to the selected face will also be selected automatically as the neutral element. Then, set the value of the draft angle in the **Angle** spinner and choose the **OK** button. Figure 6-55 shows the faces to be drafted and the face to be selected as the neutral plane. Figure 6-56 shows the resulting drafted faces.

Figure 6-57 shows the xy plane to be selected as the neutral plane and Figure 6-58 shows the resulting drafted faces.

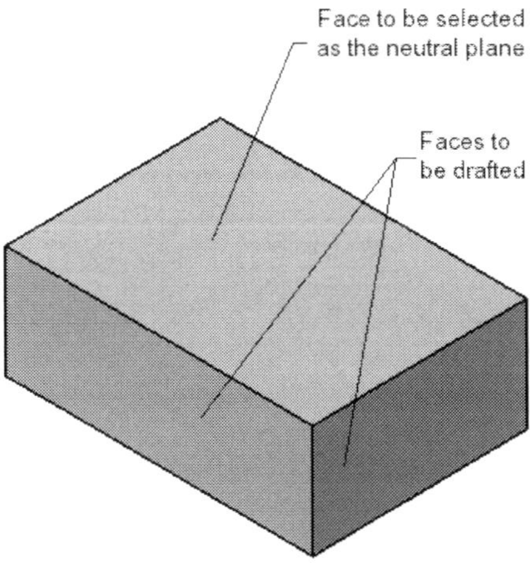

Figure 6-55 Faces to be selected

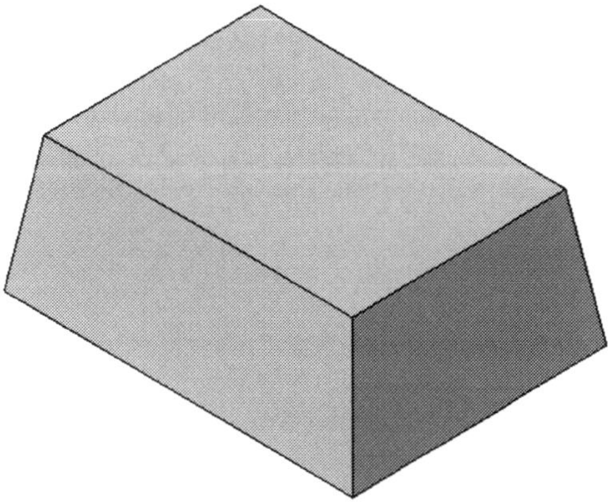

Figure 6-56 Resulting drafted faces

Tip
To add a draft to all faces which are in contact with the neutral face, instead of selecting all the faces one by one, you can select the ***Selection by neutral face*** *check box and select the neutral face.*

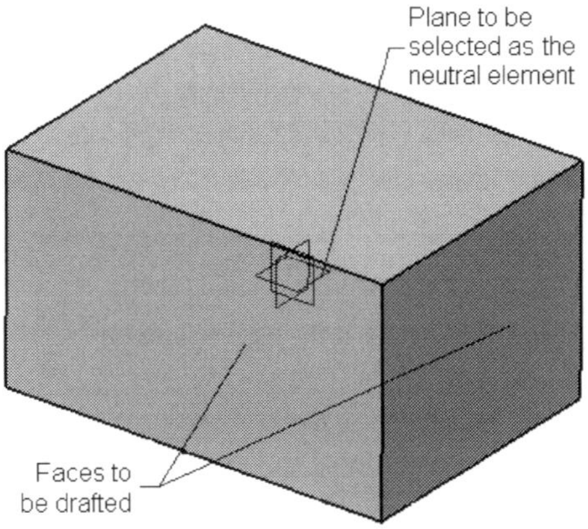

Figure 6-57 Faces and planes to be selected

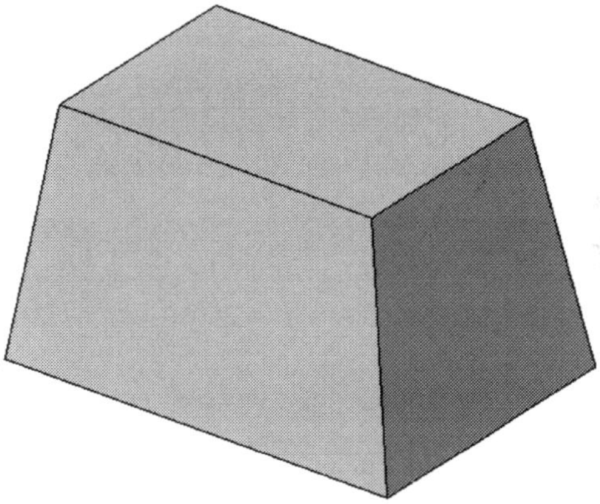

Figure 6-58 Resulting drafted faces

Defining the Parting Element while Adding Drafts to the Faces

You can also define the parting element while drafting the faces of the model. To define it, choose the **More** button from the **Draft Definition** dialog box to expand it. If you choose the **Parting = Neutral** check box from the **Parting Element** area, the neutral element will be selected as the parting element. Consider the case shown in Figure 6-57, in which a plane passing through the center of the model is selected as the neutral plane. Figure 6-59 shows the faces drafted with the **Parting = Neutral** check box selected. Note that when you select this check box, the **Draft both sides** check box will be enabled. Select this check box to add the draft to both sides of the parting element, refer to Figure 6-60.

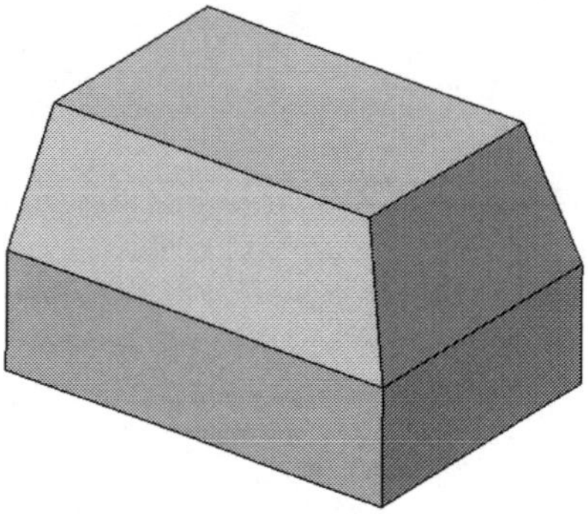

Figure 6-59 *Faces drafted with the* ***Parting= Neutral*** *check box selected*

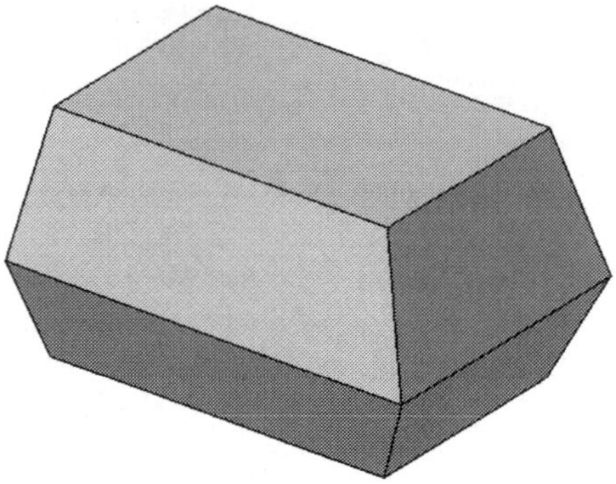

Figure 6-60 *Parting element after adding draft to both side*

You can also select a user-defined parting element other than the neutral plane. To select a user-defined parting element, select the **Define parting element** check box from the **Parting Element** area and select the parting element from the geometry area. You can also create the parting element using the options from the shortcut menu displayed on right-clicking in the **Selection** display box. Now, set the other parameters of the draft and choose the **OK** button from the **Draft Definition** dialog box. Figure 6-61 shows the faces to be drafted, neutral plane, and the parting plane. Figure 6-62 shows the resulting drafted faces.

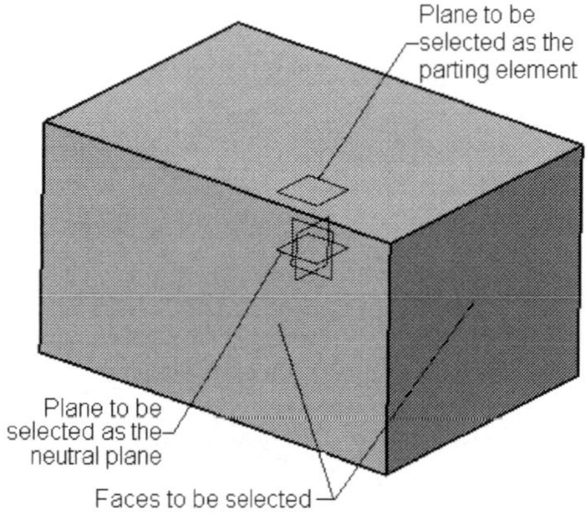

Figure 6-61 *Faces and planes to be selected*

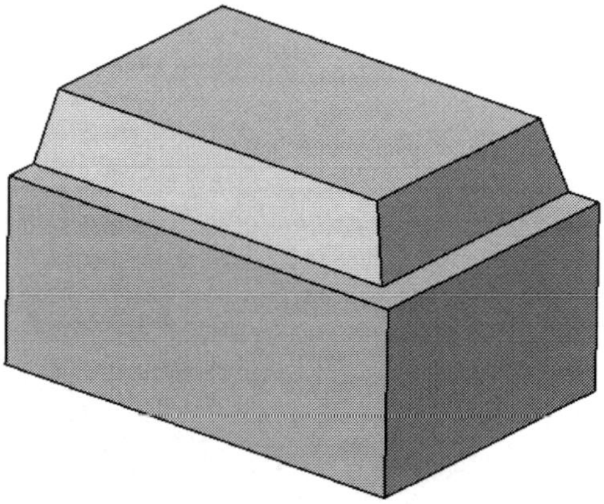

Figure 6-62 *Resulting drafted faces*

You can also define the limit elements while adding a draft to the faces of the model. To do so, click once in the **Limiting Element(s)** selection box and select the limiting elements from the geometry area. You need to make sure that if you specify two limiting elements, the feature is created in the opposite directions. Figure 6-63 shows the limiting elements to be selected and Figure 6-64 shows the resulting draft feature.

Tip

By default, the pulling direction is selected along the Z axis direction. You can also specify a user-defined pulling direction by clicking once in the ***Pulling Direction*** *selection box and then selecting the pulling direction from the geometry area.*

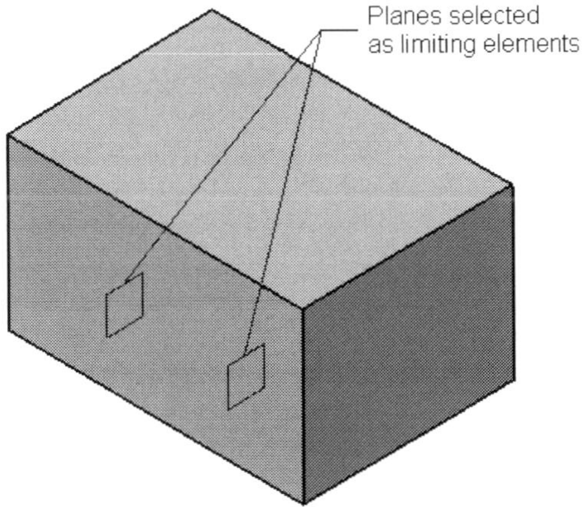

Figure 6-63 *Limiting elements to be selected*

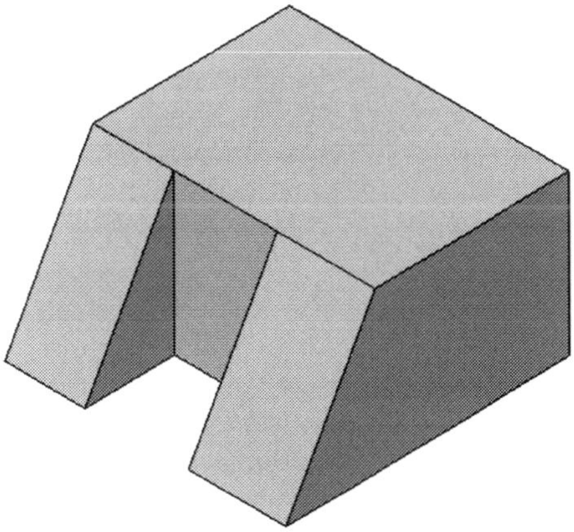

Figure 6-64 *Resulting drafted feature*

Adding Drafts Using the Reflect Line

Menubar: Insert > Dress-Up Features > Draft Reflect Line
Toolbar: Dress-Up Features > Drafts sub-toolbar > Draft Reflect Line

The **Draft Reflect Line** tool is used to create the draft feature using the silhouette lines of the selected curved face as the neutral element. To create this type of draft feature, invoke the **Draft Reflect Line** tool from the **Drafts** sub-toolbar in the **Dress-Up Features** toolbar; the **Draft Reflect Line Definition** dialog box will be displayed, as shown in Figure 6-65. Select a curved face from the geometry area. You can also filter the selection using the Selection Manager button. The faces tangent to the selected face are also selected automatically. You will notice that a pink color sketch will be created along the silhouette of the selected face. Now, expand the dialog box by using the **More** button and click on the selection box adjacent to the **Selection** option in the **Parting Element** area; you will be prompted to select the parting element. Select the plane or planar face that will be used as the parting element. Set the value of the draft angle and choose the **OK** button from the **Draft Reflect Line Definition** dialog box. Figure 6-66 shows the face to be drafted and the plane to be selected as the parting element. Figure 6-67 shows the resulting drafted feature.

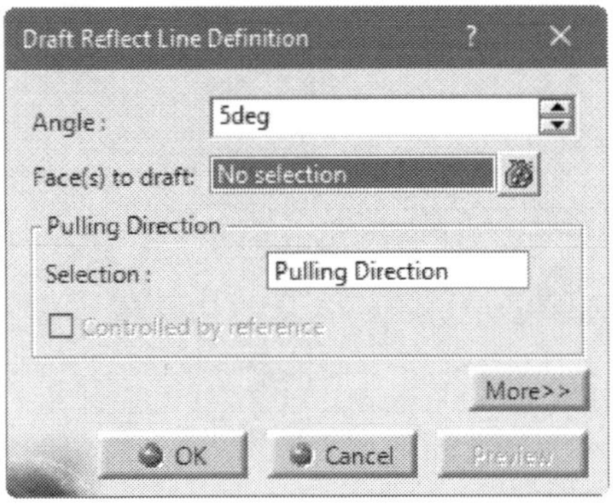

Figure 6-65 *The* ***Draft Reflect Line Definition*** *dialog box*

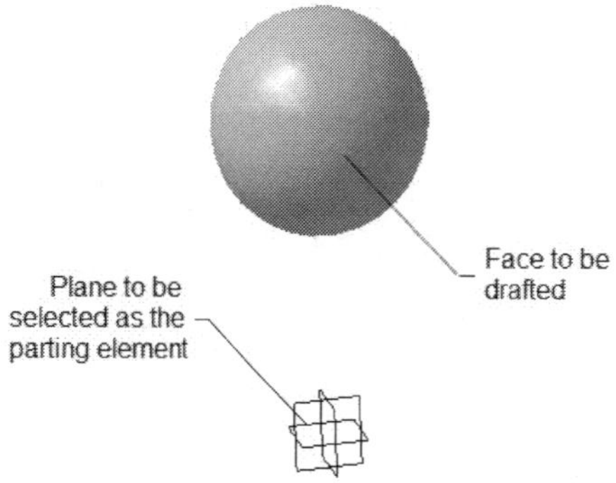

Figure 6-66 *Face to be drafted and plane to be selected*

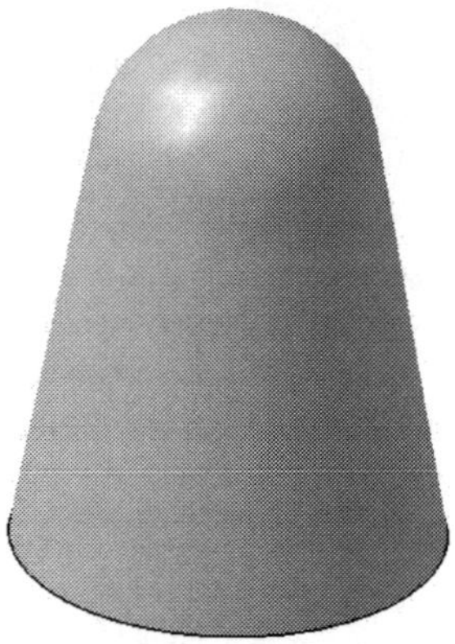

Figure 6-67 *Resulting drafted feature*

Adding a Variable Angle Draft

Menubar: Insert > Dress-Up Features > Variable Angle Draft
Toolbar: Dress-Up Features > Drafts sub-toolbar > Variable Angle Draft

To create a variable angle draft, invoke the **Variable Angle Draft** tool from the **Drafts** sub-toolbar in the **Dress-Up Features** toolbar; the **Draft Definition** dialog box will be displayed, as shown in Figure 6-68. Also, you will be prompted to select the face to draft.

Select the face on which you want to add the variable angle draft. You can select only one face for adding a draft using this tool. Click in the **Selection** selection box in the **Neutral Element** area and then define the neutral element by selecting a plane or face. You will notice that two angular dimensions are displayed attached to the end points of the selected face. One by one, select both the angles and set their values using the **Angle** spinner. You can also filter the selections using the Selection Manager button. Figure 6-69 shows the references to be selected and Figure 6-70 shows the resulting face after adding the draft.

Figure 6-68 *The* ***Draft Definition*** *dialog box*

You can also define additional points to specify other variable angles. Note that the point should only be selected on the edge from which the angle will be measured. To define an additional point, click in the **Points** selection box and then click anywhere on the edge from which the angle will be measured. If you want to define points whose distances need to be controlled, right-click in the **Points** selection box to invoke the contextual menu. Create additional points and then set the draft angle by using the options in the contextual menu.

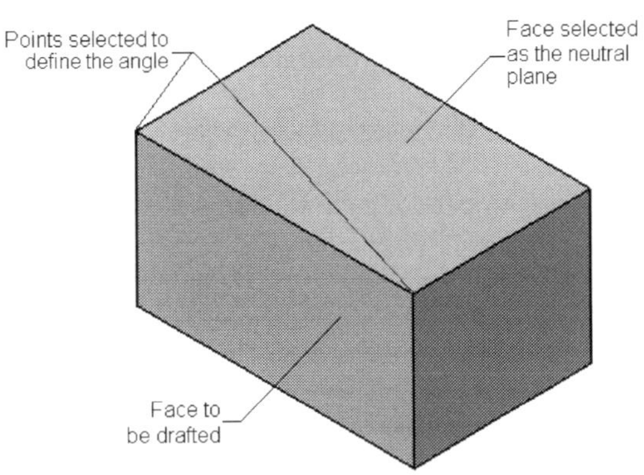

Figure 6-69 *References to be selected*

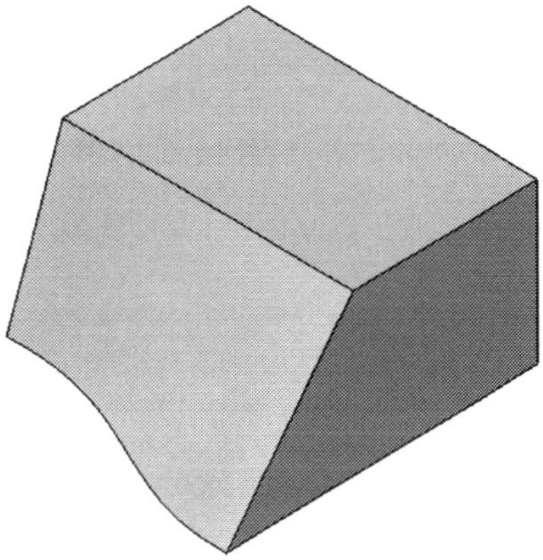

Figure 6-70 *Resulting face after adding the draft*

Creating Shell Feature

Menubar:	Insert > Dress-Up Features > Shell
Toolbar:	Dress-Up Features > Shell

The **Shell** tool is used to scoop out the material from the model and remove the selected faces, thereby resulting in a thin walled structure. To create a shell feature, invoke the **Shell** tool from the **Dress-Up Features** toolbar; the **Shell Definition** dialog box will be displayed, as shown in Figure 6-71.

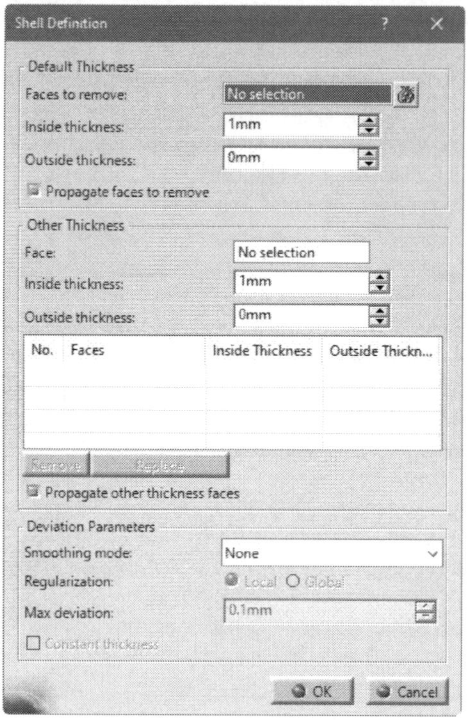

Figure 6-71 *The* ***Shell Definition*** *dialog box*

Next, you need to select the face or faces to be removed. Select them from the geometry area. When the **Propagate faces to remove** check box is selected, the faces tangent to the selected face are selected automatically. You can filter the selection using the Selection Manager button. Now, set the value of the wall thickness in the **Inside thickness** spinner from the **Default Thickness** area in the **Shell Definition** dialog box. You can also define the outside thickness of the shell using the **Outside thickness** spinner from the **Default Thickness** area. Now, choose the **OK** button from the **Shell Definition** dialog box; the shell feature will be created. Figure 6-72 shows the faces to be removed and Figure 6-73 shows the resulting shelled model. If you do not select any faces to be removed, the resulting shelled model will be a hollow model with the specified wall thickness.

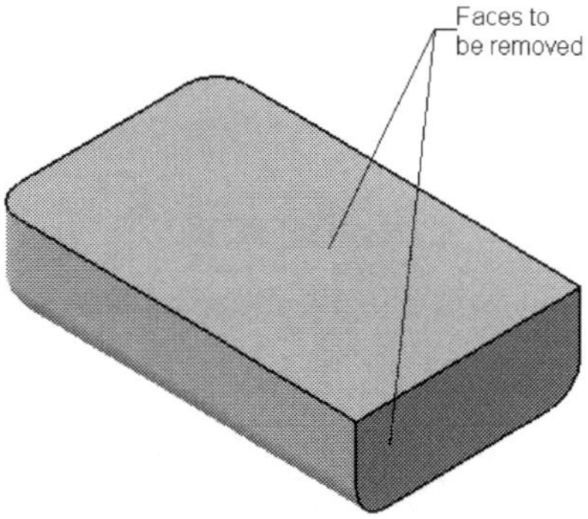

Figure 6-72 *Faces selected for removal*

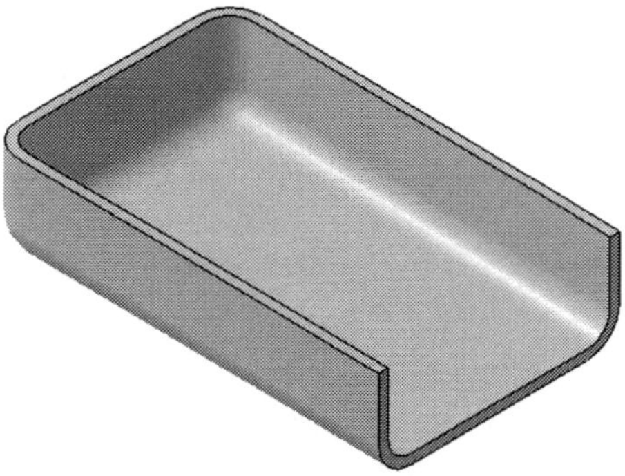

Figure 6-73 *Resulting shelled model*

Creating a Multi-thickness Shell

You can also define different shell thickness values for the faces of the shell feature. To create a multi-thickness shell feature, first select the faces to be removed and then specify the inside or outside thickness of the shell. To do so, click once on the **Face** display box in the **Other Thickness** area and then select the faces on which you need to define different shell thickness values. If the **Propagate other thickness faces** check box is selected, the other faces tangent to the other thickness faces are selected automatically. The selected faces will be highlighted in brown and the shell thickness dimensions will be attached to them. Now, select the thickness value of one of the highlighted faces from the geometry area; the selected value will be displayed in the **Inside thickness** spinner of the **Other Thickness** area of the **Shell Definition** dialog box. Modify the thickness value and repeat the process for the remaining highlighted faces. After setting all shell thickness values, choose the **OK** button from the **Shell Definition** dialog box. Figure 6-74 shows the face to be removed and the faces to define different shell thicknesses and Figure 6-75 shows the resulting shelled model.

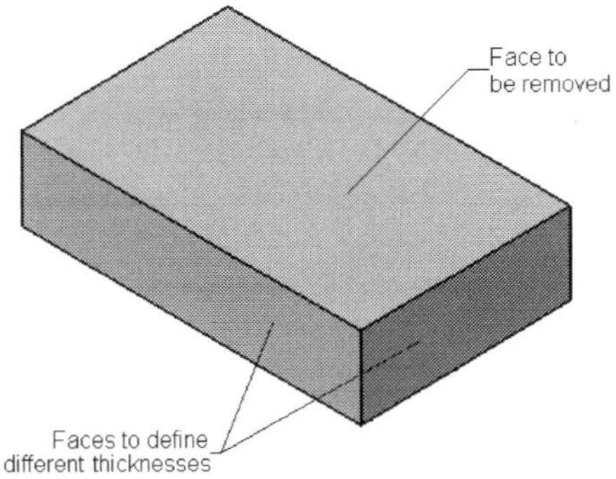

***Figure 6-74** Faces to be selected*

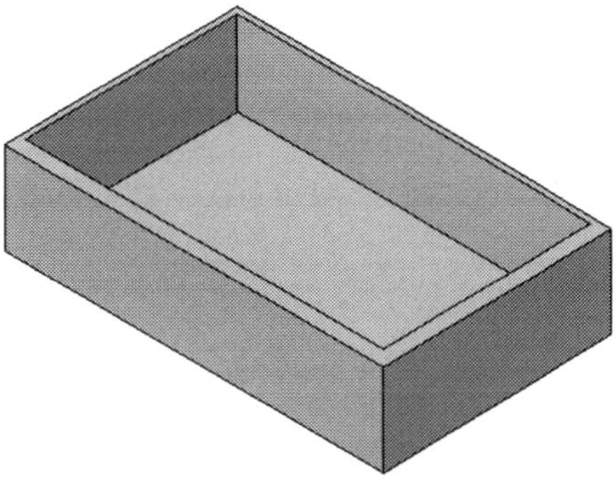

***Figure 6-75** Resulting shelled model*

Adding Thickness

Menubar:	Insert > Dress-Up Features > Thickness
Toolbar:	Dress-Up Features > Thickness

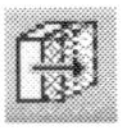

The **Thickness** tool is used to add or remove thickness to the selected faces of the part. Adding or removing thickness to a face is required sometimes before machining the part. To apply thickness to a face, choose the **Thickness** tool from the **Dress-Up Features** toolbar; the **Thickness Definition** dialog box will be displayed, as shown in Figure 6-76.

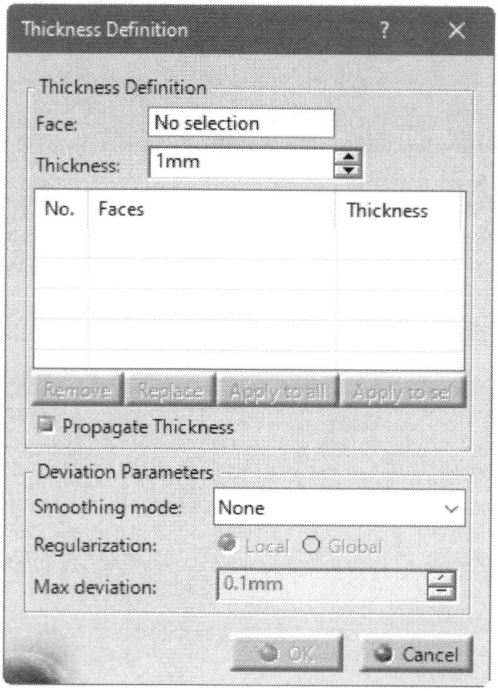

***Figure 6-76** The **Thickness Definition** dialog box*

Select the faces to apply thickness; the selected faces will be highlighted and the thickness value will be displayed in the geometry area, refer to Figure 6-77. Now, enter the value in the **Thickness** spinner in the **Default Thickness** area. When the **Propagate Thickness** check box is selected, the faces tangent to the selected face gets thicken automatically. Next, choose **OK**; the thickness value will be added to the selected faces, refer to Figure 6-78. A positive thickness value will add material to the feature whereas a negative value will remove material.

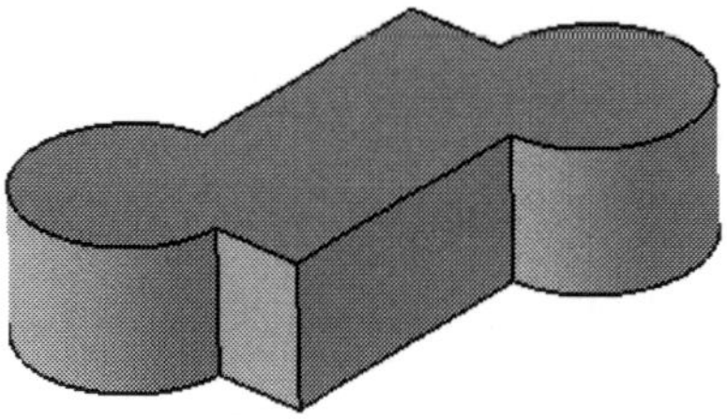

***Figure 6-77** Faces to be selected*

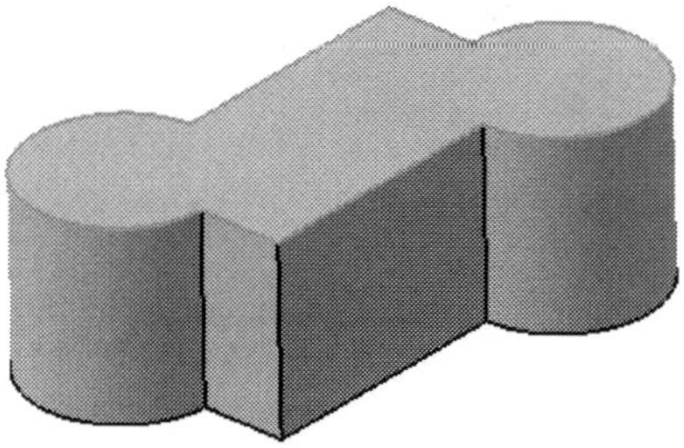

***Figure 6-78** Thickness added to the face*

Removing Faces

Menubar:	Insert > Dress-Up Features > Remove Face
Toolbar:	Dress-Up Features > Remove Face

The **Remove Face** tool is used to remove the selected faces of a solid part. Choose this tool from the **Dress-Up Features** toolbar; the **Remove Face Definition** dialog box will be displayed, as shown in Figure 6-79. Select the faces you want to remove in the **Faces to**

remove selection box; the selected faces turn purple indicating that they will be removed, refer to Figure 6-80. Select the faces to keep in the **Faces to keep** selection box; the faces get highlighted indicating that they will not be removed, refer to Figure 6-81. Click the **Show all faces to remove** check box to preview all the faces adjacent to the purple face that will be removed. Click **OK** to complete the feature, refer to Figure 6-82.

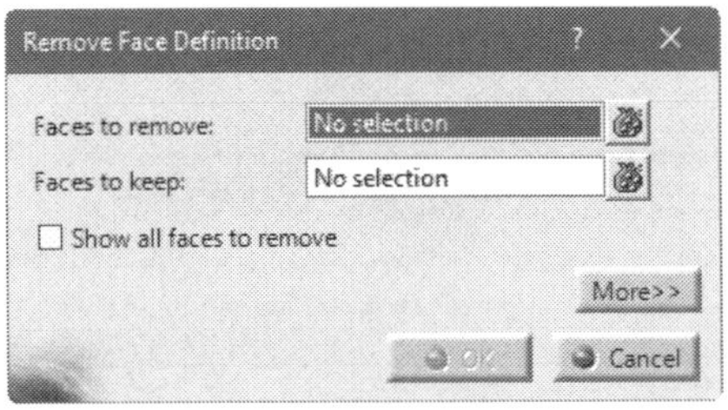

Figure 6-79 *The **Remove Face Definition** dialog box*

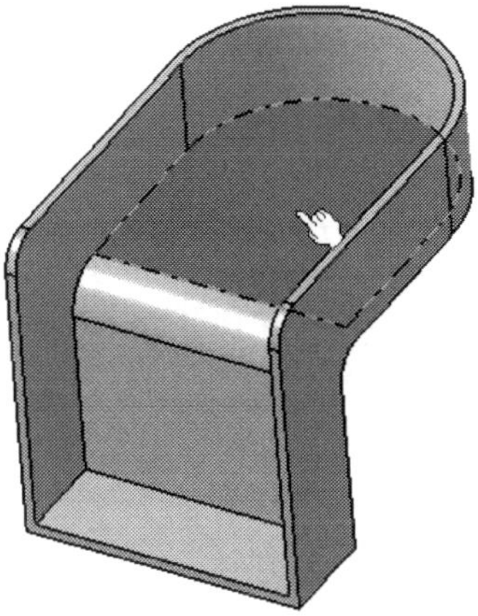

Figure 6-80 *Faces selected to remove*

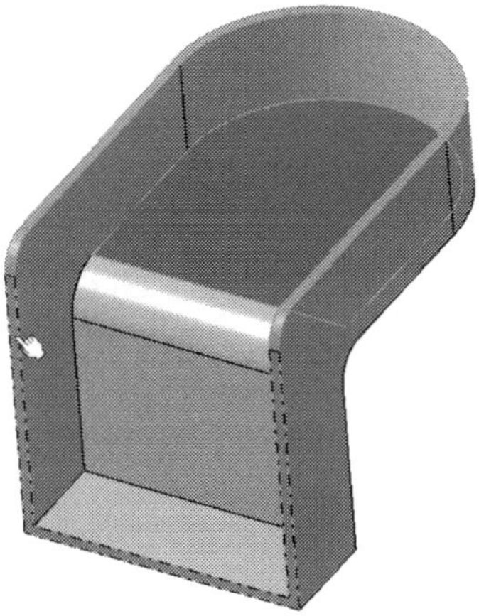

Figure 6-81 *Faces selected to keep*

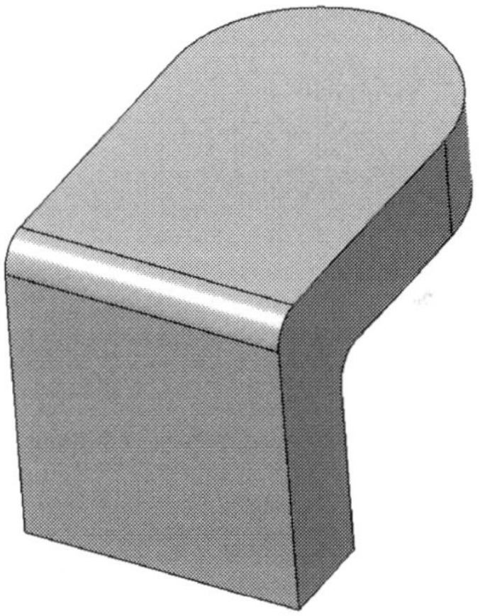

Figure 6-82 *Face removed from the solid*

Replacing Faces

Menubar:	Insert > Dress-Up Features > Replace Face
Toolbar:	Dress-Up Features > Remove face sub-toolbar >Replace Face

The **Replace Face** tool is used to replace a face or a set of tangent faces with a surface or a face belonging to the same body as the selected face. Invoke the **Replace Face** tool from the **Dress-Up Features** toolbar; the **Replace Face Definition** dialog box will be displayed, as shown in Figure 6-83. Select the surface as the replacing surface. Click on the arrow to reverse the direction of face being replaced if it is not in the desired direction. Now select the faces you want to replace in the **Face to remove** selection box; the selected faces turn purple indicating that it will be replaced, refer to Figure 6-84. Click **OK** to complete the feature, refer to Figure 6-85.

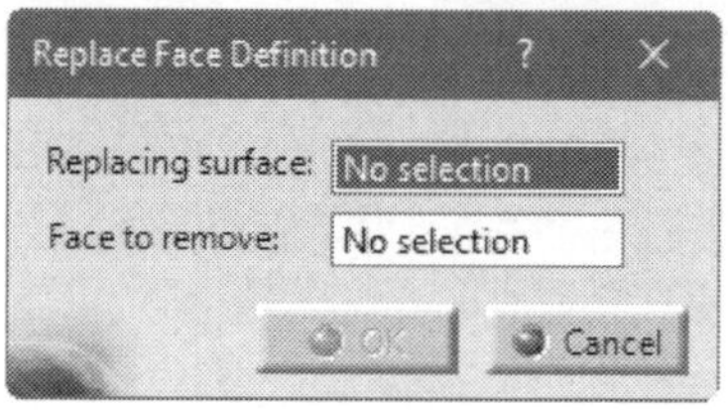

Figure 6-83 *The* ***Replace Face Definition*** *dialog box*

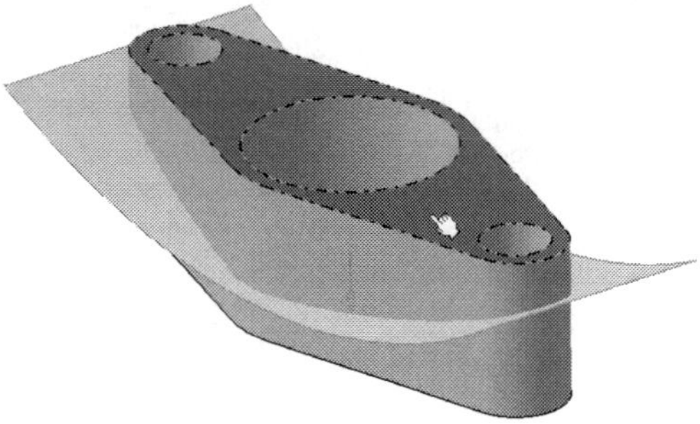

Figure 6-84 *Face selected to replace*

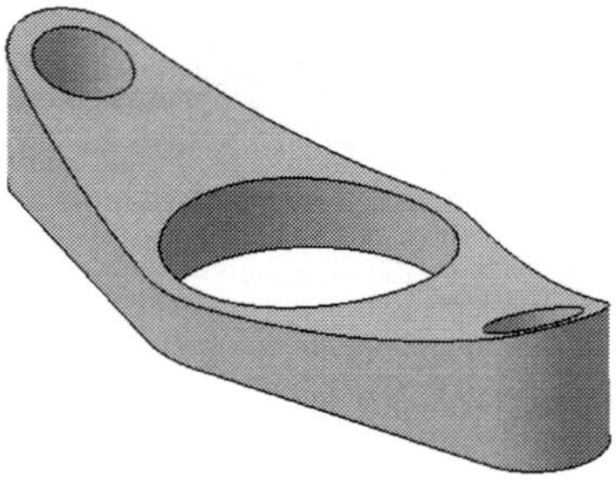

Figure 6-85 *Face replaced by the surface*

TUTORIALS

Tutorial 1

In this tutorial, you will create the model of the nozzle of a vacuum cleaner, as shown in Figure 6-86. Its views and dimensions are shown in Figure 6-87. **(Expected time: 45 min)**

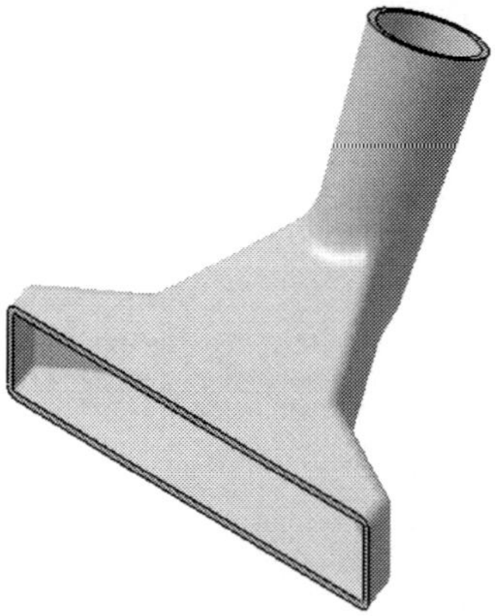

Figure 6-86 *Model for Tutorial 1*

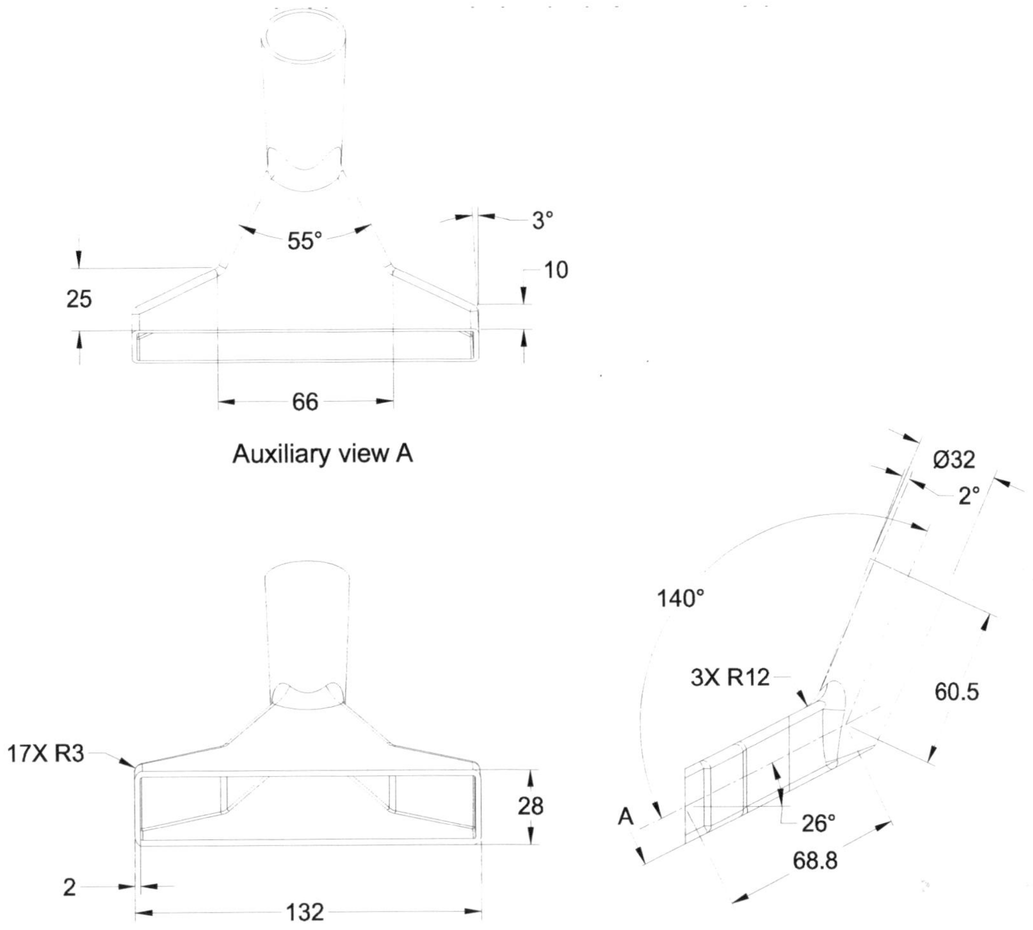

Figure 6-87 Views and dimensions for Tutorial 1

The following steps are required to complete this tutorial:

a. Start a new file in the **Part** workbench and create the base feature of the model by extruding the sketch along the selected direction.
b. Create the second feature of the model by extruding a sketch using the **Drafted Fillet Pad** tool.
c. Apply fillets to all edges of the model.
d. Shell the model using the **Shell** tool.

Creating the Base Feature of the Model

The base feature of this model will be created by extruding the sketch drawn on the zx plane. The sketch will be extruded equally on both the sides of the sketching plane using the **Mirrored extent** option.

1. Start a new file in the **Part** workbench. Select the zx plane and invoke the **Sketcher** workbench.

2. Draw the sketch, as shown in Figure 6-88, and then exit the **Sketcher** workbench.

3. Choose the **Pad** tool from the **Sketch-Based Features** toolbar; the **Pad Definition** dialog box is displayed.

4. Select the **Mirrored extent** check box and set the value of the **Length** spinner to **66**; the preview of the extruded feature is displayed in the geometry area. Choose **OK** from the dialog box.

5. Invoke the **Sketcher** workbench after selecting the upper face of the base feature as sketching plane and draw the sketch of the pocket feature, as shown in Figure 6-89, and then exit the **Sketcher** workbench.

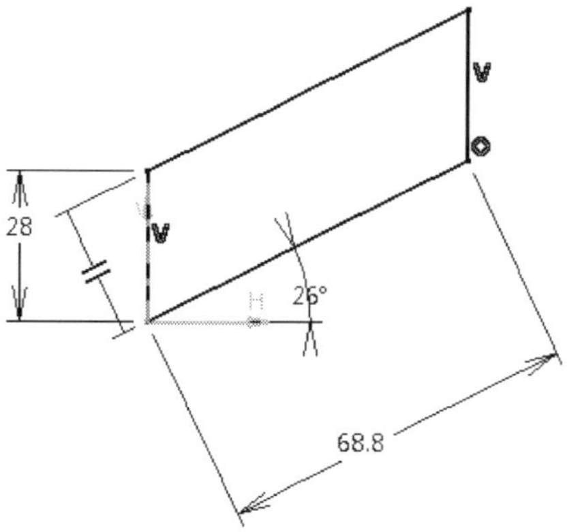

Figure 6-88 *Reference sketch*

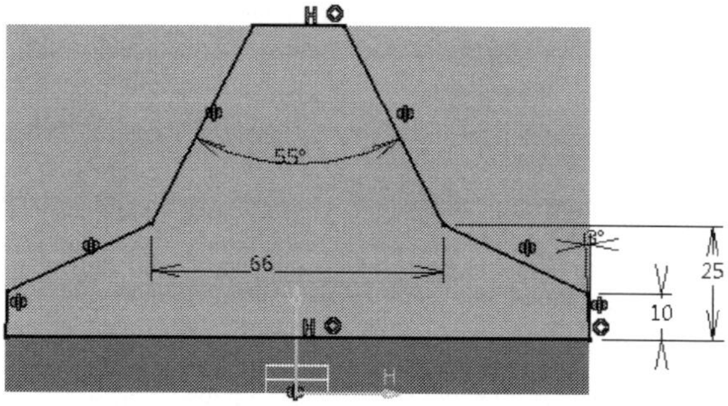

Figure 6-89 *Sketch of the pocket feature*

6. Choose the **Pocket** tool from the **Sketch-Based Features** toolbar; the **Pocket Definition** dialog box is displayed.

7. Choose the **Reverse Side** button from the **Profile/Surface** area of this dialog box and choose the **Up to next** option from the **First limit** area of this dialog box; the preview of the extruded feature is displayed in the geometry area.

8. Now, choose the **More** button to expand the **Pocket Definition** dialog box.

9. Clear the **Normal to profile** check box located in the **Direction** area and select the edge of the base feature, as shown in Figure 6-90.

10. Choose the **Preview** button form this dialog box. The preview is displayed. Choose the **OK** button from the **Pocket Definition** dialog box to complete the feature creation. The model after creating the pocket feature is shown in Figure 6-91.

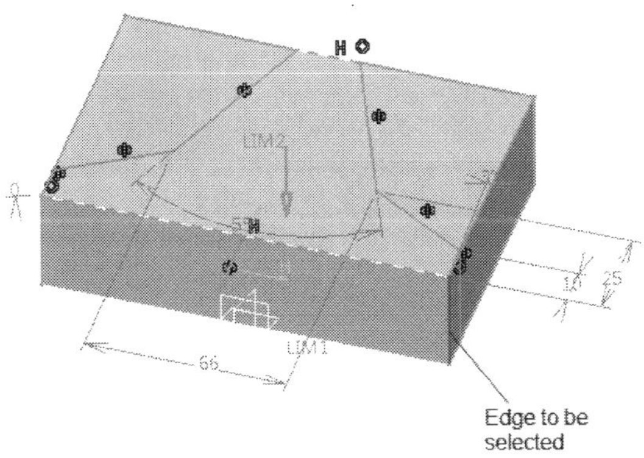

Figure 6-90 Reference edge to be selected

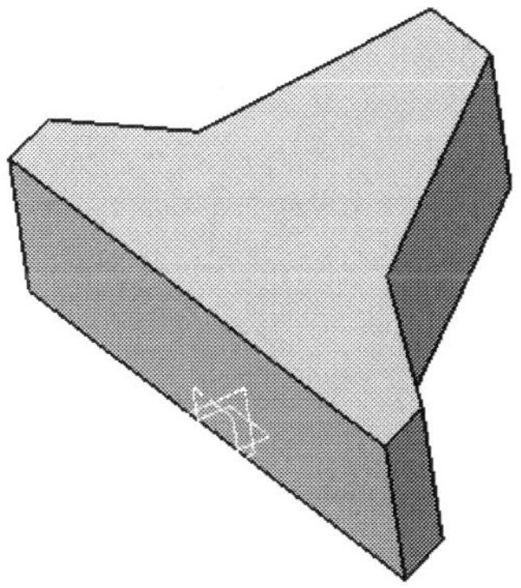

Figure 6-91 Model after creating the pocket feature

Creating the Second Feature

The second feature of this model is a drafted extruded feature created using the **Drafted Filleted Pad** tool. To create this feature, you need to create a plane. Next, you will create the sketch for the draft feature on the created plane, and then extrude the sketch.

1. Select the zx plane and invoke the **Sketcher** workbench. Draw the sketch, as shown in Figure 6-92, and then exit the **Sketcher** workbench.

2. Invoke the **Plane** tool and select the **Normal to curve** option from the **Plane type** drop-down list; you are prompted to select the reference curve.

3. Now, select the reference curve from the geometry area and then select the upper endpoint of the reference curve as the point on which the plane needs to be created, refer to Figure 6-93. The preview of the plane displayed in the geometry area is shown in Figure 6-93.

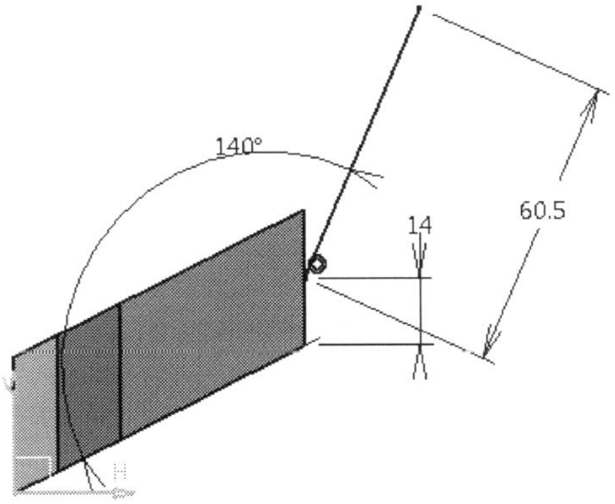

Figure 6-92 Sketch of the reference line

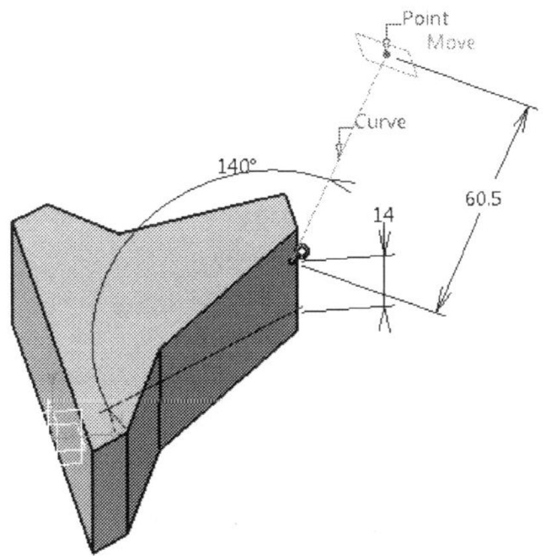

Figure 6-93 Reference line and the endpoint to be selected

4. Choose the **OK** button from the **Plane Definition** dialog box to create a plane and exit the dialog box.

5. Use the newly created plane to invoke the **Sketcher** workbench and draw the sketch of the second feature, as shown in Figure 6-94.

6. Exit the **Sketcher** workbench and invoke the **Drafted Filleted Pad** tool from the **Pads** toolbar.

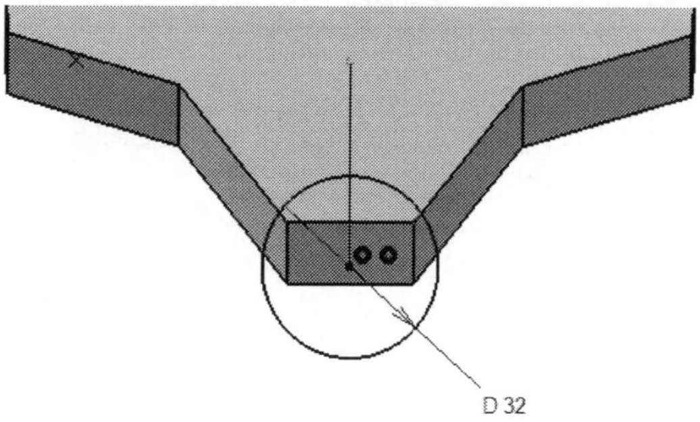

Figure 6-94 *Sketch for the second feature*

7. Set the value in the **Length** spinner to **0** and then select the xy plane from the Specification tree as the second limit.

8. Set the value of the draft angle in the **Angle** spinner to **-2deg**.

9. If the arrow is pointing upward in the preview, then choose the **Reverse Direction** button to change the direction. Clear all the check boxes in the **Fillets** area and choose the **OK** button from the **Drafted Filleted Pad Definition** dialog box.

   Now, you need to cut the extra portion of the second feature.

10. Select the bottom face of the base feature as the sketching plane to invoke the sketching environment.

11. Draw a rectangle arbitrarily to cut the portion of the drafted feature and exit the sketch, refer to Figure 6-95. Make sure that the rectangle covers both the features.

12. Invoke the **Pocket** tool from the **Sketch-Based Features** toolbar; the **Pocket Definition** dialog box is displayed.

13. Select the sketch and reverse the direction, if needed and choose the **OK** button from the **Pocket Definition** dialog box. The model after creating the second feature is shown in Figure 6-96.

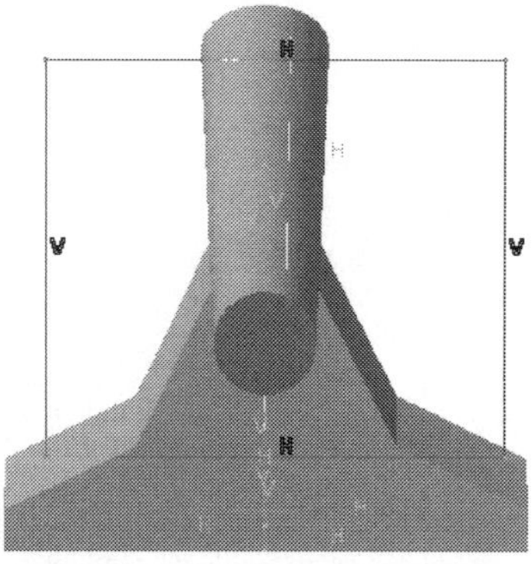

Figure 6-95 *Rectangle drawn to create a pocket*

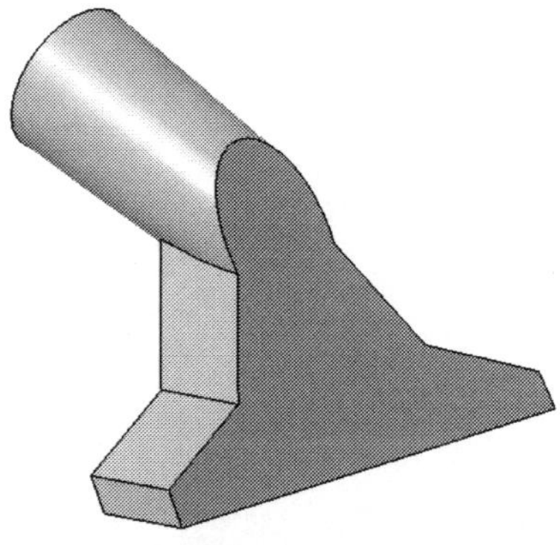

Figure 6-96 *Resulting second feature*

Filleting the Edges of the Model

You need to fillet two sets of edges of the model. Therefore, you need to apply the fillet feature twice because the two sets of edges need different fillet radii. First, you will fillet the set of edges for which the fillet radius is 3.

1. Double-click on the **Edge Fillet** tool from the **Dress-Up Features** toolbar; the **Edge Fillet Definition** dialog box is displayed.

2. Select the edges, as shown in Figure 6-97, and set value **3** in the **Radius** spinner. In this figure, the transparency of the model has been changed for better visualization. Also, model has rotated for better view of the edges to be filleted.

3. Choose the **OK** button from the **Edge Fillet Definition** dialog box. To change the display of the model, choose the **Shading with Edges without Smooth Edges** button from the **View mode** sub-toolbar in the **View** toolbar. The model after creating the first set of fillet and changing its display mode is shown in Figure 6-98.

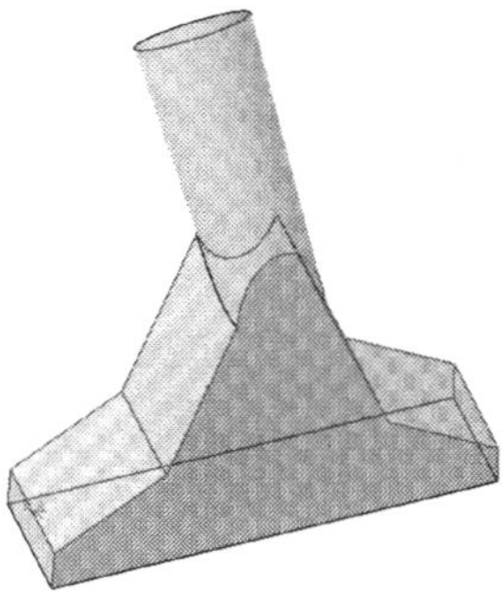

Figure 6-97 Edges selected to be filleted

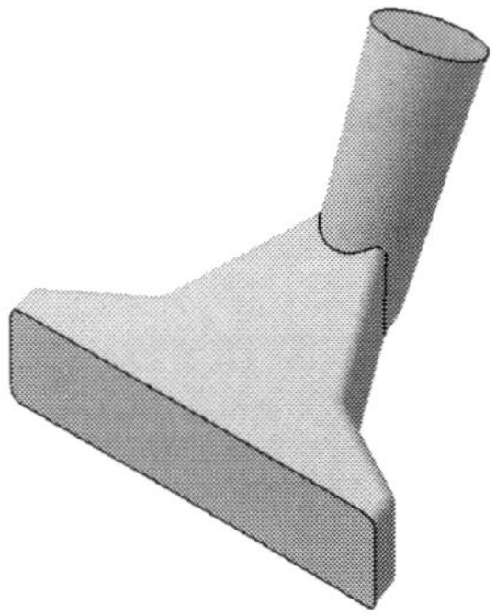

Figure 6-98 The model after creating the fillet

Next, you need to apply the fillet to the second set of edges.

4. Select the edges of the model as shown in Figure 6-99. In this figure, the transparency in the model has been changed for better visualization.

5. Set the value in the **Radius** spinner to **12** and choose the **OK** button from the **Edge Fillet Definition** dialog box. Choose the **Cancel** button from this dialog box, if it is displayed again. The model after applying the fillet to the second set of edges is shown in Figure 6-100.

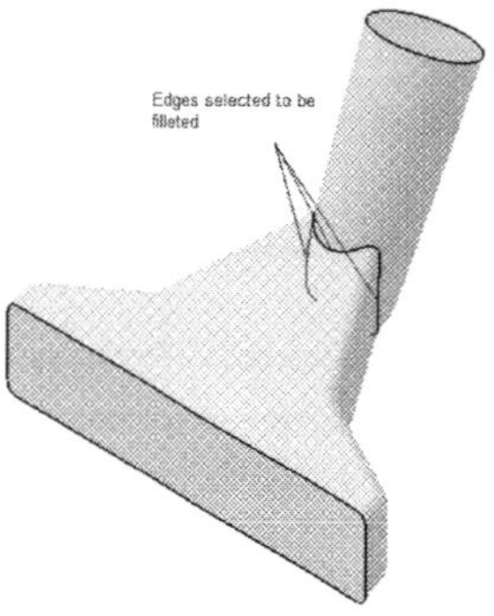

Figure 6-99 Edges be selected

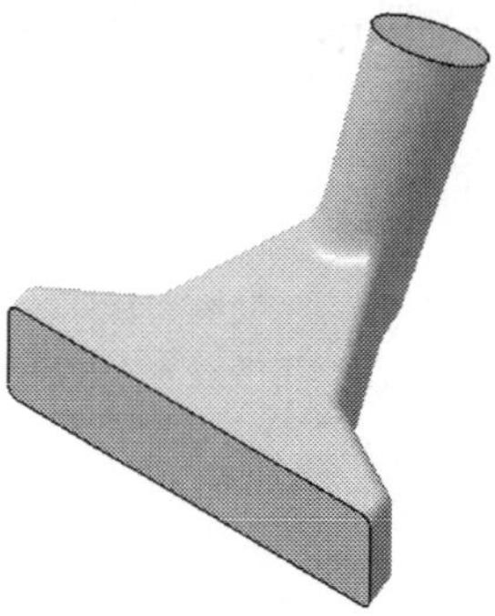

Figure 6-100 Model after creating the second fillet

Creating the Shell Feature

Finally, you need to create the shell feature. The shell feature will also be used to remove the end faces of the model leaving behind a thin wall structure.

1. Choose the **Shell** tool from the **Dress-Up Features** toolbar; the **Shell Definition** dialog box is displayed.

2. Select the faces to be removed, as shown in Figure 6-101.
3. Set the value in the **Inside thickness** spinner to **2** in the **Default Thickness** area and choose the **OK** button from the **Shell Definition** dialog box; the shell feature is created. The final model after creating the shell feature is shown in Figure 6-102.

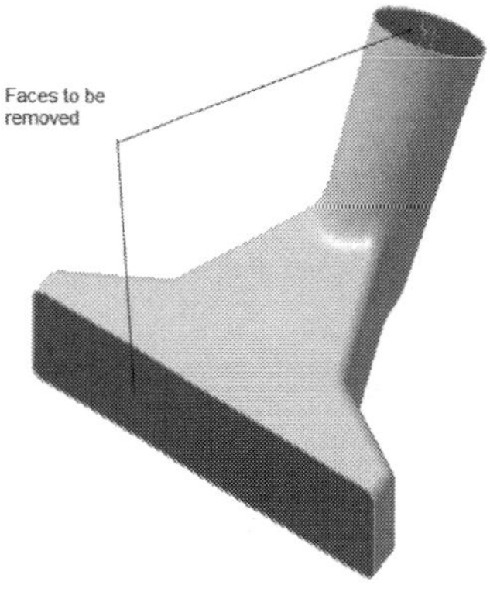

Figure 6-101 Faces to be removed

Figure 6-102 Final model after shelling

Saving and Closing the File

1. Choose the **Save** button from the **Standard** toolbar to invoke the **Save As** dialog box. Create a folder with the name *c06* in the *CATIA* folder.

2. Enter the name of the file as **c06tut1** in the **File name** edit box and choose the **Save** button; the file is saved at *C:\CATIA\c06*.

3. Close the part file by choosing **File** > **Close** from the menu bar.

Tutorial 2

In this tutorial, you will create the model of the plastic cover shown in Figure 6-103. Its views and dimensions are shown in Figure 6-104. (**Expected time: 30 min**)

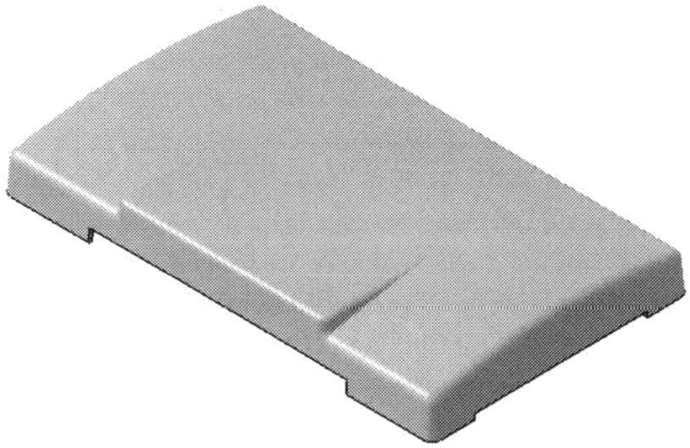

Figure 6-103 *Model for Tutorial 2*

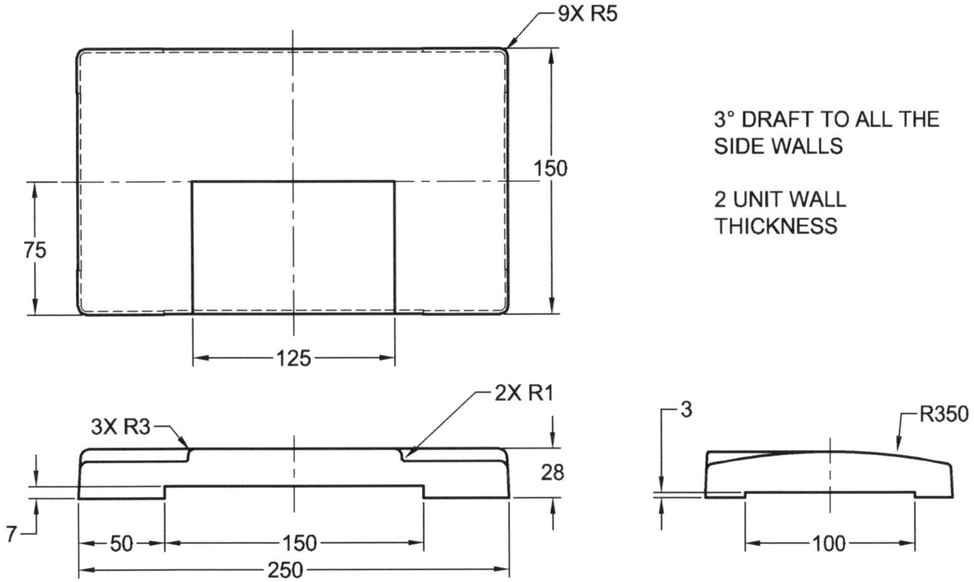

Figure 6-104 *Views and dimensions for Tutorial 2*

The following steps are required to complete this tutorial:

a. Create the base feature of the model by extruding the sketch drawn on the zx plane, equally on both sides of the sketch plane.
b. Create the second feature by extruding the sketch drawn on a plane created at an offset distance from the bottom planar face of the mode.
c. Add the draft feature to all faces of the model except the upper and lower faces.
d. Fillet the edges of the model.
e. Using the **Shell** tool, remove the bottom face of the model.
f. Create two pocket features to complete the model.

Creating the Base Feature of the Model

First, you need to create the base feature of the model by extruding the sketch drawn on the zx plane. The sketch will be extruded equally on both the sides of the sketching plane using the **Mirrored extent** option.

1. Start a new part file. Select the zx plane as the sketching plane and invoke the **Sketcher** workbench.

2. Draw the sketch of the base feature, as shown in Figure 6-105, and exit the **Sketcher** workbench.

3. Invoke the **Pad Definition** dialog box and set the value in the **Length** spinner to **125**.

4. Select the **Mirrored extent** check box and choose the **OK** button from the **Pad Definition** dialog box. The model after creating the base feature is shown in Figure 6-106.

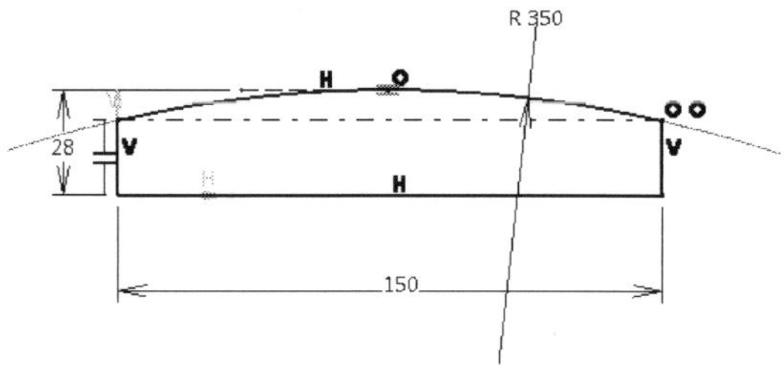

Figure 6-105 *Sketch of the base feature*

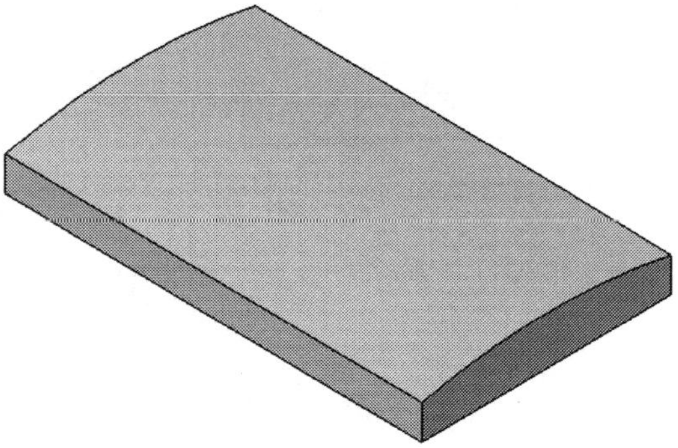

Figure 6-106 *The model after creating the base feature*

Creating the Second Feature

The second feature of the model will be created by extruding the sketch drawn on a plane created at an offset distance of 28 mm from the bottom planar face of the model.

1. Create a plane at an offset distance of 28 mm from the bottom planar face of the model. Note that the plane should be created above the model. Choose the **Reverse Direction** button, if required.

2. Invoke the **Sketcher** workbench using the newly created plane as the sketching plane.

3. Draw the sketch, as shown in Figure 6-107, and exit the **Sketcher** workbench.

4. Invoke the **Pad Definition** dialog box and choose the **Reverse Direction** button.

5. Select the **Up to next** option from the **Type** drop-down list and choose **OK**. The model after creating the second feature is shown in Figure 6-108.

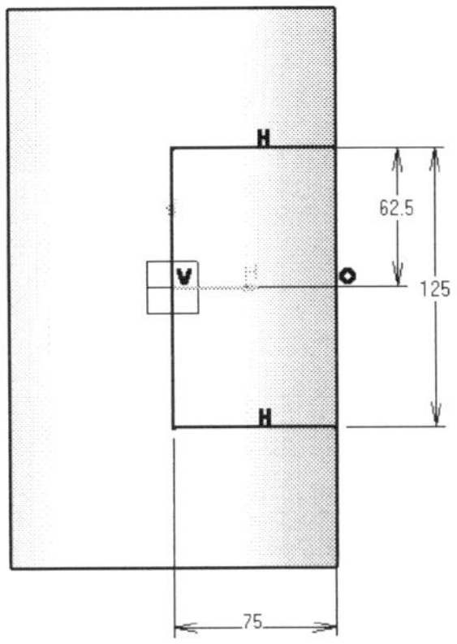

Figure 6-107 *Sketch of the second feature*

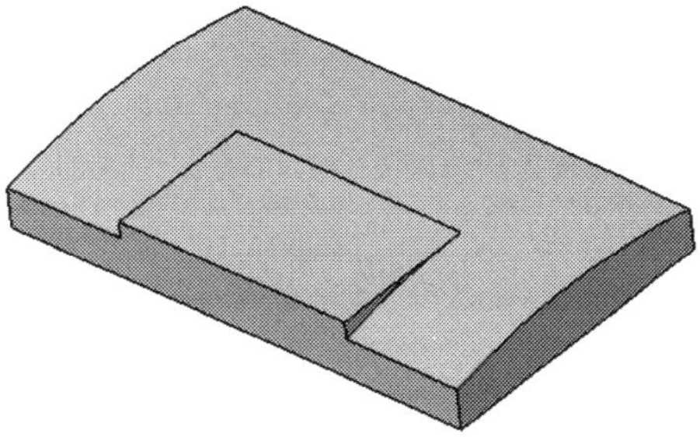

Figure 6-108 *The model after creating the second feature*

Adding the Draft to the Faces of the Model

Next, you need to add draft to the faces of the model. The draft angle is added to make sure that the component can be smoothly ejected from the die. The draft angle is one of the most important aspects of designing the components to be cast formed, or molded.

1. Choose the **Draft Angle** tool from the **Drafts** sub-toolbar in the **Dress-Up Features** toolbar; the **Draft Definition** dialog box is displayed and you are prompted to select the faces to be drafted.

2. Select all the vertical faces of the base feature and the second feature from the geometry area.

3. Click once in the **Selection** selection box in the **Neutral Element** area and select the bottom face of the base feature as the neutral element. Make sure that the pulling direction is downward. You can use the arrow displayed on the model to change the directions.

4. Set the value in the **Angle** spinner to **-3** and choose the **OK** button from the **Draft Definition** dialog box. The model after creating the draft feature is shown in Figure 6-109.

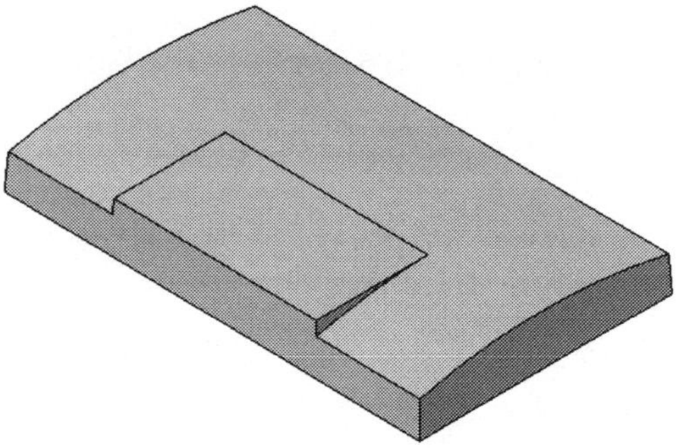

Figure 6-109 The model after drafting all vertical faces

Filleting the Edges of the Model

Next, you need to fillet the edges of the model. In this model, you need to fillet three separate set of edges using the **Edge Fillet** tool.

1. Choose the **Edge Fillet** tool from the **Dress-Up Features** toolbar; the **Edge Fillet Definition** dialog box is displayed.

2. Select the edges, as shown in Figure 6-110, and set the value in the **Radius** spinner to **3**.

3. Choose the **OK** button from the **Edge Fillet Definition** dialog box. The model after filleting the first set of edges is shown in Figure 6-111.

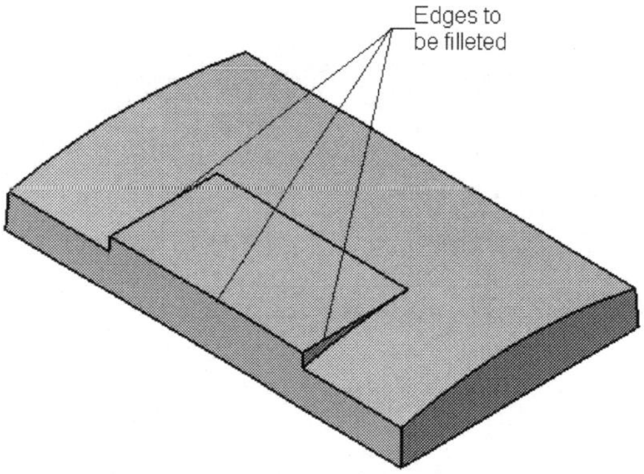

Figure 6-110 Edges to be filleted

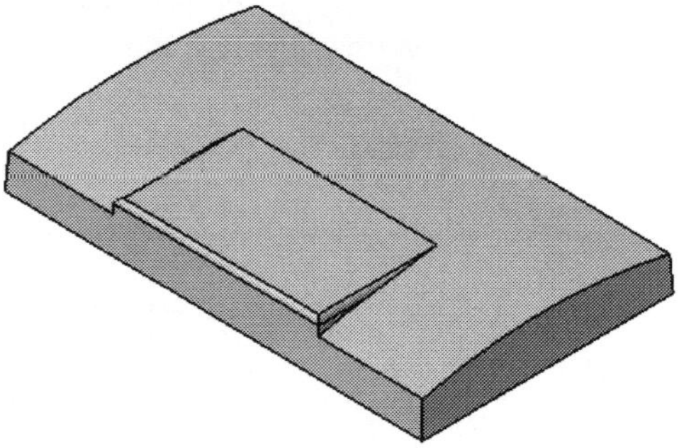

Figure 6-111 The model after filleting the first set of edges

4. Invoke the **Edge Fillet Definition** dialog box again to fillet the second set of edges.

5. Select the edges, as shown in Figure 6-112, and set the value in the **Radius** spinner to **1**.

6. Choose the **OK** button from the **Edge Fillet Definition** dialog box. The model, after filleting the second set of edges, is shown in Figure 6-113. In this figure, the model has been displayed

using the **Shading with Edges without Smooth Edges** tool from the **View mode** toolbar for better visualization.

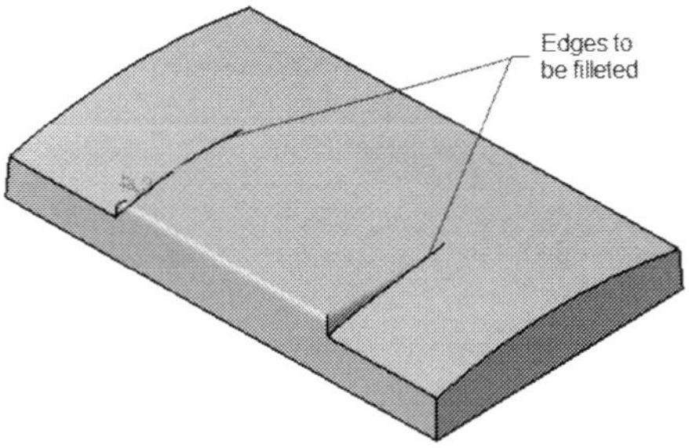

Figure 6-112 *Edge to be filleted*

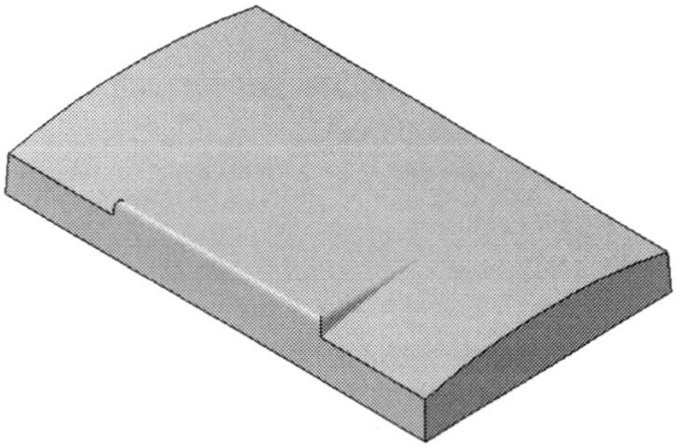

Figure 6-113 *The model after filleting the second set of edges*

7. Invoke the **Edge Fillet Definition** dialog box again to fillet the third set of edges.

8. Select all the edges of the model to be filleted except the edges shown in Figure 6-114 and set the value in the **Radius** spinner to **5**.

9. Choose the **OK** button from the **Edge Fillet Definition** dialog box. The resulting filleted model is shown in Figure 6-115.

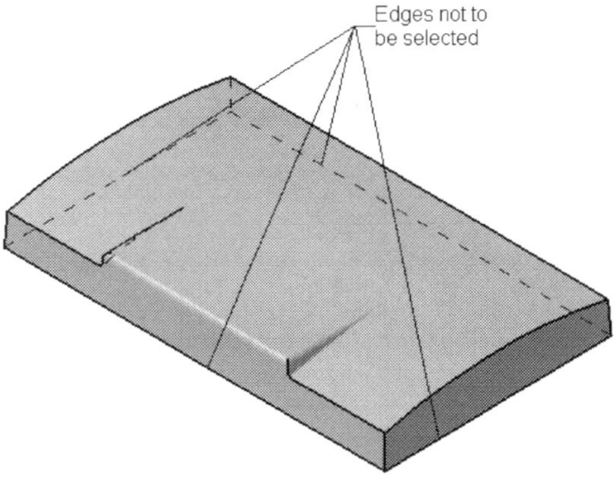

Figure 6-114 *Edges not to be selected*

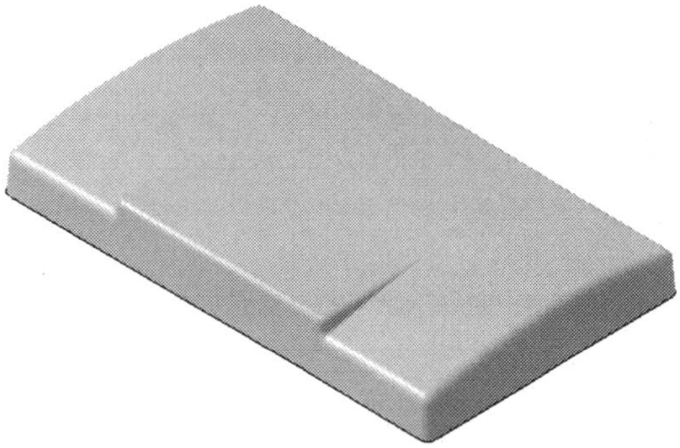

Figure 6-115 *Resulting filleted model*

Creating the Shell Feature

Finally, you need to shell the model and remove its bottom face.

1. Choose the **Shell** tool from the **Dress-Up Features** toolbar; the **Shell Definition** dialog box is displayed.

2. Select the face to be removed, as shown in Figure 6-116, and set the value in the **Inside thickness** spinner to **2** in the **Default Thickness** area.

3. Choose the **OK** button from the **Shell Definition** dialog box. The rotated view of the model after adding the shell feature is shown in Figure 6-117.

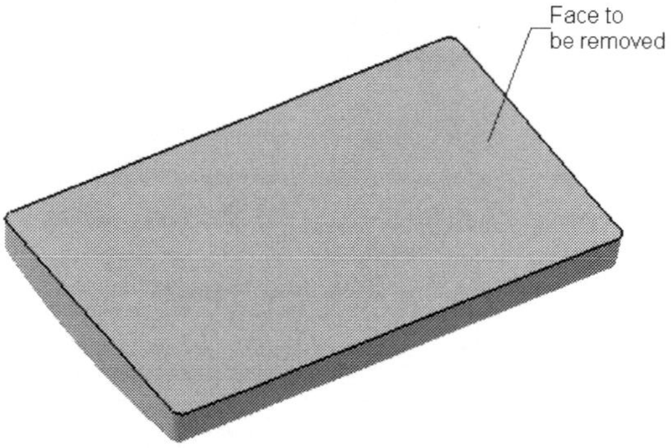

Figure 6-116 *Face to be removed*

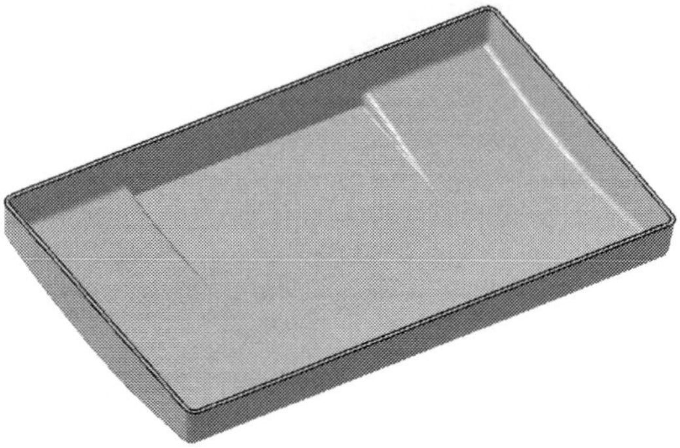

Figure 6-117 *Resulting shelled model*

4. Use the **Pocket** tool to create two pocket features at the bottom. Refer to Figure 6-104 for dimensions. Use the yz and zx planes to create the profile for these features. The final model after creating these two features is shown in Figure 6-118.

> **Tip**
> *It is recommended to shell the model after adding the draft angle and the fillet feature to maintain the draft angle and fillet curvature on the inside walls of the shelled model.*

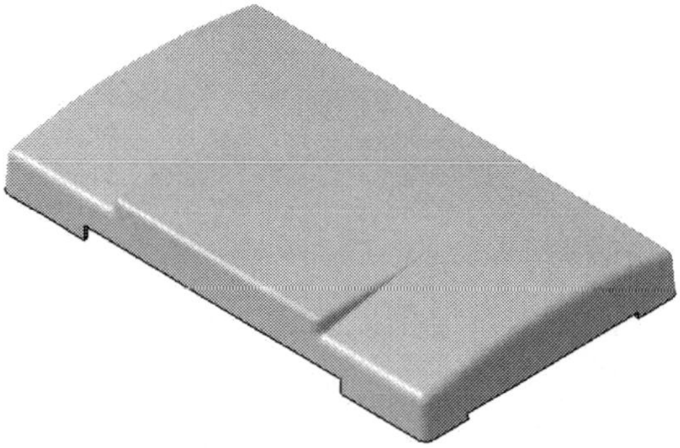

Figure 6-118 *Final model after creating the pocket features*

Saving and Closing the File

1. Choose the **Save** button from the **Standard** toolbar to invoke the **Save As** dialog box.
2. Enter the name of the file as **c06tut2** in the **File name** edit box and choose the **Save** button. The file will be saved at *C:\CATIA\c06*.
3. Close the part file by choosing **File > Close** from the menu bar.

Self-Evaluation Test

Answer the following questions and then compare them to those given at the end of this chapter:

1. The __________ tool is used to create a draft feature using the silhouette lines of the selected curved face as the neutral element.

2. The __________ tool is used to scoop out material from a model and remove the selected faces resulting in a thin-walled structure.

3. When you create a draft feature, by default the pulling direction is selected along the __________ axis.

4. The __________ tool is used to apply a fillet between the selected faces of the model.

5. __________ is defined as a process in which the sharp edges are beveled in order to reduce the area of stress concentration.

6. While creating a hole using the **Hole** tool, you can also use a hole callout to display the hole tolerance. (T/F)

7. You can create a countersink hole using the **Hole** tool. (T/F)

8. You can add user-defined thread standards for creating a threaded hole. (T/F)

9. You cannot set the limits for the fillet along the selected edge. (T/F)

10. Instead of selecting or creating a limiting element, you can specify the limit of the fillet by directly selecting the points on the edge to fillet. (T/F)

Review Questions

Answer the following questions:

1. Which of the following tools is used to taper the faces of the model?

   (a) **Draft Angle** (b) **Edge Fillet**
   (c) **Chamfer** (d) **Shell**

2. Which of the following tools is used to apply cosmetic threads on plane holes or plane cylindrical shafts?

   (a) **Extend** (b) **Thread Definition**
   (c) **Thread/Tap** (d) None of these

3. Which of the following tools is used to create a fillet feature tangent to three faces?

(a) **Face-Face Fillet** (b) **Variable Radius Fillet**
(c) **Tritangent Fillet** (d) **Edge Fillet**

4. Which of the following tabs in the **Hole Definition** dialog box is used to define the parameters to create a threaded hole?

(a) **Extension** (b) **Type**
(c) **Hole** (d) **Thread Definition**

5. Which of the following tools is used to create a variable angle draft?

(a) **Draft Angle** (b) **Draft Reflect Line**
(c) **Face-Face Fillet** (d) None of these

6. To trim the intersecting surfaces, select the __________ check box in the **Edge Fillet Definition** dialog box.

7. The __________ fillet type is used to smoothly blend the transition surfaces generated from the edges to the fillet vertex.

8. You cannot create a counterdrilled hole using the **Hole** tool. (T/F)

9. You cannot apply a different shell thickness value to the faces of the model while creating the shell feature. (T/F)

10. To create an edge fillet, choose the **Face-Face Fillet** button from the **Fillets** toolbar. (T/F)

EXERCISES

Exercise 1

Create the model shown in Figure 6-119. Its views and dimensions are shown in Figure 6-120.

(Expected time: 30 min)

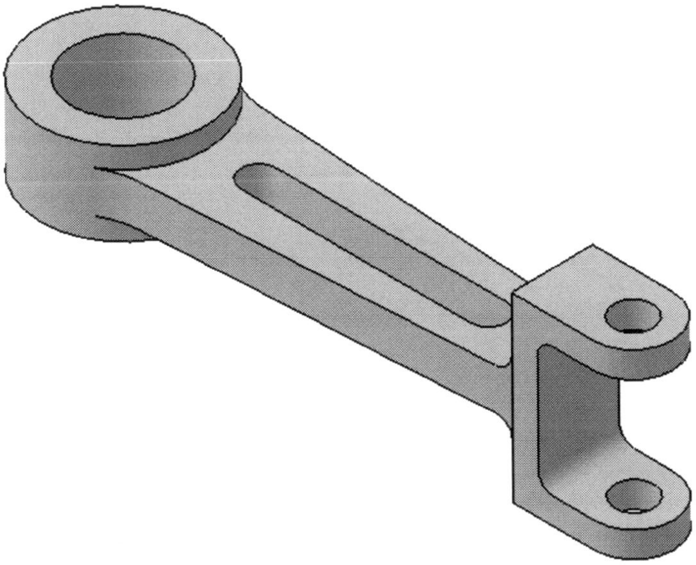

Figure 6-119 Model for Exercise 1

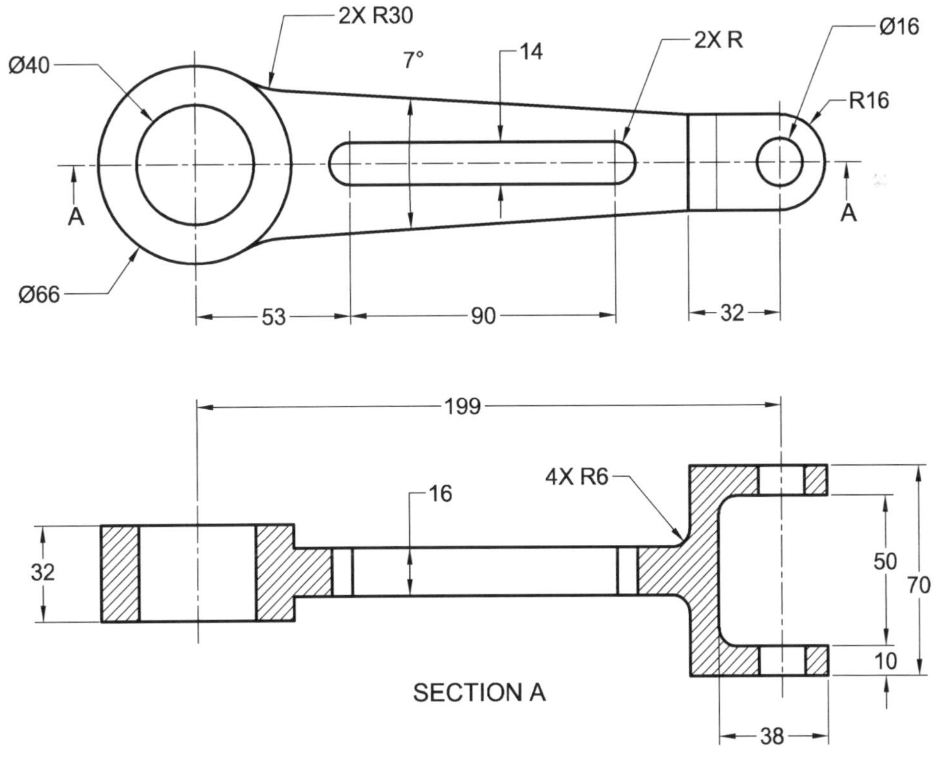

Figure 6-120 Views and dimensions for Exercise 1

Exercise 2

Create the model of the Clamp Stop shown in Figure 6-121. Its views and dimensions are shown in Figure 6-122. **(Expected time: 1 hr)**

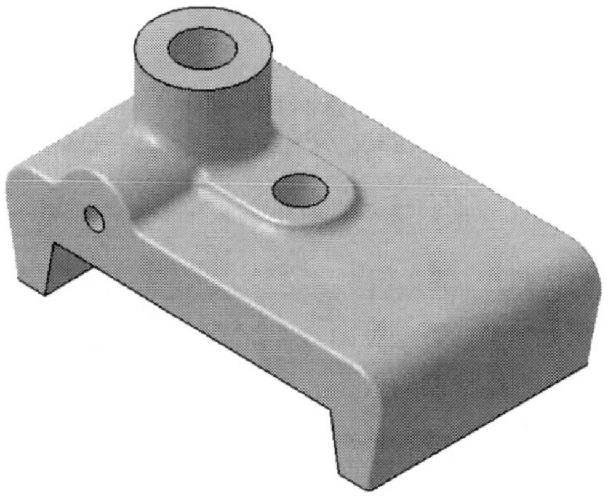

Figure 6-121 *Model for Exercise 2*

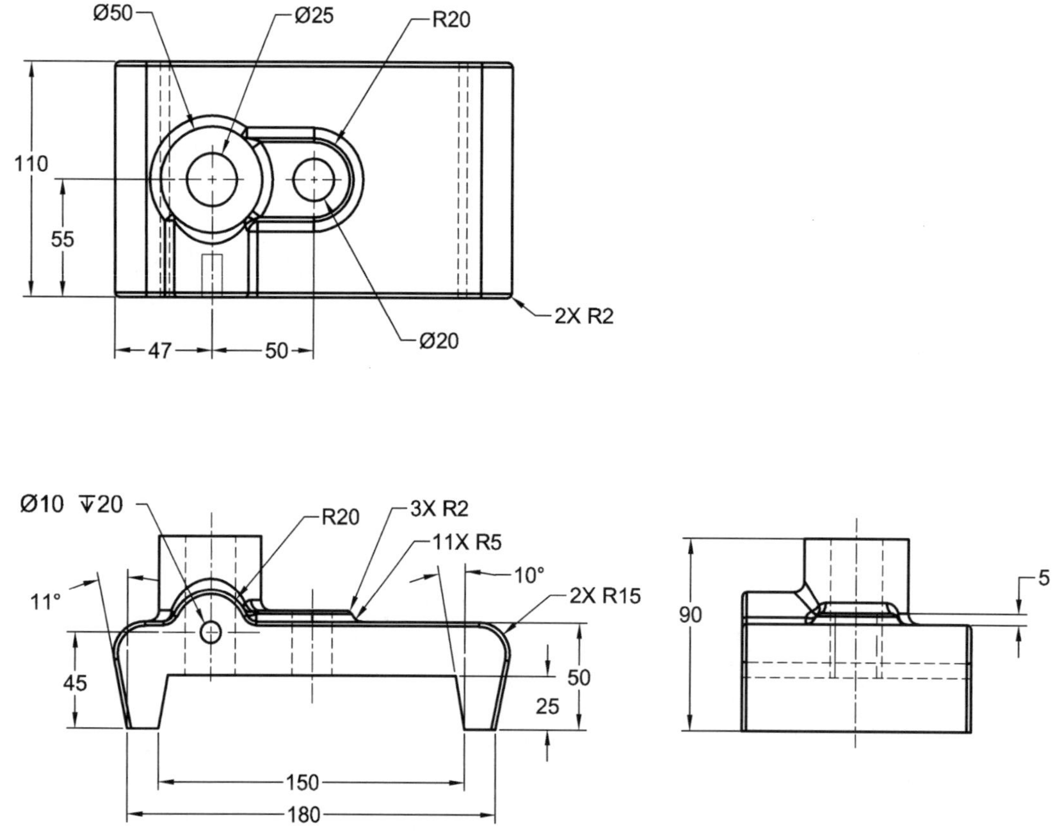

Figure 6-122 *Views and dimensions for Exercise 2*

Exercise 3

Create the model of the Machine-Block shown in Figure 6-123. Its views and dimensions are also shown in the same. Assume the missing dimensions, if any.

(Expected time: 1 hr)

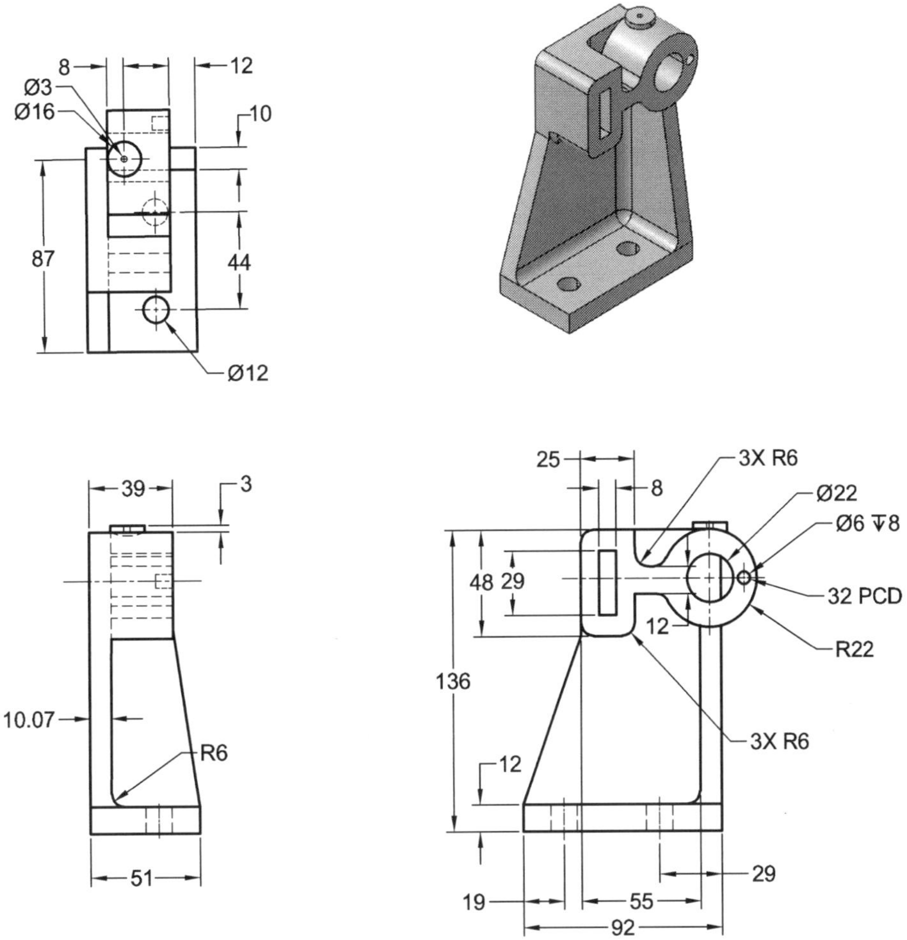

Figure 6-123 The Model, Views and dimensions for Exercise 3

Answers to Self-Evaluation Test

1. **Draft Reflect Line**, **2.** **Shell**, **3.** Z, **4.** **Face-Face Fillet**, **5.** Chamfering, **6.** T, **7.** T, **8.** T, **9.** F, **10.** T

Chapter 7

Editing Features

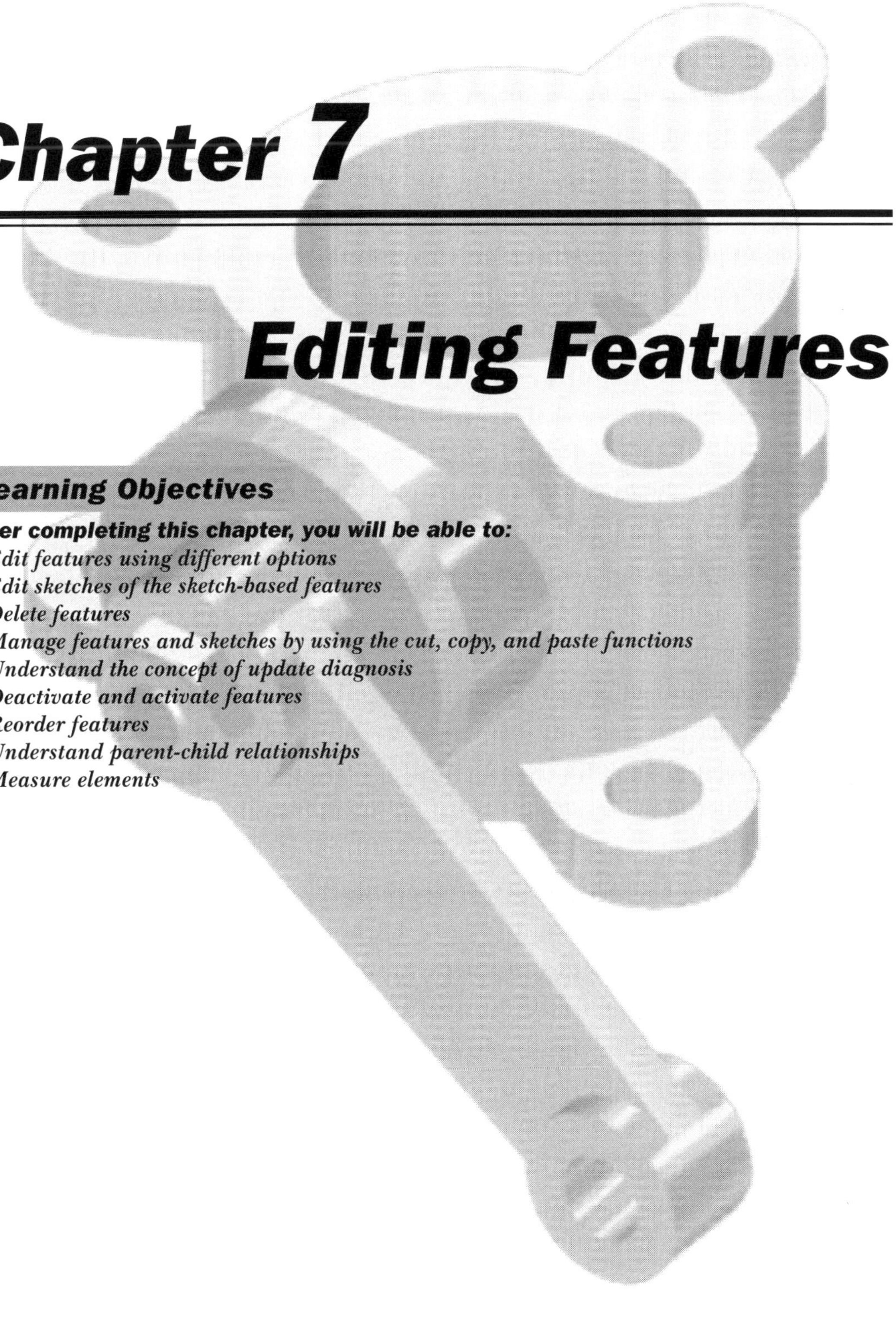

Learning Objectives

After completing this chapter, you will be able to:

- *Edit features using different options*
- *Edit sketches of the sketch-based features*
- *Delete features*
- *Manage features and sketches by using the cut, copy, and paste functions*
- *Understand the concept of update diagnosis*
- *Deactivate and activate features*
- *Reorder features*
- *Understand parent-child relationships*
- *Measure elements*

EDITING FEATURES OF A MODEL

Editing is one of the most important aspects of the product design cycle. Almost all the designs require editing, either during their creation or after they are created. As discussed earlier, CATIA V5 is a feature-based and parametric software. Therefore, the designs created in CATIA V5 are individual features integrated together to form a solid model. All these features can be edited individually. For example, a Base Plate with simple holes created using the **Hole** tool shown in Figure 7-1.

In this example, you can use a simple editing operation to replace the simple holes placed on the base plate with counterbore holes. To do so, double-click on the hole to edit; the **Hole Definition** dialog box will be displayed. Choose the **Type** tab from it and select the **Counterbored** option from the drop-down list. Choose the **OK** button from the **Hole Definition** dialog box. The simple hole will be replaced by a counterbore hole. Figure 7-2 shows the Base Plate with counterbore holes.

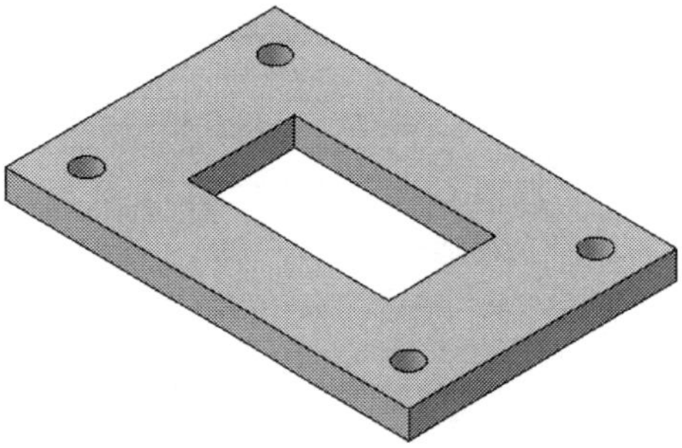

Figure 7-1 *Base Plate with simple holes*

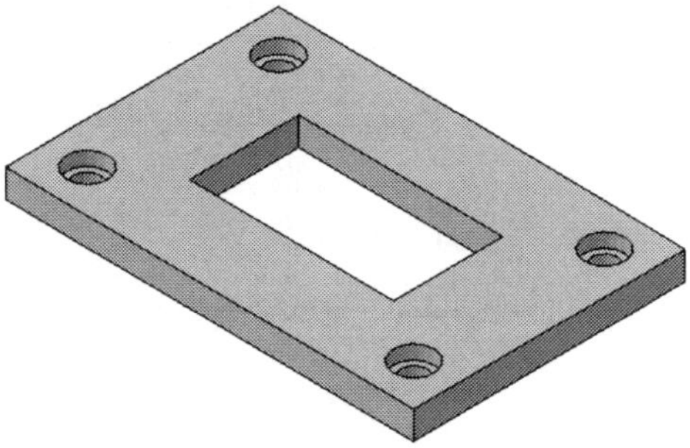

Figure 7-2 *Modified Base Plate with counterbore hole*

Similarly, you can edit the reference elements, sketches of the sketched features, and other features of the model. Remember, when you modify the reference elements, features that are dependent on them will be modified automatically. For example, consider a case in which you have created an extruded feature by extruding a sketch drawn on a plane which is created at an offset distance from the front face of the model. When you modify the offset distance, the feature created on the plane gets modified accordingly. The editing methods in CATIA V5 are discussed in the following sections.

Editing Using the Definition Option

In CATIA V5, to edit a feature by using the **Definition** option, select it from the Specification tree or from the graphics window and then right-click to invoke the contextual menu. Move the cursor to the name of the feature in the contextual menu; a cascading menu will be displayed. Choose the **Definition** option from it; the dialog box related to the selected feature will be displayed. Modify the parameters of the feature using the options in the dialog box and then choose the **OK** button; the dialog box will be closed and the feature will be modified automatically.

Editing by Double-Clicking

In CATIA V5, features are generally edited by double-clicking on it. When you double-click on a feature, a dialog box related to the selected feature will be displayed. In this dialog box, edit the required parameters and then choose **OK**; the dialog box will be closed and the feature will be modified automatically.

Editing the Sketch of a Sketch-Based Feature

You can also edit the sketches of the sketch-based features in CATIA V5. To do so, expand the Specification tree and then expand the branch of the sketch-based feature whose sketch you need to edit. Select **Sketch** and right-click to invoke the contextual menu. Move the cursor to the name of the sketch at the bottom of the contextual menu and choose the **Edit** option from the cascading menu; the **Sketcher** workbench is invoked. Alternatively, you can double-click on the sketch to invoke the **Sketcher** workbench to edit the sketch. Now, edit the sketch using the tools in the **Sketcher** workbench and then exit the sketcher workbench. The feature will be modified automatically.

Note

In case the changes are not applied immediately to the model then the model will turn red. Choose the ***Update All*** *button from the* ***Tools*** *toolbar to apply the changes to the model.*

Redefining the Sketch Plane of Sketches

You can also redefine the sketching plane of sketches of the sketch-based features to place them on some other plane. To edit a sketch, select it from the Specification tree and right-click to invoke the contextual menu. Move the cursor to the name of the sketch at the bottom of the contextual menu; a cascading menu will be displayed. Next, choose the **Change Sketch Support...** option from the cascading menu; the **Sketch Positioning** dialog box will be displayed, as shown in Figure 7-3. The name of the current sketching plane is displayed in the **Reference** selection box. Select any other plane or face as the sketching plane; the name of the selected plane or face will be displayed in the **Reference** selection box. Also, preview of the sketch drawn on the newly selected plane is displayed in the geometry area. Choose the **OK** button from the **Sketch Positioning** dialog box.

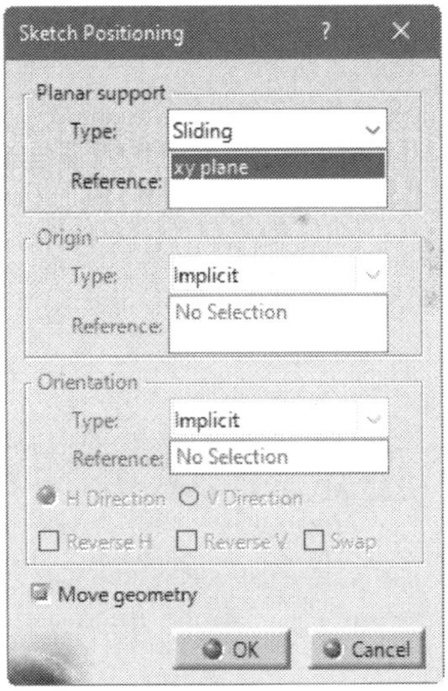

Figure 7-3 *The* ***Sketch Positioning*** *dialog box*

Figure 7-4 shows the base feature of the model created by extruding the sketch drawn on the yz plane. Figure 7-5 shows the model after redefining the sketch plane of the base feature from the yz plane to the zx plane.

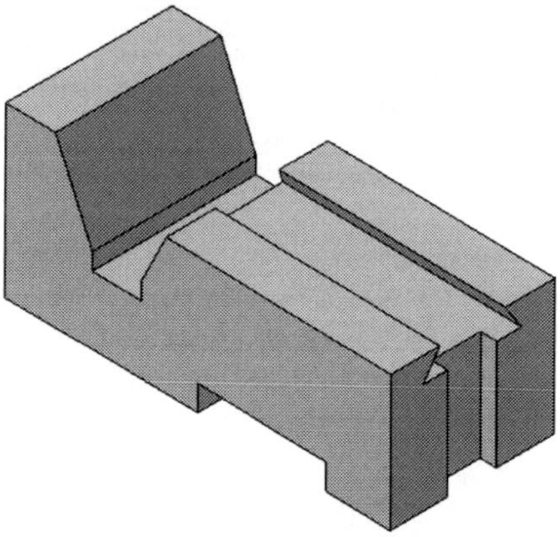

Figure 7-4 Base feature created on the yz plane

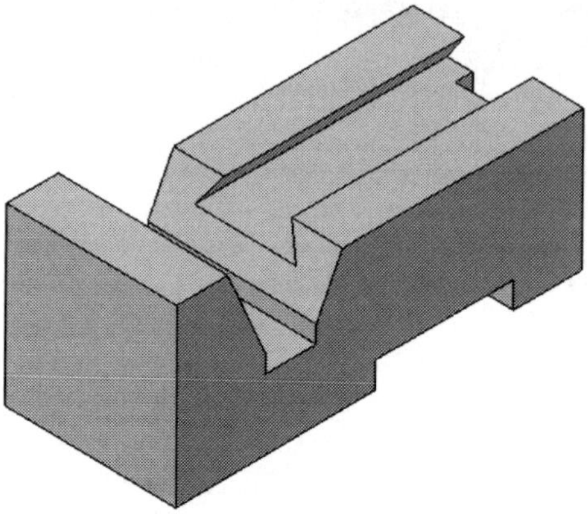

Figure 7-5 The model after redefining the sketch plane of the base feature to the zx plane

When you redefine the sketching plane of the sketch-based features, sometimes a **Warning** message box is displayed, as shown in Figure 7-6. It informs you that if you redefine the sketching plane, the sketch may become inconsistent or over-defined. Choose the **OK** button from the **Warning** message box. In this case, you may need to edit the sketch, if it becomes inconsistent or over-defined.

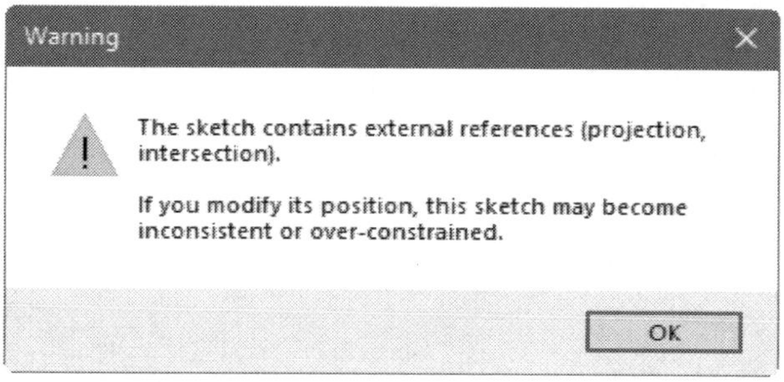

*Figure 7-6 The **Warning** message box*

Deleting Unwanted Features

In CATIA V5 you can also delete the unwanted features of a model. To do so, select the feature to be deleted from the geometry area or the Specification tree and then press the DELETE key. Alternatively, right-click on the feature to be deleted; a contextual menu will be displayed. Choose the **Delete** option from the contextual menu; the edges of the feature will be highlighted. If the feature to be deleted has a child feature referenced to it, the child feature will be highlighted in the geometry area as well as in the Specification tree. Also, the **Delete** dialog box will be displayed, as shown in Figure 7-7.

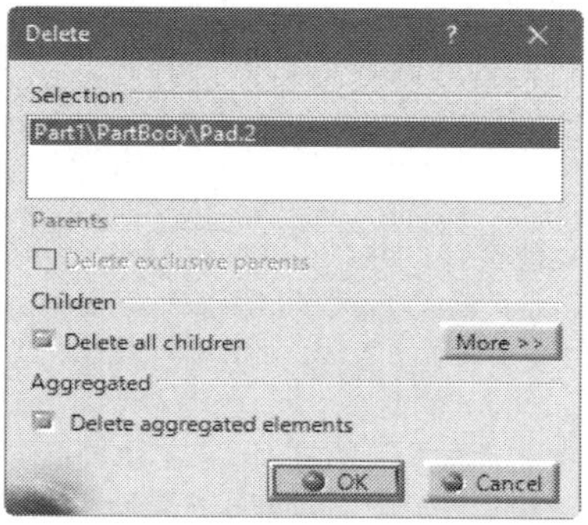

*Figure 7-7 The **Delete** dialog box*

In the **Delete** dialog box, the **Delete all children** check box is selected by default. As a result, all the features dependent on the features being deleted will also be deleted. If you clear this check box and then choose the **OK** button from the **Delete** dialog box, the part body turns red and the **Update Diagnosis** dialog box will be displayed. In case, the **Update Diagnosis** dialog box is not displayed, then choose the **Update All** button from the **Tools** toolbar to display the **Update Diagnosis** dialog box. Using this dialog box, you can modify the missing references of the features that were referenced in the deleted feature. You will learn more about the **Update Diagnosis** dialog box later in this chapter.

If you have started the file with the **Enable hybrid design** check box selected from the **New Part** dialog box, the **Delete aggregated elements** check box will be enabled. If you select this check box while deleting a feature, the sketch associated with it will also be deleted. However, if you clear this check box, the sketch associated with this feature will not be deleted.

If you have started the file with the **Enable hybrid design** check box cleared, the **Delete exclusive parents** check box will be enabled. If you select this check box while deleting a feature, the sketch associated with it will also be deleted. Also, a **Confirm Deletion** dialog box will be displayed to confirm if you really want to delete the selected features. Choose the **Yes** button to continue.

If the feature is not referenced to any other feature, then the **Delete** dialog box will not be displayed. When you select the feature and press the DELETE key, the feature will be deleted automatically.

Managing Features and Sketches by using the Cut, Copy, and Paste Functions

CATIA V5 allows you to follow the Windows functionality of Cut, Copy, and Paste to cut or copy the features and sketches and then paste them on the selected plane or face. The method of using these functionalities is same as used in the other Windows-based applications. Select the feature or sketch to cut or copy. To cut the selected item, choose the **Cut** option from the contextual menu or use the Ctrl+X keys; the selected item will be deleted. Now, choose the **Paste** option from the contextual menu or use the Ctrl+V keys to paste the deleted item on selected plane or face.

Note that if the resulting feature merges with the model, the **Warnings** message box will be displayed, as shown in Figure 7-8. This message box informs you that the operation performed on the feature is unnecessary. Choose the **Close** button to exit the message box. In this case, you need to edit the feature and flip the direction of the feature creation.

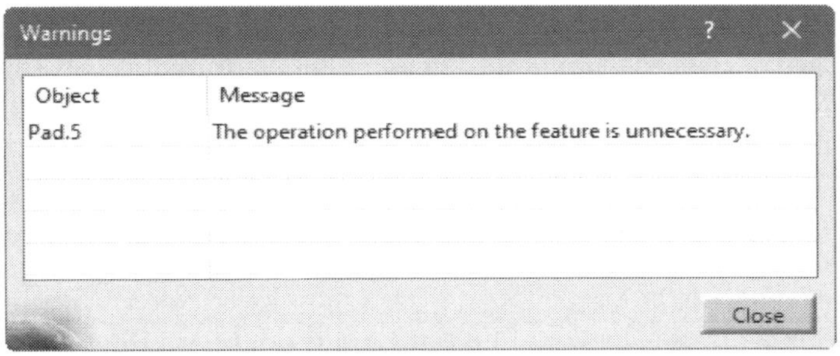

Figure 7-8 *The* ***Warnings*** *message box*

Sometimes, it is not possible to paste the feature at the desired location where it is being pasted because of some geometrical inconsistency. In such cases, the **Update Diagnosis** dialog box is displayed. This dialog box displays the errors that occur while pasting the feature. The **Update Diagnosis** dialog box is discussed later in this chapter.

If you copy and paste an item, the selected item will remain at its position and its copy will be pasted on the selected plane or face. To copy an item, select the item and choose **Copy** from the contextual menu, or use the Ctrl+C keys. Now, select the face or plane as the reference on which you need to paste the copied item and paste it using the Ctrl+V keys.

After pasting the sketch-based features, it is always recommended to edit the sketch of the copied features to properly locate it with respect to its surroundings.

Tip
For pasting a selected sketch, or a sketched based feature, or a hole created using the ***Hole*** *tool, you need to select a plane or a planar face as reference. For pasting chamfers, fillets, and drafts, you need to select an edge or edges as reference.*

Understanding the Concept of Update Diagnosis

Sometimes, after editing or modifying a feature, the **Update Diagnosis** dialog box is displayed, as shown in Figure 7-9.

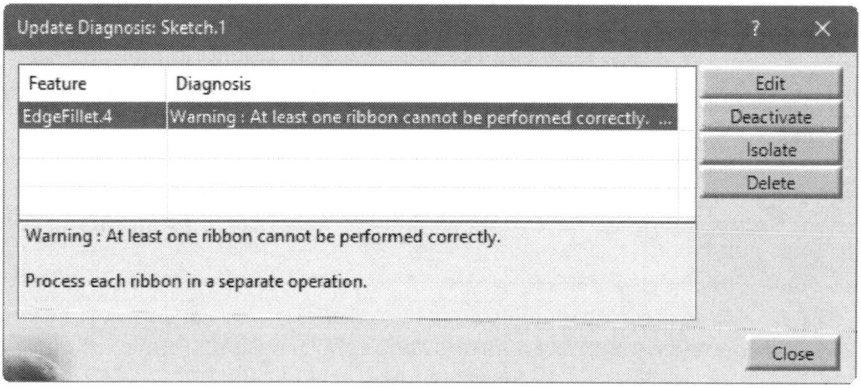

Figure 7-9 *The* ***Update Diagnosis*** *dialog box*

This dialog box displays the problems caused in the other features due to of editing of the selected feature. Therefore, you also need to edit the other features. Consider a case in which a block is created as the base feature of the model and its edges are filleted. Create another feature whose sketch is extracted from the tangential edge of the fillet. Now, according to the change in the design of the model, you need to delete the fillet. On deleting a fillet, the **Delete** dialog box will be displayed and it will prompt you to delete the child features. If you delete the child features, the feature whose sketch was being referenced to the tangential edge will also be deleted. If you do not select the option to delete the child features, the **Update Diagnosis** dialog box will be displayed. This is because the tangential edge that was selected as a reference for the sketch of this feature is deleted. Therefore, the geometry or the dimension related to that edge becomes dangling. As a result, you need to invoke the **Sketcher** workbench and redefine the sketch.

On choosing the **Edit** button from the **Update Diagnosis** dialog box, you can edit the feature or sketch that has a problem. If you choose the **Deactivate** button, the feature or sketch with a problem will be deactivated. If you choose the **Isolate** button, the sketch with a problem will be isolated and the dimensions and constraints applied to it will be deleted. If you select the **Delete** button, the feature or the sketch with the problem will be deleted.

If you choose the **Close** button without editing the feature or sketch with a problem, a yellow spiral will be displayed on the feature and the exclamation mark on the sketch will also be displayed. This indicates that one or more features in the current part body have errors. Therefore, you can easily determine the features that need to be edit.

Cut, Copy, and Paste Features and Sketches

You can also cut or copy the features and sketches from one file and paste them in another file. Consider a case in which you need to copy a fillet from an edge of the model in one file and paste it on the edge of a model in another file. In this case, the fillet that you need to copy from the geometry area and use the Ctrl+C keys to copy the selected feature. Now, choose **Window** from the menu bar and choose the name of the other file to paste the selected fillet. The other file will be displayed on the screen. Select the edge to paste the fillet and use the Ctrl+V keys to paste it.

Copying Features using Drag and Drop

CATIA V5 also allows you to use the drag and drop function of Windows to copy and paste the features within a file. To do so, press and hold the Ctrl key, select and drag the features to be copied from the geometry area or from the Specification tree to the location where you need to paste it. Release the mouse button on that location; the feature will be copied. If the sketch of the pasted item becomes inconsistent or over-defined, the **Update Diagnosis** dialog box will be displayed. In this case, you may need to edit the feature or the sketch of the pasted feature. Figure 7-10 shows the feature being dragged and Figure 7-11 shows the resulting pasted feature.

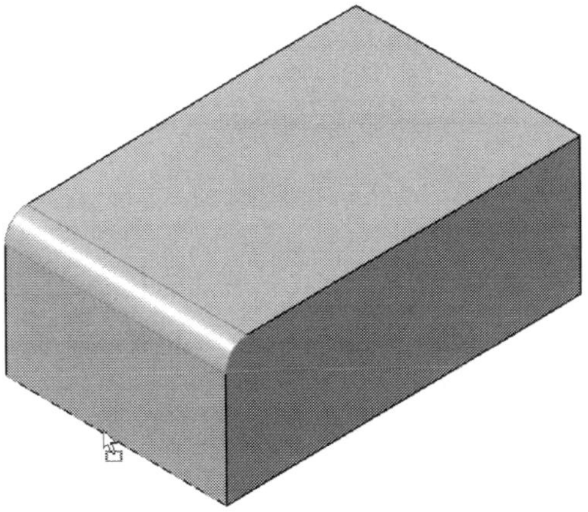

Figure 7-10 *Feature being dragged*

Figure 7-11 *Resulting pasted feature*

You can also use the drag and drop function of Windows to copy and paste the item from one document to the other. To drag and drop the feature from one file to another, first you need to tile the files vertically or horizontally by selecting the options from the **Window** menu. Next, follow the procedure discussed earlier to drag and drop the feature from one file to another. Figure 7-12 shows the fillet feature being dragged to be pasted on the edge of the model in another file.

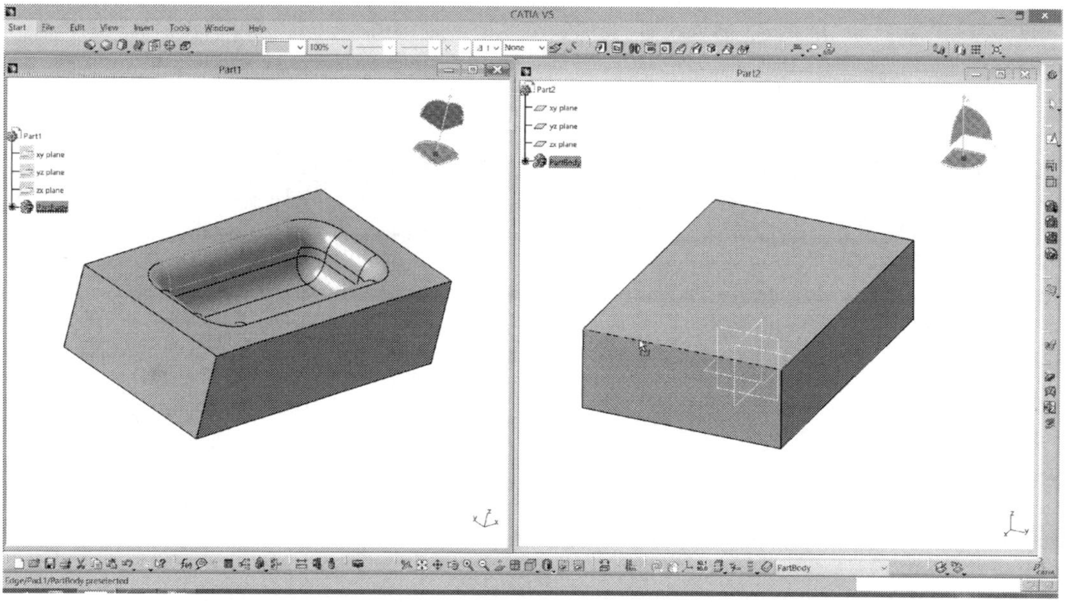

Figure 7-12 *Fillet feature being dragged to be pasted on the edge of the model in another file*

Cut, Copy, and Paste the Features using Drag and Drop function

You can also use the drag and drop function to cut and paste the feature within a file or from one file to another. To cut and paste the feature by dragging and dropping, select it and drag the cursor. Now, move it to the location where you need to place the selected feature and then release the left mouse button. The selected feature will be removed from its original location and will be pasted on the specified location.

Copying and Pasting PartBodies

You can copy the complete PartBody and paste it in another file. Any such copy will be pasted as another body. You will learn more about Partbodies later in this chapter. To copy and paste the entire part body, select PartBody from the Specification tree and use the Ctrl+C keys to copy it. Now, invoke the other document in which you need to paste the PartBody and select PartBody from the Specification tree. Use the Ctrl+V keys to paste it. On doing so, the copied PartBody will be pasted as another body.

Another method of pasting the PartBody is by using the **Paste Special** option. To use this option, copy the PartBody from one of the files and then select PartBody in the other file or in the same file in which you need to paste it. Right-click and then choose the **Paste Special** option from the contextual menu; the **Paste Special** dialog box will be displayed, as shown in Figure 7-13.

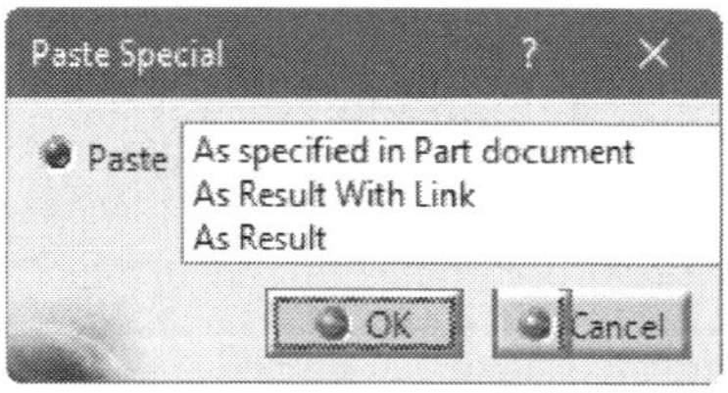

*Figure 7-13 The **Paste Special** dialog box*

If you select the **As specified in Part document** option and choose the **OK** button, the PartBody will be pasted as a new body and the feature information will also be displayed in the Specification tree.

If you choose the **As Result With Link** option, the PartBody will be pasted as a new body and instead of feature information, only a **Solid** feature will be displayed. You can add features to the **Solid** feature. If you make any change in the parent model from which it is copied, then that modification will be reflected in the pasted body also. To view the modification, you may need to update the model using the **Update All** button in the **Tools** toolbar. You can also use the Ctrl+U keys to update it.

If you choose the **As Result** option, then the PartBody will be pasted as a new body, but there will be no feature information or link between the pasted body and the parent body. Therefore, any modifications in the parent body will not be reflected in the pasted body.

Tip

*Select any one of the faces of the body that you have pasted using the **As Result With Link** option and invoke the contextual menu. Move the cursor to the name of the body at the bottom of the contextual menu and choose the **Open the Pointed Document** option from the cascading menu displayed. The file with the parent model will be opened and you can perform any modification on the parent body. After completing the modifications, save the file and exit. Now, update the model using the Ctrl+U keys.*

Deactivating Features

Sometimes, you do not want a feature to be displayed in the model or in its drawing views. Instead of deleting these features, you can deactivate them. When you deactivate a feature, it remains invisible in the model or in its drawing views. If you create an assembly using this model, the deactivated feature will not be displayed in it. Note that if you deactivate a feature, the features that are referenced to it will also be deactivated.

To deactivate a feature, select it from the Specification tree or the geometry area and right-click to invoke the contextual menu. Now, move the cursor on the name of the selected feature at the bottom of the contextual menu and choose the **Deactivate** option from the cascading menu displayed. If no other feature is referenced to the feature to be deactivated, then the **Deactivate** dialog box will be displayed, as shown in Figure 7-14. Also, the name

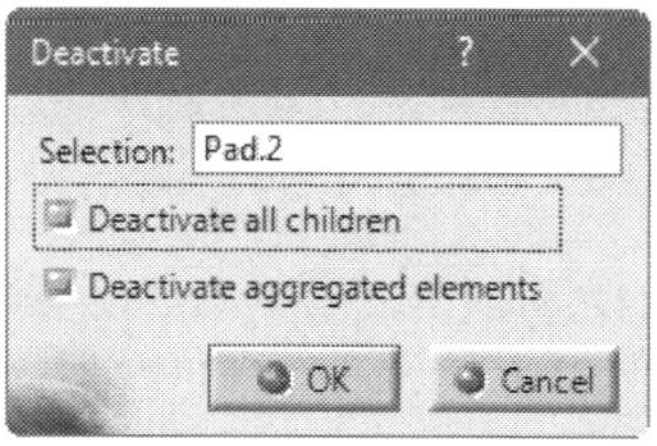

*Figure 7-14 The **Deactivate** dialog box*

of the feature to be deactivated will be displayed in the **Selection** display box. Choose the **OK** button from the **Deactivate** dialog box to deactivate the selected feature. The selected feature will get deactivated and its display will be turned off. Also, the symbol will be displayed on the deactivated feature in the Specification tree. If the feature to be deactivated has references with other features, then the **Deactivate all children** check box will be activated and selected in the **Deactivate** dialog box. The features that are referenced to it are highlighted in the Specification tree. Choose the **OK** button from the **Deactivate** dialog box to deactivate the selected and the references features.

Activating Deactivated Features

The deactivated feature can be activated by choosing the **Activate** option from the contextual menu. To resume the deactivated feature, select it from the Specification tree and invoke the contextual menu. Move the cursor on the name of the selected feature at the bottom of the contextual menu and choose the **Activate** option from the cascading menu; the **Activate** dialog box will be displayed, as shown in Figure 7-15. Choose the **OK** button from this dialog box; the deactivated feature will be activated. If the deactivated feature had some child features that were deactivated, the **Activate all children** check box will get activated and will be default selected in the **Activate** dialog box.

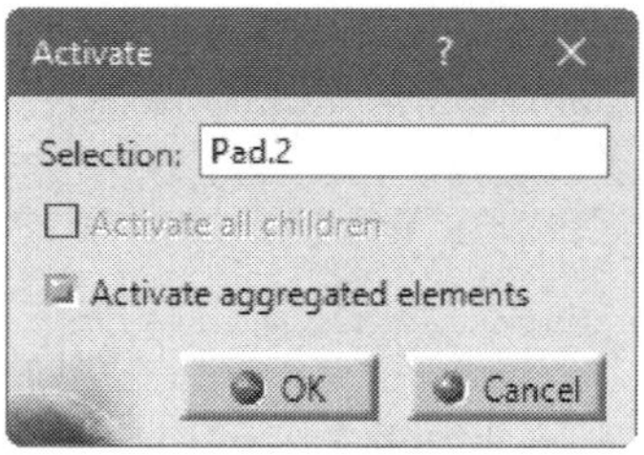

***Figure 7-15** The **Activate** dialog box*

> **Tip**
> *While working on large assemblies, it is recommended to keep only that part active on which you want to work and deactivate the rest. This will help you to get quick output. The **Deactivate** option temporarily removes the information from the RAM. As a result, the computing speed of the system increases. After completing the work, you can activate the deactivated parts. You will learn more about activation in the later chapters.*

Defining Features in Work Object

Defining a feature in the work object is a process in which you rollback a model to an earlier stage. By doing so, the features that are not defined in the current work object are suppressed. You can add new features to the model when it is in the rollback state. The newly added features are added before the features that are not defined in the current work object due to the rollback. While working with a multi featured model, if you need to edit a feature that was created at the starting of its design cycle, it is recommended that you rollback the model up to that feature. This is because after each editing operation, the time of regeneration will be minimized. To rollback a model, select the feature up to which you need to rollback and invoke the contextual menu. Choose the **Define In Work Object** option from it. To restore the model back to the normal state, select its last feature and choose the **Define In Work Object** option from the contextual menu.

Reordering Features

Reordering is defined as a process of changing the order in which the features were created in the model. Sometimes, after creating a model, it may be required to change the order in which the features in the model were created. By reordering the features, you can place them before, after, or inside another feature in the Specification tree.

Consider a case in which you have created a rectangular block with four through holes on the top face, as shown in Figure 7-16. Now, if you create a shell feature and remove the top, right, and front faces from the block, the model will appear similar to the one shown in Figure 7-17.

Tip

You can also reorder the features by dragging and dropping them within the Specification tree. To reorder the features using this method, select the feature from the Specification tree. Now drag the cursor to the feature after which you need to place the selected feature.

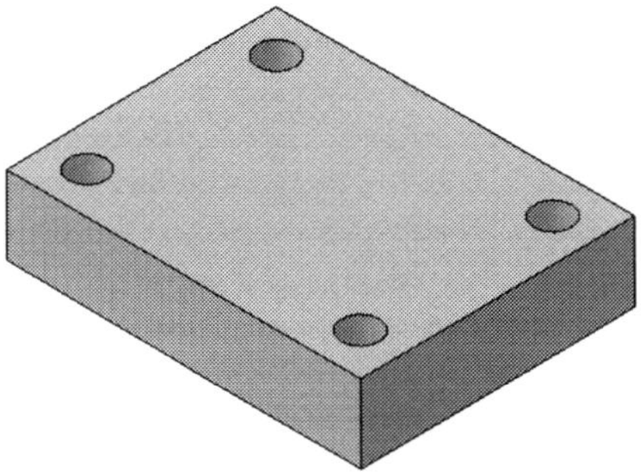

Figure 7-16 *Model with through holes on the top face*

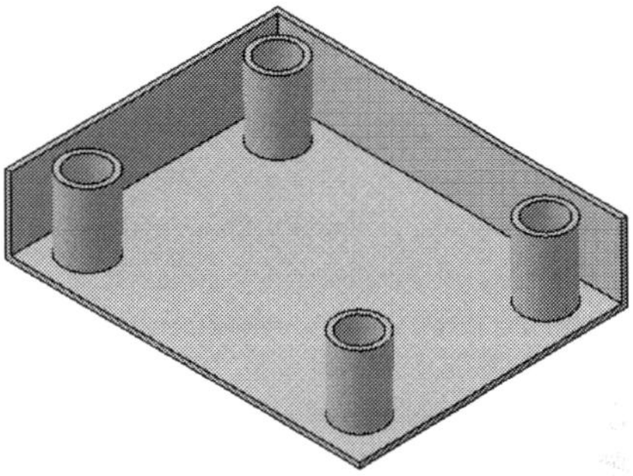

Figure 7-17 *Model after shelling*

If this is not the required result, you need to reorder the shell feature before the hole feature. Select the **Shell** feature from the Specification tree or from the geometry area and invoke the contextual menu. Next, invoke the cascading menu by moving the cursor on the name of the feature and choose the **Reorder** option; the **Feature Reorder** dialog box will be displayed, as shown in Figure 7-18. Also, you will be prompted to specify the new location of the feature. Note that the name of the selected feature will be displayed in the first display box of the **Feature Reorder** dialog box. The three options in the **Reorder** drop-down list are: **After**, **Before**, and **Inside**. Select the appropriate option from the **Reorder** drop-down list. Next, select the feature from the Specification tree. The new location will be displayed in the second display box of the **Feature Reorder** dialog box. Choose the **OK** button from the **Feature Reorder** dialog box; the feature selected first will be reordered according to option selected in the drop-down list of the **Feature Reorder** dialog box. Note that after reordering the features, you need to define the last feature in the work object because the model will automatically roll back to the reordered feature. Figure 7-19 shows the model with the shell feature reordered after the base feature. You can also select more than one feature to reorder.

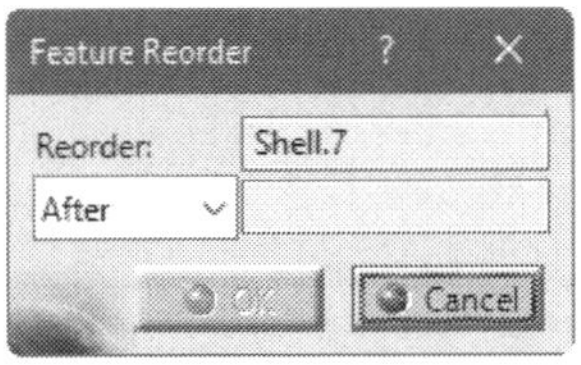

Figure 7-18 *The* ***Feature Reorder*** *dialog box*

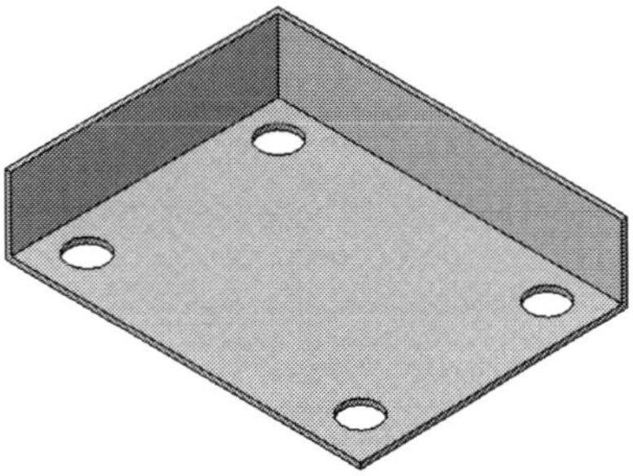

Figure 7-19 *The model after reordering the shell feature*

Understanding the Parent-Child Relationships

In CATIA V5, every model is composed of features. These features, in some way or the other, are related to each other. Generally, the base feature of a model is considered as the parent of all other features. But the sketch of the base feature is the parent of the base feature, and the sketch plane on which the sketch of the base feature is drawn is known as the parent of the sketch of the base feature. Therefore, the plane on which the sketch of the base feature is drawn is considered as the ultimate parent. The other features are the child features of the ultimate parent. Consider a case in which a feature is created by extruding a sketch drawn on the top face of the base feature. In this case, the extruded feature is a child of the sketch using which this extruded feature will be created. The base feature is the parent of the sketch of the extruded feature because the top face of the base feature was selected to draw the sketch of the extruded feature.

Remember that if parent feature is modified, the modification will be reflected in the child feature also. Similarly, if the parent feature is deleted, the **Delete** dialog box provides you with an option to delete the child features. If you do not delete them, you need to redefine the references to place the feature. To view the parent-child relationship between the features, select the feature whose parent and child features you need to view. Invoke the contextual menu and choose the **Parents/Children** option; the **Parents and Children** dialog box will be displayed, as shown in Figure 7-20.

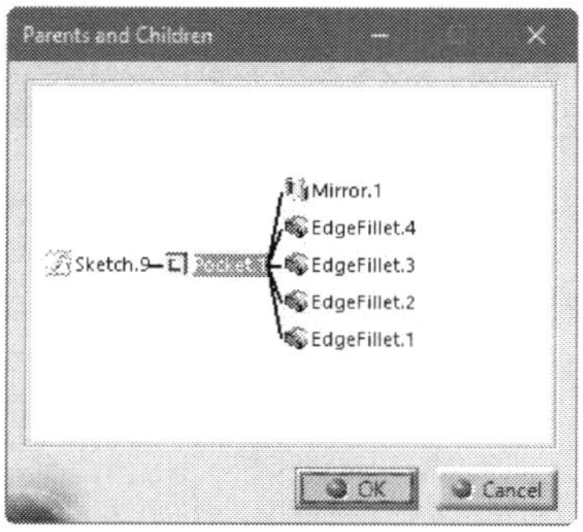

*Figure 7-20 The **Parents and Children** dialog box*

MEASURING ELEMENTS

The **Part Design** workbench of CATIA V5 also provides you with the tools to measure the distance, angle, radius, area, and inertia. To measure these elements, you can use the tools in the **Measure** toolbar shown in Figure 7-21.

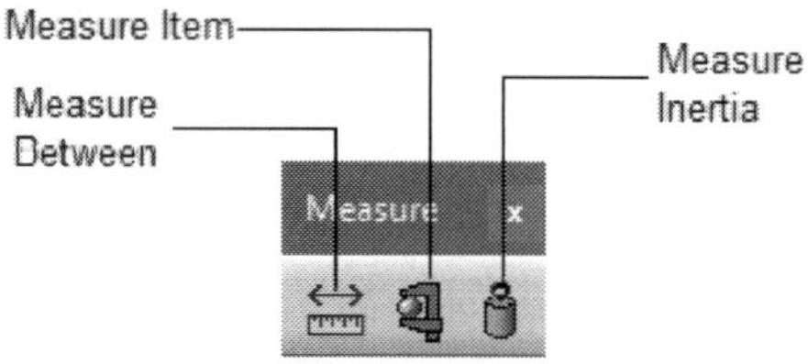

***Figure 7-21** The **Measure** toolbar*

Measuring between Elements

Toolbar:	Measure > Measure Between

To measure the distance and angle between two elements, choose the **Measure Between** tool from the **Measure** toolbar; the **Measure Between** dialog box will be displayed, as shown in Figure 7-22. Also, you will be prompted to specify the first selection item to be measured. Select the first entity to be measured; you will be prompted to specify the second selection item. Select the second element from the geometry area; a callout will be displayed attached to the selected entities. This callout will display the measurement of length and angle of the selected entities. The length and angle measurement of the selected entities will also be displayed in the **Minimum distance** and the **Angle** display boxes of the **Measure Between** dialog box. Next, choose the **OK** button from the **Measure Between** dialog box. Note that the number of measuring parameters displayed in the **Results** area of the **Measure Between** dialog box depend upon the check boxes selected in the **Measure Between Customization** dialog box. To invoke the **Measure Between Customization** dialog box, choose the **Customize** button in the **Measure Between** dialog box.

The options in the **Definition** area of the **Measure Between** dialog box are used to specify the type of measurement and elements to be selected. You will notice that by default, the **Measure between** button is chosen in the **Definition** area of this dialog box. You can measure the distance and angle between two selected entities using this option.

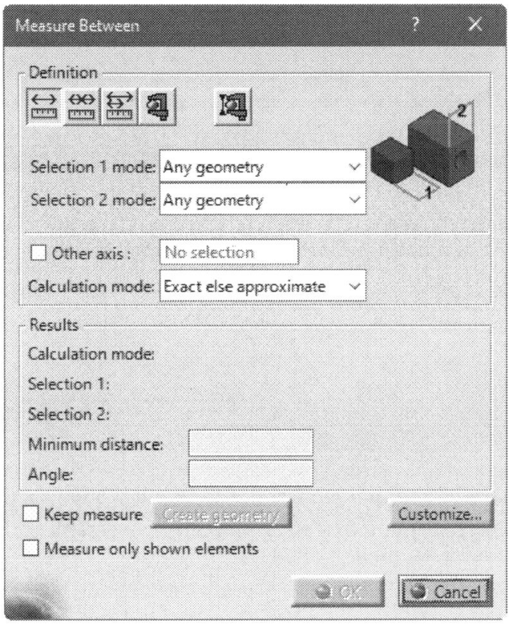

Figure 7-22 *The **Measure Between** dialog box*

If you choose the **Measure between in chain mode** button from the **Definition** area of the **Measure Between** dialog box, it will take rest of the measurements after the first measurement in the chain mode. The current measurement will be taken between the current two selections.

If you choose the **Measure between in fan mode** button, it will take rest of the measurements after the first measurement in the absolute mode.

The **Selection 1 mode** and the **Selection 2 mode** drop-down lists are used as a filter to specify the type of element that you need to select.

You can also define the mode of calculation by using the **Calculation mode** drop-down list. The **Results** area is used to display the results of the selected elements. The **Create geometry** button in this dialog box is used to create a point on the selected geometry.

If the **Keep measure** check box is selected, then a callout displaying the measurement will be attached to the selected elements and will remain displayed in the geometry area, even if you exit the **Measure Between** dialog box.

Measuring Items

Toolbar: Measure > Measure Item

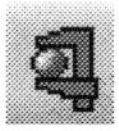

The **Measure Item** tool is used to measure length of a line, area of a planar face, radius of a circular element, and thickness of a feature. To measure elements using this tool, choose the **Measure Item** tool from the **Measure** toolbar; the **Measure Item** dialog box will be displayed, as shown in Figure 7-23.

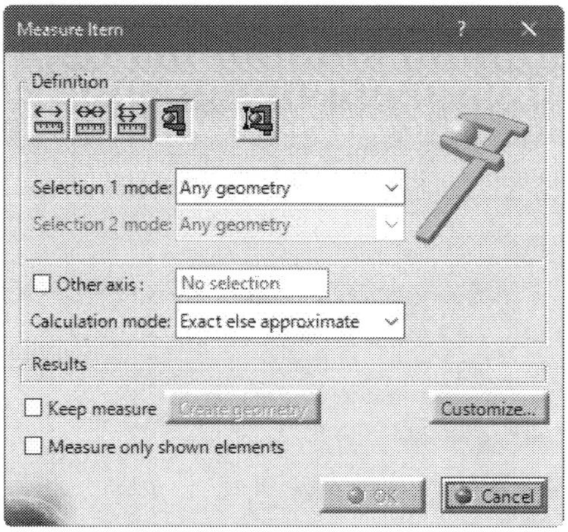

*Figure 7-23 The **Measure Item** dialog box*

Also, you will be prompted to indicate the item to be measured. Select the item; the resulting measurement will be displayed attached to the selected item. Choose the **OK** button from the **Measure Item** dialog box.

Measuring Inertia

Toolbar: Measure > Measure Inertia

The **Measure Inertia** tool is used to measure inertia of the selected body or selected face. To measure inertia, choose the **Measure Inertia** tool from the **Measure** toolbar; the **Measure Inertia** dialog box will be displayed, as shown in Figure 7-24.

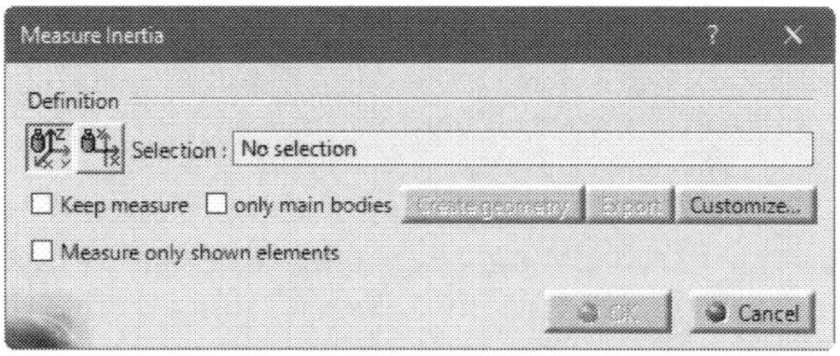

*Figure 7-24 The **Measure Inertia** dialog box*

By default, the **Measure Inertia 3D** button is chosen in the **Measure Inertia** dialog box. Next, select Part Body from the Specification tree; the **Measure Inertia** dialog box expands, as shown in Figure 7-25.

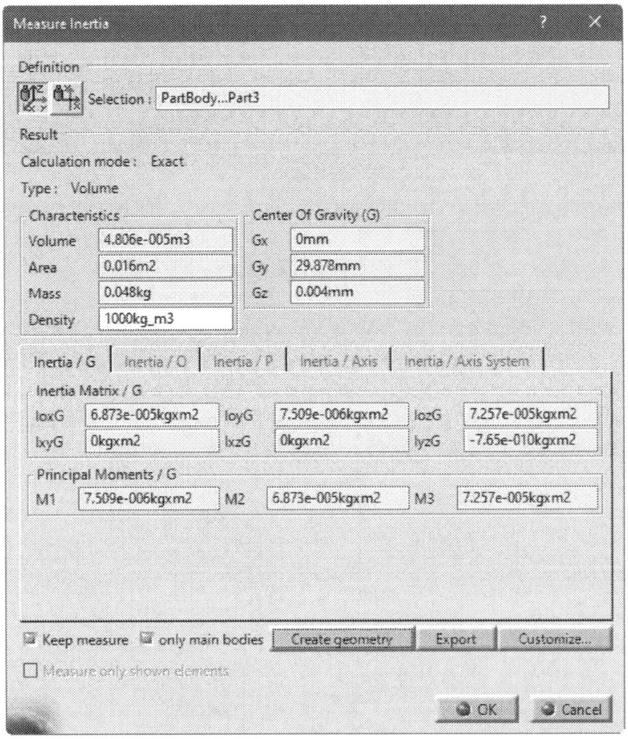

Figure 7-25 *The expanded* ***Measure Inertia*** *dialog box*

If the **only main bodies** check box is selected, the values of volume, area, mass, center of gravity, inertia, and so on of the main bodies will be displayed in the **Result** area.

When you choose the **Create geometry** button from the **Measure Inertia** dialog box, the **Creation of Geometry** dialog box is displayed, as shown in Figure 7-26. You can specify whether the type of geometry should be associative or not by selecting the corresponding radio button. You can view the center of gravity by choosing the **Center of gravity** button from the dialog box and selecting a feature/face. Also, you can create the local axis system by choosing the **Axis system** button.

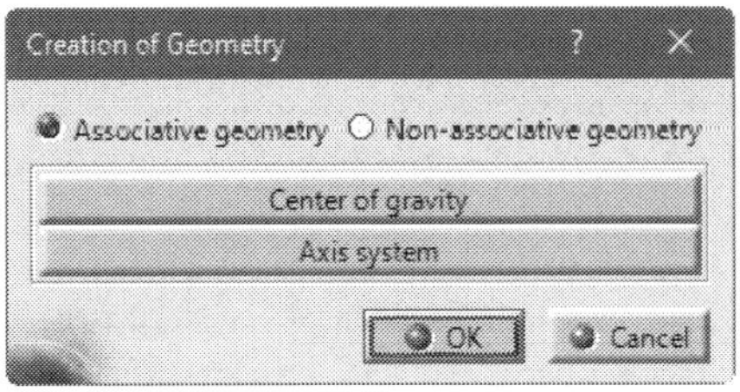

Figure 7-26 *The* ***Creation of Geometry*** *dialog box*

You can use the **Export** button from the **Measure Inertia** dialog box to save the calculated parameters in a text file.

You can also compute inertia with respect to a point, an axis, or origin. To do so, choose the **Customize** button from the dialog box; the **Measure Inertia Customization** dialog box containing a list of parameters will be displayed, as shown in Figure 7-27. Select the check box corresponding to the parameters to be measured and choose **OK**; the result will be displayed in the **Measure Inertia** dialog box. When you enter a new value in the **Density** edit box of this dialog box and then press the ENTER key, all the parameters will be recalculated and new values will get displayed in the **Result** area.

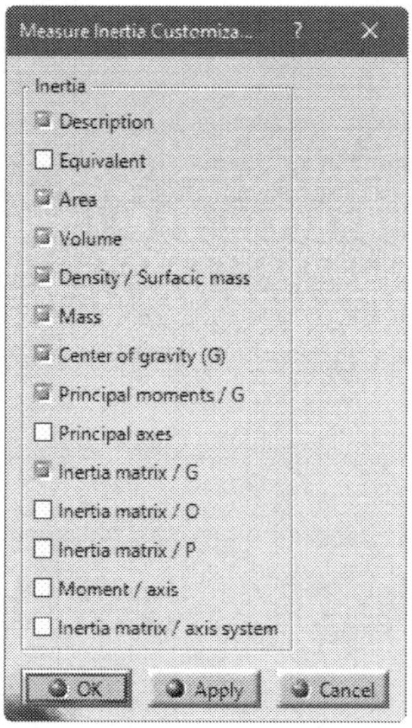

***Figure 7-27** The **Measure Inertia Customization** dialog box*

On choosing the **Measure Inertia 2D** button from the **Measure Inertia** dialog box, you can only measure the 2D inertia of the planar faces.

TUTORIALS

Tutorial 1

In this tutorial, you will create the model of a Bottom Seat, as shown in Figure 7-28. Its views and dimensions are shown in Figure 7-29. After creating this model, you will perform the following modifications.

1. Change the two holes on the front face of the model to countersunk holes.
2. Change the hole on the right face of the model to a counterbore hole.
3. Change the curved pocket feature on the upper face of the model to a rectangular slot.

The final model after modification is shown in Figure 7-30.

(Expected time: 45 min)

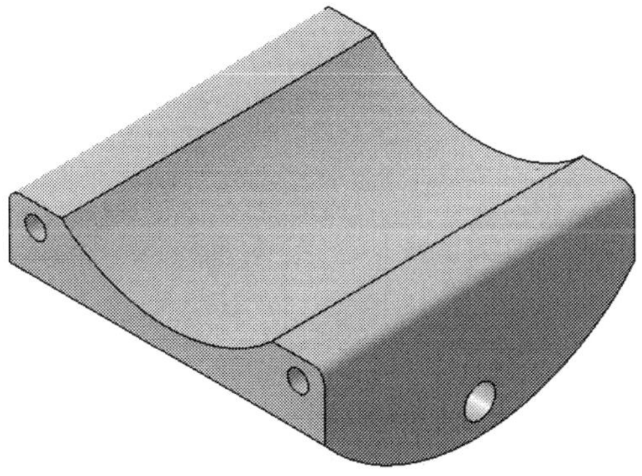

Figure 7-28 Model of the Bottom Seat for Tutorial 1

The following steps are required to complete this tutorial:

a. Start a new file in the **Part** workbench of CATIA V5 and create base feature of the model by extruding the sketch drawn on the zx plane.
b. Create a pocket feature by extruding the sketch drawn on the front face of the base feature.
c. Fillet the right and left edges of the model.
d. Create hole features on the front and right faces of the model.
e. Edit the model.

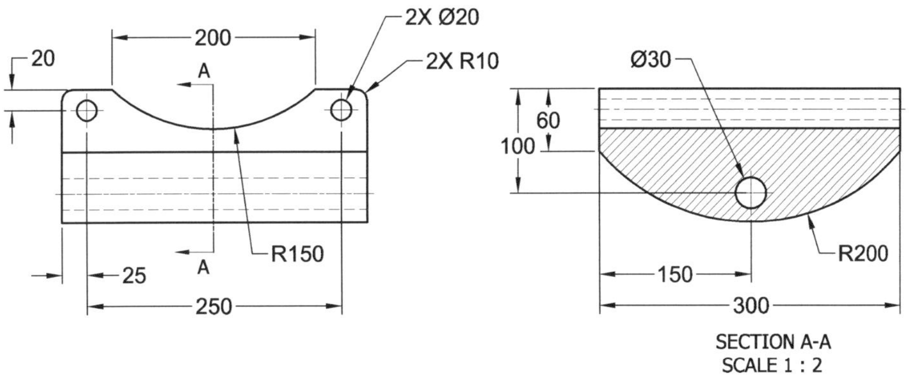

Figure 7-29 Views and dimensions of the Bottom Seat for Tutorial 1

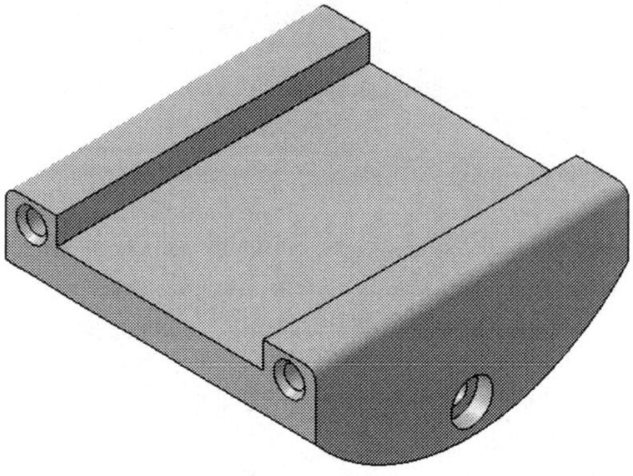

Figure 7-30 The model after modifications

Creating the Base Feature

The base feature of the model will be created by extruding the sketch drawn on the zx plane with the **Mirrored extent** check box selected.

1. Start a new file in the **Part** workbench and draw the sketch of the base feature by selecting the zx plane as the sketching plane, as shown in Figure 7-31.

2. Exit the **Sketcher** workbench and choose the **Pad** tool to invoke the **Pad Definition** dialog box.

3. In the **Pad Definition** dialog box, set the value of the **Length** spinner to **150** and select the **Mirrored extent** check box. Choose the **OK** button in this dialog box; the base feature is created, refer to Figure 7-32.

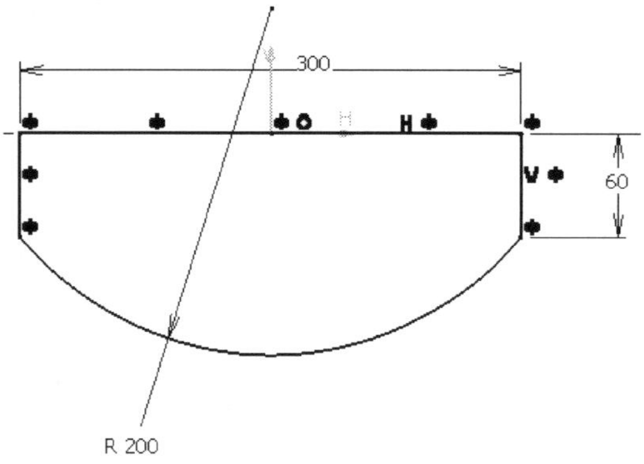

Figure 7-31 Sketch of the base feature

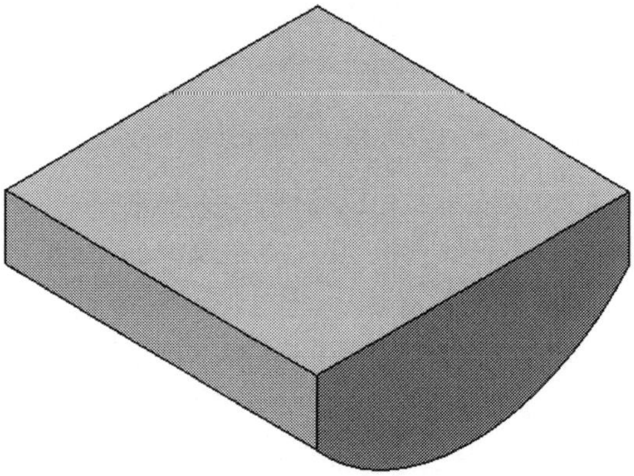

Figure 7-32 The base feature created

Creating the Pocket Feature

The second feature of this model will be created by extruding the sketch drawn on the front face of the model using the **Pocket** tool.

1. Select the front face of the model as the sketching plane and invoke the **Sketcher** workbench.

2. Draw the sketch for this feature, as shown in Figure 7-33, and exit the **Sketcher** workbench.

3. Invoke the **Pocket Definition** dialog box. In this dialog box, select the **Up to last** option from the **Type** drop-down list and choose the **OK** button; the second feature is created. The model after creating the pocket feature is shown in Figure 7-34.

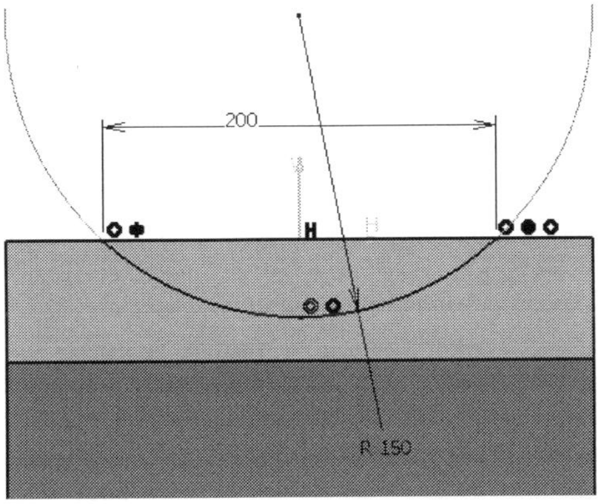

Figure 7-33 Sketch of the pocket feature

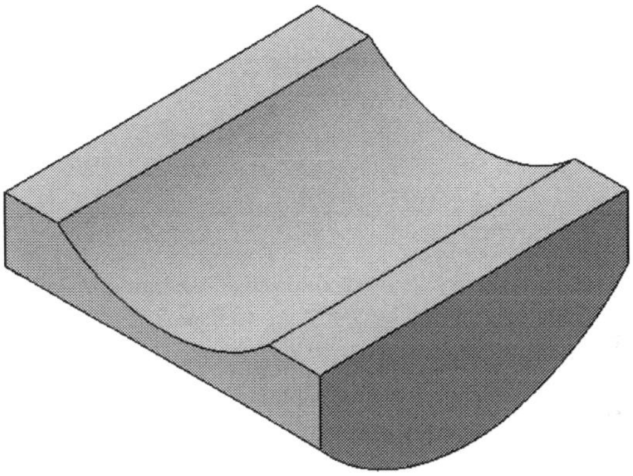

Figure 7-34 The model after creating the pocket feature

Creating the Remaining Features

1. Create rest of the features using the **Fillet** and **Hole** tools. For dimensions, refer to Figure 7-29. The model after creating all the features is shown in Figure 7-35.

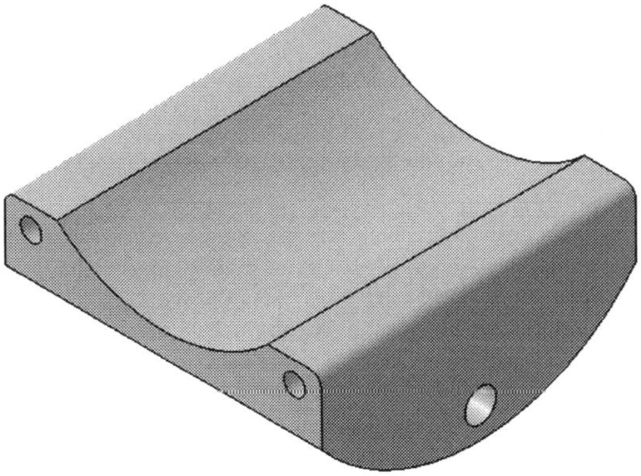

Figure 7-35 The model after creating all the features

Editing Features

After creating the model, you need to perform some editing operations on it. Like, you need to replace simple holes placed on the front face of the base feature with the countersunk holes.

1. Double-click on the right hole placed on the front face of the base feature of the model; the **Hole Definition** dialog box is displayed.

2. Choose the **Type** tab in the **Hole Definition** dialog box.

3. Select the **Countersunk** option from the drop-down list in the **Type** tab of the **Hole Definition** dialog box.

4. Select the **Depth & Angle** option from the **Mode** drop-down list.

5. Set the value of the **Depth** spinner to **5** and the **Angle** spinner to **90deg**.

6. Choose the **OK** button from the **Hole Definition** dialog box; the simple hole is converted into a countersunk hole.

7. Similarly, change the left hole placed on the front face of the base feature into a countersunk hole.

Note

In case, the changes are not applied immediately to the model then the color of the model will turn red. Choose the ***Update All*** *button from the* ***Tools*** *toolbar to apply the changes to the model.*

Next, you need to change the simple hole placed on the right face of the base feature into a counterbore hole.

8. Double-click on the simple hole placed on the right face of the base feature; the **Hole Definition** dialog box is displayed.

9. Choose the **Type** tab of the **Hole Definition** dialog box if not chosen.

10. Select the **Counterbored** option from the drop-down list in the **Type** tab of the **Hole Definition** dialog box.

11. Set the value of the **Diameter** spinner to **40** and the **Depth** spinner to **15**.

12. Choose the **OK** button from the **Hole Definition** dialog box and update the feature if not updated automatically. The simple hole placed on the right face of the base feature is converted into a counterbore hole.

    Next, you need to edit the sketch of the pocket feature of the model. According to the change in the design of the model, you need to convert the curved slot on the upper face of the model into a rectangular slot.

13. Right click on the **Pocket** feature in the Specification tree or in the geometry area; a contextual menu is displayed.

14. Move the cursor to the name of the selected feature, **Pocket1.object**, in the contextual menu; a cascading menu is displayed. Choose the **Edit Sketch.2** option from the cascading menu; the **Sketcher** workbench is invoked.

15. Modify the sketch, as shown in Figure 7-36, using the tools in the **Sketcher** workbench.

16. Exit the **Sketcher** workbench. The model after modifying the sketch is shown in Figure 7-37.

17. Create a folder with the name *c07* inside the *CATIA* folder and then save the created part file with the *c07tut1.CATPart*.

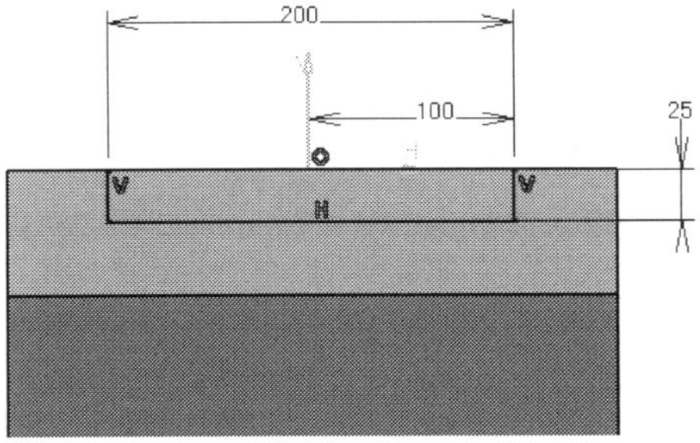

Figure 7-36 *Modified sketch of the second feature*

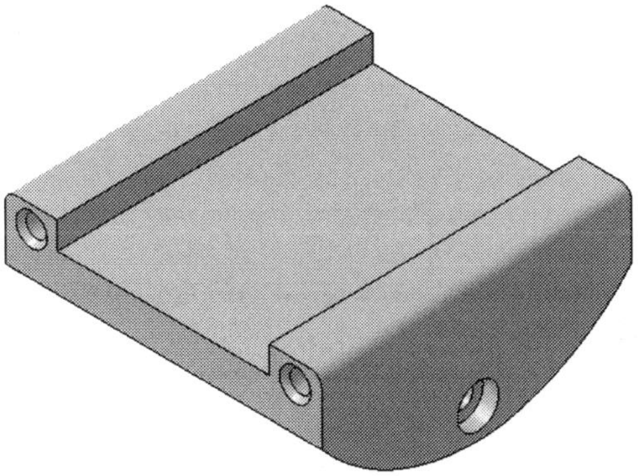

Figure 7-37 *Model after modifications*

Tutorial 2

In this tutorial, you will create the model of the Vice Jaw shown in Figure 7-38. Its views and dimensions are shown in Figure 7-39. After creating this model, you will edit some of its dimensions. Figure 7-40 shows the dimensions that need to be edited.

(Expected time: 30 min)

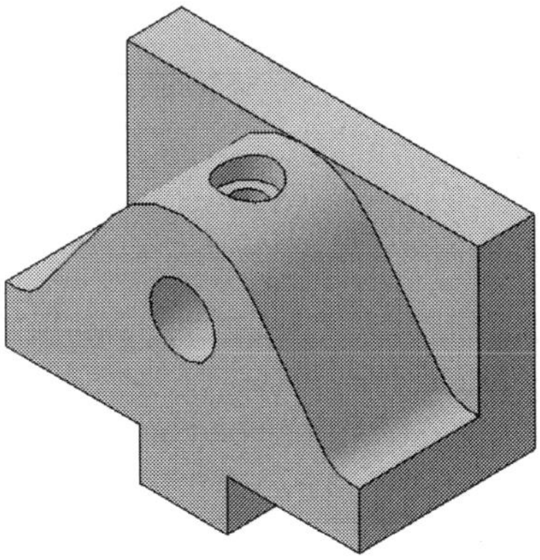

Figure 7-38 *Model of the Vice Jaw for Tutorial 2*

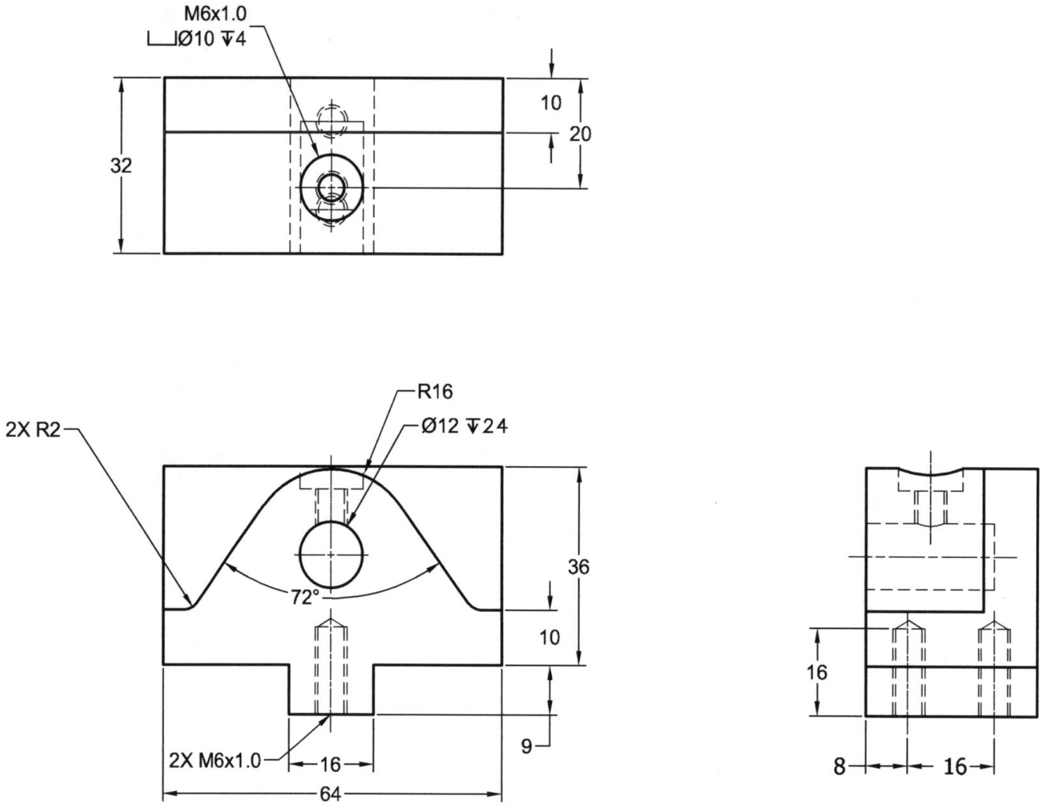

Figure 7-39 Views and dimensions of the Vice Jaw for Tutorial 2

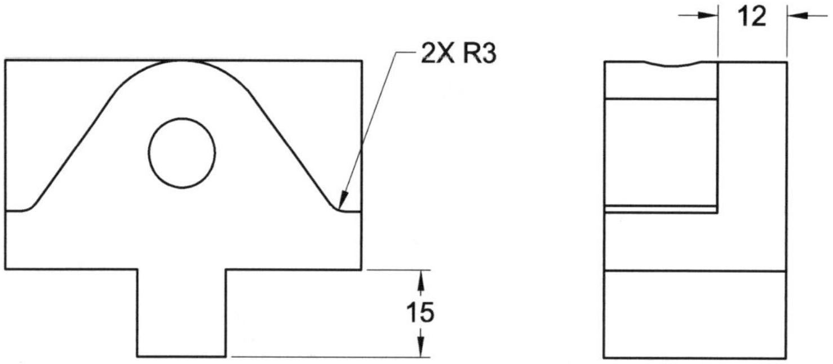

Figure 7-40 Dimensions of the Vice Jaw to be modified

The following steps are required to complete this tutorial:

a. Create the base feature of the model by extruding the sketch drawn on the zx plane.
b. Create the second feature of the model by extruding the sketch drawn on the front face of the base feature.
c. Create the third feature of the model by extruding a rectangular sketch drawn on the front face of the base feature.
d. Create holes using the **Hole** tool.
e. Add fillet to the edges of the model.
f. Modify the model.

Creating the Base Feature

The base feature of the model will be created by extruding the L-shaped sketch drawn on the zx plane with the **Mirrored extent** radio button selected.

1. Start a new part file. Select the zx plane as the sketching plane and invoke the **Sketcher** workbench.

2. Draw the sketch of the base feature, as shown in Figure 7-41, and exit the **Sketcher** workbench.

3. Invoke the **Pad Definition** dialog box and set the value in the **Length** spinner to **32**.

4. Select the **Mirrored extent** check box and choose the **OK** button in the **Pad Definition** dialog box. The model after creating the base feature is shown in Figure 7-42.

Creating the Second Feature

The second feature of this model will be created by extruding the sketch drawn on the front face of the base feature.

1. Select the front face of the base feature, refer to Figure 7-42, and invoke the **Sketcher** workbench.

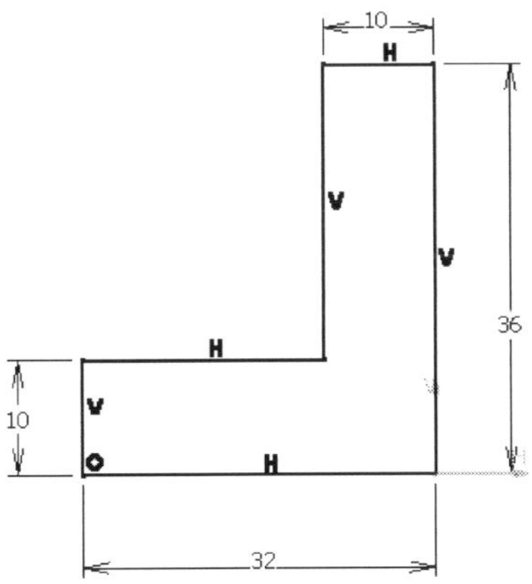

Figure 7-41 *Sketch of the base feature*

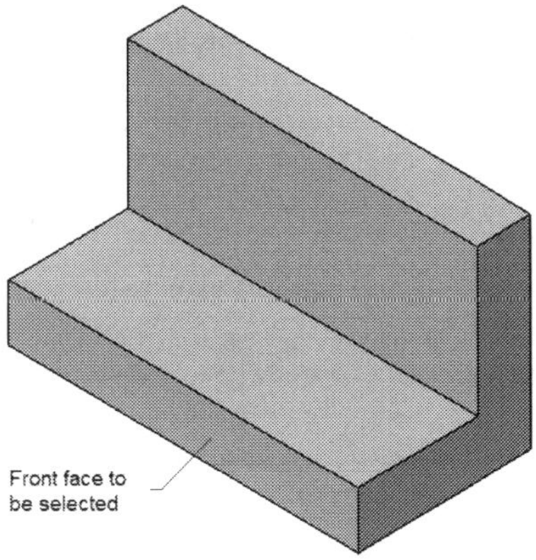

Figure 7-42 *Model after creating the base feature*

2. Draw sketch of the second feature, as shown in Figure 7-43, and exit the **Sketcher** workbench.

3. Invoke the **Pad Definition** dialog box and extrude the sketch using the **Up to next** option. Choose the **OK** button. The model after creating the second feature is shown in Figure 7-44.

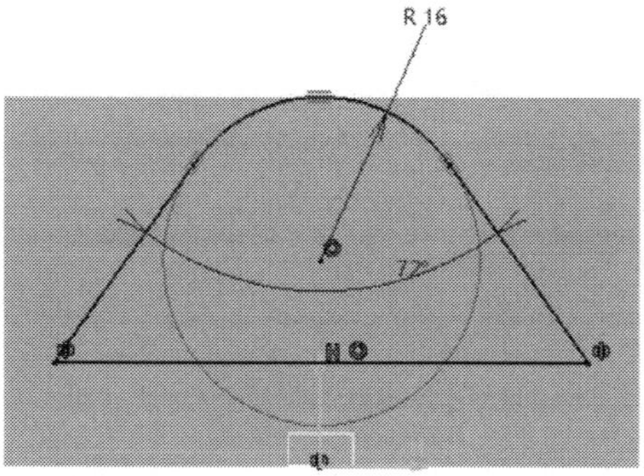

Figure 7-43 *Sketch of the second feature*

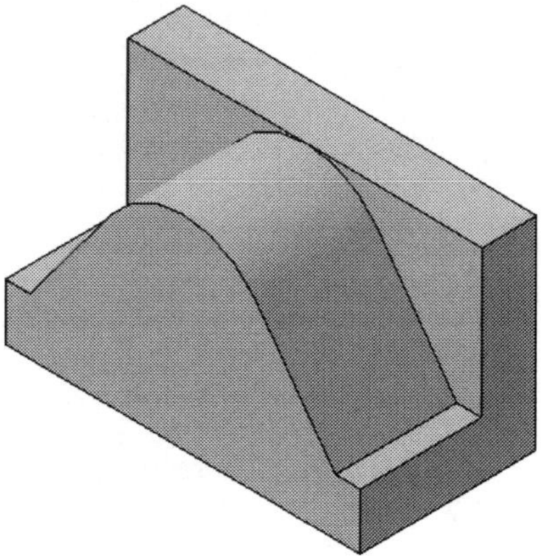

Figure 7-44 *Model after creating the second feature*

Creating the Third Feature

The third feature of the model will be created by extruding a rectangular sketch drawn on the front face of the base feature.

1. Select the front face of the base feature and invoke the **Sketcher** workbench.

2. Draw the sketch of the third feature, as shown in Figure 7-45, and exit the **Sketcher** workbench.

3. Invoke the **Pad Definition** dialog box and select the **Up to plane** option from the **Type** drop-down list. Next, choose the rear face of the model and then choose the **OK** button from the dialog box.

   The model after creating the third feature is shown in Figure 7-46.

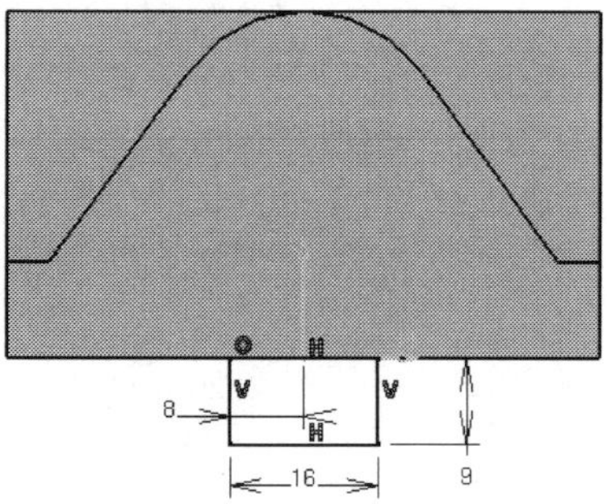

Figure 7-45 *Sketch of the third feature*

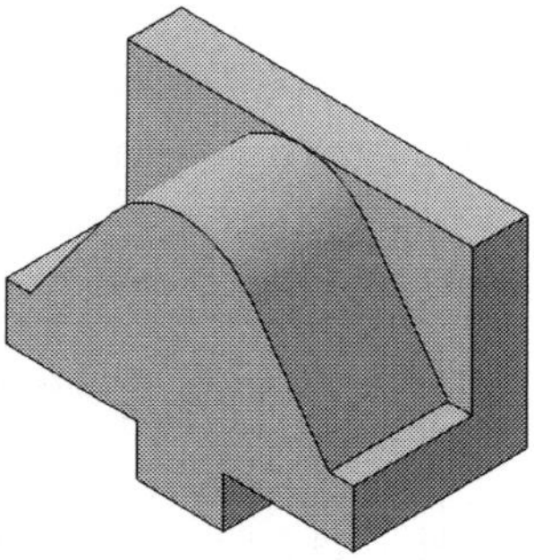

Figure 7-46 *Model after creating the third feature*

Creating Hole Features

Next, you need to create hole features using the **Hole** tool. First, you will create the counterbore hole on top of the curved face. But if you select a curved face after invoking the **Hole** tool, an arbitrary plane tangent to the selected curved face will be created which is not desirable. So, before creating the hole feature, you need to place a point on the curved face to specify the placement of the hole.

1. Select the zx plane and invoke the **Sketcher** workbench.
2. Place a point and make it coincident with the uppermost edge of the flat face on the top. Next, place a dimension between the placed point and the right edge of the base feature.
3. Set the value of the dimension placed in the previous step to **20**, as shown in Figure 7-47, and then exit the **Sketcher** workbench.
4. Select the upper curved face of the model and press and hold the Ctrl key. Now, select the point that you have placed earlier.
5. Invoke the **Hole Definition** dialog box.
6. Set the value of the **Diameter** spinner to **6** and the **Depth** spinner to **10**.

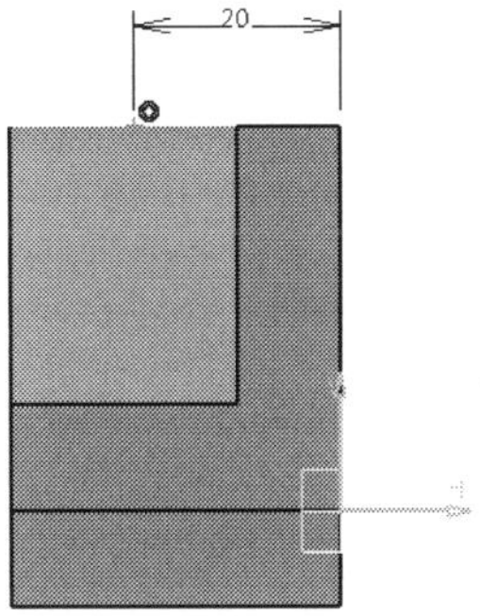

Figure 7-47 *Sketch of the **Hole** feature*

7. Choose the **Type** tab of the **Hole Definition** dialog box and select the **Counterbored** option from the drop-down list in the **Type** tab.
8. Set the value of the **Diameter** spinner to **10** and the **Depth** spinner to **4**.
9. Choose the **OK** button from the **Hole Definition** dialog box. The model after creating the counterbore hole is shown in Figure 7-48.

Creating the Remaining Features

1. Create other hole and fillet features of the model. For dimensions, refer to Figure 7-39. Note that the holes have V-bottom. The final model after creating all the features is shown in Figure 7-49.

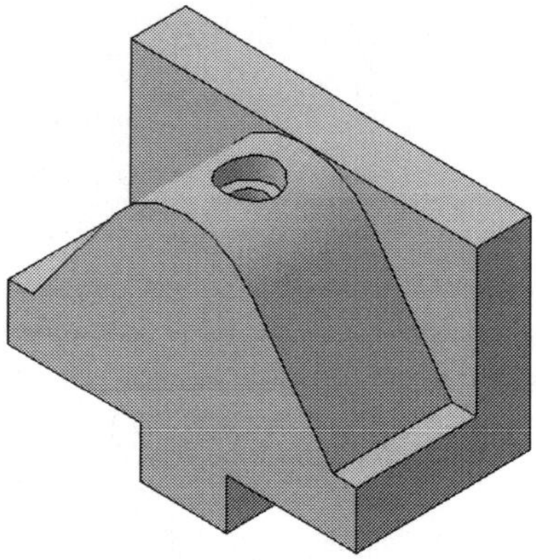

Figure 7-48 The model after creating the counterbore hole

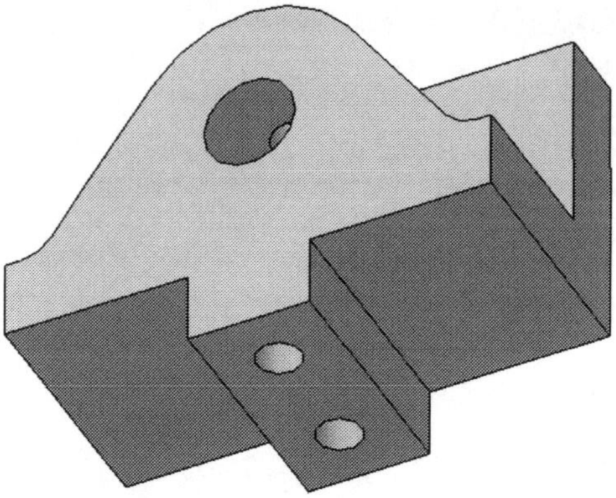

Figure 7-49 Final model after creating all the features

Modifying the Model

After creating the model, you need to modify its design to incorporate some design changes.

1. Double-click on the fillet feature from the geometry area or from the Specification tree; the **Edge Fillet Definition** dialog box is displayed.

2. Set the value of the **Radius** spinner to **3** and then choose the **OK** button from the **Edge Fillet Definition** dialog box; the radius of the fillet is modified.

   Next, you need to modify the height of the third feature.

3. Select the third feature (**Pad.3**) from the geometry area or from the Specification tree and then right-click; a cascading menu is displayed. Move the cursor to the name of the feature and choose **Edit Parameters** from the cascading menu displayed; all the dimensions of the selected feature are displayed in the geometry area.

4. Double-click on the vertical dimension **9**; the **Constraint Definition** dialog box is displayed. Set the value of the **Value** spinner to **15** and then choose the **OK** button from the dialog box.

   The color of the model changes to red and the dimensional value is modified, but the dimensional modification is not reflected in the model. Therefore, you need to update the model to see the effect of the modification.

5. Choose the **Update All** button from the **Tools** toolbar; the **Updating** message box is displayed with a progress bar displaying the updation of the model. After updating, the **Updating** dialog box is closed and the updated model is displayed in the geometry area.

6. Similarly, edit the remaining dimensions shown in Figure 7-40. The final model after the modification of dimensions is shown in Figure 7-50.

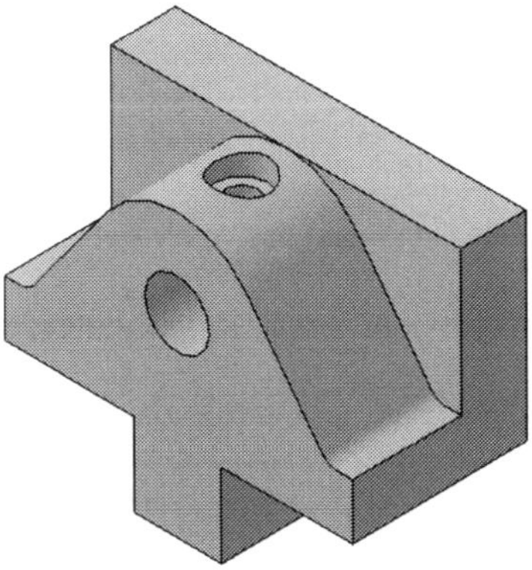

Figure 7-50 *The final model after modifying dimensions*

7. Save the file with the file name *c07tut2.CATPart* in the *c07* folder.

Tutorial 3

In this tutorial, you will create the model shown in Figure 7-51. After creating this model, you will edit its design by replacing the counterbore holes with countersunk holes. Also, replace the rectangular slot with an elongated slot. Figure 7-52 shows the edited model. The views and dimensions of the model to be created are shown in Figure 7-53. (**Expected time: 45 min**)

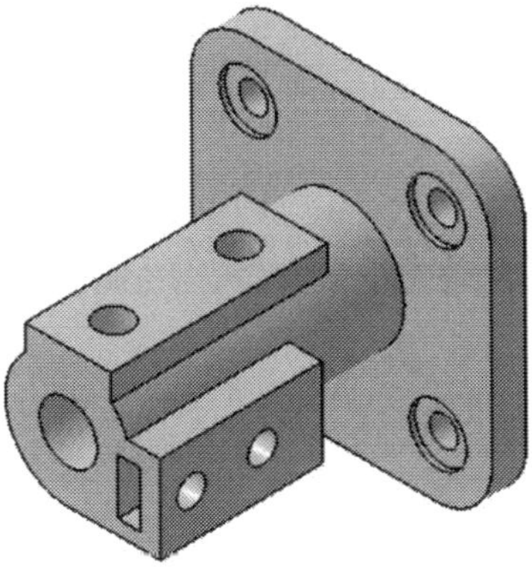

Figure 7-51 *Model for Tutorial 3*

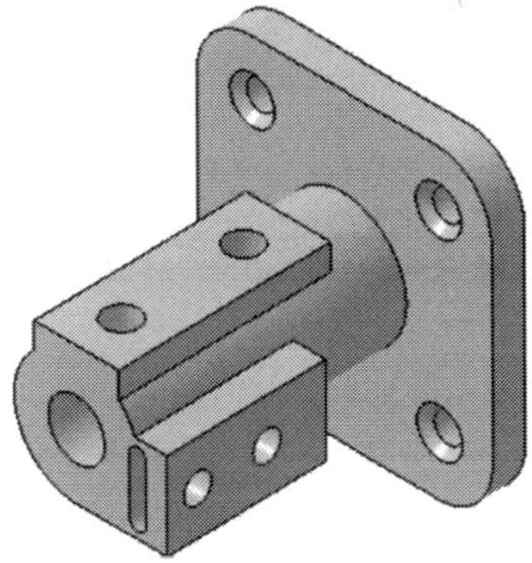

Figure 7-52 *Final edited model*

The following steps are required to complete this tutorial:

a. Start a new file in the **Part Design** workbench.
b. Create the base feature of the model by extruding the sketch drawn on the yz plane.
c. Create the second feature which is a pad feature by extruding the sketch drawn on the front face of the base feature.

d. Create the third and fourth features by extruding the sketch drawn on the front face of the second feature.
e. Create the fifth feature by using the **Pocket** tool.
f. Create the holes using the **Hole** tool and fillet the required edges.
g. Edit the design of the model.

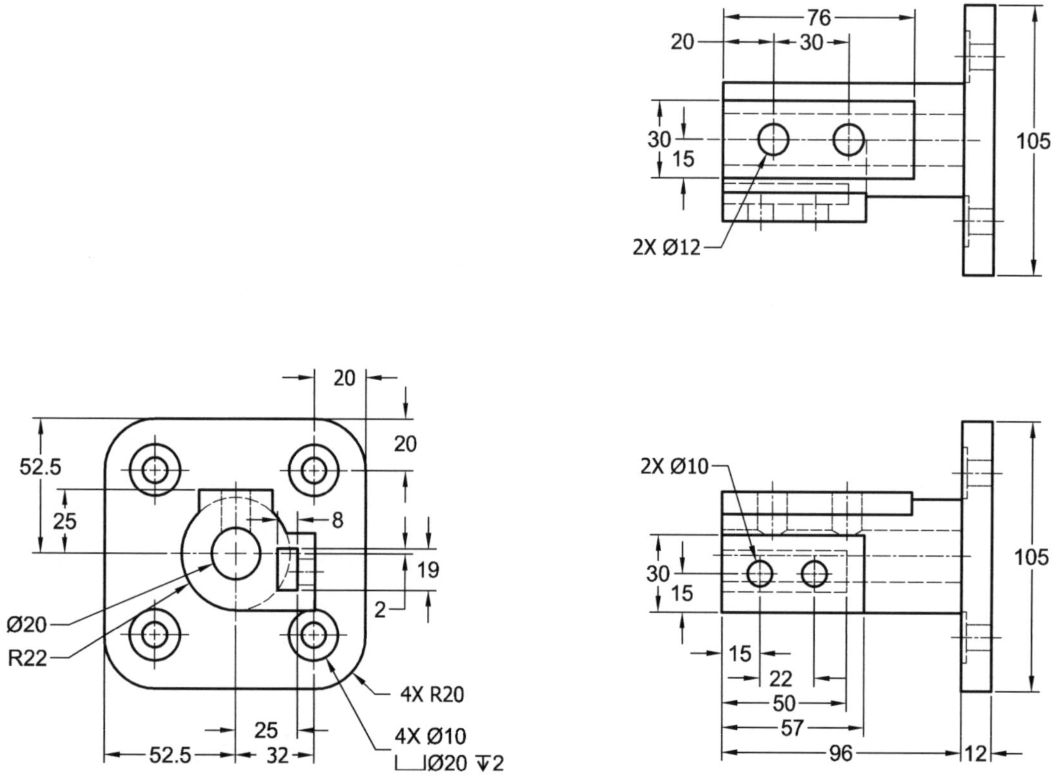

Figure 7-53 *Views and dimensions of the model for Tutorial 3*

Creating the Base Feature

1. Start a new file in the **Part Design** workbench.

2. Select the yz plane and invoke the **Sketcher** workbench.

3. Draw the sketch using the **Centered Rectangle** tool and fillet the corners of the rectangle by using the **Corner** tool. Refer to Figure 7-53 for dimensions.

4. Exit the **Sketcher** workbench and extrude the sketch up to a depth of 12 mm. The model after creating the base feature is shown in Figure 7-54.

Creating the Second Feature

The second feature of the model will be created by extruding the sketch drawn on the front face of the base feature.

1. Select the front face of the base feature as the sketching plane and invoke the **Sketcher** workbench.

2. Draw the sketch of the second feature. The sketch consists of a circular profile, refer to Figure 7-53 for dimensions.

3. Exit the **Sketcher** workbench and extrude the sketch up to a depth of 96 mm. The model after creating the second feature is shown in Figure 7-55.

Creating Rectangular Features

The third and fourth features of the model are rectangular. These features will be created by extruding the sketch drawn on the front face of the second feature. Next, you will create the fifth feature of the model which is a Pocket feature.

1. Select the front face of the second feature and invoke the **Sketcher** workbench.

2. Draw the sketch of the third feature, refer to Figure 7-53 for dimensions.

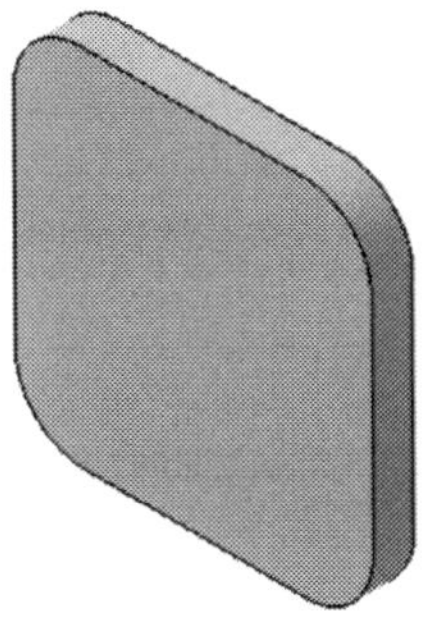

Figure 7-54 *Model after creating the base feature*

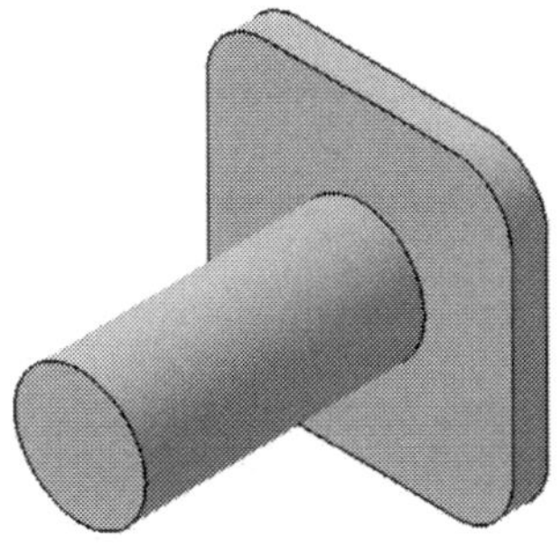

Figure 7-55 *Model after creating the second feature*

3. Exit the **Sketcher** workbench and extrude the sketch up to a depth of 76 units. Reverse the direction of extrusion, if required. The model after creating the third feature is shown in Figure 7-56.

4. Similarly, create the fourth feature by extruding the sketch drawn on the front face of the second feature to a depth of 57 units.

5. Create the fifth feature of the model by using the **Pocket** tool. Refer to Figure 7-53 for dimensions. The model after creating the fourth and fifth features is shown in Figure 7-57.

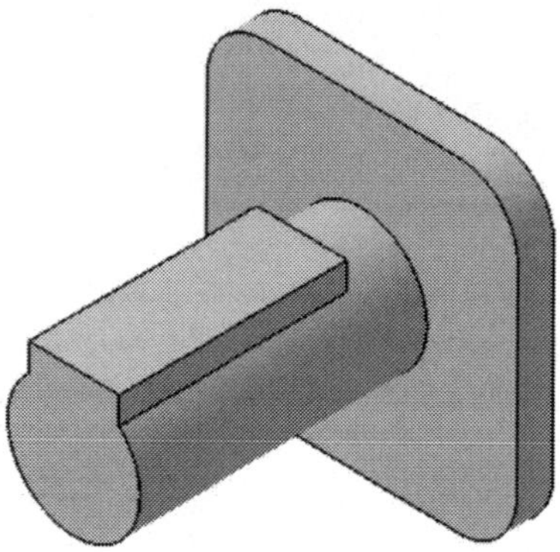

Figure 7-56 *Model after creating the third feature*

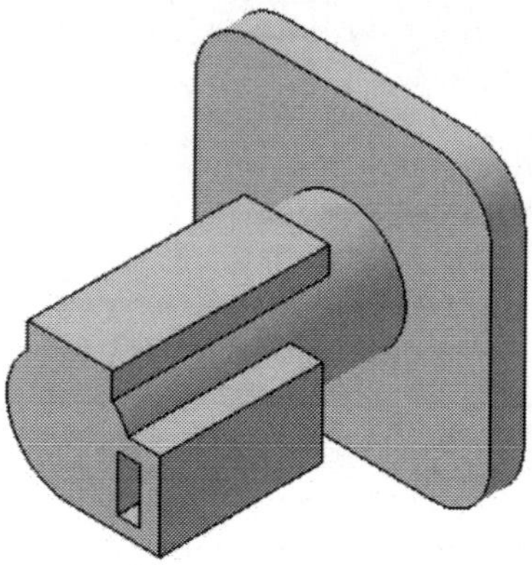

Figure 7-57 *Model after creating the fourth and fifth features*

Creating Holes

Next, you need to create three types of holes, namely simple hole, and counterbore hole.

1. Create four counterbore holes on the base feature and two holes each on the third and fourth features, as shown in Figure 7-58. Refer to Figure 7-53 for dimensions.

   The final model after creating all the features is shown in Figure 7-58.

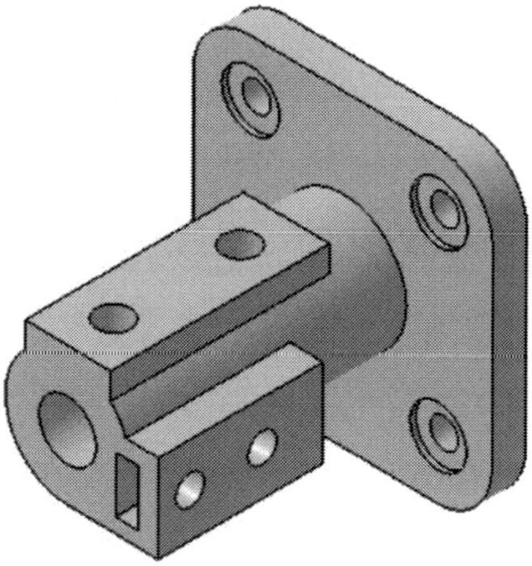

Figure 7-58 *The final model after creating all features*

Editing the Design of the Model

After creating the model, you need to perform some editing operations on it to make changes in its design. The change includes replacing the counterbore holes with the countersunk holes. You also need to replace the rectangular slot placed on the front face of the model with a rectangular elongated hole by editing the sketch of the cut feature.

1. Double-click on any one of the counterbore holes in the geometry area or Specification tree; the **Hole Definition** dialog box is displayed.

2. Choose the **Type** tab not already chosen, and select the **Countersunk** option from the drop-down list in this tab. By default, 2mm and 90 degree are displayed in the **Depth** and **Angle** spinners, respectively in the **Parameters** area.

3. Accept the default values and choose the **OK** button in the **Hole Definition** dialog box; the counterbore hole is replaced by a countersunk hole.

4. Similarly, convert the remaining three counterbore holes into countersunk holes.

   Next, you need to edit the sketch of the pocket created on the front face of the model.

5. Double-click on the sketch of the pocket feature in the Specification tree. On doing so, the **Sketcher** workbench is invoked.

6. Edit the sketch, refer to Figure 7-59, and then exit the **Sketcher** workbench. The final edited model is shown in Figure 7-60.

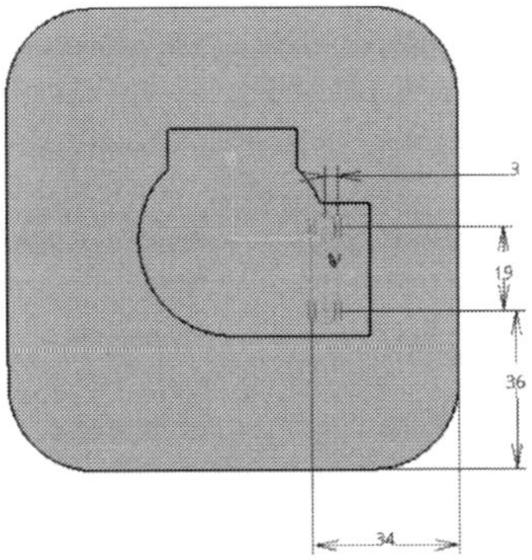

***Figure 7-59** Edited sketch of the pocket feature*

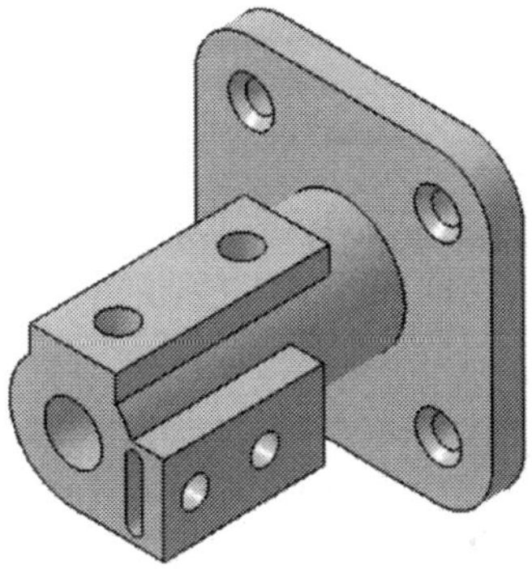

***Figure 7-60** Final edited model*

7. Save the file with the name *c07tut3.CATPart* in the *c07* folder.

Self-Evaluation Test

Answer the following questions and then compare them to those given at the end of this chapter:

1. The __________ tool is used to measure the inertia of the selected body or the selected face of the body.

2. The deactivated features can be activated by using the __________ option from the contextual menu.

3. Use the ______________ option from the contextual menu to redefine the sketch plane of a sketch for the sketch-based features.

4. When you invoke the **Measure Inertia** dialog box, the __________ button is chosen by default.

5. The _________ button in the **Measure Inertia** dialog box is used to measure the 2D inertia.

6. You cannot edit the parameters of a feature by double-clicking on it. (T/F)

7. Reordering is defined as the process of changing the position of the features in the Specification tree. (T/F)

8. Defining a model in the work object is a process in which you rollback the model to an earlier stage. (T/F)

9. You cannot redefine the sketch plane of the sketches of a sketch-based feature. (T/F)

10. You can delete unwanted features of a model. (T/F)

Review Questions

Answer the following questions:

1. Choose the _________ button from the **Measure** toolbar to measure the distance and angle between two elements.

2. When the _________ check box is selected, a callout displaying the measurement will be attached to the selected elements. This callout will remain displayed in the geometry area even if you exit the **Measure Between** dialog box.

3. The _________ tool is used to measure the length of a line, the area of a planar face, the radius of a circular element, and the thickness of a feature.

4. If you select the _________ check box from the **Delete** dialog box, all the features referenced to the feature to be deleted will also be deleted.

5. The _________ button in the **Update Diagnosis** dialog box is used to edit the feature or the sketch with errors.

6. In CATIA V5, every model consists of features. These features in some way or the other are related to each other. (T/F)

7. If the parent feature is deleted, the **Delete** dialog box is displayed. This dialog box provides you with an option to delete the child features. (T/F)

8. You cannot deactivate a feature while working in the **Part** workbench of CATIA V5. (T/F)

9. You cannot measure the distances between the selected elements in the **Part** workbench. (T/F)

10. You can use the drag and drop functionalities to cut and paste the item within a file or from one file to another. (T/F)

EXERCISES

Exercise 1

Create the model shown in Figure 7-61. Its views and dimensions are shown in Figure 7-62.

(Expected time: 45 min)

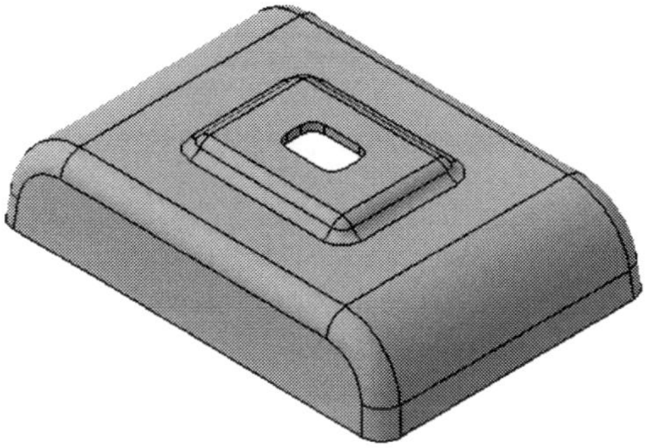

Figure 7-61 Model for Exercise 1

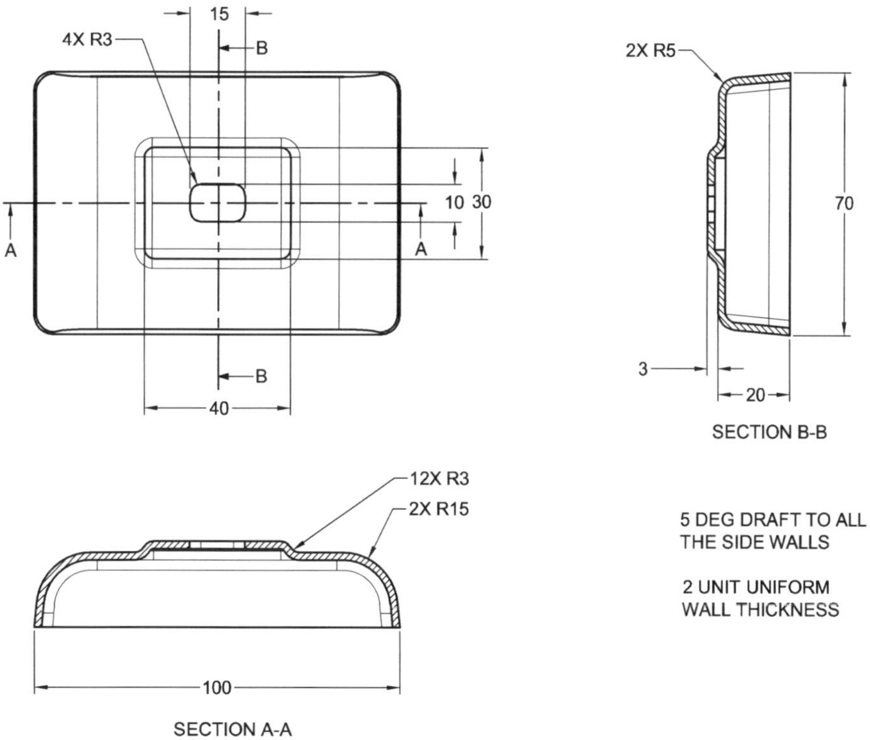

Figure 7-62 Views and dimensions for Exercise 1

Exercise 2

Create the model of the Plummer Block Casting shown in Figure 7-63. The views and dimensions of the model are shown in Figure 7-64. **(Expected time: 30 min)**

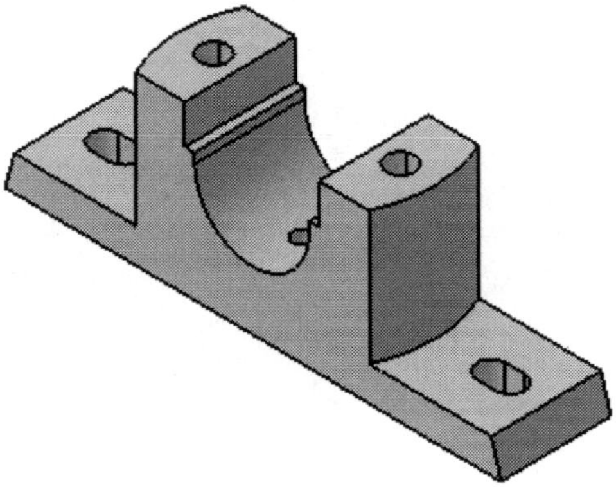

Figure 7-63 Model for Exercise 2

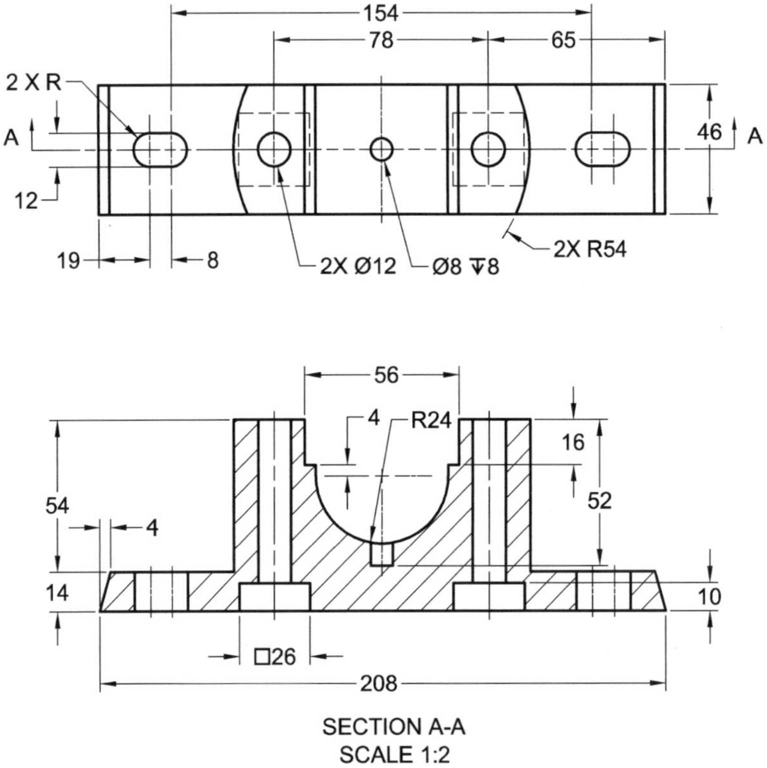

Figure 7-64 Views and dimensions for Exercise 2

Exercise 3

Create the model of the Machine Block shown in Figure 7-65. The views and dimensions of the model are shown in Figure 7-66. **(Expected time: 30 min)**

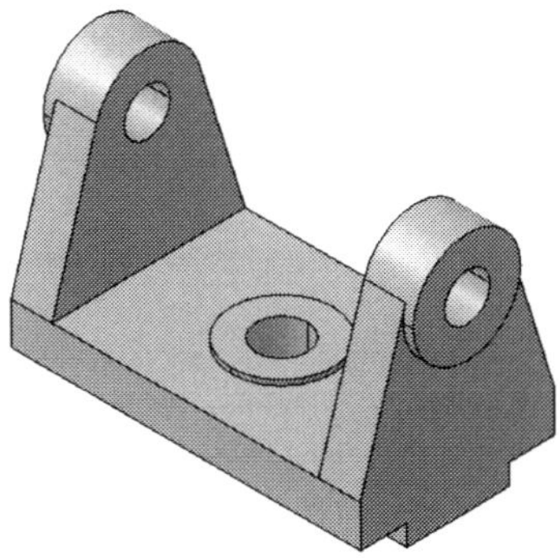

Figure 7-65 Model for Exercise 3

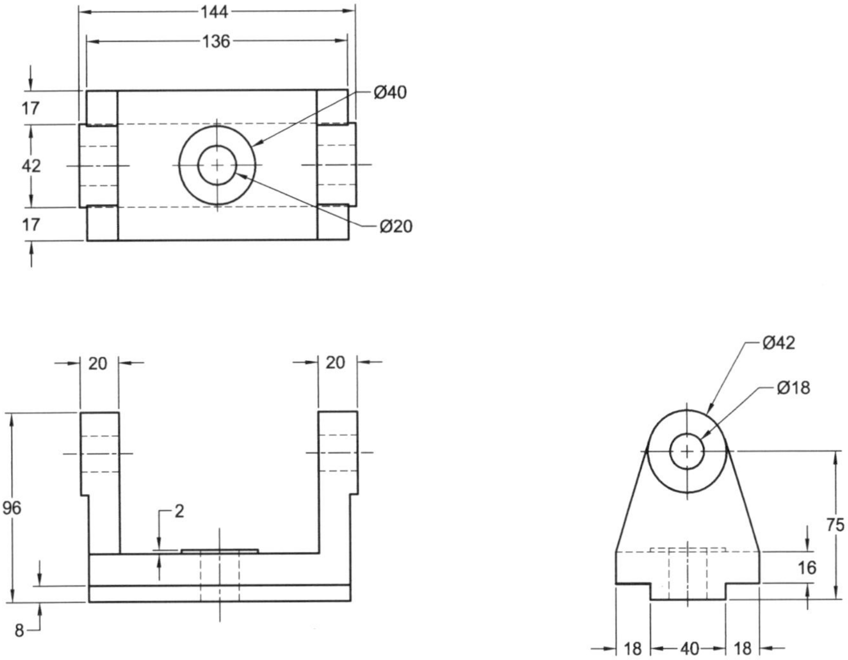

Figure 7-66 Views and dimensions for Exercise 3

Answers to Self-Evaluation Test

1. **Measure Inertia**, **2.** **Activate**, **3.** **Change Sketch Support**, **4.** **Measure Inertia 3D**, **5.** **Measure Inertia 2D**, **6.** F, **7.** T, **8.** T, **9.** F, **10.** T

Chapter 8

Transformation Features and Advanced Modeling Tools-I

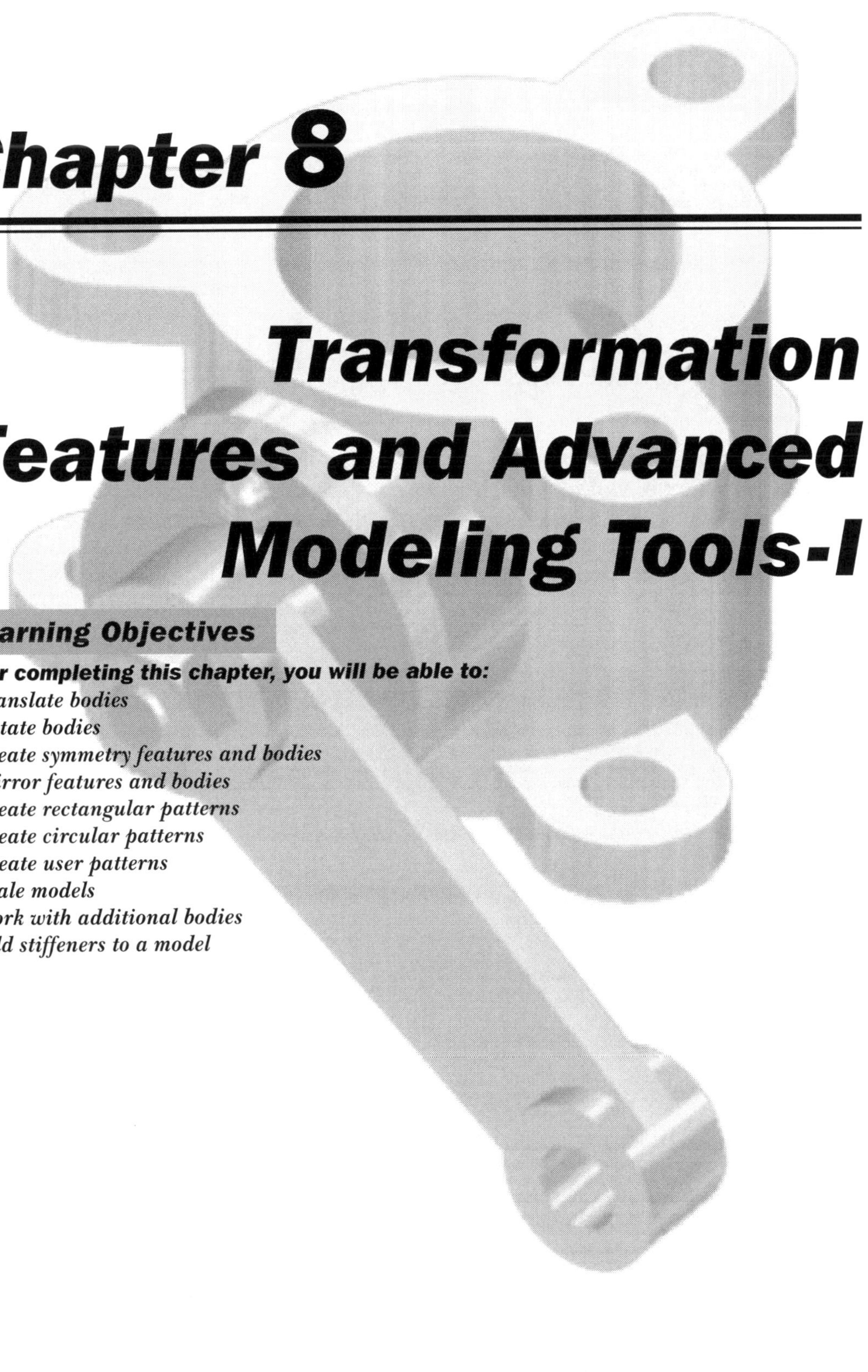

Learning Objectives

After completing this chapter, you will be able to:

- *Translate bodies*
- *Rotate bodies*
- *Create symmetry features and bodies*
- *Mirror features and bodies*
- *Create rectangular patterns*
- *Create circular patterns*
- *Create user patterns*
- *Scale models*
- *Work with additional bodies*
- *Add stiffeners to a model*

TRANSFORMATION FEATURES

In this chapter, you will learn about the transformation tools that are used to move, rotate, mirror, pattern, and scale selected features and bodies. The usages of these transformation tools are discussed next.

Translating the Bodies

Menubar: Insert > Transformation Features > Translation
Toolbar: Transformation Features > Transformations sub-toolbar > Translation

The **Translation** tool is used to move the current body by defining a specific destination. To translate a body, choose the down arrow on the right of the **Translation** tool in the **Transformation Features** toolbar; the **Transformations** sub-toolbar will be displayed, as shown in Figure 8-1.

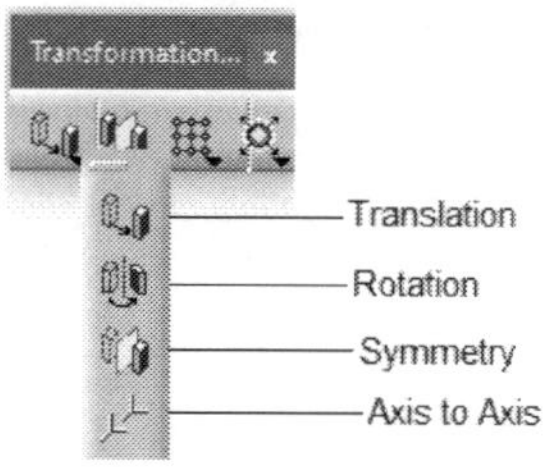

***Figure 8-1** The **Transformations** sub-toolbar*

Choose the **Translation** tool from the **Transformations** sub-toolbar; the **Translate Definition** dialog box and the **Question** message box will be displayed. The **Question** message box is shown in Figure 8-2.

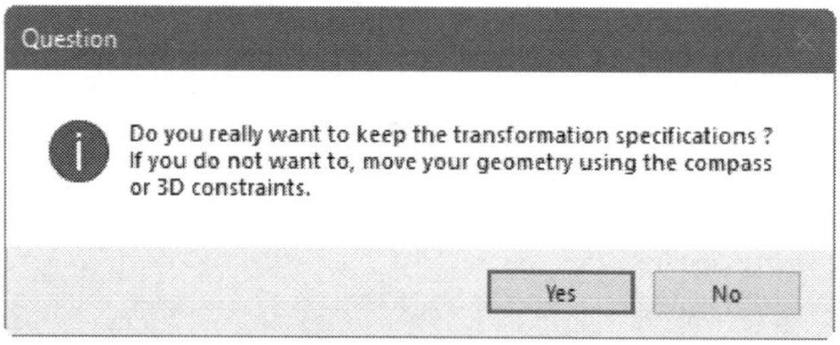

***Figure 8-2** The **Question** message box*

This message box prompts you to specify whether you really want to keep the transformation specifications or not. If not, then choose the **No** button from the message box and move the geometry using the compass or 3D constraints. Else, choose the **Yes** button from this message box; the **Translate Definition** dialog box will be activated, as shown in Figure 8-3. Also, you will be prompted to select or enter the direction because the **Direction, distance** option is selected in the **Vector Definition** drop-down list of this dialog box. Using this option, you can specify the translated distance and direction in which the body will be translated.

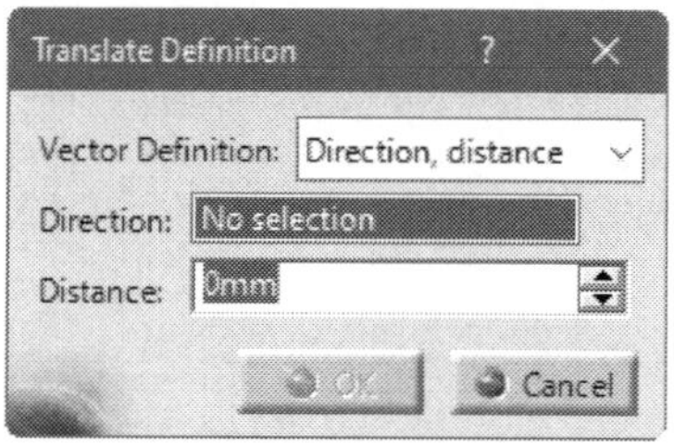

***Figure 8-3** The **Translate Definition** dialog box*

Select the translation direction from the geometry area or from the Specification tree. The direction of translation could be along a straight edge, sketched line, or plane. In case of planes, their normal direction is taken as the direction of translation.

Next, you need to specify the translation distance. Set the value of the translation distance in the **Distance** spinner; the preview of the translated body will be displayed in the geometry area. Choose the **OK** button from the **Translate Definition** dialog box.

> **Tip**
> *You can place an existing part body anywhere in the 3D space with respect to the origin by using Compass. To do so, drag the red point of the Compass, place it over the part body, and then drag the axis perpendicular to the sketch plane of the Compass; the **Move Warning** message box will be displayed informing that you can break the relations between the sketch plane and the sketch. To break the relations, choose the **OK** button from the message box; the positioning relations of the sketch will be broken and the selected part body will move from its original position. Remember that this functionality is not available after using the transformation tools such as **Mirror**, **Translation**, and **Symmetry**.*

If you select the **Point to point** option from the **Vector Definition** drop-down list, you need to specify a point on the model as the reference point and then the destination point where you want to move the body.

If you select the **Coordinates** option from the **Vector Definition** drop-down list, you need to specify only the coordinates of the destination point where you have to place the body with respect to the default coordinates of the **Part** workbench.

Figure 8-4 shows the plane to be selected as directional reference. Figure 8-5 shows the resultant translated body moved normal to the selected plane.

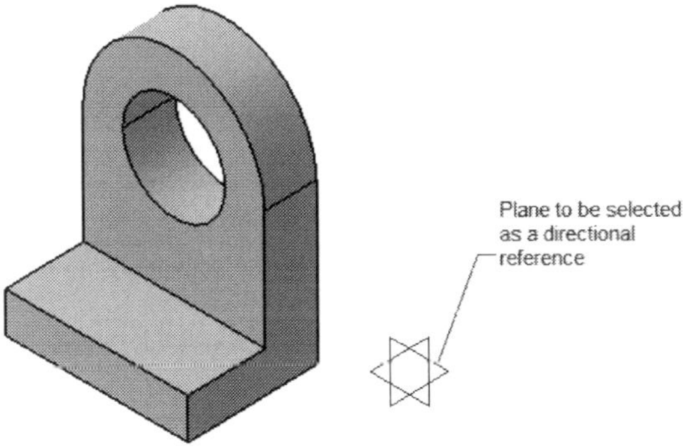

Figure 8-4 *Plane to be selected as the directional reference*

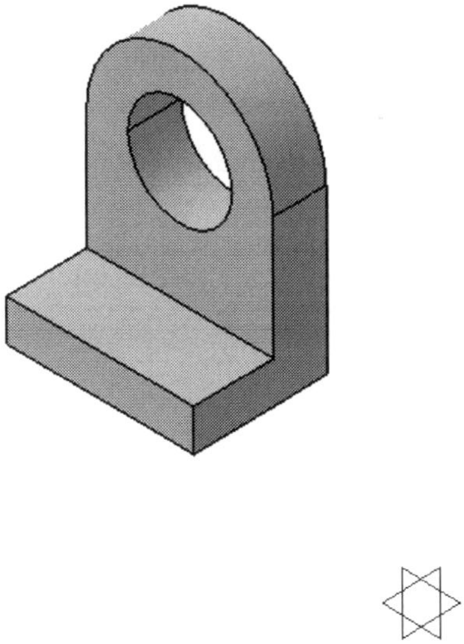

Figure 8-5 *Resultant translated body*

Rotating the Bodies

Menubar: Insert > Transformation Features > Rotation
Toolbar: Transformation Features > Transformations sub-toolbar > Rotation

The **Rotation** tool is used to rotate the body with respect to an axis of rotation. To do so, choose the **Rotation** tool from the **Transformations** sub-toolbar of the **Transformations Features** toolbar; the **Question** message box will be displayed along with the **Rotate Definition** dialog box. Choose the **Yes** button from this message box; the **Rotate Definition** dialog box will be activated, as shown in Figure 8-6.

The **Axis-Angle** option is selected by default in the **Definition Mode** drop-down list. As a result, you will be prompted to select the rotation axis. Select an edge or sketched line as the axis of rotation; the current body will be highlighted and its edges will be displayed in orange. Now, set the value of the angle of rotation in the **Angle** spinner. You can also use the **Angle** drag handle to dynamically rotate the body.

There are two more options for rotating the model in the **Definition Mode** drop-down list of the **Rotate Definition** dialog box, **Axis-Two Elements** and **Three Points**. On selecting the **Axis-Two Elements** option, you can specify the axis of rotation and then select two elements to define the angle of rotation. The angle formed between the first element, the axis of rotation, and the second element will be the angle of rotation. You can select points, linear edges/sketched lines, or planes/planar faces as the elements to define the rotation. Note that if one of the two elements is a plane, the normal of that plane will be used to determine the angle of rotation.

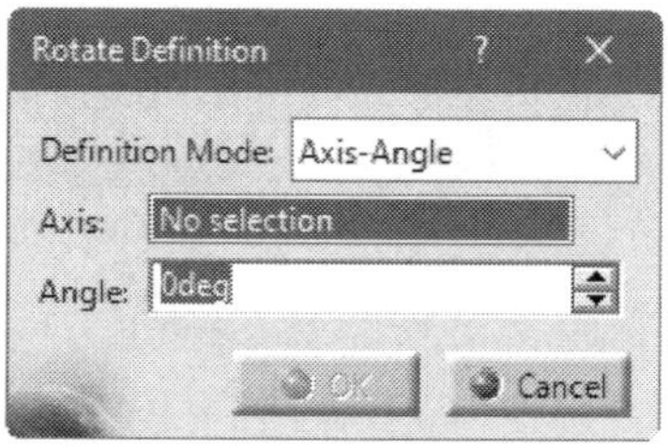

Figure 8-6 *The **Rotate Definition** dialog box*

Selecting the **Three Points** option allows you to select three points to define the rotation angle. In this case, the axis of rotation will be defined by the normal of the plane created by the three points and will pass through the second point. After setting the parameters, choose the **OK** button from the **Rotate Definition** dialog box. Figure 8-7 shows the edge to be selected as the axis of rotation and Figure 8-8 shows the resultant rotated body.

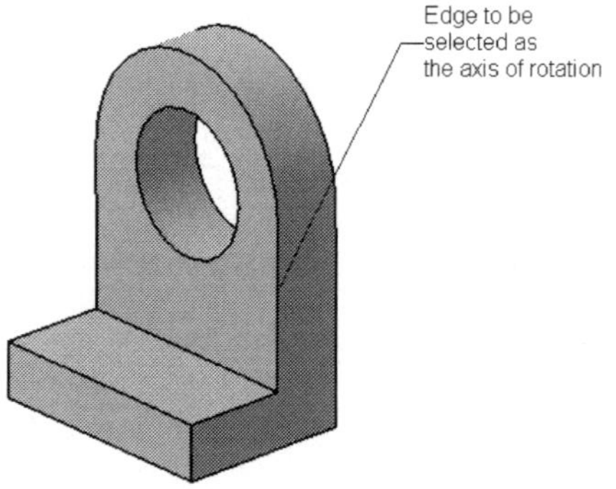

Figure 8-7 *Edge to be selected as the axis of rotation*

Figure 8-8 *Resulting rotated body*

Creating the Symmetry Features

Menubar:	Insert > Transformation Features > Symmetry
Toolbar:	Transformation Features > Transformations sub-toolbar > Symmetry

The **Symmetry** tool is used to flip the position of the body about the symmetry plane without creating its instance. To flip the position of body using this tool, choose the **Symmetry** tool from the **Transformations** sub-toolbar of the **Transformations Features** toolbar; the **Question** message box will be displayed along with the **Symmetry Definition** dialog box. Choose the **Yes** button from the **Question** message box; the **Symmetry Definition** dialog box will be activated, as shown in Figure 8-9. Also, you will be prompted to select the reference point, line, or plane.

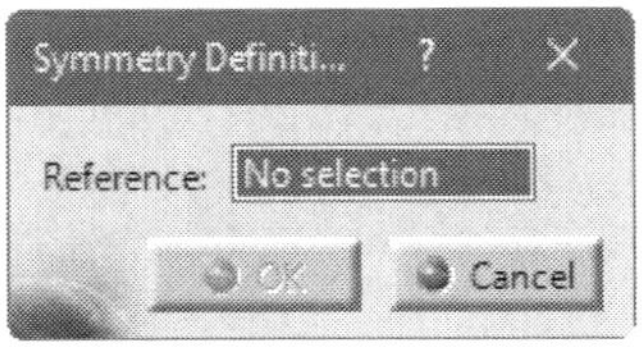

*Figure 8-9 The **Symmetry Definition** dialog box*

Select a point, line, plane, or a face that will be used as a reference for the symmetry; the preview of the symmetric body will be displayed in the geometry area. Choose the **OK** button to flip the position of the body. Figure 8-10 shows the planar face to be selected as the symmetry reference and Figure 8-11 shows the resulting symmetry feature.

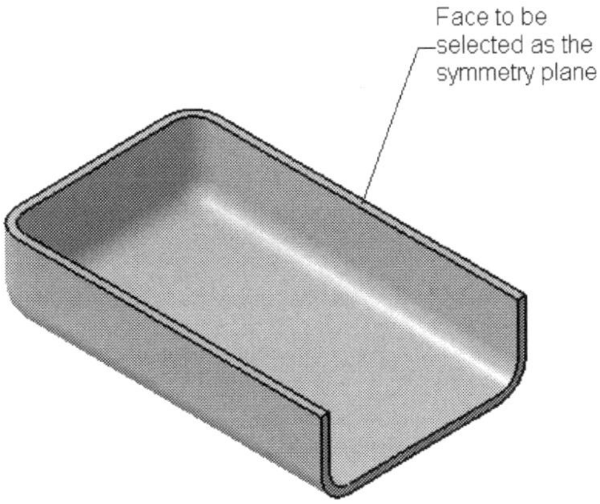

Figure 8-10 The planar face to be selected as the symmetry reference

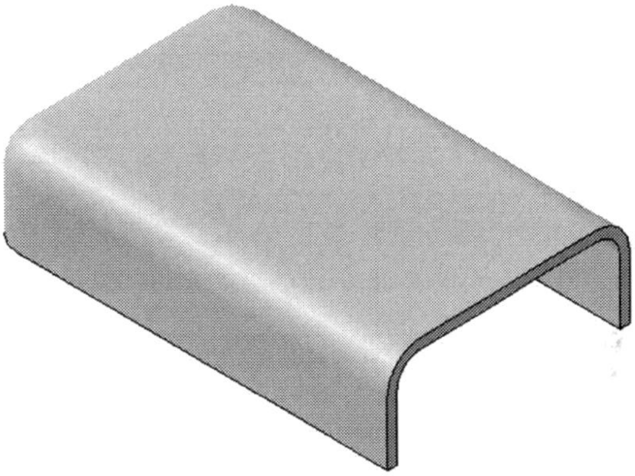

Figure 8-11 Resulting symmetry feature

Transforming the Axis System

Menubar:	Insert > Transformation Features > AxisToAxis
Toolbar:	Transformation Features > Transformations sub-toolbar > AxisToAxis

The **AxisToAxis** tool is used to transform the entire model from one axis system to the other. To do so, choose the **AxisToAxis** tool from the **Transformations** sub-toolbar; the **Question** message box will be displayed along with the **Axis To Axis Definition** dialog box. Next, choose the **Yes** button from this message box; the **Axis To Axis Definition** dialog box will be activated, as shown in Figure 8-12.

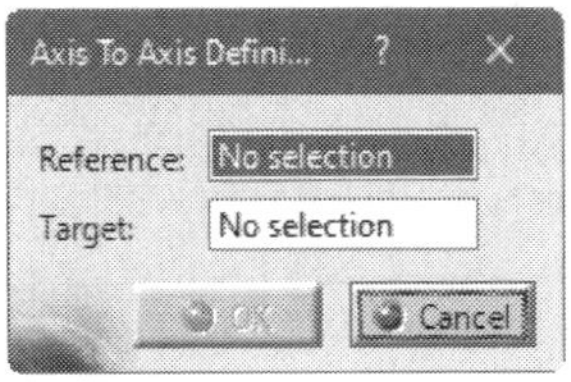

*Figure 8-12 The **Axis To Axis Definition** dialog box*

Also, you will be prompted to select the reference axis system. Select an axis system to be transformed; the name of the selected axis system will be displayed in the **Reference** display box and you will be prompted to select the target axis system. Select the axis system about which you want to transform the model; the model will get oriented with respect to the selected target axis system and the preview of the transformed element will be displayed. Choose the **OK** button from the **Axis To Axis Definition** dialog box to accept the transformation changes. Figure 8-13 shows the model with the reference axis system and the target axis system and Figure 8-14 shows the model after transforming the axis system.

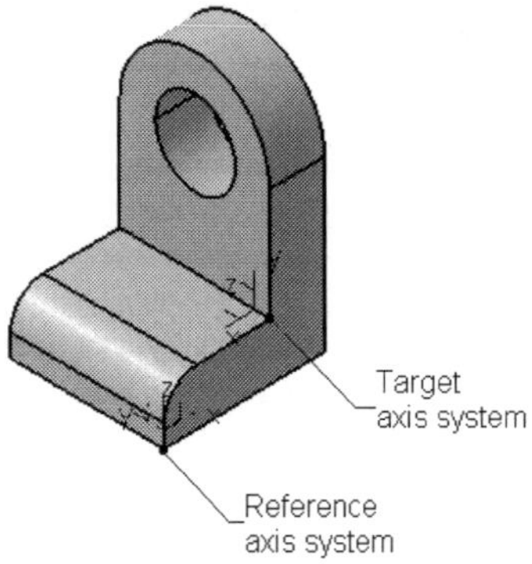

Figure 8-13 *The reference axis system and target axis system to be selected*

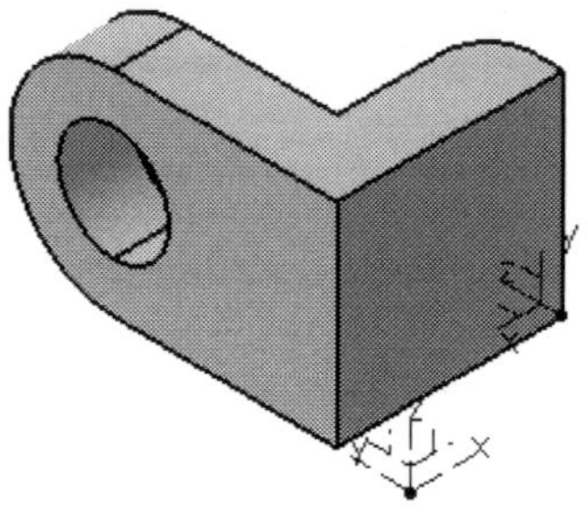

Figure 8-14 *The model after transforming the axis system*

> **Tip**
> *You can create multiple user-defined axis systems using the* ***Axis System*** *tool. This tool is available in the* ***Tools*** *toolbar.*

Mirroring the Features and Bodies

Menubar:	Insert > Transformation Features > Mirror
Toolbar:	Transformation Features > Mirror

The **Mirror** tool allows you to mirror the selected features or whole model about a mirror plane. The mirror plane could be a plane or a planar face. Note that while mirroring, the original item is retained and a mirrored copy is created. To mirror the selected feature, select it from the geometry area or from the Specification tree and then choose the **Mirror** tool from the **Transformation Features** toolbar; the **Mirror Definition** dialog box will be displayed, as shown in Figure 8-15.

Next, you need to select the mirroring element. Select a planar face or a plane as the mirroring element; the wireframe preview of the mirrored feature will be displayed in the geometry area. Choose the **OK** button from the **Mirror Definition** dialog box. Figure 8-16 shows the feature and the mirror plane to be selected. Figure 8-17 shows the resultant mirrored feature. If you select the **Keep Specifications** check box, a mirror copy of the feature will be created in the same direction as the parent object and also having same limiting specifications. This check box will only be activated when you select an object to be mirrored from the graphics window.

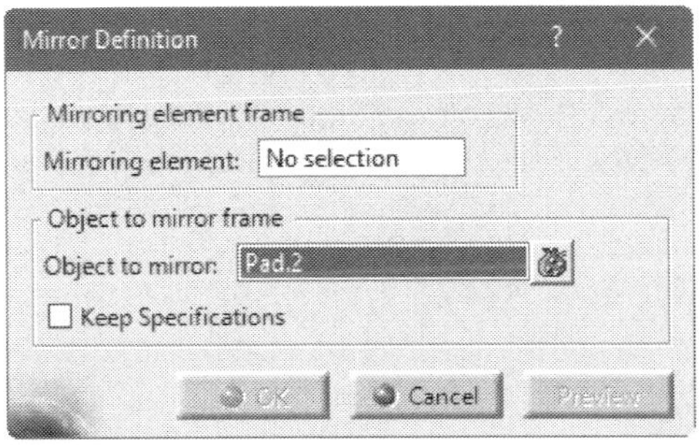

Figure 8-15 *The **Mirror Definition** dialog box*

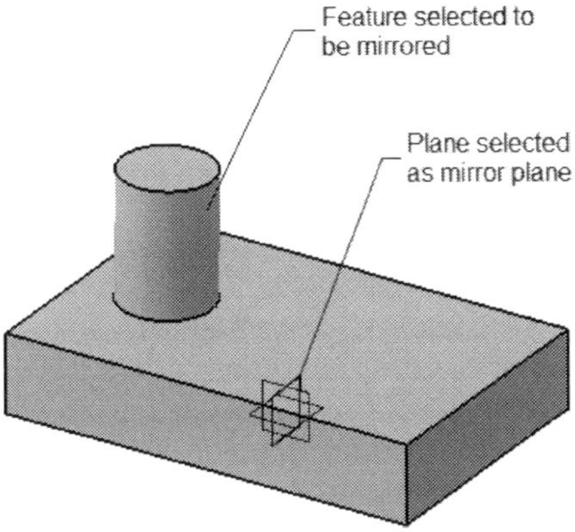

Figure 8-16 *Feature and mirror plane to be selected*

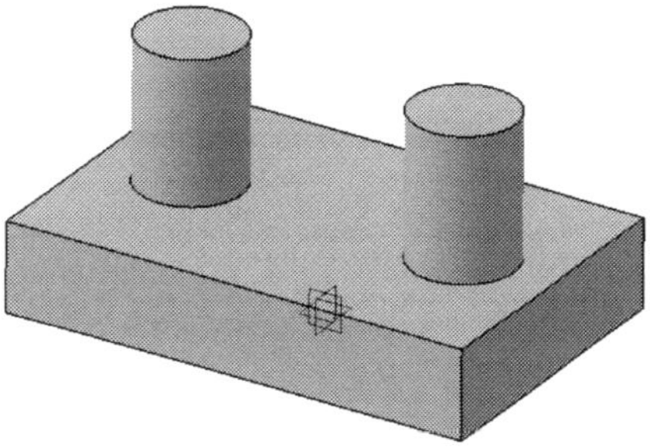

Figure 8-17 *Resultant mirrored feature*

To mirror the entire model, choose the **Mirror** tool from the **Transformation Features** toolbar; the **Mirror Definition** dialog box will be displayed. Next, you need to select the mirror plane from the geometry area, the face to be selected as the mirror plane is shown in Figure 8-18; a preview of the mirrored feature will be displayed in the geometry area and then choose the **OK** button from this dialog box. Figure 8-19 shows the resultant mirrored model.

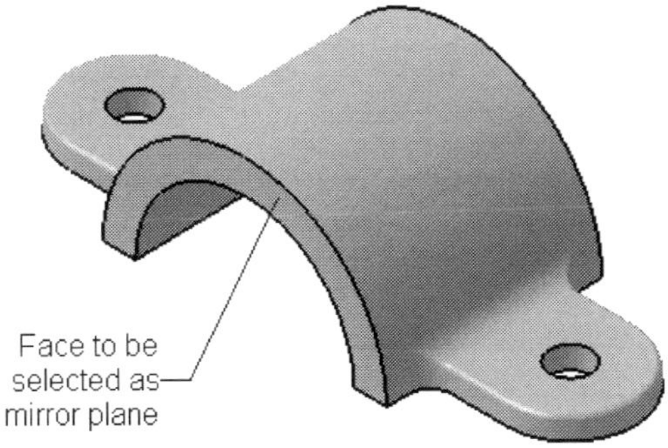

Figure 8-18 *Face to be selected as the mirror plane*

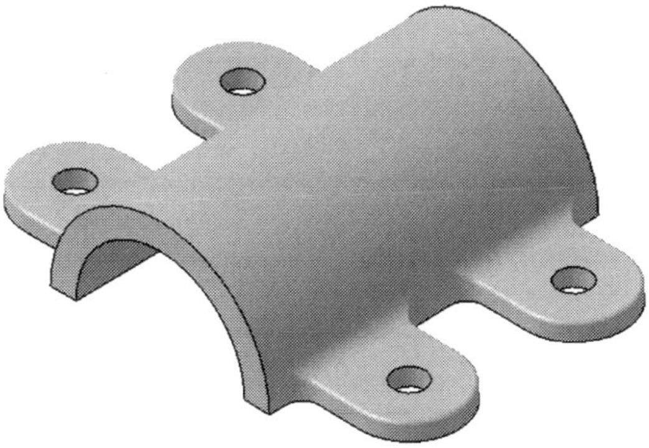

Figure 8-19 *The mirrored model*

You can create a plane or choose the datum plane about which you need to mirror the features. To create the plane, right-click in the **Mirroring Element** selection box and select the **Create Plane** option from the contextual menu; the **Plane Definition** dialog box will be displayed. Set the parameters and choose **OK**.

Note

There are some limitations of the ***Mirror*** *tool. Shelled, translated, scaled, mirrored, symmetric, and imported features and thickened, closed, and sewed surfaces cannot be mirrored by using this tool.*

Creating Rectangular Patterns

Menubar: Insert > Transformation Features > Rectangular Pattern
Toolbar: Transformation Features > Patterns sub-toolbar > Rectangular Pattern

Sometimes, you may need multiple instances of features to create a model. You can arrange these instances in a rectangular, circular, or a user-defined pattern. The tools for creating these patterns are available in the **Patterns** sub-toolbar of the **Transformation Features** toolbar, as shown in Figure 8-20.

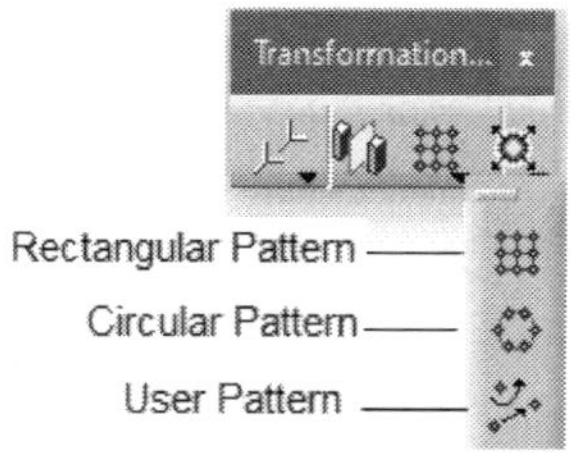

Figure 8-20 *The* ***Patterns*** *sub-toolbar*

In this section, you will learn to arrange the selected items in a rectangular pattern. Here, you can create a pattern of an existing feature as well as you can create a feature and then pattern it without exiting the pattern command. To create a rectangular pattern of an existing feature, choose the **Rectangular Pattern** tool from the **Patterns** sub-toolbar of the **Transformation Features** toolbar; the **Rectangular Pattern Definition** dialog box will be displayed, as shown in Figure 8-21. Also, you will be prompted to select the directions and specify the required parameters. Click once in the **Object** display box in the **Object to Pattern** area and select the feature to be patterned from the geometry area or from the Specification tree. If you do not select a feature, the entire body will be patterned. If the feature to be pattern is not created already, right click on the **Object** selection box in the **Object to Pattern** area of the **Rectangular Pattern Definition** dialog box; a contextual menu will be displayed. Select the desired option from this contextual menu to create an feature

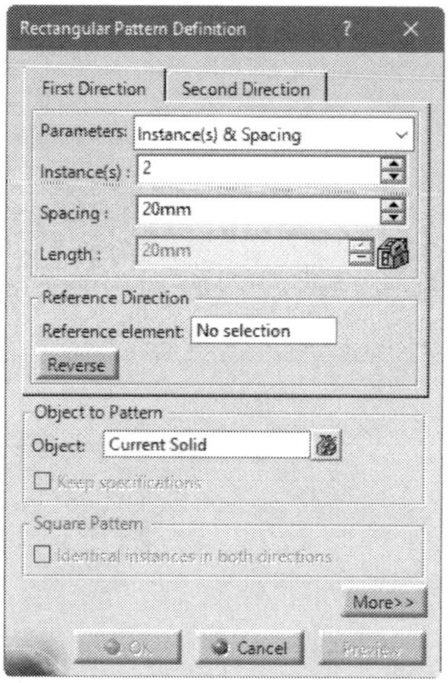

Figure 8-21 *The* ***Rectangular Pattern Definition*** *dialog box*

that is to be patterned. Now, click once in the **Reference element** selection box in the **Reference Direction** area and specify the directional reference from the geometry area. You can specify the reference direction by selecting an edge, line, or plane. The preview of the resulting pattern, with the default values, is displayed in the geometry area. You can use the **Reverse** button in the **Reference Direction** area to reverse the direction of the pattern creation. If the preview is not displayed, choose the **Preview** button.

The **Instance(s) & Spacing** option is selected by default in the **Parameters** drop-down list. Therefore, you need to specify the number of instances in the **Instance(s)** spinner and the value of the incremental distance between the instances in the **Spacing** spinner. If you select the **Instance(s) & Length** option from the **Parameters** drop-down list, you need to specify the number of instances and the total length of the pattern in the **Instance(s)** and **Length** spinners, respectively. If you select the **Spacing & Length** option, you need to specify the distance between two instances and the total length of the pattern in the **Spacing** and **Length** spinners, respectively. The **Instance(s) & Unequal Spacing** option allows you to set the number of instances in the pattern as well as the individual spacing for each instance. On selecting this option, the spacing between individual instances will be displayed with a dimension in the preview of the pattern. Double-click on the dimension in the preview to change it as per your requirement, refer to Figure 8-22.

After setting the parameters, choose the **OK** button. Figure 8-23 shows the feature and the first reference direction to be selected and Figure 8-24 shows the resulting pattern with equal spacing.

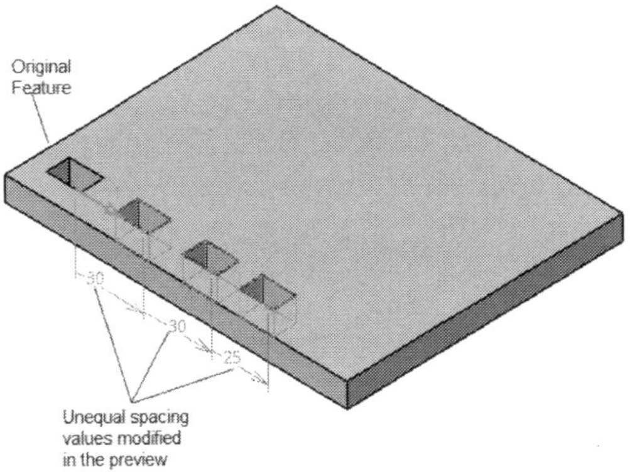

Figure 8-22 Modified values of individual spacing

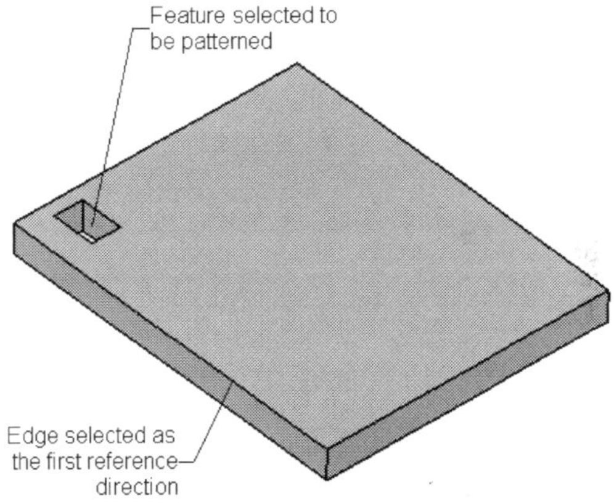

Figure 8-23 Feature to be patterned

To define the second direction of the pattern, choose the **Second Direction** tab of the **Rectangular Pattern Definition** dialog box. Click once on the **Reference element** selection box in the **Reference Direction** area and select the second directional reference from the geometry area. Now, set the values of the instances and the spacing in the **Instance(s)** and **Spacing** spinners, respectively. As you modify these values, preview of the pattern updates automatically. After setting the parameters, choose the **OK** button from the **Rectangular Pattern Definition** dialog box. Figure 8-25 shows the rectangular pattern created in two directions which are defined using the edges of the base feature.

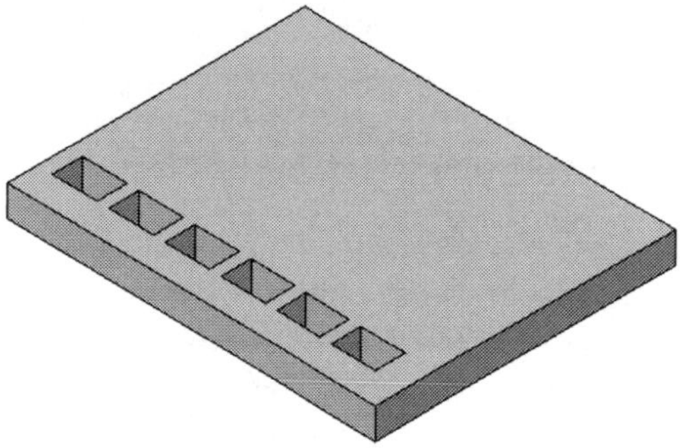

Figure 8-24 *Resulting pattern with equal spacing*

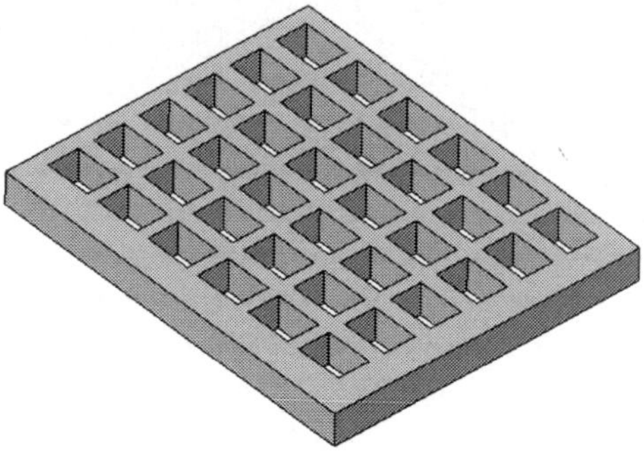

Figure 8-25 *Rectangular pattern in two directions*

You can also pattern the entire Part Body. To do so, invoke the **Rectangular Pattern Definition** dialog box and click once in the **Reference element** selection box of the **Reference Direction** area. Select the directional reference for the first direction. You will notice that by default the **Current Solid** option is selected in the **Object** display box of the **Object to Pattern** area. Therefore, it will pattern the entire Part Body. Now, define the parameters and choose the **OK** button from the **Rectangular Pattern Definition** dialog box. Figure 8-26 shows the directional references to be selected. Figure 8-27 shows the resulting pattern of the Part Body.

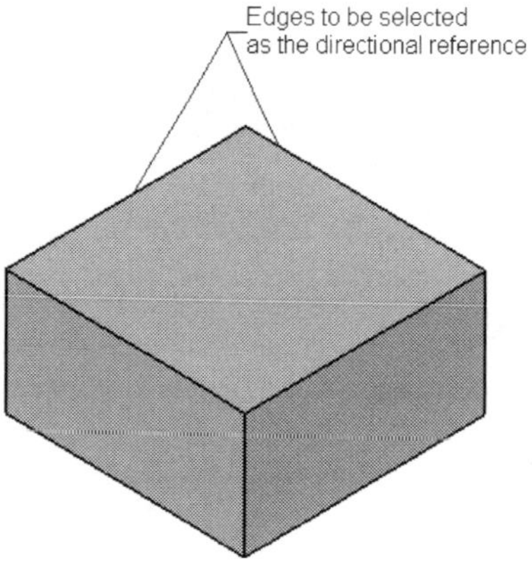

Figure 8-26 *Directional references to be selected*

Figure 8-27 *Resulting pattern of the Part Body*

The **Keep specifications** check box is used to retain the specifications of the original instance in the patterned instances also. For example, consider a case in which a pad feature is created by extruding a sketch drawn on a plane that is at an offset distance from the xy plane. This feature is extruded using the **Up to surface** option. If you pattern this feature with the **Keep specifications** check box cleared, the pattern instances will be created such that they are exactly similar to the original instance, as shown in Figure 8-28. However, if you select the **Keep specifications** check box, the resulting patterned instances will also be created same as the original feature was created, as shown in Figure 8-29.

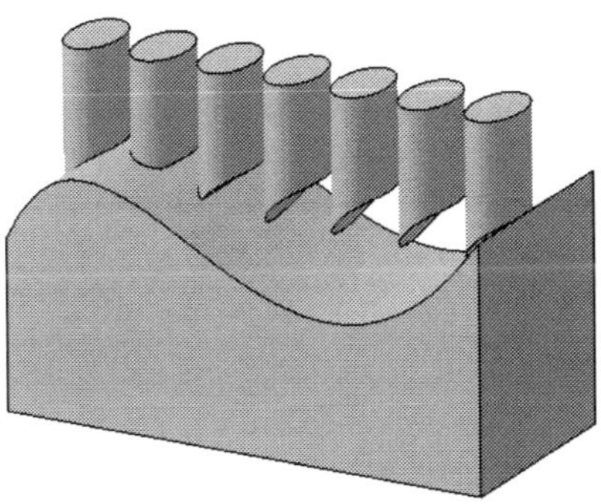

Figure 8-28 Pattern created with the ***Keep specifications*** *check box cleared*

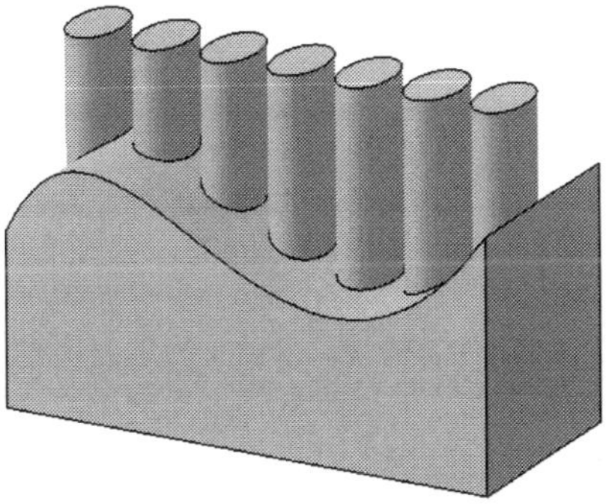

Figure 8-29 Pattern created with the ***Keep specifications*** *check box selected*

When the **Identical instances in both directions** check box is selected from the **Square Pattern** area of the **Rectangular Pattern Definition** dialog box after defining both the directions of the rectangular pattern feature, identical number of instances will be set along both the directions automatically. Note that this check box will not be activated if the **Spacing & Length** option is selected from the **Parameters** drop-down list.

While creating patterns, CATIA V5 allows you to skip some of the instances of the pattern. You will notice orange dots displayed along with the preview of the instances while creating the pattern. To skip the instances of the pattern, click on the orange dot of the instance that you need to skip; the preview of the instance will no longer be displayed. Similarly, skip the other instances. To restore the skipped instances, again click on the orange dot of the skipped instance to be retained. Figure 8-30 shows a patterned feature with all the instances retained. Figure 8-31 shows the patterned feature after skipping some of the instances.

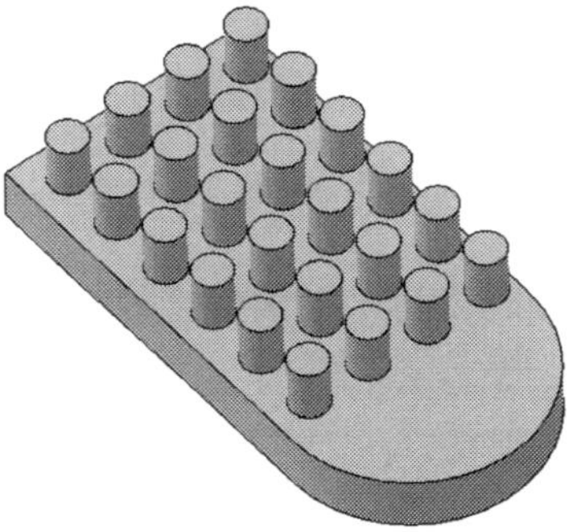

Figure 8-30 Patterned feature without skipping any instance

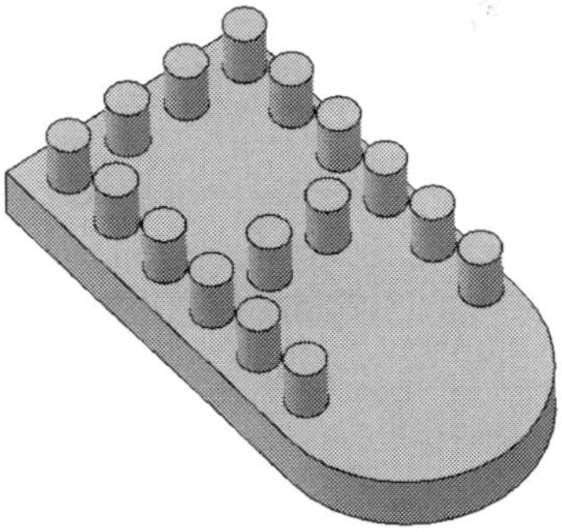

Figure 8-31 Patterned feature after skipping some of the instances

Generally, the parent feature is placed at the extreme corners of the patterned feature. You can also change its position in the pattern. To do so, expand the **Rectangular Pattern Definition** dialog box by choosing the **More** button from this dialog box. Figure 8-32 shows the expanded **Rectangular Pattern Definition** dialog box.

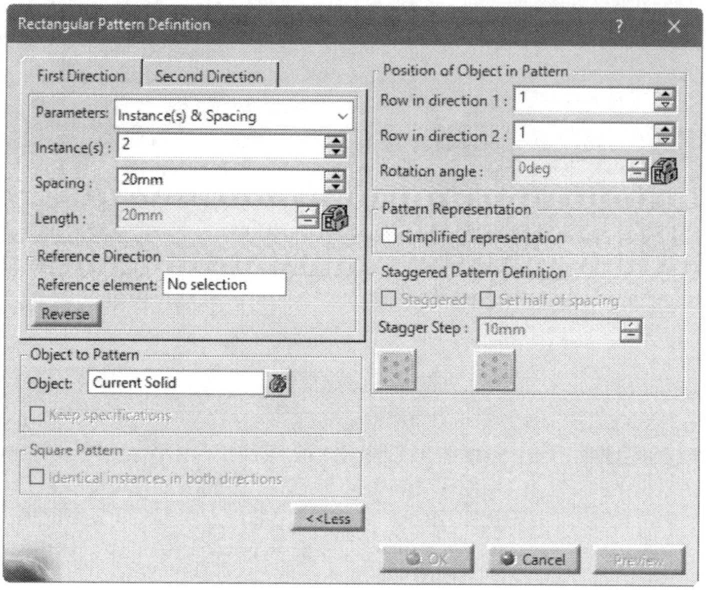

Figure 8-32 *The expanded* ***Rectangular Pattern Definition*** *dialog box*

Use the **Row in direction 1** and **Row in direction 2** spinners to set the position of the pattern instances in direction 1 and direction 2. You can also rotate it by specifying the angle of rotation in the **Rotation angle** spinner.

Figure 8-33 shows the feature to be patterned and Figure 8-34 shows the position of the parent feature in the pattern. Figure 8-35 shows the position of the pattern instances modified using the **Row in direction 1** and **Row in direction 2** spinners. Figure 8-36 shows a rotated pattern after specifying the angle of rotation in the **Rotation angle** spinner.

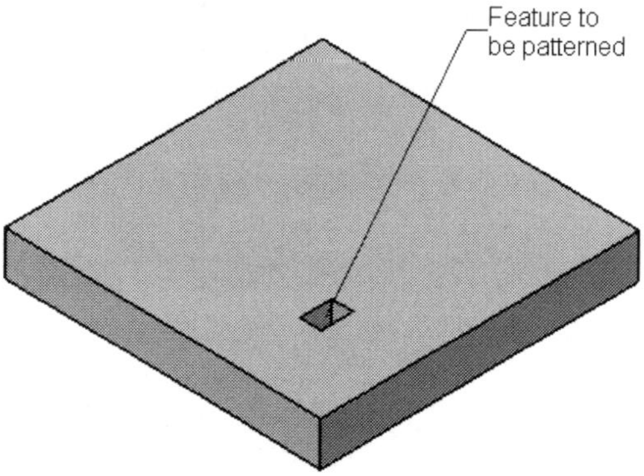

Figure 8-33 *Feature to be patterned*

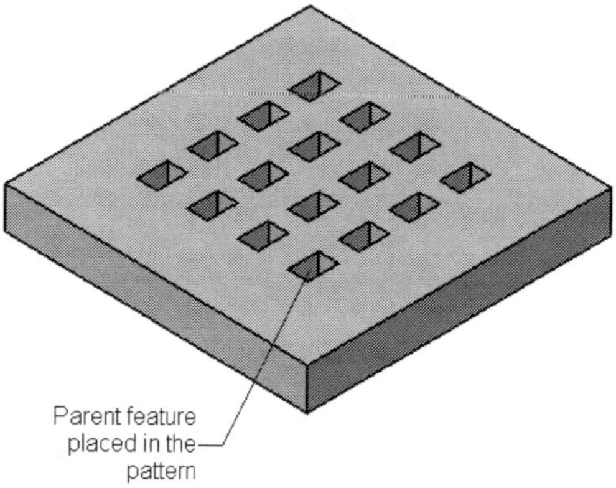

Figure 8-34 *Position of the parent feature in the pattern*

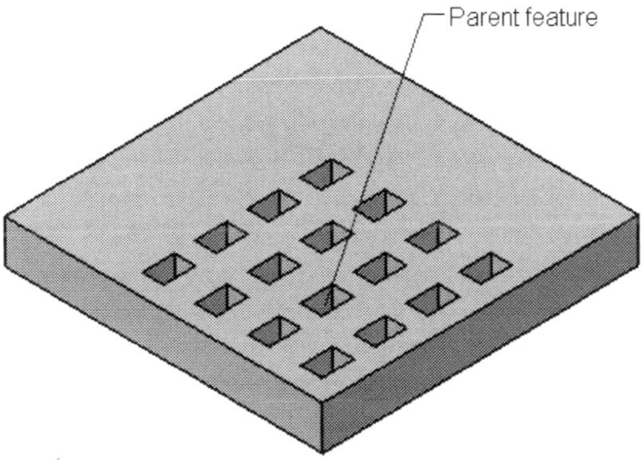

Figure 8-35 Modified position of the pattern

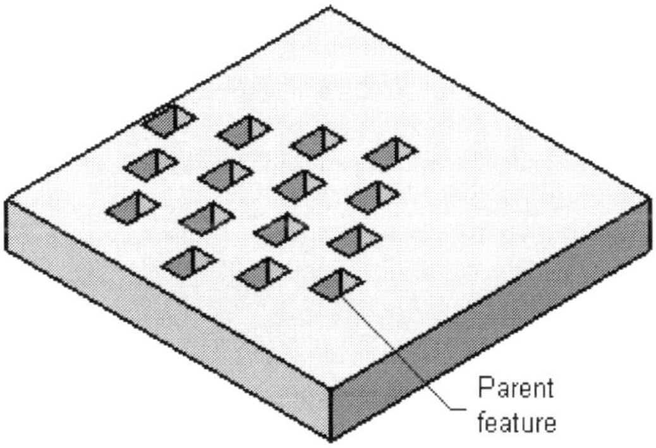

Figure 8-36 Rotated pattern

The **Simplified representation** check box is used for patterns that have a large number of instances. If this check box is selected, only the first and the last four instances of the pattern will be created and the rest of the instances will be hidden. This reduces the regeneration time. You can restore the hidden instances by clearing this check box.

The **Staggered** check box is used to create a staggered pattern in which the instances of the alternate rows are staggered by certain distance in direction 2. When you select this check box, the Config buttons below this check box are activated. You can choose any of these buttons as required. Note that the **Staggered** check box is activated only when there are more than one instance in the direction 1 and direction 2. The **Set half of spacing** check box is used to set the stagger step value to half of the spacing value in the direction 2.

Creating Circular Patterns

Menubar:	Insert > Transformation Features > Circular Pattern
Toolbar:	Transformation Features > Patterns sub-toolbar > Circular Pattern

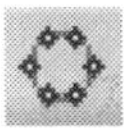

To arrange features or the current body in a circular manner, choose the **Circular Pattern** tool from the **Patterns** sub-toolbar in the **Transformation Features** toolbar; the **Circular Pattern Definition** dialog box will be displayed, as shown in Figure 8-37.

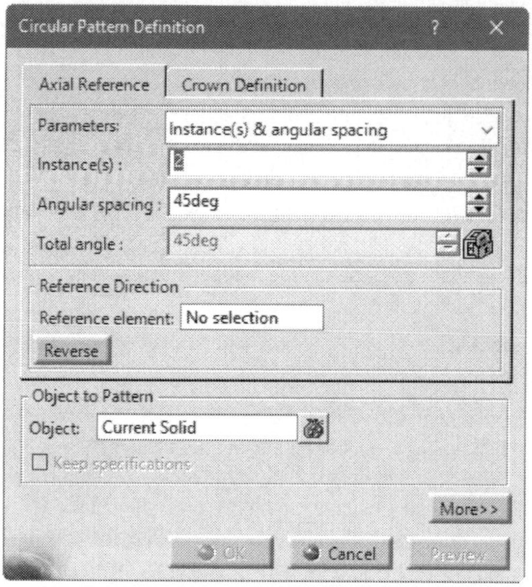

Figure 8-37 The ***Circular Pattern Definition*** *dialog box*

Additionally, you will be prompted to select the directions and specify the required parameters. Here, also you can create the feature to be patterned if it is not created already by right-clicking on the **Object** selection box as discussed earlier in the **Creating Rectangular Patterns** section. Click once in the **Object** selection box and select the feature to be patterned from the geometry area. Now, click in the **Reference element** selection box of the **Reference Direction** area and select a directional reference that will be used as the axis of the pattern. You can select the sketched line, an edge, or a circular face as the reference element. The preview of the pattern, with the default setting, is displayed in the geometry area.

You will notice that the **Instance(s) & angular spacing** option is selected by default in the **Parameters** drop-down list. Therefore, you need to set the values of the number of instances and the angular spacing in the **Instance(s)** and **Angular spacing** spinners, respectively. If you select the **Instance(s) & total angle** option, you need to set the values in the **Instance(s)** and **Total angle** spinners. On selecting the **Angular spacing & total angle** option, you need to set the values in the **Angular spacing** and **Total angle** spinners. On selecting the **Complete crown** option, you need to define only the number of instances in the **Instance(s)** spinner because on selecting this option, all instances are automatically adjusted to 360. Selecting the **Instance(s) & unequal angular spacing** option allows you to specify the number of instances and individual spacing for each instance. Note that on selecting this option, the dimension of the angular spacing for each instance is displayed in the preview of the pattern. You can double-click on the angular dimensions and set their values as per your requirement.

Figure 8-38 shows the feature to be patterned and the sketch to be selected as the reference element. Figure 8-39 shows the resultant patterned feature.

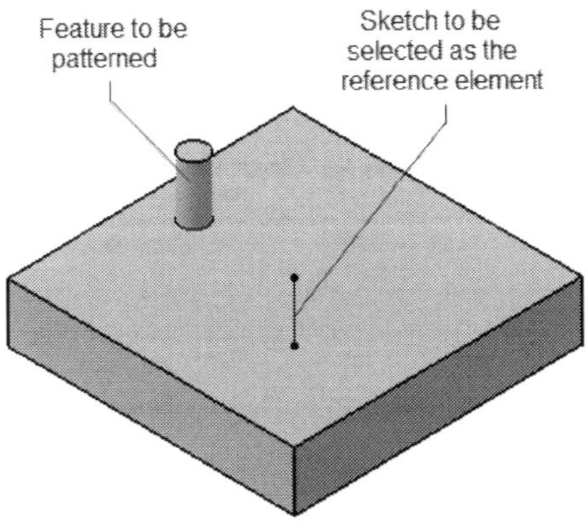

Figure 8-38 *Feature to be patterned and the reference element to be selected*

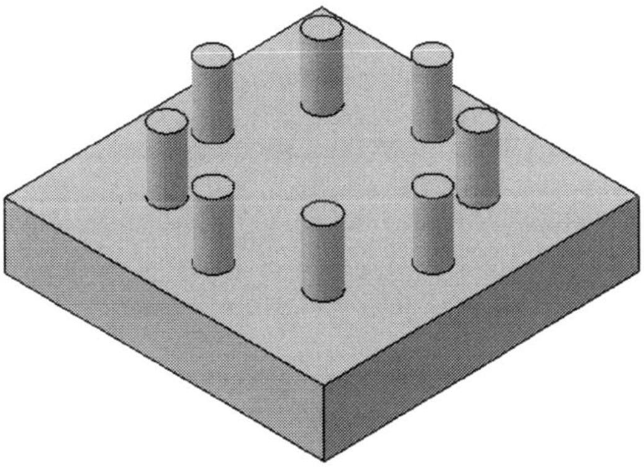

Figure 8-39 *Resulting patterned feature*

You can also create a crown-shaped circular pattern using the **Circular Pattern Definition** dialog box. To do so, choose the **Crown Definition** tab in this dialog box, refer to Figure 8-40.

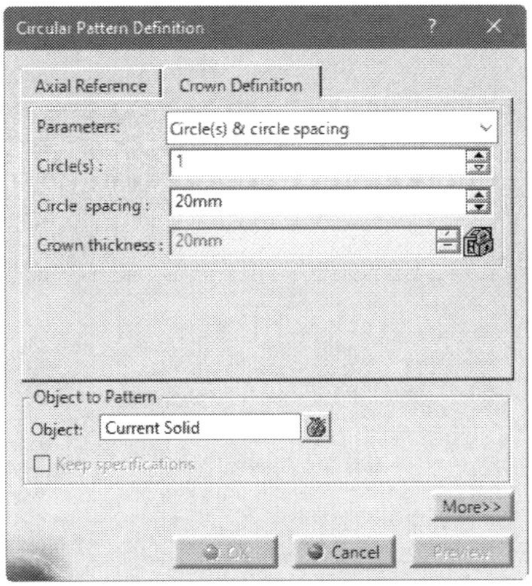

Figure 8-40 *The* ***Crown Definition*** *tab chosen in the* ***Circular Pattern Definition*** *dialog box*

By default, the **Circle(s) & circle spacing** option is selected in the **Parameters** drop-down list. Therefore, you need to set the value of the number of circles and the circle spacing in the **Circle(s)** and **Circle spacing** spinners, respectively. If you select the **Circle(s) & crown thickness** option, you need to set the values in the **Circle(s)** and **Crown thickness** spinners. If you select the **Circle spacing & crown thickness** option, you need to set the values in the **Circle spacing** and **Crown thickness** spinners. After setting all the parameters, choose **OK**.

Figure 8-41 shows various parameters used to define the crown in a circular pattern. Figure 8-42 shows the pattern after defining the crown parameters.

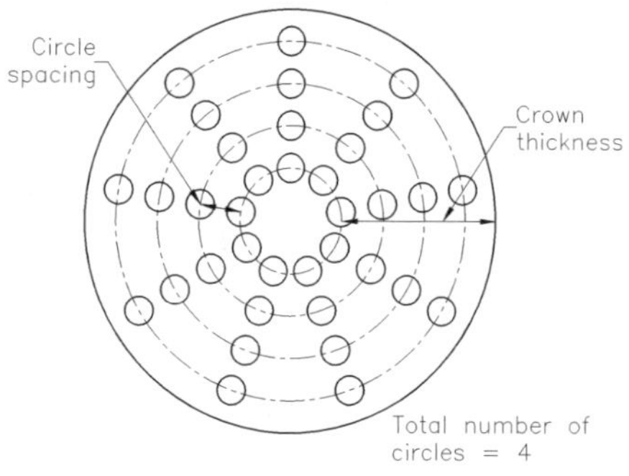

Figure 8-41 Various parameters used to define the crown in the pattern

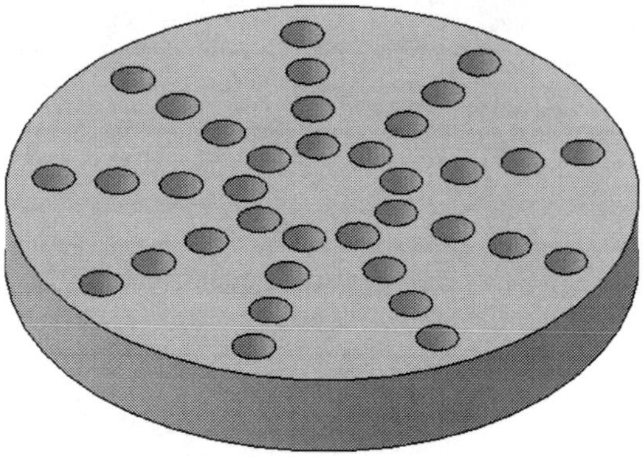

Figure 8-42 Resultant pattern after defining the crown parameters

Choose the **More** button from the **Circular Pattern Definition** dialog box to expand it. The **Radial alignment of instance(s)** check box is selected by default. Therefore, the features will be aligned radially when you create a circular pattern. If you clear this check box, the features will not be radially aligned. Figure 8-43 shows a circular pattern with the **Radial alignment of instance(s)** check box selected and Figure 8-44 shows a pattern with this check box cleared.

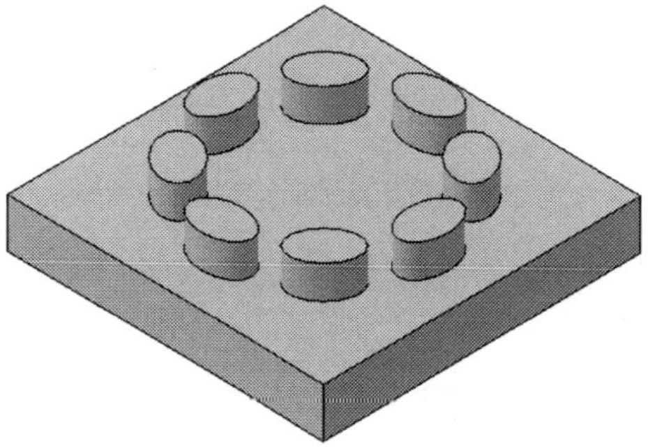

Figure 8-43 *Pattern created with the* ***Radial alignment of instance(s)*** *check box selected*

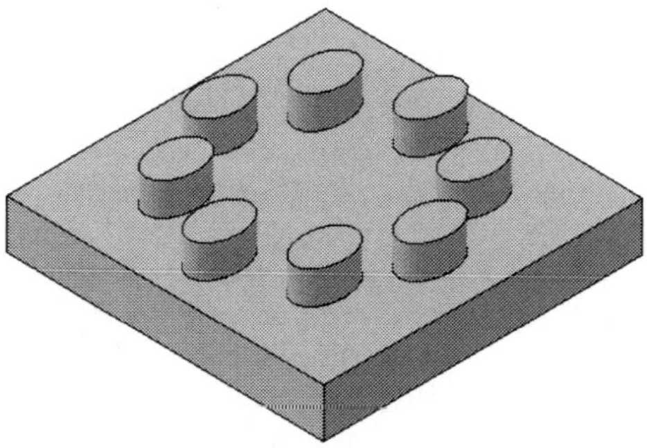

Figure 8-44 *Pattern created with the* ***Radial alignment of instance(s)*** *check box cleared*

The other options in this tool are similar to those discussed in rectangular patterns.

Creating User Patterns

Menubar: Insert > Transformation Features > User Pattern
Toolbar: Transformation Features > Patterns sub-toolbar > User Pattern

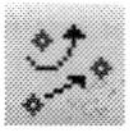

The **User Pattern** tool is used to place the instances of the selected feature in a sequence defined by reference points. These points can be defined in the sketch in the **Sketcher** workbench. To create a user pattern, select the feature to be patterned and then choose the **User Pattern** tool from the **Patterns** sub-toolbar in the **Transformation Features** toolbar; the **User Pattern Definition** dialog box will be displayed, as shown in Figure 8-45, and you will be prompted to select a sketch.

To specify the location of instances, select the sketch that includes the reference points defined in the **Sketcher** workbench; the instances of the pattern will be displayed on the points placed in this reference sketch. Choose the **OK** button from the **User Pattern Definition** dialog box. You can also specify the anchor point using the **Anchor** display box. Select any sketch point or vertex as the anchor point. On doing so, the patterned instances will be placed at an offset distance that is defined by calculating the distance and the direction between the anchor point and the selected object. Figure 8-46 shows the feature to be patterned and the reference sketch and Figure 8-47 shows the resultant pattern.

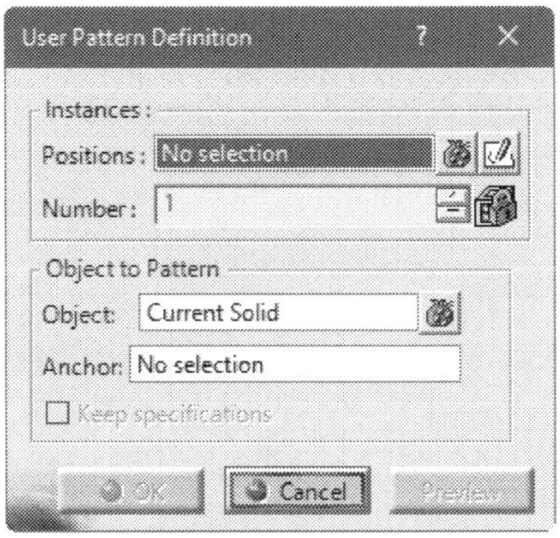

Figure 8-45 *The* ***User Pattern Definition*** *dialog box*

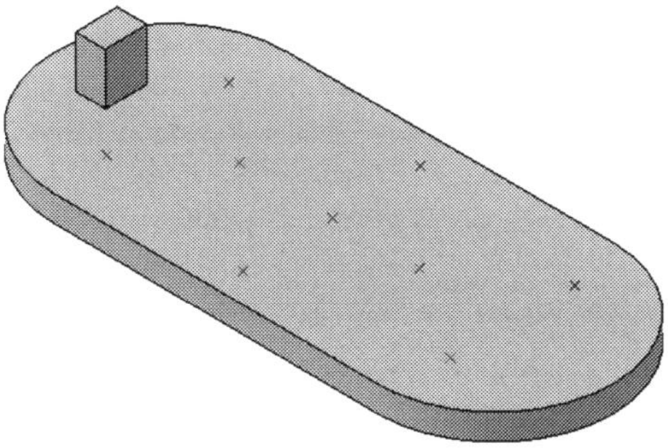

Figure 8-46 *Feature to be patterned and the reference sketch to be selected*

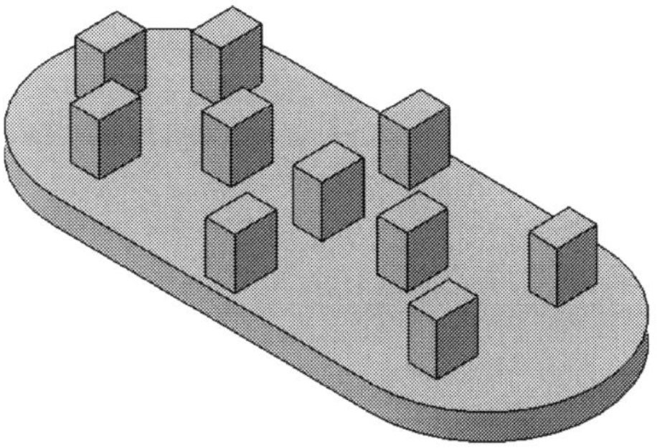

Figure 8-47 *Resultant pattern*

Note

*You can pattern multiple features using same parameters. To do so, select multiple features by pressing and holding the CTRL key by using the Selection manager button. Make sure that while patterning multiple features, the features in the **Object** selection box are sketch-based features otherwise an error message will be displayed on the screen. Note that you cannot pattern transformed and shell features.*

Uniform Scaling of Models

Menubar:	Insert > Transformation Features > Scaling
Toolbar:	Transformation Features > Scale sub-toolbar > Scaling

The **Scaling** tool is used to scale a model uniformly along the selected reference. To scale a model, choose the **Scaling** tool from the **Scale** sub-toolbar of the **Transformation Features** toolbar, as shown in Figure 8-48. On doing so, the **Scaling Definition** dialog box will be displayed, as shown in Figure 8-49. Also, you will be prompted to select a reference point, plane, or planar surface. Next, you need to set the value of the scale factor in the **Ratio** spinner; the preview of the scaled model will be displayed in the geometry area. Choose the **OK** button from the **Scaling Definition** dialog box.

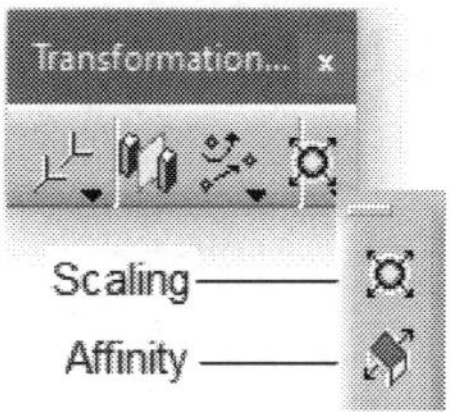

Figure 8-48 The ***Scale*** *sub-toolbar*

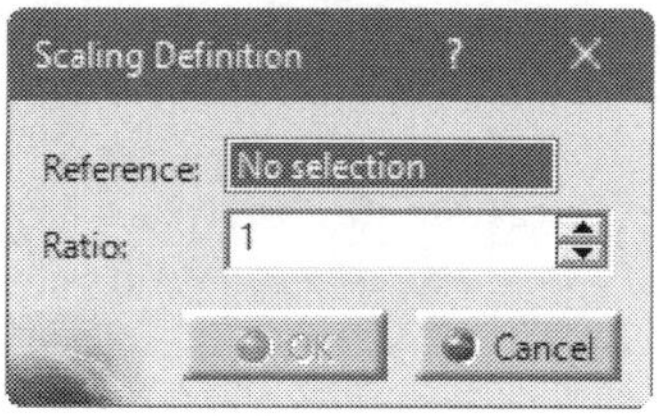

Figure 8-49 The ***Scaling Definition*** *dialog box*

If you select a plane as the reference, then the model will be scaled in the direction normal to the selected plane. However, if a point is selected as the reference, the model will be resized uniformly in three directions (x, y, and z axes).

Non-uniform Scaling of Models

Menubar: Insert > Transformation Features > Affinity
Toolbar: Transformation Features > Scale sub-toolbar > Affinity

The **Affinity** tool is used to scale a model non-uniformly. You can scale the model non-uniformly using different scale factors along the x, y, and z directions. On invoking this tool, the **Affinity Definition** dialog box will be displayed, as shown in Figure 8-50. Also, a new axis system will be displayed at the origin of the current axis system and you will be prompted to select the origin of the affinity axis system. Specify the origin, orientation of the xy plane, and the direction of the x-axis using the respective options in the **Axis system** area; you will be prompted to specify the scaling factors along the x, y, and z axes. Specify different scaling factors using the **X**, **Y**, and **Z** spinners available in the **Ratios** area and then choose the **OK** button; the model will be scaled by the specified values with respect to the specified axis system. Figure 8-51 shows the model to be scaled non-uniformly and Figure 8-52 shows the model after scaling. Here the model is scaled by the scale factor of 0.5 along the x direction, 2 along the y direction, and 1 along the z direction.

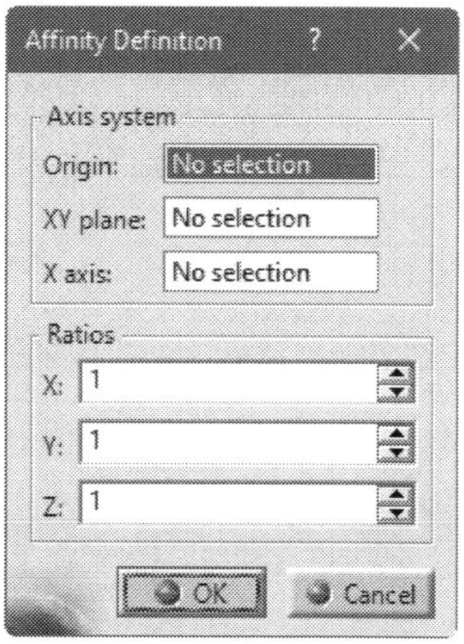

Figure 8-50 The ***Affinity Definition*** *dialog box*

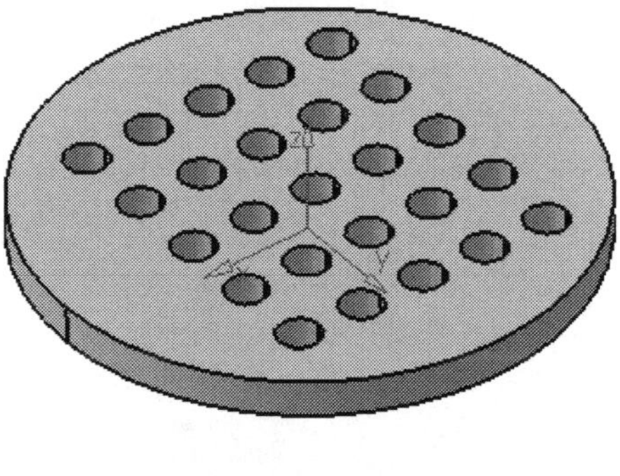

Figure 8-51 Model to be scaled non-uniformly

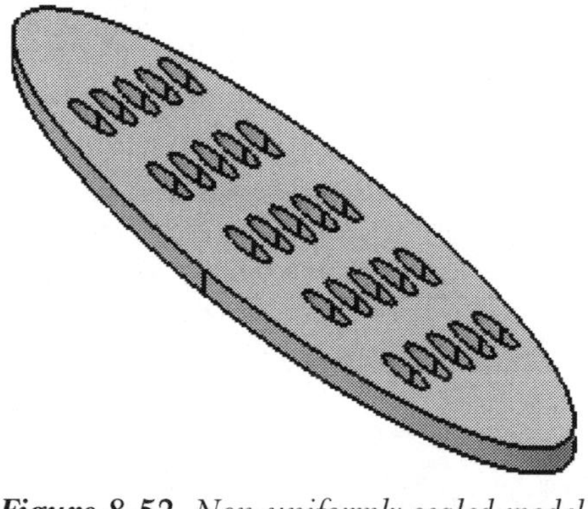

Figure 8-52 Non-uniformly scaled model

WORKING WITH ADDITIONAL BODIES

The **Part** workbench of CATIA V5 provides you with a tool to insert new bodies in the current model. You can create features in the newly created body and then perform the boolean operations on two or more than two part bodies. You will learn more about boolean operations later in this chapter. The tools that are used to insert bodies are provided in the **Insert** toolbar, as shown in Figure 8-53.

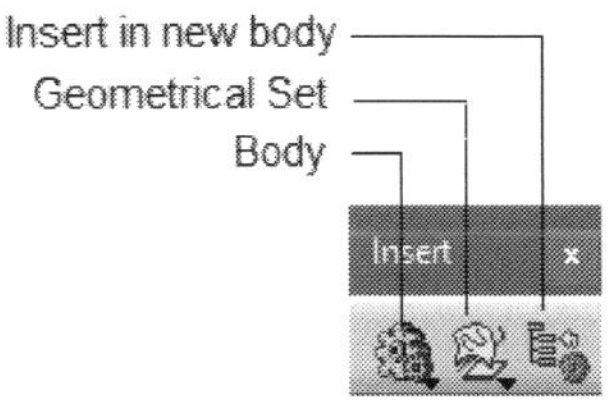

*Figure 8-53 The **Insert** toolbar*

If you want to create complicated parts such as the housing cover of a diesel pump, you need to work with multiple bodies and then combine them together to create a single part. The advantage of working with multiple bodies is that the features created on the active body will not affect any other body. Suppose, you apply the pocket feature on an active body, the material will be removed only from that body. After getting the desired shape, you can apply boolean operations on that body to create a single part. You will learn more about boolean operations and assembling of bodies later in this chapter.

Tip
*To invoke the **Insert** toolbar, move the cursor on any one of the toolbars and invoke the contextual menu. Choose **Insert** from it; the **Insert** toolbar will be displayed, as shown in Figure 8-53.*

Inserting a New Body

Menubar:	Insert > Body
Toolbar:	Insert > Bodies > Body

To insert a new body, choose the **Body** tool from the **Bodies** sub-toolbar of the **Insert** toolbar; a new body named Body.2 will be added under the current body in the Specification tree, as shown in Figure 8-54.

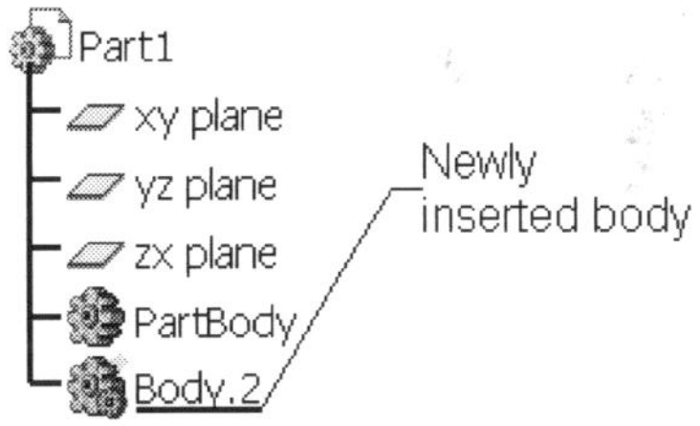

Figure 8-54 Newly inserted body

You will notice that the newly created body is underlined, indicating that the current body is active. If you add a feature to the current model, that feature will be automatically added to the active body. To add features to an inactive body, select it from the Specification tree and invoke the contextual menu. Then, choose the **Define In Work Object** option from it; the selected inactive body will be activated.

Inserting Features in the New Body

Menubar:	Insert > Insert in new body
Toolbar:	Insert > Insert in new body

The **Insert in new body** tool is used to place the selected feature inside a new body and then assemble the newly created body with the current body. To assemble a feature, select it and then choose the **Insert in new body** tool from the **Insert** toolbar; the selected feature will be placed in a new body and the newly created body will be assembled with the current body, as shown in Figure 8-55.

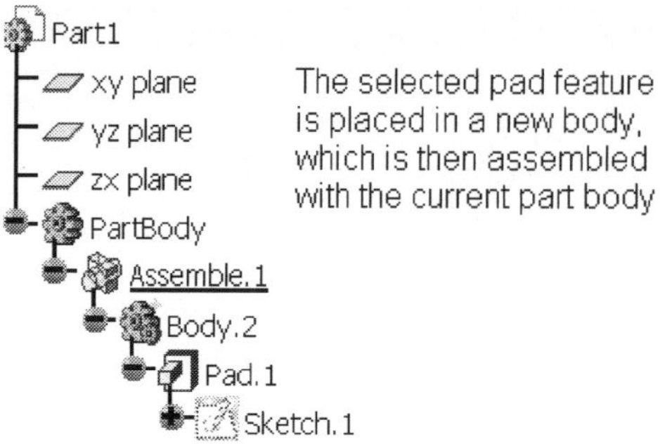

Figure 8-55 *The Specification tree*

Applying Boolean Operations to Bodies

After inserting the bodies, you can apply boolean operations, such as addition of two bodies, subtraction of one body from the other, retaining the intersected portion of two bodies, and so on. These operations are called boolean operations because they are based on boolean algebra. Boolean operations that are used to manipulate the bodies are discussed next.

Assembling Bodies

Menubar:	Insert > Boolean Operations > Assemble
Toolbar:	Boolean Operations > Assemble

The **Assemble** tool is used to assemble two selected bodies. To assemble the two bodies together, choose the **Assemble** tool from the **Boolean Operations** toolbar; the **Assemble** dialog box will be displayed, as shown in Figure 8-56 and you will be prompted to select a body (or volume) to be assembled.

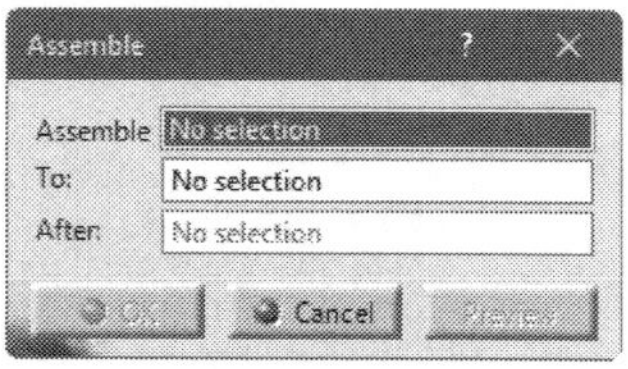

Figure 8-56 *The* ***Assemble*** *dialog box*

Select the body to assemble with the parent body if it is not selected by default. It is always recommended to select the bodies from the Specification tree. The name of the selected body will be displayed in the **Assemble** display box and the name of the parent body will be displayed in the **To** display box of the **Assemble** dialog box. After selecting the bodies, choose the **OK** button from the **Assemble** dialog box; the selected bodies will be combined with each other. Figure 8-57 shows the bodies selected to assemble and Figure 8-58 shows the resultant assembled body.

You can also create a body as a pocket feature by drawing a sketch and extruding it using the **Pocket** tool. Although the body looks like a protruded feature, it will remove the material when assembled with another body. If you assemble this pocket body with the parent body, both the bodies will be combined and the portion occupied by the pocket body will be subtracted from the parent body. This is because you have assembled a body having the properties of a cut feature with the parent body. Figure 8-59 shows the pocket body and the parent body. Figure 8-60 shows the resultant assembled body. Note that if the pocket feature is the first feature in its body, it will look similar to the protruded feature.

Note

You might have observed " + " or "-" symbol attached with the body added after the Part body. These symbols show polarity of that body. While working with add material tools, the polarity will be positive but will be negative with material removal tools.

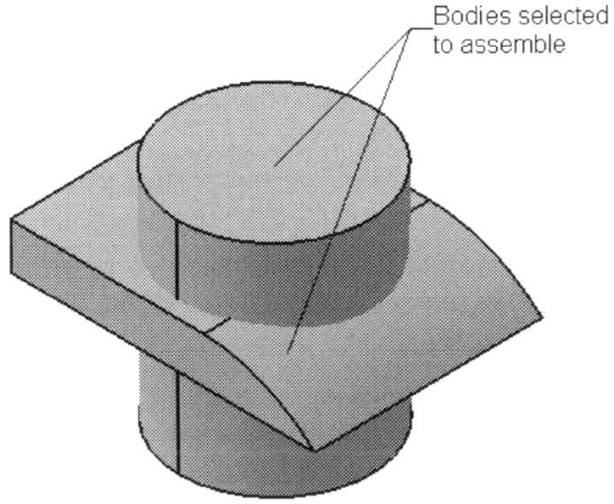

Figure 8-57 *Bodies selected to assemble*

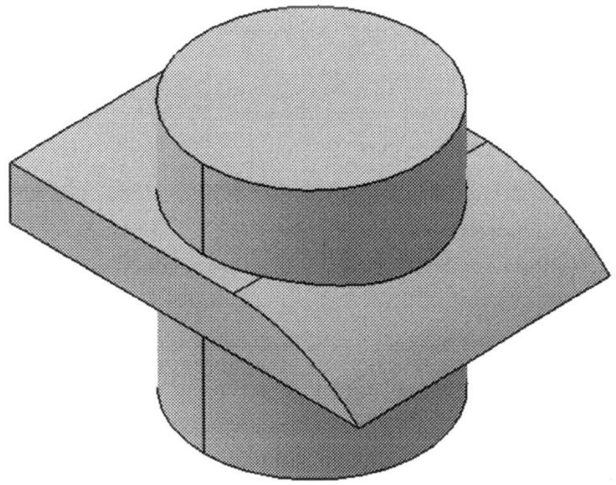

Figure 8-58 *Resultant assembled body*

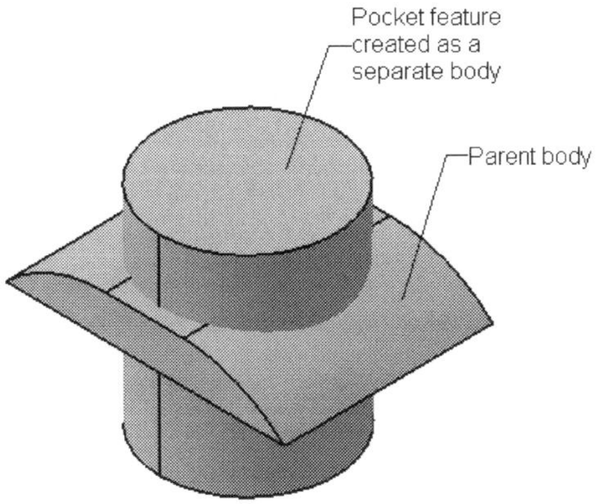

Figure 8-59 *Pocket body and parent body*

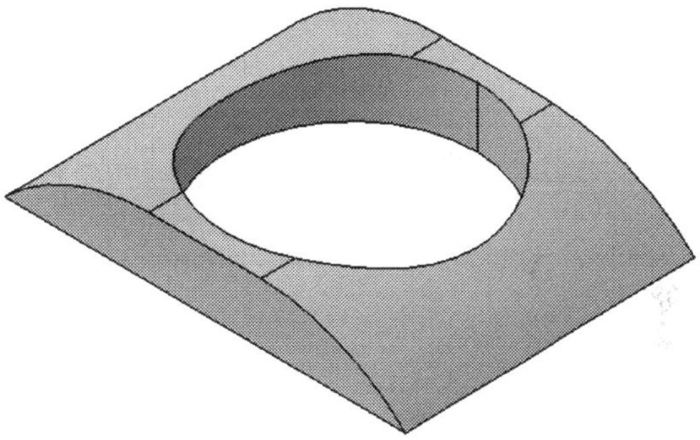

Figure 8-60 *Resultant assembled body*

Adding Bodies

Menubar:	Insert > Boolean Operations > Add
Toolbar:	Boolean Operations > Boolean Operations sub-toolbar > Add

The **Add** tool is used to add the selected bodies together. To do so, choose the down arrow at the right of the **Boolean Operations** tool in the **Boolean Operations** toolbar; the **Boolean Operations** sub-toolbar will be displayed, as shown in Figure 8-61. To invoke the **Add** tool, choose the **Add** button from the sub-toolbar; the **Add** dialog box will be displayed, as shown in Figure 8-62.

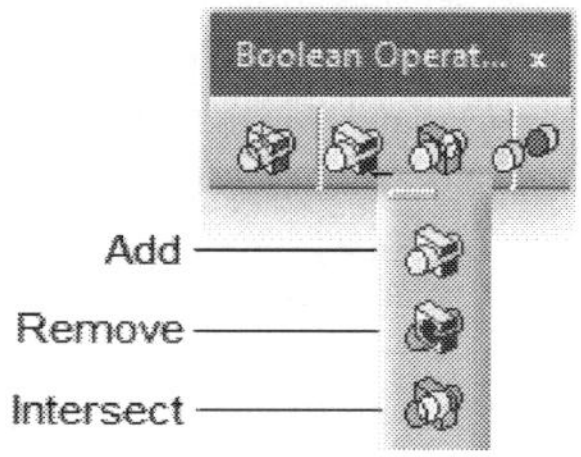

*Figure 8-61 The **Boolean Operations** sub-toolbar*

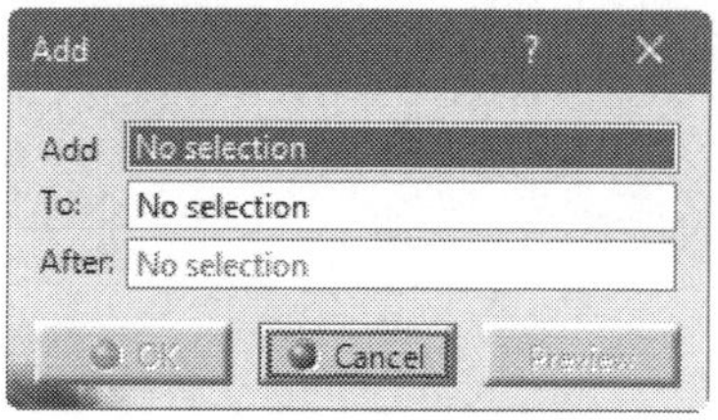

*Figure 8-62 The **Add** dialog box*

Also, you will be prompted to select the body (or volume) to be operated. Select the body to be added to the parent body and then select the parent body. Next, choose the **OK** button from the **Add** dialog box. Figure 8-63 shows the bodies to be added and Figure 8-64 shows the resultant body.

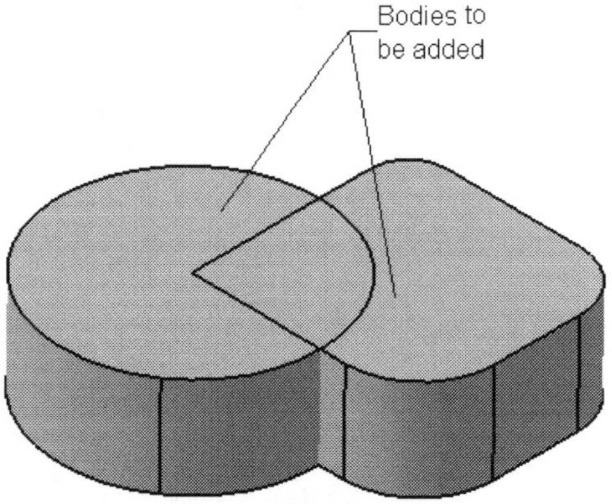

Figure 8-63 Bodies to be added

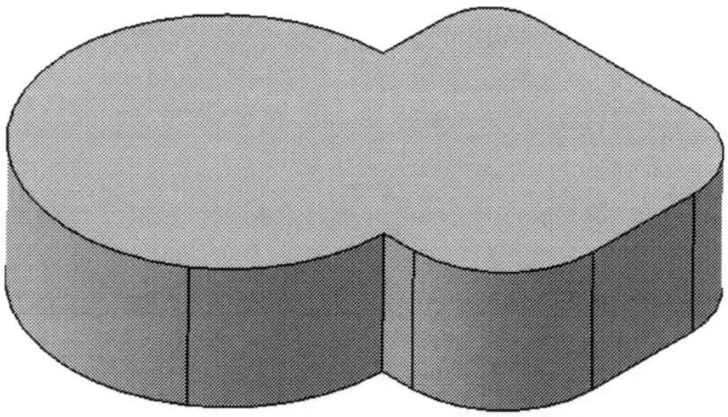

Figure 8-64 Resultant body

Subtracting Bodies

Menubar:	Insert > Boolean Operations > Remove
Toolbar:	Boolean Operations > Boolean Operations sub-toolbar > Remove

The **Remove** tool is used to subtract the selected body from another body. On invoking this tool, the **Remove** dialog box will be displayed. Select the body that you need to remove from the parent body. Note that the selected body will act as a cutting tool. Now, select the body from which the initially selected body needs to be removed. Choose the **OK** button from the **Remove** dialog box. Figure 8-65 shows the body to be removed and the parent body. Figure 8-66 shows the resultant body.

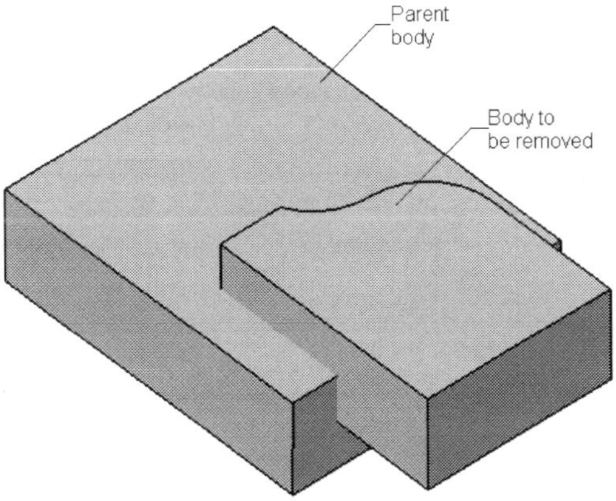

Figure 8-65 The parent body and the body to be removed

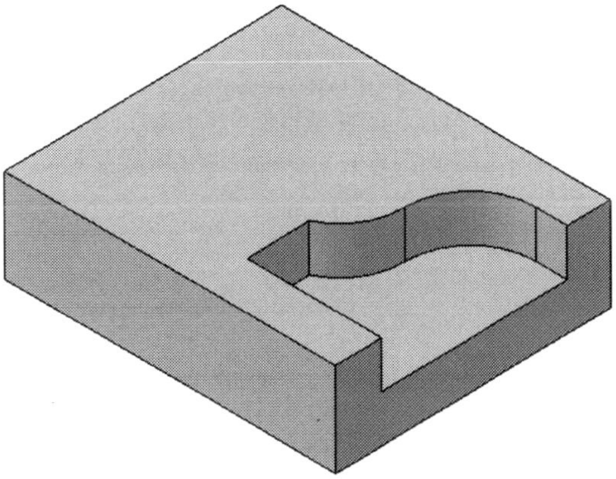

Figure 8-66 Resultant body

Intersecting Bodies

Menubar: Insert > Boolean Operations > Intersect
Toolbar: Boolean Operations > Boolean Operations sub-toolbar > Intersect

The **Intersect** tool is used to retain the common portion of two intersecting bodies and remove the other portions of the selected bodies. You can create complex geometries very easily using this tool. On invoking this tool, the **Intersect** dialog box will be displayed, as shown in Figure 8-67.

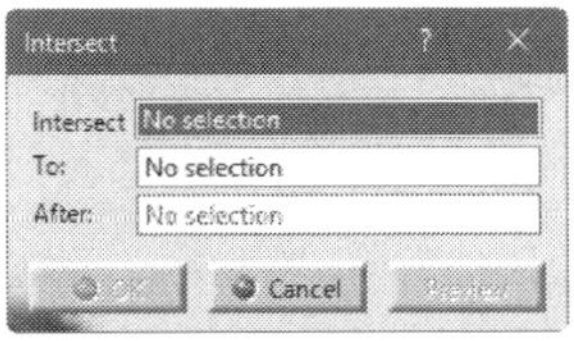

*Figure 8-67 The **Intersect** dialog box*

Select the bodies to be intersected and choose the **OK** button from the **Intersect** toolbar; the common portion between the selected bodies will be retained and other portions of the selected bodies will be removed. Figure 8-68 shows the bodies to be intersected and Figure 8-69 shows the resultant body.

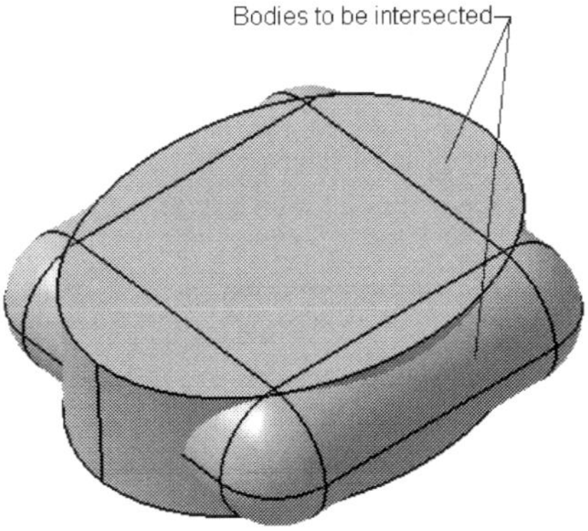

Figure 8-68 Bodies to be intersected

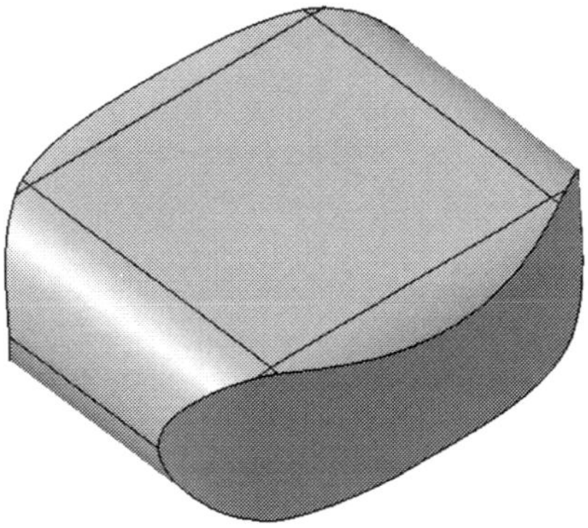

Figure 8-69 Resultant body

Trimming Bodies

Menubar:	Insert > Boolean Operations > Union Trim
Toolbar:	Boolean Operations > Union Trim

You can remove the portions of a body while performing the joining operation by using the **Union Trim** tool. On invoking this tool, you will be prompted to select the body to be trimmed. Select the existing body other than the parent body; the **Trim Definition** dialog box will be displayed, as shown in Figure 8-70. Note that the parent body cannot be trimmed.

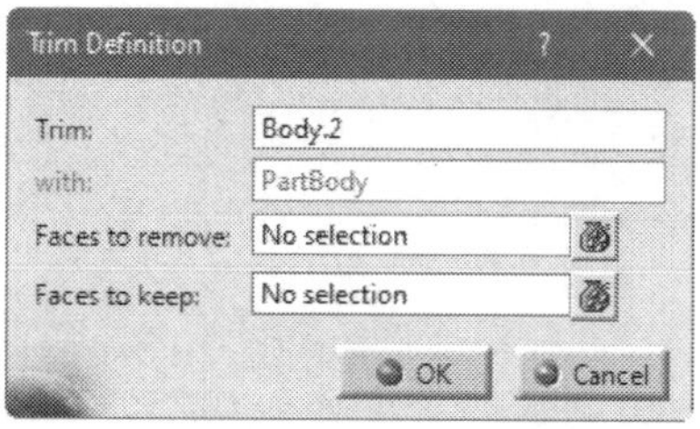

Figure 8-70 *The **Trim Definition** dialog box*

If your design consists of only two bodies, then the parent body will be automatically selected as the trimming body and the other selected body get trimmed. If your design consists of more than two bodies, and you select the second body as the body to be trimmed, it will automatically select the parent body as the trimming body, and the selected body will be trimmed. However, if you select the third body as the body to be trimmed, then you need to select a trimming body so that the selected body can be trimmed.

Next, you need to select the faces to be removed. Click once on the **Faces to remove** display box and select the face or faces to be removed. You can also select the faces that you need to retain using the **Faces to keep** display box. Then, choose the **OK** button from the **Trim Definition** dialog box. Figure 8-71 shows the body to be trimmed and the face of the body to be retained. Figure 8-72 shows the resulting trimmed body.

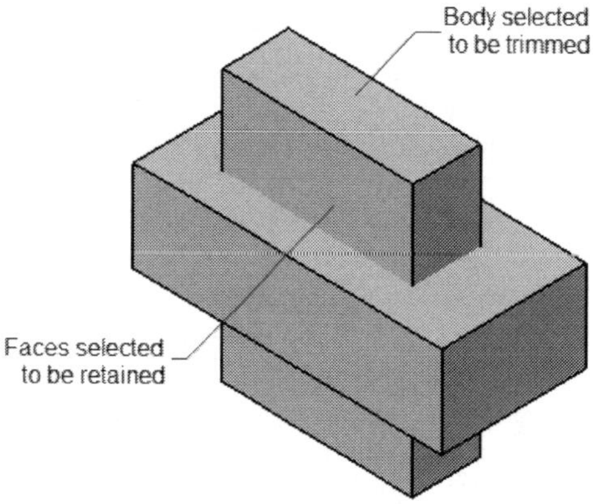

Figure 8-71 *Body to be trimmed and the face to be retained*

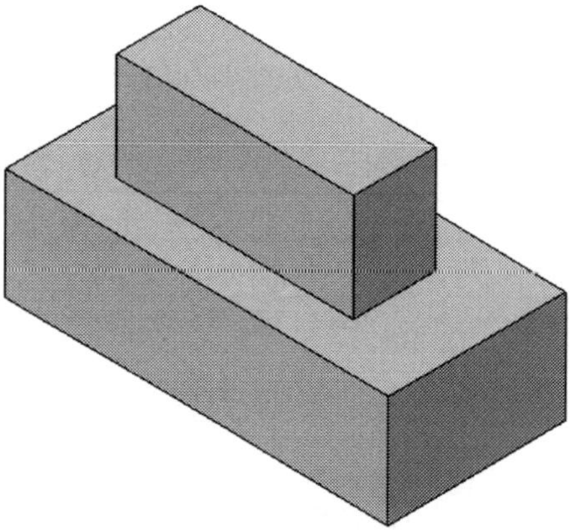

Figure 8-72 *Resultant trimmed body*

Removing Lumps

Menubar: Insert > Boolean Operations > Remove Lump
Toolbar: Boolean Operations > Remove Lump

The **Remove Lump** tool is used to remove material from a part body (model) that consists of two or more than two disjointed portions. On invoking this tool, you will be prompted to select the body to be trimmed. Select the body from the geometry area; the **Remove Lump Definition (Trim)** dialog box will be displayed, as shown in Figure 8-73.

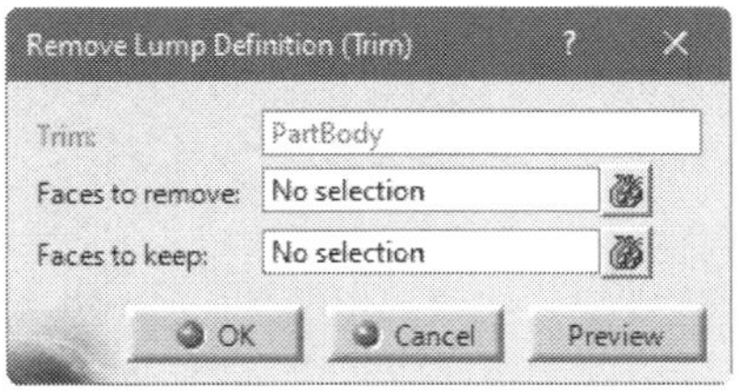

*Figure 8-73 The **Remove Lump Definition (Trim)** dialog box*

Select the face of the disjointed piece that you need to remove using the **Faces to remove** display box. You can also select the face of the disjoint piece that you need to retain using the **Faces to keep** display box. Then, choose the **OK** button from the **Remove Lump Definition (Trim)** dialog box. Figure 8-74 shows the face of the disjoint feature to be removed and Figure 8-75 shows the body after removing the lump.

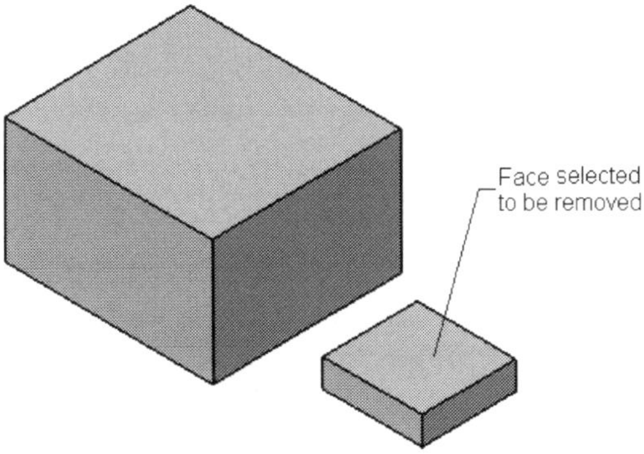

Figure 8-74 Face selected to be removed

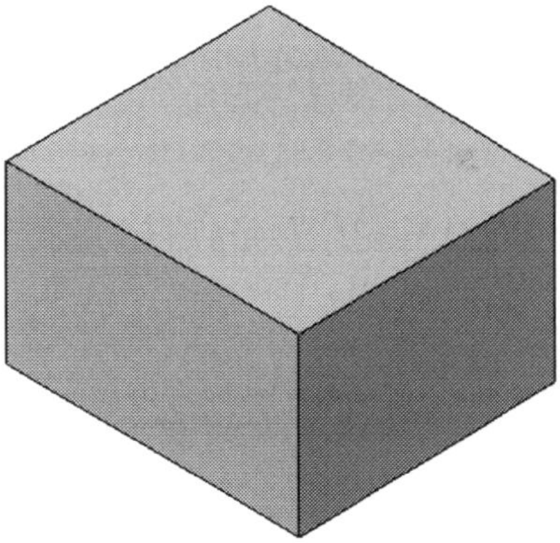

Figure 8-75 The resultant body after removing the lump

ADDING STIFFENERS TO A MODEL

Menubar: Insert > Sketch-Based Features > Stiffener
Toolbar: Sketch-Based Features > Advanced extruded features sub-toolbar > Stiffener

Stiffeners are generally added to increase the strength of components. To add a stiffener, choose the **Stiffener** tool from the **Advanced extruded features** sub-toolbar in the **Sketch-Based Features** toolbar, as shown in Figure 8-76. On choosing the **Stiffener** tool, the **Stiffener Definition** dialog box will be displayed, as shown in Figure 8-77.

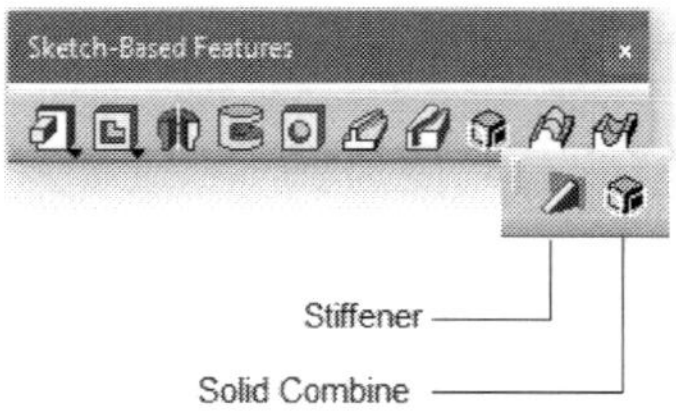

Figure 8-76 *The* ***Advanced extruded features*** *sub-toolbar*

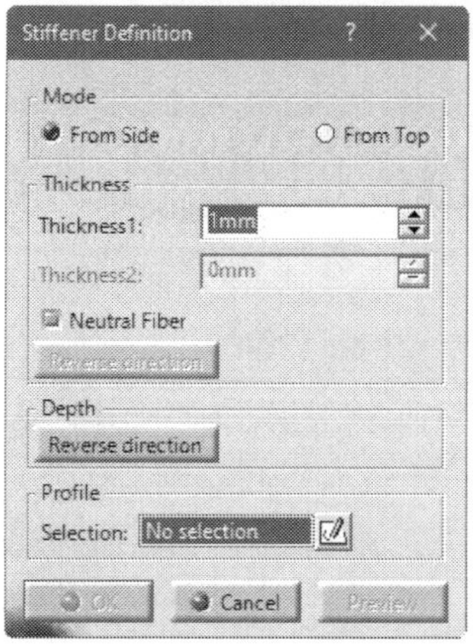

Figure 8-77 *The* ***Stiffener Definition*** *dialog box*

Next, you need to select the sketch for creating a stiffener. An open sketch is required for creating a stiffener feature, refer to Figure 8-78. If the sketch is not already drawn, choose the **Sketcher** button on the right of the **Selection** display box in the **Profile** area and then select the sketching plane to draw the sketch; the **Sketcher** workbench will be invoked. Draw the sketch and exit the **Sketcher** workbench.

Depending upon the type of sketch drawn, the **From Side** or **From Top** radio button can be selected from the **Mode** area of the **Stiffener Definition** dialog box. Set the thickness of the stiffener using the **Thickness1** spinner provided in the **Thickness** area. You will notice that the **Neutral Fiber** check box is selected by default.

While creating the stiffener, if you select the **From Side** radio button and clear the **Neutral Fiber** check box, the **Reverse Direction** button provided below this check box will be invoked. This button is used to toggle between left and right side of the sketch to create the stiffener. Figure 8-78 shows the sketch for creating the stiffener. Figure 8-79 shows the stiffener created with the **From Side** radio button selected.

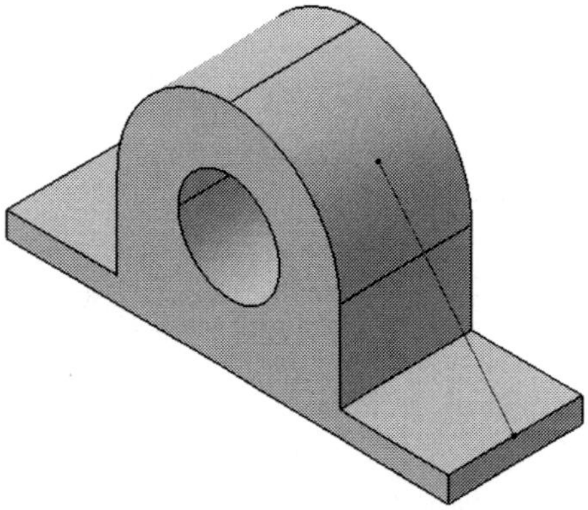

Figure 8-78 *Sketch for creating the stiffener*

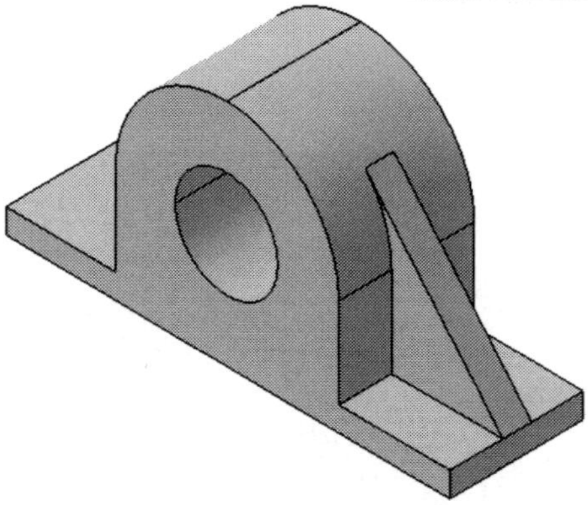

Figure 8-79 *The stiffener created with the* ***From Side*** *radio button selected*

While creating the stiffener, if you select the **From Top** radio button and clear the **Neutral Fiber** check box, the **Thickness2** spinner will be activated. You can set the value of thickness in the other direction using the **Thickness2** spinner. While creating the stiffener using this method, the endpoints of the sketched entities need not be merged with the edges of the model, refer to Figure 8-80. On doing so, the edges will be automatically extended to the nearest intersecting surface. The stiffener created by selecting the From Top radio button is shown in Figure 8-81. The **Reverse Direction** button in the **Depth** area is used to reverse the direction of the feature creation. After setting the parameters, choose the OK button from the **Stiffener Definition** dialog box.

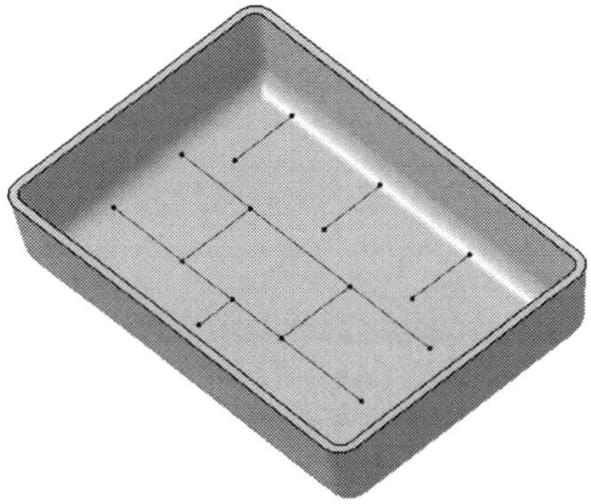

Figure 8-80 *Sketch for creating the stiffener*

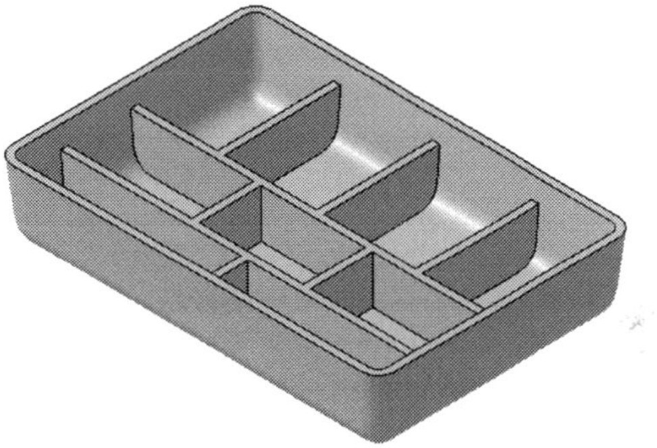

Figure 8-81 *The stiffener created with the* ***From Top*** *radio button selected*

GENERATING SOLID COMBINE

Menubar: Insert > Sketch-Based Features > Solid Combine
Toolbar: Sketch-Based Features > Advanced extruded features sub-toolbar > Solid Combine

The **Solid Combine** tool is used to generate a solid from two profiles. To invoke this tool, choose the **Solid Combine** tool from the **Advanced extruded features** sub-toolbar of the **Sketch-Based Features** toolbar; the **Combine Definition** dialog box will be displayed, as shown in Figure 8-82 and you will be prompted to select the first profile. Select any closed profile or planar surface from the graphics area; you will be prompted to select the second profile. On selecting the second profile, a preview of the solid will be displayed. This preview will be generated by calculating the intersected portion that is generated by extruding the two closed profiles. Choose the **OK** button to accept the generated solid combine. The **Normal to profile** check box in both the **First component** and **Second component** areas is selected by default. So, the resulting combine solid will be computed by calculating the portion that will intersect on virtually extruding the two profiles normal to their sketching plane. You can also generate the solid combine by virtually extruding the two profiles along the specified direction instead of normal to the sketching plane. To do so, clear both the **Normal to profile** check boxes in the **Combine Definition** dialog box; the **Direction** selection box will be activated. Select an edge, line, axis, or plane to specify the desired direction. The combine solid will be generated by virtually extruding the two profiles along the specified directional reference.

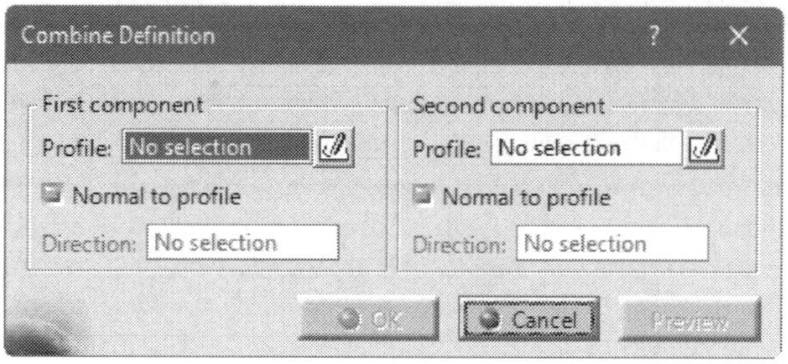

***Figure 8-82** The **Combine Definition** dialog box*

Figure 8-83 shows the two profiles to be selected for generating the solid combine and Figure 8-84 shows the resultant solid combine.

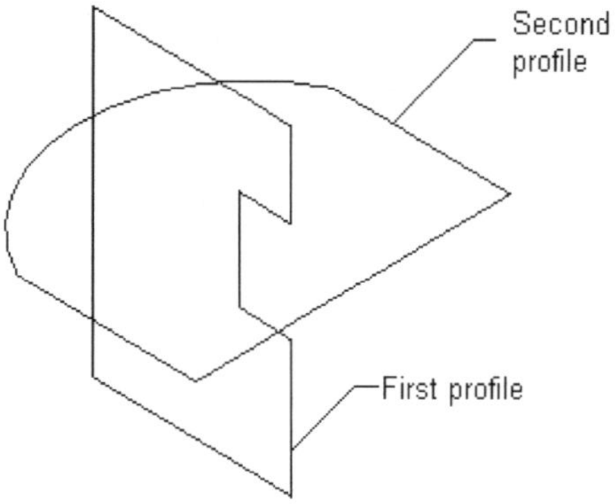

***Figure 8-83** Profiles to be selected*

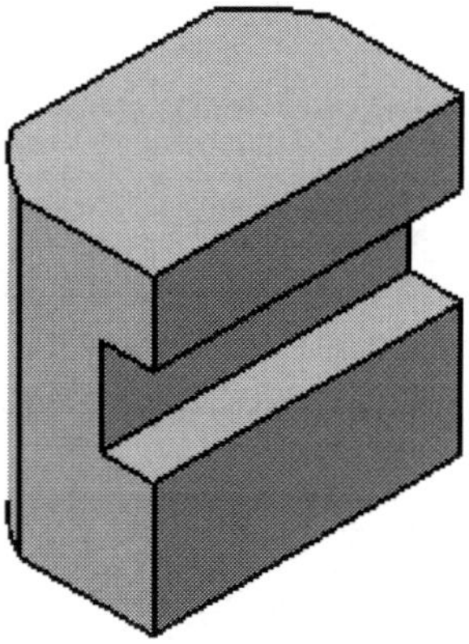

***Figure 8-84** Resultant solid combine*

TUTORIALS

Tutorial 1

In this tutorial, you will create the model of the Soap Case shown in Figure 8-85. Its orthographic views and dimensions are shown in Figure 8-86. **(Expected time: 45 min)**

***Figure 8-85** Model of the Soap Case for Tutorial 1*

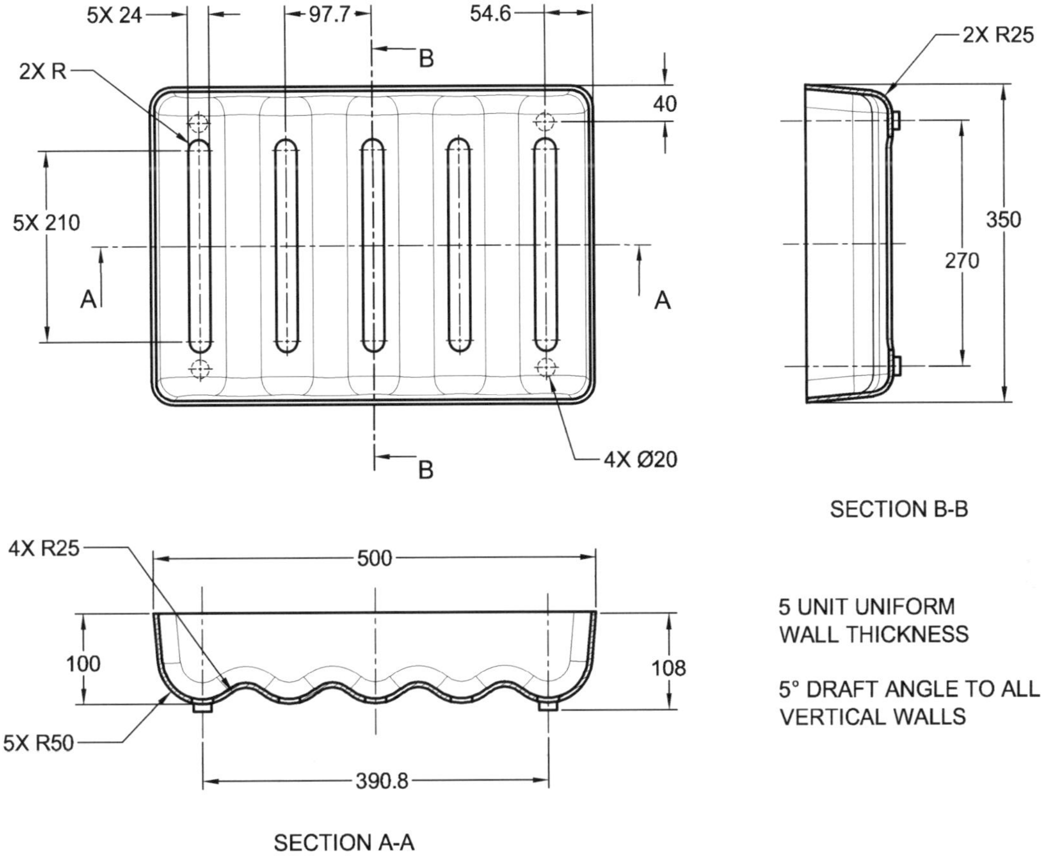

***Figure 8-86** Orthographic views and dimensions of the Soap Case for Tutorial 1*

The following steps are required to complete this tutorial:

a. Create the base feature sketch of the model.
b. Apply draft to the faces of the base feature.
c. Create a pocket feature to create the shape of the lower portion of the model.
d. Fillet the edges of the model.
e. Shell the model and remove the top face.
f. Create a sketch of pocket feature that will be used as a vent for removing water from the Soap Case.
g. Pattern the newly created pocket feature.
h. Create the standoff for the Soap Case.
i. Pattern the standoff.

Creating the Base Feature

The base feature of this model will be created by extruding the sketch drawn on the yz plane.

1. Create the sketch of the base feature on the yz plane, as shown in Figure 8-87.

2. Invoke the **Pad** tool from the **Sketch-Based Features** toolbar; the **Pad Definition** dialog box is displayed.

3. Select the **Mirrored extent** check box, set **175** in the **Length** spinner, and then choose **OK**; the base feature is created.

Applying Draft to the Base Feature

After creating the base feature of the model, you need to apply a draft to its faces.

1. Invoke the **Draft Angle** tool from the **Dress-Up Features** toolbar to display the **Draft Definition** dialog box.

2. Set the value for the draft angle to **5deg** in the **Angle** spinner.

   Now, you need to select and draft all the faces that are normal to the top face.

3. Select the **Selection by neutral face** check box from the **Draft Definition** dialog box; you are prompted to select a neutral element.

4. Select the top face of the base feature as the neutral element and click on the arrowhead if it points in the upward direction. Choose the **OK** button from the **Draft Definition** dialog box. The model after creating the draft feature is shown in Figure 8-88.

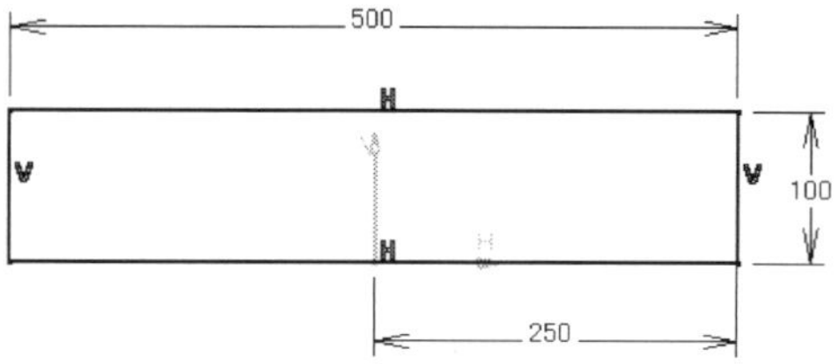

Figure 8-87 The sketch for the base feature of the model

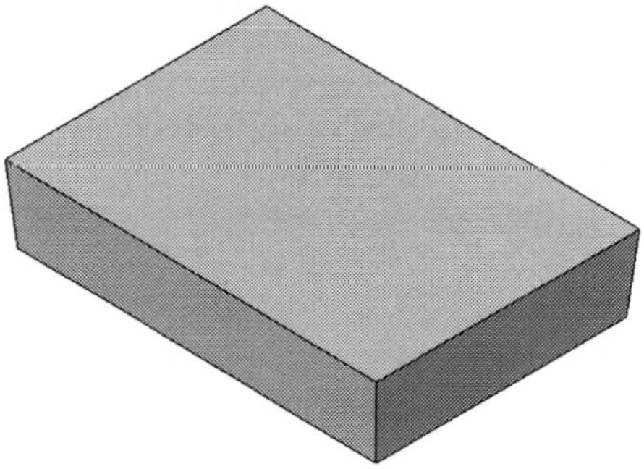

Figure 8-88 The model after creating the draft feature

Creating the Pocket Feature

Next, you need to create a pocket feature. This feature is used to shape the bottom portion of the model and is created by extruding a sketch drawn on the yz plane using the **Up to last** option in both directions.

1. Draw a sketch on the yz plane, as shown in Figure 8-89, and then exit the **Sketcher** workbench.

2. Invoke the **Pocket Definition** dialog box and choose the **Reverse Side** button.

3. Next, expand the **Pocket Definition** dialog box and then in the **First Limit** and **Second Limit** areas, select the **Up to last** option from the **Type** drop-down list and choose **OK**. The model after creating the pocket feature is shown in Figure 8-90.

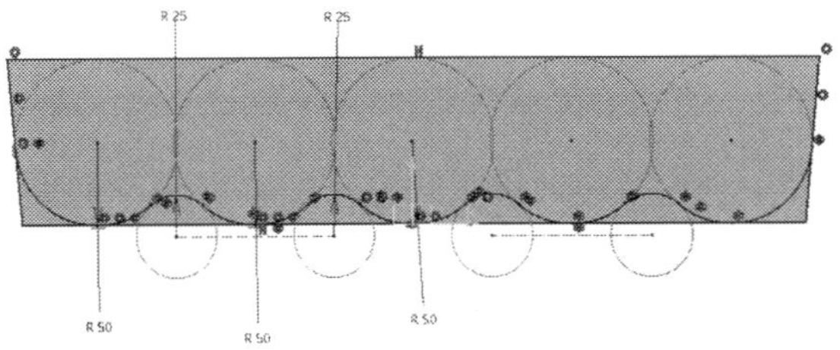

Figure 8-89 *The sketch for creating the pocket feature*

Figure 8-90 *The model after creating the pocket feature*

Filleting the Edges and Creating the Shell Feature

After creating the pocket feature, you need to fillet the edges of the model using the **Edge Fillet** tool.

1. Invoke the **Edge Fillet Definition** dialog box.

2. Select the edges to be filleted, as shown in Figure 8-91.

3. Set the value of the fillet radius in the **Radius** spinner to **25** mm and choose **OK**. The model after filleting the edges is shown in Figure 8-92.

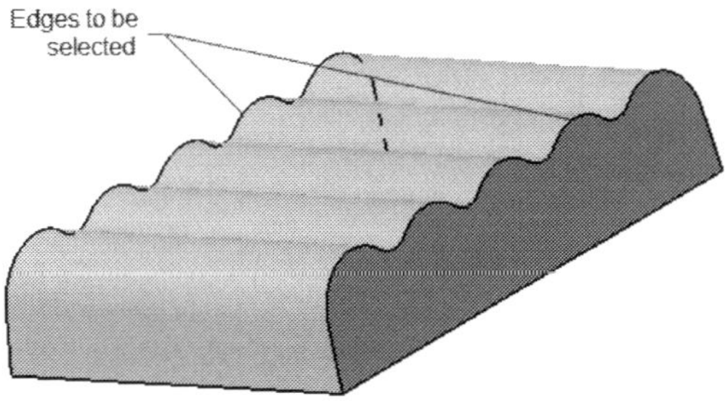

Figure 8-91 *Edges to be selected*

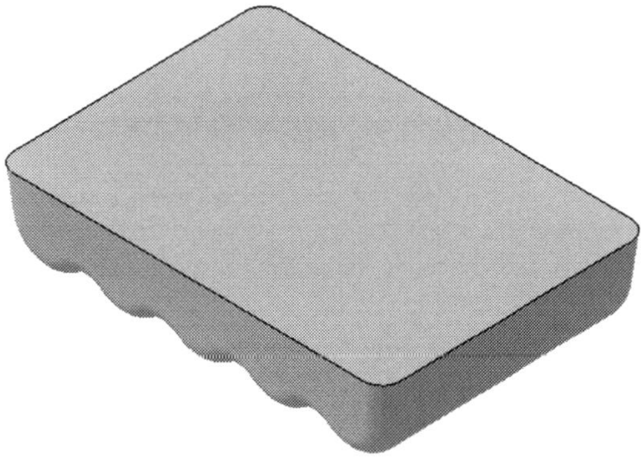

Figure 8-92 *The model after filleting the edges*

4. Now, create the shell feature by removing the top face and applying inside wall thickness of 5 units, refer to Figure 8-93.

Creating and Patterning the Pocket Feature

Now, according to the design requirement, you need to create another pocket feature and then pattern it using the **Rectangular Pattern** tool.

1. Create a sketch on the xy plane, as shown in Figure 8-94.

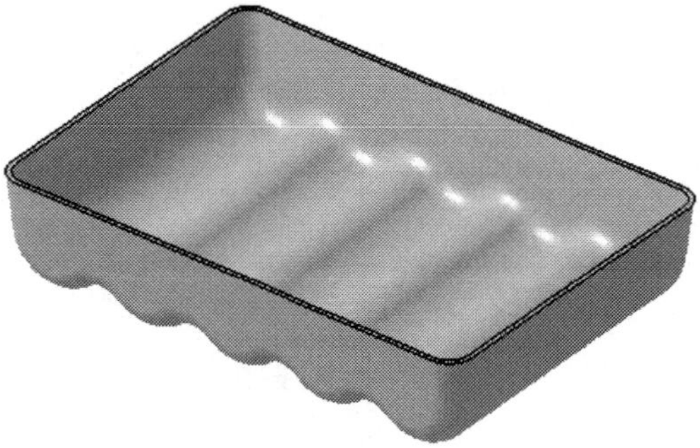

Figure 8-93 The model after creating the shell feature

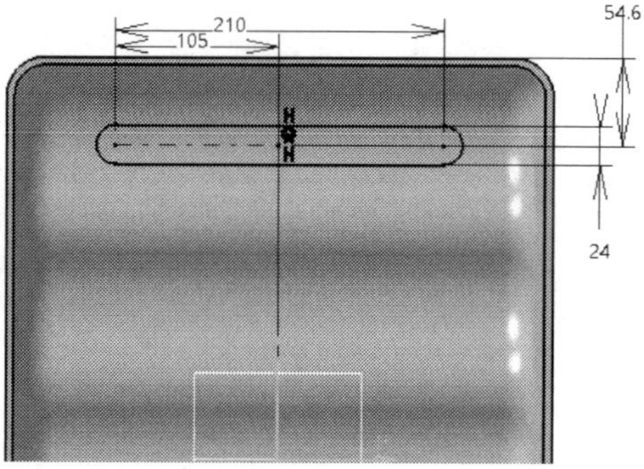

Figure 8-94 Sketch for creating another pocket feature

2. Exit the **Sketcher** workbench and invoke the **Pocket Definition** dialog box.

3. Select the **Up to next** option from the **Type** drop-down list from the **First Limit** area, and flip the direction of arrow displayed in the graphics window by choosing the **Reverse Direction** button in the dialog box if not pointing upward.

4. Click on the **More** button; the **Pocket Definition** dialog box gets expanded.

5. Select the **Up to next** option from the **Type** drop-down list in the **Second Limit** area and choose **OK** to create the pocket feature and exit the dialog box.

6. Select the pocket feature created in the previous step and then choose the **Rectangular Pattern** tool from the **Transformation Features** toolbar; the **Rectangular Pattern Definition** dialog box is displayed. Also, you are prompted to select the directions and specify the required parameters.

7. Click once in the **Reference element** selection box and select the edge, as shown in Figure 8-95, as the direction of pattern; a preview of the patterned instance with default settings is displayed in the geometry area.

   If the default direction of the pattern creation is not same as the required direction, you need to flip the direction.

8. Choose the **Reverse** button from the **Reference Direction** area of the **Rectangular Pattern Definition** dialog box; the direction of the pattern gets flipped.

9. Set the value of the **Instance(s)** spinner to **5** and the **Spacing** spinner to **97.7**. Make sure that the **Keep specifications** check box is selected to retain the specification of the original instance. Then, choose the **OK** button from the **Rectangular Pattern Definition** dialog box to accept the entered values. The model after patterning the pocket feature is shown in Figure 8-96.

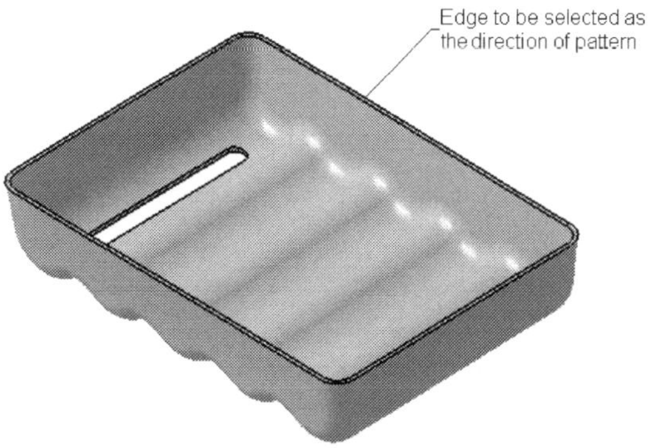

Figure 8-95 *Edge to be selected as reference*

Figure 8-96 *The model after creating the pattern*

Creating and Patterning the Standoffs

Next, you need to create the standoff that will keep the Soap Case lifted from the ground. After creating the standoff, you need to pattern it.

1. Create a sketch on the xy plane, as shown in Figure 8-97. Note that the center of the sketch is aligned with the center of the elongated hole. Exit the **Sketcher** workbench and then invoke the **Pad Definition** dialog box.

2. Select the **Up to next** option from the **Type** drop-down list in the **First Limit** area and then choose the **More** button to expand the dialog box.

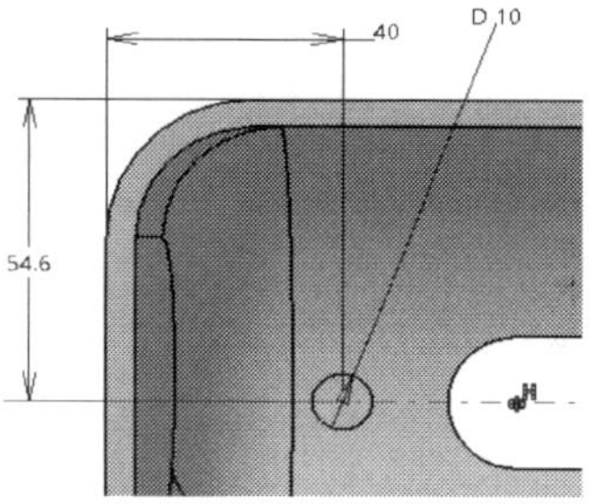

Figure 8-97 *The sketch for creating the standoff*

3. Set the value to **8** in the **Length** spinner in the **Second Limit** area and then choose the **OK** button to create the pad feature. Next, exit the dialog box.

4. Select the newly created pad feature from the Specification tree and choose the **Rectangular Pattern** button from the **Patterns** sub-toolbar in the **Transformation Features** toolbar; the **Rectangular Pattern Definition** dialog box is displayed.

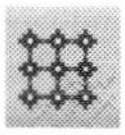

5. Click once in the **Reference element** display box and then select the first direction of the pattern creation, as shown in Figure 8-98.

6. Set the value of the **Instance(s)** spinner to **2** and the **Spacing** spinner to **390.8**; a preview of the pattern, with the default settings, is displayed in the geometry area.

7. Choose the **Reverse** button to flip the direction, if required.

8. Choose the **Second Direction** tab from the **Rectangular Pattern Definition** dialog box.

9. Click once in the **Reference element** display box and then select the second direction of the pattern creation, refer to Figure 8-98.

10. Set the value of the **Instance(s)** spinner to **2** and the **Spacing** spinner to **270**.

11. Choose the **Reverse** button to flip the direction and then choose the **OK** button from the **Rectangular Pattern Definition** dialog box.

    The model after patterning the standoff is shown in Figure 8-99.

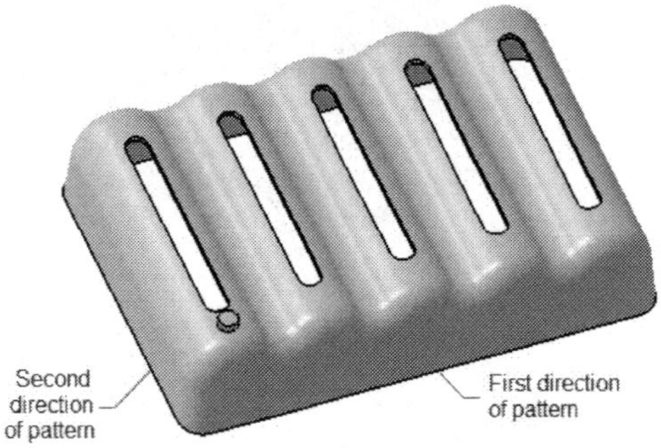

Figure 8-98 *Edges to be selected as reference*

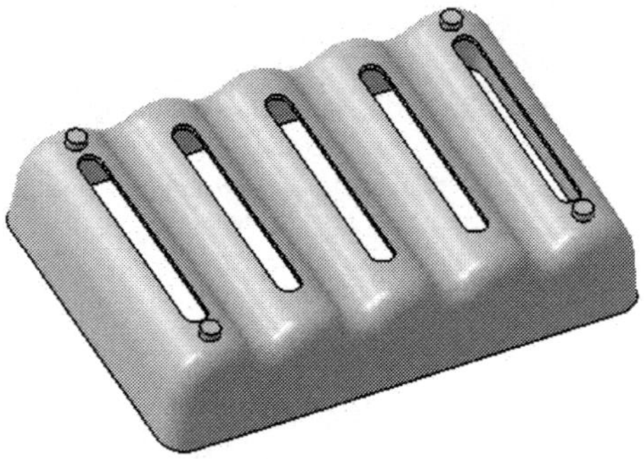

Figure 8-99 *Final model after patterning the standoff*

Saving and Closing the File

1. Choose the **Save** button from the **Standard** toolbar to invoke the **Save As** dialog box. Create the *c08* folder inside the *CATIA* folder.

2. Enter the name of the file as **c08tut1.CATPart** in the **File name** edit box and choose the **Save** button. The file is saved at *C:\CATIA\c08*.

3. Close the part file by choosing **File > Close** from the menu bar.

Tutorial 2

In this tutorial, you will create the model of the Motor Cover shown in Figure 8-100. Its views and dimensions are shown in Figure 8-101. **(Expected time: 45 min)**

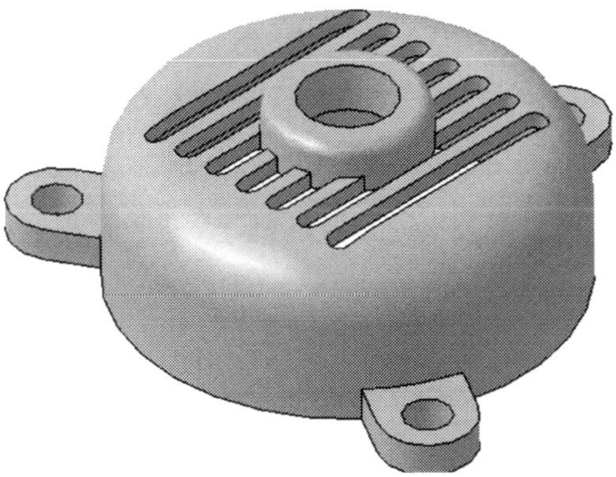

Figure 8-100 Model of the Motor Cover for Tutorial 2

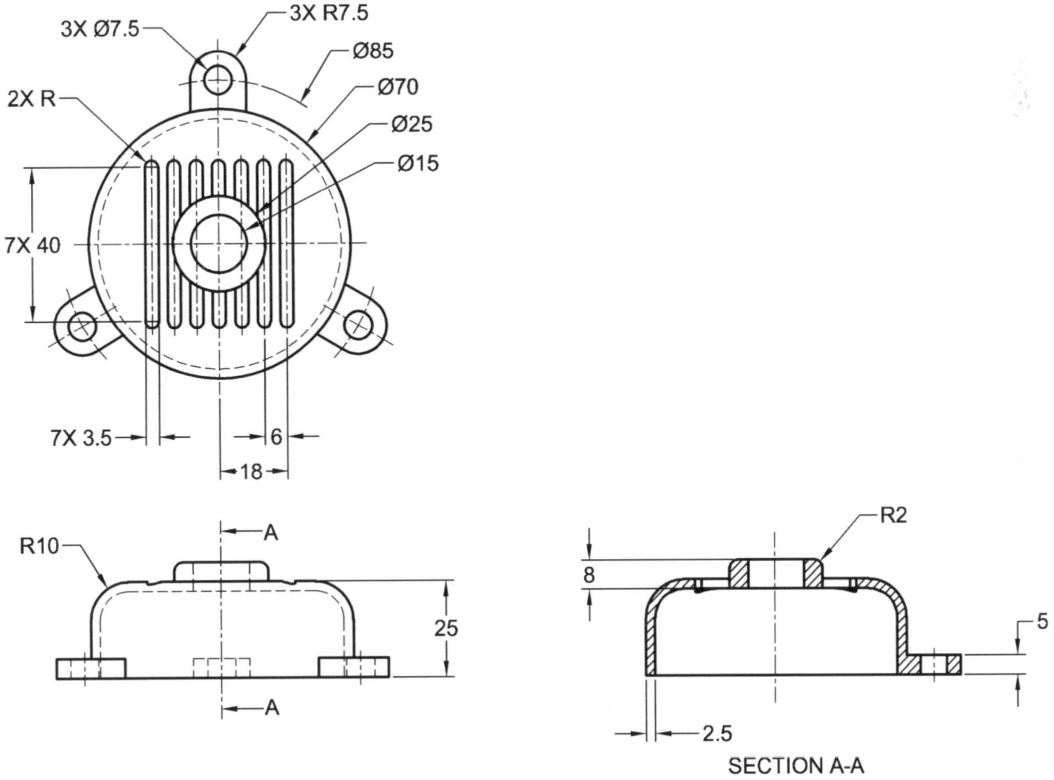

Figure 8-101 Views and dimensions of the Motor Cover for Tutorial 2

The following steps are required to complete this tutorial:

a. Create the base feature of the model by revolving the sketch drawn on the yz plane.
b. Shell the model and remove the bottom face of the model.
c. Create the pad feature on the outer periphery of the base feature.
d. Pattern the newly created pad feature using the **Circular Pattern** tool.

e. Create the pocket feature on the top face of the base feature.
f. Pattern the pocket feature using the **Rectangular Pattern** tool.
g. Create the remaining features of the model.

Creating the Base Feature

1. Draw the sketch on the yz plane, as shown in Figure 8-102, and then exit the **Sketcher** workbench. Note that a vertical axis is also drawn at the origin.

2. Invoke the **Shaft** tool from the **Sketch-Based Features** toolbar; the **Shaft Definition** dialog box and the preview of the model are displayed.

3. Choose the **OK** button from the **Shaft Definition** dialog box to create the base feature.

Shelling the Model

Next, you need to shell the model using the **Shell** tool.

1. Invoke the **Shell** tool from the **Dress-Up Features** toolbar and set the value of the **Inside thickness** spinner to **2.5** in the **Default Thickness** area of the **Shell Definition** dialog box.

2. Select the bottom face of the base feature as the face to be removed. The rotated view of the model after shelling is shown in Figure 8-103.

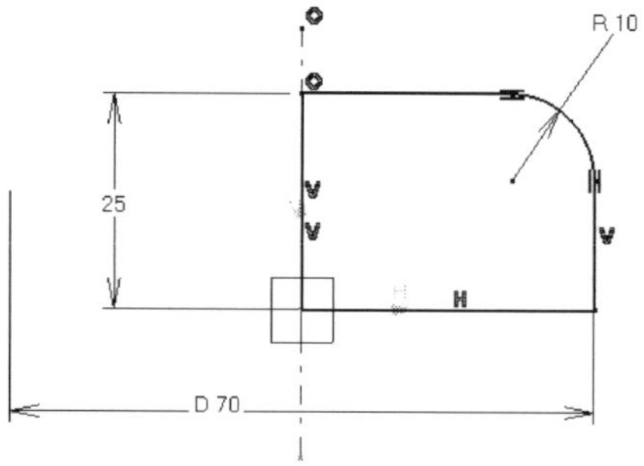

Figure 8-102 *The sketch to create the base feature of the model*

Figure 8-103 *The model after creating the shell feature*

Creating and Patterning the Pad Feature

After shelling the model, you need to create a pad feature on the outer periphery of the base feature and then pattern it using the **Circular Pattern** tool.

1. Select the bottom face of the model and invoke the **Sketcher** workbench.

2. Draw the sketch for the pad feature, as shown in Figure 8-104. Exit the **Sketcher** workbench.

3. Invoke the **Pad Definition** dialog box and create the pad feature by extruding the sketch by 5 units. If required, you can reverse the direction of extrusion using the **Reverse Direction** button and then exit the **Pad Definition** dialog box.

Next, you need to pattern the newly created pad feature using the **Circular Pattern** tool.

4. Select the newly created pad feature and invoke the **Circular Pattern** tool from the **Patterns** sub-toolbar in the **Transformation Features** toolbar; the **Circular Pattern Definition** dialog box is displayed.

5. Click once in the **Reference element** display box and select the curved face of the base feature as the reference element; a preview of the pattern is displayed with the default settings in the geometry area.

6. Select the **Complete crown** option from the **Parameters** drop-down list if not already selected.

7. Set the value of the **Instance(s)** spinner to **3** and choose the **OK** button from the **Circular Pattern Definition** dialog box. The model after patterning the pad feature is shown in Figure 8-105.

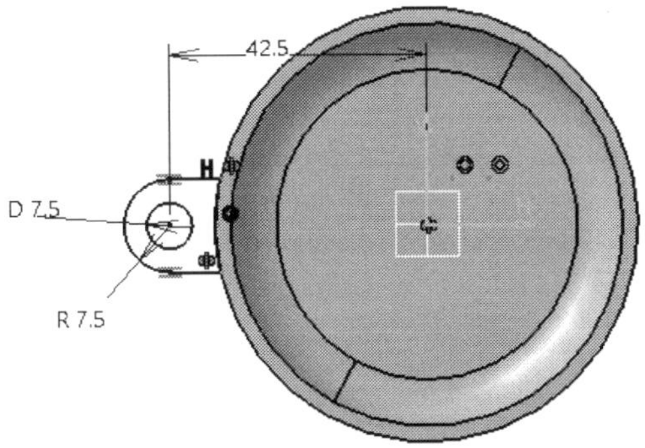

Figure 8-104 *The sketch to create the pad feature*

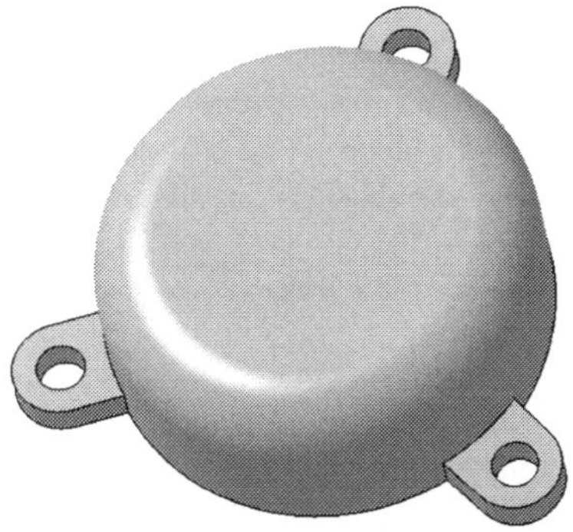

Figure 8-105 *The model after patterning the pad feature*

Creating and Patterning the Pocket Feature

Next, you need to create and pattern the pocket feature.

1. Select the top face of the base feature and invoke the **Sketcher** workbench.

2. Draw the sketch, as shown in Figure 8-106, and exit the **Sketcher** workbench. Note that, in this figure, the constraints have been hidden for better visualization.

3. Invoke the **Pocket** tool and extrude the sketch by 5 units; the pocket feature is created.

   Next, you need to pattern the pocket feature.

4. Select the pocket feature and invoke the **Rectangular Pattern** tool from the **Patterns** sub-toolbar in the **Transformation Features** toolbar; the **Rectangular Pattern Definition** dialog box is displayed.

5. Click once in the **Reference element** display box and select the yz plane as the reference element. The preview of the pattern with the default settings is displayed in the geometry area.

6. Set the value of the **Instance(s)** spinner to **7** and the **Spacing** spinner to **6** and choose the **Reverse** button, if required.

7. Select the **Keep specifications** check box and choose the **OK** button from the **Rectangular Pattern Definition** dialog box. The model after patterning the pocket feature is shown in Figure 8-107.

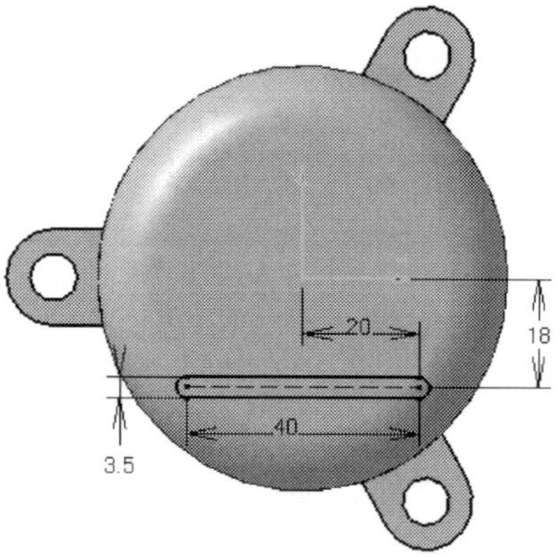

Figure 8-106 *The sketch for creating the pocket feature*

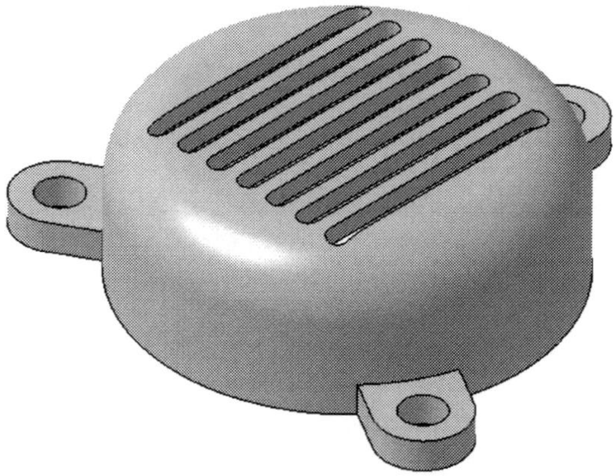

Figure 8-107 *The model after patterning the pocket feature*

Creating Remaining Features

1. Now, draw a circle of 25 unit diameter on the internal flat face of the component in the **Sketcher** workbench and extrude it by 8 units.

2. Next, apply the fillet and pocket features to the model. The model after creating all features is shown in Figure 8-108.

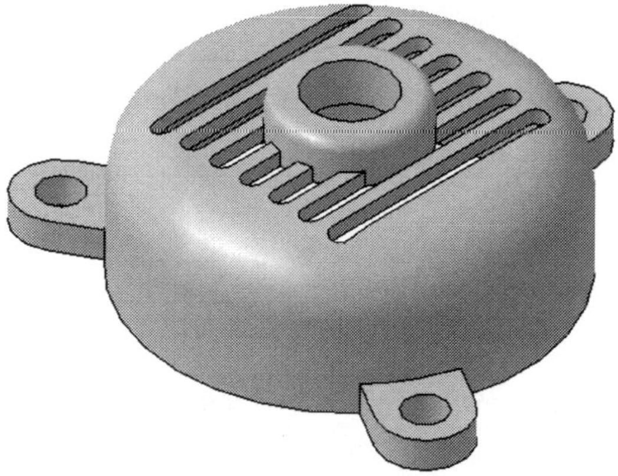

Figure 8-108 *Final model of the Motor Cover*

Saving and Closing the File

1. Choose the **Save** button from the **Standard** toolbar to invoke the **Save As** dialog box.

2. Save the file with the name *c08tut2.CATPart* at *C:\CATIA\c08*.

3. Close the part file by choosing **File > Close** from the menu bar.

Tutorial 3

In this tutorial, you will create the model shown in Figure 8-109, by using boolean operations. Its views and dimensions are shown in Figure 8-110. **(Expected time: 45 min)**

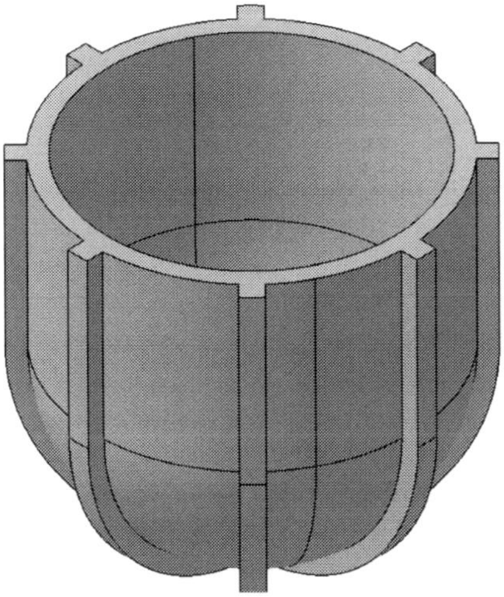

Figure 8-109 *Model for Tutorial 3*

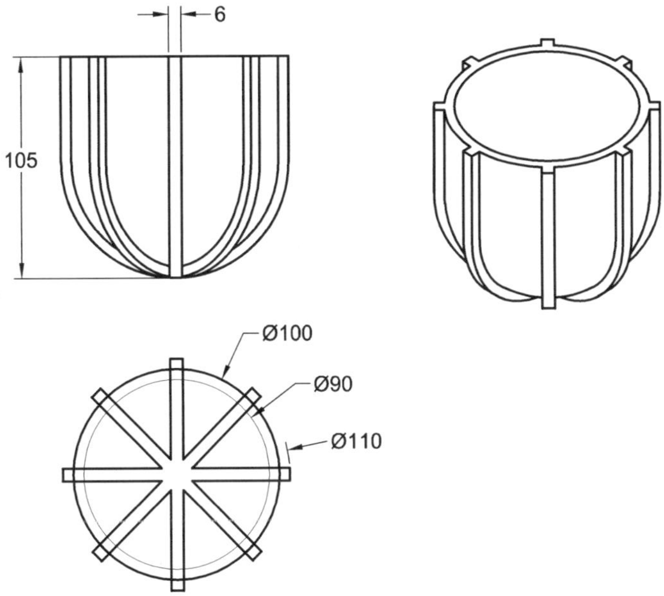

Figure 8-110 *Views and dimensions for Tutorial 3*

The following steps are required to complete this tutorial:

a. Create standard tree structure for boolean operations.
b. Create the base feature of the model by revolving the sketch drawn along the yz plane.
c. Create hollowness by removing core from the base feature of the model.
d. Add thickness to copy of the base feature and create rib plates.
e. Create intersection between rib and plate bodies.
f. Add rib body to the result.

Creating Tree Structure and Adding New Bodies

You need to create standard tree structure which you can use to create any model by using boolean method.

1. Choose the **Body** tool from the **Bodies** sub-toolbar of the **Insert** toolbar; a new body named **Body.2** is added under the current body in the Specification tree. Similarly, add two more bodies.

   Next, you need to rename the bodies.

2. Right-click on the **Body.2** and choose the **Properties** option from the contextual menu; the **Properties** dialog box is displayed. Next, choose the **Features Properties** tab. Enter **Result** in the **Feature Name** edit box and choose the **OK** button from the dialog box. Similarly, rename **Body.3** and **Body.4** as **Material_addition** and **Material_removal** respectively.

3. Choose the **Add** tool from the **Boolean Operations** toolbar; the **Add** dialog box is displayed.

4. Select **Material_addition** from the Specification tree; **Material_addition** gets displayed in the **Add** display box. Next, select **Result** from the Specification tree; **Result** gets displayed in the **To** display box.

5. Choose the **OK** button from the dialog box.

6. Choose the **Remove** tool from the **Boolean Operations** toolbar; the **Remove** dialog box is displayed.

7. Select **Material_removal** from the Specification tree; **Material_removal** gets displayed in the **Remove** display box. Next, select **Result** from the Specification tree; **Result** gets displayed in the **From** display box.

8. Choose the **OK** button. The expanded **Result** node is shown in Figure 8-111.

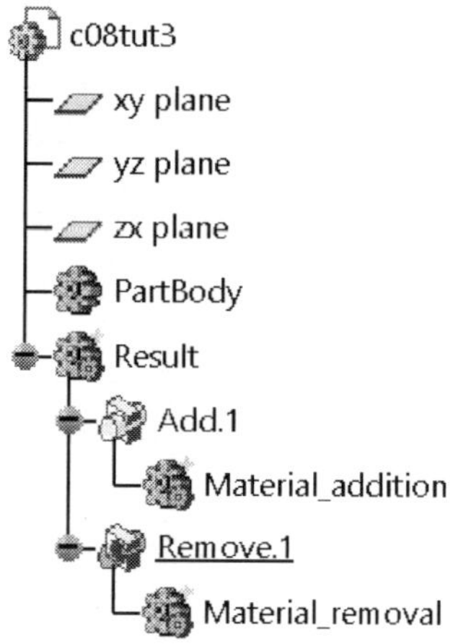

***Figure 8-111** The expanded **Result** node*

Creating the Base Feature

After creating the standard tree structure, you need to create the base feature in new body.

1. Add new body and rename it as **Base_feature** as done earlier. Now select **Base_feature** and invoke the contextual menu.

2. Choose the **Define In Work Object** option from the contextual menu; the selected body gets activated.

3. Draw the sketch on the yz plane, as shown in Figure 8-112, and then exit the **Sketcher** workbench. Note that vertical axis should drawn at the origin.

4. Invoke the **Shaft** tool from the **Sketch-Based Features** toolbar; the **Shaft Definition** dialog box along with preview of the model is displayed.

5. Choose the **OK** button from the **Shaft Definition** dialog box with the default settings; the base feature is created as shown in Figure 8-113.

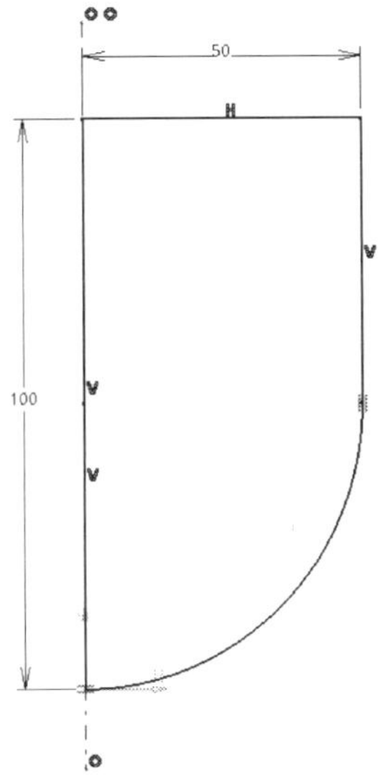

Figure 8-112 *The sketch to create the base feature of the model*

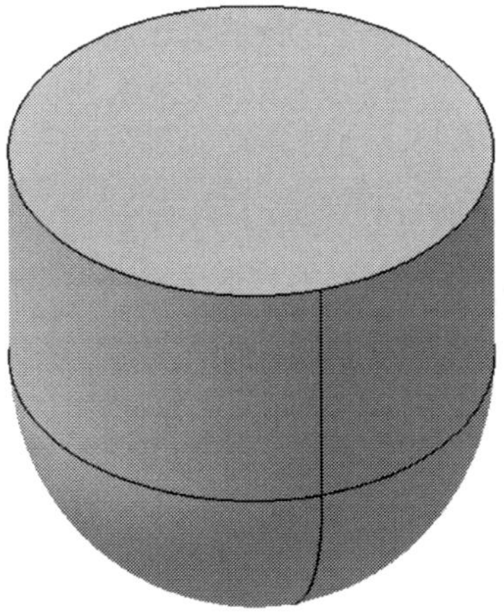

Figure 8-113 *The model after creating the base feature*

Copying the Base Feature

After creating the base feature, you need to copy it.

1. Right-click on **Base_feature** in the Specification tree and select the **Copy** option from the contexual menu.

2. Again right-click on **Base_feature** and select the **Paste Special** option from the contextual menu; the **Paste Special** dialog box is displayed.

3. Select **As Result with Link** from the dialog box; the copy of the **Base_feature** feature along with **Solid.1** gets added to the Specification tree.

4. Now, choose the **Add** tool from **Boolean Operations**; the **Add** dialog box is displayed.

5. Select the copied **Base_feature**; the selected body gets displayed in the **Add** display box. Next select **Material_addition**; the selected body gets displayed in the **To** display box. Choose the **OK** button from the dialog box.

Creating Hollowness

Next, you need to remove material to create hollowness by using the same base feature.

1. Create a copy of the base feature and rename it as **Core** as done earlier.

2. Choose the **Only Current Body** tool from the **View** sub-toolbar of the **Tools** toolbar if not selected. Now select the **Thickness** option from the **Dress-Up Features** toolbar; the **Thickness Definition** dialog box is displayed.

3. Select the cylindrical face from the graphics window and enter **-5** in **Thickness** edit box. Choose the **OK** button.

4. Right-click on **Core** and select the **Add** option from the **Core object** cascading menu; the **Add** dialog box is displayed with **Core** displayed in the **Add** display box.

5. Select **Material_removal** from the Specification tree; the selected body gets displayed in the **To** display box.

6. Choose the **OK** button. Now activate the **Result** body to get the resultant model, as shown in Figure 8-114.

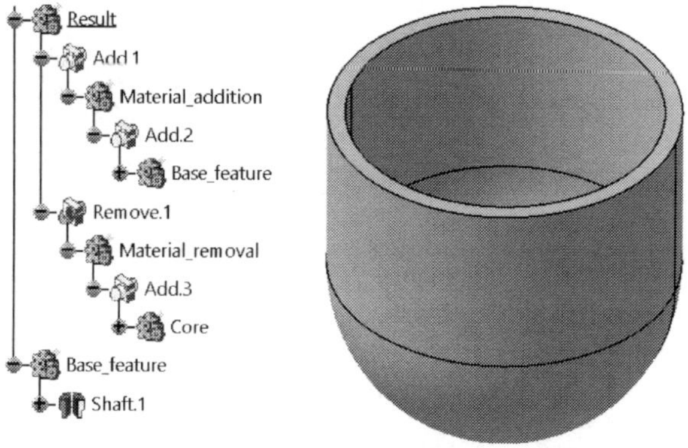

Figure 8-114 *The model after removing core from the base feature*

Creating Required Features for Rib

For creating ribs of the model, you need to add thickness to the newly copied base feature and then create ribs in the new body. After that you need to intersect these two bodies.

1. Create a copy of the base feature and rename the copied feature as Rib as done earlier.

2. Choose the **Only Current Body** tool from the **View** sub-toolbar of the **Tools** toolbar if not chosen by default. Next, select the **Thickness** option from the **Dress-Up Features** toolbar; the **Thickness Definition** dialog box is displayed.

3. Select the cylindrical face from graphics window and enter **5** in the **Thickness** edit box and choose the **OK** button.

   Next, you need create ribs in the new body.

4. Choose the **Body** tool from the **Bodies** sub-toolbar of the **Insert** toolbar; the new body gets added to the Specification tree. Rename this body as **Rib_plates**.

5. Now right-click on **Rib_plates** and choose the **Define In Work Object** option to activate the body.

6. Click on **Only Current Body** option in the **View** sub-toolbar of the **Tools** toolbar; all the features are now visible in graphics window and you can use these features as a reference.

7. Now, select the top face of **Base_feature**, as shown in Figure 8-115 and create the sketch, as shown in Figure 8-116.

8. Exit the Sketcher Workbench and choose the **Pad** tool from the **Sketch-Based Features** toolbar; the **Feature Error** dialog box is displayed.

9. Select the **Yes** button from the dialog box; the **Pad** dialog box is displayed with the preview of the model.

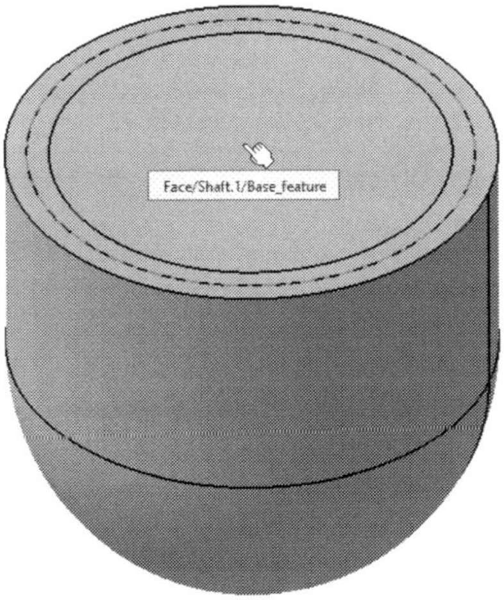

Figure 8-115 *The face selection for creating sketch*

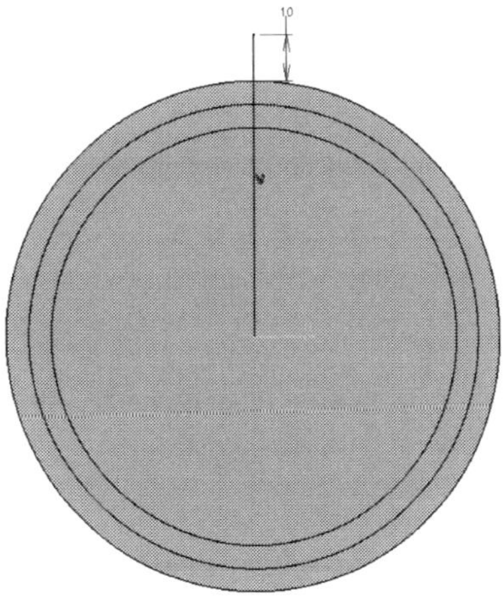

Figure 8-116 *The sketch to create the pad feature*

10. Select the **Thick** check box in the **Profile/Surface** area; the **Pad Definition** dialog box gets expanded. Select the **Neutral Fiber** check box and enter **6** in the **Thickness1** edit box of the **Thin Pad** area.

11. Enter **110** in the **Length** edit box of the **Second Limit** area and choose the **OK** button.

12. Select the pad feature and invoke the **Circular Pattern** tool from the **Patterns** sub-toolbar in the **Transformation Features** toolbar; the **Circular Pattern Definition** dialog box is displayed.

13. Click once in the **Reference element** display box and select the cylindrical face of **Base_feature** as the reference element.

14. Select **Complete crown** from the **Parameters** drop-down list if not already selected.

15. Set the value of the **Instance(s)** spinner to **8** and choose the **OK** button. Ribs are created, as shown in Figure 8-117.

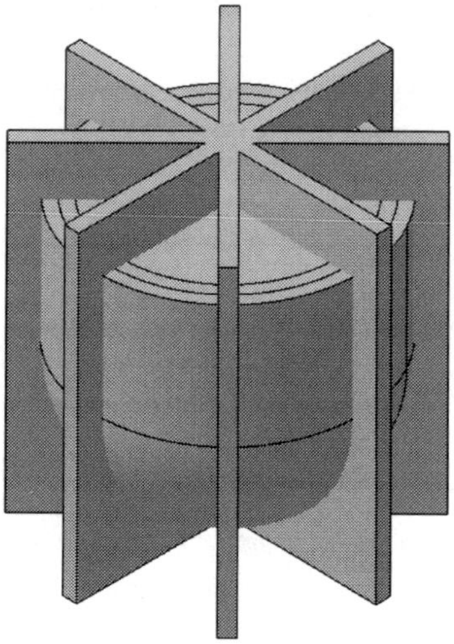

Figure 8-117 *The model after patterning the pad feature*

Creating Intersection of the Rib and Plate Bodies

1. Right-click on the **Rib_plates** and choose the **Intersect** option from the **Rib_plates object** cascading menu; the **Intersect** dialog box is displayed with **Rib_plates** in the **Intersect** display box.

2. Select **Rib** from the Specification tree; **Rib** gets displayed in the **To** display box. Choose the **OK** button from the dialog box.

3. Choose the **Only Current Body** tool in the **View** sub-toolbar of the **Tools** toolbar; the resultant model gets displayed, as shown in Figure 8-118.

4. Right-click on **Rib** and choose the **Add** option from the **Rib object** cascading menu; the **Add** dialog box is displayed with **Rib** in the **Add** display box.

5. Select **Material_addition** from the Specification tree; **Material_addition** gets displayed in the **To** selection box. Choose the **OK** button from the **Add** dialog box. Now activate **Result** to see the resultant model, as shown in Figure 8-119.

Figure 8-118 *The resultant model*

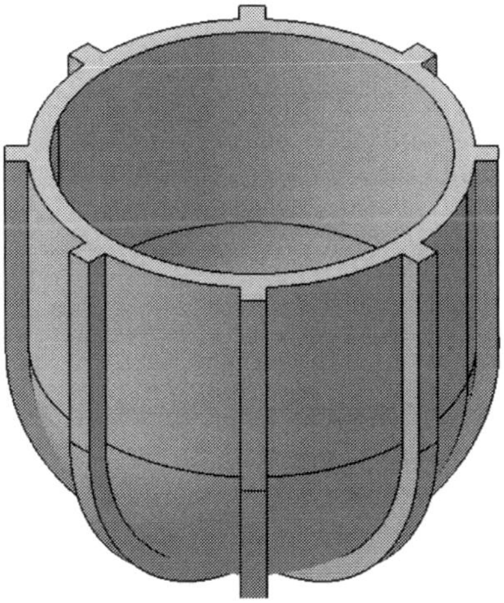

Figure 8-119 *The final model after adding ribs*

Saving and Closing the File

1. Choose the **Save** button from the **Standard** toolbar to invoke the **Save As** dialog box.
2. Save the file with the name *c08tut3.CATPart* at *C:\CATIA\c08*.
3. Close the part file by choosing **File > Close** from the menu bar.

Self-Evaluation Test

Answer the following questions and then compare them to those given at the end of this chapter:

1. Choose the _________ button from the **Insert** toolbar to insert a new body in the current file.
2. The _________ tool is used to add stiffeners to strengthen the component.
3. The _________ tool is used to remove material from the part body that consists of two or more than two disjointed portions.
4. The _________ check box is used to automatically align the pattern instances when you create a circular pattern.
5. Choose the _________ tab in the **Circular Pattern Definition** dialog box to define the crown parameters of the circular pattern.
6. The **Translation** tool is used to move the current body by defining a specified destination. (T/F)

7. A parent-child relationship is developed between the first instance and the patterned instances of the pattern. (T/F)

8. The **Symmetry** tool allows you to mirror the selected features or the whole body along a mirror element. (T/F)

9. The **User Pattern** tool is used to arrange the features of the current body in a sequence defined by the reference points in the **Sketcher** workbench. (T/F)

10. Choose the **Assemble** button from the **Boolean Operations** toolbar to assemble two bodies together. (T/F)

Review Questions

Answer the following questions:

1. Which of the following tools is used to retain the common portion of two intersecting bodies and remove the remaining portion of the selected bodies?

   (a) **Add** (b) **Union Trim**
   (c) **Shell** (d) **Intersect**

2. Which of the following tools is used to place the selected feature inside a new body and then assemble the newly created body with the current body?

   (a) **Mirror** (b) **Assemble**
   (c) **Subtract** (d) None of these

3. Which of the following tools is used to flip the position of the body with respect to the symmetry plane without creating another instance of the body?

   (a) **Intersect** (b) **Rectangular Pattern**
   (c) **Symmetry** (d) **Mirror**

4. Which of the following tools is used to insert a body in the Specification tree?

   (a) **Add** (b) **Intersect**
   (c) **Body** (d) None of these

5. Which of the following dialog boxes is used to rotate a selected body?

   (a) **Rotation Definition** (b) **Translation Definition**
   (c) **Rotational Pattern Definition** (d) None of these

6. The **Assemble** tool is used to assemble two selected bodies. (T/F)

7. The **Add** tool is used to add two selected bodies. (T/F)

8. The **Scaling** tool is used to scale the model with respect to the selected reference. (T/F)

9. You cannot add a body inside a body. (T/F)

10. You can place the part body anywhere in the 3D space with respect to the origin by using the Compass. (T/F)

EXERCISES

Exercise 1

Create the model of the Bracket shown in Figure 8-120. Its views and dimensions are shown in the same figure. **(Expected time: 45 min)**

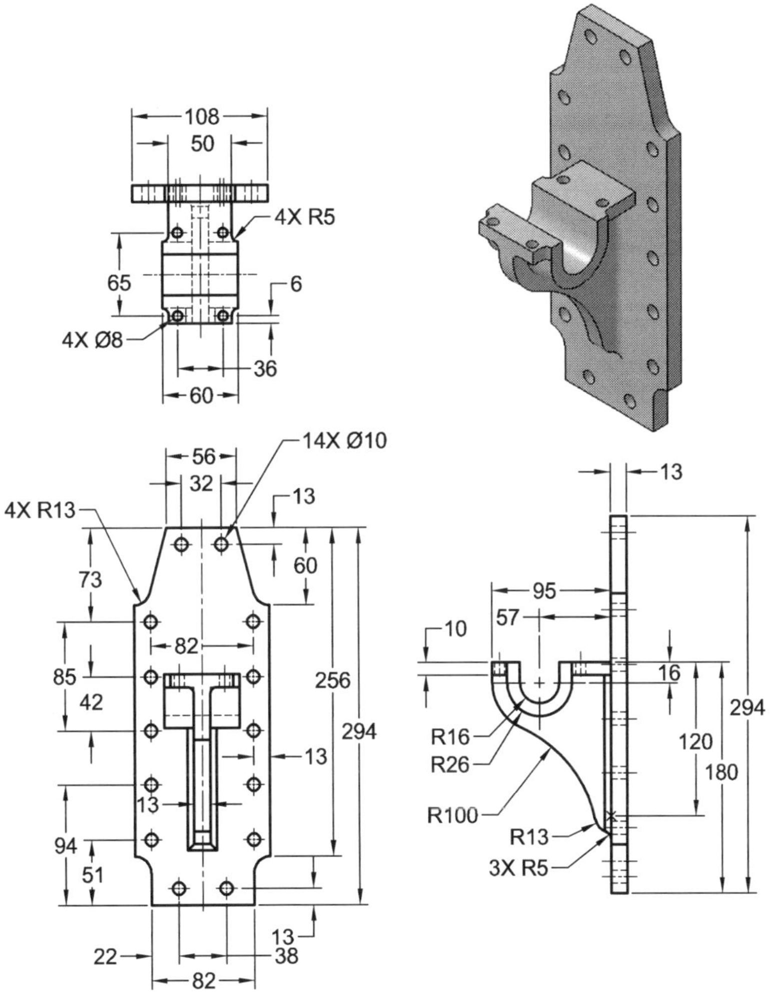

Figure 8-120 The Bracket with its views and dimensions

Exercise 2

Create the model of the Machine-Block shown in Figure 8-121. Its views and dimensions are shown in Figure 8-122. **(Expected time: 30 min)**

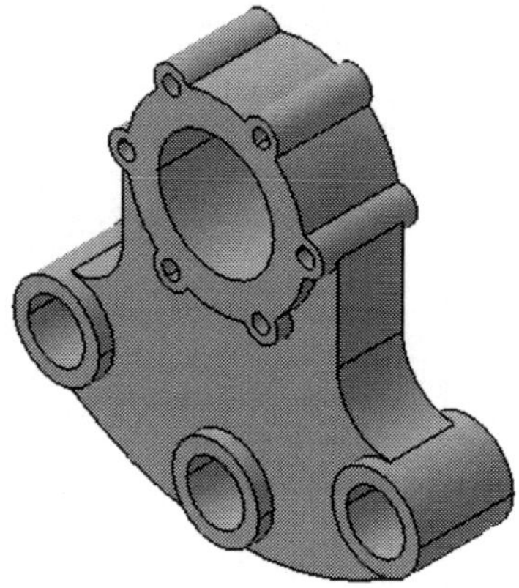

Figure 8-121 *Solid model for Exercise 2*

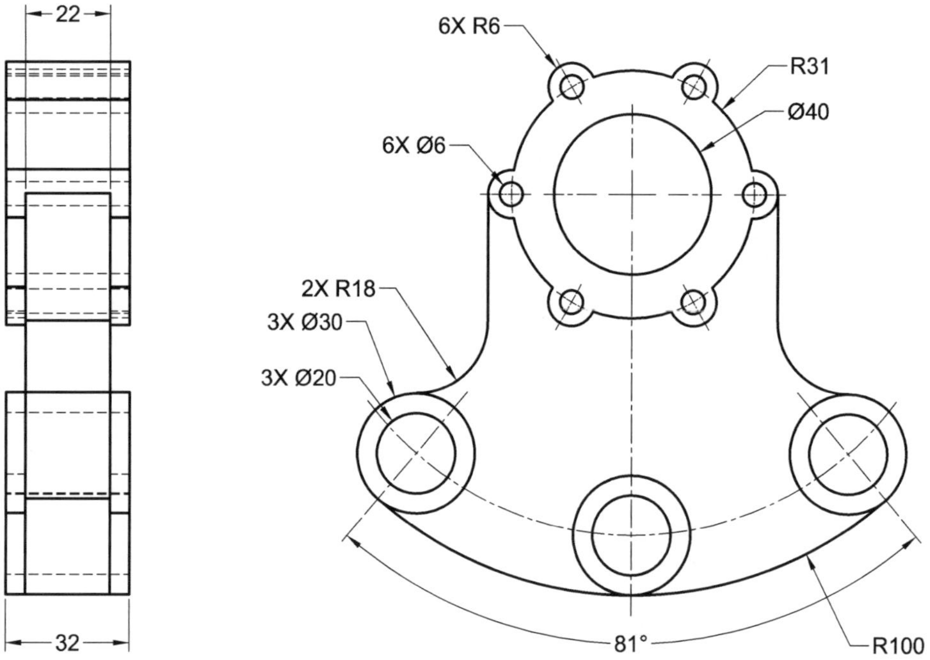

Figure 8-122 *Views and dimensions of the model for Exercise 2*

Exercise 3

Create the model shown in Figure 8-123. Its views and dimensions are shown in the same figure. **(Expected time: 30 min)**

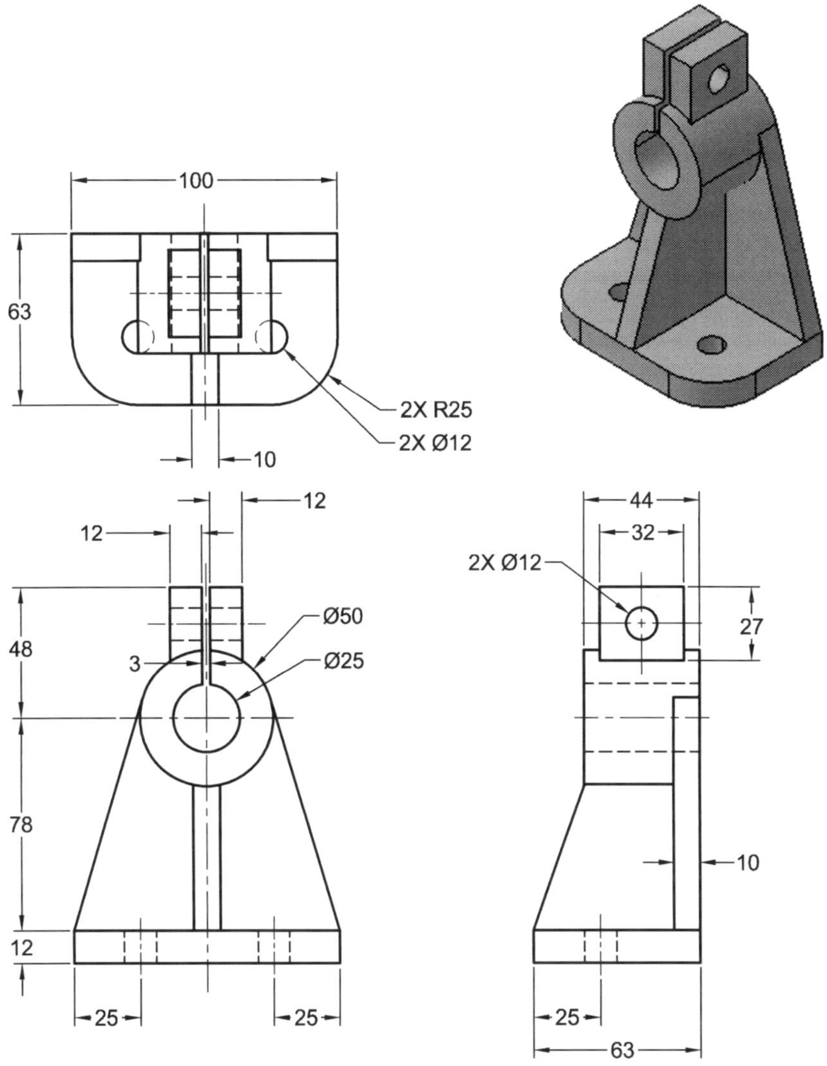

Figure 8-123 The views and dimensions of the model for Exercise 3

Answers to Self-Evaluation Test

1. Body, **2.** Stiffener, **3.** Remove Lump, **4.** Radial alignment of instance(s), **5.** Crown Definition, **6.** T, **7.** T, **8.** F, **9.** T, **10.** T

Chapter 9

Advanced Modeling Tools-II

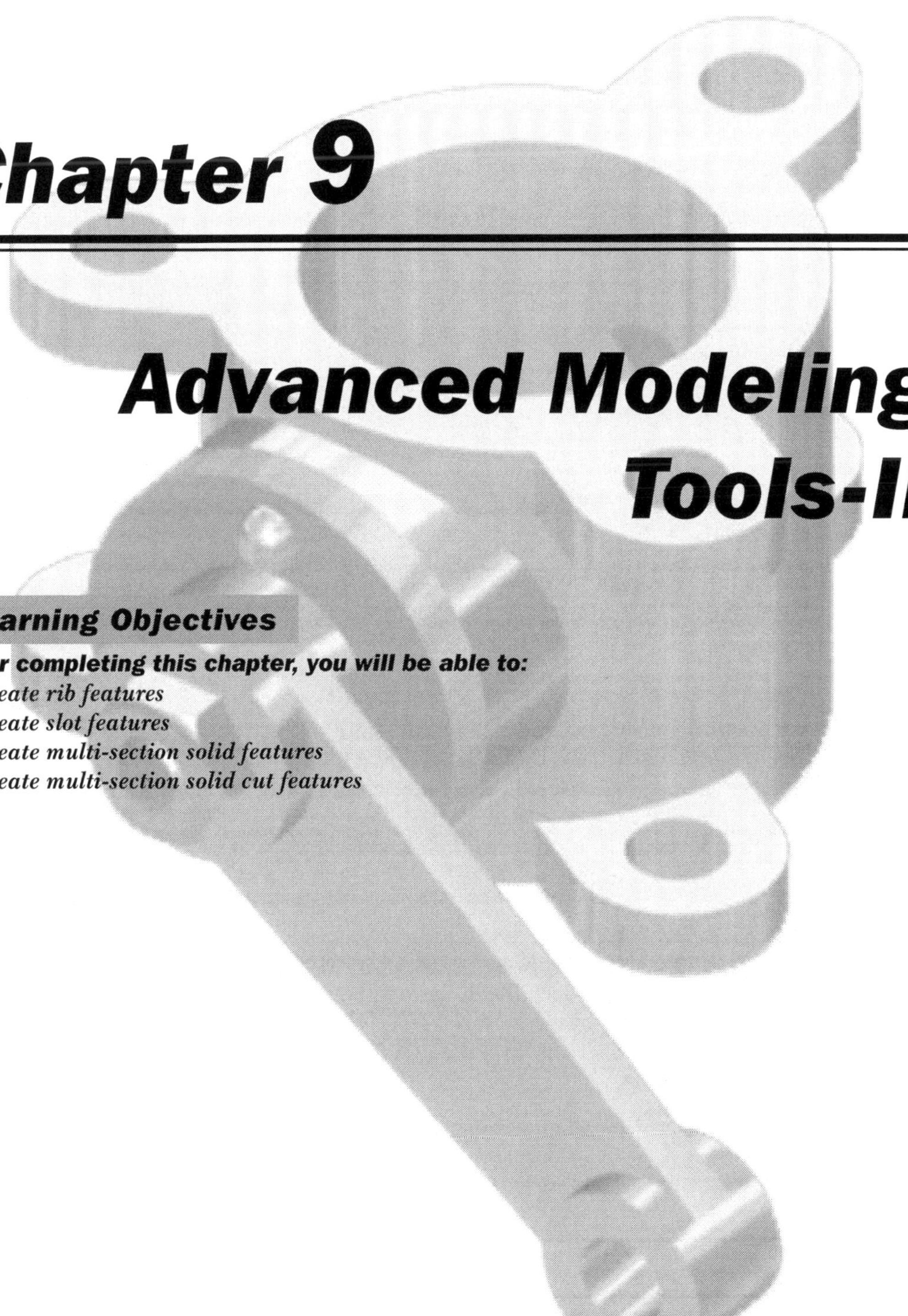

Learning Objectives

After completing this chapter, you will be able to:

- *Create rib features*
- *Create slot features*
- *Create multi-section solid features*
- *Create multi-section solid cut features*

ADVANCED MODELING TOOLS

In this chapter, you will learn about some more advanced modeling tools that will help you in creating complex features of models. The tools to be discussed in this chapter are **Rib**, **Slot**, **Multi-sections Solid**, and **Removed Multi-sections Solid**.

Creating Rib Features

Menubar: Insert > Sketch-Based Features > Rib
Toolbar: Sketch-Based Features > Rib

One of the most important advanced modeling tools is the **Rib** tool. This tool is used to sweep an open or a closed profile along an open or a closed center curve. A profile is the cross-section for the rib feature and the center curve is the course taken by the profile while creating the rib feature. To create a rib feature, invoke the **Rib** tool from the **Sketch-Based Features** toolbar; the **Rib Definition** dialog box will be displayed, as shown in Figure 9-1. Also, you will be prompted to define the profile of the rib feature.

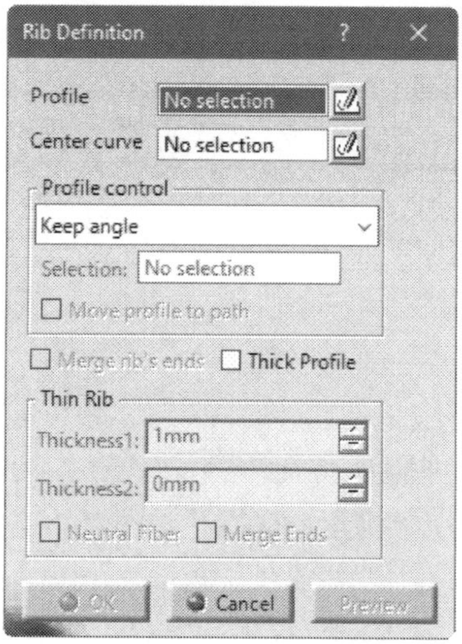

***Figure 9-1** The **Rib Definition** dialog box*

Select the sketch of the profile from the geometry area. Make sure that the profile is a closed sketch. Creating a rib feature with an open profile is discussed later in this chapter. On selecting the sketch, the name of the selected profile will be displayed in the **Profile** display box and you will be prompted to define the center curve. You can either select an open sketch, a closed sketch, or an edge as the center curve. Select the sketch; the name of the selected center curve will be displayed in the **Center curve** display box. Also, the preview of the feature is displayed in the graphics window. Choose the **OK** button from the **Rib Definition** dialog box.

Figure 9-2 shows a closed profile and the open center curve to be selected and Figure 9-3 shows the resulting rib feature. It is not necessary that the profile curve and the center curve must coincide. You can also create a rib feature using an unattached profile and a center curve. Figure 9-4 shows an unattached profile and the center curve and Figure 9-5 shows the resulting rib feature. Figure 9-6 shows a closed profile and a closed center curve and Figure 9-7 shows the resulting rib feature.

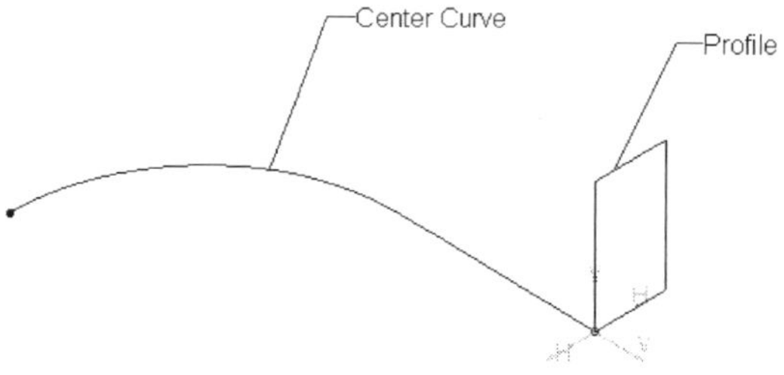

***Figure 9-2** Closed profile and the center curve to be selected*

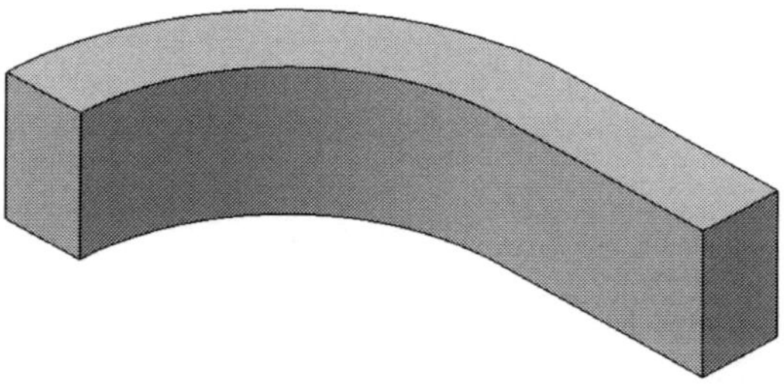

***Figure 9-3** Resulting rib feature*

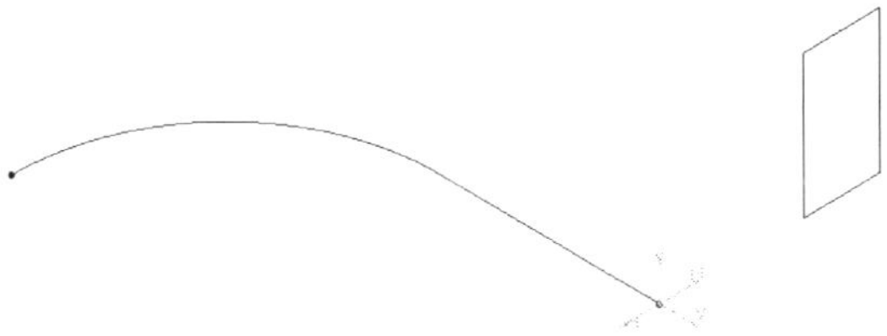

Figure 9-4 An unattached profile and the center curve

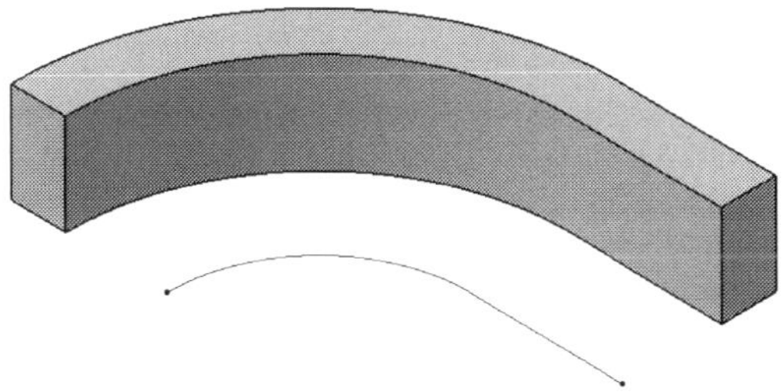

Figure 9-5 Resulting rib feature

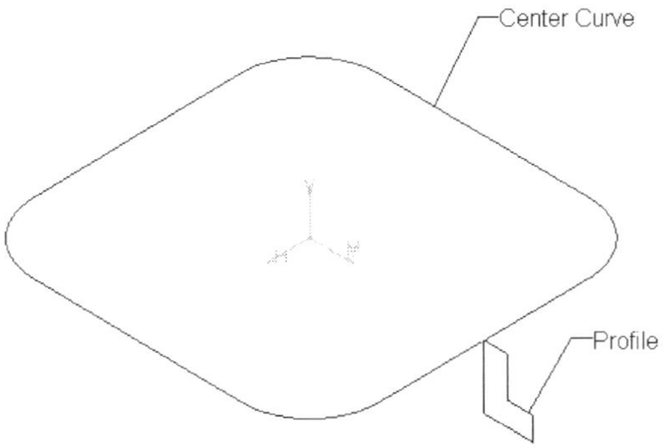

Figure 9-6 Closed profile and closed center curve

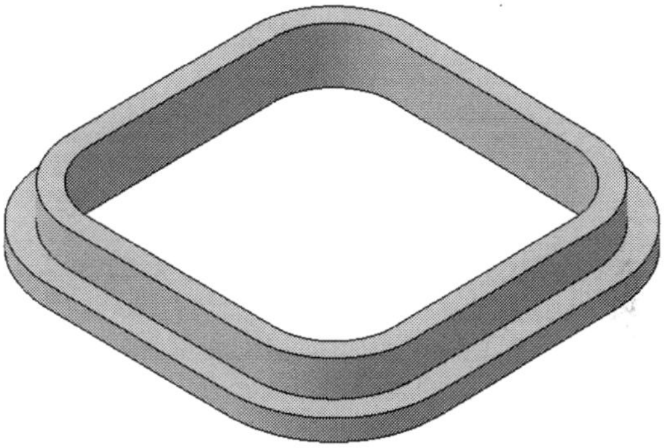

Figure 9-7 Resulting rib feature

Defining the Pulling Direction

The **Keep angle** option is selected by default in the drop-down list in the **Profile control** area. As a result, the profile of the rib feature will be normal throughout the center curve. You can also define a pulling direction while creating the rib feature. In this case, the end section of the rib feature will be normal to the selected pulling direction reference. To create such a sweep feature, select the **Pulling direction** option from the drop-down list in the **Profile control** area; you will be prompted to define the pulling direction. Select a plane or planar face; the name of the selected reference will be displayed in the **Selection** display box. The **Reference surface** option from the drop-down list in the **Profile control** area is used to select a reference surface along which the rib feature will be created. This option is discussed in detail later in this chapter. Figure 9-8 shows the profile, center curve, and pulling direction reference to be selected. Figure 9-9 shows the rib feature created with the **Keep angle** option selected from the drop-down list in the **Profile control** area. Figure 9-10 shows the rib feature created by selecting the **Pulling direction** option from the **Profile control** drop-down list.

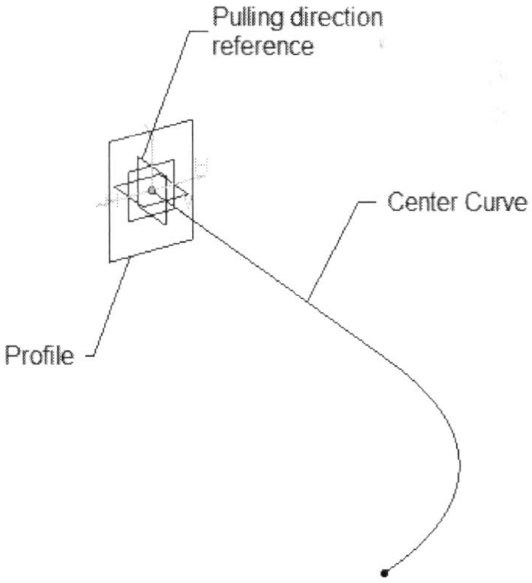

Figure 9-8 The profile, center curve, and pulling direction to be selected

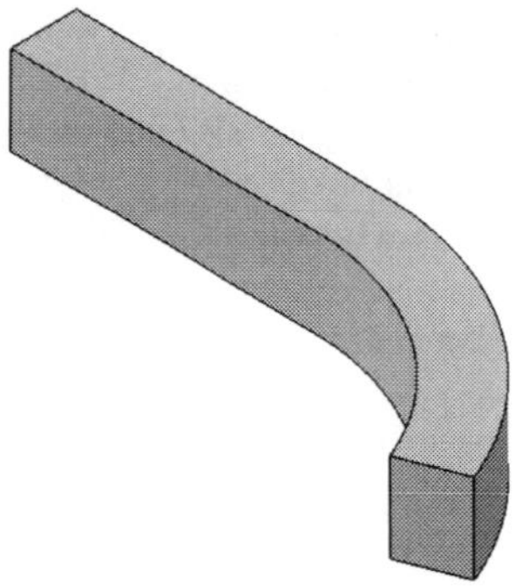

Figure 9-9 *The rib feature created with the* ***Keep angle*** *option selected*

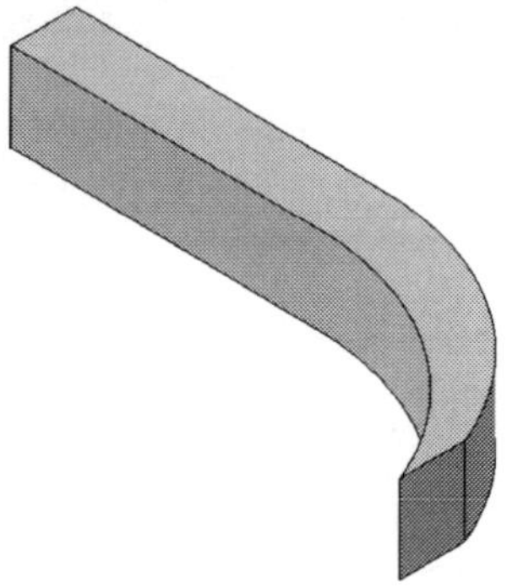

Figure 9-10 *The rib feature created with the* ***Pulling direction*** *option selected*

Moving the Center of the Profile to the Path

The **Move profile to path** check box in the **Rib Definition** dialog box will be available when the **Pulling Direction** or **Reference Surface** option is selected from the **Profile control** drop-down list. On selecting this check box, the profile will be swept along the center curve such that the origin of the profile lies on the center curve. Note that the term origin refers to the first point selected on the profile to draw the curve and not the sketch origin. Figure 9-11 shows the profile, center curve, and pulling direction to be specified for creating a rib feature by selecting the **Move profile to path** check box and Figure 9-12 shows the resulting rib feature.

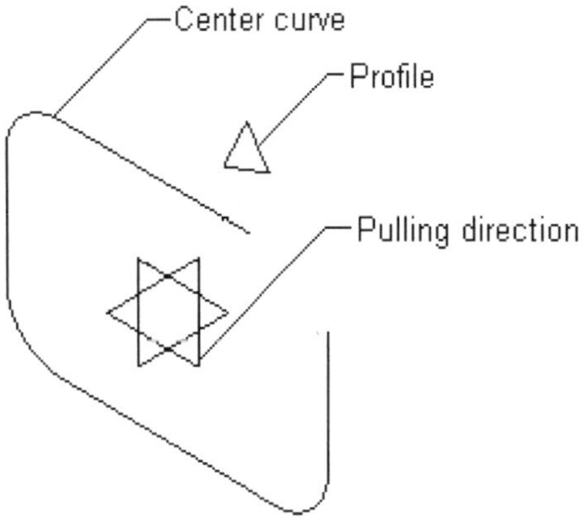

Figure 9-11 *Center curve, profile, and pulling direction to be selected*

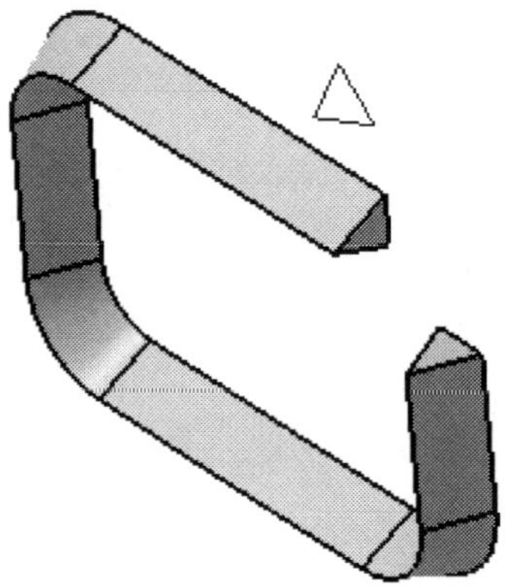

Figure 9-12 *The rib feature created with the* ***Move profile to path*** *check box selected*

With the **Move profile to path** check box selected, you can also sweep the same profile along more than one center curve. Figure 9-13 shows a profile, two center curves, and pulling direction selected for creating a rib feature and Figure 9-14 shows the resulting rib feature.

Note

If you are using more than one center curve for creating the rib feature, the center curves should be drawn in a single sketch. Also on selecting a sketch with more than one center curve, the ***Feature Definition Error*** *box will be displayed, informing that the selected sketch contains several profiles. Choose the* ***Yes*** *button from the error box to continue.*

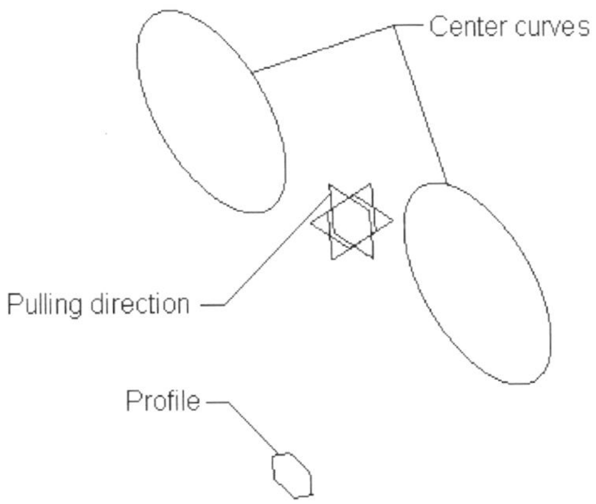

Figure 9-13 *Center curve, profile, and pulling direction to be selected*

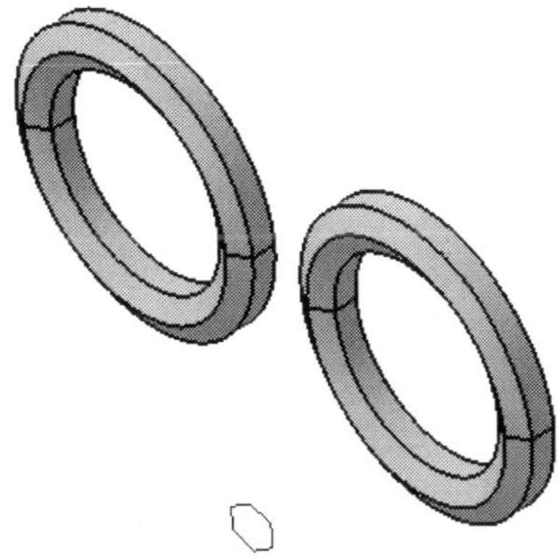

Figure 9-14 *The resulting rib feature with the **Move profile to path** check box selected*

Merging the End Faces of the Rib

The **Merge rib's ends** check box is used only when you have at least one feature already created in the current body. This option is used to trim or extend the ends of the rib feature to merge with the nearest end face of an existing feature. Figure 9-15 shows the profile and the center curve to create the rib feature. Figure 9-16 shows the rib feature with the **Merge rib's ends** check box cleared and Figure 9-17 shows the rib feature with the **Merge rib's ends** check box selected.

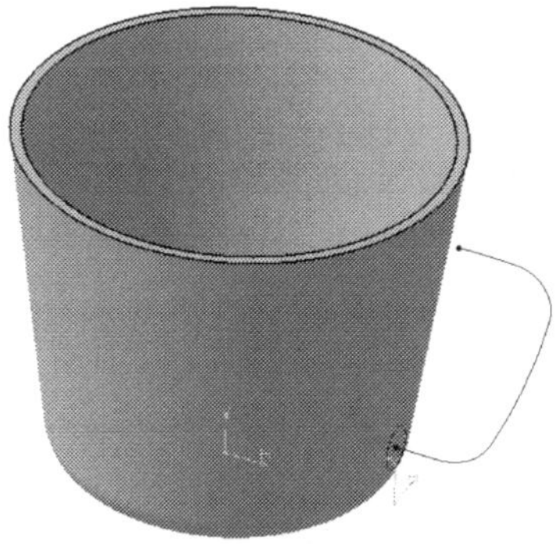

Figure 9-15 *The profile and the center curve*

Figure 9-16 *Rib feature with the **Merge rib's ends** check box cleared*

Figure 9-17 *Rib feature with the **Merge rib's ends** check box selected*

Creating Thin Rib Features

You can also create a thin rib feature by selecting the **Thick Profile** check box. On selecting this check box, the options in the **Thin Rib** area will be enabled, as shown in Figure 9-18. The value in the **Thickness1** spinner specifies the inner thickness and the value in **Thickness2** spinner specifies the outer thickness. You can specify the thickness of the rib feature by using these options in this area. A thin sweep feature is shown in Figure 9-19.

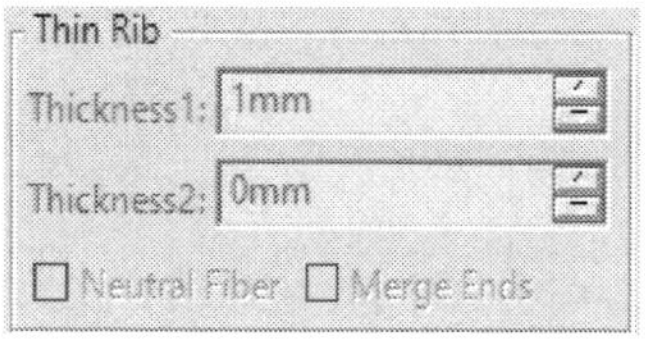

*Figure 9-18 The **Thin Rib** area of the **Rib Definition** dialog box*

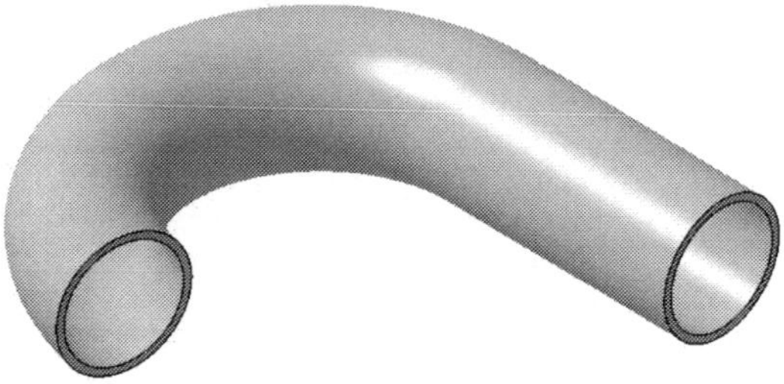

Figure 9-19 A thin sweep feature

Defining the Reference Surface

The **Rib** tool also provides you an option to define the reference surface for creating the rib feature. Consider a case in which you need to create a rib by sweeping a rectangular section along a center curve drawn on a multi-section solid feature, refer to Figure 9-20. You will learn more about the multi-solid feature later in this chapter. In this case, select the **Reference surface** option from the drop-down list in the **Profile control** area; you will be prompted to define the reference surface. Select the surface that you want to set as the reference; the preview of the rib feature, aligned to the selected surface, will be displayed in the geometry area. Choose the **OK** button from the **Rib Definition** dialog box. Figure 9-20 shows the profile, center curve, and reference surface to be selected. Figure 9-21 shows the rib feature created with the **Keep angle** option selected from the drop-down list in the **Profile control** area and Figure 9-22 shows the rib feature created after defining the reference surface.

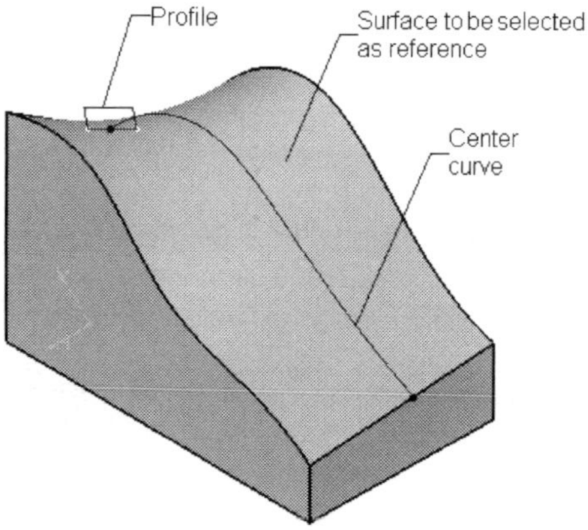

Figure 9-20 The profile, center curve, and reference surface to be selected

Tip

*To extract a sketch at an intersection of the sketch plane and a planar or non planar face, invoke the **Sketcher** workbench using that plane, and then choose the **Intersect 3D element** button from the **Operation** toolbar or the **3D Geometry** toolbar. Next, select the element from which you need to extract the geometry.*

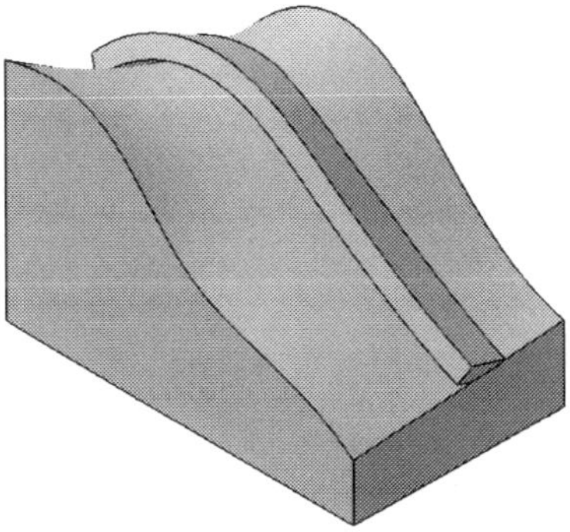

Figure 9-21 *Rib feature created on selecting the* ***Keep Angle*** *option from the drop-down list in the* ***Profile control*** *area*

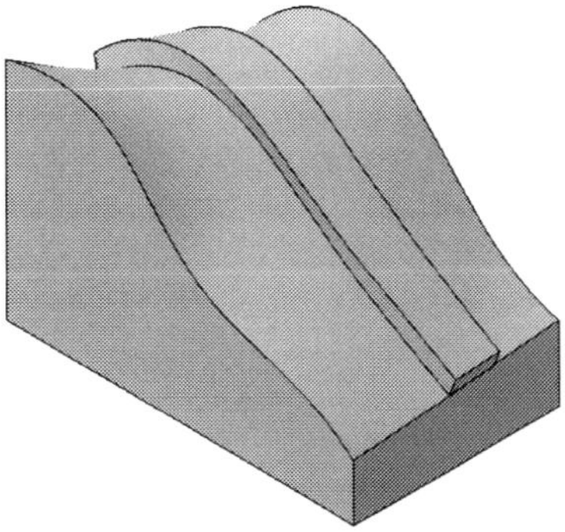

Figure 9-22 *Rib feature created after defining the reference surface*

Creating Rib Features by Using Open Profile and Open Center Curve

You can also create a rib feature using an open profile and an open center curve. But it is only possible if at least one feature already exists in the current body and the resulting rib feature is not crossing the boundary defined by the existing feature. To create this type of rib feature, draw the open sketches of the profile and the center curve, refer to Figure 9-23. Next, select the sketches to create the rib feature, as shown in Figure 9-24.

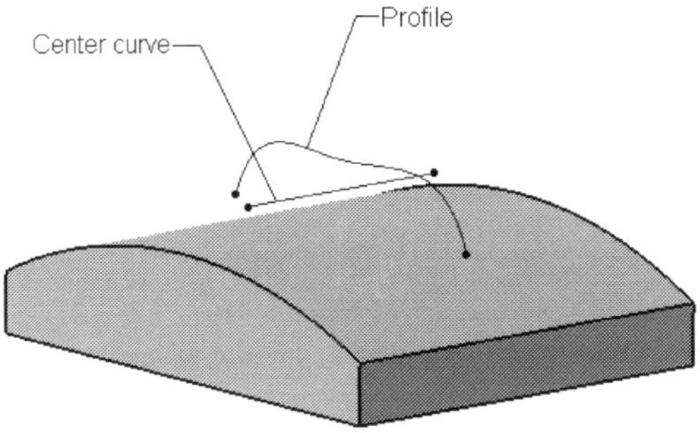

Figure 9-23 *The profile and the center curve*

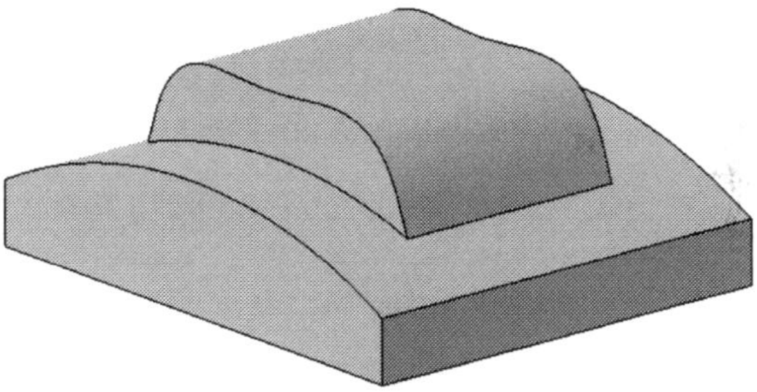

Figure 9-24 *Resulting rib feature*

Creating Slot Features

Menubar:	Insert > Sketch-Based Features > Slot
Toolbar:	Sketch-Based Features > Slot

The **Slot** tool is used to remove material by sweeping a profile along the center curve. To create this feature, choose the **Slot** tool from the **Sketch-Based Features** toolbar; the **Slot Definition** dialog box will be displayed, as shown in Figure 9-25.

Select the profile from the geometry area and then select the center curve; the preview of the slot feature will be displayed in the geometry area. Choose the **OK** button from the **Slot Definition** dialog box. Other options in this tool are the same as those discussed while creating the rib feature. Figure 9-26 shows the profile and the center curve. Figure 9-27 shows the resulting slot feature. Figure 9-28 shows the slot feature created with the **Merge slot's ends** check box selected.

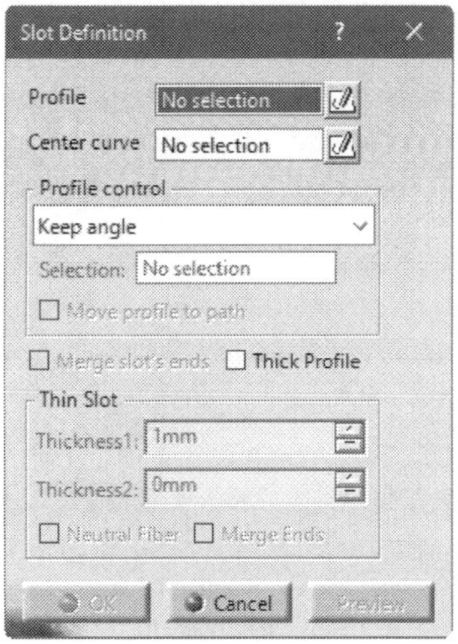

Figure 9-25 *The* ***Slot Definition*** *dialog box*

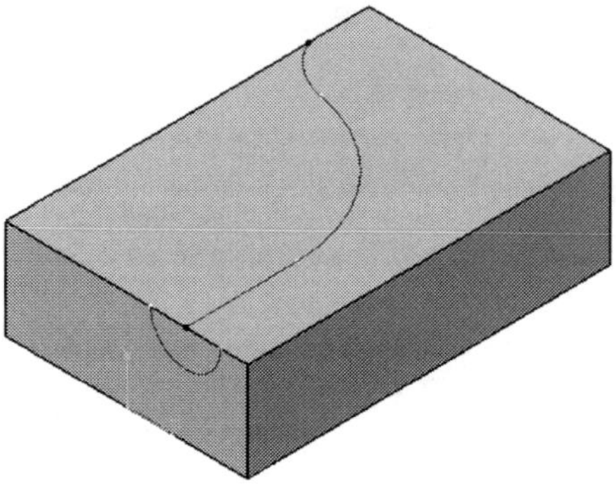

Figure 9-26 *Profile and center curve*

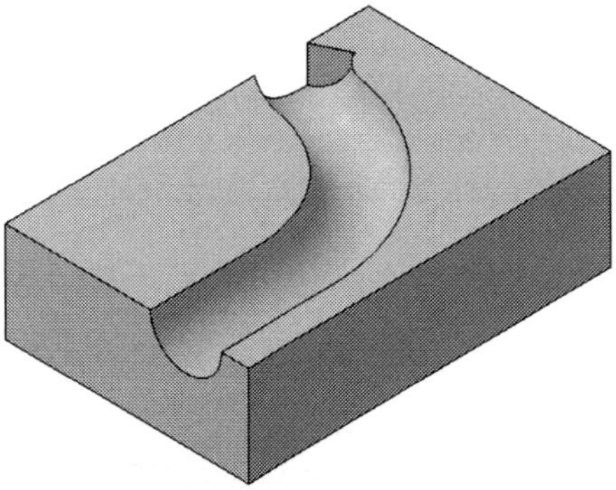

Figure 9-27 *Resulting slot feature*

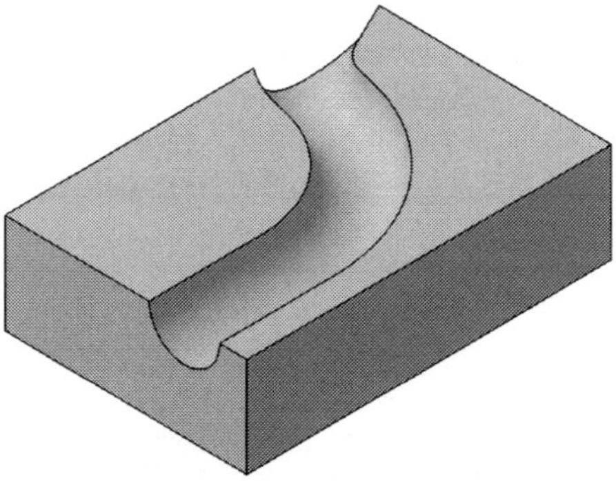

Figure 9-28 *The slot feature created with the* ***Merge slot's ends*** *check box selected*

Creating Multi-Section Solid Features

Menubar:	Insert > Sketch-Based Features > Multi-sections Solid
Toolbar:	Sketch-Based Features > Multi-sections Solid

The **Multi-sections Solid** tool is used to create a feature by blending more than one similar or dissimilar geometries together to get a free-form shape. These similar or dissimilar geometry may or may not be parallel to each other. Note that the sketches used to create the multi-sections solid features must be closed. To create a multi-sections solid feature, choose the **Multi-sections Solid** tool from the **Sketch-Based Features** toolbar; the **Multi-Sections Solid Definition** dialog box will be displayed, as shown in Figure 9-29, and you will be prompted to select a curve. Select a closed sketch, planar face, or a non-planar face. The name of the selected section will be displayed in the **Multi-Sections Solid Definition** dialog box and the **Section** and **Closing Point** callouts will be attached to the selected sections. Also, you will be prompted to select a new curve, support surface, or closing point. Select the second section from the geometry area. You will learn more about the tangent surface and the closing points later in this chapter.

If the closing point of the second selected section is not specified at the same location as that in the first section, then you need to replace the position of the closing point. Select the closing point that you need to replace and right-click to invoke the contextual menu. Select the **Replace** option from it. Now, select an appropriate vertex that can be used as the closing point. Also, you need to match the direction of arrow of second selected section with the first selected section in clockwise or anti-clockwise direction.

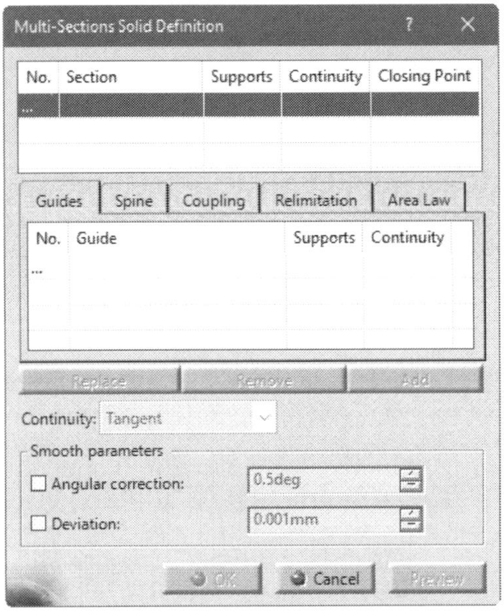

Figure 9-29 *The **Multi-Sections Solid Definition** dialog box*

To add more sections, right-click in the geometry area and then choose the **Add Section** option from the contextual menu. You can select and add as many sections as you need. Now, choose the **Preview** button from the **Multi-Sections Solid Definition** dialog box. If the preview of the multi-sections solid is not the required one, you may need to change the closing point or direction of the closing point. You will learn more about the direction of the closing point later in this chapter. After setting the parameters, choose the **OK** button. Figure 9-30 shows the sections to be selected to create the multi-sections solid feature. Figure 9-31 shows the preview of the resulting multi-section solid feature.

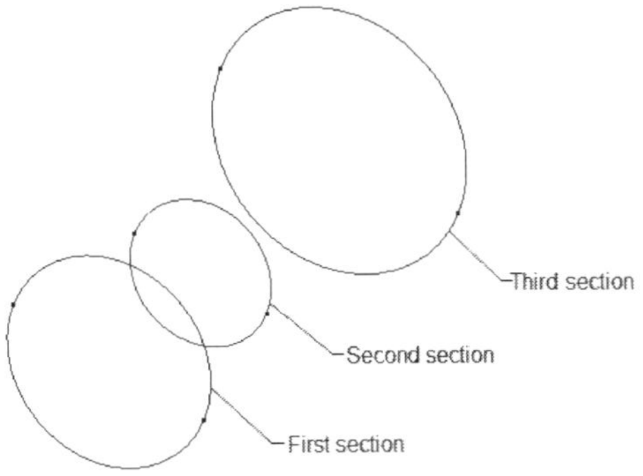

Figure 9-30 *Sections to be selected to create a multi-section solid*

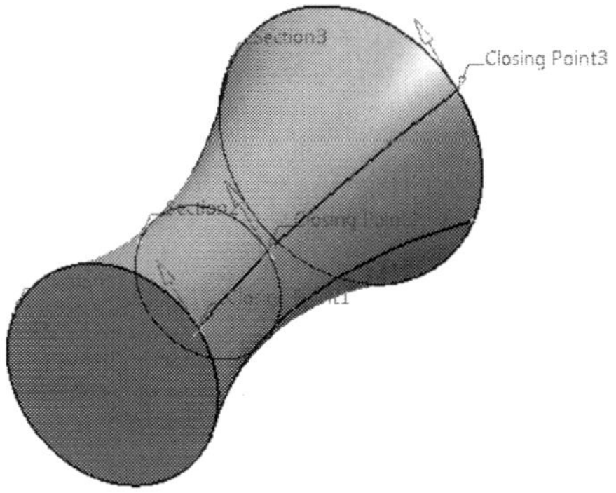

Figure 9-31 *Preview of the resulting multi-section solid feature*

Tip

On selecting a section for creating multi-sections solid, you will notice that a closing point is automatically selected, and the ***Closing Point*** *callout is attached to it. The point from which you start drawing a sketch is selected as the closing point by default. To flip the direction of the closing point, click on the arrow attached to it.*

While selecting the section, if you find that a closing point is not at proper location relative to closing point of other sections, then define a new position of the closing point by selecting a vertex immediately after specifying the section.

Creating a Multi-Section Solid Feature with Unequal Number of Vertices

You can also create a multi-section solid feature between the sections with unequal number of vertices. While creating such a multi-section solid feature, and also to have a better control over the blending points, you need to set the coupling options. To do so, choose the **Coupling** tab and select the **Ratio** option from the **Sections coupling** drop-down list. For example, consider a case in which you need to create a multi-section solid feature by blending a triangular and a rectangular section. To create this feature, invoke the **Multi-Sections Solid Definition** dialog box and select the triangular and rectangular sections. Set the closing points, if required.

Now, choose the **Coupling** tab in the **Multi-Sections Solid Definition** dialog box, as shown in Figure 9-32. Then click on the three dots displayed under the **No.** Column head and choose the **Add** button in the **Coupling** tab; the **Coupling** dialog box will be displayed, as shown in Figure 9-33, and you will be prompted to select the coupling point. Select a vertex of the triangle as the first coupling point. Next, you need to select the second coupling point. Note that you need to select the coupling points in the same sequence, in which the sections were selected. Select the corresponding vertex of the rectangle as the second coupling point. Figure 9-34 indicates the points to be selected for defining the couplings.

The **Display coupling curves** check box is by default selected and it is used to display the coupling curve. After the selection, the **Coupling** dialog box will disappear. Now, right-click in the geometry area to invoke the contextual menu and then choose the **Add Coupling** option from it; the **Coupling** dialog box will be displayed. Select the vertices for the second coupling, refer to Figure 9-34.

Similarly, add the third coupling. Now, choose the **Preview** button from the **Multi-Sections Solid Definition** dialog box. If the preview of the multi-sections solid feature is not the same as the required one, then you may need to edit the closing points or its direction. After setting all parameters, choose the **OK** button from the **Multi-Sections Solid Definition** dialog box.

Figure 9-34 shows the sections, closing points, and couplings for creating a multi-sections solid feature between a triangular section and a rectangular section. Figure 9-35 shows the resulting multi-sections solid feature.

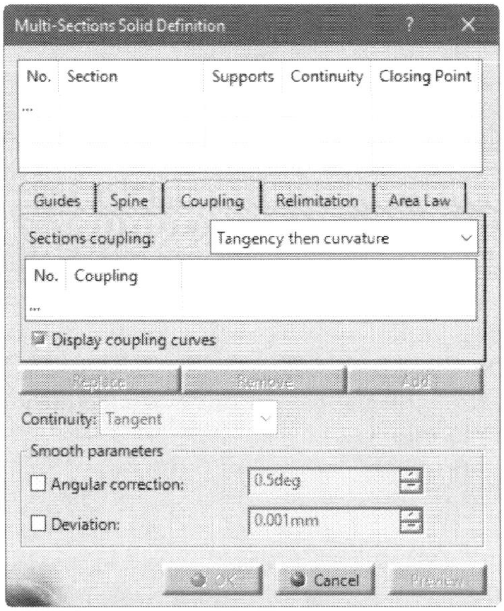

***Figure 9-32** The **Multi-Sections Solid Definition** dialog box with the **Coupling** tab chosen*

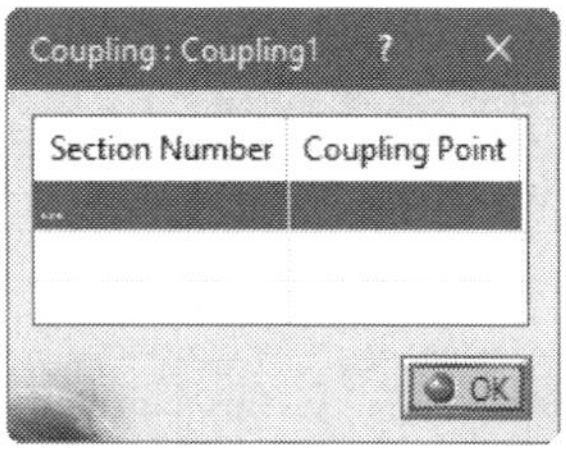

***Figure 9-33** The **Coupling** dialog box*

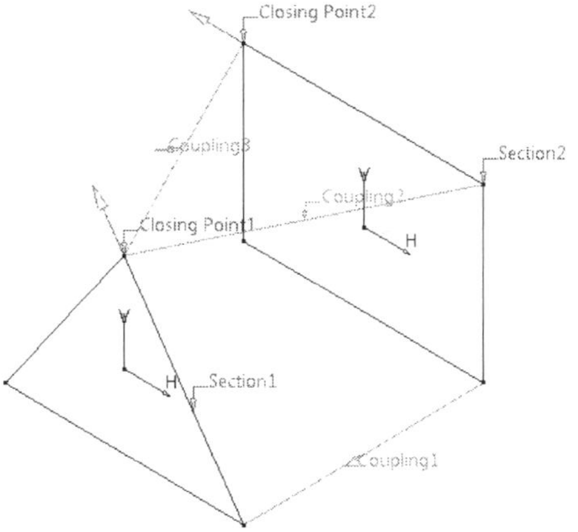

***Figure 9-34** Sections, closing points, and couplings*

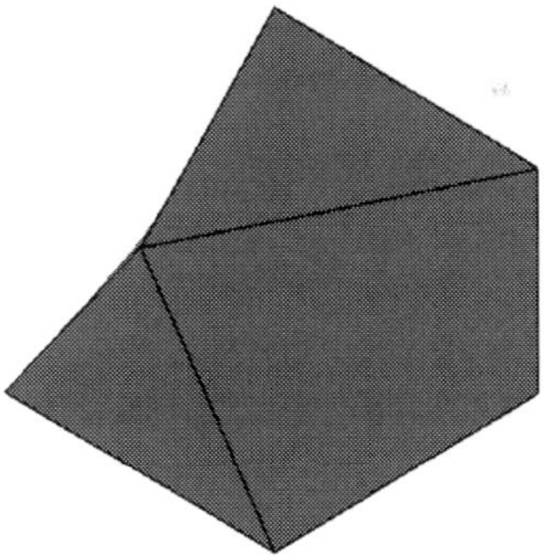

***Figure 9-35** Resulting multi-sections solid feature*

Note

*On selecting the **Tangency** option from the **Section coupling** drop-down list, the path of propagation will be tangential to the sections. On selecting the **Tangency then curvature** option, the path of propagation will be tangent to sections for some distance and thereafter it will follow a curved course. On selecting the **Vertices** option, the sections are coupled according to their vertices. The common point in these three options is that you cannot use these options if the number of vertices on sections is not the same.*

Creating a Multi-Section Solid Feature between a Circular and a Polygonal Section

Consider a case in which you need to blend a circular section with a pentagonal section. You can easily create a multi-section solid feature between these two sections by selecting the **Ratio** option from the **Sections coupling** drop-down list in the **Coupling** tab. However, to have a better control over the blending, you need to place five points on the periphery of the circular section. This is done to make the closing point and the coupling points on the circular section equal to the number of vertices on the pentagon section. Now, invoke the **Multi-Sections Solid Definition** dialog box and then select the pentagonal and circular sections. Set the closing points for polygon and circle by selecting a vertex from the polygon and a point at similar side on the circle. Choose the **Coupling** tab and define the couplings, as shown in Figure 9-36. After specifying all couplings, choose the **OK** button from the **Multi-Sections Solid Definition** dialog box. Figure 9-37 shows preview of the resulting multi-sections solid feature.

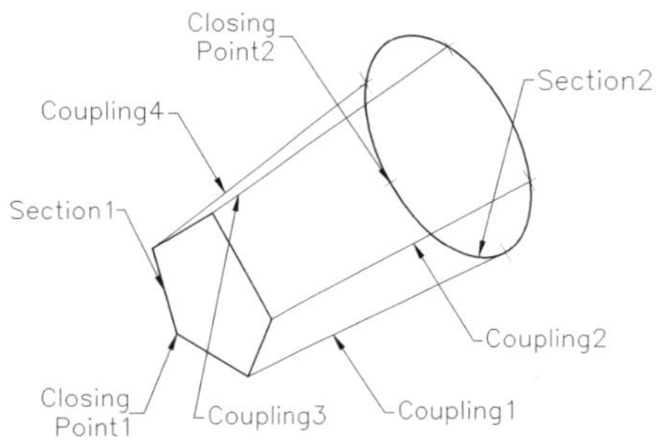

Figure 9-36 Sections, closing points, and couplings

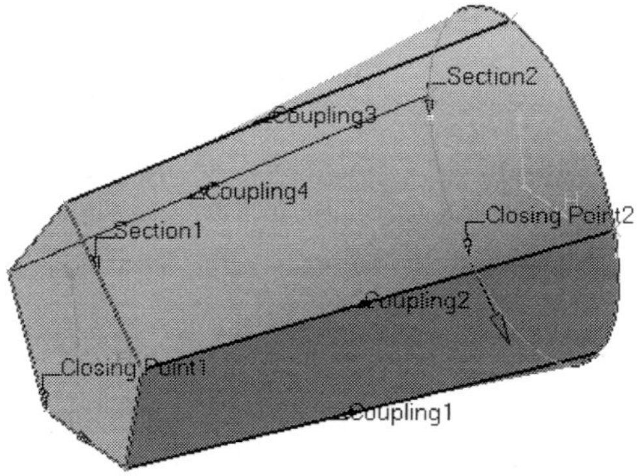

Figure 9-37 The preview of the resulting multi-sections solid feature

Creating a Multi-Section Solid Feature along a Spine

You can also define a spine while creating the multi-sections solid features. This option is used to create a multi-sections solid feature by blending two or more than two sections. The path of transition of the multi-sections solid feature is defined by a spine which could be a sketched element or an edge. To create a multi-section solid feature using this option, choose the **Spine** tab in the **Multi-Sections Solid Definition** dialog box, as shown in Figure 9-38.

By default, the **Computed spine** check box is selected in this tab. Therefore, the path of transition to be followed by the multi-section solid feature is calculated automatically. Click once in the **Spine** display box and select the spine from the geometry area. You will notice that the **Computed spine** check box is cleared automatically. Choose the **OK** button from the **Multi-Sections Solid Definition** dialog box. Figure 9-39 shows the sections and the spine to be selected. Figure 9-40 shows the multi-section solid feature created with computed spine and Figure 9-41 shows the multi-section solid feature created after selecting the spine.

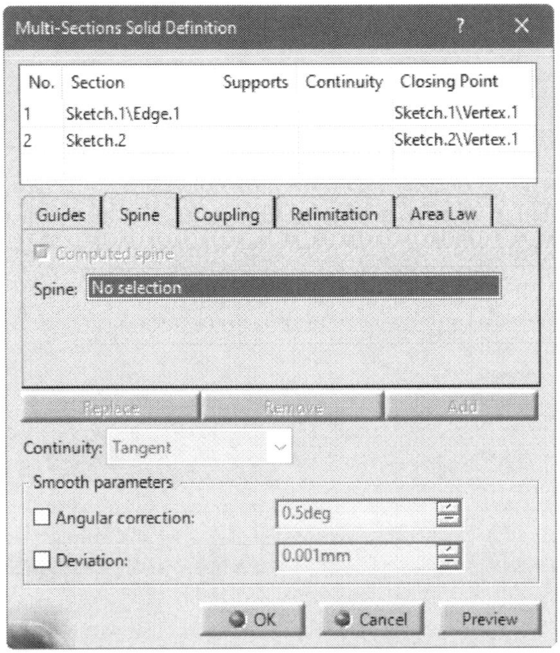

Figure 9-38 *The **Multi-Sections Solid Definition** dialog box with the **Spine** tab chosen*

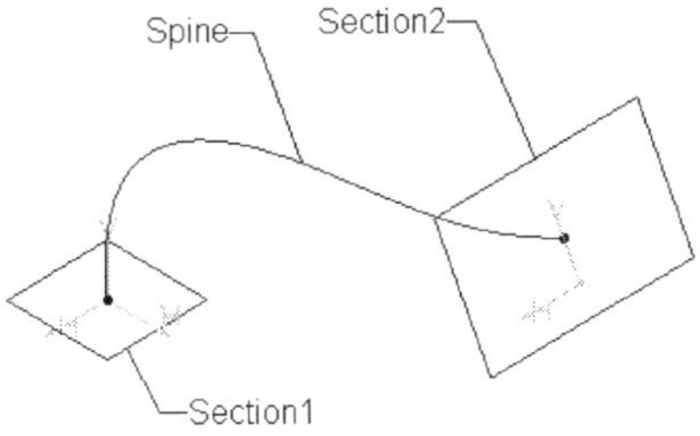

Figure 9-39 *The sections and the spine to be selected*

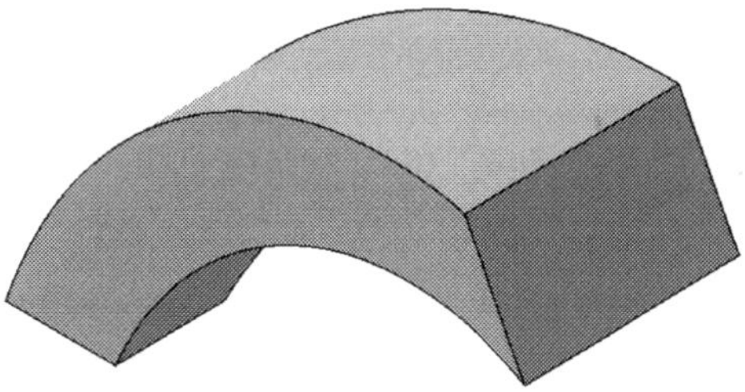

Figure 9-40 *Multi-section solid feature created with computed spine*

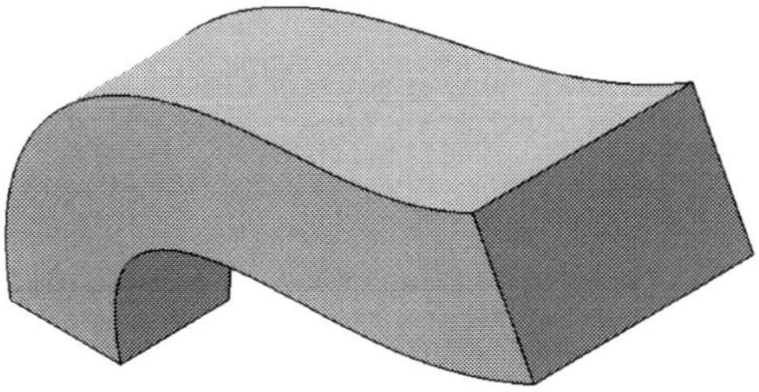

Figure 9-41 *Multi-section solid feature after selecting the user defined spine*

Creating a Multi-Section Solid Feature with Guides

You can also create the guide curves to define the variations in the section of multi-section solid along the spine. While drawing the sketch of the guide, make sure that the sketch of the guide curve intersects the sections of the multi-section solid. After selecting the sections for the multi-section solid feature, you can define the guide curves. To do so, click once in the display box provided in the **Guides** tab. Then, select the guide curves from the geometry area; the names of the selected guide curves will be displayed in the geometry area. Choose the **OK** button from the **Multi-Sections Solid Definition** dialog box. Figure 9-42 shows the sections and guides for a multi-section solid feature. Figure 9-43 shows the resulting multi-section solid feature.

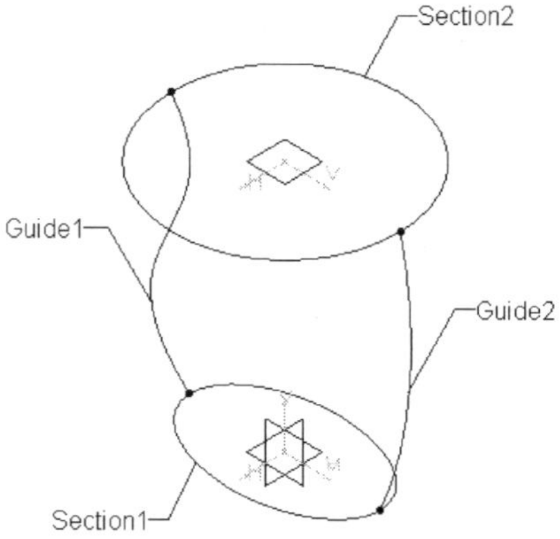

Figure 9-42 *The sections and guides for a multi-section solid feature*

Figure 9-43 *Resulting multi-section solid feature*

Creating a Multi-Section Solid Feature with Relimitations

You will notice that the propagation of the multi-section solid feature is extended up to the end points of spine or guide or both, if these entities are extended beyond the specified section profiles. You can limit the propagation of the multi-section solid feature up to the specified start and end sections by using the check boxes in the **Relimitation** tab. To create a multi-section solid feature with relimitation, first specify the sections and then the spine or the guide curve, or both. Next, choose the **Relimitation** tab from the **Multi-Sections Solid Definition** dialog box, refer to Figure 9-44. By default, the **Relimited on start section** and **Relimited on end section** check boxes are selected in the **Relimitation** tab. As a result, the multi-section solid feature will be created only up to the start and end sections even though the spine curve or the guide curve is created beyond the start and end sections. If you clear the **Relimited on start section** check box, the multi-section solid feature will start from the start of the spine curve or the guide curve.

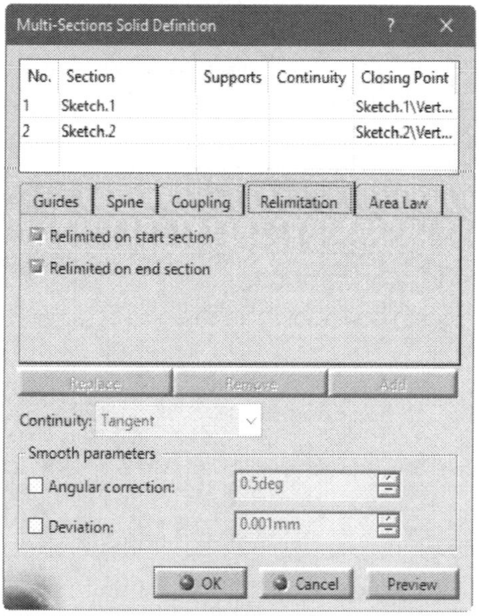

Figure 9-44 *The **Multi-Sections Solid Definition** dialog box displayed after choosing the **Relimitation** tab*

Similarly, if you clear the **Relimited on end section** check box, the multi-section solid feature will be created up to the end point of the spine curve or guide curve.

Figure 9-45 shows the sections and the spine to be selected and Figure 9-46 shows the multi-section solid feature created with the **Relimited on start section** and **Relimited on end section** check boxes selected. Figure 9-47 shows the resulting multi-section solid feature created with the **Relimited on start section** and **Relimited on end section** check boxes cleared.

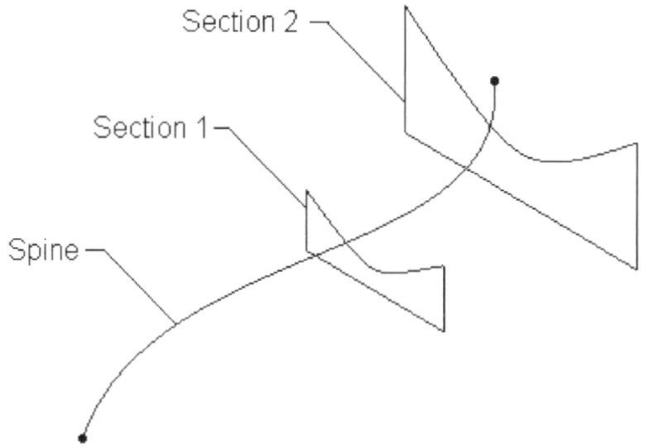

Figure 9-45 *The sections and the spine to be selected*

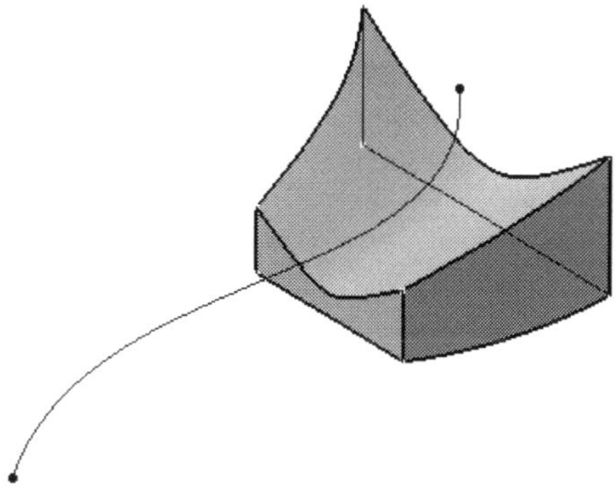

Figure 9-46 *The multi-section solid feature created by using the relimitation check boxes*

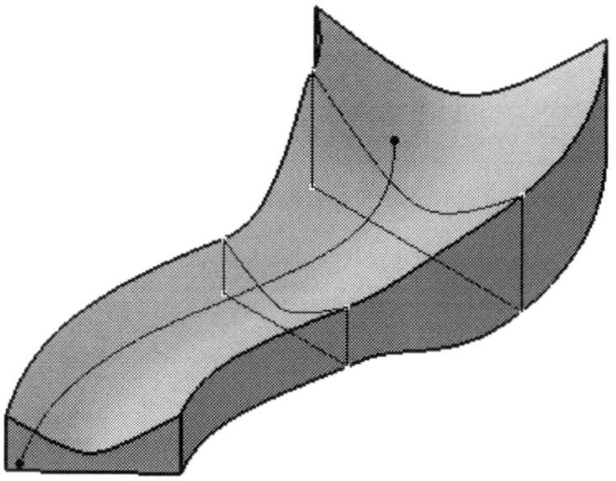

Figure 9-47 *The multi-section solid feature created after clearing the relimitation check boxes*

If you are creating a multi-section solid feature without relimitations and the length of the spine curve is shorter than the guide curve, the multi-section solid feature will stop up to the end of the spine curve. And, if the length of the guide curve is shorter than the spine curve, the multi-section solid feature will stop at the end of the guide curve. Figure 9-48 shows the sections, spine, and guide curve for creating the multi-section solid feature. Figure 9-49 shows the resulting multi-section solid feature without relimitation at the start and end sections. At the start section (Section 1), the spine is shorter than the guide curve, so the resulting feature is created only up to the end of the spine. On the other hand, at the end section (Section 2), the guide curve is shorter than the spine curve, so the resulting feature is created only up to the end of the guide curve.

Note

*The **Angular correction** and **Deviation** check boxes in the **Smooth parameters** area of the **Multi-Sections Solid Definition** dialog box are used to control the smoothness with which the sections follow the guide curves. This smoothness is controlled by allowing the angular deviation in tangency with the spine curve, perpendicularity with the guide curves, and linear deviation from the guide curves.*

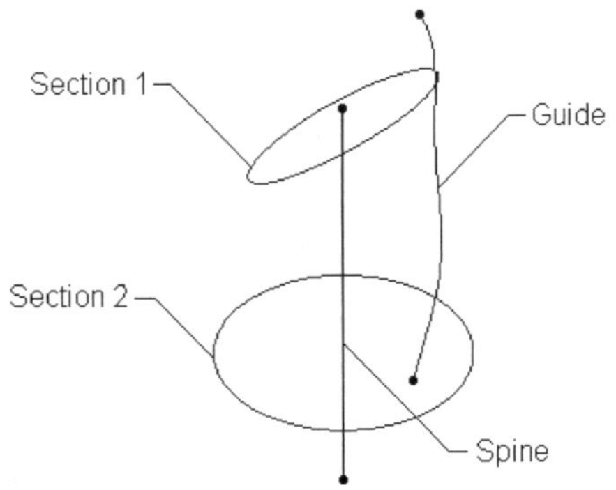

Figure 9-48 *The sections, spine, and guide curve for creating the multi-section solid feature*

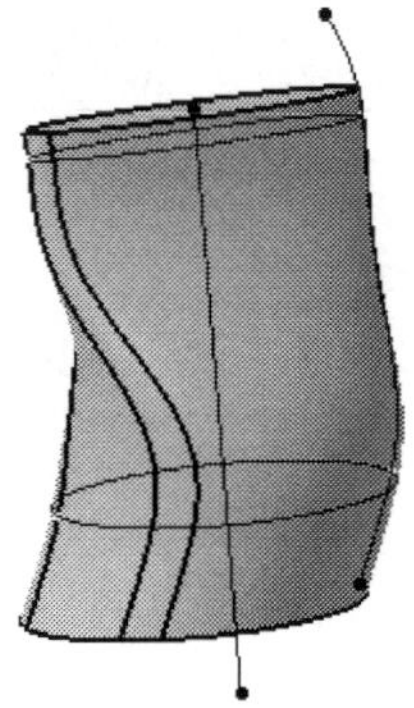

Figure 9-49 *The multi-section solid feature created without relimitation at the start and end sections*

Creating the Multi-Section Solid Cut Feature

Menubar: Insert > Sketch-Based Features > Removed Multi-sections Solid
Toolbar: Sketch-Based Features > Removed Multi-sections Solid

The **Removed Multi-sections Solid** tool is used to remove material by blending two or more than two sections. To remove material, invoke this tool, the **Removed Multi-Sections Solid Definition** dialog box will be displayed, as shown in Figure 9-50.

Select the sections and then define the closing points, if required. You can also select guide, spine, or vertices to define the path of propagation. The procedure for creating this feature is the same as that discussed for creating the multi-sections solid feature. After setting all parameters, choose the **OK** button from the **Removed Multi-Sections Solid Definition** dialog box. Figure 9-51 shows the sections and the spine selected to create the multi-section solid cut feature. Figure 9-52 shows the resulting multi-section solid cut feature.

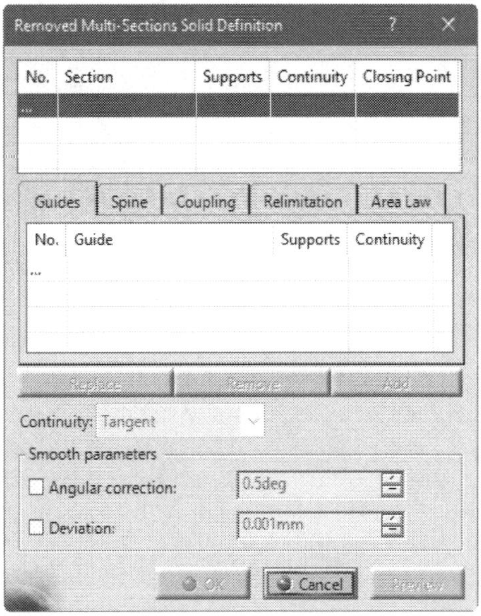

Figure 9-50 *The **Removed Multi-Sections Solid Definition** dialog box*

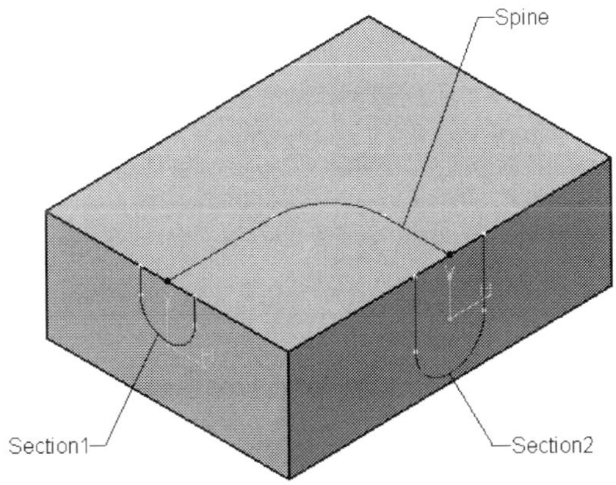

Figure 9-51 *Sections and spine selected to create the multi-section solid cut feature*

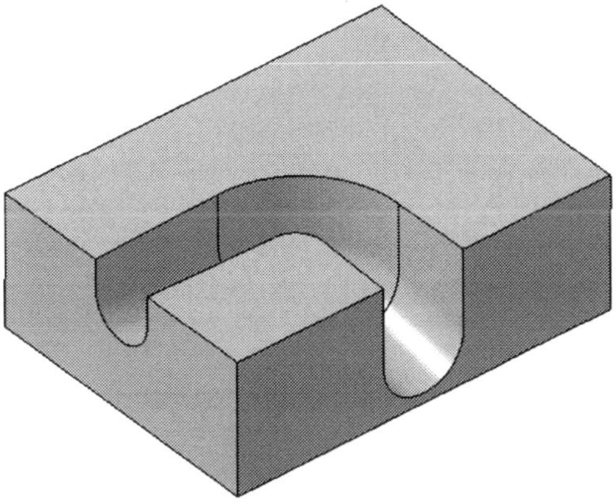

Figure 9-52 *Resulting multi-section solid cut feature*

TUTORIALS

Tutorial 1

In this tutorial, you will create the model of the Upper Housing shown in Figure 9-53. Its orthographic views and dimensions are shown in Figure 9-54.

(Expected time: 1 hr)

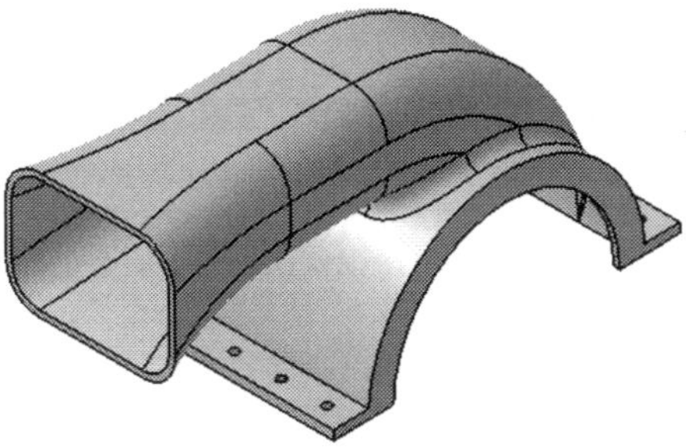

Figure 9-53 *Model of the Upper Housing for Tutorial 1*

The following steps are required to complete this tutorial:

a. Create the base feature of the model.
b. Create the rib feature.
c. Create the multi-section solid feature.
d. Fillet the edges of the model.
e. Shell the model by removing the bottom and left planar faces of the model.
f. Create other features to complete the model.

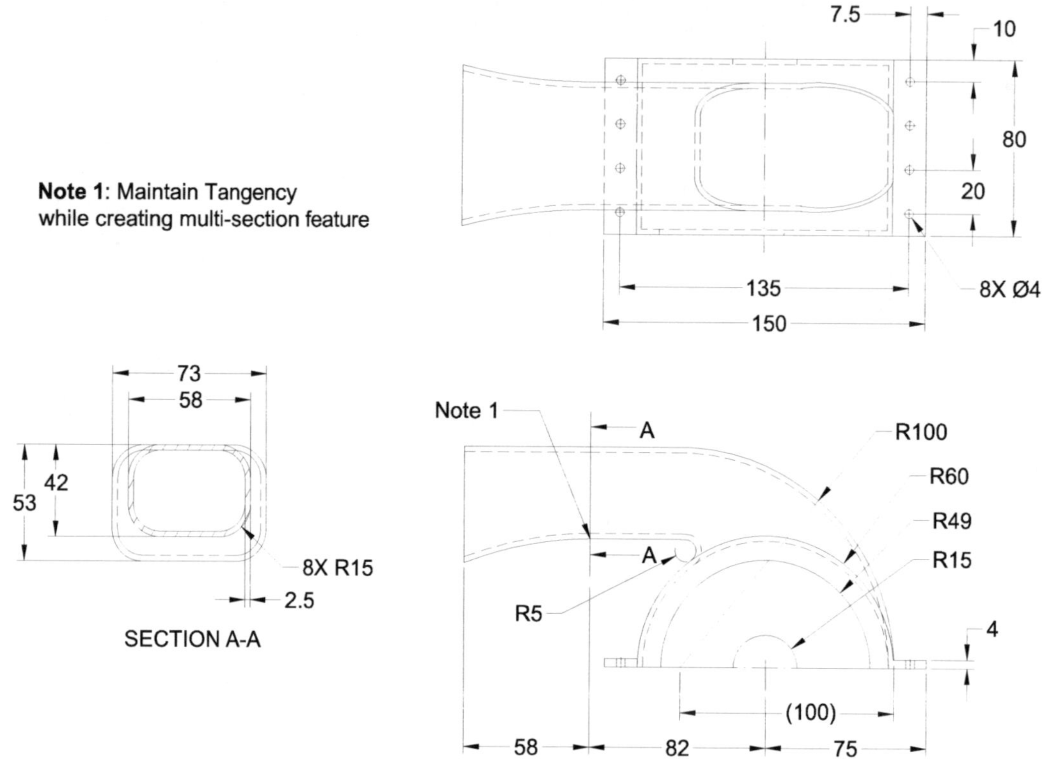

Figure 9-54 Views and dimensions of the Upper Housing

Creating the Base Feature

1. Create a semicircular sketch on the zx plane, refer to Figure 9-55.

2. Next, invoke the **Pad** tool and create the base feature by extruding the semicircular sketch, drawn on the zx plane to both sides of the sketching plane. For dimensions and parameters of the pad feature, refer to Figure 9-54. Preview of the resulting base feature is shown in Figure 9-55.

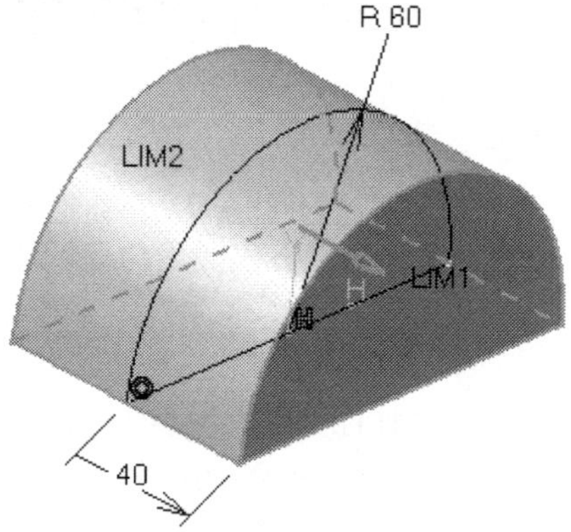

Figure 9-55 Preview of the base feature

Creating the Rib Feature

After creating the base feature, you need to create the rib feature of the model. To create it, first you need to sketch the center curve and the profile.

1. Select the zx plane as the sketching plane and draw the sketch of the center curve, as shown in Figure 9-56. Exit the **Sketcher** workbench.

Next, you need to draw the sketch for the profile of the rib feature at the endpoint of the sketch drawn earlier (center curve), and the profile should be normal to the center curve.

2. Invoke the **Plane Definition** dialog box by choosing the **Plane** tool from the **Reference Element** toolbar and then select the **Normal to curve** option from the **Plane** type drop-down list.

3. Select the sketch of the center curve, if it is not already selected and then select the endpoint of the horizontal line of the center curve.

4. Choose the **OK** button from the **Plane Definition** dialog box.

5. Draw the sketch of the profile on the newly created plane, as shown in Figure 9-57.

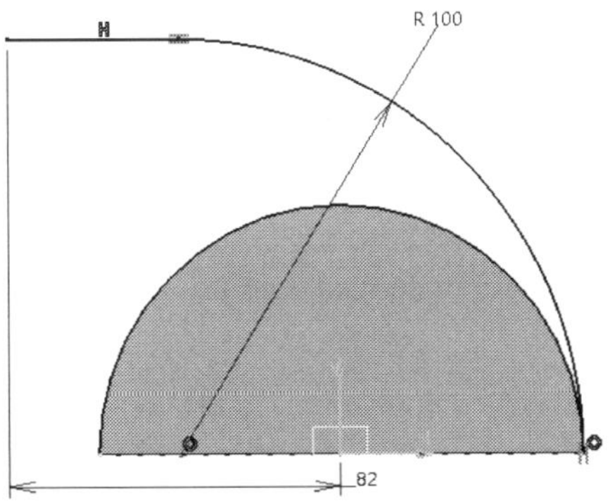

Figure 9-56 *Sketch of the center curve*

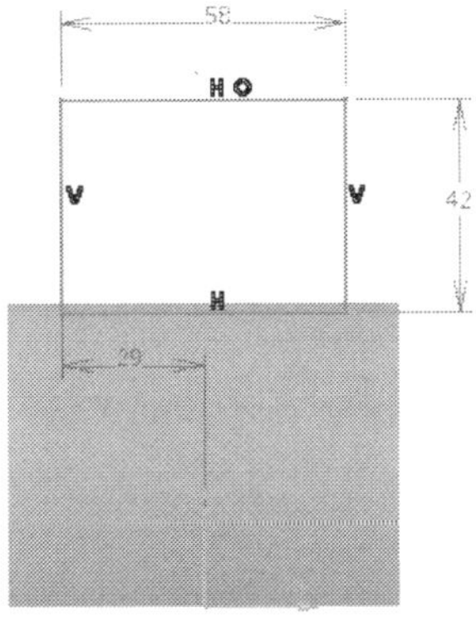

Figure 9-57 *Sketch of the profile*

6. Exit the **Sketcher** workbench. The model after drawing the sketch is shown in Figure 9-58.

7. Invoke the **Rib** tool from the **Sketch-Based Features** toolbar; the **Rib Definition** dialog box is displayed and you are prompted to define the profile.

8. Select the sketch of the profile from the geometry area if not already selected; you are prompted to define the center curve.

9. Select the sketch of the center curve from the geometry area; the wireframe preview of the rib feature is displayed.

10. Choose the **OK** button from the **Rib Definition** dialog box. The model after creating the rib feature is shown in Figure 9-59.

Creating the Multi-section Solid Feature

After creating the rib feature, you need to create the multi-section solid feature by lofting the two sections. The first section of the multi-section solid feature will be the front planar face of the rib and the other section will be drawn on a plane at an offset distance from the front planar face of the rib feature.

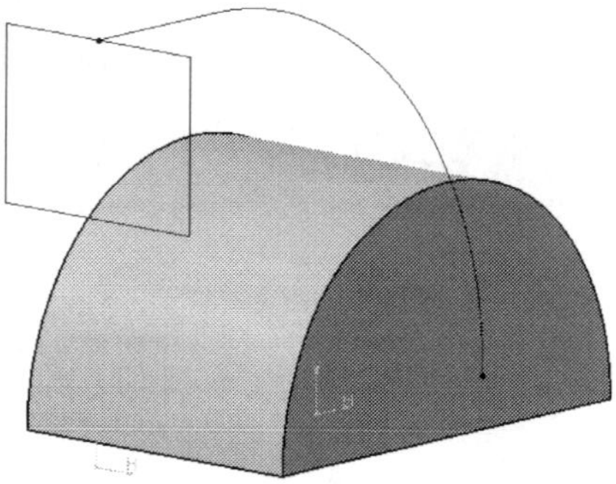

Figure 9-58 Sketches for the rib feature

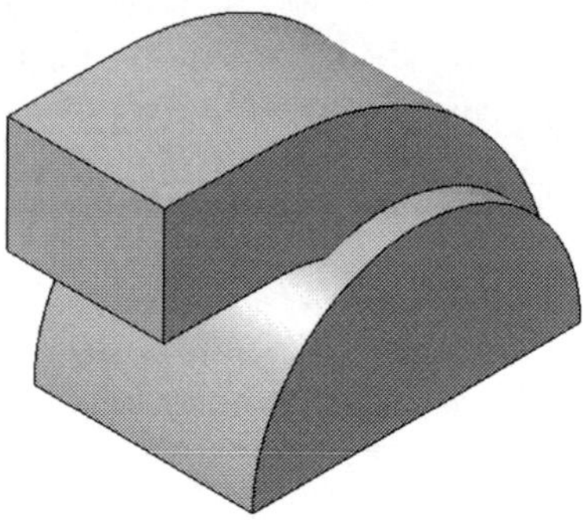

Figure 9-59 The model after creating the rib feature

1. Create a plane using the **Plane** tool from the **Reference Element** toolbar at an offset distance of 58 mm from the front planar face of the rib feature. Make sure that the plane is created in the direction opposite to that of the propagation of the rib feature.

2. Select the newly created plane and invoke the **Sketcher** workbench.

3. Draw the sketch of the second section of the multi-section solid feature, which is a rectangle, as shown in Figure 9-60. In this figure, the model is oriented in 3D space for better understanding of the position of the sketch and its dimensions.

4. Exit the **Sketcher** workbench and then press the ESC key to remove the sketch from the selection set.

5. Invoke the **Multi-sections Solid** tool from the **Sketch-Based Features** toolbar; the **Multi-Sections Solid Definition** dialog box is displayed and you are prompted to select a curve. 

6. Select the front planar face of the rib feature, as shown in Figure 9-61; you are prompted to select a new curve. Note that after selecting the front planar face as a section, a surface is automatically generated on the selected planar face. You can hide this surface after creating the multi-section solid.

7. Select the newly created curve and relocate the closing point if required, as shown in Figure 9-61.

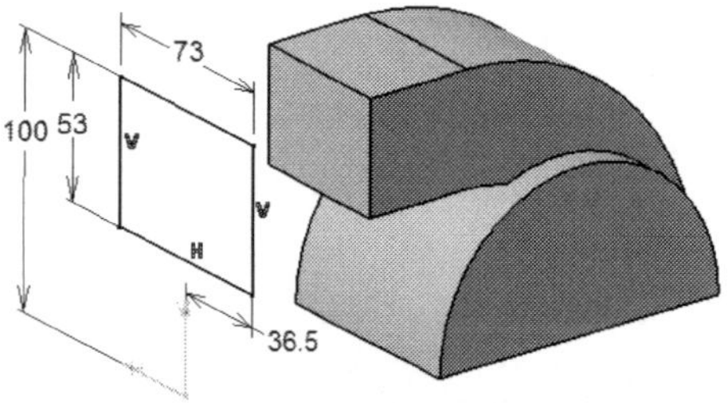

Figure 9-60 Second section of the multi-section solid feature

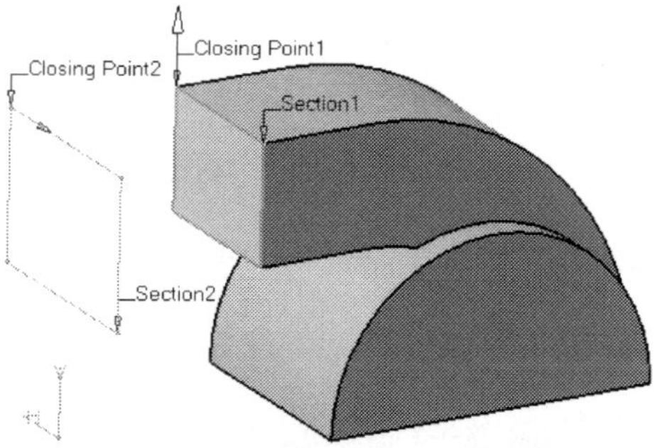

Figure 9-61 Newly created curve and the relocated closing point

Note

Consider a case in which the closing points in the first rectangular section and the second rectangular section are the upper left vertex and the lower left vertex, respectively. In order to create a multi-section solid feature without a twist, you need to replace any one of the closing points with the respective closing points. To replace the closing point of the second section, select the callout of the closing point and invoke the contextual menu by right-clicking on it. Choose the ***Replace*** *option from it. Then, select the required vertex of the sketch to define the closing point on that vertex.*

8. Flip the direction of any one of the closing points if they point in the opposite direction.

9. Choose the **OK** button from the **Multi-Sections Solid Definition** dialog box.

10. Create a fillet of radius 15 units at the four tangent edges of the multi-section solid feature, refer to Figure 9-62. Note that while filleting the tangent edges, the **Warning** message box is displayed, informing that the tangent design is detected. Choose the **Close** button to exit this message box.

11. After filleting the edges, the automatically generated surface becomes visible, as shown in Figure 9-62. You can hide this surface by selecting it and then invoking the contextual menu. Finally, choose the **Hide/Show** option from the menu.

12. Create a fillet of radius 5 units on the edges formed between the pad feature and the rib feature.

Shelling the Model

1. Invoke the **Shell** tool from the **Dress-Up Features** toolbar; the **Shell Definition** dialog box is displayed and you are prompted to select the face to be removed. Select the front and bottom planar faces of the model.

2. Set the value in the **Inside thickness** spinner to **2.5** in the **Default Thickness** area and choose the **OK** button from the **Shell Definition** dialog box. The model after shelling is shown in Figure 9-63.

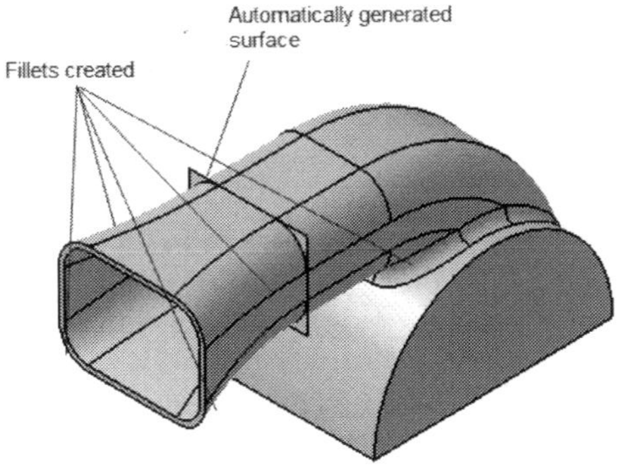

Figure 9-62 *Model after filleting the edges*

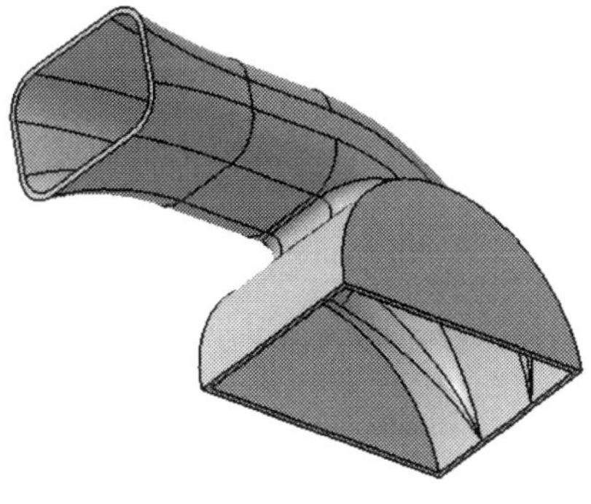

Figure 9-63 *Model after shelling*

Creating the Remaining Features

1. Create a pocket feature on the right planar face of the model by extruding a semicircle of radius 49 units by using the **Up to plane** option from the **Type** drop-down list and zx plane in the **Limit** area of the **Pocket Definition** dialog box.

2. Similarly, create another pocket feature at the left planar face of the model by extruding a semicircle of radius 15 unit by using the **Up to next** option from the **Type** drop-down list of the **Pocket Definition** dialog box.

3. Create the mounting plate by extruding the sketch drawn on extreme right planer face up to the extreme left planar face. Refer to Figure 9-64 for creating the sketch.

4. Mirror the mounting plate about the yz plane.

5. Create a simple hole and then pattern it. Refer to Figure 9-54 for the dimensions and position of holes. The final model after creating all features is shown in Figure 9-65.

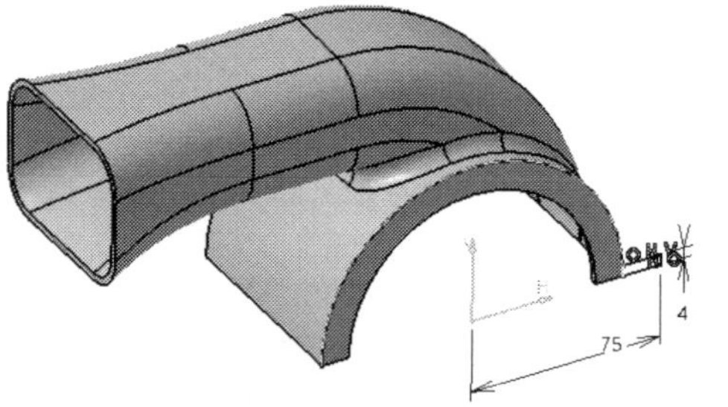

Figure 9-64 Sketch for creating the mounting plate

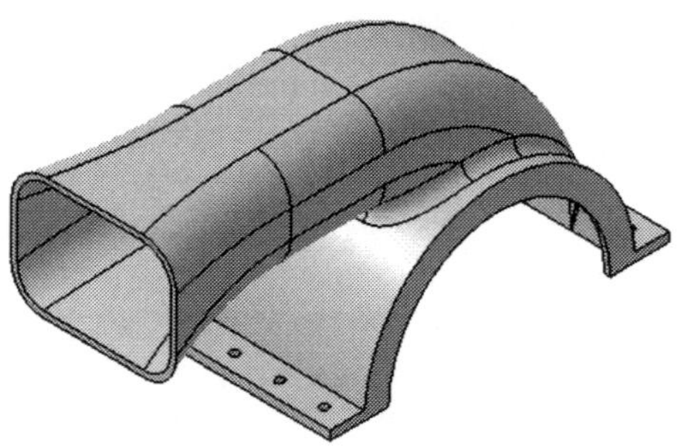

Figure 9-65 Final model of the Upper Housing

Saving and Closing the File

1. Choose the **Save** button from the **Standard** toolbar; the **Save As** dialog box is invoked. Create the *c09* folder inside the *CATIA* folder.

2. Enter **c09tut1** in the **File name** edit box and choose the **Save** button. The file is saved at *C:\CATIA\c09*.

3. Close the part file by choosing **File > Close** from the menu bar.

Tutorial 2

In this tutorial, you will create the model of the Helical gear shown in Figure 9-66. Its views and dimensions are shown in Figure 9-67. **(Expected time: 1 hr)**

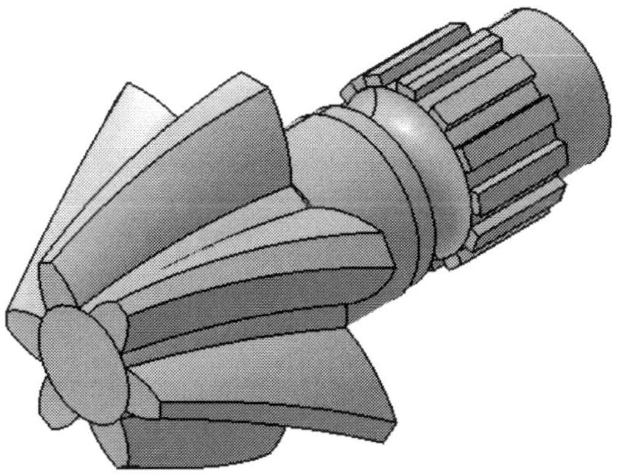

Figure 9-66 Model of the Helical gear for Tutorial 2

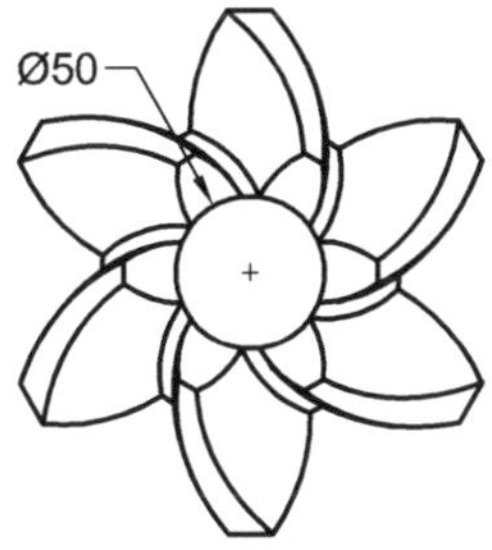

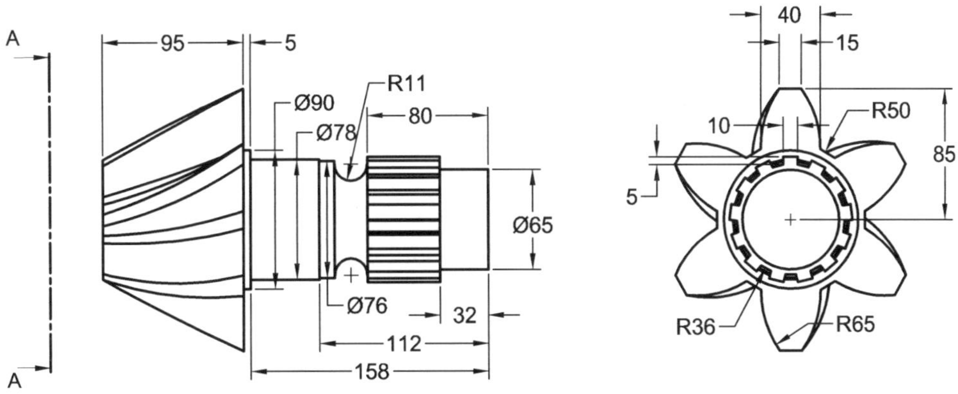

Create multi-section solid feature with 3 sketches size 100%, 75%, 50% with rotation angle of 15° between each sketch

The planes for drawing the first, second and third sections are at the offset distance of 163, 210.5 and 258 units, respectively from the right side of the base feature.

Figure 9-67 Views and dimensions for Tutorial 2

The following steps are required to complete this tutorial:

a. Create the base feature of the model.
b. Create a pad feature on the right of the base feature and then pattern it using the **Circular Pattern** tool.
c. Draw the sketch of the first section of the gear tooth on a plane that is at an offset distance from the right most planar face of the base feature.
d. Draw the sketch of the second section of the gear tooth on a plane that is at an offset distance from the right most planar face of the base feature. This sketch is the 75% scaled sketch of the first section and is rotated at an angle of 15°.
e. Draw the sketch of the third section of the gear on the left most planar face of the base feature.
f. Create a multi-section solid feature by using three sections.
g. Create the pattern of the gear tooth by using the circular pattern tool.

Creating the Base Feature

1. Draw the sketch of the base feature on the zx plane, as shown in Figure 9-68. In this figure, the constraints have been hidden for better visualization.

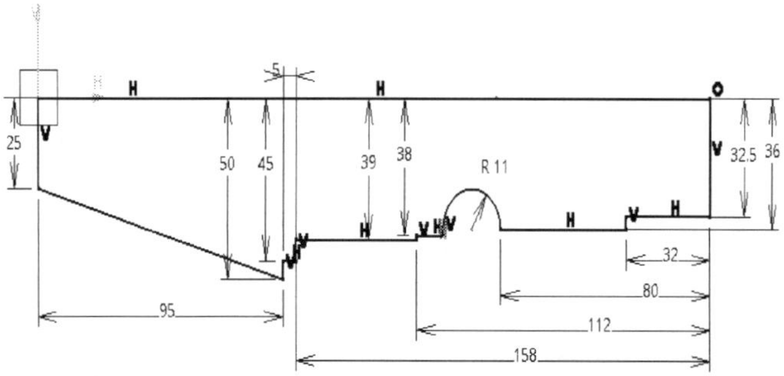

Figure 9-68 *The sketch for creating the Shaft feature*

2. Invoke the **Shaft** tool and revolve the sketch. The model after creating the base feature is shown in Figure 9-69.

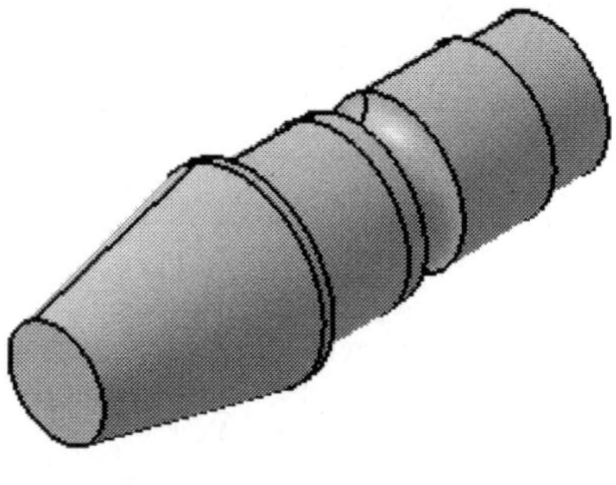

Figure 9-69 *Base feature of the model*

Creating the Pad Feature

1. Create the sketch of the pad feature on the planar face of the base feature, as shown in Figure 9-70. In this figure, the constraints have been hidden and the model has been oriented in 3D space for better visualization of dimensions and position of sketch.
2. Invoke the **Pad** tool from the **Sketch-Based Feature** toolbar and extrude the sketch by 48 units. Use the **Reverse Direction** button, if required.

3. Invoke the **Circular Pattern** tool from the **Patterns** toolbar and create 13 instances using the **Complete crown** option. The model after creating the circular pattern of the pad feature is shown in Figure 9-71.

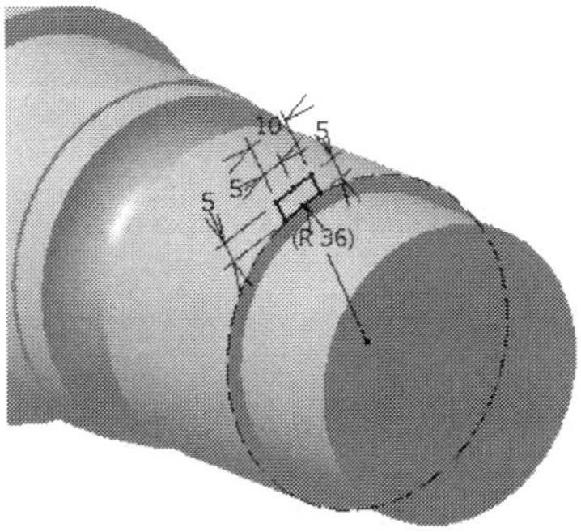

Figure 9-70 *The sketch for creating the pad feature*

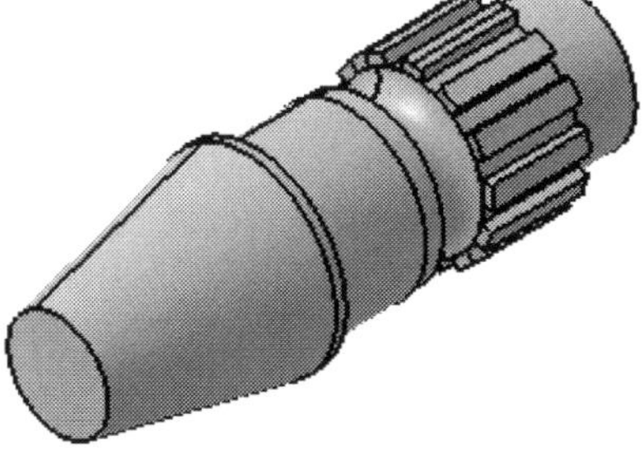

Figure 9-71 *The model after creating the circular pattern of the pad feature*

Creating Sections for the Multi-section Solid Feature

Next, you need to create the multi-section solid feature. You will create this feature by adding the material uniformly through three sections. Therefore, first you need to create the sections of the multi-section solid feature.

1. Create a plane at an offset of 163 units from the right most planar face of the base feature, refer to Figure 9-67.

2. Draw the sketch on the newly created plane, as shown in Figure 9-72. Exit the **Sketcher** workbench.

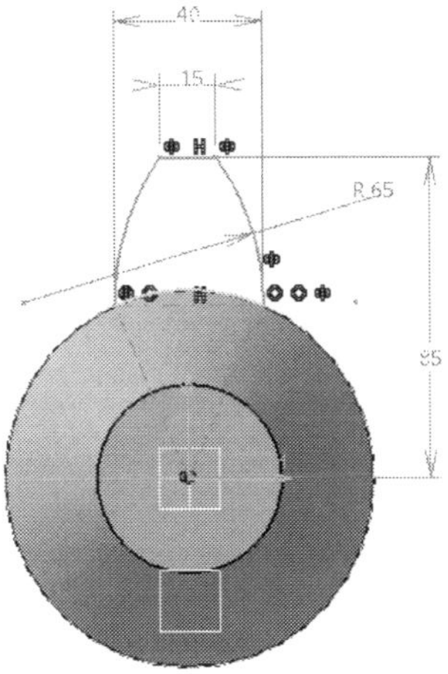

Figure 9-72 *Sketch created on the front planar face*

3. Create another plane at an offset of 210.5 units from the right most planar face of the base feature of the model.

4. To draw the second sketch of the multi-section solid feature, select the newly created plane and invoke the **Sketcher** workbench.

5. Extract the geometry of the previously drawn sketch by using the **Project 3D Elements** button of the **Operation** toolbar.

6. Select the projected elements of the sketch and invoke the contextual menu by right-clicking in the graphics area. Next, choose **Mark.1 object > Isolate** from it; a warning message box may be displayed informing that reroute is needed.

7. Choose **OK** from this message box to exit it.

8. Select all the isolated entities and choose the **Scale** tool from the **Transformation** toolbar; the **Scale Definition** dialog box is displayed and you are prompted to specify the scaling center point.

9. Click at the origin to specify the scaling center point.

10. Clear the **Duplicate mode** check box in this dialog box if selected.

11. Set **0.75** in the **Value** spinner of the **Scale Definition** dialog box and press the ENTER key.

12. Select all entities of the sketch and then invoke the **Rotate** tool from the **Transformation** sub-toolbar; the **Rotation Definition** dialog box is displayed and you are prompted to select the rotation center point.

13. Click at the origin to specify the rotation center point; you are prompted to select a point to define the reference line for the angle.

14. Move the cursor horizontally toward the left with respect to the specified point and then click to specify the reference line for the angle.

15. Clear the **Duplicate mode** check box in the **Rotation Definition** dialog box, if it is already selected.

16. Next, set **15** in the **Value** spinner of the **Rotation Definition** dialog box and then press the ENTER key; the profile thus created is rotated by 15 degree. Figure 9-73 shows the second section after scaling and rotating its profile.

    After rotating the sketch, you need to reconstrain the sketch by applying the **Fix** constraints, to make it fully defined.

***Figure 9-73** Second section for the multi-section solid feature*

17. Exit the **Sketcher** workbench. On doing so, the second section of the multi-section solid feature is created.

18. Similarly, create the third section on the left most planar face of the base feature of the model. While creating the third section, project the first section and reduce its size to 50% and then rotate it by 30 degree from the horizontal. After scaling and rotating the feature, the profile of the third section will be created, as shown in Figure 9-74.

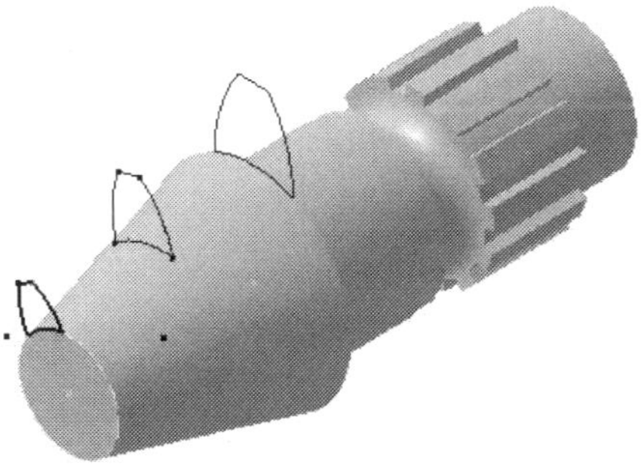

Figure 9-74 Third section for the multi-section solid feature

19. After rotating the sketch, you need to reconstrain the sketch by applying the **Fix** constraints, to make it fully defined.

20. Exit the **Sketcher** workbench.

Creating the Multi-section Solid Feature

1. Invoke the **Multi-sections Solid** tool from the **Sketch-Based Features** toolbar; the **Multi-Sections Solid Definition** dialog box is displayed. 

2. Select the three sections in the descending order of their sizes to create the multi-section solid feature.

3. If required, set the closing points of the sections and flip their directions, refer to Figure 9-75.

4. Choose the **Preview** button from the **Multi-Sections Solid Definition** dialog box; a preview of the resultant model is displayed. If the preview displayed is different from the required shape then you need to reset the closing points and their directions.

5. Choose **OK** from the **Multi-Sections Solid Definition** dialog box. The final resulting feature is shown in Figure 9-76.

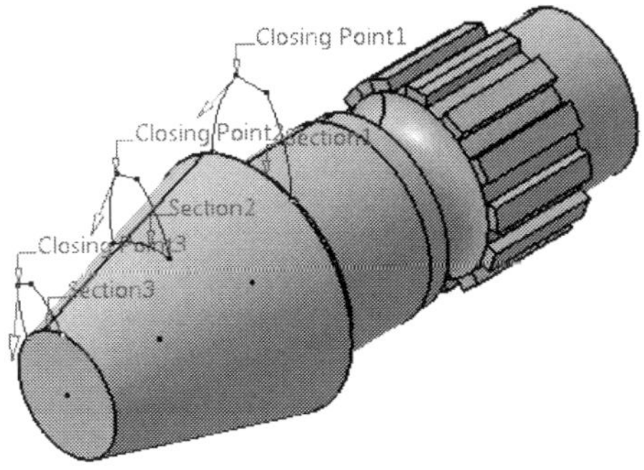

Figure 9-75 Closing points and their directions

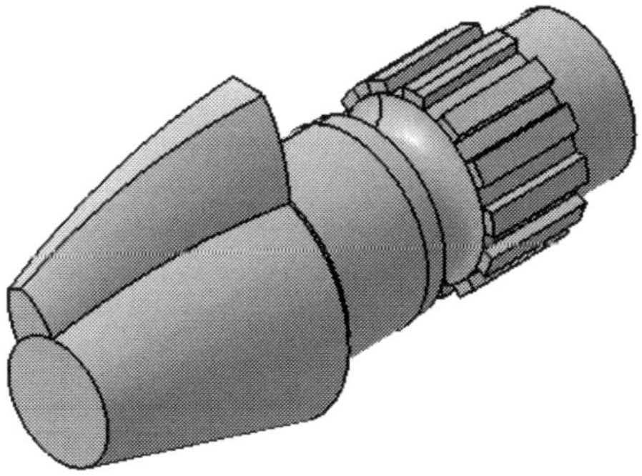

Figure 9-76 Preview of the resultant feature

6. Invoke the **Circular Pattern** tool from the **Patterns** sub-toolbar and create 6 instances of newly created multi-section solid feature using the **Complete crown** option. To define the reference element, select the circular edge of the front face. The model after creating the circular pattern of the multi-section solid feature is shown in Figure 9-77.

Note

If tangent configurations have been created during boolean operation; a warning message box is displayed while creating the multi-section solid feature. Choose the ***Close*** *button to close the message box*

Saving and Closing the File

1. Choose the **Save** button from the **Standard** toolbar to invoke the **Save As** dialog box.

2. Enter **c09tut2** in the **File name** edit box and choose the **Save** button. The file is saved at *C:\CATIA\c09*.

3. Close the part file by choosing **File > Close** from the menu bar.

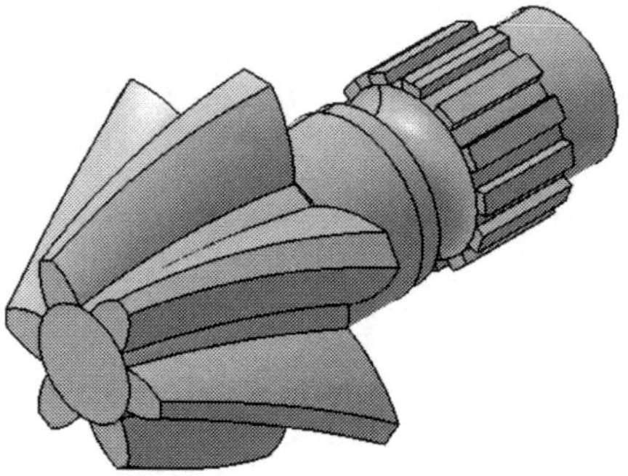

Figure 9-77 *Final model of the Helical gear*

Tutorial 3

In this tutorial, you will create the model of the Mouse Cover shown in Figure 9-78. Its views and dimensions are shown in Figure 9-79. **(Expected time: 30 min)**

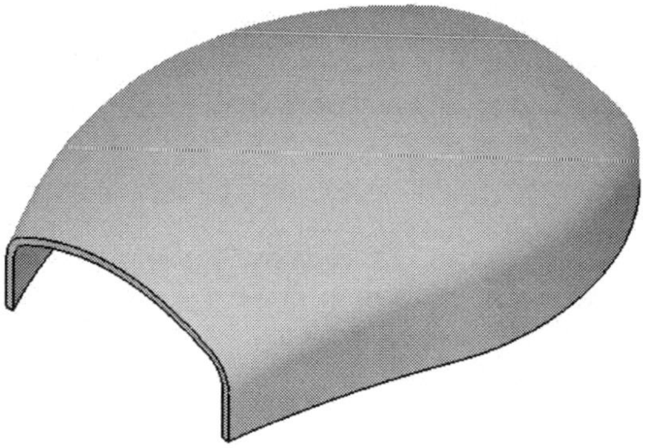

Figure 9-78 *Model of the Mouse Cover for Tutorial 3*

The following steps are required to complete this tutorial:

a. Draw sketches for creating a multi-section solid feature with guides. The sketches include two sections and three guides.
b. Create the base feature by creating multi-sections solid with guides.
c. Fillet the edges of the base feature.
d. Shell the model by removing the front and bottom faces of the model.

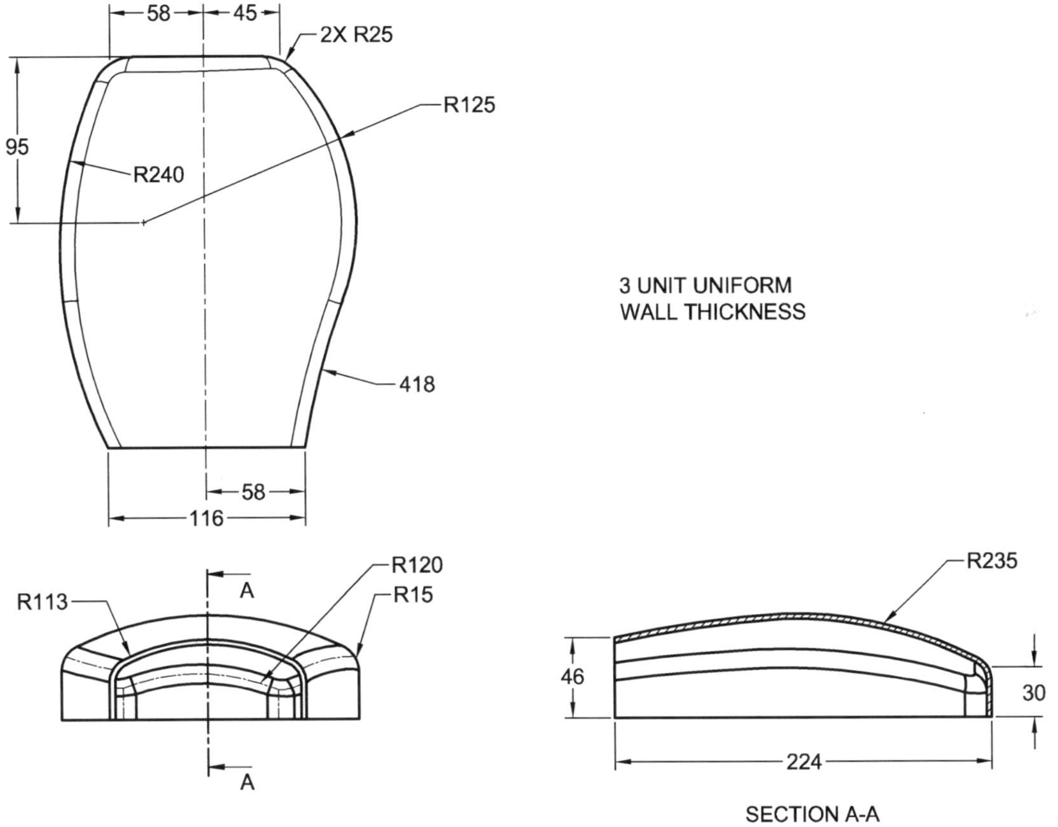

Figure 9-79 Orthographic views and dimensions for Tutorial 3

Drawing the Sketches for the Base Feature

The base feature of the model will be created by creating a multi-section solid having two sections. Each section plane will be at an offset distance from one another and the profile of the multi-section solid feature will be controlled by the guides. Therefore, before proceeding further, you need to draw the sketches of three guides and two sections.

1. Select the xy plane and invoke the **Sketcher** workbench.

2. Draw the sketch of the first guide, as shown in Figure 9-80, and exit the **Sketcher** workbench.

3. Select the xy plane again and invoke the **Sketcher** workbench.

4. Draw the sketch of the second guide, as shown in Figure 9-81, and exit the **Sketcher** workbench.

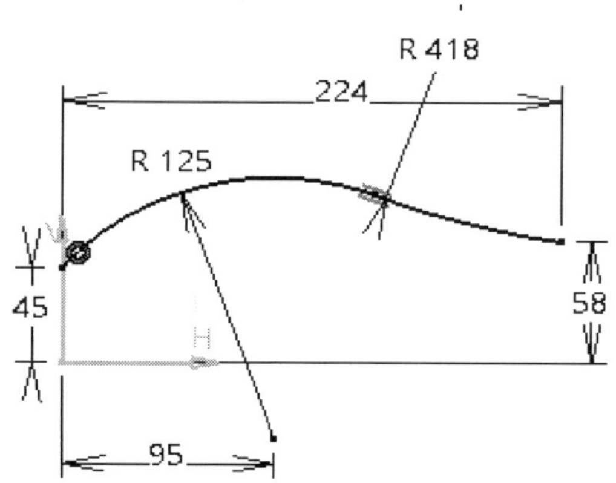

Figure 9-80 Sketch of the first guide

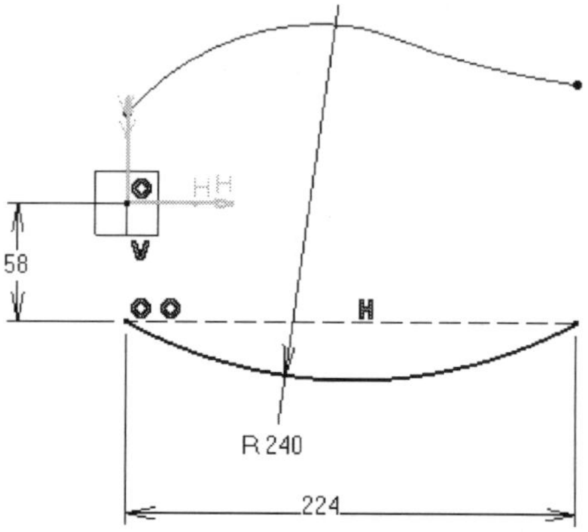

Figure 9-81 Sketch of the second guide

5. Select the zx plane and invoke the **Sketcher** workbench.

6. Draw the sketch of the third guide, as shown in Figure 9-82, and exit the **Sketcher** workbench.

7. Select the yz plane and invoke the **Sketcher** workbench.

8. Draw the sketch of the first section, as shown in Figure 9-83, and exit the **Sketcher** workbench. Note that the ends of the section are coincident with the ends of the guide curves.

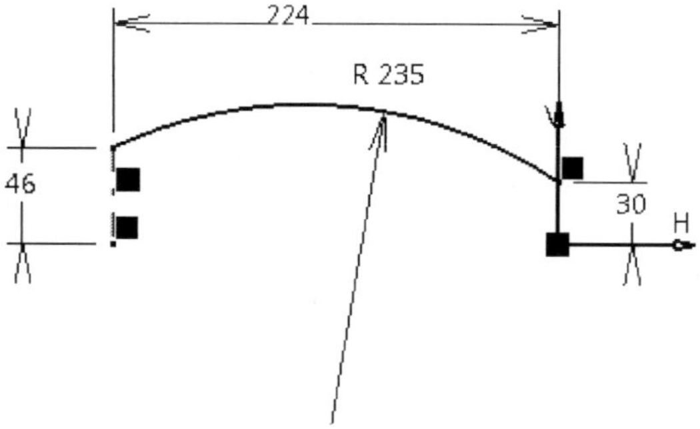

Figure 9-82 Sketch of the third guide

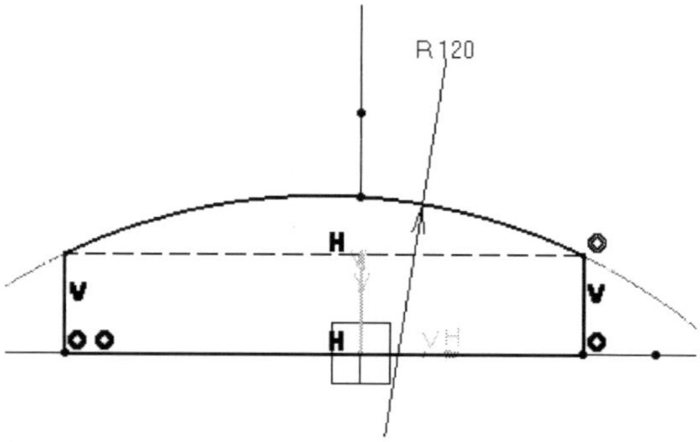

Figure 9-83 Sketch of the first section

9. Invoke the **Plane Definition** dialog box and then select the **Through three points** option from the **Plane type** drop-down list.

10. Select the three open endpoints of the curve and then choose **OK**; a plane passing through all the selected points is created.

11. Select the newly created plane as the sketching plane and invoke the **Sketcher** workbench.

12. Draw the sketch of the second section, as shown in Figure 9-84, and exit the **Sketcher** workbench. The model after drawing all sketches is shown in Figure 9-85.

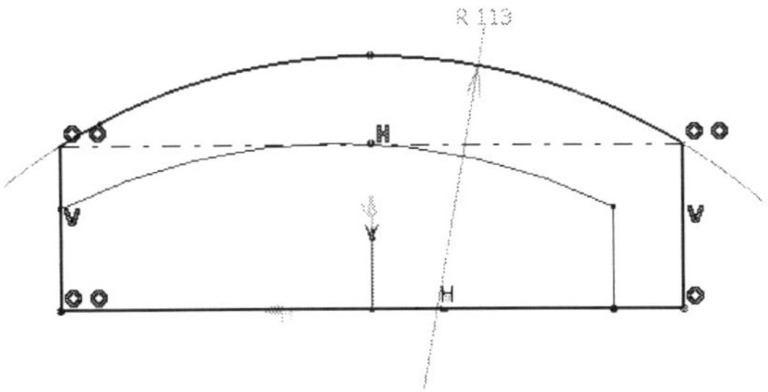

Figure 9-84 Sketch of the second section

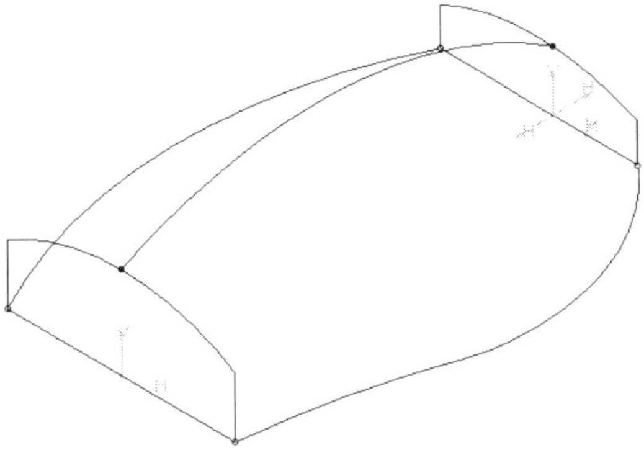

Figure 9-85 Model after drawing all sketches

Creating the Base Feature

After drawing the sketches of the base feature, you need to create the base feature using these sketches.

1. Invoke the **Multi-sections Solid** tool from the **Sketch-Based Features** toolbar; the **Multi-Sections Solid Definition** dialog box is displayed. 

2. Select the first section from the geometry area, refer to Figure 9-86.

3. Select the second section from the geometry area, refer to Figure 9-86.

   After selecting both sections, you need to make sure that the closing points are located at the same position on both sections. Otherwise, you may need to replace the closing points.

   After selecting the sections and setting the closing points, you need to select the guides.

4. Click on the three dots displayed under the **No** column head in the **Guides** tab.

5. Select all three guide curves from the geometry area, refer to Figure 9-86.

6. Choose the **OK** button from the **Multi-Sections Solid Definition** dialog box. The model after creating the base feature is shown in Figure 9-87.

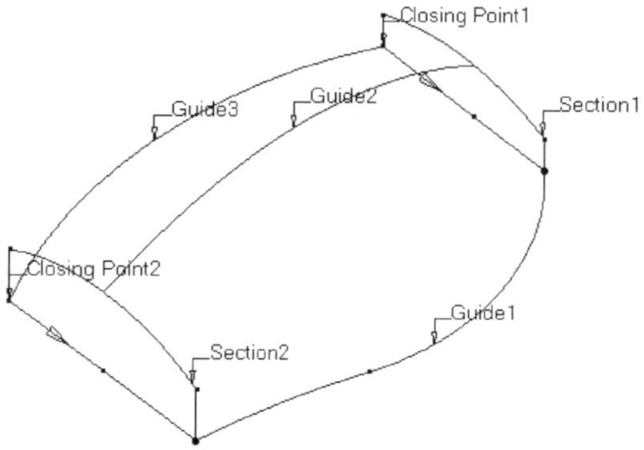

Figure 9-86 Sections and guides to be selected

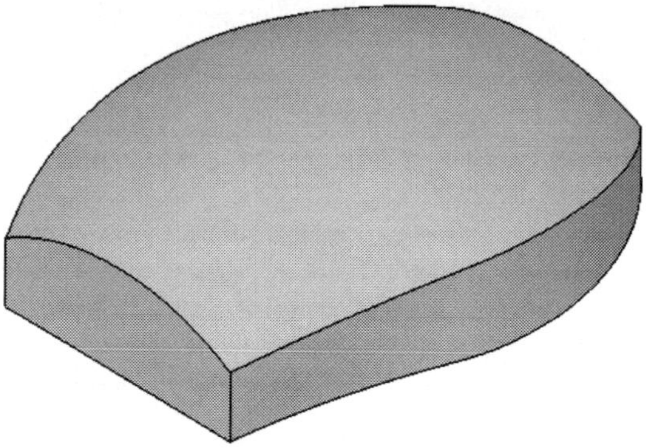

Figure 9-87 Base feature of the model

Filleting the Edges of the Model

After creating the base feature of the model, you need to fillet its edges.

1. Double-click on the **Edge Fillet** button in the **Dress-Up Features** toolbar to invoke the **Edge Fillet Definition** dialog box.

2. Select the edges, as shown in Figure 9-88, and set the value of the radius to **25** in the **Radius** spinner.

3. Choose the **OK** button from the **Edge Fillet Definition** dialog box; the edges are filleted, as shown in Figure 9-89, and the dialog box is displayed again.

4. Now, select the edges shown in Figure 9-90 and set the value of the radius to **15** in the **Radius** spinner. Next, choose the **OK** button from the **Edge Fillet Definition** dialog box. The model after filleting the edges is shown in Figure 9-91. Exit the **Edge Fillet Definition** dialog box.

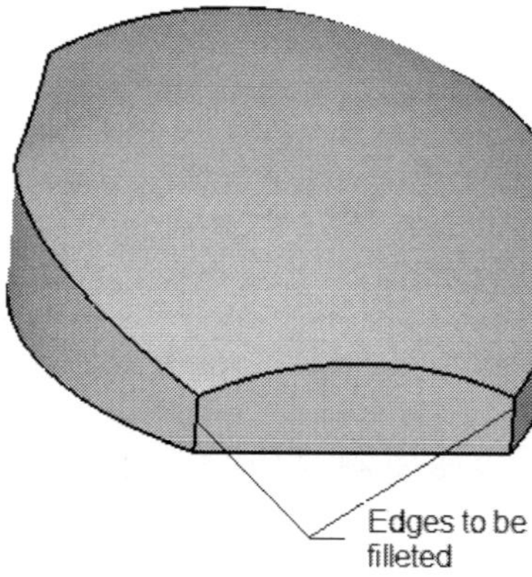

Figure 9-88 Edges to be filleted

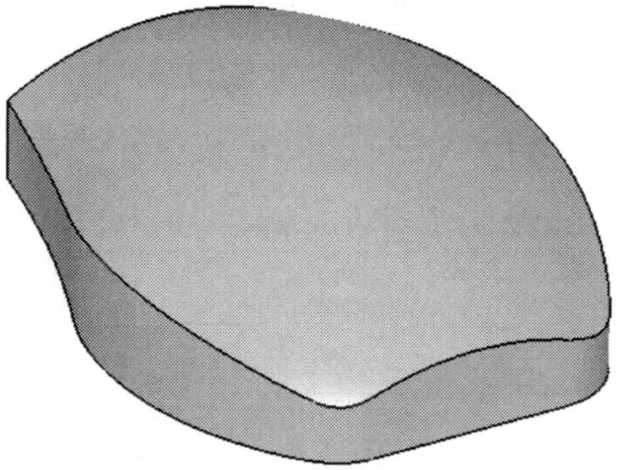

Figure 9-89 Model after filleting the edges

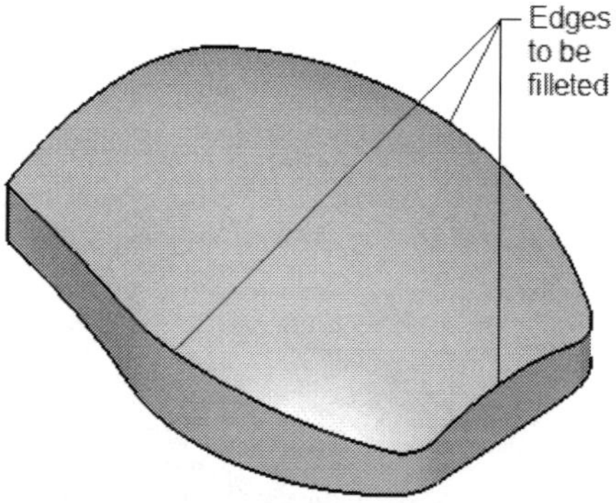

Figure 9-90 Edges to be filleted

5. Using the **Shell** tool, shell the model by 3 units by removing its front and bottom faces. The final model is shown in Figure 9-92.

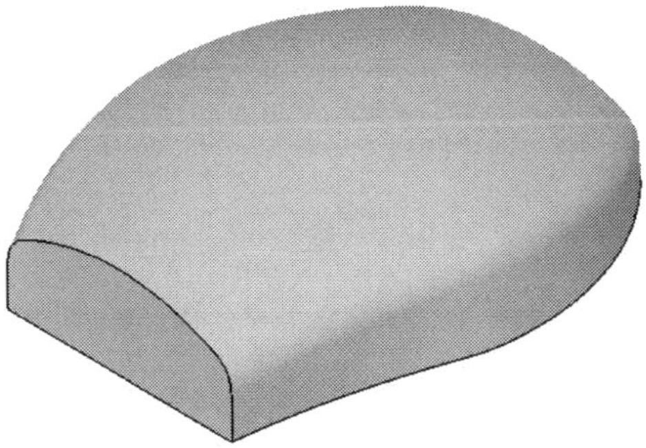

Figure 9-91 *The model after filleting the selected edges*

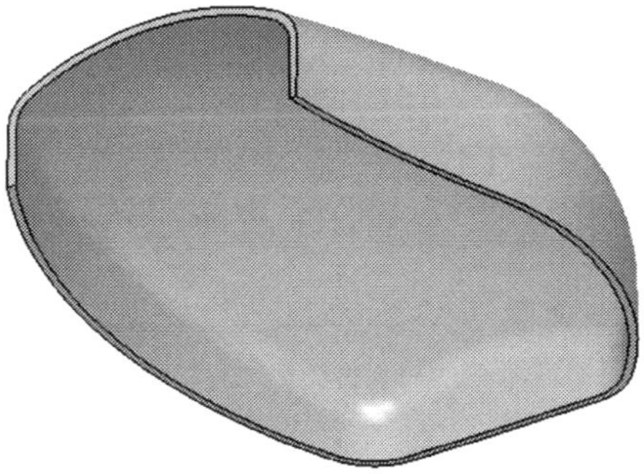

Figure 9-92 *Final model of the Mouse Cover*

Saving and Closing the File

1. Choose the **Save** button from the **Standard** toolbar to invoke the **Save As** dialog box.
2. Enter **c09tut3** in the **File name** edit box and choose the **Save** button. The file will be saved at *C:\CATIA\c09*.
3. Close the part file by choosing **File > Close** from the menu bar.

Tutorial 4

In this tutorial, you will create the model shown in Figure 9-93. Its views and dimensions are shown in the same figure. The hidden lines in the top and side views are suppressed for clarity.

(Expected time: 45 min)

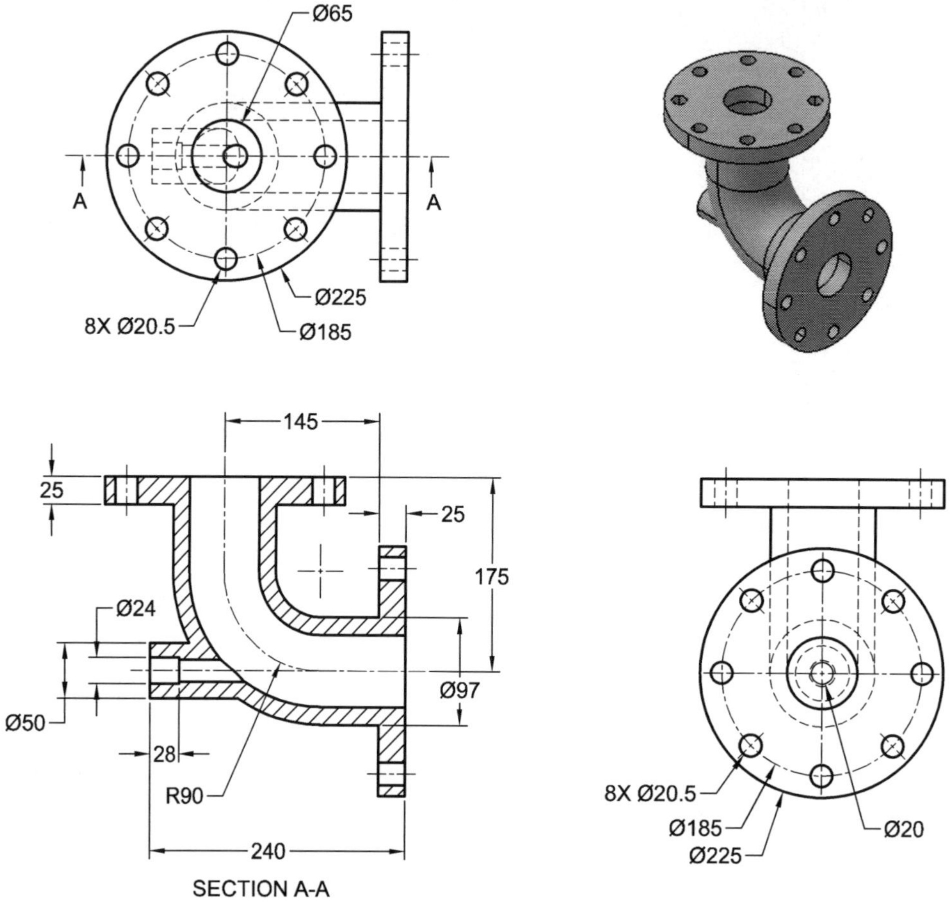

Figure 9-93 Views and dimensions of the model for Tutorial 4

The following steps are required to complete this tutorial:

a. Create the path for the rib feature.
b. Create the profile of the rib feature.
c. Create the rib feature.
d. Create the remaining features.
e. Save and close the file.

Creating the Path for the Rib Feature

As discussed earlier, the base feature of the model is a rib feature. To create the rib feature, you first need to create its path on the yz plane.

1. Draw the sketch of the path for the rib feature on the yz plane and add the required constraints and dimensions to the sketch, as shown in Figure 9-94. Exit the sketching environment and change the view to isometric.

Creating the Profile of the Rib Feature

After creating the path for the rib feature, you need to create the profile of the rib feature. To create the profile, you need to create a reference plane normal to the path. The newly created plane will be selected as the sketching plane for creating the profile of the rib feature.

1. Invoke the **Plane Definition** dialog box by choosing the **Plane** tool from the **Reference Elements** toolbar and then select the **Normal to curve** option from the **Plane type** drop-down list.

2. Select the sketch and create the plane at the endpoint of its horizontal entity, as shown in Figure 9-95.

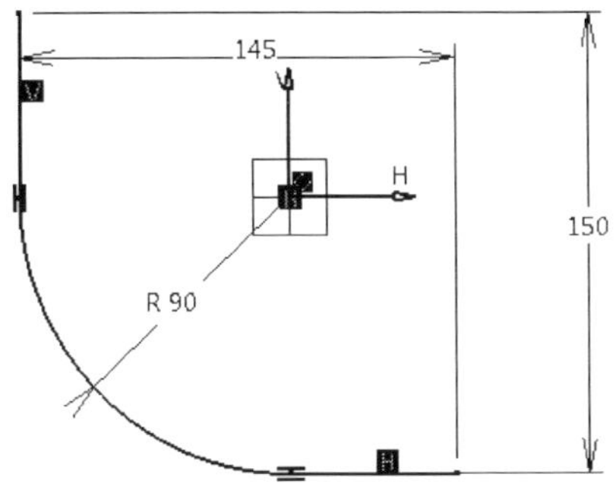

Figure 9-94 Sketch of the path

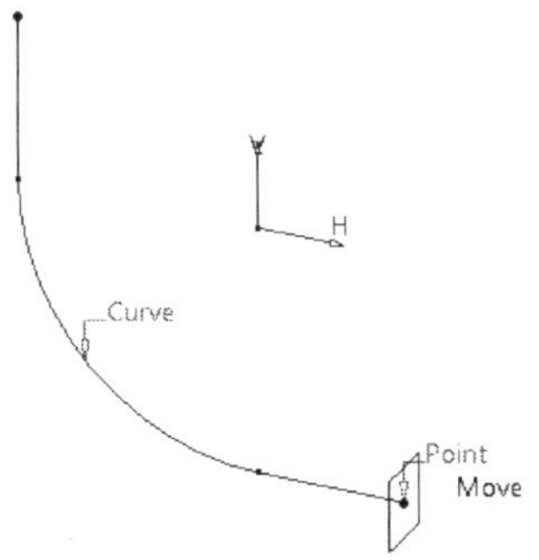

Figure 9-95 Plane created normal to the path

3. Invoke the **Sketcher** workbench by selecting the newly created plane as the sketching plane.

4. Draw the sketch of the profile of the rib feature by using the **Circle** tool. The center of the circle is at the end point of the curve on the new plane and the diameter of the circle is 97 mm.

5. Exit the **Sketcher** workbench after drawing the profile of the rib feature.

Creating the Rib Feature

After creating the profile of the rib feature, you need to create the rib feature by using the **Thick profile** check box in the **Rib Definition** dialog box.

1. Choose the **Rib** tool from the **Sketch-Based Features** toolbar; the **Rib Definition** dialog box is displayed and you are prompted to select a profile.

2. Select the circle as the profile for the rib feature; you are prompted to select the center curve.

3. Select the path created for the rib feature as the center curve. Next, select the **Thick profile** check box; the **Thin Rib** area gets activated. Then, select the **Neutral Fiber** check box.

4. Enter **16** in the **Thickness1** spinner and choose the **OK** button; the rib feature is created, as shown in Figure 9-96.

Creating Remaining Features

Next, you need to create the remaining features to complete the model.

1. Create the flanges on both the ends of the rib feature using the **Pad** tool, refer to Figure 9-96.

2. Create a hole on the flange using the **Hole** tool and then create a circular pattern of the hole feature using the **Circular pattern** tool.

3. Create a plane at an offset distance of 240 mm from the right face of the model, and then create a circle of diameter 50 and extrude it by using the **Up to next** option from the **Pad** tool.

4. Create a counterbore hole using the **Hole** tool on the extruded feature created in the previous step. The final solid model is shown in Figure 9-97.

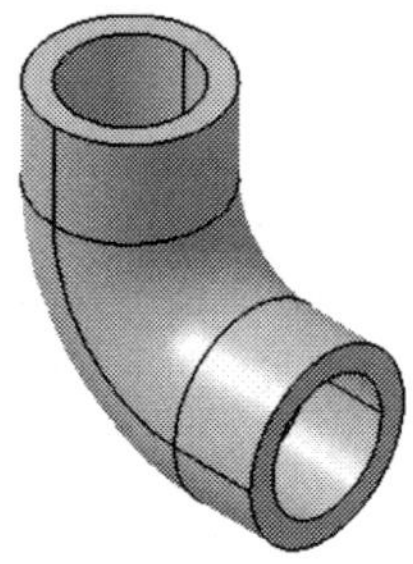

Figure 9-96 *Base feature of the model*

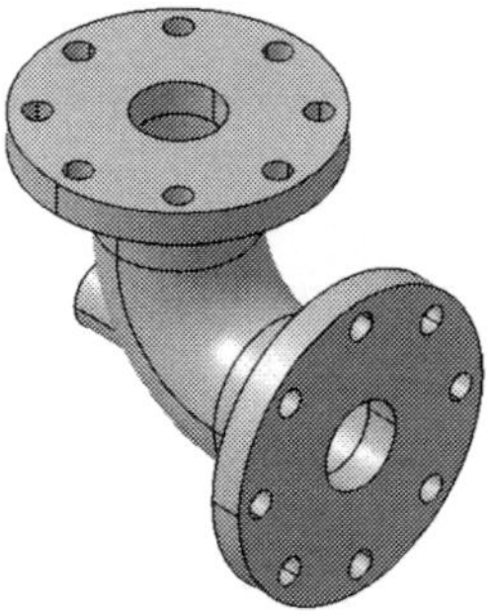

Figure 9-97 *The final model*

Saving and Closing the File

1. Choose the **Save** button from the **Standard** toolbar to invoke the **Save As** dialog box.

2. Enter **c9tut4** in the **File name** edit box and choose the **Save** button; the file is saved at *C:\CATIA\c09*.

3. Close the part file by choosing **File > Close** from the menu bar.

Self-Evaluation Test

Answer the following questions and then compare them to those given at the end of this chapter:

1. The rib feature is created using the __________ dialog box.

2. The __________ tool is used to remove the material by sweeping a section along with a center curve.

3. The __________ check box is selected to merge the ends of the rib feature with the surrounding surfaces.

4. To replace the closing point of a section, select the point to be replaced, right-click to invoke the contextual menu, and then choose the __________ option from it.

5. To create a thin rib feature, select the __________ check box from the **Rib Definition** dialog box.

6. The **Rib** tool is used to sweep an open or a closed profile along an open or a closed center curve. (T/F)

7. You cannot define the pulling direction while creating a rib feature. (T/F)

8. The **Rib** tool also provides you with an option to define the reference surface for creating the rib feature. (T/F)

9. You cannot define a spine while creating a multi-section feature. (T/F)

10. You can also create a multi-section solid feature by using the sections with uneven number of vertices. (T/F)

Review Questions

Answer the following questions:

1. Which of the following dialog boxes is used to remove the material by sweeping a section along the center curve?

   (a) **Cut sweep** (b) **Remove rib**
   (c) **Slot** (d) None of these

2. Which of the following tabs of the **Multi-Sections Solid Definition** dialog box is used to define the guide curves for creating a loft feature?

   (a) **Spine** (b) **Guide curves**
   (c) **Guides** (d) **Coupling**

3. Which of the following check boxes in the **Rib Definition** dialog box is used to create a thin rib feature?

   (a) **Thin** (b) **Thick Profile**
   (c) **Merge rib's ends** (d) None of these

4. Which of the following dialog boxes is used to remove the material by blending the sections together?

   (a) **Multi-Sections Solid Definition** (b) **Remove Multi-Sections Solid Definition**
   (c) **Slot** (d) None of these

5. Which of the following check boxes should be selected to add the material of the thin rib feature on both sides of the sketch?

   (a) **Thickness2** (b) **Neutral Fibre**
   (c) **Thickness1** (d) None of these

6. You can create a rib feature by using an open profile and an open center curve. (T/F)

7. If you clear the **Relimited on start section** check box, the multi-section solid feature will be created from the start point of the spine curve or the guide curve. (T/F)

8. The **Intersect 3D element** tool is used to extract the sketch formed by the intersection of the planar or non-planar face with the sketch plane. (T/F)

9. When you create a multi-section solid feature, the point from which you start drawing the sketch is selected by default as the closing point of the section. (T/F)

10. The **Multi-Sections Solid Definition** dialog box is used to remove the material by creating a multi-sections solid feature. (T/F)

EXERCISES

Exercise 1

Create the model of the Angle Flange shown in Figure 9-98. Its orthographic views and dimensions are shown in Figure 9-99. **(Expected time: 45 min)**

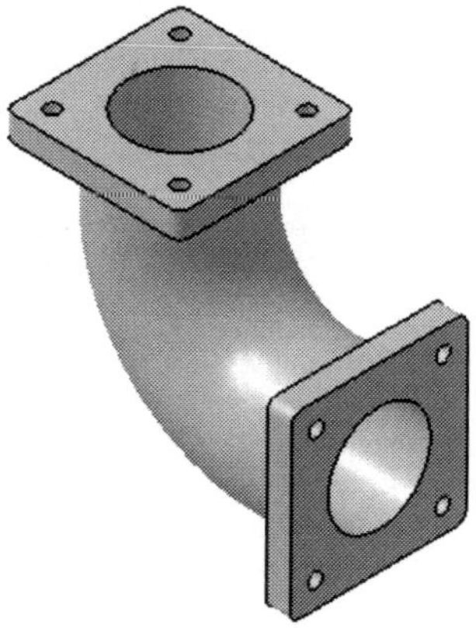

Figure 9-98 *Model for Exercise 1*

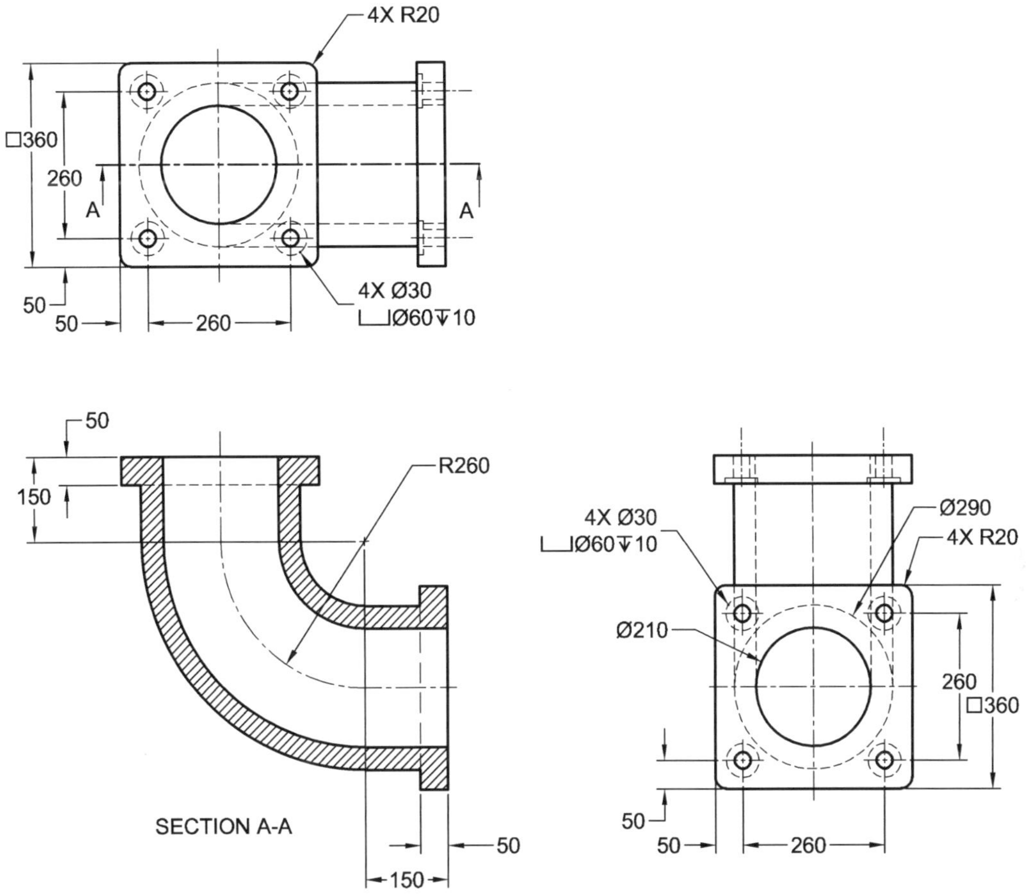

Figure 9-99 *Views and dimensions for Exercise 1*

Exercise 2

Create the model of the Carburetor Cover shown in Figures 9-100 and 9-101. Its orthographic views and dimensions are shown in Figure 9-102. **(Expected time: 45 min)**

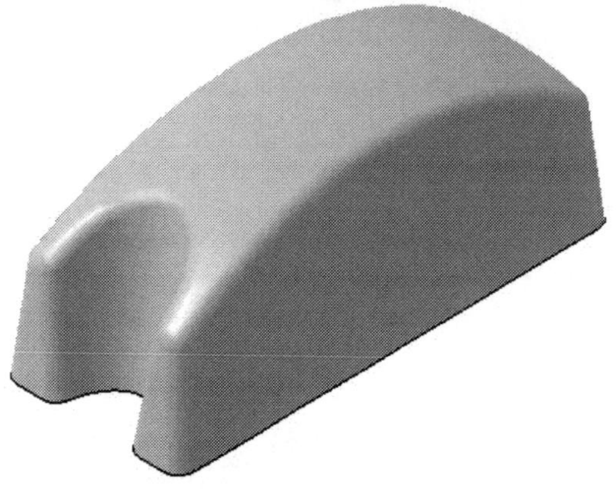

Figure 9-100 Model of the Carburetor Cover for Exercise 2

Figure 9-101 Rotated view of the Carburetor Cover for Exercise 2

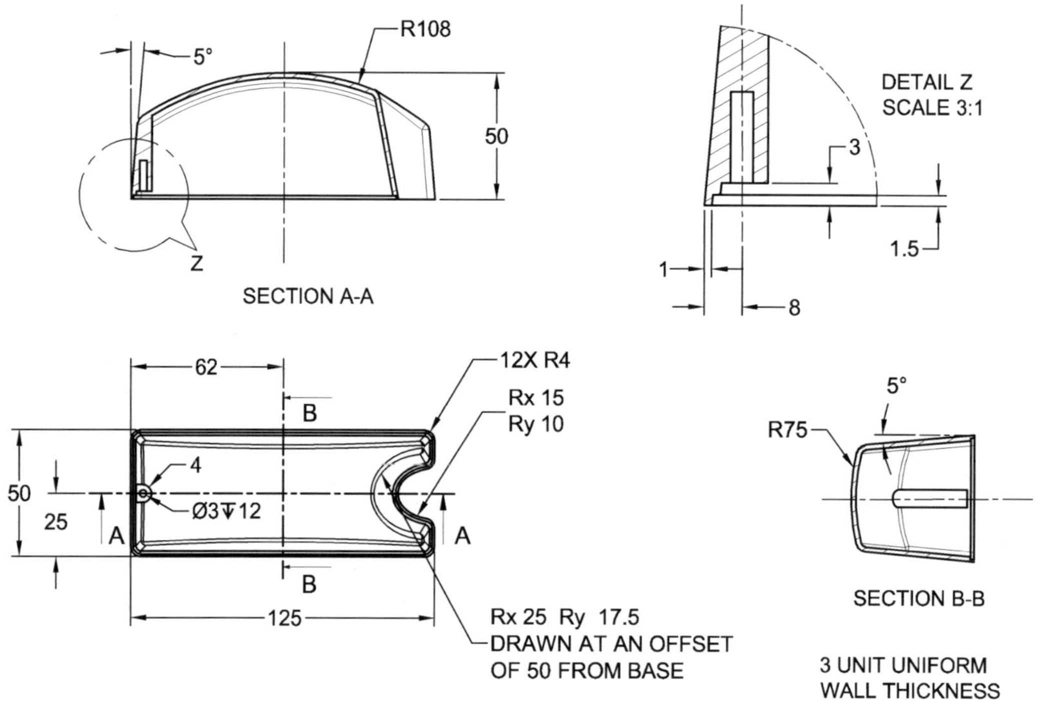

Figure 9-102 Views and dimensions for Exercise 2

Answers to Self-Evaluation Test

1. **Rib Definition**, **2.** **Slot**, **3.** **Merge rib's ends**, **4.** **Replace**, **5.** **Thick Profile**, **6.** T, **7.** F, **8.** T, **9.** F, **10.** T

Chapter 10

Working with the Wireframe and Surface Design Workbench

Learning Objectives

After completing this chapter, you will be able to:

- *Create wireframe geometry*
- *Create extruded surfaces*
- *Create revolved surfaces*
- *Create spherical surfaces*
- *Create offset surfaces*
- *Create swept surfaces*
- *Create fill surfaces*
- *Create multi-sections surfaces*
- *Create blended surfaces*
- *Split surfaces*
- *Trim surfaces*
- *Join surfaces*

NEED OF SURFACE MODELING

The product and industrial designers give special importance to product styling and provide a unique shape to the components. Generally, this is done to make the product look attractive and presentable. Most of the times, the shape of the product is managed using the surface modeling techniques. Surface models are three-dimensional models with no thickness and do not have mass properties. CATIA V5 provides a number of surface modeling tools to create complex three-dimensional surface models. Various workbenches in CATIA V5 with surface creation tools are:

1. Wireframe and Surface Design
2. Generative Shape Design
3. FreeStyle

In this chapter, you will learn about the surface modeling tools in the **Wireframe and Surface Design** workbench.

WIREFRAME AND SURFACE DESIGN WORKBENCH

The **Wireframe and Surface Design** workbench provides tools to create wireframe construction elements during preliminary design and enrich an existing 3D mechanical part design with wireframe and basic surface features.

Starting the Wireframe and Surface Design Workbench

Start a new session of CATIA V5 and close the new product file, which is open by default. Next, choose **Wireframe and Surface Design** from **Start > Mechanical Design** in the menu bar to start a new file in the **Wireframe and Surface Design** workbench.

CREATING WIREFRAME ELEMENTS

On invoking the **Wireframe and Surface Design** workbench, the **Wireframe** toolbar is displayed, as shown in Figure 10-1. The tools in this toolbar are used to create wireframe geometries. These geometries aid in creating surfaces. The sketches drawn in the sketcher workbench can also be used to create surfaces. The tools for constructing wireframe geometries are **Line**, **Polyline**, **Circle**, **Spline**, **Helix**, and so on. Some of these tools are discussed next.

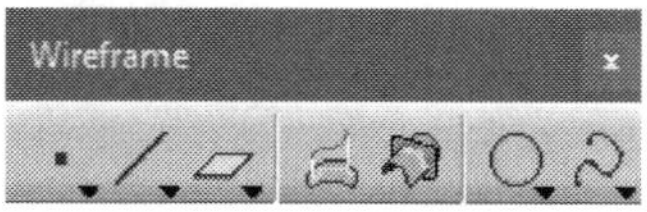

Figure 10-1 *The **Wireframe** toolbar*

Creating Circles

Menubar:	Insert > Wireframe > Circle
Toolbar:	Wireframe > Circle-Corner-Connect sub-toolbar > Circle

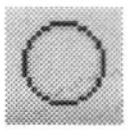

In the **Wireframe and Surface Design** workbench, the **Circle** tool is used to create arcs and circles. To create arcs and circles, choose the **Circle** tool from the **Circle-Corner-Connect** sub-toolbar of the **Wireframe** toolbar; the **Circle Definition** dialog box will be displayed, as shown in Figure 10-2. In this dialog box, the **Center and radius** option is selected by default in the **Circle type** drop-down list. As a result, you are prompted to select the center point. You can select a predefined point or create a new one by choosing any one of the options from the contextual menu which is displayed when you right-click in the **Center** display box of the **Circle Definition** dialog box. As soon as you specify the center point, you are prompted to select the support surface. Select a plane as the support surface and then specify the required radius value in the **Radius** spinner. You can set the angular limits of the arc from the **Circle Limitations** area. You can also create a circle by choosing the **Whole Circle** button from the **Circle Limitations** area of the **Circle Definition** dialog box and then choosing the **OK** button from the dialog box. To project the arc on the selected surface or plane, choose the **Geometry on support** check box. Finally, choose the **OK** button to complete the arc.

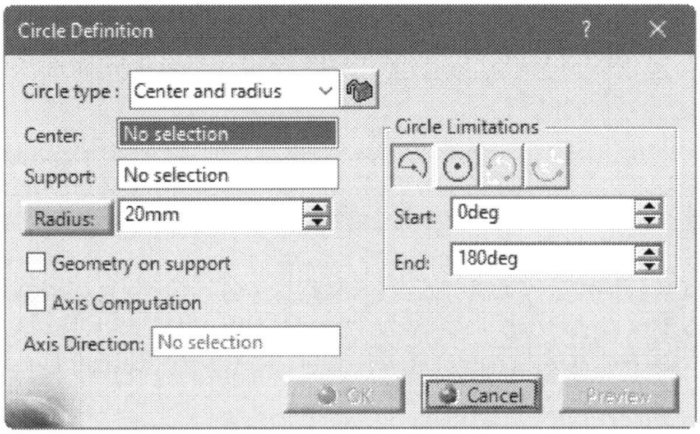

***Figure 10-2** The **Circle Definition** dialog box*

Note

*To create axis of a circle or arc, you need to select the **Axis Computation** check box. After selecting this check box, you can define the axis direction.*

Creating Splines

Menubar:	Insert > Wireframe > Spline
Toolbar:	Wireframe > Curves sub-toolbar > Spline

The **Spline** tool is used to draw a spline in three-dimensional space by selecting the connecting points. This tool is available in the **Curves** sub-toolbar of the **Wireframe** toolbar, as shown in Figure 10-3.

To draw a spline, choose the **Spline** tool from the **Curves** sub-toolbar; the **Spline Definition** dialog box will be displayed, as shown in Figure 10-4, and you will be prompted to select a point. You can select a predefined point, or right-click in the display box under the **Points** column head of the dialog box to create a new point using the options from the contextual menu. After selecting a predefined point or creating a new point, you will be prompted to select another point. In this way, you can specify a number of points to draw spline.

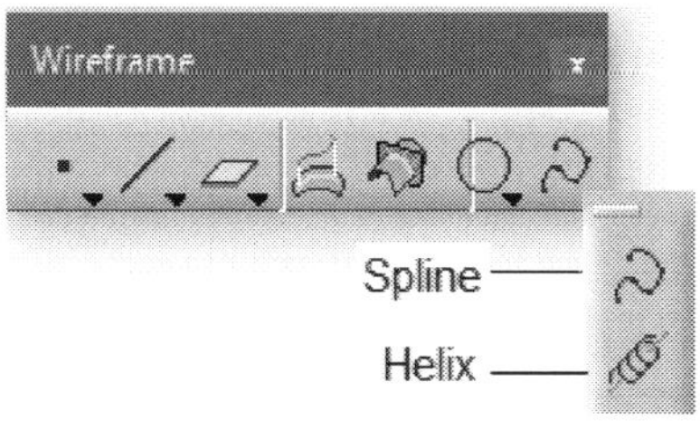

***Figure 10-3** The **Curves** sub-toolbar in the **Wireframe** toolbar*

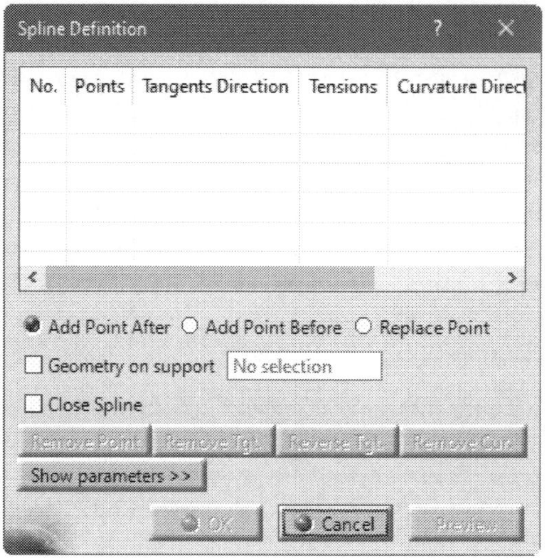

Figure 10-4 *The* ***Spline Definition*** *dialog box*

After specifying all the points, select the **Geometry on support** check box from the **Spline Definition** dialog box; you will be prompted to select a support element. Select a plane or surface on which you want to create the spline. Remember that after specifying the support element, all the specified points should lie completely on the selected support element. Choose the **OK** button from the **Spline Definition** dialog box to create the spline and exit the dialog box.

If you need to create a spline in 3D space, the **Geometry on Support** check box must be cleared in the **Spline Definition** dialog box.

To replace a wrongly selected point for defining the spline, select it from the **Points** column, and then select the **Replace Point** radio button. Next, select the replacement point for the selected point and choose **OK**. You can create a closed spline by selecting the **Close Spline** check box.

Creating a Helix

Menubar:	Insert > Wireframe > Helix
Toolbar:	Wireframe > Curves sub-toolbar > Helix

The **Helix** tool is used to create a helical curve. On invoking this tool, the **Helix Curve Definition** dialog box will be displayed, as shown in Figure 10-5, and you will be prompted to select the helix starting point. Select a predefined point or create a new point by using the options from the contextual menu that is displayed on right clicking in the **Starting Point** display box of the **Helix Curve Definition** dialog box.

On selecting the point, you will be prompted to select a line as the helix axis. Select a predefined line or draw a new line by using the options from the contextual menu that is displayed on right-clicking in the **Axis** display box. On doing so, the preview of the helix curve, with the default values, will be displayed in the drawing area. You can also use the axis of the coordinates system as the axis of helix. You can create helical curves by using different options available in the **Helix Type** drop-down list in the **Type** area. These options are discussed next.

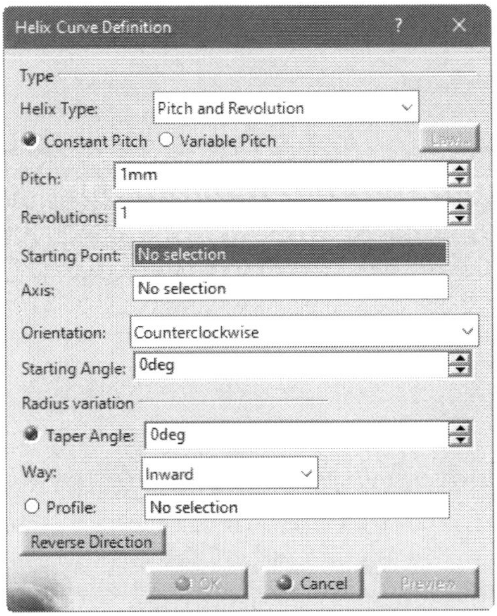

***Figure 10-5** The **Helix Curve Definition** dialog box*

Pitch and Revolution

The **Pitch and Revolution** option in **Helix Type** drop-down list in **Type** area is selected by default. As a result, the **Pitch** and **Revolutions** spinners are displayed. You can specify the pitch and number of revolutions of helical curve in the **Pitch** and **Revolutions** spinners, respectively.

By using the **Orientation** drop-down list, you can control the rotation of the helix either in clockwise direction or counterclockwise direction. You can specify the start angle of the helical curve using the **Starting Angle** spinner in the **Type** area.

If you need to create a helix of varying pitch, select the **Variable Pitch** radio button from the **Type** area and specify the start value, end value, and number of revolutions in their respective spinners. Note that the **Variable Pitch** radio button will be active only when the **Pitch and Revolution** option is selected in **Helix Type** drop-down list in the **Type** area.

Height and Revolution

The **Height and Revolution** option from the **Helix Type** drop-down list is used to specify the total height and number of revolutions of helical curve. You can specify the height and revolutions of the helical curve in the **Height** and **Revolutions** spinners, respectively.

Height and Pitch

When you select the **Height and Pitch** option from the **Helix Type** drop-down list in the **Type** area, the **Height and Pitch** spinners are displayed. You can specify the height and pitch of the helical curve in the **Height** and **Pitch** spinners, respectively.

You can also add a taper angle to the helix by setting a value in the **Taper Angle** spinner in the **Radius variation** area of the dialog box. Figure 10-6 shows a helix without a taper angle. Figure 10-7 shows a helix with a taper angle.

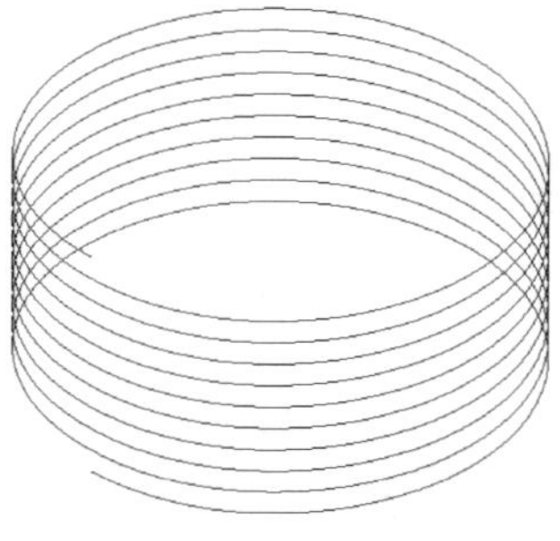

Figure 10-6 *The helix without a taper angle*

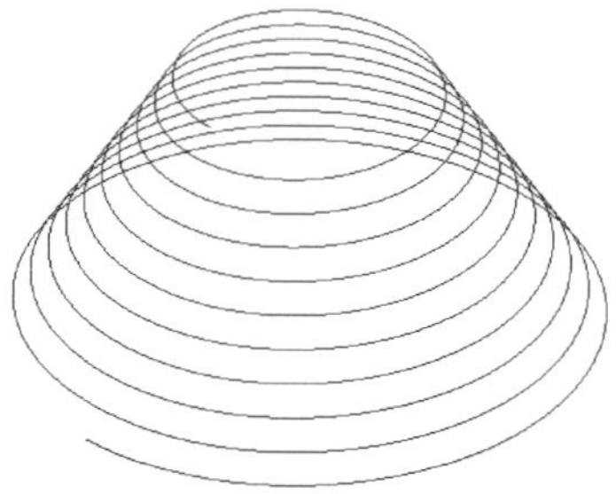

Figure 10-7 *The helix with a taper angle*

You can create taper toward the selected axis or away from the selected axis by selecting the **Inward** or **Outward** option from the **Way** drop-down list. To create a helix curve on a profile, select the **Profile** radio button, you will be prompted to select a planar curve as a helix of variable radius. Select the planar curve, as shown in Figure 10-8; the preview of a resultant profile will be displayed. Note that the selected profile curve must pass through the selected starting point.

If you choose the **Law** button from the **Helix Curve Definition** dialog box, the **Law Definition** dialog box will be displayed, as shown in Figure 10-9. You can use this dialog box to define the evolution of the pitch along the helix by specifying the law type for the helix curve by selecting the **Constant** or **S type** radio button from the **Law type** area. By default, the **S type** radio button is selected in this area. As a result, you can specify only the start value using the **Start value** spinner in the **Law Definition** dialog box. If you select the **S type** radio button, the **End value** spinner will be activated and then you can specify both the start and end values.

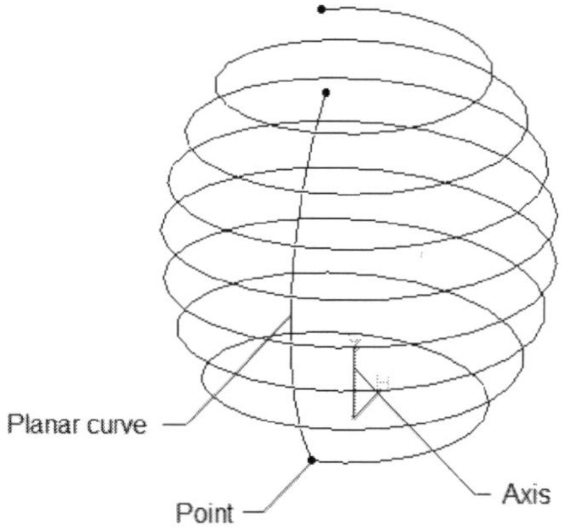

Figure 10-8 *Selected planar curve and the preview of the resultant profile*

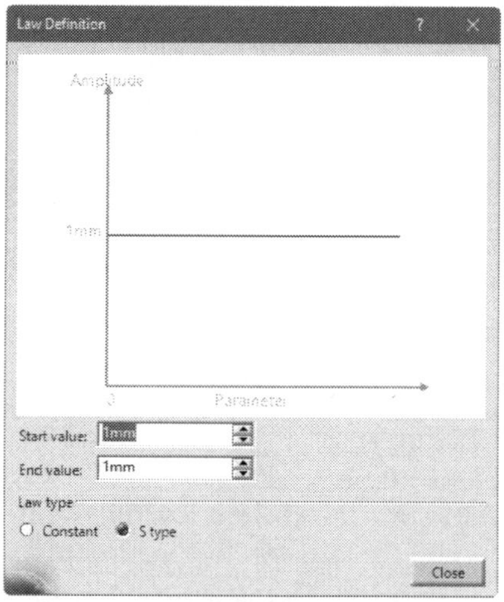

Figure 10-9 *The* ***Law Definition*** *dialog box*

Note that the **Law** button will be activated only when the **Variable Pitch** radio button is selected in the **Type** area.

CREATING SURFACES

The tools provided in the **Wireframe and Surface Design** workbench to create simple and complex surfaces are discussed next.

Creating Extruded Surfaces

Menubar: Insert > Surfaces > Extrude
Toolbar: Surfaces > Extrude

The extruded surfaces are created by extruding a profile and specifying the extrusion depth and direction vector. The basic parameters required for creating an extruded surface are profile, direction for extrusion, and extrusion limits. To create an extruded surface, first you need to draw the profile to be extruded using the **Sketcher** workbench or the tools in the **Wireframe** toolbar. Once you have drawn the profile, choose the **Extrude** tool from the **Surfaces** toolbar; the **Extruded Surface Definition** dialog box will be displayed, as shown in Figure 10-10.

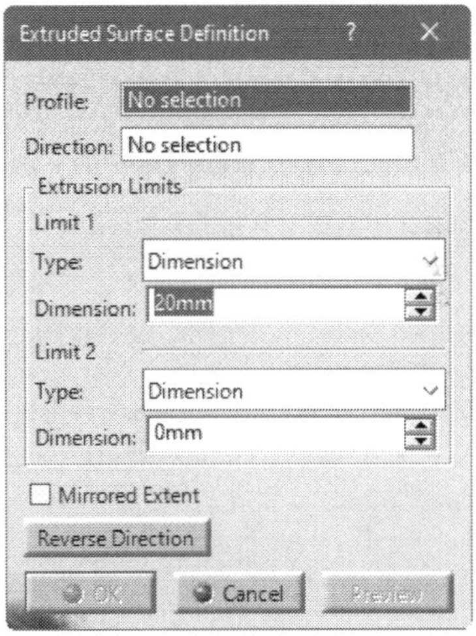

Figure 10-10 *The* ***Extruded Surface Definition*** *dialog box*

If the profile is created in the sketching environment and selected before invoking the **Extrude** tool, a preview of the extruded surface will be displayed in the geometry area. If you select a profile that is created using the tools from the **Wireframe** toolbar or an edge from a solid/existing surface, you will be prompted to specify the direction for extrusion. You can select any plane, planar face, line, edge, or axis to specify the direction for extrusion. By default, the **Dimension** option is selected in the **Type** drop-down list in both the **Limit 1** and **Limit 2** areas of the **Extruded Surface Definition** dialog box. You can specify the extrusion values for **Limit 1** and **Limit 2** in the **Dimension** spinners of the **Extrusion Limits** area. You can also select the **Up-to element** option from the **Type** drop-down list to specify the limiting element for feature termination. Figure 10-11 shows the profile to be extruded and Figure 10-12 shows the resultant extruded surface. You can flip the direction of extrusion of a surface by choosing the **Reverse Direction** button. The **Mirrored Extent** check box helps you extrude the surface on both sides of the sketch by the same distance.

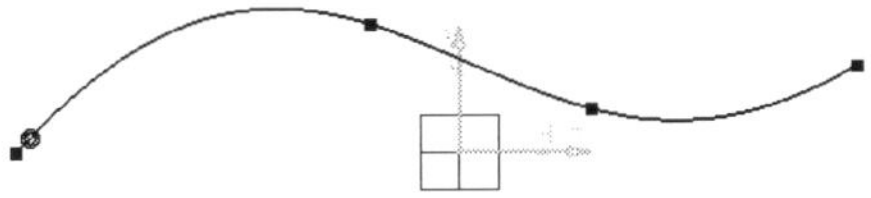

Figure 10-11 *The profile to be extruded*

Figure 10-12 *The resultant extruded surface*

> **Tip**
> *You can also select an edge of an existing surface or a solid body as the profile to create an extruded surface.*

If you select a sketch having multiple entities to create an extruded surface, choose **OK** from the **Extruded Surface Definition** dialog box; the **Multi-Result Management** dialog box will be displayed. You can select the surface bodies that you need to retain using the three options in this dialog box. By default, the **keep only one sub-element using a Near/far** option is selected in this dialog box. As a result, the extruded surface nearest or farthest to the reference element is retained. If you select the **keep only one sub-element using an Extract** option, the **Extract Definition** dialog box will be displayed and you will be prompted to select the elements to be extracted. The **Propagation type** drop-down list provides options to select the elements to be extracted easily. The selected reference elements will be extracted from the surface and retained. On selecting the **keep all the sub-elements** option, each individual surface will be retained.

Creating Revolved Surfaces

Menubar: Insert > Surfaces > Revolve
Toolbar: Surfaces > Revolve

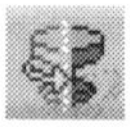

The revolved surfaces are created by revolving the given profile about a revolution axis. To do so, first create the profile and the revolution axis around which the profile will be revolved. Choose the **Revolve** tool from the **Surfaces** toolbar; the **Revolution Surface Definition** dialog box will be displayed, as shown in Figure 10-13. Next, select the profile to be revolved. By default, the axis you sketched using the axis tool in the sketcher workbench is selected as the axis for revolution. You can also select axis of revolution. Now, select the required option from the **Type** drop-down list and set the required angular limits in the **Angle 1** and **Angle 2** spinners. You can also use the graphic handles to define the start and end limits. Figure 10-14 shows a profile and an axis of revolution to create the revolved surface. The resultant surface created by revolving the sketch through an angle of 180 degrees is shown in Figure 10-15.

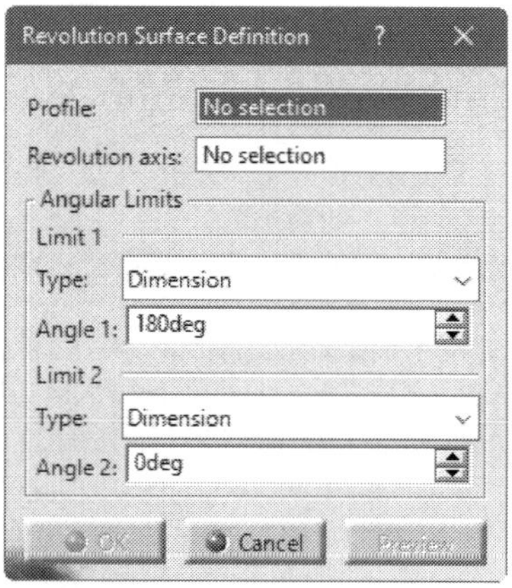

Figure 10-13 *The* ***Revolution Surface Definition*** *dialog box*

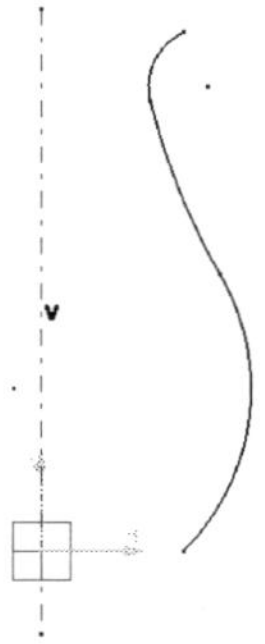

Figure 10-14 *The profile and the axis of revolution*

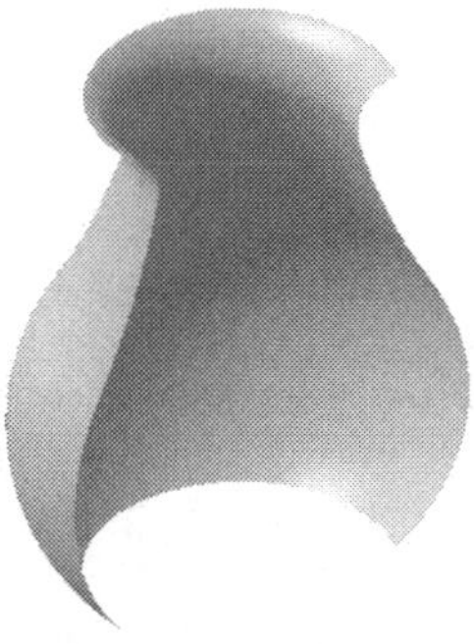

Figure 10-15 *Revolved surface created by revolving the sketch through 180 degrees*

Creating Spherical Surfaces

Menubar: Insert > Surfaces > Sphere
Toolbar: Surfaces > Sphere

The spherical surfaces are created by using the **Sphere** tool from the **Surfaces** toolbar. When you invoke this tool, the **Sphere Surface Definition** dialog box will be displayed, as shown in Figure 10-16. You need to select the center point and an axis system as the sphere axis. You can select an existing point as the center point or create a point by using the options from the contextual menu, which will be displayed on right-clicking in the **Center** selection box. The **Default (Absolute)** axis system is automatically selected. You can also select any previously created axis system. The preview of the spherical surface will be displayed in the geometry area. You can vary the angle values using the options in the **Sphere Limitations** area or by directly dragging the limiting arrows in the geometry area. Figure 10-17 shows the spherical surface created by specifying the angles and by defining the origin as the center. This surface has the default axis system and sphere limitation values.

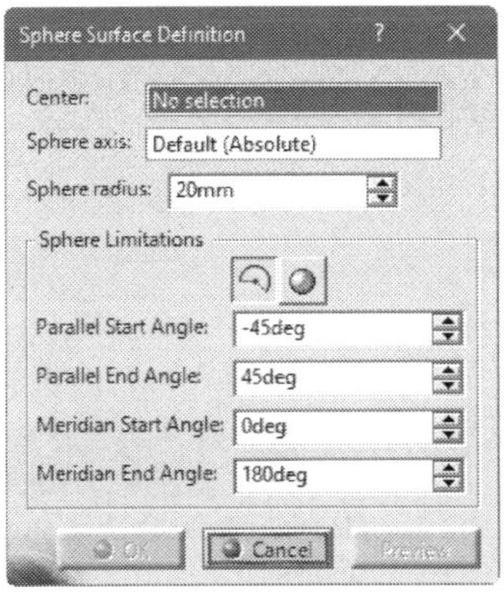

*Figure 10-16 The **Sphere Surface Definition** dialog box*

Figure 10-17 A spherical surface

Tip
*You can create a complete sphere using the **Create the whole sphere** button from the **Sphere Limitations** area of the **Sphere Surface Definition** dialog box.*

Creating Cylindrical Surfaces

Menubar: Insert > Surfaces > Cylinder
Toolbar: Surfaces > Cylinder

The cylindrical surfaces are created by using the **Cylinder** tool. Choose this tool from the **Surfaces** toolbar, the **Cylinder Surface Definition** dialog box will be displayed, as shown in Figure 10-18, and you will be prompted to select the center of the cylinder. You can select an existing point or create a new point by using the options in the contextual menu, which is displayed on right-clicking in the **Point** selection box. After specifying the center of the cylinder, you are prompted to specify the direction for the cylinder. Select a plane, normal to which the cylinder will be extruded. You can also select a direction vector from the contextual menu, which can be invoked by right-clicking in the **Direction** selection box. You can also select an existing line or edge from the geometry area for defining the direction of cylinder creation. Set the parameters using the spinners in the **Parameters** area of the **Cylinder Surface Definition** dialog box. Choose **OK** to create the cylindrical surface.

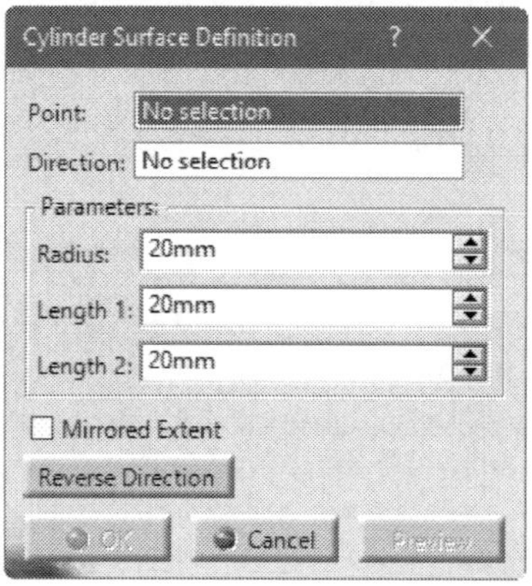

***Figure 10-18** The **Cylinder Surface Definition** dialog box*

Creating Offset Surfaces

Menubar:	Insert > Surfaces > Offset
Toolbar:	Surfaces > Offset

The **Offset** tool is used to create a surface at an offset distance from a reference surface. To do so, choose the **Offset** tool from the **Surfaces** toolbar; the **Offset Surface Definition** dialog box will be displayed, as shown in Figure 10-19, and you will be prompted to select a reference surface.

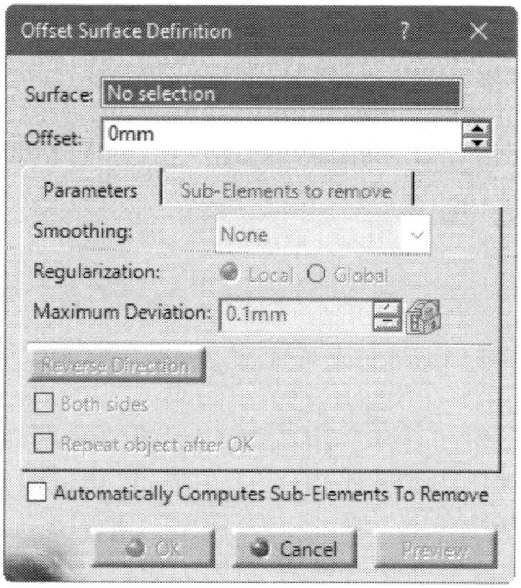

***Figure 10-19** The **Offset Surface Definition** dialog box*

Select the reference surface from the geometry area and specify the offset value in the **Offset** spinner. Choose the **Reverse Direction** button in the dialog box to reverse the offset direction. The **Both sides** check box is used to create a surface on both sides of the selected reference surface. The **Repeat object after OK** check box is used to create multiple offset surfaces. Select the **Repeat object after OK** check box and choose **OK** from the **Offset Surface Definition** dialog box; the **Object Repetition** dialog box will be displayed, as shown in Figure 10-20.

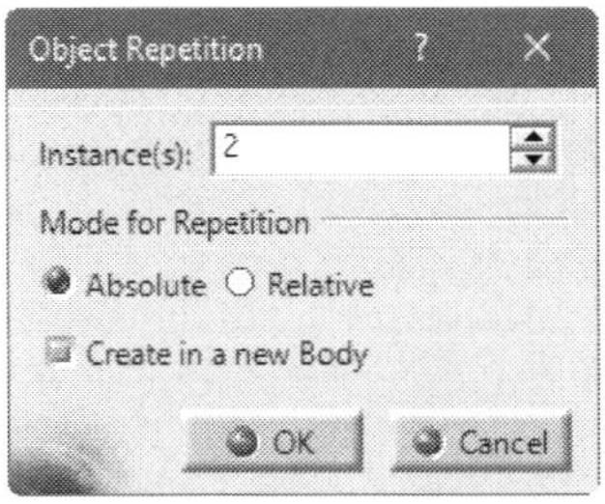

***Figure 10-20** The **Object Repetition** dialog box*

In this dialog box, specify the required number of instance(s). Choose the **OK** button to create the offset surfaces. Figure 10-21 shows a reference surface and an offset surface.

***Figure 10-21** The reference and offset surfaces*

The options for smoothening the complex offset surfaces are provided in the **Smoothing** drop-down list of the **Parameters** tab in the **Offset Surface Definition** dialog box. By default, the **None** option is selected in the **Smoothing** drop-down list. This ensures that a uniform smoothening is applied throughout the offset surface. If you select the **Automatic** option, additional smoothening will be applied to those areas of the surface that cannot be offset because of geometric conditions. Therefore, the offset surface that cannot be created with the **None** option can be completed by using the **Automatic** option. While creating an offset surface by using the **Automatic** option, sometimes the **Warning** message is displayed informing that some faces cannot be offset accurately, so a local smoothening needs to be applied to those faces. Choose the **Close** button from this dialog box to continue.

Note
Sometimes, for complex reference surfaces, the offset surface may not be created. In such cases, you need to reduce the offset value or modify the initial geometry.

Creating Sweep Surfaces

Menubar:	Insert > Surfaces > Sweep
Toolbar:	Surfaces > Sweep

The **Sweep** tool is used to create surfaces by sweeping a profile along a guide curve in the **Wireframe and Surfaces Design** workbench of CATIA V5. To create a sweep surface, first you need to draw two separate sketches, a profile and a guide curve. Next, choose the **Sweep** tool from the **Surfaces** toolbar; the **Swept Surface Definition** dialog box will be displayed, as shown in Figure 10-22, and you will be prompted to select a profile. Select the profile from the geometry area; you will be prompted to select a guide curve. Select the guide curve from the geometry area. Next, choose the **OK** button from the **Swept Surface Definition** dialog box. Figure 10-23 shows a profile and a guide curve. Figure 10-24 shows the resulting swept surface.

> **Tip**
> *Sometimes, the swept surface may not be created as the geometry created forms a cusp. In such a case, you need to reduce the curvature of the guide curves.*

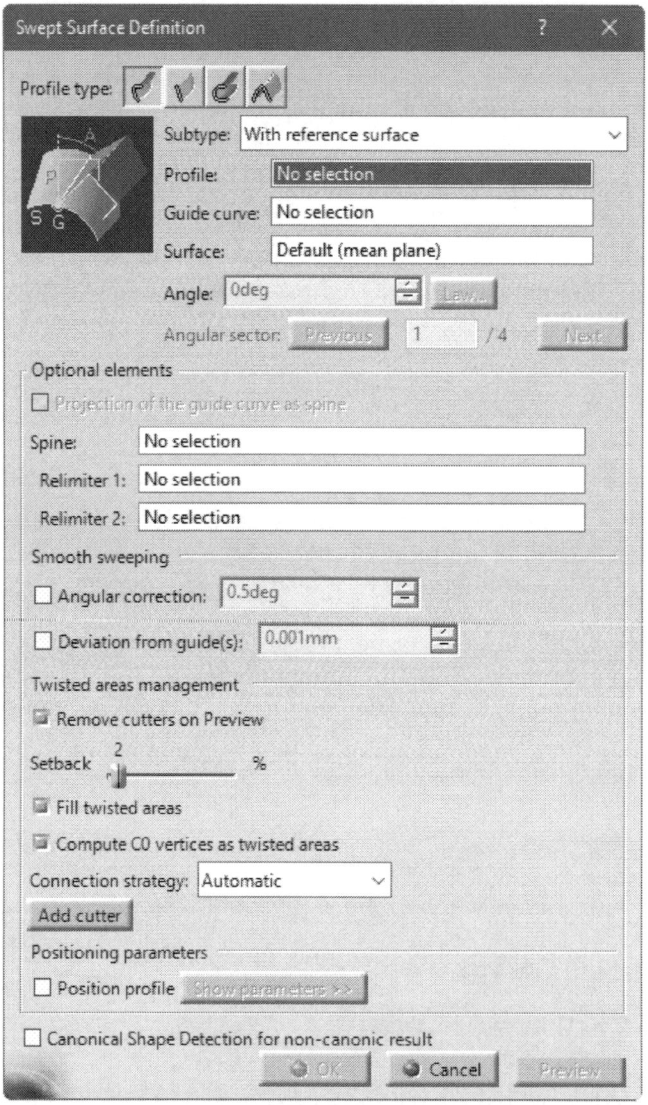

Figure 10-22 *The* ***Swept Surface Definition*** *dialog box*

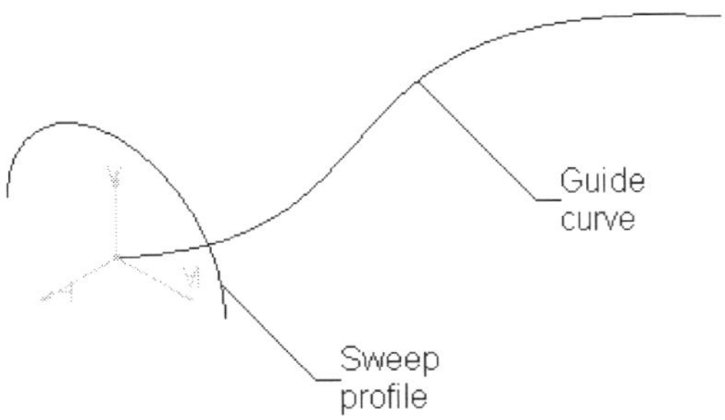

Figure 10-23 The sweep profile and the guide curve

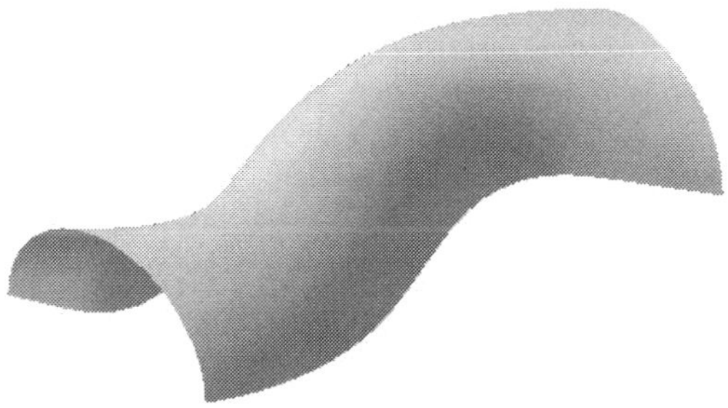

Figure 10-24 The resulting swept surface

Various other methods to create swept surfaces are discussed in the next section.

Creating a Swept Surface with Two Guide Curves

You can also create a swept surface using two guide curves. To do so, draw a profile and two guide curves as separate sketches. Next, invoke the **Swept Surface Definition** dialog box. Now, choose the **Explicit** button from the **Profile type** area and select the **With two guide curves** option from the **Subtype** drop-down list; you will be prompted to select a profile. After you have selected the profile, you will be prompted to select a guide curve. Select the first and second guide curves. Next, select the anchor points for the respective guide curves. The intersection point between the profile and the first guide curve gives the first anchor point. The second anchor point is defined by the intersection point of the plane passing through the first anchor point and normal to the profile at the second guide curve. Choose the **Preview** button from the **Swept Surface Definition** dialog box to preview the surface created. Choose the **OK** button from the **Swept Surface Definition** dialog box. Figure 10-25 shows a profile and guide curves. The preview of the swept surface created using two guide curves is shown in Figure 10-26.

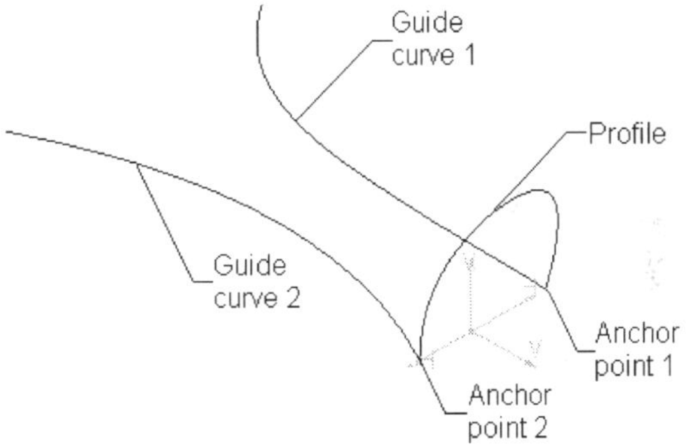

Figure 10-25 The sweep profile and the guide curves

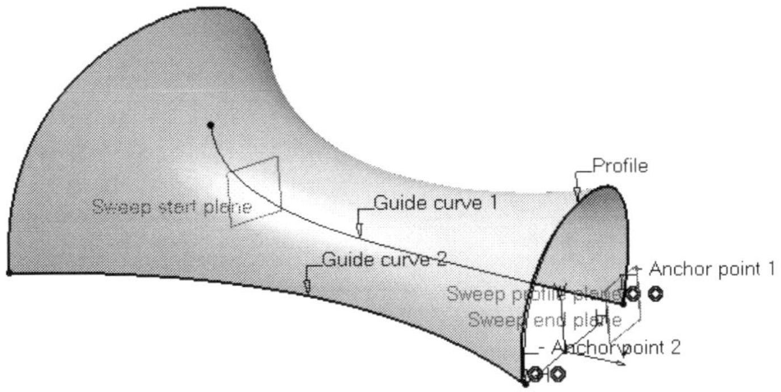

Figure 10-26 The preview of the resulting swept surface

Creating a Swept Surface with Pulling Direction

You can also create a swept surface by defining a pulling direction. To do so, create a profile and a guide curve as separate sketches. Next, invoke the **Swept Surface Definition** dialog box. Now, choose the **Explicit** button from the **Profile type** area and select the **With pulling direction** option from the **Subtype** drop-down list; you will be prompted to select a profile. Select the profile from the graphics area. On doing so, you will be prompted to select the guide curve. On selecting the guide curve, you will be prompted to select a pulling direction. Select the required pulling direction. Now, set the angle of rotation in the **Angle** spinner and set the angle sector in the **Angular sector** edit box and choose the **OK** button from the dialog box.

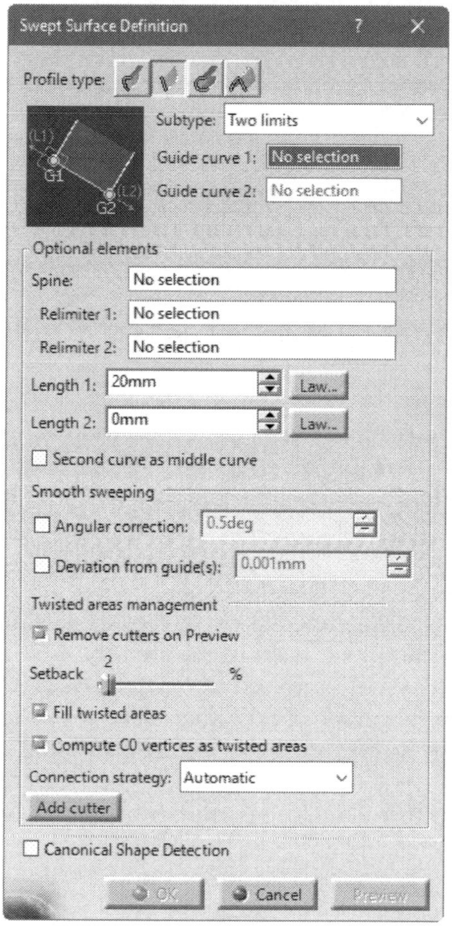

*Figure 10-27 The **Swept Surface Definition** dialog box*

Creating a Swept Surface with Two Limits

In CATIA V5, you can create a swept surface by defining two limit curves. The limit curves can be in the same or different planes. To create a swept surface with two limits, you need to draw two limit curves. Next, invoke the **Swept Surface Definition** dialog box. Choose the **Line** button from the **Profile type** area in the dialog box; you will be prompted to select the first guide curve. On selecting the guide curve, you will be prompted to select the second guide curve. Select the second guide curve and choose the **Preview** button to display the swept surface created between the limiting curves. The parameters in the **Swept Surface Definition** dialog box after selecting the **Line** button are shown in Figure 10-27. Note that in the **Optional elements** area, guide curve 1 is selected by default in the **Spine** display box. You can also select another curve to be defined as the spine. Choose the **OK** button to create the swept surface. Figure 10-28 shows the guide curves to be selected and Figure 10-29 shows the resulting swept surface.

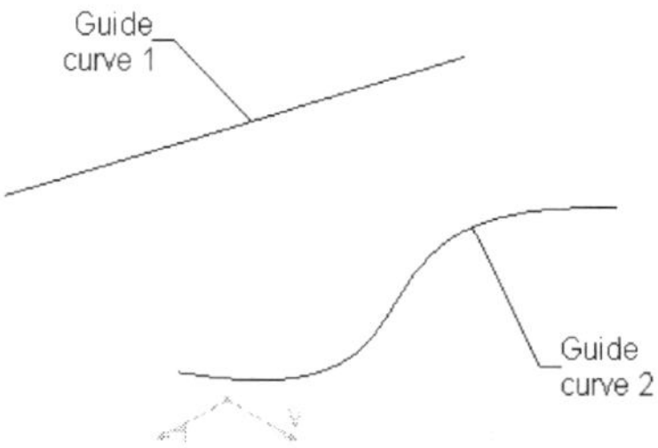

***Figure 10-28** The guide curves to be selected*

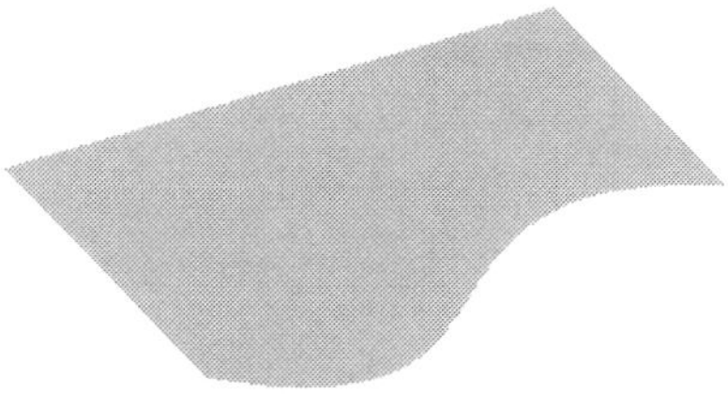

***Figure 10-29** The resulting swept surface*

Creating a Swept Surface with Limit and Middle curves

You can also create a swept surface by defining a limit and a middle curves. To create a swept surface with limit and middle curves, you need to create two curves as separate sketches. Next, invoke the **Swept Surface Definition** dialog box. Choose the **Line** button from the **Profile type** area and select the **Limit and middle** option from the **Subtype** drop-down list; you will be prompted to select the first guide curve. On selecting the first guide curve, you will be prompted to select the second guide curve which will act as the middle guide curve. Select the second guide curve and choose the **Preview** button to display the resultant swept surface. Choose the **OK** button from this dialog box to create the swept surface.

Creating a Swept Surface with Reference Surface

You can create a swept surface using a reference surface. To create a swept surface with a guide curve and a reference surface, you need to create a guide curve. To do so, invoke the **Swept Surface Definition** dialog box. Choose the **Line** button from the **Profile type** area and select the **With reference surface** option from the **Subtype** drop-down list; you will be prompted to select a guide curve. On selecting the guide curve, you will be prompted to select the reference surface. Select the reference surface. Next, set value of the extrusion length in the **Length 1** and **Length 2** spinners and choose the **Preview** button to display the resultant swept surface. Choose the **OK** button from this dialog box to create the swept surface.

Creating a Swept Surface with Reference Curve

You can create a swept surface using a reference curve. To do so, you need to create two guide curves as separate sketches. Next, invoke the **Swept Surface Definition** dialog box. Choose the **Line** button from the **Profile type** area and select the **With reference curve** option from the **Subtype** drop-down list; you will be prompted to select a guide curve. On selecting the guide curve, you will be prompted to select the reference curve. Select the required reference curve. Next, set the value of length in the **Length 1** and **Length 2** spinners and choose the **Preview** button to display the swept surface. Choose the **OK** button to create the swept surface.

Creating a Swept Surface with Tangency Surface

You can create a swept surface using a guide curve and a tangency surface. To do so, you need to create a surface and a guide curve separately. Next, invoke the **Swept Surface Definition** dialog box. Choose the **Line** button from the **Profile type** area and select the **With tangency surface** option from the **Subtype** drop-down list; you will be prompted to select a guide curve. On selecting the guide curve, you will be prompted to select the tangency surface, refer to Figure 10-30. Select the surface and choose the **Preview** button to display the swept surface, as shown in Figure 10-30. Choose the **OK** button to create the swept surface.

Creating a Swept Surface with Two Tangency Surface

You can create a swept surface using two tangency surfaces. To do so, you need to create two surfaces separately. Next, invoke the **Swept Surface Definition** dialog box. Choose the **Line** button from the **Profile type** area and select the **With two tangency surfaces** option from the **Subtype** drop-down list; you will be prompted to select a guide curve as spine. On selecting the spine, you will be prompted to select the first tangency surface, on selecting the first tangency surface you will be prompted to select the second tangency surface. Select the surface and choose the **Preview** button to display the swept surface, as shown in Figure 10-31. In case there are more than one solution, select the required surface from the graphics window and choose the **OK** button to create the swept surface.

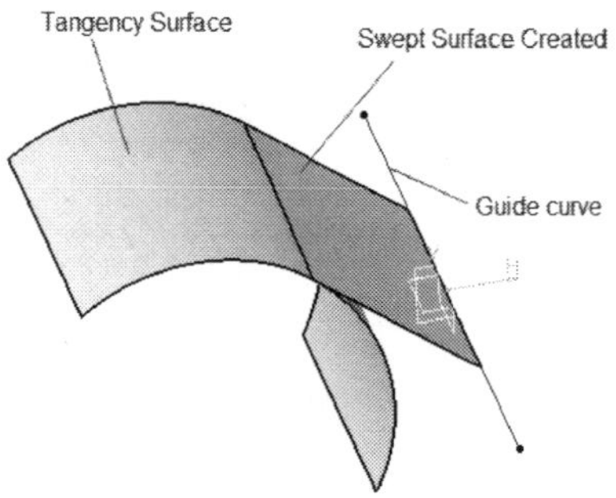

Figure 10-30 The tangency surface and the guide curves to be selected

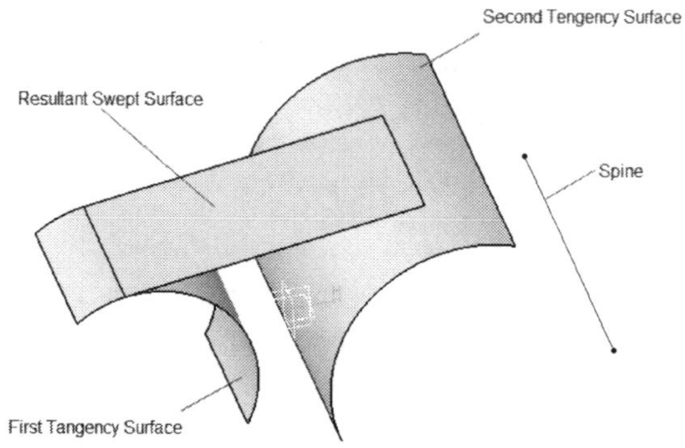

Figure 10-31 The two tangency surface and the spine to be selected

Creating a Swept Surface with Draft Direction

You can create a swept surface using a draft direction. To do so, create a guide curve. Next, invoke the **Swept Surface Definition** dialog box. Choose the **Line** button from the **Profile type** area and select the **With draft direction** option from the **Subtype** drop-down list; you will be prompted to select a guide curve. On selecting the guide curve, you will be prompted to select the draft direction. Select the required draft direction from the **Draft Direction** selection box or from the graphics window. Now, select the **Square** or the **Cone** radio button from the **Draft computation mode** area. Next, set the angle in the **Angle** spinner of this dialog box and select the draft length type from the **Length type 1** and **Length type 2** selection areas, set the value of extrusion of the draft length in the **Length 1** and **Length 2** spinners. Next, choose the **OK** button to create the swept surface. Figure 10-32 shows a draft direction and guide curve. Figure 10-33 shows the resulting swept surface.

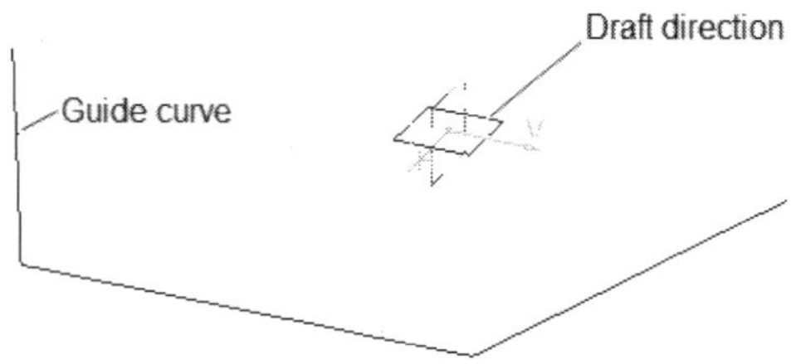

Figure 10-32 The sweep profile guide curve

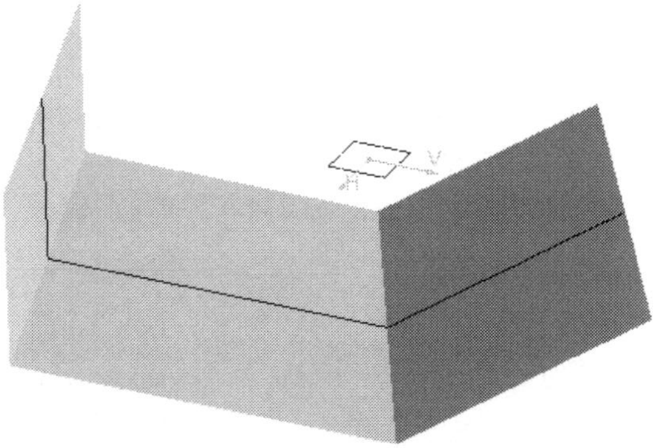

Figure 10-33 The resultant draft direction swept surface

Creating a Circular Swept Surface with Three Curves

You can also create a swept surface by using three guide curves. This helps you to create a circular swept surface. To create this type of surface, first you need to draw three guide curves as separate sketches. After drawing the curves, invoke the **Swept Surface Definition** dialog box and then choose the **Circle** button from the **Profile type** area; the parameters in the **Swept Surface Definition** dialog box will change, as shown in Figure 10-34, and you will be prompted to select the first guide curve that will define the first extremity of the circular arc. Also, make sure the **Three guides** option is selected in the **Subtype** drop-down list. Select the first guide curve; you will be prompted to select the second guide curve. Select the second guide curve; you will be prompted to select the guide curve that defines the second extremity of the circular arc. Select the guide curve and choose the **OK** button from the dialog box; the circular swept surface will be created. Figure 10-35 shows the guide curves and Figure 10-36 shows the resulting swept surface. If a message box informing that one guide curve has been relimited is displayed, choose the **Close** button from the message box to exit.

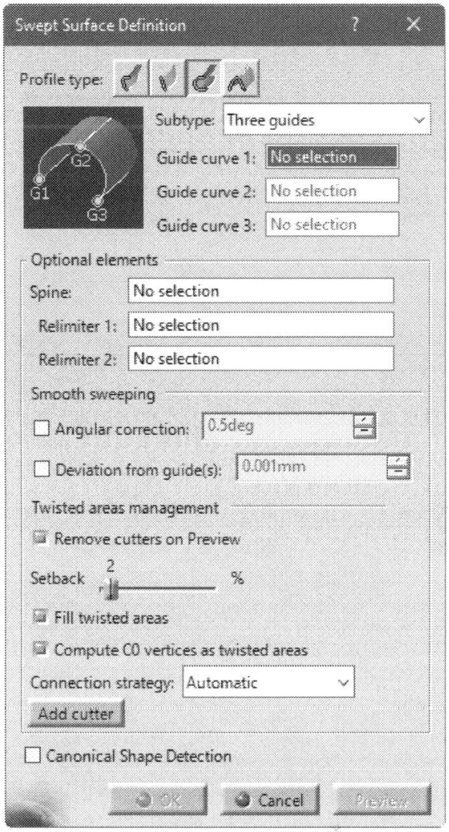

Figure 10-34 *The* ***Swept Surface Definition*** *dialog box*

Note

The options in the ***Twisted areas management*** *area are used to control the generation of any twisted areas in the swept surface.*

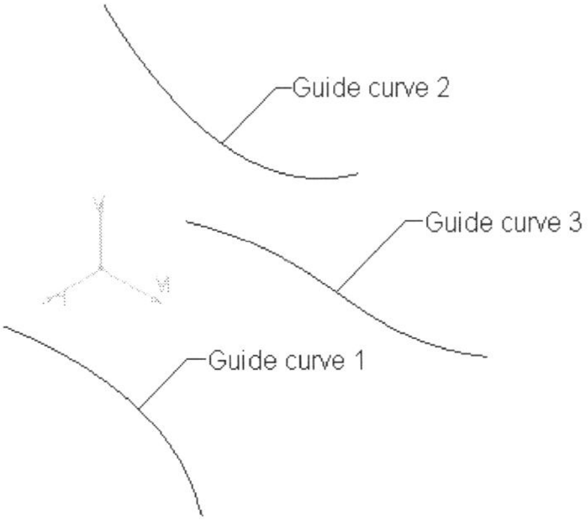

Figure 10-35 *The guide curves*

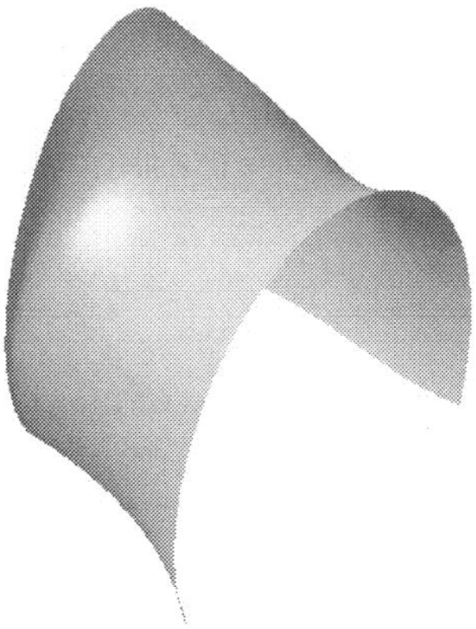

Figure 10-36 *The resulting swept surface*

Creating a Swept Surface with Two Guides and Radius

You can also create a swept surface by using two guide curves and a radius. This helps you to create a circular swept surface of specified radius and two selected guides lying on it. To create this type of surface, first you need to create two guide curves as separate sketches. After creating the curves, invoke the **Swept Surface Definition** dialog box and then choose the **Circle** button from the **Profile type** area and select the **Two guides and radius** option from the **Subtype** drop-down list; you will be prompted to select the first guide curve that will define the first extremity of the circular arc. Select the first guide curve and you will be prompted to select the second guide curve. Select the second guide curve and specify the value of the radius in the **Radius** spinner and choose the **OK** button from the dialog box.

Creating a Swept Surface with Center curve and Two Angles

You can also create a swept surface by using two angles and a center curve. To create this type of surface, first you need to create a center curve and a reference curve. After drawing the curves, invoke the **Swept Surface Definition** dialog box, and then choose the **Circle** button from the **Profile type** area. Select the **Center and two angles** option from the **Subtype** drop-down list; you will be prompted to select the center curve that will define the center of the circular curve. Select the center curve; you will be prompted to select the reference curve. Next, select the reference curve and specify the value of the rotation angles in the **Angle 1** and **Angle 2** spinners. Next, choose the **OK** button. Figure 10-37 shows a center curve and reference curve. Figure 10-38 shows the resultant swept surface. If a message box informing that one guide curve has been relimited is displayed, choose the **Close** button from the message box to exit.

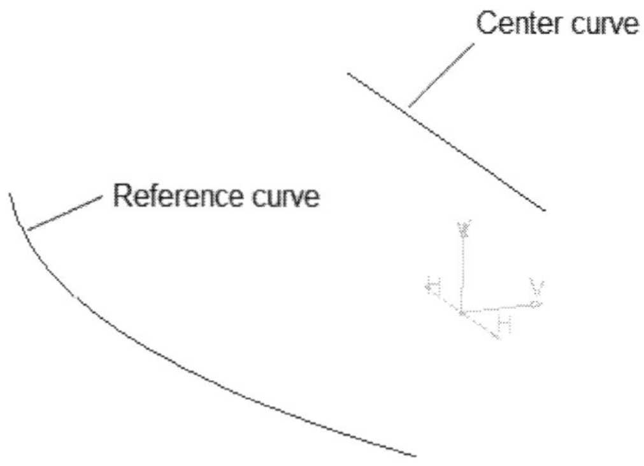

Figure 10-37 *Center and reference curves*

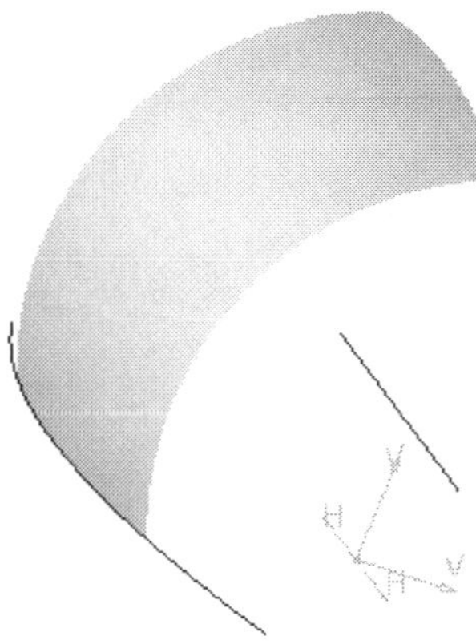

Figure 10-38 *The resultant swept surface*

Creating a Swept Surface with Center curve and Radius

You can also create a swept surface by using radius and a center curve. To create this type of surface, first you need to create a center curve. After creating the curve, invoke the **Swept Surface Definition** dialog box, and then choose the **Circle** button from the **Profile type** area and select the **Center and radius** option from the **Subtype** drop-down list; you will be prompted to select the center curve that will define the center of the circular curve. Select the center curve and specify the value of the radius in the **Radius** spinner. Choose the **Preview** button to display the swept surface. Choose the **OK** button from the dialog box to create the swept surface.

Creating a Swept Surface with Two Guides and Tangency Surface

You can also create a swept surface by using two limit guide curves and a tangency surface. To create a swept surface with two guides and a tangency curve, first you need to create a tangency surface and a limit curve tangent to the tangency surface. Next, you need to create another limit curve for the circular surface to be created. After creating the curve, invoke the **Swept Surface Definition** dialog box, and then choose the **Circle** button from the **Profile type** area and select the **Two guides and tangency surface** option from the **Subtype** drop-down list; you will be prompted to select the first guide curve. Select the first limit curve which is having tangency with the tangency surface and you will be prompted to select the tangency surface. Select the tangency surface; you will be prompted to select the limit curve. Select the limit curve to set the limit of the circular surface that is to be created. Next, choose the **OK** button to create the swept surface. Figure 10-39 shows a tangency surface and limit curves. Figure 10-40 shows the resultant swept surface. In case there are more than one solution, select the required surface from the graphics window and choose the **OK** button from the dialog box.

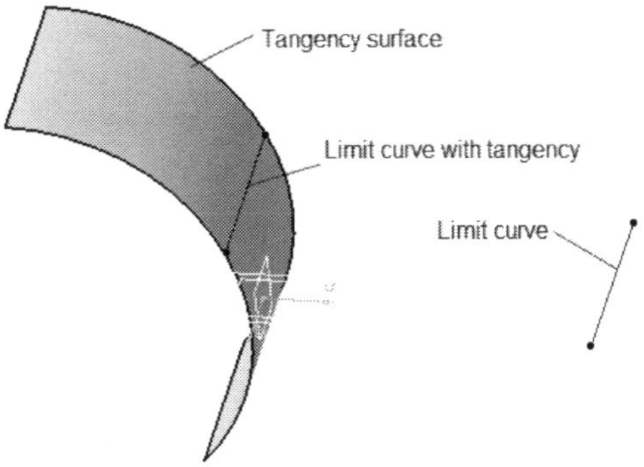

Figure 10-39 *The tangency surface and limit curves*

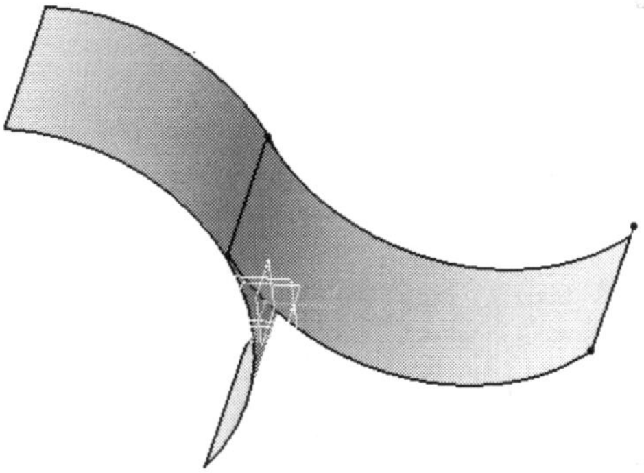

Figure 10-40 *The resultant swept surface*

Creating a Swept Surface with One Guide and Tangency Surface

You can also create a swept surface by using a guide curve and a tangency surface. To create a swept surface with a guide curve and a tangency curve, first you need to create a tangency surface and a guide curve. After creating the curve and the surface, invoke the **Swept Surface Definition** dialog box, and then choose the **Circle** button from the **Profile type** area. Select the **One guide and tangency surface** option from the **Subtype** drop-down list and you will be prompted to select the first guide curve. Select the guide curve and you will be prompted to select the tangency surface. Next, select the tangency surface and specify the value of radius of the circular surface to be created in the **Radius** spinner. Note that the value of the radius entered must be greater than the distance between the tangency surface and the guide curve. Choose the **Preview** button to display the swept surface. In case there are more than one solution, select the required surface from the graphics window and choose the **OK** button from the dialog box.

Creating a Swept Surface with Limit Curve and Tangency Surface

You can also create a swept surface by using a tangency surface and a guide curve as starting limit of the circular swept surface to be created. To create a swept surface with a limit curve and a tangency surface, first you need to create a tangency surface and a limit curve. Note that the limit curve must be tangent to the tangency surface. After creating the curve and the surface, invoke the **Swept Surface Definition** dialog box and then choose the **Circle** button from the **Profile type** area. Select the **Limit curve and tangency surface** option from the **Subtype** drop-down list and you will be prompted to select the limit curve. Select the limit curve and you will be prompted to select the tangency surface. Next, select the tangency surface and specify the value of radius of the circular surface to be created in the **Radius** spinner. Next, choose the **OK** button to create the swept surface. Figure 10-41 shows a tangency surface and limit curve. Figure 10-42 shows the resultant swept surface. In case there are more than one solution, select the required surface from the graphics window and choose the **OK** button from the dialog box.

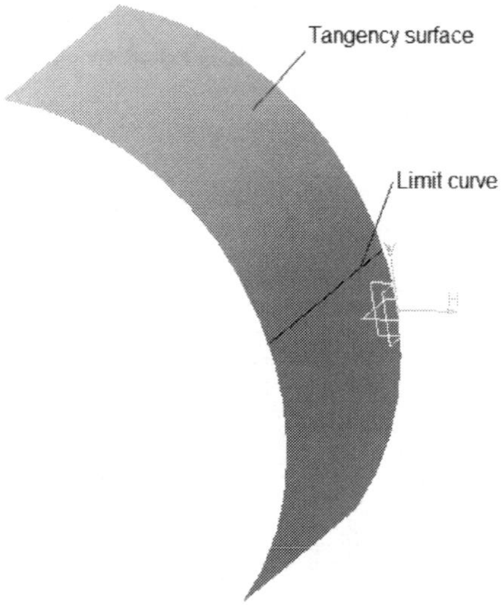

***Figure 10-41** The limit curve and tangency surface*

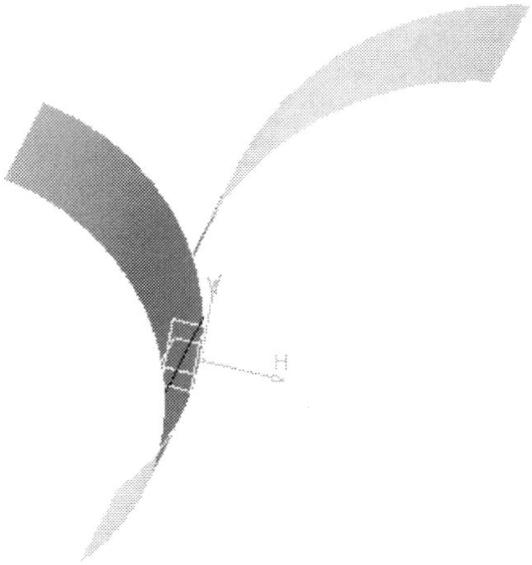

***Figure 10-42** The resultant swept surface*

Creating a Conical Swept Surface with Two Guide Curves

To create a conical swept surface by using two guide curves, first you need to draw two guide curves as separate sketches. Next, you need to create a curve to specify tangency element. Now, invoke the **Swept Surface Definition** dialog box, and then choose the **Conic** button from the **Profile type** area; the options in the **Swept Surface Definition** dialog box will change, as shown in Figure 10-43, and you will be prompted to select the guide curve1. Make sure the **Two guide curves** option is selected in the **Subtype** drop-down list. Select the guide curve1; you will be prompted to define the tangency. Select the curve to define the tangency; you will be prompted to define the guide curve2. Select the guide curve2; you will be again prompted to define tangency. Select the middle curve to define the tangency and choose the **OK** button from the dialog box; the conic swept surface will be created. Figure 10-44 shows the guide curves and Figure 10-45 shows the resulting swept surface.

You can define the conic parameters by using the **Angle** and **Parameter** spinners.

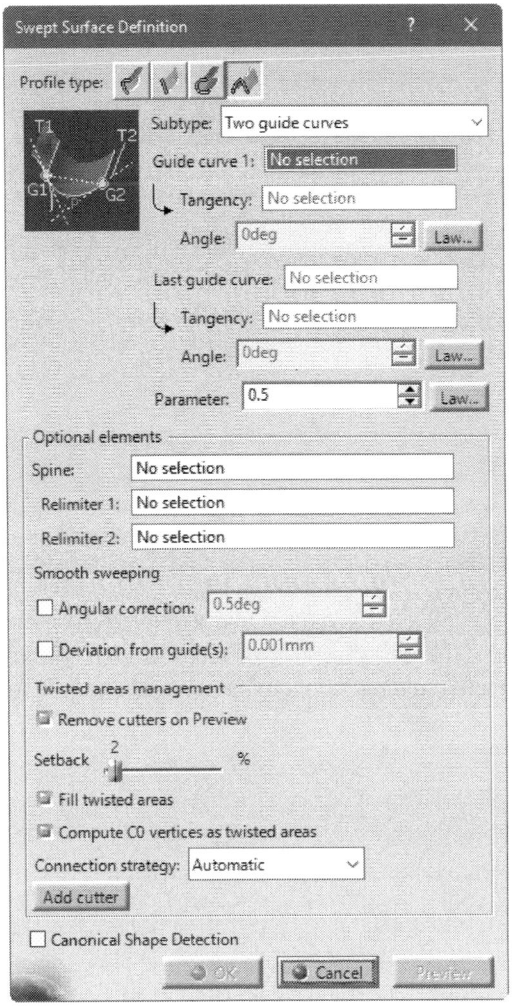

Figure 10-43 *The* ***Swept Surface Definition*** *dialog box*

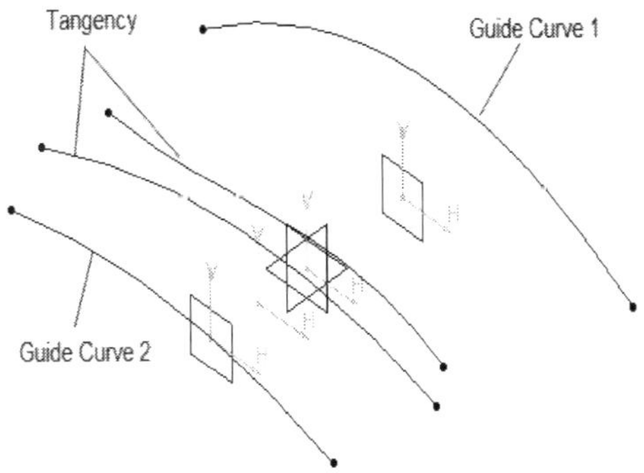

Figure 10-44 *The guide curves*

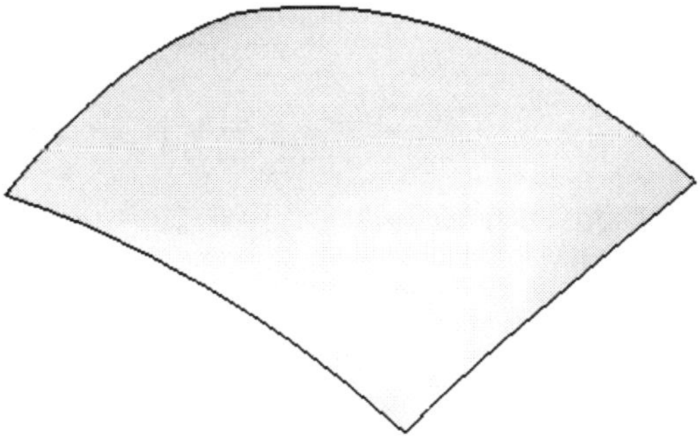

Figure 10-45 *The resulting swept surface*

Creating a Conical Swept Surface with Three Guide Curves

To create a conical swept surface by using three guide curves, first you need to create three guide curves as separate sketches. Next, you need to create a surface or a curve to specify tangency element. Now, invoke the **Swept Surface Definition** dialog box, and then choose the **Conic** button from the **Profile type** area and select the **Three guide curves** option from the **Subtype** drop-down list; you will be prompted to select the first guide curve. Select the first guide curve; you will be prompted to define the tangency. Select a tangent surface or plane to define the tangency; you will be prompted to define the second guide curve. Select the second guide curve, you will be prompted to define the third guide curve. Select the third guide curve; you will be again prompted to define the tangency. Select the tangency surface or a plane to define the tangency of the third guide curve. Next, choose the **OK** button to create the swept surface. Figure 10-46 shows a tangency surface and guide curves. Figure 10-47 shows the resultant swept surface.

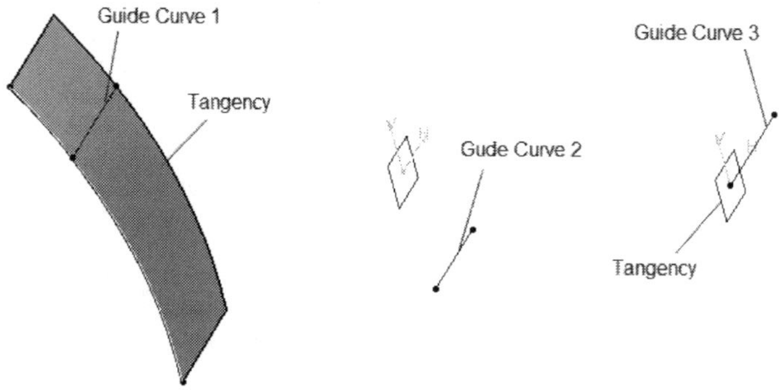

Figure 10-46 *The guide curves and tangency surface*

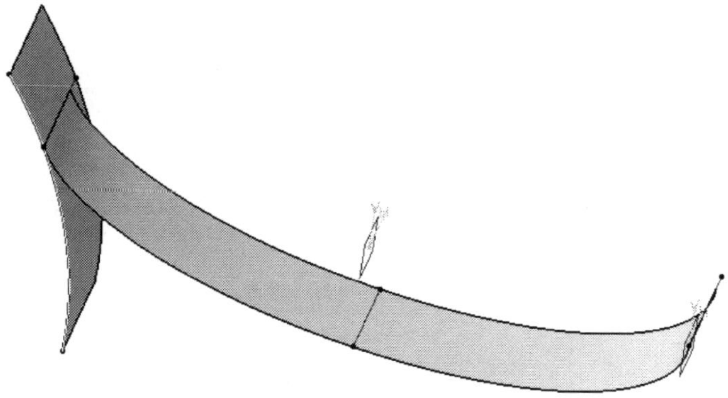

Figure 10-47 *The resultant swept surface*

Creating a Conical Swept Surface with Four Guide Curves

To create a conical swept surface by using four guide curves, first you need to create four guide curves as separate sketches. Next, you need to create a surface or a curve to specify tangency element. Now, invoke the **Swept Surface Definition** dialog box, and then choose the **Conic** button from the **Profile type** area and select the **Four guide curves** option from the **Subtype** drop-down list; you will be prompted to select the first guide curve. Select the first guide curve; you will be prompted to define the tangency. Select the tangency surface or a plane to define the tangency; you will be prompted to define the second guide curve. Select the second guide curve; you will be prompted to define the third guide curve. Select the third guide curve; you will be prompted to select the last guide curve. Select the guide curve and choose the **OK** button to create the swept surface. Figure 10-48 shows a tangency surface and guide curves. Figure 10-49 shows the resultant swept surface.

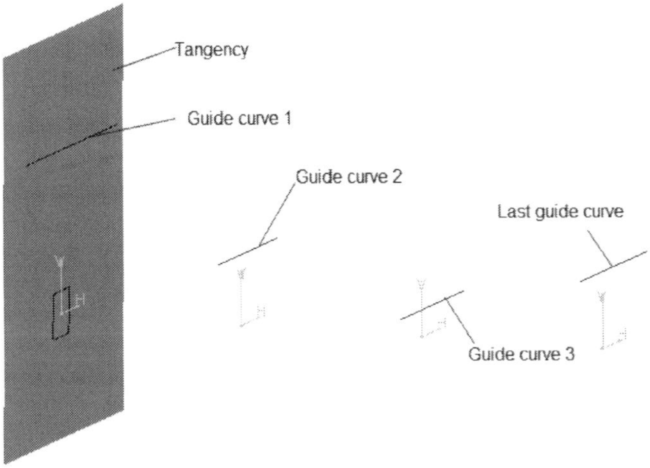

Figure 10-48 *The guides and tangency surface to be selected*

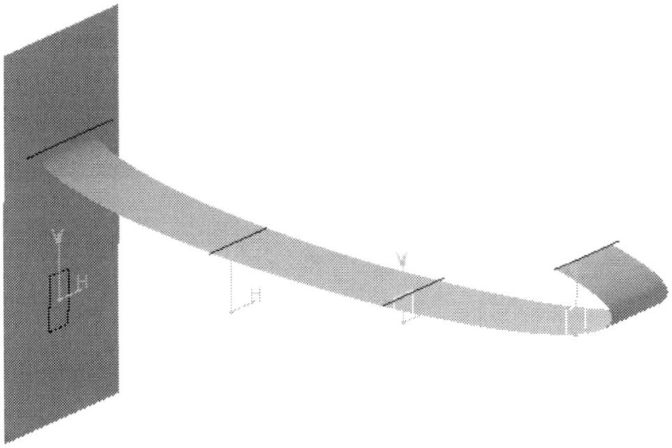

Figure 10-49 *The resultant swept surface*

Creating a Conical Swept Surface with Five Guide Curves

To create a conical swept surface by using five guide curves, first you need to draw five guide curves as separate sketches. Now, invoke the **Swept Surface Definition** dialog box, and then choose the **Conic** button from the **Profile type** area and select the **Five guide curves** option from the **Subtype** drop-down list; you will be prompted to select the first guide curve. Select the first guide curve; you will be prompted to define the second guide curve. Select the second guide curve, you will be prompted to define the third guide curve. Select the third guide curve; you will be prompted to define the fourth guide curve. Select the fourth guide curve, you will be prompted to select the last guide curve. Select the last guide curve and choose the **OK** button to create the swept surface. Figure 10-50 shows guide curves. Figure 10-51 shows the resultant swept surface.

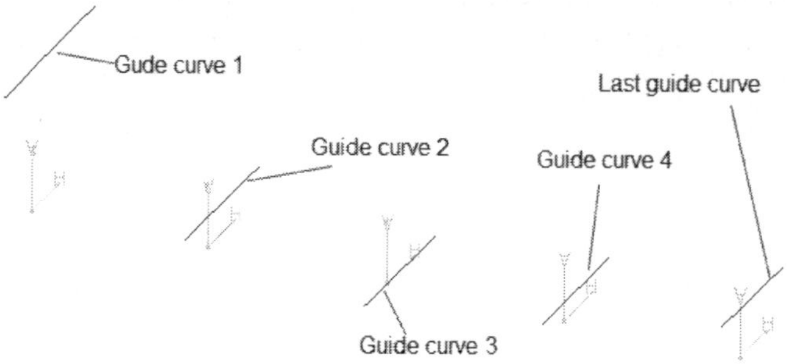

Figure 10-50 *The five guide curves to be selected*

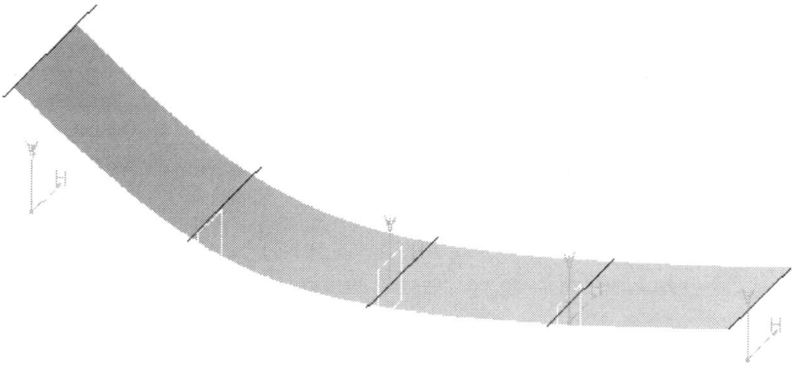

Figure 10-51 *The resultant swept surface*

Creating Fill Surfaces

Menubar:	Insert > Surfaces > Fill
Toolbar:	Surfaces > Fill

The **Fill** tool is used to create fill surfaces between a number of boundary segments. The boundary segments may be planar or non-planar, but there should not be a gap greater than 0.1 mm between the consecutive boundary segments. To create a fill surface, you will need boundary curves. Choose the **Fill** tool from the **Surfaces** toolbar; the **Fill Surface Definition** dialog box will be displayed, as shown in Figure 10-52. Also, you will be prompted to select the first curve. Select curves or surface edges in a chain to form the outer closed boundary in the **Outer Boundaries** tab. A flag note will be displayed in the geometry area showing whether the boundary is a closed contour or not. If the selected curves or surface edges do not form a closed boundary a flag note will display the gap between the open ends of the boundary. Once you have selected the boundary curves, choose the **Preview** button to preview the fill surface. Figure 10-53 shows the boundary curves selected to create the fill surface and Figure 10-54 shows the preview of the fill surface. You can select support surfaces with the corresponding curve to ensure continuity between the fill and support surfaces.

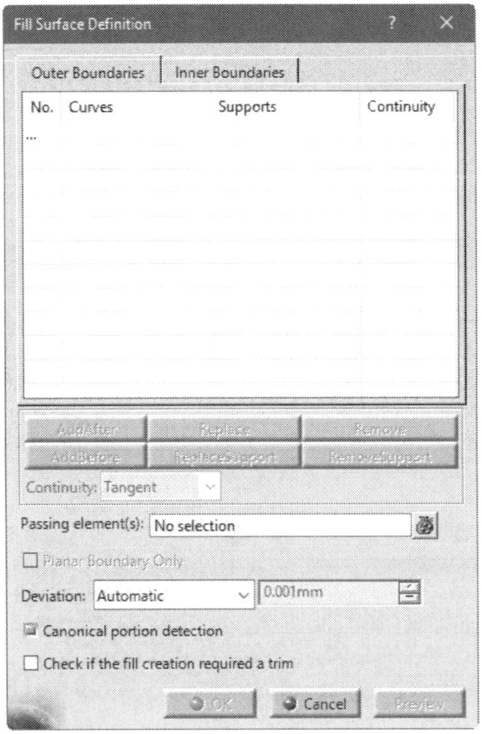

***Figure 10-52** The **Fill Surface Definition** dialog box*

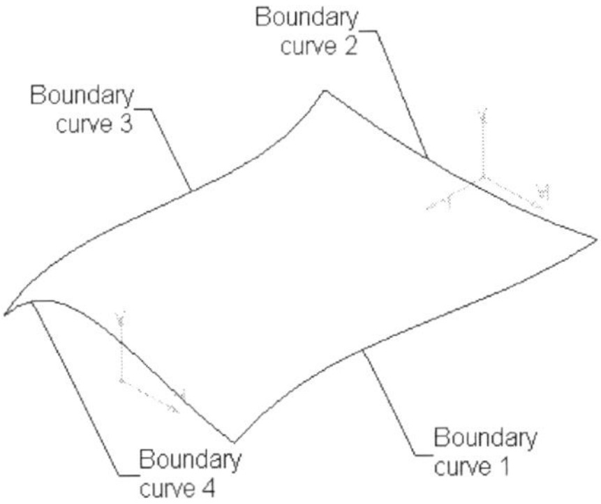

***Figure 10-53** The boundary curves*

***Figure 10-54** The resulting fill surface*

To create a fill surface between nested boundaries, first select curves or surface edges that form an outer closed boundary in the **Outer Boundaries** tab and then select the curves or surface edges that form an inner closed boundary in the **Inner Boundaries** tab. Next, choose the **OK** button to create a fill surface between outer and inner boundaries, refer to Figures 10-55 and 10-56. If required, choose the **Add New Boundary** button to add a new boundary or **Remove Current Boundary** to remove the selected boundary in the **Inner Boundaries** tab. Choose the **Previous** or **Next** button, if you want to navigate between all inner curves and their supports. If the boundary selected is the last one, the **Next** button will be disabled. Similarly, when the boundary selected is the first one, the **Previous** button will be disabled. You can also navigate between boundaries by using the **Inner Boundary No:** spinner.

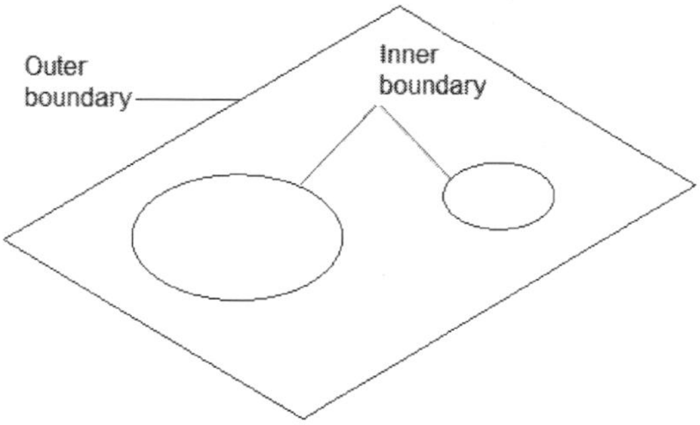

Figure 10-55 The inner and outer boundary curves

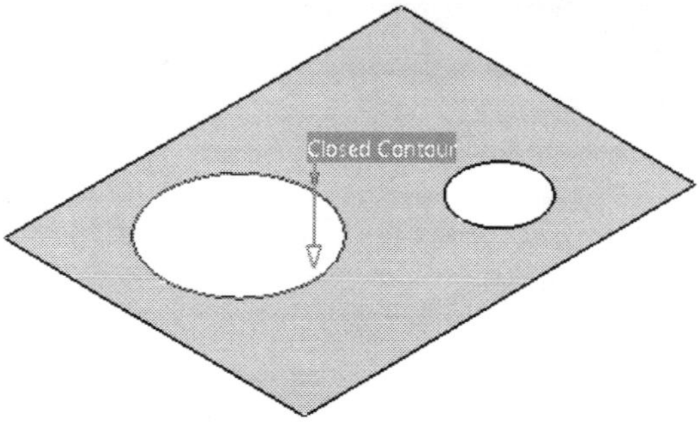

Figure 10-56 The resulting fill surface

You can select the type of continuity for support surface by selecting the options from the **Outer Boundaries** and **Inner Boundaries** tabs. Choose the type of continuity for the selected from the **Continuity** drop-down list. You can also create a fill surface that passes through the selected points and curves. To do so, select the points and curves in the **Passing element(s)** selection box and then choose **Preview** to preview the created fill surface. You can select **Planar Boundary Only** to fill only planar boundaries, when the boundary is defined by one curve on one surface.

By default, the **Canonical portion detection** check box is selected in the **Fill Surface Definition** dialog box. As a result, canonical portion detection automatically computes the information of the planer, cylindrical and conical surfaces, if they exist in the fill surface. When the **Check if Fill creation required a trim** check box is selected while creating the intersected filled surface, refer to Figure 10-57; a warning message box will be displayed, as shown in Figure 10-58.

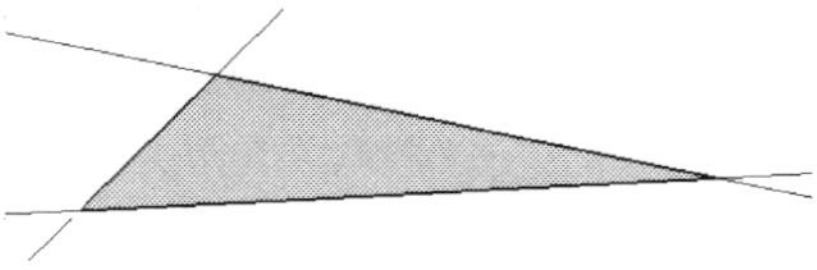

Figure 10-57 The intersected filled surface

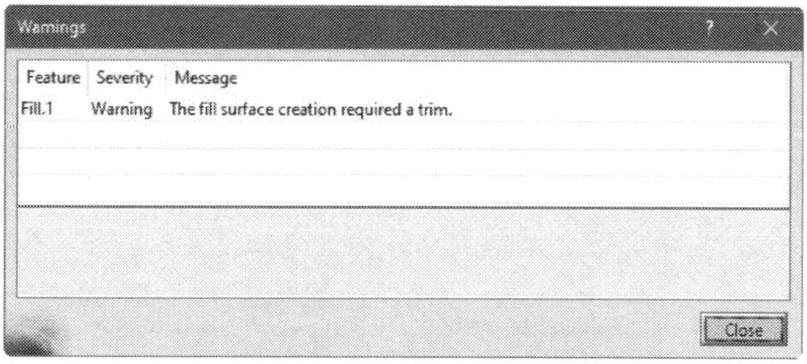

Figure 10-58 The warning message box

Note that in this dialog box, you can define deviation with the help of the **Deviation** drop-down list. Select the **None** option if you do not want to define the deviation. Select the **Automatic** option to define deviation automatically with a default value of 0.001 mm. Select the **Manual** option to define the deviation value manually. When this option is selected, the spinner next to it gets activated and you can specify the value of deviation.

Creating Multi-Sections Surfaces

Menubar: Insert > Surfaces > Multi-Sections Surface
Toolbar: Surfaces > Multi-Sections Surface

The **Multi-Sections Surface** tool is used to create lofted multi-section surfaces. The surface is created between the sections along the computed or user-defined spine. You can create a multi-section surface between two or more than two sections. To create a multi-section surface, first you need to create sections and guide curves as separate sketches. Next, choose the **Multi-Sections Surface** tool from the **Surfaces** toolbar; the **Multi-Sections Surface Definition** dialog box will be displayed, as shown in Figure 10-59, and you will be prompted to select a curve. Select the first section curve; you will be prompted to select a new curve or a tangent surface. Select the second section curve from the geometry area. Next, click on the three dots in the display box of the **Guides** tab; you will be prompted to select a curve. One by one, select the guide curves that were drawn earlier. Choose the **OK** button to exit the **Multi-Sections Surface Definition** dialog box and create the multi-section surface. Figure 10-60 shows sections and guide curves to create the multi-section surface and Figure 10-61 shows preview of the resulting surface.

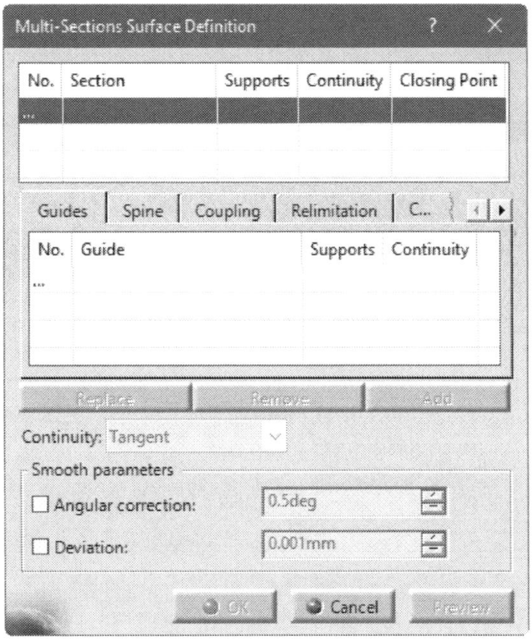

*Figure 10-59 The **Multi-Sections Surface Definition** dialog box*

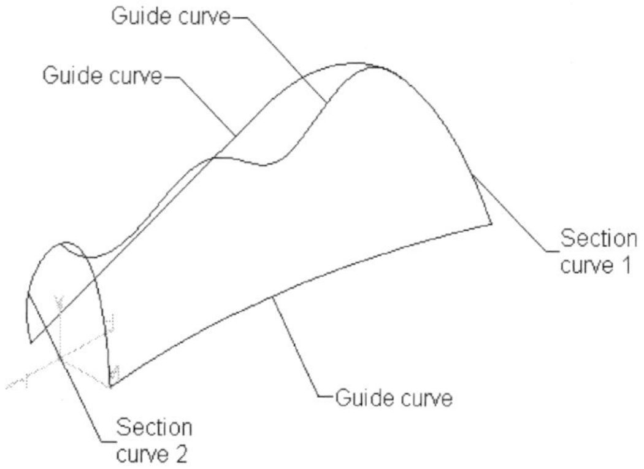

Figure 10-60 Sections and guide curves

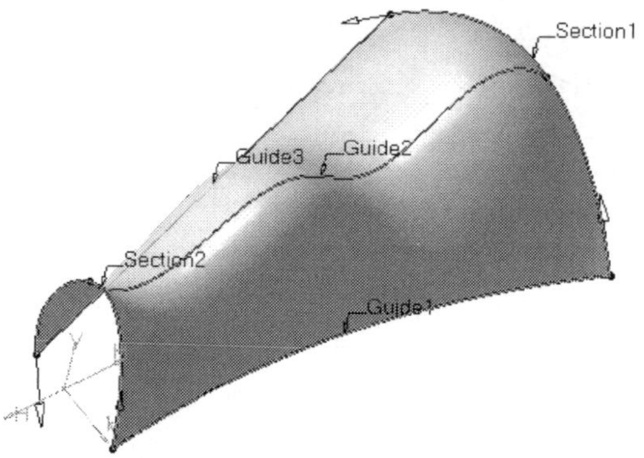

Figure 10-61 Preview of the resulting multi-sections surface

Note

While selecting the section curve, make sure the direction arrow associated with each section curve points in the same direction. Else, the surface will result in a cusp and will not be created. In some cases, when arrows point in the opposite direction, a twisted surface may be formed.

Creating Blended Surfaces

Menubar: Insert > Surfaces > Blend
Toolbar: Surfaces > Blend

The **Blend** tool is used to create a surface by blending two curves. These curves can be sketched curves, wireframe geometries, or edges of existing surfaces. If you select support surfaces with curves, the resulting blended surface will be tangent to the support surfaces. To create a blended surface, draw some curves and create support surfaces. Choose the **Blend** tool from the **Surfaces** toolbar; the **Blend Definition** dialog box will be displayed, as shown in Figure 10-62. Also, you will be prompted to select the first curve and first support.

Select the curve and support. Next, you will be prompted to select the second curve and second support. Select them and choose the **OK** button from the **Blend Definition** dialog box. Figure 10-63 shows curves and support surfaces to create the blended surface and Figure 10-64 shows the preview of the resulting blended surface.

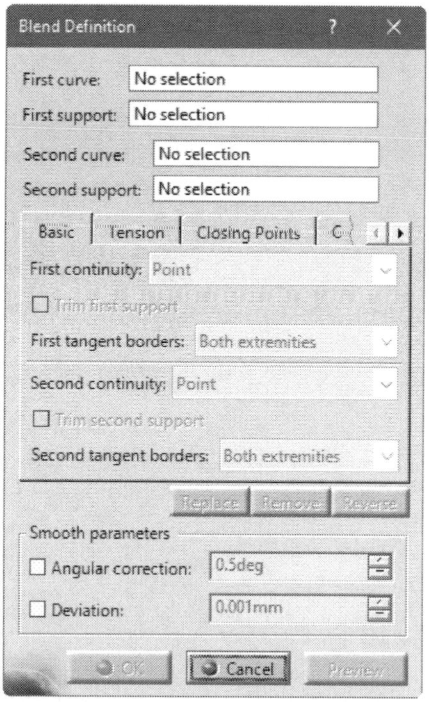

*Figure 10-62 The **Blend Definition** dialog box*

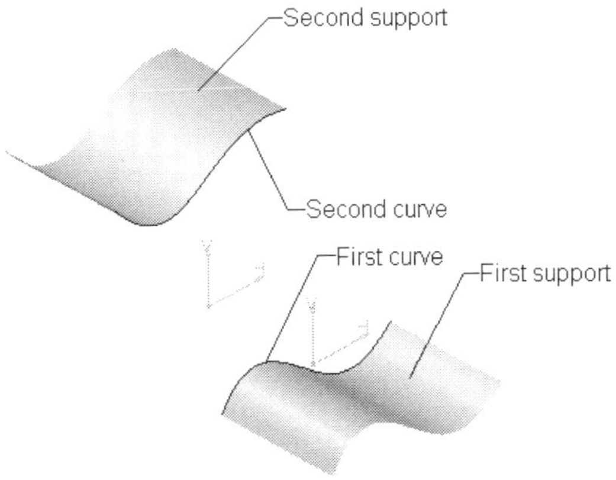

Figure 10-63 The curves and the support surfaces

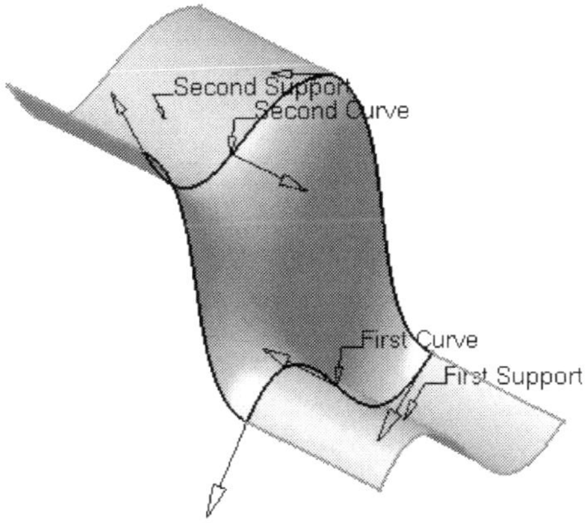

Figure 10-64 Preview of the resulting blended surface

OPERATIONS ON SHAPE GEOMETRY

Generally, surface models are a combination of various surfaces. You can perform various operations such as join, trim, split, or translate to manage multiple surfaces. For performing such operations, CATIA V5 provides a number of operation and tools. Some of these tools are discussed in this chapter and the remaining tools will be discussed in the next chapter.

Joining Surfaces

Menubar:	Insert > Operations > Join
Toolbar:	Operations > Join-Healing sub-toolbar > Join

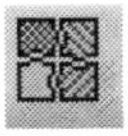

Generally, most of the surface models comprise of various individual surfaces connected to each other. To use the surface model for creating a solid model, first you need to join all the individual surfaces to form a single surface. You can also join individual curves that are connected to each other to form a single curve using this tool. You will learn more about curves in the later chapters.

Choose the **Join** tool from the **Join-Healing** sub-toolbar in the **Operations** toolbar; the **Join Definition** dialog box will be displayed, as shown in Figure 10-65. Select the surfaces to be joined from the geometry area; their names will be displayed in the **Elements To Join** display box. Keep the other default settings as they are and choose the **OK** button; the resultant join surface will be formed by joining all selected surfaces. Figure 10-66 shows the surfaces to be joined.

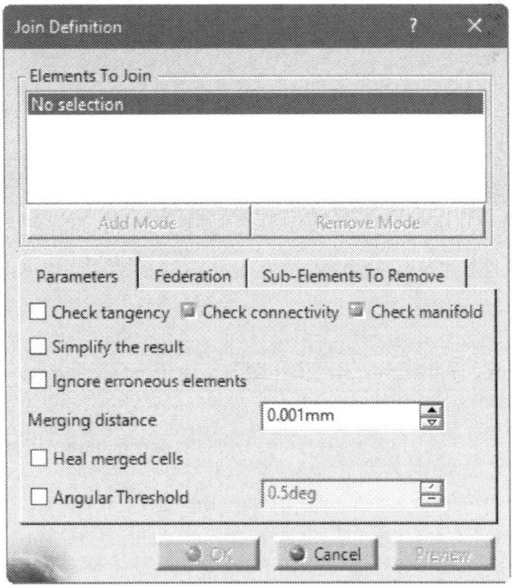

*Figure 10-65 The **Join Definition** dialog box*

Figure 10-67 shows the resulting single surface after joining. Note that if you need to heal the merged cells, select the **Heal merged cells** check box in the **Join Definition** dialog box.

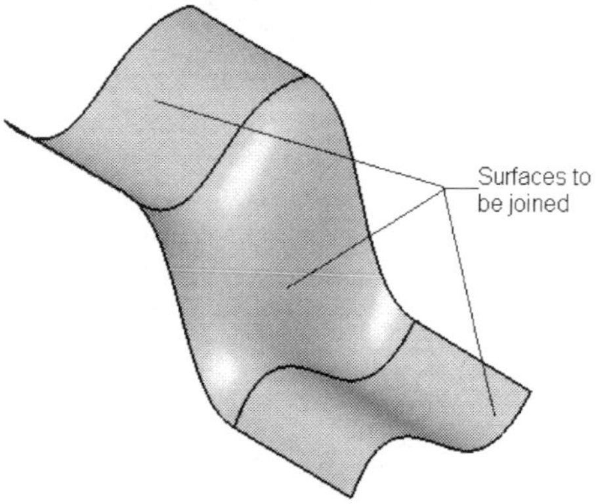

Figure 10-66 Surfaces to be joined

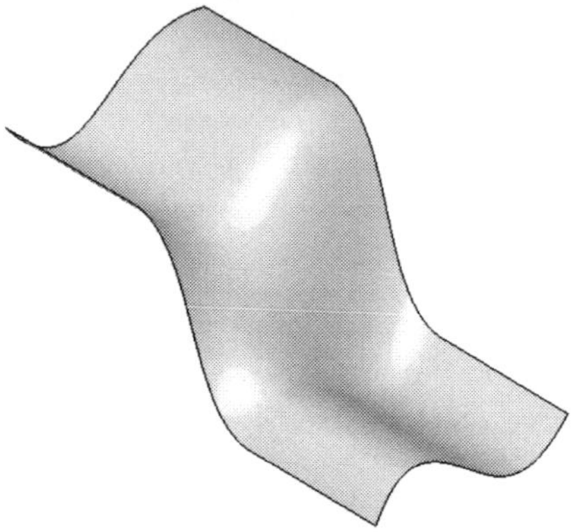

Figure 10-67 Resulting joined surface

Splitting Surfaces

Menubar:	Insert > Operations > Split
Toolbar:	Operations > Split-Trim sub-toolbar > Split

The **Split** tool is used to split a surface or a wireframe element by using a cutting element. A wireframe element can be split using a point, another wireframe element, or a surface. A surface can be split using another surface or wireframe element. To understand the usage of this tool, consider the case of two intersecting surfaces, as shown in Figure 10-68. Choose the **Split** tool from the **Split-Trim** sub-toolbar in the **Operations** toolbar; the **Split Definition** dialog box will be displayed, as shown in Figure 10-69, and you will be prompted to select the curve or surface to be split. Select a surface for splitting it and then select another surface as the cutting element; a preview of the split surface will be displayed. In the preview, the portion of the surface which you selected as the first surface will be retained and the other portion will be removed. Choose the **OK** button to split the surface. Figure 10-70 shows the split surface. You can choose the **Other side** button from the **Split Definition** dialog box to reverse the side of the surface to be removed. You can also retain both sides of the split surface by selecting the **Keep both sides** check box in the **Split Definition** dialog box. On selecting the **Keep both sides** check box, the surface will be split by the cutting element and will be divided into two separate entities.

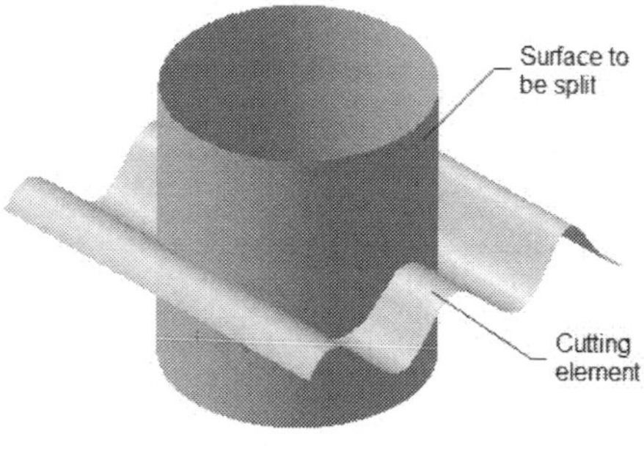

Figure 10-68 The split surface and the cutting surfaces

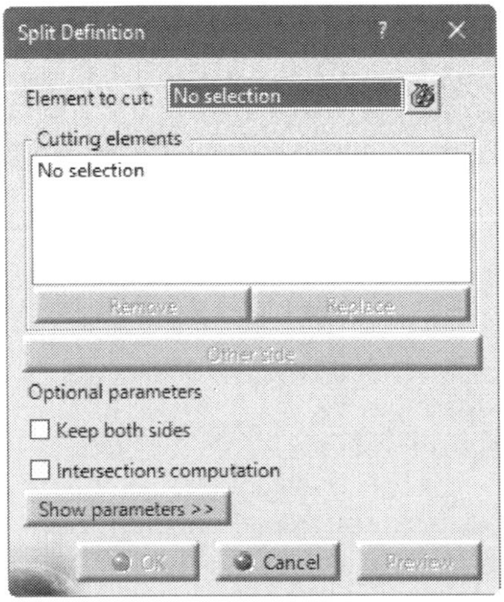

*Figure 10-69 The **Split Definition** dialog box*

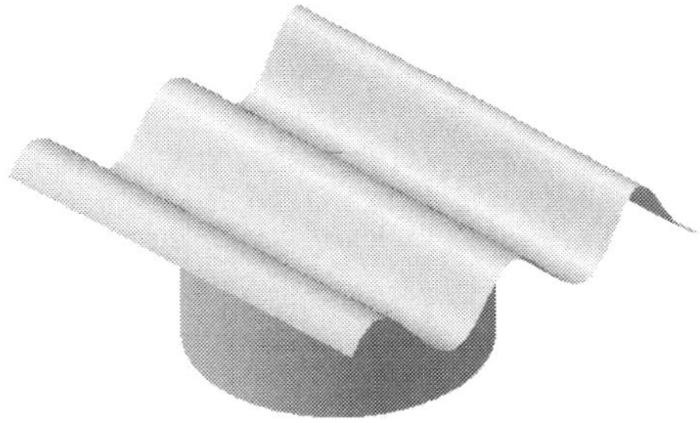

Figure 10-70 The resulting split surface

You can select more than one surface or wireframe elements to be split. For a multiple selection, you need to choose the **Elements to cut** button provided on the right of the **Element to cut** display box. On doing so, the **Elements to cut** dialog box is displayed. Select multiple elements that you need to split. After selecting multiple entities, choose the **Close** button from the **Elements to cut** dialog box. Click once in the **Cutting elements** display box and select the cutting elements. Now, choose the **OK** button from the **Split Definition** dialog box.

Trimming Surfaces

Menubar:	Insert > Operations > Trim
Toolbar:	Operations > Split-Trim sub-toolbar > Trim

The **Trim** tool is used to trim two intersecting surfaces or curves with respect to each other. On invoking this tool, the **Trim Definition** dialog box will be displayed, as shown in Figure 10-71, and you will be prompted to select an element to trim/remove. Select a surface; you will be prompted to select another element to trim/remove. Select other surface as the second element to be trimmed; a preview of the surface to be trimmed and the elements to be retained will be displayed, as shown in Figure 10-72. You can choose the **Other side / next element** and **Other side / previous element** buttons to reverse the sides of the surfaces selected to be trimmed. Next, choose the **OK** button to trim the selected surfaces. Figure 10-73 shows the resulting trimmed surfaces after reversing the sides of the selected surfaces.

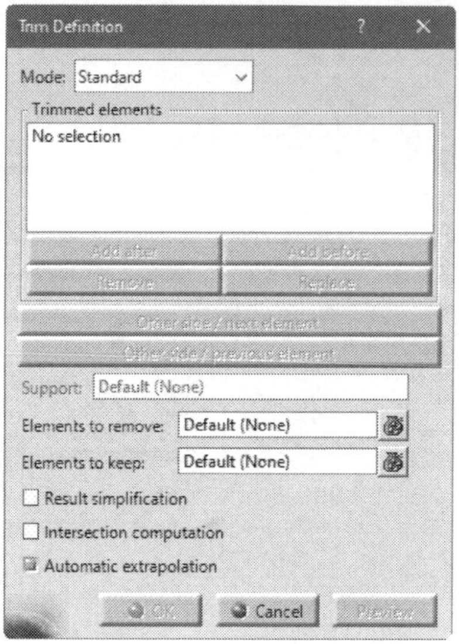

Figure 10-71 *The **Trim Definition** dialog box*

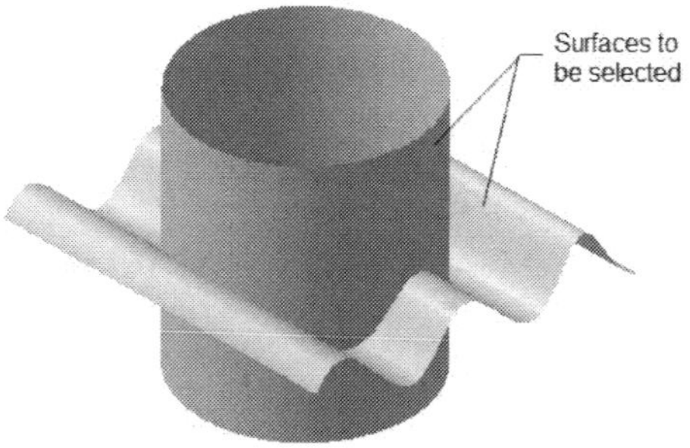

Figure 10-72 *The surface to be trimmed*

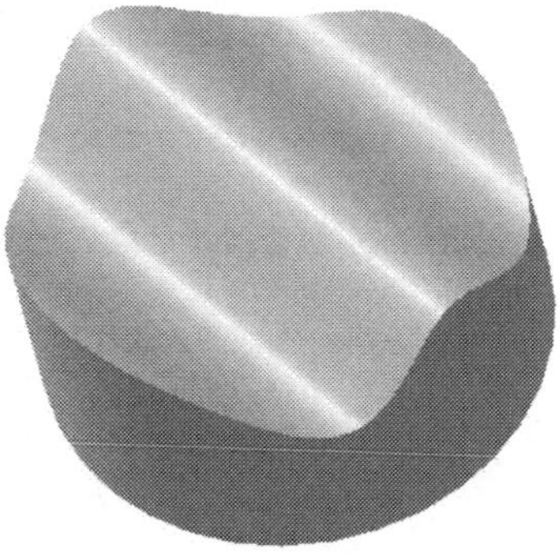

Figure 10-73 *The resulting trimmed surfaces*

TUTORIALS

Tutorial 1

In this tutorial, you will create the model shown in Figure 10-74. Its views and dimensions are shown in Figure 10-75. **(Expected time: 45 min)**

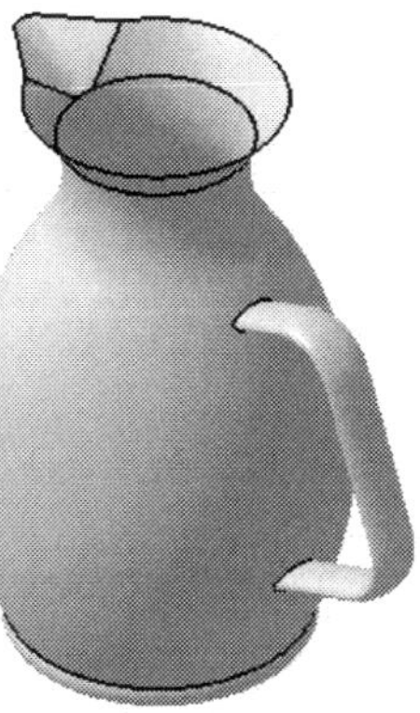

Figure 10-74 *The isometric view of the model*

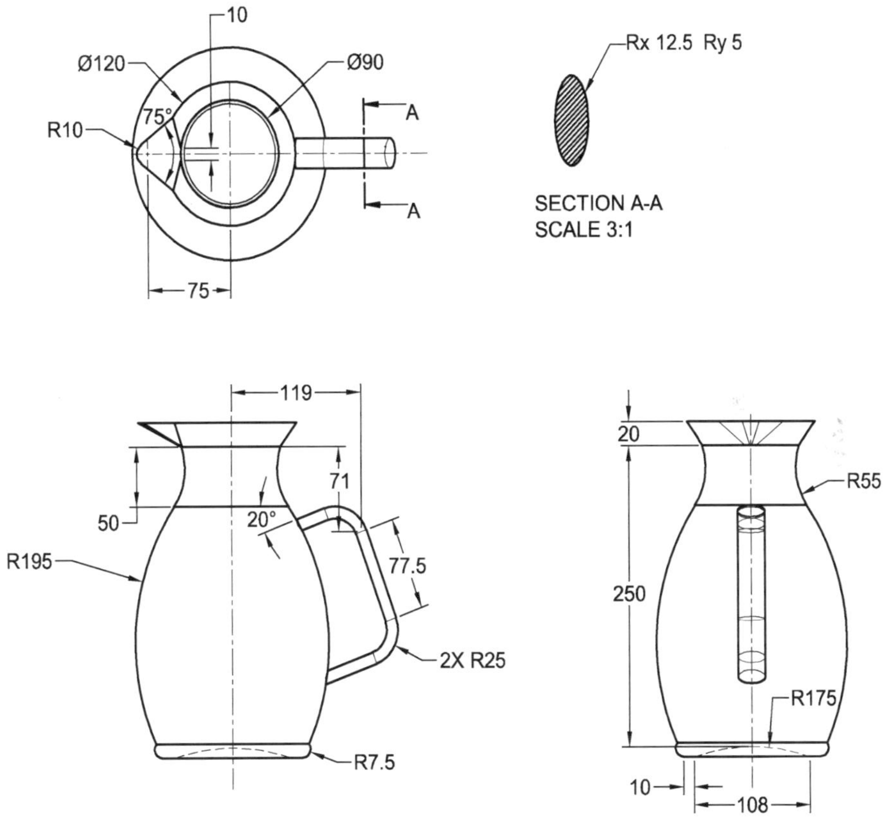

Figure 10-75 *Orthographic views and dimensions of the model*

The following steps are required to complete this tutorial:

a. Start CATIA V5 and then start a new file in the **Wireframe and Surface Design** workbench.
b. Draw sketches for the multi-sections surface.
c. Create multi-sections surface.
d. Draw a sketch to create the revolved surface.
e. Create the revolved surface.
f. Draw the sketch to create the sweep profile.
g. Create the sweep surface.
h. Split the sweep surface with the revolved surface.
i. Join the surfaces.

Starting a New Part File

1. Start CATIA V5 and close the default product file. The start screen of CATIA V5 is displayed. Choose **Start > Mechanical Design > Wireframe and Surface Design** from the menu bar to display the **New Part** dialog box. Enter **c10tut1** in the **Enter part name** edit box and choose **OK** from the **New Part** dialog box to start a new file in the **Wireframe and Surface Design** workbench.

Drawing the Sketch for the Base Surface

1. Invoke the **Sketcher** workbench by selecting the xy plane as the sketching plane and then draw a circle with its center as the origin and diameter 90 mm.

2. Place two points on the circle, as shown in Figure 10-76. These points will be used to create the guide curve and define the closing point for creating the loft surface later in this tutorial.

3. Exit the **Sketcher** workbench and create a plane at an offset of 20 mm from the xy plane in upper direction.

4. Invoke the **Sketcher** workbench again by selecting **Plane.1** as the sketching plane.

5. Draw the sketch, as shown in Figure 10-77, and exit the **Sketcher** workbench.

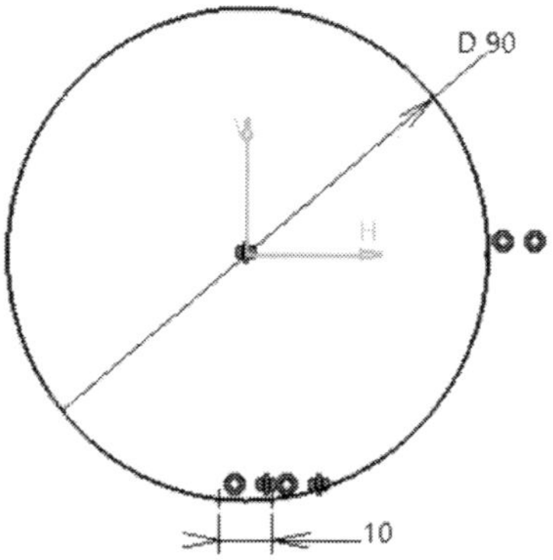

***Figure 10-76** Points placed on a circle for defining the closing point of the loft surface*

Next, you need to draw a line joining the two points in the two sections.

6. Choose the **Line** button from the **Wireframe** toolbar; the **Line Definition** dialog box is displayed. Select two points, as shown in Figure 10-78, and then choose the **OK** button from the **Line Definition** dialog box; a line joining the points is created.

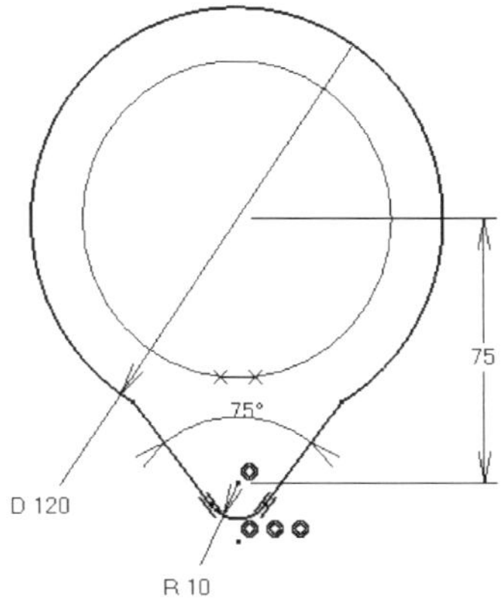

Figure 10-77 *The sketch for the second section of the base surface*

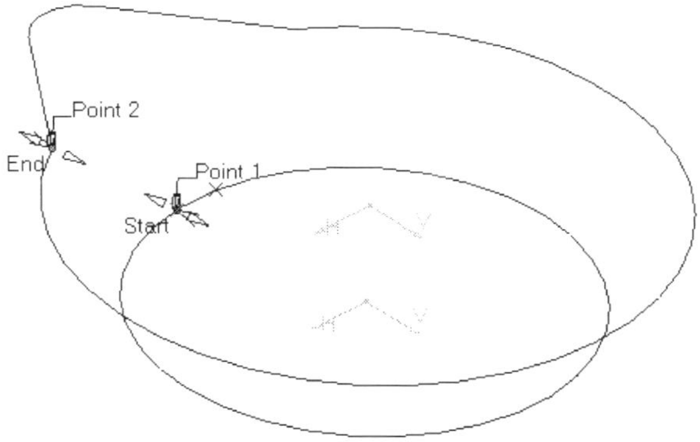

Figure 10-78 *Points selected to create a line*

Creating the Base Surface

In this section, you will create a base surface by using the **Multi-Sections Surface** tool.

1. Choose the **Multi-Sections Surface** tool from the **Surfaces** toolbar; the **Multi-Sections Surface Definition** dialog box is displayed and you are prompted to select a curve.

2. Select the sketch that was drawn first as Section 1, refer to Figure 10-79. The name of the selected sketch is displayed in the display box of the **Multi-Sections Surface Definition** dialog box. By default, a closing point is created. Move the cursor on the text in the **Closing Point** column of the dialog box and right-click; a contextual menu is invoked.

3. Choose the **Replace** option from the contextual menu and select a point on Section 1 in the geometry area, refer to Figure 10-79.

4. Choose the **Add** button from the **Multi-Sections Surface Definition** dialog box to select the second section; you are prompted to select a curve.

5. Select the second sketch (second section) from the geometry area and replace the default closing point with the required closing point, refer to Figure 10-79.

6. Next, click on the three dots in the display box of the **Guides** tab to specify the guide; you are prompted to select a curve.

7. Select the line created to specify the guide curve for creating a multi-section surface, as shown in Figure 10-79.

8. Choose the **OK** button from the **Multi-Sections Surface Definition** dialog box to complete the creation of the multi-section surface. Figure 10-80 shows the resulting multi-section surface.

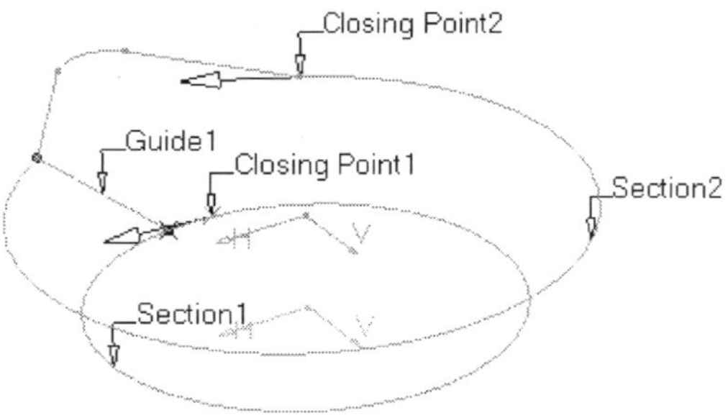

Figure 10-79 *The sketch showing the position of the closing points*

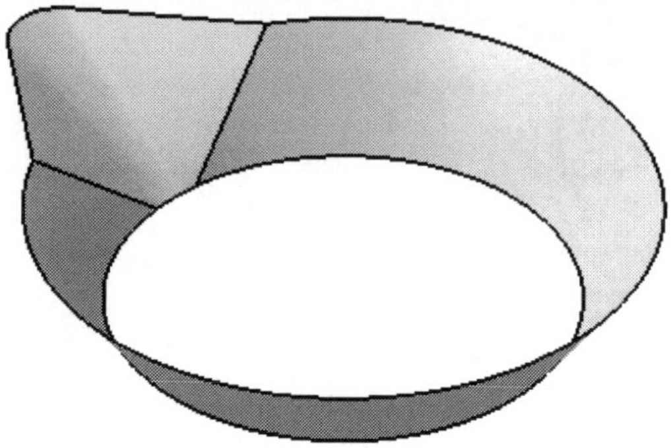

Figure 10-80 *The resulting multi-section surface*

Creating the Revolved Surface

Next, you need to create the revolved surface.

1. Invoke the **Sketcher** workbench by selecting the yz plane as the sketching plane.

2. Choose the **Axis** button from the **Profile** toolbar and draw a vertical axis passing through the origin.

3. Next, draw the sketch, as shown in Figure 10-81, and exit the **Sketcher** workbench.

4. Choose the **Revolve** tool from the **Surfaces** toolbar; the **Revolution Surface Definition** dialog box is displayed. 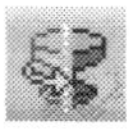

5. Select the profile to be revolved if not already been selected. The axis drawn in the sketcher workbench is automatically selected as the revolution axis and a preview of the revolved surface with default angle limits is displayed in the geometry area.

6. Set the value in the **Angle 1** spinner to **360** and choose the **OK** button from the **Revolution Surface Definition** dialog box. Figure 10-82 shows the model after creating the revolved surface and hiding the sketch.

Creating the Sweep Surface

Next, you need to create a surface by sweeping profile along a guide to create the handle of the jug.

1. Invoke the **Sketcher** workbench by selecting the yz plane as the sketching plane.

2. Draw the sketch for the guide curve, as shown in Figure 10-83.

3. Exit the **Sketcher** workbench. Click anywhere in the geometry area to exit the current selection set.

4. Create a plane normal to the guide at the upper endpoint.

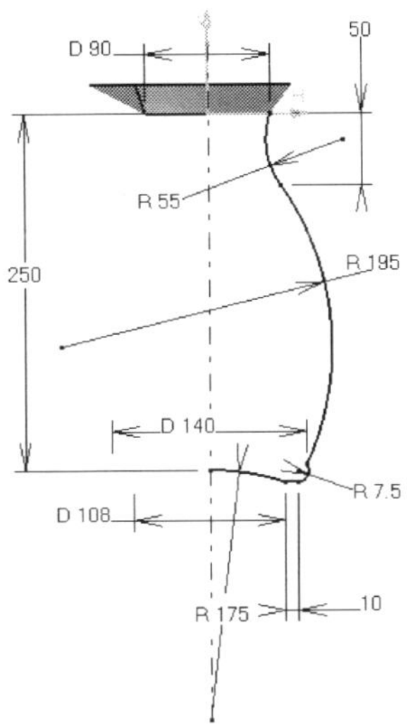

Figure 10-81 The sketch for creating the revolved surface

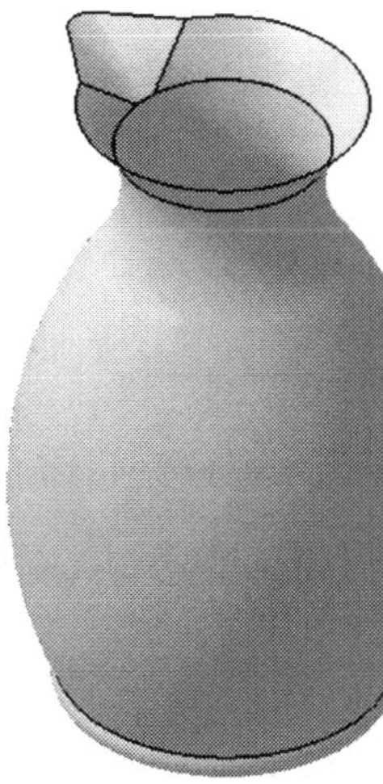

Figure 10-82 The model after creating the revolved surface

5. Invoke the **Sketcher** workbench by selecting the **Plane. 2** as the sketching plane.

6. Draw an ellipse for the profile of the sweep surface, as shown in Figure 10-84.

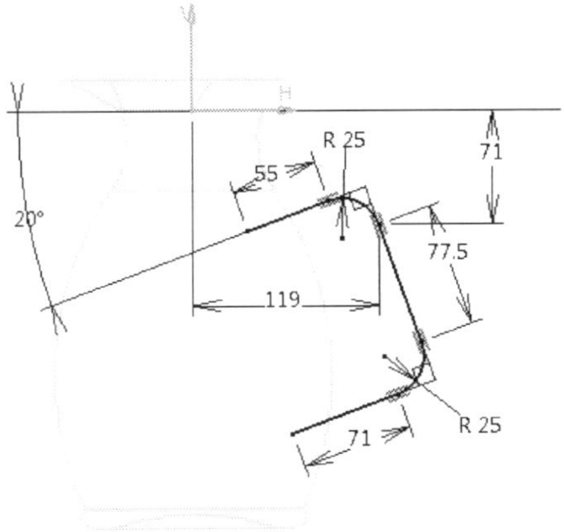

Figure 10-83 The guide for the sweep surface

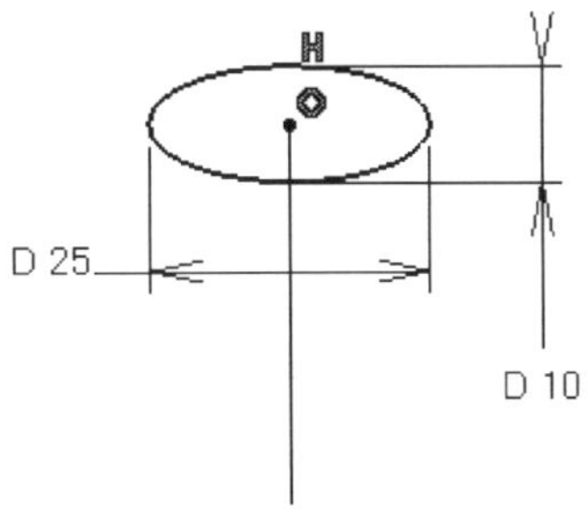

Figure 10-84 The profile for the sweep surface

7. Exit the **Sketcher** workbench. Click anywhere in the geometry area to exit the current selection set.

8. Choose the **Sweep** tool from the **Surfaces** toolbar; the **Swept Surface Definition** dialog box is displayed. In this dialog box, choose the **Explicit** button from the **Profile type** area, if it is not selected by default; you are prompted to select a profile.

9. Select the profile (ellipse) from the geometry area if not selected, as shown in Figure 10-85; you are prompted to select a guide curve.

10. Select the guide curve from the geometry area, refer to Figure 10-85, and choose the **OK** button from the dialog box to complete the creation of the sweep surface.

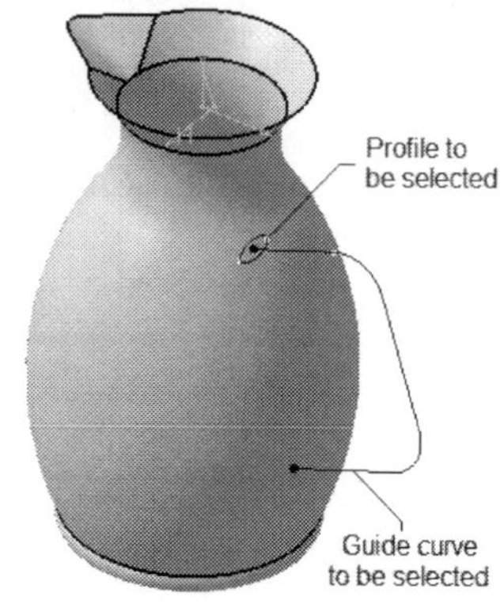

Figure 10-85 *The profile and the guide curve to be selected*

Splitting and Joining the Sweep Surface

The sweep surface is extended beyond the revolved surface. Therefore, you need to remove the unwanted portion of the swept surface which is inside the jug.

1. Choose the **Split** tool from the **Split-Trim** sub-toolbar in the **Operations** toolbar; the **Split Definition** dialog box is displayed and you are prompted to select the curve or the surface to be split.

2. Select the sweep surface from the geometry area.

3. Now, select the revolved surface as the cutting element.

4. Choose the **OK** button from the **Split Definition** dialog box to complete the split operation. After making the required modifications, the model will look similar to the one shown in Figure 10-86.

   On rotating the view of the model, you will see that the unwanted portion of the swept surface has been removed.

5. Next, choose the **Join** tool from the **Operation** toolbar; you are prompted to select the surfaces to be joined.

6. Select the Multi-sections surface, Revolute surface and Splitted surface and clear the **Check connectivity** check box, if selected by default.

7. Choose the **OK** button; the selected surfaces are joined, refer to Figure 10-86.

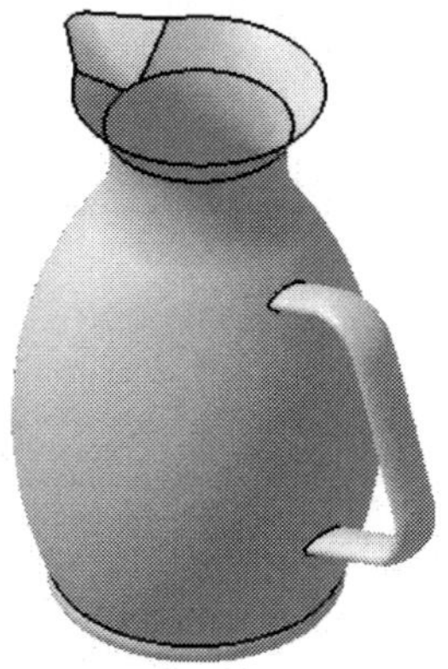

Figure 10-86 *The isometric view of the model after splitting and joining the swept surface*

Saving the File

1. Choose the **Save** button from the **Standard** toolbar; the **Save As** dialog box is displayed. Create the *c10* folder at *C:/CATIA*.

2. Enter **c10tut1** in the **File name** edit box. Choose the **Save** button from the **Save as** dialog box. The file is saved at *C:/CATIA/c10*.

3. Close the part file by choosing **File > Close** from the menu bar.

Tutorial 2

In this tutorial, you will create the model shown in Figure 10-87. Its drawing views and dimensions are shown in Figure 10-88. **(Expected time: 45 min)**

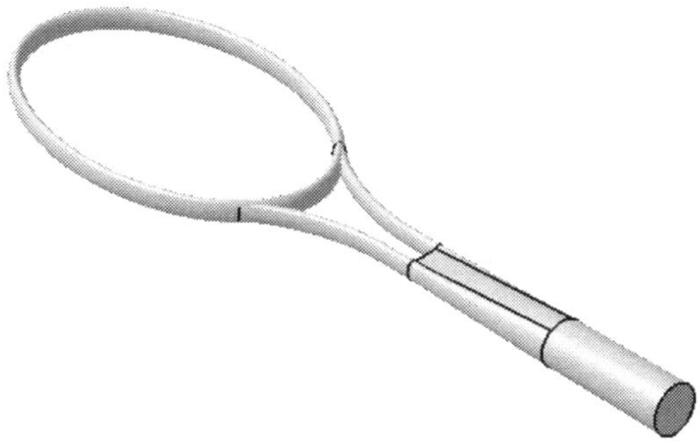

Figure 10-87 *The isometric view of the model*

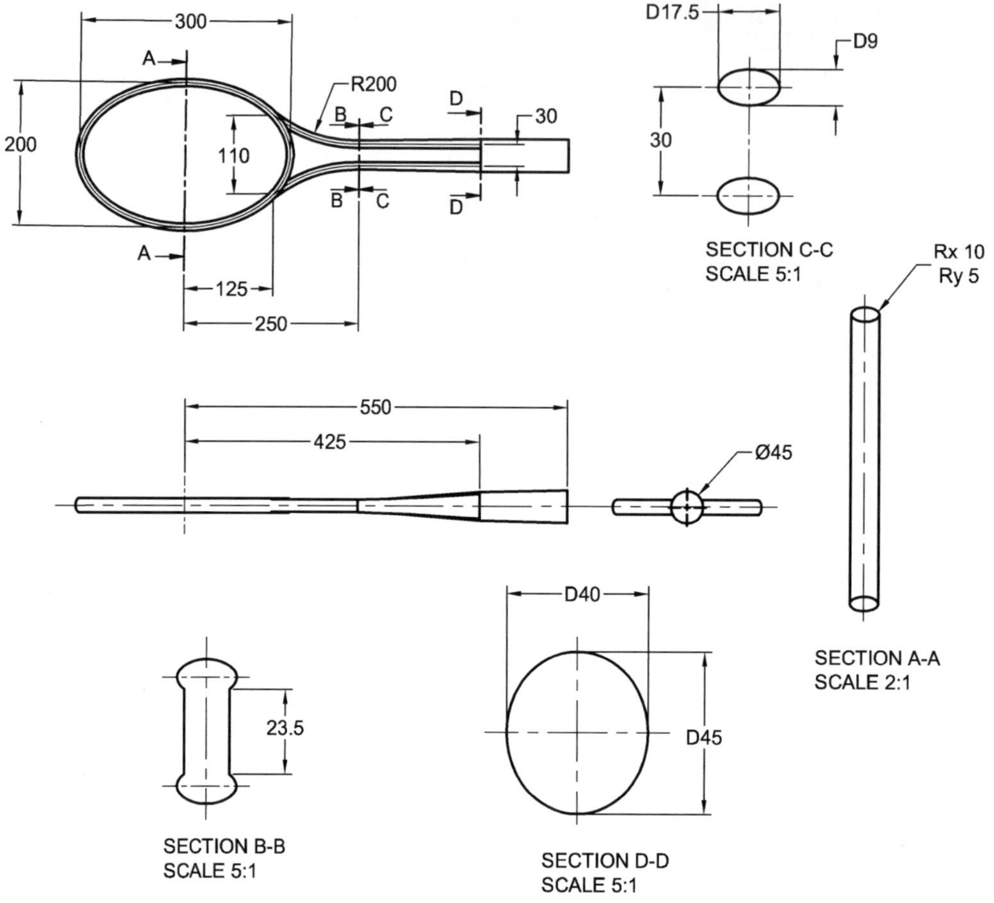

Figure 10-88 *The views and dimensions of the model*

The following steps are required to complete this tutorial:

a. Start a new file in the **Wireframe and Surface Design** workbench.
b. Draw the sketch for the base surface.
c. Create a sweep surface as the base feature.
d. Create the second sweep surface.
e. Create the symmetry feature of the second swept surface.
f. Create the multi-sections surface.
g. Create the blended surface.
h. Create fill surfaces.
i. Join the surfaces.

Drawing the Sketch for the Base Surface

First, you need to draw the profile and the guide curve for creating the sweep surface.

1. Start a new part file in the **Wireframe and Surface Design** workbench.

2. Invoke the **Sketcher** workbench by selecting the xy plane as the sketching plane.

3. Draw an ellipse for the guide curve, as shown in Figure 10-89, and exit the **Sketcher** workbench. If the sketch is selected, then you need to click anywhere on the screen to deselect it.

4. Invoke the **Sketcher** workbench by selecting the zx plane as the sketching plane.

5. Draw an ellipse for the sweep profile, as shown in Figure 10-90, and exit the **Sketcher** workbench.

Creating the Base Surface

After drawing the profile and the guide curve, you need to create a sweep surface by sweeping the profile along the guide curve.

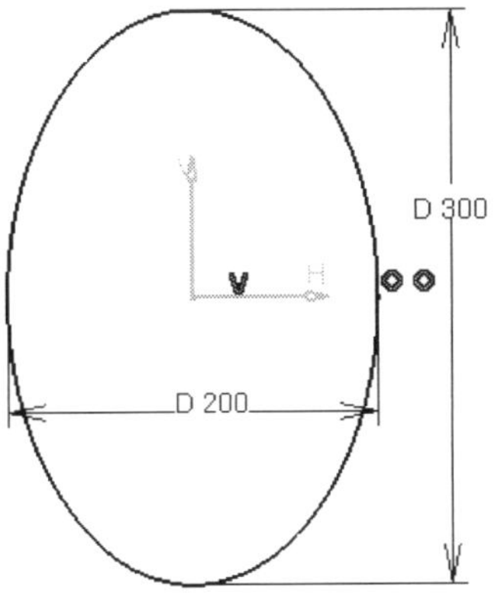

Figure 10-89 *The guide curve*

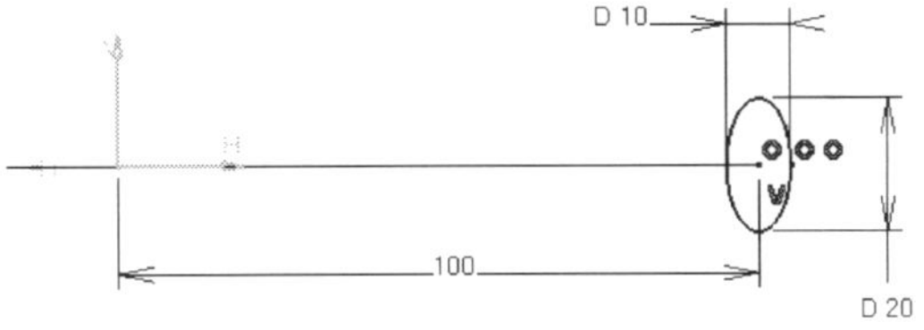

Figure 10-90 *The sweep profile*

1. Choose the **Sweep** tool from the **Surfaces** toolbar; the **Swept Surface Definition** dialog box is displayed.

2. In this dialog box, make sure the **Explicit** button is chosen in the **Profile type** area. You are prompted to select a profile.

3. Select the profile from the geometry area to create a sweep surface, as shown in Figure 10-91; you are prompted to select a guide curve.

4. Now, select the guide curve from the geometry area, as shown in Figure 10-91, and then choose the **OK** button. The resulting sweep surface after hiding the sketches is shown in Figure 10-92.

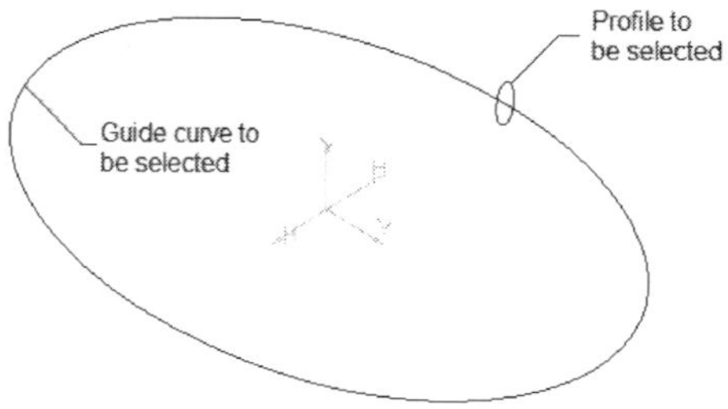

Figure 10-91 The profile and the guide curve to be selected

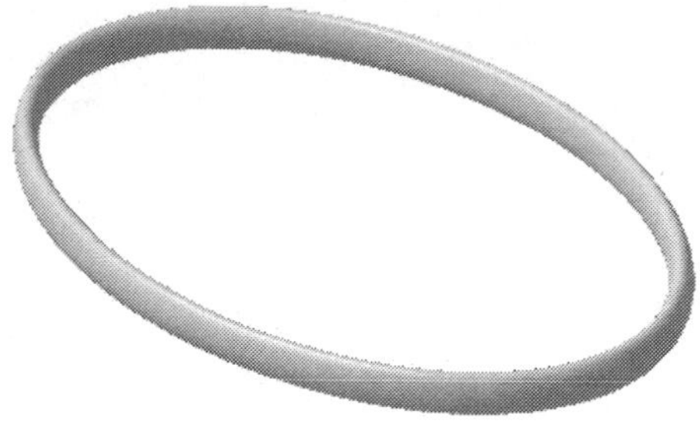

Figure 10-92 The resulting sweep surface

Drawing Sketches for the Second Sweep Feature

The second feature is also a sweep surface. Therefore, you need a profile and a guide curve to create this surface.

1. Invoke the **Sketcher** workbench by selecting the xy plane as the sketching plane.
2. Draw the guide curve, as shown in Figure 10-93.
3. Exit the **Sketcher** workbench and click anywhere in the geometry area to clear the current selection set.
4. Create a plane normal to the guide curve at the free endpoint of the straight line.
5. Invoke the **Sketcher** workbench by selecting **Plane.1** as the sketching plane.
6. Draw the sketch of the sweep profile coinciding with the guide curve, as shown in Figure 10-94.
7. Apply the Coincident constraint between the center of the ellipse and the endpoint of the guide curve, if it is not applied.
8. Exit the **Sketcher** workbench and click anywhere in the geometry area to remove the current selection set.

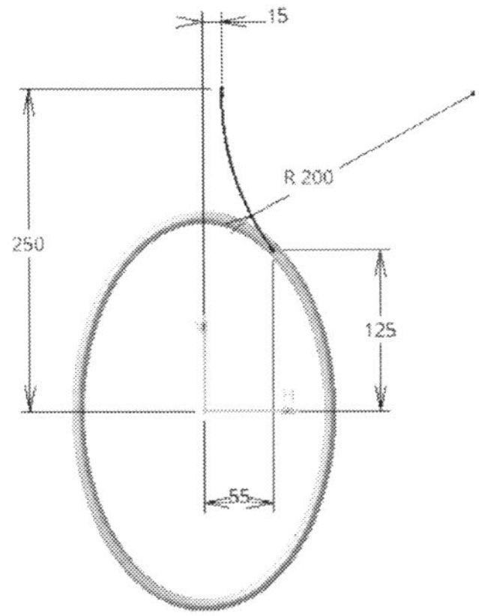

Figure 10-93 *The sketch of the guide curve for creating the second sweep feature*

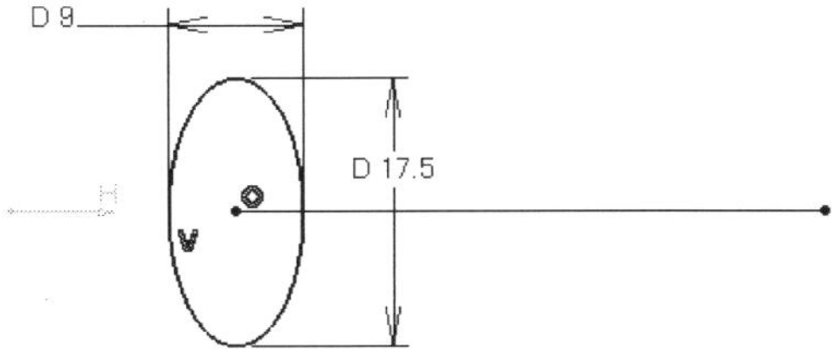

Figure 10-94 *The sketch of the sweep profile for creating the second sweep feature*

Creating the Second Sweep Feature

After drawing the sketches for the second sweep feature, you need to create the swept surface.

1. Choose the **Sweep** tool from the **Surfaces** toolbar; the **Swept Surface Definition** dialog box is displayed.

2. In this dialog box, the **Explicit** button is chosen by default in the **Profile type** area. If this button is not chosen by default, choose it; you are prompted to select a profile.

3. Select the profile to be swept from the geometry area, as shown in Figure 10-95; you are prompted to select a guide curve.

4. Select the guide curve from the geometry area and choose the **OK** button from the **Swept Surface Definition** dialog box. The model after creating the second sweep surface looks similar to the one shown in Figure 10-96.

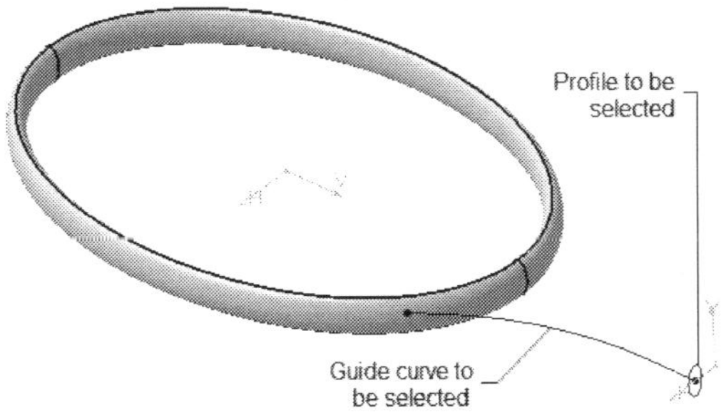

Figure 10-95 *The profile and the guide curve to be selected*

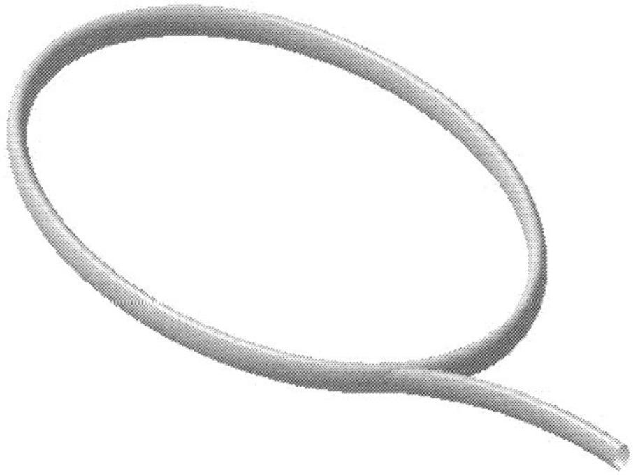

Figure 10-96 *The model after creating the second sweep feature*

5. Create a mirror copy of the second sweep feature. To do so, select **Sweep.2** from the Specification tree and then invoke the **Symmetry** tool by choosing **Insert > Operations >**

Symmetry from the menu bar. On doing so, the **Symmetry Definition** dialog box is displayed and you are prompted to select the reference point, line, or plane.

6. Select the yz plane from the Specification tree; a preview of the symmetry surface is displayed in the geometry area.

7. Choose the **OK** button from the **Symmetry Definition** dialog box to complete the symmetric feature. Figure 10-97 shows the model after creating the symmetry feature.

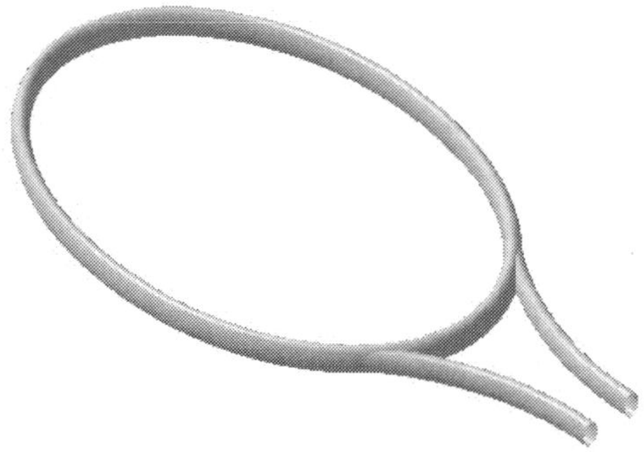

Figure 10-97 *The model after creating the symmetry feature*

Creating the Multi-Sections Surface

The next feature to be created is a multi-sections surface. For creating this surface, you need to draw two sections.

1. Invoke the **Sketcher** workbench by selecting **Plane.1** as the sketching plane.

2. Choose the **Normal View** tool from the **View** toolbar to flip the view direction, if required.

3. Click on the **Project 3D Elements** button from the **3D Geometry** sub-toolbar in the **Operation** toolbar and select the elliptical edges of the second sweep and symmetry feature. 

4. Complete the sketch for the first section, as shown in Figure 10-98, and exit the **Sketcher** workbench.

5. Create a plane at an offset of 425 mm from the zx plane.

6. Invoke the **Sketcher** workbench by using the newly created plane.

7. Draw the sketch for the second section, as shown in Figure 10-99. Note that the sketch consists of an ellipse and four points.

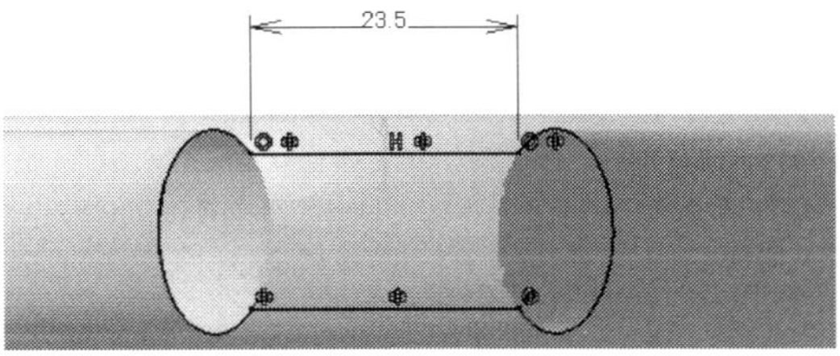

Figure 10-98 The first section for creating the multi-sections surface

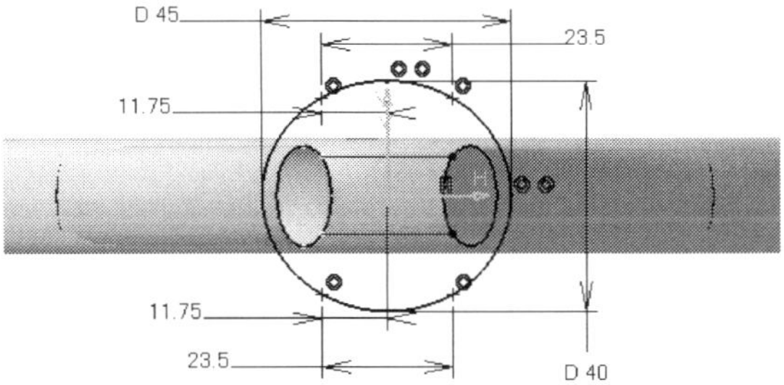

Figure 10-99 The second section for creating the multi-sections surface

8. Exit the **Sketcher** workbench and click anywhere in the geometry area to remove the current selected set. Change the view to the isometric view.

9. Choose the **Multi-Sections Surface** button from the **Surfaces** toolbar; the **Multi-Sections Surface Definition** dialog box is displayed and you are prompted to select a curve. 

10. Select the first section from the geometry area, refer to Figure 10-100; the closing point for the first section is selected by default.

11. Next, select the second section from the geometry area, refer to Figure 10-100; the closing point for the second section is selected by default.

12. Replace the closing points and set their directions, refer to Figure 10-101.

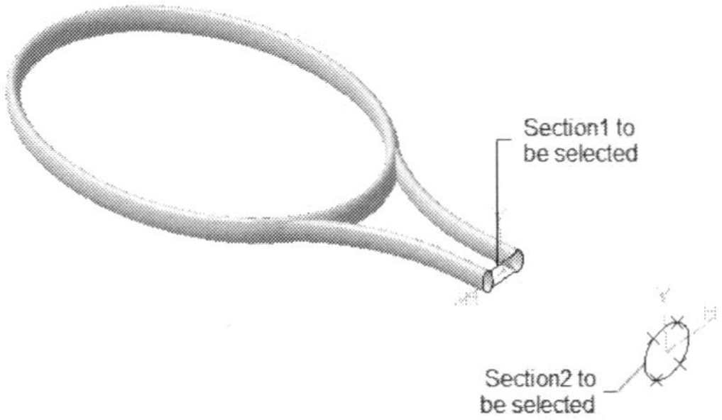

Figure 10-100 The sections to be selected for creating the multi-sections surface

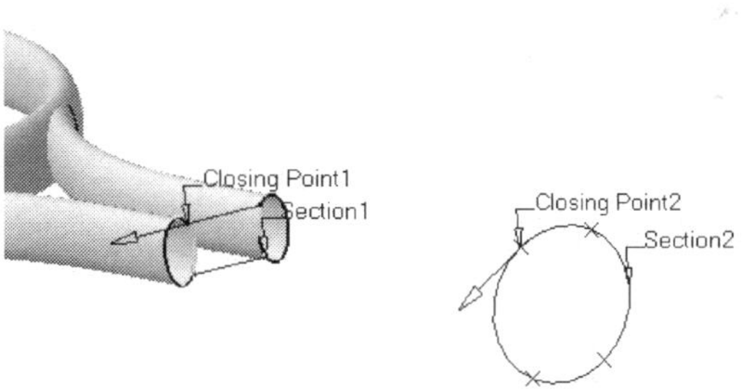

Figure 10-101 Specifying the closing points and their directions

13. Choose the **Coupling** tab from the **Multi-Sections Surface Definition** dialog box; you are prompted to add, remove or edit the coupling, or select a point to add the coupling.

14. Choose the **Add** button; the **Coupling** dialog box is displayed and you are prompted to select the coupling point.

15. Select the first coupling point on the first section, refer to Figure 10-102; the selected coupling point is displayed in the **Coupling** dialog box.

16. Select the second coupling point on the second section, refer to Figure 10-102; the coupling is created and displayed in the geometry area.

17. Activate the Coupling display box by clicking in the **Coupling** column, and then choose the **Add** button from the **Multi-Sections Surface Definition** dialog box; the **Coupling** dialog box is displayed again and you are prompted to select a coupling point.

18. Similarly, add the second, third, and fourth couplings on the sections, as shown in Figure 10-103.

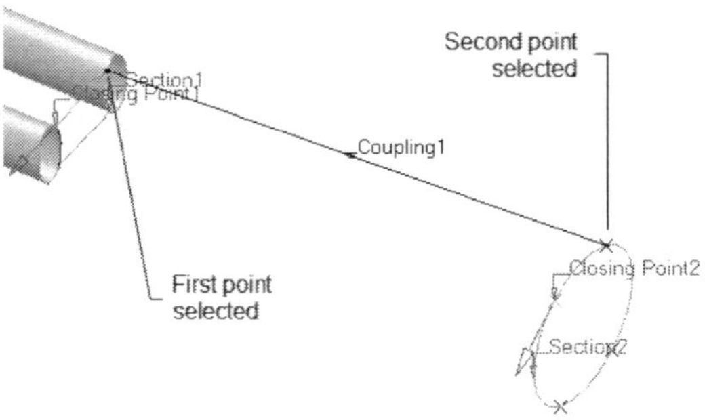

Figure 10-102 The points selected for creating the coupling

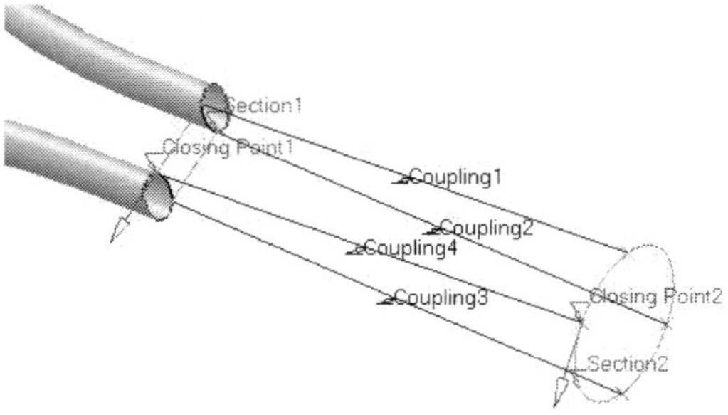

Figure 10-103 Couplings for creating the multi-sections surface

19. Choose the **OK** button from the **Multi-Sections Surface Definition** dialog box to create the multi-sections surface. Figure 10-104 shows the model after creating the multi-sections surface and hiding the sketches and the plane.

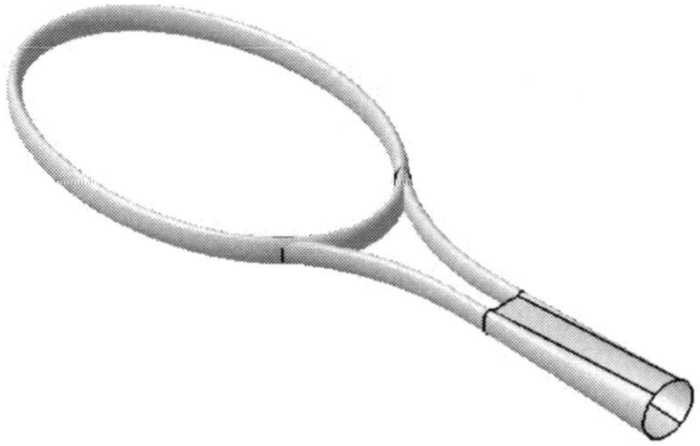

Figure 10-104 The model after creating the multi-sections surface

Creating the Blended Surface

Next, you need to create a blended surface.

1. Create a plane at an offset distance of 550 mm from the zx plane and invoke the **Sketcher** workbench by selecting the newly created plane.

2. Draw the sketch of the blended section, as shown in Figure 10-105. In this figure, the display of the previously created surfaces has been turned off.

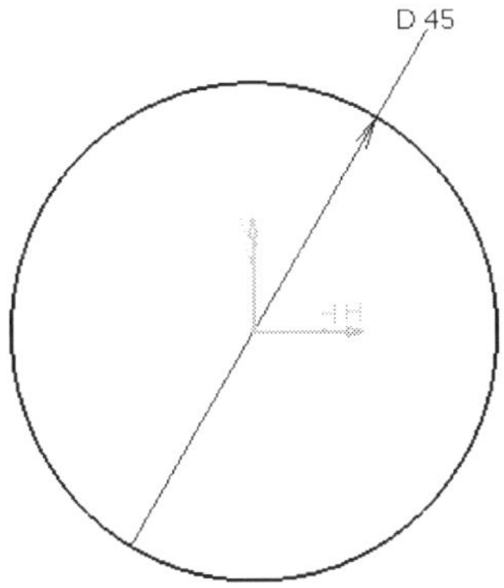

Figure 10-105 The sketch of the blended section

3. Exit the **Sketcher** workbench and click in the geometry area to remove the current selection set.

4. Choose the **Blend** tool from the **Surfaces** toolbar; the **Blend Definition** dialog box is displayed and you are prompted to select the first curve or the first support. 

5. Select the elliptical edge, as shown in Figure 10-106.

6. Next, select the sketch of the blended section from the geometry area, refer to Figure 10-106.

7. Choose the **OK** button from the **Blend Definition** dialog box. The model with the blended surface is displayed, as shown in Figure 10-107.

Creating the Fill Feature and Joining Surface

Next, you need to create the fill surface to close the open end of the blended feature.

1. Choose the **Fill** button from the **Surfaces** toolbar; the **Fill Surface Definition** dialog box is displayed.

2. Select the sketch shown in Figure 10-106.

3. Choose the **OK** button from the **Fill Surface Definition** dialog box to create a surface and exit the dialog box.

4. Next, choose the **Join** tool from the **Operation** toolbar; you are prompted to select the surface to be joined.

5. Select the surfaces from the geometry area or from the Specification tree; the names of the selected surfaces are displayed in the **Elements To Join** display box.

6. Choose the **OK** button; the selected surfaces are joined.

The final model is shown in Figure 10-108.

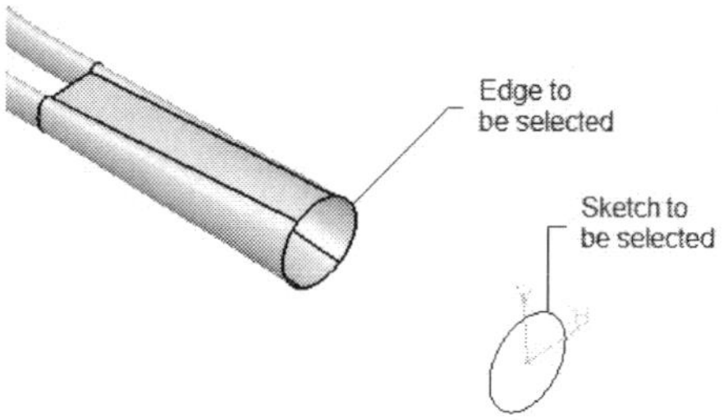

Figure 10-106 *The edge and the sketch to be selected for creating the blended surface*

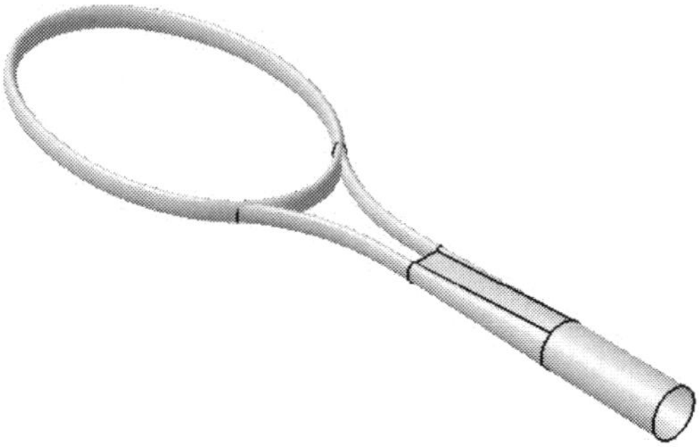

Figure 10-107 *The model after creating the blended surface*

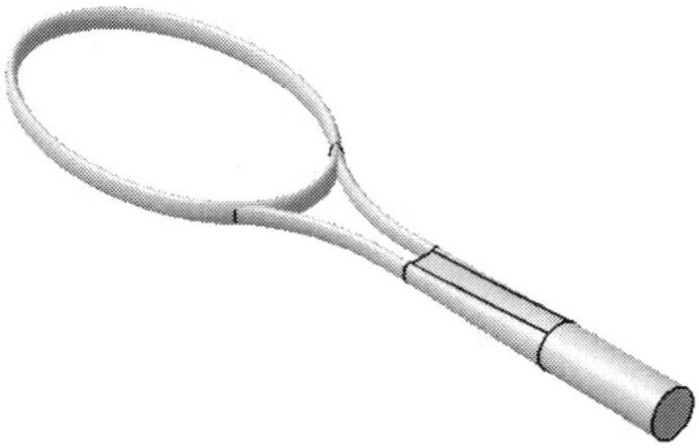

Figure 10-108 *The final model*

Saving the File

1. Once the model is complete, you need to save the file. Choose the **Save** button from the **Standard** toolbar; the **Save As** dialog box is displayed.

2. Enter **c10tut2** in the **File name** edit box. Choose the **Save** button from the **Save as** dialog box. Next, save the file at *C:\CATIA\c10*.

3. Close the part file by choosing **File > Close** from the menu bar.

Self-Evaluation Test

Answer the following questions and then compare them to those given at the end of this chapter:

1. In **Wireframe and Surface Design** workbench, the _________ tool is used to extrude a closed or open profile up to the defined limits.

2. The _________ tool is used to create a feature by revolving a profile about an axis.

3. The _________ tool is used to create a spherical surface by defining angular limits.

4. The _________ tool is used to create a cylindrical surface by defining the center point and the direction.

5. The _________ tool is used to create a surface by sweeping a profile along a guide curve.

6. You cannot use the **Multi-Sections Surface** tool to create a surface using more than two sections. (T/F)

7. The **Blend** tool is used to create a surface using only two sections. (T/F)

8. The **Join** tool is used to join individual surfaces to form a single surface. (T/F)

9. The **Split** tool is used to split a surface or a curve by using a cutting element. (T/F)

10. The **Trim** tool is used to trim only the surface element. (T/F)

Review Questions

Answer the following questions:

1. In CATIA V5, which of the following tools is used to create an offset surface?

   (a) **Extrude** (b) **Revolve**
   (c) **Offset** (d) **Sweep**

2. Which of the following options in the **Subtype** drop-down list of the **Swept Surface Definition** dialog box is used to create a surface using two guide curves?

   (a) **With reference surface** (b) **With pulling direction**
   (c) **With two guide curves** (d) None of these

3. Which of the following tools is used to create a blended surface between two curves?

   (a) **Revolve** (b) **Multi-Sections Surface**
   (c) **Blend** (d) **Split**

4. Which of the following tools is used to trim two surfaces with respect to each other?

   (a) **Circle** (b) **Trim**
   (c) **Offset** (d) **Split**

5. The _________ tool is used to create a helix shaped wireframe element.

6. The _________ tool is used to create a fill surface.

7. The _________ tool is used to create a circular sweep surface using three guide curves.

8. The __________ tool is used to create a surface by offsetting a reference surface.

9. You can invoke the **Object Repetition** dialog box by selecting the __________ check box in the **Offset Surface Definition** dialog box.

10. You can select the edge of an existing surface or a solid body as a profile to create an extruded surface. (T/F)

EXERCISES

Exercise 1

In this exercise, you will create the surface model shown in Figure 10-109. Its orthographic views and dimensions are shown in Figure 10-110. **(Expected time: 30 min)**

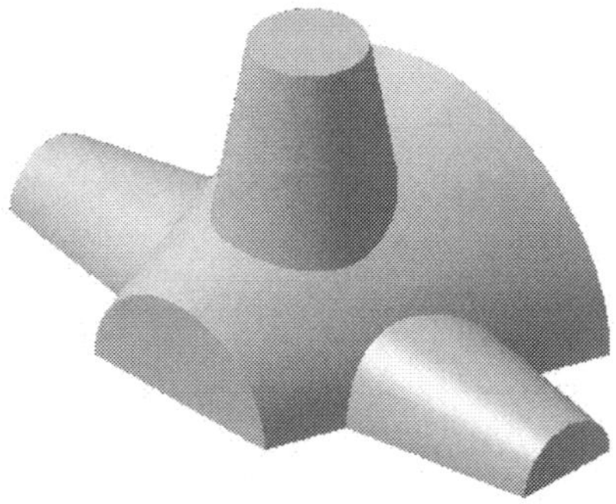

Figure 10-109 *The isometric view of the model*

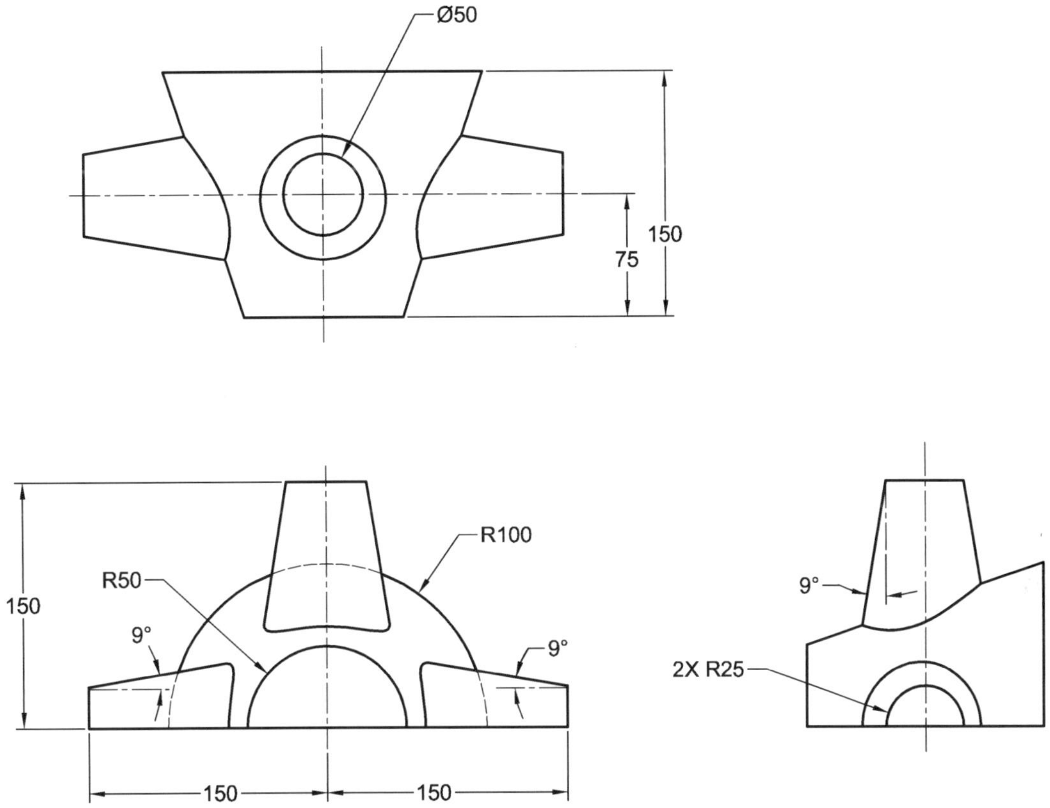

Figure 10-110 *Orthographic views and dimensions of the model*

Exercise 2

In this exercise, you will create the surface model shown in Figure 10-111. Its orthographic views and dimensions are shown in Figure 10-112. **(Expected time: 45 min)**

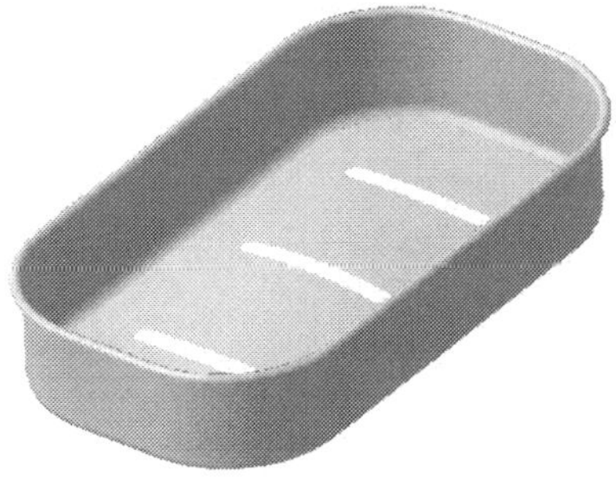

Figure 10-111 *The isometric view of the model*

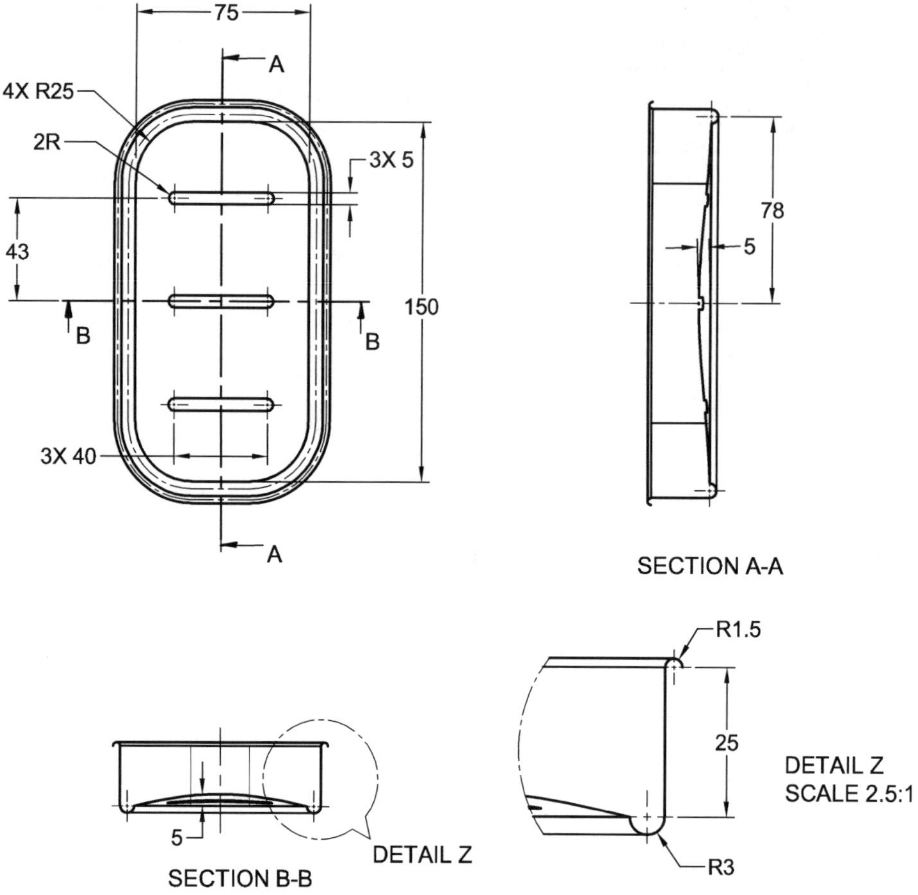

Figure 10-112 Orthographic views and dimensions of the model

Answers to Self-Evaluation Test

1. **Extrude**, **2.** **Revolve**, **3.** **Sphere**, **4.** **Cylinder**, **5.** **Sweep**, **6.** F, **7.** T, **8.** T, **9.** T, **10.** F

Chapter 11

Editing and Modifying Surfaces

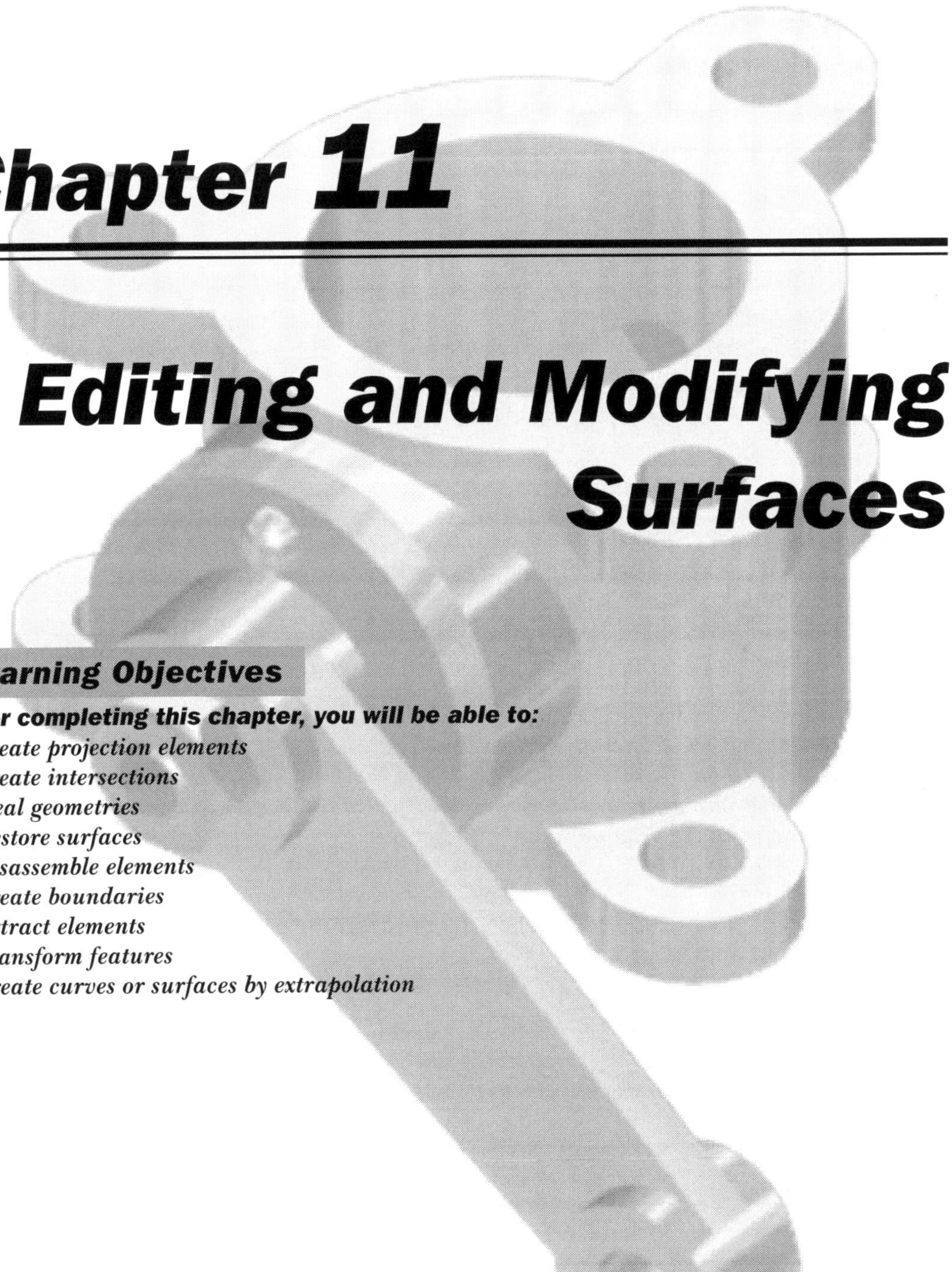

Learning Objectives

After completing this chapter, you will be able to:

- *Create projection elements*
- *Create intersections*
- *Heal geometries*
- *Restore surfaces*
- *Disassemble elements*
- *Create boundaries*
- *Extract elements*
- *Transform features*
- *Create curves or surfaces by extrapolation*

SURFACE OPERATIONS

In the previous chapter, you learned to work with various surface creation tools. CATIA V5 also provides you with tools to modify surfaces. These are known as surface operation tools and are used regularly while creating surface models. These tools enhance the surface creation capabilities and help to create complex geometries and save modeling time.

Various surface operation tools are discussed in the following sections:

Creating Projection Curves

Menubar:	Insert > Wireframe > Projection
Toolbar:	Wireframe > Projection

The **Projection** tool is used to project one or more elements on a support which could be a surface, a plane, or a wireframe geometry. In case of wireframe geometries, you can only project a point on them. The projection could be normal to the support or along a specified direction. To understand the use of the **Projection** tool, create an extruded surface and a curve, as shown in Figure 11-1.

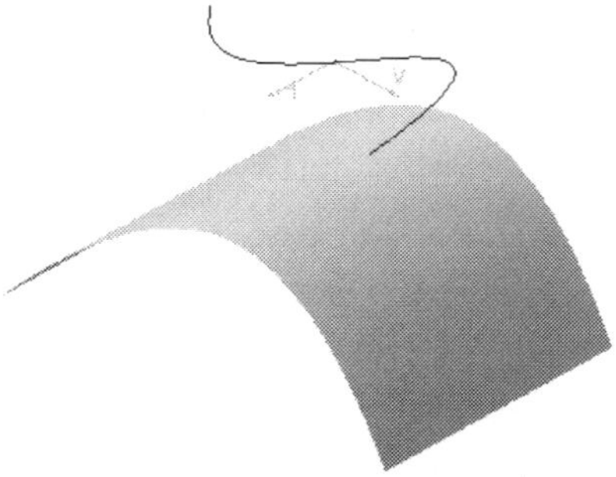

Figure 11-1 The curve and the support surface

Next, choose the **Projection** tool from the **Wireframe** toolbar; the **Projection Definition** dialog box will be displayed, as shown in Figure 11-2. By default, the **Normal** option is selected in the **Projection type** drop-down list of this dialog box. This option is used to project the element normal to the selected support. Select the curve to be projected from the geometry area; you will be prompted to select the support surface or plane. Select the surface and choose the **OK** button from the dialog box to create the projected element on the selected support surface. Figure 11-3 shows the resulting projected element. If you select the **Along a direction** option from the **Projection type** drop-down list the projection will be created along the specified direction.

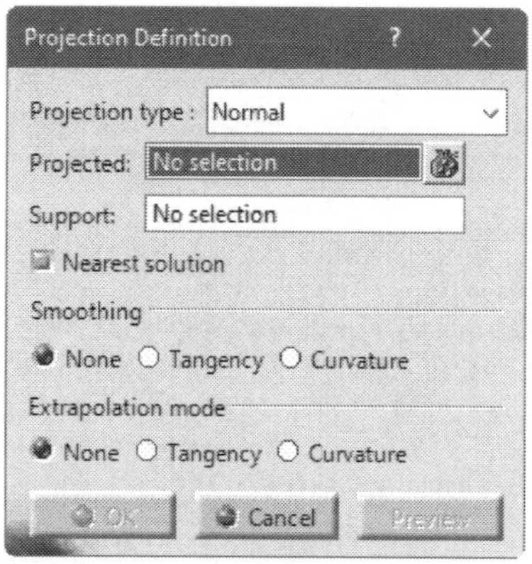

*Figure 11-2 The **Projection Definition** dialog box*

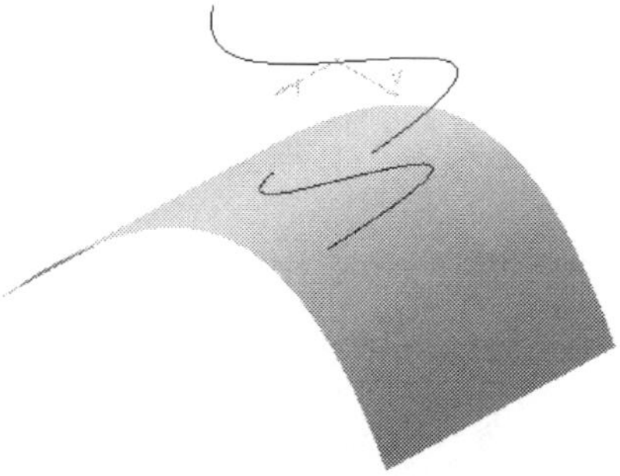

Figure 11-3 The resulting projected element

While projecting entities on the curved face of the cylinder, two projection solutions are possible: on the curved portion near the curve and on the other side of the cylinder. In such cases, select the **Nearest solution** check box from the **Projection Definition** dialog box to project the selected entities on the curved portion nearest to them.

You can also select multiple entities to be projected on the surface in a single operation. You can select entities from the drawing area or the Specification tree. To select multiple entities, choose the Selection Manager button next to the **Projected** selection box; the **Projected** dialog box will be displayed, as shown in Figure 11-4. Next, select multiple entities from the drawing area; the selected entities will be listed in the **Projected** dialog box. You can remove or replace the selected entities by choosing the **Remove** or **Replace** button in the **Projected** dialog box. After doing the necessary action, close the **Projected** dialog box and then select the support surface from the drawing area. Next, choose the **OK** button from the **Projection Definition** dialog box; the selected elements will be projected on the surface, as shown in Figure 11-5. The projection is identified as **Multi Output.xxx(Project)** and the projected elements are grouped and displayed under it in the Specification tree.

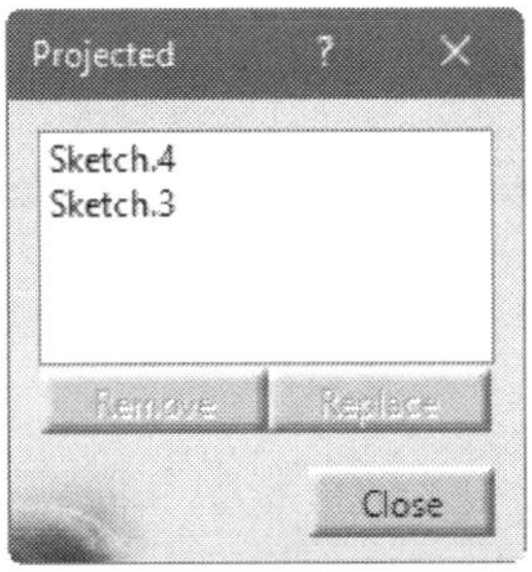

Figure 11-4 *The **Projected** dialog box*

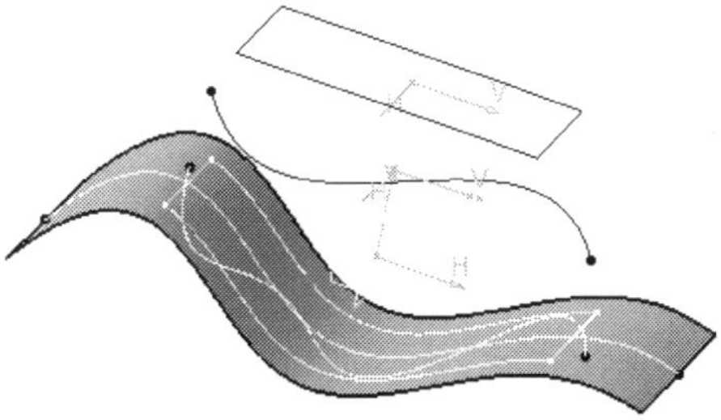

Figure 11-5 *The resulting projected elements using multi-output*

Note

*The Selection Manager option is also available in the **Intersection Definition**, **Symmetry Definition**, **Scaling Definition**, **Affinity Definition**, **Axis to Axis Definition**, **Split Definition**, **Extrapolate Definition**, and **Translation Definition** dialog boxes.*

Creating Intersection Elements

Menubar:	Insert > Wireframe > Intersection
Toolbar:	Wireframe > Intersection

The **Intersection** tool is used to create intersecting elements using two curves, a curve and a surface or two surfaces, a solid and a surface. Depending upon the elements that were selected while performing the intersection, the result of the intersection can be a point, a curve, a contour, or a surface. On choosing this tool from the **Wireframe** toolbar, the **Intersection Definition** dialog box will be displayed, as shown in Figure 11-6, and you will be prompted to select the elements to be intersected.

The process of intersection of different elements is discussed next.

Intersection of Two Curves

When two curves intersect, the result will be a point or a curve, depending on how they intersect. If the intersecting curves overlap, the result will be a curve. If they cross each other, the result will be points at every intersection of the curve.

To create intersecting elements, choose the **Intersection** tool from the **Wireframe** toolbar and then select the curves that need to be intersected.

If the curves overlap and you choose the **OK** button from the **Intersection Definition** dialog box, a **Warnings** message window will be displayed. This message window warns that the input elements are tangent and you need to modify the inputs, if necessary. If you select the **Points** radio button from the **Curves Intersection With Common Area** area before choosing the **OK** button, then some points will be created as a result of the intersection, no matter whether the curves are overlapping, tangent, or crossing.

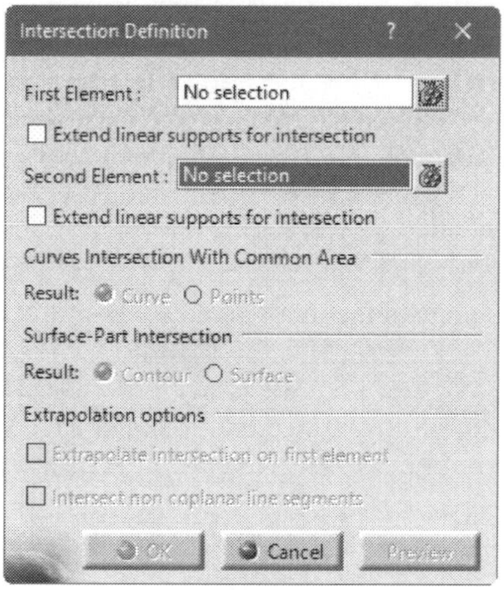

Figure 11-6 The ***Intersection Definition*** *dialog box*

The **Extend linear supports for intersection** check boxes available below the **First Element** and **Second Element** display boxes, respectively are used to extend the selected curves to the nearest intersection.

After setting all the parameters, choose the **OK** button from the **Intersection Definition** dialog box. Figure 11-7 shows the elements selected for intersection and the resulting intersected elements created using different options.

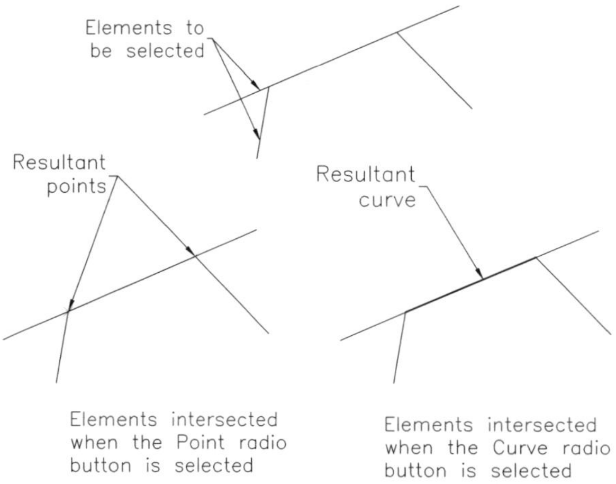

Figure 11-7 *Elements to be intersected and the resulting intersections*

Intersection of Two Surfaces

When you intersect two surfaces, the result will be the formation of a curve at their intersection. Figure 11-8 shows the surfaces to be intersected and Figure 11-9 shows the resulting intersection curve. Note that although the resulting curve looks like a combination of multiple curves, yet it is actually a single curve.

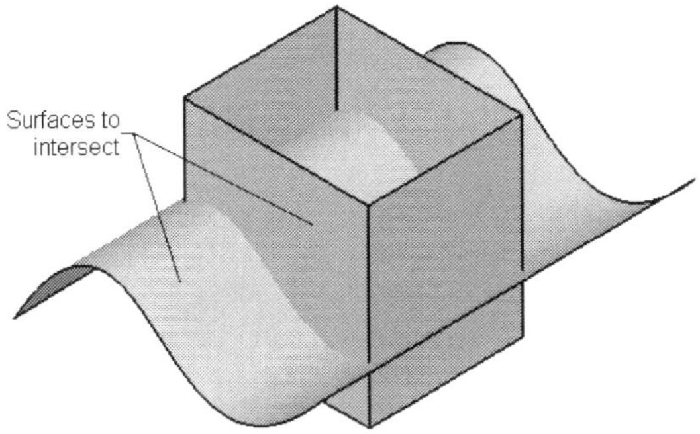

Figure 11-8 Surfaces to be intersected

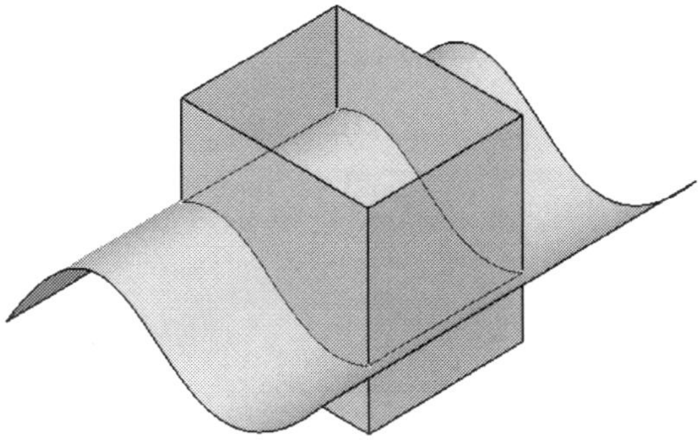

Figure 11-9 Resulting intersection curve

You will notice that the **Extrapolate intersection on first element** option in the **Extrapolation options** area will become active after you select two surfaces. This option is used to extrapolate the intersection curve on the first element.

Intersection of a Solid and a Surface

When a solid and a surface intersect, the result could be a contour or a surface, depending upon the option selected in the **Surface-Part Intersection** area of the **Intersection Definition** dialog box.

To intersect a solid with a surface, choose the **Intersection** tool from the **Wireframe** toolbar; the **Intersection Definition** dialog box will be displayed. Select a solid as the first element to be intersected; you will be prompted to select the second element to be intersected. Select a surface as the second element to be intersected. By default, the **Contour** radio button is selected in the **Surface-Part Intersection** area of the **Intersection Definition** dialog box. If you select the **Surface** radio button, the resulting intersection will be a surface created by the intersection of a solid and a surface. After selecting the desired option, choose **OK** from the dialog box to create the intersection curve.

Healing Geometries

Menubar:	Insert > Operations > Healing
Toolbar:	Operations > Join-Healing sub-toolbar > Healing

The **Healing** tool is used to fill a small gap that is left after joining two or more surfaces. This tool is available in the **Join-Healing** sub-toolbar, as shown in Figure 11-10. This sub-toolbar can be invoked by choosing the down arrow on the right of the **Join** tool in the **Operations** toolbar.

To understand the function of this tool, create surfaces with a small gap between them. Choose the **Healing** tool from the **Join-Healing** sub-toolbar; the **Healing Definition** dialog box will be displayed, as shown in Figure 11-11. Also, you will be prompted to select the elements to be healed. Select them from the geometry area.

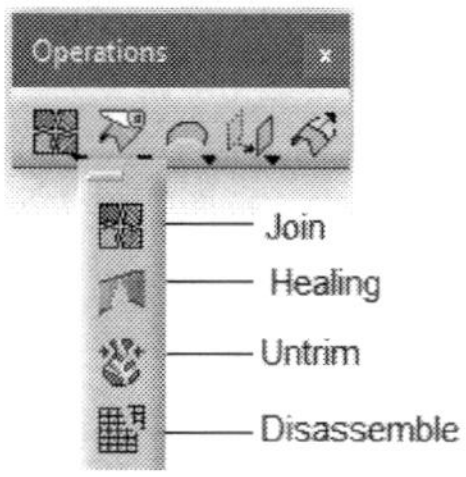

Figure 11-10 *The* ***Join-Healing*** *sub-toolbar*

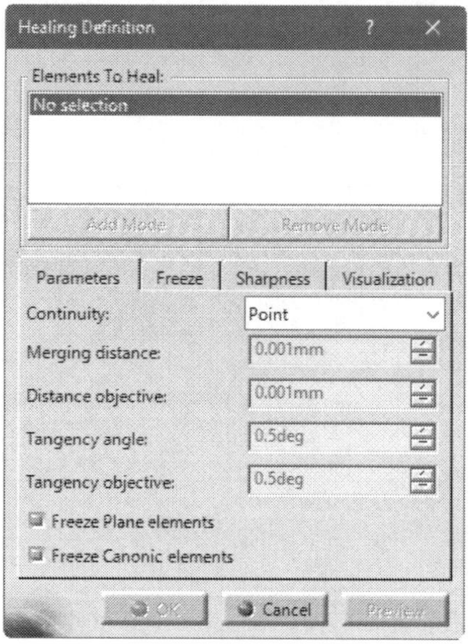

Figure 11-11 *The* ***Healing Definition*** *dialog box*

Next, in the **Merging distance** spinner, set the value greater than the existing value of gap between the two elements to be healed. Set the value in this spinner as per the requirement.

Choose the **Preview** button from the **Healing Definition** dialog box; the resulting healed surface will be displayed with the edges highlighted in green. Next, choose **OK** from the dialog box to accept the healing surface created.

Figure 11-12 shows the surfaces with a small gap between them. Figure 11-13 shows the healed surface. In this case, the **Merging distance** value is increased. But you must remember that if the value of the merging distance is increased too much, the original surfaces will deviate from their actual positions to create the required surface.

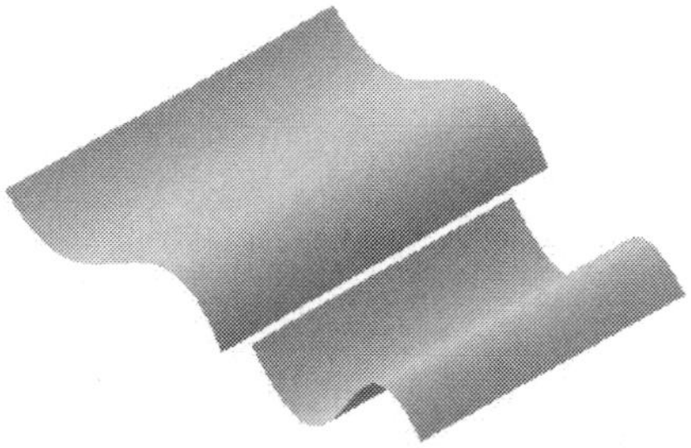

Figure 11-12 *The surfaces to be healed*

Figure 11-13 *The resulting healed surface*

If the two surfaces to be healed have distance between them greater than the value specified in the **Merging distance** spinner of the dialog box then on choosing the **OK** button from the **Healing Definition** dialog box, the surfaces will not be healed and the **Multi-Result Management** dialog box will be displayed, as shown in Figure 11-14. In this dialog box, you can specify the surface to be retained by selecting the required radio button from the **Pointing elements** tab. By default, the **keep only one sub-element using a Near/Far**, radio button is selected in this tab. If you choose the **OK** button, the **Near/Far Definition** dialog box will be displayed with the **Near** radio button selected by default, as shown in Figure 11-15. Also, you will be prompted to select a reference element. Select a reference element and choose the **OK** button; the surface near the reference element will be retained. If you select the **Far** radio button in this dialog box, the surface which is far away from the reference element will be retained.

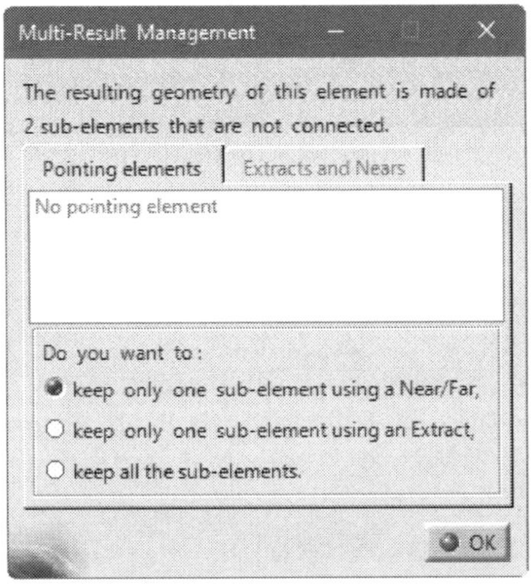

Figure 11-14 *The* ***Multi-Result Management*** *dialog box*

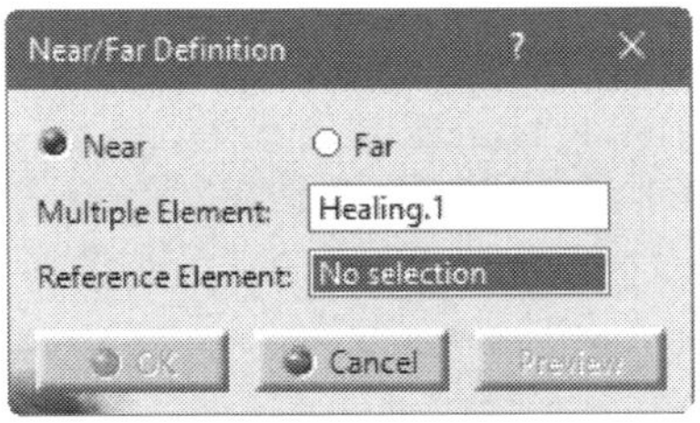

Figure 11-15 *The* ***Near/Far Definition*** *dialog box*

If you select the **keep only one sub-element using an Extract** radio button from the **Multi-Result Management** dialog box, the **Extract Definition** dialog box will be displayed and you will be prompted to select the elements to be extracted. The **Propagation type** drop-down list in this dialog box will provide you options to select the reference elements easily. The selected reference elements will be extracted from the surface and retained. On selecting the **keep all the sub-elements** option from the **Multi-Result Management** dialog box, each individual surface will be retained.

When the **Freeze Plane elements** or **Freeze Canonic elements** check boxes are selected, no selected plane/canonic element is affected by the healing operation.

Disassembling Elements

Menubar: Insert > Operations > Disassemble
Toolbar: Operations > Join-Healing sub-toolbar > Disassemble

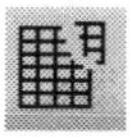

The **Disassemble** tool is used to break the sketched entities into individual curves at points where the tangency or point continuity changes. You can also disassemble a surface into individual surfaces using this tool.

To disassemble elements, choose the **Disassemble** tool from the **Join-Healing** sub-toolbar; the **Disassemble** dialog box will be displayed, as shown in Figure 11-16. Also, you will be prompted to select the elements to be disassembled. You can disassemble them completely or partially. The **All Cells** option in the **Disassemble mode** area of the **Disassemble** dialog box is selected by default. As a result, you can disassemble the elements completely. To disassemble the elements into their domains partially, select the **Domains Only** option in the **Disassemble mode** area. You can also select an edge of a surface or a solid element to disassemble. If you disassemble a solid body, it will split into individual surfaces. Note that after applying this tool, the duplicate entities will be created over the original entities.

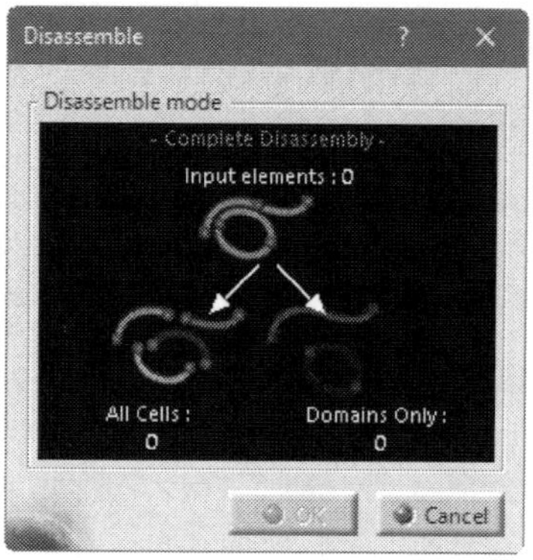

*Figure 11-16 The **Disassemble** dialog box*

Untrimming a Surface or a Curve

Menubar: Insert > Operations > Untrim
Toolbar: Operations > Join-Healing sub-toolbar > Untrim

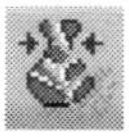

The **Untrim** tool is used to restore the portion of a surface or a curve that is removed by using the **Split** tool. To untrim a surface, choose the **Untrim** tool from the **Join-Healing** sub-toolbar; the **Untrim** dialog box will be displayed, as shown in Figure 11-17, and you will be prompted to select the element. Select the element from the geometry area; the **Warning** message box will be displayed, as shown Figure 11-18. Choose the **OK** button to exit it.

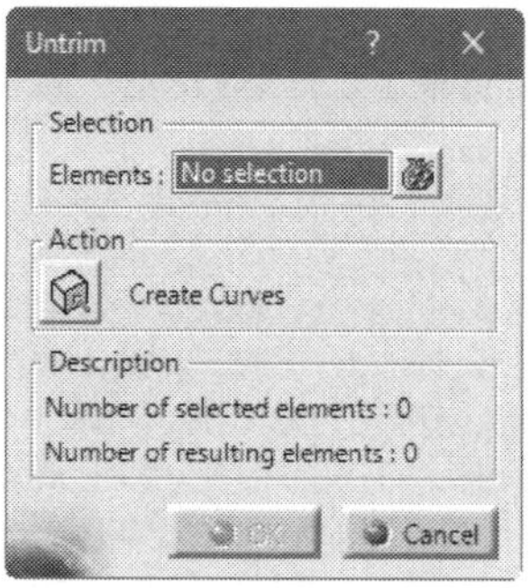

*Figure 11-17 The **Untrim** dialog box*

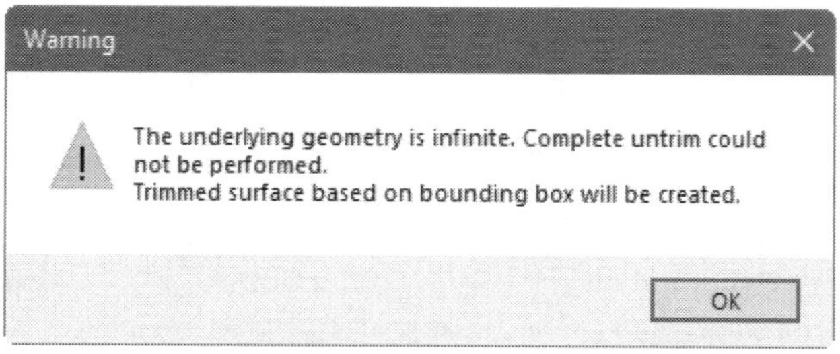

*Figure 11-18 The **Warning** message box*

The number of selected elements and the number of resulting elements will be displayed in the **Untrim** dialog box. Choose the **OK** button from it; a new surface will be created and the **Surface Untrim.xxx** feature will be added in the Specification tree. Figure 11-19 shows the trimmed surface and Figure 11-20 shows the untrimmed surface.

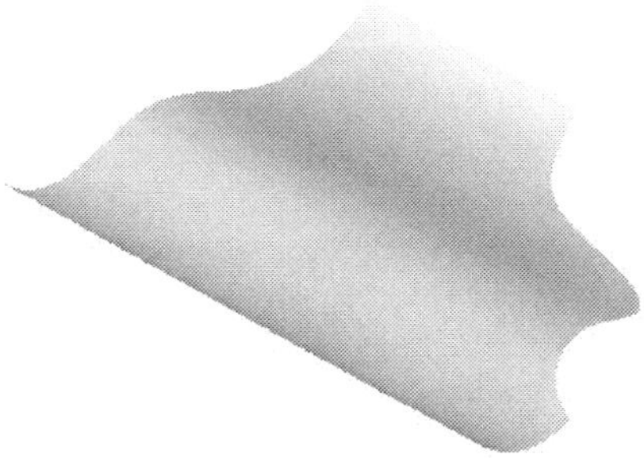

Figure 11-19 *The trimmed surface*

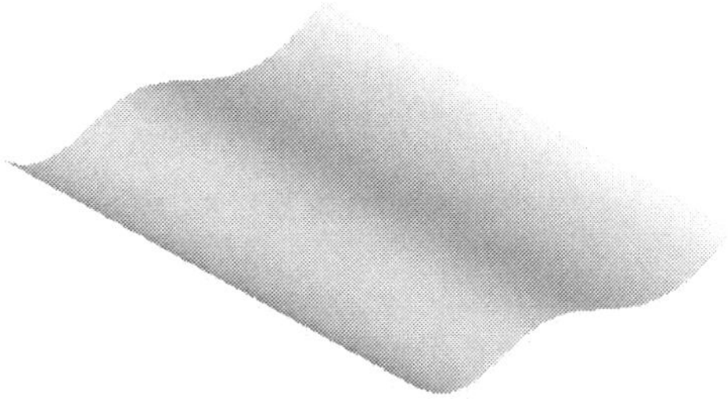

Figure 11-20 *The untrimmed surface*

Creating Boundary Curves

Menubar: Insert > Operations > Boundary
Toolbar: Operations > Extracts sub-toolbar > Boundary

The **Boundary** tool is used to extract boundary curves from surfaces. You can select a surface or an edge to create a boundary. To do so, choose the **Boundary** button from the **Extracts** sub-toolbar in the **Operations** toolbar; the **Boundary Definition** dialog box will be invoked, as shown in Figure 11-21.

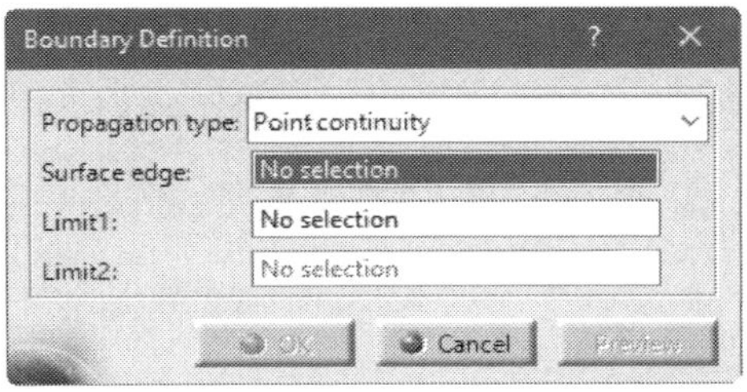

Figure 11-21 *The* ***Boundary Definition*** *dialog box*

On invoking this dialog box, you will be prompted to select the propagation type and an edge. Retain the default settings for the propagation type and select an edge of a surface from the geometry area to extract boundary curves from the surface. Note that, if you select a surface, the **Propagation type** option will not be available in the **Boundary Definition** dialog box.

In this dialog box, the **Propagation type** drop-down list provides you four options to create boundary curves: the complete boundary, point continuity, Tangency community, and No propagation Options. The **Complete boundary** option is used to create all boundaries associated with the surface whose edge is selected. Figure 11-22 shows the selected edge and preview of the resultant boundary curve to be created using the **Complete boundary** option from the **Propagation type** drop-down list. The **Point continuity** option is used to create boundary with point continuity. Figure 11-23 shows preview of the boundary curve with the **Point continuity** option selected.

If you select the **Tangent continuity** option, all the edges tangent to the selected edge will automatically be selected to create the boundary curve. Figure 11-24 shows preview of the boundary curve with the **Tangent continuity** option selected. The **No propagation** option is used to create boundary with the selected edge only. Figure 11-25 shows preview of the boundary curve with the **No propagation** option selected.

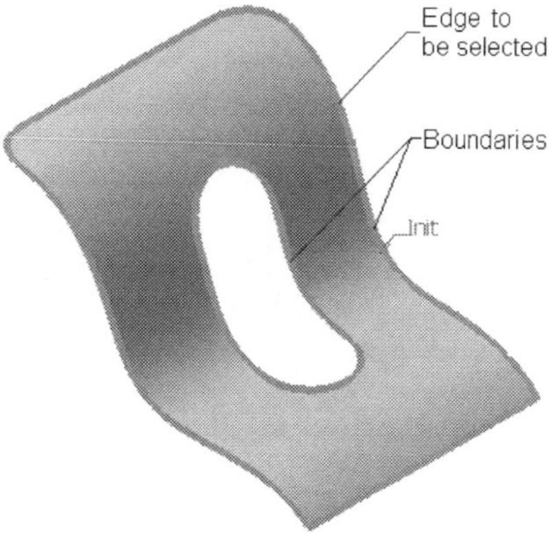

Figure 11-22 *Preview of the boundary curve on using the **Complete boundary** option*

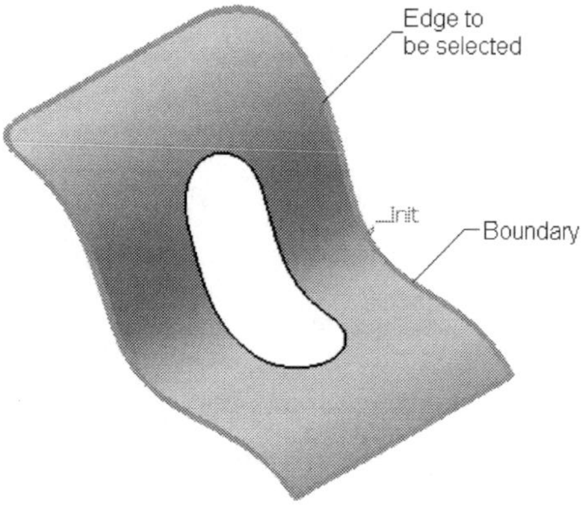

Figure 11-23 *Preview of the boundary curve on using the **Point continuity** option*

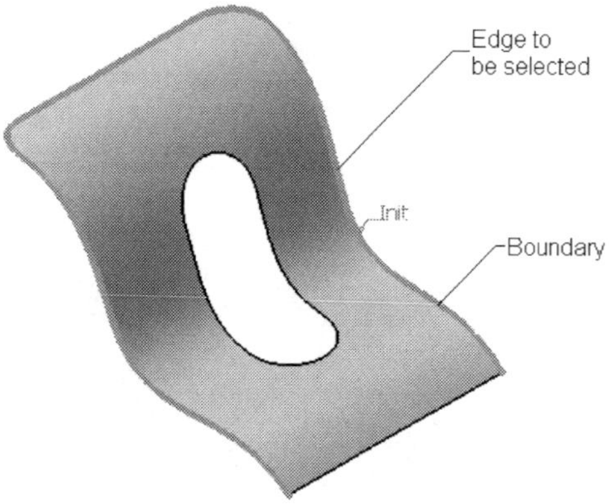

Figure 11-24 *Preview of the boundary curve with the **Tangent continuity** option selected*

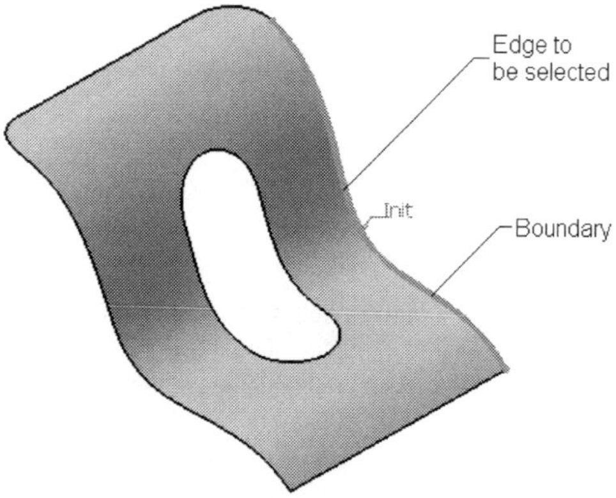

Figure 11-25 *Preview of the boundary curve with the **No propagation** option selected*

Depending on the propagation type selected, the boundary created is highlighted in green color. You can also limit the propagation of a boundary along an edge by defining the limits. You can select a predefined point or create a point by choosing an option from the shortcut menu displayed on right-clicking in the geometry area. Choose the **OK** button from the **Boundary Definition** dialog box to complete the boundary operation.

Tip

You can also extract boundaries from a solid body. The procedure to do so is same as the one discussed while extracting boundaries from a surface.

Extracting Geometry

Menubar:	Insert > Operations > Extract
Toolbar:	Operations > Extracts sub-toolbar > Extract

The **Extract** tool is used to create geometry by extracting an edge or a surface from an existing surface or a solid geometry. To extract geometry from a solid geometry, create a solid geometry in the **Part Design** workbench and then switch back to the **Wireframe and Surface Design** workbench using the **Start** menu. Next, choose the **Extract** tool from the **Extracts** sub-toolbar; the **Extract Definition** dialog box will be displayed, as shown in Figure 11-26, and you will be prompted to select the element to be extracted.

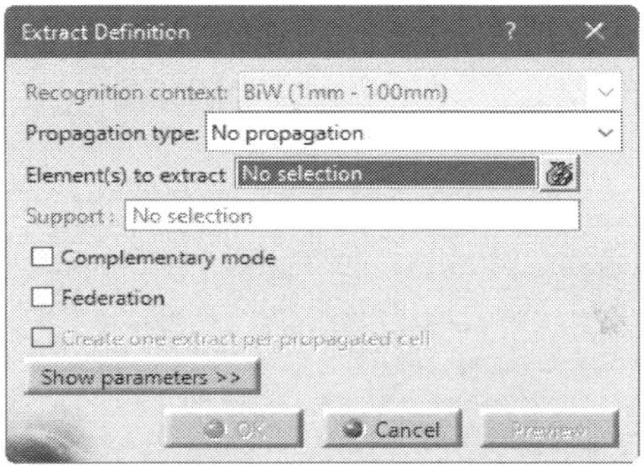

*Figure 11-26 The **Extract Definition** dialog box*

In the **Propagation type** drop-down list, the **No propagation** option is selected by default. You can select the required propagation type from this drop-down list. Next, select an element from the graphics window; the selected element will be displayed in green. Select the **Complementary mode** check box to select the complementary portion of the specified element. On choosing the **Show parameters** button, the dialog box will expand and will show three more options, namely **Distance Threshold**, **Angular Threshold**, and **Curvature Threshold**. These options are applicable only for curves. After selecting the required geometry and setting other options, choose the **OK** button from the **Extract Definition** dialog box; the selected element will be extracted.

Transformation

Transformation is used to change the physical position of the geometry through different processes such as translating, rotating, scaling, and so on. You can apply transformations by using the tools available in the **Transformations** sub-toolbar. This sub-toolbar is invoked by clicking on the down arrow on the right of the **Translate** tool in the **Operations** toolbar. Figure 11-27 shows the **Transformations** sub-toolbar.

The use of the tools in **Transformations** sub-toolbar is discussed next.

Translating Elements

Menubar:	Insert > Operations > Translate
Toolbar:	Operations > Transformations sub-toolbar > Translate

The **Translate** tool is used to translate one or more than one element in the specified direction. You can translate points, lines, surfaces, and solid bodies by using this tool. To translate an element, choose the **Translate** tool from the **Transformations** sub-toolbar; the **Translate Definition** dialog box will be displayed, as shown in Figure 11-28. Also, you will be prompted to select the element to translate.

Select the element to translate from the geometry area. If you want to select more elements, choose the button on the right of the **Element** display box; the **Elements** dialog box will be displayed. Select the required entities from the geometry area, and then close the **Elements** dialog box; you will be prompted to select or choose the direction. Select the translation direction and set the translation value in the **Distance** spinner.

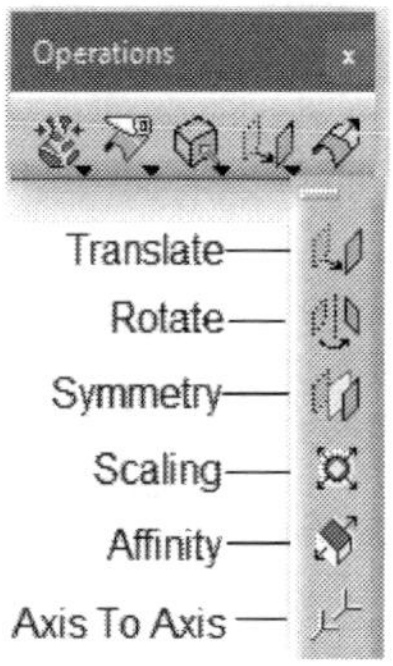

Figure 11-27 *The* ***Transformations*** *sub-toolbar*

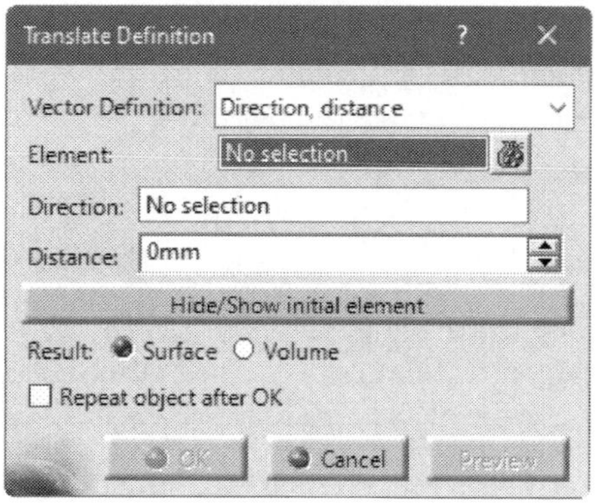

Figure 11-28 *The* ***Translate Definition*** *dialog box*

The **Hide/Show initial element** button switches the display of the selected element. Using this button, you can turn on or off the display of the selected elements.

You will notice that the **Result** area is disabled by default. The options in this area are available only when you start the file in the conventional design mode by clearing the **Enable hybrid design** check box from the **New Part** dialog box. With the **Surface** radio button selected, the resulting translation will be a surface body. If you select the **Volume** radio button, the resulting translation will be a volume body. After setting the parameters, choose the **OK** button from the **Translate Definition** dialog box.

If the **Repeat object after OK** check box is selected, and you choose the **OK** button from the **Translate Definition** dialog box; the **Object Repetition** dialog box will be displayed. Set the number of instances in the **Instance(s)** spinner. The **Create in a new Body** check box is selected by default. As a result, the copied objects will be created in the new body. Clear this check box and choose the **OK** button from the **Object Repetition** dialog box. If the **Create in a new Body** check box is selected, then all the instances other than the first instance, will be placed under the **Ordered Geometrical Set.xxx** in the Specification tree. Figure 11-29 shows the surface to be selected and the directional reference and Figure 11-30 shows the resulting translated feature.

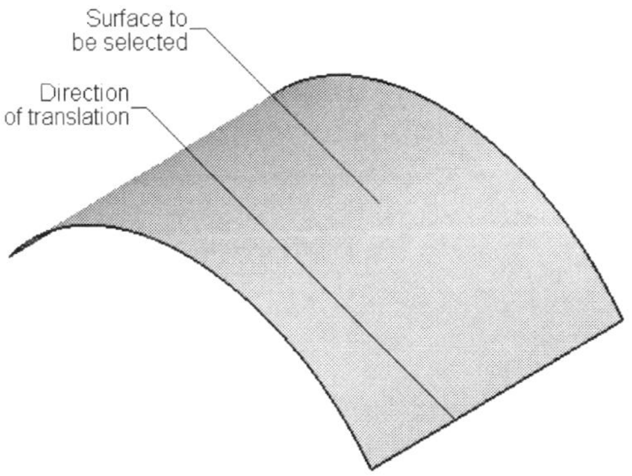

Figure 11-29 *Surface and directional reference*

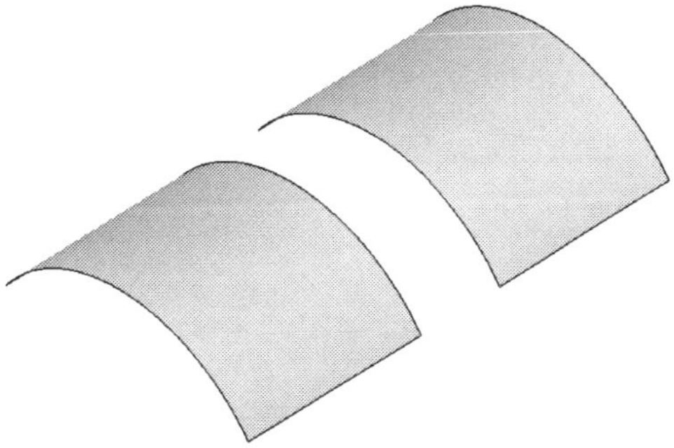

Figure 11-30 *Resulting translated surface*

> **Tip**
> *While working in the hybrid design mode, you need to create a new geometric set to activate the option in the* ***Result*** *area of the* ***Translate Definition*** *dialog box.*
>
> *To create a new geometric set, choose* ***Insert > Geometric Set*** *option from the menubar; the* ***Insert Geometrical Set*** *dialog box will be displayed. Specify the name of the geometric set and choose the* ***OK*** *button; a new geometric set will be created and displayed in the Specification tree. From now onward, all the surface and reference features will be saved in the newly created geometric set.*

On selecting the **Point to point** option from the **Vector Definition** drop-down list, you need to specify the start and end points to define the translation distance. If you select the **Coordinates** option, you need to specify the coordinates of the destination point.

Rotating Elements

Menubar: Insert > Operations > Rotate
Toolbar: Operations > Transformations sub-toolbar > Rotate

The **Rotate** tool is used to rotate one or more elements about a specified axis. You can rotate points, lines, surfaces, and solid bodies by using this tool. To do so, choose the **Rotate** tool from the **Transformation** sub-toolbar; the **Rotate Definition** dialog box will be displayed, as shown in Figure 11-31.

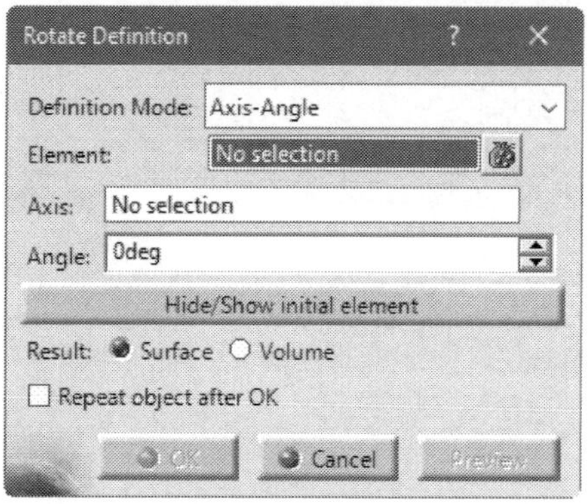

***Figure 11-31** The **Rotate Definition** dialog box*

Select the rotation method from the **Definition Mode** drop-down list and then select the element that you need to rotate; the name of the selected element will be displayed in the **Element** selection box and you will be prompted to select the rotation axis. Select it from the geometry area and set the value of the rotation angle in the **Angle** spinner; the preview of the rotated element will be displayed. The **Hide/Show initial element** button is used to hide the parent element. Select the **Repeat object after OK** check box to create multiple instances of a selected element. After setting all parameters, choose the **OK** button from the **Rotate Definition** dialog box. Figure 11-32 shows the surface and the axis of rotation to be selected and Figure 11-33 shows the resulting rotated surface with repeated instances.

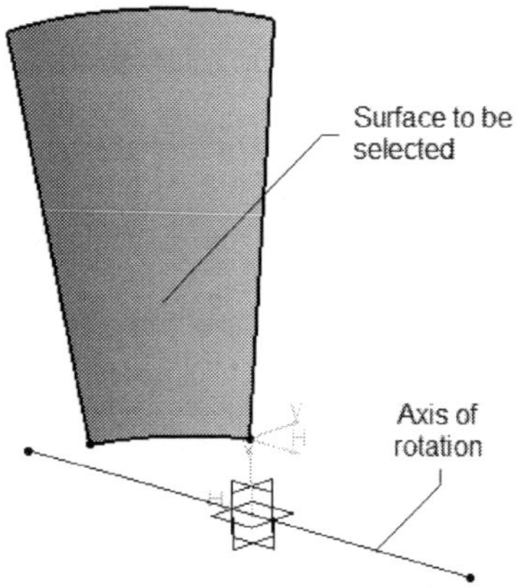

***Figure 11-32** The surface and the axis of rotation*

***Figure 11-33** Resulting rotated surface with repeated instances*

Creating Symmetry Elements

Menubar: Insert > Operations > Symmetry
Toolbar: Operations > Transformations sub-toolbar > Symmetry

The **Symmetry** tool is used to mirror a selected element about a selected reference. On invoking this tool, the **Symmetry Definition** dialog box will be displayed, as shown in Figure 11-34.

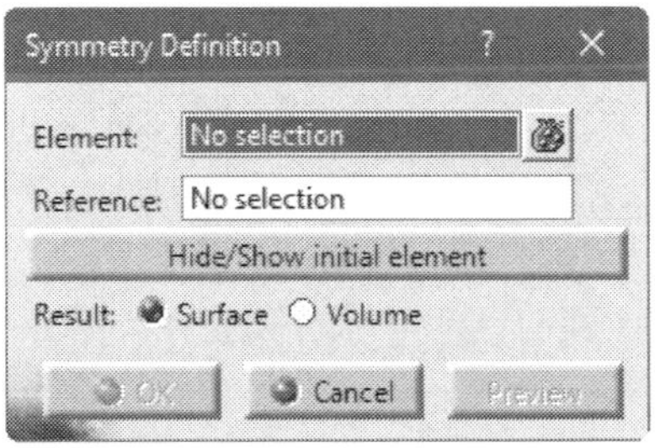

*Figure 11-34 The **Symmetry Definition** dialog box*

Select the element that you need to mirror from the geometry area. You can also mirror an axis system. On selecting the element to be mirrored, you will be prompted to select a reference point, line, or plane. Select the reference element from the geometry area; the preview of the mirrored element will be displayed in the geometry area. Choose the **OK** button from the **Symmetry Definition** dialog box. Figure 11-35 shows the surface to be mirrored and the reference element to be selected. Figure 11-36 shows the resulting mirrored surface.

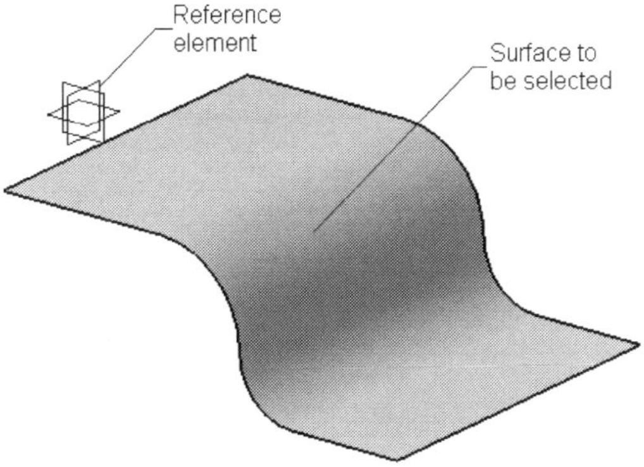

Figure 11-35 *The surface and the reference element*

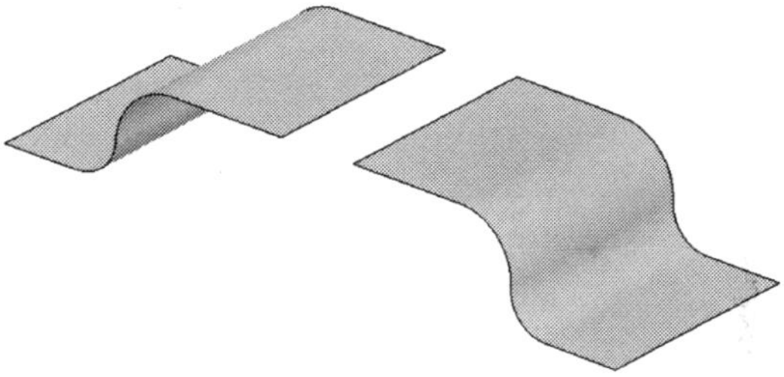

Figure 11-36 *Resulting mirrored surface*

Note

*To create an axis system, choose **Insert > Axis System** from the menubar; the **Axis System Definition** dialog box will be displayed. Using the options in this dialog box, you can create the axis system.*

Scaling the Elements

Menubar: Insert > Operations > Scaling
Toolbar: Operations > Transformations sub-toolbar > Scaling

The **Scaling** tool is used to scale elements with respect to a selected reference element. To do so, choose the **Scaling** tool from the **Transformations** sub-toolbar; the **Scaling Definition** dialog box will be displayed, as shown in Figure 11-37.

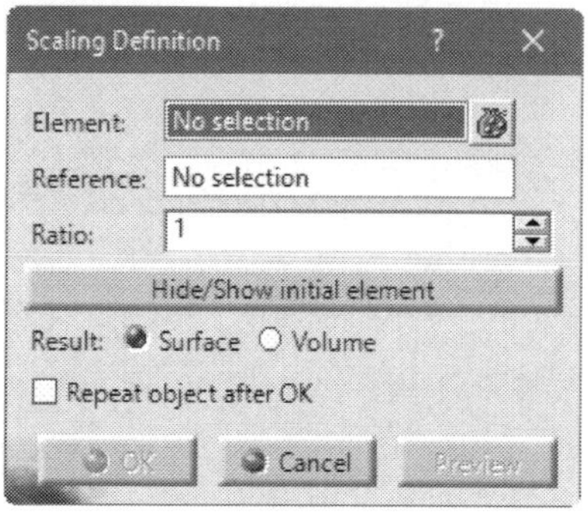

Figure 11-37 *The* ***Scaling Definition*** *dialog box*

Select the element to be scaled from the geometry area and then select a reference that will be fixed while scaling. Set the value of the scale factor in the **Ratio** spinner, and then choose the **OK** button from the **Scaling Definition** dialog box. Figure 11-38 shows the surface and the reference to be selected and Figure 11-39 shows the resulting scaled surface.

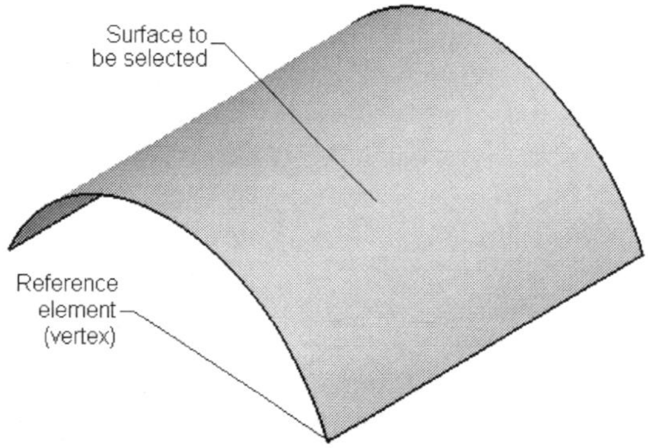

Figure 11-38 *The surface and the reference element*

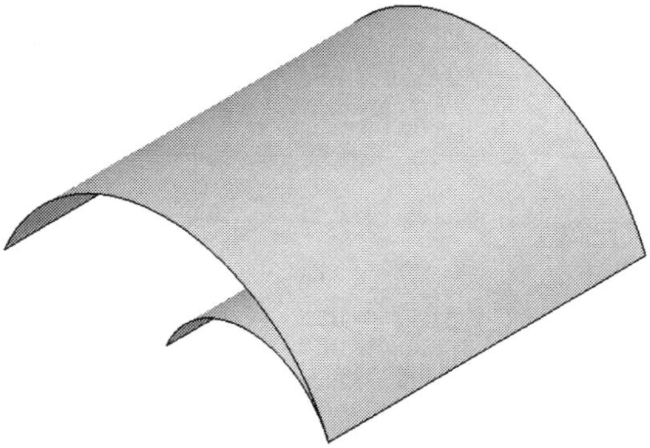

Figure 11-39 *Resulting scaled surface*

Non-uniform Scaling of Elements

Menubar: Insert > Operations > Affinity
Toolbar: Operations > Transformations sub-toolbar > Affinity

The **Affinity** tool is used to scale elements non-uniformly. To do so, invoke this tool from the **Transformations** sub-toolbar; the **Affinity Definition** dialog box will be displayed, as shown in Figure 11-40. By default, a coordinate system will be displayed at the origin, and you will be prompted to select the element to be transformed by affinity. Select the element to be transformed from the geometry area; the preview of the element with the default settings, will be displayed. By default, the origin is selected as the reference on which the coordinate system will be displayed. You can also define a user-defined reference using the options from the **Axis system** area. The **X**, **Y**, and **Z** scale factors are set to 1 by default. Set the individual scaling parameters along the X, Y, and Z axes in the spinners provided in the **Ratios** area. After setting all parameters, choose the **OK** button from the **Affinity Definition** dialog box. Figure 11-41 shows the surface selected to be scaled and Figure 11-42 shows the resulting non-uniformly scaled surface.

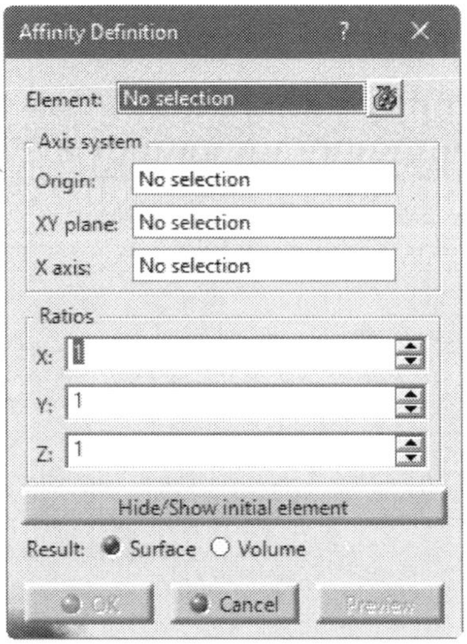

*Figure 11-40 The **Affinity Definition** dialog box*

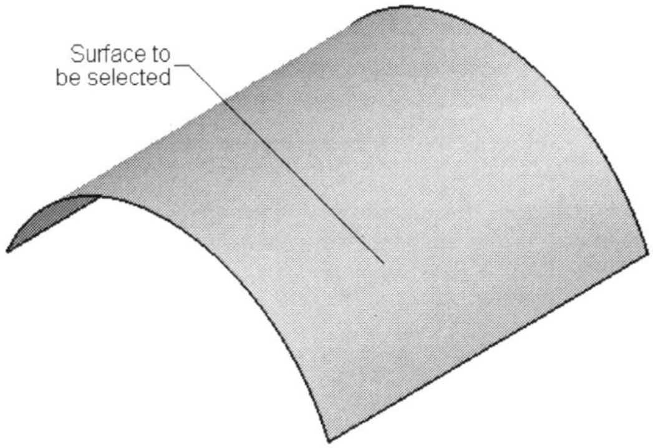

Figure 11-41 Surface selected to be scaled

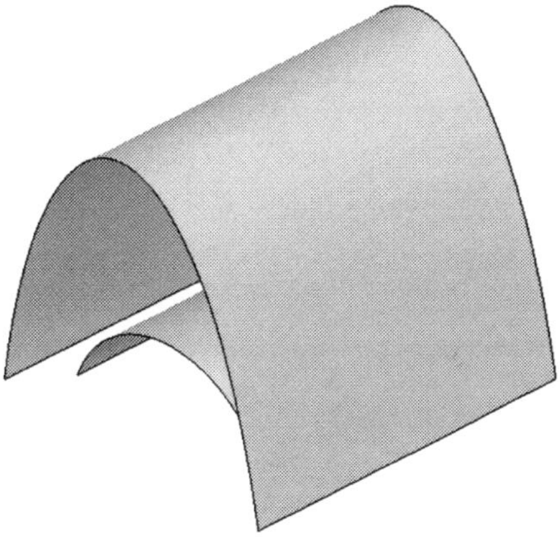

Figure 11-42 Resulting non-uniformly scaled surface

Transforming an Element from Axis to Axis

Menubar: Insert > Operations > Axis To Axis
Toolbar: Operations > Transformations sub-toolbar > Axis To Axis

The **Axis To Axis** tool is used to transform an element from one axis system to the other. On invoking this tool, the **Axis To Axis Definition** dialog box will be displayed, as shown in Figure 11-43, and you will be prompted to select an element to be transformed from one axis system to another. Select the element to be transformed from the geometry area.

Note that you can also select an axis system to be transformed. On selecting the element to be transformed, you will be prompted to select a reference axis. After selecting the reference axis, you will be prompted to select the target axis. After doing so, the preview of the transformed element will be displayed. Choose the **OK** button from the **Axis To Axis Definition** dialog box. Figure 11-44 shows the surface, reference axis, and target axis to be selected and Figure 11-45 shows the resulting transformed surface.

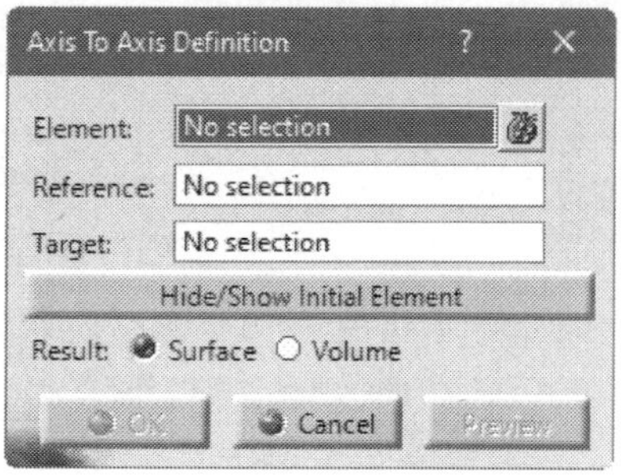

Figure 11-43 The ***Axis To Axis Definition*** *dialog box*

To create an axis system while performing a transformation operation, right-click in the required field to invoke the shortcut menu. Next, choose **Create Axis System** from the shortcut menu.

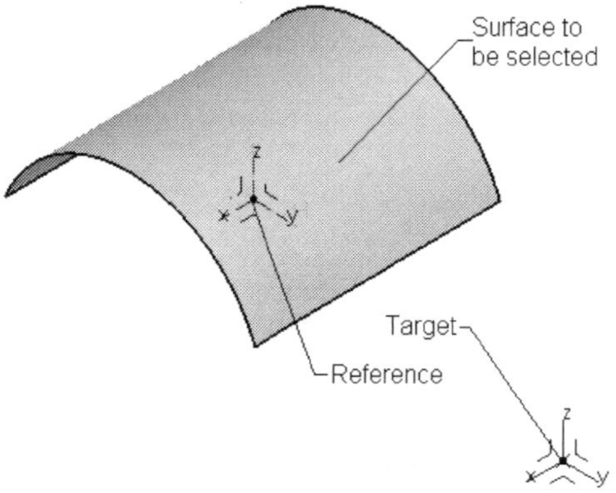

Figure 11-44 Surface and axes to be selected

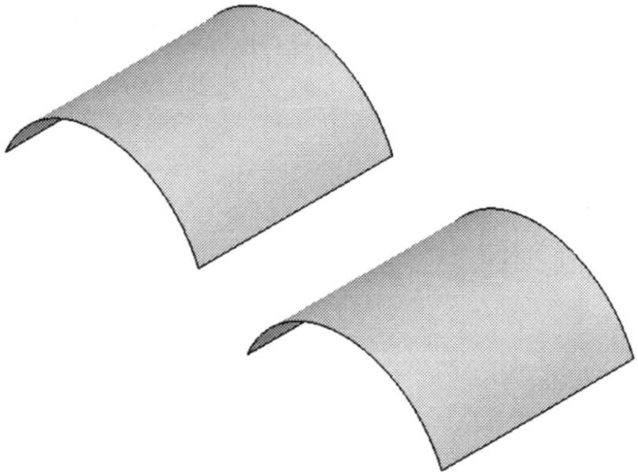

Figure 11-45 Resulting transformed surface

Extrapolating Surfaces and Curves

Menubar: Insert > Operations > Extrapolate
Toolbar: Operations > Extrapolate

The **Extrapolate** tool is used to extend a surface or a curve to a desired distance while maintaining the tangency or the curvature continuity. On invoking this tool, the **Extrapolate Definition** dialog box will be displayed, as shown in Figure 11-46. Also, you will be prompted to select a boundary or a curve.

Select an edge of the surface body as the boundary and the surface to extrapolate will be selected by default; the preview of the extrapolated surface will be displayed. You can select the feature termination condition from the **Type** drop-down list. By default, the **Length** option is selected in the **Type** drop-down list. Set the value of the length of extrapolation in the **Length** spinner. On selecting the **Up to element** option from the **Type** drop-down list as the feature termination condition, the **Up to** selection box will be activated. Now, you can select the element upto which you need to extrapolate from the geometry area, the name of selected element will be displayed in the **Up to** selection box.

You can also define the continuity of the extrapolated surface with the reference surface. By default, the tangent continuity is selected. To define the curvature continuity, select the **Curvature** option from the **Continuity** drop-down list of the **Extrapolate Definition** dialog box. The **Tangent** option is selected in the Extremities drop-down list by default. This option allows the extreme ends of the extrapolated surface to be tangent to the extreme ends of the reference surface. If you select the Normal option, the extreme edges of the extrapolated surface will be normal to the selected boundary.

You can also define the propagation using the options in the **Propagation mode** drop-down list. The **Assemble result** check box is selected by default. Therefore, the extrapolated surface will be merged with the reference surface. If you clear this check box, the extrapolated surface will be created as a separate surface.

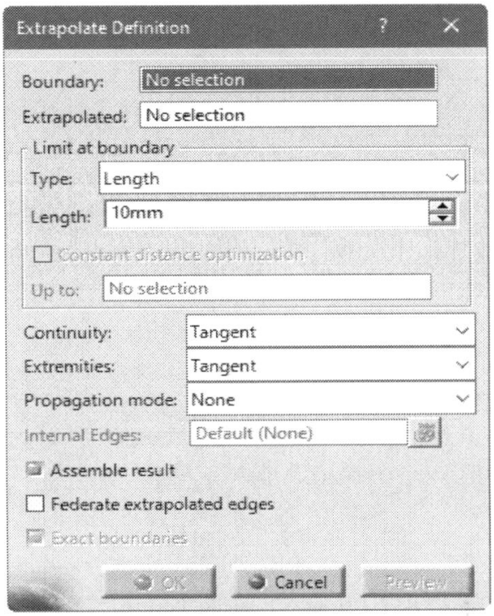

Figure 11-46 The ***Extrapolate Definition*** *dialog box*

The **Constant distance optimization** check box is used to extrapolate the surface with a constant length and create a surface without deformation. Note that this check box will not be available when the **Extend extrapolated edges** check box is selected and the **Up-to-element** option is selected in the **Type** drop down list. The **Internal Edges** selection box enables you to determine a privileged direction for the extrapolation. In this case, you can only select edges in contact with the boundary. Note that this option will not be available with **Curvature** option selected in the **Continuity** drop-down list.

On selecting the **Federate extrapolated edges** check box, you can reconnect the features based on the elements of the extrapolated surface. This option is especially useful if you work within an ordered geometrical set environment. Note that this check box will only be activated when the **Tangent** option is selected in the **Continuity** and **Extremities** drop-down lists. After setting all parameters, choose the **OK** button from the **Extrapolate Definition** dialog box.

Figure 11-47 shows the boundary and the surface to be selected and Figure 11-48 shows the resulting extrapolated surface.

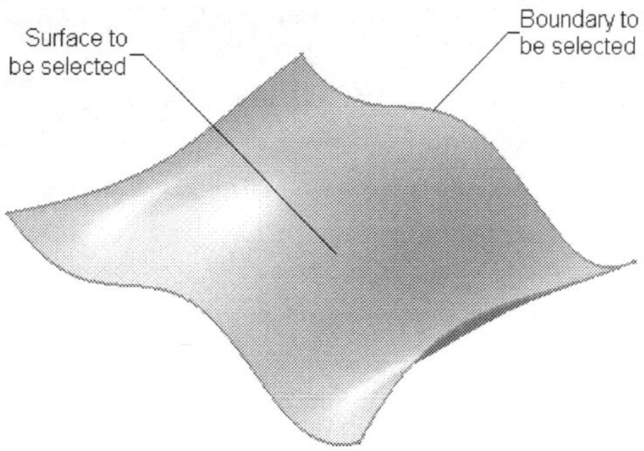

Figure 11-47 *Boundary and surface to be selected*

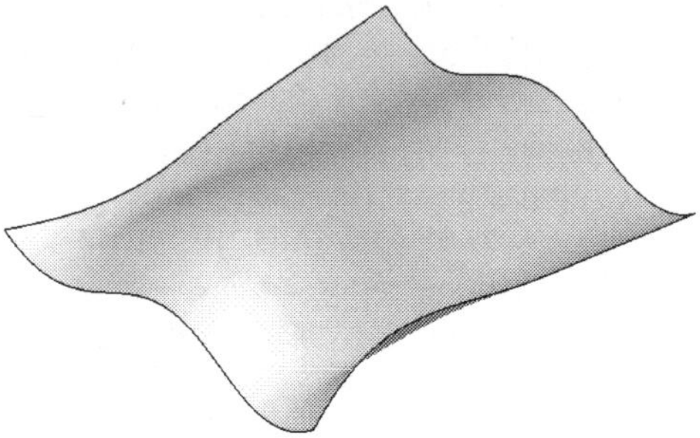

Figure 11-48 *Resulting extrapolated surface*

To extrapolate a curve, invoke the **Extrapolate Definition** dialog box and select the endpoint of the curve as the boundary along which you need to extrapolate it. On doing this, you will notice that the curve gets selected in the **Extrapolated** selection box as the element to be extrapolated. Set the length of extrapolation in the **Length** spinner and choose the **OK** button from the **Extrapolate Definition** dialog box. Figure 11-49 shows the boundary and the curve to be selected and Figure 11-50 shows the resulting extrapolated curve.

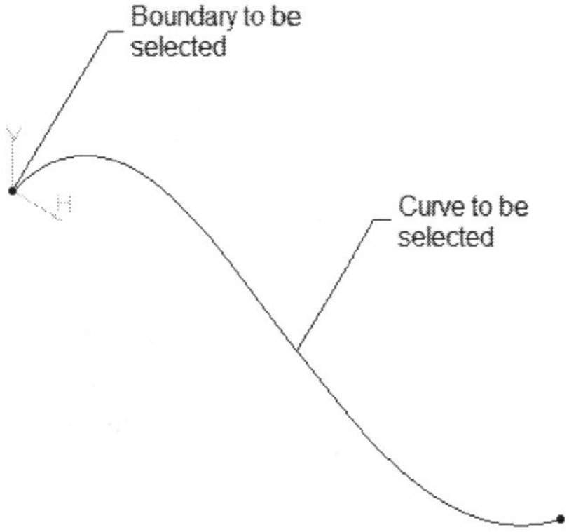

Figure 11-49 *Boundary and curve to be selected*

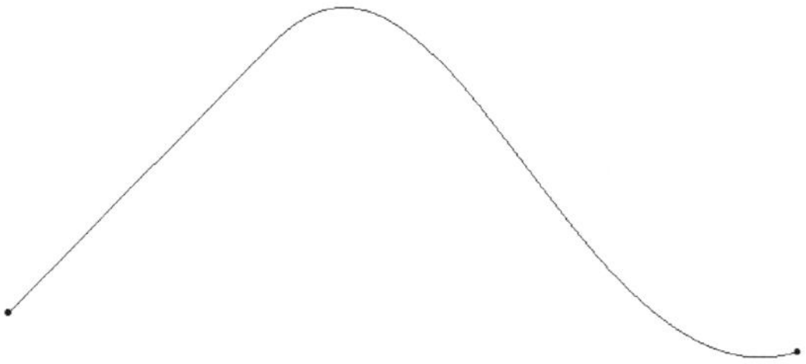

Figure 11-50 *Resulting extrapolated curve*

Splitting a Solid Body with a Surface

Menubar:	Insert > Surface-Based Features > Split
Toolbar:	Surface-Based Features (Extended) > Split

The **Split** tool is used to divide a solid body into two or more than two parts using a surface or a plane. To split a solid body, select a surface that will be used as the splitting tool and then switch back to the **Part** workbench by choosing **Start > Mechanical Design > Part Design** from the menubar. Now, invoke the **Surface-Based Features** toolbar, if not displayed by default. Figure 11-51 shows the **Surface-Based Features (Extended)** toolbar.

Choose the **Split** tool from the **Surface-Based Features (Extended)** toolbar; the **Split Definition** dialog box will be displayed, as shown in Figure 11-52. By default, the **None** option is selected in the **Extrapolation type** drop-down list. If you select the **Tangent** option, the body will get split tangentially from the ends of the splitting surface. When you select the **Curvature** option, the splitting surface will split the body with respect to the curvature of the splitting surface.

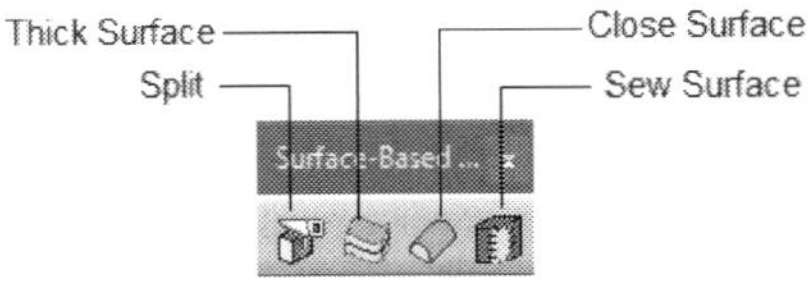

Figure 11-51 *The* ***Surface-Based Features (Extended)*** *toolbar*

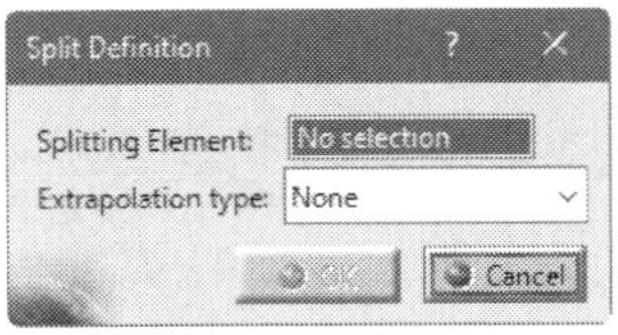

Figure 11-52 *The* ***Split Definition*** *dialog box*

Select the splitting surface from the geometry area; the name of the selected surface will be displayed in the **Splitting Element** display box. Also, an arrow will be displayed on the splitting surface. Click on the arrow to flip the direction in which you need to retain the solid body. Next, choose the **OK** button from the **Split Definition** dialog box to split the solid body. Figure 11-53 shows the body to be split and the splitting surface and Figure 11-54 shows the resulting solid body after hiding the splitting surface.

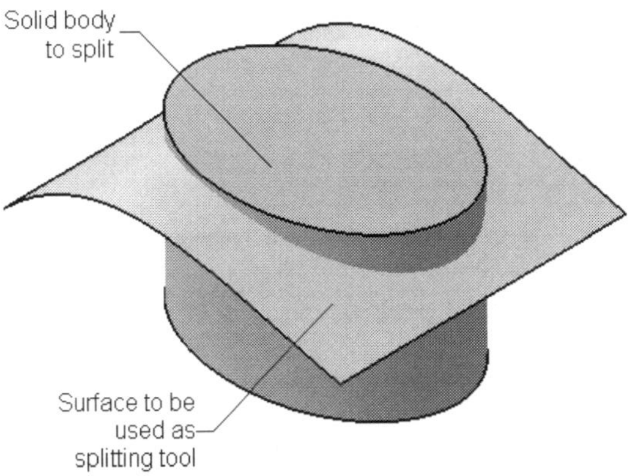

Figure 11-53 *Body to be split and the splitting surface*

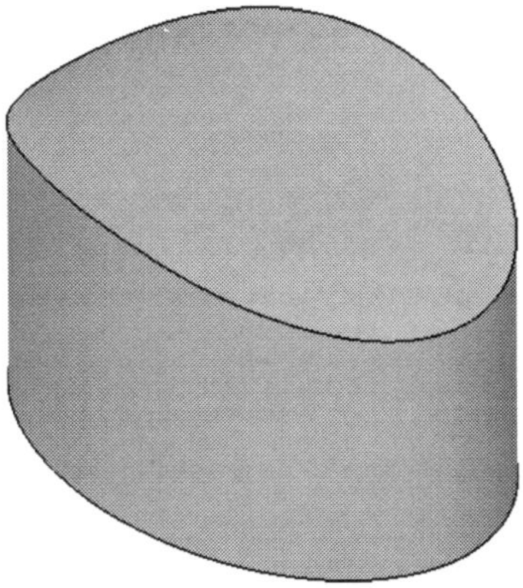

Figure 11-54 *Resulting solid body after hiding the splitting surface*

SOLIDIFYING SURFACE MODELS

After creating the surface model, you need to apply thickness to convert it into a solid model. In this section, you will learn about the tools that will be used to perform the surface based operations in the **Part** workbench.

Adding Thickness to a Surface

Menubar: Insert > Surface-Based Features > Thick Surface
Toolbar: Surface-Based Features (Extended) > Thick Surface

After creating surfaces and joining them together, you may need to thicken them. To do this, first you need to invoke the **Part Design** workbench and then choose the **Thick Surface** tool from the **Surface-Based Features (Extended)** toolbar. On doing so, the **ThickSurface Definition** dialog box will be displayed, as shown in Figure 11-55.

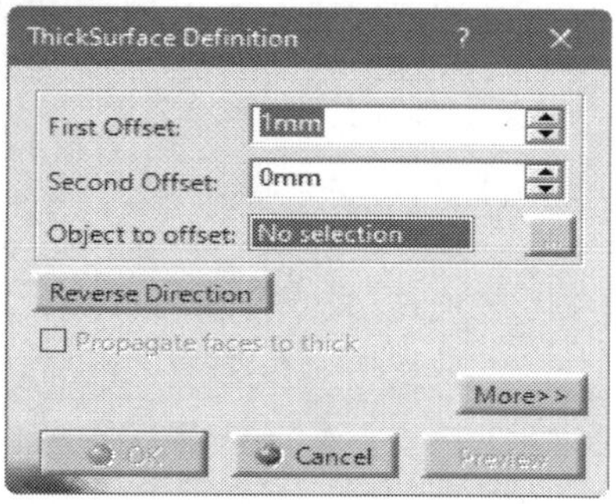

Figure 11-55 *The* ***ThickSurface Definition*** *dialog box*

Select the surface to be thickened from the geometry area; the preview of the thickened surface with the default settings will be displayed. Set the value of the thickness in the **First Offset** spinner. You can also specify the value on the other side of the surface using the **Second Offset** spinner. You can flip the direction of material addition using the **Reverse Direction** button. Choose the **Preview** button from the **ThickSurface Definition** dialog box to display preview of the thickened surface. After setting all parameters, choose the **OK** button from the **ThickSurface Definition** dialog box. Figure 11-56 shows the surface to be thickened and Figure 11-57 shows the solid model created by thickening the surface. In Figure 11-57, the resulting solid body is shown after hiding the splitting surface.

Figure 11-56 *Surface to be thickened*

Figure 11-57 *Resulting solid model*

Creating a Solid Body from a Closed Surface Body

Menubar: Insert > Surface-Based Features > Close Surface
Toolbar: Surface-Based Features (Extended) > Close Surface

The **Close Surface** tool is used to convert a closed surface body into a solid body. To create a solid body, you need to ensure that all the surfaces forming a closed body are joined, thus resulting in a single surface. The surface may not be closed at its ends, as shown in Figure 11-58. The **Close Surface** tool is used to convert closed surface body into a solid body keeping both the ends as limit, as shown in Figure 11-59.

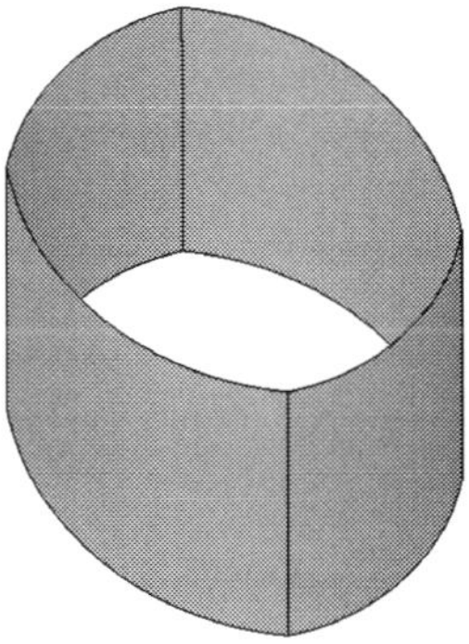

Figure 11-58 Surface with open ends

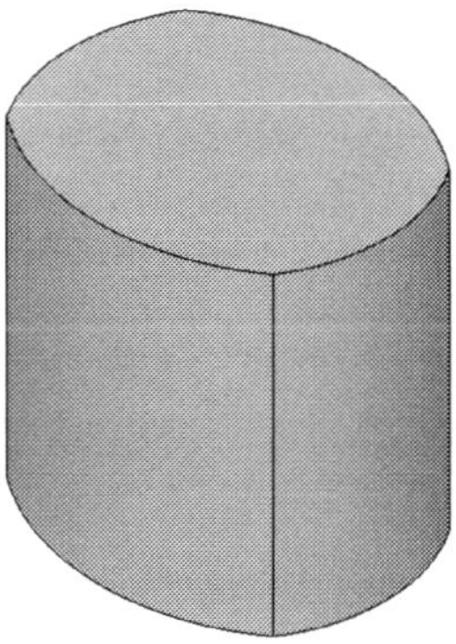

Figure 11-59 Resulting solid body

On invoking the **Close Surface** tool, the **CloseSurface Definition** dialog box will be displayed, as shown in Figure 11-60. Select the surface to be closed from the geometry area and choose the **OK** button. You may need to hide the surface body after creating the solid body.

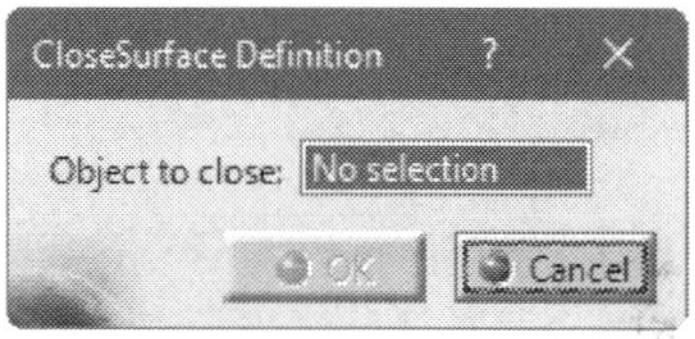

*Figure 11-60 The **CloseSurface Definition** dialog box*

Sewing a Surface to a Solid Body

Menubar:	Insert > Surface-Based Features > Sew Surface
Toolbar:	Surface-Based Features (Extended) > Sew Surface

The **Sew Surface** tool is used to sew a surface to the faces of a solid body, thus merging the faces with the surface. This may result in material addition or material removal. In some cases, it may result in both, depending on the surface selected. To understand the use of this tool, create a solid body, as shown in Figure 11-61, and then create a surface, as shown in Figure 11-62.

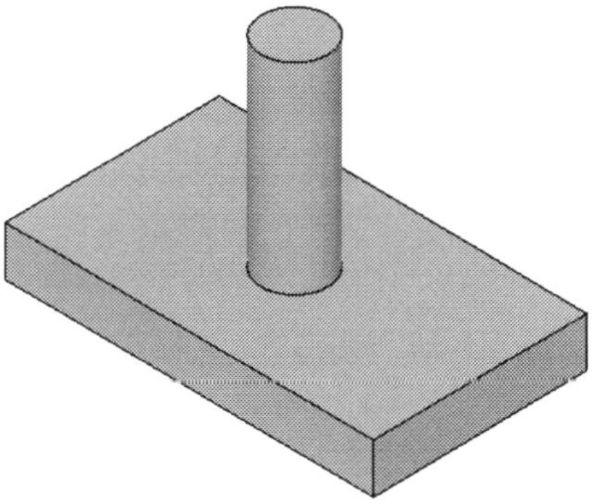

Figure 11-61 Solid body

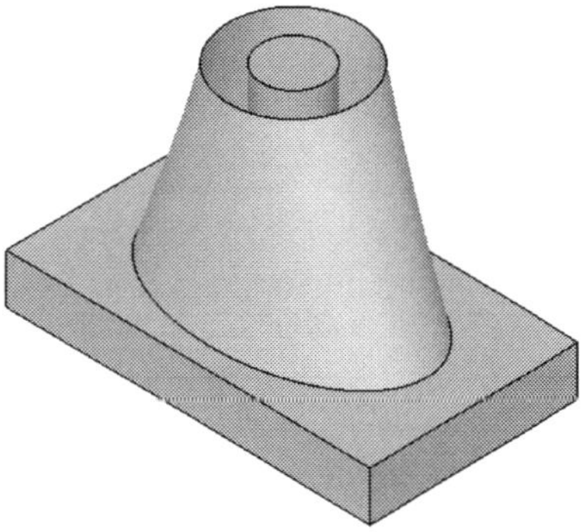

Figure 11-62 Surface created on the solid body

Next, switch to the **Part** workbench and then choose the **Sew Surface** tool from the **Surface-Based Features (Expanded)** toolbar; the **Sew Surface Definition** dialog box will be displayed, as shown in Figure 11-63.

Click in the **Object to sew** display box and select the surface, as shown in Figure 11-64. If the arrow points outward, click on it to flip the direction. Now, click once in the **Faces to remove** display box and select the face of the solid body from the geometry area. Next, choose the **OK** button from the **Sew Surface Definition** dialog box; the material will be added to the solid body. The resulting model after hiding the surface body is shown in Figure 11-65. Figure 11-66 shows another face to be removed and the surface to be sewed. Figure 11-67 shows the resulting solid body.

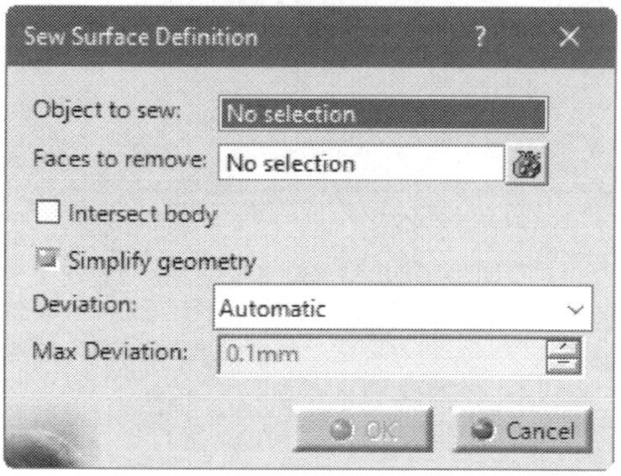

***Figure 11-63** The **Sew Surface Definition** dialog box*

If the surface boundary entirely lies on the solid, you need to select the **Simplify geometry** check box to sew the surface. However, if the surface boundary crosses the solid, you need to select the **Intersect body** check box to sew the surface.

Note that in the **Sew Surface Definition** dialog box, you can define deviation from the **Deviation** drop-down list. By default, the **Automatic** option is selected in this drop-down list. As a result, the **Max Deviation** spinner remains inactive with a default value of 0.1 mm. When you select the **Manual** option, the **Max Deviation** spinner gets activated prompting you to specify maximum value of deviation in this spinner.

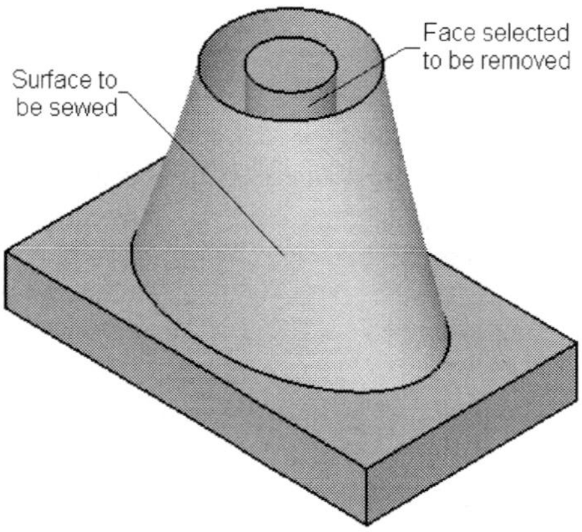

***Figure 11-64** Surface to be sewed and the face to be removed*

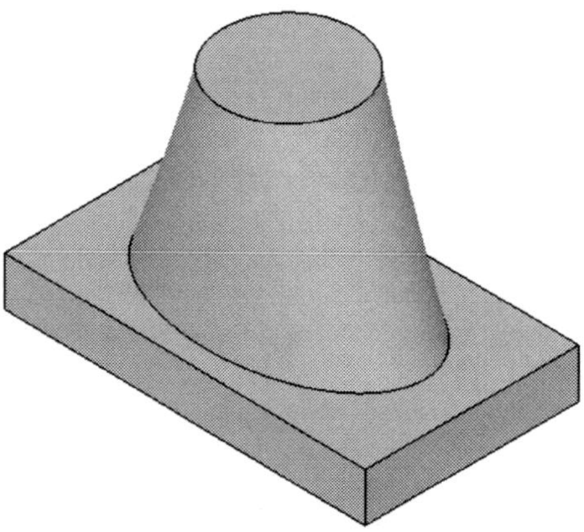

***Figure 11-65** Resulting solid body*

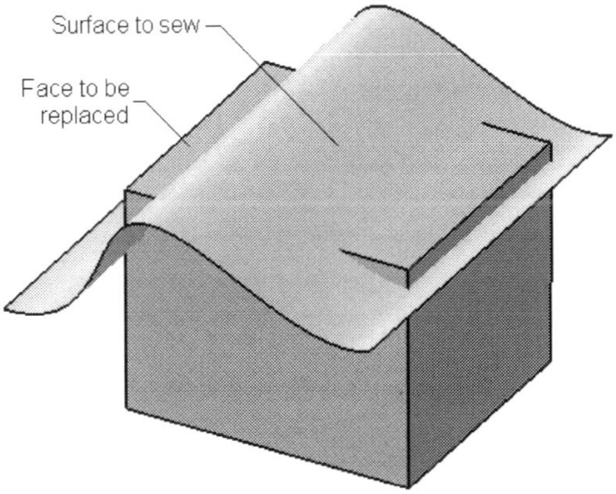

Figure 11-66 Surface to be sewed and the face to be removed

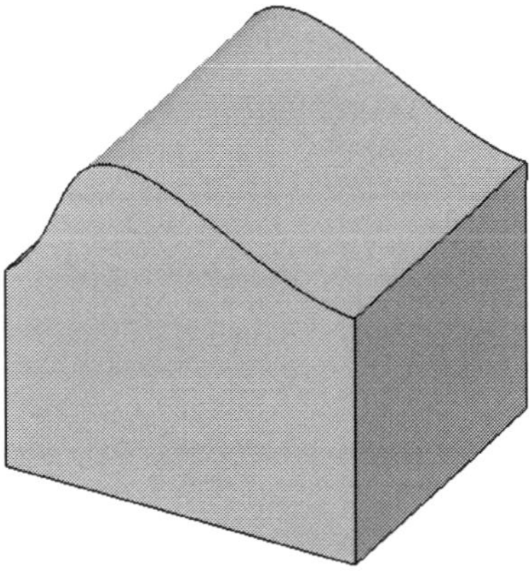

Figure 11-67 Resulting solid body

TUTORIALS

Tutorial 1

In this tutorial, you will create the model of the back cover of a toy monitor shown in Figure 11-68. To create this model, you will use the tools in the **Wireframe and Surface Design** workbench and the **Part Design** workbench. After creating the surface model, you will convert it into a solid body. The orthographic views and dimensions of the model are shown in Figure 11-69.

(Expected time: 1hr)

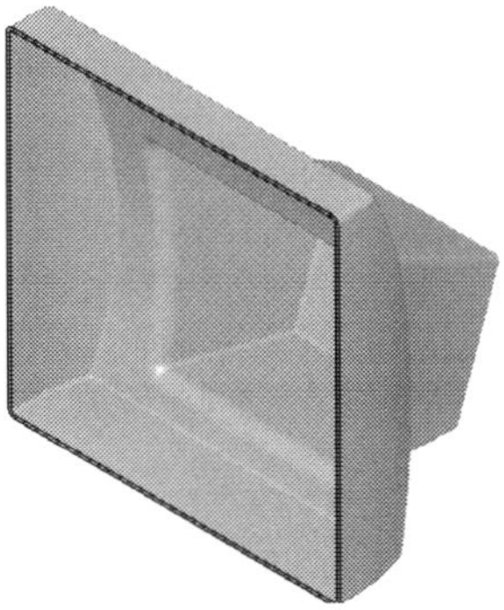

Figure 11-68 Back cover of a toy monitor

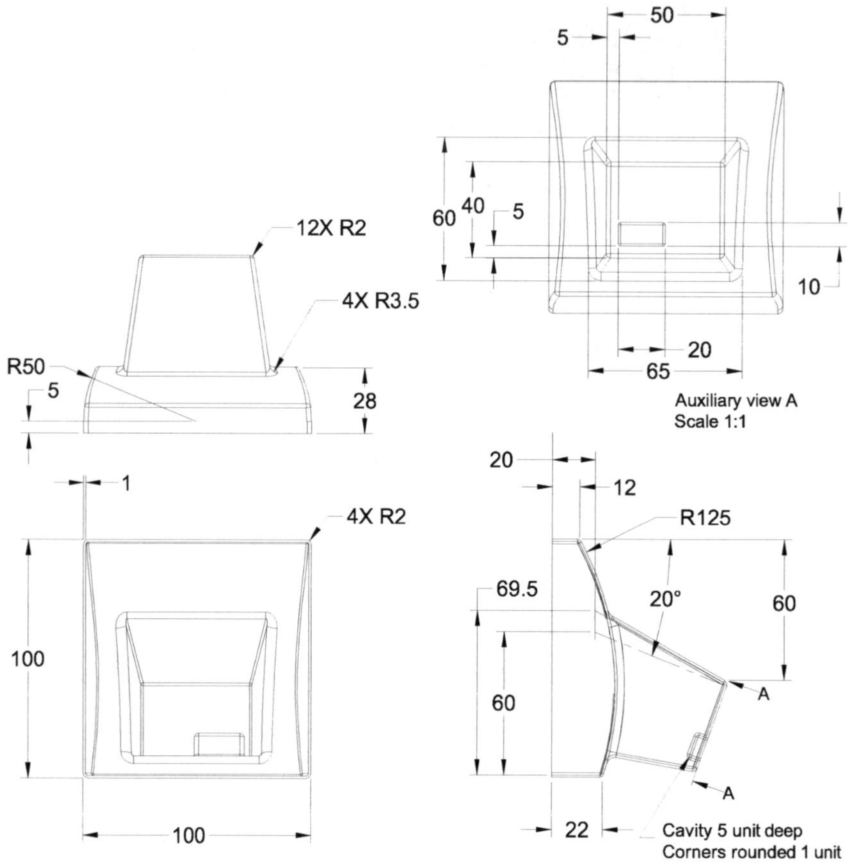

Figure 11-69 Orthographic views and dimensions of the model for Tutorial 1

The following steps are required to complete this tutorial:

a. Start a new file in the **Wireframe and Surface Design** workbench and create the base surface of the model by extruding the sketch drawn on the zx plane.
b. Create other required surfaces for the basic structure of the model.
c. Trim unwanted surfaces.
d. Invoke the **Part Design** workbench and convert the surface model into a solid model.
e. Fillet edges and shell the model.

Starting a New File in the Wireframe and Surface Design Workbench

To start this tutorial, first you need to start a new file in the **Wireframe and Surface Design** workbench.

1. Choose **Start > Mechanical Design > Wireframe and Surface Design** from the menubar to invoke the **New Part** dialog box. In this dialog box, enter **c11tut1** in the **Enter part name** edit box and then choose the **OK** button; a new file is started in the **Wireframe and Surface Design** workbench.

Creating Surfaces

You will create base surface of this model by extruding the sketch drawn on the zx plane.

1. Select the zx plane from the Specification tree and invoke the **Sketcher** workbench.

2. Draw the sketch of the base surface, as shown in Figure 11-70, and exit the **Sketcher** workbench.

3. Choose the **Extrude** tool from the **Surfaces** toolbar; the **Extruded Surface Definition** dialog box is displayed.

4. Select the **Mirrored Extent** check box and set the value for **Limit 1** in the **Dimension** spinner to **50**. Choose **OK** from the **Extruded Surface Definition** dialog box. The model after creating the base surface is shown in Figure 11-71.

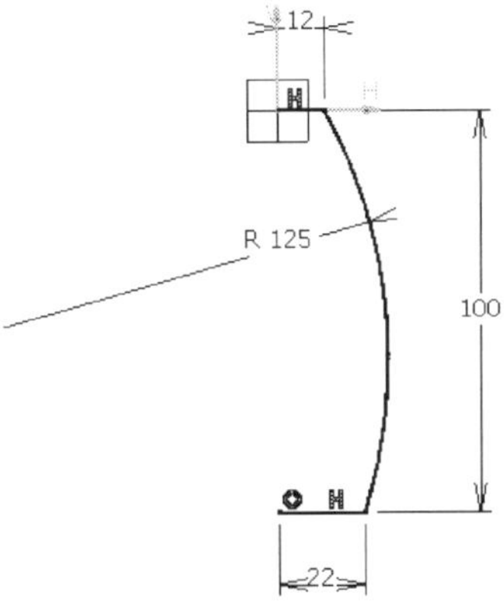

Figure 11-70 Sketch of the base surface

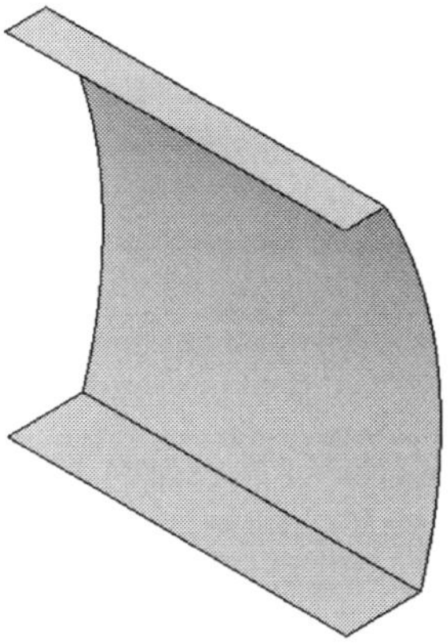

Figure 11-71 Resulting base surface

Next, you need to draw a reference sketch to create two planes, and also the sketches that will be used to create a multi-sections surface.

5. Select the zx plane and invoke the **Sketcher** workbench. Draw the reference sketch, as shown in Figure 11-72, and exit the **Sketcher** workbench.

6. Using the **Plane** tool, create a plane normal to the curve by selecting the curve and its endpoint, as shown in Figure 11-73.

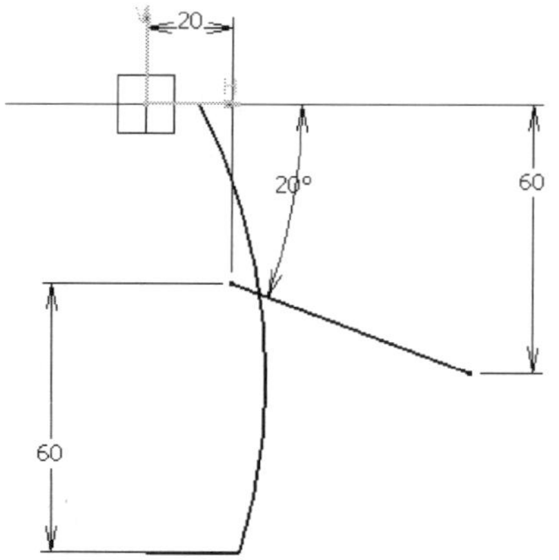

Figure 11-72 *Reference sketch*

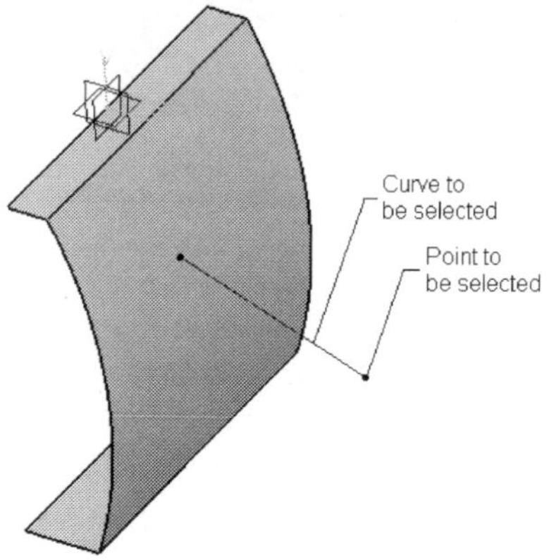

Figure 11-73 *Point and curve to be selected*

7. Invoke the **Sketcher** workbench by using the newly created plane as the sketching plane

8. Draw the sketch, as shown in Figure 11-74, and exit the Sketcher workbench. Click anywhere on the screen to exit the selection.

9. Now, using the **Plane** tool, create another plane normal to the curve by selecting the curve and its endpoint, as shown in Figure 11-75.

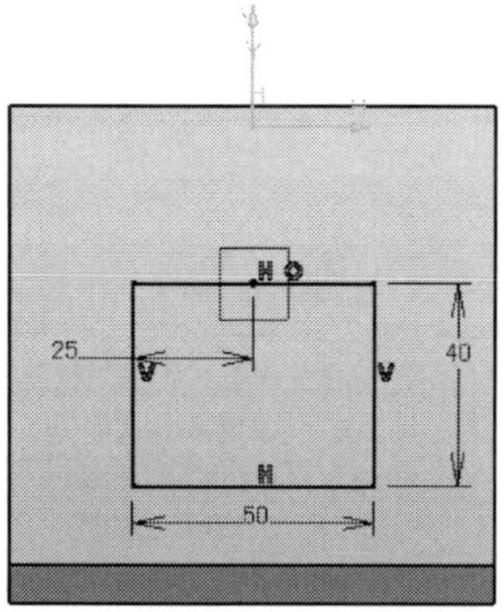

Figure 11-74 *Sketch drawn on the new plane*

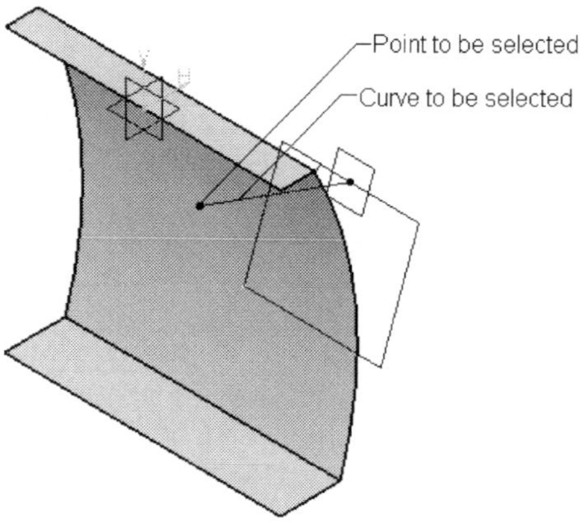

Figure 11-75 *Curve and its endpoint selected*

10. Again invoke the **Sketcher** workbench using the newly created plane as the sketching plane.

11. Draw the sketch, as shown in Figure 11-76, and exit the **Sketcher** workbench.

Next, you need to create the multi-section surface.

12. Choose the **Multi-Sections Surface** tool from the **Surfaces** toolbar; the **Multi-Sections Surface Definition** dialog box is displayed.

13. Select both the rectangular sketches. Next, set the closing points, if required.

14. Choose the **Coupling** tab and then select the **Vertices** option from the **Sections coupling** drop-down list.

15. Choose the **OK** button from the **Multi-Sections Surface Definition** dialog box. After the surface has been generated, hide all the sketches and planes. The model after creating the multi-section surface is shown in Figure 11-77.

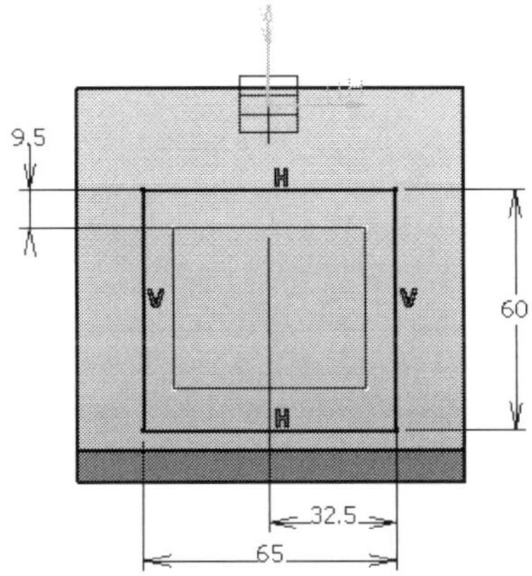

Figure 11-76 *Sketch drawn on the new plane*

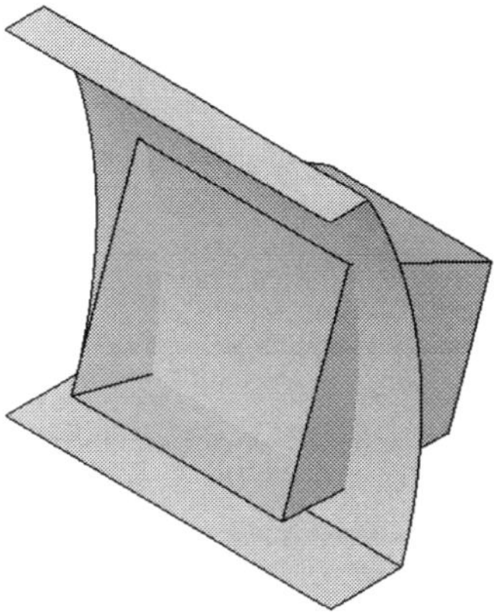

Figure 11-77 *Resulting multi-section surface*

Next, you need to create surface on the right of the base surface and mirror it about the zx plane.

16. Select the top face of the base surface and invoke the **Sketcher** workbench.

17. Draw the sketch, as shown in Figure 11 -78, and exit the **Sketcher** workbench.

18. Invoke the **Extruded Surface Definition** dialog box and then set **100** for **Lim 1** and **0** for **Lim 2** in the respective **Dimension** spinners; a preview of the extrusion is displayed. If required, flip the direction by choosing the **Reverse Direction** button from this dialog box. Next, choose **OK** to exit the dialog box.

    The model after creating the extruded surface and hiding the sketch is shown in Figure 11-79. After creating the extruded surface, you need to mirror it about the zx plane.

19. Choose the **Symmetry** tool from the **Transformations** sub-toolbar in the **Operations** toolbar; the **Symmetry Definition** dialog box is displayed and you are prompted to select the element to be transformed by symmetry.

20. Select the extruded surface as the element to be mirrored, as shown in Figure 11-80; you are prompted to select a reference point, line, or plane.

21. Select the zx plane as the reference plane, refer to Figure 11-80; a preview of the mirrored surface is displayed in the geometry area. Choose the **OK** button from the **Symmetry Definition** dialog box to create the mirror feature. The model after mirroring the extruded surface is shown in Figure 11-81.

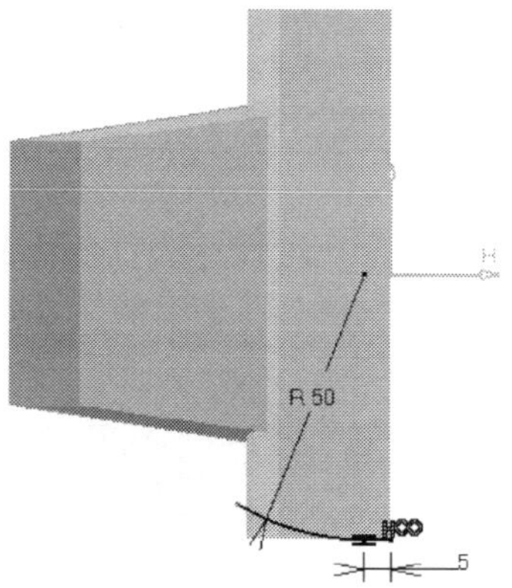

Figure 11-78 *Sketch drawn*

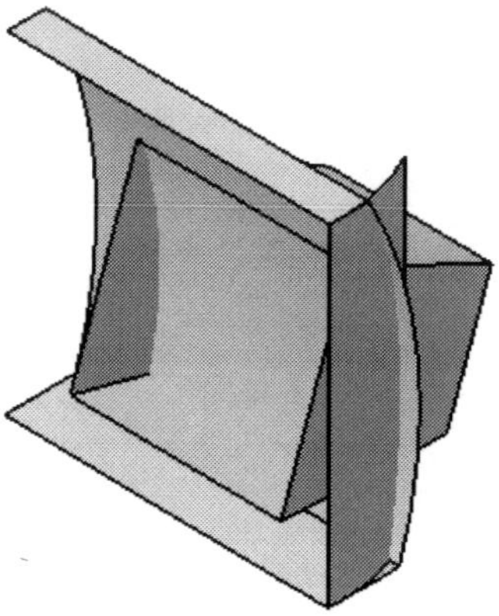

Figure 11-79 *The extruded surface*

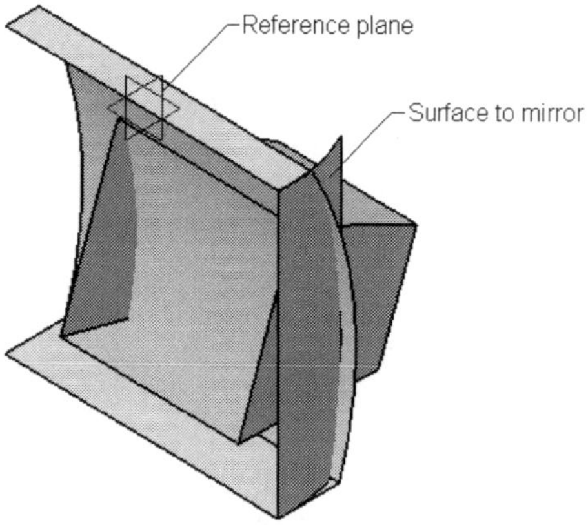

Figure 11-80 *Surface and reference to be selected*

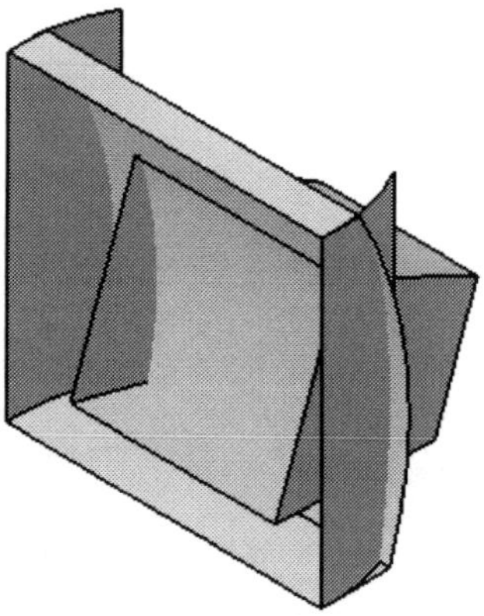

Figure 11-81 *The mirrored surface*

Trimming Surfaces

After creating the surfaces, you need to trim their unwanted portions using the **Trim** tool.

1. Choose the **Trim** tool from the **Split-Trim** sub-toolbar in the **Operations** toolbar; the **Trim Definition** dialog box is invoked.

2. Select the right extruded surface as the first element and the base surface as the second element, as shown in Figure 11-82; names of both the selected surfaces are displayed in the **Trimmed elements** area of the **Trim Definition** dialog box. Also, preview of the trimmed surfaces is displayed in the geometry area, refer to Figure 11-82.

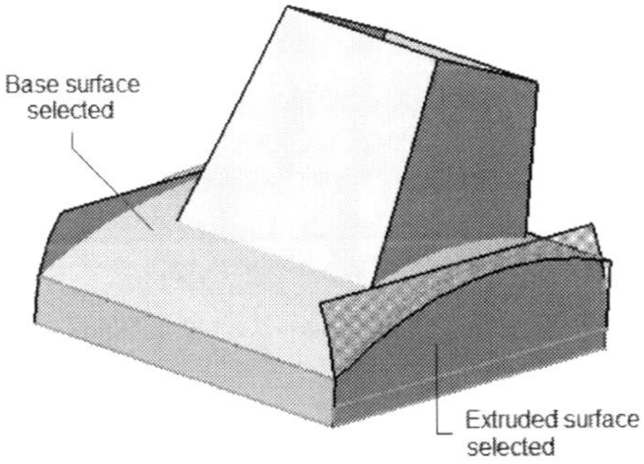

Figure 11-82 Surfaces selected for trimming

If the surface displayed in the preview is not the required one, you need flip the surface.

3. Choose the **Other side / next element** button. Similarly, if the surface displayed to be trimmed from the second surface is not the desired one, choose the **Other side /previous element** button from the **Trim Definition** dialog box.

4. Choose the **OK** button from the **Trim Definition** dialog box to trim the surface and exit the dialog box.

5. Similarly, trim the left extruded surface.

   After trimming the left extruded surface, you need to trim the unwanted portion of the base surface and the lofted surface.

6. Invoke the **Trim Definition** dialog box, and then select the base surface and the lofted surface as the surfaces to be trimmed.

7. Choose the **Other side /next element** button to flip the trimming side of the first element, and the **Other side /previous element** button of the second element, if required.

8. Choose the **OK** button from the **Trim Definition** dialog box to trim the surface and exit the dialog box. The model after trimming the surfaces is shown in Figure 11-83.

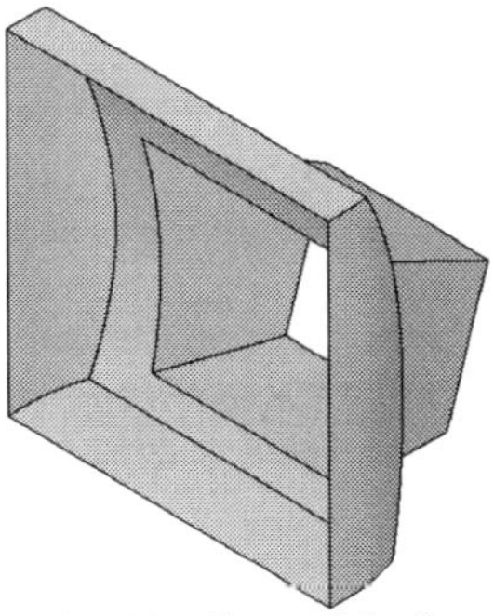

Figure 11-83 The model after trimming the surfaces

Solidifying the Surface Model

Next, you need to solidify the model using the **Close Surface** tool. For this, you need to invoke the **Part Design** workbench.

Note that a closed surface model is required to solidify the model. If the end edges of the surface form a flat patch, you do not need to create a flat patched surface to make it a closed surface and to solidify it.

1. Choose **Start > Mechanical Design > Part Design** from the menubar; the **Part Design** workbench is invoked.

2. Invoke the **Surface-Based Features (Extended)** toolbar if not invoked by default, and then choose the **Close Surface** tool from it; the **CloseSurface Definition** dialog box is displayed.

3. Select the surface from the geometry area; the name of the surface is displayed in the **Object to close** display box.

4. Choose the **OK** button from the **CloseSurface Definition** dialog box; the surface body is solidified. Next, you need to hide the surface body.

5. Select **Trim.3** from the Specification tree and then choose the **Hide/Show** button from the **View** toolbar; the display of the surface body is turned off.

   The model after solidifying and hiding the surface body is shown in Figure 11-84.

6. Add other features such as pocket, fillet, and a shell to the model using the tools in the **Part Design** workbench. For dimensions, refer to Figure 11-69. The final model is shown in Figure 11-85.

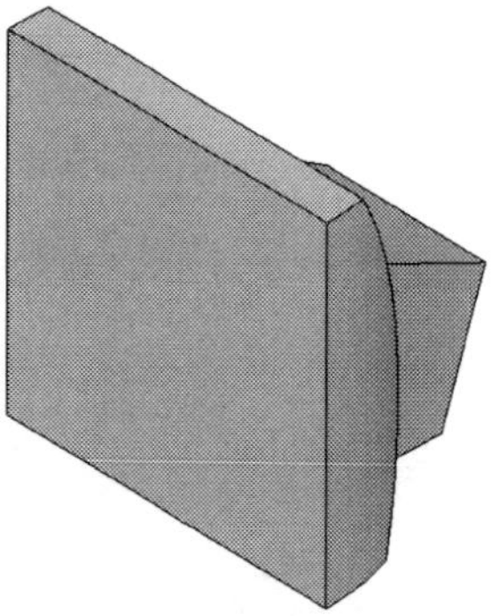

Figure 11-84 *The model after solidifying the surface body*

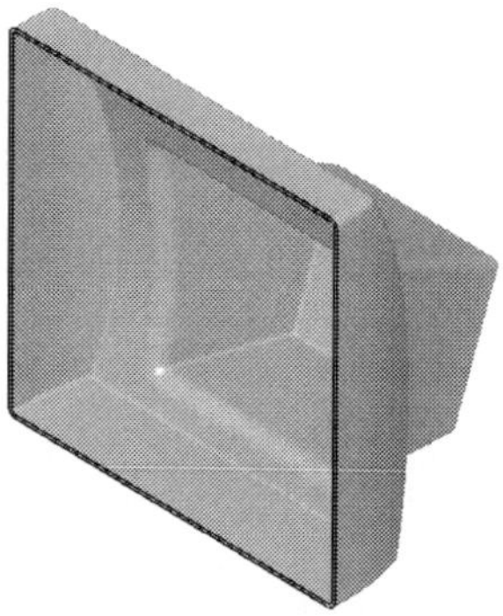

Figure 11-85 *Final model for Tutorial 1*

Saving and Closing the File

1. Choose the **Save** button from the **Standard** toolbar to invoke the **Save As** dialog box. Create a folder *c11* in the *CATIA* folder.

2. Enter **c11tut1.CATPart** in the **File name** edit box if not displayed, and choose the **Save** button. The file is saved at *C:\CATIA\c11*.

3. Close the part file by choosing **File > Close** from the menubar.

Tutorial 2

In this tutorial, you will create the model of a Hair Dryer Cover shown in Figure 11-86 using the tools in the **Wireframe and Surface Design** workbench and the **Part Design** workbench. The views and dimensions of the model are shown in Figure 11-87.

(Expected time: 1hr)

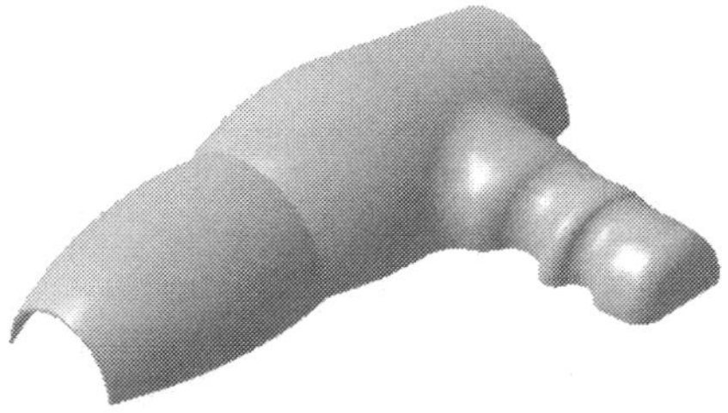

Figure 11-86 *Model of the Hair Dryer Cover*

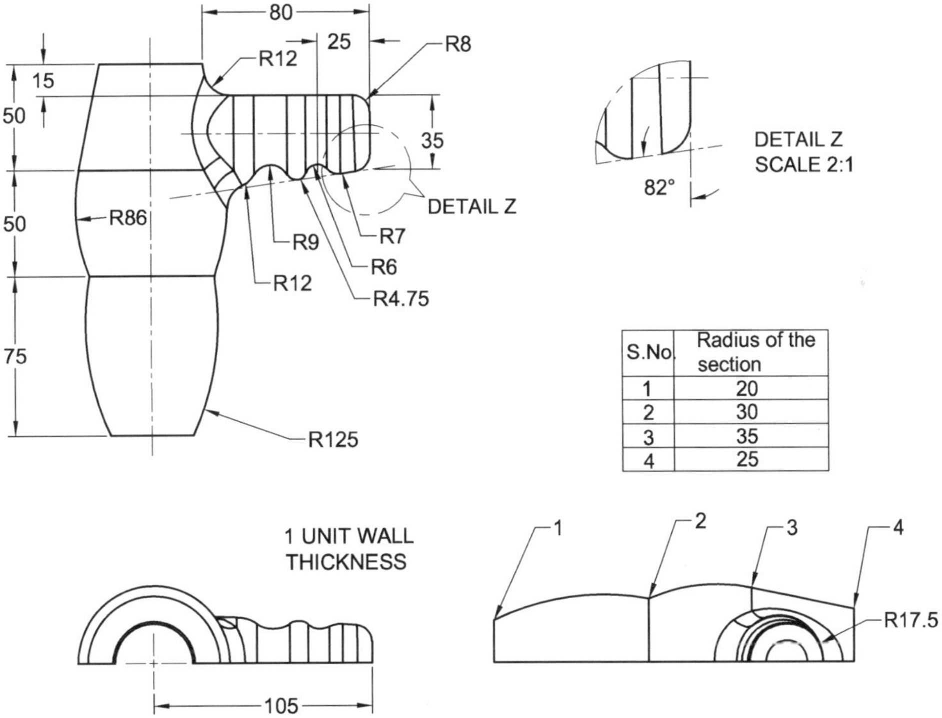

Figure 11-87 *Orthographic views and dimensions of the model*

The following steps are required to complete this tutorial:

a. Start a new file in the **Wireframe and Surface Design** workbench and using the **Multi-Sections Surface** tool, create the base surface which forms body of the Hair Dryer Cover.
b. Create the swept surface that forms the handle of the cover.
c. Trim the surfaces and create a flat boundary surface to close the surface body.
d. Join the surfaces together.
e. Invoke the **Part Design** workbench and solidify the surface body using the **Close Surface** tool.
f. Create the remaining features of the model using the **Edge Fillet** and **Shell** tools.

Creating the Base Surface

You will create the base (multi-section) surface of the model by blending four sections (sketches) drawn on different planes. These planes will be placed at an offset from each other. The resulting multi-section surface will be guided by using the guide curves throughout its path of transition. To create this surface, first you need to create three new planes at an offset distance. After creating the planes, you need to draw semicircular sketches (sections) on these planes. After drawing all the sections, you need to draw the sketches of guide curves on the xy plane.

1. Start a new file in the **Wireframe and Surface Design** workbench.

2. Create three planes at an offset distance of 75, 125, and 175 from the yz plane. Note that to get proper orientation in the isometric view, you need to create three planes in a direction opposite to the default direction.

3. Draw semicircular sketches on the respective planes, as shown in Figure 11-88. For dimensions, refer to the table given in Figure 11-87.

4. Draw two guide curves as two sketches on the xy plane by using the **Three Point Arc Starting With Limits** and **Line** tools. Figure 11-88 shows the model after drawing all sketches.

   Next, you need to create the base surface of the model using the **Multi-Sections Surface** tool.

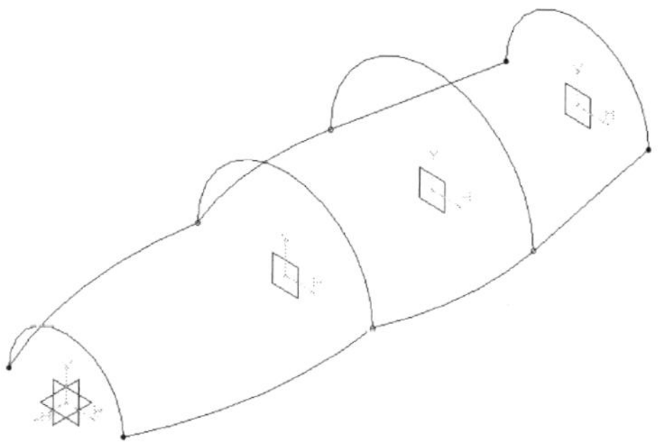

Figure 11-88 Sketches of sections and guide curves

5. Invoke the **Multi-Sections Surface Definition** dialog box. Select all the semicircular sketches. Set the directions of the arrowhead, if required, refer to Figure 11-89.

   Next, you need to select guide curves.

6. Click once on the three dots in the display box of the **Guides** tab and select the sketches of the guide curves in the geometry area, refer to Figure 11-89.

7. Choose the **OK** button from the **Multi-Sections Surface Definition** dialog box. Hide all the sketches and planes after the surface has been generated. Figure 11-90 shows the resulting base surface.

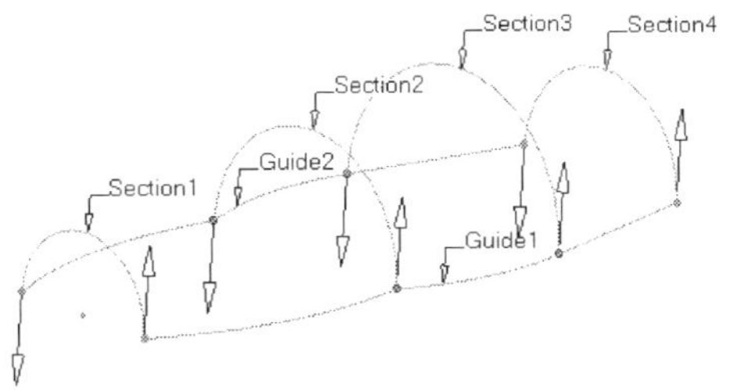

Figure 11-89 *Selection of sections and guide curves*

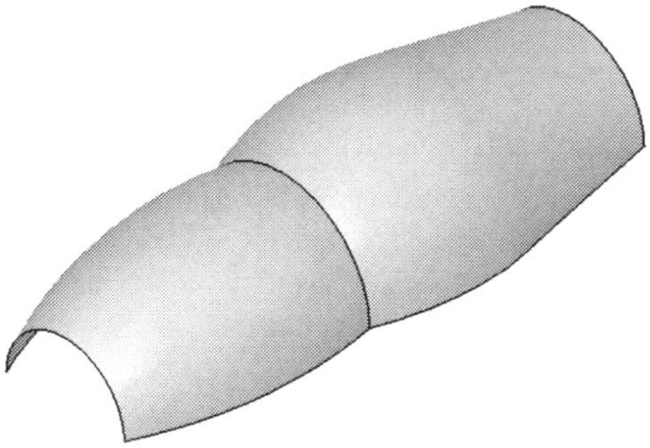

Figure 11-90 *Resulting base surface*

Creating the Swept Surface

Next, you need to create a swept surface that will form the handle of the Hair Dryer Cover. The swept surface will be created by sweeping a section along two guide curves.

1. Select the xy plane and invoke the **Sketcher** workbench.

2. Draw the sketch of the guide curve, as shown in Figure 11-91, and exit the **Sketcher** workbench.

Next, you need to draw the sketch of the second guide curve.

3. Select the xy plane as the sketching plane and invoke the **Sketcher** workbench.

4. Draw the sketch of the second guide curve, as shown in Figure 11-92, and exit the **Sketcher** workbench.

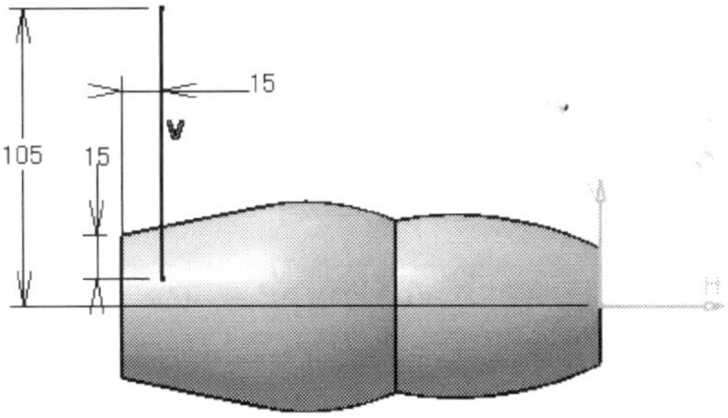

Figure 11-91 *Sketch of the first guide curve*

After drawing both the guide curves, you need to draw the sketch of the section. The sketch of the section will be drawn on a plane at an offset distance of 105 from the zx plane.

5. Create a plane at an offset distance of 105 from the zx plane.

6. Invoke the **Sketcher** workbench using the newly created plane as the sketching plane.

7. Draw a semicircular sketch for the section of the sweep surface, as shown in Figure 11-93, and exit the **Sketcher** workbench.

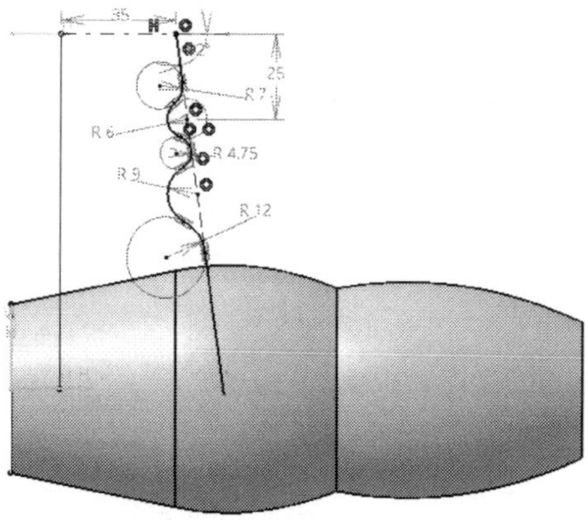

Figure 11-92 *Sketch of the second guide curve*

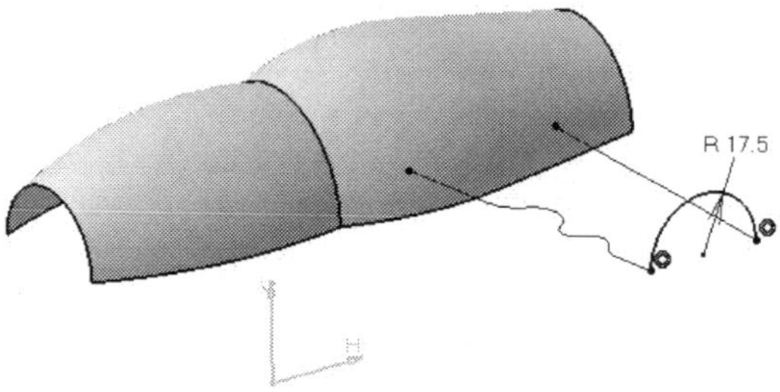

Figure 11-93 *Profile, guide curves, and anchor points to be selected*

Next, you need to create the swept surface.

8. Choose the **Sweep** tool from the **Surfaces** toolbar; the **Swept Surface Definition** dialog box is displayed. 

9. Select the **With two guide curves** option from the **Subtype** drop-down list in this dialog box; you are prompted to select a profile.

10. Select the required profile from the geometry area, refer to Figure 11-94; you are prompted to select a guide curve.

11. Select the sketch of a guide curve from the geometry area, refer to Figure 11-94; you are prompted to select the second guide curve.

12. Select the second guide curve from the geometry area, refer to Figure 11-94.

13. By default, the **Two points** option is selected in the **Anchoring type** drop-down list in the **Swept Surface Definition** dialog box. As a result, you can specify two anchoring points for creating the sweep surface.

14. Select the first guide anchor point from the geometry area, refer to Figure 11-94; you are prompted to select the second guide anchor point.

15. Select the second guide anchor point from the geometry area.

    In some cases, CATIA itself takes the anchor points by default. You need not to select the anchor points.

16. Choose the **OK** button from the **Swept Surface Definition** dialog box. Figure 11-95 shows the resulting sweep surface.

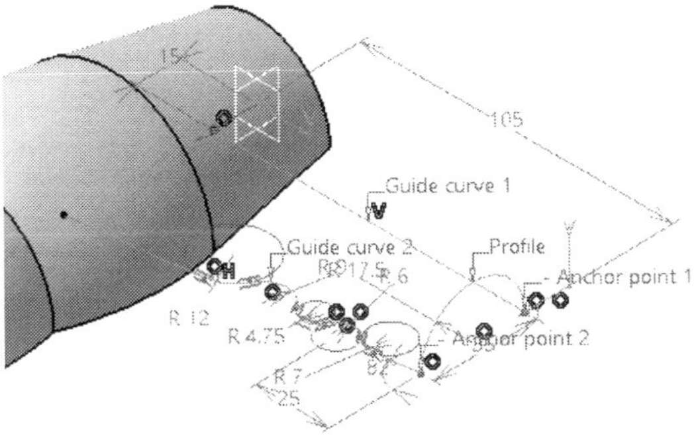

Figure 11-94 *Entities to be selected to create a sweep surface*

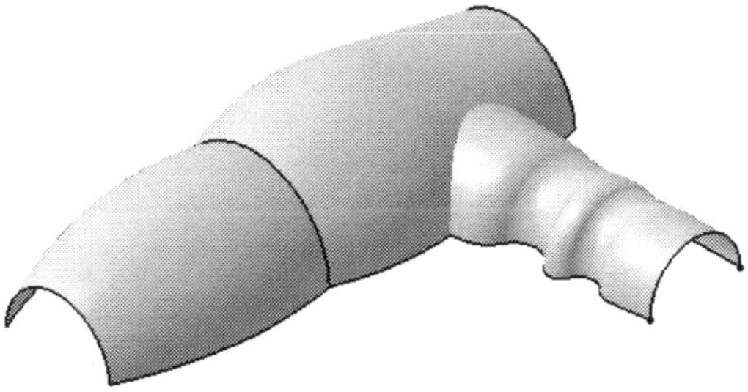

Figure 11-95 *Resulting sweep surface*

Trimming Unwanted Surfaces

After creating the base surface and the swept surface if you rotate the view of the surface model, you will notice that both the surfaces are intersecting each other. Figure 11-96 shows the intersecting surfaces of the model. In this section, trim their unwanted portions.

1. Invoke the **Trim Definition** dialog box, and select the sweep surface and then the multi-section surface from the geometry area; a preview of the trimmed surfaces is displayed in the geometry area, refer to Figure 11-96.

   The default trimmed surfaces are as required, therefore, you do not need to change the default settings in the **Trim Definition** dialog box.

2. Choose the **OK** button from the **Trim Definition** dialog box. Figure 11-97 shows the trimmed surface.

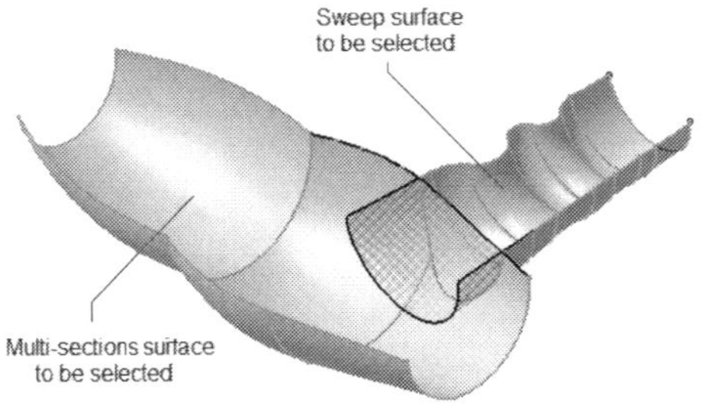

Figure 11-96 *Intersecting surfaces*

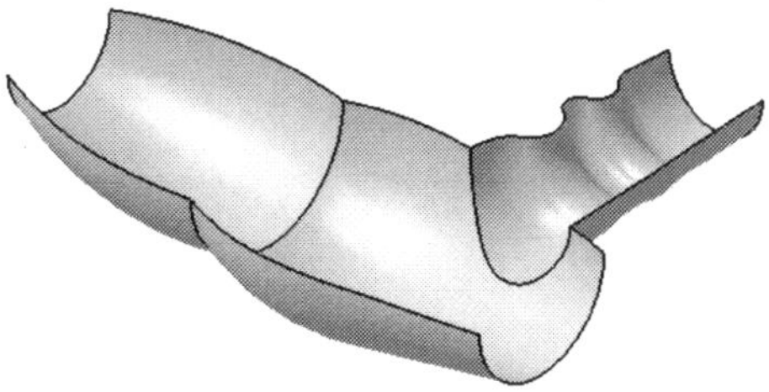

Figure 11-97 *The trimmed surface*

Creating Flat Boundary Surfaces

After trimming the unwanted portion of the surface, you need to create flat boundaries for creating a closed surface body.

1. Choose the **Line** tool from the **Wireframe** toolbar; the **Line Definition** dialog box is displayed and you are prompted to select the first element.

2. Select the right endpoint of the semicircular edge on the front of the base surface; you are prompted to select the second point.

3. Select the left endpoint of the same edge and choose the **OK** button from the **Line Definition** dialog box; a line is drawn between the selected points.

   Next, you need to create a flat boundary surface.

4. Choose the **Fill** tool from the **Surfaces** toolbar; the **Fill Surface Definition** dialog box is displayed and you are prompted to select the curves. 

5. Select the curves, as shown in Figure 11-98 and choose the **OK** button from the **Fill Surface Definition** dialog box. Figure 11-99 shows the resulting filled surface.

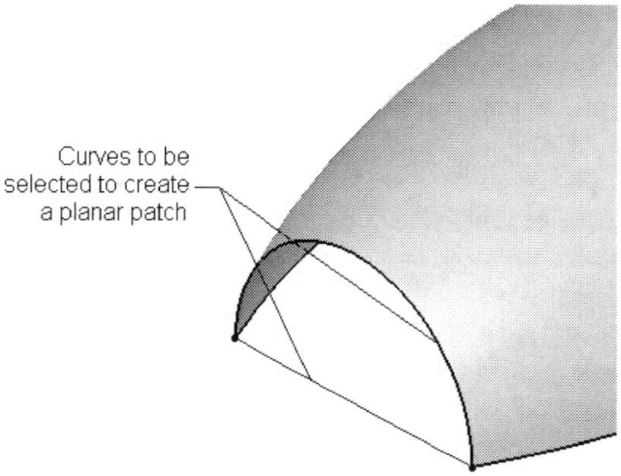

Figure 11-98 *Curves to be selected*

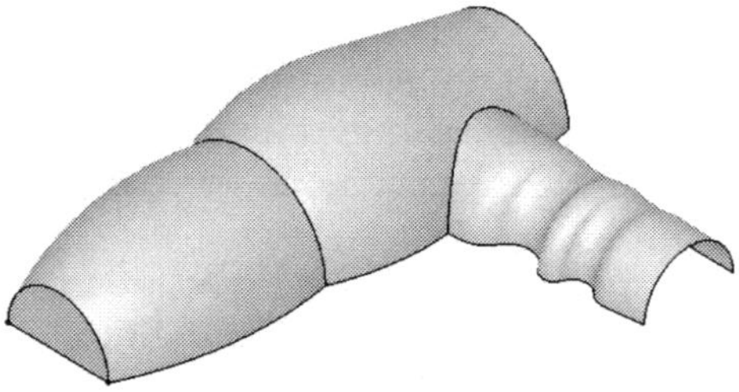

Figure 11-99 *Resulting filled surface*

6. Similarly, create other flat boundary surfaces to create a closed surface body. Figure 11-100 shows the surface model after creating all the flat boundary surfaces.

Note

If you face a problem while creating the above-mentioned flat boundary, you can leave the surfaces open. They will automatically be closed when you solidify the model.

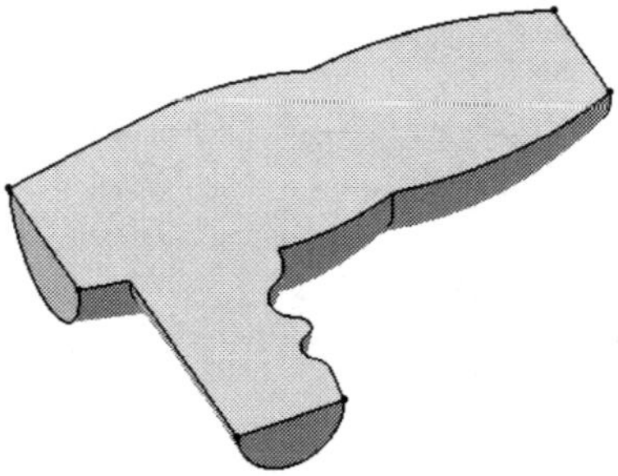

Figure 11-100 *The model after creating all flat boundary surfaces*

Joining Surfaces

After creating all the surfaces, you need to join them together using the **Join** tool.

1. Choose the **Join** tool from the **Join-Healing** sub-toolbar in the **Operations** toolbar; the **Join Definition** dialog box is displayed.

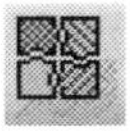

2. Select all the surfaces from the geometry area or from the Specification tree; the names of the selected surfaces are displayed in the **Elements To Join** display box.

3. Choose the **OK** button from the **Join Definition** dialog box.

Solidifying the Surface Body

After joining all the surfaces to each other, you need to solidify the closed surface body by invoking the **Part Design** workbench.

1. Choose **Start > Mechanical Design > Part Design** from the menubar; the **Part Design** workbench is invoked.

2. Choose the **Close Surface** tool from the **Surface-Based Features (Extended)** toolbar; the **CloseSurface Definition** dialog box is displayed.

3. Select **Join.1** from the Specifications tree; the name of the selected surface is displayed in the **Object to close** display box of the **CloseSurface Definition** dialog box.

4. Choose the **OK** button from the **CloseSurface Definition** dialog box; the solid body is created.

5. Hide the joined surface; the solid body is displayed in the geometry area, as shown in Figure 11-101.

Apply fillets and shell the model. The model after creating all the features is shown in Figure 11-102.

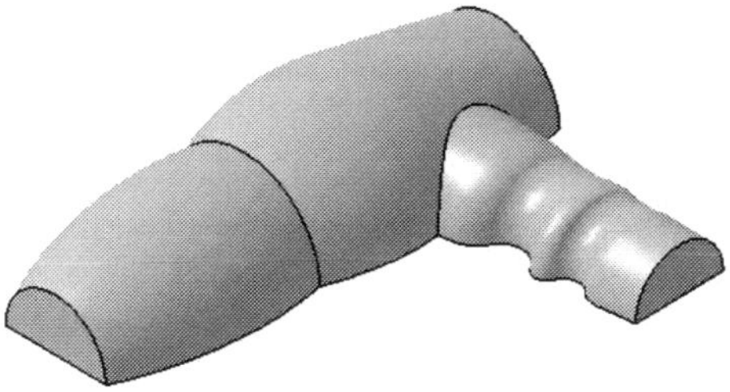

Figure 11-101 *Resulting solid body*

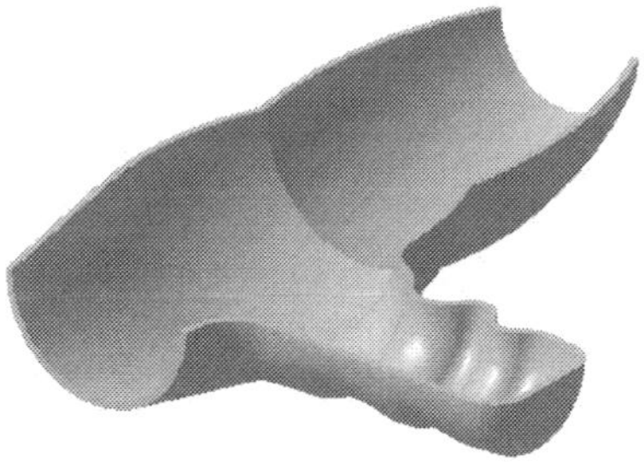

Figure 11-102 *Final model of the Hair Dryer Cover*

Saving and Closing the File

1. Choose the **Save** button from the **Standard** toolbar to invoke the **Save As** dialog box.

2. Enter **c11tut2** in the **File name** edit box and choose the **Save** button. The file will be saved at *C:\CATIA\c11*.

3. Close the part file by choosing **File > Close** from the menubar.

Self-Evaluation Test

Answer the following questions and then compare them to those given at the end of this chapter:

1. The __________ tool is used to transform an element from one axis system to another axis system.

2. To thicken a surface, choose the __________ button from the **Surface-Based Features (Extended)** toolbar in the **Part Design** workbench.

3. The __________ tool is used to scale elements with respect to a selected reference element.

4. The __________ tool is used to extract curves from surfaces.

5. The __________ tool is used to fill small gaps left between the joined surfaces.

6. The **Projection** tool is used to project one or more elements on a support, which could be a surface, a plane, or a wireframe geometry. (T/F)

7. The **Extrapolate** tool is used to extend a surface or a curve to a desired distance by maintaining the tangency or the curvature continuity. (T/F)

8. The **Disassemble** tool is used to trim curves and surfaces. (T/F)

9. The **Sew Surface** tool is used to sew a selected surface to the faces of a solid body. (T/F)

10. You cannot join the surfaces that are connected with each other. (T/F)

Review Questions

Answer the following questions:

1. Which dialog box is used to scale elements non-uniformly?

   (a) **Scaling Definition** (b) **Affinity Definition**
   (c) **Rotate Definition** (d) None of these

2. Which dialog box is used to mirror a selected element along a reference?

   (a) **Translate Definition** (b) **Split Definition**
   (c) **Symmetry Definition** (d) None of these

3. Which tool is used to convert a closed surface body into a solid body?

   (a) **Sew Surface** (b) **Split**
   (c) **Close Surface** (d) None of these

4. Which option in the **Extremities** drop-down list of the **Extrapolate Definition** dialog box is selected by default?

   (a) **Normal** (b) **Tangent**
   (c) **Symmetry Definition** (d) None of these

5. Which of the following tools is used to restore the portion of the surface or the curve removed using the **Split** tool?

   (a) **Untrim** (b) **Symmetry**
   (c) **Trim** (d) **Split**

6. You can non-uniformly scale entities using the **Scale** tool. (T/F)

7. The **Symmetry** tool is used to mirror a selected element about a selected reference. (T/F)

8. The **Split** tool is used to divide a solid body into two or more than two parts using a surface. (T/F)

9. You cannot define the continuity of an extrapolated surface with a reference surface. (T/F)

10. You can flip the direction of the material addition by using the **Reverse Direction** button from the **ThickSurface Definition** dialog box. (T/F)

EXERCISE

Exercise 1

Create the model of the cover shown in Figure 11-103. Its views and dimensions are shown in Figure 11-104. **(Expected time: 30 min)**

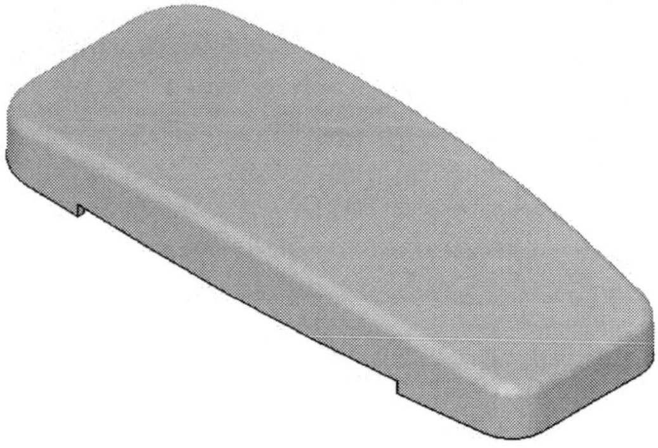

Figure 11-103 Model of the cover

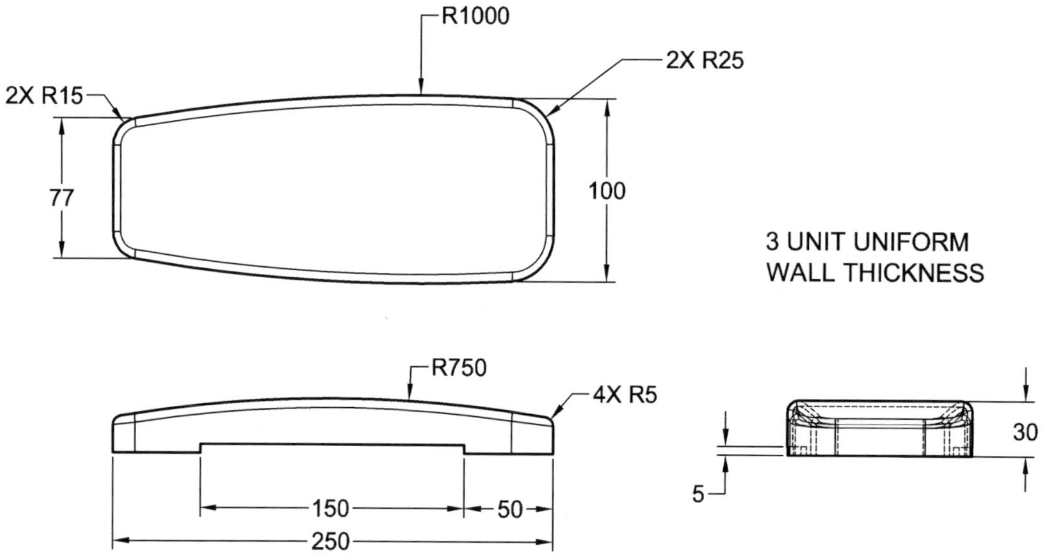

Figure 11-104 Orthographic views and dimensions of the cover

Answers to Self-Evaluation Test

1. **Axis To Axis**, **2.** **Thick Surface**, **3.** **Scaling**, **4.** **Boundary**, **5.** **Healing**, **6.** T, **7.** T, **8.** F, **9.** T, **10.** F

Chapter 12

Assembly Modeling

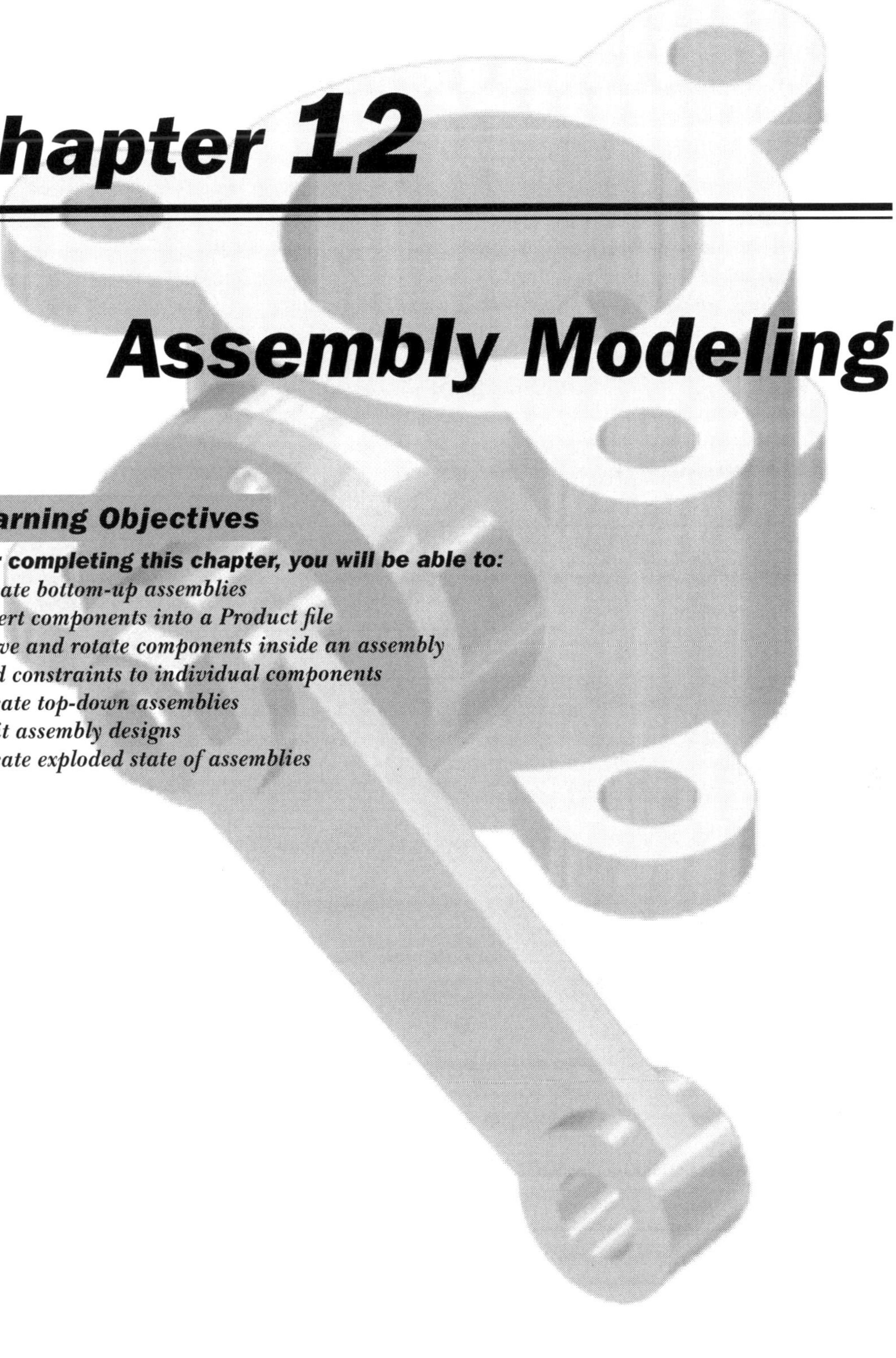

Learning Objectives

After completing this chapter, you will be able to:

- *Create bottom-up assemblies*
- *Insert components into a Product file*
- *Move and rotate components inside an assembly*
- *Add constraints to individual components*
- *Create top-down assemblies*
- *Edit assembly designs*
- *Create exploded state of assemblies*

ASSEMBLY MODELING

Assembly modeling is the process of creating designs with two or more components assembled together at their respective work positions. The components are brought together and assembled in the **Assembly Design** workbench by applying suitable parametric assembly constraints to them. The assembly constraints allow you to restrict the degrees of freedom of the components at their respective work positions. The assembly files in CATIA V5 are called product files. There are two methods to invoke the **Assembly Design** workbench of CATIA V5. The primary method to start a new product file is by choosing **Start > Mechanical Design > Assembly Design** from the menubar. The other method of invoking the **Assembly Design** workbench is by choosing **File > New** from the menubar. On doing so, the **New** dialog box will be displayed. In this dialog box, select **Product**, as shown in Figure 12-1.

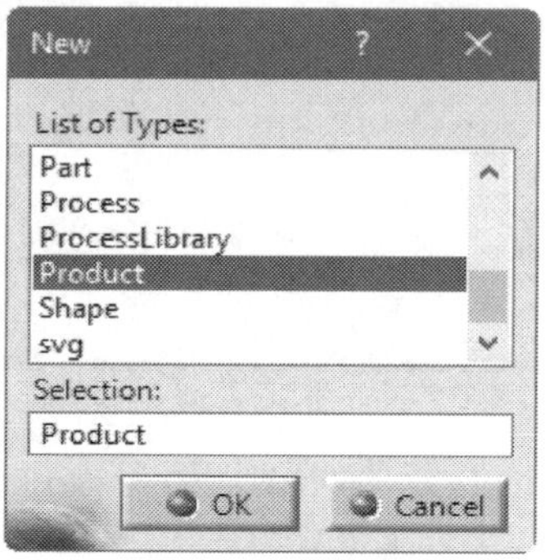

***Figure 12-1** The **Product** option selected from the **New** dialog box*

On invoking the **Assembly Design** workbench, a new file is started. The screen display of CATIA V5 after starting the new file in the **Assembly Design** workbench is shown in Figure 12-2. You will also notice that the toolbars related to the assembly will be displayed. The tools in these toolbars will be discussed later in this chapter.

***Figure 12-2** Screen display after starting a new file in the **Assembly Design** workbench*

Types of Assembly Design Approaches

In CATIA V5, you can create assembly models by adopting two types of design approaches. The first design approach is the bottom-up approach and the second is the top-down approach. Both these design approaches are discussed next.

Bottom-Up Assembly

The bottom-up assembly design approach is the most preferred approach for creating assembly models. In this approach, the components are created in the **Part Design** workbench as the **.CATPart* file. After creating the part files, the product (**.CATProduct*) file is created and all the previously created components are inserted in it using the tools provided in the **Assembly Design** workbench. After inserting all the components into the product file, constraints are applied to position them properly in the 3D space.

Adopting the bottom-up approach gives the user an opportunity to pay more attention to the details of the components while they are being designed individually. Because other components are not present in the same window, it becomes easier to maintain a relationship between the features of the current component. This approach is preferred for large assemblies especially those having intricate individual components.

Top-Down Assembly

In the top-down assembly design approach, components are created inside the **Assembly Design** workbench. Therefore, there is no need to create separate part files for the components. This design approach is completely different from the bottom-up design approach. Here you have to start the product file first and thereafter all the components will be created one by one. Note that even though the components are created inside the product file, they are saved as individual part files, which can be separately opened later.

Adopting the top-down assembly design approach gives the user a distinctive advantage of using the geometry of one component to define the geometry of another component. In these cases, the construction and assembly takes place simultaneously. As a result, the user can view the development of the product in real time. This design approach is highly preferred while working on a conceptual design or a tool design where a reference of the previously created parts is required to develop a new part.

Note

An assembly can also be created by combining both the top-down and bottom-up assembly design approaches.

CREATING BOTTOM-UP ASSEMBLIES

As mentioned earlier, while creating an assembly using the bottom-up approach, the components are created in separate part files and are then inserted into the assembly file. They are assembled at their working position by applying assembly constraints to them. To create an assembly using this approach, it is recommended to insert the first component and fix its position after properly orienting it in the 3D space. The other components can be inserted and positioned with reference to the first component. The method used for placing components inside the product file is discussed next.

Inserting Components in a Product File

Menubar:	Insert > Existing Component
Toolbar:	Product Structure Tools > Existing Component

To insert the existing component in the product file, choose the **Existing Component** tool from the **Product Structure Tools** toolbar; you will be prompted to select a component into which the existing component to be inserted. Select **Product1** from the Specification tree; the **File Selection** dialog box will be displayed. Browse to the location where part files are saved and double-click on the component to be inserted; the component will be inserted in the current product file and the name of the component will appear in the Specification tree. Note that when you insert a component into an assembly, a default part number will be assigned to the inserted component. This part number can be changed by the user. The process of changing the part number of a component is discussed later in this chapter. After inserting the component, you need to fix it by using the **Fix Component** constraint. The method of applying the **Fix Component** constraint to the components is discussed later in the chapter.

The above procedure needs to be repeated for inserting the next component. When you insert components with the same part number, the **Part number conflicts** dialog box is displayed, as shown in Figure 12-3.

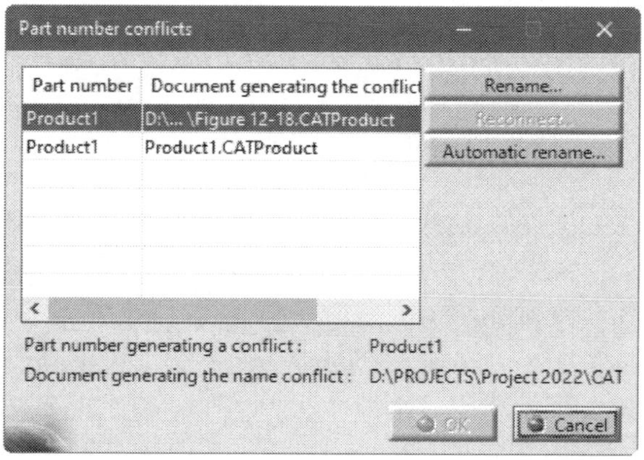

Figure 12-3 *The* ***Part number conflicts*** *dialog box*

This dialog box is displayed because there is a clash between the part numbers of the previously inserted component and the currently inserted component. Note that in the display box of the dialog box, the part number of both components is same, but the names of the files are different. You can change the part number by using the options in this dialog box.

There are two active buttons on the right of the display box of this dialog box: **Rename** and **Automatic rename**. If after selecting the component to be renamed, you choose the **Automatic rename** button, the part number of the selected component will be renamed from **Part1** to **Part1.1**. Choose the **OK** button from the **Part number conflicts** dialog box to insert the second component into the Product file. Follow the same procedure to rename the part number while inserting other components. Note that while inserting the third component, the first time when you rename the component using the **Automatic rename** option, the part number is changed to **Part1.1**. Because this part number is already assigned to the second component, the **Part number conflicts** dialog box will again be displayed after choosing the **OK** button. It shows the conflict between the second and third components. You need to choose the **Automatic rename** button again to change the part number of the third component. Now, the third component will be renamed from **Part1.1** to **Part1.1.2**. If you choose the **OK** button the third component will be inserted into the product file. This means if you insert the nth component, the **Automatic rename** button has to be used n-1 times. In this way, the part number of every new component keeps on changing in a similar fashion, and the same is represented in the Specification tree, as shown in Figure 12-4.

If you choose the **Rename** button from the **Part number conflicts** dialog box, the **Part Number** dialog box will be displayed, as shown in Figure 12-5.

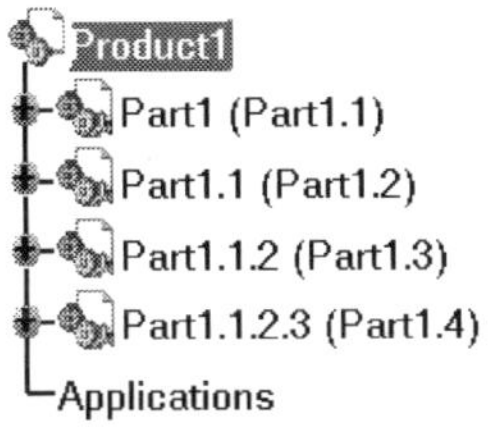

Figure 12-4 The Specification tree showing four components

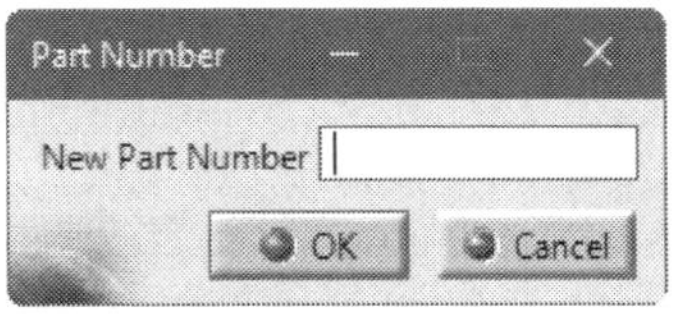

*Figure 12-5 The **Part Number** dialog box*

In this dialog box, you can enter the new part number for the selected component. After entering the new part number, choose the **OK** button to exit the **Part Number** dialog box. Now, choose the **OK** button from the **Part number conflicts** dialog box to insert the component in the product file. Ideally, the part number entered should be the same as the file name. If you enter the same part number for two different components, it will not be accepted by the software, and the **Part number conflicts** dialog box will be displayed again. Again, choose the **Rename** button and enter a unique name for that part such that it does not conflict with any other part number. The advantage of using this option is that the user can enter the desired part number, which can be useful especially when the individual components are referred to in an assembly using the number coding. Figure 12-6 shows the Specification tree with individual part numbers.

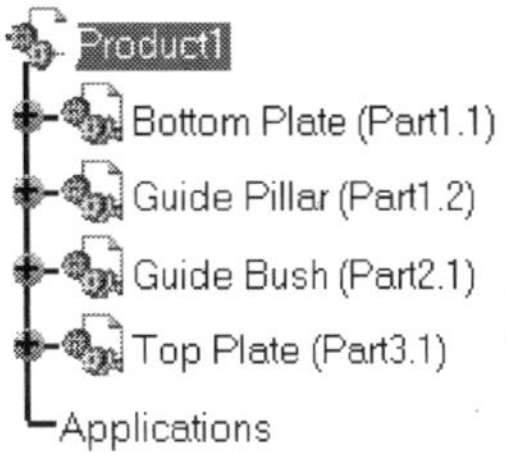

Figure 12-6 The Specification tree showing four components with unique part numbers

Note that in the Specification tree, the part numbers of each component are suffixed by the instance number displayed within parenthesis. This instance number, generated by the software itself, is unique for each component.

When a component is inserted into a product file, its placement in the 3D space depends on the location of its default planes. The default planes of the component placed will become the default plane of the product file. When more than one component is inserted into the product file, the default planes of all the components are placed one over the other. Therefore, they appear as one set of default planes. When the components are moved away, the default planes of each component are distinctly visible. You will learn more about moving the components later in this chapter.

Note

If the default planes of the inserted components are not visible, this means their visibility is turned off in the part file. Therefore, you need to turn on the visibility of the reference planes in the part file to display them in the assembly file.

Tip

You can also insert components in the product file by using the ***Copy*** *and* ***Paste*** *methods. To insert components using these methods, open the part file of the component. Select the name of the component from the top of the Specification tree and choose* ***Copy*** *from the contextual menu. Now switch to the product file and select the name of the assembly on the top of the Specification tree. Invoke the contextual menu and choose* ***Paste*** *from it; the component will be placed in the assembly.*

Moving Individual Components

Generally, the components when inserted in a product file are overlapped by the other components placed earlier. As a result, their visualization is hampered and it becomes difficult to apply constraints to them. Therefore, it is necessary to reposition the components in the 3D space such that they are distinctly visible and the mating references are accessible in the assembly. CATIA V5 allows you to move and rotate the individual unconstrained components inside the product file without affecting the position and location of the other components. The reorientation of the component can be carried out using three different methods, which are discussed next.

Moving and Rotating Components by Using the Manipulation Tool

Menubar:	Edit > Move > Manipulate
Toolbar:	Move > Manipulation

The **Manipulation** tool is used to move or rotate a component freely by dragging the cursor. To translate or rotate a component, choose the **Manipulation** tool from the **Move** toolbar; the **Manipulation Parameter** dialog box will be displayed, as shown in Figure 12-7.

This dialog box contains buttons that are arranged in three rows. The currently active button is displayed on top of the dialog box. The buttons in the first row are used to translate the component along a particular direction. There are four buttons in this row, which are discussed below:

The **Drag along X axis** button is the first button and it is chosen by default. This button is used to translate the selected component along the X-axis of the assembly coordinate system. To move the component, select it and then drag it to the desired location and then release the left mouse button.

The **Drag along Y axis** button is used to translate the component along the Y-axis of the assembly coordinate system. It works similar to the button discussed above. After choosing the **Drag along Y axis** button, select the component to be moved and then drag it.

Figure 12-7 *The* ***Manipulation Parameter*** *dialog box*

The **Drag along Z axis** button is used to translate the component along the Z-axis of the assembly coordinate system.

The **Drag along any axis** button is used to translate the component along an axis defined by selecting any line, a line of a component, or a line of a component to be dragged.

The buttons in the second row of the **Manipulation Parameter** dialog box are used to move the selected component along a particular plane. These buttons are called the planar translation buttons and are discussed next.

The **Drag along XY plane** button is used to translate the selected component parallel to the xy plane of the assembly coordinate system.

The **Drag along YZ plane** button is used to translate the selected component parallel to the yz plane of the assembly coordinate system.

The **Drag along XZ plane** button is used to translate the selected component parallel to the xz plane of the assembly coordinate system.

The **Drag along any plane** button is used to move the selected component parallel to a specified plane. After choosing the **Drag along any plane** button, you need to select a plane for the planar translation. This plane can be a construction plane, planar face, or a surface. After selecting the plane, select the component to move and then drag it on the selected plane.

The buttons in the third row of the **Manipulation Parameter** dialog box are used to rotate the selected component around an axis. These buttons are called the rotation buttons and are discussed next.

The **Drag around X axis** button is used to rotate the selected component around the X-axis of the assembly coordinate system.

The **Drag around Y axis** button is used to rotate the selected component around the Y-axis of the assembly coordinate system.

The **Drag around Z axis** button is used to rotate the selected component around the Z-axis of the assembly coordinate system.

The **Drag around any axis** button is used to rotate the selected component around a specified axis. After choosing the **Drag around any axis** button, you need to select a line to define the rotation axis. This line can be an edge of the component or an axis of a cylindrical feature. After selecting the axis for rotation, select the component to rotate and then drag it.

The **With respect to constraints** check box is selected to move or rotate the components within the available degrees of freedom after applying constraints. You will learn more about applying constraints later in this chapter.

Moving Components by Using the Snap Tool

Menubar:	Edit > Move > Snap
Toolbar:	Move > Snap sub-toolbar > Snap

The **Snap** tool is used to move a component in an assembly by snapping the geometric element of the first component on another component or on the same component. The movement of the component depends on the selection of the geometric elements. The element selected first will move to snap the second element. For example, if you first select a line and then a point, the line will be reoriented in such a way that it passes through the selected point.

To move a component using the **Snap** tool, choose the **Snap** tool from the **Snap** sub-toolbar in the **Move** toolbar; you will be prompted to select the first geometric element on a component, which can be an axis system, point, line, or plane. Select a suitable geometrical element that will be projected on the next selection. In Figure 12-8, the upper right edge of the left component is selected as the first geometrical element.

On selecting the suitable geometric element, you will be prompted to select the second geometric element on the same component or another component, which can be a point, line or plane. Select the component to which the first selection is to be snapped. In Figure 12-8, the upper left edge of the right component is selected as the second geometric element. On selecting the second element to be snapped, the first selection will be snapped to the second selection, and a green arrow will be displayed at the snapping location, as shown in Figure 12-9. You can click on the arrow to reverse the snapping direction, else click anywhere in the geometry area to exit the tool. Figure 12-10 shows the components after the snapping direction is reversed.

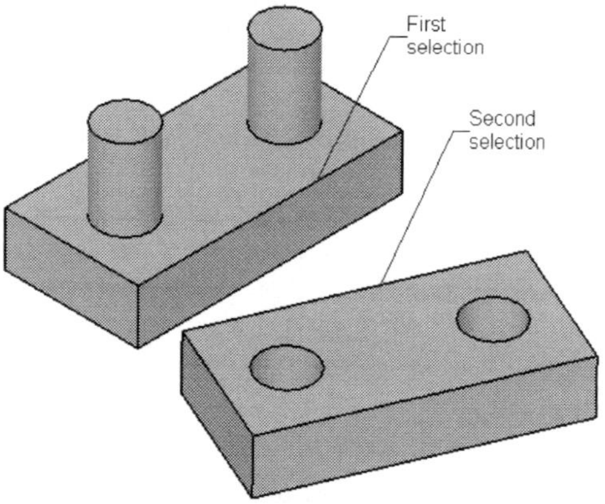

Figure 12-8 Geometric elements selected to be snapped

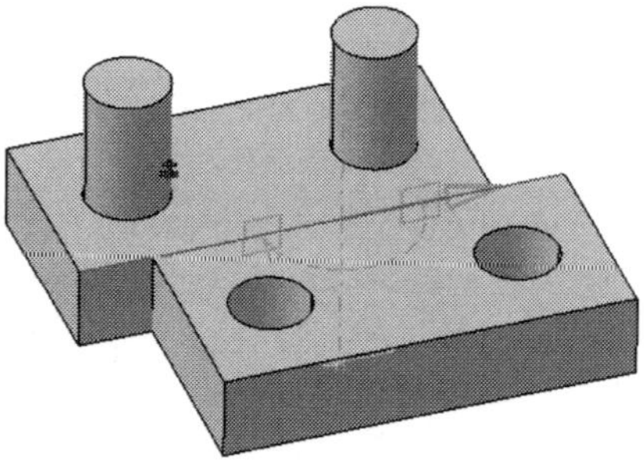

Figure 12-9 Position of the components after snapping

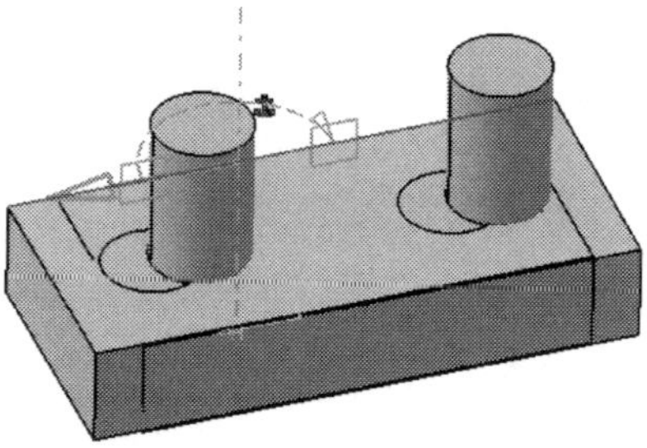

Figure 12-10 Position of the components after the snapping direction is reversed

In some cases, the arrows are not displayed when you snap the two elements, such as snapping a point to a point, point to a surface, point to a cylindrical surface or a planar surface, and so on.

Moving Components by Using the Smart Move Tool

Menubar:	Edit > Move > Smart Move
Toolbar:	Move > Snap sub-toolbar > Smart Move

The **Smart Move** tool works as a multipurpose tool.This tool has the capability of manipulating and snapping components and applying constraints to them, if required. To move components, choose the down arrow besides the **Snap** tool and choose the **Smart Move** tool from the sub-toolbar; the **Smart Move** dialog box will be displayed. Choose the **More** button to expand the **Smart Move** dialog box, refer to Figure 12-11.

If the **Automatic constraint creation** check box is selected, a permanent constraint will be applied between the selected elements of the components to be snapped. Also, the same will be displayed in the Specification tree. If you clear this check box, the components will be repositioned but no permanent constraint will be applied. You will learn more about applying constraints later in this chapter.

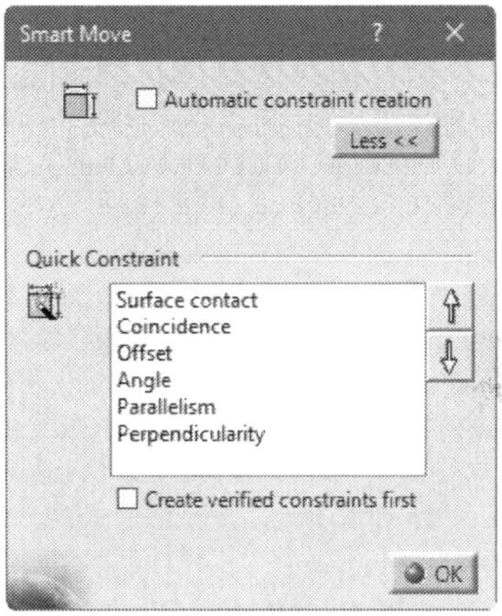

Figure 12-11 The expanded ***Smart Move*** *dialog box*

The **Quick Constraint** area displays the constraints in a hierarchical order. These constraints will be applied to the components while they are being snapped. If more than one constraints are applied to the current selection set, then priority will be given to the constraint that is on the top of the hierarchy. To change the position of a constraint, select it and choose the Up or Down arrows on the right side of the **Quick Constraint** area.

After specifying the options in this dialog box, select the first geometric element on a component. The element can be a point, line, axis, plane, planar face, or circular face. Now, select the second geometric element on the other component; the suitable constraint will be applied between the two components, depending on the current selection set. A green arrow may be displayed at the constraint location. You can click on it to reverse the orientation of the mating components. The component from which the first selection is made will move to snap the component on which the second element is selected. After the components are reoriented, choose the **OK** button from the **Smart Move** dialog box. Figure 12-12 shows two cylindrical surfaces to be selected and Figure 12-13 shows the concentric constraint applied between the two surfaces.

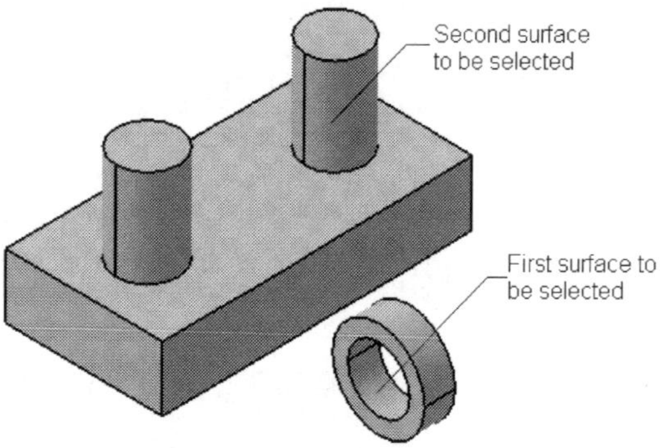

Figure 12-12 Surfaces to be selected

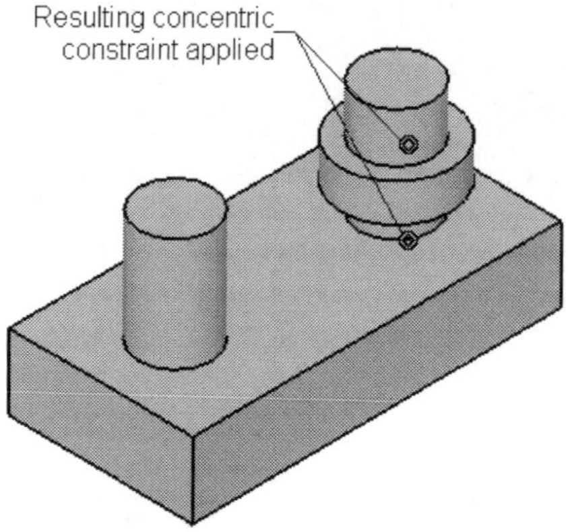

Figure 12-13 Concentric constraint applied

The **Smart Move** tool can be invoked along with a viewer, which makes the selection of geometric elements easier. To invoke the viewer, first select the components that need to be moved and then choose the **Smart Move** tool from the **Snap** sub-toolbar. This time the **Smart Move** dialog box will be displayed with a viewer on its top. Partial view of the **Smart Move** dialog box is shown in Figure 12-14. The component selected first is displayed in the viewer. The part number of the displayed component is displayed on its top. Only one component is displayed in the viewer and its geometric element can be easily selected as you can zoom in and rotate the component in it. This is very helpful especially when there are a number of components in the geometry area and some of the components are fully or partially placed inside another component. After selecting the geometric element of the first component, choose the **Next component** button from the **Smart Move** dialog box. Note that this button will not be available if you select only one component before invoking the **Smart Move** dialog box. Now, the other component is displayed in the viewer and you can select its geometric element. Note that while you zoom or rotate the component in the viewer, the actual orientation of the component in the geometry area will not change. After the selections are made from the viewer, the components are reoriented in the geometric area. Now, choose the **OK** button to close the **Smart Move** dialog box.

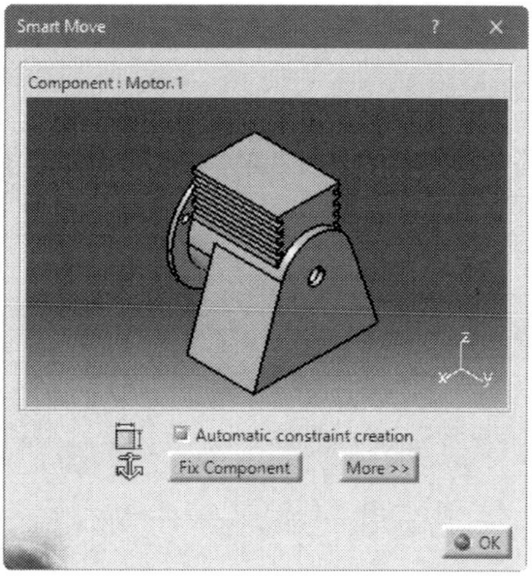

*Figure 12-14 Partial view of the **Smart Move** dialog box with the viewer*

Manipulating Components by Using the Compass

The orientation of the components can also be manipulated using the compass at the top right corner of the geometry area. To move a component using the compass, you first need to associate it with the component that needs to be moved or rotated. To associate the compass to a part, move the cursor over the red square displayed on the base of the compass. When the selection cursor is replaced by the move cursor represented by four directional arrows, hold the left mouse button, and drag the compass on the surface of the component to be manipulated. Once the compass is moved, a gray dot appears at its original location. To associate the compass with a component, place it on the surface of the component by releasing the left mouse button. The gray dot is no longer displayed. To move the component, place the cursor on any of the straight edges of the compass. The edges are highlighted and the cursor is replaced by the hand symbol. Press and hold the left mouse button and drag the cursor along the highlighted edge to move the component in that direction. After moving it, release the left mouse button. Similarly, to rotate the component, place the cursor over any of the circular edges of the manipulator and when the hand symbol is displayed, drag the cursor along that edge.

Once the manipulation is over, you need to place the compass back at its original position. To do so, move the cursor on the red square on the base of the compass. Once the move cursor symbol is displayed, drag the cursor anywhere in the geometry area away from all the components in the assembly and release the left mouse button.

Note

When the compass is placed back at its original location, the orientation of the compass remains the same as it was after manipulating the component. This may lead to confusion. To bring the compass back to its default orientation, place the compass over a perfectly horizontal surface and then place the compass back at its default location.

Applying Constraints

After placing the components in the product file, you need to assemble them. By assembling the components, you will constrain their degree of freedom. As mentioned earlier, the components are assembled using the constraints. Constraints help you to precisely place and position the components with respect to other components and surroundings in the assembly. If all degrees of freedom of every component of the assembly are restricted, it is called a fully constrained assembly. Else, it is called a partially constrained assembly. If a mechanism has to be created after assembling the components, some degrees of freedom of the assembly is needed to be kept free intentionally, so that movements can be achieved in that direction. Various types of constraints in CATIA V5 are discussed below.

Fix Component Constraint

Menubar:	Insert > Fix
Toolbar:	Constraints > Fix Component

The Fix Component constraint is used to fix the location of the selected component in the 3D space. Once the orientation of the component is fixed, it cannot be changed. To apply this constraint, choose the **Fix Component** tool from the **Constraints** toolbar; you will be prompted to select the component to be fixed. You can select the component from the geometry area or from the Specification tree. Once you select the component, an anchor symbol ( ) will be displayed on the component. Now, other components can be constrained with respect to the fixed component. While doing so, the orientation of the fixed component will not be altered and other components will be reoriented to apply the constraints. It is always advisable to fix the base component at its default location so that it can be used as a reference for the other components. On applying this constraint, the **Fix Component** constraint is displayed in the Specification tree. To view the applied constraints, expand the **Constraints** node in the Specification tree.

Coincidence Constraint

Menubar:	Insert > Coincidence
Toolbar:	Constraints > Coincidence Constraint

The Coincidence constraint is applied to coincide the central axis of the cylindrical features that are selected from two different components. This option can also be used to apply the Coincident constraint between edges, points, planes, or planar faces. To apply this constraint, choose the **Coincidence Constraint** tool from the **Constraints** toolbar; the **Assistant** dialog box will be displayed providing information about the selected constraint. You can select the **Do not prompt in the future** check box, if you do not want to display this dialog box again. Close the **Assistant** dialog box. Now, move the cursor over a cylindrical surface to display the central axis. When the preview of the central axis appears, as shown in Figure 12-15, click to select it. Similarly, select the central axis of the second component, as shown in Figure 12-16.

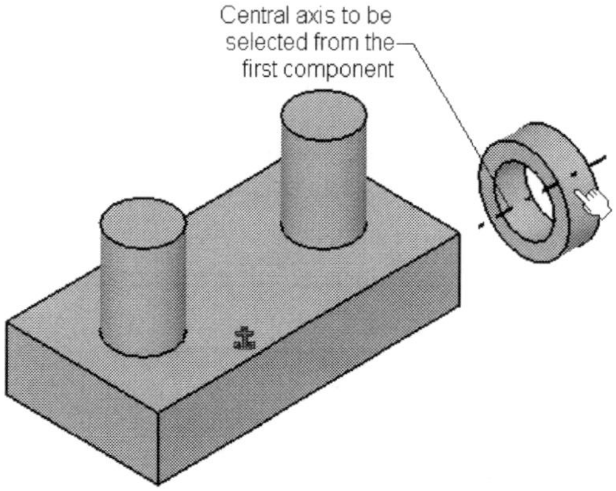

Figure 12-15 *Selecting the central axis of the first component*

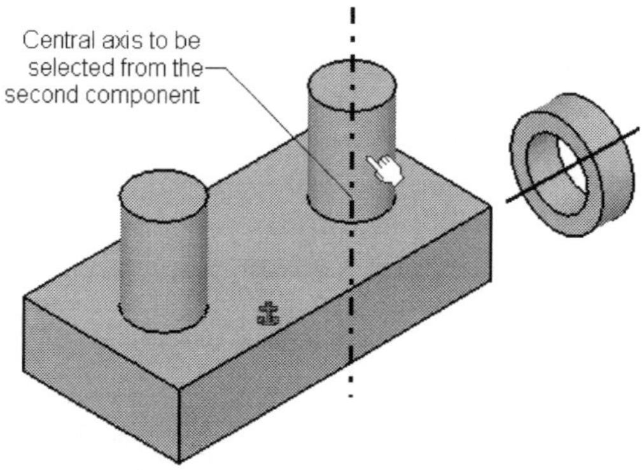

Figure 12-16 *Selecting the central axis of the second component*

Once the two axes are selected, the Coincidence constraint will be applied between them and the coincidence symbol will be displayed, as shown in Figure 12-17. You will notice that although the coincidence constraint is applied between the two components, the components are not assembled according to the constraint applied. Instead, a line connecting the two constraints is displayed. To position the components, choose the **Update All** button from the **Update** toolbar or press the CTRL+U keys. Now, the components will be placed such that the two selected cylindrical surfaces become concentric, as shown in Figure 12-18. Click once in the geometry area to remove the constraint from the current selection set. The symbol of the constraint is displayed in green on the assembled components.

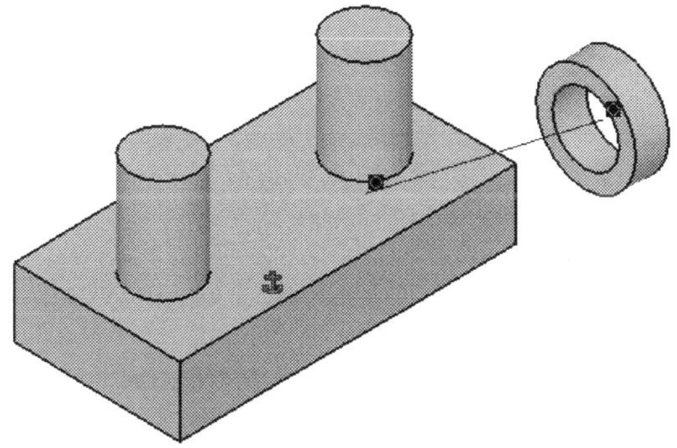

Figure 12-17 *Coincidence constraint applied between two components*

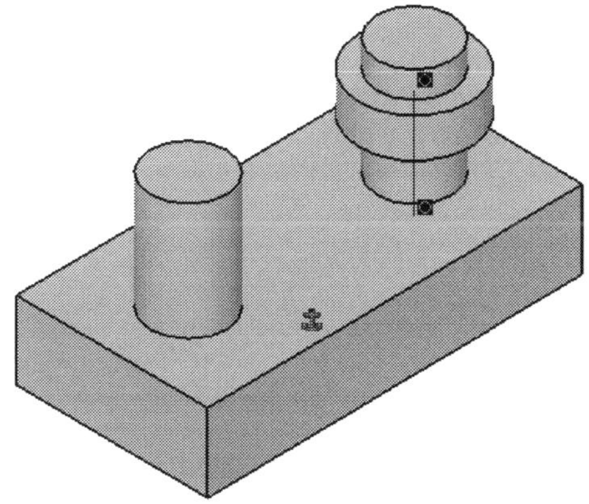

Figure 12-18 *Position of the components after updating the file*

Contact Constraint

Menubar:	Insert > Contact
Toolbar:	Constraints > Contact Constraint

The Contact constraint is applied to make a surface to surface contact between two selected elements from two different components. The elements to be selected can be planes, planar faces, cylindrical faces, spherical faces, conic faces, or circular edges. To apply this constraint, choose the **Contact Constraint** tool from the **Constraints** toolbar; you will be prompted to select the first geometric element of the Contact constraint. Select the element from the first component; you will be prompted to select the geometric element to place in contact with the first selection. Select the element from the second component; the Contact constraint will be applied between the two elements and the component will be placed accordingly after updating them. Also, a symbol of the Contact constraint (▣) will be displayed on the assembled components. Figure 12-19 shows the faces to be selected and Figure 12-20 shows the resulting components after applying the Contact constraint.

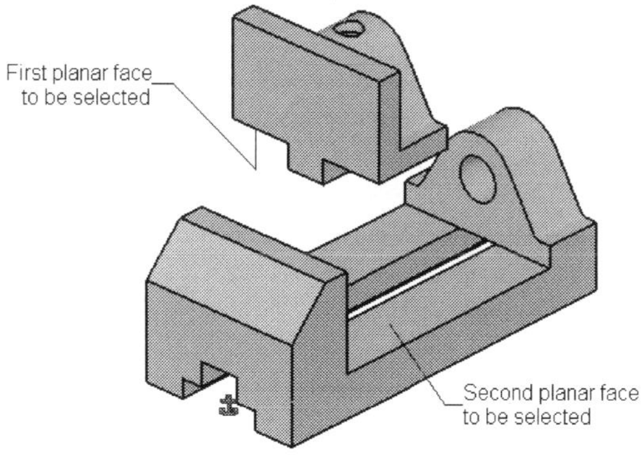

Figure 12-19 *Planar faces to be selected*

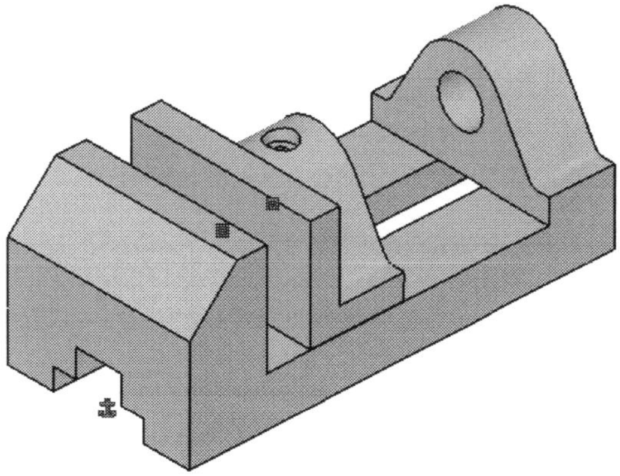

Figure 12-20 *Position of components after the Contact constraint is applied and updated*

Note

If you apply the Contact constraint between two cylindrical surfaces or a cylindrical surface and a planar face or plane, the ***Constraint Properties*** *dialog box will be displayed. You can set the orientation of the constraint using the options in the* ***Orientation*** *drop-down list. After setting the orientation, choose* ***OK*** *from the* ***Constraint Properties*** *dialog box.*

Offset Constraint

Menubar: Insert > Offset
Toolbar: Constraints > Offset Constraint

The Offset constraint is used to place the selected elements at an offset distance from each other. It also makes the two planar faces parallel to each other. To apply this constraint, choose the **Offset Constraint** tool from the **Constraints** toolbar; you will be prompted to select the first geometric element for the Offset constraint. Select a planar face, circular face, plane, axis, or a point from the geometry area; you will be prompted to select the second geometric element. Select a planar face of another component; the **Constraint Properties** dialog box will be displayed, as shown in Figure 12-21.

If you select two planar faces, then two arrows will be displayed on them. The arrows represent the orientation of the planes with respect to each other. The planes can face in the same direction or opposite direction. To flip the direction or arrows, click on any one of the arrows, or use the options in the **Orientation** drop-down list. The options in the **Constraint Properties** dialog box are discussed next.

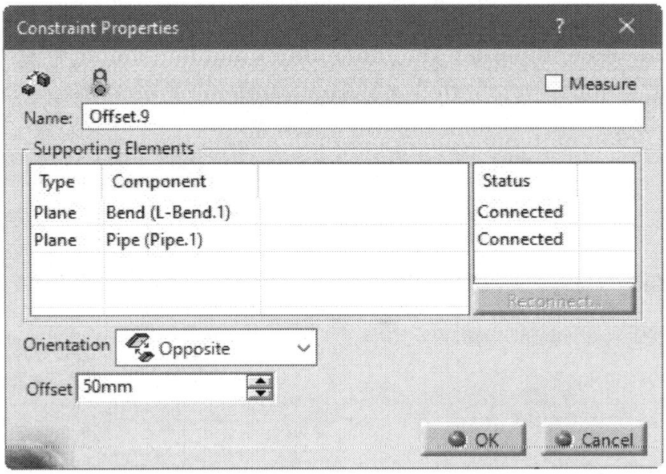

Figure 12-21 *The* ***Constraint Properties*** *dialog box*

The **Name** edit box displays the default name assigned to the constraint applied. You can change the default name and enter a new name in this edit box. If the **Measure** check box on the top right corner of the dialog box is selected, the present distance between the two selected surfaces will be measured from the geometry, and they will be assigned the same value as the offset distance. Note that this value will be displayed as a driven value in a bracket.

The **Supporting Elements** area displays the type of geometrical element selected for applying the constraint, the name of the component on which the geometrical elements are present, and the status of the constraint. The status should display **Connected**. If it displays **Disconnected**, you need to choose the **Reconnect** button, and select the geometric element again.

The **Orientation** drop-down list has three options namely, **Undefined**, **Same**, and **Opposite**. If the **Undefined** option is selected, then the software automatically orients the component in the same or in the opposite direction, depending on the orientation of the planes. Otherwise, you can select the required option from the drop-down list. In the **Offset** spinner, you need to enter the required offset distance between the planes. After setting all parameters, choose the **OK** button from the **Constraint Properties** dialog box. Select the **Update** button to place the components defined by the constraint. Now, the two selected planes will be placed parallel to each other and will have the specified separation between them. Figure 12-22 shows the faces to be selected. The orientation arrows are shown in Figure 12-23. Figure 12-24 shows the orientation of the selected planes after updating. After the **Offset** constraint is applied, the constraint symbol is displayed as the offset distance between the two selected faces.

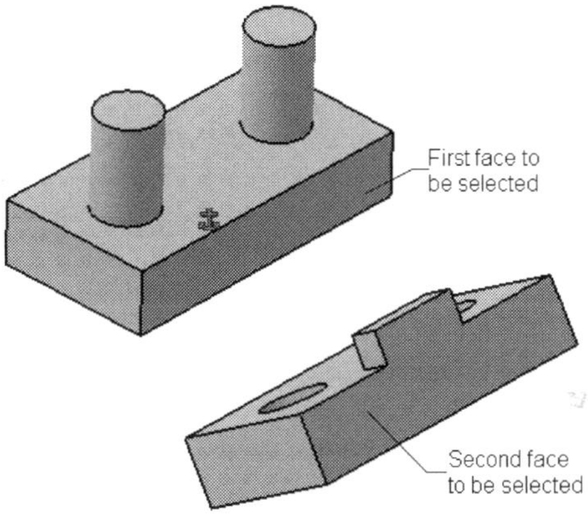

Figure 12-22 *Faces to be selected*

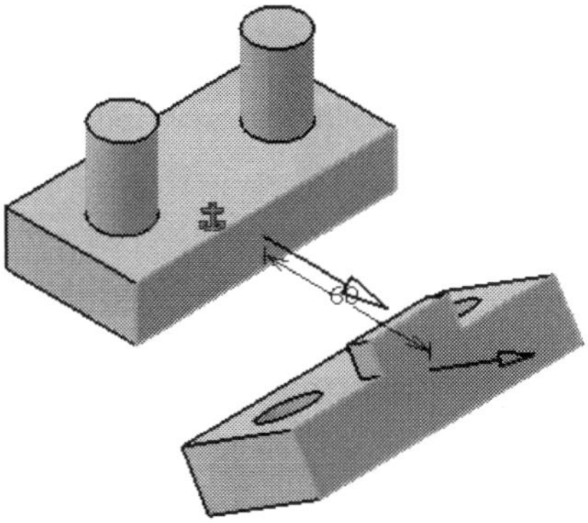

Figure 12-23 *Arrows pointing in the same direction*

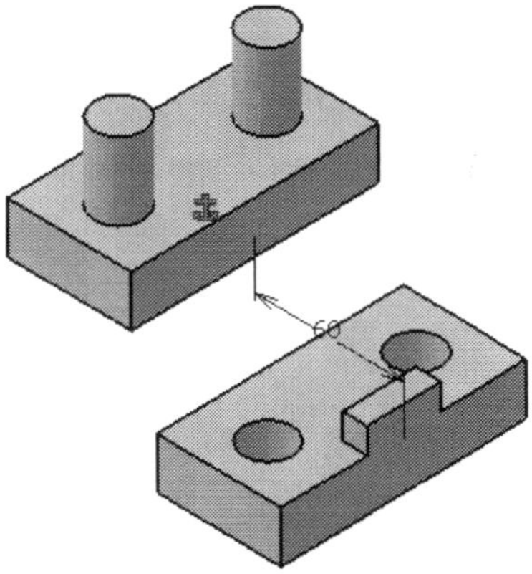

Figure 12-24 *Orientation of the selected planes after updating the file*

Angle Constraint

Menubar:	Insert > Angle
Toolbar:	Constraints > Angle Constraint

The Angle constraint is used to position two geometric constraints at a particular angle with respect to each other. You can also use this tool to make two selected elements parallel or perpendicular to each other. To apply this constraint, choose the **Angle Constraint** tool from the **Constraints** toolbar. Now, select the two planar faces from the two different components that you need to place at some angle from each other. You can also select a plane, circular face, or an edge as the geometric element. Once the selection is complete, the **Constraint Properties** dialog box will be displayed, as shown in Figure 12-25. In this dialog box, the **Angle** radio button is selected by default. If required, you can select the **Parallelism** radio button or the **Perpendicularity** radio button. If the **Perpendicularity** radio button is selected, the angle between the faces is automatically set to 90-degree and the display of the **Angle** spinner is turned off. If the **Parallelism** radio button is selected, then the **Orientation** drop-down list will be displayed. From this drop-down list, select required orientation option.

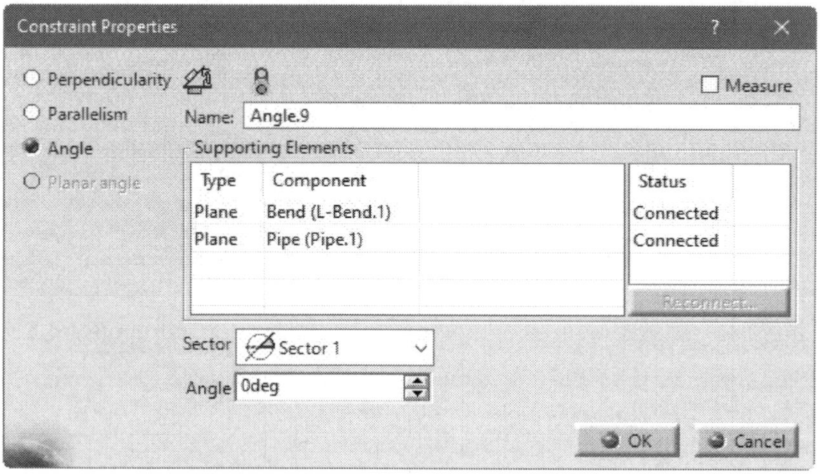

Figure 12-25 *The* ***Constraint Properties*** *dialog box for the Angle constraint*

If the **Angle** radio button is selected, then the **Angle** spinner and the **Sector** drop-down list will be displayed. Before specifying the angle, you need to select the appropriate sector, in which it will be applied. Select the sector from the **Sector** drop-down list. The selected sector will also be displayed in the geometry area. After setting all parameters, choose the **OK** button from the **Constraint Properties** dialog box. Figure 12-26 shows the faces to be selected and Figure 12-27 shows the orientation of the faces after applying the Angle constraint and updating them.

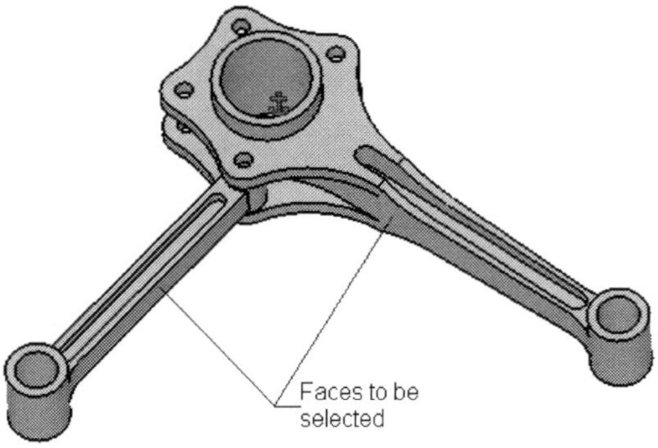

Figure 12-26 *Faces to be selected*

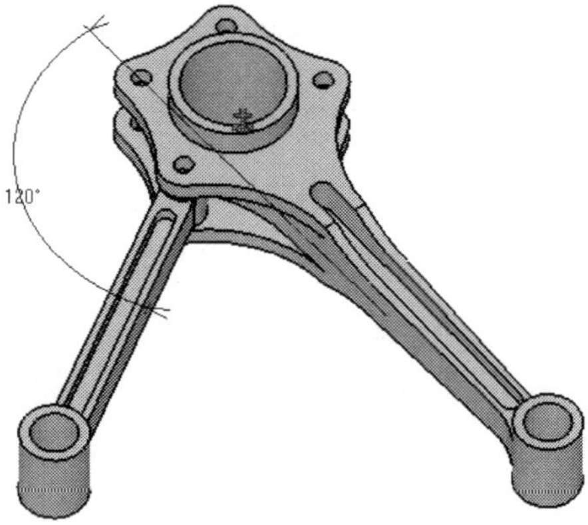

Figure 12-27 *The orientation of the faces after applying the Angle constraint*

After the constraint is applied, the angle value is displayed attached to the selected faces. In case the Perpendicularity constraint has been applied, the perpendicular (⊥) symbol appears between the selected faces. If the Parallelism constraint is applied, the parallel (⊣⊢) symbol is displayed between the selected faces. The name of the resulting constraint is also displayed in the Specification tree.

Fix Together Constraint

Menubar:	Insert > Fix Together
Toolbar:	Constraints > Fix Together

The Fix Together constraint is used to fix the position of the selected components with respect to each other. Once the selected components are fixed together, they can be moved as a single component such that the position of one component remains the same with respect to another component. To apply this constraint, choose the **Fix Together** tool from the **Constraints** toolbar; the **Fix Together** dialog box will be displayed. Now, select the components to be linked; the part numbers of the selected components will be displayed in the **Fix Together** dialog box, as shown in the Figure 12-28. If you need to remove a particular component from the list, click on its part number in the **Components** display box and then choose the **OK** button from the **Fix Together** dialog box to apply the Fix Together constraint.

Tip

1. After applying the Fix Together constraint to move the linked components by using the ***Manipulation*** *tool, select the* ***With respect to constraints*** *check box from the dialog box displayed. If you move the linked components using the compass, hold the SHIFT key and then move the component. Else, the components will move separately.*

2. You can move the components involved in fixed together. To do so, choose the required option from the ***General*** *tab from* ***Tool > Options > Mechanical Design > Assembly Design*** *from the menu bar.*

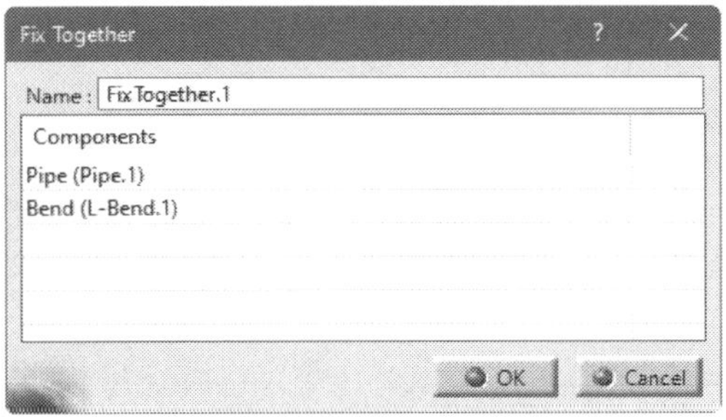

Figure 12-28 The ***Fix Together*** *dialog box*

Quick Constraint

Menubar:	Insert > Quick Constraint
Toolbar:	Constraints > Quick Constraint

In CATIA V5, there is an option in which the software applies the most appropriate constraint to the entities in the current selection set. To apply constraints using this method, choose the **Quick Constraint** tool from the **Constraints** toolbar. The possibility of applying constraints depends on the priority of constraints in the **Quick Constraints** priority list. You can invoke this list to set the priority by choosing **Tools > Options** from the menu bar; the **Options** dialog box will be displayed. Choose **Mechanical Design > Assembly Design** from the left of the **Options** dialog box. Next, choose the **Constraints** tab; the **Quick Constraint** priority list will be displayed in the **Quick Constraint** area.

To use the **Quick Constraint** tool, select the two geometric elements to be constrained. The software will automatically apply the most appropriate constraint between the selected geometric elements. This saves time in assembling components. However, you need to be careful while selecting the geometric elements because the orientation of the components depends on it.

Reuse Pattern

Menubar:	Insert > Reuse Pattern
Toolbar:	Constraints > Reuse Pattern

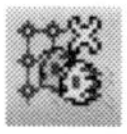

Sometimes, while assembling components, you may need to assemble more than one instance of the component in a specified arrangement. Consider the case of a flange coupling where you need to assemble eight instances of nuts and bolts to fasten the coupling. This is a very tedious and time-consuming process. Therefore, to reduce the time in the assembly design cycle, CATIA V5 provides you with the **Reuse Pattern** tool to insert and constrain multiple copies of a component over an existing pattern. The pattern can be rectangular, circular, or a user pattern.

The first step of using this tool is to insert the first instance of the component and constrain it with any instance of the pattern in the other component. Figure 12-29 shows a Plate with holes created using the circular pattern and a Pin that needs to be placed in each instance of the hole. After inserting the Plate and the Pin into the product file, constrain the Pin to any one instances of the hole on the Plate, as shown in Figure 12-30. To assemble the Pin with a hole in the Plate, apply the Coincidence constraint to the central axis of the Pin and the instance of the hole on the Plate. Next, apply the Contact constraint between the bottom face of the head of the Pin and the top face of the Plate.

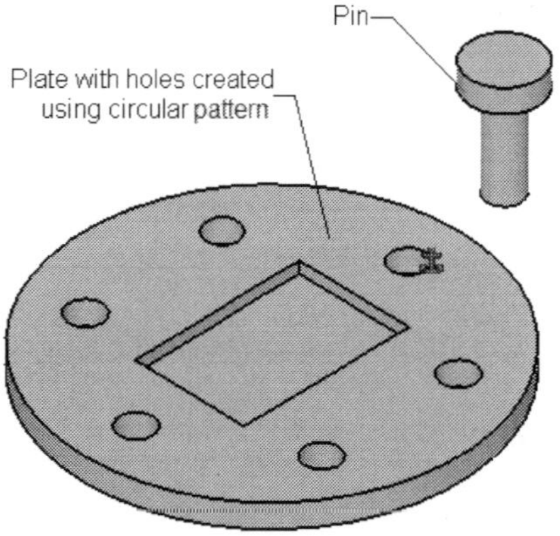

Figure 12-29 Plate having patterned holes and Pin to be placed in the hole instances

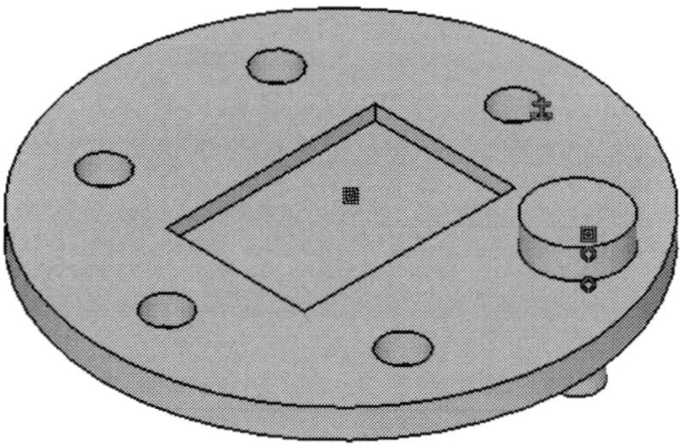

Figure 12-30 The Pin assembled to one of the instances of the patterned hole

Next, you need to select the constraint that associates the position of the Pin with the pattern instance. Therefore, in this case, the Coincidence constraint needs to be selected from the Specification tree. After selecting the constraint, choose the **Reuse Pattern** tool from the **Constraints** toolbar; preview of Pins assembled with all the instances of the hole will be displayed in the geometry area and the **Instantiation on a pattern** dialog box will be displayed, as shown in Figure 12-31. Note that the **Pattern** area of the dialog box indicates the name of the pattern, the number of instances to be created, and the name of the component, in which the pattern has been created. The name of the component to be repeated is displayed in the **Component to instantiate** area. The **Keep Link with the pattern** check box, at the top of the dialog box, is selected by default. This makes the newly created instances of the pattern associative with the pattern geometry.

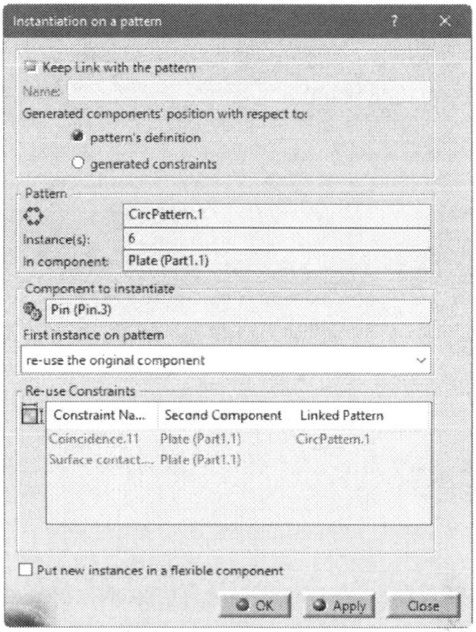

***Figure 12-31** The **Instantiation on a pattern** dialog box*

The **First instance on pattern** drop-down list has three options. These options are used to define the first instance of the component to be duplicated, and are discussed below:

re-use the original component
This option is used to retain the original component at its location, generate the instances, and populate only the vacant locations.

create a new instance
This option is used to place the new instance of the component on all the patterned instances, including the original instance. As a result, there will be two overlapping instances at the location of the original component.

cut and paste the original component
This option is used to remove the original component from its location and place the new instance of the component at all locations.

By default, the **pattern's definition** radio button is selected in the **Generated components' position with respect to** area of the **Instantiation on a pattern** dialog box. This option facilitates placing and constraining of all instances based on the selected reference pattern. If you select the **generated constraints** radio button, the constraints applied to the parent instance are also applied individually to all pattern instances.

After setting all options, choose the **OK** button from the **Instantiation on a pattern** dialog box. Figure 12-32 shows the resulting assembly after using the **Reuse Pattern** tool. List of instances created is displayed in the Specification tree shown in Figure 12-33. Note that the part number of every instance remains the same, but the instance number displayed in parenthesis is different. A new node called **Assembly features** will be created in the Specification tree and the **Reused Circular Pattern.1** is displayed under it.

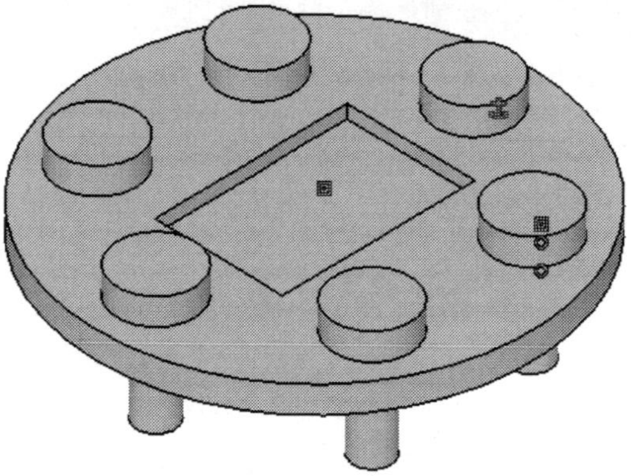

Figure 12-32 The assembly after the selected component is patterned

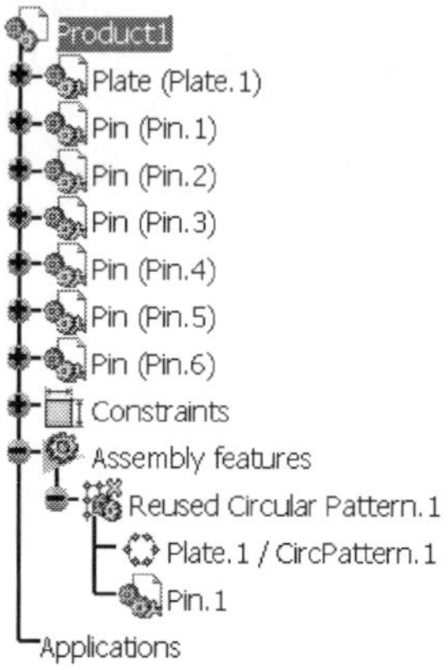

Figure 12-33 The Specification tree after creating the component pattern

Note

*You can choose the **Reuse Pattern** tool without selecting any constraint. On doing so, the **Instantiation on a pattern** dialog box with no entity selected will be displayed. Now, select the pattern instance and the component to instantiate from the Specification tree or from the geometry area.*

Inserting Existing Components with Positioning

Menubar: Insert > Existing Component With Positioning
Toolbar: Product Structure Tools > Existing Component With Positioning

The **Existing Component With Positioning** tool is used to insert, position, and apply constraints to a component in a single operation. It is an enhanced form of the **Insert Existing Component** tool. To insert a existing component, choose the **Existing Component With Positioning** tool from the **Product Structure Tools** toolbar. Now, click on the **Product1** in the Specification tree; the **File Selection** dialog box will be displayed. Select the part to be inserted and choose the **Open** button; the **Smart Move** dialog box along with the viewer, will be displayed. Use the **Smart Move** dialog box to position and constraint the newly inserted component. You need to make sure that the **Automatic constraint creation** check box is selected. Else, the only component will be placed and the constraint will not be applied. The applied constraint is displayed in the Specification tree. By using this tool, you can save the assembly creation time.

Note

*By using the **Existing component** and **Existing Component With Positioning** tools, you can insert an existing product file into the currently active product file as a subassembly. You can use individual parts of the subassembly to apply constraints. The parts and the subassemblies are associative with the parent product file. Therefore any modifications done in part or subassembly will be visible in the product file.*

CREATING TOP-DOWN ASSEMBLIES

As discussed earlier, in the top-down assembly design approach, all the components of the assembly are created inside the **Assembly Design** workbench. To create the components, you need to invoke the **Part Design** workbench within the **Assembly Design** workbench.

Creating Base Part in the Top-Down Assembly

Menubar:	Insert > New Part
Toolbar:	Product Structure Tools > Part

To start working on the top-down assembly, start a new product file. Click on **Product1** in the Specification tree and choose the **Part** tool from the **Product Structure Tools** toolbar. On doing so, a new component named **Part1** will be displayed in the Specification tree and a default name will be assigned to it. Once a new part is inserted into the product file, the geometry area will display the default planes. These planes belong to the new part and can be used to draw sketches and create features. By default, the origin of these planes is placed over the origin of the assembly coordinate system.

To change the name of the part, choose the **Properties** option from the contextual menu invoked by right-clicking on the part name displayed in the Specification tree. On doing so, the **Properties** dialog box will be displayed. The **Product** tab is chosen by default in the **Properties** dialog box. Specify the name of the part in the **Instance name** and **Part Number** edit boxes. After making changes, choose the **OK** button. The part names are modified in the Specification tree.

To create the model, you need to invoke the **Part Design** workbench. Click on the plus sign (+) displayed on the left of part name in the Specification tree to expand it. Now, double-click on the part name that is displayed inside the expanded Specification tree to invoke the **Part Design** workbench. The fully expanded Specification tree is shown in Figure 12-34. Now, you can create a part using tools in this workbench.

After the part is completed, double-click on **Product1** in the Specification tree to switch back to the **Assembly Design** workbench. Now, you can move and apply constraints to the base component.

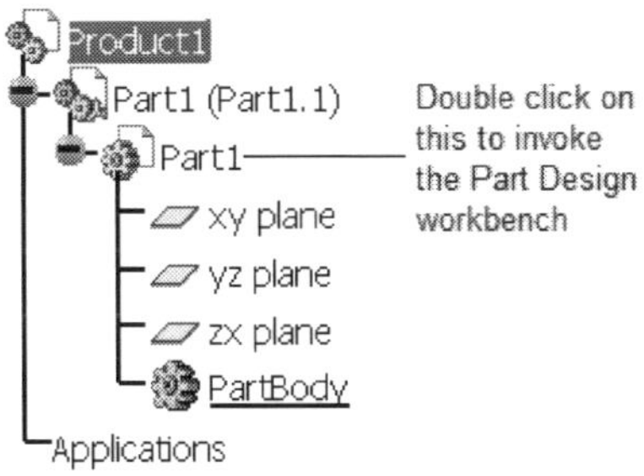

Figure 12-34 *The fully expanded Specification tree after inserting a part in the Product file*

Creating Subsequent Components in the Top-Down Assembly

After the base component is created inside a product file, you need to create other components of the assembly. The process of creating subsequent components is similar to that of creating the base component. Click on **Product1** in the Specification tree and then choose the **Part** button from the **Product Structure Tools** toolbar; if a geometry already exists under the **Product1** node then the **New Part: Origin Point** dialog box will be displayed which prompts you to define a new origin point for the new part. Select the **No** button to define the default origin point for the assembly as the origin part for the new part or select the **Yes** button to define the origin point of the assembly as a pre defined point or a vertex of the existing component. A new part is created and its name is displayed in the Specification tree. Now, invoke the **Part Design** workbench.

While creating the subsequent components using the top-down approach, you can refer to the geometry of components already created in the assembly to extract the geometry of the sketches of the current component. You can also refer to the geometry of the already created components, while creating features of the current component. For example, you can sketch a circle and then apply the tangent constrain to it with an edge of another part in the assembly or extrude a sketch up to a surface that belongs to another part. To retain this kind of relation between the external references, you need to activate the Keep link with selected object option. Choose **Tools > Options** from the menu bar to invoke the **Options** dialog box. Select the **Infrastructure** option from the left pane of the dialog box to expand the branch and select the **Part Infrastructure** option. Now, choose the **General** tab if it is not chosen by default to display the entries under it. In the **External References** area, select the **Keep link with selected object** check box. Now, choose the **OK** button from the **Options** dialog box. After activating this option, the relations with the external references will be maintained.

You can select a planar surface of the base component or the default planes as the sketching plane to draw sketches for creating the features of the new part. The edges of the base feature can be used to constraint the sketch and its surfaces can work as limits while creating the extruded feature. Once the part creation is complete, double-click on **Product1** to switch back to the **Assembly Design** workbench.

Similarly, you can create more parts in the current product file. Note that if you reorient a part that has some relation with external references, then the product file needs to be updated to reestablish the relation. Figure 12-35 shows two parts created inside the assembly file. The cylinder is extruded up to the surface that belongs to the base part. Figure 12-36 shows the Up to surface relation still maintained even after moving the cylinder base along the Z-direction.

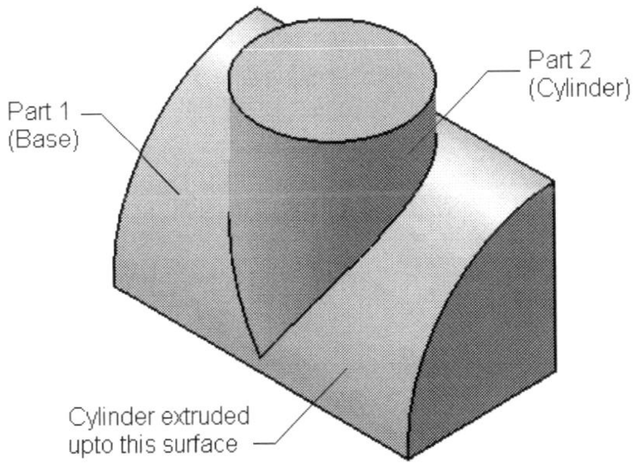

Figure 12-35 Two different parts created in a product file and the cylinder extruded up to the surface

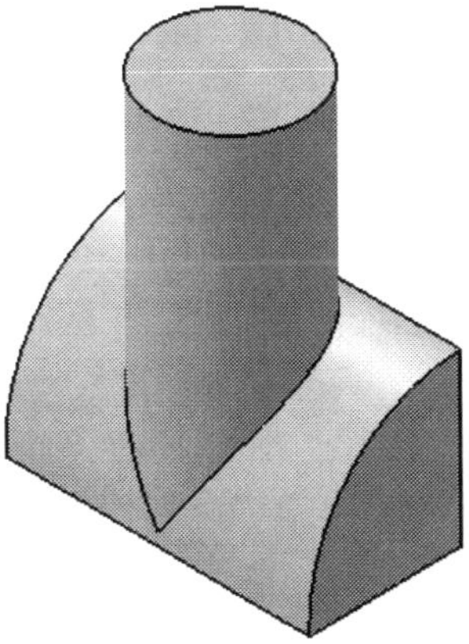

Figure 12-36 The Up to surface relation maintained even after moving the base downward

Creating Subassemblies in the Top-Down Assembly

While creating complicated assemblies, you may need to have subassemblies inside an assembly. While working in the top-down approach, you can directly create a subassembly inside the product file. In CATIA V5, there are two types of subassemblies that can be created in the **Assembly Design** workbench: **Product** and **Component**. Both these subassemblies are discussed next.

Product Subassemblies

Menubar:	Insert > New Product
Toolbar:	Product Structure Tools > Product

If you create a subassembly using the **Product** tool, the resulting subassembly and the parts created within it are saved as separate product and part files within the folder in which the main assembly file is saved. This gives the benefit of managing the subassembly or the part files individually. You can also open these files separately and work on the design changes. This results in greater flexibility. Once the modifications are made and the files are saved, the changes are automatically reflected in the main assembly file. To create a subassembly inside the product file, select **Product1** from the Specification tree and choose the **Product** tool from the **Product Structures Tools** toolbar. The new subassembly named **Product2** is displayed in the Specification tree. The name of this subassembly can also be modified using the method similar to that discussed while renaming parts. Because the newly created subassembly is already highlighted, choose the **Part** tool to insert a new part inside it. In this way, you can create more subassemblies inside the main assembly or inside the subassembly itself.

Note

You can activate a subassembly by double-clicking on the name of the subassembly in the Specification tree. To switch back to the main assembly, double-click on the name of the main assembly in the Specification tree.

Component Subassemblies

Menubar:	Insert > New Component
Toolbar:	Product Structure Tools > Component

If you create a subassembly using the **Component** tool, the resulting subassembly becomes an integral part of the main assembly file and will not be saved as a separate product file. However, the individual parts are saved as separate part files. To make any modification in the subassembly, you need to access it from the main assembly because the subassembly file is not saved separately. To create this type of subassembly inside the product file, select **Product1** from the Specification tree and choose the **Component** tool from the **Product Structures Tools** toolbar. The newly created subassembly named **Product3** will be displayed in the Specification tree, as shown in Figure 12-37. Now, you can rename this file if required and create parts inside it.

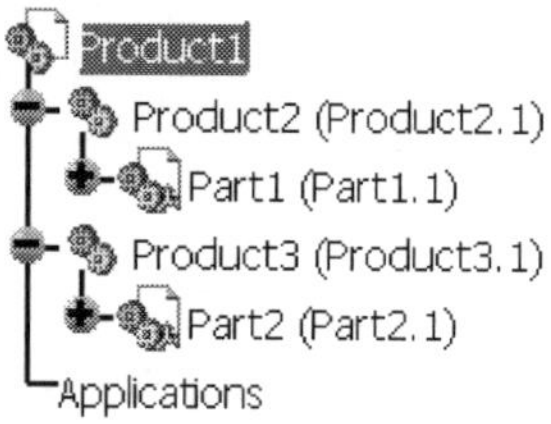

Figure 12-37 *The Specification tree showing products and components*

Note

Once all the parts and subassemblies are created inside the product file then you need to save them. These parts and subassemblies will be automatically saved as separate files inside the folder in which the main product file is saved, and so there is no need to save them separately. The file names will be the same as those given to parts and subassemblies in the ***Properties*** *dialog box.*

EDITING ASSEMBLIES

After creating the assembly, you may need to modify the parts, subassembly, or the applied constraints. You may also need to replace the existing part with another part. These editing operations are discussed next.

Deleting Components

While working in the **Assembly Design** workbench, you may need to delete some of the constituent parts and subassemblies. To delete a part or a subassembly, right-click on its name in the Specification tree and choose the **Delete** option from the contextual menu displayed. You can also delete a part or subassembly by selecting it from the Specification tree and pressing the DELETE key. If there are some relations associated with the selected part, the **Delete** dialog box will be displayed, as shown in the Figure 12-38.

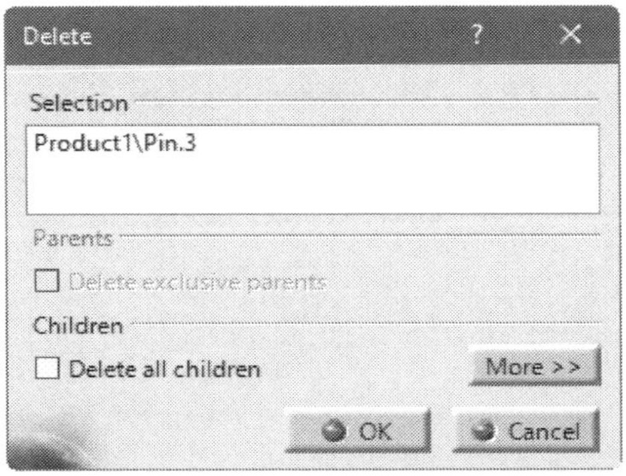

***Figure 12-38** The **Delete** dialog box*

The Display box of the **Delete** dialog box will display the names of the parts to be deleted and the name of the assembly to which the parts belong. Choose the **OK** button to complete the deletion process. The associated constraints now become inconsistent and a yellow error symbol is displayed in the Specification tree. These constraints have to be deleted separately.

If you select the **Delete all children** check box in the **Delete** dialog box, all relations associated with the selected part will be deleted along with it. Similarly, the subassemblies can also be deleted from the main assembly.

Note
In a product file with more than one subassembly, you cannot delete the currently activated subassembly. To do so, first activate any other subassembly and then delete the required subassembly.

Replacing Components

Menubar:	Edit > Components > Replace Component
Toolbar:	Product Structure Tools > Replace Component

In CATIA V5, you can replace an existing component with another component inside an assembly. If the new component being placed has the same basic geometry that the original component has, then it will be placed exactly at the same location where the original component was placed. Otherwise, the replaced component will be placed arbitrarily in space with no association with the location where the original component was present.

To replace a component, select it from the Specification tree and choose the **Replace Component** tool from the **Product Structure Tools** toolbar; the **File Selection** dialog box will be displayed. Select the component and choose the **Open** button from the **File Selection** dialog box; the **Impacts On Replace** dialog box will be displayed. In this dialog box, the **Yes** radio button is selected by default. As a result, all the instances of the selected component will be replaced. If you select the **No** radio button from the **Impacts On Replace** dialog box, then only the selected components will be replaced. Choose the **OK** button from this dialog box to replace the existing component with the selected component. Note that the constraints that were applied earlier on the previous component now become inconsistent. You can reattach these constraints, or delete them and apply new ones. The process of reattaching a constraint is discussed later in this chapter. Figures 12-39 and 12-40 show the original and the replaced components, respectively. Note that you cannot undo a Replace Component operation. Therefore, you need to be careful while performing it. If you need to undo the Replace Component operation, use the same command again to replace the new component with the previous component.

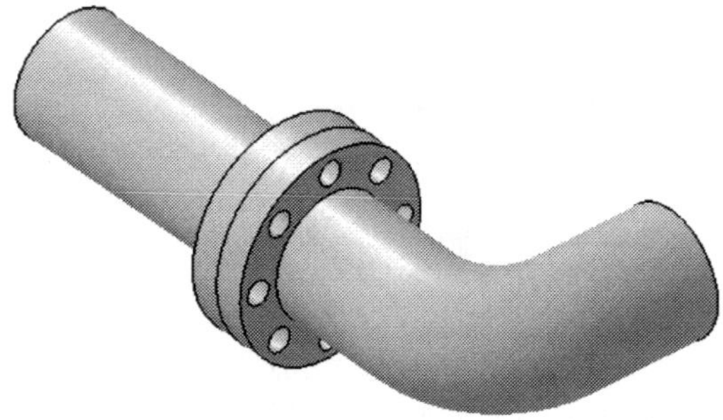

Figure 12-39 *The original component*

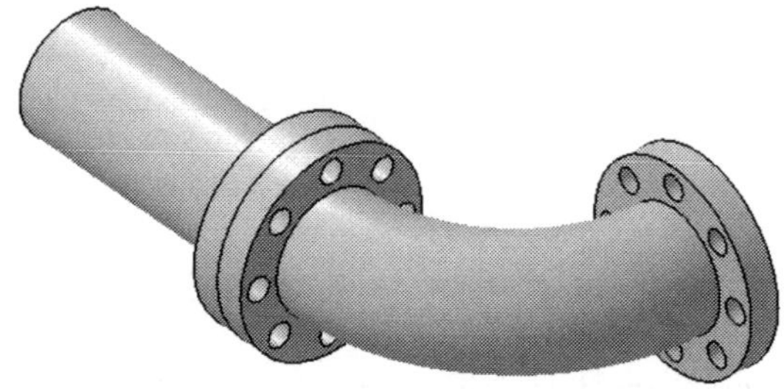

Figure 12-40 *The replaced component*

Editing Components Inside an Assembly

You can also edit features and modify sketches of the parts of the assembly within the **Assembly Design** workbench. For this, you need to activate it by invoking the **Part Design** workbench. To invoke the **Part Design** workbench for editing a part, click on the plus sign (+) displayed on the left of the part name to expand it in the Specification tree. Now, double click on the part name inside the expanded branch to expand it as well as invoke the **Part Design** workbench. In the **Part Design** workbench, you can make the modifications to the features and sketches of the part. After you have made all changes, double-click on the **Product** name to return back to the **Assembly Design** workbench. Note that in the **Part Design** workbench, all parts of the assembly are visible but changes are made only to the part that is active. Similarly, you can also edit the components in the subassemblies. After performing modifications on the components, it is recommended to press the CTRL+U keys on the keyboard to update the files.

Editing Subassemblies Inside an Assembly

You can also edit subassemblies that are placed inside the main assembly. To edit a subassembly, double-click on its name in the Specification tree; the subassembly will be activated. You can insert or remove the components, or you can edit the constraints applied to the components of the subassembly. After making the necessary changes, double-click on the main assembly to switch back to it.

Note

The changes made to a part or subassembly inside the product file are also reflected in its respective parts and product files. Therefore, the changes made at one place will take place wherever these parts and subassemblies have been used.

Tip

*On double-clicking on the sketch of the feature of a component in the Specification tree, the **Sketcher** workbench is invoked directly. Modify the sketch and exit the **Sketcher** workbench. You will switch to the **Part Design** workbench. To switch back to the **Assembly Design** workbench, double-click on **Product1** in the Specification tree.*

Editing Assembly Constraints

In an assembly, the constituent parts are positioned at their respective locations using the constraints. Sometimes, you need to replace the existing constraint with another constraint or to change the entities to which the constraints are applied. The methods to modify the constraints are discussed below.

Editing the Constraint Definition

All assembly constraints need to be associated with entities of two different components. These entities can be planes, surfaces, axes, edges, and so on. In the **Offset** and **Angle** constraints, some numeric values which define the offset distance and the rotation angle are also specified. These associated entities and the numerical values can be modified by editing the definition of the constraint. The definition of a constraint can be accessed by double-clicking on its name in the Specification tree or its symbol from the graphics area. On doing so, the **Constraint Definition** dialog box will be displayed. Choose the **More** button from this dialog box to expand the dialog box, as shown in the Figure 12-41.

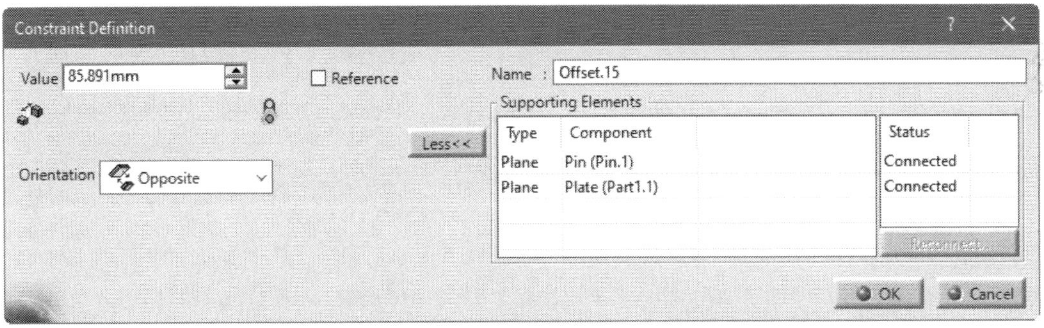

*Figure 12-41 The **Constraint Definition** dialog box*

The **Value** spinner is available on the left of this dialog box. You can modify the displayed value to change the offset distance. The appropriate option from the **Orientation** drop-down list can be selected to change the position of the faces between which the **Offset** constraints have been created. In the expanded region of the dialog box, the name of the constraint is displayed. If required, you can enter a new name for the constraint and the same will be displayed in the Specification tree.

The **Supporting Elements** area displays the type of entities and their corresponding components between which the constraint is applied. Ideally, the **Status** area should display **Connected**. This means that the association of the constraint with the entity exists. If the status shows **Disconnected**, then it would imply that the association between the entities has been broken and so the constraint has become inconsistent. In this case, the constraint has to be reconnected with the proper entity. The **Reconnect** button is used to select the entity on which the constraint will be connected. This button can also be used to replace an existing element with another element, even if the status is connected. By default, the **Reconnect** button is disabled. To enable it, select the reference to be replaced from the **Supporting Elements** area and then choose the **Reconnect** button; you will be prompted to select the element. Select the element from the geometry area; the new selection will be displayed in the dialog box. Choose the **OK** button to complete the constraint editing operation. Next, update the model to incorporate the changes. Figure 12-42 shows the Offset constraint applied between the faces of two components. Figure 12-43 shows the position of the components after the Offset constraint is reconnected to another surface and the offset distance is modified. You can modify any constraint in a similar manner by editing its definition.

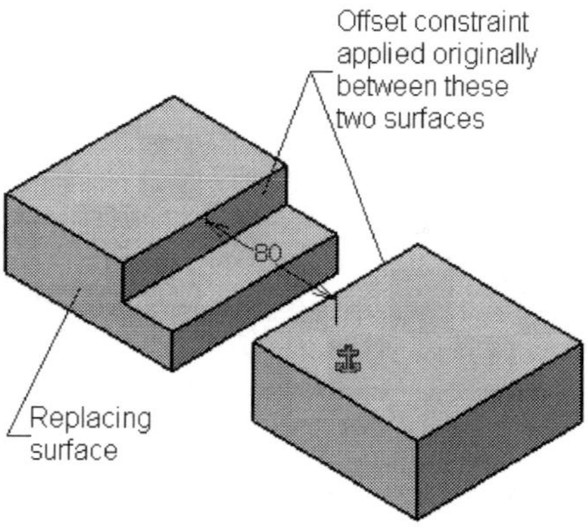

Figure 12-42 The associated and replacing surfaces for the Offset constraint

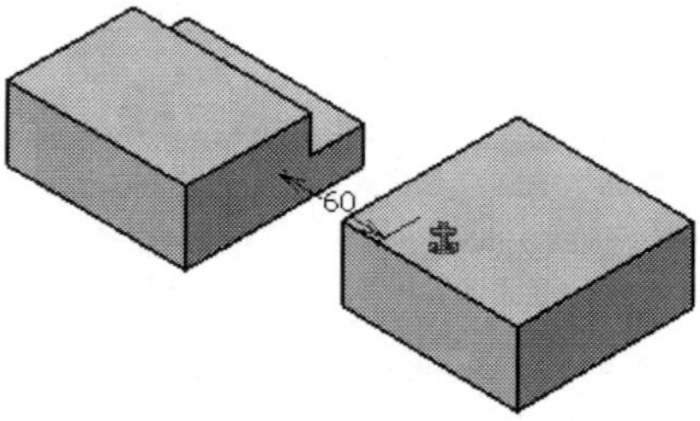

Figure 12-43 The components after editing the offset constraint and updating it

Replacing a Constraint

Toolbar: Constraints > Change Constraint

To replace an existing constraint with another constraint, select it from the Specification tree or from the geometry area. Next, choose the **Change Constraint** tool from the **Constraints** toolbar to display the **Possible Constraints** dialog box. This dialog box displays all possible constraints that can be used to replace the selected constraint. Select the appropriate constraint and choose the **OK** button from the **Possible Constraints** dialog box. The previously applied constraint will now be replaced by the new constraint. You can change the definition of the replaced constraint as per your requirement. After making the changes, update the assembly to bring the newly applied constraint into effect. Figure 12-44 shows the Contact constraint to be replaced by the Offset constraint and Figure 12-45 shows the components after replacing the Contact constraint by the Offset constraint, which is applied between the flange of the pipes.

Simplifying the Assembly

While working on large assemblies consisting of a large number of parts and subassemblies, you may face a difficulty while managing the components of the assembly. Therefore, it is recommended to hide some of the parts to improve the visibility of the other parts. While working with a large assembly, you may experience some difficulty in updating it because all parts of the assembly are rebuilt during updating. Therefore, it is recommended to deactivate the parts that are not required at that particular stage of the design cycle. This reduces the regeneration time of the assembly. The process of Hiding and deactivating the components is discussed next.

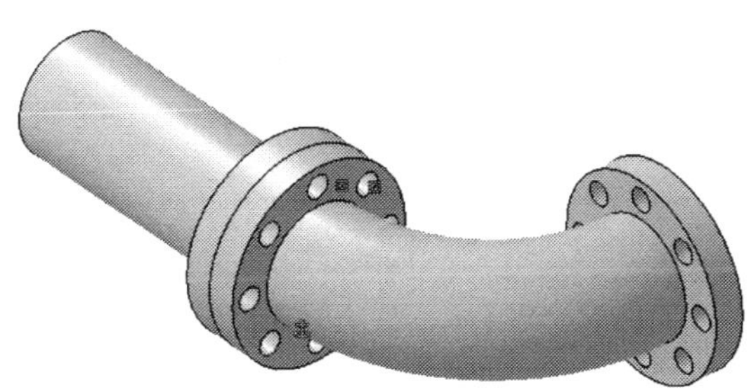

Figure 12-44 The Contact constraint to be replaced by the Offset constraint

Figure 12-45 The components after replacing the Contact constraint by the Offset constraint

Hiding a Component

Menubar:	View> Hide/Show > Hide/Show
Toolbar:	View> Hide/Show

The **Hide/Show** tool is used to turn off the display of the selected components of the assembly. But the component exists in the hierarchy of the assembly and participates in the assembly updating. The symbol of the hidden component is displayed in light gray color in the Specification tree. The **Hide/Show** tool can also be accessed from the contextual menu by right-clicking on the name of the component to be hidden.

Deactivating a Component

When you deactivate a component, it is removed temporarily from the assembly. This substantially decreases the regeneration time of the model. To deactivate a part, invoke the contextual menu by right-clicking on the name of the component in the Specification tree. Now place the cursor over the instance name to open the cascading menu and choose the **Activate/ Deactivate Component** option. On doing so, a red symbol () will be displayed on the left of the name of the deactivated component in the Specification tree. Follow the same procedure to activate the component and make it visible. Note that once a component is deactivated, the constraints associated with it become inconsistent and a yellow symbol is displayed against them in the Specification tree. These constraints are no longer displayed in the geometry area.

Interference Detection

Menubar:	Analyze > Clash
Toolbar:	Space Analysis > Clash

It is recommended to check the interference and clearance between the components of the assembly to make sure that the components do not interfere with each other and the right type of fit is maintained between the mating parts. The interference is detected using the **Clash** tool invoked by choosing the **Clash** button from the **Space Analysis** toolbar. On invoking this tool, the **Check Clash** dialog box will be displayed, as shown in Figure 12-46.

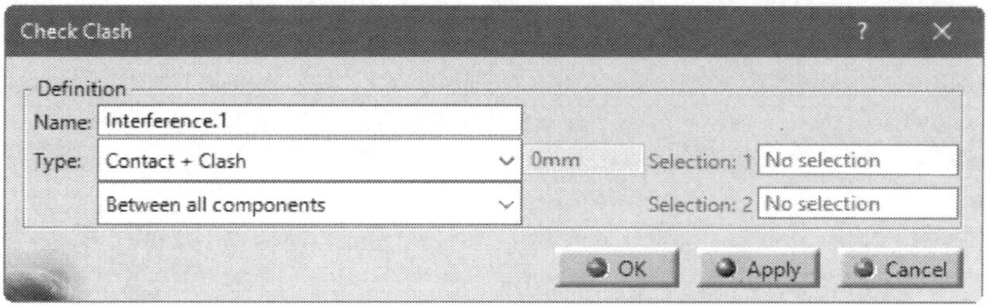

*Figure 12-46 The **Check Clash** dialog box*

There are two drop-down lists in the **Type** area. From the upper drop-down list, select the type of analysis that you need to perform. From the lower drop-down list, you can select the option for defining the selection of components between which the interference will be calculated. To perform the analysis, choose the **Apply** button. On doing so, the **Check Clash** dialog box will expand and display the result of the analysis. The result display area serially lists the name of components and the type of interference between them, which can be clearance, contact, or clash. On selecting a particular interference from the list area, the corresponding interference value will be displayed. The **Preview** window will be displayed to show the location of the interference. Figure 12-47 shows the **Preview** window displaying the location and value of the interference between the two components.

There are three tabs on the top of the **List** area. They are used to change the display format of the list. You can also use the drop-down lists of the **Filter list** area to display specific type of interferences in the list area. After checking the required interferences, choose the **OK** button to close the **Check Clash** dialog box.

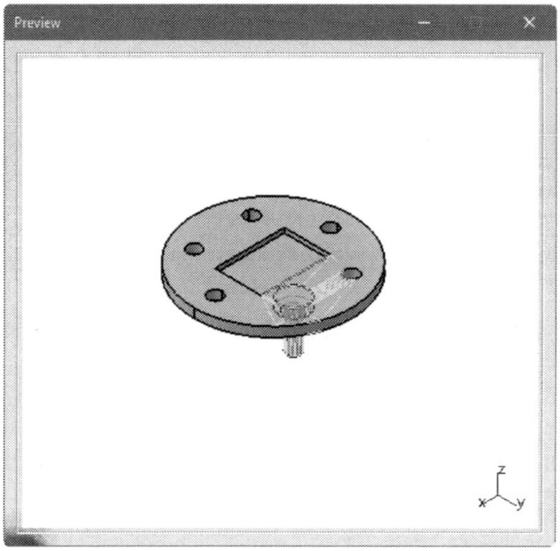

*Figure 12-47 The **Preview** window*

Sectioning an Assembly

Menubar: Analyze > Sectioning
Toolbar: Space Analysis > Sectioning

Sometimes, it is required to section an assembly model to view its cross-section. This is required to analyze the clearance and interference of internal parts, which may not be visible from outside. To section an assembly model, choose the **Sectioning** tool from the **Space Analysis** toolbar; the **Sectioning Definition** dialog box will be displayed as shown in Figure 12-48. Also, the 2D representation of the section view is represented in another window along with the default window. The assembly window displays the sectioning plane.

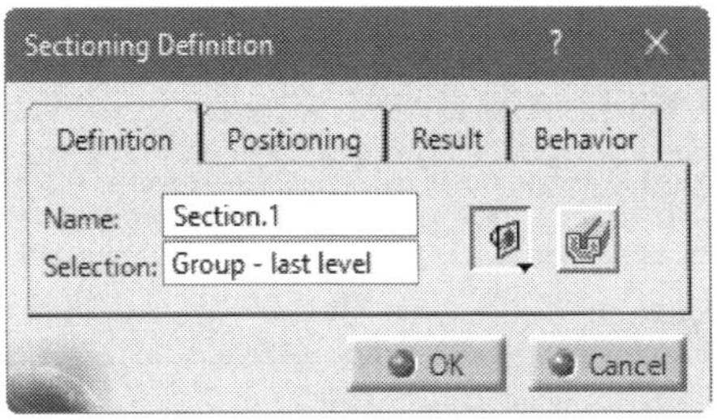

*Figure 12-48 The **Sectioning Definition** dialog box*

By default, the sectioning plane will be coincident to the yz plane. To change the position of the sectioning plane, place the cursor over it; a bidirectional green arrow will be displayed. This arrow will be normal to the sectioning plane. Drag the green arrow to reposition the plane. You can also use the red compass to rotate the sectioning plane in the same way as it is used for reorienting parts. The size of the sectioning plane can be modified by dragging its edges. You can also position it using the options in the **Positioning** tab of the **Sectioning Definition** dialog box. Figures 12-49 and 12-50 show 2D sectional view generated by sectioning the model and the plane used to create the section, respectively.

When the **Sectioning Definition** dialog box is invoked, the **Definition** tab is chosen by default. Under this tab, there are two buttons. The left button is used to select the type of sectioning required. Choose the down arrow displayed on the right of this button to open the flyout. This flyout displays three sectioning options. The section can be created using a plane or a slice of the model can be generated by sectioning it between two planes. Also, a portion of the model can be sectioned out by placing it in a bounding box. The next is the **Volume Cut** button. If this button is chosen, the solid section view of the assembly is displayed in the assembly model window. If you need to view only the 2D section view of some selected parts, then click in the **Selection** display box and select the required parts from the Specification tree. Selecting the same part again will remove it from the current selection set. To view the section of the whole assembly again, remove all parts from the current selection set and select the product, if it is not available in the current selection set. After viewing the required sectional view, choose the **OK** button to exit the **Sectioning Definition** dialog box. The cross-section that is generated after sectioning the model is now displayed in the geometry area and the corresponding section name is displayed in the Specification tree inside the **Applications** node. Now, close the window with the section view and maximize the product file. Hide the section, if you do not want it to be displayed.

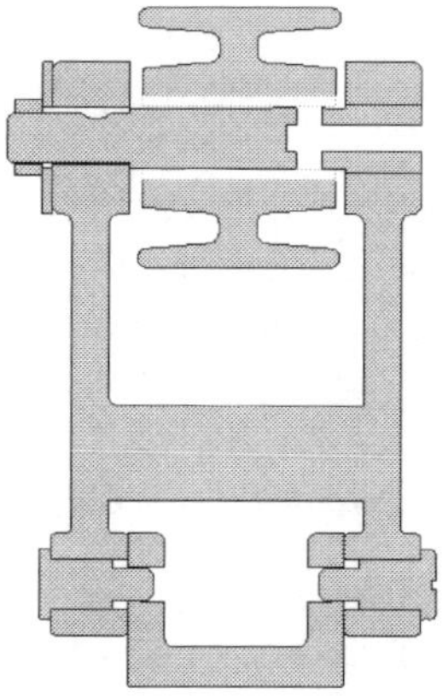

Figure 12-49 *2D section view of the complete assembly*

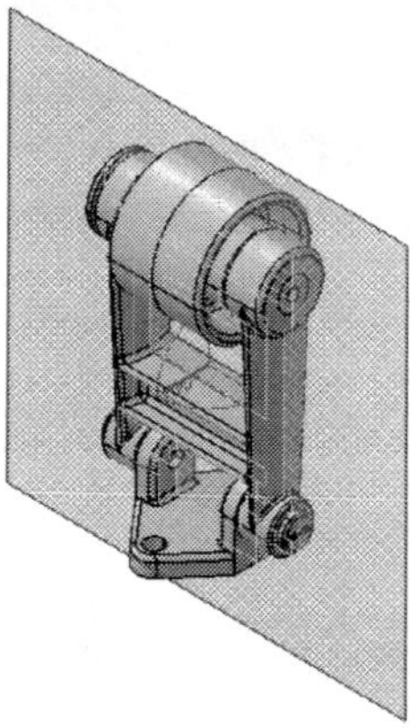

Figure 12-50 *The sectioning plane*

Exploding an Assembly

Menubar:	Edit > Move > Explode in assembly design
Toolbar:	Move > Explode

Generally, an assembly model consists of a large number of parts. Some of the parts are assembled inside the other parts. Therefore, these parts are not visible and the user is unable to see all components present in the assembly. To resolve this problem, the assembly is exploded such that all components are moved from their original position to a location where they are clearly visible. To explode an assembly, choose the **Explode** tool from the **Move** toolbar; the **Explode** dialog box will be displayed, as shown in the Figure 12-51. Make sure **Product1** is activated in the Specification tree before invoking the **Explode** tool. If there are multiple assemblies in the product file, you can select any one of them to explode.

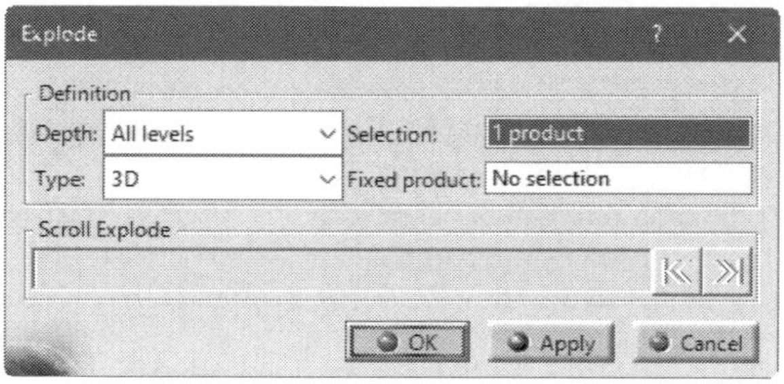

Figure 12-51 *The* ***Explode*** *dialog box*

You can set the parameters for exploding the assembly in the **Explode** dialog box. The options in this dialog box are discussed next.

The **Depth** drop-down list has two options. If the **First level** option is selected from this drop-down list, the parts of the subassembly are not exploded. Rather, the subassembly will be treated as a single component. The components of the subassembly will be exploded only if the **All levels** option is selected from the **Depth** drop-down list. The **Selection** display box displays the number of products that have been selected for explosion. The **Fixed product** display box is used to select a part of the assembly that needs to be fixed while exploding the assembly. All other parts will be moved with respect to it. In the **Type** drop-down list, there are three options. By default, the **3D** option is selected which enables the assembly model to explode in the **3D** space and the components to be placed arbitrarily in it. The assembled view of the Belt Tightener assembly is shown in Figure 12-52. Figure 12-53 shows the position of the components after the assembly is exploded using the **3D** option.

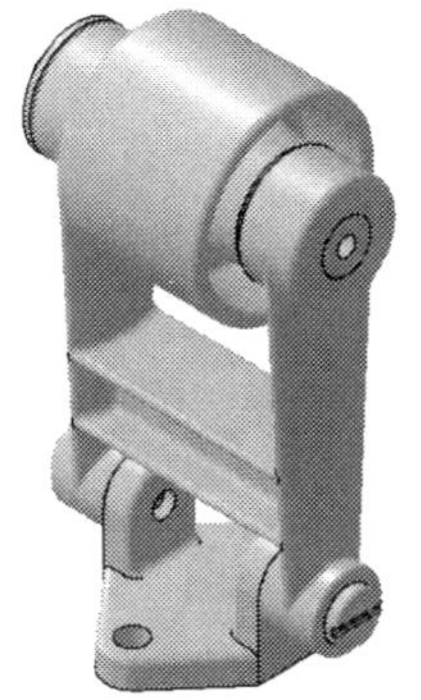

Figure 12-52 *The Belt Tightener in the assembled state*

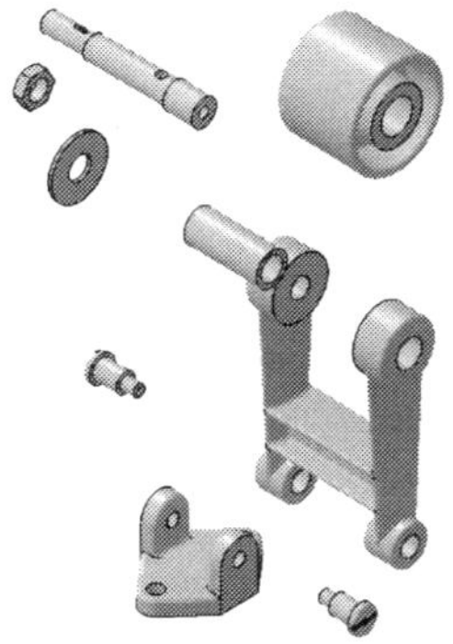

Figure 12-53 *Components after 3D explosion of the assembly*

If you select the **2D** option from the **Type** drop-down list, the components will be exploded and placed parallel to the viewing plane. Figure 12-54 shows the Belt Tightener assembly exploded using the **2D** option, with the front plane parallel to the screen. Figure 12-55 shows the top view of the same exploded assembly.

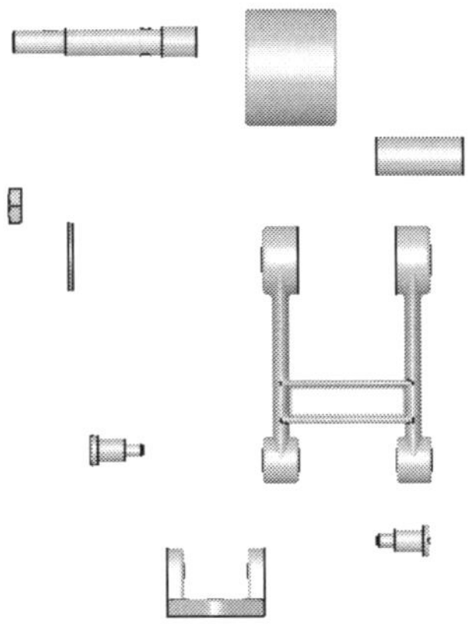

Figure 12-54 *Front view of the exploded Belt Tightener assembly exploded using the **2D** option*

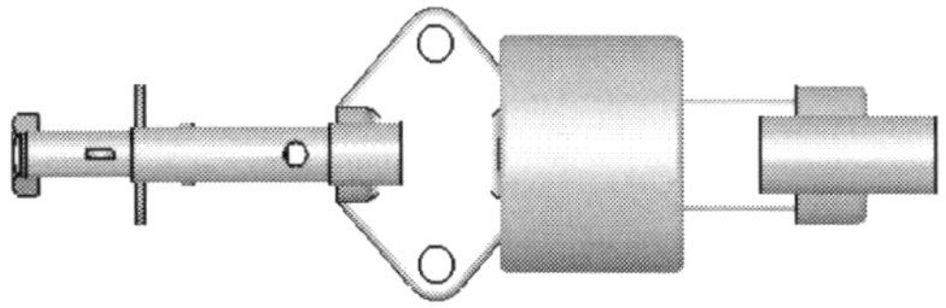

Figure 12-55 *Top view of the exploded Belt Tightener assembly*

The third option in the **Type** drop-down list is **Constrained**. You can select this option to explode the assembly in such a way that some of the constraints applied to the parts are maintained. This results in a more organized explosion, as shown in the Figure 12-56.

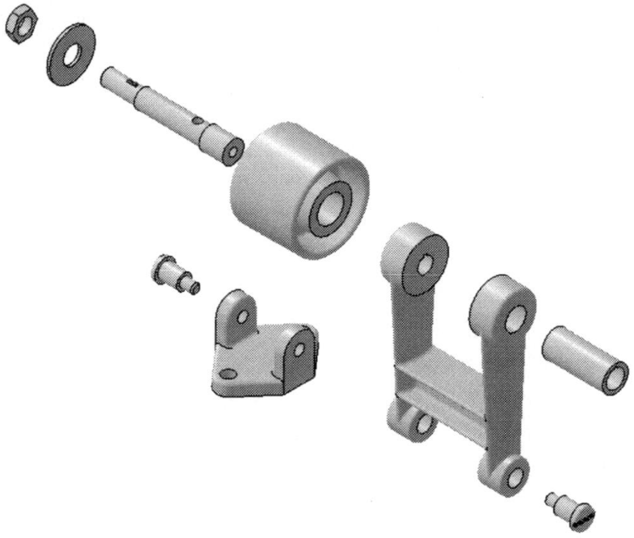

Figure 12-56 *The exploded assembly after selecting the* ***Constrained*** *option from the* ***Type*** *drop-down list*

After all the selections are made in the **Explode** dialog box, choose the **Apply** button; the assembly will be exploded and the **Information Box** will be displayed. This box informs you that the exploded parts can now be moved using the 3D compass. Move the components to arrange them in a more realistic manner, if required. Choose the **OK** button to close the **Information Box**. You can clear the **Show this message next time** check box to prevent the **Information Box** from appearing every time you explode a model. You can use the scroller in the **Scroll Explode** area to change the distance between parts. Finally, choose the **OK** button from the **Explode** dialog box to close it and then choose **Yes** from the **Warning** message box. The exploded assembly is shown in the geometry area.

To switch back to the assembled view, choose the **Update** button from the **Update** toolbar. Note that the components which are not fully constrained will not be updated.

Creating Scenes of an Assembly

Menubar:	Insert > Create Enhanced Scene
Toolbar:	Scenes > Enhanced Scene

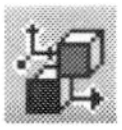

The **Enhanced Scene** tool allows you to create multiple assembly configurations of different types. It also makes the visualization of arranged assemblies on drawings easier and also reduces the drafting time. To create a new scene of an assembly, choose the **Enhanced Scene** tool from the **Scenes** toolbar; the **Enhanced Scene** dialog box will be displayed, as shown in Figure 12-57. Note that **Product1** is activated in the Specification tree before invoking the **Enhanced Scene** dialog box. Now, clear the **Automatic naming** check box; the **Name** edit box gets activated. Enter the desired name in the **Name** edit box and choose the **OK** button; a warning message box will be displayed. Now, choose the **Close** button; the Enhanced Scene environment will be invoked. In this environment, you can create different scenes by using tools such as **Explode**, **Apply Scene on Assembly**, and **Apply Assembly on scene** available in the **Enhanced Scenes** toolbar. Choose the **Save Viewpoint** and then the **Exit Scene** tool in this toolbar to exit the environment. The name of the scene created is displayed in the Specification tree inside the **Applications** node.

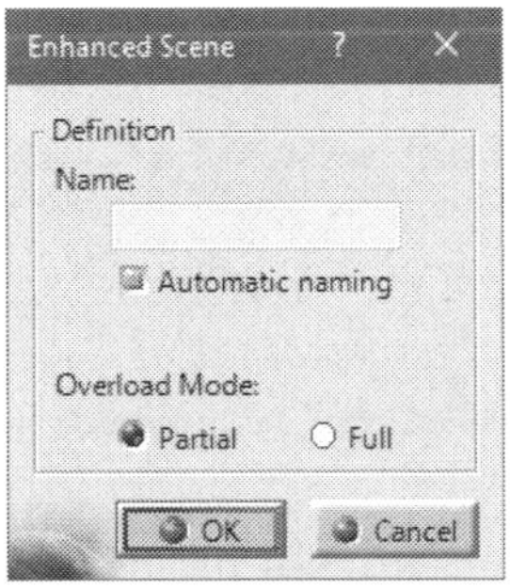

Figure 12-57 *The* ***Enhanced Scene*** *dialog box*

TUTORIALS

Tutorial 1

In this tutorial, you will create all the components of the Blower assembly shown in Figure 12-58 and then assemble them together. After creating it, you will generate the exploded view. The exploded view of the Blower assembly is shown in Figure 12-59. The dimensions of all the components are shown in Figures 12-60 through 12-65. **(Expected time: 2.5 hr)**

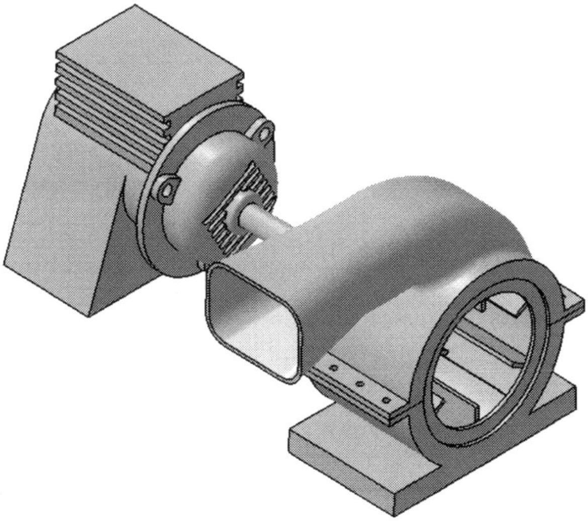

Figure 12-58 *The Blower assembly*

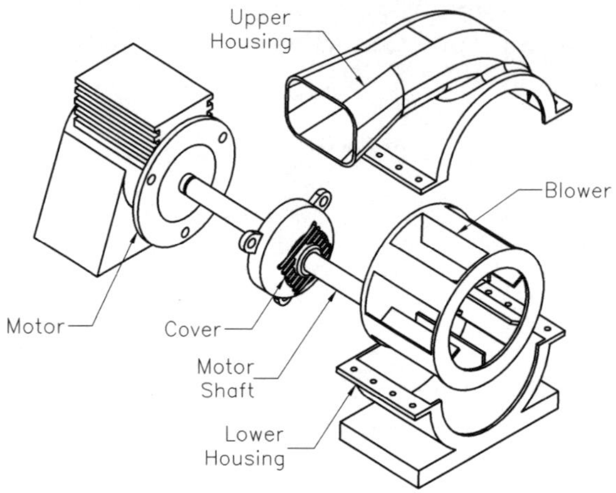

Figure 12-59 Exploded view of the Blower assembly

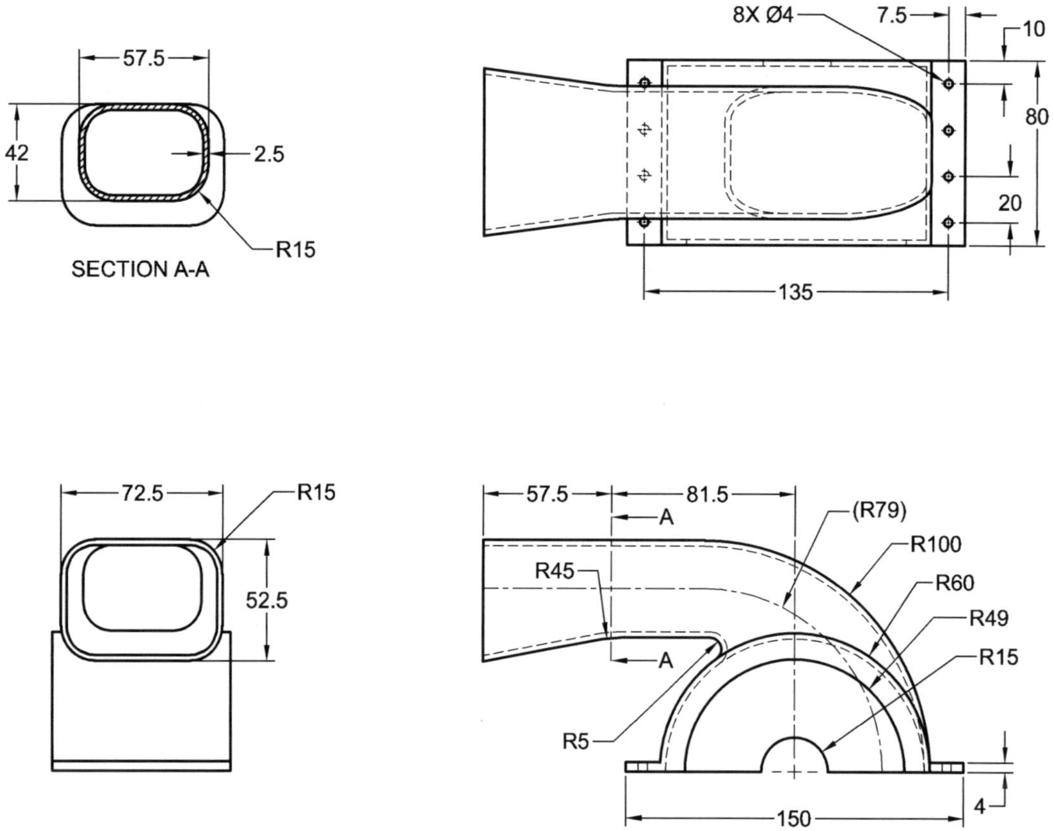

Figure 12-60 Orthographic views and dimensions of the Upper Housing

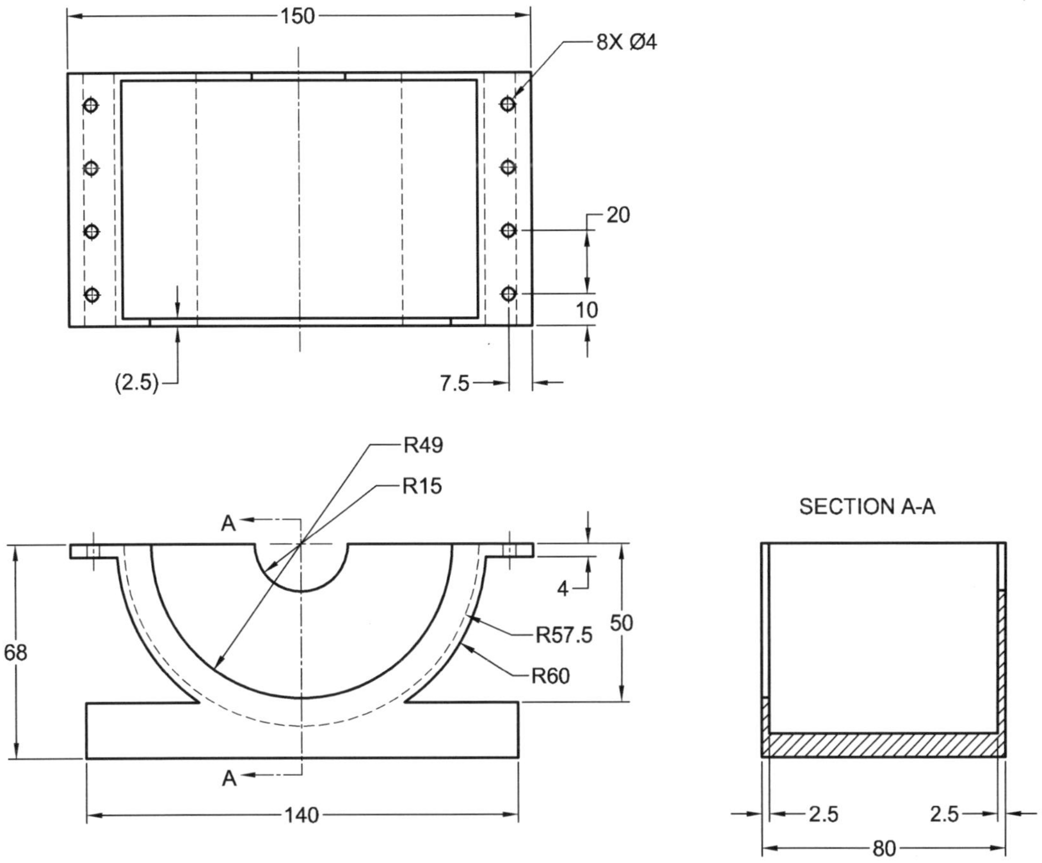

***Figure 12-61** Orthographic views and dimensions of the Lower Housing*

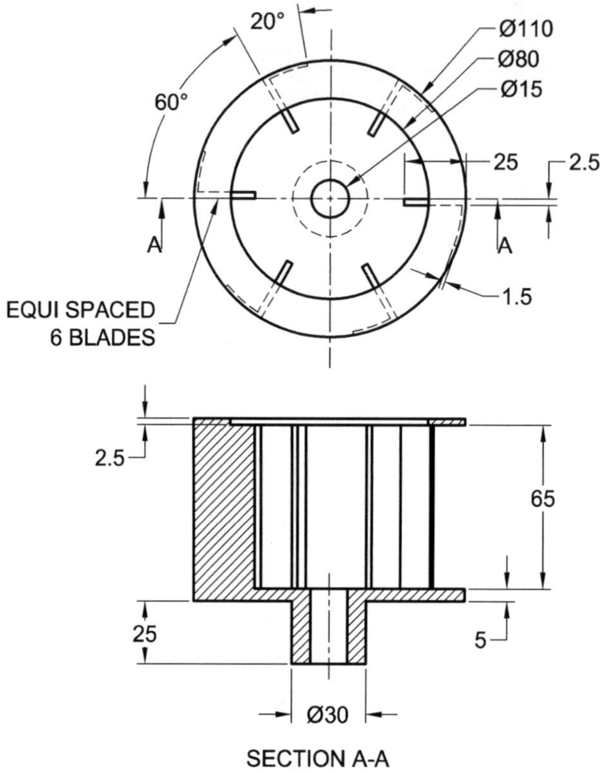

Figure 12-62 Orthographic views and dimensions of the Blower

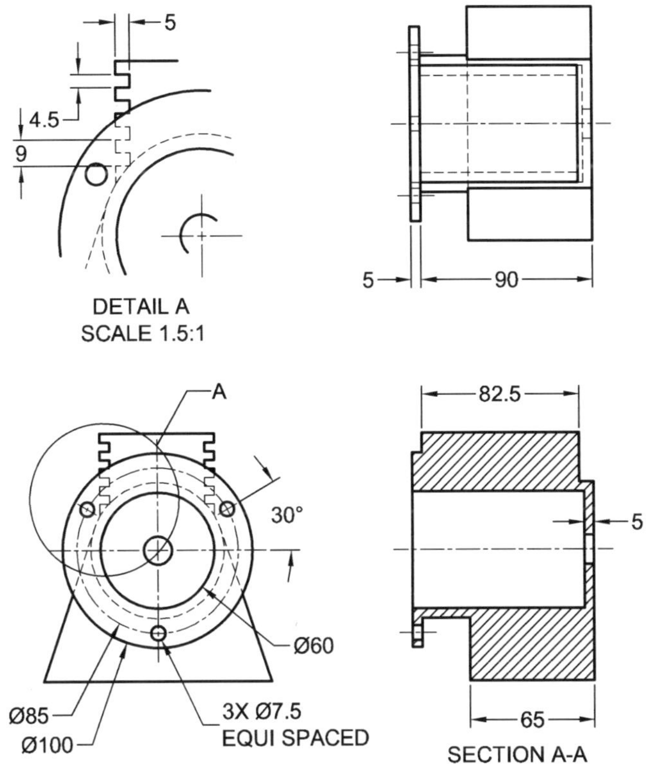

Figure 12-63 *Orthographic views and dimensions of the Motor*

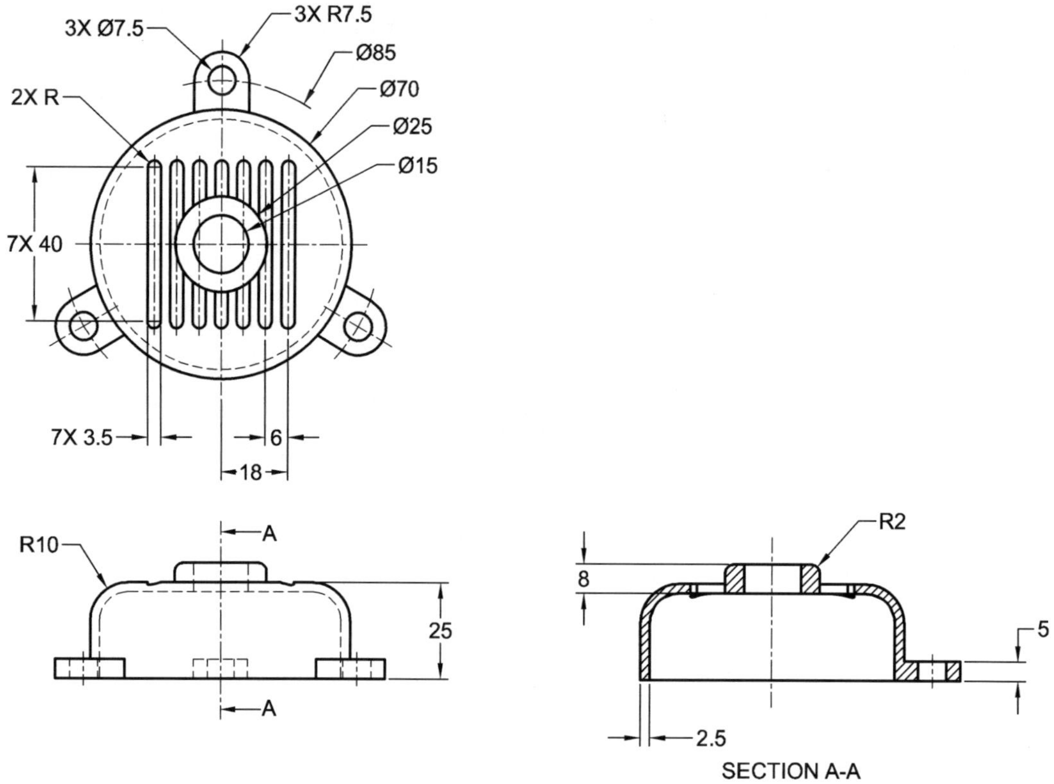

Figure 12-64 *Views and dimensions of the Cover*

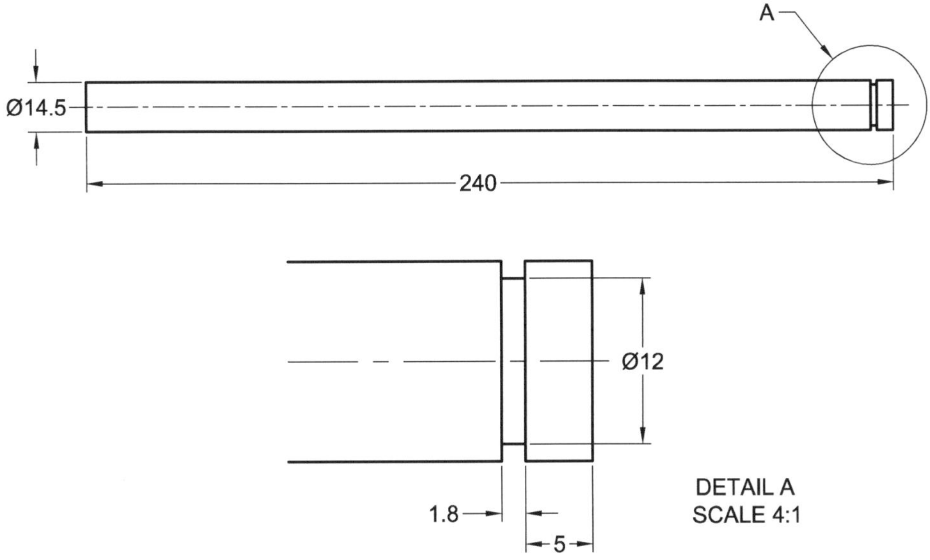

Figure 12-65 *Views and dimensions of the Motor Shaft*

The following steps are required to complete this tutorial:

a. Create all components of the assembly as separate part files in the **Part Design** workbench.
b. Start a new file in the **Assembly Design** workbench.
c. Insert the Lower Housing into the assembly as the base component, set its orientation, and apply the Fix constraint to it at its default location.
d. Insert the Upper Housing into the assembly and place it over the Lower Housing by applying proper constraints.
e. Hide the Upper Housing. Insert and place the blower inside the Lower Housing.
f. Insert and constrain the Motor, the Motor Shaft, and the Cover.
g. Turn on the display of the Upper Housing.
h. Create exploded state of the assembly.
i. Save the assembly file.

Before creating components for this tutorial, create a folder with the name *Blower Assembly* at *C:\CATIA\c12*. You need to save the parts of the Blower assembly in this folder. Also, you need to change the part number of every component before saving it. To change the part number, select the part name in the Specification tree and right-click. Then, choose the **Properties** option from the contextual menu; the **Properties** dialog box will be displayed. Specify the name for the part in the **Part Number** edit box in the **Product** tab.

Note

While performing this tutorial, the orientation of the components should be the same, as shown in figures of this tutorial.

Creating Components of the Assembly

The Blower assembly will be created using the bottom-up approach. You have already learned that in the bottom-up assemblies, all the parts of the assembly are first created as individual part files and then inserted into the assembly file.

1. Create all the parts of the assembly and save them as separate part files at *C:\CATIA\c12\ Blower Assembly*.

2. Close all the part files if they are opened.

Starting a New File in the Assembly Workbench

All the components that you have created above need to be assembled in an assembly file. The assembly file has a file extension *.CATProduct*. You need to start a new file in the **Assembly Design** workbench to assemble the parts.

1. Choose the **New** button from the **Standard** toolbar; the **New** dialog box is displayed.

2. Choose the **Product** option from the **List of Types** list box.

3. Choose the **OK** button from the **New** dialog box; a new file is started in the **Assembly Design** workbench and **Product1** is displayed on the top of the Specification tree. In case, the product file starts in an environment other than the assembly environment, then choose **Start > Mechanical Design > Assembly Design** from the menu bar.

Inserting the First Component and Fixing it

After the new product file is started, you can insert the base component into the assembly. In this case, the Lower Housing is the base component. After inserting the Lower Housing, you need to set the orientation of the Lower Housing and then fix its location.

1. Choose the **Existing Component** tool from the **Product Structure Tools** toolbar. 

2. Select **Product1** from the Specification tree; the **File Selection** dialog box is displayed. From this dialog box, browse to the location of the file of the Lower Housing and open it.

   The Lower Housing is displayed in the geometry area and its name appears in the Specification tree. If the current orientation of the isometric view of the placed component is not as the required one, you need to reset it. The orientation of the model can be set by using the **Snap** tool. If the current orientation of the isometric view of the placed component is as per your requirement, you can skip step 3 and 4.

3. Choose the **Snap** tool from the **Move** toolbar and then select the first and second elements, as shown in Figure 12-66. The orientation of the Lower Housing is changed and a flip arrow is displayed on it. 

4. Click anywhere in the geometry area to exit the **Snap** tool. Set the orientation of the view of the assembly to Isometric; the orientation of Lower Housing is modified as shown in Figure 12-67.

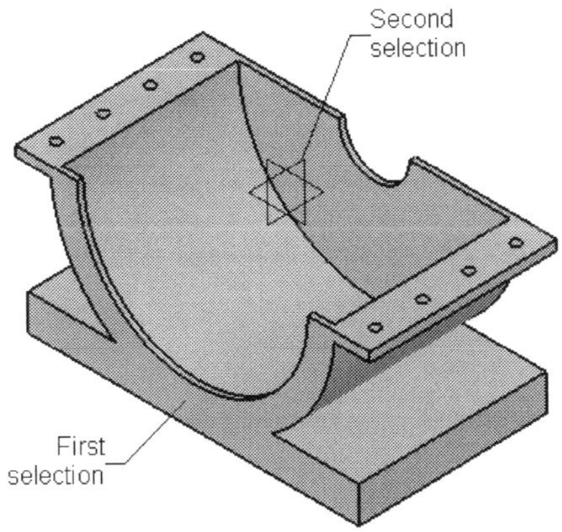

Figure 12-66 First and second elements to be selected

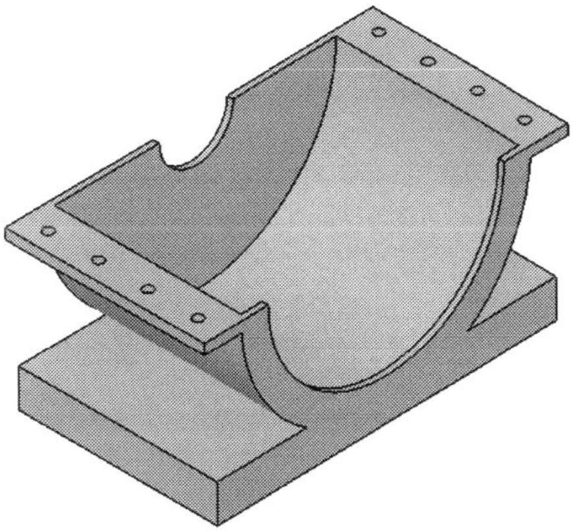

Figure 12-67 The Lower Housing after modifying its orientation

Next, you need to apply the Fix constraint to lock the position of the lower housing.

5. Choose the **Fix Component** tool from the **Constraints** toolbar and select the Lower Housing from the geometry area or from the Specification tree.

The symbol of the Fix constraint is displayed on the Lower Housing in the geometry area. Figure 12-68 shows the Lower Housing after making it fixed at its default location.

Inserting the Upper Housing and Constraining it

You can insert different parts into the assembly as per your requirement. However in this tutorial, the Upper Housing will be the second component to be inserted into the Blower assembly.

1. Insert the Upper Housing in a similar way, as discussed earlier.

Note

The part number of all the components were modified before saving the part files. Therefore, the ***Part Number Conflicts*** *dialog box will not be displayed for any component.*

The Upper Housing will be placed at its default location, refer to Figure 12-69. You need to apply constraints to place it properly over the Lower Housing. The first constraint that will be applied to the Upper Housing is the Contact constraint. This constraint will be applied between the upper face of the Lower Housing and the lower face of the Upper Housing.

2. Choose the **Contact Constraint** tool from the **Constraints** toolbar if the **Assistant** message box is displayed, then select the **Do not prompt in the future** check box and choose the **Close** button. Select the two faces, as shown in Figure 12-69.

You need to rotate the view of the assembly to select the surface that is not visible in the current display.

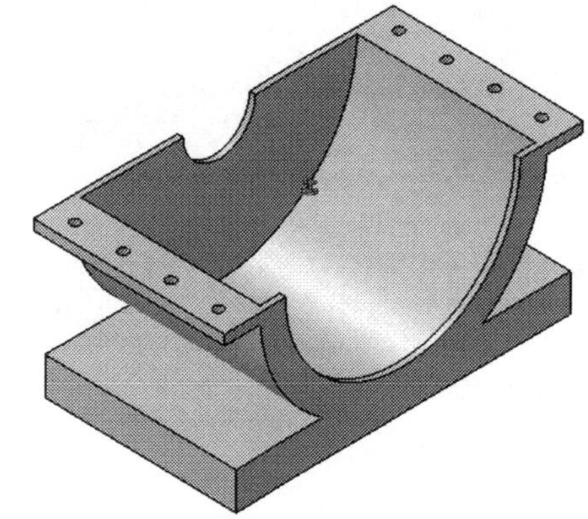

Figure 12-68 Lower Housing after it is fixed at its default location

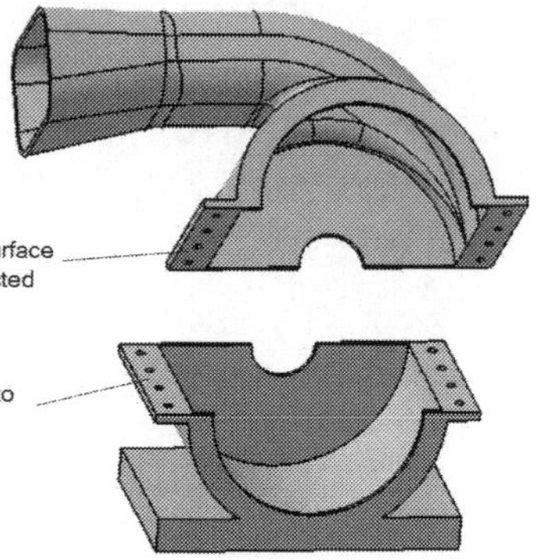

Figure 12-69 The rotated view of upper housing and the surfaces to be selected to apply the Contact constraint

3. Choose the **Update All** button if active. If this button is not active, the assembly does not need updating.

   The Contact constraint is applied between the two surfaces and name of the constraint is displayed in the Specification tree under the **Constraints** node.

   Next, you need to apply the Coincidence constraint between the cylindrical surfaces of the two components to make them concentric.

4. Choose the **Manipulation** tool from the **Move** toolbar and select the **Drag around Z axis** button. Next, select the Upper housing part to change its orientation.

5. Choose the **Coincidence Constraint** tool from the **Constraints** toolbar. Now, click on the two cylindrical surfaces to select the central axes of these surfaces, as shown in the Figure 12-70. 

6. Choose the **Update All** button from the **Update** toolbar to reorient the Upper Housing. Alternatively, you can use the CTRL+U keys to update the assembly.

Note

*To confirm the presence of the free degree of freedom, double-click on **Upper Housing** in the Specification tree; the Upper Housing is activated. Now, choose **Analyze > Degree(s) of freedom** from the menu bar; the **Degrees of Freedom Analysis** dialog box is displayed along with a set of arrows in the X direction. This set of arrows shows the free direction of the degree of freedom. Choose the **Close** button from the **Degrees of Freedom Analysis** dialog box and then double-click on **Product1** in the Specification tree to activate the **Constraints** toolbar.*

Next, you need to align the right face of the Upper Housing with the right face of the Lower Housing. This can be done by applying the Offset constraint with 0 offset.

7. Choose the **Offset Constraint** tool from the **Constraints** toolbar, and then select the faces shown in Figure 12-71 to apply the constraint between them. 

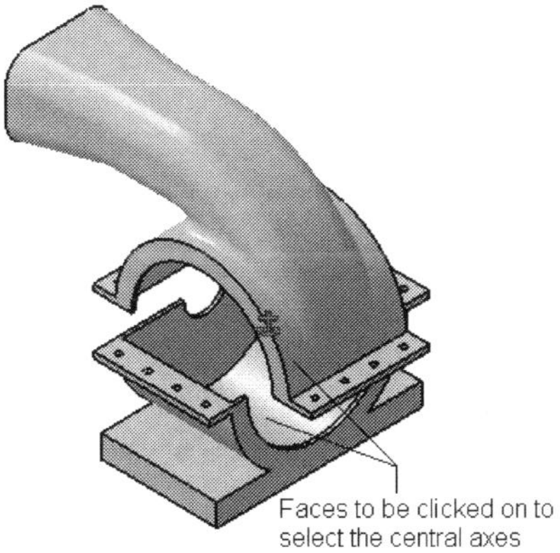

Figure 12-70 Surfaces on which you need to click to select the central axes

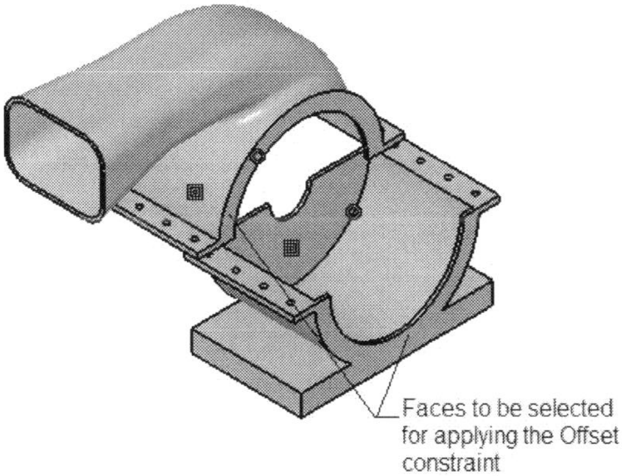

Figure 12-71 The rotated view of upper housing and the surfaces to be selected for applying the Offset constraint

On doing so, the **Constraint Properties** dialog box is displayed. Now, make sure that the **Orientation** is set to **Same** and the **Offset** value is set to **0**.

8. Choose the **OK** button, and then choose the **Update All** button to apply the Offset constraint. The Upper Housing is fully constrained.

Tip

*To check whether a part is fully constrained, try to reorient it using the **Manipulation** tool with the **With respect to constraints** check box selected. If the part is fully constrained, it will not move or rotate in any direction.*

Assembling the Blower after Hiding the Upper Housing

The Blower needs to be assembled in between the Upper and Lower Housings. To ease the process of assembling the Blower, you need to hide the Upper Housing.

1. Right-click on the name of the Upper Housing in the Specification tree and choose the **Hide/Show** option to turn off the display of the selected component.

2. Now, insert the Blower in the assembly as discussed earlier. Choose the **Coincidence Constraint** tool and click on the faces shown in Figure 12-72 to select the central axes of these faces.

3. Update the assembly, if required. If the orientation of the Blower is not as shown in Figure 12-73, orient it using the **Manipulation** tool and update again.

   Next, you need to place the left face of the Blower at an offset distance of 3.175 from the inner left face of the Lower Housing using the Offset constraint.

4. Choose the **Offset Constraint** tool and apply the Offset constraint between the faces shown in Figure 12-72. Enter **3.175** in the **Offset** spinner and make sure that the **Orientation** is set

to **Opposite** in the **Constraint Properties** dialog box. Then, choose the **OK** button. Next, update the model to bring the Blower to its proper position.

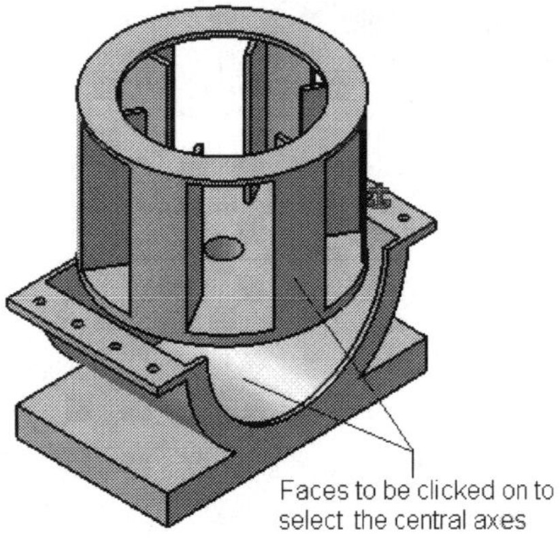

Figure 12-72 Faces to be clicked on to select the central axes

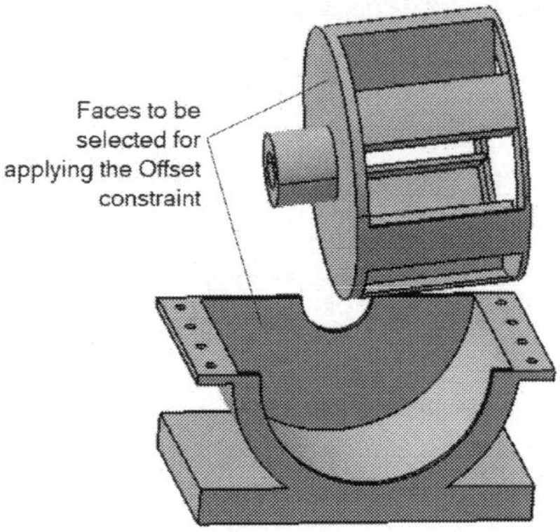

Figure 12-73 The rotated view of Blower and the faces to be selected for applying the Offset constraint

Assembling the Motor Shaft

Next, you need to assemble the Motor Shaft.

1. Insert the Motor Shaft into the assembly file. By default, it is placed in the middle of the existing assembly, as shown in Figure 12-74. You need to move it out of the assembly to get a better view of the shaft.

2. Choose the **Manipulation** tool from the **Move** toolbar. Next, choose the **Drag along X axis** button and drag the Motor Shaft to move it out of the assembly and rotate it if required, as shown in Figure 12-75. Choose **OK** and exit the **Manipulation Parameters** dialog box.

3. Choose the **Offset Constraint** tool from the **Constraints** toolbar and then select the faces shown in Figure 12-75; the **Constraint Properties** dialog box is displayed. Set the **Orientation** to **Same** and the **Offset** value to **0**. Then, choose **OK** to apply the constraint and exit the dialog box. Update the model to place the Motor Shaft at its proper location.

Next, you need to apply the Coincidence constraint between the axes of the Motor Shaft and the Blower hub. You will use the **Quick Constraint** tool to apply this constraint.

4. Choose the **Quick Constraint** tool from the **Constraints** toolbar and move the cursor over the Motor Shaft. The axis of the shaft will be displayed as a center line. Select the axis by clicking over the center line. The axis will now be highlighted in orange. Similarly, select the axis of the Blower hub. The Coincidence constraint will be automatically applied between the two selected axes. Update the model to place the Motor Shaft inside the Blower hub, as shown in Figure 12-76.

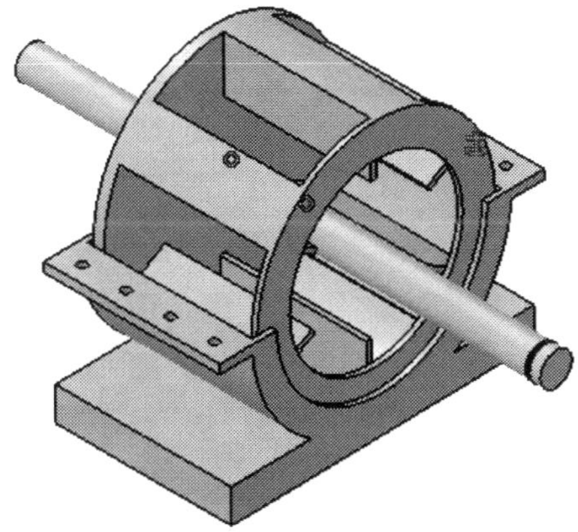

Figure 12-74 Motor Shaft inserted at its default location

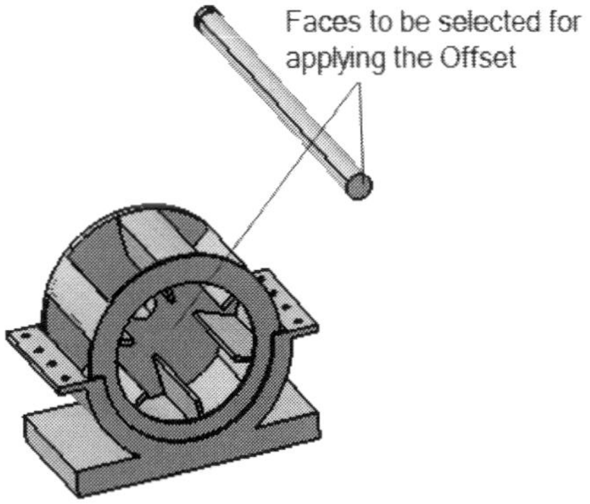

Figure 12-75 Faces to be selected for the Offset constraint

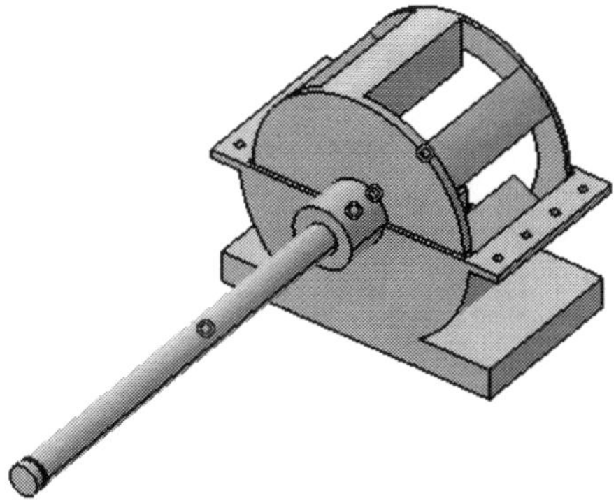

Figure 12-76 Position of the Motor Shaft with respect to the Blower shown from the back side

Assembling the Motor

Next, you need to assemble the Motor with the Motor Shaft.

1. Insert the Motor in the Blower assembly. By default, it is placed in such a way that its body overlaps existing assembly parts. Therefore, you need to use the **Manipulation** tool to move it out into the open space.

2. Now apply the Offset constraint between the faces shown in Figure 12-77. Set the **Orientation** to **Same** and the **Offset** value to **0**.

3. Apply the Coincidence constraint between the axis of the shaft and the axis of the hole on the back side of the Motor, refer to Figure 12-78. After applying both the constraints, update the model.

   You will notice that the base of the Motor and the Lower Housing appear to be parallel, but there is no constraint applied to both the faces. Therefore, you need to apply the **Angle** constraint to these two faces.

4. Choose the **Angle Constraint** tool from the **Constraints** toolbar and select the faces, as shown in Figure 12-77. Now, select the **Parallelism** radio button to make the selected faces parallel and set the **Orientation** to **Same**. Next, choose **OK** from the **Constraint Properties** dialog box.

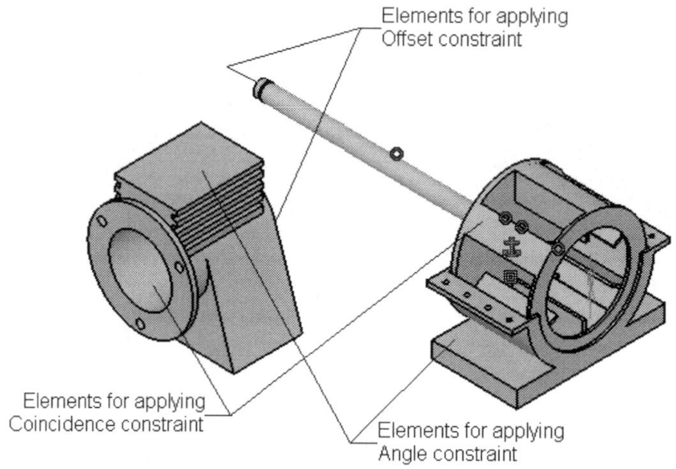

***Figure 12-77** Elements to be selected for applying various constraints*

Assembling the Cover and Turning On the Display of the Upper Housing

The last component to be assembled is the Cover. After assembling all the components, you need to turn on the display of the hidden Upper Housing.

1. Insert the Cover into the Blower assembly. By default, it will be placed inside the blower. Use the **Manipulation** tool to move the Cover away from the assembly. Next, you need to apply constraints to the Motor Cover.

2. Apply the Contact constraint between the front face of the Motor and the bottom face of the Cover, refer to Figure 12-78.

3. Apply the Coincidence constraint between the central hub of the Cover and cylindrical face of the Motor.

4. Apply another Coincidence constraint between one of the screw holes in the Cover and Motor. Various faces to be selected for applying these three constraints are shown in Figure 12-78. After applying all the three constraints, update the model to properly orient the cover in the Blower assembly.

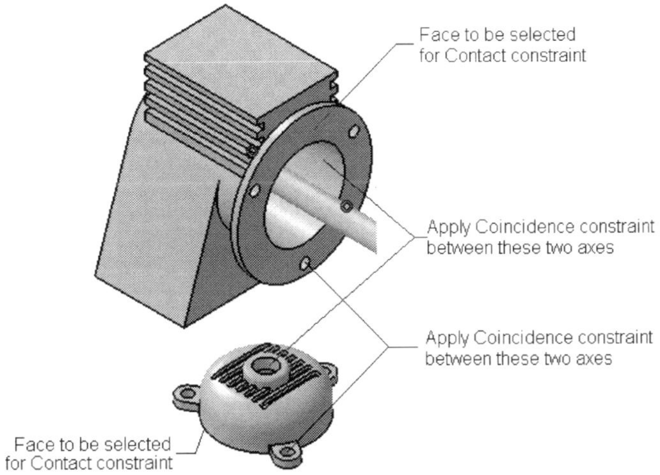

Figure 12-78 *Various faces to be selected for applying the three constraints*

Next, you need to turn on the display of the Upper Housing.

5. Select the Upper Housing from the Specification tree and choose the **Hide/Show** option from the contextual menu; the Upper Housing is displayed in the geometry area.

6. Select the **Constraints** node from the Specification tree and then hide it. The final Blower assembly is shown in Figure 12-79.

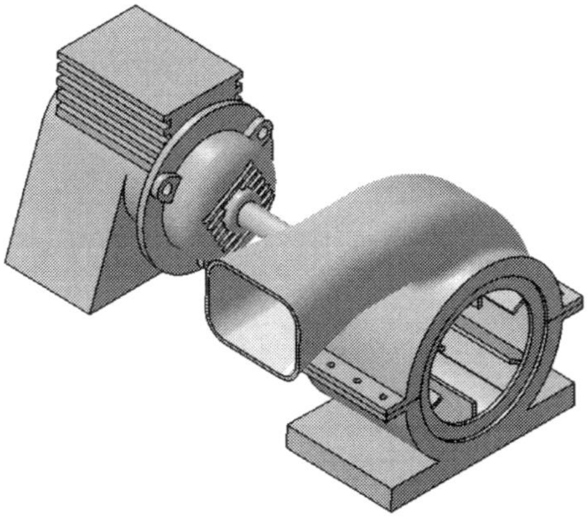

Figure 12-79 *The final Blower assembly*

Creating the Exploded State of the Assembly

After creating the assembly, you can create its exploded state. In the exploded state, all parts of the assembly are distinctly visible.

1. Select **Product1** from the Specification tree and choose the **Explode** tool from the **Move** toolbar; the **Explode** dialog box is displayed.

2. Select the **All levels** option from the **Depth** drop-down list and select **2D** from the **Type** drop-down list. Click in the **Fixed product** display box and then select the Lower Housing as the product to remain fixed while exploding the assembly.

3. Choose the **Apply** button from the **Explode** dialog box to generate an exploded view; the **Information Box** message box is displayed. Choose the **OK** button from this message box to close it. Exploded view of the Blower assembly is shown in Figure 12-80.

Note
*If you have cleared the **Show this message next time** check box in the **Information Box** message box earlier, this message box will not be displayed.*

4. Choose the **OK** button from the **Explode** dialog box and then choose **Yes** from the **Warning** dialog box. The exploded state of the assembly is displayed in the geometry area.

5. Press the CTRL+U keys to switch back to the assembled mode.

Saving the File

1. Choose the **Save** button from the **Standard** toolbar; the **Save As** dialog box is displayed.

2. Save the file at *C:\CATIA\c12\Blower Assembly*.

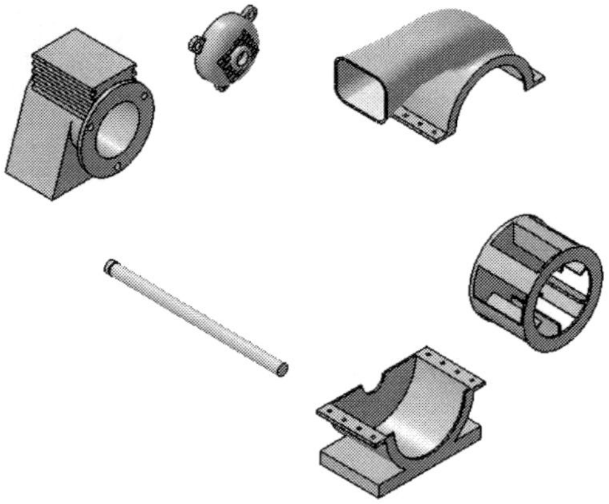

Figure 12-80 *Exploded view of the Blower assembly*

Tutorial 2

In this tutorial, you will create some components of the Press Tool Base assembly shown in Figure 12-81 using the top-down assembly approach. The exploded state of this assembly is shown in Figure 12-82. The dimensions of all components are shown in Figures 12-83 and 12-84.

(Expected time: 45 min)

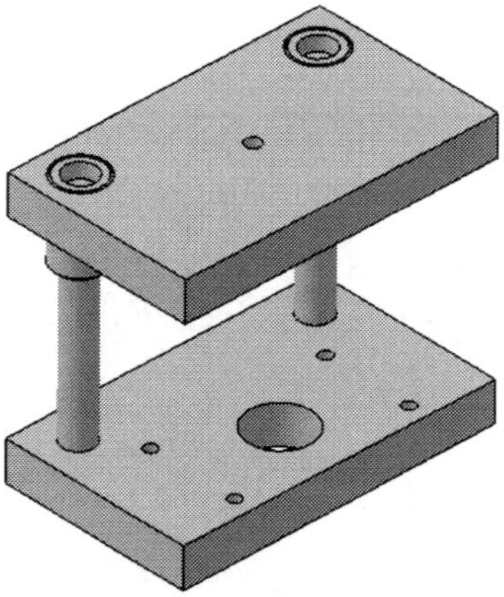

Figure 12-81 *The Press Tool Base assembly*

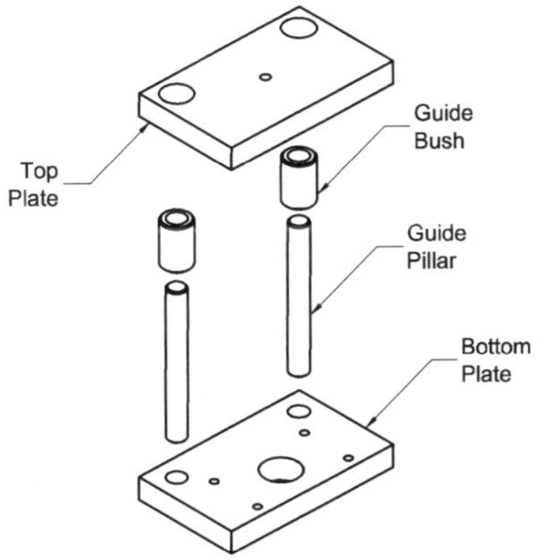

Figure 12-82 The exploded state of the Press Tool Base assembly

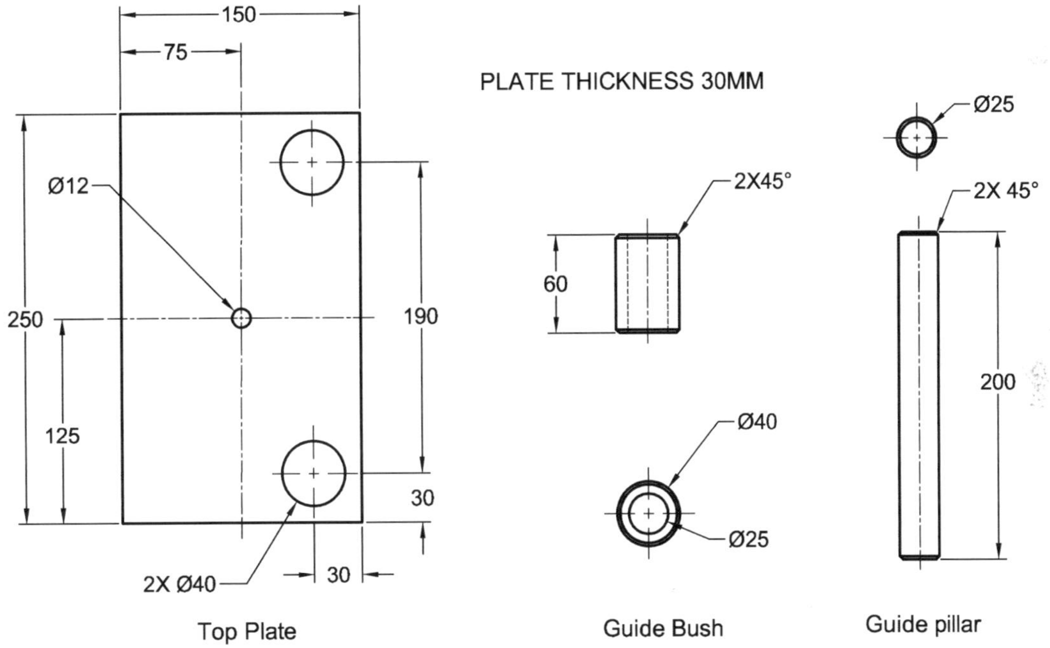

Figure 12-83 Orthographic views and dimensions of the Top Plate, Guide Pillar, and Guide Bush

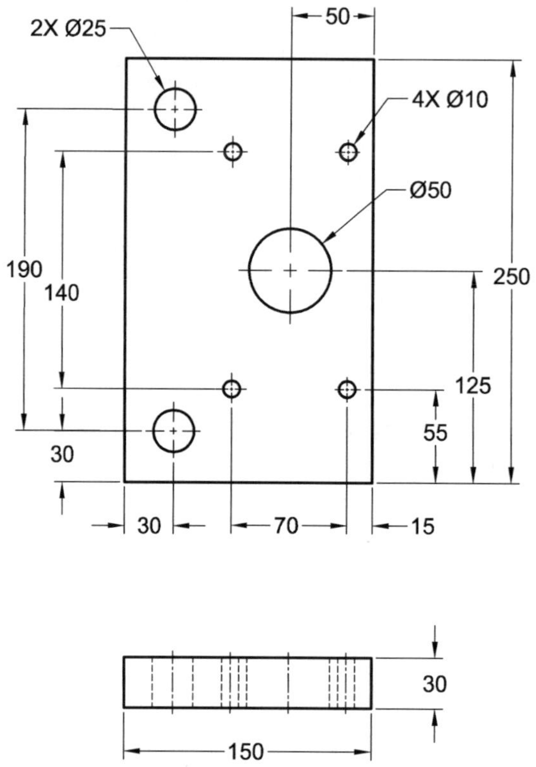

***Figure 12-84** Views and dimensions of the Bottom Plate*

Note

The assembly shown in Figure 12-81 is not a complete assembly of the Press Tool Base and has been created only to explain the procedure followed while creating it using the top-down assembly approach.

The following steps are required to complete this tutorial:

a. Start a new product file.
b. Create a new part inside the assembly. Modify its name and create features of the base component. In this assembly, the Bottom Plate will be the base component.
c. Create the Guide Bush and Guide Pillar as subsequent components inside the product file.
d. Create duplicates of Guide Pillar and Guide Bush that are to be duplicated using the **Reuse Pattern** tool.
e. Create the Top Plate.
f. Finally, save the Product file. The Part files will be saved automatically.

Before you start creating the assembly using the top-down assembly design approach, create the *Press Tool Base* folder *at C:\CATIA\c12\Press Tool*. You need to save the product file of the Press Tool assembly in this folder. All the part files will also be automatically saved in the same folder.

Starting a New Product File

1. Choose the **New** button from the **Standard** toolbar and select the **Product** option from the **List of Types** area of the **New** dialog box. Next, choose the **OK** button to start a new Product file.

2. Invoke the contextual menu by right-clicking on **Product1** in the Specification tree. Select **Properties** from the contextual menu to display the **Properties** dialog box.

3. Choose the **Product** tab if not already chosen, and enter the name **Press Tool Base** in the **Part Number** edit box. Next, choose the **OK** button from the **Properties** dialog box; the name **Press Tool Base** is displayed at the top of the Specification tree.

   When the product file is saved, the name Press Tool Base is automatically assigned to the product file.

Creating a New Part Inside the Assembly

1. Select **Press Tool Base** from the Specification tree and choose the **Part** tool from the **Product Structure Tools** toolbar; a new part named as **Part1** is added under the **Press Tool Base** product file in the specification tree. Also, the default planes of the added part are displayed in the geometry area.

2. Set the part number and the instance name of the new part as Bottom Plate in the **Properties** dialog box. This dialog box is invoked by right-clicking on **Part1** in the Specification tree.

3. Choose **Tools > Options** from the menu bar to invoke the **Options** dialog box. Click on **Infrastructure** on the left of this dialog box to expand this branch. Now, select **Part Infrastructure** from the **Infrastructure** branch. Select the **Keep link with selected object** check box if not selected. Choose the **OK** button from the **Options** dialog box.

Creating Features of the Bottom Plate

After a new part file is created inside the assembly, you need to invoke the **Part Design** workbench to create the features. Since the Bottom Plate is the base component, you will now create its features.

1. Click on the plus sign on the left of **Bottom Plate** in the Specification tree to expand it. Now, double-click on **Bottom Plate** displayed inside the expanded branch to invoke the **Part Design** workbench.

2. Create the Bottom Plate using the part modeling tools. For dimensions, refer to Figure 12-84. The final model of the Bottom Plate is shown in Figure 12-85.

Note

Create the Bottom Plate on the XY-plane. Create holes of diameter 10 and 25 as separate features, and then pattern them separately, refer to Figure 12-85.

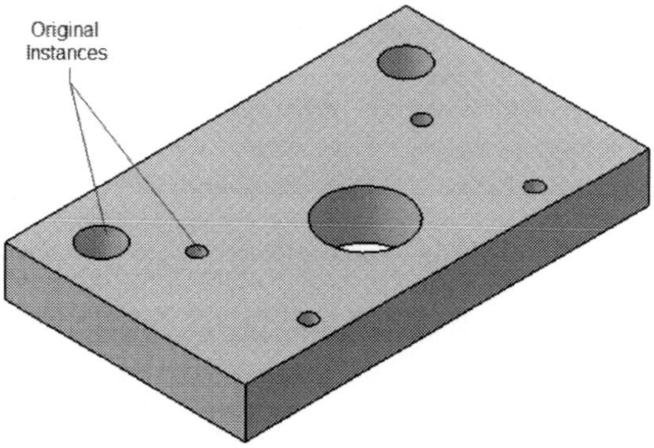

Figure 12-85 *The final model of the Bottom Plate*

3. Double click on **Press Tool Base** in the Specification tree for switching to the **Assembly Design** workbench.

4. Apply the Fix constraint to the Bottom Plate using the **Fix Component** tool.

Creating the Guide Pillar

Next, you need to create the Guide Pillar by referring to the geometry of the Base Plate.

1. Choose the **Part** button from the **Product Structure Tools** toolbar and select **Press Tool Base** from the Specification tree to insert the second part in the assembly. On doing so, the **New Part: Origin Point** dialog box is displayed.

2. Choose the **No** button from the **New Part: Origin Point** dialog box to place the origin of the second part over the origin of the assembly coordinate system.

3. Rename the second component to **Guide Pillar** and expand its branch. Next, double-click on the **Guide Pillar** displayed inside the expanded branch to invoke the **Part Design** workbench.

4. For creating the Guide Pillar, you need to take the reference of the geometry of the Bottom Plate. Select the bottom face of the Bottom Plate as the sketching plane and invoke the **Sketcher** workbench.

5. Select the circular edge of the original instance having a diameter of 25 and choose the **Project 3D Elements** button from the **Operation** toolbar. The geometry is extracted from the selected edge and is projected over the sketching plane.

Note that there is no need to provide any dimension to the circle. The size of the extracted circle is the same as that of the edge of the hole from which it is extracted.

6. Exit the **Sketcher** workbench and extrude the sketch, refer to Figure 12-83, for dimensions.

7. Hide the Bottom Plate and apply chamfer on both the ends of the Guide Pillar. This will complete the feature creation of the Guide Pillar.

8. Turn on the display of the Bottom Plate and switch back to the **Assembly Design** workbench. The Guide Pillar is created in the assembly, as shown in Figure 12-86.

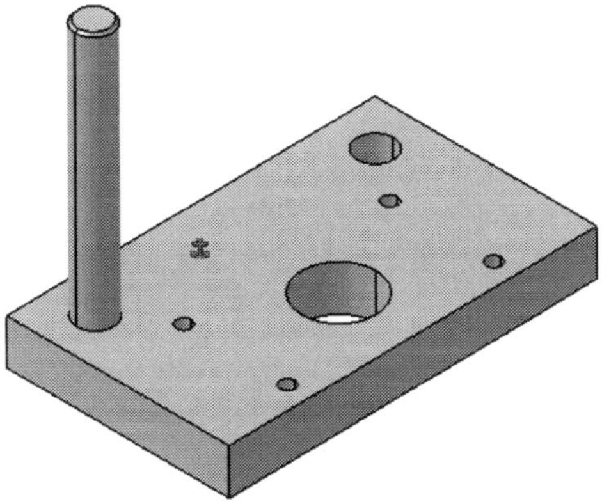

Figure 12-86 *The Guide Pillar created in the assembly*

Creating the Guide Bush

It is evident from Figure 12-81 that the Guide Bush will be placed over the Guide Pillar. Therefore, the geometry of the Guide Pillar will be used to create the Guide Bush.

1. Start another Part file inside the assembly file and rename it as Guide Bush. Invoke the **Part Design** workbench.

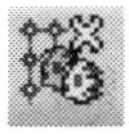

2. Select the top face of the Guide Pillar as the sketching plane and draw two concentric circles. Make the inner circle coincident with the outer edge of the Guide Pillar and apply dimension to the outer circle, as shown in Figure 12-87.

3. Exit the **Sketcher** workbench and extrude the sketch to 60 units; reverse the direction of extrusion if required. Now, apply chamfers at both the outer edges of the Guide Bush. This completes the feature creation of the Guide Bush, refer to Figure 12-88.

4. Return to the **Assembly Design** workbench by double clicking on the **Press Tool Base** in the Specification tree.

Creating the Second Set of Guide Bush and Guide Pillar

In the Press Tool Base assembly, Guide Bush and Guide Pillar are used in pairs. Therefore to create the second set of the Guide Bush and Guide Pillar, you need to use the **Reuse Pattern** tool. The advantage of using this tool is that the second set of components will be placed at desired location without applying any constraints between them.

1. Expand the branch of the **Bottom Plate** in the Specification tree to display the **Rectangular Pattern** used for creating holes.

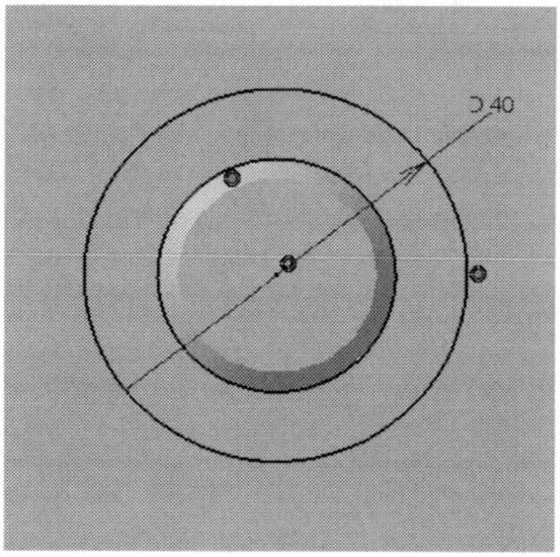

Figure 12-87 *Sketch of the Pad feature for creating the Guide Bush*

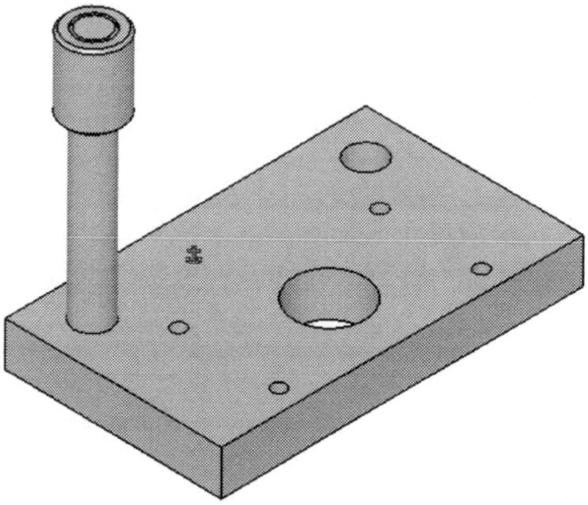

Figure 12-88 *Final Guide Bush*

2. Press and hold the CTRL key. Select **RectPattern.1** (used for creating a pattern of 25 units diameter hole) and **Guide Pillar** from the Specification tree.

3. Choose the **Reuse Pattern** tool from the **Constraints** toolbar; the **Instantiation on a pattern** dialog box and the **Warning** message box is displayed. Choose the **OK** button from the message box. If another message box is displayed, then choose the **OK** button again to exit. Next, choose the **OK** button in the **Instantiation on a pattern** dialog box to create the second Guide Pillar.

4. Using the **Reuse Pattern** tool, select **RectPattern.1** (used for creating a pattern of 25 units diameter hole) and **Guide Bush** from the Specification tree to assemble the second Guide Bush over the newly placed Guide Pillar.

   Once both the components are duplicated, you can close the expanded branch of the Specification tree. Figure 12-89 shows the assembly model after placing the second set of Guide Pillar and Guide Bush.

Creating the Top Plate

The Top Plate is the last component to be created in the Press Tool Base assembly. You will use the reference of the geometries of the Bottom Plate and Guide Bush to draw the sketch for the Pad feature of the Top Plate.

1. Start another part inside the assembly file and rename it as Top Plate. Invoke the **Part Design** workbench.

2. Select the top face of the Guide Pillar as the sketching plane and invoke the **Sketcher** workbench.

3. Extract four side edges of the Bottom Plate and the outer circular edge of the Guide Bush using the **Project 3D Elements** tool, as shown in Figure 12-90.

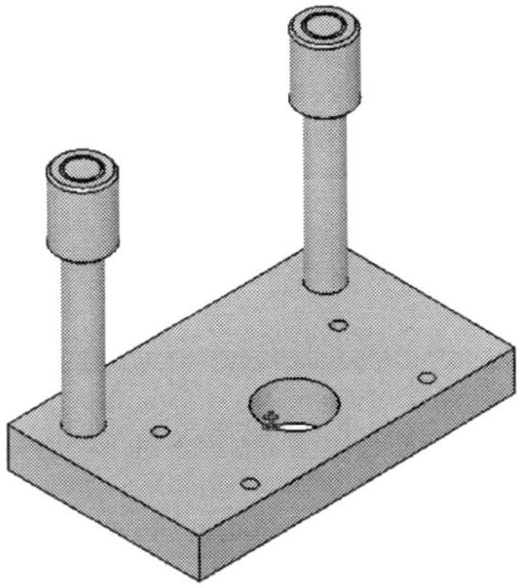

Figure 12-89 *The assembly after placing the second set of the Guide Pillar and Guide Bush*

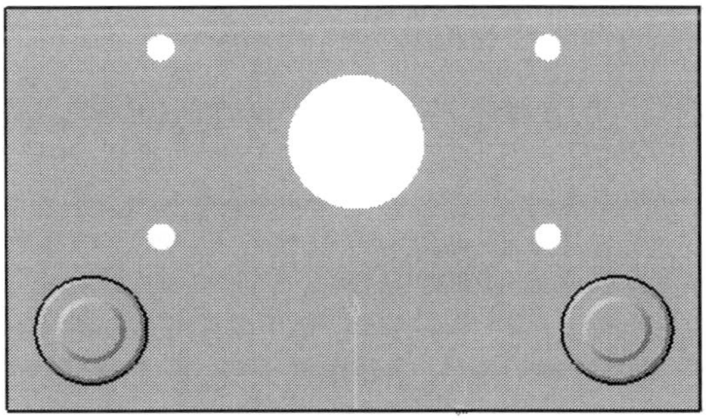

Figure 12-90 *Sketch of the Pad feature for creating the Top Plate*

4. Exit the **Sketcher** workbench and extrude the sketch by 30 units by flipping its direction of extrusion. Now, create a hole on the top face of the Top Plate. For position of the hole, refer to Figure 12-83.

   This completes the creation of Top Plate.

5. Switch back to the **Assembly Design** workbench. The final assembly is shown in Figure 12-91.

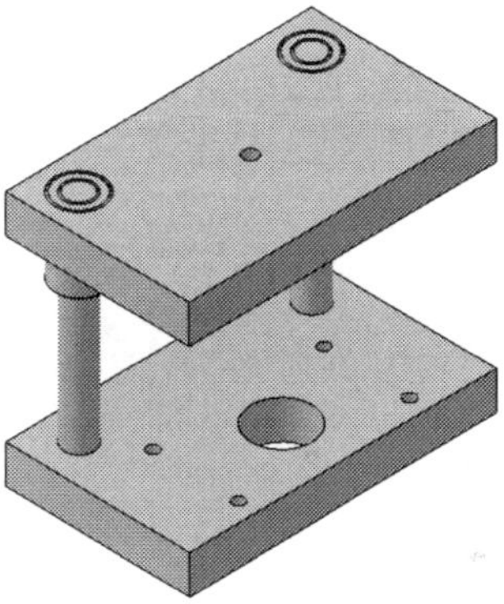

Figure 12-91 *The final Press Tool Base assembly*

Saving the Assembly File

1. Make sure that you are in the **Assembly Design** workbench and choose the **Save** button from the **Standard** toolbar; the **Save As** dialog box is displayed.

2. Browse to the *Press Tool Base* folder that you have created in the beginning of this tutorial.

3. Choose the **Save** button from the **Save As** dialog box; the **Save As** confirmation box is displayed, as shown in Figure 12-92.

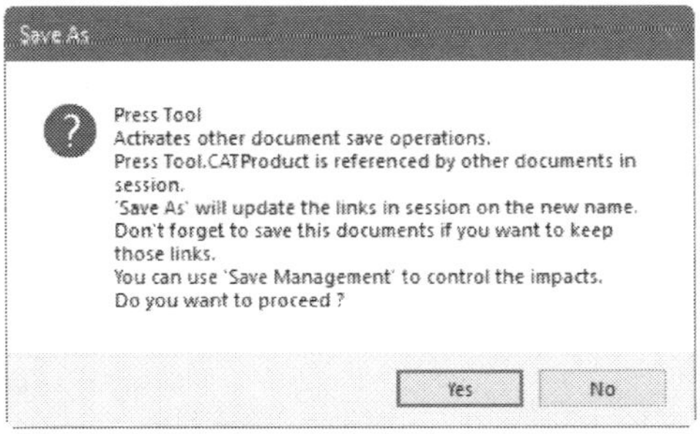

Figure 12-92 *The* ***Save As*** *confirmation box*

4. Choose the **Yes** button from this confirmation box to save the assembly file along with all the Part files.

5. Close the assembly file by choosing **File > Close** from the menu bar.

Self-Evaluation Test

Answer the following questions and then compare them to those given at the end of this chapter:

1. The _________constraint is used to place two planar faces at a specified angle with respect to each other.

2. You can place multiple instances of a component over a predefined pattern by using the _________tool.

3. The _________option is selected from the **Type** drop-down list in the **Explode** dialog box to explode all the components of an assembly along the 2D plane currently parallel to the screen.

4. The edges of a component can be snapped on to the edge of another component by using the _________ tool.

5. If two parts with same name are inserted into an assembly file, then the _________ dialog box is displayed.

6. In the bottom-up assembly design approach, all the parts are created in separate part files and then inserted into the product file. (T/F)

7. While creating a top-down assembly in CATIA V5, all the individual parts created in the assembly need to be saved separately after saving the product file. (T/F)

8. The Angle constraint can be used to make two surfaces parallel to each other. (T/F)

9. The **Manipulation** tool is used to move and rotate a part of an assembly. (T/F)

10. You cannot select a component to fix it at its original position while exploding an assembly model. (T/F)

Review Questions

Answer the following questions:

1. Which of the following tools is used to calculate the interference between two mating components?

   (a) **Measure** (b) **Clash**
   (c) **Smart Move** (d) **Snap**

2. Which of the following buttons is used to replace a constraint with another constraint?

   (a) **Reuse Pattern** (b) **Replace Component**
   (c) **Change Constraint** (d) **Replace Constraint**

3. Which button is used to apply the most appropriate constraint to the current selection set?

   (a) **Quick Constraint** (b) **Change Constraint**
   (c) **Contact Constraint** (d) None of these

4. Which of the following dialog boxes is displayed if the **Reuse Pattern** button is chosen?

   (a) **Instantiation on a pattern** (b) **Reuse pattern**
   (c) **Constraint properties** (d) **Constraint definition**

5. Which of the following tools is used to fix the position of a part in 3D space?

   (a) **Contact Constraint** (b) **Fix Component**
   (c) **Coincidence Constraint** (d) **Angle Constraint**

6. The _________ constraint is used to make two cylindrical surfaces concentric.

7. The _________ tool is used to turn off the display of the selected components.

8. The _________ tool can be used to move a component and also apply a constraint to it.

9. The cross-section of an assembly model can be viewed using the _________ tool.

10. An existing part can be inserted into a Product file using the ________ button.

EXERCISE

Exercise 1

Create the assembly of the Radial Engine shown in Figure 12-93. The assembly in the exploded state is shown in Figure 12-94. Note that this exploded view is provided only for your understanding and has not been generated using CATIA V5. The dimensions of various parts of this assembly model are given in Figure 12-95 through Figure 12-100.

(Expected time: 3 hr 30 min)

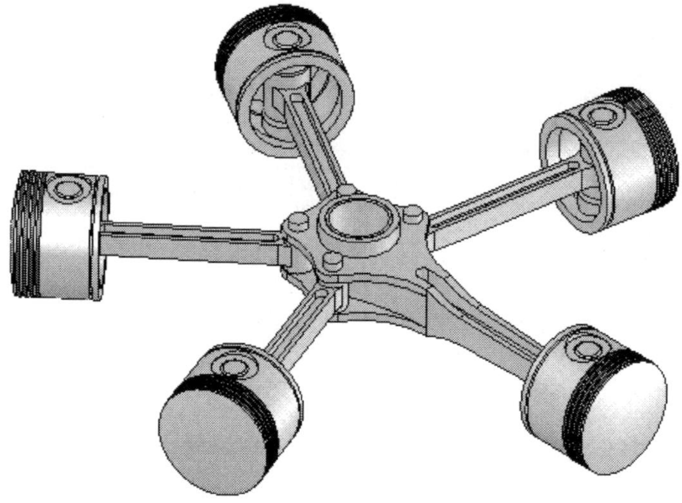

Figure 12-93 *The Radial Engine assembly*

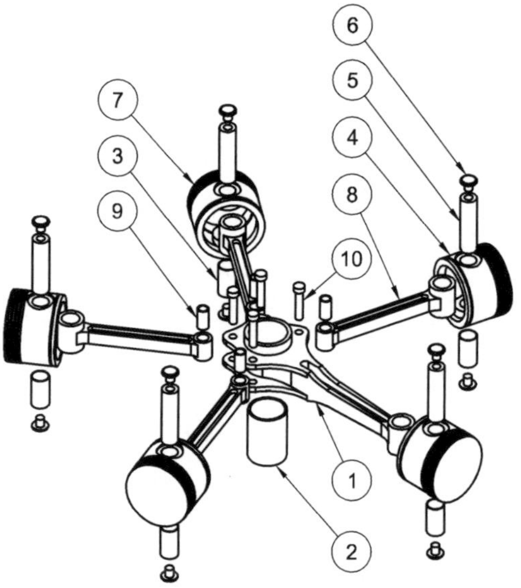

ITEM NO.	PART NO.	QTY.
1	Master Rod	1
2	Master Rod Bearing	1
3	Rod Bush Upper	5
4	Piston	5
5	Piston Pin	5
6	Piston Pin Plug	10
7	Piston Ring	20
8	Articulated Rod	4
9	Rod Bush Lower	4
10	Link Pin	4

Figure 12-94 *Exploded view of the assembly*

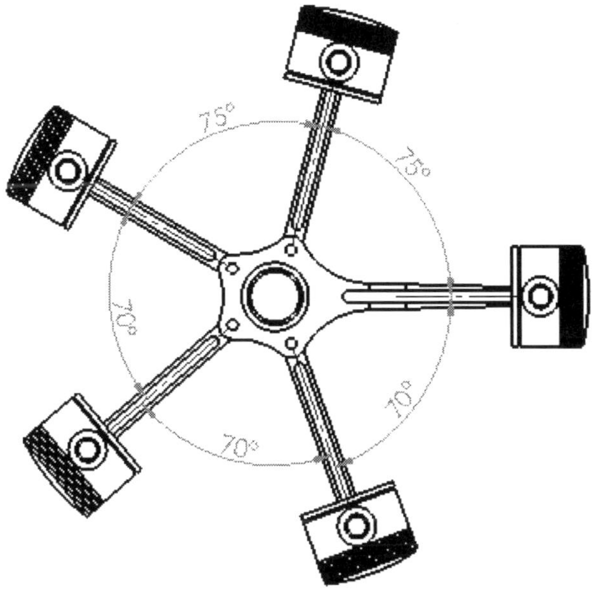

Figure 12-95 Positioning of the articulated Rods

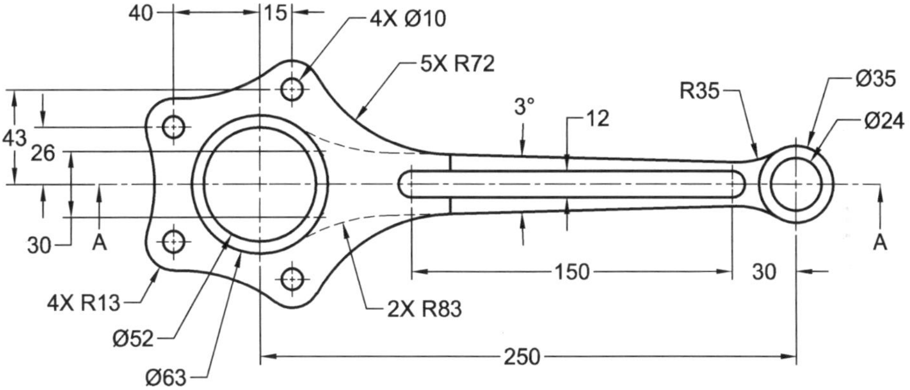

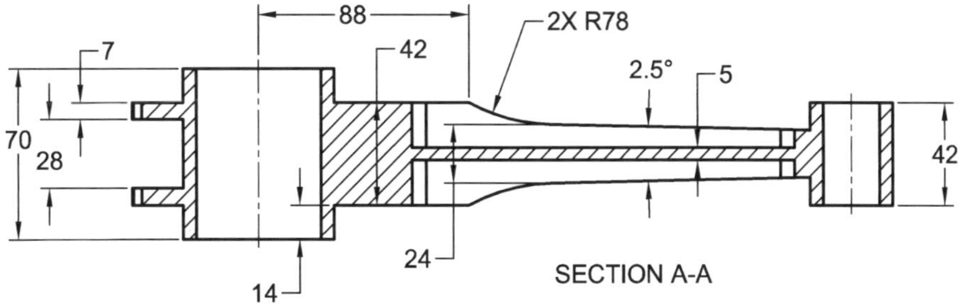

Figure 12-96 Orthographic views and dimensions of the Master Rod

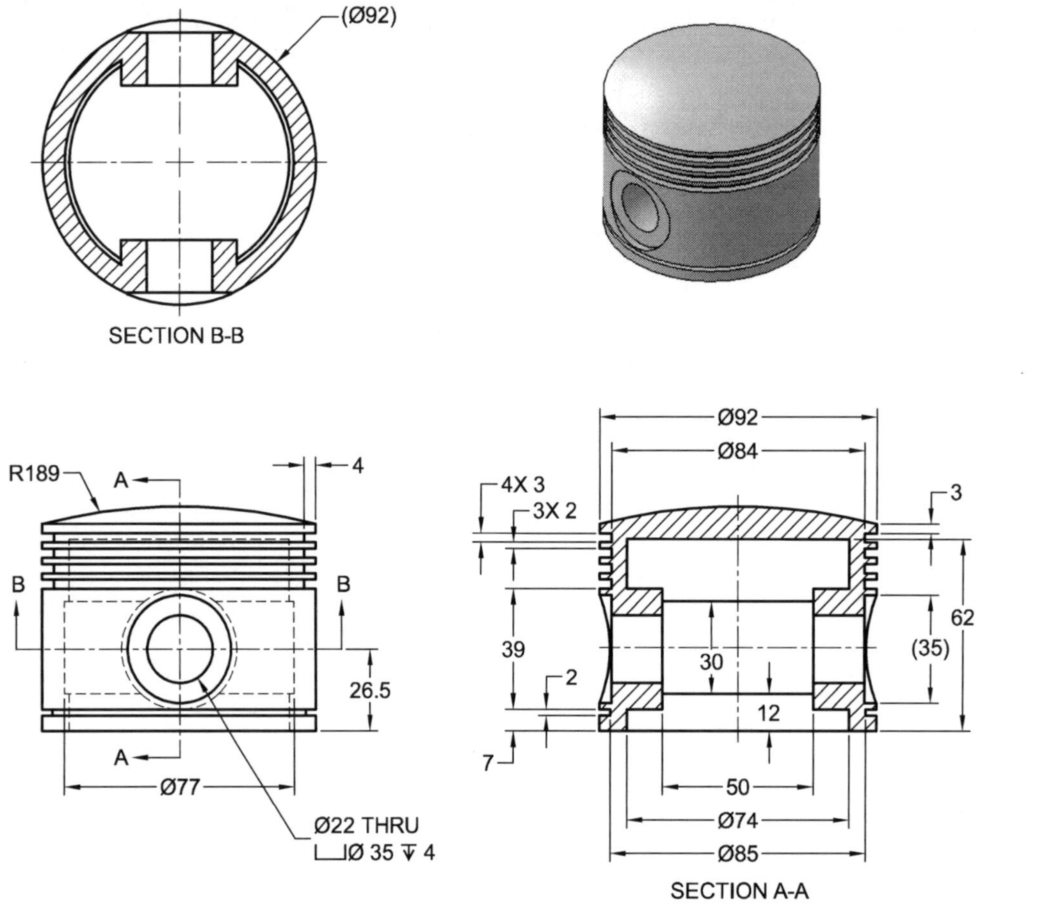

Figure 12-97 *Orthographic views and dimensions of the Piston*

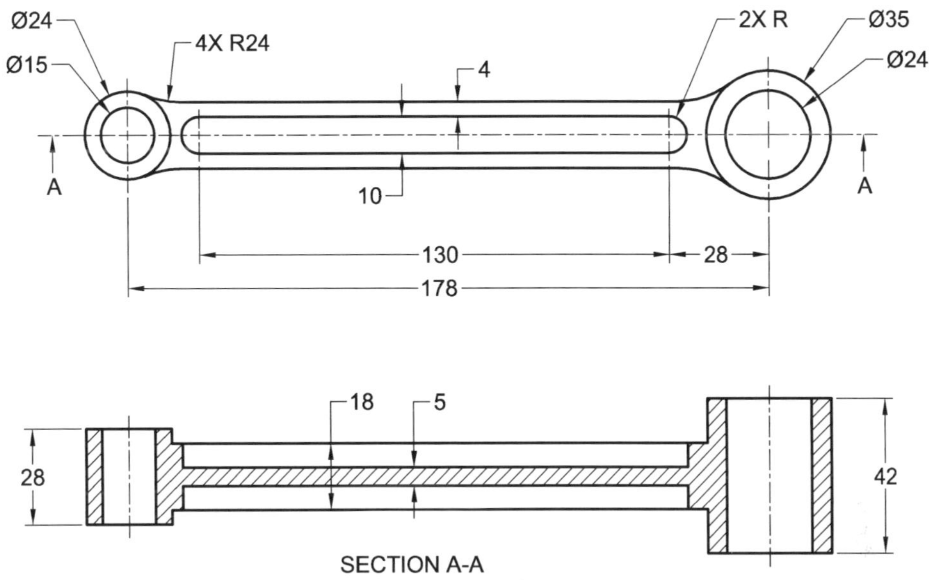

Figure 12-98 *Orthographic views and dimensions of the Articulated Rod*

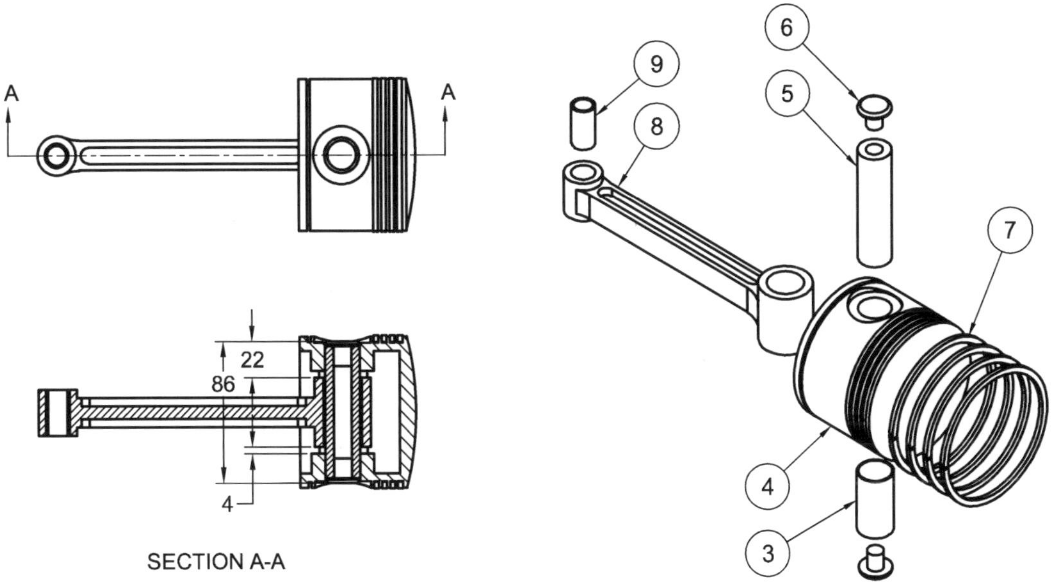

Figure 12-99 *Views and dimensions of Piston Articulated Rod Subassembly*

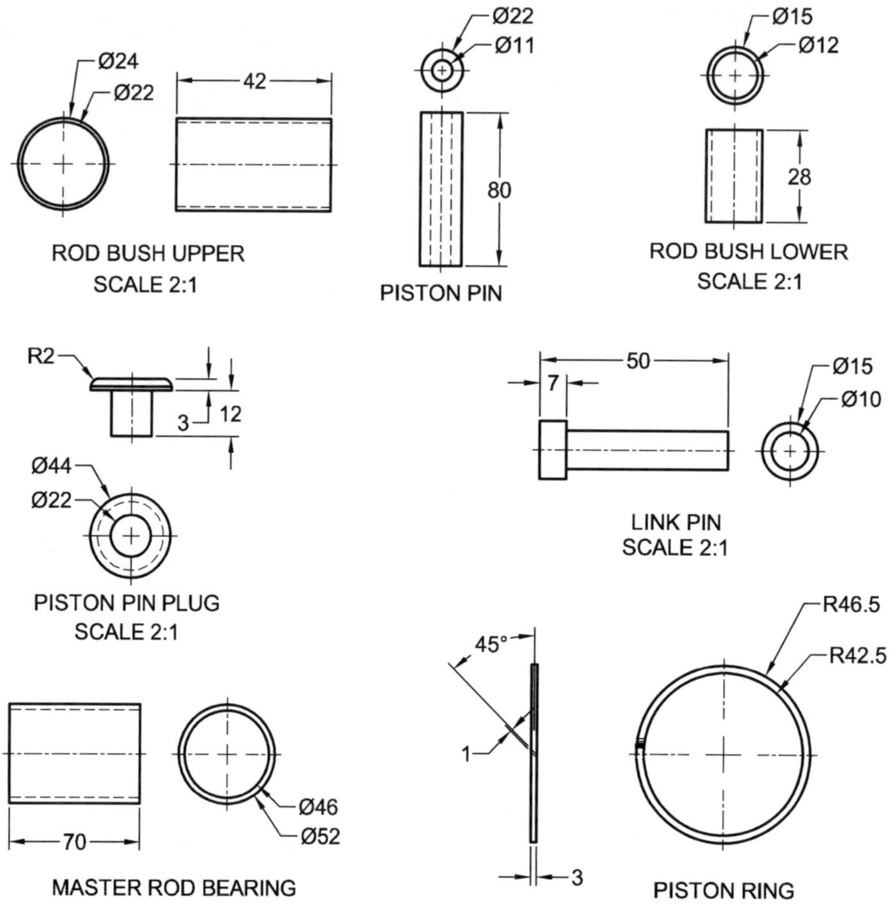

Figure 12-100 *Views and dimensions of other components*

Answers to Self-Evaluation Test

1. Angle, **2. Reuse Pattern**, **3. 2D**, **4. Snap**, **5. Part number conflicts**, **6.** T, **7.** F, **8.** T, **9.** T, **10.** F

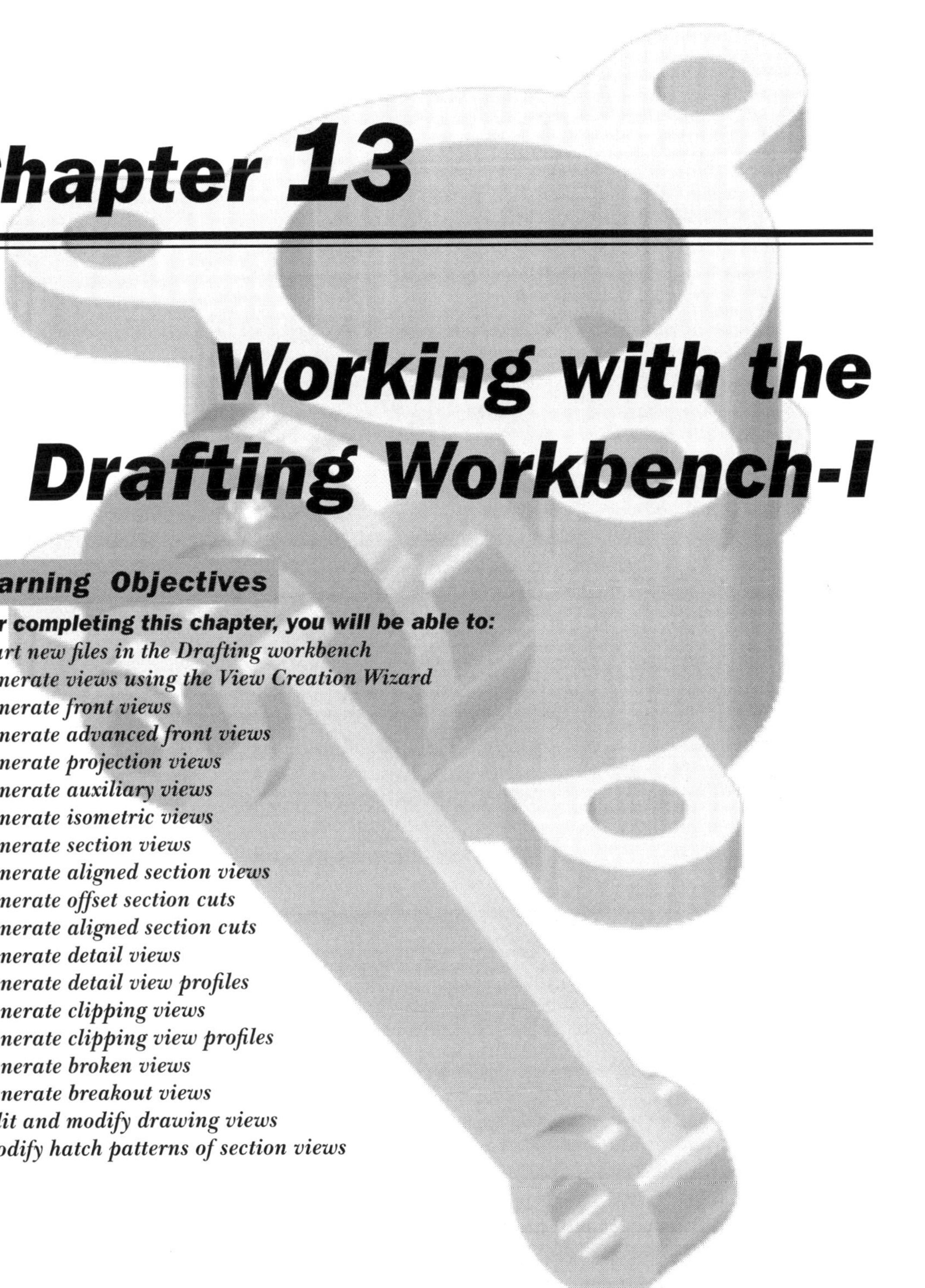

Chapter 13

Working with the Drafting Workbench-I

Learning Objectives

After completing this chapter, you will be able to:

- *Start new files in the Drafting workbench*
- *Generate views using the View Creation Wizard*
- *Generate front views*
- *Generate advanced front views*
- *Generate projection views*
- *Generate auxiliary views*
- *Generate isometric views*
- *Generate section views*
- *Generate aligned section views*
- *Generate offset section cuts*
- *Generate aligned section cuts*
- *Generate detail views*
- *Generate detail view profiles*
- *Generate clipping views*
- *Generate clipping view profiles*
- *Generate broken views*
- *Generate breakout views*
- *Edit and modify drawing views*
- *Modify hatch patterns of section views*

THE DRAFTING WORKBENCH

2D drawing is the most important part of the manufacturing systems because they are required by a machinist for manufacturing on the shop floor or in the tool room. CATIA V5 provides you with the **Drafting** workbench which is a specialized environment for generating 2D drawing views. This workbench provides all tools required to generate drawing views, modify, and apply dimensions and add annotations. In other words, you can get the final shop floor drawing using this workbench of CATIA V5. There are two types of drafting techniques in CATIA V5: Generative drafting and Interactive drafting. Generative drafting is a technique of generating the drawing views using a solid model or an assembly model. Interactive drafting is a technique, in which the sketcher tools are used to draw the 2D drawing views.

In this chapter, you will learn how to generate the drawing views of parts and assemblies. One of the major advantages of working in CATIA V5 is its bi-directional associative nature. This property ensures that the modifications made in the model in any one of the workbenches (**Part**, **Assembly Design**, or **Drafting**) are reflected in other two workbenches.

Starting a New File in the Drafting Workbench

To generate drawing views, you first need to start a new file in the **Drafting** workbench. There are two methods of starting a new file in the **Drafting** workbench, which are discussed next.

Starting a New File in the Drafting Workbench Using the New Tool

Menubar:	File > New
Toolbar:	Standard > New

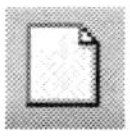

To start a new file in the **Drafting** workbench using the **New** tool, choose the **New** tool from the **Standard** toolbar; the **New** dialog box will be displayed. Select **Drawing** from the **List of Types** drop-down list and choose the **OK** button; the **New Drawing** dialog box will be displayed, as shown in Figure 13-1. The options in this dialog box are discussed next.

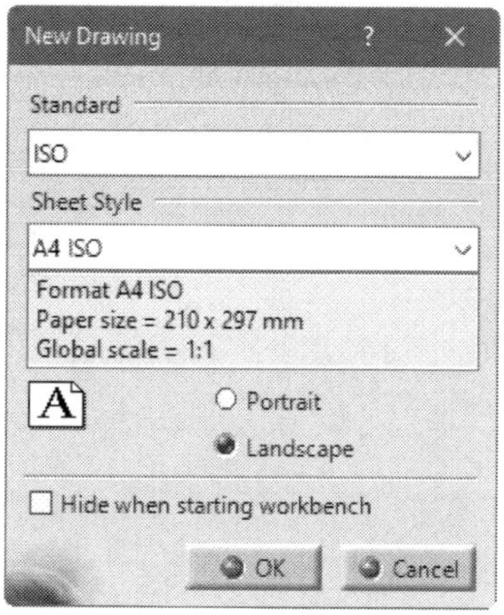

Figure 13-1 The New Drawing dialog box

Standard

The options in the **Standard** drop-down list are used to define the dimensioning standard. By default, the **ISO** standard is selected as the dimensioning standard. You can also select other standards such as **JIS**, **ASME_3D**, **ASME**, **ANSI**, and so on from the drop-down list.

Sheet Style

The **Sheet Style** drop-down list is used to define the sheet format, which changes with the change in the drawing standards. By default, only the ISO formats are available and the **A0 ISO** format is selected in the **Sheet Style** drop-down list. This is because the **ISO** standard is selected in the **Standard** drop-down list.

You can also modify the default orientation of the sheet using the two radio buttons provided in the **Sheet Style** area. By default, the **Landscape** radio button is selected. As a result, the resulting sheet will be in the landscape orientation. If you select the **Portrait** radio button, the resulting sheet will be in the portrait orientation.

After setting all parameters in the **New Drawing** dialog box, choose the **OK** button; a new file will start in the **Drawing** workbench. The initial screen of CATIA V5 when a new file is opened in the **Drafting** workbench is shown in Figure 13-2.

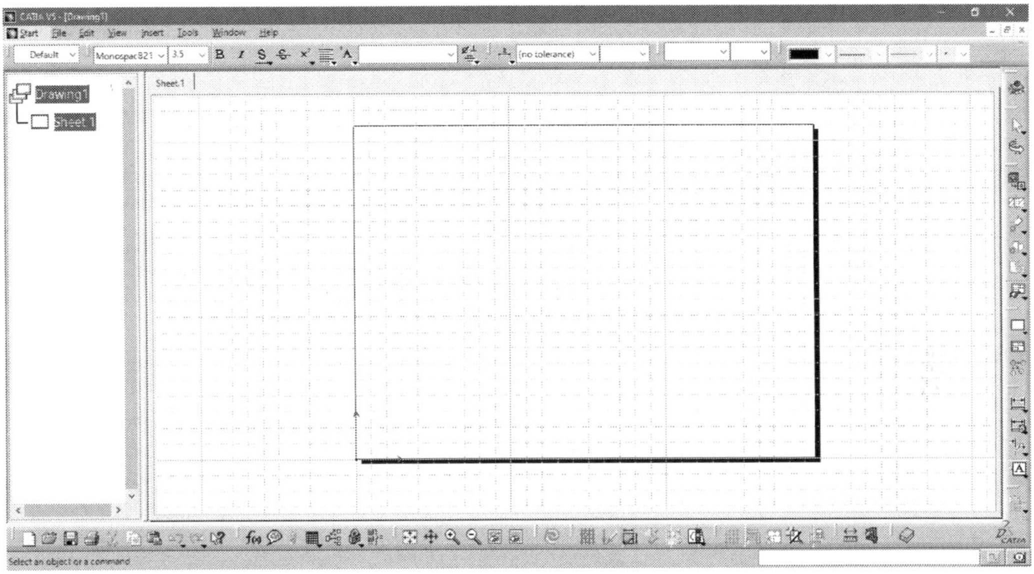

***Figure 13-2** The initial screen after starting a new file in the **Drafting** workbench*

Starting a New File in the Drafting Workbench Using the Start Menu

Menubar: Start > Mechanical Design > Drafting

You can also start a new file in the **Drafting** workbench using the **Start** menu. Note that this option is recommended only if the part or assembly file is already open in a separate window and you want to generate drawing views for that part or assembly. To start a new file in the **Drafting** workbench using this option, choose **Start > Mechanical Design > Drafting** from the menubar; the **New Drawing Creation** dialog box will be displayed, as shown in Figure 13-3. The options in this dialog box are discussed next.

Note
*The **New Drawing Creation** dialog box will not be displayed if any part or assembly file is not opened in the current window.*

Empty sheet

The **Empty sheet** option is used to start an empty sheet without any view. After starting a new file using this option, you need to manually generate the views.

All Views

The **All Views** option is used to generate the front, right, left, top, bottom, rear, and isometric views automatically after starting a new file in the **Drafting** workbench.

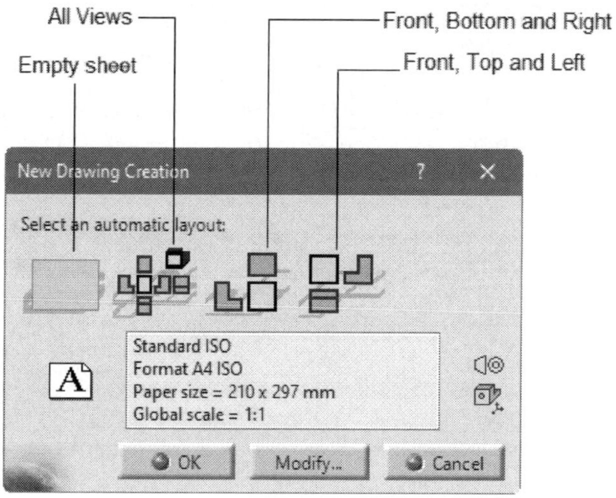

*Figure 13-3 The **New Drawing Creation** dialog box*

Front, Bottom and Right

The **Front, Bottom and Right** option is used to generate the front, bottom, and right views automatically after starting a new file in the **Drafting** workbench.

Front, Top and Left

The **Front, Top and Left** option is used to generate the front, top, and left views automatically after starting a new file in the **Drafting** workbench.

The standard of dimensioning, the format of the drawing sheet, and the default scale of the views are displayed in the **New Drawing Creation** dialog box. The views are generated using the first angle projection from this dialog box. This book will follow the third angle projection standard. Setting the option to generate the views in the third angle projection is discussed later in this chapter.

To modify the dimensioning standard, format of sheet, and the default view scale, choose the **Modify** button from the **New Drawing Creation** dialog box; the **New Drawing** dialog box will be displayed. Modify the setting and choose the **OK** button to apply the settings; the **New Drawing Creation** dialog box will be activated. Choose the **OK** button from the dialog box to start a new drawing sheet.

TYPES OF VIEWS

In CATIA V5, you can generate various types of drawing views. Generally, first you generate the front view and then use this view to generate the remaining views. The views that you can generate using the tools in the **Drafting** workbench are discussed next.

Front View

The front view is the base view that is generated on the drawing sheet in the **Drafting** workbench. You can set the orientation of the front view using the knob. Other views are generated from the front view.

Projected View

The projected view is generated by projecting lines parallel to an existing view, which is called the parent view. The resulting view will be an orthographic view.

Section View

A section view is generated by chopping a part of an existing view using a plane and then viewing the model in the direction normal to the section plane. In CATIA V5, the section plane is defined using one or more sketched line segments.

Aligned Section View

An aligned section view is the section of the features that are created at a certain angle to the main section planes. Align sections straighten these features by revolving them about an axis normal to the view plane. Remember that the axis about which the feature is straightened should lie on the cutting planes.

Auxiliary View

An auxiliary view is generated by projecting the lines normal to a specified edge in an existing view.

Detail View

A detail view is used to display the details of a portion of an existing view. The portion that you have selected will be magnified and placed as a separate view. You can control the magnification of the detail view.

Clipping View

A clipping view is used to crop an existing view enclosed in a closed sketch associated to it. The portion of the view that lies inside the associated sketch is retained and the remaining portion is removed.

Broken View

A broken view is used to display a component by removing a portion of it from between, keeping the ends of the drawing view intact. This type of view is used to display the components whose length to width ratio is very high. This means that either the length is very large as compared to the width, or the width is very large as compared to the length. The broken view will break the view along the horizontal or vertical direction, such that the drawing view fits the area you require.

Breakout View

A breakout section view is used to remove a part of the existing view and display the area of the model or the assembly that lies behind the removed portion. This type of view is generated using a closed sketch that is associated with the view.

GENERATING DRAWING VIEWS

In CATIA V5, there are two methods of generating the views, after you start an empty file in the **Drafting** workbench. The first method is by using wizards and other is by generating each view one after another. Both these methods of generating the drawing views are discussed in this chapter. Before you start generating the drawing views, you need to set the standard of the views generation to the third angle projection. To set this option, select **Sheet.1** from the Specification tree provided on the left of the drawing sheet. Invoke the contextual menu and choose the **Properties** option; the **Properties** dialog box will be displayed and the **First angle standard** radio button will be selected in the **Projection Method** area. Select the **Third angle standard** radio button and choose the **OK** button from the **Properties** dialog box.

Generating Views Automatically

You can generate the drawing views automatically using the tools from the **Wizard** sub-toolbar in the **Drafting** workbench. To generate the views automatically, open the part or the assembly whose drawing views you want to generate in a separate window. Now, invoke the **Drafting** workbench window and choose the down arrow provided on the right of the **View Creation Wizard** tool in the **Views** toolbar; the **Wizard** sub-toolbar will be displayed, as shown in Figure 13-4. The tools in this toolbar are used to automatically generate the views. The method to generate views using these tools is discussed next.

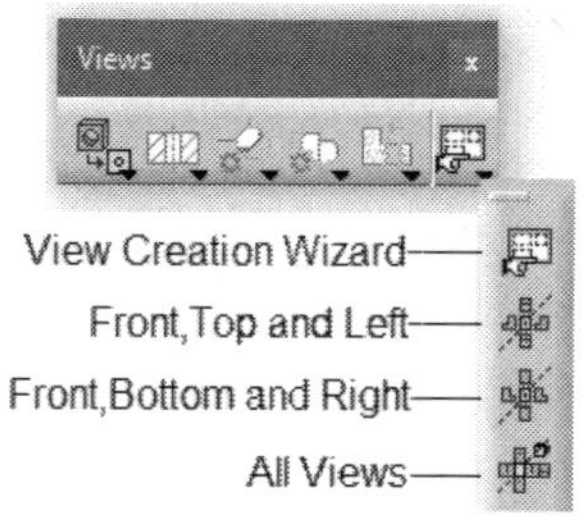

Figure 13-4 *Tools in the **Wizard** sub-toolbar*

Generating Views Using the View Creation Wizard

Menubar:	Insert > Views > Wizard > Wizard
Toolbar:	Views > Wizard sub-toolbar > View Creation Wizard

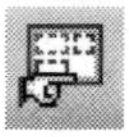

Before generating views using the **View Creation Wizard** tool, you need to make sure that the part or the assembly for which the views need to be generated is open in the CATIA V5 window. When you invoke this tool, the **View Wizard** dialog box will be displayed, as shown in Figure 13-5. This dialog box provides you with various options to generate drawing views, depending on predefined configurations.

If you choose the **Configuration 1 using the 3rd angle projection method** button from the **View Wizard** dialog box, the top, front, and right views will be displayed in the preview area of this dialog box.

If you choose the **Configuration 2 using the 3rd angle projection method** button from the **View Wizard** dialog box, the top, front, and left views will be displayed in the preview area of this dialog box.

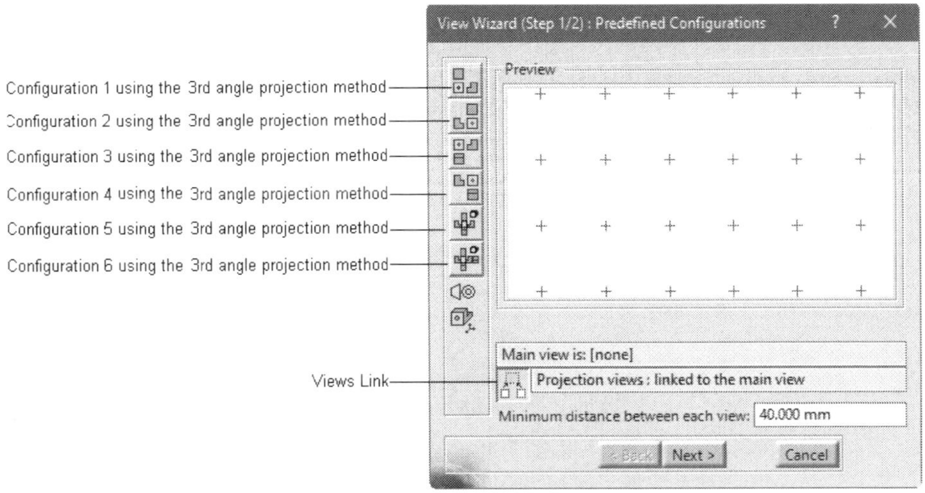

Figure 13-5 *The* ***View Wizard*** *dialog box*

If you choose the **Configuration 3 using the 3rd angle projection method** button, the front, right, and bottom views will be displayed in the preview area of this dialog box.

If you choose the **Configuration 4 using the 3rd angle projection method** button, the front, left, and bottom views will be displayed in the preview area of this dialog box.

If you choose the **Configuration 5 using the 3rd angle projection method** button, the front, top, left, right, bottom, and isometric views will be displayed in the preview area of this dialog box.

If you choose the **Configuration 6 using the 3rd angle projection method** button, the front, top, left, right, bottom, rear, and isometric views will be displayed in the preview area of this dialog box.

On selecting a configuration, you will notice that the **Front** view is displayed inside a green box. This indicates that the **Front** view is the main view. By default, the **Views Link** button is chosen in the **View Wizard** dialog box. As a result, a relation between the front view and the other views will be maintained. If you move the front view after placing it, the other views will also move with respect to it. If you move any other view, it will move with respect to the front view. If you deselect the **Views Link** button, the relation between the front view and the other views will not exist.

After selecting the configuration type, you need to choose the **Next** button from the **View Wizard** dialog box to invoke the second step of view generation and to arrange the views in the **Preview** area. To move a view, select it, press and hold the left mouse button, and drag the selected view; the view will snap to a point in the **Preview** area. Next, release the left mouse button to place the view at the desired location.

The **Clear Preview** button in the **View Wizard** dialog box is used to clear all views from the **Preview** area so that you can define them again. To define new views, use the buttons in the left of the **View Wizard** dialog box and place the new views at the desired location. You can also define the minimum distance between views using the **Minimum distance between each view** edit box.

After setting all parameters, choose the **Finish** button from the **View Wizard** dialog box; you will be prompted to select a reference plane on a 3D geometry. Choose **Window > "Name of the part or product file"** from the menubar; the part or the product document will be invoked. Next, you need to select a plane or planar face from the Specification tree or from the geometry area that will be used to specify the front view. When you move the cursor on a plane or a planar face, a preview of the front view is displayed in the **Oriented Preview** area at the lower right corner of the geometry area, as shown in Figure 13-6.

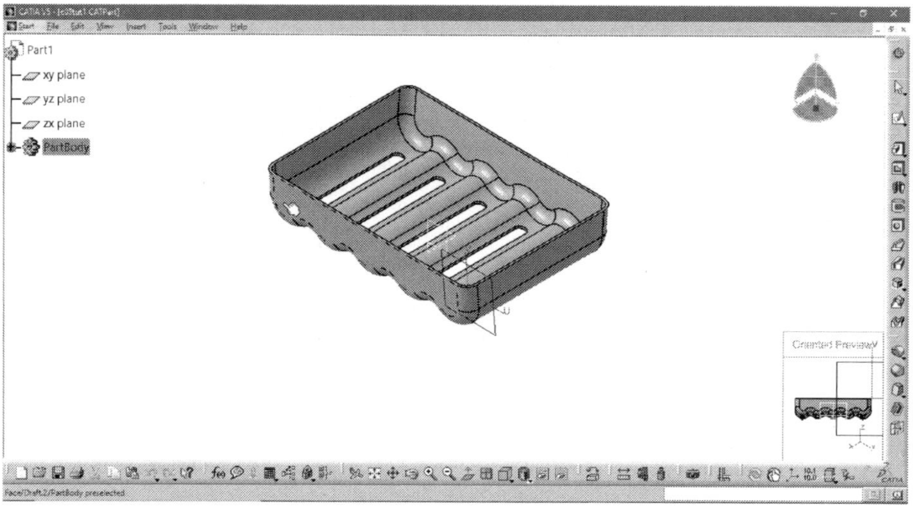

***Figure 13-6** Preview of the front view at the lower-right corner of the geometry area*

Select an appropriate plane or planar face from the geometry area or from the Specification tree; the document you are working with in the **Drafting** workbench will be invoked, and the preview of the **Front** view with other empty views will be displayed in the geometry area. You can also display the preview of the other views that are empty in the geometry area while displaying the preview of the front view. To do so, move the cursor over an empty view; the preview of that view will be displayed in the geometry area. Also, a blue colored knob will be displayed at the upper right corner of the geometry area. Using the controls on this knob, you can orient the base view. The functions of these controls (buttons) on the knob are shown in Figure 13-7.

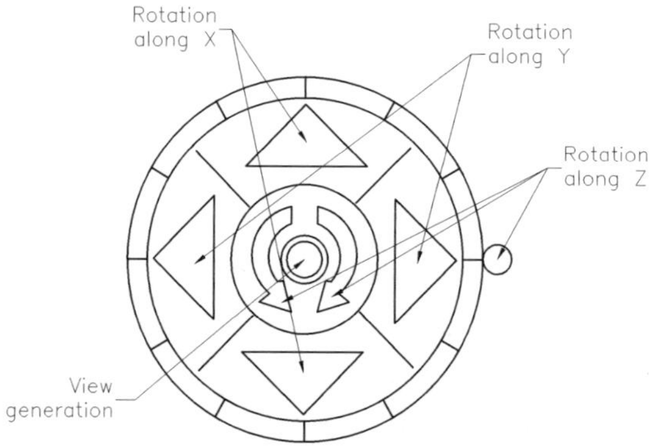

***Figure 13-7** Controls provided on the blue knob*

After setting the orientation, press the view generation button on the center of the knob or click anywhere in the geometry area. After placing the views, you may need to move the views to place them at the desired location. To move a view, press and hold the left mouse button on the boundary of the view to be moved and then drag the cursor to move it. Next, release the left mouse button to place it at the desired location. Figure 13-8 shows the views after moving them to a new location.

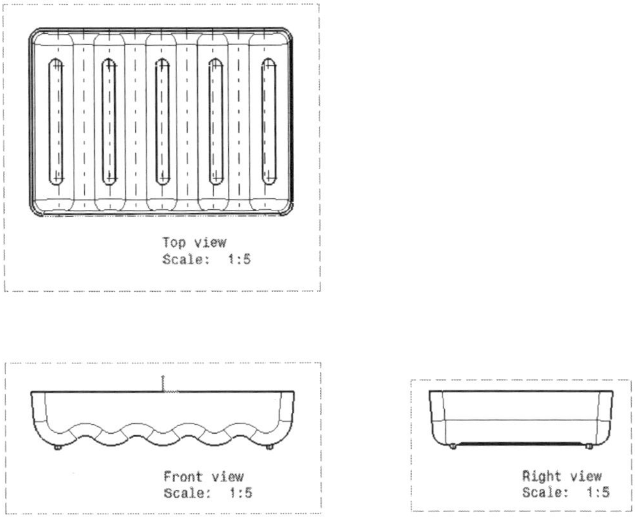

Figure 13-8 *Views generated using the* ***View Wizard*** *dialog box*

> **Tip**
> *To set the global scale factor of the sheet after it is invoked, select* ***Sheet.1*** *from the Specification tree. Invoke the contextual menu, and then choose the* ***Properties*** *option from it; the* ***Properties*** *dialog box will be displayed. Set the value of the scale in the* ***Scale*** *edit box and choose the* ***OK*** *button from the* ***Properties*** *dialog box.*

Generating the Front, Top, and Left Views

Menubar:	Insert > Views > Wizard > First Config
Toolbar:	Views > Wizard sub-toolbar > Front, Top and Left

The **Front, Top and Left** tool is used to generate the front, top, and left views. To generate the views using this tool, choose the **Front, Top and Left** tool from the **Wizard** sub-toolbar in the **Views** toolbar; you will be prompted to select a reference plane on a 3D geometry. Choose **Window > "Name of the File"** from the menubar. Now, select the plane or the planar face that you need to place parallel to the screen in the front view. Set the orientation of the front view using the knob in the graphics area of the drawing sheet, if required. Click anywhere in the geometry area to generate the views.

Generating the Front, Bottom, and Right Views

Menubar:	Insert > Views > Wizard > Second Config
Toolbar:	Views > Wizard sub-toolbar > Front, Bottom and Right

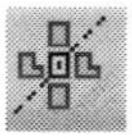

The **Front, Bottom and Right** tool is used to generate the front, bottom, and right views at a time. To generate the views using this tool, choose the **Front, Bottom and Right** tool from the **Wizard** sub-toolbar in the **Views** toolbar; you will be prompted to select a reference plane on the 3D geometry. Choose **Window > "Name of the File"** from the menubar. Now, select the plane or the planar face that you need to place parallel to the screen in the front view. Set the orientation of the front view using the knob in the graphics area of the drawing sheet, if required. Click anywhere in the geometry area to generate the views.

Generating All Views

Menubar:	Insert > Views > Wizard > Third Config
Toolbar:	Views > Wizard sub-toolbar > All Views

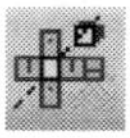

The **All Views** tool is used to generate the front, top, bottom, right, left, rear, and isometric views by using a single tool. To generate all these views using this tool, choose the **All Views** tool from the **Wizard** sub-toolbar in the **Views** toolbar; you will be prompted to select a reference plane on the 3D geometry. Choose **Window > "Name of the File"** from the menubar. Now, select the plane or the planar face that you need to place parallel to the screen in the front view. Set the orientation of the front view using the knob in the graphics area of the drawing sheet, if required. Click anywhere in the geometry area to generate the views.

Generating Individual Drawing Views

In the previous section, you learned how to generate various drawing views automatically. In this section, you will learn how to generate the drawing views individually. For example, first you will generate the front view. Next, you will generate the other projected views with respect to the front view. You can also generate the section views, detail views, broken views, and so on.

The tools to generate the drawing views are grouped in the **Projections** sub-toolbar. To view these tools, click on the down arrow on the right of the **Front View** tool in the **Views** toolbar; the **Projections** sub-toolbar will be displayed with the tools, as shown in Figure 13-9.

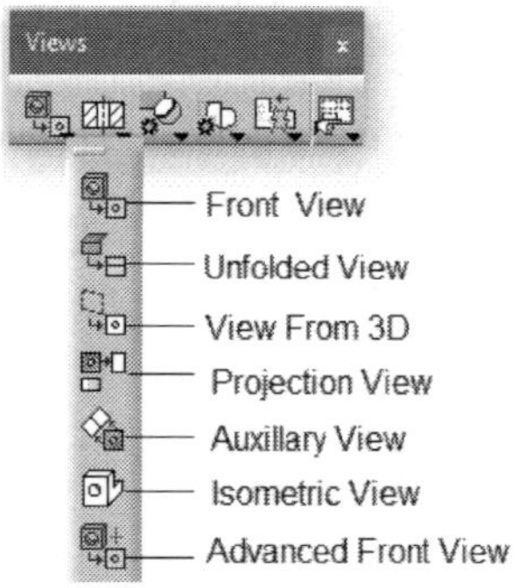

Figure 13-9 *Tools in the* ***Projections*** *sub-toolbar*

The procedure to generate the individual drawing views is discussed next.

Generating the Front View

Menubar: Insert > Views > Projections > Front View
Toolbar: Views > Projections sub-toolbar > Front View

The front view is the main view for generating any other views in the **Drafting** workbench of CATIA V5. To generate the front view, first you need to make sure that the model from which the view needs to be generated is opened in another window. It is recommended that before generating any view, tile the document window horizontally or vertically. Next, choose the **Front View** tool from the **Projections** sub-toolbar in the **Views** toolbar; you will be prompted to select a reference plane on a 3D geometry.

From the model, select a reference plane or a planar face that you need to place parallel to the screen in the front view; the preview of the front view will be displayed on the drawing sheet. Maximize the drawing sheet window and set the orientation of the front view using the knob displayed at the upper right corner of the drawing sheet. Next, click anywhere in the geometry area to complete the generation of the front view. You will notice that the front view is displayed in an orange dotted rectangle and the name **Front View** is highlighted in blue background in the Specification tree. This implies that the front view is the active view on the drawing sheet. Figure 13-10 shows the front view generated using the **Front View** tool.

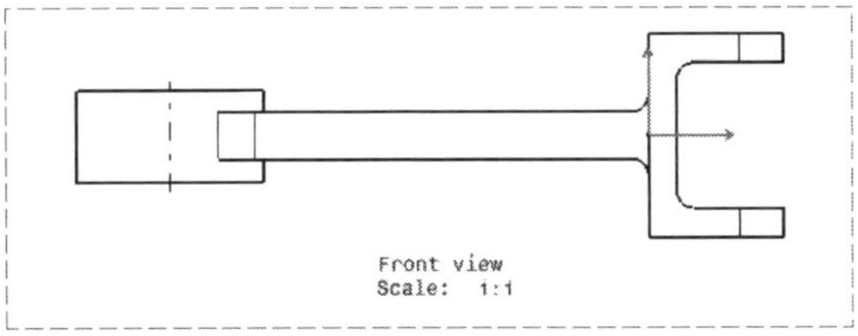

Figure 13-10 *The front view generated using the* ***Front View*** *tool*

Tip
When you generate the view by selecting the reference plane or face from the tiled window, a preview of the front face is not displayed at the bottom right corner of the part window. To preview the front view in the tiled window, you first need to click on the title bar of the part window to activate it and then move the cursor on the reference plane or face to preview the front view on the bottom right corner of the part window.

Generating the Advanced Front View

Menubar: Insert > Views > Projections > Advanced Front View
Toolbar: Views > Projections sub-toolbar > Advanced Front View

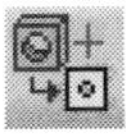

The **Advanced Front View** tool is used to generate the front view. The only difference between this tool and the **Front View** tool is that in this tool, you can specify the name and the scale of the front view while generating it. To generate a front view using this tool, choose the **Advanced Front View** tool from the **Projections** sub-toolbar in the **Views** toolbar; the **View Parameters** dialog box will be displayed, as shown in Figure 13-11. Specify the name and scale of the front view to be generated in this dialog box.

After specifying the name and scale of the front view, choose the **OK** button from the **View Parameters** dialog box. The remaining procedure is similar to the one explained in the **Front View** tool. Figure 13-12 shows the front view generated using the **Advanced Front View** tool.

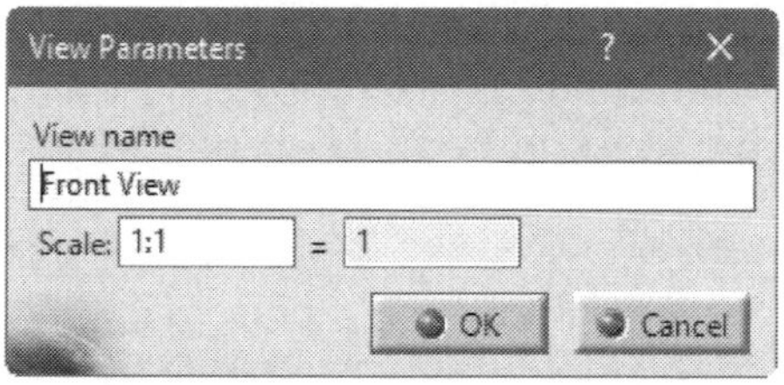

Figure 13-11 *The* ***View Parameters*** *dialog box*

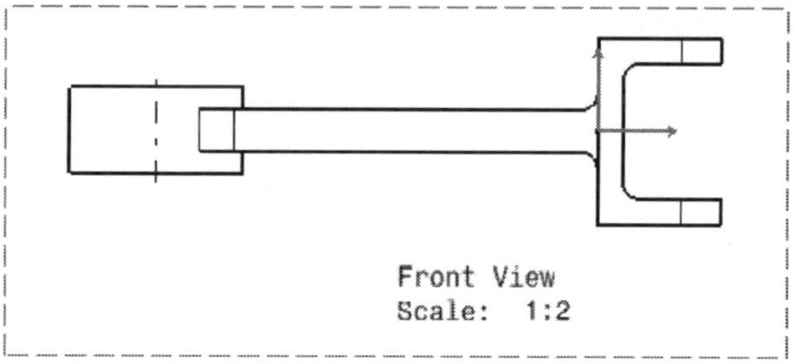

Figure 13-12 *The front view generated using the* ***Advanced Front View*** *tool*

Generating Projection Views

Menubar:	Insert > Views > Projections > Projection
Toolbar:	Views > Projections sub-toolbar > Projection View

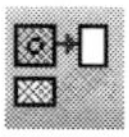

After generating the front view, you need to generate the projected views. As mentioned earlier, the projected views are generated by projecting lines horizontally or vertically from an existing view. To generate a projected view, choose the **Projection View** tool from the **Projections** sub-toolbar in the **Views** toolbar. The projected views will be generated using the front view as the parent view because it is set active by default. Move the cursor in the direction normal to the one of the four edges of the front view boundary; the preview of the projection view will be displayed. Click at the appropriate location to place the projected view on the drawing sheet. Figure 13-13 shows the projected view generated from the parent front view.

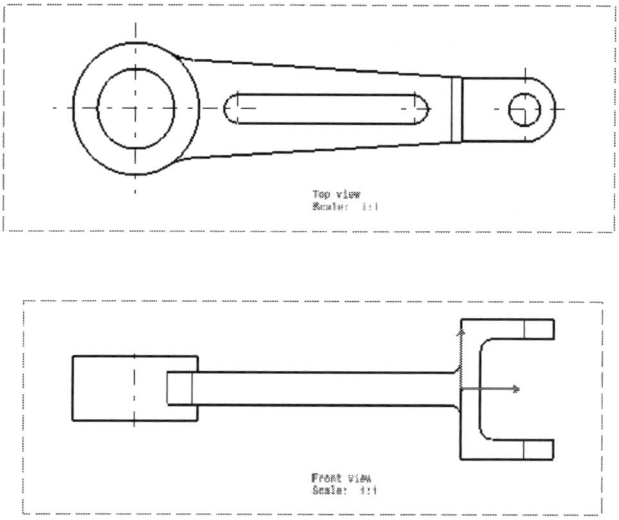

Figure 13-13 *Projected view (Top view) generated from the front view using the* ***Projection View*** *tool*

Tip

1. *After the projected view is generated, it is displayed inside a blue dotted rectangle indicating that the view is not active.*

2. *If you want to generate a projected view from a view other than the front view, first you need to activate the view. To activate the view, select it from the Specification tree or from the geometry area and choose the* ***Activate View*** *option from the contextual menu. The activated view will now be displayed in a red dotted rectangle.*

3. *You can also activate a view by double-clicking on the view in the Specification tree or on the boundary of the view in the geometry area.*

Generating Auxiliary Views

Menubar:	Insert > Views > Projections > Auxiliary
Toolbar:	Views > Projections sub-toolbar > Auxiliary View

This tool is used to generate an auxiliary view by projecting lines normal to the specified edge of an existing view. To generate an auxiliary view, choose the **Auxiliary View** tool from the **Projections** sub-toolbar in the **Views** toolbar; you will be prompted to select the start point or a linear edge to define the orientation. You can define the reference of the auxiliary view by selecting two points to define a line or by selecting an inclined edge. Select the inclined edge normal to which you need to place the view. Move the cursor on the drawing sheet and click to specify the placement of the inclined line. Now, move the cursor and define a point on the drawing sheet to place the auxiliary view. Figure 13-14 shows the reference to be selected for generating the auxiliary view. Figure 13-15 shows the resulting auxiliary view placed above the parent view (front view). Figure 13-16 shows the auxiliary view placed below the parent view (front view).

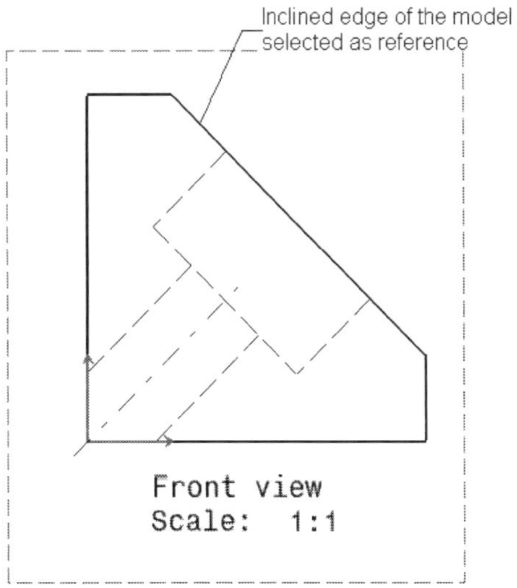

Figure 13-14 *Reference to be selected for generating the auxiliary view*

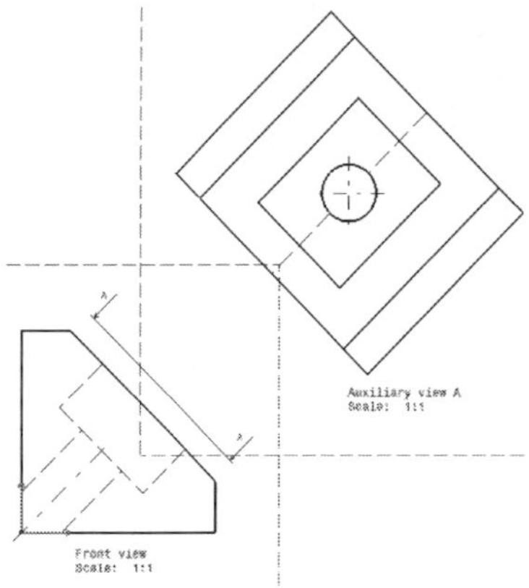

Figure 13-15 Auxiliary view placed above the parent view (Front View)

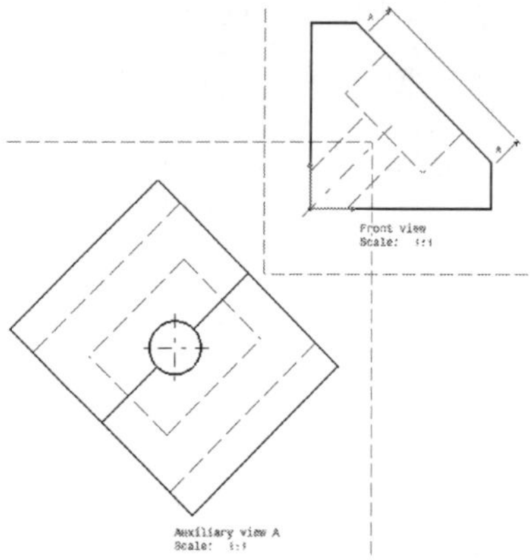

Figure 13-16 Auxiliary view placed below the parent view (Front View)

Generating Isometric Views

Menubar: Insert > Views > Projections > Isometric
Toolbar: Views > Projections sub-toolbar > Isometric View

The **Isometric View** tool is used to generate an isometric view. To do so, choose the **Isometric View** tool from the **Views** toolbar; you will be prompted to select a reference plane on a 3D geometry. Select the reference planar face or a plane from the model; the preview of the isometric view will be displayed on the drawing sheet. Click anywhere on the screen to generate the isometric view. Its orientation will depend on the orientation of the view of the model in the part or assembly file. Figure 13-17 shows an isometric view generated using the **Isometric View** tool.

In CATIA V5, you can change the orientation of the part by using the 3D Orienter, as shown in Figure 13-18. The arrows in the 3D Orienter are used to orient the object by rotating it along the direction of the selected arrows. A green knob is available on the periphery of the 3D Orienter. You can also rotate the object using the knob also. When you right-click on the knob; a shortcut menu is displayed, as shown in Figure 13-19. You can use the **Free hand rotation** option from the shortcut menu to rotate the object freely. You can specify the settings for the rotational increment of the object in the **Increment Setting** dialog box that is displayed on choosing the **Set increment** option from the shortcut menu.

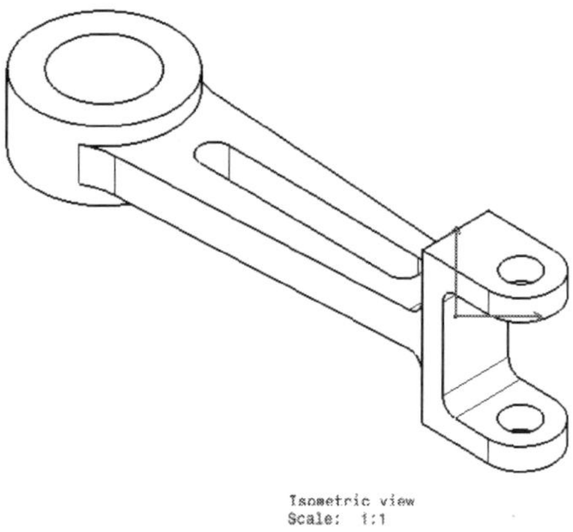

*Figure 13-17 Isometric view generated using the **Isometric View** tool*

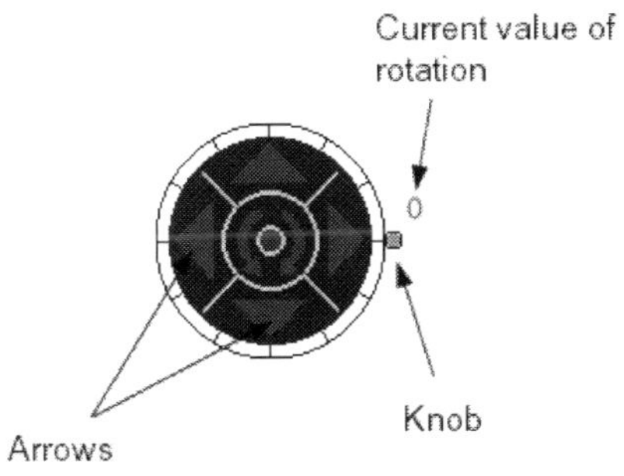

Figure 13-18 The 3D Orienter

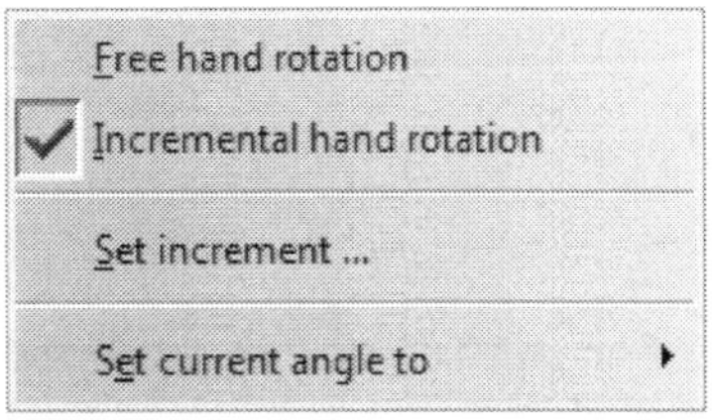

Figure 13-19 The shortcut menu

Tip

1. To display the hidden lines of a view, select the view and choose the ***Properties*** *option from the contextual menu; the* ***Properties*** *dialog box is displayed. Select the* ***Hidden Lines*** *check box from the* ***Dress-up*** *area and choose the* ***OK*** *button from the* ***Properties*** *dialog box.*

2. When you generate a projected or an aligned view, its position is aligned to the parent view or the selected reference. To place the view in some other locations such that it is not aligned to the parent view, select the view to be moved and invoke the contextual menu. Choose ***View Positioning > Position Independently of Reference View*** *from it. Now, you can move the projected or the auxiliary view independently.*

Generating Offset Section Views

Menubar:	Insert > Views > Sections > Offset Section View
Toolbar:	Views > Sections sub-toolbar > Offset Section View

As mentioned earlier, section views are generated by chopping a portion of the existing view using a cutting plane defined by sketched lines and then viewing the parent view in a direction normal to the cutting plane.

The **Offset Section View** tool is used to generate a full section view, as shown in Figure 13-20, an offset section view, as shown in Figure 13-21, and a half section view, as shown in Figure 13-22.

A full section view is defined using a single line segment, but an offset section view or a half section view is defined using multiple line segments. To generate a section view, first activate the view that needs to be defined as the parent view. Choose the **Offset Section View** tool from the **Sections** sub-toolbar that is invoked by choosing the down arrow provided on the right of the **Offset Section View** tool in the **Views** toolbar, as shown in Figure 13-23.

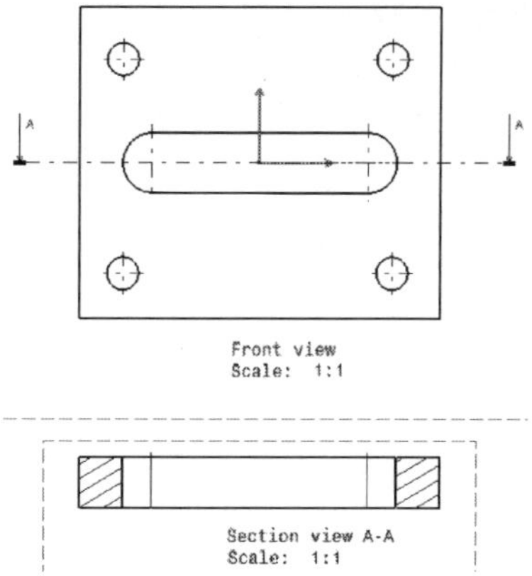

Figure 13-20 Full section view

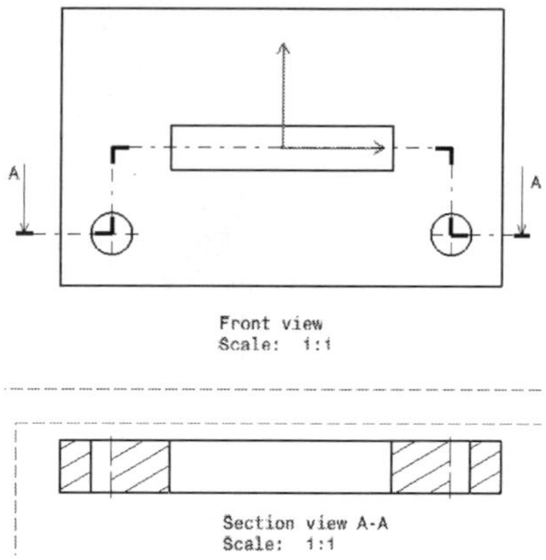

Figure 13-21 Offset section view

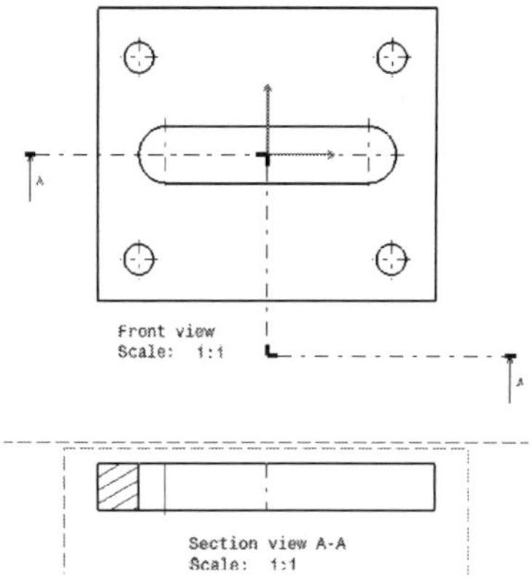

Figure 13-22 Half section view

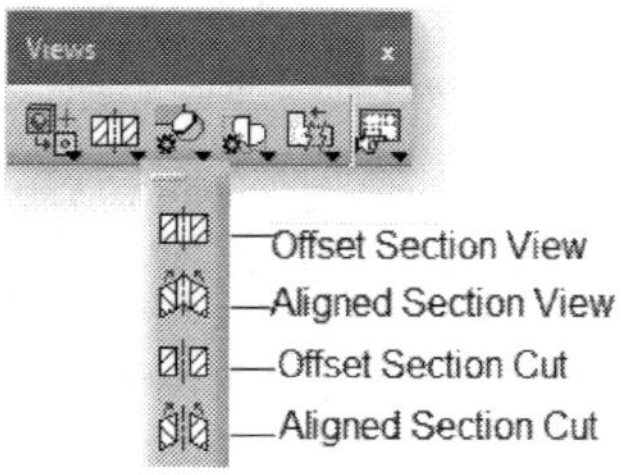

*Figure 13-23 The **Sections** sub-toolbar*

When you choose the **Offset Section View** tool, you will be prompted to select the start point, a circular edge, or an axis. Draw the line to define the section plane. At the end of the section line, double-click to exit the line creation. Move the cursor to define the position of the section view and click in the drawing area to define its placement.

When you generate a section view and move the cursor to place it, you will observe that the view is aligned to the direction of the arrows on the section line. If you need to remove this alignment, move the view to the desired location and select a point on the drawing sheet to place it.

Note

The default hatch pattern in the section view depends on the material assigned to the model. Also, the default spacing of the hatch may not be as per the requirement. Therefore, you may need to decrease the spacing of the hatch pattern. You will learn more about editing the hatch pattern later in this chapter.

Creating the Partial Section View

If the section lines do not cut through the model, the resulting view will be a partial section view, where the portion of the view that is sectioned using the section line is retained, and the remaining portion will be removed, as shown in Figure 13-24.

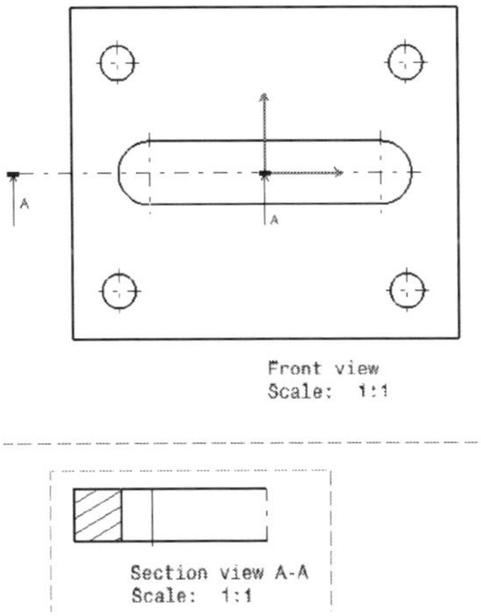

Figure 13-24 The partial section view

Generating Aligned Section Views

Menubar: Insert > Views > Sections > Aligned Section View
Toolbar: Views > Sections sub-toolbar > Aligned Section View

The **Aligned Section View** tool is used to generate a section view of the component, in which at least one of the features is at an angle. In the aligned section view, the sectioned portion revolves about an axis normal to the view such that it is straightened. Figure 13-25 shows the concept of an aligned section view. Note that the inclined feature sectioned in this view is straightened. As a result, the section view is longer than the parent view.

Activate the view to create the aligned section view. Choose the **Aligned Section View** tool from the **Sections** sub-toolbar in the **Views** toolbar. Draw the sketch that defines the section plane. Note that the resulting view will be projected normal to the line drawn first in the section sketch. After sketching the section plane, the aligned section view will be attached to the cursor. Place the view at an appropriate location on the drawing sheet. Figure 13-26 shows the aligned section view in which the inclined line in the section sketch is drawn first and Figure 13-27 shows the aligned section view in which the vertical line in the section sketch is drawn first.

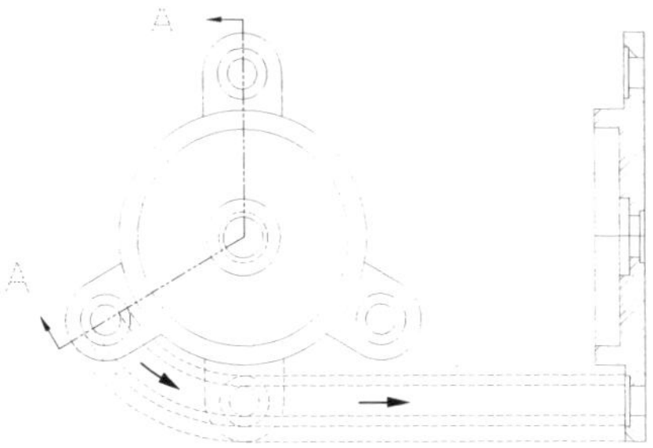

Figure 13-25 Aligned section view

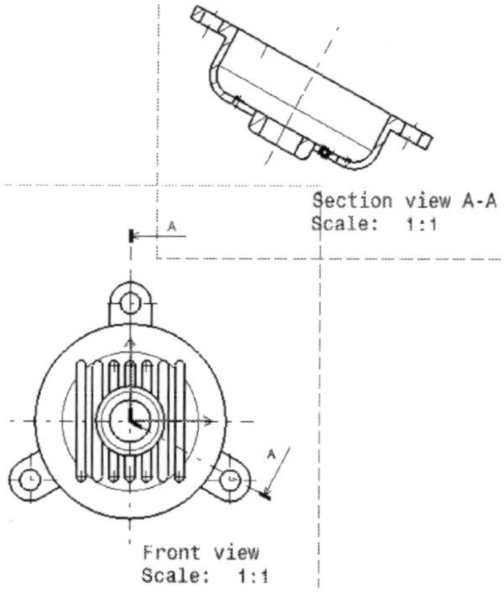

***Figure 13-26** Aligned section view with the inclined line drawn first*

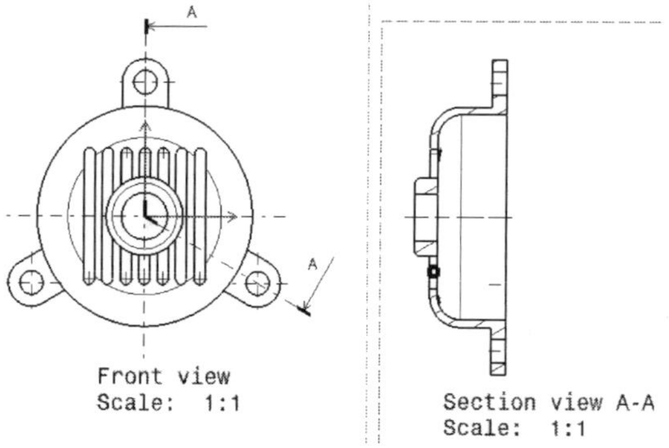

***Figure 13-27** Aligned section view with the vertical line drawn first*

Generating the Offset Section Cut

Menubar: Insert > Views > Sections > Offset Section Cut
Toolbar: Views > Sections sub-toolbar > Offset Section Cut

The **Offset Section Cut** tool is used to generate the section view in which only the section surface is displayed. Entities of the view, other than those sectioned, are not displayed in it. The procedure of generating the offset section cut is the same as that of generating the offset section view. Figure 13-28 shows an offset section cut.

Generating the Aligned Section Cut

Menubar: Insert > Views > Sections > Aligned Section Cut
Toolbar: Views > Sections sub-toolbar > Aligned Section Cut

The **Aligned Section Cut** tool is used to generate aligned section cut in which only the sectioned surface is displayed. Entities of the view which are not sectioned are not displayed in it. The procedure of generating the aligned section cut is the same as that of generating the aligned section view. Figure 13-29 shows an aligned section cut.

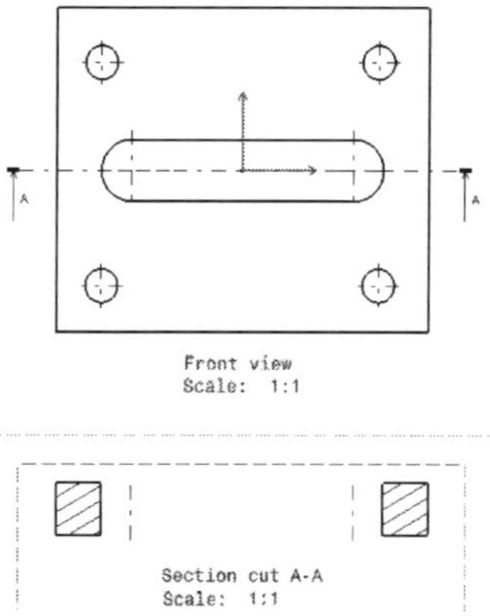

***Figure 13-28** Offset section cut*

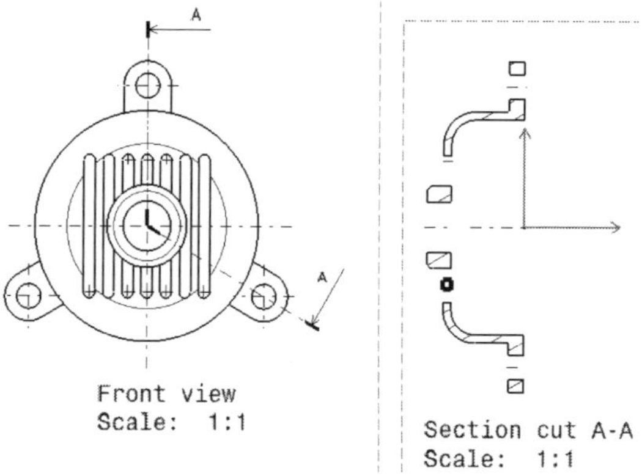

***Figure 13-29** Aligned section cut*

> **Tip**
> *To edit the section sketch used to generate the section view, double-click on the section line; the* ***Profile Edition*** *environment will be invoked where you can edit the section sketch. You can also flip the direction of section arrows using the* ***Invert Profile Direction*** *button in this environment. After editing the profile, choose the* ***End Profile Edition*** *button to return to the* ***Drafting*** *workbench.*

Generating Detail Views

Menubar:	Insert > Views > Details > Detail
Toolbar:	Views > Details sub-toolbar > Detail View

The detail view is used to display the details of a portion of an existing view. You can select the portion whose detailing has to be shown in the parent view. The portion that you select will be magnified and placed as a separate view. You can control the magnification of the detail view. To generate a detail view, first you need to activate the view from which you will generate the detail view. Next, choose the down arrow on the right of the **Detail View** tool in the **Views** toolbar, as shown in Figure 13-30, to invoke the **Details** sub-toolbar. Next, choose the **Detail View** tool from the sub-toolbar.

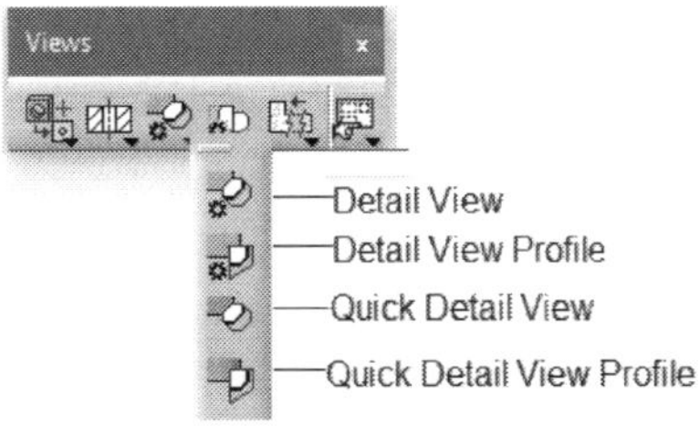

Figure 13-30 *Tools in the* ***Details*** *sub-toolbar*

When you invoke the **Detail View** tool, the **Tools Palette** dialog box will be displayed and you will be prompted to select a point or click to define the circle center. Define the center point of the circle on the view whose detail needs to be generated. Move the cursor and specify the radius of the circle. You can also set the value of the radius in the **Tools Palette** dialog box.

Move the cursor to place the view. The preview of the detail view will be attached to the cursor and its scale will also be mentioned on the preview. Click on the drawing sheet to define the placement point for placing the drawing view. To edit the scale of the detail view, select the view and invoke the shortcut menu. Choose the **Properties** option from the shortcut menu; the **Properties** dialog box will be displayed. Set the scale of the view in the **Scale** edit box and choose the **OK** button from the **Properties** dialog box. Figure 13-31 shows the detail view.

Generating Detail View Profiles

Menubar:	Insert > Views > Details > Sketched Detail Profile
Toolbar:	Views > Details sub-toolbar> Detail View Profile

The **Detail View Profile** tool is used to generate the magnified portion of the view defined by a polygonal profile. To generate the detail view using this tool, choose the **Detail View Profile** tool from the **Details** sub-toolbar in the **Views** toolbar. Draw a closed polygon; the preview of the detail view will be attached to the cursor. Place the view at the desired location on the drawing sheet. Figure 13-32 shows the detail view generated using this tool.

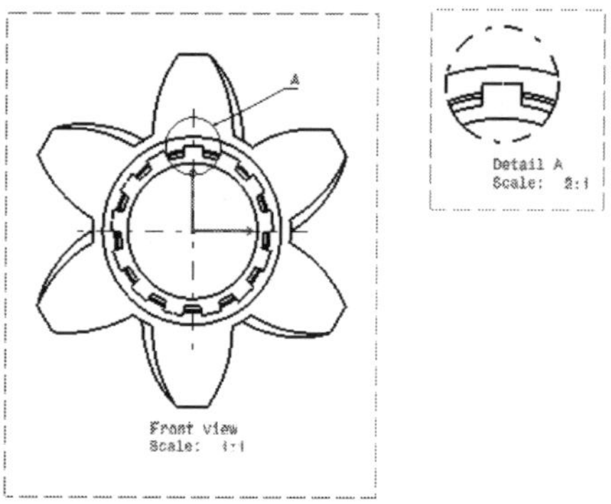

Figure 13-31 Detail view

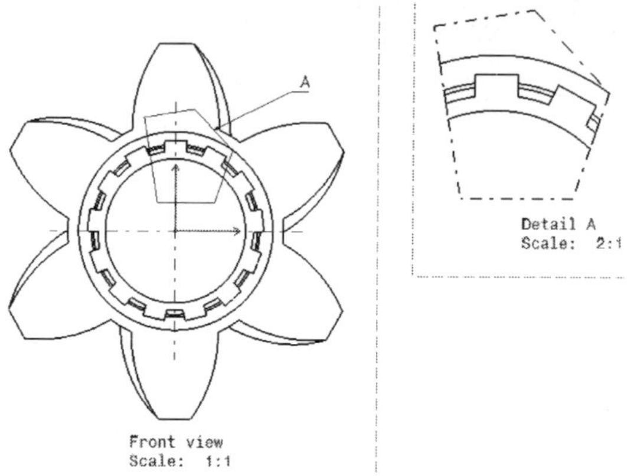

Figure 13-32 Detail view profile

Generating Clipping Views

Menubar: Insert > Views > Clippings > Clipping
Toolbar: Views > Clippings sub-toolbar > Clipping View

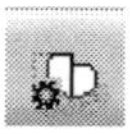

The **Clipping View** tool is used to crop an existing view by using a circular profile. The portion of the view that lies inside it is retained and the remaining portion is removed. To crop a view, activate it and choose the **Clipping View** tool from the **Clippings** sub-toolbar, as shown in Figure 13-33. This sub-toolbar is invoked by choosing the down arrow on the right of the **Clipping View** tool in the **Views** toolbar.

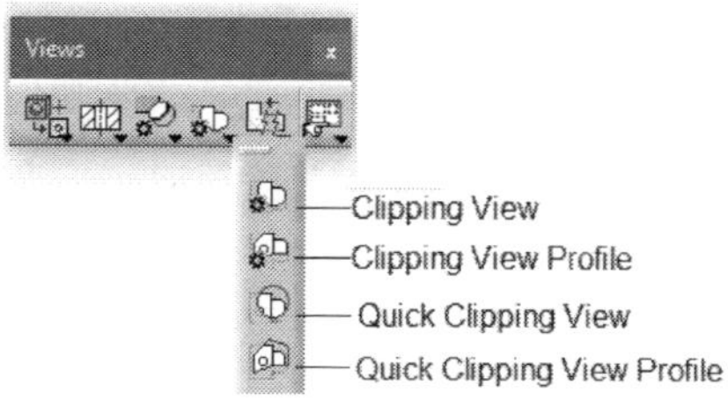

*Figure 13-33 Tools in the **Clippings** sub-toolbar*

When you invoke this tool, the **Tool Palette** toolbar will be displayed and you will be prompted to select a point or click to define the center of the circle. Click on the drawing sheet to define the center point of the circle. Now, move the cursor and specify its radius. You can also specify the radius in the **Tools Palette** toolbar. The portion of the view inside the circular profile will be retained and the remaining portion will be removed, as shown in Figure 13-34.

Generating Clipping View Profiles

Menubar: Insert > Views > Clippings > Sketched Clipping Profile
Toolbar: Views > Clippings sub-toolbar > Clipping View Profile

The **Clipping View Profile** tool is used to crop an existing view using a polygonal profile. The portion of the view that lies inside the polygonal profile is retained and the remaining portion is removed. To crop a view, activate it and choose the **Clipping View Profile** tool from the **Clippings** sub-toolbar in the **Views** toolbar; you will be prompted to click on the first point. Draw the profile of the closed polygon. After specifying the end point of the profile; the line creation will be exited and the portion of the view inside the polygonal profile will be retained and the remaining portion will be removed, as shown in Figure 13-35.

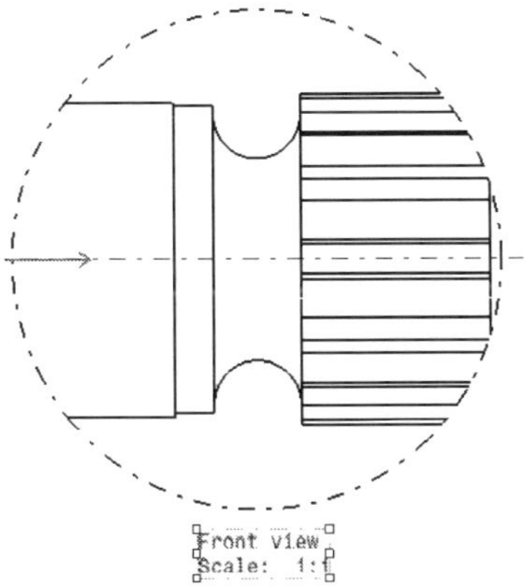

Figure 13-34 Clipping view

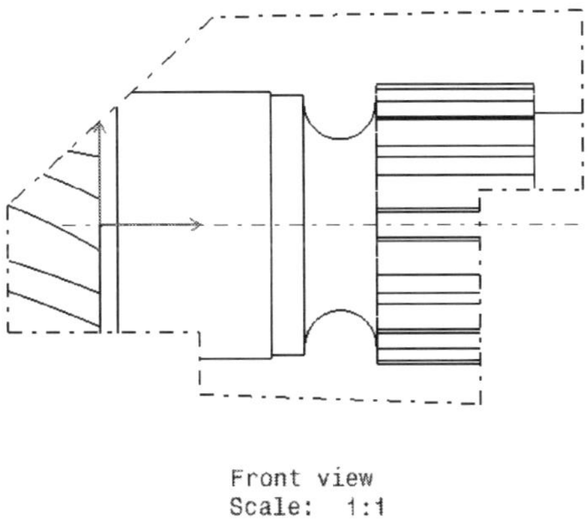

Figure 13-35 Clipping view profile

Generating Broken Views

Menubar:	Insert > Views > Break view > Broken View
Toolbar:	Views > Break view sub-toolbar > Broken View

The **Broken View** tool is used to display a component by removing a portion of it while keeping the ends of the view intact. This type of view is used for displaying those components whose length to width ratio is very high. This means the length is very large as compared to the width, or the width is very large as compared to the length. The broken view breaks the view along the horizontal or vertical direction such that the drawing view fits into the required area. To generate a broken view, activate it, and then choose the **Broken View** tool from the **Break view** sub-toolbar, refer to Figure 13-36. On doing so, you will be prompted to select a point inside the view to indicate the position of the first breakout line.

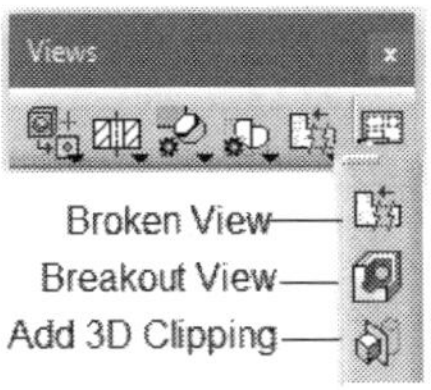

*Figure 13-36 Tools in the **Break view** sub-toolbar*

Select a point; the preview of the breakout line will be displayed on the drawing sheet. Move the cursor and click to specify its orientation, which can be horizontal or vertical; the preview of the second breakout line will be attached to the cursor and you will be prompted to select a point inside the view to indicate its position. Move the cursor and click on a location where you need to place the second breakout line; you will be prompted to click on the sheet to generate the broken view. Click anywhere on the drawing sheet to generate the broken view. Figure 13-37 shows the front view that needs to be broken. Figure 13-38 shows the resultant broken view. You can also break an isometric view, as shown in Figure 13-39. If you break a 3D view placed horizontally, the parts of the view will lose their alignment.

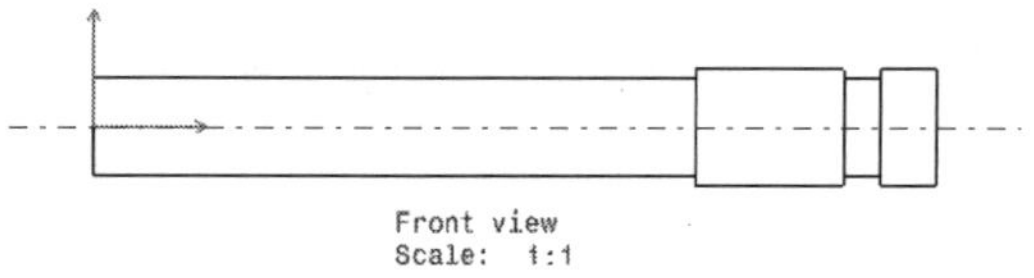

Figure 13-37 *The front view that needs to be broken*

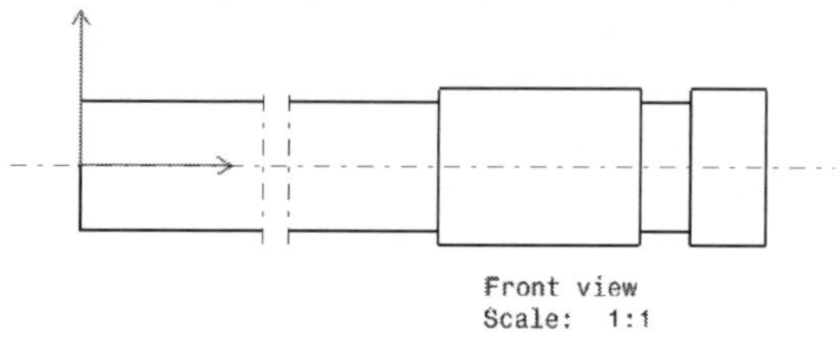

Figure 13-38 *Resultant broken view*

To change the linetype of the breakout line, as shown in Figures 13-39 and 13-40, select the breakout line and choose the **Properties** option from the contextual menu; the **Properties** dialog box will be displayed. In this dialog box, select the required linetype from the **Linetype** drop-down list and choose the **OK** button.

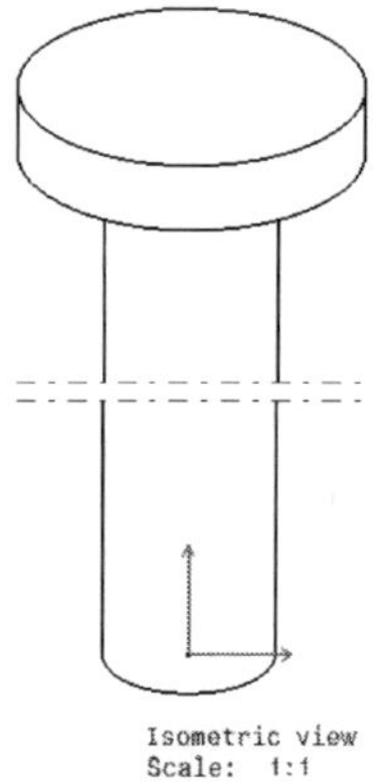

Figure 13-39 *A broken isometric view*

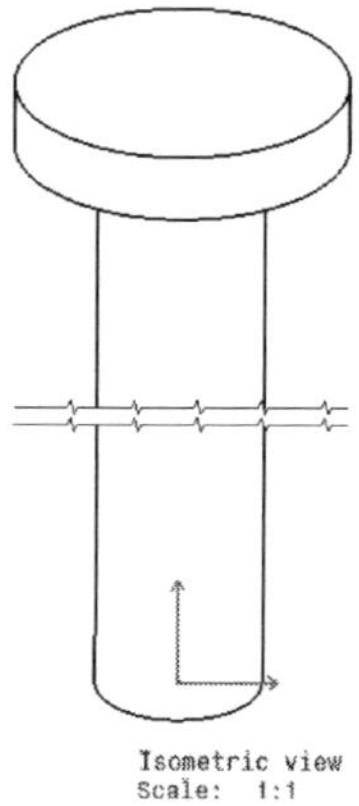

Figure 13-40 *Zigzag broken lines*

Tip

To unbreak the broken view, select the view or the break lines and invoke the contextual menu. Choose ***"Callout (Broken View)" > Unbreak*** *from the contextual menu.*

Generating Breakout Views

Menubar: Insert > Views > Break view > Breakout View
Toolbar: Views > Break view sub-toolbar > Breakout View

The **Breakout View** tool is used to remove material from an existing view and then display the area of the model or assembly that lies behind the removed portion. This type of view is generated by using the closed polygonal sketch drawn on the view. To generate a breakout view, activate the view, and then choose the **Breakout View** tool from the **Break view** sub-toolbar in the **Views** toolbar. Next, draw the polygon profile, as shown in Figure 13-41.

After drawing the polygon profile, the **3D Viewer** window will be displayed, as shown in Figure 13-42, and you will be prompted to move the plane or select its position with respect to an element. Move the plane that is highlighted in the preview window to define the depth of the cut. You can also click once in the **Reference element** display box and select a 2D element from the drawing sheet that will be selected to define the depth of the cut. The reference to be selected could be a plane, face, line, edge, center line, or center mark. The name of the selected reference will be displayed in the **Reference element** display box and its preview will also be displayed in the **Preview** area of the **3D Viewer** window.

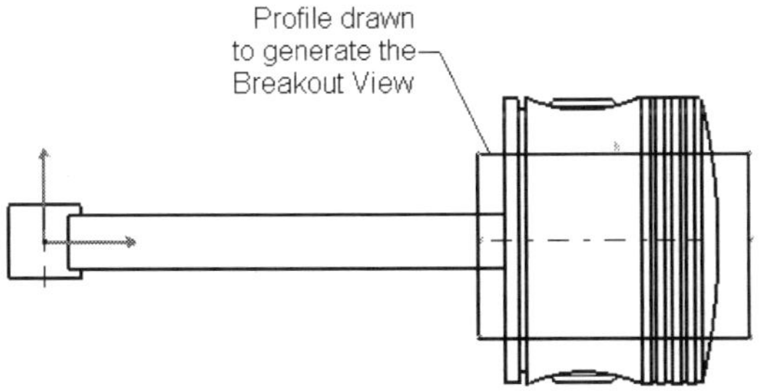

Figure 13-41 *Polygon profile drawn to generate the Breakout view*

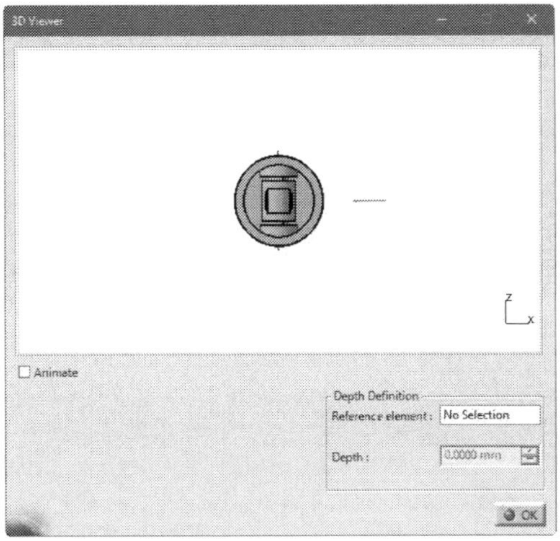

Figure 13-42 *The **3D Viewer** window*

You can set the value of the offset distance of the plane using the **Depth** spinner. If you select the **Animate** check box from the **3D Viewer** window, the view of the model in the **Preview** area will rotate with respect to the view, in the drawing area, in which you place the cursor while selecting the reference. You can use the mouse buttons to zoom and rotate the view of the model in the preview window. After setting all parameters, choose the **OK** button from the **3D Viewer** window. Figure 13-43 shows a breakout view.

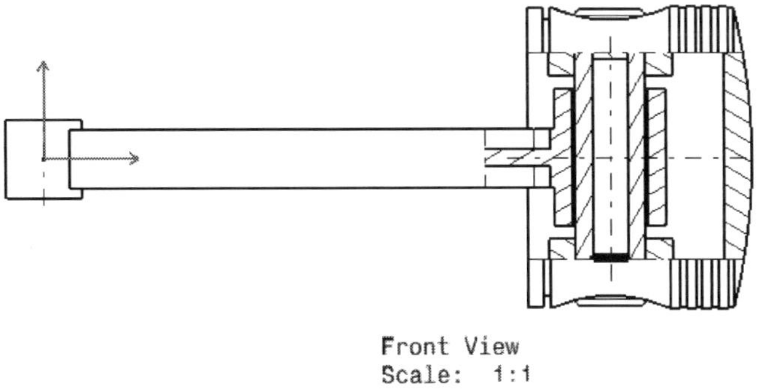

Figure 13-43 *Breakout view*

GENERATING AN EXPLODED VIEW

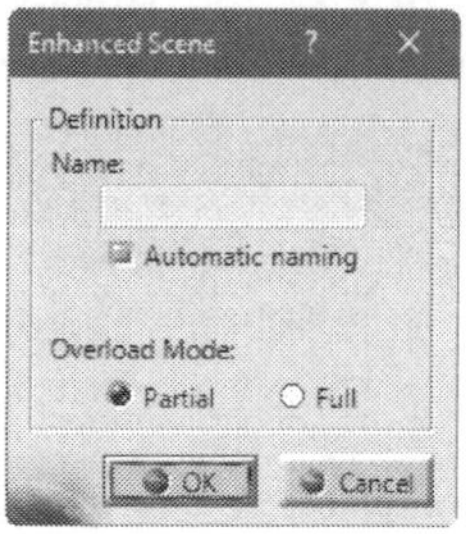

***Figure 13-44** The **Enhanced Scene** dialog box*

You can also generate an exploded view of an assembly in the **Drafting** workbench of CATIA V5. To generate the exploded view of an assembly, first you need to create a scene in the **DMU Navigator** workbench. This scene will be used for generating the exploded view. To create the scene, open the product file of the assembly that you need to explode. Next, choose **Start > Digital Mockup > DMU Navigator** from the menubar; the **DMU Navigator** workbench will be invoked. Choose the **Enhanced Scene** tool from the **DMU Review Creation** toolbar to display the **Enhanced Scene** dialog box, as shown in Figure 13-44. Alternatively, you can invoke this dialog box from the **Assembly Design** workbench by selecting the **Enhanced Scene** tool from the **Scenes** toolbar.

Clear the **Automatic naming** check box if it is selected by default. Enter the name of the scene as **Exploded View** in the **Name** edit box. Now, select the **Full** radio button from the **Overload Mode** area and choose the **OK** button from the **Enhanced Scene** dialog box. The scene environment will be invoked and the background color will be changed to dark green. Also, the **Warning** message window will be displayed warning you that any moves performed in the scene context may invalidate the analysis data results. Choose the **Close** button from this message window. The **Enhanced Scenes** toolbar will also be displayed in the geometry area.

Choose the **Explode** tool from the **Enhanced Scenes** toolbar; the **Explode** dialog box will be displayed. Explode the assembly using this dialog box. Exploding of the assemblies is discussed in Chapter 12. After creating the exploded view, choose the **Exit Scene** tool from the **Enhanced Scenes** toolbar; the assembly will switch back to the assembled mode. Now, you can save it. Next, start a new file in the **Drafting** workbench. Choose the **Isometric View** tool from the **Projections** toolbar and invoke the **Assembly** window. Expand the **Applications** branch of the Specification tree and then expand the **Scenes** branch. Select the **Exploded View** from the **Scenes** branch. Now, select any one of the faces or edges of the assembly; the drawing window will be invoked and the preview of the exploded view will be displayed. Click anywhere on the drawing sheet to generate the drawing view. Figure 13-45 shows the drawing view of an exploded assembly.

Note

*After generating an exploded view of an assembly, it is very important to draw the explode line sketch. In the **Drafting** workbench, the sketching tools provided in the **Geometry Creation** toolbar are used to draw it manually.*

Tip

*You can also generate an exploded view directly from an assembly. To do so, create the exploded view in the **Assembly** workbench and then start a new file in the **Drafting** workbench. Next, choose the **Isometric View** tool from the **Projections** toolbar and then switch to the assembly window. In the assembly window, select any one of the faces or edges of the assembly. As soon as you select a face or an edge, the drawing window will be invoked automatically and a preview of the exploded view will be attached with the cursor. Next, click anywhere on the drawing sheet to specify the placement point for the exploded view.*

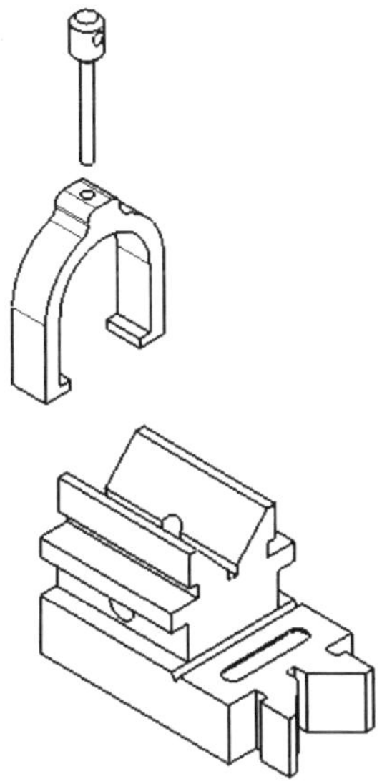

Figure 13-45 Drawing view of the exploded assembly

WORKING WITH INTERACTIVE DRAFTING IN CATIA V5

As mentioned earlier, you can also sketch the 2D drawing views in the **Drafting** workbench of CATIA V5. In technical terms, sketching 2D drawings is known as interactive drafting. Before drawing, it is recommended that you insert an empty view. To place an empty view, choose **Insert > Drawing > New View** from the menubar. Click on the drawing sheet to place the empty view. An empty view will be placed on that drawing sheet. The view will be enclosed in a dotted rectangle displayed in orange, which implies it is active. Now, use the sketching tools in the **Drafting** workbench to draft the views. Similarly, to draw another drawing view, it is recommended to place another empty view and activate it.

EDITING AND MODIFYING DRAWING VIEWS

In CATIA V5, you can edit and modify the drawing views. For example, you can change the view scale, delete, or rotate it. All these operations are discussed next.

Changing the Scale of Drawing Views

In the **Drafting** workbench of CATIA V5, you can change the scale of a drawing view by selecting it and invoking the contextual menu. Choose the **Properties** option; the **Properties** dialog box will be displayed. Set the scale of the drawing view in the **Scale** edit box and choose the **OK** button. If you change the scale of the main view, the scale of the dependent views will not be changed. To change the scales of the dependent views, you need to change them manually using the above mentioned procedure.

Modifying the Project Plane of the Parent View

In CATIA V5, you have been provided with a new functionality to modify the projection plane that is used to generate the parent view. As discussed earlier, a parent view is generated by using the **Front View** or **Advanced Front View** tool. To modify the projection plane, right-click on the boundary of the parent view and invoke the contextual menu. Move the cursor on **Front view**

object (or the name of the object, if it is changed) in the contextual menu to invoke a cascading menu. Now, choose the **Modify Projection Plane** option from this cascading menu; you will be prompted to select a reference plane on a 3D geometry. Invoke the part file and select the face or the plane that you need to select as the projection plane; the drawing file will be invoked and the parent view with the modified orientation will be displayed on the original view. You can also use the knob provided at the upper right corner to change the orientation of the view. Click once on the drawing area to accept the orientation of the model view. You will notice that the orientation of all views referred to the parent view is not changed. To update the orientation of all reference views, choose the **Update current sheet** button from the **Update** toolbar.

Deleting Drawing Views

You can delete the unwanted views from the drawing sheet using the Specification tree or directly from the drawing sheet. Select the view to be deleted from the Specification tree or select the boundary of the view from the drawing sheet and invoke the contextual menu. Choose the **Delete** option from it; the selected view will be deleted. You can also delete the drawing view by selecting it and then using the DELETE key. If you delete the parent view, the child views related to the parent views will not deleted.

Rotating Drawing Views

CATIA V5 allows you to rotate a drawing view in the 2D plane. To do so, select the view and invoke the contextual menu. Choose the **Properties** option from the menu; the **Properties** dialog box will be displayed. Set the angle of rotation in the **Angle** edit box provided in the **Scale and Orientation** area of the dialog box. Next, choose the **OK** button to apply the changes on the selected view and exit the **Properties** dialog box.

Hiding Drawing Views

To hide a view, select it from the drawing sheet or from the Specification tree and invoke the contextual menu. Choose the **Hide/Show** option; the display of the selected view will be turned off.

MODIFYING THE HATCH PATTERN OF SECTION VIEWS

As discussed earlier, when you generate a section view of an assembly or a component, a hatch pattern based on the material assigned to the components in the part document is applied to the component(s). If the material is not applied to the model, then it applies the default hatch pattern. To modify the default hatch pattern, double-click on it or select it and invoke the contextual menu. Choose the **Properties** option from the contextual menu; the **Properties** dialog box will be displayed, as shown in Figure 13-46.

To modify the hatch spacing, change the value in the **Pitch** spinner. You can use the other options in this dialog box to modify the hatch parameters.

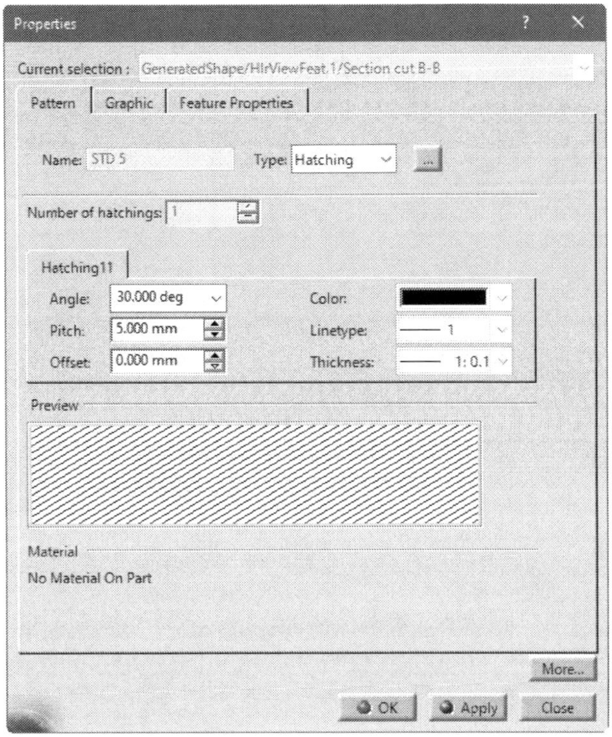

*Figure 13-46 The **Properties** dialog box*

TUTORIALS

Tutorial 1

In this tutorial, you will generate the drawing views of the model of the Motor Cover created in Tutorial 2 of Chapter 8. You will generate the front view, top view, aligned section view, isometric view, and detail view from the top view. You will use the A4 standard size sheet and the third angle projection. The drawing sheet after generating all views is shown in Figure 13-47.

(Expected time: 30 min)

The following steps are required to complete this tutorial:

a. Copy the part document of Tutorial 2 of Chapter 8 to the folder of the current chapter.
b. Start a new file in the **Drafting** workbench with the standard A4 sheet size.
c. Set the projection standard to third angle.
d. Generate the front view.
e. Generate the top view by using the **Projection View** tool.
f. Generate the aligned section view from the top view after activating it.
g. Generate the detail view.
h. Generate the isometric view.
i. Save and close the file.

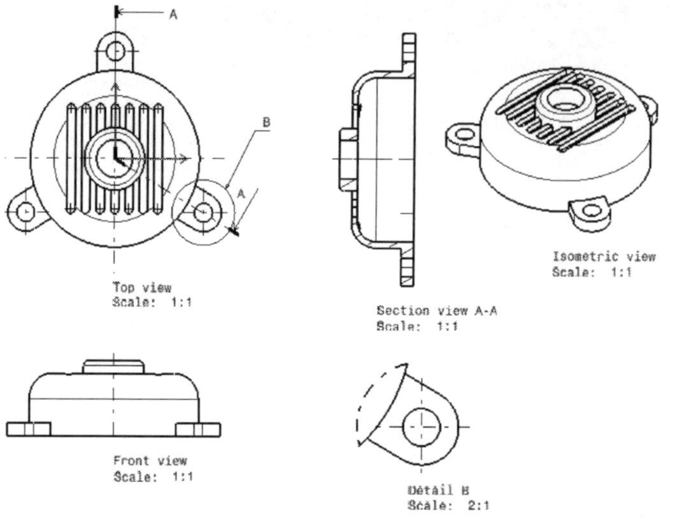

Figure 13-47 *Drawing views to be generated*

Copying and Opening the Part Document

1. Create a folder with the name *c13* in the *CATIA* folder and then copy *c8tut2.CATPart* from *C:\CATIA\c08* to *c13* folder.

2. Start CATIA V5 and open the part document from *C:\CATIA\c13\c08tut2*.

Starting a New File in the Drafting Workbench

1. Invoke the **New** dialog box to start a new file. Select the **Drawing** option from the **New** dialog box and then choose the **OK** button to display the **New Drawing** dialog box.

2. Select the **A4 ISO** option from the **Sheet Style** drop-down list in the **New Drawing** dialog box if not selected. Choose the **OK** button from the **New Drawing** dialog box; a new file is started in the **Drafting** workbench.

Generating the Front View

After starting new file in the **Drafting** workbench, you need to turn off the display of grid and set the standard of projection to the third angle. After making all settings, you will generate the front view.

1. Choose the **Sketcher Grid** button from the **Visualization** toolbar to turn off the grid display if not already turned off. 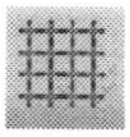

2. Right-click on **Sheet.1** in the Specification tree and choose the **Properties** option from the contextual menu; the **Properties** dialog box is displayed.

3. Select the **Third angle standard** radio button and choose the **OK** button from the **Properties** dialog box.

After setting all the options, you need to generate the front view of the model.

4. Choose the **Front View** tool from the **Projections** sub-toolbar in the **Views** toolbar; you are prompted to select a reference plane on a 3D geometry. 

   Note that the **Front View** tool will only be activated if a part or product file is open in the background.

5. Choose **Window > Tile Vertically** from the menu bar to tile the window vertically.

6. Select the yz plane from the Specification tree in the Part window; a preview of the view is displayed in the drawing view and the knob is also displayed at the top right of the drawing sheet.

7. Maximize the drawing window. Fit the display area of the drawing sheet to the screen using the **Fit All In** tool.

8. Press and hold the left mouse button on the green dashed rectangular frame in which the preview of the front view is enclosed, and then drag the cursor close to the lower-left corner of the drawing sheet, refer to Figure 13-48.

9. Click on the drawing sheet to generate the front view. The drawing sheet after generating the front view is shown in Figure 13-48.

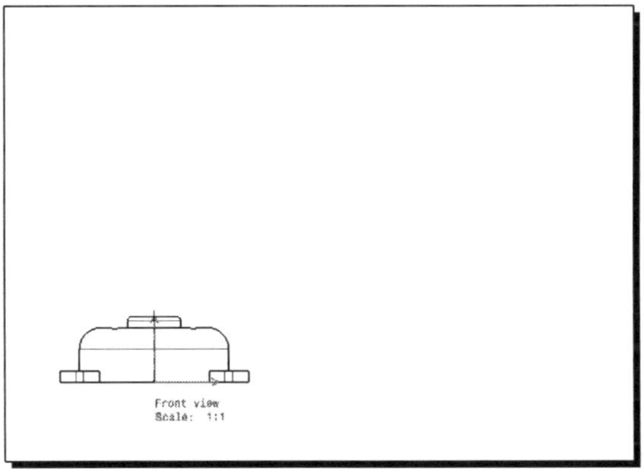

Figure 13-48 *The drawing sheet after generating the front view*

After generating the front view, you will notice that the front view is enclosed in a red frame. To improve the display of views on the drawing sheet, you need to turn off the display of the frame.

10. Right-click on the view in the drawing sheet or in the Specification tree, and then choose **Properties** from the contextual menu displayed; the **Properties** dialog box is displayed. Select the **Axis** check box from the **Dress-up** area available under the **View** tab from the dialog box.

11. Clear the **Display View Frame** check box and choose the **OK** button from the **Properties** dialog box. The display of the frame is turned off.

Generating the Top View

Next, you need to generate the top view using the **Projection View** tool. You will generate the top view with the help of the front view.

1. Choose the **Projection View** tool from the **Projections** sub-toolbar in the **Views** toolbar; a preview of the projected view is attached to the cursor.

2. Move the cursor close to the top left corner of the drawing sheet, refer to Figure 13-49, and click to generate the top view.

3. Next, invoke the **Properties** dialog box for the top view and then select the **Center Line** check box from it. To turn off the display of the view frame, clear the **Display View Frame** check box and then choose the **OK** button in the **Properties** dialog box. The drawing sheet after generating the top view is shown in Figure 13-49.

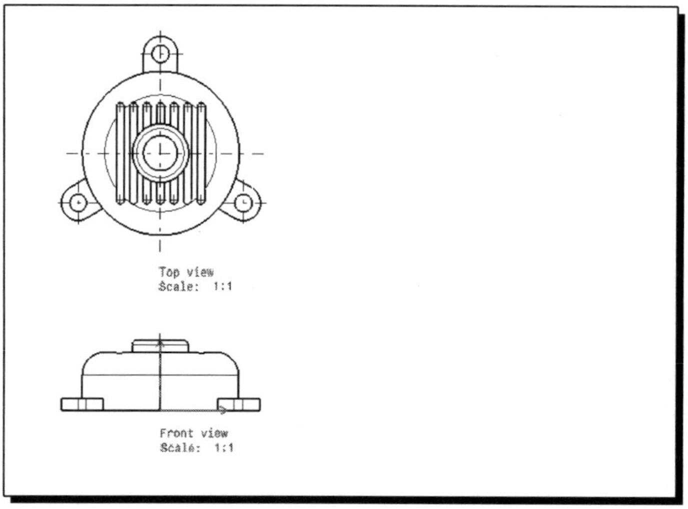

***Figure 13-49** The drawing sheet after generating the top view*

Generating the Aligned Section View

Next, you need to generate the aligned section view from the top view. Therefore, first you need to activate the top view.

1. Select **Top view** from the Specification tree and then right-click; a contextual menu is displayed. Choose the **Activate View** option from it to activate the top view. Next, click on the drawing sheet; the **Top view** in the Specification tree is highlighted with blue background indicating that the top view is the active view. Also, the axis system is displayed on the top view in the drawing sheet indicating that the selected view is the active view.

2. Choose the **Aligned Section View** tool from the **Sections** sub-toolbar in the **Views** toolbar.

Next, you need to define the section plane by sketching the section lines.

3. Draw sketch for section lines in the top view, as shown in Figure 13-50. You need to make sure that the vertical line of the sketch is drawn first. Double-click on the endpoint of the line drawn last to exit section line creation; preview of the aligned section view is attached to the cursor.

   Note that in the preview, it may not look like an aligned section view.

4. Move the cursor to the right of the top view, refer to Figure 13-51.

5. Click on the drawing sheet to generate the aligned section view. Turn off the display of the view frame.

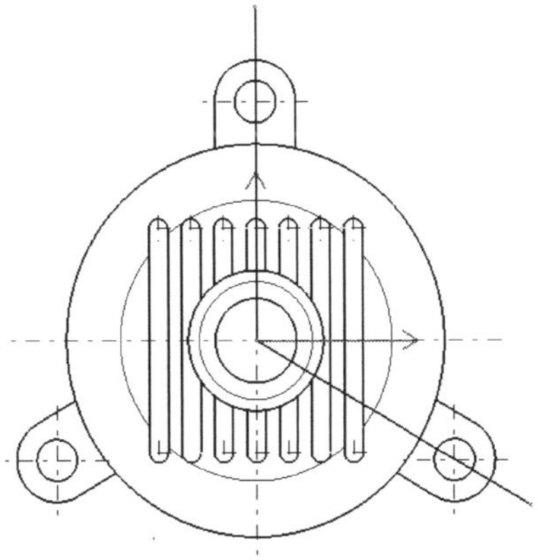

Figure 13-50 *Sketch for the section lines*

The drawing sheet after generating the aligned section view is shown in Figure 13-51.

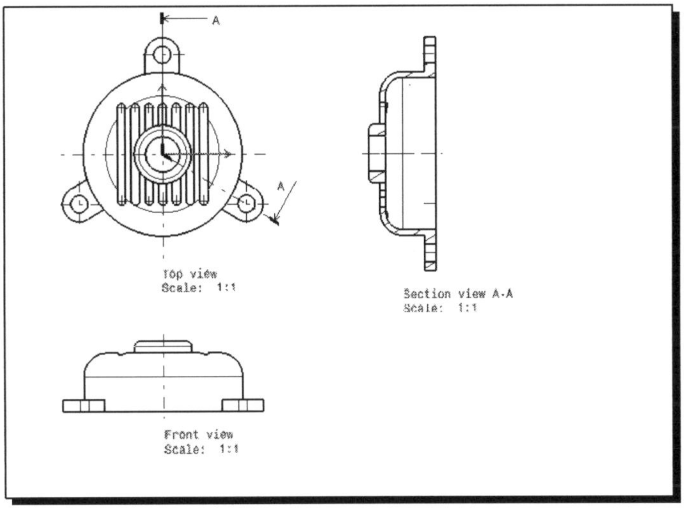

Figure 13-51 *The drawing sheet after generating the aligned section view*

Generating the Detail View

After generating the front, top, and aligned section views, you need to generate the detail view from the top view using the **Detail View** tool.

1. Choose the **Detail View** tool from the **Details** sub-toolbar in the **Views** toolbar; you are prompted to select a point to define the center of the circle.

2. Draw a circle, as shown in Figure 13-52.

3. Move the cursor in the drawing area; the detail view gets attached to the cursor. Click on the drawing sheet below the aligned section view to generate the detail view. Turn off the display of the view frame.

   The drawing sheet after generating the detail view is shown in Figure 13-53.

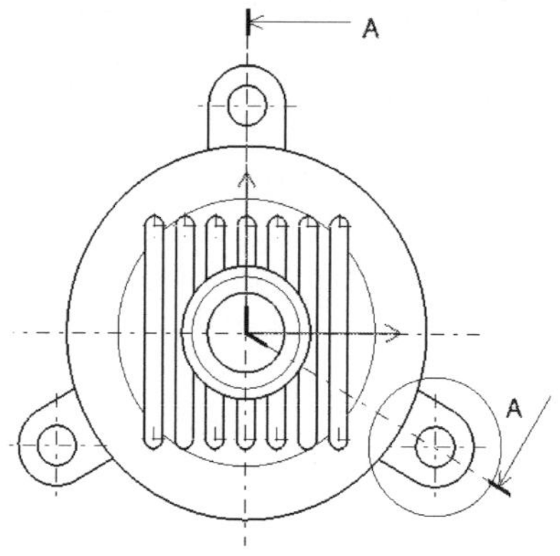

***Figure 13-52** Sketch for the detail view*

Generating the Isometric View

The last view that you need to generate is the isometric view. The view will be generated by using the **Isometric View** tool.

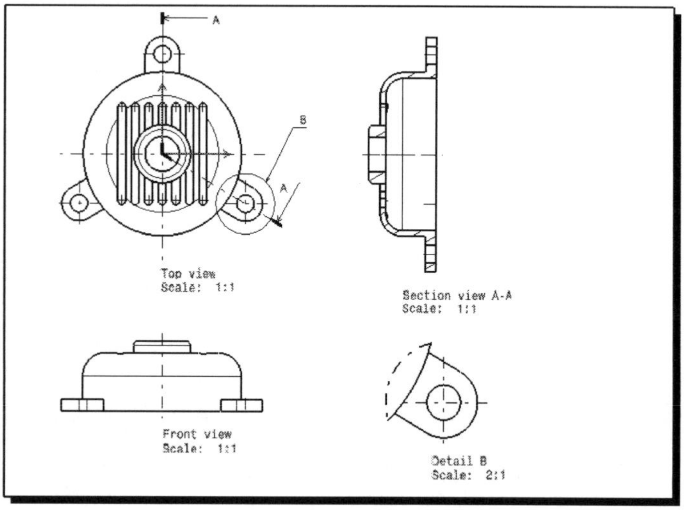

***Figure 13-53** The drawing sheet after generating the detail view*

1. Choose the **Isometric View** tool from the **Projections** sub-toolbar in the **Views** toolbar; you are prompted to select a reference from the 3D geometry.

2. Choose **Window > Tile Vertically** from the menubar.

3. Select any one of the default planes in the Specification tree of the part file. Make sure the part file is displayed in isometric view.

   The preview of the isometric view with the knob is displayed on the drawing sheet.

4. Maximize the drawing file window and drag the preview of the isometric view close to the upper right corner of the drawing sheet by holding the view frame.

5. Click on the drawing sheet to generate the isometric view. Turn off the display of the view frame.

   The final drawing sheet after generating all the drawing views is shown in Figure 13-54.

Saving and Closing the File

1. Choose the **Save** button from the **Standard** toolbar to invoke the **Save As** dialog box and browse to *C:\CATIA\c13*.

2. Enter the name of the file as **c13tut1.CATDrawing** in the **File name** edit box and choose the **Save** button. The file is saved at *C:\CATIA\c13*.

3. Close the part file and the drawing file by choosing **File > Close** from the menubar.

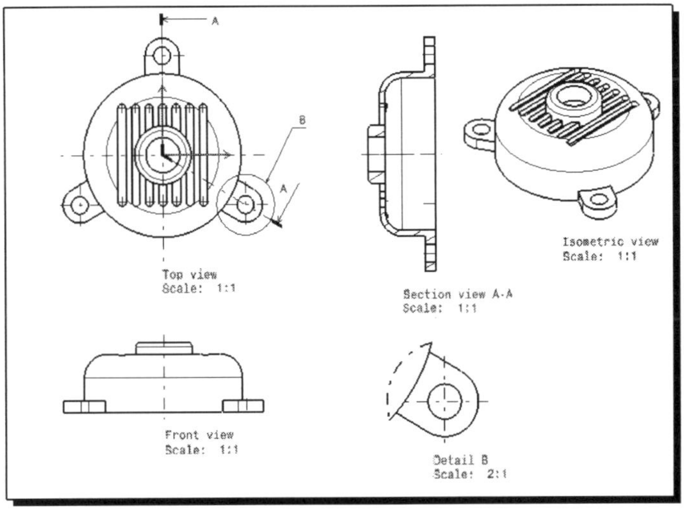

Figure 13-54 *The final drawing sheet with all the views*

Tutorial 2

In this tutorial, you will create the Bench Vice assembly and then generate the front, top, right, and isometric views of the Bench Vice assembly, as shown in Figure 13-55. Figures 13-56 through Figures 13-59 show the views and dimensions of the components of the Bench Vice assembly. Orthographic views and BOM of the Bench Vice Assembly is shown in Figure 13-60.

(Expected time: 2 hr)

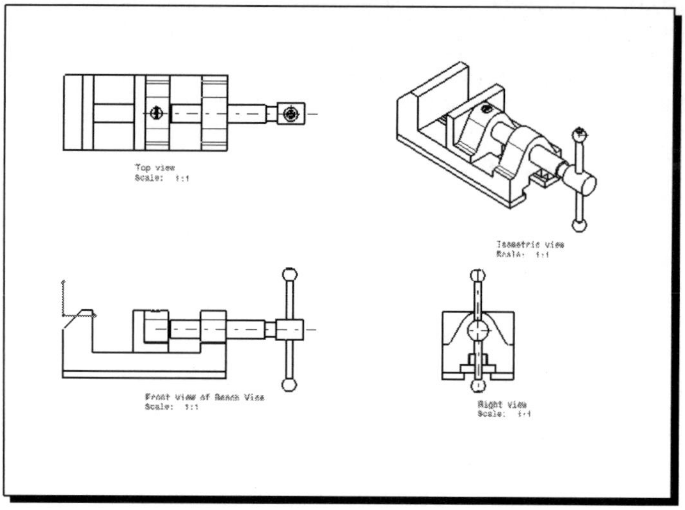

Figure 13-55 Drawing view of the Bench Vice assembly

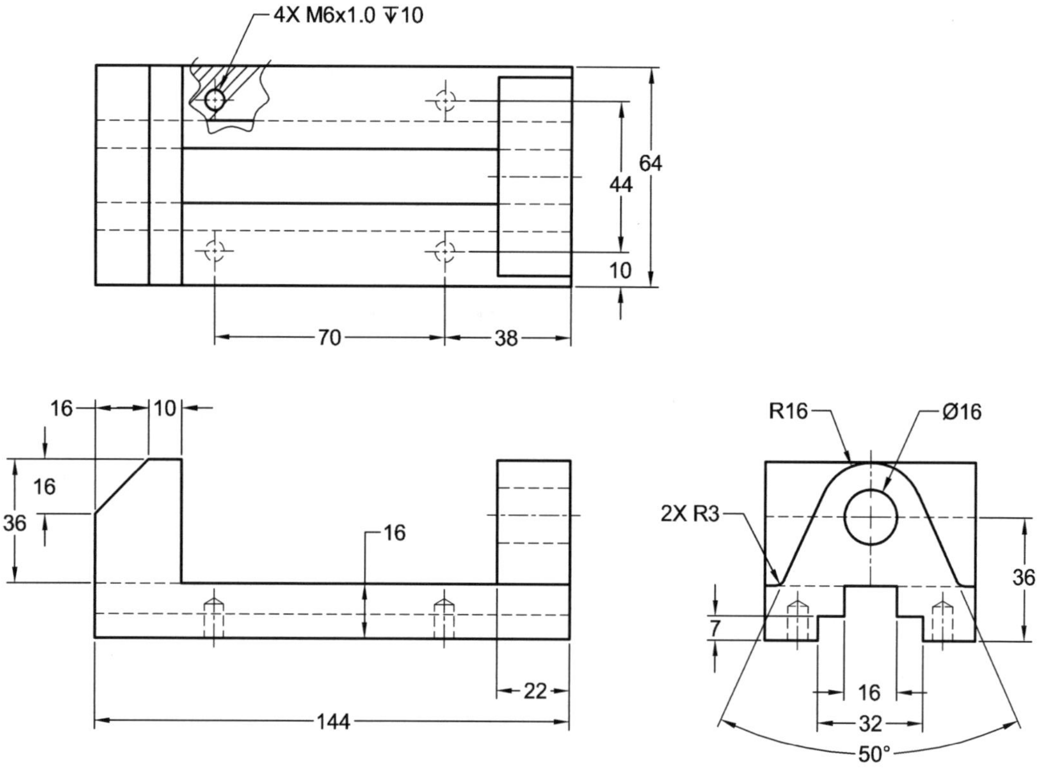

Figure 13-56 Orthographic views and dimensions of the Base

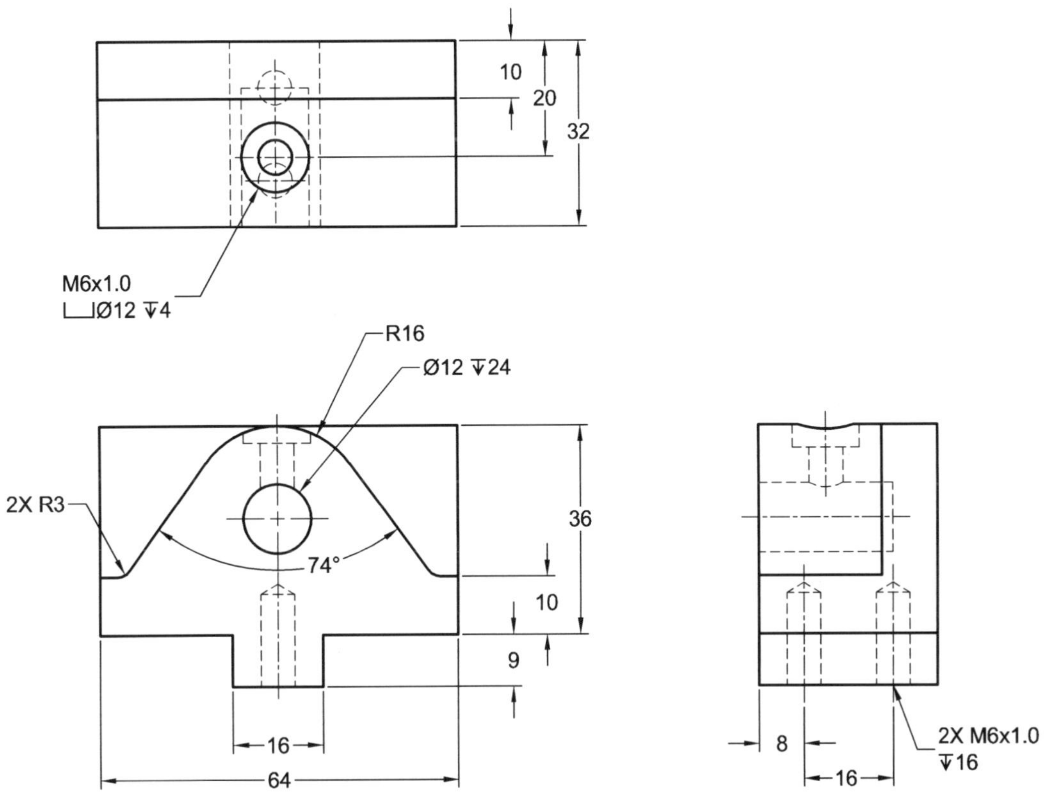

Figure 13-57 Orthographic views and dimensions of the Vice Jaw

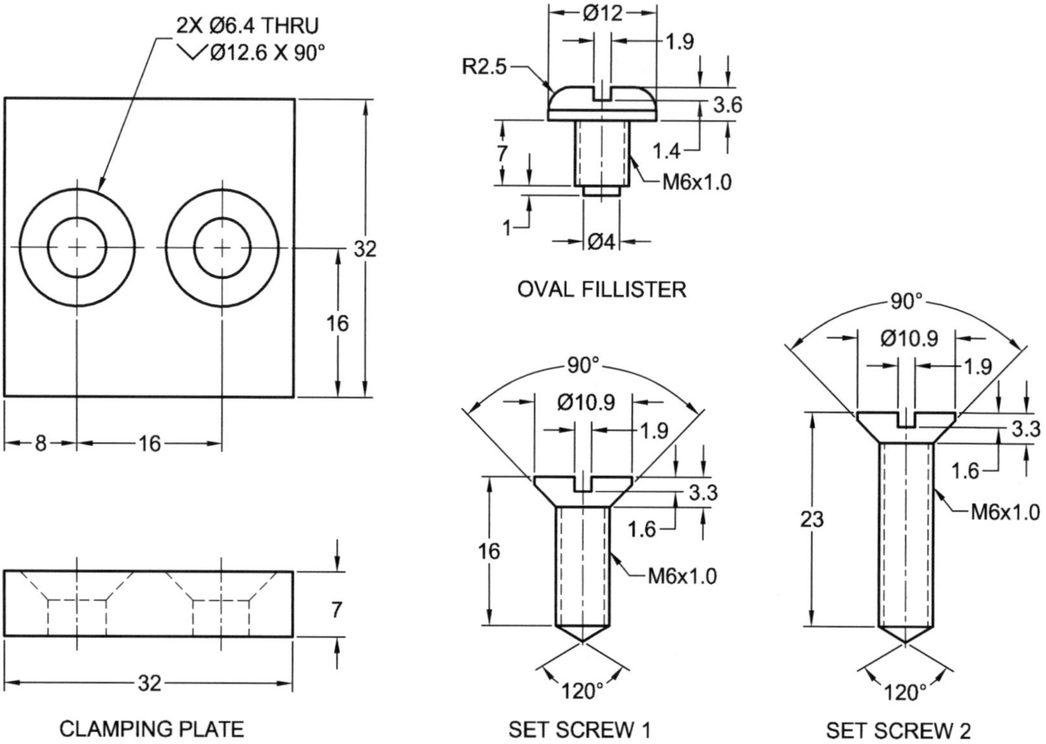

***Figure 13-58** Orthographic views and dimensions of the Clamping plate, Oval fillister and Set screw*

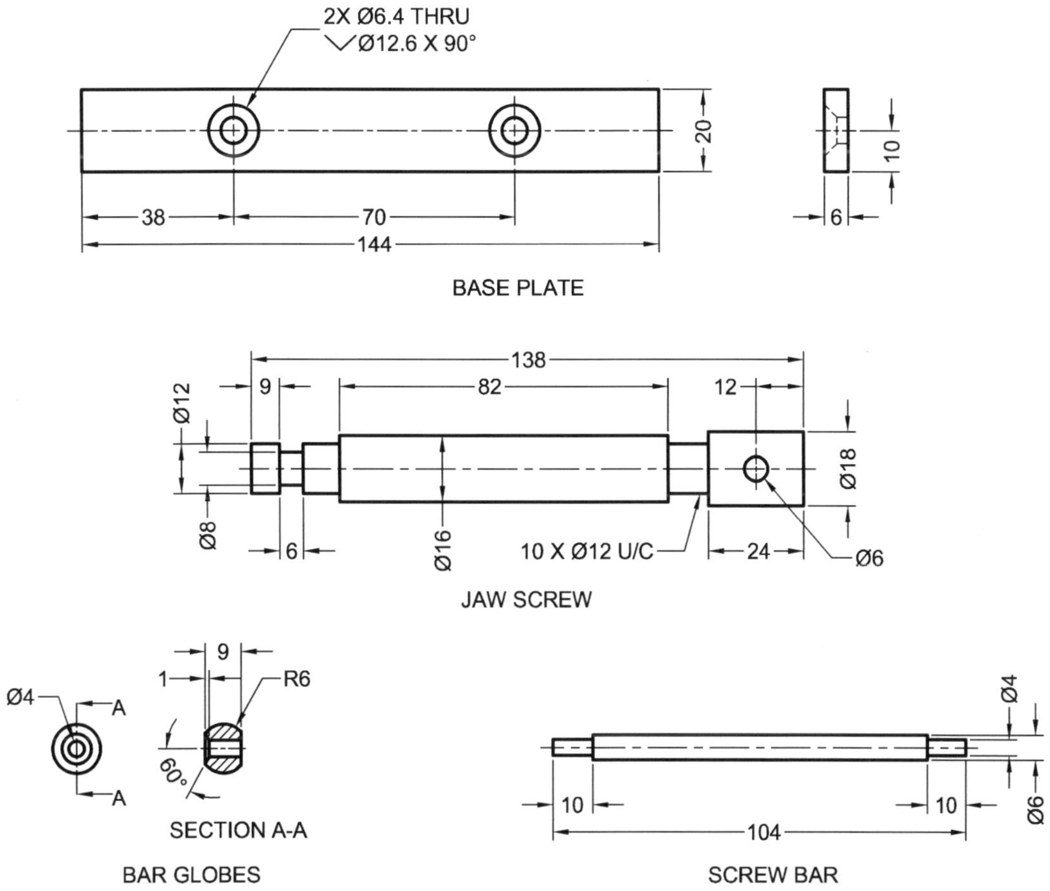

Figure 13-59 *Orthographic views and dimensions of the Jaw screw, Bar globe and Screw bar*

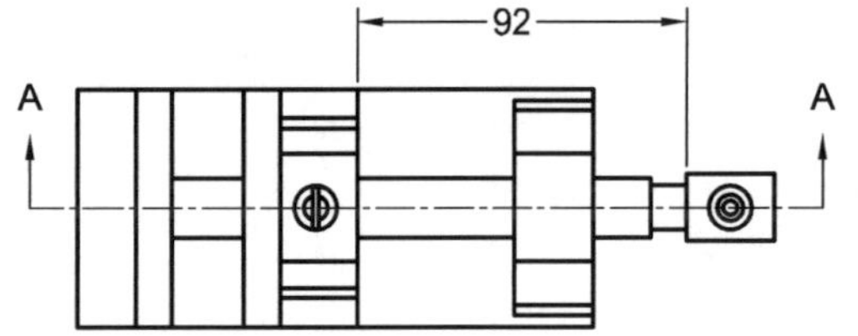

ITEM NO.	PART NO.	DESCRIPTION	QTY.
1	Vice Body		1
2	Vice Jaw		1
3	Jaw Screw		1
4	Clamping Plate		1
5	Base Plate		2
6	Screw Bar		1
7	Bar Globes		2
8	Oval Fillister		1
9	Set Screw1		4
10	Set Screw2		2

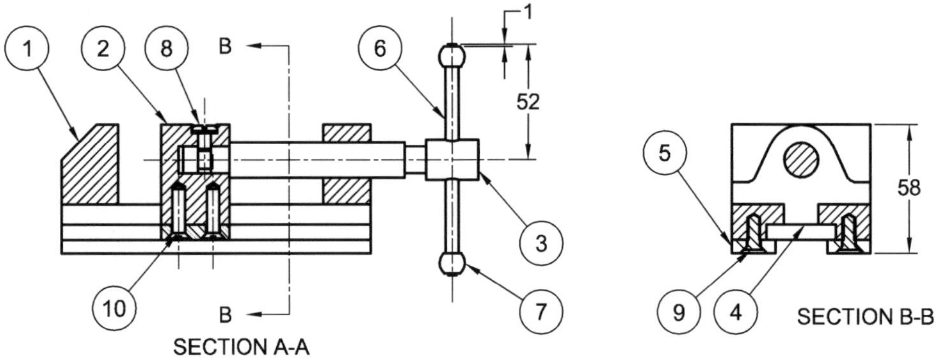

***Figure 13-60** Orthographic views and BOM of the Bench Vice Assembly*

The following steps are required to complete this tutorial:

a. Create all components of the Bench Vice assembly and assemble them.
b. Start a new file in the **Drafting** workbench using the A2 size sheet.
c. Set the standard of the projection to third angle.
d. Generate the front view, refer to Figure 13-61.
e. Generate the top view, refer to Figure 13-62.
f. Generate the right view, refer to Figure 13-63.
g. Generate the isometric view, refer to Figure 13-64.
h. Save and close the file.

Creating and Assembling Components

1. Create all components of the Bench Vice assembly. Create a folder with the name *Bench Vice* inside the folder *c13* and save all components in it.

2. Assemble all components in the **Assembly Design** workbench and save the assembly file.

3. Close all part files, but keep the assembly file open.

Starting a New File in the Drawing Workbench

Next, you need to start a new file in the **Drafting** workbench. Since the assembly file of the Bench Vice is already opened in the CATIA V5 window, you will use the **Start** menu to start a new file in the **Drafting** workbench.

1. Choose **Start > Mechanical Design > Drafting** from the menubar to display the **New Drawing Creation** dialog box.

2. Choose the **Empty sheet** button and then the **OK** button from the **New Drawing Creation** dialog box; a new file is started in the **Drafting** workbench.

   Next, you need to change the sheet size and the projection standard to generate the drawing views of the model in third angle projection standard.

3. Choose **File > Page Setup** from the menubar to display the **Page Setup** dialog box.

4. Select **A2 ISO** from the **Sheet Style** drop-down list and choose the **OK** button from the **Page Setup** dialog box.

5. Use the **Fit All In** tool to fit the display area of the sheet into the screen.

6. Right-click on **Sheet.1** in the Specification tree and choose the **Properties** option from the contextual menu; the **Properties** dialog box is displayed.

7. Select the **Third angle standard** radio button and choose the **OK** button from the **Properties** dialog box.

Generating the Front View

After setting all parameters, you need to generate the front view of the Bench Vice assembly.

1. Choose the **Advanced Front View** tool from the **Projections** sub-toolbar in the **Views** toolbar; the **View Parameters** dialog box is displayed.

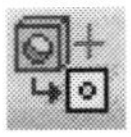

2. In the **View Parameters** dialog box, enter the name of the view as **Front view of Bench Vice** in the **View name** edit box.

3. Choose the **OK** button from the **View Parameters** dialog box; you are prompted to select a reference plane on a 3D geometry.

4. Choose **Window > Tile Vertically** from the menubar; multiple windows are displayed on the screen.

5. Select the front planar face of the Base from the **Product** workbench; the preview of the front view is displayed on the drawing sheet. Also, the knob is displayed on the top right of the drawing sheet.

6. Maximize the drawing window and fit the view of the sheet into the screen using the **Fit All In** tool.

7. Drag the view close to the lower-left corner of the drawing sheet, refer to Figure 13-61.

8. Click on the drawing sheet to generate the front view.

9. Invoke the **Properties** dialog box for the front view and then select the **Center Line** and **Axis** check boxes from it. To turn off the display of the view frame, clear the **Display View Frame** check box, and then choose the **OK** button from the **Properties** dialog box. The drawing sheet after generating the front view is shown in Figure 13-61.

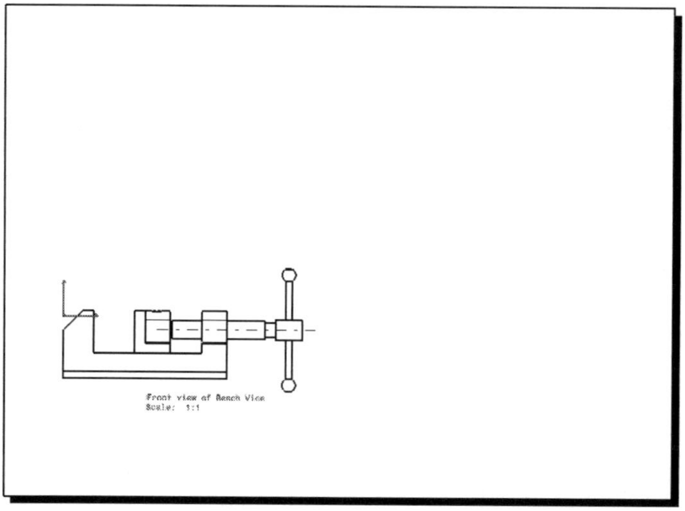

Figure 13-61 *The drawing sheet after generating the front view*

Generating the Top View

Next, you need to generate the top view of the Bench Vice assembly.

1. Choose the **Projection View** tool from the **Projections** sub-toolbar in the **Views** toolbar and then place the cursor on the front view; a preview of the projected view is attached to the cursor.

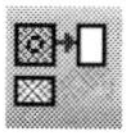

2. Move the cursor close to the top-left corner of the drawing sheet and then click; the top view is generated, refer to Figure 13-62.

3. Turn off the display of the view frame.

   The drawing sheet after generating the top view is shown in Figure 13-62.

Generating the Right View

After generating the front and top views, you need to generate the right view of the Bench Vice assembly.

1. Choose the **Projection View** tool from the **Views** toolbar and place the cursor on the right of the front view; a preview of the right view is attached to the cursor.

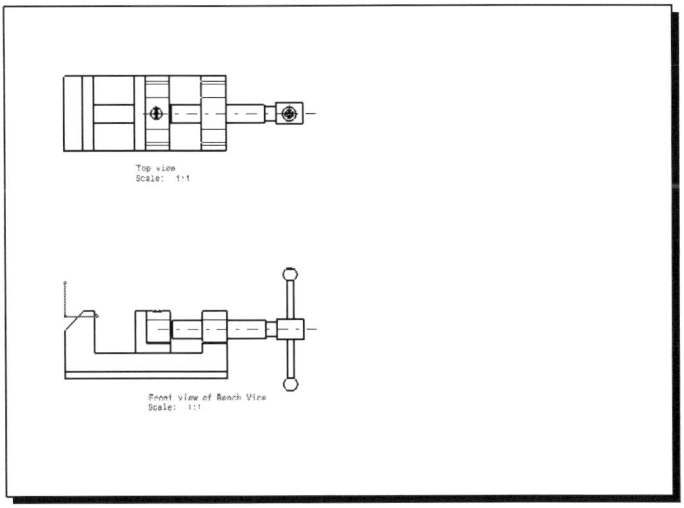

Figure 13-62 *The drawing sheet after generating the top view*

2. Move the cursor further on the right of the front view and click; the right view is generated, refer to Figure 13-63.
3. Turn off the display of the view frame.

The drawing sheet after generating the right view is shown in Figure 13-63.

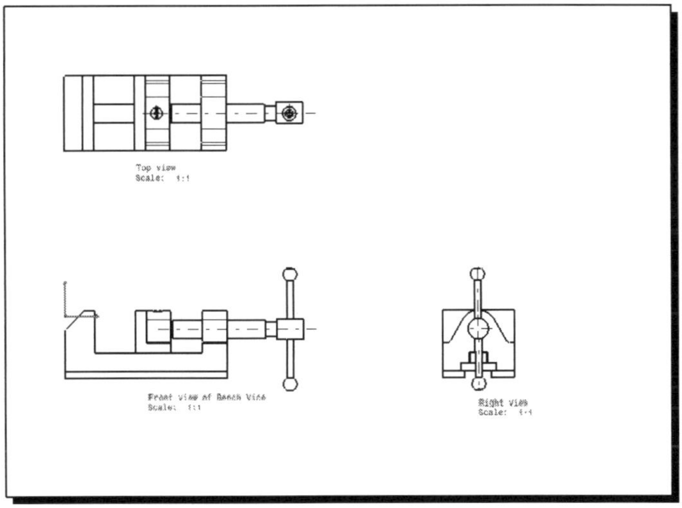

Figure 13-63 *The drawing sheet after generating the right view*

Generating the Isometric View

The last view that you need to generate is the isometric view. This view can be generated by using the **Isometric View** tool.

1. Choose the **Isometric View** tool from the **Projections** sub-toolbar in the **Views** toolbar; you are prompted to select a reference plane from the 3D geometry.

2. Choose **Window > Tile Vertically** from the menubar.

3. Select any of the faces of the Bench Vice assembly from the geometry view; a preview of the isometric view with the knob is displayed on the drawing sheet.

4. Maximize the drawing file window. Drag the preview of the isometric view close to the upper right corner of the drawing sheet and then click; the isometric view is generated.

5. Turn off the display of the view frame.

   The final drawing sheet after generating all drawing views is shown in Figure 13-64.

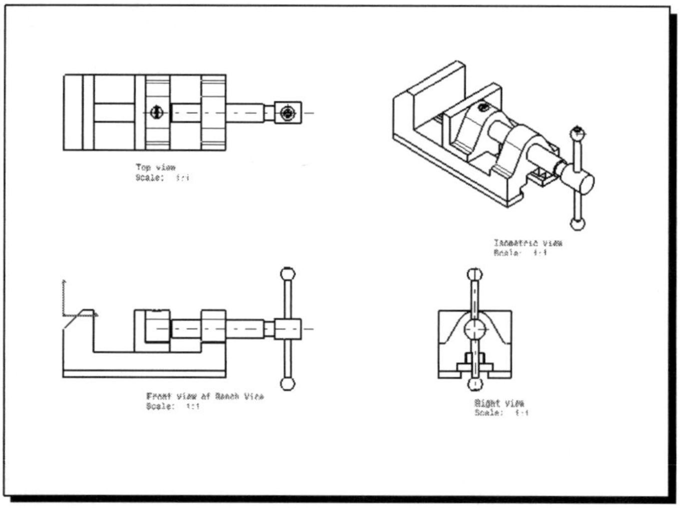

Figure 13-64 *Final drawing sheet after generating all drawing views*

Saving and Closing the File

1. Choose the **Save** button from the **Standard** toolbar to invoke the **Save As** dialog box.

2. Enter the name of the file as **c13tut2.CATDrawing** in the **File name** edit box and save the file at *C:\CATIA\c13\Bench Vice*.

3. Close all files by choosing **File > Close** from the menubar.

Self-Evaluation Test

Answer the following questions and then compare them to those given at the end of this chapter:

1. The __________ tool is used to display a component by removing a portion of it while keeping the ends of the view intact.

2. The _________ tool is used to generate a view by projecting lines normal to the specified edge of an existing view.

3. The _________ tool is used to generate the magnified portion of the view defined by a polygonal profile.

4. The _________ tool is used to generate a full section view.

5. Choose the **New button** from the **Standard** toolbar to start a new file in the **Drafting** workbench by using the **New** tool. (T/F)

6. The **All Views** option in the **New Drawing Creation** dialog box is used to automatically generate only the front, right, and left views. (T/F)

7. The **Breakout View** tool is used to remove material from an existing view and then display the area of the model or the assembly behind the removed portion. (T/F)

8. The **Clipping View** tool is used to crop an existing view by using a polygonal profile. (T/F)

9. If a section line does not cut through a model, the resulting section view will be a partial section view. (T/F)

10. In the **Drafting** workbench, you can generate the drawing views automatically using the tools in the **Wizard** toolbar. (T/F)

Review Questions

Answer the following questions:

1. Which of the following tools is used to generate an aligned section view when only the sectioned surface is displayed?

   (a) **Aligned Section Cut** (b) **Offset Section Cut**
   (c) **Aligned Section View** (d) None of these

2. Which of these dialog boxes is used to edit the hatch pattern of a section view?

   (a) **Hatch Pattern** (b) **Hatch**
   (c) **Properties** (d) None of these

3. Which of the following dialog boxes is displayed if you start a new file in the **Drawing** workbench using the **Start** menu?

   (a) **Drawing** (b) **New Drawing Creation**
   (c) **New Drawing** (d) None of these

4. Which of the following tools is used to generate a projected view?

   (a) **Project View** (b) **Projection View**
   (c) **Isometric View** (d) None of these

5. Which of the following options in the **New Drawing Creation** dialog box is used to generate the front, top, and left views?

   (a) **All Views** (b) **Front, Bottom and Right**
   (c) **Front, Top and Left** (d) None of these

6. You can invoke the **Profile Edition** environment for editing the section line by double-clicking on the profile. (T/F)

7. In the **Drafting** workbench of CATIA V5, you cannot change the scale of drawing views. (T/F)

8. You can delete unwanted views from a drawing sheet. (T/F)

9. You can modify the hatch spacing by changing the value in the **Pitch** spinner. (T/F)

10. The **Activate View** option is used to activate an inactive view. (T/F)

EXERCISE

Exercise 1

Create the components of the V-Block assembly as shown in Figure 13-65, and generate the front, top, right, detail, and isometric views, as shown in Figure 13-66. The views and dimensions of the components of the V-Block are shown in Figures 13-67 and 13-68.

(Expected time: 1 hr)

Figure 13-65 *Components of the V-Block assembly*

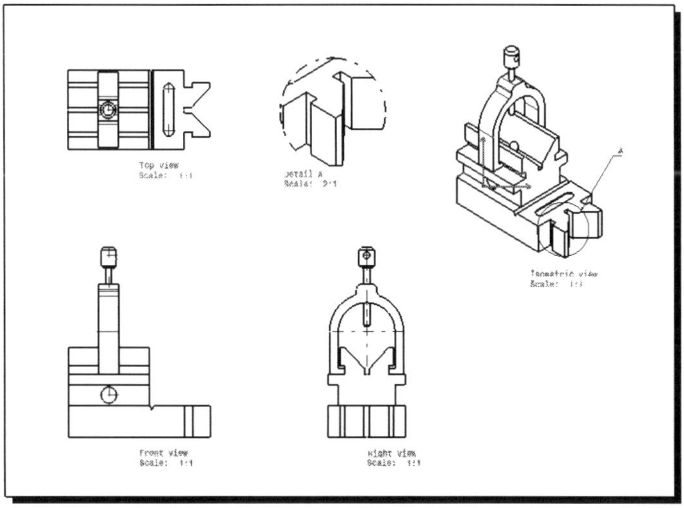

Figure 13-66 *Drawing views of the V-Block assembly*

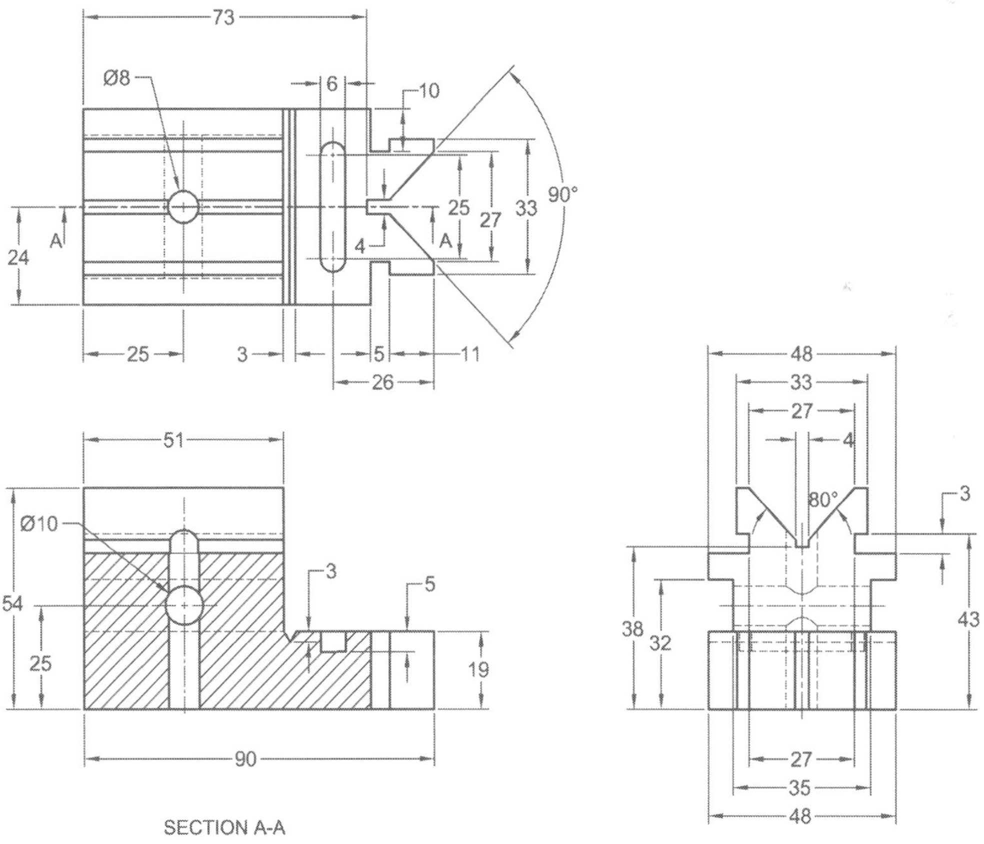

Figure 13-67 *Views and dimensions of the V-Block body*

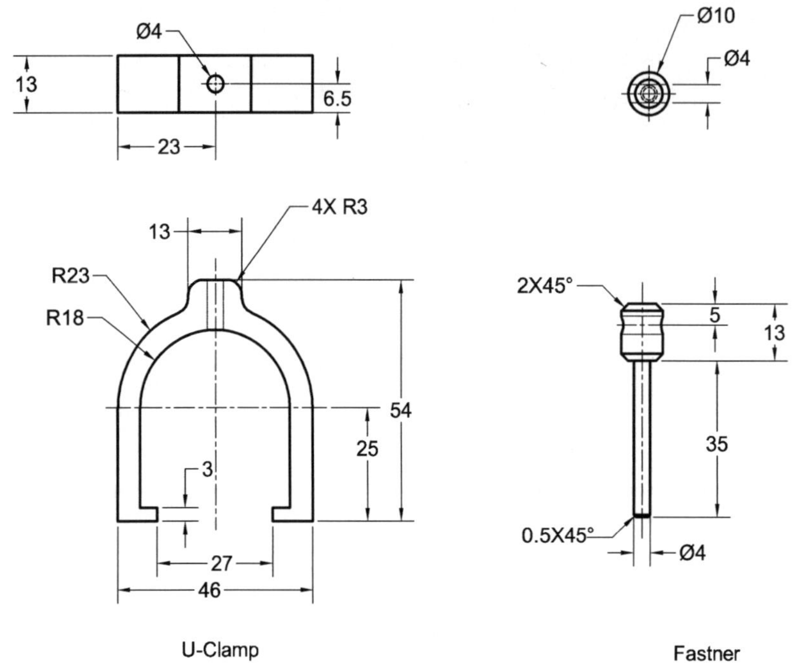

Figure 13-68 *Views and dimensions of the U-Clamp and Fastener*

Answers to Self-Evaluation Test

1. **Broken View**, **2.** **Auxiliary View**, **3.** **Detail View Profile**, **4.** **Offset Section View**, **5.** T, **6.** F, **7.** T, **8.** F, **9.** T, **10.** T

Chapter 14

Working with the Drafting Workbench-II

Learning Objectives

After completing this chapter, you will be able to:

- *Insert additional sheets in the current drawing file*
- *Insert frames and title blocks*
- *Add annotations to the drawing views*
- *Edit annotations*
- *Generate Bill of Material (BOM)*
- *Generate balloons*

INSERTING SHEETS IN THE CURRENT FILE

Menubar: Insert > Drawing > Sheets > New Sheet
Toolbar: Drawing > Sheets sub-toolbar > New Sheet

You can insert additional sheets to the current drafting file by using the **New Sheet** tool. This is a good practice when you need to generate the drawing views of all the components of an assembly and also its other views such as the isometric view or a view with the Bill of Material (BOM) and balloons in a single drawing file. You will learn more about BOM and balloons later in this chapter.

It is easier to manage all the drawings at the same time because a multi sheet drawing file will act as a single storage space for the drawings of all the components of that assembly.

To insert a new drawing sheet, choose the **New Sheet** tool from the **Sheets** sub-toolbar in the **Drawing** toolbar, as shown in Figure 14-1; a new sheet is added to the current drawing file, as shown in Figure 14-2.

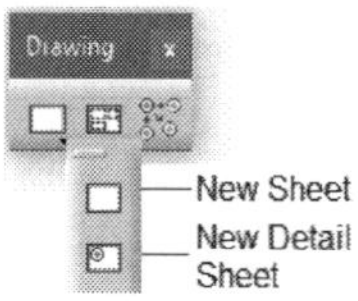

*Figure 14-1 Tools in the **Sheets** sub-toolbar*

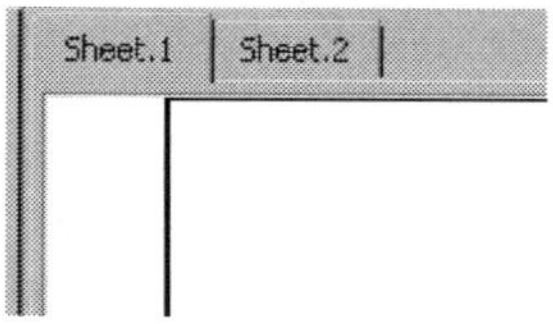

Figure 14-2 Partial view of the drawing file with a new sheet added to it

The newly inserted sheet becomes the active sheet. To activate the previous sheet, choose the **Sheet.1** tab from the top of the drawing sheet or double-click on **Sheet.1** in the Specification tree. You can also choose the **Activate Sheet** option from the contextual menu to activate the previous sheet.

You can also reorder the sequence of the sheets. To do so, right-click on the sheet that you want to reorder in the Specification tree and choose **Sheet.x object > Reorder** from the contextual menu; the cursor will change into a down-arrow when you move it on the name of the sheet in the Specification tree. Next, click on the sheet below which you want to place the selected sheet in the Specification tree; the sheet will get reordered. To move a selected sheet up in the Specification tree, press and hold the CTRL key once the cursor changes into a down arrow; the cursor will change into an up arrow. Next, click on the sheet above which you want to place the selected sheet in the Specification tree and then release the CTRL key.

You can rename a sheet from the Specification tree. To do so, select it from the Specification tree and then right-click; a contextual menu will be displayed. Next, select the **Properties** option from it to display the **Properties** dialog box. Change the name of the sheet in the **Name** edit box and then choose **OK** to apply the changes and exit the dialog box.

To delete a drawing sheet, select the drawing sheet to be removed from the Specification tree and invoke the contextual menu. Choose the **Delete** option from the contextual menu; the **Confirm Delete** message box will be displayed. Choose the **OK** button from the message box; the selected sheet will be deleted.

INSERTING THE FRAME AND THE TITLE BLOCK

To insert the frame and the title block in the drawing sheet, you first need to set the drawing sheet to the edit background mode. To set this mode, choose **Edit > Sheet Background** from the menu bar; the background editing mode will be invoked and the color of the sheet will automatically be changed to gray. There are two methods of inserting the frame and the title block. Both the methods are discussed next.

Automatic Insertion of the Frame and the Title Block

Menubar: Insert > Drawing > Frame and Title Block
Toolbar: Drawing > Frame and Title Block

This method is used to automatically insert the frame and the title block from those available in the system by default. To insert frame and title block, choose the **Frame and Title Block** tool from the **Drawing** toolbar; the **Manage Frame And Title Block** dialog box will be displayed, as shown in Figure 14-3.

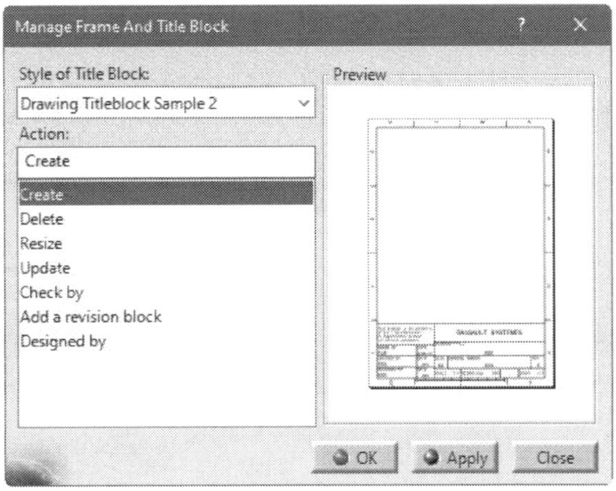

*Figure 14-3 The **Manage Frame And Title Block** dialog box*

The options available in this dialog box are discussed next.

Style of Title Block

The **Style of Title Block** drop-down list is used to set the style of the frame and the title block. By default, the previous style of title block is selected in this drop-down list. The other four styles in this drop-down list are **Drawing Titleblock PlyBook**, **Drawing Titleblock Sample1**, **Drawing Titleblock Sample2**, and **Drawing Titleblock Sample Enovia 1**. You can select the type of style from this drop-down list according to your need.

Action

The options in the **Action** list box are used to define the type of action that you want to perform. The options in this list box are discussed next.

Create

The **Create** option in the **Action** list box is selected by default. This option is used to create the frame and the title block on the current drawing sheet. The preview of the frame and the title block to be created is shown in the preview area of the **Manage Frame And Title Block** dialog box. Figure 14-4 shows an A2 ISO sheet with the frame and the title block inserted using this option.

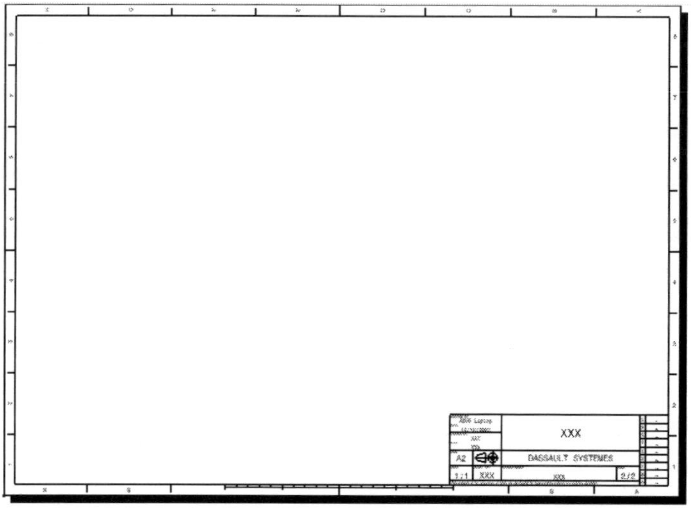

***Figure 14-4** The drawing sheet after adding the frame and the title block to the drawing sheet*

Delete

The **Delete** option in the **Action** list box is used to delete the existing frame and the title block.

Resize

The **Resize** option is used to resize the frame and the title block, if you have modified the size of the sheet after placing them. After modifying the sheet size, the frame and the title block will not adjust to the sheet boundaries automatically. Therefore, to resize them, choose **Resize** from the **Action** list box and then choose the **OK** button from the **Manage Frame And Title Block** dialog box. On doing so, the frame and the title block will adjust to the new sheet size.

Update

The **Update** option in the **Action** list box is used to update the frame and the title block. This is done, in case you have modified some of the parameters of the sheet, such as the projection standard or sheet scale, or you have inserted new sheets. After performing the modifications, you need to update the frame and the title block using this option.

Check by

The **Check by** option is used to specify the name of the person who has checked the drawing sheet. To specify the name, choose this option and then choose the **OK** button from the **Manage Frame And Title Block** dialog box; the **Controller's name** dialog box will be displayed. Specify the name of the controller in the edit box and choose the **OK** button. The name specified in this dialog box will be displayed in the **CHECKED BY** section of the title block in the drawing sheet.

Add a revision block

The **Add a revision block** option is used to add revision blocks of the drawing to the drawing sheets. To add revision blocks, choose this option and then choose the **OK** button from the **Manage Frame And Title Block** dialog box; the **Reviewer's name** dialog box will be displayed. Specify the name of the reviewer and choose the **OK** button; the **Description** dialog box will be displayed. Specify the description in the edit box available in this dialog box and choose the **OK** button. The information specified in these dialog boxes will be displayed in the revision block. Similarly, you can add other revision blocks to form a complete revision table.

After inserting the frame and the title block, you can specify the required parameters in the columns of the title block. To do so, double-click on the default parameter of the title block; the **Text Editor** dialog box will be displayed. In text box of this dialog box, enter the text that you want to display in the drawing, and then choose the **OK** button from the **Text Editor** dialog box.

Next, you need to exit the background editing mode. To do so, choose **Edit > Working Views** from the menu bar.

Creating the Frame and the Title Block Manually

You can also create frames and title blocks manually. To do so, you need to invoke the background editing mode first. To invoke this mode, choose **Edit > Sheet Background** from the menu bar. Now, using the sketching tool in this mode, draw the sketch of the frame and title block. A drawing sheet after drawing sketch for the frame and the title block using the sketching tools, is shown in Figure 14-5.

Adding Text in the Title Block

Menubar:	Insert > Annotations > Text > Text
Toolbar:	Annotations > Text sub-toolbar > Text

After drawing the frame and title block, you can add text to the title block by invoking the background editing mode. To do so, choose the **Text** tool from the **Text** sub-toolbar in the **Annotations** toolbar; you will be prompted to indicate the text anchor point. Select a point on the drawing sheet to place the text; the **Text Editor** dialog box will be displayed, as shown in Figure 14-6.

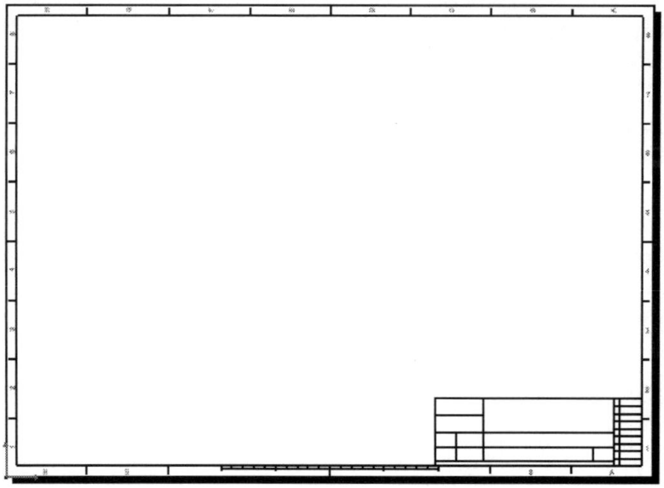

Figure 14-5 The drawing sheet after drawing the frame and the title block

*Figure 14-6 The **Text Editor** dialog box*

Enter the text in the text box in the **Text Editor** dialog box and choose the **OK** button; the text will be displayed at the selected point in the drawing sheet. To relocate the text, select and drag it to a new location.

To modify the font, style, text height, and so on, select the text and choose the **Properties** option from the contextual menu; the **Properties** dialog box will be displayed. Using the options available in this dialog box, you can modify the font, style, text height, and other properties of the text. After setting the parameters, choose the **OK** button from the **Properties** dialog box.

Inserting the Object

You can insert a company logo in the title block. To do so, invoke the background editing mode and then choose **Insert > Object** from the menu bar. On doing so, the **Insert Object** dialog box will be displayed. Select the **Create from File** radio button to insert a graphic created earlier. Choose the **Browse** button from the **Insert Object** dialog box; the **Browse** dialog box will be displayed. Browse to the graphic file and double-click on it; the complete path name of the selected file will be displayed in the **File** edit box of the **Insert Object** dialog box. Choose the **OK** button from the **Insert Object** dialog box; the graphic will be placed aligned to the lower left corner of the drawing sheet. You can move it by dragging. You can also resize it by holding and dragging one of its corners. Note that the aspect ratio of the image will change on resizing.

Inserting the Picture

You can also insert an image in the title block. To do so, invoke the background editing mode and then choose **Insert > Picture** from the menu bar. On doing so, the **File selection** dialog box will be displayed. Next, browse to the image and double-click on it. Now, click at the location on the sheet where you want to place the image; the image will be placed at the selected location. You can move it by dragging. You can also resize it by holding it from one of the corners and dragging. For example you can also insert a log as a picture in the title block.

After adding the text and logo, choose **Edit > Working Views** from the menu bar to exit the sheet editing mode. On doing so, the color of the sheet will automatically change from back to white. The drawing sheet after completing the frame and the title block is displayed in Figure 14-7.

Figure 14-7 *The drawing sheet after completing the frame and the title block*

ADDING ANNOTATIONS TO THE DRAWING VIEWS

After generating the drawing views, you need to generate the dimensions in the drawing views and add other annotations such as notes, surface finish symbols, geometric tolerance, and so on. Two types of annotations can be generated in the drawing views. The first type is the generative annotation in which you can generate the dimensions that were added while creating the part in the **Part Design** workbench. The second type of annotations are those that are added manually to the geometry of drawing views, such as reference dimensions, notes, surface finish symbols, and so on. Both these types of annotations and the procedures to add them are discussed next.

Generating Dimensions

The dimensions applied to the sketches and features of the part can be generated and displayed in the drawing views using the tools discussed next.

Generating All Dimensions Together

Menubar: Insert > Generation > Generate Dimensions
Toolbar: Generation > Dimension Generation sub-toolbar > Generate Dimensions

The **Generate Dimensions** tool is used to generate all the dimensions of the model in its drawing views. To generate dimensions, first invoke the **Options** dialog box and then select the **Drafting** sub-node of the **Mechanical Design** node on the left pane of this dialog box. Now, choose the **Generation** tab and then select the **Filters before generation** check box, if it is not chosen by default. Choose **OK** to apply changes and exit the **Options** dialog box. Next, choose the **Generate Dimensions** tool from the **Dimension Generation** sub-toolbar in the **Generation** toolbar; the **Dimension Generation Filters** dialog box will be displayed, as shown in Figure 14-8. The options in this dialog box are discussed next.

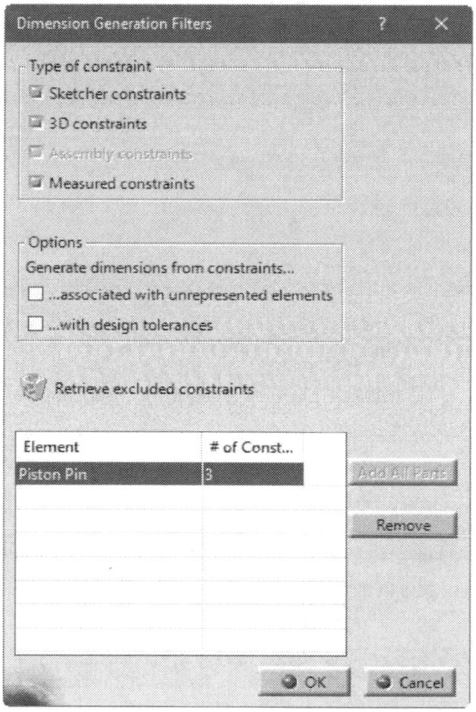

Figure 14-8 The ***Dimension Generation Filters*** *dialog box*

Type of constraint Area

The options in the **Type of constraint** area are used to define the type of constraint that you need to generate. By default, all the check boxes are selected in this area. While generating the dimensions of the assembly drawing views, the **Assembly constraints** check box will also get activated. Clear the check box of the type of constraint that you do not want to generate.

Options Area

The **associated with unrepresented elements** check box provided in the **Options** area is also used to generate those dimensions whose references are not displayed in the drawing view. These dimensions include the dimensions that are referenced to a plane, axis, reference sketch, and so on.

The **with design tolerances** check box is selected to generate the dimensions with design tolerances. Tolerance is also applied to the generated dimensions.

Retrieve excluded constraints

After deleting some of the generated dimensions if you need to restore them, invoke the **Dimensions Generation Filters** dialog box and choose the **Retrieve excluded constraints** button. Now, choose the **OK** button from the **Dimensions Generation Filters** dialog box. The dimensions, that were deleted earlier, will be restored in their respective views.

Add All Parts

The **Add All Parts** button is activated only while generating dimensions of the drawing views of an assembly. By default, only the name of the assembly is displayed in the **Element** column provided on the left of this button. If you choose this button, then the name of all parts of the assembly are displayed in this column. Also, the dimensions of all components will be generated in the assembly drawing views. To remove a part from the selection set of the **Element** column, select it and choose the **Remove** button provided below the **Add All Parts** button.

After setting all the parameters, choose the **OK** button from the **Dimension Generation Filters** dialog box. If the **Generated Dimension Analysis** dialog box is displayed, choose the **OK** button from it; all dimensions of the model will be generated.

> **Tip**
> *By default, dimensions are generated in all views. To generate dimensions in a particular view, select the view and choose the **Generate Dimensions** tool.*

Generating Dimensions Step by Step

Menubar:	Insert > Generation > Generate Dimensions Step by Step
Toolbar:	Generation > Dimension Generation sub-toolbar > Generate Dimensions Step by Step

To generate dimensions step by step, choose the **Generate Dimensions Step by Step** tool from the **Dimension Generation** sub-toolbar in the **Generation** toolbar. If the **Dimension Generation Filters** dialog box is displayed, set the options in this dialog box and choose the **OK** button to exit it; the **Step by Step Generation** dialog box will be displayed, as shown in Figure 14-9.

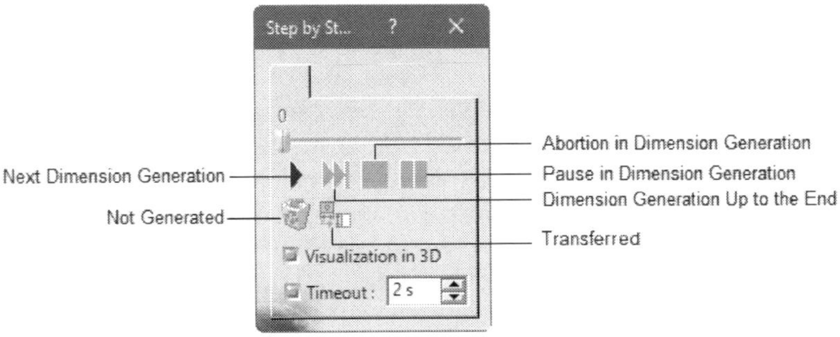

*Figure 14-9 The **Step by Step Generation** dialog box*

Choose the **Next Dimension Generation** button from the **Step by Step Generation** dialog box. It will start generating dimensions one after the other. To generate all the dimensions at once, choose the **Dimension Generation Up to the End** button from this dialog box. If you need to abort the dimension generation, choose the **Abortion in Dimension Generation** button.

If you do not need to generate a particular dimension then choose the **Pause in Dimension Generation** button immediately after it is displayed in the drawing sheet. Now, choose the **Not Generated** button from the **Step by Step Generation** dialog box. Choose the **Next Dimension Generation** button to resume generation of the remaining dimensions.

To switch the dimensions from one view to another, pause the dimension generation process immediately after the dimension to be transferred is displayed on the drawing sheet. Now, choose the **Transferred** button and select the view on which you need to place the transferred dimension. After transferring it, continue generating the dimensions by choosing the **Next Dimension Generation** button.

The **Visualization in 3D** check box, which is selected by default, is used to display the dimensions in the solid model as they are generated in the drawing views. You can visualize the dimensions in the solid model also, if you have tiled the windows of the part and drawing.

The **Timeout** check box is used to specify the time gap between the generation of the consecutive dimensions after the **Next Dimension Generation** button is chosen. The time gap can be specified using the spinner on the right of this check box. If you clear this check box, only a single dimension will be generated on choosing the **Next Dimension Generation** button. Choose this button again to generate the next dimension.

Tip

You can also modify the dimensions of a solid model by modifying the dimensions generated in the ***Drafting*** *workbench. To do so, invoke the* ***Options*** *dialog box by choosing* ***Tools > Options*** *from the menu bar. Next, select* ***Mechanical Design > Drafting*** *node from the left pane of the* ***Options*** *dialog box, if it is not already selected. Now, choose the* ***Administration*** *tab from the right pane of this dialog box. Next, clear the* ***Prevent dimensions from driving 3D constraint*** *check box, if it is selected. Choose* ***OK*** *to apply the changes and exit the* ***Options*** *dialog box. Now, if you modify the generated dimensions in the* ***Drafting*** *workbench and update the solid model, the modified dimensions will be applied to the solid model.*

Adding Reference Dimensions

Menubar:	Insert > Dimensioning > Dimensions > Dimensions
Toolbar:	Dimensioning > Dimensions sub-toolbar > Dimensions

In the **Drafting** workbench of CATIA V5, you can use the **Dimensions** tool to create reference dimensions. To do so, invoke the **Options** dialog box by choosing **Tools > Options** from the menu bar. Select **Drafting**, if not already selected, from the **Mechanical Design** option on the left pane of the **Options** dialog box. Now, choose the **Dimension** tab from the right pane of this dialog box and then select the **Dimension following the cursor (CTRL toggles)** check box if not already selected. Choose the **OK** button to exit. Now, to create reference dimensions, choose the **Dimensions** tool from the **Dimensions** sub-toolbar in the **Dimensioning** toolbar; the **Tools Palette** toolbar will be displayed, as shown in Figure 14-10.

Specify the type of dimension that you need to add using this **Tools Palette** toolbar. Now, select the geometrical element or elements to which you need to add the dimension; the dimension will be attached to the cursor. Move the cursor and click on the drawing sheet to place the dimension.

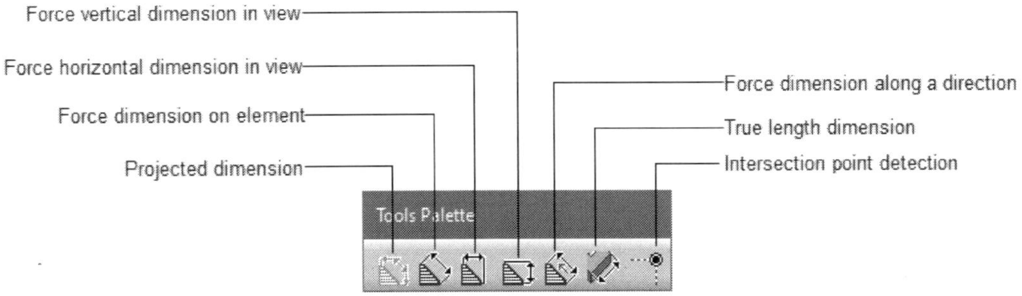

*Figure 14-10 The **Tools Palette** toolbar*

The procedure to create dimensions using the **Tools Palette** toolbar is same as the one discussed for dimensioning the entities in the **Sketcher** workbench. The remaining dimensioning tools and the procedure to use them are discussed next.

Adding Chamfer Dimensions

Menubar:	Insert > Dimensioning > Dimensions > Chamfer Dimensions
Toolbar:	Dimensioning > Dimensions sub-toolbar > Chamfer Dimensions

To create a chamfer dimension, choose the **Chamfer Dimensions** tool from the **Dimensions** sub-toolbar in the **Dimensioning** toolbar; the **Tools Palette** toolbar will be displayed, as shown in Figure 14-11.

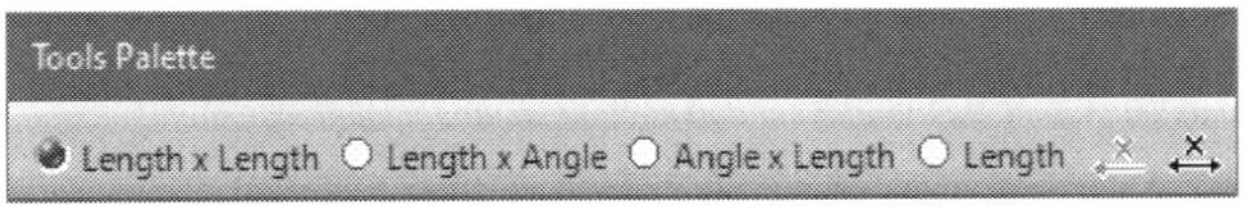

*Figure 14-11 The **Tools Palette** toolbar*

The options in this **Tools Palette** toolbar are used to define the form of the chamfer dimension. The **Length x Length** radio button is used to create the chamfer dimension using the chamfer lengths in the X and Y directions. The **Length x Angle** radio button is used to create the chamfer dimension by defining the length and angle of the chamfer. The **Angle x Length** radio button is used to create the chamfer by defining the angle first and then the length. The **Length** radio button is used to create the chamfer by defining the length of one side of the chamfer. After selecting the form of chamfer from the **Tools Palette** toolbar, move the cursor to the chamfered edge. You will notice that **2 1 3** and **3 1 2** digits swap their positions as you move the cursor to the upper or lower portion of the chamfer. When you move the cursor close to the upper portion of the chamfered edge, **2 1 3** will be displayed. This implies that the chamfered edge will be selected first and then the upper edge will be selected as the dimensional reference. The chamfer dimension will be created with reference to the upper edge.

If you move the cursor close to the lower portion of the chamfered edge, then **3 1 2** will be displayed. This implies that the chamfered edge will be selected first and then the edge close to the digit **2** will be selected as the dimensional reference. The resulting chamfered dimension will be created with reference to that edge. Figures 14-12 through 14-15 show selection sequences and resulting chamfer dimensions.

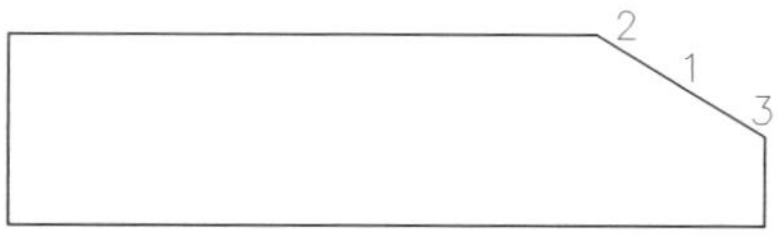

***Figure 14-12** Selection sequence*

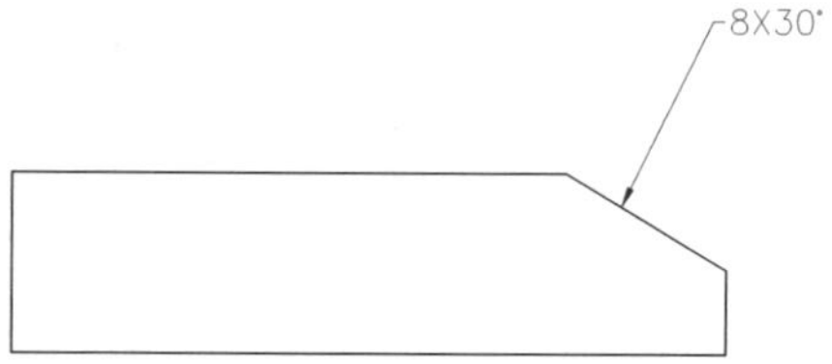

***Figure 14-13** Resulting chamfer dimension*

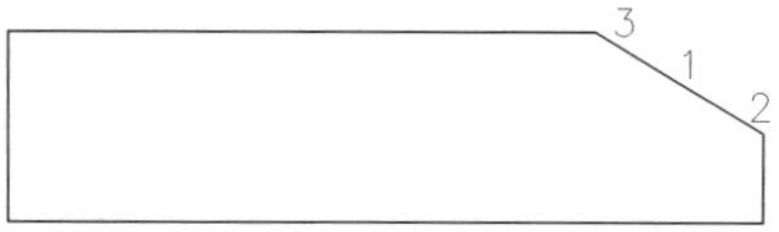

***Figure 14-14** Selection sequence*

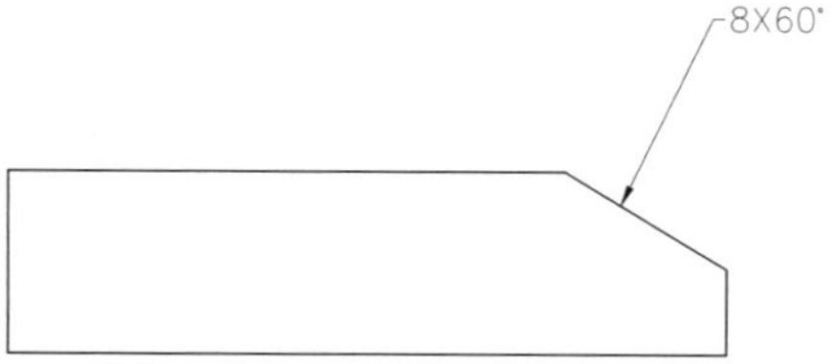

***Figure 14-15** Resulting chamfer dimension*

Select the chamfered edge and move the cursor to an appropriate location. Click at a point on the drawing sheet to place the chamfer dimension.

Adding Datum Features

Menubar:	Insert > Dimensioning > Tolerancing > Datum Feature
Toolbar:	Dimensioning > Tolerancing sub-toolbar > Datum Feature

You can add the datum feature symbol to the drawing views using the **Datum Feature** tool. The datum feature symbols are used as the datum references while adding geometric tolerances to the drawing view. To add the datum feature symbol, choose the **Datum Feature** tool from the **Tolerancing** sub-toolbar in the **Dimensioning** toolbar; you will be prompted to select an element or click the leader anchor point. Select the reference element on which you need to attach the datum feature; its preview will be attached to the cursor. Move the cursor to the desired location and click on the drawing sheet to place the datum feature; the **Datum Feature Creation** dialog box will be displayed, as shown in Figure 14-16. Choose the **OK** button from this dialog box; the dialog box will be closed and the datum feature will be displayed attached to the selected edge and placed at the specified location.

***Figure 14-16** The **Datum Feature Creation** dialog box*

Adding Geometric Tolerance to the Drawing Views

Menubar:	Insert > Dimensioning > Tolerancing > Geometrical Tolerance
Toolbar:	Dimensioning > Tolerancing sub-toolbar > Geometrical Tolerance

In shop floor drawings, you need to provide various other parameters along with the dimensions and the dimensional tolerance. These parameters can be a geometric condition, material condition, and so on. All these type of parameters are defined using the geometric tolerance. To add the geometric tolerance to the drawing views, choose the **Geometrical Tolerance** tool from the **Tolerancing** sub-toolbar in the **Dimensioning** toolbar. Select the element to add the geometric tolerance; the preview of the geometric tolerance will get attached to the cursor. Move the cursor to an appropriate location and click on the drawing sheet to place the tolerance on that location; the **Geometrical Tolerance** dialog box will be displayed, as shown in Figure 14-17.

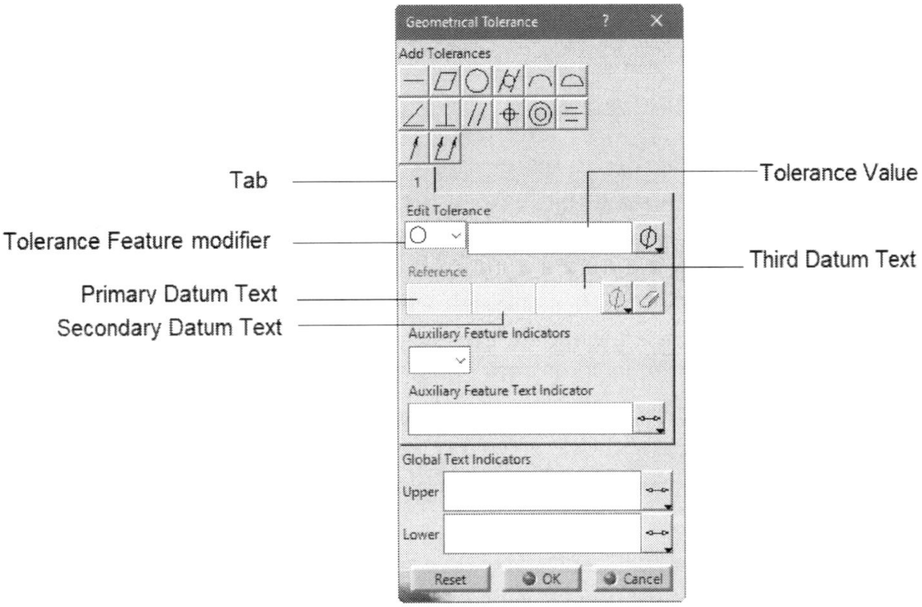

***Figure 14-17** The **Geometrical Tolerance** dialog box*

Add Tolerances

The options in the **Add Tolerances** area are used to specify the geometrical symbol for the tolerance and the value of tolerance. To add a tolerance value to the current tolerance feature, select the required tolerance indicator from the **Add Tolerances** area. You can add as many tolerances as required. This tolerance indicator gets added below the previous one in the drawing area.

Tab

Each line of the geometric tolerance on the drawing is represented by a tab in this dialog box. The options available under a tab are discussed below.

Edit Tolerance

The options in the **Edit Tolerance** area are used to edit the tolerance value. You can replace the selected tolerance indicator with another one from the Tolerance Feature modifier drop-down list available in the **Edit Tolerance** area. You can also select **Remove Line** from the Tolerance Feature modifier drop-down list to remove the selected line. This option is unavailable if the selected tolerance line is the only one in the list. Specify the tolerance value in the Tolerance Value edit box available in the **Edit Tolerance** area of this dialog box and select the required engineering symbol from the drop-down next to the Tolerance Value edit box.

Reference

The options in the **Reference** area of the **Geometrical Tolerance** dialog box are used to define reference for applying tolerance. After specifying the geometric condition and the value of tolerance, define the primary reference for applying the tolerance in the **Primary Datum Text** edit box. Once you specify the primary reference, the **Secondary Datum Text** edit box will be invoked. Specify the second reference in this edit box. After specifying the secondary reference, the **Third Datum Reference** edit box will be invoked. This edit box is used to define the third reference.

Auxiliary Feature Indicators

Select a local indicator from the **Auxiliary Feature Indicators** drop-down list. Enter indication text in the **Auxiliary Feature Text Indicator** edit box. You can also select engineering symbols from the drop-down available next to this edit box.

Global Text Indicators

The options in the **Global Text Indicators** area are used to enter text above and below the geometrical tolerance value. In this area, the **Upper** Text and **Lower** Text edit boxes are used to enter text above and below the geometrical tolerance value, respectively. The Insert Symbol flyout is used to insert symbols.

Select the **Reset** button to set the parameters of the **Geometrical Tolerance** dialog box to default. After setting all the parameters, choose the **OK** button. Figure 14-18 shows a drawing after adding datum features and tolerances.

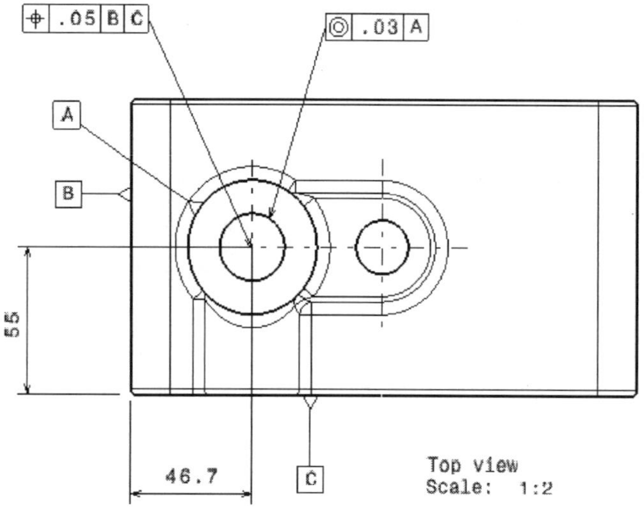

Figure 14-18 *The drawing view after adding the datum features and tolerances*

Adding Surface Finish Symbols

Menubar:	Insert > Annotations > Symbols > Roughness Symbol
Toolbar:	Annotations > Symbols sub-toolbar > Roughness Symbol

The **Roughness Symbol** tool is used to add roughness symbol to the drawing views. To do so, choose this tool from the **Symbols** sub-toolbar in the **Annotations** toolbar, as shown in Figure 14-19.

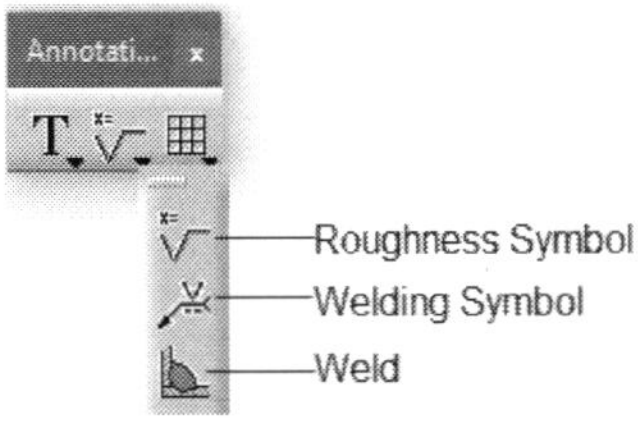

Figure 14-19 *Tools in the* ***Symbols*** *sub-toolbar*

On choosing the **Roughness Symbol** tool, you will be prompted to click the roughness anchor. Click on the surface from where you want to place the roughness symbol; a preview of the symbol will be displayed attached to the selected point and also the **Roughness Symbol** dialog box will be displayed, as shown in Figure 14-20.

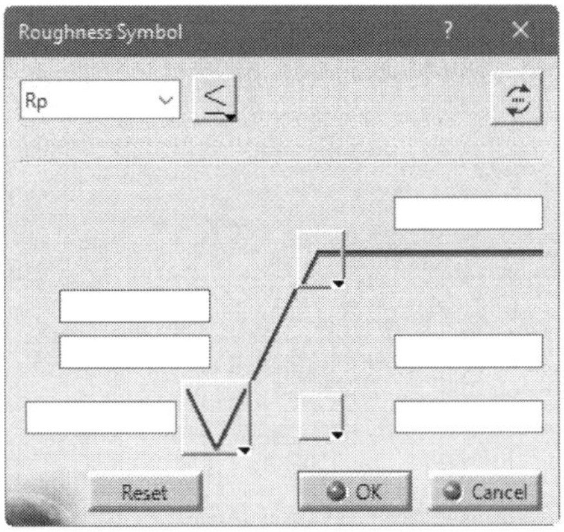

Figure 14-20 *The* ***Roughness Symbol*** *dialog box*

Select the prefix of the roughness symbol from the **Prefix** drop-down list in this dialog box. Specify the equality symbol using the **Inequality symbol** flyout on the right of the **Prefix** drop-down list. You can choose the type of rugosity and surface texture using the **Rugosity type** and **Surface texture/All surfaces around** flyouts. You can also set the value of the surface roughness by using the edit boxes provided in the **Roughness Symbol** dialog box. You can also invert the surface roughness symbol using the **Invert** button. After setting the required parameters, choose the **OK** button.

Adding Welding Symbols

Enhanced

Menubar: Insert > Annotations > Symbols > Welding Symbol
Toolbar: Annotations > Symbols sub-toolbar > Welding Symbol

To add the welding symbol, choose the **Welding Symbol** tool from the **Symbols** sub-toolbar in the **Annotation** toolbar; you will be prompted to select the first element or indicate the leader anchor point. Select the first element on which you need to add the welding symbol from the drawing sheet; you will be prompted to select the second element. Select the second element from the drawing sheet; a preview of the welding symbol will be attached to the cursor. Click on some appropriate location on the drawing sheet; the **Welding Symbol** dialog box will be displayed, as shown in Figure 14-21.

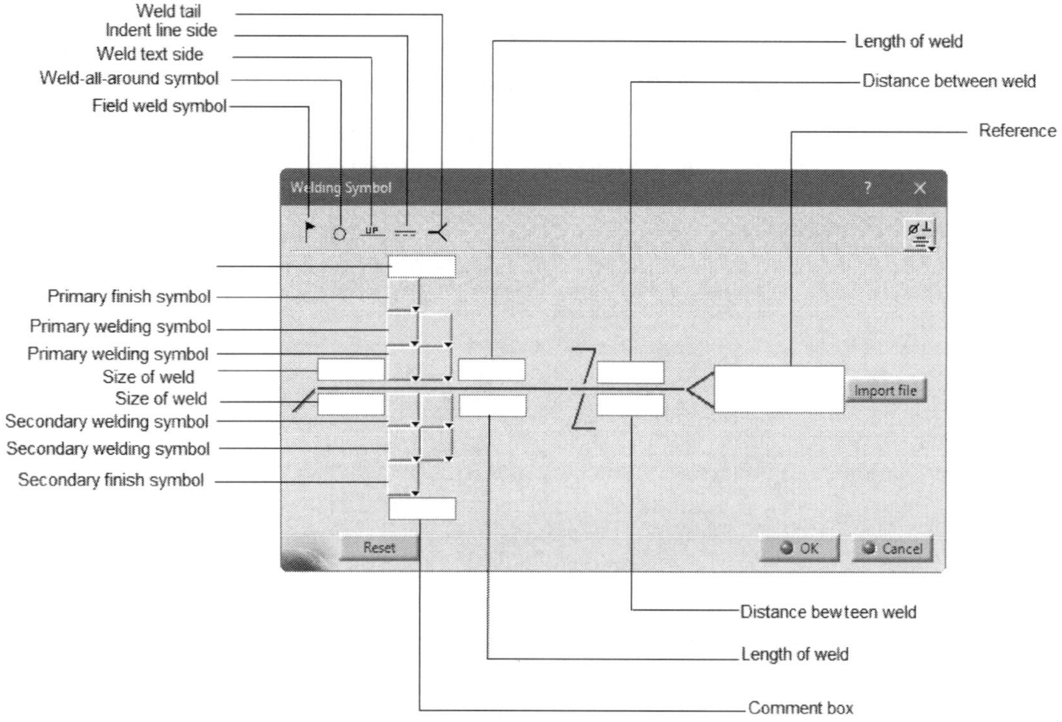

***Figure 14-21** The **Welding Symbol** dialog box*

The buttons on the upper row of the **Welding Symbol** dialog box are used to define the conditions of welding such as field weld symbol, weld all-around, side of the weld text, side of indent line, and the welding symbol tail. The comment boxes in this dialog box are used to add comments about the weld.

Set the size of the weld in the **Size of weld** edit box. You can also define the parameters for the other side of the weld if you weld the component on both sides by setting the value of the size in the lower **Size of weld** edit box. All the parameters regarding the size, length, and other properties on one side can also be defined for the other side.

The **Primary finish symbol** button is used to define the type of finish for the weld. To do so, choose this button; a flyout will be displayed. Select the type of primary finish symbol for the weld.

The **Primary welding symbol** button is used to define the type of weld such as the square-groove weld, v-groove weld, fillet weld, and so on. To define the type of weld, choose this button; a flyout will be displayed. Select the type of weld from this flyout.

The **Primary complementary symbol** button is used to define the profile of the weld such as straight, convex, concave, and so on. To define the profile of the weld, choose this button; a flyout will be displayed. Select the type of weld profile symbol from this flyout.

The **Secondary finish symbol** button is used to define the type of finish for the weld. To do so, choose this button; a flyout will be displayed. Select the type of primary finish symbol for the weld.

The **Secondary welding symbol** button is used to define the type of weld such as the square-groove weld, v-groove weld, fillet weld, and so on. To define the type of weld, choose this button; a flyout will be displayed. Select the type of weld from this flyout.

The **Secondary complementary symbol** button is used to define the profile of the weld such as straight, convex, concave, and so on. To define the profile of the weld, choose this button; a flyout will be displayed. Select the type of weld profile symbol from this flyout.

The **Length of weld** edit box is used to specify the length of the weld. You can also define the reference of the weld in the **Reference** edit box. The **Import file** button on the left of the **Reference** edit box is used to browse the text file to insert the reference.

The Symbols button at the right corner of the dialog box is used to specify the geometrical symbols. To add a geometrical symbol, choose this button; a flyout will be displayed. Select the required symbols from the flyout.

After setting all the parameters, choose the **OK** button from the **Welding Symbol** dialog box. Figure 14-22 shows a welding symbol attached to a drawing view.

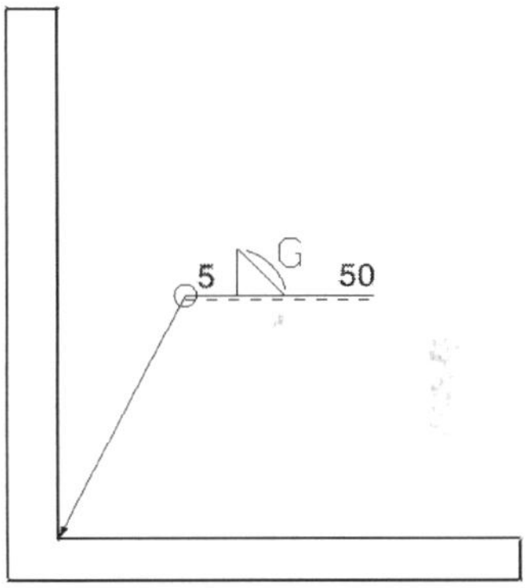

***Figure 14-22** The drawing view after adding the welding symbol*

Applying Weld

Menubar:	Insert > Annotations > Symbols > Weld
Toolbar:	Annotations > Symbols sub-toolbar > Weld

To apply a weld, choose the **Weld** tool from the **Symbols** sub-toolbar in the **Annotations** toolbar; you will be prompted to select the first edge. Select the first edge on which you need to apply the weld; you will be prompted to select the second edge. Select the second edge from the drawing sheet; the **Welding Editor** dialog box will be displayed, as shown in Figure 14-23.

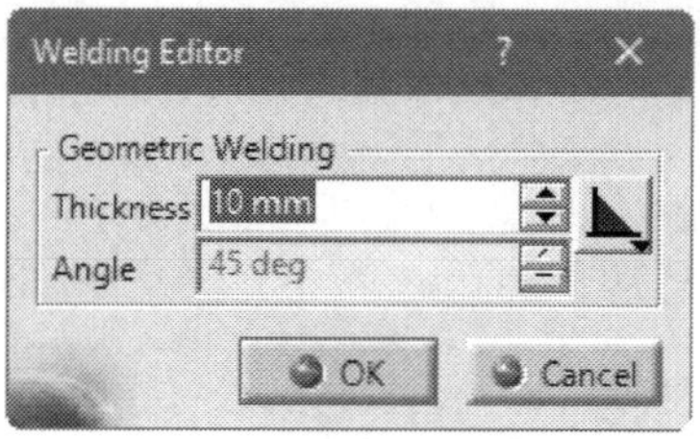

*Figure 14-23 The **Welding Editor** dialog box*

A preview of the weld is also displayed on the drawing sheet. Depending on the entities selected, the system will choose the most appropriate type of weld. You can also select the weld type to be applied from the flyout that is displayed when you choose the **Change Type** button on the right of the **Thickness** spinner. Set the value of thickness using the **Thickness** spinner. If required, set the value of the angle using the **Angle** spinner.

After setting all the parameters, choose the **OK** button from the **Welding Editor** dialog box. Figure 14-24 shows a drawing view after applying the weld.

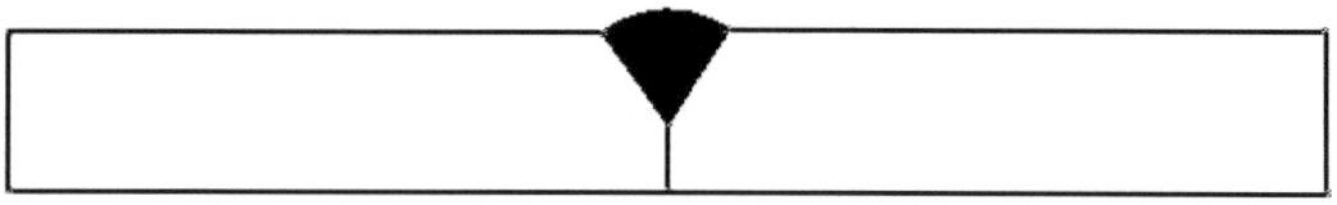

Figure 14-24 The drawing view after applying the welding symbol

EDITING ANNOTATIONS

You can edit the annotations added to the drawing views by double-clicking on them. On doing so their respective dialog boxes will be displayed. You can edit the parameters of the annotations using these dialog boxes.

GENERATING THE BILL OF MATERIAL (BOM)

After generating the drawing views of an assembly, it is very necessary to generate the Bill of Material (BOM). The BOM is a table that provides you information related to the number of components in an assembly, their name, quantity, and so on. You can add the BOM to the drawing sheet to display the part list of the components used in the assembly. The BOM placed on the drawing sheet is parametric in nature. Therefore, if you add or delete a part from the assembly, the change will be reflected in the BOM on the drawing sheet. Before generating the BOM, you need to apply numbers to the components of the assembly. This will help in providing the serial number to the component, which will, in turn, help to number the components while generating balloons. To number the components, switch to the assembly file from which the drawing views are generated. Choose the **Generate Numbering** tool from the **Product Structure Tools** toolbar. Now, select **Product** from the Specification tree; the **Generate Numbering** dialog box will be displayed, as shown in Figure 14-25.

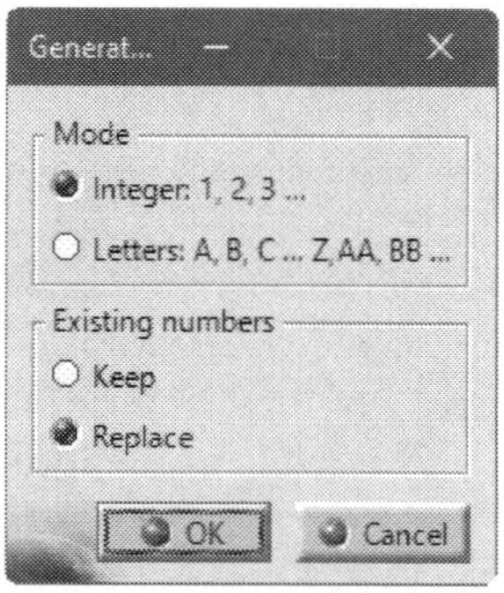

*Figure 14-25 The **Generate Numbering** dialog box*

The **Integer** radio button is selected by default in the **Mode** area of the **Generate Numbering** dialog box. To generate the numbering in alphabets, select the **Letters** radio button.

The options in the **Existing numbers** area are used to specify whether you need to keep or replace the existing numbers. Choose the **OK** button from the **Generate Numbering** dialog box to generate the numbers and exit the dialog box.

After generating the numbers, you need to edit the format of the BOM in the assembly environment. This is because, by default, the number column is not available in the BOM format. To edit the BOM format, choose **Analyze > Bill of Material** from the menu bar; the **Bill of Material : Product1** dialog box will be displayed. Choose the **Define formats** button from this dialog box; the **Bill of Material : Define formats** dialog box will be displayed. Select the required property from the **Hidden properties** list box and choose the **Show properties** button; the selected property will be displayed at the bottom in the **Displayed properties** list box. You can reposition this property to the top in the **Displayed properties** list box. To do so, select the required property from the **Displayed properties** list box and choose the **Change order** button. Next, select a property available on the top in the **Displayed properties** list box. As soon as you select a property, the property to be repositioned will move on the top of the selected property. Similarly, if you want to hide any unwanted property displayed in the **Displayed properties** list box, select the property to be hidden from the **Displayed properties** list box and then choose the **Hide properties** button.

After setting the properties, choose the **OK** button from the **Bill of Material : Define formats** dialog box and then choose the **OK** button from the **Bill of Material : Product1** dialog box. Save the assembly file and switch back to the drawing window.

To generate a BOM, choose **Insert > Generation > Bill of Material > Bill of Material** from the menu bar; you will be prompted to click at a location to insert the Bill of Material. Click on the drawing sheet; the BOM and the recapitulation list will be placed at the selected point. Figure 14-26 shows the drawing sheet after generating the BOM.

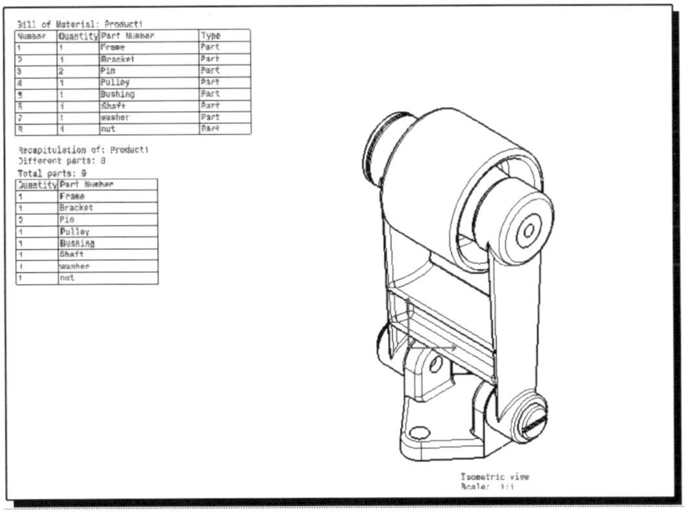

Figure 14-26 *The drawing sheet after generating the BOM*

GENERATING BALLOONS

Menubar: Insert > Generation > Balloon generation
Toolbar: Generation > Dimension Generation sub-toolbar > Generate Balloons

The **Generate Balloons** tool is used to generate balloons that are attached to the drawing view of an assembly. The naming of the balloons depends on the sequence of the parts in the BOM. To generate balloons, you need to make sure that numbering of the components in the assembly is already done. Numbering the components of the assembly has been discussed earlier while generating the BOM. Now, choose the **Generate Balloons** tool from the **Dimension Generation** sub-toolbar in the **Generation** toolbar. Balloons are automatically attached to the components of the assembly in the active drawing view. By default, they are placed arbitrarily on the drawing sheet. You need to move them manually and place them at an appropriate location. Figure 14-27 shows a drawing sheet after generating balloons.

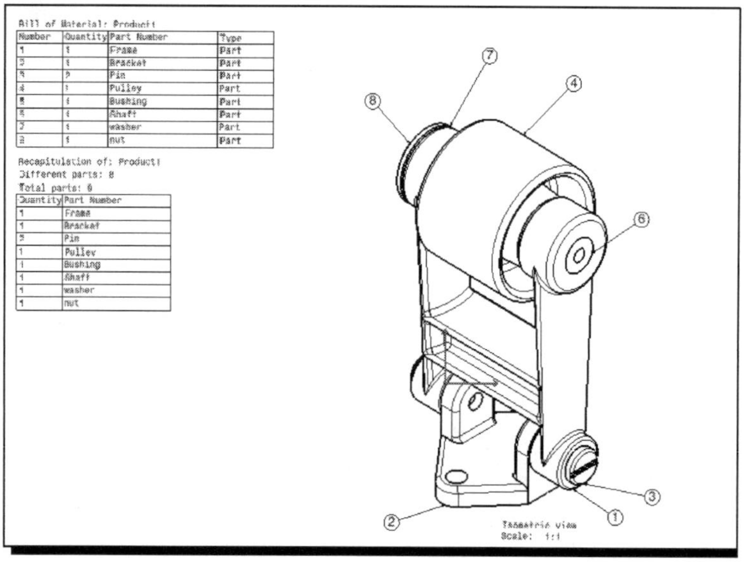

Figure 14-27 *The drawing sheet after generating balloons*

TUTORIALS

Tutorial 1

In this tutorial, you will create the model shown in Figure 14-28. After creating it, you need to generate the front, top, and isometric views in the **Drafting** workbench. You also need to generate the dimensions in the drawing views. The views and dimensions of the model are shown in Figure 14-29.
(Expected time: 45 min)

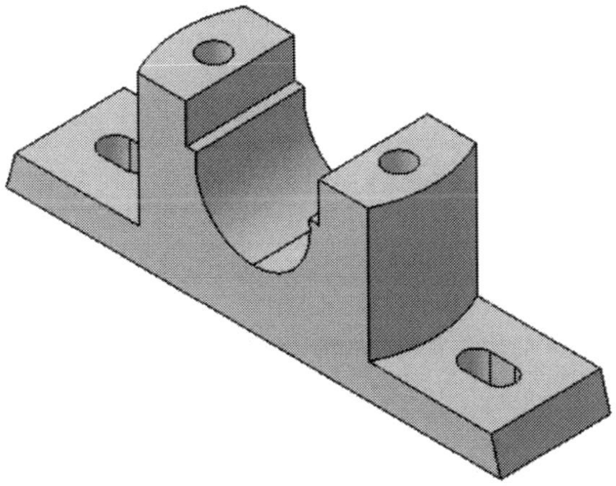

Figure 14-28 *Model for Tutorial 1*

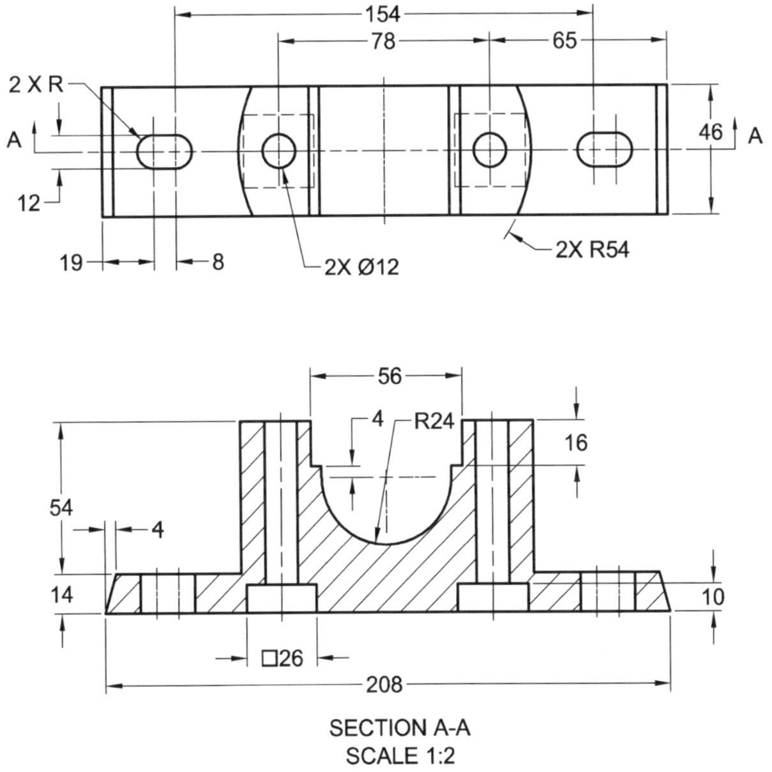

Figure 14-29 *Views and dimensions for Tutorial 1*

The following steps are required to complete this tutorial:

a. Create the model in the **Part** workbench and save the file in the *c14* folder.
b. Start a new file in the **Drafting** workbench with a standard A2 sheet.
c. Set the projection standard to the third angle.

d. Create a standard title block and frame in the background editing mode.
e. Generate the front view.
f. Generate the top view.
g. Generate the isometric view.
h. Generate the dimensions.
i. Arrange the dimensions and delete the unwanted ones.

Creating the Model

In this section, you need to create the model in the **Part** workbench.

1. Start a new file in the **Part** workbench.

2. Create the model shown in Figure 14-28. Refer to Figure 14-29 for dimensions.

3. Create a folder with the name *c14* at *C:/CATIA* and save the model in it with the name *c14tut1*.

Starting a New File in the Drafting Workbench

After creating and saving the model, you need to start a new file in the **Drafting** workbench.

1. Choose **File > New** from the menu bar to display the **New** dialog box. Select **Drawing** from the **New** dialog box and choose the **OK** button; the **New Drawing** dialog box is displayed.

2. Select the **A2 ISO** option from the **Sheet Style** drop-down list. Choose the **OK** button from the **New Drawing** dialog box; a new file gets started in the **Drafting** workbench.

3. Set the projection mode to the third angle projection using the **Properties** dialog box.

Creating the Title Block and the Frame

After starting a new file in the **Drafting** workbench, you need to create the title block and the frame by invoking the background editing mode.

1. Choose **Edit > Sheet Background** from the menu bar to invoke the background editing mode. On doing so, the color of the sheet changes to gray.

2. Choose the **Frame and Title Block** tool from the **Drawing** toolbar; the **Manage Frame And Title Block** dialog box is displayed.

3. In the dialog box, select the **Drawing Titleblock Sample 2** option from the **Style of Title Block** drop-down list.

   The **Create** option is selected by default in the **Action** list box. You need to use this option to create the feature and the title block.

4. Choose the **OK** button from the **Manage Frame And Title Block** dialog box.

5. Double-click on the **DASSAULT SYSTEMS** in the Title Block; the **Text Editor** dialog box is displayed. Enter **CADCIM Technologies** in the dialog box. Choose the **OK** button; **DASSAULT SYSTEM** is renamed as CADCIM Technologies.

6. Similarly edit the other text in the title block. The drawing sheet after creating the title block and the frame is shown in Figure 14-30.

   After creating the title block, you need to exit the background editing mode.

7. Choose **Edit > Working Views** from the menu bar; you return to the working views mode.

Generating the Front View

After creating the title block and frame, you need to generate the front view.

1. Choose the **Front View** tool from the **Projections** sub-toolbar in the **Views** toolbar; you are prompted to select a reference plane on a 3D geometry. 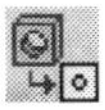

2. Choose **Window > c14tut1.CATPart** from the menu bar; the part file is displayed.

3. Select the yz plane (of the front planar face) from the Specification tree; the drawing file is invoked.

   A preview of the front view is displayed and a knob appears on the view to set the orientation of the front view.

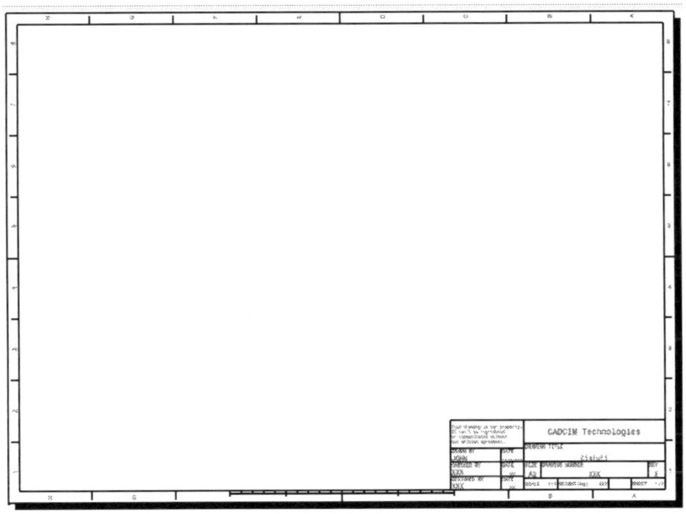

Figure 14-30 The drawing sheet after creating the title block and the frame

4. Move the cursor on the frame that shows the preview of the front view; the cursor is replaced by a hand cursor. Press and hold the left mouse button and drag the cursor close to the lower left corner of the drawing sheet.

5. Click anywhere on the drawing sheet to generate the front view. The drawing sheet after generating the front view is shown in Figure 14-31.

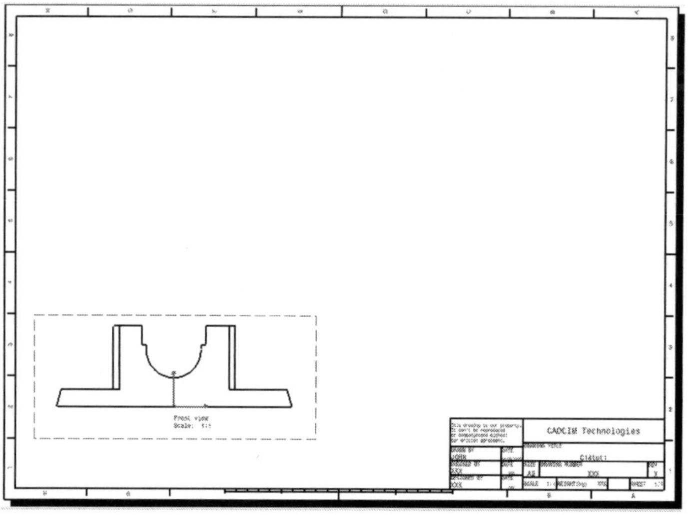

***Figure 14-31** The drawing sheet after generating the front view*

Generating the Top View

After generating the front view, you need to generate the top view by using the **Projection View** tool.

1. Choose the **Projection View** tool from the **Projections** sub-toolbar in the **Views** toolbar. 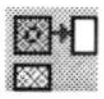

2. Move the cursor vertically upward in the front view; a preview of the projected view is displayed attached to the cursor.

3. Click on the drawing sheet to place the top view.

You can move the view labels by dragging them, if required. The drawing sheet after generating the top view is shown in Figure 14-32.

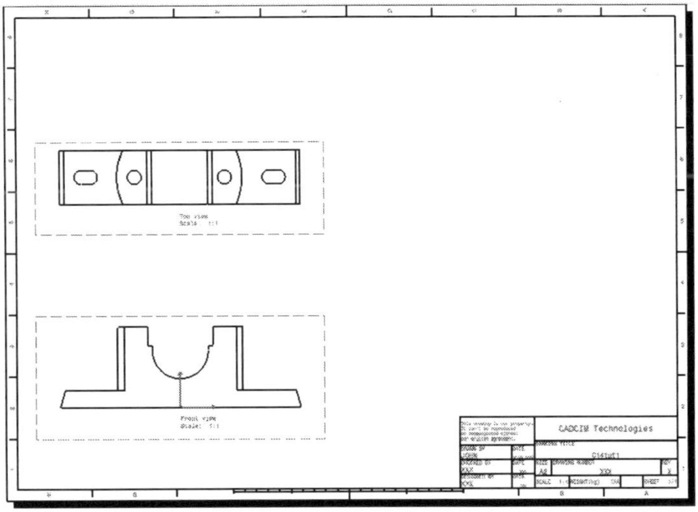

Figure 14-32 The drawing sheet after generating the top view

Generating the Isometric View

Next, you need to generate the isometric view.

1. Choose the **Isometric View** tool from the **Projections** sub-toolbar in the **Views** toolbar; you are prompted to select a reference plane on a 3D geometry. 

2. Choose **Window > c14tut1.CATPart** from the menu bar; the part file is displayed.

3. Select the front face of the model from the geometry area; the drawing sheet is displayed. Also, a preview of the isometric view of the model is displayed on the drawing sheet along with a knob which can be used to orient the isometric view.

   If the isometric view of the model is placed outside the drawing sheet, you may need to move it inside.

4. Drag the view to the desired location by holding its frame, refer to Figure 14-33.

5. Click anywhere on the drawing sheet to generate the isometric view. The drawing sheet after generating the isometric view is shown in Figure 14-33.

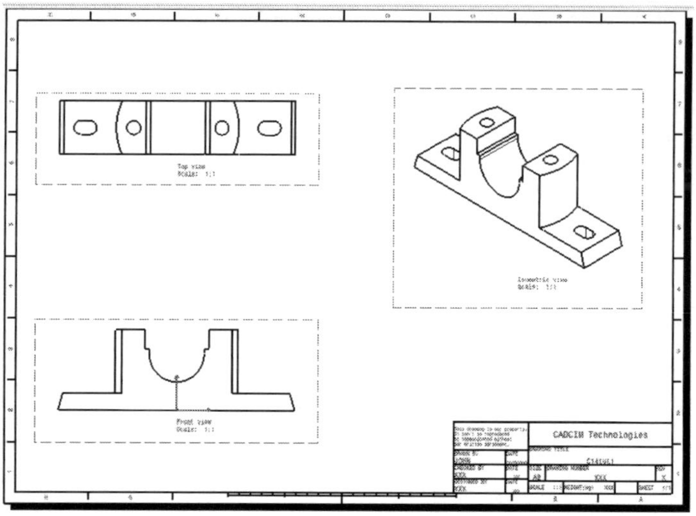

Figure 14-33 *The drawing sheet after generating the isometric view*

Generating Dimensions

After generating the drawing view, you need to generate the dimensions. Before generating them, you need to turn on the display of the hidden lines of the top and the front views to define some of the dimensions from the hidden lines. You also need to turn on the display of the center lines. After this, you need to turn off the display of the view frames.

1. Hold the CTRL key down and select **Front view** and **Top view** from the Specification tree. Right-click to invoke the contextual menu.

2. Choose the **Properties** option; the **Properties** dialog box is displayed.

3. Select the **Hidden Lines** and **Center Line** check boxes from the **Dress-up** area and choose the **OK** button from the **Properties** dialog box.

   Next, you need to turn off the display of the view frames.

4. Hold the CTRL key down and select **Front view**, **Top view**, and **Isometric view** from the Specification tree.

5. Invoke the contextual menu and choose the **Properties** option from it.

6. Clear the **Display View Frame** check box from the **Visualization and Behavior** area and then choose the **OK** button from the **Properties** dialog box. The drawing sheet after modifying the display is shown in Figure 14-34. Alternatively, you can turn off the display of the view frame by deactivating the **Display View Frame as Specified for Each View** button from the **Visualization** toolbar.

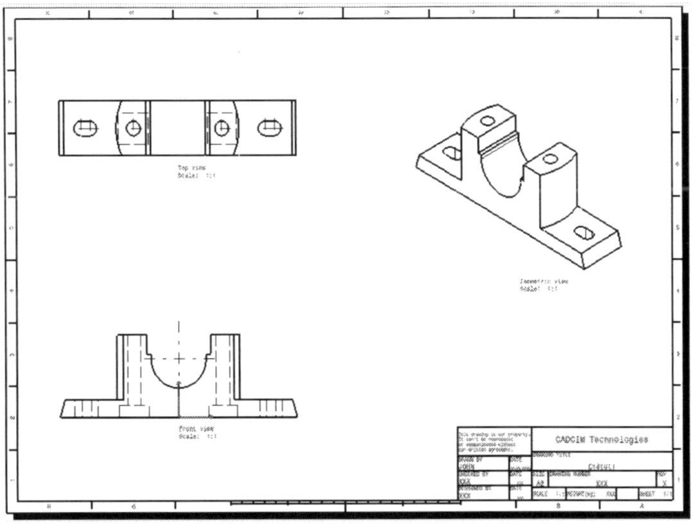

Figure 14-34 *Drawing sheet after modifying the display*

Now, you need to generate the dimensions.

Note

Before generating the dimensions, it is recommended to change some settings. To do so, choose ***Tools > Options*** *from the menu bar; the* ***Options*** *dialog box will be displayed. Select* ***Mechanical Design > Drafting*** *from the left pane of the dialog box. Now, choose the* ***Generation*** *tab and select the* ***Filters before generation*** *check box if not already selected. Choose* ***OK*** *to apply the changes and exit the dialog box.*

7. Choose the **Generate Dimensions** tool from the **Dimension Generation** sub-toolbar in the **Generation** toolbar; the **Dimension Generation Filters** dialog box is displayed. 

8. Select the ...**associated with unrepresented elements** check box from the **Options** area of the **Dimension Generation Filters** dialog box and choose the **OK** button; the **Generated Dimension Analysis** dialog box is displayed. Choose **OK** from it to generate dimension and exit the dialog box.

   On exiting the dialog box, all the dimensions are generated in the front and top views, as shown in Figure 14-35.

   After generating dimensions, you may need to delete the repeated dimensions. Also, the dimensions are placed randomly on the drawing sheet, and therefore, you need to arrange them properly.

9. Delete the repeated dimensions by selecting the dimensions and pressing the DELETE key

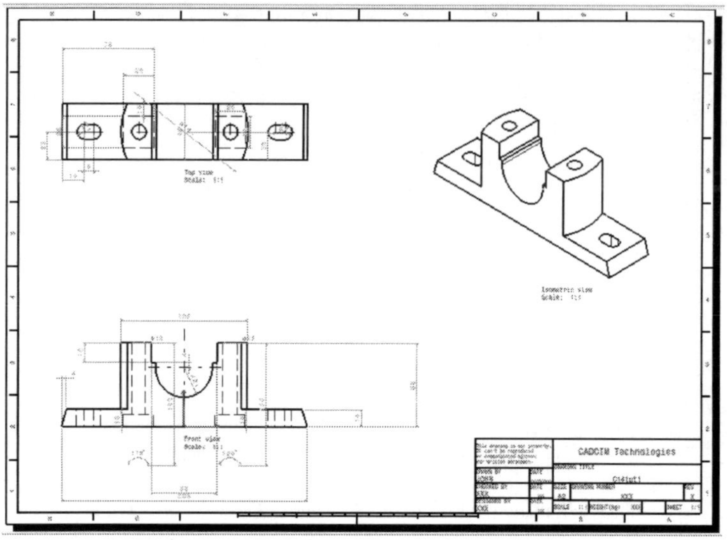

Figure 14-35 *The drawing sheet after generating dimensions*

10. Select the dimensions one by one and drag them to the desired location, refer to Figure 14-36.

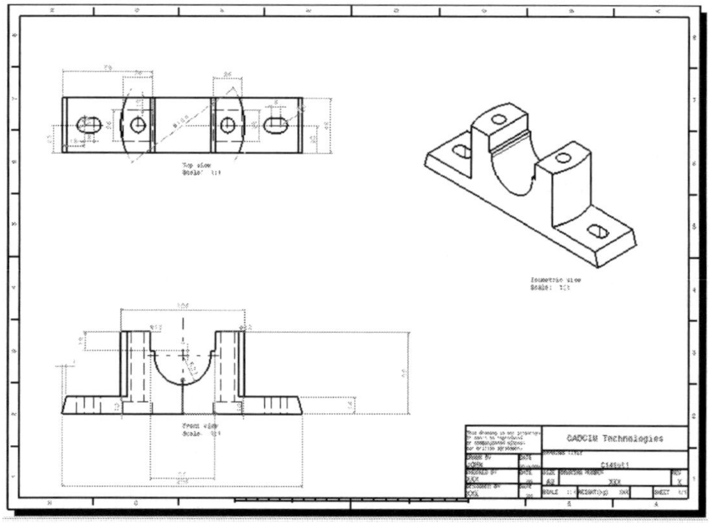

Figure 14-36 *Drawing sheet after deleting the repeated dimensions*

The final drawing sheet is shown in Figure 14-37.

Saving and Closing the File

1. Choose the **Save** button from the **Standard** toolbar to invoke the **Save As** dialog box.

2. Enter the name of the file as **c14tut1.CATDrawing** in the **File name** edit box and choose the **Save** button. The file will be saved at *C:\CATIA\c14*.

3. Close the part file by choosing **File > Close** from the menu bar.

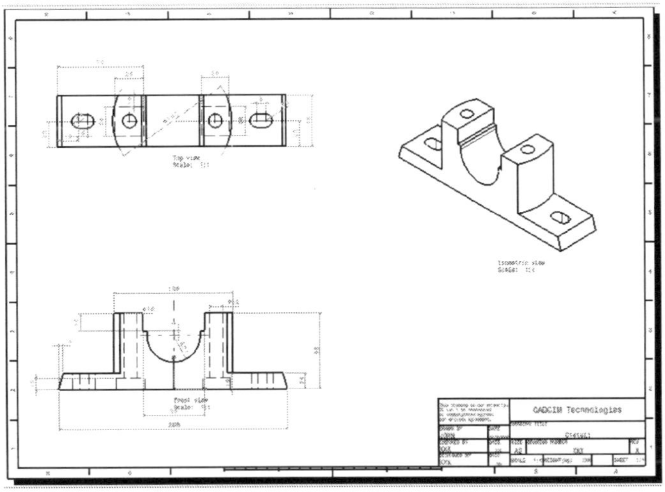

Figure 14-37 *Final drawing sheet*

Tutorial 2

In this tutorial, you will generate the front, top, right, and isometric views of the V-Block assembly. You can download V-Block assembly file from *www.cadcim.com*. The complete path for downloading the files is as follows: *Textbooks > CAD/CAM > CATIA > CATIA V5-6R 2022 for Designers > Input File*. Next, you also need to generate the BOM and balloons. **(Expected time: 30 min)**

The following steps are required to complete this tutorial:

a. Download and open the V-Block assembly file.
b. Start a new file in the **Drafting** workbench.
c. Generate front view.
d. Generate top view.
e. Generate right view.
f. Generate isometric view.
g. Generate BOM.
h. Generate balloons.

Downloading and Opening the Assembly Document

1. Download the *V-Block* assembly file CADCIM website.

2. Start CATIA V5 and open the assembly file that you have downloaded.

Starting a New File in the Drafting Workbench

1. Choose **File > New** from the menu bar to display the **New** dialog box. Select **Drawing** and then choose the **OK** button from the **New** dialog box; the **New Drawing** dialog box is displayed.

2. Select the **A2 ISO** option from the **Sheet Style** drop-down list and then choose the **OK** button from the **New Drawing** dialog box.

3. Set the projection standard to the third angle.

Creating the Title Block and the Frame

After starting a new file in the **Drafting** workbench, you need to create the title block and frame by invoking the background editing mode.

1. Choose **Edit > Sheet Background** from the menu bar to invoke the background editing mode. On doing so, the color of the sheet turns gray.

2. Choose the **Frame and Title Block** tool from the **Drawing** toolbar; the **Manage Frame And Title Block** dialog box is displayed.

3. Select the **Drawing Titleblock Sample 2** option from the **Style of Title Block** drop-down list and then choose the **OK** button from the dialog box.

   After creating the title block, you need to exit the background editing mode.

4. Choose **Edit > Working Views** from the menu bar.

Generating Drawing Views

Next, you need to generate front, top, right drawing views of the V-Block assembly.

1. Generate front, top, and right views of the V-Block assembly.

   Next, you need to generate an exploded view of the of the V-Block assembly. To generate the exploded view of an assembly, first you need to create a scene in the **DMU Navigator** workbench. This scene will be used for generating the exploded view. To create the scene, open the product file of the assembly that you need to explode.

2. Choose **Start > Digital Mockup > DMU Navigator** from the menubar; the **DMU Navigator** workbench will be invoked.

3. Choose the **Enhanced Scene** tool from the **DMU Review Creation** toolbar to display the **Enhanced Scene** dialog box.

4. Clear the **Automatic naming** check box if selected by default. Enter **Exploded View** as the name of the scene in the **Name** edit box.

5. Now, select the **Full** radio button from the **Overload Mode** area and choose the **OK** button from the **Enhanced Scene** dialog box. The scene environment is invoked and the background color turns dark green. Also, the **Warning** message window warns you that any move performed in the scene context may invalidate the analysis data results.

6. Choose the **Close** button from this message window.

7. Choose the **Explode** tool from the **Enhanced Scenes** toolbar; the **Explode** dialog box is displayed.

8. Select **All levels** from the **Depth** drop-down list and **3D** from the **Type** drop-down list. Choose the **OK** button from the dialog box.

9. Choose the **Exit Scene** tool from the **Enhanced Scenes** toolbar; the assembly switches to the assembled mode.

10. Next, start a new file in the **Drafting** workbench. Choose the **Isometric View** tool from the **Projections** toolbar and invoke the **Assembly** window.

11. Expand the **Applications** branch of the Specification tree and then expand the **Scenes** branch. Select the **Exploded View** from the **Scenes** branch.

12. Now, select any one of the faces or edges of the assembly; the drawing window is invoked and a preview of the exploded view is displayed. Click anywhere on the drawing sheet to generate the drawing view. The drawing sheet after generating the views is shown in Figure 14-38.

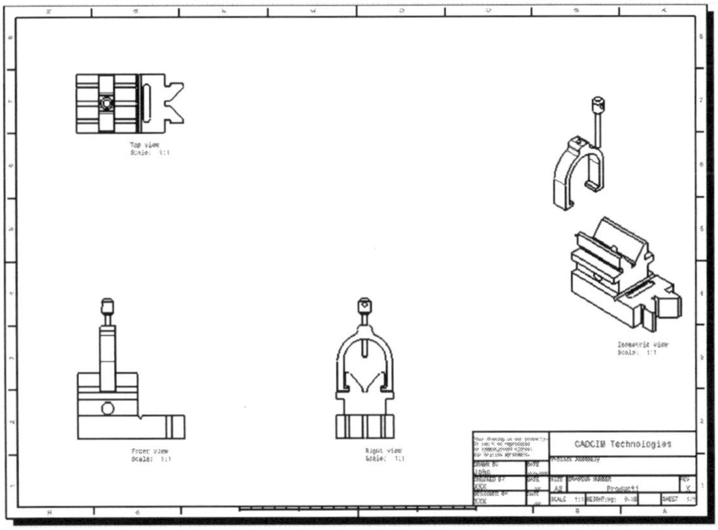

Figure 14-38 *The drawing sheet after generating front, top, right, and isometric views*

Generating BOM and Balloons

Before generating the BOM and balloons, you need to number the components of the assembly and then modify the BOM format.

1. Choose **Window > V-Block Assembly.CATProduct** from the menu bar; the assembly window is invoked.

   Numbering the components of the assembly will help you generate balloons and number the components in the BOM.

2. Choose the **Generate Numbering** tool from the **Product Structure Tools** toolbar; you are prompted to select a component to manage its representation.

3. Select **Product1** from the Specification tree; the **Generate Numbering** dialog box is displayed.

4. Choose the **OK** button from the **Generate Numbering** dialog box.

   After numbering the components of the assembly, you need to redefine the BOM format.

5. Choose **Analyze > Bill of Material** from the menu bar; the **Bill of Material : Product1** dialog box is displayed.

6. Choose the **Define formats** button from the **Bill of Material : Product1** dialog box; the **Bill of Material : Define formats** dialog box is invoked.

7. In the **Bill of Material : Define formats** dialog box, select **Number** from the **Hidden properties** list box and choose the **Show Properties** button on the left of the **Hidden properties** list box; the **Number** property is displayed in the **Displayed properties** selection area.

   You will notice that the **Number** property is displayed at the end in the **Displayed properties** list. You need to change the sequence of the properties by moving the **Number** property to the top.

8. Select the **Number** property from the **Displayed properties** list box and choose the **Change order** button located on the right of the **Displayed properties** list box. 

9. Select the **Quantity** property from the **Displayed properties** list box; the **Number** property is moved to the top.

   Next, you need to remove some of the properties from the **Displayed properties** list box.

10. Select **Nomenclature** from the **Displayed properties** list box and choose the **Hide properties** button on the right; the **Nomenclature** property moves into the **Hidden properties** list box. 

11. Similarly, remove the **Revision** property from the **Displayed properties** list box.

12. Choose the **OK** button from the **Bill of Material : Define formats** dialog box and then choose the **OK** button from the **Bill of Material : Product1** dialog box.

13. Choose **Window > Drawing** from the menu bar to invoke the drawing window.

    Next, you need to generate balloons in the isometric view.

14. In the Specification tree, double-click on the **Isometric view** to activate it.

15. Choose the **Generate Balloons** tool from the **Dimension Generation** sub-toolbar in the **Generation** toolbar; the balloons are generated and displayed in the isometric view.

16. Move the balloons to appropriate locations if they are placed arbitrarily.

    Next, you need to generate the BOM.

17. Choose **Insert > Generation > Bill of Material > Bill of Material** from the menu bar; you are prompted to click at a location to insert the Bill of Material.

18. Click near the upper right corner of the drawing sheet to place the BOM, refer to Figure 14-39.

19. Move the BOM into the drawing sheet if placed outside the drawing sheet. Turn off the display of the view frames of all the views.

    The final drawing sheet after generating the BOM and balloons is shown in Figure 14-39.

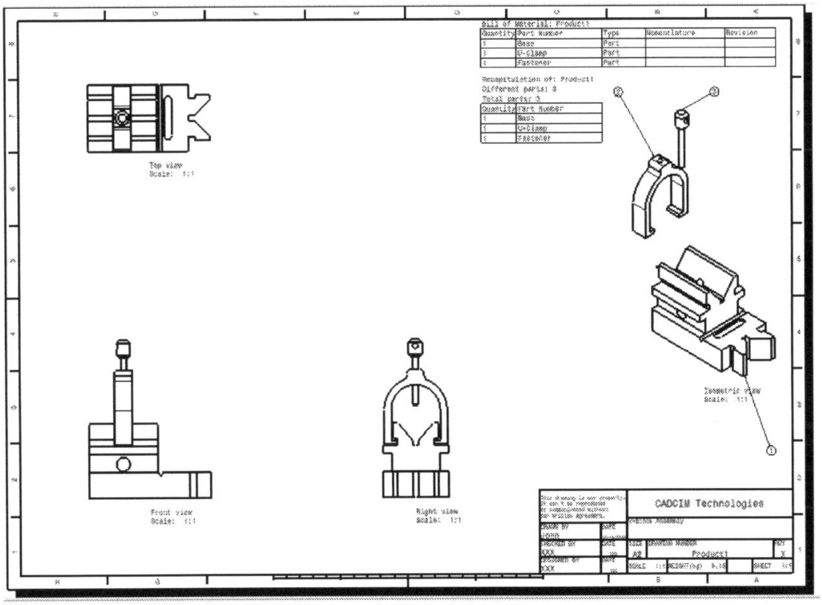

Figure 14-39 *The final drawing sheet*

20. Save the drawing file with the name *c14tut2.CATDrawing* at *C:\CATIA\c14*.

Self-Evaluation Test

Answer the following questions and then compare them to those given at the end of this chapter:

1. The _________ tool is used to add a welding symbol to the drawing views.

2. The _________ check box in the **Dimension Generation Filters** dialog box is used to generate the dimensions with design tolerances.

3. You can set the size of the weld by specifying a value in the _________ edit box of the **Welding creation** dialog box.

4. The _________ option in the **Action** list box of the **Manage Frame And Title Block** dialog box is used to delete existing frame and title block.

5. The options in the _________ area of the **Geometrical Tolerance** dialog box are used to specify the geometrical condition for the tolerance and the value of the tolerance.

6. You cannot insert additional sheets to the current drafting file. (T/F)

7. The **Style of Title block** drop-down list in the **Manage Frame And Title Block** dialog box is used to set the style of the frame and the title block. (T/F)

8. The options in the **Reference** area of the **Geometrical Tolerance** dialog box are used to define the reference for applying the tolerance. (T/F)

9. You cannot apply the datum feature symbol to the drawing views. (T/F)

10. The naming of the balloons depends upon the sequence of the parts in the BOM. (T/F)

Review Questions

Answer the following questions:

1. The _________ button in the **Geometrical Tolerance** dialog box is used to reset all the parameters of the geometrical tolerance to default values.

2. The _________ edit box in the **Welding creation** dialog box is used to specify the length of the weld.

3. You can exit the background editing mode after inserting the frame and the title block by choosing the _________ option from the menu bar.

4. The _________ option in the **Action** list box of the **Manage Frame And Title Block** dialog box is used to resize the frame and the title block.

5. The _________ tool is used to generate the dimensions step by step.

6. To switch the dimensions from one view to another, choose the **Transfer** button from the **Step By Step Generation** dialog box. (T/F)

7. The **Upper Text** edit box in the **Geometrical Tolerance** dialog box is used to specify the text above the geometrical tolerance value. (T/F)

8. You can add the surface finish symbols by choosing the **Roughness Symbol** button from the **Symbols** sub-toolbar in the **Annotations** toolbar. (T/F)

9. You cannot insert a logo in the title block. (T/F)

10. You cannot generate a BOM in the **Drafting** workbench of CATIA V5. (T/F)

EXERCISE

Exercise 1

Generate the front, top, right, and isometric views of the Blower assembly created in Tutorial 1 of Chapter 12. The drawing view will be generated with a scale factor of 1:5. After generating the drawing views, you need to generate balloons and BOM, as shown in Figure 14-40.

(Expected time: 30 min)

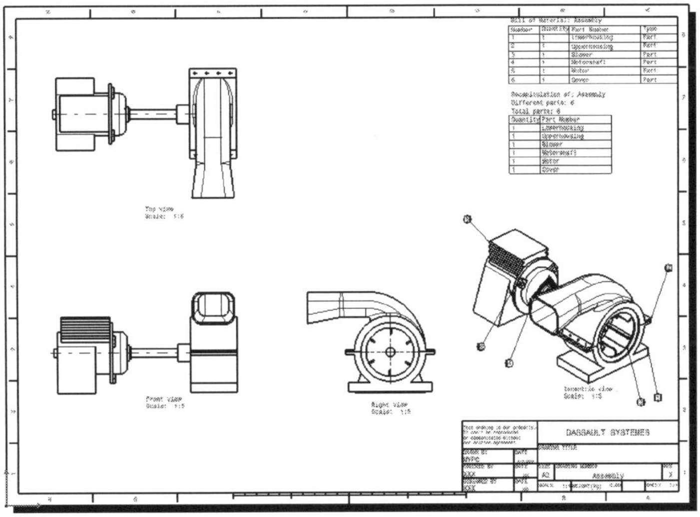

Figure 14-40 *Drawing views for Exercise 1*

Answers to Self-Evaluation Test

1. Welding Symbol, **2.** with design tolerances, **3.** Size of weld, **4.** Delete, **5.** Tolerance, **6.** F, **7.** T, **8.** T, **9.** F, **10.** T

Chapter 15

Working with Sheet Metal Components

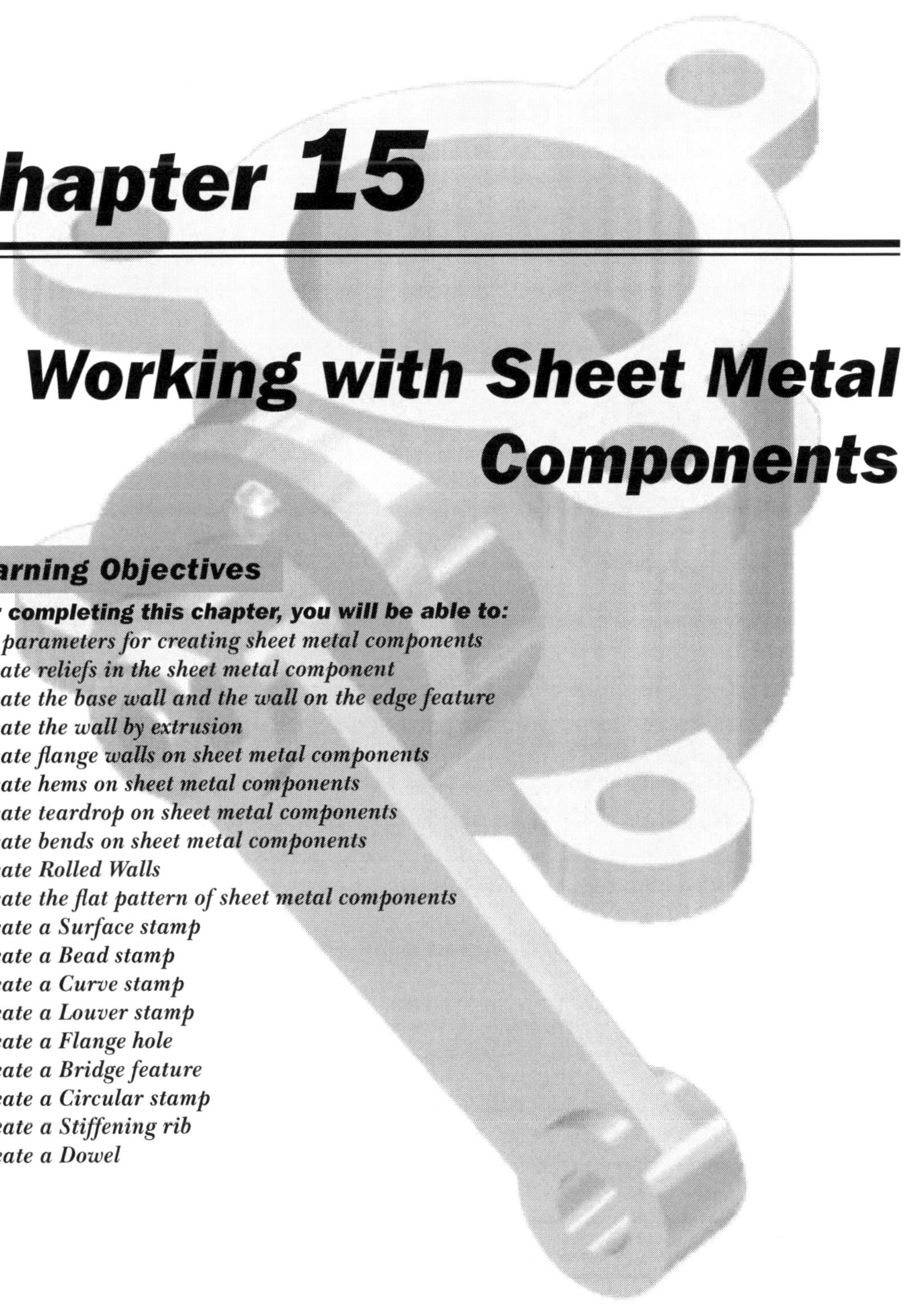

Learning Objectives

After completing this chapter, you will be able to:

- *Set parameters for creating sheet metal components*
- *Create reliefs in the sheet metal component*
- *Create the base wall and the wall on the edge feature*
- *Create the wall by extrusion*
- *Create flange walls on sheet metal components*
- *Create hems on sheet metal components*
- *Create teardrop on sheet metal components*
- *Create bends on sheet metal components*
- *Create Rolled Walls*
- *Create the flat pattern of sheet metal components*
- *Create a Surface stamp*
- *Create a Bead stamp*
- *Create a Curve stamp*
- *Create a Louver stamp*
- *Create a Flange hole*
- *Create a Bridge feature*
- *Create a Circular stamp*
- *Create a Stiffening rib*
- *Create a Dowel*

THE SHEET METAL COMPONENT

A component whose thickness is less than 12 mm is called a sheet metal component. The sheet metal fabrication is a chip-less process and an easy way to create components by using manufacturing processes such as bending, stamping, and so on.

A sheet metal component of uniform thickness is shown in Figure 15-1. It is not possible to machine such a thin component. After creating a sheet metal component, you need to flatten it in order to find the strip layout. Based on the layout detail, you can design punch and die. Figure 15-2 shows a flattened view of the sheet metal component shown in Figure 15-1.

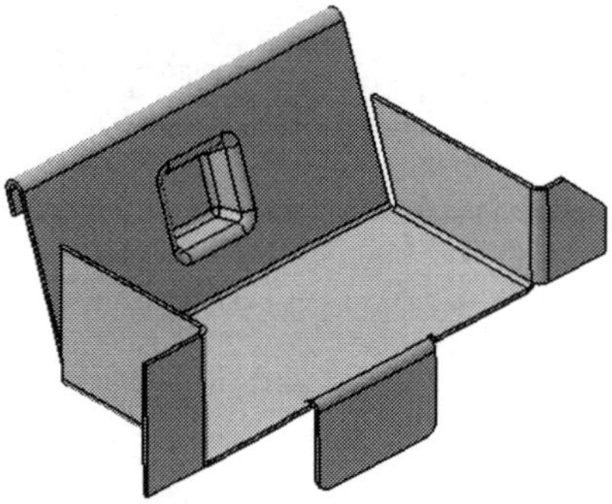

Figure 15-1 *Sheet metal component*

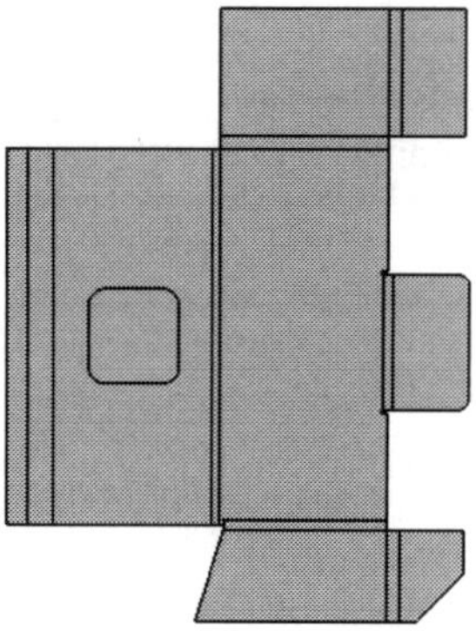

Figure 15-2 *Flattened view of the sheet metal component*

In CATIA V5, you can create sheet metal components in a separate environment called the **Generative Sheetmetal Design**. In this environment, parts are created in **.CATPart* file format.

Starting a New File in Generative Sheet Metal Workbench

To start a new file in the **Generative Sheetmetal Design** workbench, first close the existing file and then choose **Start > Mechanical Design > Generative Sheetmetal Design** from the menu bar; the **New Part** dialog box will be displayed. Choose the **OK** button from the dialog box; a new file will open in the **Generative Sheetmetal Design** workbench. Figure 15-3 shows the initial screen appearance of the **Generative Sheetmetal Design** workbench.

Tip

For the ease of locating and invoking tools, you can create customized toolbars and add the frequently used tools in the customized toolbar. To do so, choose ***Tools > Customize*** *from the menu bar; the* ***Customize*** *dialog box will be displayed. Choose the* ***Toolbars*** *tab and then the* ***New*** *button; the* ***New Toolbar*** *dialog box will be displayed. Enter* ***Generative Sheet Metal Design*** *as the name of the toolbar in the* ***Toolbar Name*** *edit box and then choose* ***OK****; the toolbar will be added to the* ***Toolbars*** *list box. Also, a new toolbar will be displayed in the graphics window with the name* ***Generative Sheet Metal Design****. Next, you need to add tools to the new toolbar. To do so, choose the* ***Commands*** *tab from the* ***Customize*** *dialog box. From the* ***Categories*** *list box, select the* ***All Commands*** *option; all the commands will be displayed in the* ***Commands*** *list box of the* ***Customize*** *dialog box. Select the tool that you want to add and then drag it to the* ***Generative Sheet Metal Design*** *toolbar.*

Figure 15-3 The initial screen appearance of the ***Generative Sheetmetal Design*** *workbench*

SETTING SHEET METAL PARAMETERS

Menubar:	Insert > Sheet Metal Parameters
Toolbar:	Walls > Sheet Metal Parameters

After invoking the **Generative Sheetmetal Design** workbench, you need to set Sheet Metal parameters based on your design requirement. To do so, choose the **Sheet Metal Parameters** tool from the **Walls** toolbar; the **Sheet Metal Parameters** dialog box will be displayed, as shown in Figure 15-4. The parameters in this dialog box are applicable to all the additional features. The options in the **Sheet Metal Parameters** dialog box are discussed next.

Parameters Tab

The options in this tab are used to set the parameters related to the sheet thickness and bend radius. These options are discussed next.

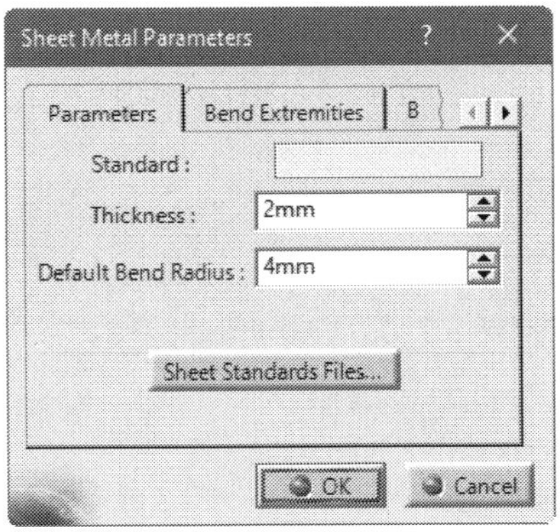

Figure 15-4 The ***Sheet Metal Parameters*** *dialog box with the* ***Parameters*** *tab chosen*

Thickness

The **Thickness** spinner is used to specify the sheet thickness. The value that you set in this spinner will be displayed as the default value while creating the sheet metal part.

Default Bend Radius

The **Default Bend Radius** spinner is used to set the inner radius of the bend.

Bend Extremities Tab

Whenever you bend a sheet metal component or create a flange such that the bend does not extend throughout the length of the edge, a groove is added at the end of the bend so that the walls of the sheet metal part do not intersect when folded or unfolded. This groove is known as relief.

Choose the **Bend Extremities** tab to define the type of relief, as shown in Figure 15-5. Select an option from the drop-down list under this tab to define the relief type. Alternatively, choose the down arrow on the button available underneath; a flyout will be displayed with different types of relief. Select the required relief from the flyout. The types of relief in CATIA V5 are discussed next.

Figure 15-5 *The* ***Sheet Metal Parameters*** *dialog box with the* ***Bend Extremities*** *tab chosen*

Minimum with no relief

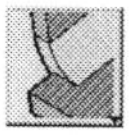

This option is selected by default, and it does not provide any relief to the common area between the supporting walls of the sheet metal part, refer to Figure 15-6.

Square relief

This option provides a square relief between the supporting walls of the sheet metal part, as shown in Figure 15-7. You can modify the L1 and L2 values in their respective spinners.

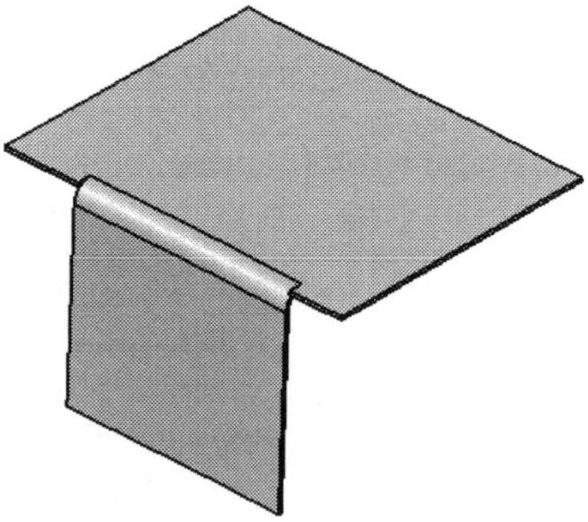

Figure 15-6 *Sheet metal part with no relief*

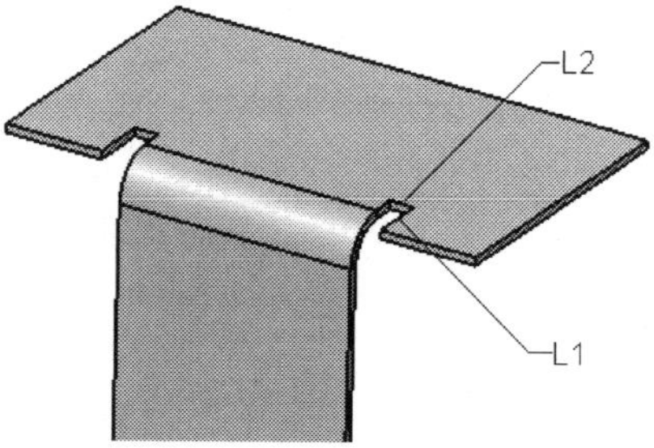

Figure 15-7 *Sheet metal part with square relief*

Round relief

This option provides a round relief between the supporting walls of the sheet metal part, as shown in Figure 15-8. You can modify the L1 and L2 values in their respective spinners.

Linear

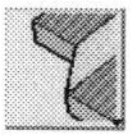

The **Linear** option provides a linear relief between the supporting walls of the sheet metal part, as shown in Figure 15-9.

Tangent

The **Tangent** option provides a tangent relief between the supporting walls of the sheet metal part, as shown in Figure 15-10.

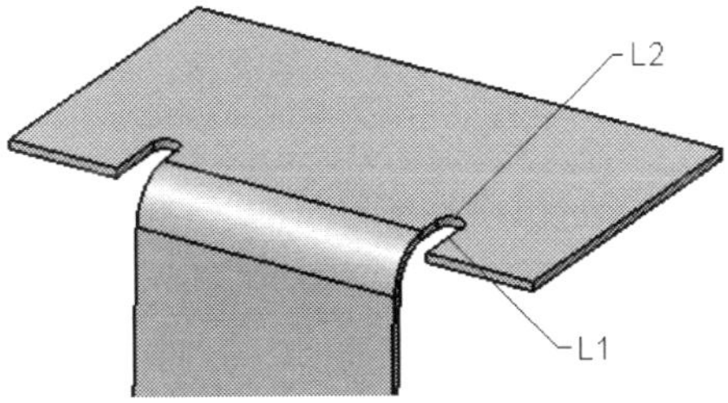

Figure 15-8 Sheet metal with round relief

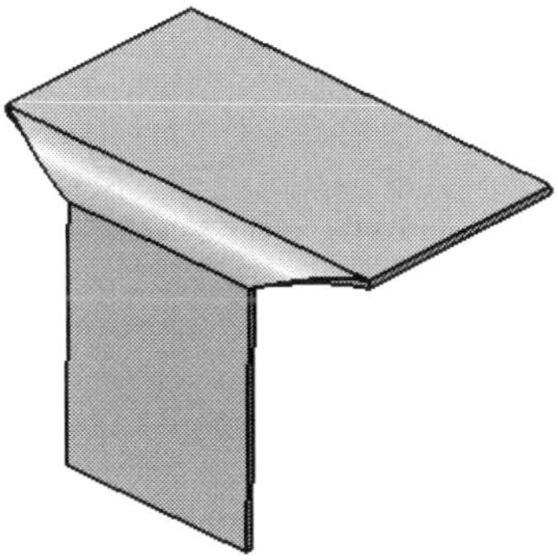

Figure 15-9 Sheet metal with linear relief

Maximum

On selecting the **Maximum** option, the bend is calculated between the extreme edges of two supporting walls, as shown in Figure 15-11.

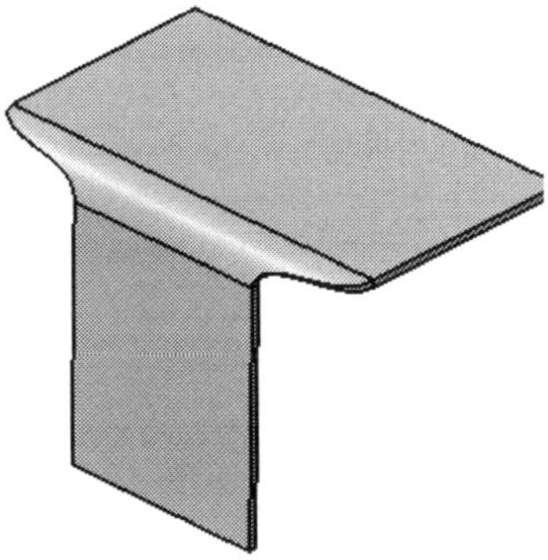

Figure 15-10 *Sheet metal part with tangent relief*

Figure 15-11 *Sheet metal part with maximum relief*

Closed

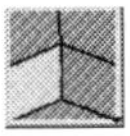

The **Closed** option provides relief to the intersection between the bends of two supporting walls.

Flat joint

The **Flat joint** option provides flat relief to the intersection between the bends of two supporting walls, as shown in Figure 15-12.

Figure 15-12 *Sheet metal part with flat joint relief*

Note

You can also first draw the sketch of the base feature in the ***Sketcher*** *workbench and then set the parameters in the* ***Sheet Metal Parameters*** *dialog box.*

Bend Allowance Tab

The option available in the **Bend Allowance** tab is discussed next.

K Factor

K factor is the ratio between the distance from the neutral bend line and the upper surface of the sheet metal part to the total thickness of the part.

Note

*1. You can change the values of parameters in the **Sheet Metal Parameters** dialog box any time during the design.*

*2. The tools in this workbench will be available only after setting the parameters in the **Sheet Metal Parameters** dialog box.*

INTRODUCTION TO SHEET METAL WALLS

A wall refers to any section in a sheet metal design. There are two types of walls in CATIA V5 and the methods to create them are discussed next.

Creating the Base Wall

Menubar:	Insert > Walls > Wall
Toolbar:	Walls > Wall

The first feature created while designing a sheet metal part is called the base wall. The other features are added to this base wall. Remember that the parameters of the base wall and additional features are based on the parameters set in the **Sheet Metal Parameters** dialog box. The **Wall** tool is used to create the base feature of a sheet metal component. To create a wall, choose the **Wall** tool from the **Walls** toolbar; the **Wall Definition** dialog box will be displayed, as shown in Figure 15-13. You can draw the sketch of the base feature before invoking the **Wall Definition** dialog box.Alternatively, you can draw the sketch of the base feature after invoking the **Wall Definition** dialog box. To do so, choose the **Sketch** button on the right of the Profile display box in the **Wall Definition** dialog box; you will be prompted to select the sketching plane. Select the sketching plane and draw the sketch of the base feature in the **Sketcher** workbench, as shown in Figure 15-14. Next, exit the **Sketcher** workbench to redisplay the **Wall Definition** dialog box.

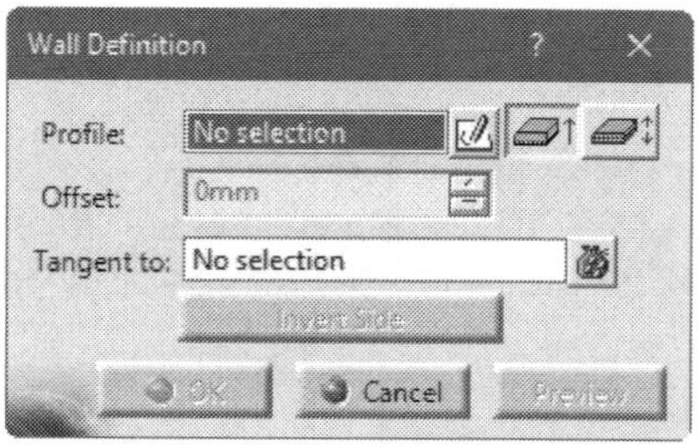

*Figure 15-13 The **Wall Definition** dialog box*

If the sketch is already drawn, select the sketch from the geometry area; a preview of the wall will be displayed. You can flip the direction of the material by choosing the **Invert Side** button from the dialog box. The **Sketch at extreme position** button on the right of the **Profile** display box in the **Wall Definition** dialog box is chosen by default. This button allows you to set the sketch at the extreme position of the wall thickness. To set the sketch at the middle position of the wall thickness, choose the **Sketch at middle position** button on the right of the **Sketch at extreme position** button. You can also create the wall at an offset distance from the sketch plane. To do so, specify the required offset value in the **Offset** edit box. By default, this edit box is inactive, it becomes active on selecting the sketch. By default, the value is set to 0 mm. Select the **Invert Side** button to invert the side of extrusion. This button is available only when the **Sketch at extreme position** button is selected. Next, choose the **OK** button to create the base wall. Figure 15-15 shows the resulting base wall.

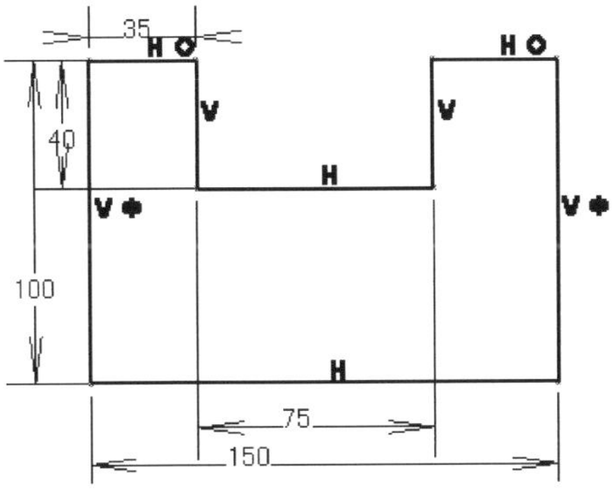

Figure 15-14 *The sketch of the base feature*

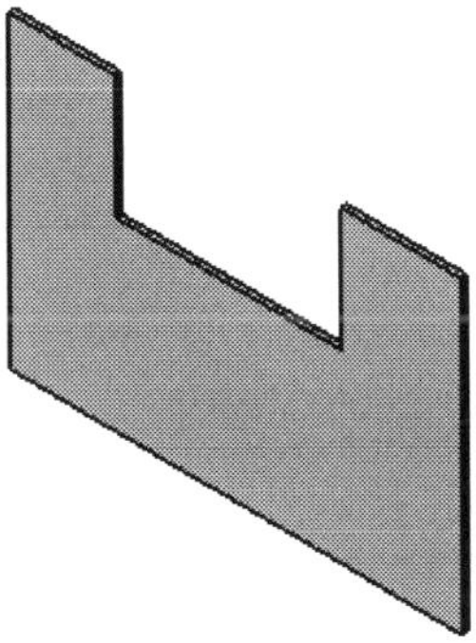

Figure 15-15 *The resulting base feature*

Creating a Wall On Edge

Menubar: Insert > Walls > Wall On Edge
Toolbar: Walls > Wall On Edge

The **Wall On Edge** tool is used to create a wall on the edges of an existing wall. To do so, choose the **Wall On Edge** tool from the **Walls** toolbar; the **Wall On Edge Definition** dialog box will be displayed, as shown in Figure 15-16. Also, you will be prompted to select the edge on which the wall is to be created. In the **Wall On Edge Definition** dialog box, the **Type** drop-down list has two options: **Automatic** and **Sketch Based**. These options are discussed next.

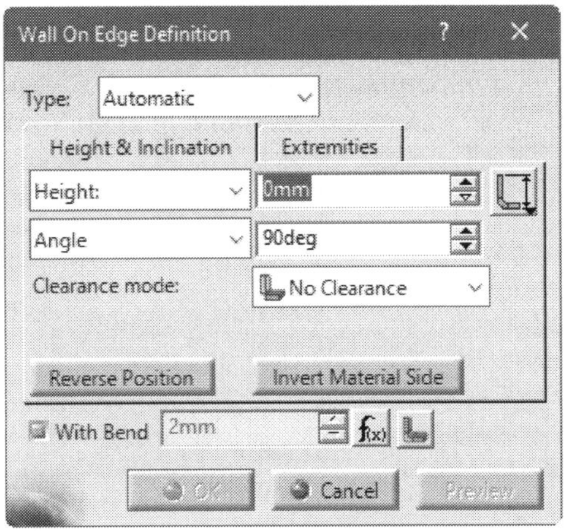

Figure 15-16 *The* ***Wall On Edge Definition*** *dialog box*

Automatic

The **Automatic** option allows you to select an existing edge to create a wall. By default, this option is selected in the **Type** drop-down list. If you select an existing edge of the wall, a preview of the new wall will be displayed. You can also select multiple edges to create walls. The options that will be displayed on selecting the **Automatic** option from the **Type** drop-down list are discussed next.

Height & Inclination Tab

By default, the **Height** option is selected in the first drop-down list under the **Height & Inclination** tab. Set the value of the height in the spinner. To define the height, click on the down arrow available on the **Length type** button; a flyout will be displayed with four options. These options are discussed next.

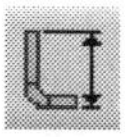
This option allows you to specify the vertical distance from the lower face of the base wall to the extreme edge of the wall created.

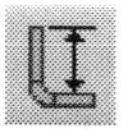
This option allows you to specify the vertical distance from the upper face of the base wall to the extreme edge of the wall created.

This option is used to specify the height of the flange from the start of the bend to the extreme edge of the wall created.

This option allows you to specify the slant distance from the extreme edge of the vertical wall to the apparent intersection of the vertical and horizontal walls.

This option allows you to specify the slant distance from the extreme edge of the vertical wall to the upper face of the apparent intersection of the vertical and horizontal walls.

If you need the wall on the edge to be created upto a pre-defined plane, select the **Up To Plane/Surface** option from the first drop-down list and then select the plane from the geometry area or the Specification tree. You can also create a new plane. To do so, right-click in the **Up To Plane/Surface** display box; a contextual menu will be displayed. Choose the **Create Plane** option from the contextual menu and create a plane as discussed earlier. While creating the wall on edge using the **Up To Plane/Surface** option, if you select the plane from the geometry area, the **Offset** spinner will be available in the dialog box. You can use this spinner to terminate the wall at a certain offset from the selected plane.

To create a wall at an angle with respect to the support wall, select the **Angle** option, if it is not already selected, from the second drop-down list and set the inclination in the **Angle** spinner.

If you need to create an inclined wall with respect to the existing face or plane, select the **Orientation plane** option from the second drop-down list. Select the plane as discussed earlier. If you need to change the inclination of the wall with respect to the selected plane, set the angle in the **Rotation angle** spinner.

Clearance mode

The wall can be created at an offset from the selected edge by using the options in the **Clearance mode** drop-down list. The options in this drop-down list are **No Clearance**, **Monodirectional**, and **Bidirectional**. On selecting the **No Clearance** option, the wall will be created without clearance between the base wall and the side wall, as shown in Figure 15-17. When you select the **Monodirectional** option, the wall will be created with the horizontal clearance, as shown in Figure 15-18. You can set the value for the horizontal clearance in the **Clearance value** spinner.

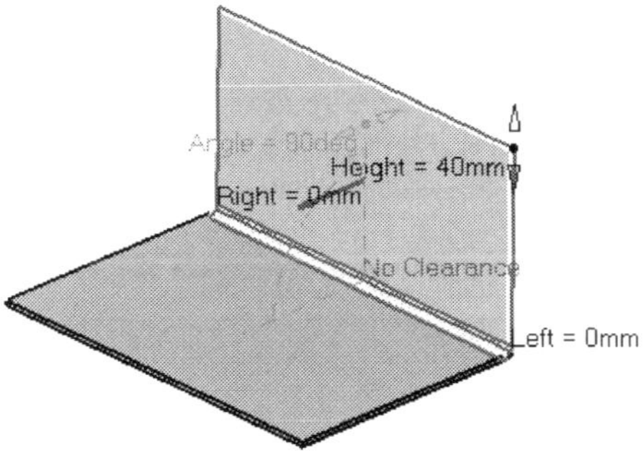

Figure 15-17 Side wall with no clearance

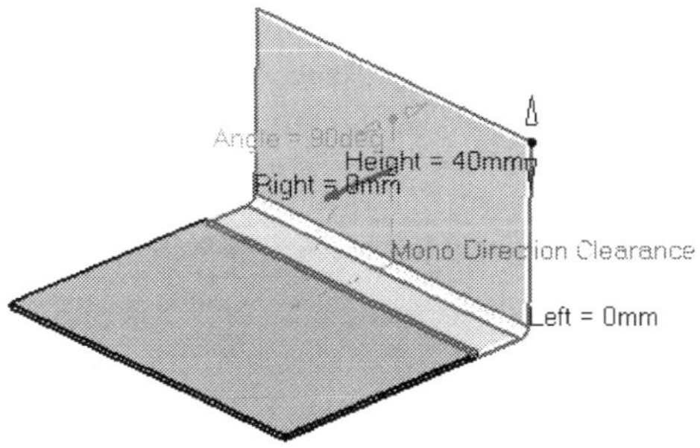

Figure 15-18 Side wall with the horizontal clearance

On selecting the **Bidirectional** option, the **Feature Definition Warning** message box will be displayed. If you choose the **Yes** button, the **Clearance value** spinner will not be available and the clearance value will be the same as set for the bend radius in the **Sheet Metal Parameters** dialog box. When you choose the **No** button from the message box, the **Clearance value** spinner will be available and you can change its value. The **Bidirectional** option provides clearance in the horizontal and vertical directions, as shown in Figure 15-19.

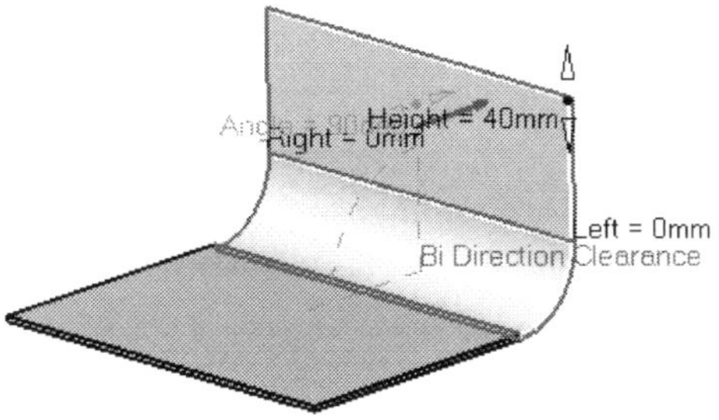

Figure 15-19 Side wall with the bidirectional clearance

To change the direction of the wall on the edge, choose the **Reverse Position** button from the **Wall On Edge Definition** dialog box. To change the material side of the wall, choose the **Invert Material Side** button from the dialog box. Note that the **Invert Material Side** button will not be available, if the **Bidirectional** option is selected in the **Clearance mode** drop-down list.

Extremities Tab

The options in the **Extremities** tab are used to define the length of the wall with respect to a reference plane/wall. The options and the procedure to set the limits are discussed next.

Left limit

Select the predefined plane or the wall that you need to set as the left limit from the geometry area; the name of the selected plane will be displayed in the **Left limit** display box. You can also create a plane as discussed earlier.

Left offset

The **Left offset** spinner is used to set the offset distance from the plane that is selected as the left limit in the **Left limit** display box. If there is no selection in the **Left limit** display box, the endpoint on the left side of the edge is selected as the left limit by default.

Right limit

Select the predefined plane or the wall that you need to set as the right limit from the geometry area; the name of the selected plane will be displayed in the **Right limit** display box. You can also create a plane as discussed earlier.

Right offset

The **Right offset** spinner is used to set the offset distance from the plane that is selected as the right limit in the **Right limit** display box. If there is no selection in the **Right limit** display box, the endpoint on the right side of the edge is selected as the right limit by default.

With Bend

The **With Bend** check box is used to create a wall with or without a bend. Select the **With Bend** check box, if it is not already selected; the wall will be created with a bend. To change the type of bend relief, choose the **Bends parameters** button on the right of the **With Bend** check box; the **Bend Definition** dialog box will be displayed. In this dialog box, you can change the type of relief, as discussed earlier. The **Left Extremity** and **Right Extremity** tabs in this dialog box can be used to specify the type of relief for the left and right ends of the resulting wall.

If you select the **Monodirectional** option from the **Clearance mode** drop-down list and set the clearance value more than the bend radius in the **Clearance value** spinner, the **Extrapolation mode** button will be enabled next to the **Bends parameters** button. Click on the down arrow of the **Extrapolation mode** button; a flyout will be displayed with two buttons. If you choose the first button, the width of the clearance will be equal to the width of the base feature, as shown in Figure 15-20. On choosing the second button, the width of the clearance will be equal to the width of the wall on the edge feature, as shown in Figure 15-21. Choose the **OK** button; the resultant side wall will be created, as shown in Figure 15-22.

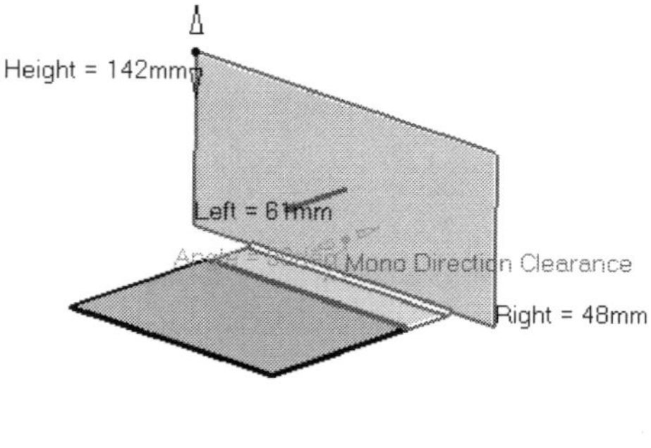

Figure 15-20 *Preview of the wall to be created when the first button is chosen from the* ***Extrapolation mode*** *flyout*

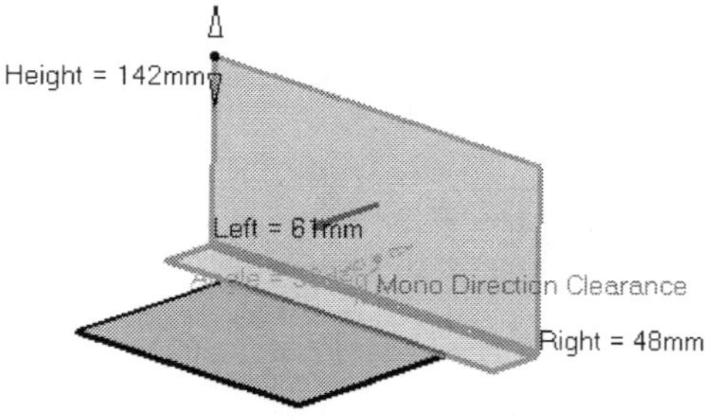

Figure 15-21 *Preview of the wall to be created when the second button is chosen from the* ***Extrapolation mode*** *flyout*

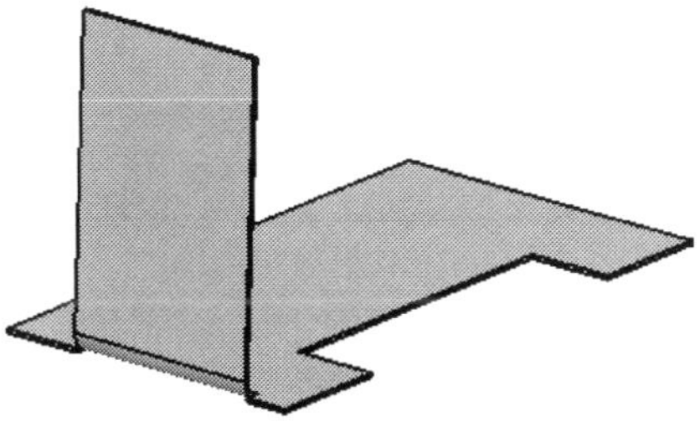

Figure 15-22 *Wall created by selecting the* ***Automatic*** *option*

Tip
On selecting the ***Swap limits*** *symbol displayed on the left of the spinners in the* ***Extremities*** *tab of the* ***Wall On Edge Definition*** *dialog box, the limits on the left and right sides of the wall will be interchanged.*

Sketch Based

Select the **Sketch Based** option in the **Type** drop-down list; the **Wall On Edge Definition** dialog box will be modified, as shown in Figure 15-23. The options in this dialog box are discussed next.

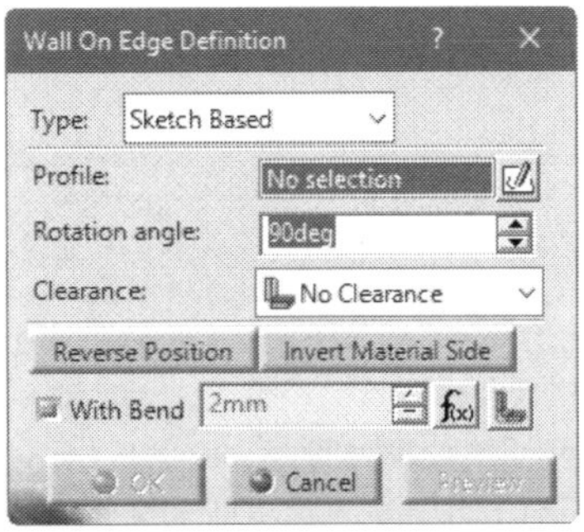

Figure 15-23 *The* ***Wall On Edge Definition*** *dialog box with the* ***Sketch Based*** *option selected*

Profile

Select the edge of the base wall feature on which you need to create a wall, as shown in Figure 15-24. Next, choose the **Sketch** button on the right of the **Profile** display box; you will be prompted to select the sketching plane. Select the face adjacent to the selected edge; the **Sketcher** workbench will be invoked. The selected face will act as the sketching plane of the wall feature. Create a closed sketch line in the **Sketcher** workbench, as shown in Figure 15-24. Remember that any sketch line of the profile and the selected edge must coincide with each other. Next, exit the **Sketcher** workbench. The preview of the feature will be displayed in the geometry area. Choose the **OK** button; the wall on the edge will be created, as shown in Figure 15-25.

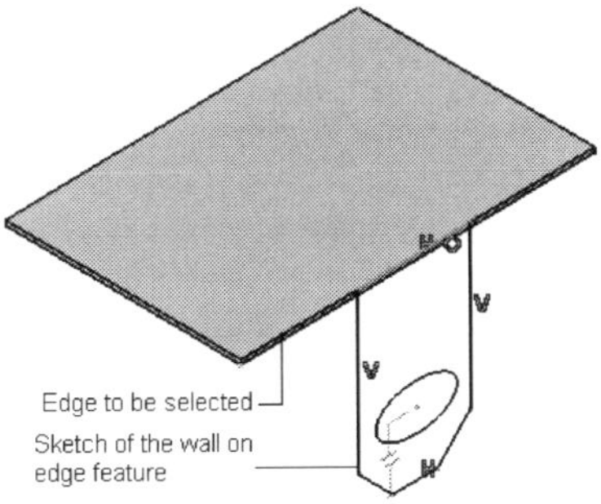

Figure 15-24 *Edge to be selected and sketch for the wall on edge feature*

Figure 15-25 *Wall on edge created using the* ***Sketch Based*** *option*

Rotation angle

You can also create a wall inclined at an angle to the selected edge. To do so, set the angular value in the **Rotation angle** spinner. The default value in the **Rotation angle** spinner is **90deg**.

The other options in this dialog box have been discussed earlier.

Note

If you have created a wall without a bend, you can create the bend later using the ***Bend*** *tool. This tool will be discussed later in this chapter.*

CREATING EXTRUSIONS

Menubar:	Insert > Walls > Extrusion
Toolbar:	Walls > Extrusion

The **Extrusion** tool is used to extrude an open sketch drawn on an existing wall. To extrude the sketch, choose the **Extrusion** tool from the **Walls** toolbar; the **Extrusion Definition** dialog box will be displayed, as shown in Figure 15-26. Also, you will be prompted to select a profile. Select the existing open profile (open loop sketch) from the geometry area to extrude it. If there is no existing profile in the geometry area, then you need to create a sketch using the **Sketch** button available on the right of the **Profile** display box, as discussed earlier. The options in the **Extrusion Definition** dialog box are discussed next.

By default, the **Limit 1 dimension** option is selected in the **Sets First limit** drop-down list and the **Limit 2 dimension** option is selected in the **Sets Second limit** drop-down list of the **Extrusion Definition** dialog box. Therefore, the wall will extrude normal to the base wall.

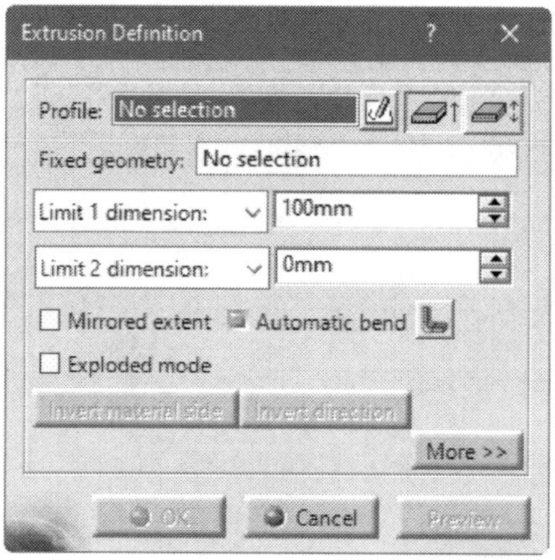

Figure 15-26 The ***Extrusion Definition*** *dialog box*

You can specify the length of the wall to be extruded in the limit 1 and limit 2 directions by using the length1 and length2 spinners that are available on the right of **Sets first limit** and **Sets Second limit** drop-down lists, respectively.

To extrude the sketch in the limit 1 direction upto a selected plane or surface, select the **Limit 1 up to plane** or **Limit 1 up to surface** option from the **Sets first limit** drop-down list. You can also extrude the sketch in the limit 2 direction upto a selected plane or surface. To do so, select the **Limit 2 up to plane** or **Limit 2 up to surface** option from the **Sets Second limit** drop-down list. Next, select the plane or surface from the geometry area.

The options in the second drop-down list are used to extrude the wall on the other side of the base wall.

The **Automatic bend** check box in the **Extrusion Definition** dialog box is selected by default. This check box allows you to create bends on the sharp vertices of a sketch. To extrude a feature by the same value in both the directions, select the **Mirrored extent** check box. But note that on selecting this check box, the **Sets first limit** and **Sets Second limit** drop-down lists will not be available. On selecting the **Exploded mode** check box, you can explode the extrusion into elementary features. The **Sketch at extreme position** button located on the right of the **Profile** display box is chosen by default. As a result, the material will be added to either sides of the sketch. If you choose the **Sketch at middle position** button, the sketch will be considered as a neutral plane and the material will be added to both sides of the sketch. You can change the direction of the material by choosing the **Invert material side** button from the dialog box. To preview the feature, you can choose the **Preview** button from this dialog box. To specify the K factor for the bend, you need to expand the **Extrusion Definition** dialog box by choosing the **More** button.

Figure 15-27 shows the sketch of the extrusion wall and Figure 15-28 shows the resulting extruded wall.

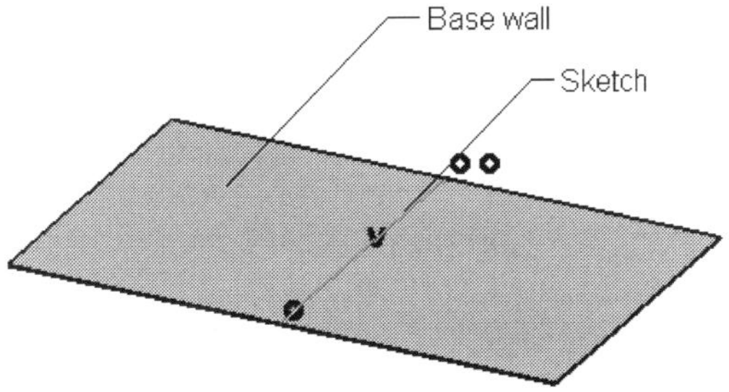

Figure 15-27 *Sketch of the extrusion wall*

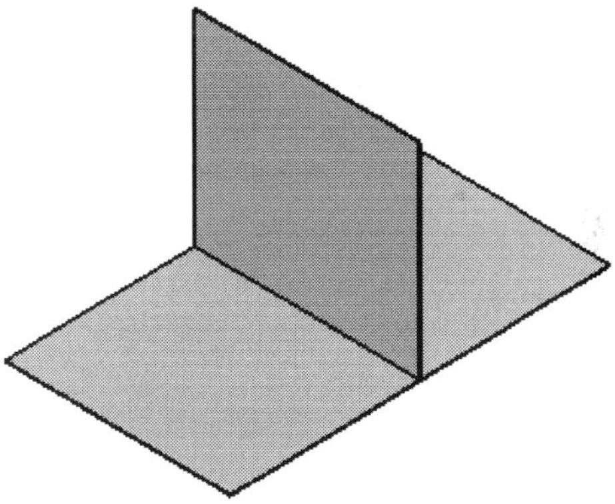

Figure 15-28 *Resulting extruded wall*

Note

You need to create the bend after extruding the profile. The procedure to do so will be discussed later in this chapter.

CREATING SWEPT WALLS

Swept walls are created by sweeping a profile along the selected edge. Different types of swept walls are **Flange**, **Hem**, **Tear Drop**, and **User Flange**. These walls are discussed next.

Creating Flanges on the Sheet Metal Component

Menubar: Insert > Walls > Swept Walls > Flange
Toolbar: Walls > Swept Walls sub-toolbar > Flange

Flange is the bend section of a Sheet Metal. To create a flange feature, choose the **Flange** tool from the **Swept Walls** sub-toolbar of the **Walls** toolbar, refer to Figure 15-29; the **Flange Definition** dialog box will be displayed, as shown in Figure 15-30. Also, you will be prompted to select the guide element. Select an existing edge from the wall; a preview of the flange will be displayed with the default values. The options in the **Flange Definition** dialog box are discussed next.

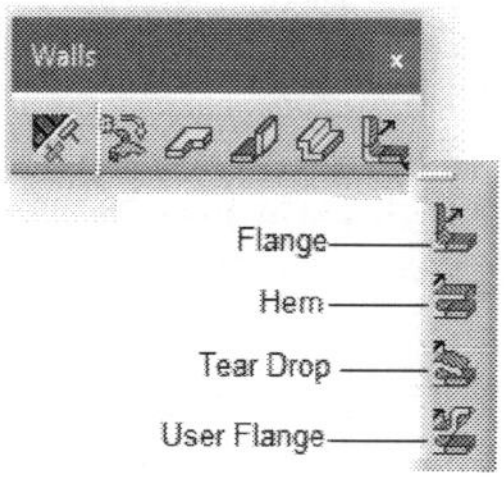

*Figure 15-29 Tools in the **Swept Walls** sub-toolbar*

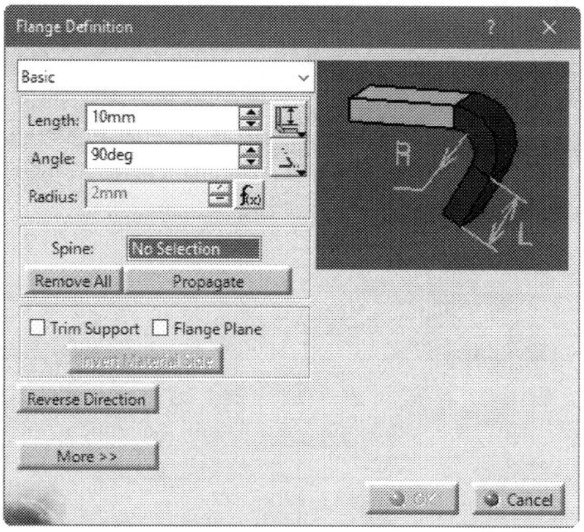

*Figure 15-30 The **Flange Definition** dialog box*

Flange type

There are two options in the **Flange type** drop-down list and these are discussed next.

Basic

The **Basic** option in the **Flange Type** drop-down list is selected by default and in this case, the width of the flange created will be equal to the width of the selected edge. The options available on selecting the **Basic** option from the **Flange type** drop-down list are discussed next.

Length

Set the length of the flange in the **Length** spinner. To change the length type of the flange, choose the down arrow on the right side of the **Length type** button; a flyout will be displayed with four options. The options in the flyout have been discussed earlier.

Angle

You can also create a flange wall inclined at an angle to the selected edge. To do so, set the angular value in the **Angle** spinner. To change the angle type of the flange, choose the down arrow on the right of the **Angle type** button; a flyout will be displayed with two buttons. If you choose the **Inner Angle type** button from the flyout, the angle that you define will be considered as the included angle from the face. If you choose the **Outer Angle type** button, the angle that you define will be considered as the excluded angle.

Radius

Set the bend radius of the flange in the **Radius** spinner.

Spine

The **Spine** display box displays the name of the selected edge of the existing wall. To remove the selected entity from the **Spine** display box, choose the **Remove All** button available below this display box.

Trim Support

By default, this check box is clear. As a result, a flange will be created outward from the selected edge, as shown in Figure 15-31. When you select the **Trim Support** check box, the flange will be created inward from the selected edge at a distance equal to the radius of the flange, as shown in Figure 15-32. Also, the portion of the wall beyond the radius will be trimmed.

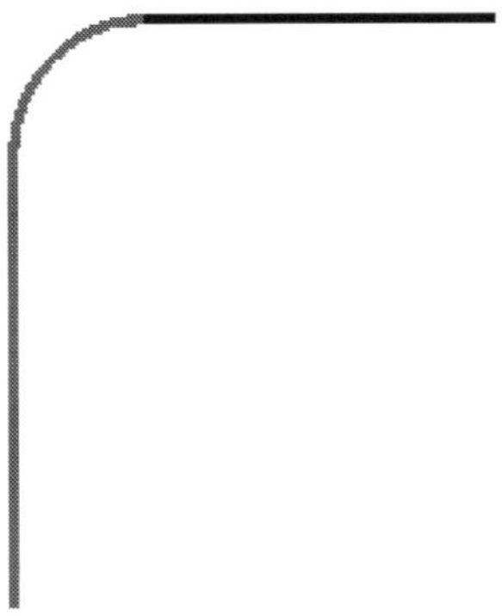

Figure 15-31 *Flange created with the* ***Trim Support*** *check box cleared*

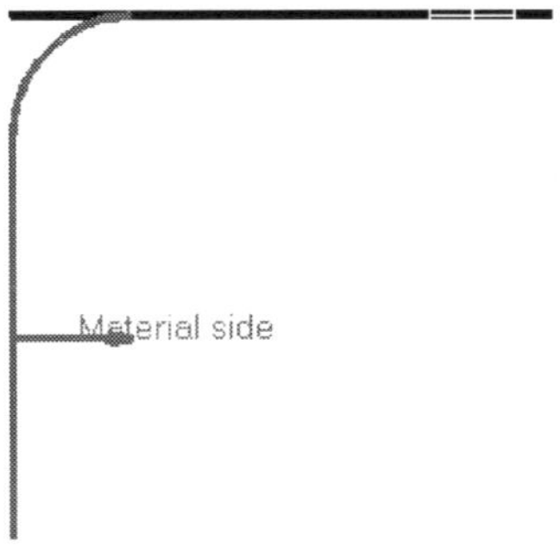

Figure 15-32 *Flange created with the* ***Trim Support*** *check box selected*

Flange Plane

If you need to create a flange at an angle similar to that of an existing plane, then select the **Flange Plane** check box and select the required plane; a preview of the flange orienting to the angle of the selected plane will be displayed.

Choose the **Propagate** button to select the edges that are connected tangentially.

Choose the **Invert Material Side** button to change the material side of the flange. Note that the **Invert Material Side** button will be activated when the **Trim Support** or **Flange Plane** check box is selected.

Relimited

To create the flange up to a specified limit, select the **Relimited** option from the **Flange type** drop-down list. Next, define the limits in the **Limit 1** and the **Limit 2** display box. You can select the predefined limits from the geometry area or create the new one by right-clicking in the **Limit 1** and **Limit 2** display box as discussed earlier.

The rest of the options are the same as discussed in the **Basic** option. Figure 15-33 shows the flange feature created by selecting the **Basic** option. Figure 15-34 shows the flange feature created by selecting the **Relimited** option.

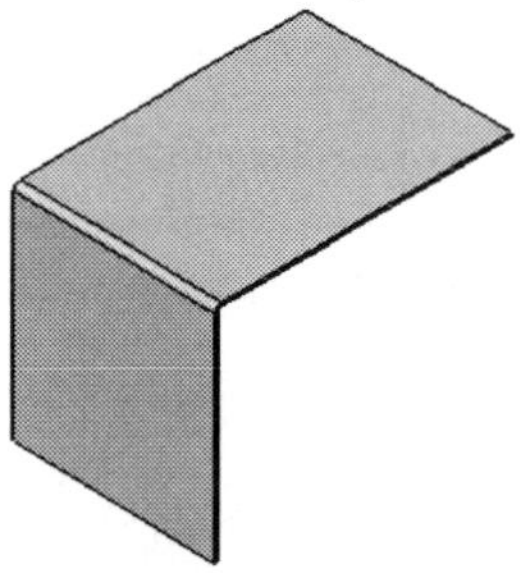

***Figure 15-33** The flange feature created by selecting the **Basic** option*

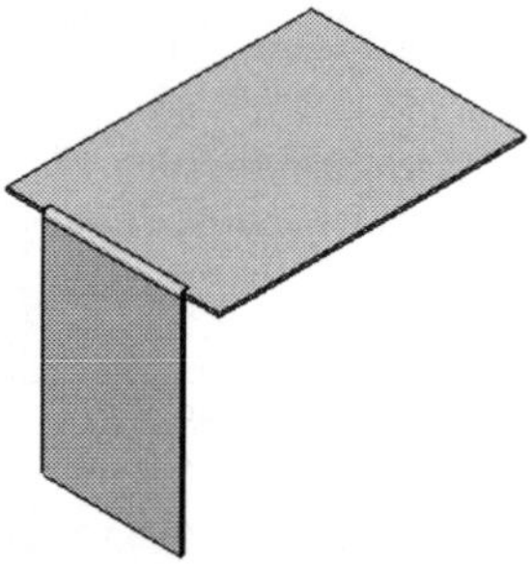

***Figure 15-34** The flange feature created by selecting the **Relimited** option*

Creating Hems on the Sheet Metal Component

Menubar:	Insert > Walls > Swept Walls > Hem
Toolbar:	Walls > Swept Walls sub-toolbar > Hem

Hem is a rounded face created on the sharp edges of a sheet metal component. It is used to reduce sharpness in the sheet metal components. This makes the sheet metal components easy to handle and assemble. To create a hem, choose the **Hem** tool from the **Swept Walls** sub-toolbar; the **Hem Definition** dialog box will be displayed, as shown in Figure 15-35. Also, you will be prompted to select a guide element. Select the edge on which you want to create the hem as the guide element; the preview of the hem will be displayed with the default values. The options in the **Hem Definition** dialog box are discussed next.

By default, the **Basic** option is selected in the **Flange type** drop-down list. Therefore, the width of the hem created will be equal to the width of the selected edge. On selecting the **Relimited** option from the **Flange type** drop-down list, you need to select the limits from the geometry area. Next, you need to set the length and radius of the hem in their respective spinners. The selected guide element will be displayed in the **Spine** display box. To remove the selected guide element, choose the **Remove All** button.

To select the edges that are connected tangentially, choose the **Propagate** button. To change the material side of the flange, choose the **Reverse Direction** button. After setting the required parameters, choose the **OK** button to create the hem and exit the dialog box. Figure 15-36 shows the hem created on the side wall.

Creating a Tear Drop on the Sheet Metal Component

Menubar:	Insert > Walls > Swept Walls > Tear Drop
Toolbar:	Walls > Swept Walls sub-toolbar > Tear Drop

A tear drop is similar to a hem with the only difference that the flat wall of the tear drop will be inclined at an angle. To create a tear drop, choose the **Tear Drop** tool from the **Swept Walls** sub-toolbar; the **Tear Drop Definition** dialog box will be displayed, as shown in Figure 15-37. The options in this dialog box are discussed next.

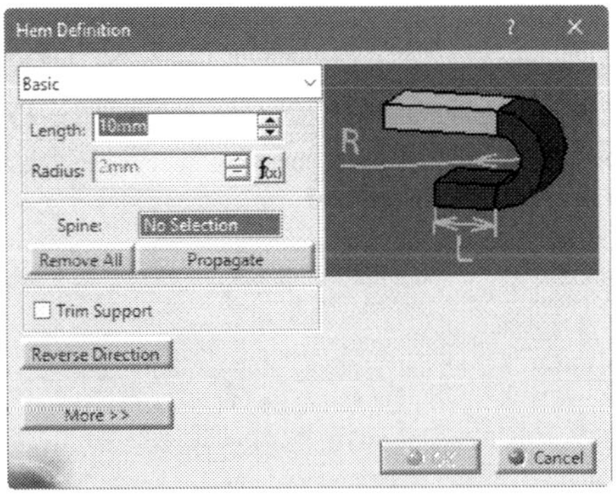

*Figure 15-35 The **Hem Definition** dialog box*

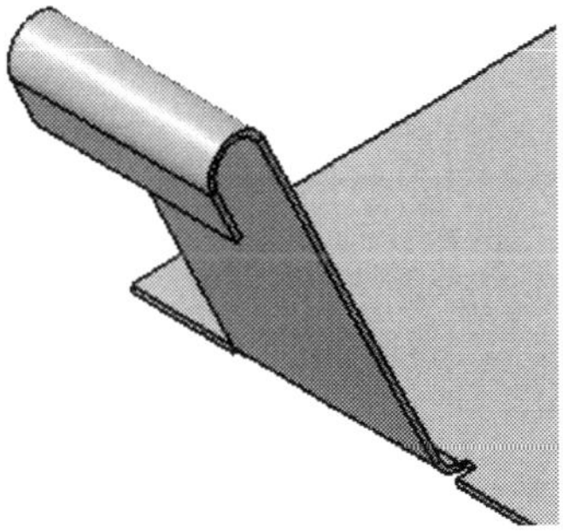

Figure 15-36 The hem created on the side wall

By default, the **Basic** option is selected in the **Flange type** drop-down list. As a result, the width of the tear drop feature created will be equal to the width of the selected edge. If you select the **Relimited** option from the **Flange type** drop-down list, you need to select the limits from the geometry area. Next, you need to set the length of the flat wall and radius of the tear drop in their respective spinners.

Select the edge on which you need to create the tear drop. The selected edge will be displayed in the **Spine** display box. To remove the selected edge, choose the **Remove All** button. To select the edges that are connected tangentially, choose the **Propagate** button. To change the material side of the flange, choose the **Reverse Direction** button. After setting the required parameters, choose the **OK** button to create the final feature. Figure 15-38 shows the tear drop feature created on the side wall.

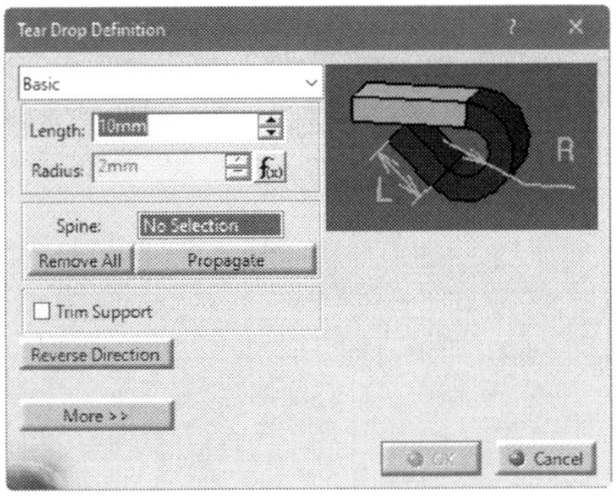

*Figure 15-37 The **Tear Drop Definition** dialog box*

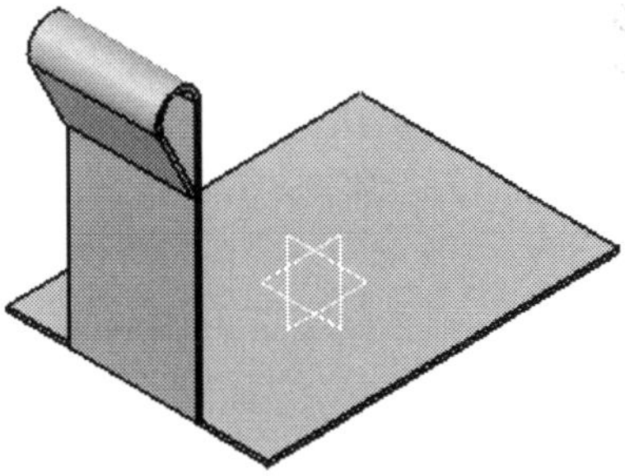

Figure 15-38 The tear drop feature created on a side wall

Creating a User Flange on the Sheet Metal Component

Menubar:	Insert > Walls > Swept Walls > User Flange
Toolbar:	Walls > Swept Walls sub-toolbar >User Flange

User flanges are created by sweeping an open sketch along an edge. To create a user flange, choose the **User Flange** tool from the **Swept Walls** sub-toolbar; the **User-Defined Flange Definition** dialog box will be displayed, as shown in Figure 15-39. The options in this dialog box are discussed next.

By default, the **Basic** option is selected in the **Flange type** drop-down list, so the width of the flange created will be equal to the width of the selected edge. If you select the **Relimited** option from the **Flange type** drop-down list, you will need to select the limits from the geometry area. Next, select the profile from the geometry area; the selected profile will be displayed in the **Profile** display box. Note that the profile must be tangent to the supporting wall. Next, select an edge along which the profile has to be swept. To remove the selected edge from the **Spine** selection box, choose the **Remove All** button. To select the edges that are connected tangentially, choose the **Propagate** button. After setting the required parameters, choose the **OK** button to create the feature.

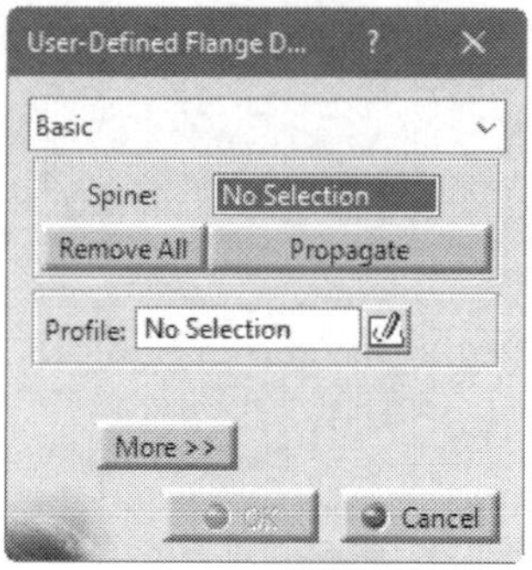

***Figure 15-39** The **User-Defined Flange Definition** dialog box*

Figure 15-40 shows the profile and the edge selected for creating the user flange and Figure 15-41 shows the resulting user flange feature.

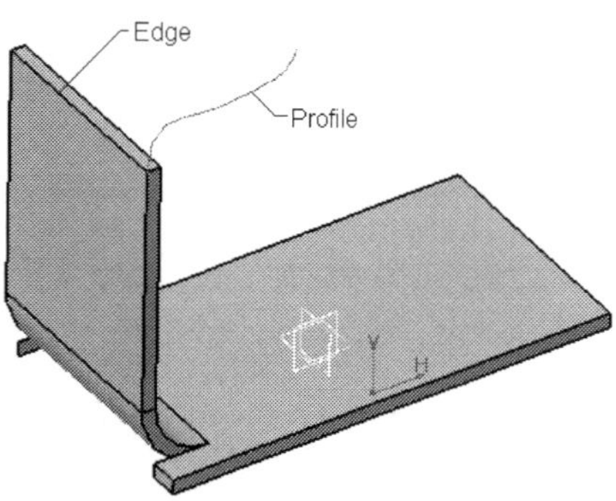

***Figure 15-40** The profile and the edge selected for creating the user flange*

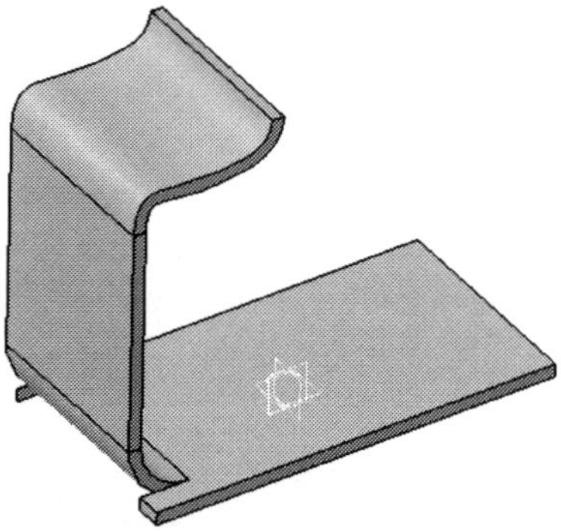

***Figure 15-41** The resulting user flange feature*

CREATING A BEND

Menubar:	Insert > Bending > Bend
Toolbar:	Bending > Bends sub-toolbar > Bend

This tool is used to create a bent face at the intersection of two walls. To create a bent face, choose the **Bend** tool from the **Bends** sub-toolbar of the **Bending** toolbar, refer to Figure 15-42; the **Bend Definition** dialog box will be displayed. The procedure to create a bend, and the options in the **Bend Definition** dialog box are discussed next.

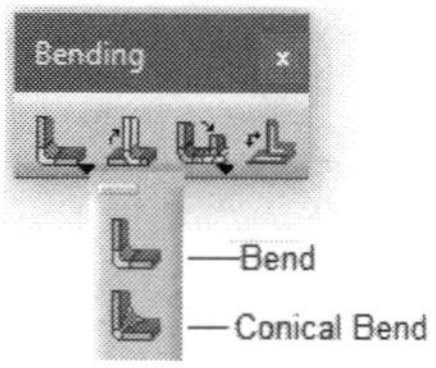

***Figure 15-42** Tools in the **Bends** sub-toolbar*

Select two walls in succession from the geometry area; the selected walls will be displayed in the **Support 1** and **Support 2** display boxes. The bend radius and the angle value that were set in the **Sheet Metal Parameters** dialog box will be displayed as the default values in the **Radius** and **Angle** spinners of the **Bend Definition** dialog box, respectively.

To change the type of relief provided to a bend, choose the **More** button; the **Bend Definition** dialog box will be expanded, as shown in Figure 15-43. The **Left Extremity** tab in the dialog box is chosen by default. The options in this tab are used to provide relief to the left side of the wall. Select the down arrow on the **Select the extremity type** button; a flyout will be displayed with various relief options. Select the type of relief from the options in the flyout. Next, choose the **Right Extremity** tab and select the type of relief to be provided on the right side of the wall.

Choose the **Preview** button to see the preview of the feature, as shown in Figure 15-44. Next, choose the **OK** button; the final bend feature will be created, as shown in Figure 15-45.

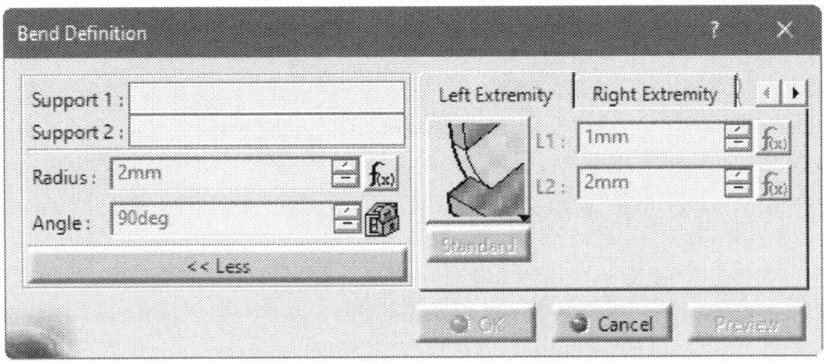

*Figure 15-43 The expanded **Bend Definition** dialog box*

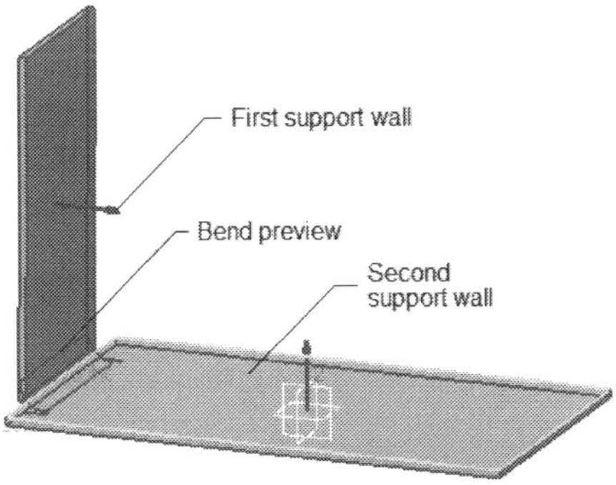

Figure 15-44 Preview of the bend feature

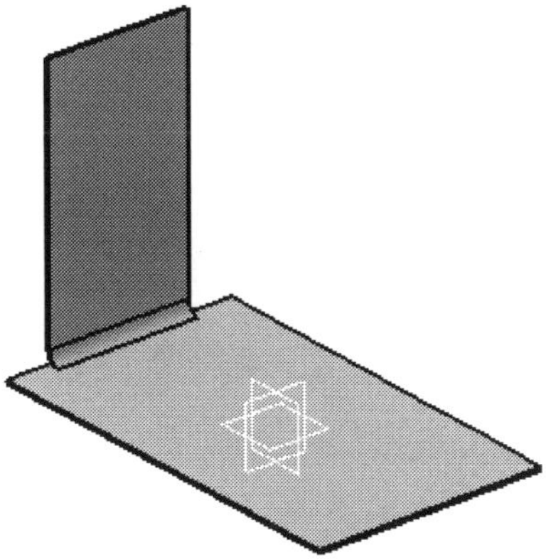

Figure 15-45 The resulting bend feature

Tip
*You can create a bend even if the two support walls are not intersecting provided the distance between the two walls is equal to or less than the default bend radius set in the **Sheet Metal Parameters** dialog box.*

Creating a Conical Bend

Menubar: Insert > Bending > Conical Bend
Toolbar: Bending > Bends sub-toolbar > Conical Bend

The **Conical Bend** tool is used to create a bend of variable radius at the intersection of two walls. The bend created using this tool will have a conical shape and varying radius from left to right of the wall, or vice-versa. To create a conical bent face, choose the

Conical Bend tool from the **Bending** toolbar; the **Bend Definition** dialog box will be displayed, as shown in Figure 15-46.

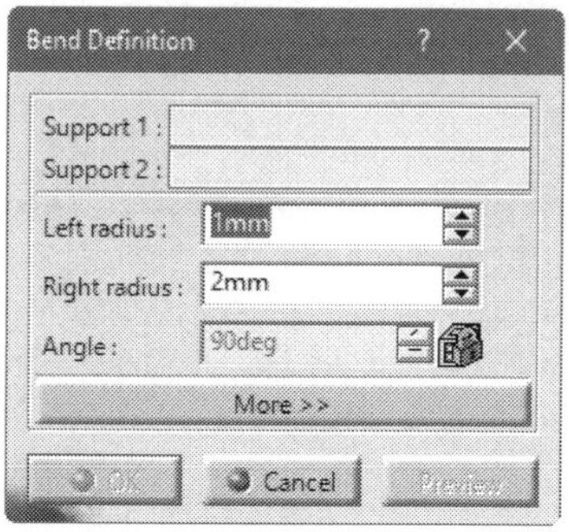

*Figure 15-46 The **Bend Definition** dialog box*

Choose two support walls in succession from the geometry area; the selected walls will be displayed in the **Support 1** and **Support 2** display boxes, respectively. Set the radius values in the **Left radius** and **Right radius** spinners, respectively. To add or modify the type of relief in the bend, choose the **More** button; the **Bend Definition** dialog box will be expanded. The **Left Extremity** tab is chosen by default. Now, select the type of relief from the flyout of the **Select the extremity type** button. Next, choose the **Right Extremity** tab and select the type of relief for the right side of the wall.

You can see the preview of the sheet metal feature by choosing the **Preview** button from the **Bend Definition** dialog box, as shown in Figure 15-47. Next, choose the **OK** button; the resulting conical bend feature will be created, as shown in Figure 15-48.

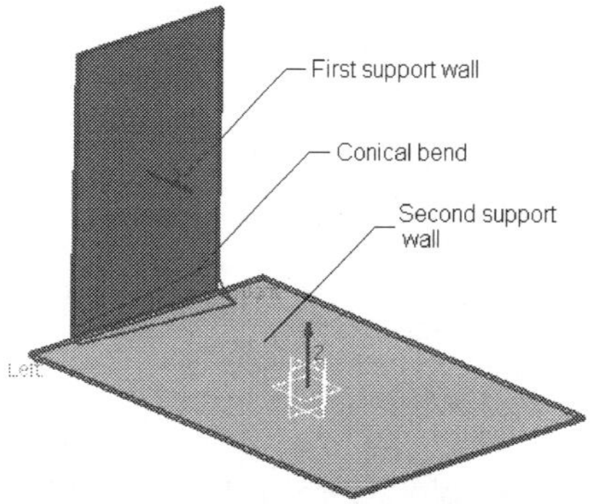

Figure 15-47 Preview of the conical bend feature

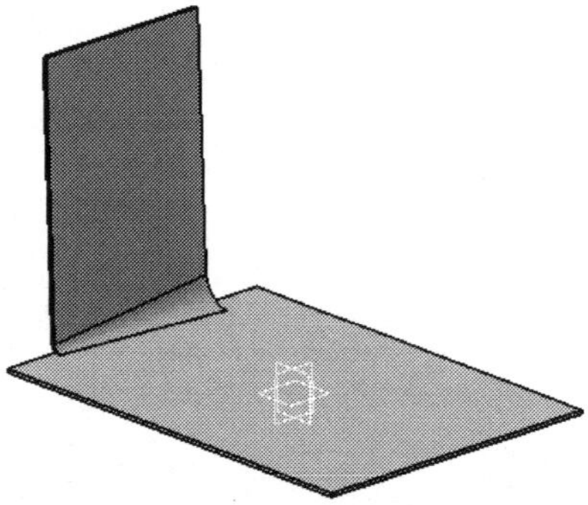

Figure 15-48 Resulting conical bend feature

BEND FROM FLAT

Menubar: Insert > Bending > Bend From Flat
Toolbar: Bending > Bend From Flat

CATIA V5 allows you to bend a part of a sheet metal component by using the **Bend From Flat** tool. Remember that in such cases the sheet metal part will be folded along the different lines. To invoke this tool, choose the **Bend From Flat** tool from the **Bending** toolbar; the **Bend From Flat Definition** dialog box will be displayed, as shown in Figure 15-49. The options in the **Bend From Flat Definition** dialog box are discussed next.

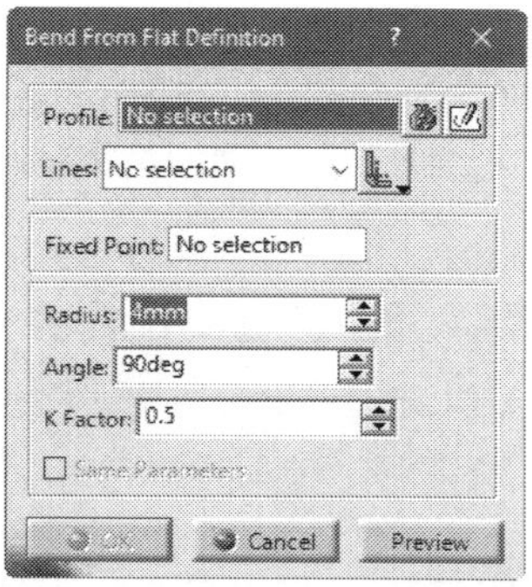

*Figure 15-49 The **Bend From Flat Definition** dialog box*

Profile

Select the sketch from the geometry area and it will be displayed in the **Profile** display box. The selected sketch can contain more than one line, but these should be noncontinuous ones. You can also create the profile by using the **Sketch** button in the **Bend From Flat Definition** dialog box.

Lines

If the selected profile has more than a line, then all lines will be listed in the **Lines** drop-down list. Select the line along which you need to create the bend from the **Lines** drop-down list. Next, select the down arrow on the button at the right of the **Lines** drop-down list; a flyout will be displayed with five buttons. The buttons in the flyout are discussed next.

Axis

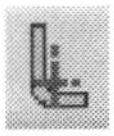

When you choose this button, the bend line will be placed at the centre of the bend and the bend will be created equally in both the directions of the bend line, as shown in Figure 15-50.

BTL (Bent Tangent Line) Base Feature

When you choose this button, the bend line will be placed at the start of the bend, as shown in Figure 15-51.

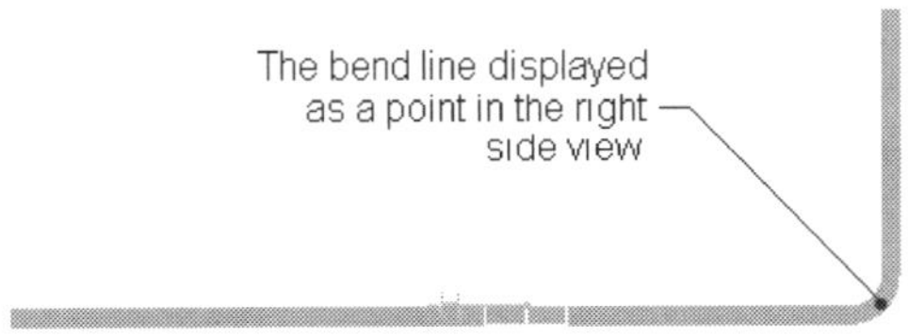

*Figure 15-50 Bend created using the **Axis** button*

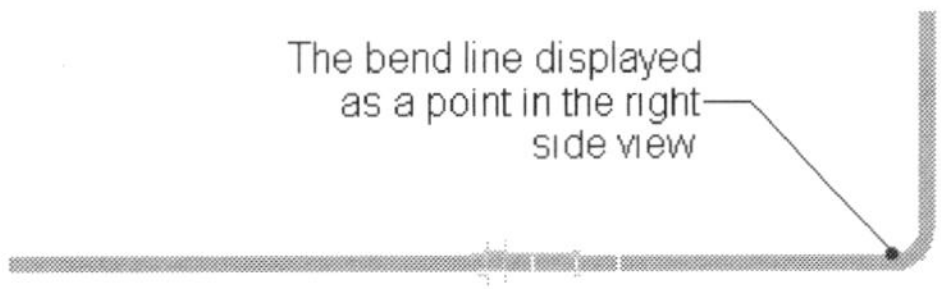

*Figure 15-51 Bend created using the **Bent Tangent Line** button*

IML (Inner Mold Line)

When you choose this button, the bend line will be placed at the apparent intersection of the inner surface of both the horizontal and vertical walls, as shown in Figure 15-52.

OML (Outer Mold Line)

On choosing this button, the bend line will be placed at the apparent intersection of the outer surface of both the horizontal and vertical walls, as shown in Figure 15-53.

BTL Support

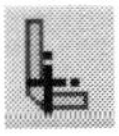

When you choose this button, the bend line will be placed at the end of the bend, Figure 15-54 shows the proposed bend in the top view. Figure 15-55 shows the preview of the bend in the left side view.

Fixed Point

Once you select the sketch, the fixed point is automatically set on the face where the sketch is positioned. This means that no portion of the wall will move when the bend is created. You can change the fixed point by selecting another point on the wall. The fixed point is generally the start point of the sketch. To create a new fixed point, right-click on the **Fixed Point** display box, select the **Create Point** option, and define a point.

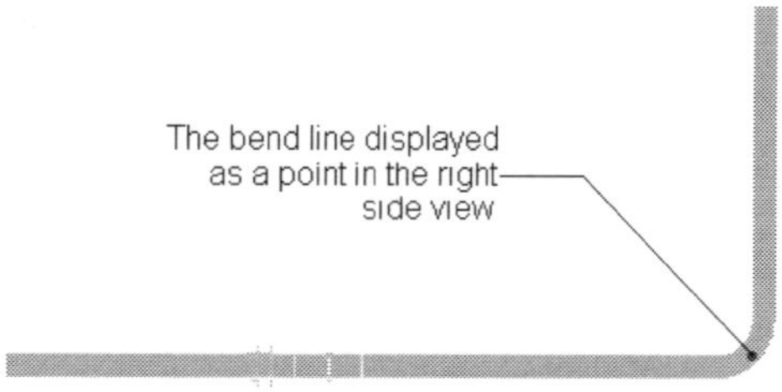

***Figure 15-52** Bend created using the **Inner Mold Line** button*

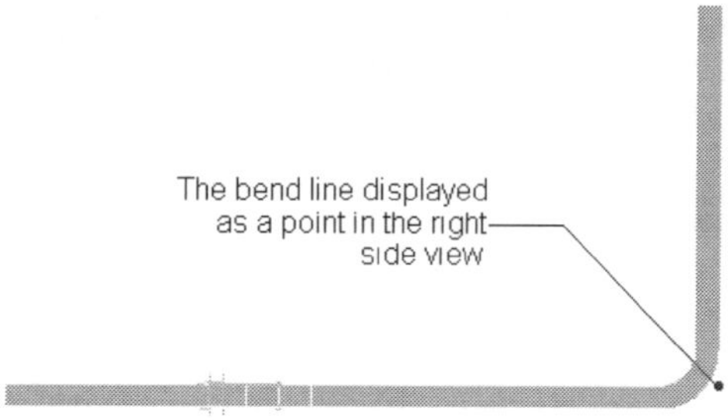

***Figure 15-53** Bend created using the **Outer Mold Line** button*

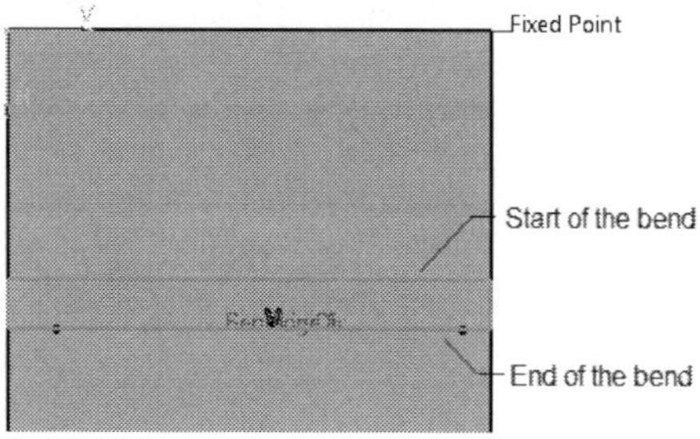

***Figure 15-54** Proposed bend in the top view*

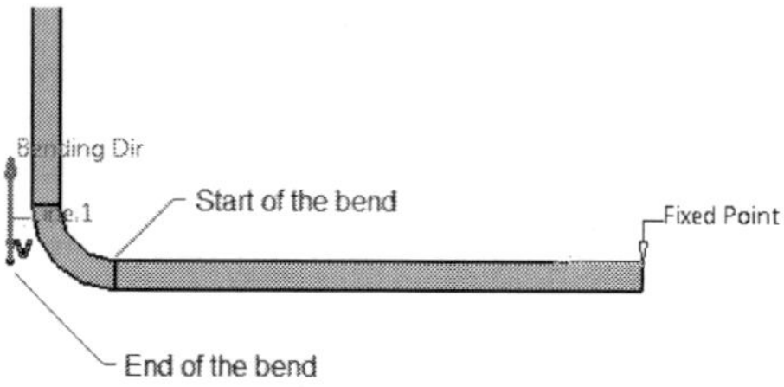

***Figure 15-55** Preview of the bend in the left side view*

Radius

By default, the value of the bend radius is the value that was set earlier in the **Sheet Metal Parameters** dialog box. You can also change the radius value in the **Radius** spinner. To do so, right-click on the **Radius** spinner and then choose **Formula > Deactivate** from the contextual menu displayed.

Angle

The **Angle** spinner is used to set the angle of the fold for the sheet metal component. The default value in this edit box is 90-degree. You can set a value below 180-degree in this spinner. Figure 15-56 shows the sheet metal part folded by an angle of 90-degree along Line 1 and 45-degree along Line 2. To bend a component using two lines, first you need to create a sketch consisting of two lines in the same sketching plane. Next, invoke the **Bend From Flat Definition** dialog box. In this dialog box, specify the angle value for these lines in their respective **Angle** spinners. Note that if you select **Line 1** in the **Lines** drop-down list, then you can specify the angle value only for Line1. You can repeat the same procedure for specifying the angle value for Line 2. To reverse the bending direction, click on the blue colored arrow in the preview. You can also apply the first line parameter values to all lines by selecting the **Same Parameters** check box and then choosing the **OK** button from the **Bend From Flat Definition** dialog box.

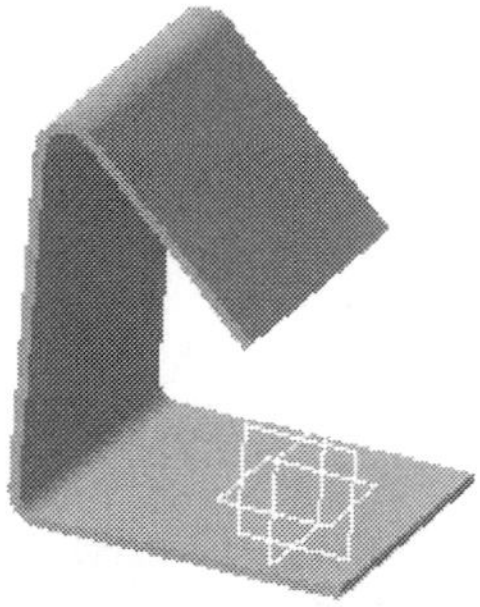

***Figure 15-56** Sheet metal part after folding*

CREATING ROLLED WALLS

Rolled walls are created by rolling a sheet metal onto a shape. Different types of rolled walls are **Hopper**, **Free Form surface,** and so on. These walls are discussed next.

Creating a Hopper Wall

Menubar:	Insert > Rolled Walls > Hopper
Toolbar:	Rolled Walls > Hopper

The **Hopper** tool is used to create a surfacic hopper or a canonic hopper. To create a surfacic hopper, first you need to create sections. The planes of the sections can be parallel or non-parallel. Note that the first profile and the second profile should be similar and should not have sharp edges. Next, choose the **Hopper** tool from the **Rolled Walls** toolbar; the **Hopper** dialog box will be displayed, as shown in Figure 15-57. In this dialog box, select the **Surfacic Hopper** option from the **Hopper Type** drop-down list. Next, right-click in the **Selection** display box in the **Surface** area; a contextual menu will be displayed. Next, choose the **Create Multi-sections Surface** option; the **Multi-Sections Surface Definition** dialog box will be displayed. Select the cross-sections and add couplings as discussed in the Chapter10 under the Creating Multi-Sections Surfaces heading, as shown in Figure 15-58. If you want to offset a surfacic hopper, specify the required offset value in the **Offset** edit box. Next, choose the **OK** button; a preview of the unfolded hopper will be displayed.

Next, click in the **Reference wire** display box to activate it and select an edge on the hopper surface as the reference wire, refer to Figure 15-59. Click in the **Invariant point** display box to activate it and select a point on the surface as the invariant point. Note that the point should lie on the surface as well as the selected reference wire, refer to Figure 15-59. Click in the **Tear wires** display box and then select an edge on the surface. Next, choose the **OK** button; the hopper will be created, as shown in Figure 15-60.

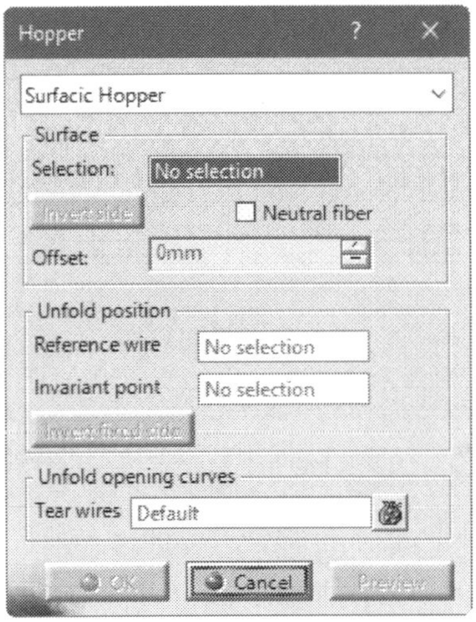

Figure 15-57 The ***Hopper*** *dialog box*

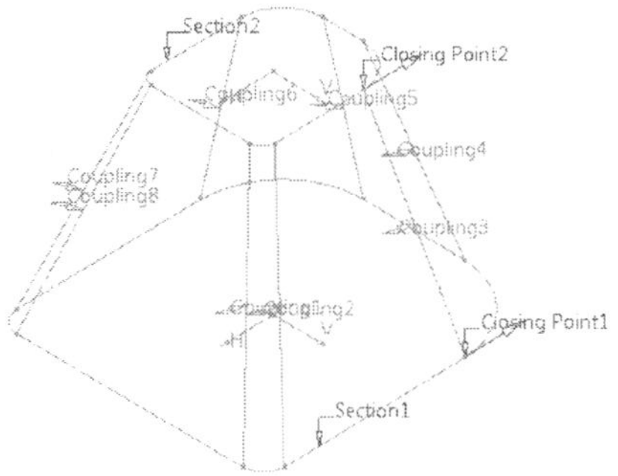

Figure 15-58 Sections and Couplings

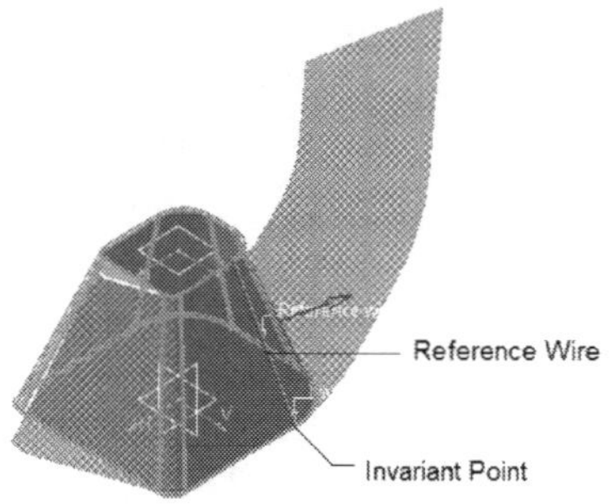

Figure 15-59 Reference wire and the invariant point to be selected

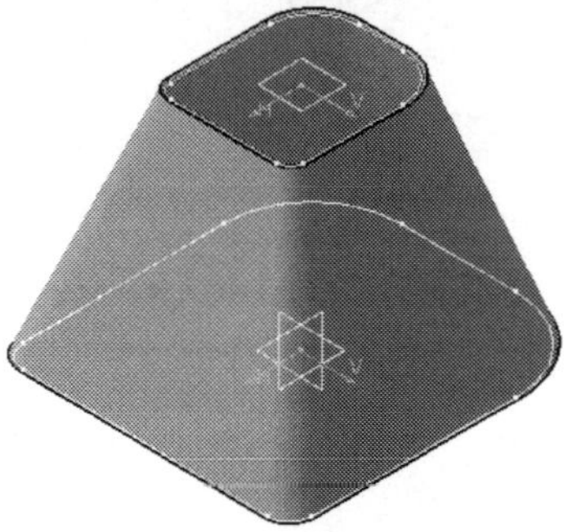

Figure 15-60 Resulting surfacic hopper

To create a canonic hopper, choose the **Canonic Hopper** option from the **Hopper Type** drop-down list; the **Hopper** dialog box will be modified, as shown in Figure 15-61. Select the first and second profile, as shown in Figure 15-62. Note that the first profile and the second profile should be similar and consist of arcs and straight lines. Also, the centers of the arcs of the two profiles should coincide when projected onto each other. Next, you need to specify an opening line of the hopper which is drawn by using the first point and the second point, refer to Figure 15-62. Note that the first point of the opening line should lie on the first profile and the second point on the second profile. You can choose the **Invert fixed side** button to change the fixed side of the unfolded view of the hopper. The **Invert material side** button is used to invert the side on which the material is to be added. Next, choose the **OK** button; the canonic hopper will be created. To create a lofted bend, you need to select the **Generate lofted bend** check box from the **Lofted bend** area. On selecting this check box, the **Approximate by** drop-down list gets displayed. You can define the approximation of lofted bend by selecting the options available in the **Approximate by** drop-down list.

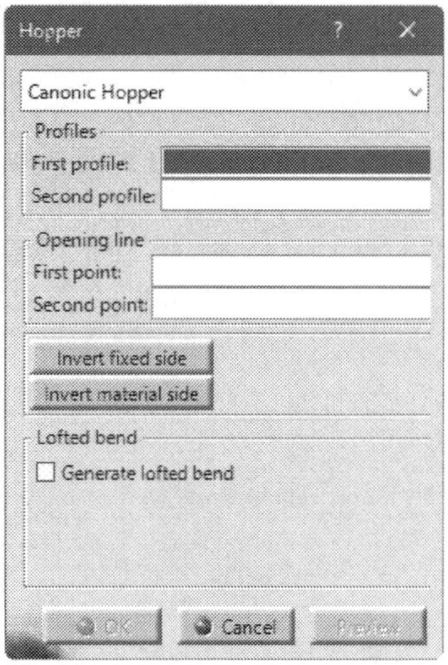

*Figure 15-61 The **Hopper** dialog box to create a canonic hopper*

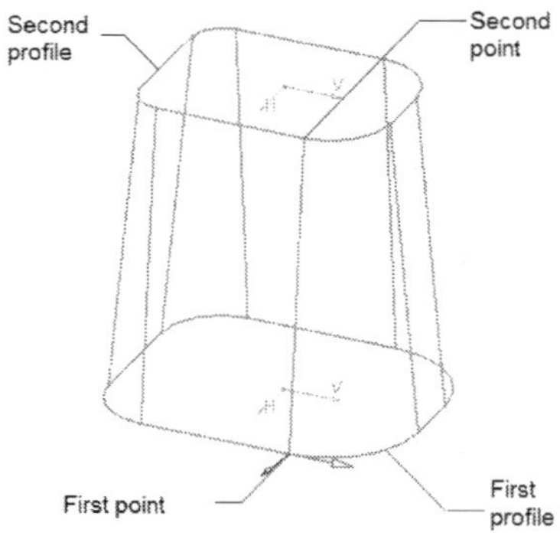

Figure 15-62 Profiles and the points to be selected to create a canonic hopper

Creating a Rolled Wall

Menubar:	Insert > Rolled Walls > Rolled Wall
Toolbar:	Rolled Walls > Rolled Wall

The **Rolled Wall** tool is used to create rolled sheet metal components such as pipes, open pipes, and so on. To create a rolled wall, first you need to create a circle or an arc. Next, choose the **Rolled Wall** tool from the **Rolled Walls** toolbar; the **Rolled Walls** dialog box will be displayed. Choose the **Dimension** option from the **Type** drop-down list and then select the sketch; a preview of the rolled wall will be displayed, as shown in Figure 15-63. You can specify its length in the **Length** spinner. You can specify the reference point to unfold the rolled wall by using the options from the **Sketch Location** drop-down list. After specifying the parameters in the dialog box, choose the **OK** button; the rolled wall will be created, as shown in Figure 15-64.

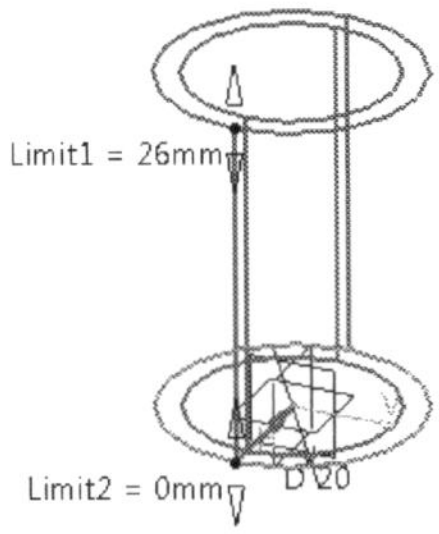

Figure 15-63 *Preview of the rolled wall feature*

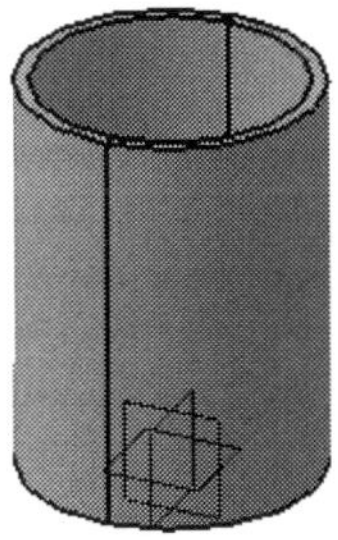

Figure 15-64 *Resultant rolled wall*

FOLDING AND UNFOLDING SHEET METAL PARTS

Sometimes, you need to perform operations like cutout, hole, and so on across the bend face of a wall, as shown in Figure 15-65. You cannot perform these operations in bend condition. Therefore, you need to unfold the sheet metal part, add features, and fold it again. The folding and unfolding operations can be carried out by using the tools available in the **Folding/Unfolding** sub-toolbar in the **Bending** toolbar refer to Figure 15-66. These tools are discussed next.

Figure 15-65 *Cutout operation performed across the bend face*

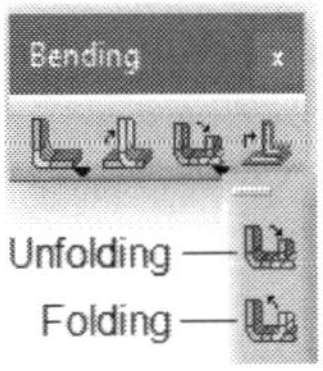

Figure 15-66 *Tools in the **Folding/Unfolding** sub-toolbar*

Unfolding Sheet Metal Parts

Menubar:	Insert > Bending > Unfolding
Toolbar:	Bending > Folding/Unfolding sub-toolbar > Unfolding

The **Unfolding** tool is used to unfold the bent face of a sheet metal part. To invoke this tool, choose the **Unfolding** tool from the **Folding/Unfolding** sub-toolbar; the **Unfolding Definition** dialog box will be displayed, as shown in Figure 15-67.

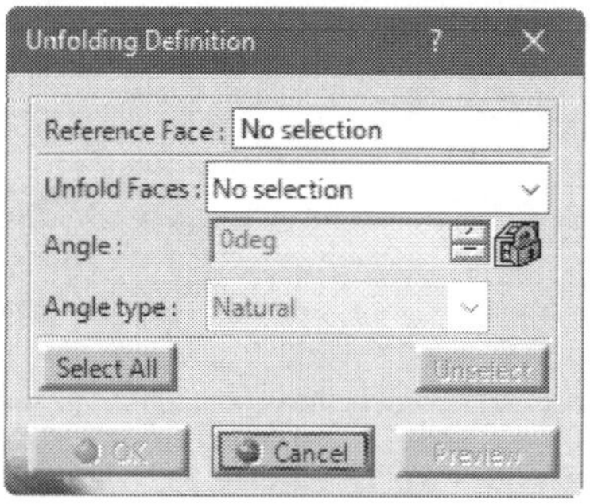

*Figure 15-67 The **Unfolding Definition** dialog box*

On invoking this dialog box, you will be prompted to select a reference face. Select a face of the wall from the geometry area; it will be displayed in the **Reference Face** display box. Next, you will be prompted to select a face to unfold. If you need to unfold a particular bend, select it from the drawing area, as shown in Figure 15-68. To unfold all bends, choose the **Select All** button. By choosing the **Unselect** button you can clear the selected bends. Next, choose the **OK** button to unfold the part, as shown in Figure 15-69.

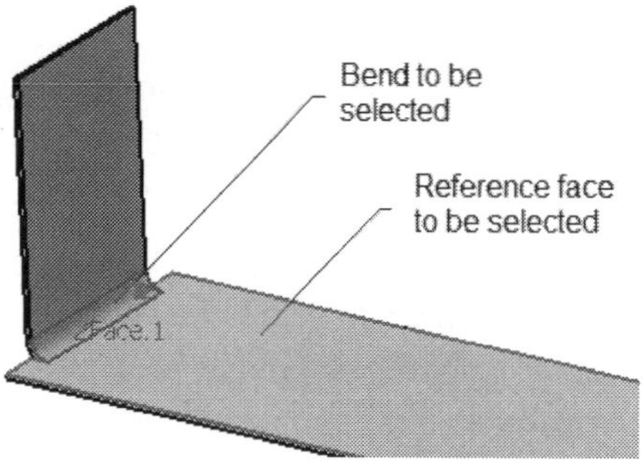

Figure 15-68 The reference face and the bend to be selected

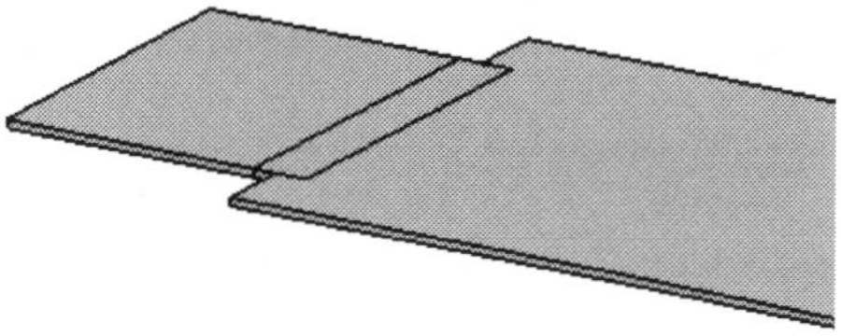

Figure 15-69 The unfolded part

Folding Unfolded Parts

Menubar: Insert > Bending > Folding
Toolbar: Bending > Folding/Unfolding sub-toolbar > Folding

The **Folding** tool is used to fold the wall that is unfolded by using the**Unfolding** tool, back to its original position. To do so, choose the **Folding** tool from the **Folding/Unfolding** sub-toolbar; the **Folding Definition** dialog box will be displayed, as shown in Figure 15-70. Also, you will be prompted to select a reference face. Select a face of the wall from the geometry area; name of the selected face will be displayed in the **Reference Face** display box. Next, select the bend section; name of the selected bend section will be displayed in the **Fold Faces** drop-down list. The **Angle** spinner is used to specify the angle upto which the unfolded wall has to be folded. By default, the angular value is the difference between 180-degree and the bend angle of the original part. Select an option from the **Angle type** drop-down list to specify the angle type of the resulting folded part. The options in the **Angle type** drop-down list are discussed next.

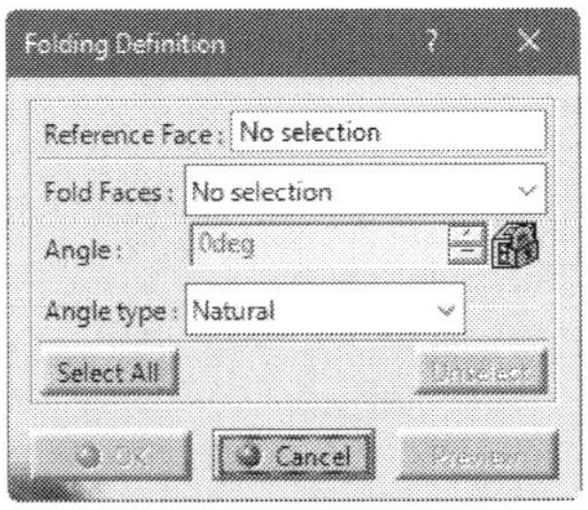

*Figure 15-70 The **Folding Definition** dialog box*

Natural

By default, this option is selected and the wall will be folded back to its original position. So, the **Angle** spinner will not be available.

Defined

Select this option if you need to fold the wall to an angle other than its original position. Specify the angle in the **Angle** spinner.

Spring back

Select this option if you need to fold the wall at an angle with respect to its original position. Specify the angle in the **Angle** spinner.

By choosing the **Select All** button, you can select all the unfolded walls. By choosing the **Unselect** button, you can deselect the selected unfolded walls. Choose the **Preview** button; a preview of the folding feature will be displayed, as shown in Figure 15-71. Next, choose the **OK** button to create the final feature, as shown in Figure 15-72.

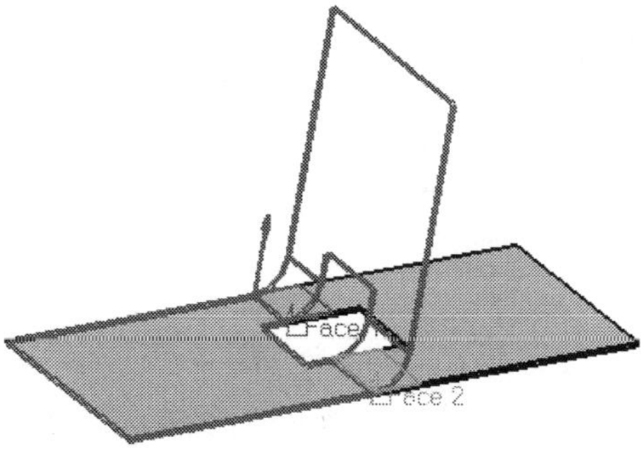

Figure 15-71 Preview of the folding feature

Figure 15-72 Resulting folded feature

Mapping the Geometry

Menubar:	Insert > Bending > Point or Curve Mapping
Toolbar:	Bending > Point or Curve Mapping

The **Point or Curve Mapping** tool is used to create a geometrical object, such as a point or a curve, and recognize it as a sheet metal element. This tool is specially used for generating a logo, defining an area for chemical milling, or creating a cut on components. To create a geometrical object, draw the required sketch in the **Sketcher** environment and then invoke the **Unfold object definition** dialog box by choosing the **Point or Curve Mapping** tool from the **Bending** toolbar, refer to Figure 15-73.

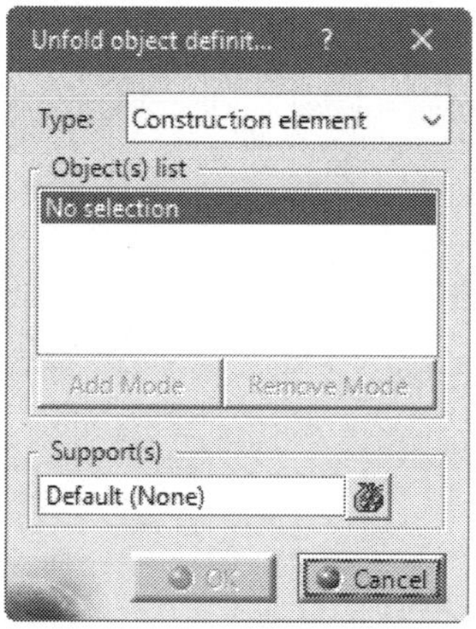

*Figure 15-73 The **Unfold object definition** dialog box*

Next, select the sketch from the geometry area; the selected sketch will be displayed in the **Object(s) list** display box and its preview will be displayed in the geometry area. Figure 15-74 shows the sketch for mapping and Figure 15-75 shows the preview of the mapping. The **Add Mode** and **Remove Mode** buttons are used to add and remove elements from the **Object(s) list** display box, respectively. The **Support(s)** display box is used to select the supported wall with respect to the sketch.

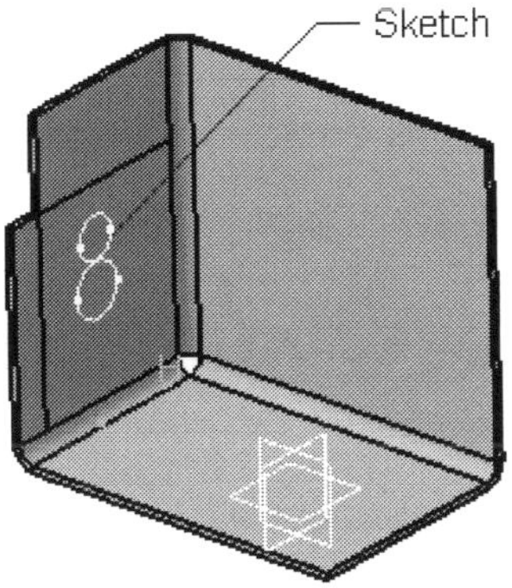

Figure 15-74 *Sketch for mapping*

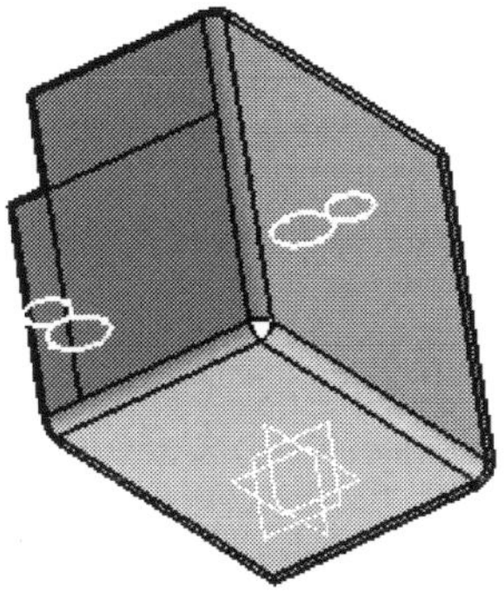

Figure 15-75 *Preview of the mapping*

CREATING FLAT PATTERNS OF SHEET METAL COMPONENTS

Menubar:	Insert > Views > Fold/Unfold
Toolbar:	Views > Fold/Unfold sub-toolbar > Fold/Unfold

The flattened view of a sheet metal component plays an important role during the manufacturing stage. The flattened view of the sheet metal component is required so that it can be cut in exact size and shape and various operations can be performed on it.

To create a flat pattern of a folded sheet metal component, choose the **Fold/Unfold** button from the **Fold/Unfold** sub-toolbar of the **Views** toolbar (see Figure 15-76); the flat pattern will be created. As this is a toggle button, you can choose it again to fold back the sheet metal components. Note that you cannot make any modifications in the unfolded sheet metal components. Figure 15-77 shows a sheet metal component and Figure 15-78 shows the flat pattern of the same component.

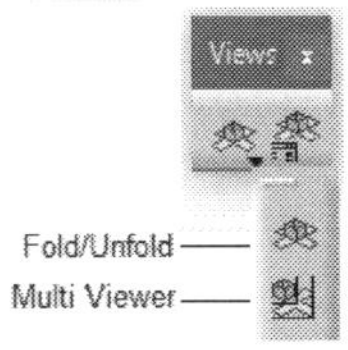

Figure 15-76 *Tools in the* ***Fold/Unfold*** *sub-toolbar*

Figure 15-77 *Sheet metal component*

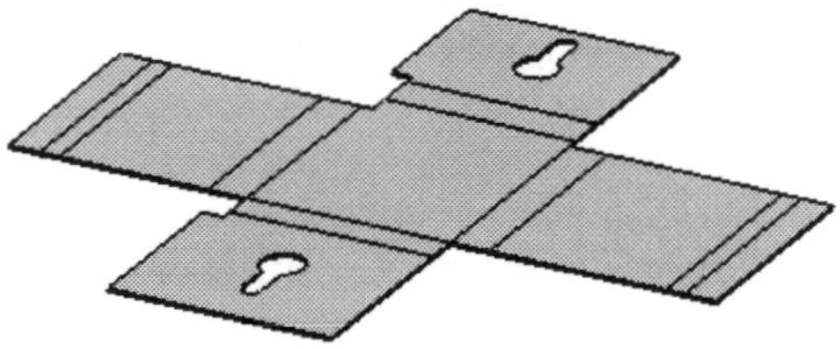

Figure 15-78 *Flat pattern of the sheet metal component*

VIEWING A SHEET METAL COMPONENT IN MULTIPLE WINDOWS

Menubar:	Insert > Views > Multi Viewer
Toolbar:	Views > Fold/Unfold sub-toolbar > Multi Viewer

The **Multi Viewer** tool is used to unfold a sheet metal component and display it in another window. To do so, choose the **Multi Viewer** button from the **Views** toolbar; the part, which is currently the folded part, will be displayed as the unfolded part in the second window. To view both the windows on the same screen, choose **Window > Tile Horizontally** from the menu bar. If you modify any one view, the modification will be reflected in the other view also.

USING VIEWS MANAGEMENT

Menubar:	Insert > Views > Views Management
Toolbar:	Views > Views Management

The **Views Management** tool is used to manage the activation or deactivation of views. On invoking this tool from the **Views** toolbar, the **Views** dialog box will be displayed, as shown in Figure 15-79. Alternatively, right-click on the **PartBody** feature from the Specification tree; a contextual menu will be displayed. Choose **PartBody object > Views** from the contextual menu to display the **Views** dialog box.

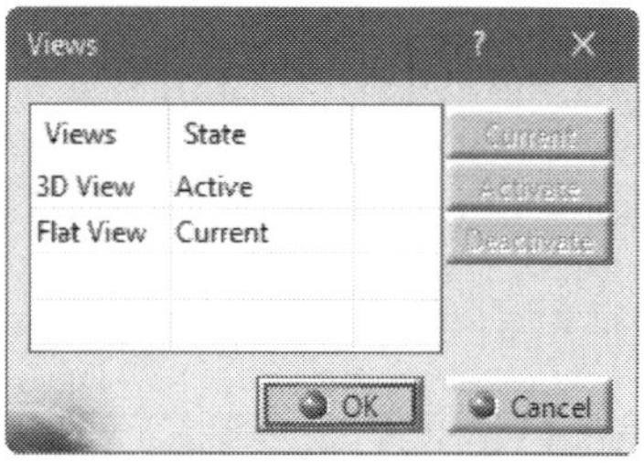

Figure 15-79 *The* ***Views*** *dialog box*

In this dialog box, the state of the **3D View** is **Current** which means the part is folded currently. To make the state of the **Flat View** as **Current**, select it from the dialog box; the **Current** and **Deactivate** buttons will be enabled in the dialog box. Next, choose the **Current** button; the part is displayed as flat pattern. If you want to make the folded view again as the current view, first select the **3D View** and then choose the **Current** button. You can activate and deactivate the views by choosing the **Activate** and **Deactivate** buttons from this dialog box.

STAMPING

Stamping is a metal working process in which a sheet is punched in a press tool. This process is used to create stamping features such as blanking, piercing, and forming on the sheet metal part. CATIA V5 provides a set of tools to design a sheet metal part that can be created by the stamping process. These tools are grouped in the **Cutting/Stamping** sub-toolbar, as shown in Figure 15-80. These tools are discussed next.

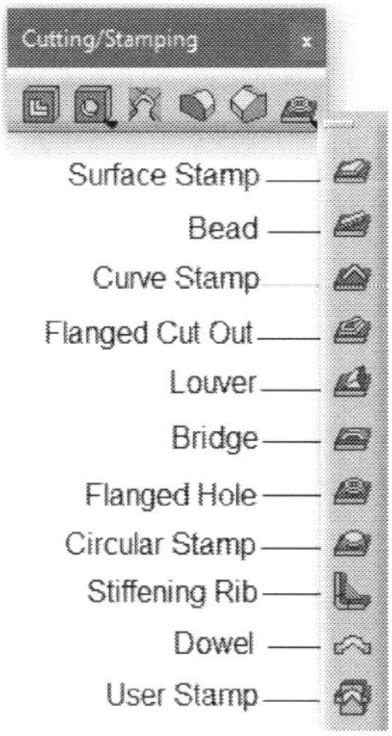

Figure 15-80 Tools in the ***Cutting/Stamping*** *sub-toolbar*

Note

You can create holes, fillets, and chamfer features using the tools in the ***Cutting/Stamping*** *toolbar. The procedure to create these features is similar to that of part modeling.*

Creating a Surface Stamp

Menubar: Insert > Stamping > Surface Stamp
Toolbar: Cutting/Stamping > Stamping sub-toolbar > Surface Stamp

In CATIA V5, you can emboss an area enclosed by a closed profile using the **Surface Stamp** tool. To do so, create a close sketch on a wall, as shown in Figure 15-81. Next, choose the **Surface Stamp** tool from the **Cutting/Stamping** toolbar; the **Surface Stamp Definition** dialog box will be displayed, as shown in Figure 15-82. Select the profile from the geometry area, if it is not already selected; the selected profile will be displayed in the **Profile** display box. The other options in the **Surface Stamp Definition** dialog box are discussed next.

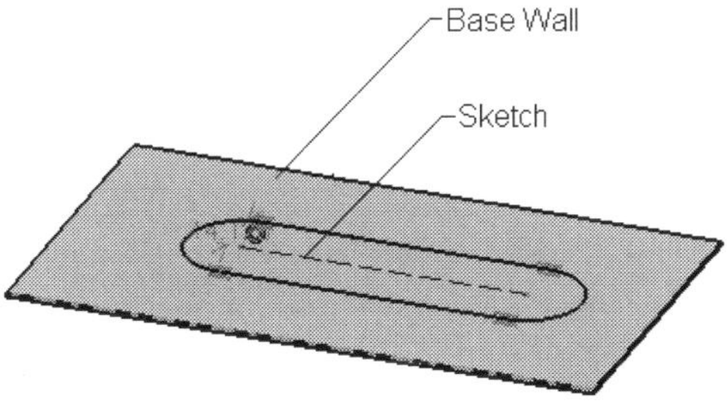

Figure 15-81 Sketch of the surface stamp feature

Definition Type Area

The options in the **Definition Type** area are discussed next.

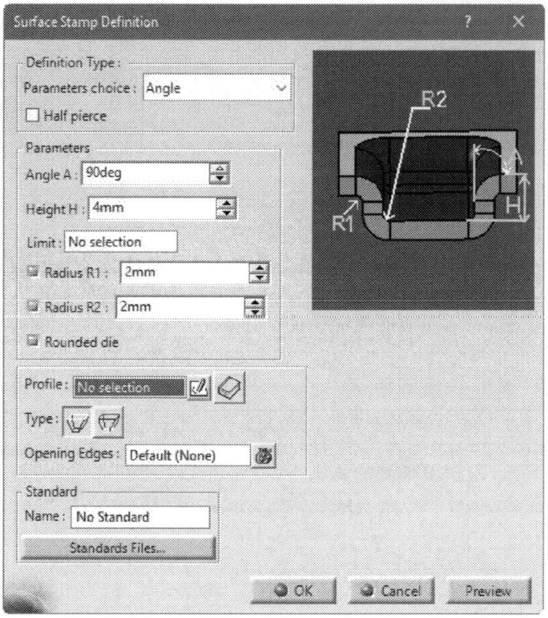

*Figure 15-82 The **Surface Stamp Definition** dialog box*

Parameters choice

The three options available in the **Parameters choice** drop-down list are **Angle**, **Punch & Die**, and **Two profiles**. By default, the **Angle** option is selected in the **Parameters choice** drop-down list.

Half pierce

If you select this check box, the stamping will be created such that the height is equal to the half of the sheet thickness. However, if needed you can define the height upto the value of sheet thickness. But the **Angle A**, **Radius R1**, **Radius R2**, **Rounded die**, and **Opening Edges** options will not be active in the dialog box.

Parameters Area

When the **Angle** option is selected in the **Parameters choice** drop-down list, the following options will become available in the **Parameters** area:

Angle A

Set the angular value of the stamp feature in this spinner. The angle will be provided between the wall of the surface stamp and the face where the profile is drawn. The default value in the **Angle A** spinner is **90deg**.

Height H

The height of the stamp feature is specified in the **Height H** spinner. Note that this spinner will not be available if you specify the height using the **Limit** option.

Limit

You can also specify the height of the stamp feature by specifying the limits. The limits specified can be a surface or plane. To create a plane, right-click in the **Limit** display box and then choose the **Create Plane** option from the contextual menu.

Radius R1, Radius R2, and Rounded Die

Select the **Radius R1** and **Radius R2** check boxes to create fillets on the top and bottom surfaces of the stamp, respectively, refer to Figure 15-83. Set the fillet radii in the **Radius R1** and **Radius R2** spinners. If you select the **Rounded die** check box, the side edges of the stamp feature will be filleted. Figure 15-83 shows the surface stamp feature with the **Rounded die** check box cleared. Figure 15-84 shows the surface stamp feature with the **Rounded die** check box selected.

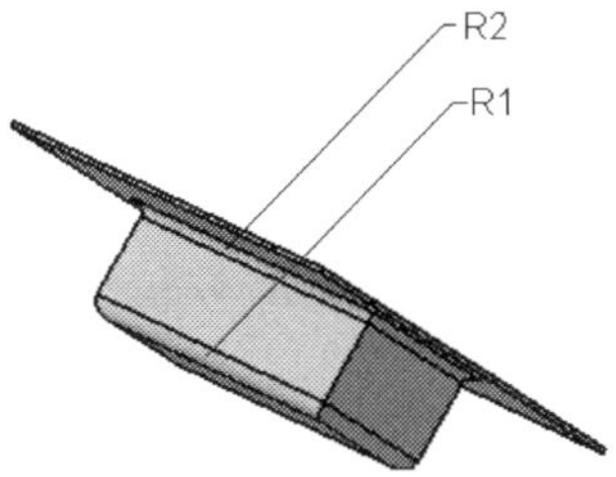

Figure 15-83 *Stamp feature without the side edges filleted*

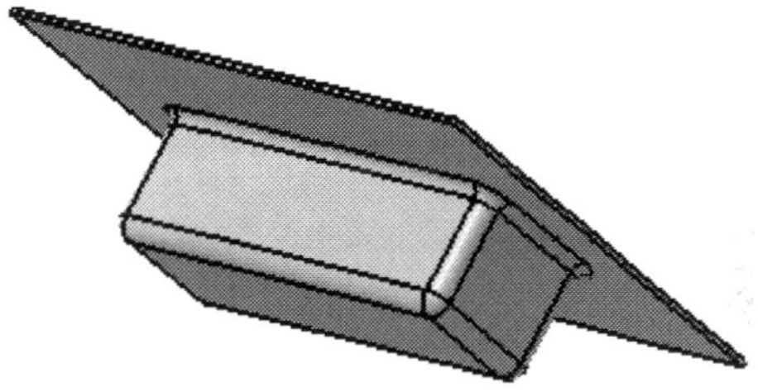

Figure 15-84 *Stamp feature with the side edges filleted*

Profile Area

The options in the **Profile** area are discussed next

Profile Display box

The **Profile** display box displays the name of the profile selected for the surface stamp. You can sketch the profile of the surface stamp by using the **Sketch** button available on the right of the **Profile** display box.

Type buttons

On choosing the **Upward sketch profile** button in this area, the dimensions of the top face of the stamp feature will be equal to the sketch selected initially, as shown in Figure 15-85. On choosing the **Downward sketch profile** button, the dimensions of the bottom of the stamp feature will be equal to the sketch selected initially, as shown in Figure 15-86.

Opening Edges

You can also create a stamp feature with an open wall. To do so, click in the **Opening Edges** display box and select a sketch entity from the profile; the stamp feature with an open wall will be created at the selected edge, as shown in Figure 15-87. To see the preview of the resulting feature, choose the **Preview** button.

After specifying the required parameters, choose the **OK** button from the **Surface Stamp Definition** dialog box; the surface stamp feature will be created, as shown in Figure 15-88.

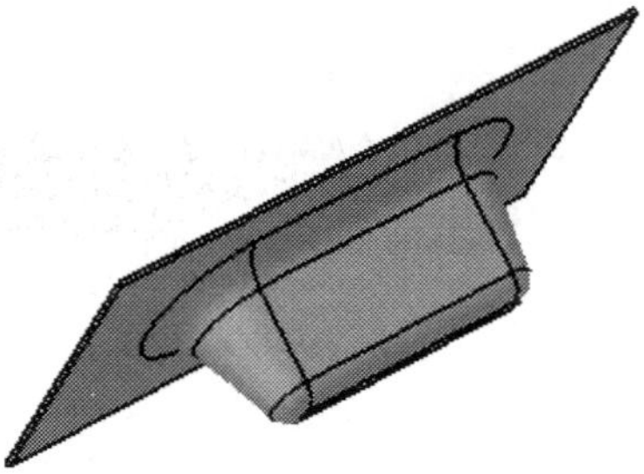

*Figure 15-85 Stamp feature on choosing the **Upward sketch profile** button*

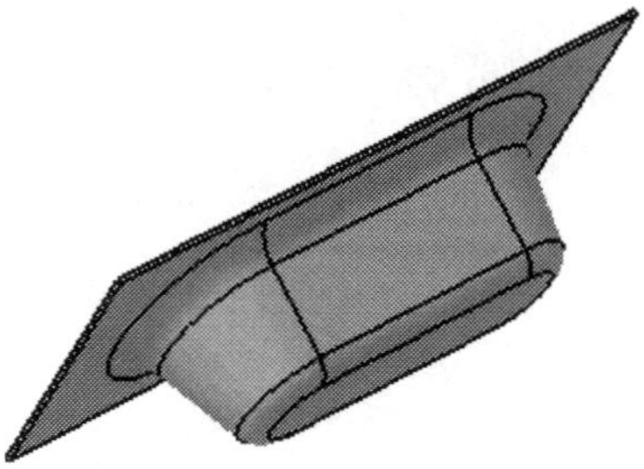

*Figure 15-86 Stamp feature on choosing the **Downward sketch profile** button*

Figure 15-87 The stamp feature with the open wall

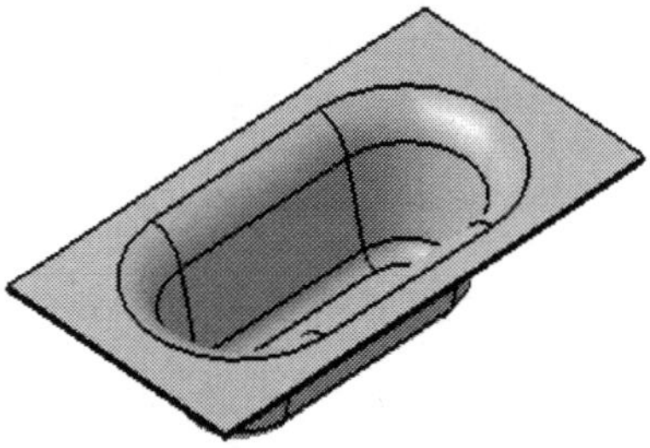

Figure 15-88 The surface stamp feature

Creating a Bead Stamp

Menubar:	Insert > Stamping > Bead
Toolbar:	Cutting/Stamping > Stamping sub-toolbar > Bead

The **Bead** tool is used to create an embossed or an engraved bead on a sheet metal component along an open sketch. If the sketch has multiple entities, they should be tangentially connected. Else, you will get an error message. Choose the **Bead** tool from the **Cutting/Stamping** toolbar; the **Bead Definition** dialog box will be displayed, as shown in Figure 15-89.

Also, you will be prompted to select a profile. Select it from the geometry area. Next, set the values of the section radius and the end radius in their respective spinners. Note that the value of the end radius should be greater or equal to section radius. Set the height of the stamp in the **Height H** spinner. Various parameters of the bead feature are shown in Figure 15-90. If you clear the **Radius R** check box, the resulting bead feature will be created without a fillet, as shown in Figure 15-91. You can change the direction of the bead by clicking on the orange arrowhead in the geometry area. To see the preview of the feature, choose the **Preview** button. Next, choose the **OK** button; the bead feature will be created.

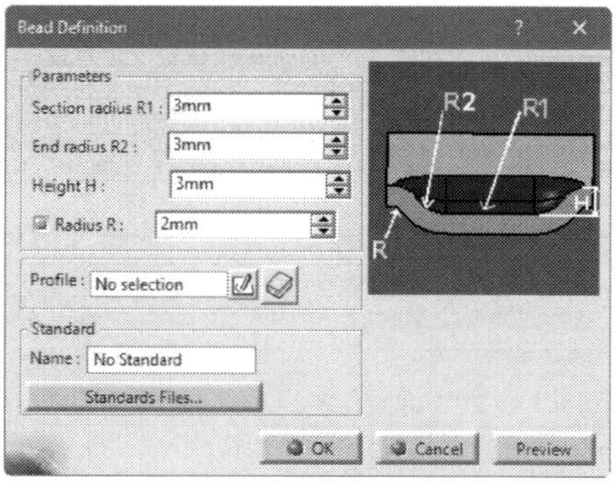

*Figure 15-89 The **Bead Definition** dialog box*

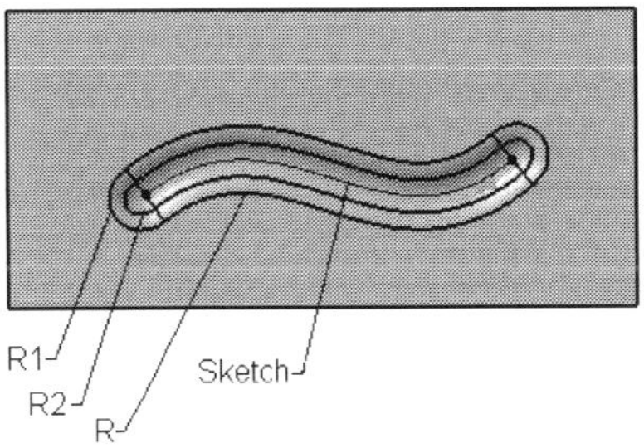

Figure 15-90 Parameters of the bead feature

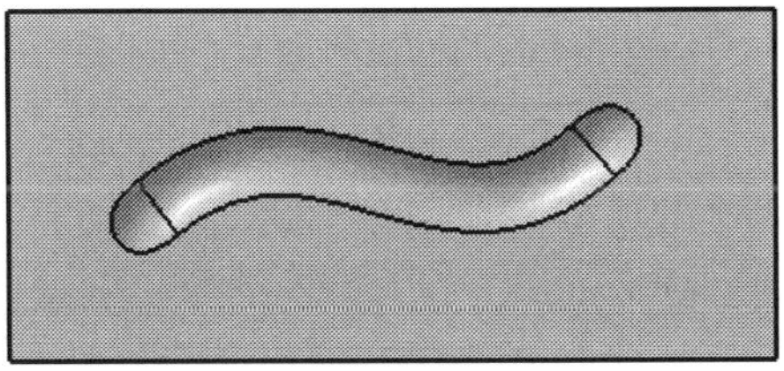

Figure 15-91 Bead feature without fillet

Creating a Curve Stamp

Menubar:	Insert > Stamping > Curve Stamp
Toolbar:	Cutting/Stamping > Stamping sub-toolbar > Curve Stamp

The **Curve Stamp** tool is used to create an embossed feature along the entities that are not connected tangentially. To create a curve stamp feature, create an open or a closed profile on the wall of the sheet metal component, as shown in Figure 15-92. Next, choose the **Curve Stamp** tool from the **Cutting/Stamping** toolbar; the **Curve stamp definition** dialog box will be displayed, as shown in Figure 15-93. The options in this dialog box are discussed next.

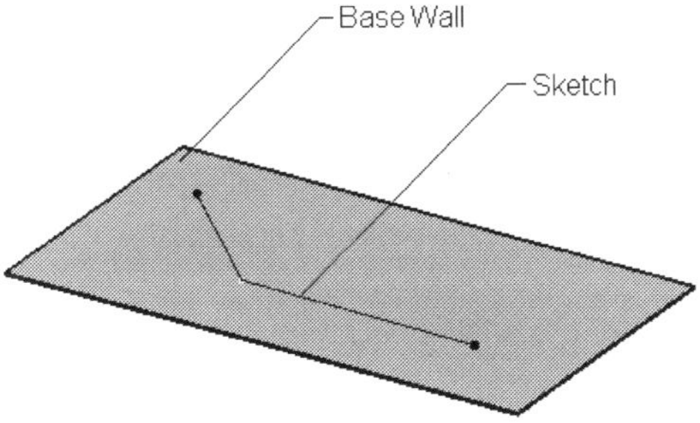

Figure 15-92 Sketch of a curve stamp

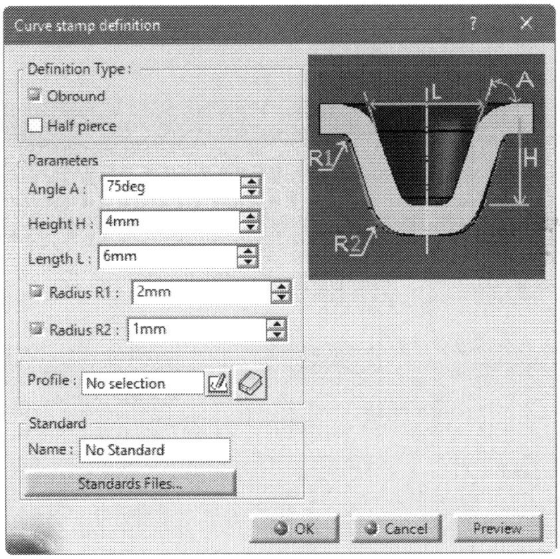

*Figure 15-93 The **Curve stamp definition** dialog box*

Definition Type Area

The options in this area are discussed next.

Obround

By selecting the **Obround** check box, you can round-off the end of the edges of the curve stamp, as shown in Figure 15-94. Figure 15-95 shows the curve stamp feature with the **Obround** check box cleared.

Half pierce

If you select the **Half pierce** check box, you cannot specify the height of the feature in the **Height H** spinner more than the specified thickness of the sheet metal, also the **Angle** spinner and the **Radius R1** and **Radius R2** check boxes gets deactivated.

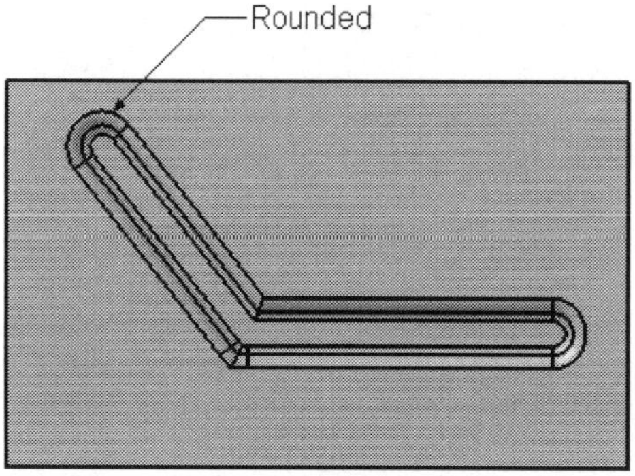

***Figure 15-94** Curve stamp feature with the **Obround** check box selected*

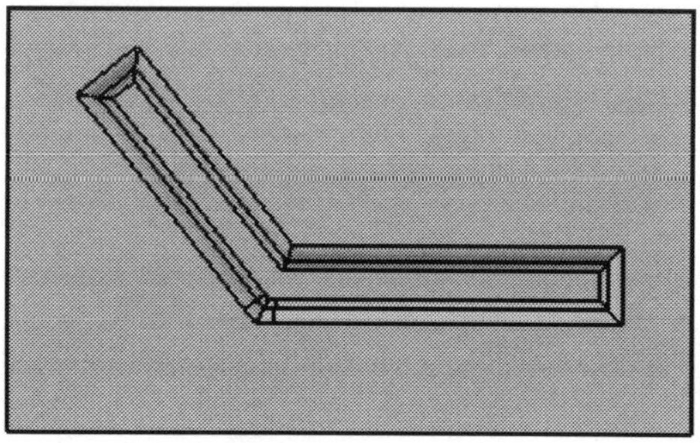

***Figure 15-95** Curve stamp feature with the **Obround** check box cleared*

Parameters Area

Set the angular value in the **Angle A** spinner to provide a draft to the stamp wall. The angular value can range from 0.1 to 90-degree.

Next, set the height and length of the stamp feature in the **Height H** and **Length L** spinners, respectively. The **Radius R1** spinner is used to create a fillet on the surface where you have created the profile and the **Radius R2** spinner is used to create a fillet on the other surface, where the stamp ends. To activate the **Radius R1** and **Radius R2** spinners, select their corresponding check boxes.

Profile

If the sketch is already drawn, select it from the geometry area; the preview of the curve stamp will be displayed. If the sketch is not drawn, choose the **Sketch** button on the right of the **Profile** display box, select the surface of the base feature, and then create a sketch.

To see the preview of the stamp feature, choose the **Preview** button. Choose the **OK** button to create a curve stamp feature.

Creating a Flanged Cut Out Stamp

Menubar:	Insert > Stamping > Flanged Cut Out
Toolbar:	Cutting/Stamping > Stamping sub-toolbar > Flanged Cut Out

To create a flanged cut out stamp feature, create a closed profile on a wall, as shown in Figure 15-96. Next, choose the **Flange Cut Out** tool from the **Cutting/Stamping** toolbar; the **Flanged cutout Definition** dialog box will be displayed, as shown in Figure 15-97. Also, you will be prompted to select the profile. Select the closed profile from the geometry area, if it is not already selected. Set the height of the flange cutout feature in the **Height H** spinner. The height will be specified normal to the face on which the sketch is drawn. Next, set the angle in the **Angle A** spinner. If you select the **Radius R** check box, the edges of the profile will be filleted. You can set the radius of the fillet in the **Radius R** spinner. Choose the **Preview** button; the preview of the flange cutout stamp will be displayed. Choose the **OK** button; the flanged cut out feature will be created, as shown in Figure 15-98.

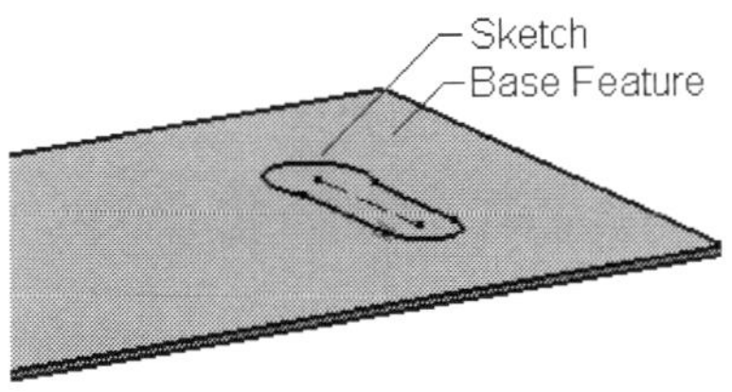

Figure 15-96 *Sketch of the flanged cutout stamp feature*

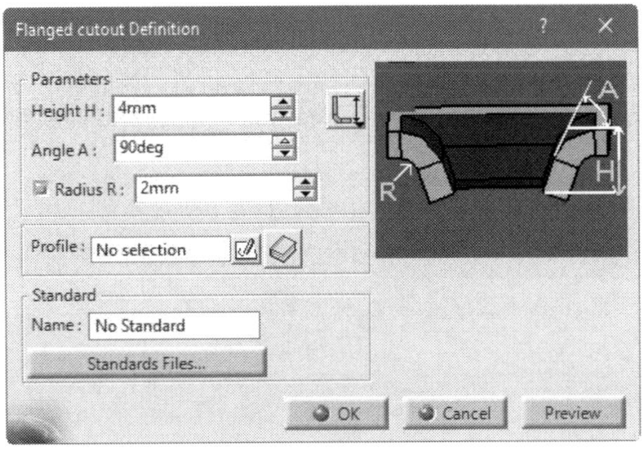

Figure 15-97 *The* ***Flanged cutout Definition*** *dialog box*

Figure 15-98 *The flanged cutout feature*

Creating a Louver Stamp

Menubar:	Insert > Stamping > Louver
Toolbar:	Cutting/Stamping > Stamping sub-toolbar > Louver

Louvers are created in a sheet metal part to provide openings for the purpose of ventilation. Generally, the rectangular pattern of louvers is created on the top face of a cover. To create a louver stamp feature, create a closed sketch on the base feature. Next, choose the **Louver** tool from the **Cutting/Stamping** toolbar; the **Louver Definition** dialog box will be displayed, as shown in Figure 15-99.

Select the sketch from the geometry area, if it is not already selected. Next, set the height of the louver in the **Height H** spinner. If you need to create the side walls of the louver at an angle to the face on which it is placed, set the angular value in the **Angle A1** spinner. By default, the angular value is 0deg. The angle between the top face of the louver and the plane where the opening is provided is specified in the **Angle A2** spinner. By default, its value is 90deg. Specify the opening edge of the louver from the sketch. The selected entity will be displayed in the **Opening line** display box. Next, choose **OK** to create the louver feature. Figure 15-100 shows the sketch to create the louver feature and Figure 15-101 shows the resulting louver feature.

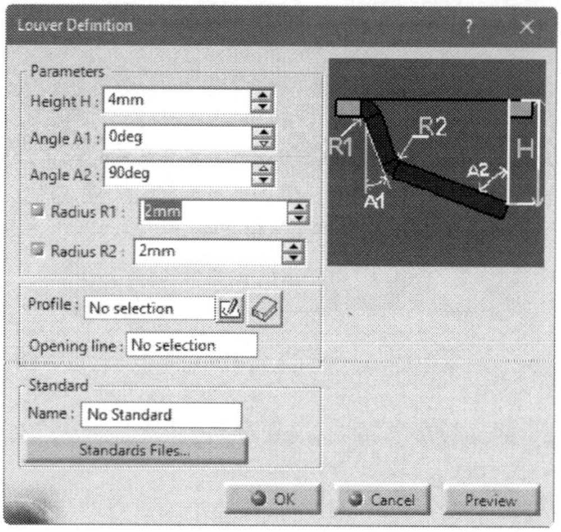

*Figure 15-99 The **Louver Definition** dialog box*

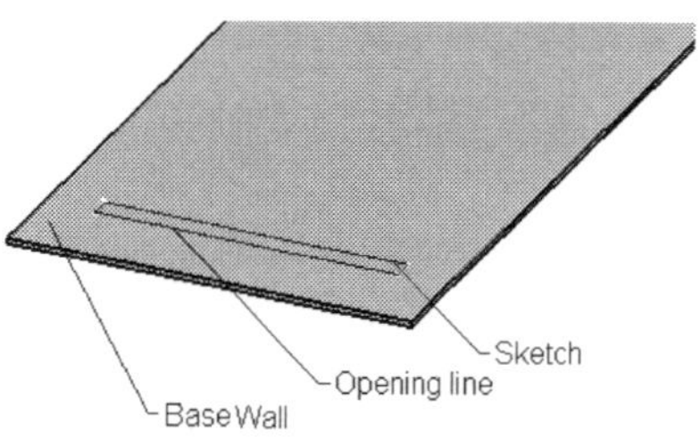

Figure 15-100 Sketch for the louver feature

Figure 15-101 Resulting louver feature

Creating a Bridge Stamp

Menubar:	Insert > Stamping >Bridge
Toolbar:	Cutting/Stamping > Stamping sub-toolbar > Bridge

The **Bridge** tool is used to create the features used for holding a component. To create a bridge feature, you need to create a point on the base wall, as shown in Figure 15-102. Select the point and the surface on which the bridge feature has to be created. Next, choose the **Bridge** tool from the **Stamping** sub-toolbar; the **Bridge Definition** dialog box will be displayed, as shown in Figure 15-103. The options in this dialog box are discussed next.

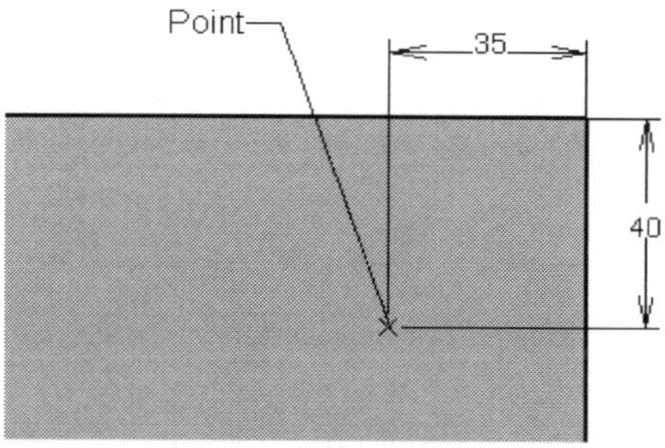

Figure 15-102 Dimensioned sketch point

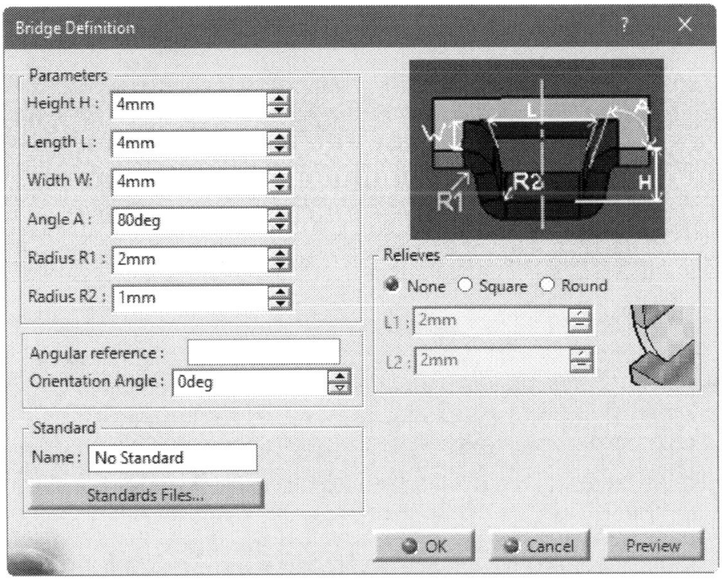

Figure 15-103 The ***Bridge Definition*** *dialog box*

Parameters Area

The options in the **Parameters Area** are discussed next.

Height H and Length L

Set the height and length of the bridge in the **Height H** and **Length L** spinners, respectively.

Width W

Set the value for the width of the bridge in the **Width W** spinner.

Angle A

If you need the walls of the bridge at an angle, set the value in the **Angle A** spinner.

Radius R1 and Radius R2

Set the values of the outer bend radius and inner bend radius of the bridge in the **Radius R1** and **Radius R2** spinners, respectively.

Angular reference

Select the angular reference about which the bridge will rotate from the geometry area.

Orientation Angle

If you need the bridge feature to be created at an angle with respect to the horizontal plane, set the value in the **Orientation Angle** spinner. Figure 15-104 shows the bridge at an angle of 90-degree and Figure 15-105 shows the bridge at an angle of 45-degree.

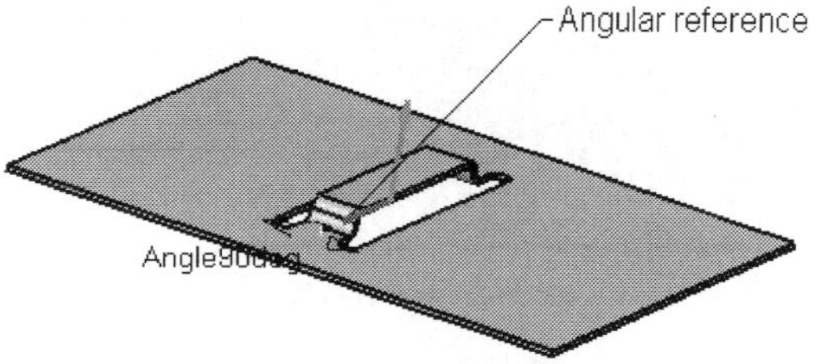

Figure 15-104 *Bridge at an angle of 90-degree*

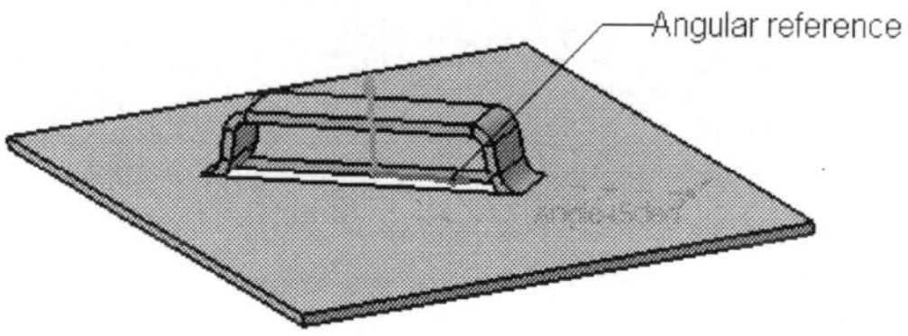

Figure 15-105 *Bridge at an angle of 45-degree*

Relieves Area

There are three options in the **Relieves** area. They are discussed next.

If you select the **None** radio button in the **Relieves** area, the bridge will not have any relief. If you want to create a square relief, select the **Square** radio button. On doing so, the **L1** and **L2** spinners will be activated in the **Relieves** area. Set the limits in the **L1** and **L2** spinners. Similarly, you can create a round relief by selecting the **Round** radio button.

Note that, you can flip the direction of the bridge feature by clicking on the orange arrowhead displayed in the preview of the bridge feature.

Creating a Flanged Hole Stamp

Menubar:	Insert > Stamping > Flanged Hole
Toolbar:	Cutting/Stamping > Stamping sub-toolbar > Flanged Hole

To create a flanged hole at a specified position, first you need to create a point on a wall to define the position of the hole, and then choose the **Flanged Hole** tool from the **Stamping** sub-toolbar. Next, select the point and the surface on which the hole has to be created; the **Flanged Hole Definition** dialog box will be displayed, as shown in Figure 15-106.

You will notice that the **Major Diameter** option is selected by default in the **Parameters choice** drop-down list in the **Definition Type** area. Next, set the values for the height of the flanged hole, fillet radius, angle of the flanged hole, and the diameter of the flanged hole in the **Height H**, **Radius R**, **Angle A**, and **Diameter D** spinners, respectively. The location of the flanged hole can be specified by the point selected on the surface. If you do not want a protruded hole, select the **Without cone** radio button in the **Definition Type** area. Figure 15-107 shows the point placed on a wall. Figure 15-108 shows the flanged hole created with the **With cone** radio button selected.

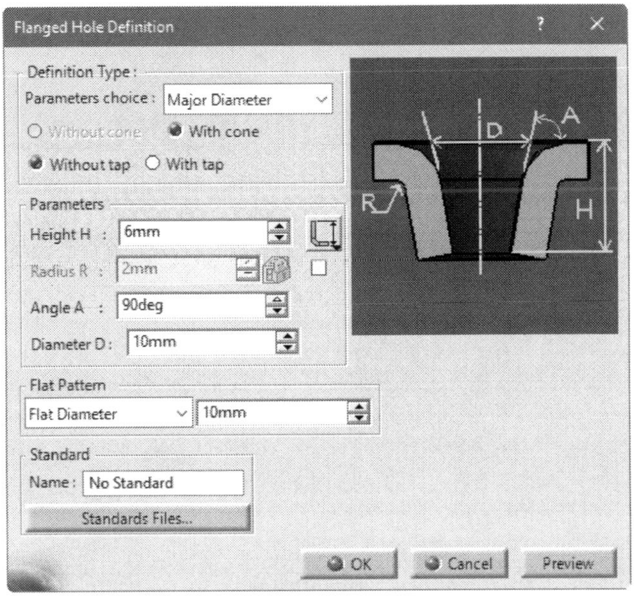

Figure 15-106 *The **Flanged Hole Definition** dialog box*

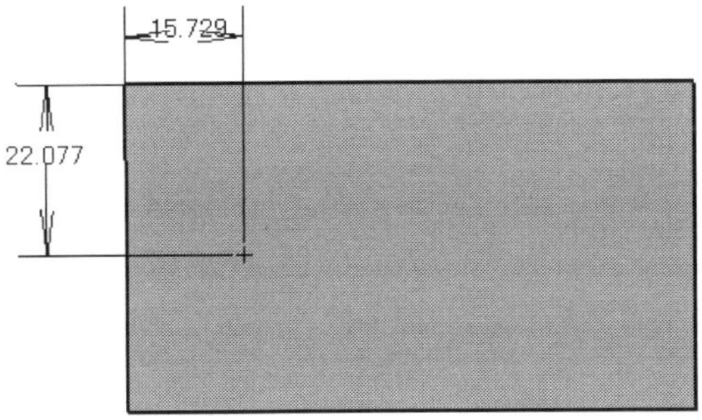

Figure 15-107 *Dimensioned sketch point*

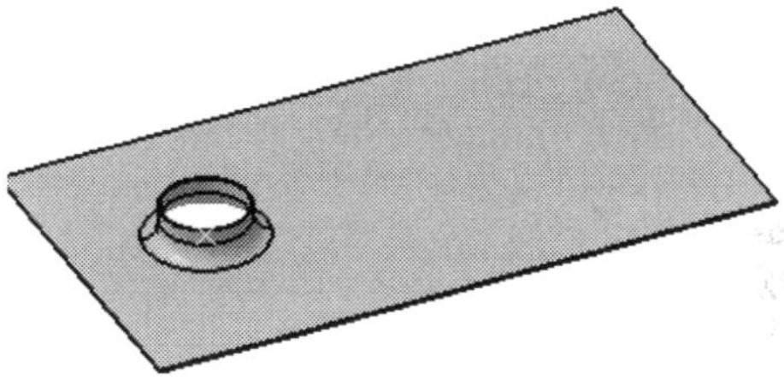

Figure 15-108 *Flanged hole feature created with the **With cone** radio button selected*

To create a tap, select the **With tap** radio button, if it is not selected, from the **Definition Type** area; the **Flanged Hole Definition** dialog box will be modified, refer to Figure 15-109.

In the **Parameters** area, set the values for the height of the flanged tap, fillet radius, angle of the flanged tap, and the diameter of the flanged tap in the **Height H**, **Radius R**, **Angle A**, and **Diameter D** spinners, respectively.

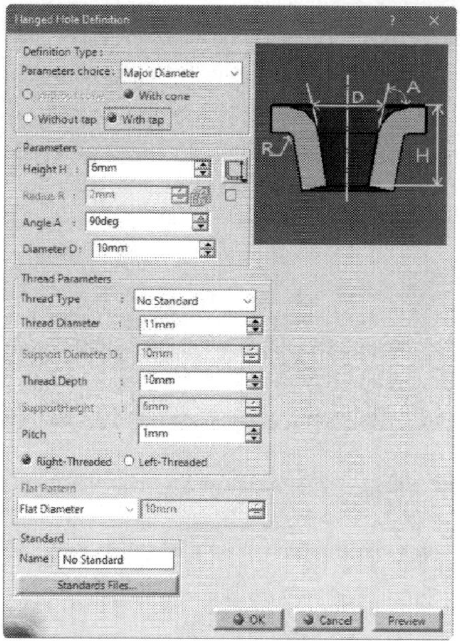

***Figure 15-109** The **Flanged Hole Definition** dialog box with the **With tap** radio button selected*

In the **Thread Parameters** area, by default the **No Standard** option is selected in the **Thread Type** drop-down list. Therefore, you need to manually specify the parameters to define the thread.

Set the value of the thread diameter in the **Thread Diameter** spinner. By default, this value is based on the diameter value specified in the **Parameters** area. Set the thread depth in the **Thread Depth** spinner. Also, set the pitch value in the **Pitch** spinner. By default, the **Right-Threaded** radio button is selected. To create a left hand thread, select the **Left-Threaded** radio button. After setting the parameters, choose the **OK** button from the **Flanged Hole Definition** dialog box; a threaded hole will be created. Note that the thread will not be displayed in the hole because it will be a cosmetic thread. However, when you generate the drawing view, the thread convention will be displayed in it.

To create standard threaded holes, select the **Metric Thin Pitch** or **Metric Thick Pitch** option from the **Thread Type** drop-down list in the **Thread Parameters** area, refer to Figure 15-109; the **Thread Diameter** spinner will be replaced by the **Thread Description** drop-down list. You can select the standard size of the thread from this drop-down list. In this case, you need to specify only the thread depth and the hole depth. The other parameters such as the hole diameter, thread diameter, and thread pitch are automatically defined on the basis of the selected standard size of the thread from the **Thread Description** drop-down list.

If the value of hole diameter matches the standard value, then the relevant thread parameters will be applied automatically as per the standards selected from the **Thread Type** drop-down list in the **Thread Parameters** area. If the value of the hole diameter does not match any standard value, the **Flanged Hole Warning** message box will be displayed. Choose the **OK** button from this message box; the hole diameter and the other thread parameters will be set in accordance with the smallest standard value of the diameter available in the **Thread Description** drop-down list.

Creating a Circular Stamp

Menubar:	Insert > Stamping > Circular Stamp
Toolbar:	Cutting/Stamping > Stamping sub-toolbar > Circular Stamp

The **Circular Stamp** tool is used to create a circular stamp by specifying the parameters of the punch. To create a circular stamp, first you need to create a point on the wall to position it. Next, choose the **Circular Stamp** tool from the **Cutting/Stamping** toolbar and then select the point and the surface on which the hole has to be created; the **Circular Stamp Definition** dialog box will be displayed, as shown in Figure 15-110.

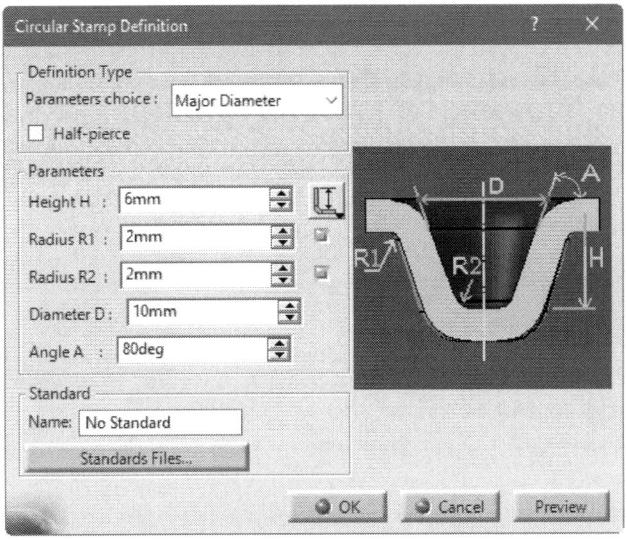

*Figure 15-110 The **Circular Stamp Definition** dialog box*

In the **Circular Stamp Definition** dialog box, the **Major Diameter** option in the **Parameters choice** drop-down list under the **Definition Type** area is selected by default. If you select the **Half-pierce** check box, you cannot specify the height of the feature in the **Height H** spinner more than the sheet metal thickness. Also, on selecting this check box, the **Radius R1**, **Radius R2**, and **Angle A** spinners will not be enabled. Figure 15-111 shows the feature created with the **Half-pierce** check box selected and Figure 15-112 shows the feature created with the **Half-pierce** check box cleared.

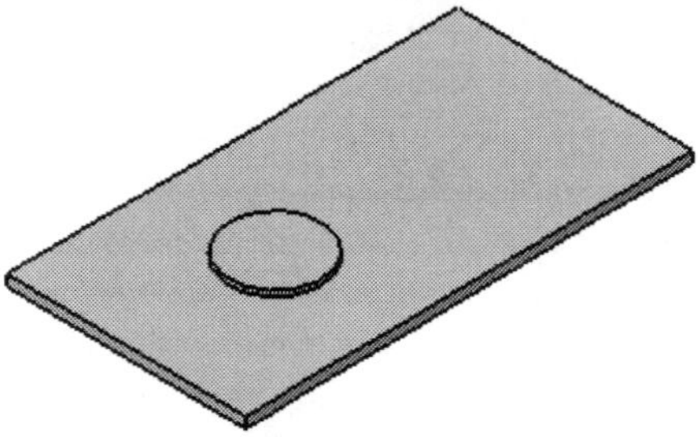

Figure 15-111 *Feature created with the* ***Half-pierce*** *check box selected*

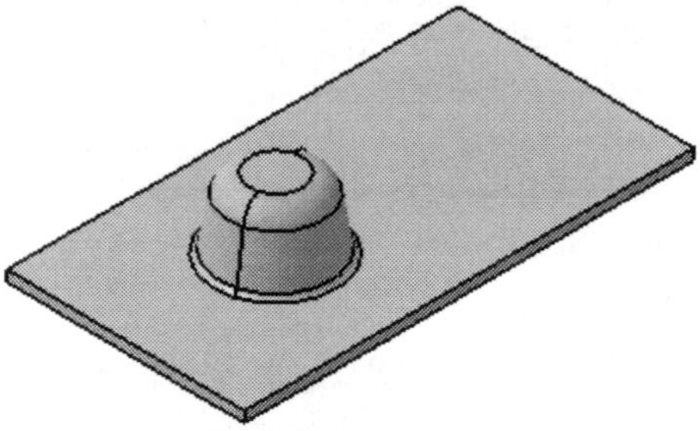

Figure 15-112 *Feature created with the* ***Half-pierce*** *check box cleared*

Note

To create a circular stamp feature at a particular point, first you need to define a point to specify its position before invoking the ***Circular Stamp Definition*** *dialog box.*

Creating a Stiffening Rib Stamp

Menubar: Insert > Stamping > Stiffening Rib
Toolbar: Cutting/Stamping > Stamping sub-toolbar > Stiffening Rib

The **Stiffening Rib** tool is used to create a stiffening rib feature by specifying geometrical parameters. To create a stiffening rib, choose the **Stiffening Rib** tool from the **Stamping** sub-toolbar; you will be prompted to select a position on the cylindrical surface for the stiffening rib. While selecting the position for the stiffening rib, the external face of the cylindrical surface should be selected, as shown in Figure 15-113. On selecting the position for the stiffening rib, the preview of the stiffening rib will be displayed in the geometrical area and the **Stiffening Rib Definition** dialog box will also be displayed.

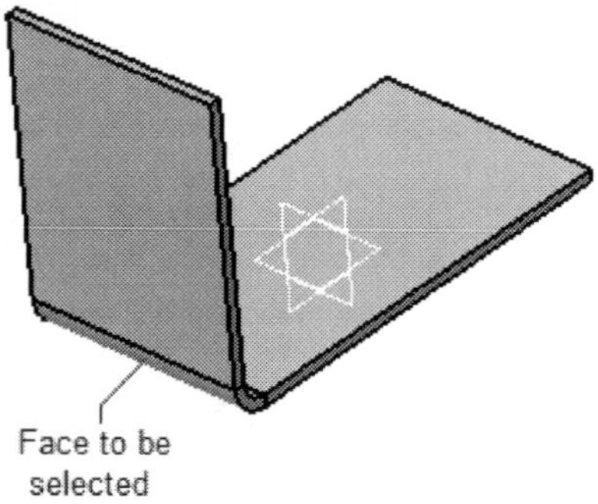

Figure 15-113 *The external face of the cylindrical surface to be selected*

The options in the **Stiffening Rib Definition** dialog box are used to specify the length, radius, and angle for the stiffening rib in their respective spinners. Choose the **OK** button to create the rib feature, as shown in Figure 15-114.

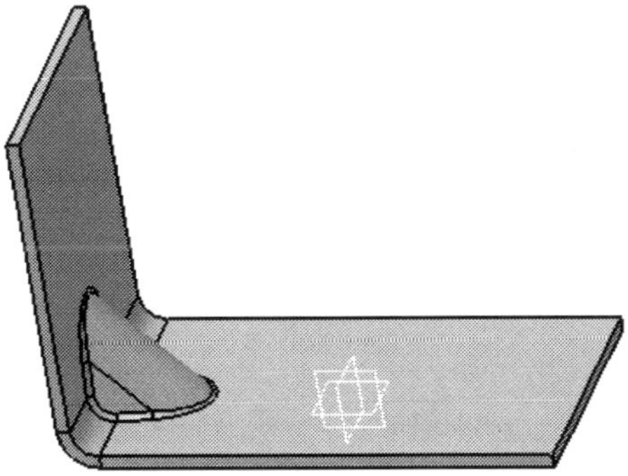

Figure 15-114 Model after creating the stiffening rib feature

Creating a Dowel Stamp

Menubar:	Insert > Stamping > Dowel
Toolbar:	Cutting/Stamping > Stamping sub-toolbar > Dowel

To create a dowel, choose the **Dowel** tool from the **Stamping** sub-toolbar; you will be prompted to select a face, a plane, or a point for positioning the dowel. Select the face of the feature for positioning the dowel; the **Dowel Definition** dialog box will be displayed, as shown in Figure 15-115.

You can change the position of the dowel by using the **Positioning Sketch** button in the **Dowel Definition** dialog box. You can also increase or decrease the diameter of the dowel by using the **Diameter D** spinner in the dialog box. Choose the **OK** button to create the dowel feature, as shown in Figure 15-116.

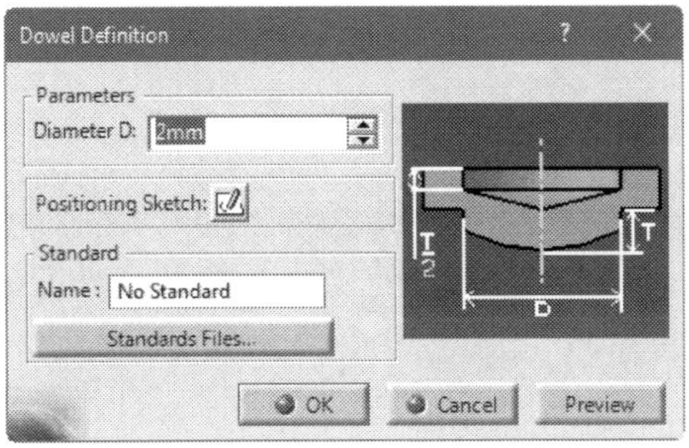

*Figure 15-115 The **Dowel Definition** dialog box*

Figure 15-116 Model after creating the dowel feature

TUTORIALS

Tutorial 1

In this tutorial, you will create the sheet metal component of a Holder Clip, as shown in Figure 15-117. The flat pattern of the component is shown in Figure 15-118. The dimensions of the component are shown in Figure 15-119. The thickness of the sheet is 1 mm. After creating the sheet metal component, create its flat pattern. **(Expected time: 45 min)**

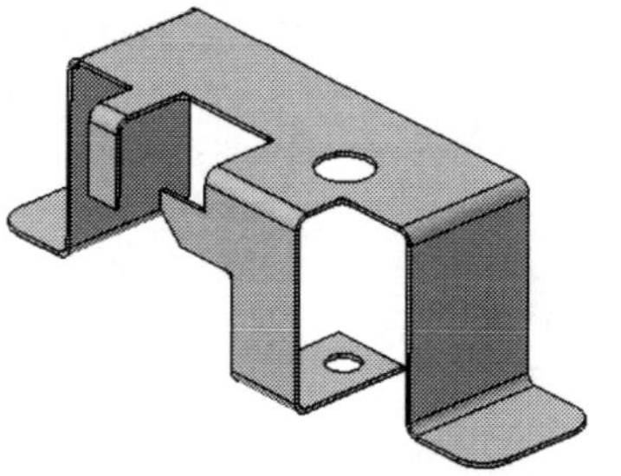

Figure 15-117 *Sheet metal component of the Holder Clip*

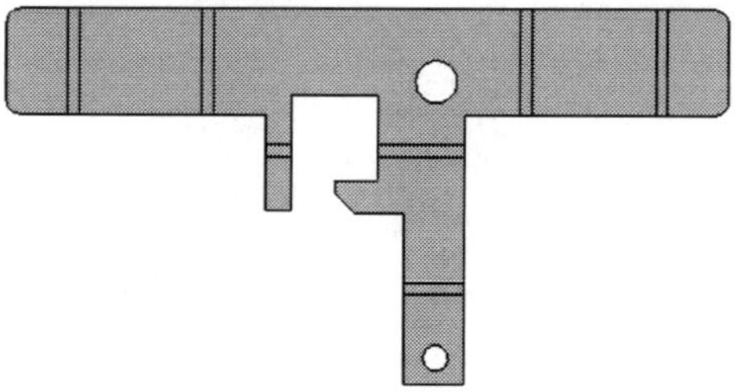

Figure 15-118 *Flat pattern of the component*

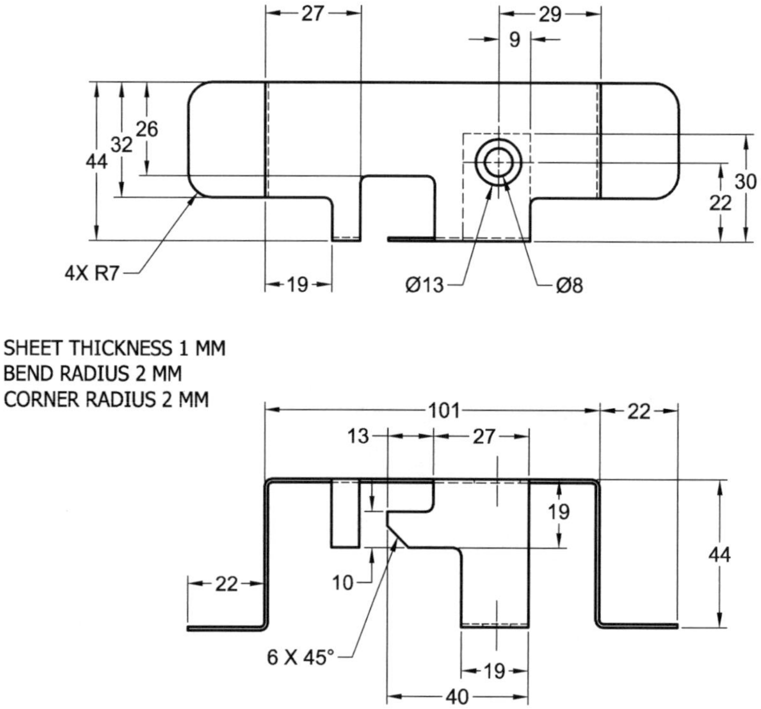

Figure 15-119 *Top and front views of the Holder Clip*

The following steps are required to complete this tutorial:

a. Start a new sheet metal file and then draw the sketch of the base wall of the sheet metal component.

b. Set parameters in the **Sheet Metal Parameters** dialog box and convert the sketch into a sheet metal face.

c. Add the flange on the right and left faces of the top feature.
d. Add flanges to the right and left ends of the flanges of the top feature.
e. Create a wall on edge feature.
f. Add another flange on the edge of the wall on edge feature.
g. Create the hole and corner features. Finally, create the flat pattern.
h. Save the sheet metal component file.

Starting a New Part File

1. To start a new file in the **Generative Sheetmetal Design** workbench, choose **Start > Mechanical Design > Generative Sheetmetal Design** from the menu bar; the **New Part** dialog box is displayed. Choose the **OK** button from the **New Part** dialog box; a new file gets started in the **Generative Sheetmetal Design** workbench.

Setting the Parameters for the Sheet Metal Component

1. Choose the **Sheet Metal Parameters** tool from the **Walls** toolbar; the **Sheet Metal Parameters** dialog box is displayed.

2. Set **1mm** in the **Thickness** spinner and **2mm** in the **Default Bend Radius** spinner in the **Sheet Metal Parameters** dialog box. Then, choose the **OK** button.

Drawing the Sketch for the Top Face

1. Invoke the **Sketcher** workbench by selecting the xy plane as the sketching plane and draw the sketch for the base wall, as shown in Figure 15-120.

2. Exit the **Sketcher** workbench.

Converting the Sketch into a Base Wall

1. Choose the **Wall** tool from the **Walls** toolbar; the **Wall Definition** dialog box is displayed.

2. Select the sketch from the geometry area, if it is not already selected; preview of the base feature is displayed in the geometry.

3. Retain rest of the settings and choose the **OK** button to get the final feature of the base wall, as shown in Figure 15-121.

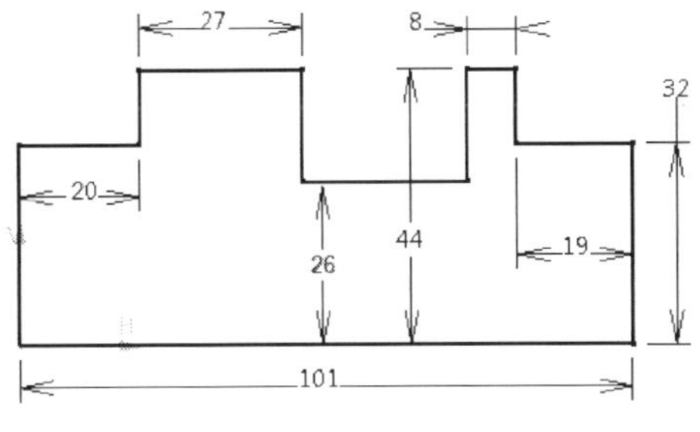

***Figure 15-120** Sketch for the base wall of the Holder Clip*

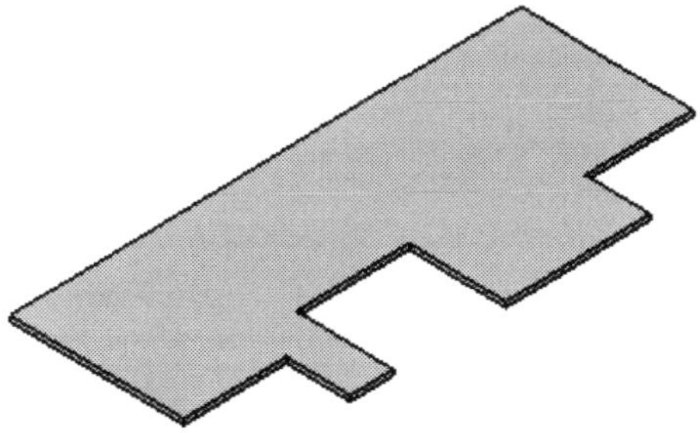

***Figure 15-121** Base wall of the Holder Clip*

Creating the Flanges (side wall) on the Base Wall

1. Choose the **Flange** tool from **Walls > Swept Walls** sub-toolbar; the **Flange Definition** dialog box is displayed.

2. Select the edge on which you need to create the flange, as shown in Figure 15-122; the name of the selected edge is displayed in the **Spine** selection box.

3. Select the **Basic** option from the **Flange type** drop-down list, if it is not already selected.

4. Set the value **44** in the **Length** spinner. Choose the down arrow on the **Length type** button; a flyout is displayed. Choose the **Outer Length type** button (the third button in the flyout from the left).

5. Set the value **90deg** in the **Angle** spinner if it is not set by default. Choose the **Reverse Direction** button, if the preview of the flange is in upward direction.

6. Select the **Trim Support** check box and choose the **OK** button to create the flange. Figure 15-123 shows the resultant flange.

7. Follow the procedure explained in the last steps and create another flange of same length on the opposite edge of the component. The sheet metal component after creating the flange is shown in Figure 15-124.

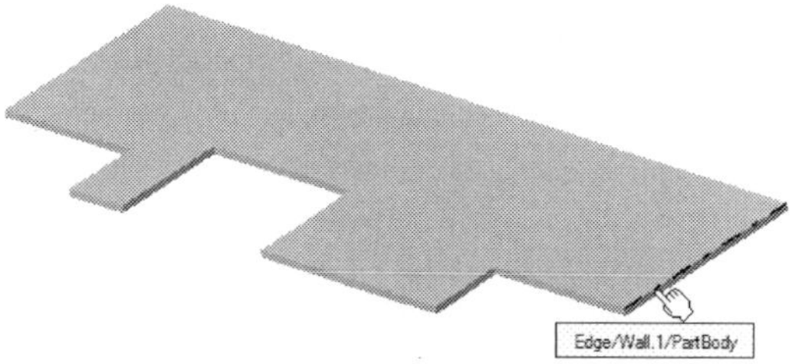

Figure 15-122 *Selecting the edge for creating the flange*

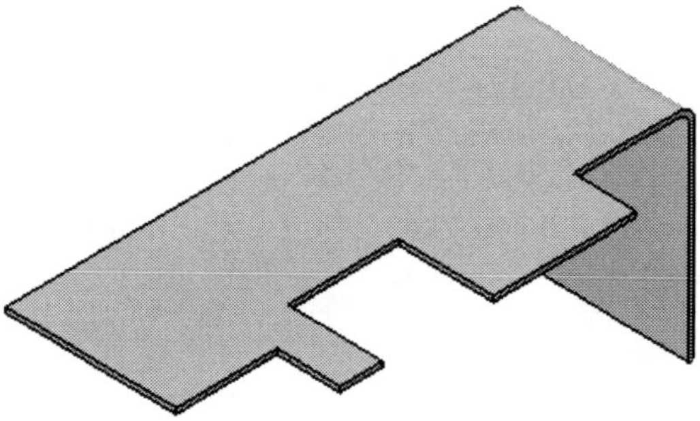

Figure 15-123 *Sheet metal component after creating the flange*

Figure 15-124 *Sheet metal component after creating the flanges*

Creating Flanges on the Side Walls

1. Again invoke the **Flange Definition** dialog box and set the value **22** in the **Length** spinner; **90deg** in the **Angle** spinner.
2. Select the edge of the base wall, as shown in Figure 15-125.
3. Choose the **Reverse Direction** button, if required and select the **Trim Support** check box, if it is not selected by default. Choose the **OK** button to create the flange. The sheet metal component after creating the flange is shown in Figure 15-126.

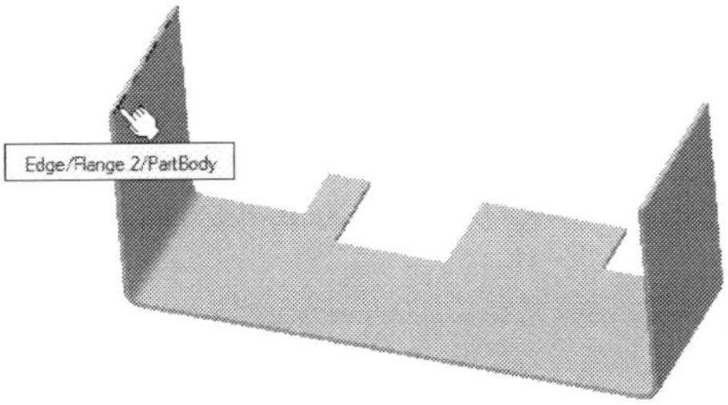

Figure 15-125 *Selecting the edge for creating the flange*

Figure 15-126 *Sheet metal component after creating the flange*

4. Similarly, create another flange of same length on the opposite edge, as shown in Figure 15-127.

Creating a Wall on the Edge

1. Choose the **Wall On Edge** tool from the **Walls** toolbar; the **Wall On Edge Definition** dialog box is displayed. Select the **Sketch Based** option from the **Type** drop-down list in the dialog box.
2. Select the lower edge of the base wall, as shown in Figure 15-128.

Figure 15-127 *Sheet metal component after creating the flange*

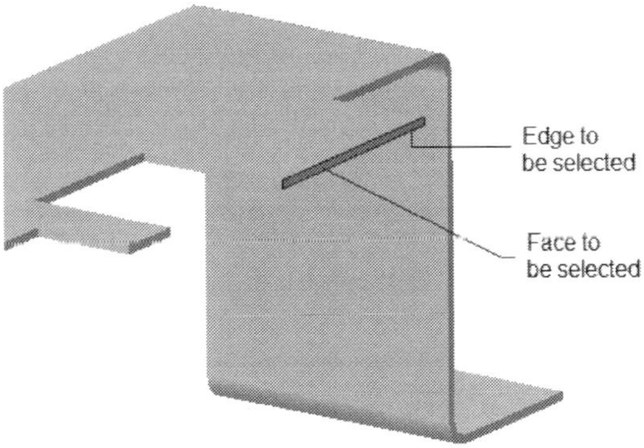

Figure 15-128 *The lower edge and the planar face to be selected for creating the wall*

3. Choose the **Sketch** button on the right of the **Profile** display box; you are prompted to select the sketching plane.

4. Select the face representing the thickness of the base wall, refer to Figure 15-128; the **Sketcher** workbench is invoked.

5. Draw the sketch for the wall on the edge feature in the **Sketcher** workbench, as shown in Figure 15-129.

6. Exit the **Sketcher** workbench; preview of the resulting wall is displayed.

7. Set the view to front and check the direction of wall creation. If the direction arrow is facing away from the existing feature, choose the **Invert Material Side** button.

8. Accept the remaining default options and choose the **OK** button to create a wall on the edge. Figure 15-130 shows the resultant wall on the edge.

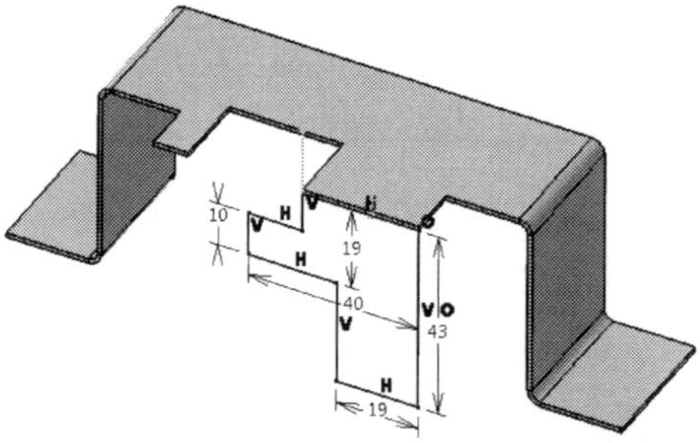

Figure 15-129 *Sketch for the wall on the edge feature*

Figure 15-130 *Sheet metal component after creating the wall on the edge*

Creating other Flanges

1. Choose the **Flange** tool from the **Swept Walls** sub-toolbar; the **Flange Definition** dialog box is displayed.

2. Set **19** in the **Length** spinner and **90deg** in the **Angle** spinner. Select the edge of the base wall, as shown in Figure 15-131.

3. Ensure that the **Trim Support** check box is selected and then choose the **OK** button. The sheet metal component after creating the flange is shown in Figure 15-132.

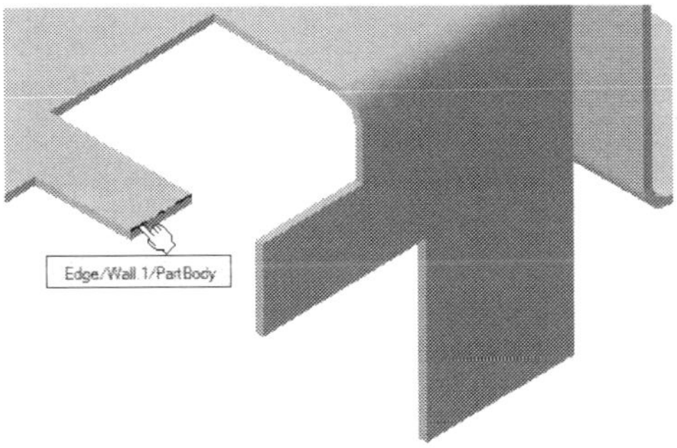

Figure 15-131 *Selecting an edge*

Figure 15-132 *The flange created*

4. Next, create another flange of 30 mm length similar to the one created earlier, refer to Figure 15-133 and Figure 15-134. The sheet metal component after creating the flange is shown in Figure 15-134.

Creating the Remaining Features

1. Create holes using the **Hole** tool from **Cutting/Stamping > Holes** sub-toolbar.

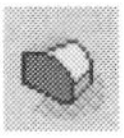

2. Create fillets by using the **Corner** tool from the **Cutting/Stamping** toolbar. For dimensions, refer to Figure 15-119.

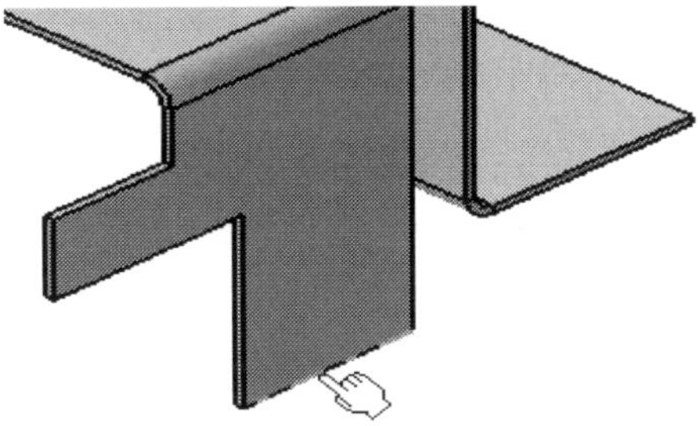

Figure 15-133 *Selecting an edge*

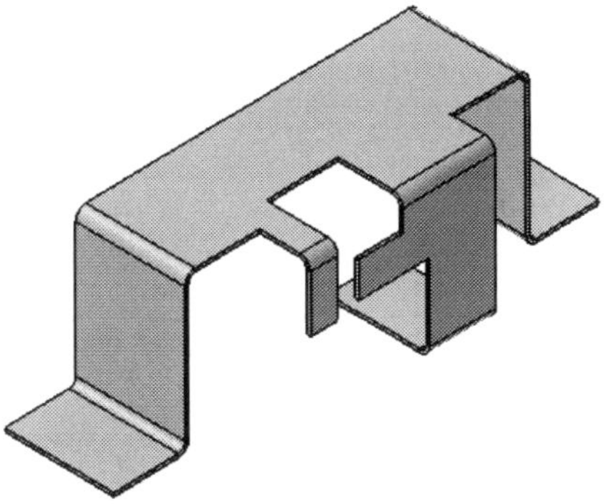

Figure 15-134 *The flange created*

3. Invoke the **Chamfer** tool from the **Cutting/Stamping** toolbar and create chamfers on walls. Figure 15-135 shows the sheet metal component after creating holes, fillets, and chamfer. For the dimension of holes, fillets, and chamfer, refer to Figure 15-119. The final sheet metal component of the Holder Clip is shown in Figure 15-135.

Creating the Flat Pattern

1. Choose the **Fold/Unfold** button from **Views > Fold/Unfold** sub-toolbar; the flat pattern of the sheet metal part is created, as shown in Figure 15-136.

Figure 15-135 Final model of the Holder Clip

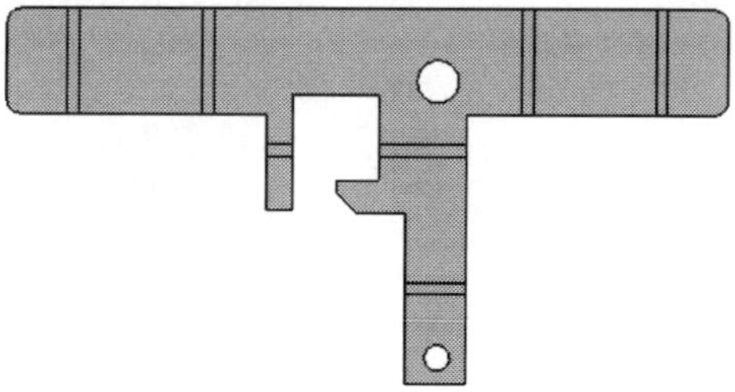

Figure 15-136 Flat pattern of the Holder Clip

2. Create the folder with the name *c15* at *C:\CATIA*.

3. Save the sheet metal components with the name *c15tut1* at *C:\CATIA\c15*.

Tutorial 2

In this tutorial, you will create the sheet metal component shown in Figure 15-137. Its dimensions are given in the same figure. The flat pattern of the component is shown in Figure 15-138. The thickness of the sheet is 1 mm and bend radius is 2 mm. **(Expected time: 45 min)**

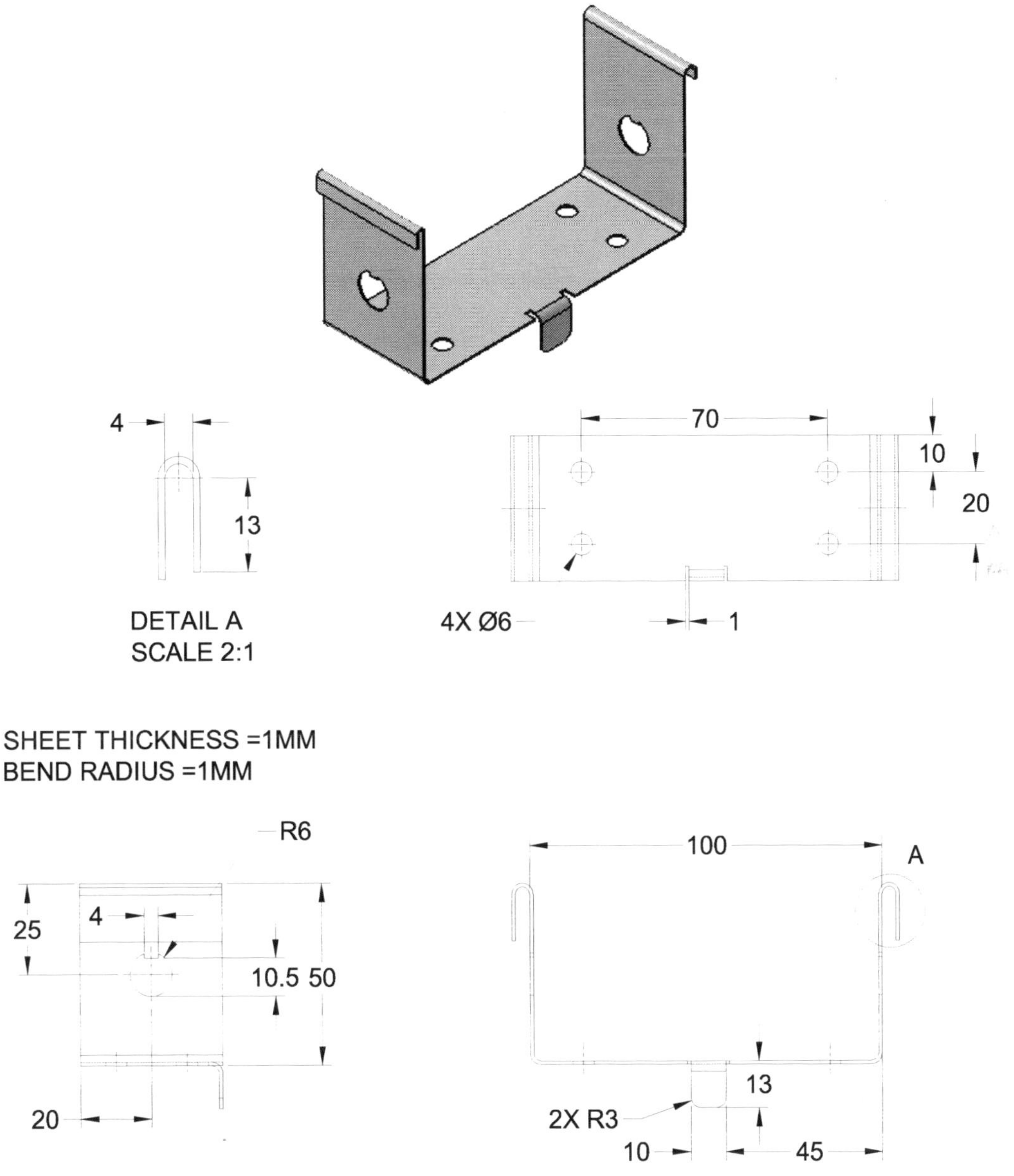

Figure 15-137 Orthographic views and dimensions of the sheet metal component

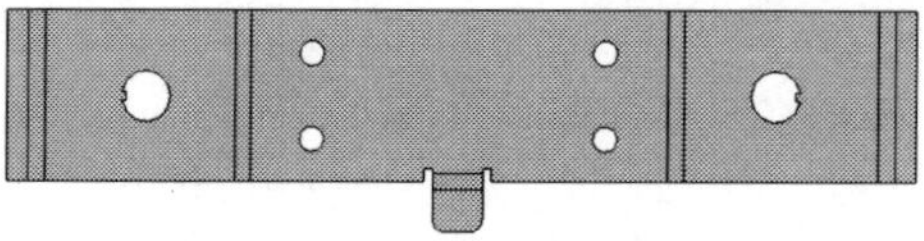

Figure 15-138 Flat pattern of the sheet metal component

The following steps are required to complete this tutorial:

a. Start a new file in the **Generative Sheetmetal Design** workbench.
b. Draw the sketch of the base feature and then convert it into a base wall by using the **Wall** tool.
c. Create one hole and then pattern it to create the remaining three instances.
d. Create flanges on the left and right edges of the base wall.
e. Create hems on the two flanges and then create the cutout feature by using the **Cut Out** tool.
f. Create the wall on the edge of the front face of the base.
g. Create the flat pattern.
h. Save the model.

Starting a New File and Setting Sheet Metal Part Parameters

1. Choose **Start > Mechanical Design > Generative Sheetmetal Design** from the menu bar to start a new file in the **Generative Sheetmetal Design** workbench.

2. Choose the **Sheet Metal Parameters** tool from the **Walls** toolbar; the **Sheet Metal Parameters** dialog box is displayed.

3. Set **1mm** in the **Thickness** spinner and **2mm** in the **Default Bend Radius** spinner.

4. Choose the **Bend Extremities** tab and then select the **Square relief** option from the drop-down list displayed below it. Set **1mm** and **1mm** in the **L1** and **L2** spinners, respectively. Next, choose the **OK** button from the **Sheet Metal Parameters** dialog box to exit it.

Drawing the Sketch for the Base Feature

1. Invoke the **Sketcher** workbench by selecting the xy plane as the sketching plane.

2. Draw the sketch for the base wall, as shown in Figure 15-139, and exit the **Sketcher** workbench.

Converting the Sketch into a Base Wall

1. Choose the **Wall** tool from the **Walls** toolbar; the **Wall Definition** dialog box is displayed. Select the sketch from the geometry area, if it is not already selected; the name of the selected sketch is displayed in the **Profile** display box. Choose **OK** to create the base wall.

2. Choose the **Hole** tool from **Cutting/Stamping > Holes** sub-toolbar and create a hole at the lower left corner of the base wall. For dimensions of the hole, refer to Figure 15-137.

3. Create the rectangular pattern of the hole. Change the current view to the isometric view. The base of the sheet metal component after creating the hole feature and the pattern feature is shown in Figure 15-140.

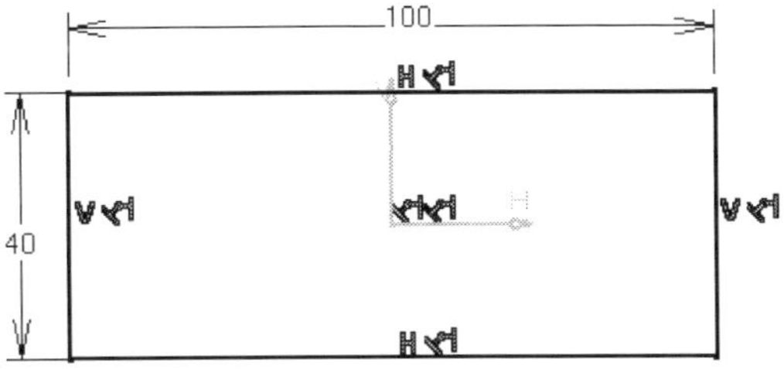

Figure 15-139 *Sketch for the base wall*

Figure 15-140 *Base wall with the hole feature*

Creating Two Flanges

1. Choose the **Flange** tool from **Walls > Swept Walls** sub-toolbar; the **Flange Definition** dialog box is displayed.
2. Select the edge of the base wall, as shown in Figure 15-141. You will notice that the preview of the flange is displayed. Reverse the direction if required.

3. Set **50** in the **Length** spinner. Choose the down arrow on the **Length type** button; a flyout is displayed. Choose the **Outer Length type** button.
4. Set **90deg** in the **Angle** spinner if not set by default.
5. Select the **Trim Support** check box and choose the **Reverse Direction** button if the preview of the flange is in opposite direction. Next, choose **OK**; a flange is created.
6. Similarly, create a flange on the other side of the base feature, as shown in Figure 15-142.

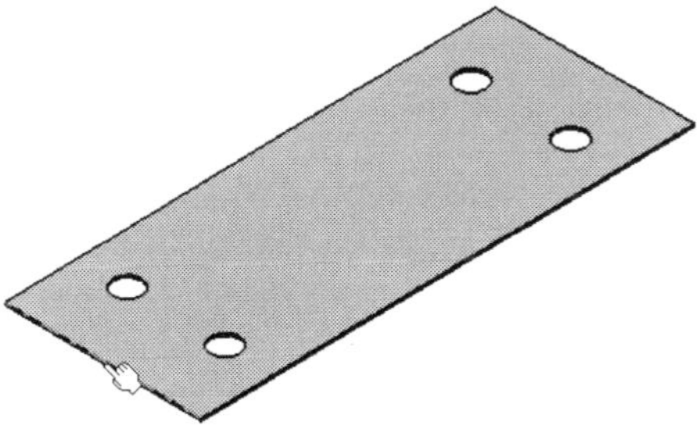

Figure 15-141 *Selecting the edge for creating a flange*

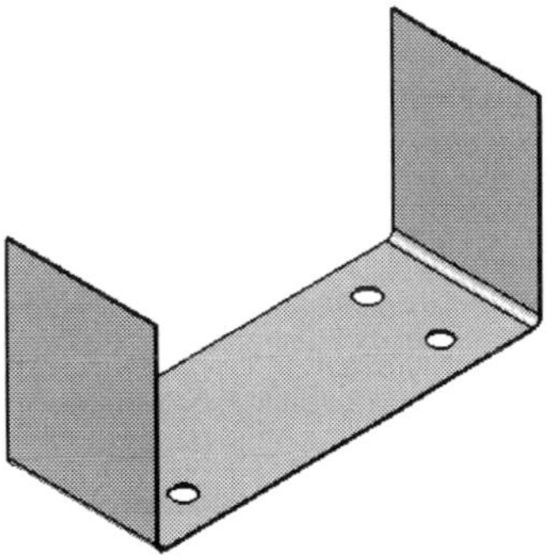

Figure 15-142 *Model after creating two flanges*

Creating the Hems

1. Choose the **Hem** button from the **Swept Walls** sub-toolbar; the **Hem Definition** dialog box is displayed. Also, you are prompted to select an edge to create the hem. 

2. Select the inner edge of the right flange; the name of the selected edge is displayed in the **Spine** display box.

3. Select the **Basic** option from the **Flange type** drop-down list.

4. Set **13** in the **Length** spinner and **2** in the **Radius** spinner.

   Note that to set the radius, right-click in the **Radius** spinner; a shortcut menu will be displayed and select **Formula > Deactivate.**

5. Ensure that the **Trim Support** check box is selected in the **Hem** dialog box and then choose the **OK** button to create the hem.

6. Similarly, create a hem on the other flange, as shown in Figure 15-143.

Creating a Cut Out Feature

Next, you need to create a cutout feature. This feature can be created by using the **Cut Out** tool.

1. Define a new sketching plane on the outer face of the right flange. Draw the sketch by referring to the dimensions given in Figure 15-137.

2. Exit the **Sketcher** workbench.

3. Choose the **Cut Out** tool from the **Cutting/Stamping** toolbar; the **Cutout Definition** dialog box is displayed.

4. Select the **Up to last** option from the **Type** drop-down list in the **End Limit** area.

5. Accept the remaining default parameters in the **Cutout Definition** dialog box and then choose the **OK** button; a cutout feature is created, as shown in Figure 15-144.

***Figure 15-143** Model after creating two hems*

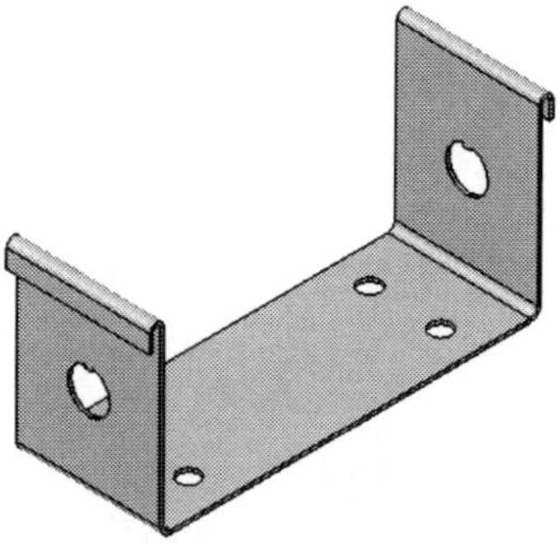

***Figure 15-144** Sheet metal after creating a cutout feature*

Creating the Side Wall with Corner and Filleting Edges

1. Choose the **Wall on Edge** tool from the **Walls** toolbar; the **Wall On Edge Definition** dialog box is displayed.

2. Choose the **Height & Inclination** tab in the **Wall On Edge Definition** dialog box if it is not already chosen, and then set **13mm** in the **Height** spinner and **90deg** in the **Angle** spinner. Select the **With Bend** check box, if it is not already selected.

3. Choose the **Extremities** tab in the dialog box and then set **-42mm** in both the **Left offset** and **Right offset** spinners.

4. Select the front edge of the base feature to create a wall on the edge; a preview of the wall is displayed.

5. Flip the direction of the wall if required. Ensure that the material is added toward the base feature. If it is not so, choose the **Invert Material Side** button and choose **OK**; the side wall feature is created. Next, create the corner feature using the dimensions given in Figure 15-137.

6. Create fillets by using the **Corner** tool. For dimensions of the fillets, refer to Figure 15-137. The final sheet metal component is shown in Figure 15-145.

Creating the Flat Pattern and Saving the File

1. Choose the **Unfold/Fold** button from **Views > Fold/Unfold** sub-toolbar; the flat pattern of the sheet metal component is created, as shown in Figure 15-146.

2. Save the sheet metal component with the name *c15tut2* at *C:\CATIA\c15*.

Figure 15-145 *Final sheet metal component*

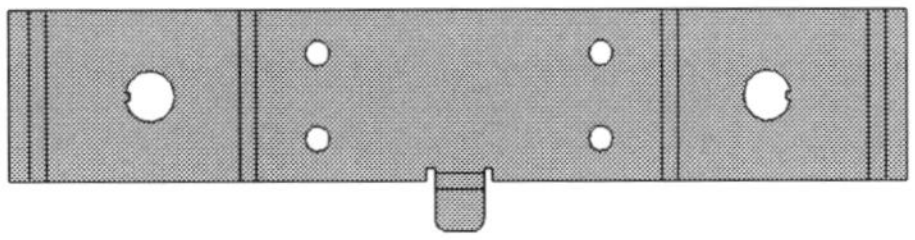

Figure 15-146 *Flat pattern of the sheet metal component*

Tutorial 3

In this tutorial, you will create the sheet metal component of a slide cover, as shown in Figure 15-147. The dimensions for the slide cover are given in Figure 15-148. The thickness of the sheet metal component is 1 mm and bend radius is 2 mm. **(Expected time: 30 min)**

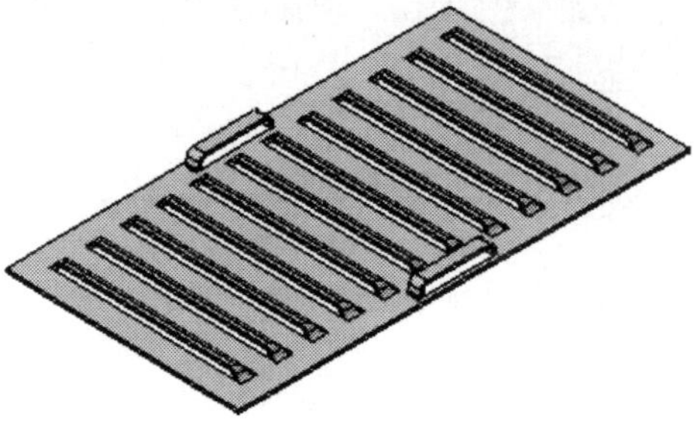

***Figure 15-147** Model for Tutorial 3*

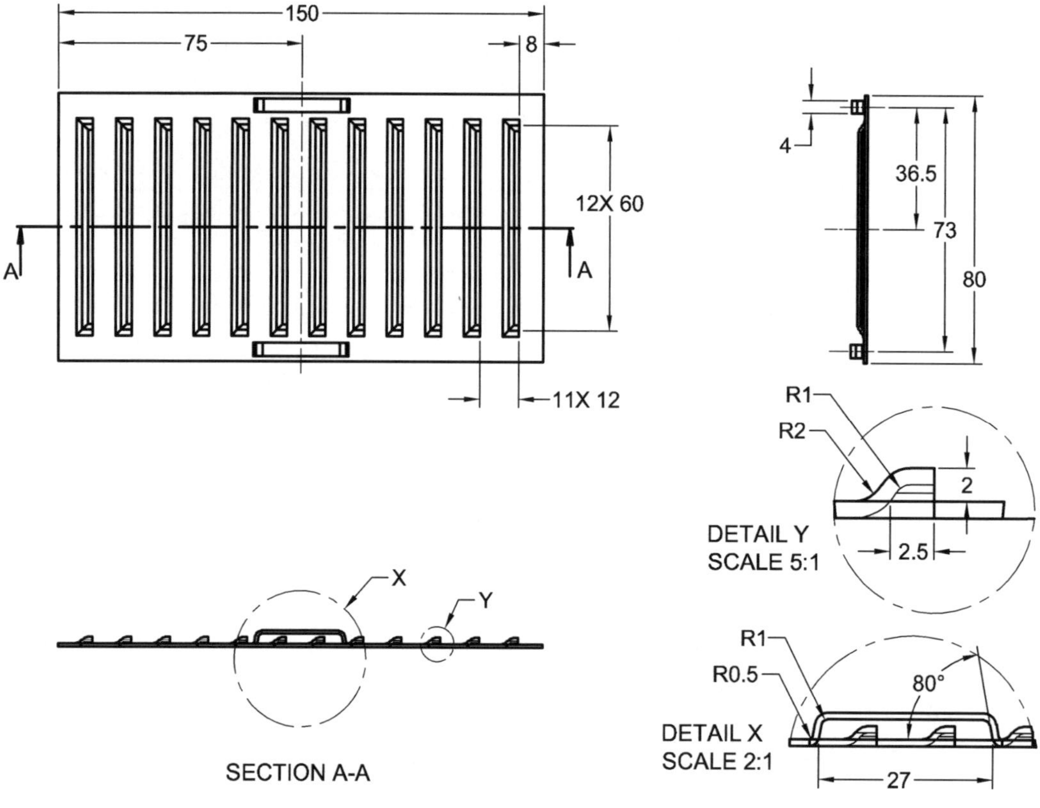

***Figure 15-148** Orthographic views and dimensions of the model for Tutorial 3*

The following steps are required to complete this tutorial:

a. Start a new file in the **Generative Sheetmetal** Design workbench.
b. Draw the sketch for the base feature and then convert it into a base wall by using the **Wall** tool.
c. Create the louver feature and then pattern it.
d. Create handles.
e. Save the model.

Starting a New Part File

1. Start a new file in the **Generative Sheetmetal Design** workbench.

2. Choose the **Sheet Metal Parameters** tool from the **Walls** toolbar; the **Sheet Metal Parameters** dialog box is displayed.

3. Set the value **1** in the **Thickness** spinner and the value 2 in the **Default Bend Radius** spinner in the **Parameters** tab of the **Sheet Metal Parameters** dialog box, if not set by default. Next, choose **OK**.

Drawing the Sketch of the Base Wall

1. Invoke the **Sketcher** workbench by selecting the xy plane as the sketching plane.

2. Draw the sketch of the base feature, as shown in Figure 15-149, and then exit the **Sketcher** workbench.

Converting the Sketch into a Base Wall

1. Choose the **Wall** tool from the **Walls** toolbar; the **Wall Definition** dialog box is displayed.

2. Select the sketch from the geometry area if not already selected.

3. Choose the **OK** button to create the base wall, as shown in Figure 15-150.

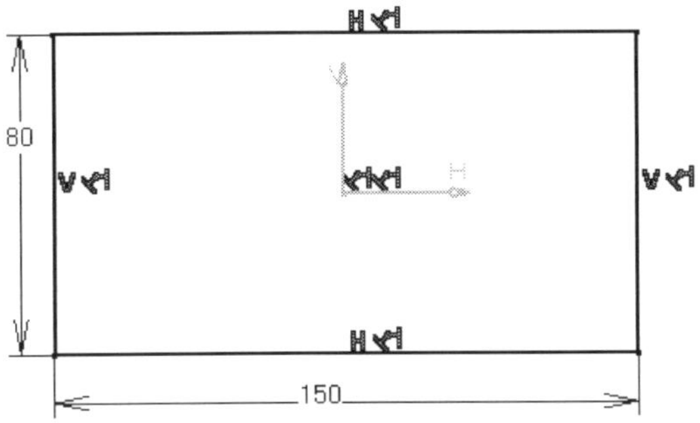

Figure 15-149 *Sketch of the base wall*

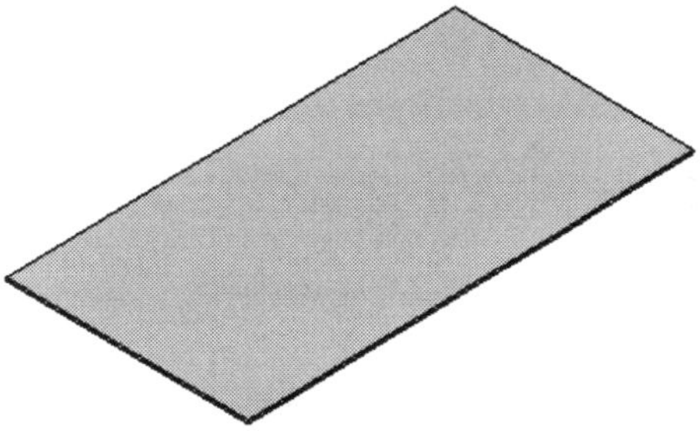

Figure 15-150 *The resultant base wall*

Drawing the Sketch of Ventilation Slots

1. Choose the **Sketch** tool from the **Sketcher** toolbar and then select the bottom face of the base wall as the sketching plane; the **Sketcher** workbench is invoked.

2. Draw the sketch for the louver feature, as shown in Figure 15-151, and then exit the **Sketcher** workbench.

Creating Ventilation Slots

1. Choose the **Louver** tool from **Cutting/Stamping > Stamping** sub-toolbar; the **Louver Definition** dialog box is displayed. Set the value **2** in the **Height** spinner in the dialog box.

2. Set **0deg** in the **Angle A1** spinner and **90deg** in the **Angle A2** spinner. Next, set **2** and **1** in the **Radius R1** and **Radius R2** spinners, respectively.

3. Select the sketch from the geometry area, if it is not already selected; the name of the sketch is displayed in the **Profile** display box.

4. Select the line nearest to the edge as the open end of the louver, refer to Figure 15-152.

5. Choose the **OK** button; the louver stamp feature is created, as shown in Figure 15-152.

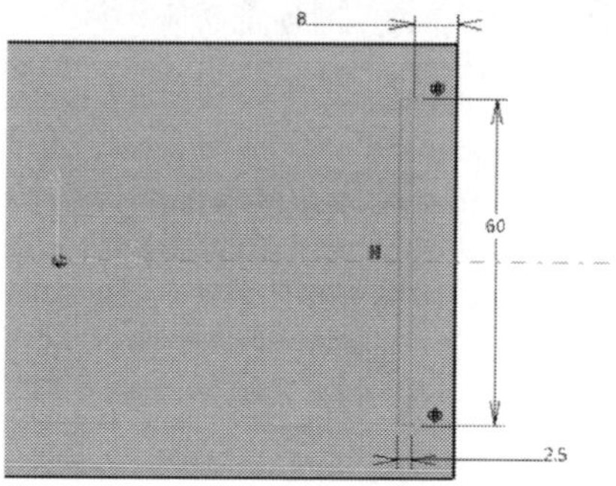

Figure 15-151 *Sketch of the louver stamp*

Creating the Pattern of the Louver Feature

1. Select the louver feature from the Specification tree and choose the **Rectangular Pattern** tool from the **Transformations > Pattern** sub-toolbar; the **Rectangular Pattern Definition** dialog box is displayed.

2. Set 12 in the Instances and Spacing spinners.

3. Select the base wall as the reference element.

4. Choose the **Reverse** button to flip the direction of the pattern feature if required.

5. Choose the **OK** button; the pattern of the louver feature is created, as shown in Figure 15-153.

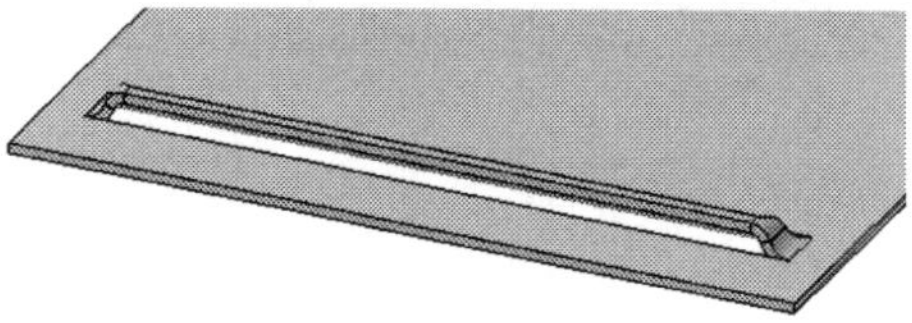

Figure 15-152 *The louver stamp feature created*

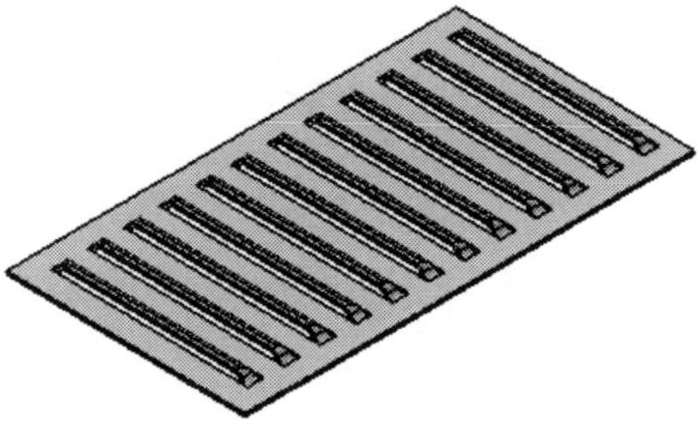

Figure 15-153 *Pattern of the louver feature*

Creating the Handles and Saving the File

1. Invoke the sketching environment to create a point on the bottom face of the base wall, as shown in Figure 15-154, and then exit the **Sketcher** workbench.

2. Press and hold the CTRL key, select the created point and the bottom face of the sheet metal part. Then, choose the **Bridge** tool from **Cutting/Stamping > Stamping** sub-toolbar; the **Bridge Definition** dialog box is displayed.

3. Set **4** in the **Height H** spinner and **27** in the **Length L** spinner.

4. Next, set **4** in the **Width W** spinner and **80deg** in the **Angle A** spinner.

5. Set **0.5** in the **Radius R1** spinner and **1** in the **Radius R2** spinner.

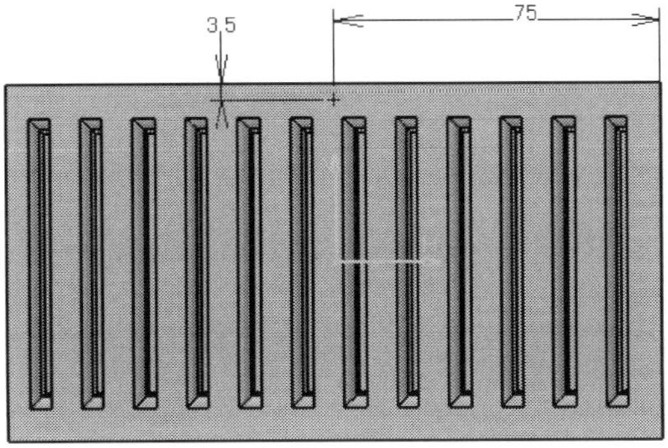

Figure 15-154 *Creating a point for the handle (Bridge feature)*

6. Flip the direction of the bridge feature if the bridge feature is created in the opposite direction, and then choose **OK**; the Bridge feature is created, as shown in Figure 15-155.

7. Create the mirror feature of the handle created about the zx plane. The final model of Tutorial 3 is shown in Figure 15-156.

8. Save the sheet metal component with the name *c15tut3* at *C:\CATIA\c15*.

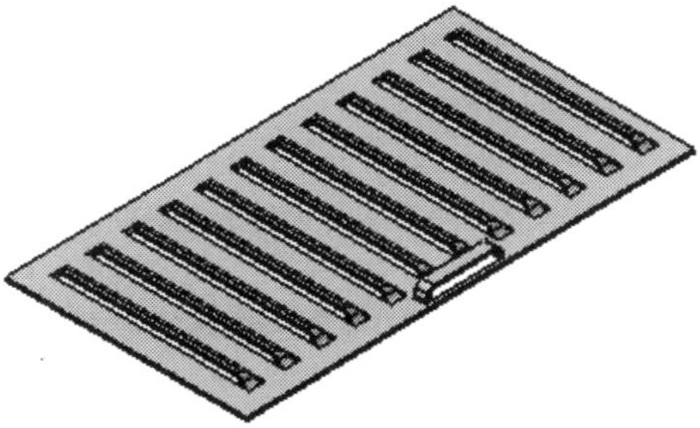

Figure 15-155 *The Bridge feature created*

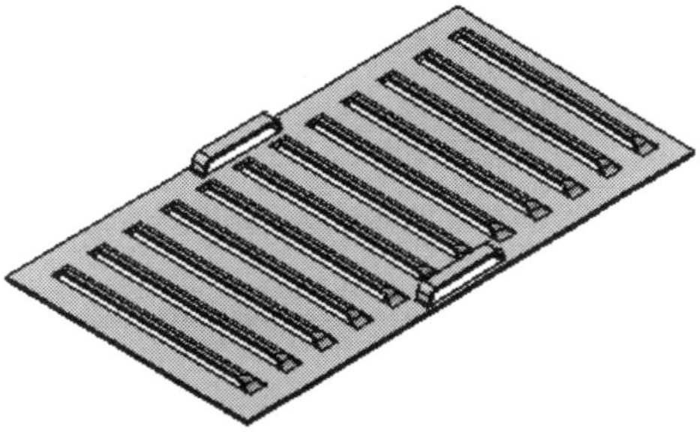

Figure 15-156 *The final model for Tutorial 3*

Tutorial 4

In this tutorial, you will create a sheet metal component, as shown in Figure 15-157. Its dimensions are shown in Figure 15-158. The thickness of the sheet metal component is 1 mm and bend radius is 2 mm. **(Expected time: 45 min)**

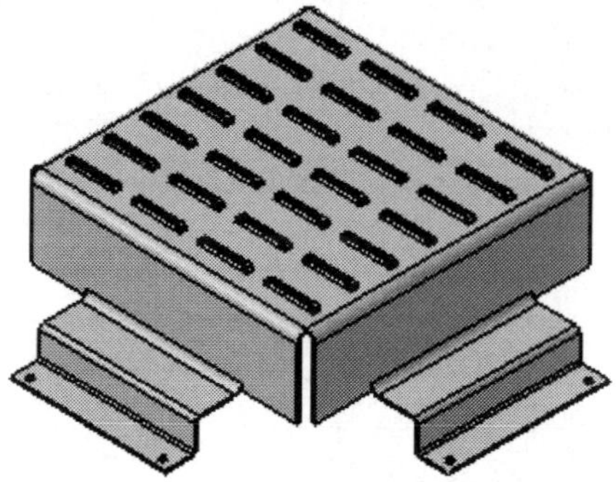

Figure 15-157 *Model for Tutorial 4*

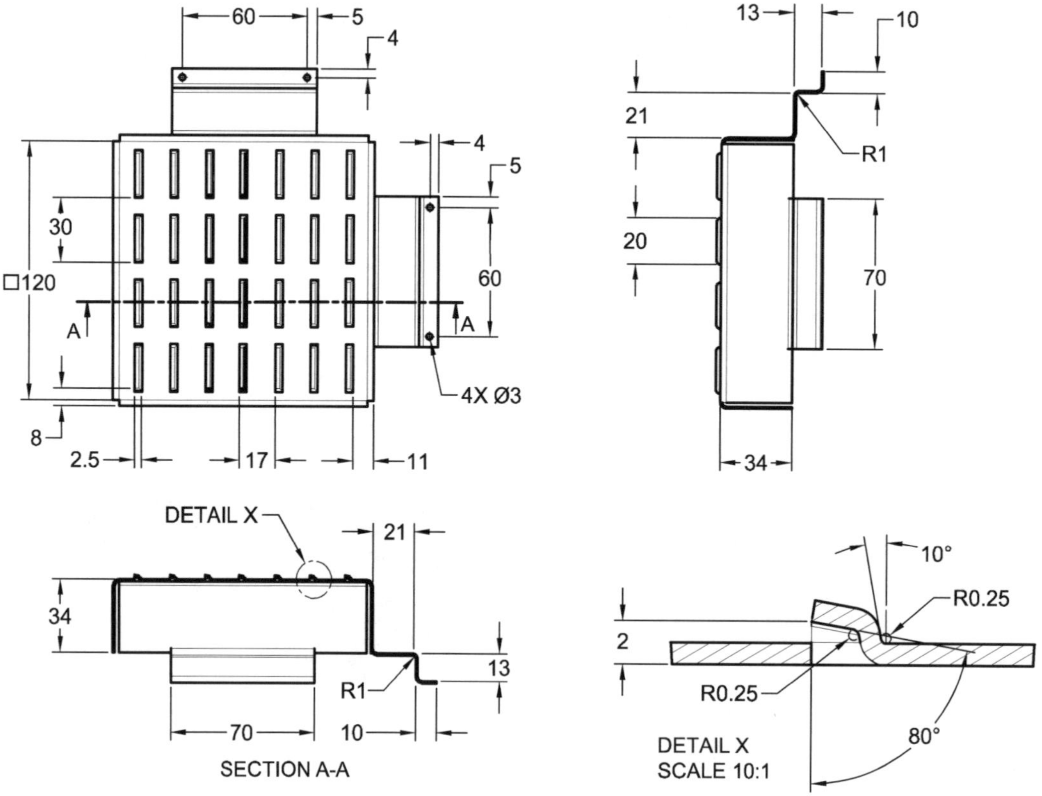

Figure 15-158 *Views and dimensions of the sheet metal component for Tutorial 4*

The following steps are required to complete this tutorial:

a. Start a new file in the **Generative Sheetmetal Design** workbench.
b. Draw the sketch of the base feature and convert it into a base wall by using the **Wall** tool.
c. Create the flanges on the four edges of the base wall.
d. Create walls on the flanges.
e. Create hole features.

f. Create the louver on the top face and pattern it.
g. Create the flat pattern and then save the model.

Starting a New File and Setting Sheet Metal Parameters

1. Start a new file in the **Generative Sheetmetal Design** workbench.

2. Choose the **Sheet Metal Parameters** tool from the **Walls** toolbar; the **Sheet Metal Parameters** dialog box is displayed.

3. Set the value **1** in the **Thickness** spinner and the value **2** in the **Default Bend Radius** spinner in the **Parameters** tab of the **Sheet Metal Parameters** dialog box. Next, choose **OK**.

Drawing the Sketch of the Base Wall

1. Invoke the **Sketcher** workbench by selecting the xy plane.

2. Draw the sketch of the base wall, as shown in Figure 15-159, and then exit the **Sketcher** workbench.

Converting the Sketch into a Base wall

1. Choose the **Wall** tool from the **Walls** toolbar; the **Wall Definition** dialog box is displayed.

2. Select the sketch from the geometry area, if it is not already selected.

3. Choose the OK button; the base wall is created, as shown in Figure 15-160.

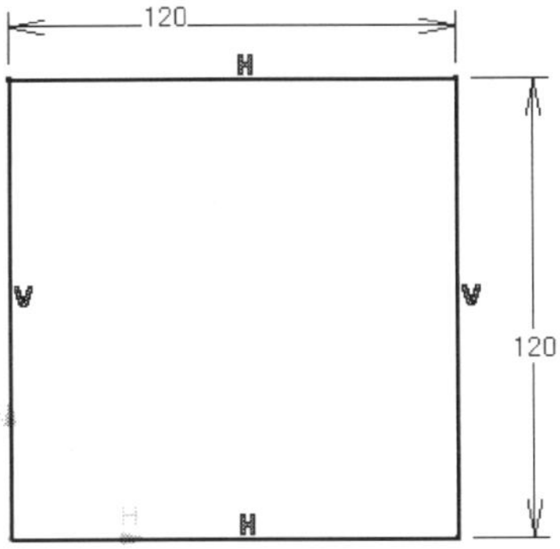

Figure 15-159 *Sketch for the base wall feature*

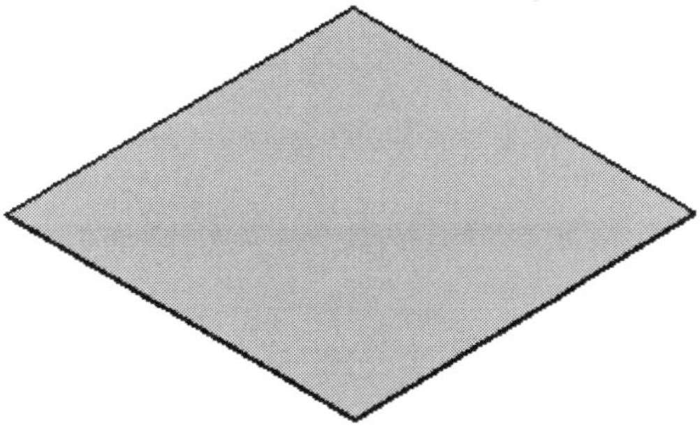

Figure 15-160 *The base wall feature*

Creating the Flange on the Edge of the Base wall

1. Select the edge on which the flange has to be created, as shown in Figure 15-161.

2. Choose the **Flange** tool from **Walls > Swept Walls** sub-toolbar; the **Flange Definition** dialog box is displayed.

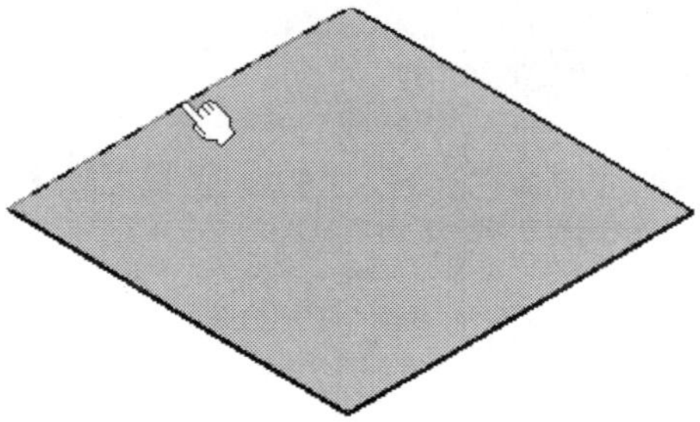

Figure 15-161 Edge selected for creating the flange

3. Select the **Basic** option from the **Flange type** drop-down list, if it is not already selected.

4. Set the value **34** in the **Length** spinner. Choose the down arrow on the **Length type** button; a flyout is displayed. Choose the **Outer Length type** button.

5. Set **90deg** in the **Angle** spinner and **2mm** in the **Radius** spinner, if required.

6. Choose the **Reverse Direction** button, if required, and then choose the **OK** button. Figure 15-162 shows the resultant flange.

7. Similarly, create three more flanges on the other edges of the base wall. Figure 15-163 shows the Sheet metal component after creating other flanges.

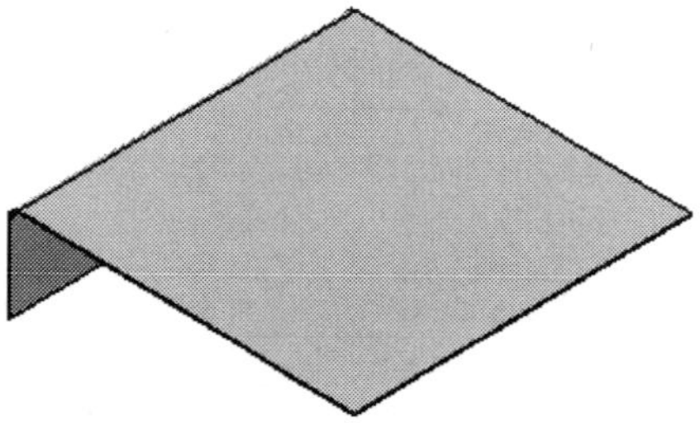

Figure 15-162 Resulting flange feature

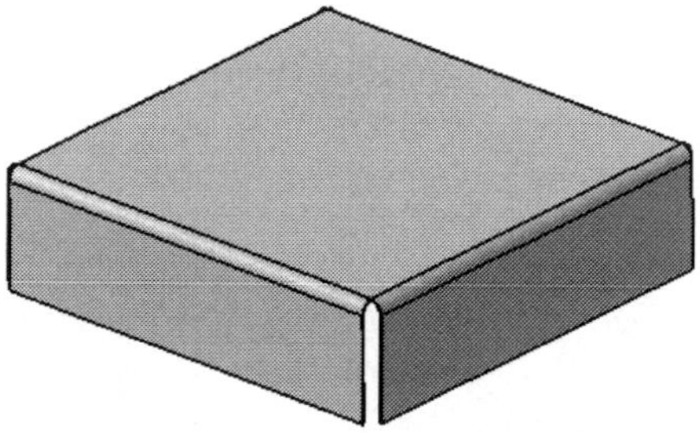

Figure 15-163 Component with all flanges

Creating the Wall on the Flange

1. Choose the **Wall On Edge** tool from the **Walls** toolbar; the **Wall On Edge Definition** dialog box is displayed. Select the **Automatic** option from the **Type** drop-down list.

2. Select an edge of the flange, as shown in Figure 15-164.

3. Set **21** in the **Height** spinner and **90deg** in the **Angle** spinner. Also, select the **With Bend** check box.

4. Choose the **Extremities** tab. Set the value **-25** in both the **Left offset** and **Right offset** spinners.

5. Choose the **OK** button; the side wall feature is created, as shown in Figure 15-165.

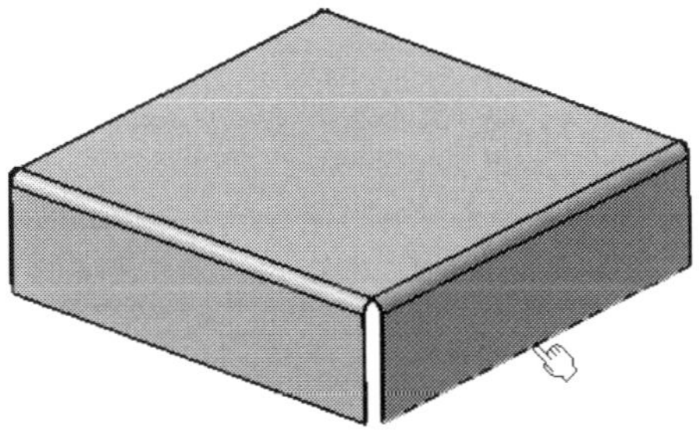

Figure 15-164 Selecting the edge for creating the wall on the flange

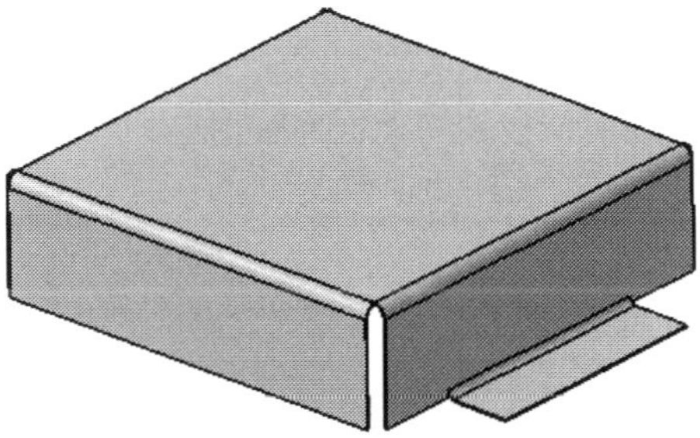

Figure 15-165 The side wall feature created

Creating the Flange on the Wall

1. Choose the **Flange** tool from **Walls > Swept Walls** sub-toolbar; the **Flange Definition** dialog box is displayed. Select the **Basic** option from the **Flange type** drop-down list.

2. Select the edge of the wall created in the previous step, as shown in Figure 15-166.

3. Set 13 in the Length spinner, 90deg in the Angle spinner, and 1 in the Radius spinner.

4. Choose the **Reverse Direction** button if required. Choose the **OK** button; a flange is created, as shown in Figure 15-167.

Figure 15-166 Selecting the edge for creating the flange

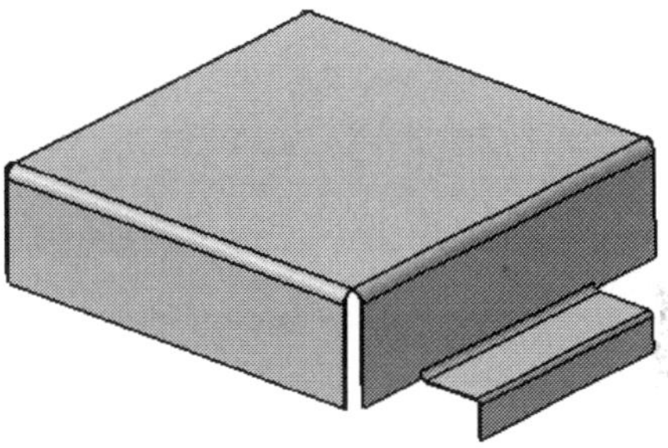

Figure 15-167 The flange created

5. Next, create a flange of 10 mm on the vertical flange created in the previous step. Make sure that the new flange is created in horizontal direction and also away from the vertical wall.

6. Similarly, create the wall on the edge and the flange features on the other side of the Sheet metal component, refer to Figure 15-168. The dimensions of the wall and flange features are the same as in the previous case.

Creating the Hole Features

1. Create holes by invoking the **Hole** tool from **Cutting/Stamping > Stamping** sub-toolbar, refer to Figure 15-168. For dimensions of holes, refer to Figure 15-158.

Creating the Louver

1. Select the bottom face of the base wall as the sketching plane and draw the sketch of the louver, as shown in Figure 15-169.

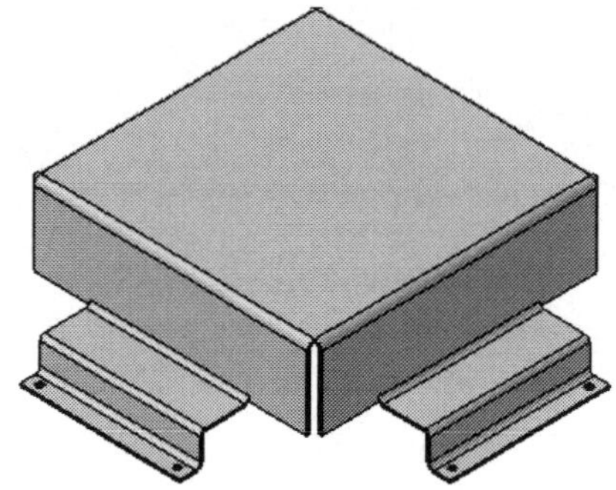

Figure 15-168 The component after creating the hole features

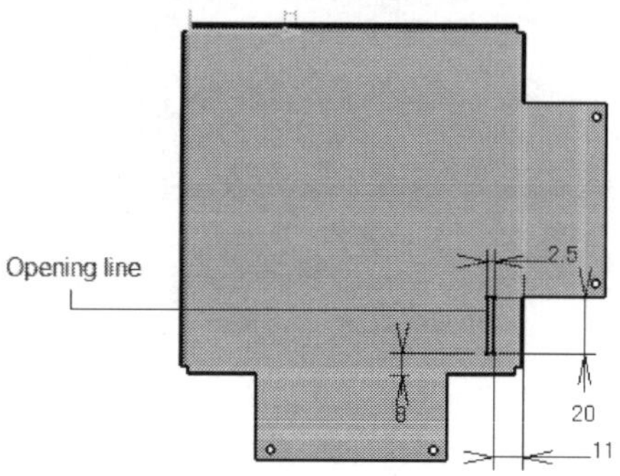

Figure 15-169 Sketch of the Louver

2. Choose the **Louver** tool from the **Stamping** toolbar. Set **2** in the **Height** spinner, **10deg** in the Angle A1 spinner, and 80deg in the Angle A2 spinner.

3. Set **0.25mm** in both the **Radius R1** and **Radius R2** spinners. Select the sketch from the geometry area.

4. Select the line that is near to the hole feature as the open end of the louver feature.

5. Choose the **OK** button; the louver feature is created.

6. Create the pattern feature of the louver, as shown in Figure 15-170. Refer to Figure 15-158 for dimensions.

Creating the Flat Pattern and Saving the File

1. Choose the **Fold/Unfold** button from **Views > Fold/Unfold** sub-toolbar; the Sheet metal component is unfolded, as shown in Figure 15-171. To fold it back, choose the Fold/Unfold button again.

2. Save the Sheet metal component with the name *c15tut4* at *C:\CATIA\c15*.

Figure 15-170 Pattern of the louver feature created

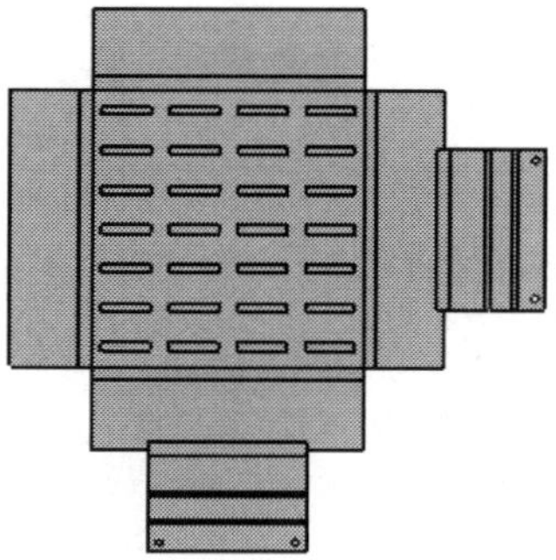

Figure 15-171 Flat pattern of the component

Self-Evaluation Test

Answer the following questions and then compare them to those given at the end of this chapter:

1. You can unfold a sheet metal component in a separate window by choosing the _________ button from the **Fold/Unfold** sub-toolbar in the **Views** toolbar.

2. _________ features are created in a sheet metal part for the purpose of ventilation.

3. You can fillet corners of a wall by using the _________ tool.

4. In the **Generative Sheet Metal Design** environment, parts are created in **.CATPart* file format. (T/F)

5. Choose **Start > Mechanical Design > Part Design** from the menu bar to invoke the **Generative Sheetmetal Design** workbench. (T/F)

6. The **Linear** option provides a round relief between the supporting walls of a sheet metal. (T/F)

7. The **Extrusion** tool is applicable only on open profiles. (T/F)

8. A tear drop is similar to a hem with the only difference that when you create a tear drop, the flat wall is inclined at an angle. (T/F)

9. You can fold as well as unfold a sheet metal component by using the **Folded/Unfolded** tool. (T/F)

10. You can create a surface stamp feature by using an open profile. (T/F)

Review Questions

Answer the following questions:

1. Which of the following buttons is not available in the flyout that is displayed on the right of the **Lines** drop-down list in the **Bend From Flat Definition** dialog box?

   (a) **Axis** (b) (**IML**) Inner Mold Line
   (c) (**OML**) Outer Mold Line (d) (**MML**) Middle Mold Line

2. The _________ tool is used to create a bent face between the intersection of two walls.

3. The_________ walls are created by sweeping a profile along the selected edge.

4. The _________ feature is created for holding a sheet metal component.

5. The _________ option provides a tangent relief between the supporting walls of a sheet metal component.

6. You can change the values of parameters in the **Sheet Metal Parameters** dialog box at any time. (T/F)

7. You can apply material to a sheet metal component using the **Sheet Metal Parameters** dialog box. (T/F)

8. To create a flanged hole, you need to create a point on a wall. (T/F)

9. For creating the Bead feature, if the sketch has multiple entities, they should be tangentially connected. (T/F)

10. Flange is the bend section of a sheet metal. (T/F)

EXERCISE

Exercise 1

Create the sheet metal component shown in Figure 15-172. The dimensions of the model are shown in Figure 15-173. **(Expected time: 30 min)**

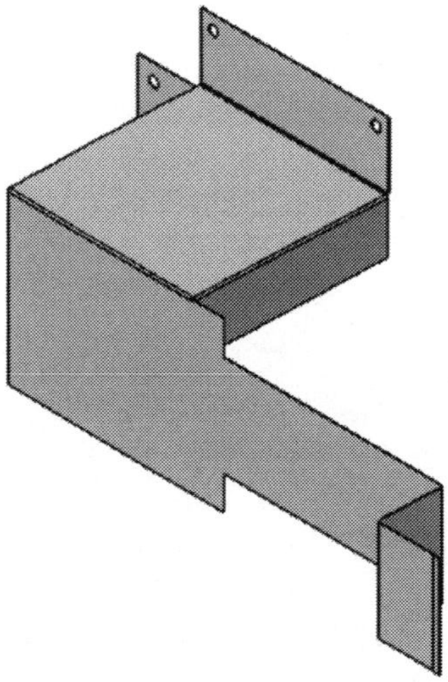

Figure 15-172 *Sheet metal part for Exercise 1*

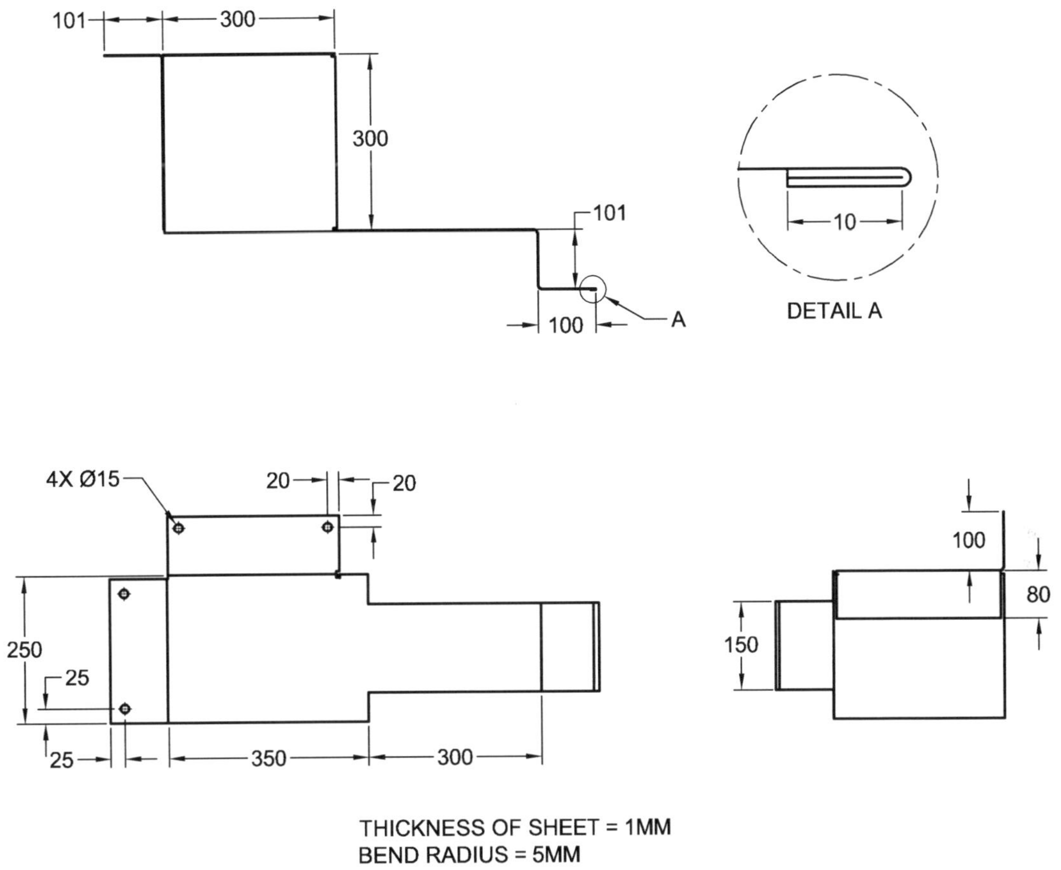

Figure 15-173 Orthographic views and dimensions of the component

Answers to Self-Evaluation Test

1. **Multi Viewer**, **2.** **Louver**, **3.** **Corner**, **4.** T, **5.** F, **6.** F, **7.** T, **8.** T, **9.** T, **10.** F

Chapter 16

DMU Kinematics

Learning Objectives

After completing this chapter, you will be able to:

- *Create the Revolute joint and simulate it*
- *Create the Prismatic joint*
- *Create the Cylindrical joint*
- *Create the Screw joint*
- *Create the Rigid joint*
- *Create the Spherical joint*
- *Create a mechanism using a combination of joints*
- *Create the Planar joint*
- *Create the Point Curve joint*
- *Create the Slide Curve joint*
- *Create the Roll Curve joint*
- *Create the Point Surface joint*
- *Create the Universal joint*
- *Create the CV joint*
- *Create the Gear joint*
- *Create the Rack joint*
- *Create the Cable joint*
- *Convert assembly constraints into joints*

INTRODUCTION TO DMU KINEMATICS

DMU Kinematics is a workbench that is used to create and edit different kinematic mechanisms. Also, it is used to study and check the working of kinematic mechanisms. In this workbench, you can simulate and analyze the mechanisms dynamically. Therefore, it is easy and convenient to check the limits and interferences of different parts of the mechanism.

In **DMU Kinematics**, minimum two parts are required to create a mechanism. You can keep one part of the mechanism fixed and move the other part with respect to the fixed part to analyze various functions of the mechanism.

You can design mechanisms by creating joints manually using the tools available in the **Kinematics Joints** sub-toolbar in the **DMU Kinematics** toolbar. Alternatively, you can create them automatically by using assembly constraints. In this chapter, you will learn about both the methods used for designing a mechanism.

DESIGNING A MECHANISM

As already stated, the **DMU Kinematics** workbench allows you to create a mechanism. The procedure to create a mechanism involves the following five major steps: invoking the **DMU Kinematics** workbench, placing / creating parts in this workbench, starting a new mechanism, making a part of the mechanism fixed (stationary), and creating joints. These steps are discussed next.

1. The first step to create a mechanism is to invoke the **DMU Kinematics** workbench. To invoke this workbench, choose **Start > Digital Mockup > DMU Kinematics** from the menu bar.

2. After invoking the **DMU Kinematics** workbench, you need to place the parts required for creating the mechanism in this workbench. You can do so by placing an existing part or by creating a new part as you did in the assembly environment. To place a part in the **DMU Kinematics** workbench, choose **Insert > Existing Component** from the menu bar and select **Product1** from the Specification tree; the **File Selection** dialog box will be displayed. In this dialog box, browse to the location of the desired part and place it in the **DMU Kinematics** workbench. To create a new part in the **DMU Kinematics** workbench, choose **Insert > New Part** from the menu bar and select **Product1** from the Specification tree; the new **Part** node will be added to the Specification tree. Now, create a new part as you did in the top-down assembly design approach in the **Assembly Design** workbench.

> **Tip**
> *If the parts placed in the **DMU Kinematics** workbench overlap each other, you need to move them apart for better visualization. To do so, drag the red point of the Compass and place it over the part to be moved. Next, drag any axis of the Compass and move the part to the desired position.*

3. Now, you need to start a new mechanism. To do so, choose **Insert > New mechanism** from the menu bar; the **Mechanism.1** node will be created under the **Mechanisms** node, this, in turn, is a sub node of the **Applications** node in the Specification tree. The **Mechanisms** node is called the parent mechanism node. One product can have multiple mechanisms. Therefore, a parent mechanism can have multiple child mechanisms. These nodes are known

as child mechanisms. For creating a mechanism, you need to create joints and commands, define all degrees of freedoms (DOFs) for each part, and make one part fixed. On doing so, all joints and commands used for creating the mechanism will be listed automatically under their respective nodes as the **Joints** and **Commands** sub nodes. Also, all undefined DOFs will be displayed beside the child mechanism node, see Figure 16-1.

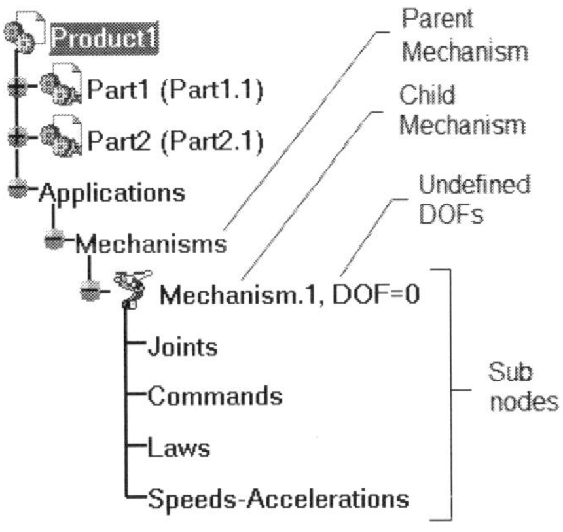

Figure 16-1 *The Specification tree showing all joints and commands*

4. To simulate a mechanism, you need to keep one part of the mechanism fixed (Stationary). To fix a part, choose the **Fixed Part** tool from the **DMU Kinematics** toolbar available in this workbench; the **New Fixed Part** dialog box will be displayed, as shown in Figure 16-2. Also, you will be prompted to select the part that you want fixed. Select the part in the drawing area; the selected part will become fixed. Now, all other parts in the **DMU Kinematics** workbench can be simulated with respect to this fixed part.

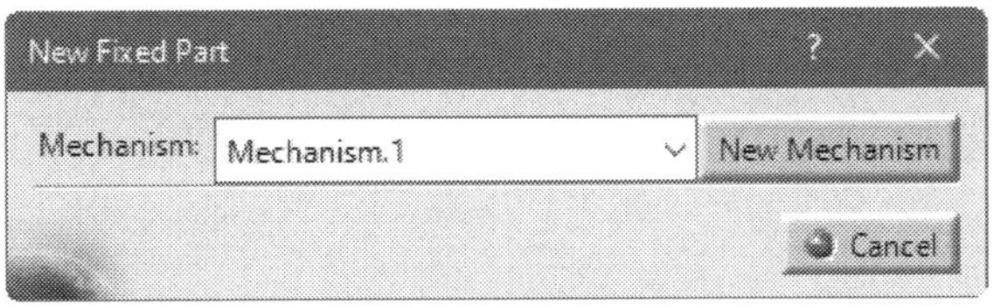

Figure 16-2 *The* ***New Fixed Part*** *dialog box*

5. For designing a mechanism, you need to apply different mechanical relations between parts. In CATIA V5, you can apply these mechanical relations by using tools from the **Kinematics Joints** sub-toolbar. To view the **Kinematics Joints** sub-toolbar, choose the down-arrow in the **Revolute Joint** tool in the **DMU Kinematics** toolbar.

   The procedure to create different types of joints using the tools available in the **Kinematics Joints** sub-toolbar is discussed next.

Creating the Revolute Joint

Menubar:	Insert > New Joint > Revolute
Toolbar:	DMU Kinematics > Kinematics Joints sub-toolbar > Revolute Joint

The **Revolute Joint** tool is used to create a joint between two parts such that one part rotates with respect to the fixed part about a common axis. For creating a Revolute joint, you need to specify one axis and one planar face (plane) from each part as well as the offset distance between the two specified planes. To create a Revolute joint between two parts, insert the parts in the **DMU Kinematics** workbench, as shown in Figure 16-3. Start a new mechanism and keep one of the parts fixed as explained earlier. For the parts shown in Figure 16-3, the column will be the fixed part. Now, to create the Revolute joint between the parts, choose the **Revolute Joint** tool from the **Kinematics Joints** sub-toolbar in the **DMU**

Kinematics toolbar; the **Joint Creation: Revolute** dialog box will be displayed, as shown in Figure 16-4. Also, you will be prompted to select the first line. You can select any axis, edge, or sketch line as the first line. Select the center axis of one part, refer to Figure 16-5; you will be prompted to select the second line. Select the center axis of another part, refer to Figure 16-5; you will be prompted to select the first plane. Select the planar face of one part; you will be prompted to select the second plane. Select the planar face of another part, refer to Figure 16-5. By default, the **Null Offset** radio button is selected in the **Joint Creation: Revolute** dialog box. As a result, both the selected planar faces will become coplanar. You can also specify an offset distance between the selected planar faces. To do so, select the **Offset** radio button from the **Joint Creation: Revolute** dialog box; the **Offset** spinner will be activated and the default distance between the two selected planar faces will be displayed in it. Now, set the required offset distance in this spinner. Next, select the **Angle Driven** check box and then choose the **OK** button; the Revolute joint will be created, as shown in Figure 16-6. Also, the **Information** message box will be displayed, as shown in Figure 16-7. This message box informs whether or not the mechanism can be simulated, depending upon the mechanical relations applied through joints.

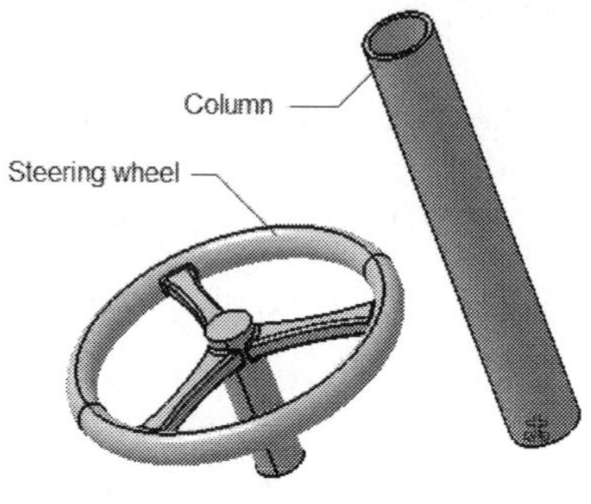

Figure 16-3 *Parts placed in the* ***DMU Kinematics*** *workbench*

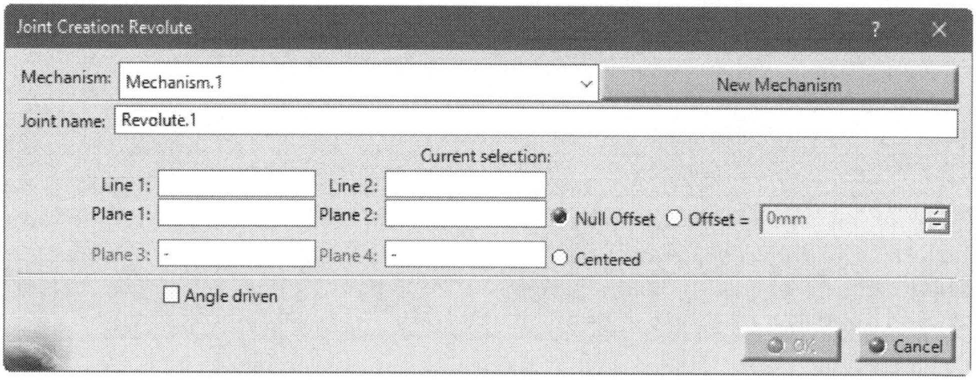

Figure 16-4 *The* ***Joint Creation: Revolute*** *dialog box*

Once the joint has been created, you will notice that the subnode **Revolute.1** has been created under the **Joints** node, and the subnode **Command.1** has been created under the **Commands** node in the Specification tree. You can use this command to simulate the mechanism while manipulating the angle.

After creating the joints, you can also modify the specified offset distance between the planes, refer to Figure 16-6. To do so, double-click on the offset distance displayed in the graphic area; the **Constraint Definition** dialog box will be displayed. Set the required value in the **Value** spinner and choose **OK** from this dialog box; the new offset distance will be applied.

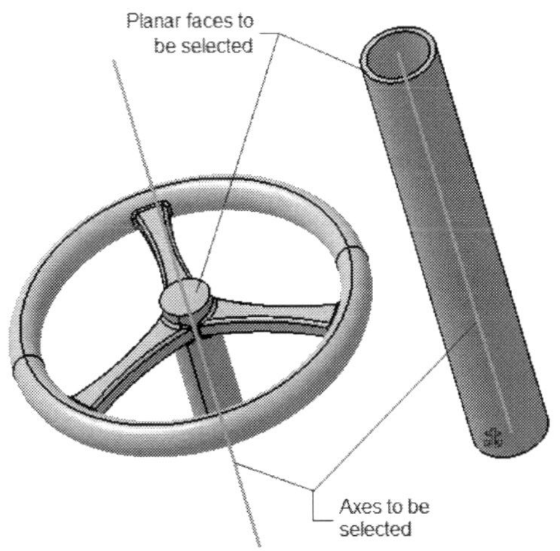

Figure 16-5 *Axes and planar faces to be selected for creating the Revolute joint*

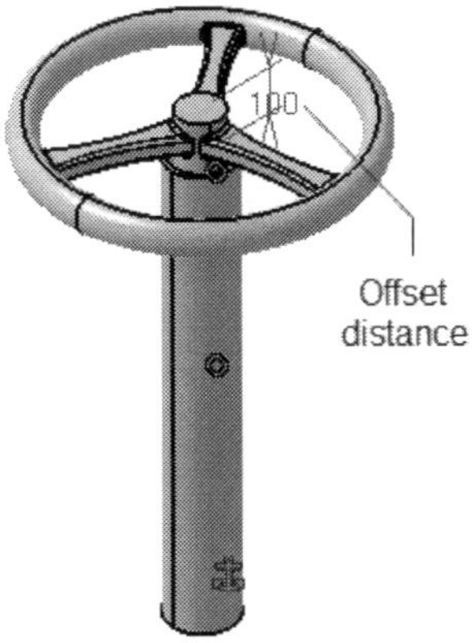

Figure 16-6 *Component after creating a Revolute joint*

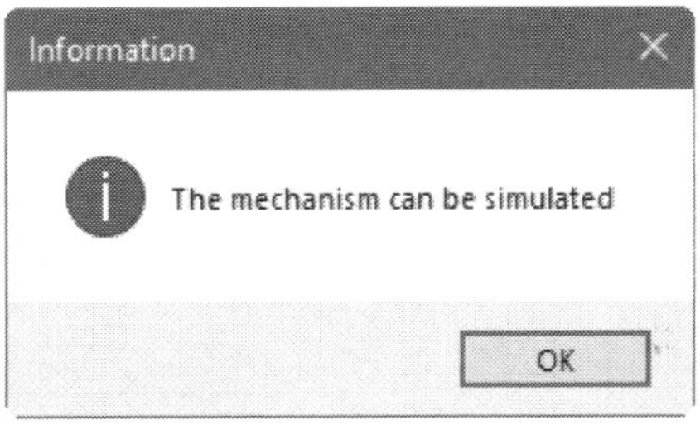

Figure 16-7 *The **Information** message box*

Note

*To simulate a mechanism, you need to have at least one command. You can create a command by selecting the **Angle Driven** or **Length Driven** check box available in the dialog boxes of corresponding joints.*

The procedure to simulate this mechanism is discussed next.

Simulating the Mechanism

To simulate the mechanism, choose the **Simulation with Commands** tool from the **Simulation** sub-toolbar in the **DMU Kinematics** toolbar; the **Kinematics Simulation** dialog box will be displayed. Choose the **More** button in this dialog box to expand it, if it is not expanded. The expanded **Kinematics Simulation** dialog box is shown in Figure 16-8. Now, move the cursor over the part that is not fixed; green arrow heads will be displayed over the component, as shown in Figure 16-9. Press and hold the left mouse button and drag the cursor; the part will move according to the movement of the cursor.

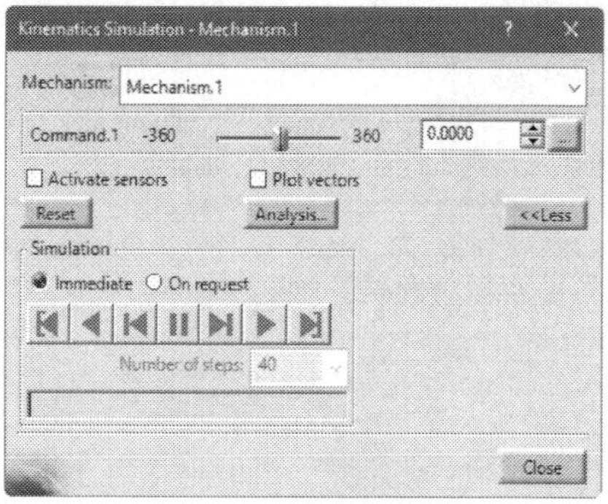

Figure 16-8 The expanded ***Kinematics Simulation*** *dialog box*

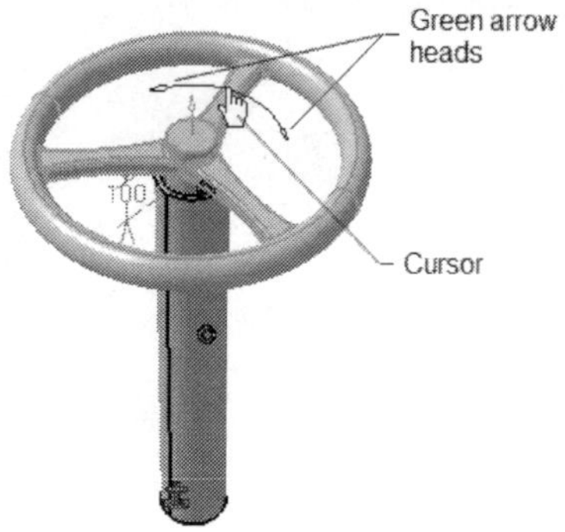

Figure 16-9 Green arrow heads displayed over the part

The areas in the Kinematics Simulation dialog box are discussed next.

Mechanism Area

The **Mechanism** drop-down list in this area displays all the mechanisms created in the current file. Select the required mechanism from this drop-down list to simulate.

Command Area

In the **Command** area, the Command slider bar, the Command spinner, and the Swatch button are available, as shown in Figure 16-10. The Command slider bar or the Command spinner is used to set the value for the simulation of mechanism. The default lower and higher limits of the Command slider bar for the rotational movement of a part are -360 to 360 degrees, respectively, refer to Figure 16-10. If you create a joint by selecting the **Length driven** check box, then the default limits for the translational movement of the part will be -100 to 100. You can change these default values as per your need. To do so, choose the Swatch button; the **Slider** dialog box will be displayed, as shown in Figure 16-11.

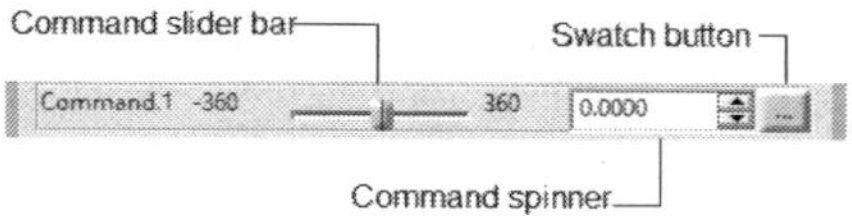

Figure 16-10 The ***Command*** *area of the* ***Kinematics Simulation*** *dialog box*

Using this dialog box, you can specify the lower and higher limits of the Command slider bar in the **Lowest Value** and **Highest Value** edit boxes, respectively. The value specified in the **Spin box increments** edit box can be used as the incremental value of the Command spinner of the **Kinematics Simulation** dialog box.

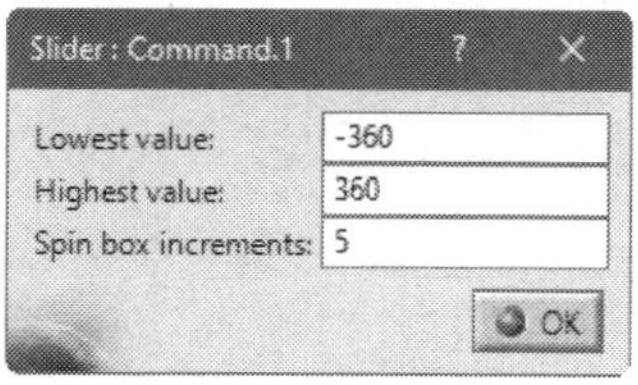

Figure 16-11 The ***Slider*** *dialog box*

Simulation Area

When the **Immediate** radio button is selected in this area, the movement of the part will be displayed instantaneously in the graphic area as soon as you move the Command slider bar or change the value in the Command spinner. Note that the options in the **Simulation** area remain inactive when the **Immediate** radio button is selected. However, if the **On request** radio button is selected in the **Simulation** area, the buttons below this option will get activated. These buttons will be used to play the simulation. If you choose the **Play forward** button without changing the value in the Command spinner or the Command bar, the **Simulation Information** message box will be displayed, as shown in Figure 16-12, informing that you need to change the command value before playing the simulation.

To play the simulation, change the command value by using the Command slider bar or the Command spinner and then choose the **Play forward** button; the simulation will run according to the specified command value. Also, the remaining buttons in the **Simulation** area get activated. To play the simulation in the reverse order, choose the **Play Back** button. If you want to set the simulation at the end position, choose the **End** button. Similarly, to set the simulation at the start position, choose the **Start** button. You can choose the **Pause** button to pause the simulation anytime while it is running. The **Number of steps** edit box is used to make the simulation smoother. More the value in this edit box, more smooth and slow will be the simulation. Choose the **Step forward** button to increment the simulation by one step. Choose the **Step back** button to decrement the simulation by one step.

Note

*If you want to create a new mechanism, choose the **New Mechanism** button from the **Joint Creation** dialog box; the **Mechanism Creation** dialog box will be displayed, as shown in Figure 16-13. You can specify a user-defined name in the **Mechanism name** edit box and then choose the **OK** button. On doing so, the new mechanism with the specified name will be displayed under the **Mechanisms** node in the Specification tree.*

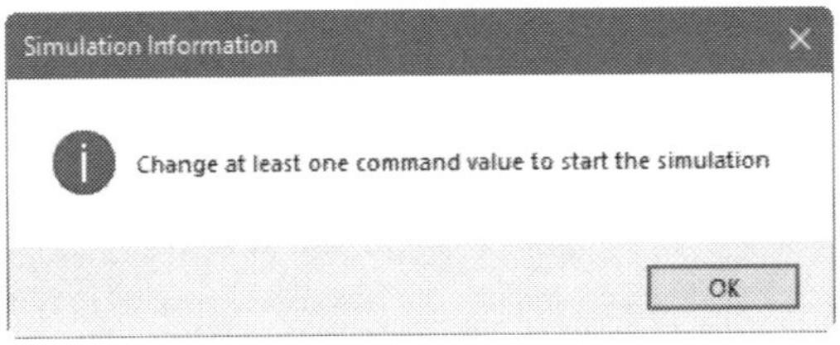

*Figure 16-12 The **Simulation Information** message box*

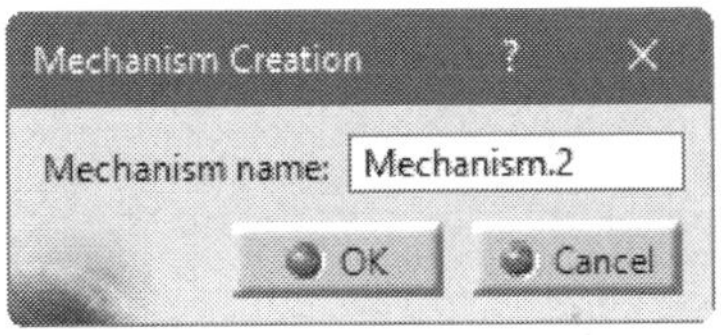

*Figure 16-13 The **Mechanism Creation** dialog box*

Tip

*It is recommended that you update the mechanism after creating, modifying, or editing it. To do so, choose the **Update Positions** button from the **Kinematic Update** toolbar; the **Update Mechanism** dialog box will be displayed. Select the required mechanism and choose **OK** from this dialog box; the mechanism will get updated.*

Creating the Prismatic Joint

Menubar:	Insert > New Joint > Prismatic
Toolbar:	DMU Kinematics > Kinematics Joints sub-toolbar > Prismatic Joint

The Prismatic joint allows you to slide a part in linear direction with respect to the fixed part at a common plane. For defining a Prismatic joint, you need to specify one edge and one planar face or datum plane from each part. To create a Prismatic joint between two parts, insert the parts in the **DMU Kinematics** workbench, as shown in Figure 16-14. Next, start a new mechanism and make one of the parts fixed. For the parts shown in Figure 16-14, Part 1 will be the fixed part.

Now, to create the prismatic joint between the parts, choose the **Prismatic Joint** tool from the **Kinematics Joints** sub-toolbar in the **DMU Kinematics** toolbar; the **Joint Creation: Prismatic** dialog box will be displayed, as shown in Figure 16-15. Also, you will be prompted to select the first line. You can select any edge, axis, or sketch line as the first line. Select an edge of one part; you will be prompted to select the second line. Select an edge of another part; the selected lines will become coaxial and one selected edge will slide over another selected edge. After selecting the edges, you will be prompted to select the first plane. You can select any plane or planar face. Select a plane from one of the parts; you will be prompted to select the second plane. Select a plane from another part; the selected planes will become coplanar and one of the planes will slide over the other selected plane during the simulation.

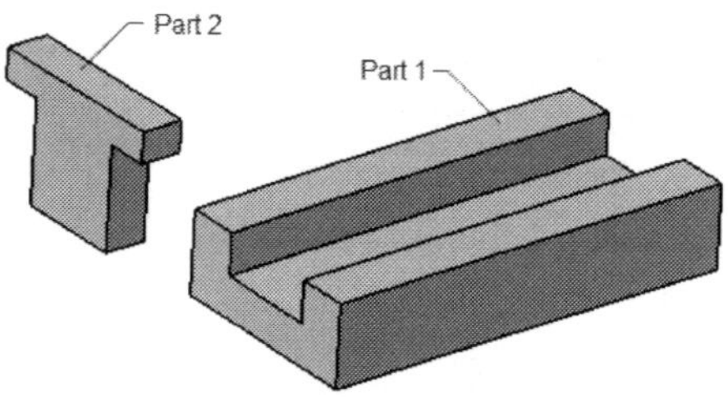

Figure 16-14 *Parts inserted in the* ***DMU Kinematics*** *workbench*

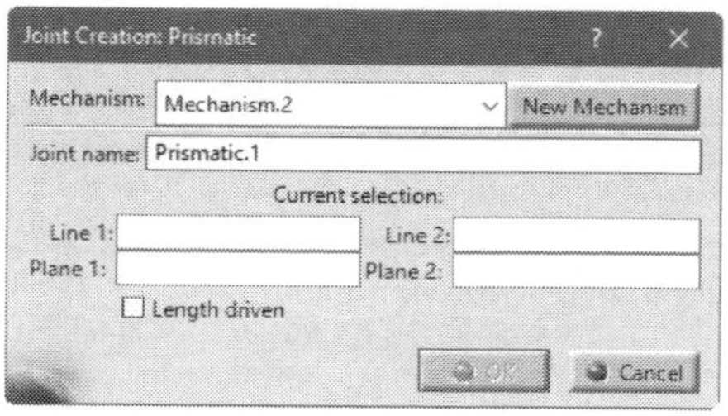

Figure 16-15 *The* ***Joint Creation: Prismatic*** *dialog box*

Figure 16-16 shows the edges and planes to be selected for creating the Prismatic joint between Part 1 and Part 2. To drive Part 2 in the linear direction, select the Length driven check box in the Joint Creation: Prismatic dialog box. On doing so, a command will be created. Now, choose the OK button to create a joint and exit the dialog box; the Prismatic. 1 and Command. 1 sub nodes will be created in the Specification tree. Next, update the mechanism to create the prismatic joint, as shown in Figure 16-17.

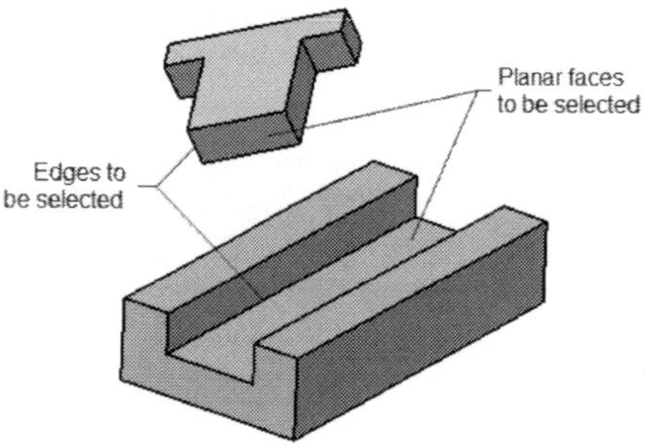

Figure 16-16 *Edges and planar faces selected for creating the Prismatic joint*

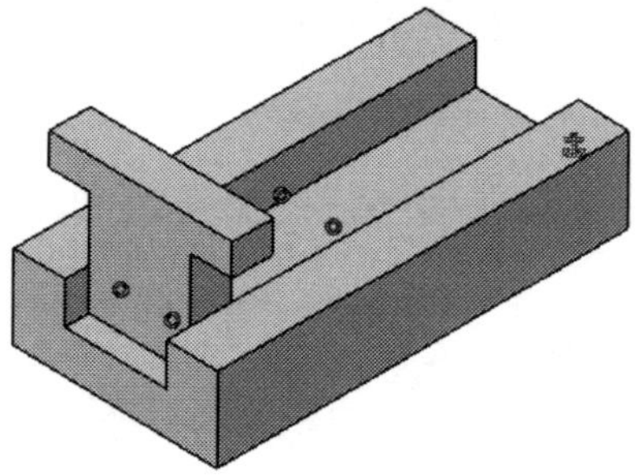

Figure 16-17 *The updated mechanism of the Prismatic joint*

Creating the Cylindrical Joint

Menubar: Insert > New Joint > Cylindrical
Toolbar: DMU Kinematics > Kinematics Joints sub-toolbar > Cylindrical Joint

The Cylindrical joint allows you to slide a part in linear direction as well as rotate it angularly with respect to the fixed part. Note that on applying this joint, the linear and rotational movements become independent of each other. For defining a Cylindrical joint, you need to specify one axis from each part. To create a Cylindrical joint between two parts, insert the parts in the **DMU Kinematics** workbench, as shown in Figure 16-18. Start a new mechanism and keep one of the parts fixed. For the parts shown in Figure 16-18, Part 1 will be the fixed part.

Now, to create the cylindrical joint between the parts, choose the **Cylindrical Joint** tool from the **DMU Kinematics** toolbar; the **Joint Creation: Cylindrical** dialog box will be displayed, as shown in Figure 16-19. Also, you will be prompted to select the first line. You can select any edge, axis, or sketch line as the first line. Select an axis from one part; you will be prompted to select the second line. Select an axis from another part; the selected axes will become coaxial such that one selected part can slide as well as rotate with respect to the fixed part during the simulation.

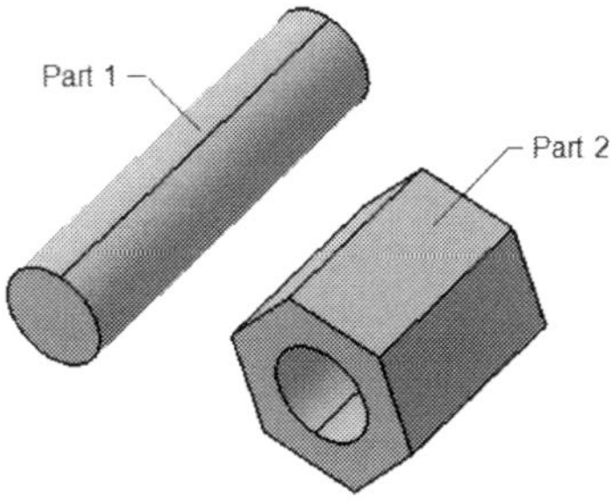

Figure 16-18 *Parts inserted in the* ***DMU Kinematics*** *workbench*

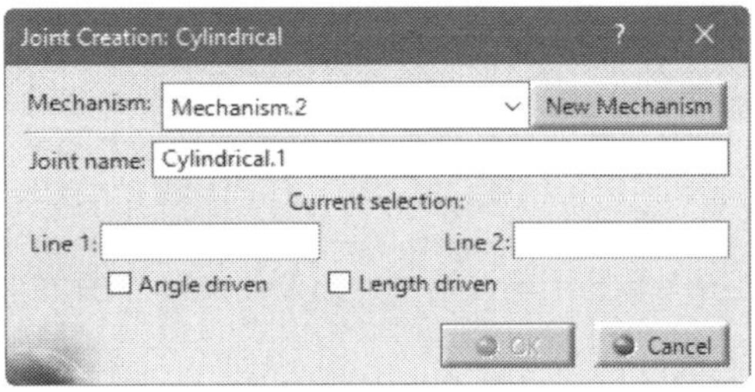

Figure 16-19 *The* ***Joint Creation: Cylindrical*** *dialog box*

Figure 16- 20 shows the axes to be selected for creating the Cylindrical joint between Part 1 and Part 2. Note that the **Length driven** and **Angle driven** check boxes in this dialog box can be used to create commands. Select the **Length driven** check box to create a command that can be used to slide one part over another. Similarly, select the **Angle driven** check box to create a command that can be used to rotate a part about its axis. If you select any one of these check boxes and create a joint, some undefined degrees of freedom (DOFs) will be generated in the mechanism. The undefined DOFs will be displayed beside the child mechanism node in the Specification tree. In this case, it will be displayed as **DOF=1** beside the **Mechanism.1** node in the Specification tree. Select both these check boxes and choose **OK**; the **Cylindrical. 1**, **Command. 1**, and **Command. 2** nodes will be created in the Specification tree. Update the created mechanism. The updated mechanism is shown in Figure 16-21.

Now, simulate the mechanism, as discussed earlier in the section Creating the Revolute Joint.

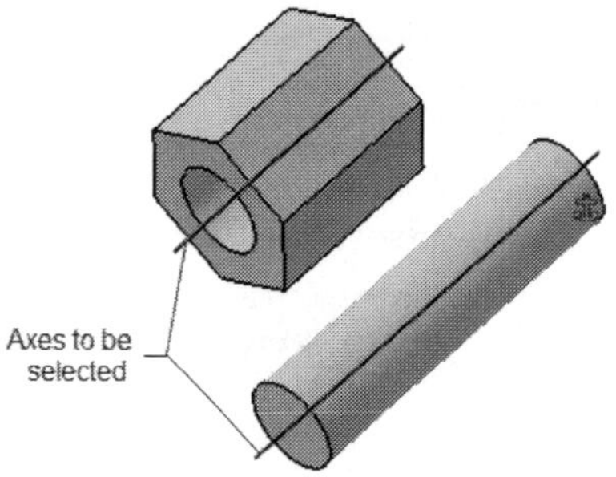

Figure 16-20 *Axes to be selected for creating the Cylindrical joint*

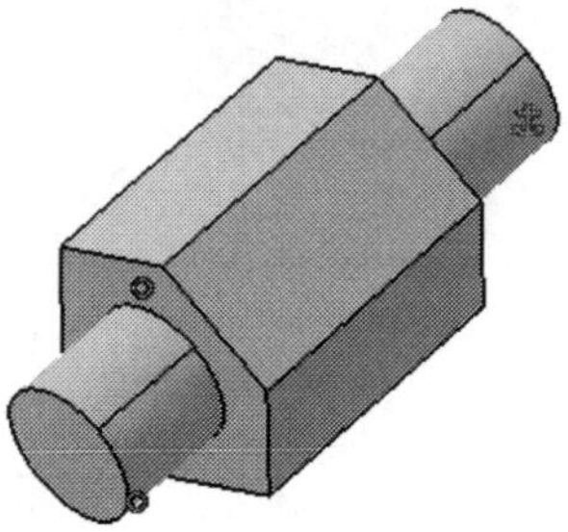

Figure 16-21 *The updated mechanism of the Cylindrical joint*

Creating the Screw Joint

Menubar: Insert > New Joint > Screw
Toolbar: DMU Kinematics > Kinematics Joints sub-toolbar > Screw Joint

A Screw joint is a joint in which one part moves linearly while rotating about its own axis with respect to the fixed part. The linear and rotational movements of a part depend on the specified pitch value. In other words, in a Screw joint, the translational and rotational movements are related to each other through the pitch value. Therefore, this mechanism is very much similar to the Nut-Bolt mechanism. For defining a Screw joint between two parts, first you need to specify one axis from each part and then the pitch value. To do so, insert two parts in the **DMU Kinematics** workbench, as shown in Figure 16-22. Next, start a new mechanism and make one of the parts fixed. In Figure 16-22, Part 1 will be the fixed part.

Tip
Generally, each part of a mechanism has six degrees of freedom (DOFs). Of these, three are rotational DOFs (about the X, Y, and Z axes) and the rest are linear (or translational) DOFs (along X, Y, and Z axes).

Choose the **Screw Joint** tool from the **DMU Kinematics** toolbar; the **Joint Creation: Screw** dialog box will be displayed, as shown in Figure 16-23. Also, you will be prompted to select the first line. You can select any edge, axis, or sketch line as the first line. Select an axis from one part; you will be prompted to select the second line. Select an axis from the other part; the selected axes will become coaxial. Next, specify the pitch value in the **Pitch** edit box. Figure 16-24 shows the axes to be selected for creating the Screw joint between Part 1 and Part 2.

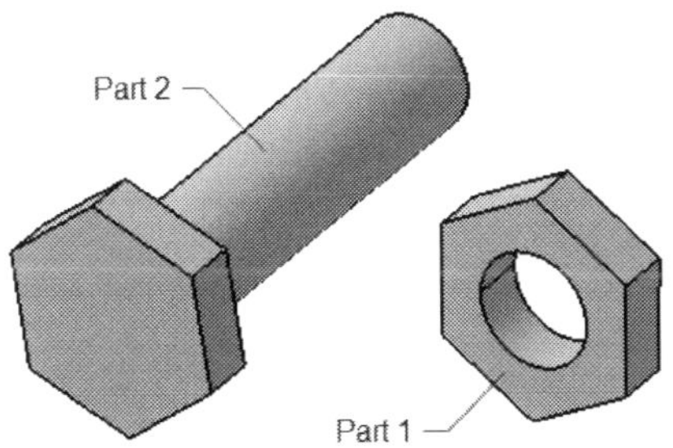

Figure 16-22 Parts inserted in the ***DMU Kinematics*** *workbench*

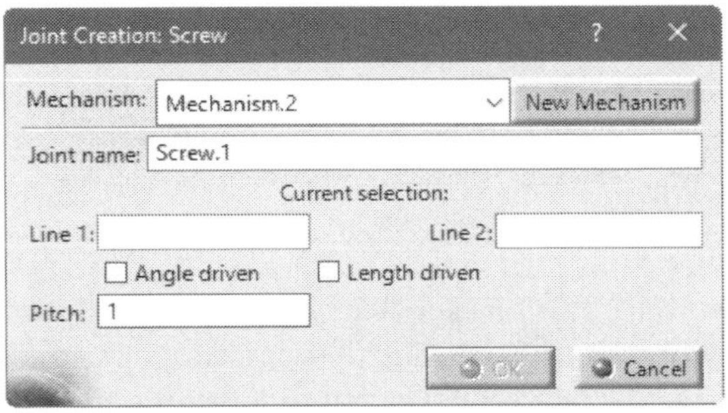

Figure 16-23 The ***Joint Creation: Screw*** *dialog box*

The **Length driven** and **Angle driven** check boxes in this dialog box act as toggle check boxes. Therefore, you can select only one of these check boxes at a time. If you select the **Length driven** check box, the part will be driven linearly and the revolution per unit length will be controlled by the specified pitch value. Similarly, if you select the **Angle driven** check box, the part will be driven rotationally and the linear distance per revolution will be controlled by the specified pitch value. Select the required check box and choose **OK**; the **Screw.1** and **Command.1** nodes will be created in the Specification tree. Next, update the created mechanism. The updated mechanism is shown in Figure 16-25.

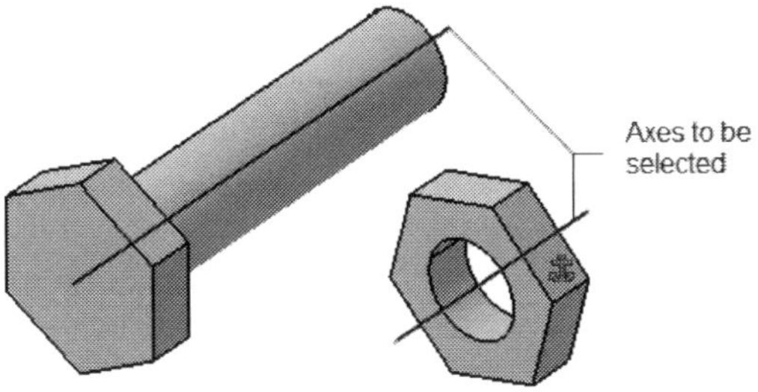

Figure 16-24 Axes selected for creating the Screw joint

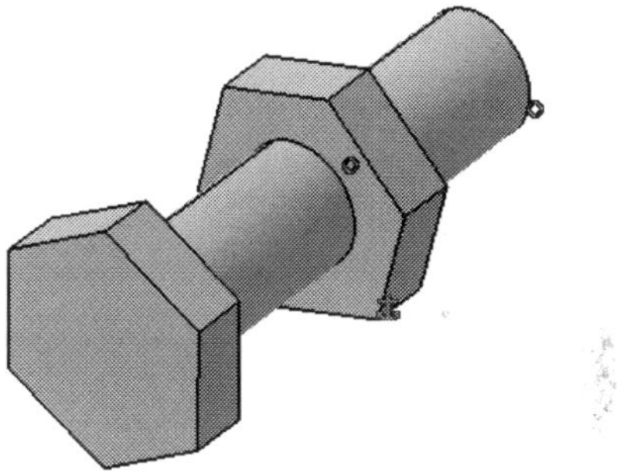

Figure 16-25 The updated mechanism of the Screw joint

Note

In the Cylindrical joint, two separate command nodes can be created in the Specification tree, one for driving the translational movement and other for driving the rotational movement. But in case of the Screw joint, only one command can be created because translational and rotational movements are inter-dependent and linked through the pitch value.

Creating the Rigid Joint

Menubar: Insert > New Joint > Rigid
Toolbar: DMU Kinematics > Kinematics Joints sub-toolbar > Rigid Joint

The Rigid joint is used to fix two parts rigidly. As a result, all DOFs between the selected parts get eliminated and they start working as a single component during the simulation.

To create a Rigid joint between two parts, insert the parts in the **DMU Kinematics** workbench and start a new mechanism. Next, choose the **Rigid Joint** tool from the **DMU Kinematics** toolbar; the **Joint Creation: Rigid** dialog box will be displayed, as shown in Figure 16-26. Also, you will be prompted to select the first part. Select a part from the geometry area; you will be prompted to select the second part. Select another part and choose the **OK** button; the selected parts will be attached rigidly to each other.

Note that, the **Length driven** and **Angle driven** check boxes are not available in the **Joint Creation: Rigid** dialog box. As a result, the command required to create the joint will be created in the Specification tree. The Rigid joint is helpful for creating complex mechanisms using a combination of joints as it can constrain all DOFs between any two parts in a mechanism, if required. You will learn more about creating mechanisms using a combination of joints later in this chapter.

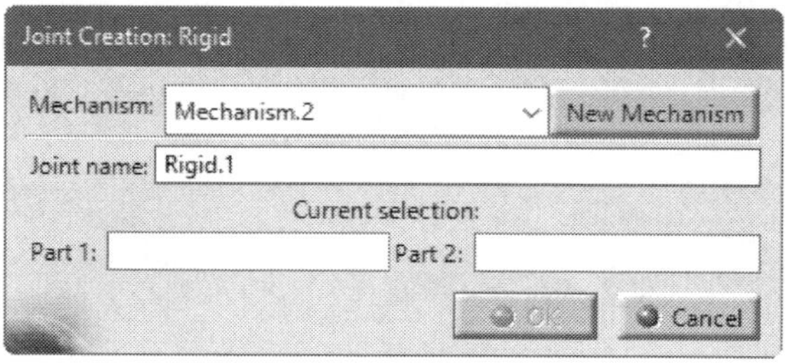

Figure 16-26 The ***Joint Creation: Rigid*** *dialog box*

Creating the Spherical Joint

Menubar: Insert > New Joint > Spherical
Toolbar: DMU Kinematics > Kinematics Joints sub-toolbar > Spherical Joint

The Spherical joint is used to create a joint between two parts such that one point from each part always remains in touch with the other. Therefore, to create a spherical joint between two parts, you need to specify one point from each part. The joint thus created will generate three undefined rotational DOFs and restrict the other three DOFs at a common point. To create this joint, insert two parts in the **DMU Kinematics** workbench, as shown in Figure 16-27. Then, start a new mechanism and make one of the parts fixed. For the parts shown in Figure 16-27, Part 1 will be the fixed part.

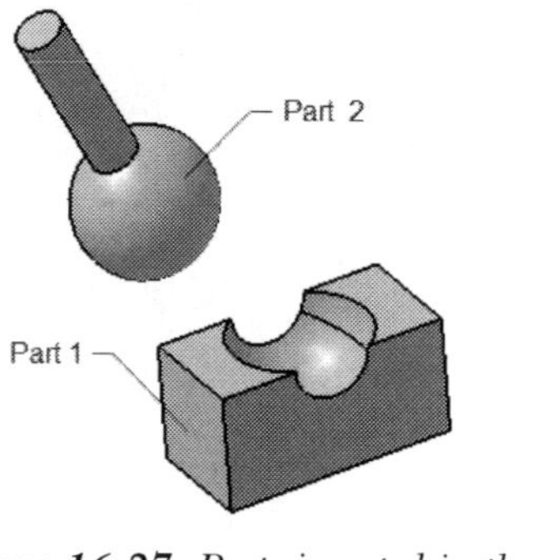

Figure 16-27 *Parts inserted in the* ***DMU Kinematics*** *workbench*

Now, choose the **Spherical Joint** tool from the **DMU Kinematics** toolbar; the **Joint Creation: Spherical** dialog box will be displayed, as shown in Figure 16-28, and you will be prompted to select the first point. You can select any vertex, sketch point, or center point to define the first point. Select a point from one part; you will be prompted to select the second point. Select a point from another part and choose the **OK** button to create a joint and exit the dialog box. Note that after creating this joint, both the selected points will coincide with each other. Figure 16-29 shows the center points to be selected for creating Spherical joint between Part 1 and Part 2. Figure 16-30 shows the updated mechanism of the Spherical joint. Note that in the **Joint Creation: Spherical** dialog box, the **Length driven** and **Angle driven** check boxes are not available. As a result, after creating this joint, no command will be created and **DOF=3** will be displayed beside the child mechanism node in the Specification tree. Here, **DOF=3** implies that there are only three undefined rotational DOFs and all the remaining DOFs are translational which are constrained at a common point.

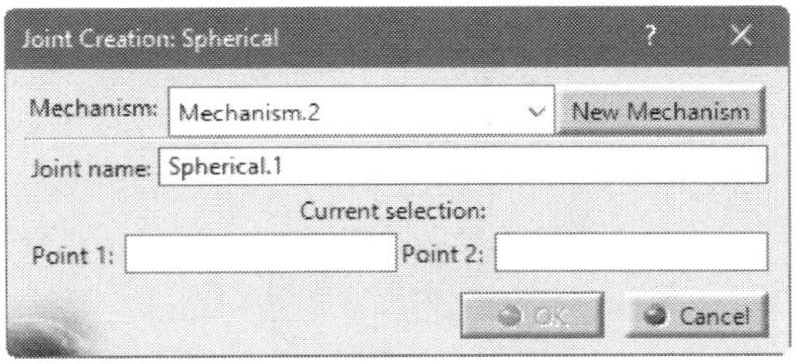

Figure 16-28 The ***Joint Creation: Spherical*** *dialog box*

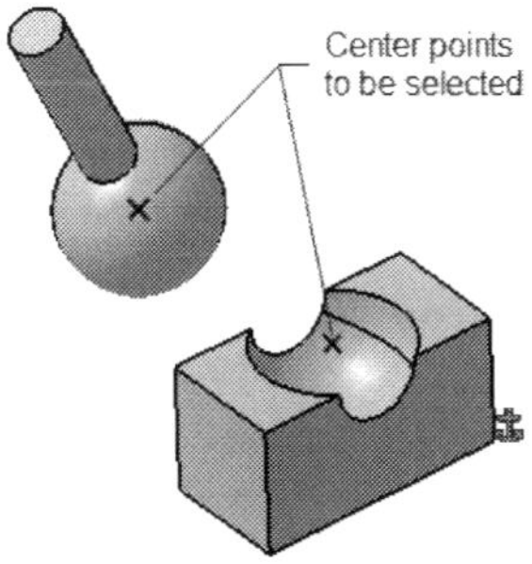

Figure 16-29 Points to be selected for creating the Spherical joint

This joint can be simulated using a combination of other joints. The creation of mechanism using combination of joints is discussed next.

Creating a Mechanism Using a Combination of Joints

To create a complex mechanism, you need to create multiple joints to define all its degrees of freedom (DOFs). To define these DOFs according to the requirement of the mechanism, you need to analyze the movement of each part of the mechanism, and then create joints such that the joints created will facilitate the required movement of the parts of the mechanism.

Consider a mechanism in which a spherical part rotates inside a semi-spherical hole, maintaining the concentricity between the centers of both the parts. To create this mechanism, first you need to create the Spherical joint so as to maintain the concentricity, refer to Figure 16-30. After creating the Spherical joint, you will notice that **DOF=3** is displayed beside the **Mechanism.1** node in the Specification tree. This indicates that three undefined DOFs are available in the **Mechanism.1** mechanism. As it is a Spherical joint, the three undefined DOFs are rotational. Now, insert a sketch line as Part 3 in the mechanism, as shown in Figure 16-31. Next, create a Revolute joint between Part 2 and Part 3 by invoking the **Joint Creation: Revolute** dialog box. To create the Revolute joint, select the central axis and the sketch line, and then the planar face of Part 2 and the plane perpendicular to the sketch line, as shown in Figure 16-32. Remember that you need to select the **Angle driven** check box in the **Joint Creation: Revolute** dialog box for creating a command. Next, update the mechanism.

Figure 16-30 The updated mechanism of the Spherical joint

After creating this joint, you will notice that the **Command.1** node is created in the Specification tree. Still, you cannot simulate the mechanism because three rotational DOFs in the mechanism are still undefined. To define these DOFs, create a rigid joint between Part 3 and Part 1; the **Information** message box will be displayed informing that this mechanism can be simulated because all DOFs are defined and command is created. In this way, you can create different mechanisms with the help of multiple joints.

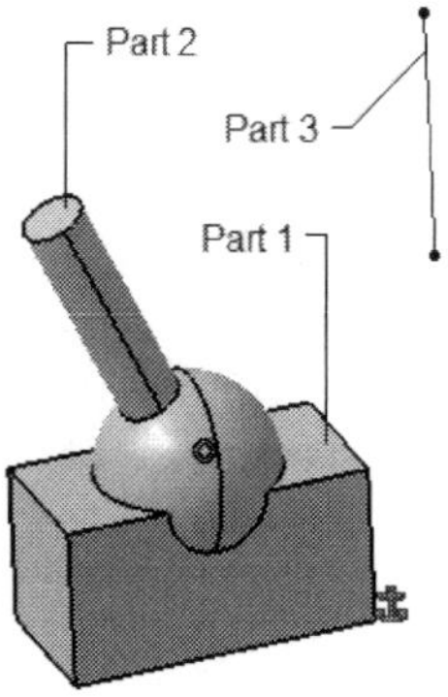

Figure 16-31 Part 3 inserted in the mechanism

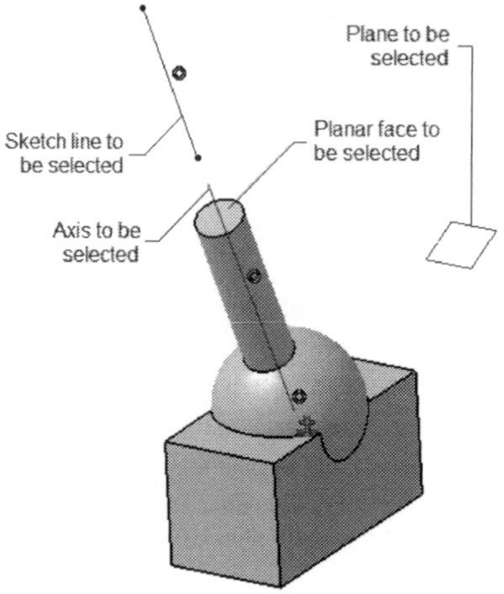

Figure 16-32 Selection of entities for creating a Revolute joint

Creating the Planar Joint

Menubar:	Insert > New Joint > Planar
Toolbar:	DMU Kinematics > Kinematics Joints sub-toolbar > Planar Joint

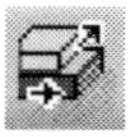

The Planar joint makes two selected planar faces (planes) coplanar. As a result, you can slide one part over the coplanar face of the other part. For creating a Planar joint between two parts, you need to specify one planar face (plane) from each part. To do so, insert two parts in the **DMU Kinematics** workbench, as shown in Figure 16-33. Start a new mechanism and make one of these parts fixed. In Figure 16-33, Part 1 will be the fixed part.

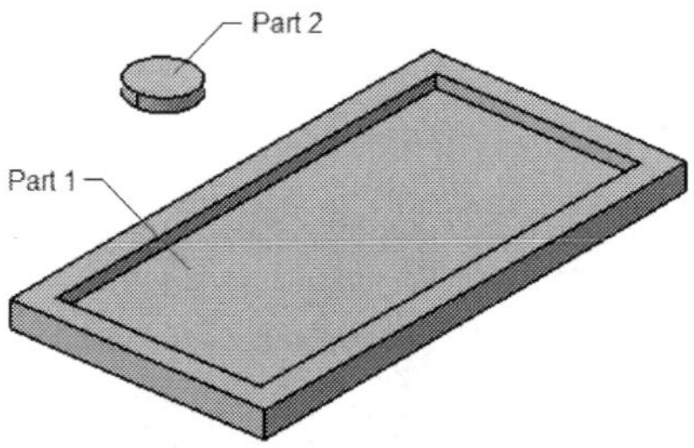

Figure 16-33 Parts inserted in the ***DMU Kinematics*** *workbench*

Now, to create the planar joint between two parts, choose the **Planar Joint** tool from the **DMU Kinematics** toolbar; the **Joint Creation: Planar** dialog box will be displayed, as shown in Figure 16-34, and you will be prompted to select the first plane. You can select any planar face or plane. Select a planar face from one part; you will be prompted to select the second plane. Select a planar face from another part; the selected planar faces will become coplanar. Choose the **OK** button to create a joint and exit the dialog box. Next, update the mechanism. Figure 16-35 shows the planar faces to be selected for creating the Planar joint between Part 1 and Part 2.

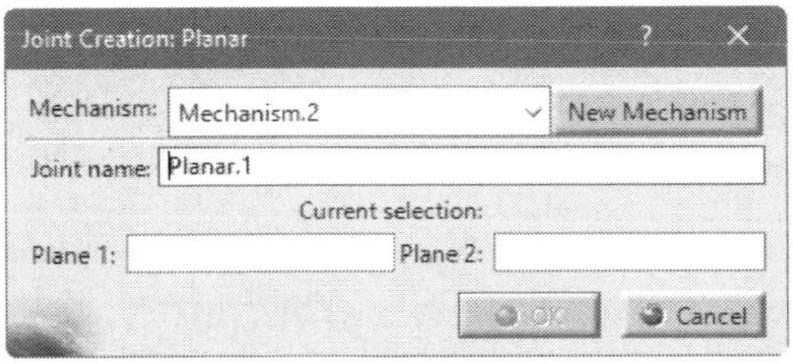

Figure 16-34 *The* ***Joint Creation: Planar*** *dialog box*

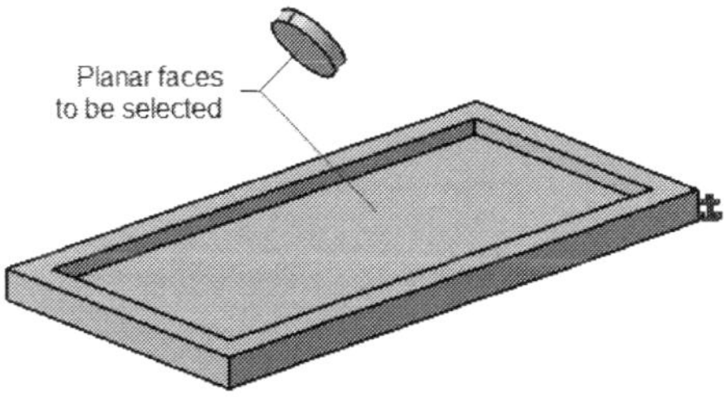

Figure 16-35 *Planar faces to be selected*

After creating the joint, you will notice that **DOF=3** is displayed beside the child mechanism node in the Specification tree. This informs that there are three undefined DOFs in the mechanism. Out of these three undefined DOFs, two are translational and one is rotational. You can use this joint to constrain the DOFs. Note that in this dialog box, the **Length driven** and **Angle driven** check boxes are not available. As a result, no command will be created. The simulation of the mechanism with the combination of Planar joint and Roll Curve joint will be discussed under the topic 'Creating the Roll Curve Joint' later in this chapter.

Creating the Point Curve Joint

Menubar:	Insert > New Joint > Point Curve
Toolbar:	DMU Kinematics > Kinematics Joints sub-toolbar > Point Curve Joint

The Point Curve Joint makes a point and a curve coincident. As a result, the selected point and curve always remain in contact with each other during simulation. This joint can be used to slide a point over the curve. This joint is very useful for creating a mechanism similar to the Cam and Follower mechanism. For defining the Point Curve Joint, you need to select a curve from one part and a point from another part. To create a Point Curve joint, insert two parts in the **DMU Kinematics** workbench, as shown in Figure 16-36. Then, start a new mechanism and make one of the parts fixed. For the parts shown in Figure 16-36, Part 1 will be the fixed part. Note that you can apply the Point Curve Joint between two parts only when a sketch point, vertex, or pointer of one part touches with the edge or sketch curve of another part. To do so, assemble two parts in the assembly environment such that a point of one part touches with the curve of another part. After assembling the parts, return to the **DMU Kinematics** workbench. In Figure 16-37, Part 1 and Part 2 are assembled such that the pointer of Part 1 touches the Sketch curve of Part 2 at a common point.

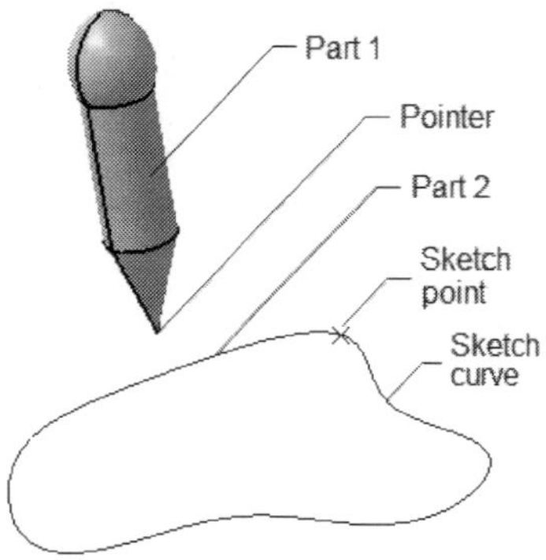

***Figure 16-36** Parts inserted in the **DMU Kinematics** workbench*

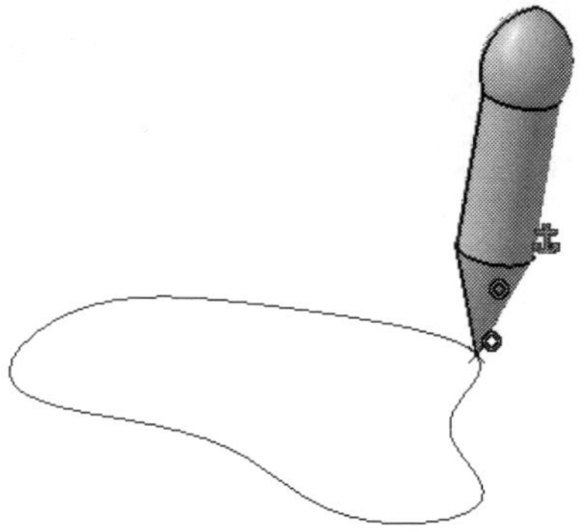

***Figure 16-37** The Part 1 and Part 2 assembled in the assembly environment*

Now, to create the joint between the parts, choose the **Point Curve Joint** tool from the **DMU Kinematics** toolbar; the **Joint Creation: Point Curve** dialog box will be displayed, as shown in Figure 16-38. Also, you will be prompted to select the first curve. You can select any edge, sketch line, or sketch curve as the first curve. Select a curve from Part 2 of the mechanism; you will be prompted to select the first point. You can select any vertex or sketch point as the first point. Select a point from Part 1 of the mechanism. Figure 16-39 shows the sketch curve and the pointer to be selected for creating the Point Curve Joint.

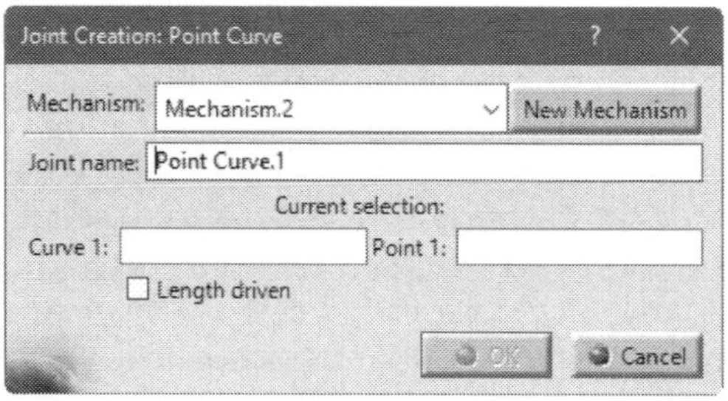

***Figure 16-38** The **Joint Creation: Point Curve** dialog box*

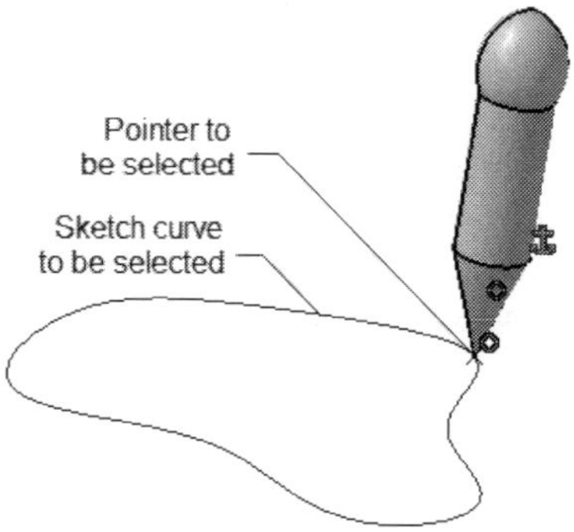

***Figure 16-39** Sketch curve and pointer to be selected for creating a Point Curve joint*

As the **Length driven** check box is available in the **Joint Creation: Point Curve** dialog box, you can create a command by selecting it. If you create a Point curve joint without selecting this check box, four undefined DOFs will be created in the mechanism and only two translational DOFs will be constrained. If you select this check box, three rotational DOFs will remain undefined.

Creating the Slide Curve Joint

Menubar:	Insert > New Joint > Slide Curve
Toolbar:	DMU Kinematics > Kinematics Joints sub-toolbar > Side Curve Joint

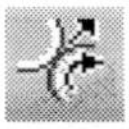

The Slide Curve joint is used to create a joint between two curves such that one curve slides over the other (fixed) curve while maintaining the tangency. For defining the Slide

Curve joint between two parts, you need to select one curve from each part. To create a Slide Curve joint, insert two parts in the **DMU Kinematics** workbench, as shown in Figure 16-40. Then, start a new mechanism and make one of the parts fixed. In Figure 16-40, Part 1 will be the fixed part. Note that you can apply the Slide Curve joint between two parts only when a curve from each part touches one another tangentially. To ensure this, assemble the parts in the assembly environment such that the curves from each part touch one another tangentially. In Figure 16-41, Part 1 and Part 2 are placed in the assembly environment with the help of the **Snap** button such that the Edge 1 and Edge 2 are tangentially connected with each other at a common point.

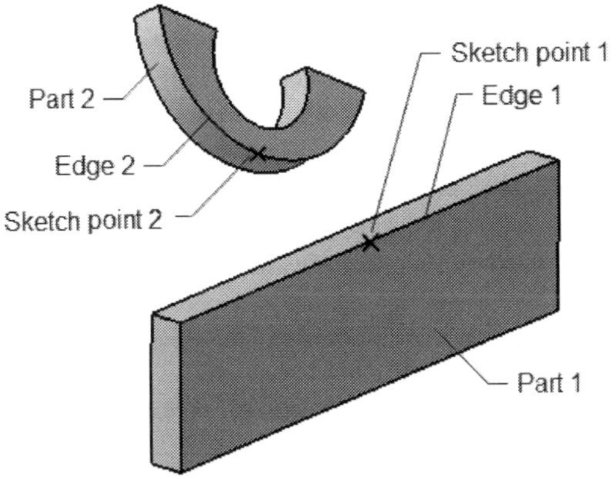

Figure 16-40 *Parts inserted in the* ***DMU Kinematics*** *workbench*

Now, to create the joint, choose the **Slide Curve Joint** tool from the **DMU Kinematics** toolbar; the **Joint Creation: Slide Curve** dialog box will be displayed, as shown in Figure 16-42. Also, you will be prompted to select the first curve. You can select any edge, sketch line, or sketch curve as the first curve. Select an edge from one of the parts; you will be prompted to select the second curve. Select an edge from another part as the second curve. Now, choose the **OK** button to create a joint and exit the dialog box. For the parts shown in Figure 16-40, Edge 1 and Edge 2 are selected for creating the Slide Curve joint. Note that after creating the joint, three undefined DOFs will be created in the mechanism which are rotational DOFs.

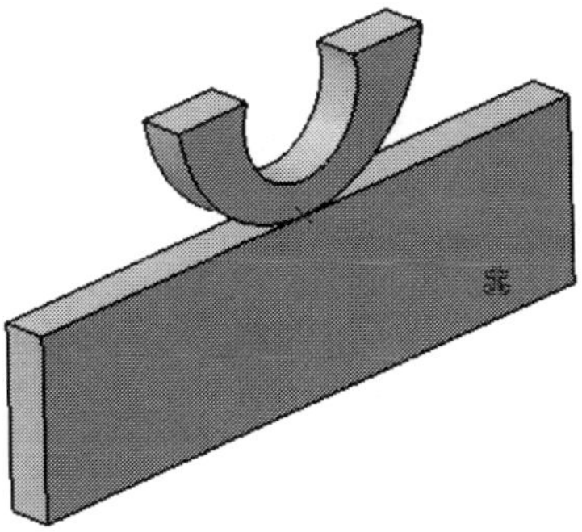

Figure 16-41 *The Edge 1 and Edge 2 are tangentially connected with each other at a common point*

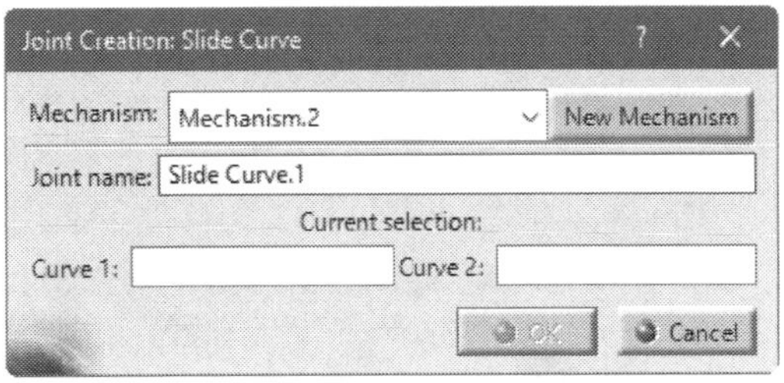

Figure 16-42 *The* ***Joint Creation: Slide Curve*** *dialog box*

Creating the Roll Curve Joint

Menubar:	Insert > New Joint > Roll Curve
Toolbar:	DMU Kinematics > Kinematics Joints sub-toolbar > Roll Curve Joint

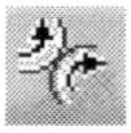

The Roll Curve joint is used to create a joint between two curves such that one curve rolls over the other (fixed curve) while maintaining tangency. For defining a Roll Curve joint between two parts, you need to select one curve from each part. To create the Roll Curve joint, insert two parts in the **DMU Kinematics** workbench, as shown in Figure 16-43. Then, start a new mechanism and fix one of the parts. For the parts shown in Figure 16-43, Part 1 will be the fixed part. Note that you can apply the Roll Curve joint between two parts only when a curve from each part touch one another tangentially at a common point. To do so, assemble the parts in the assembly environment such that the curve from each part touches one another tangentially at a common point. In Figure 16-44, Part 1 and Part 2 are assembled in the assembly environment using the **Coincidence constraint** button. In these parts, Edge 1 and Edge 2 are tangentially connected with each other at a common point where Sketch point 1 touches Sketch point 2. Next, choose the **Roll Curve Joint** tool from the **DMU Kinematics** toolbar; the **Joint Creation: Roll Curve** dialog box will be displayed, as shown in Figure 16-45, and you will be prompted to select the first curve. You can select any edge, sketch curve, or sketch line as the first curve. Select a curve from one of these parts; you will be prompted to select the second curve. Select a curve from another part. Next, you can select the **Length driven** check box in the **Joint Creation: Roll Curve** dialog box to create a command, and then choose the **OK** button to create a joint and exit the dialog box.

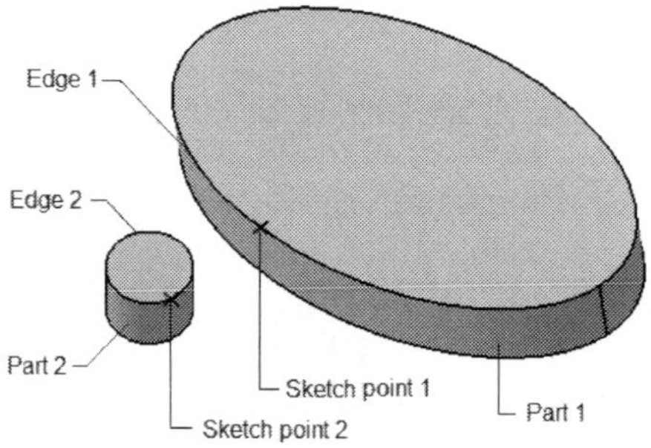

Figure 16-43 *Parts inserted in the* ***DMU Kinematics*** *workbench*

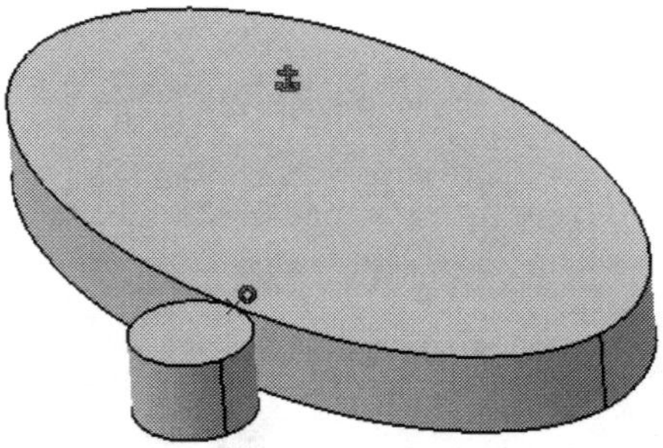

Figure 16-44 *Parts assembled in the assembly environment*

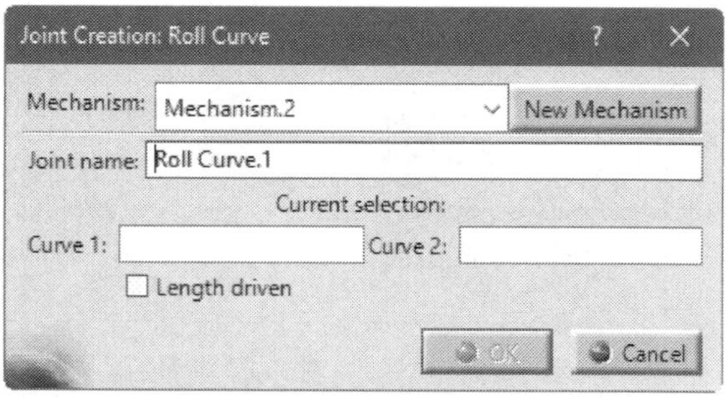

Figure 16-45 *The* ***Joint Creation: Roll Curve*** *dialog box*

For the parts shown in Figure 16-43, Edge 1 and Edge 2 are selected for creating the Roll Curve joint. Note that if you create this joint without selecting the **Length driven** check box from the **Joint Creation: Roll Curve** dialog box, two undefined DOFs will be created in the mechanism. One of these DOFs will be translational and another will be rotational. If you select the **Length driven** check box from the **Joint Creation: Roll Curve** dialog box during creation of joint, only one undefined rotational DOF will be created in the mechanism. If you define this DOF properly with the help of other joint, you can simulate the mechanism. For example, in the parts shown in Figure 16-44, if you apply the Planar joint between Part 1 and Part 2, all DOFs will be defined and the **Information** message box will be displayed informing that the mechanism can be simulated.

Creating the Point Surface Joint

Menubar:	Insert > New Joint > Point Surface
Toolbar:	DMU Kinematics > Kinematics Joints sub-toolbar > Point Surface Joint

A Point Surface joint is a joint in which a point of one part slides over the surface of another part. Therefore, for creating a Point Surface joint, you need to select a surface from one part and a point from another part. To create a Point Surface joint, insert two parts in the **DMU Kinematics** workbench, as shown in Figure 16-46. Then, start a new mechanism and make one of these parts fixed. For the parts shown in Figure 16-46, Part 1 will be the fixed part. Note that you can apply the Point Surface joint between two parts only when a point of one part touches the surface of another part. To do so, assemble two parts in the assembly environment such that a point from one part touches the surface of another part at a common point. In Figure 16-47, Part 1 and Part 2 are assembled such that the pointer and the surface touch each other at a common point.

Choose the **Point Surface Joint** tool from the **DMU Kinematics** toolbar; the **Joint Creation: Point Surface** dialog box will be displayed, as shown in Figure 16-48, and you will be prompted to select the first surface. You can select any planar face, plane, or surface as the first surface. Select a surface from one of these parts; you will be prompted to select the first point. You can select any vertex, sketch point, or pointer as the first point. Select a point from another part. Next, choose **OK** to create a joint and exit the dialog box. Figure 16-46 shows the pointer and the surface to be selected for creating the Point Surface joint between Part 1 and Part 2.

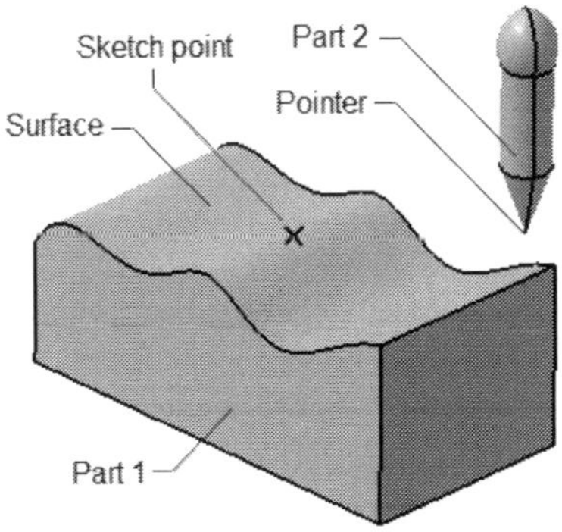

Figure 16-46 *Parts inserted in the* ***DMU Kinematics*** *workbench*

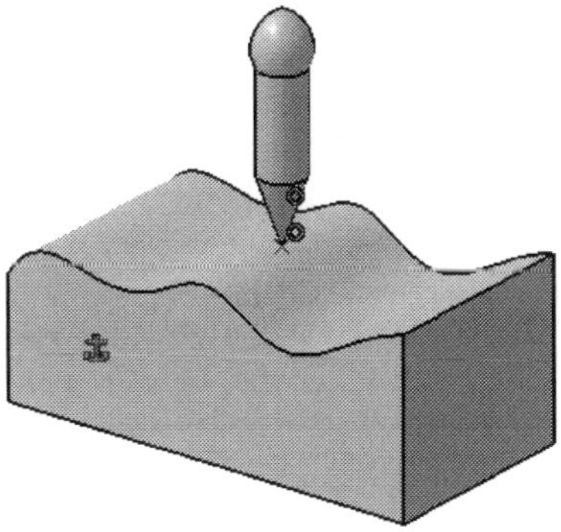

Figure 16-47 *Parts assembled in the assembly environment*

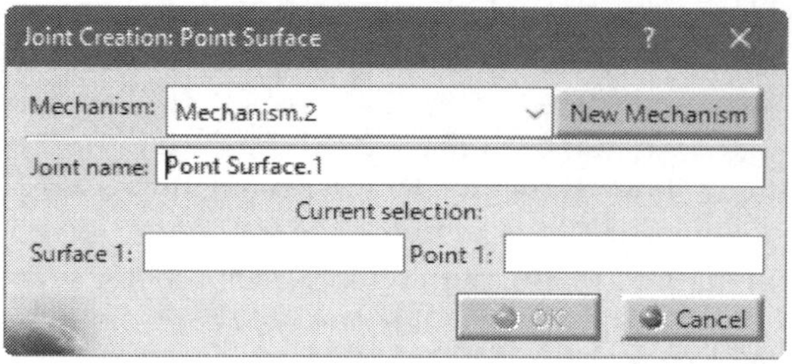

*Figure 16-48 The **Joint Creation: Point Surface** dialog box*

After creating the joint, you will notice that **DOF=5** will be displayed beside the child mechanism node in the Specification tree. This indicates that there are five undefined DOFs in this mechanism. Out of these five undefined DOFs, three are rotational and two are translational. Using this joint, you can constrain only one translational DOF. As a result, the selected point will always remain in contact with the surface. Note that in the **Joint Creation: Point Surface** dialog box, the **Length driven** and **Angle driven** check boxes are not available. As a result, no command will be created.

Creating the Universal Joint

Menubar:	Insert > New Joint > Universal
Toolbar:	DMU Kinematics > Kinematics Joints sub-toolbar > Universal Joint

In universal joints, two different parts are joined such that both the parts rotate about their own axis while keeping their axes at an angle in the same plane. This mechanism works in the same way as the universal joints used in automobiles. For creating a Universal joint between two parts, you need to select an axis from each part. These selected axes must be coincident with each other in the same plane. To create a Universal joint, insert the two parts in the **DMU Kinematics** workbench and place them as per your requirement, as shown in Figure 16-49. In this figure, the edges of the parts have been hidden for better visualization and also, the parts have been placed in such a way that Plane 1, Plane 2, Axis 1, and Axis 2 are in the same plane. Also note that the parts of the Universal joint rotate about their own axes in the mechanism. Therefore, it is recommended that you do not keep any of these parts fixed.

Start a new mechanism and then choose the **Universal Joint** tool from the **DMU Kinematics** toolbar; the **Joint Creation: U Joint** dialog box will be displayed, as shown in Figure 16-50. Also, you will be prompted to select the first line. You can select any axis, edge, or sketch line as the first line. Select an axis from one of the parts; you will be prompted to select the second line. Select an axis from another part. In Figure 16-49, Axis 1 and Axis 2 will be selected for creating the Universal joint between Part 1 and Part 2. You can select any radio button from the **Cross-pin axis direction** area to specify one of the selected axes as the axis normal to the direction of spin. Next, choose the **OK** button from the dialog box to create a joint and exit it. Note that after creating this joint, two undefined DOFs will be created in the mechanism. Out of these, one will be rotational and the other will be translational DOF. To simulate the mechanism, you need to define these DOFs according to your requirement.

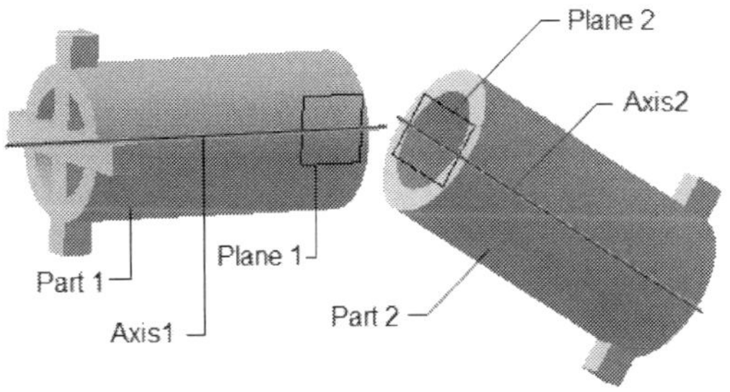

Figure 16-49 Parts inserted in the ***DMU Kinematics*** *workbench*

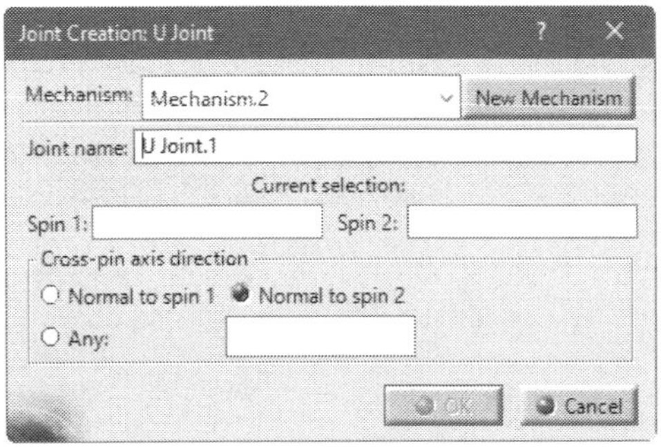

Figure 16-50 The ***Joint Creation: U Joint*** *dialog box*

Creating the CV Joint

Menubar: Insert > New Joint > CV
Toolbar: DMU Kinematics > Kinematics Joints sub-toolbar > CV Joint

The CV joint is used to create two universal joints among three parts. For creating a CV joint among three different parts, an axis from each part must lie on the same plane and one of the axes must coincide with the other two axes. To create a CV joint, insert three parts in the **DMU Kinematics** workbench and place them as per the requirement, see in Figure 16-51. In this figure, the edges of parts are hidden for better visualization, and also they are placed in such a way that Plane 1, Plane 2, Plane 3, Axis 1, Axis 2, and Axis 3 are on the same plane. Note that the parts of the CV joint rotate about their own axes in the mechanism. Therefore, it is recommended that you do not keep any of these parts fixed.

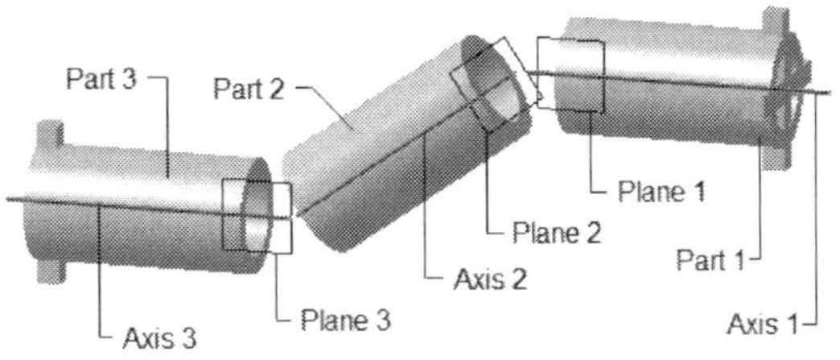

Figure 16-51 Parts inserted in the ***DMU Kinematics*** *workbench*

Next, start a new mechanism and then choose the **CV Joint** tool from the **DMU Kinematics** toolbar; the **Joint Creation: CV Joint** dialog box will be displayed, as shown in Figure 16-52. Also, you will be prompted to select the first line. You can select any axis, edge, or sketch line as the first line. Select an axis from one of these parts; you will be prompted to select the second line. Select an axis from another part; you will be prompted to select the direction of the cross-pin axis. Select the axis of the third part. Figure 16-51 shows the Axis 1, Axis 2, and Axis 3 to be selected for creating the CV joint between Part 1, Part 2, and Part 3. Next, choose the **OK** button from the **Joint Creation: CV Joint** dialog box to create the joint and exit the dialog box. Note that after creating this joint, two Universal joints will be created. The first one will be created between Part 1 and Part 2 and the second one will be created between Part 2 and Part 3. Also, four undefined DOFs will be created in the mechanism.

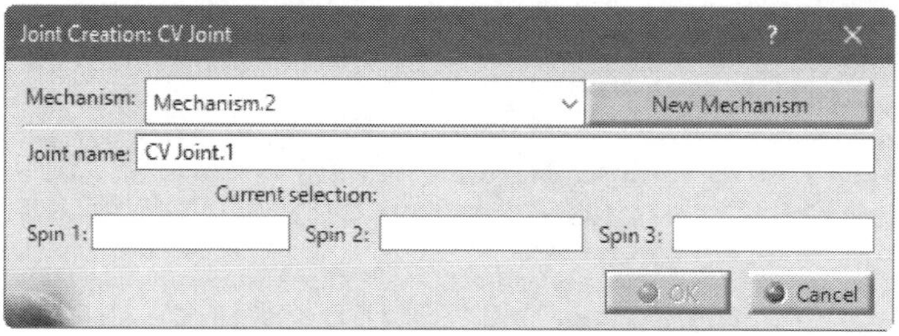

*Figure 16-52 The **Joint Creation: CV Joint** dialog box*

Creating the Gear Joint

Menubar: Insert > New Joint > Gear
Toolbar: DMU Kinematics > Kinematics Joints sub-toolbar > Gear Joint

The Gear joint is used to create a joint similar to the gear mechanism. For creating a Gear joint, you need to specify two Revolute joints and a gear ratio. In this joint, you can rotate two parts at a time with respect to the fixed part. To create the Gear joint, insert three parts in the **DMU Kinematics** workbench and create two Revolute joints. In Figure 16-53, three parts are inserted in the **DMU Kinematics** workbench. Among these parts, Part 1 is the fixed part, the first Revolute joint is created between Shaft 1 and Part 2, and the second Revolute joint is created between Shaft 2 and Part 3. Next, choose the **Gear Joint** tool from the **DMU Kinematics** toolbar; the **Joint Creation:Gear** dialog box will be displayed, as shown in Figure 16-54. Also, you will be prompted to select or create the first joint, and the **Create** button besides the **Revolute Joint 1** display box in the **Joint Creation: Gear** dialog box will be activated. If you choose this button, the **Joint Creation: Revolute** dialog box will be displayed. Use this dialog box to create the Revolute joint, if it is not already created. Else, you can select the existing Revolute joint from the Specification tree. Select the first Revolute joint; you will be prompted to select or create the second Revolute joint. Select the second Revolute joint for the mechanism; the **Define** button will be activated in the **Joint Creation: Gear** dialog box.

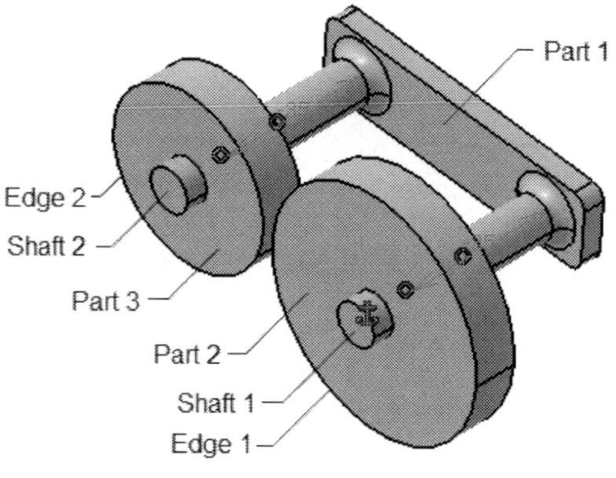

Figure 16-53 The Revolute joints created between Shafts and Parts

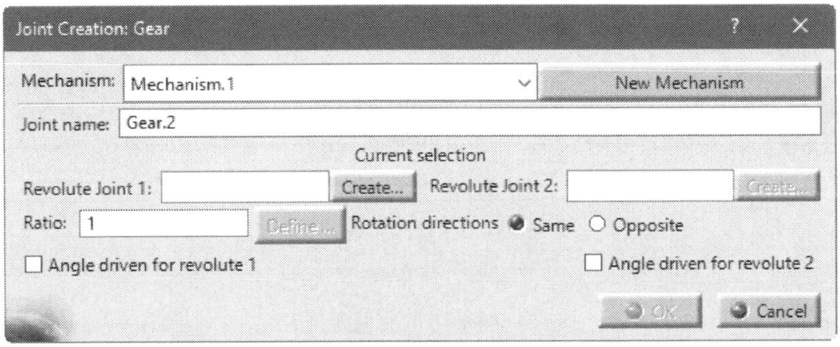

Figure 16-54 *The* ***Joint Creation: Gear*** *dialog box*

You can either specify the gear ratio manually in the **Ratio** edit box or you can define it automatically. To define the gear ratio automatically, choose the **Define** button from the **Joint Creation: Gear** dialog box; the **Gear Ratio Definition** dialog box will be displayed, as shown in Figure 16-55. In this dialog box, you can specify the gear ratio by selecting circular edges from the parts. To do so, select the circular edge from the first part; its radius will be displayed in the **Radius 1** display box of the **Gear Ratio Definition** dialog box. Similarly, select the circular edge from the second part; its radius will be displayed in the **Radius 2** display box. On doing so, the gear ratio will be automatically calculated and displayed in the **Ratio** display box of the **Gear Ratio Definition** dialog box. Choose **OK** from the **Gear Ratio Definition** dialog box; the **Joint Creation: Gear** dialog box will be re-displayed.

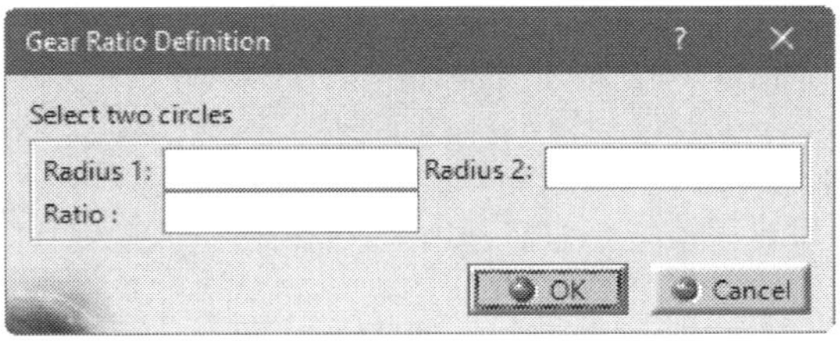

Figure 16-55 *The* ***Gear Ratio Definition*** *dialog box*

By default, the **Same** radio button is chosen in the **Joint Creation: Gear** dialog box. As a result, both the parts will rotate in the same direction during the simulation. If you select the **Opposite** radio button, both the parts will rotate in the opposite direction during the simulation. You can specify the driving gear to drive the mechanism by selecting the required check box from the **Joint Creation: Gear** dialog box. If you select the **Angle driven for revolute 1** check box, a command will be created and the first Revolute joint will drive the mechanism. If you select the **Angle driven for revolute 2** check box, the second Revolute joint will drive the mechanism. Select the required check box and choose **OK** to create the mechanism and exit the dialog box. By default, the driven Revolute joint rotates in clockwise direction during the simulation. To change the direction of revolution, double-click on the **Gear** node below the **Joint** node in the Specification tree; the **Joint Creation: Gear** dialog box will be displayed again. Also, an arrow pointing in clockwise direction will be displayed over the driving Revolute joint, as shown in Figure 16-56. In this figure, all edges have been hidden for better visualization. To change the direction, click on the displayed arrow head in the graphic area and choose **OK**. Note that in the re-displayed dialog box, you can edit the parameters of the joint, if required.

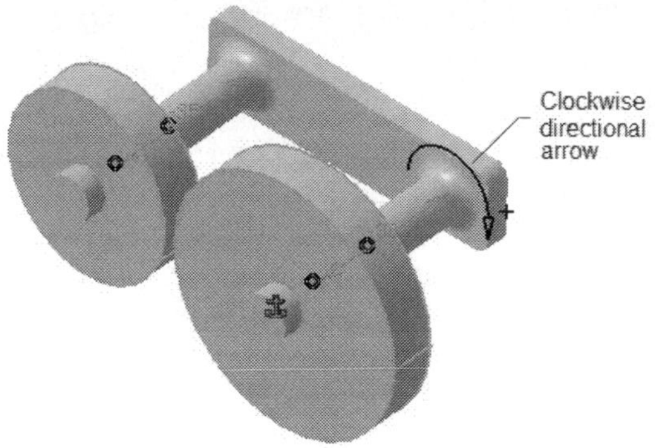

Figure 16-56 Clockwise directional arrow displayed in the graphics area

Creating the Rack Joint

Menubar:	Insert > New Joint > Rack
Toolbar:	DMU Kinematics > Kinematics Joints sub-toolbar > Rack Joint

The Rack joint is used to create a mechanism similar to the rack and pinion mechanism used in automobiles. In this joint, you can slide one part by rotating the other part or vice versa. In this mechanism, the relative motion of parts depends upon the specified rack ratio. For creating a Rack joint, you need to specify one Prismatic joint, one Revolute joint, and rack ratio for the parts used in creating the mechanism. To create a Rack joint, insert three parts in the **DMU Kinematics** workbench, and create one Prismatic joint and one Revolute joint. In Figure 16-57, three parts: Part 1, Part 2, and Part 3 are inserted in the **DMU Kinematics** workbench and Part 1 is the fixed part. Next, create a Prismatic joint between Slot 1 and Part 2, and one Revolute joint between Shaft 1 and Part 3.

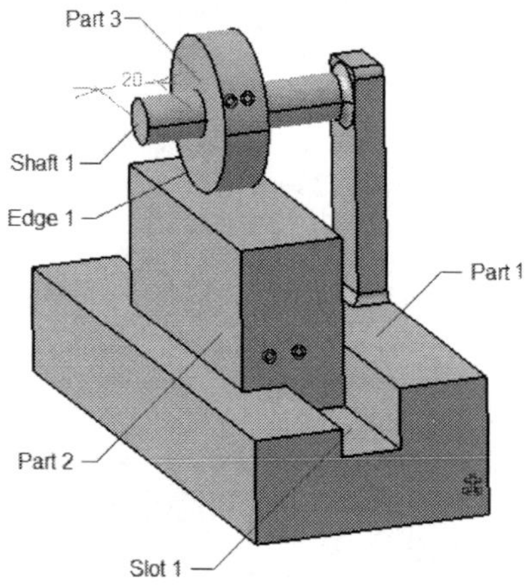

Figure 16-57 Parts inserted in the ***DMU Kinematics*** *workbench*

Now, choose the **Rack Joint** tool from the **Kinematics Joints** toolbar; the **Joint Creation: Rack** dialog box will be displayed, as shown in Figure 16-58. Also, you will be prompted to select or create the first joint, and the **Create** button beside the **Prismatic Joint** display box in the **Joint Creation: Rack** dialog box will be activated. If you have not already created the Prismatic joint, choose this button; the **Joint Creation: Prismatic** dialog box will be displayed. Use this dialog box to create the Prismatic joint. Else, select the already created Prismatic joint from the Specification tree. Now, select the Prismatic joint; you will be prompted to select or create the second joint. As the prismatic joint, select the Revolute joint for the mechanism; the **Define** button will be activated in the **Joint Creation: Rack** dialog box.

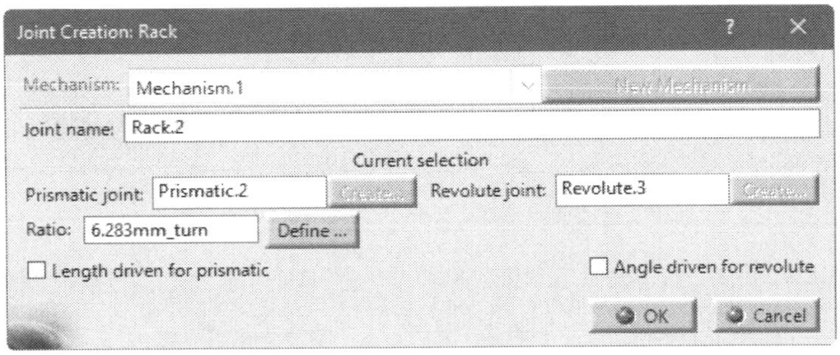

Figure 16-58 *The* ***Joint Creation: Rack*** *dialog box*

Next, specify the rack ratio manually in the **Ratio** edit box of the **Joint Creation: Rack** dialog box. You can also define it automatically. To do so, choose the **Define** button; the **Rack Ratio Definition** dialog box will be displayed, as shown in Figure 16-59. In this dialog box, you can specify the rack ratio by selecting the circular edge. To do so, select the circular edge from the part; its radius will be displayed in the **Radius** display box of the **Rack Ratio Definition** dialog box. Also, the rack ratio will automatically be calculated and displayed in the **Ratio** display box of the **Rack Ratio Definition** dialog box. After defining the gear ratio, choose **OK** to accept the rack ratio and exit the dialog box. For the parts shown in Figure 16-57, the existing Prismatic and Revolute joints are selected for defining the Rack joint and for defining the Rack ratio, Edge 1 has to be selected.

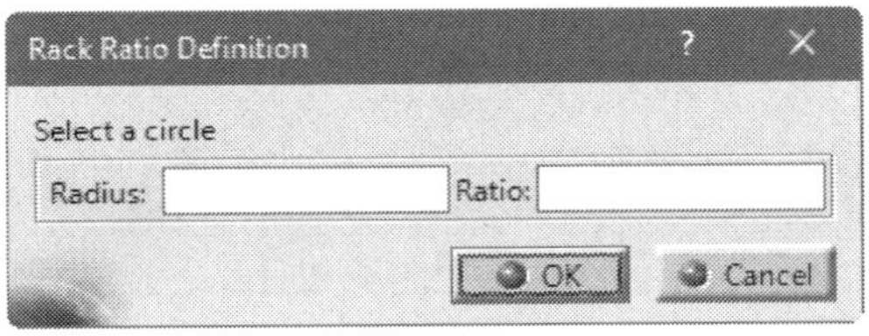

Figure 16-59 *The* ***Rack Ratio Definition*** *dialog box*

You can drive the mechanism by driving the Prismatic Joint or the Revolute Joint. To drive the mechanism using the Prismatic Joint, choose the **Length driven for prismatic** check box from the **Joint Creation: Rack** dialog box. To drive the mechanism using the Revolute joint, choose the **Angle driven for revolute** check box. Next, choose **OK** to create the mechanism and exit the dialog box.

Note

If the Revolute joint of the Rack joint is created using a negative offset value between the two specified planes, then while simulating the mechanism, the relative motion between the Revolute joint and the Prismatic joint will be in opposite direction and vice versa.

Creating the Cable Joint

Menubar:	Insert > New Joint > Cable
Toolbar:	DMU Kinematics > Kinematics Joints sub-toolbar > Cable Joint

The Cable joint is used to create a bond between two parts by using two Prismatic joints. As a result, while driving one part in the linear direction, the other part will also move linearly depending on the specified DOF and with respect to the fixed part. To create a Cable joint, insert three parts in the **DMU Kinematics** workbench, as shown in Figure 16-60, and then create two Prismatic joints. Figure 16-60 shows three parts inserted in the DMU Kinematics workbench with Part1 fixed. One Prismatic joint is created between Track 1 and Part 2 and another Prismatic joint is created between Track 2 and Part 3.

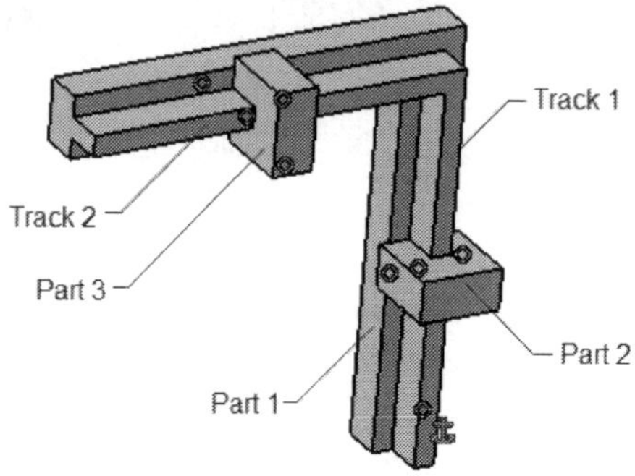

Figure 16-60 *Parts inserted in the* ***DMU Kinematics*** *workbench*

Next, choose the **Cable Joint** tool from the **DMU Kinematics** toolbar; the **Joint Creation: Cable** dialog box will be displayed, as shown in Figure 16-61. Also, you will be prompted to select or create the first joint, and the **Create** button beside the **Prismatic joint 1** display box in the **Joint Creation: Cable** dialog box will get activated. If you choose this button, the **Joint Creation: Prismatic** dialog box will be displayed. Use this dialog box to create the Prismatic joint if not already created. Else, you can select the existing Prismatic joint from the Specification tree. Select the Prismatic joint; you will be prompted to select or create the second joint. Select another Prismatic joint from the Specification tree. To drive the mechanism using the first selected Prismatic joint, choose the **Length driven for prismatic 1** check box from the **Joint Creation: Cable** dialog box. Similarly, to drive the mechanism using the second selected Prismatic joint, choose the **Length driven for prismatic 2** check box. Note that you can specify the cable ratio in the **Ratio** edit box of the **Joint Creation: Cable** dialog box. After defining the cable ratio, choose the **OK** button to create the mechanism and exit the dialog box.

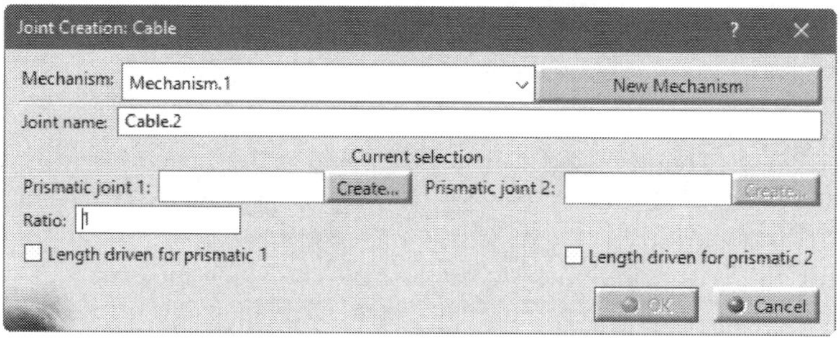

Figure 16-61 *The* ***Joint Creation: Cable*** *dialog box*

Note that, so far, you designed the mechanisms using the tools in the **DMU Kinematics** toolbar. However, you can also design mechanisms using assembly constraints. These assembly constraints automatically get converted into joints. In the next section, you will learn how assembly constraints can be used as joints in a mechanism.

CONVERTING ASSEMBLY CONSTRAINTS INTO JOINTS

Toolbar: DMU Kinematics > Assembly Constraints Conversion

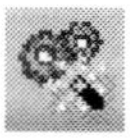

In CATIA V5, you can also create joints by using assembly constraints. To do so, open an existing assembly in the **DMU Kinematics** workbench and choose the **Assembly Constraints Conversion** tool from the **DMU Kinematics** toolbar; the **Assembly Constraints Conversion** dialog box will be displayed. Choose the **More** button from this dialog box; the **Assembly Constraints Conversion** dialog box will expand, as shown in Figure 16-62. Start a new mechanism, as discussed earlier, by choosing the **New mechanism** button, if it has not already been started. Once the mechanism has been started, you will notice that a pair of parts is highlighted in the graphic area and the part numbers are also displayed in the **Product 1** and **Product 2** display boxes, respectively. The constraints applied to these parts will be listed in the **Constraints list** list box. If you select a constraint from this list, the name of the possible joint that may be created using the selected constraint will be displayed in the **Resulting type** display box. Also, all possible commands that can be created for the joint will be available in the **Add command** drop-down list. Select the required option from this drop-down list to create a command and choose the **Create Joint** button from the **Assembly Constraints Conversion** dialog box; the joint will be specified and listed in the **Joints list** list box, which is on the right pane of this dialog box. Similarly, create other joints for the highlighted pair; the highlighted pair will be resolved. Note that the **Unresolved pairs** area in the upper part of **Assembly Constraints Conversion** dialog box shows the total numbers of pairs available in the assembly and the number of unresolved pairs. After resolving the first active pair, choose the **Step forward** button; the next pair will be activated and highlighted in the graphics window. Similarly, resolve this active pair. In this way, you can resolve all active pairs of the assembly.

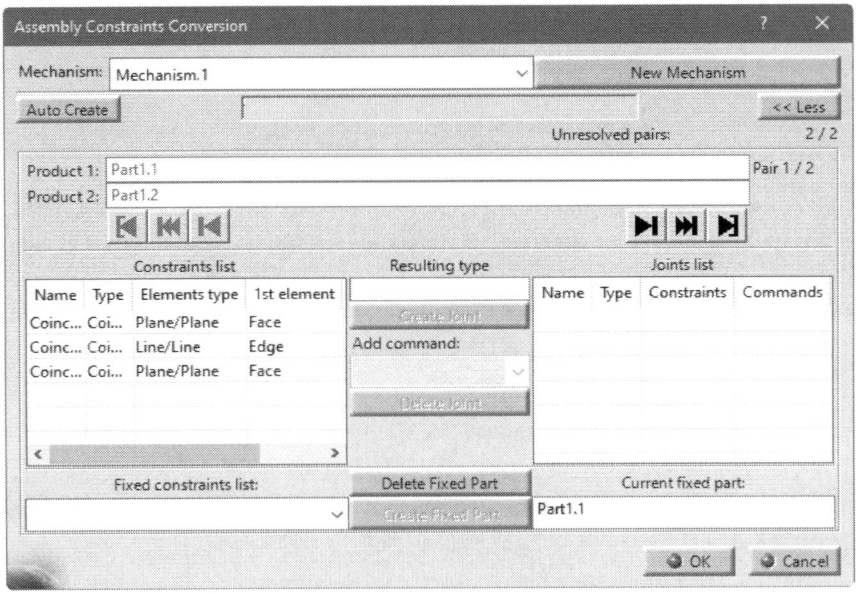

*Figure 16-62 The expanded **Assembly Constraints Conversion** dialog box*

You can also delete the unwanted joint from the active pair. To do so, select the unwanted joint from the **Joints list** list box at the right pane of the dialog box; the **Delete Joint** button will be activated. Now, choose this button to delete the specified joint. You can directly activate the next unresolved pair by choosing the **Go to next unresolved** button. To activate the last pair, choose the **Go to the last pair** button. Similarly, you can use the buttons, **Step backward**, **Go to previous unresolved**, and **Go to the first pair**, to perform their respective actions.

You can make any of the parts fixed for the mechanism from the parts that are fixed in the **Assembly Design** workbench. To do so, select the part from the **Fixed constraints list** drop-down list and choose the **Create Fixed Part** button; the selected part will be displayed in the **Current fixed part** display box.

Note that instead of specifying constraint for creating joints, you can directly create joints for the active pair by choosing the **Auto Create** button.

After specifying all the required joints and fixing a part for the mechanism, choose the OK button.

TUTORIALS

Tutorial 1

In this tutorial, you will create a mechanism for the Bench Vice assembly created in Tutorial 2 of Chapter 13. Refer to Figure 16-63. **(Expected time: 1 hr)**

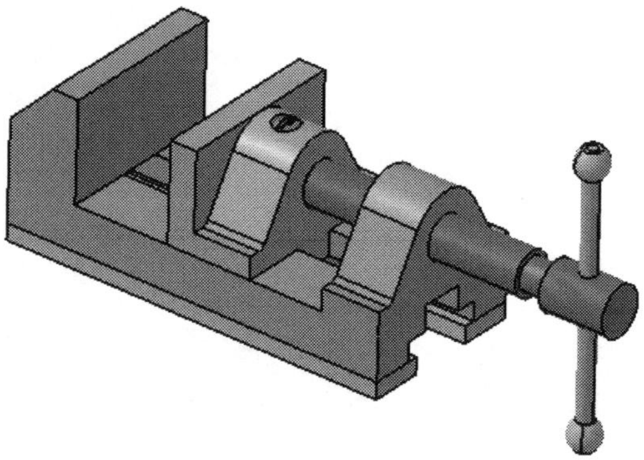

Figure 16-63 *The Bench Vice assembly*

The following steps are required to complete this tutorial:

a. Copy the *Bench Vice* folder from the *c13* folder to the *c16* folder and open the assembly file.
b. Invoke the **DMU Kinematics** workbench and save the file using the **Save As** tool.
c. Delete constraints, start a new mechanism, and fix a part.
d. Apply Rigid joints to the mechanism.
e. Apply the Revolute joint to the mechanism.
f. Apply the Planar joint to the mechanism.
g. Apply the Screw joint to the mechanism.
h. Simulate the mechanism.
i. Save and close the file.

Copying and Opening the Bench Vice Assembly

1. Create a folder with the name *c16* in the *CATIA* folder. Copy the *Bench Vice* folder from the *c13* folder and paste it in the *c16* folder.

2. Open the assembly file of Bench Vice.

If the assembly is opened in the **Assembly Design** workbench, then follow both the steps given in the next section. If the assembly is opened in the **DMU Kinematics** workbench, then follow only the second step in the next section.

Invoking the DMU Kinematics Workbench

1. Choose **Start > Digital Mockup > DMU Kinematics** from the menu bar.

2. Choose **File > Save As** from the menu bar; the **Save As** dialog box is displayed. Save the file at *CATIA\c16\Bench Vice* with the name *mechanism_1*.

Deleting Constraints, Starting a New Mechanism, and Fixing a Part in the DMU Kinematics Workbench

In this tutorial, you will create joints manually. While creating joints, the assembly constraints will be created automatically. Therefore, it is recommended that you delete those constraints before proceeding further in the tutorial.

1. Right-click on the **Constraints** node in the Specification tree; a contextual menu is displayed.

2. Choose the **Delete** option from the contextual menu; all constraints are deleted.

3. Next, choose **Insert > New Mechanism** from the menu bar; the **Mechanisms** sub-node under the **Applications** node is created in the Specification tree.

4. Now, choose the **Fixed Part** tool from the **DMU Kinematics** toolbar; the **New Fixed Part** dialog box is displayed and you are prompted to select the fixed part.

5. Select the **Base** part from the Specification tree or the graphics window; the selected part gets fixed in the mechanism.

Applying Rigid Joints to the Mechanism

To simplify the mechanism, it is recommended that you apply rigid joints between the parts that have no relative motion. As a result, these parts will act as a rigid body in the mechanism.

1. Choose the **Rigid Joint** tool from **DMU Kinematics > Kinematics Joints** sub-toolbar; the **Joint Creation: Rigid** dialog box is displayed.

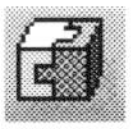

2. Select the Base and then the Base Plate on the left side; the **OK** button is enabled in the **Joint Creation: Rigid** dialog box.

3. Choose the **OK** button; a Rigid joint is created between the Base and the Base Plate on the left side.

4. Similarly, apply the Rigid joint between the Base and the Base Plate on the right side.

5. Now, apply the Rigid joint between the Base Plate on the left side and its Set Screws.

6. Similarly, apply the Rigid joint between the Base Plate on the right side and its Set Screws.

7. Next, apply the Rigid joint between the Screw Bar and the Bar Globes.

8. Similarly, apply the Rigid joint between the Sliding Jaw and the Oval Fillister.

9. Apply the Rigid joint between the Sliding Jaw and the Holding Plate.

10. Apply the Rigid joint between the Holding Plate and its Set Screws.

11. Finally, apply the Rigid Joint between the Jaw Screw and the Screw Bar.

Applying the Revolute Joint to the Mechanism

When you rotate the Screw Bar in the Bench Vice assembly, the Jaw Screw pushes the Sliding Jaw translationally. Therefore, you need to create a joint between the Jaw Screw and the Sliding Jaw. The joint to be created must facilitate the rotary motion of the Jaw Screw inside the Sliding Jaw. Also, it should facilitate the linear motion of the Jaw Screw and the Sliding Jaw. If you create a Revolute joint between the Sliding Jaw and the Jaw Screw, it will fulfill the above requirements.

1. Choose the **Revolute Joint** tool from **DMU Kinematics > Kinematics Joints** sub-toolbar; the **Joint Creation: Revolute** dialog box is displayed and you are prompted to select the first line.

2. Select the central axis of the Jaw Screw, as shown in Figure 16-64; you are prompted to select the second line.

3. Select the central axis of a hole from the Sliding Jaw, refer to Figure 16-64; you are prompted to select the first plane.

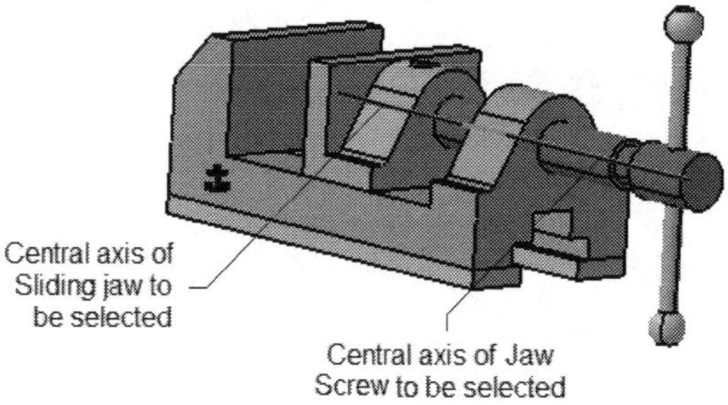

Figure 16-64 *Axes to be selected for creating the Revolute joint*

4. Select the front face of the Sliding Jaw, refer to Figure 16-65; you are prompted to select the second face.

5. Select the face of the Jaw Screw that is close to the face selected earlier, refer to Figure 16-66. Note that in this figure, the remaining parts have been hidden for better visualization.

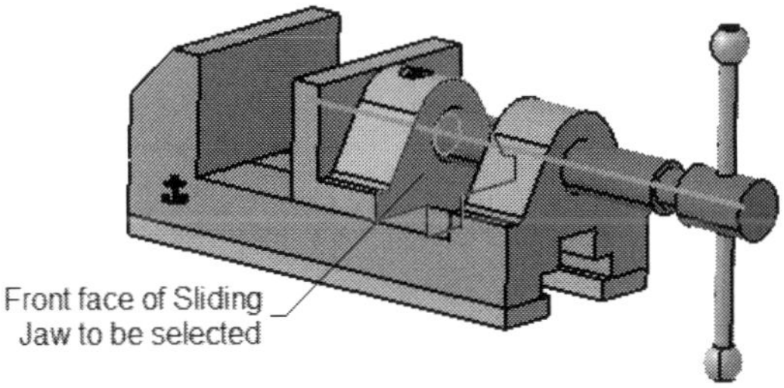

Figure 16-65 Front face of the Sliding Jaw to be selected

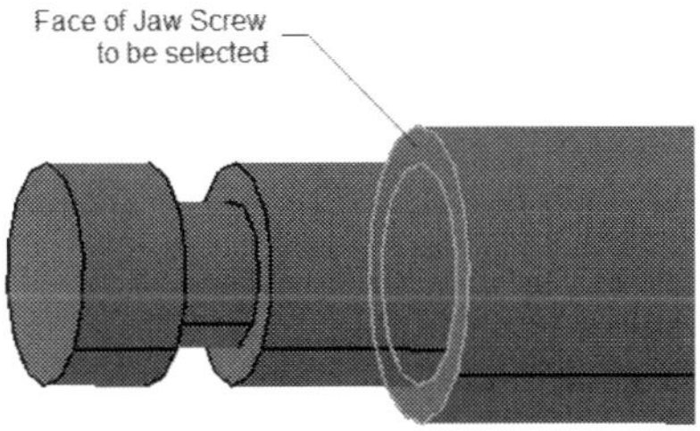

Figure 16-66 Face of the Jaw Screw to be selected

6. Next, select the **Offset** radio button in the **Joint Creation: Revolute** dialog box; the offset distance between the selected planes is displayed in the **Offset** spinner.

7. Enter **1.5** in the **Offset** spinner of the **Joint Creation: Revolute** dialog box.

8. Choose **OK** to create the joint and exit the dialog box.

Applying the Planar Joint to the Mechanism

In the Bench Vice assembly, the Sliding Jaw slides over the Base. So, you need to create a planar joint between the Sliding Jaw and the Base.

1. Choose the **Planar Joint** tool from **DMUKinematics > Kinematics Joints** sub-toolbar; the **Joint Creation: Planar** dialog box is displayed and you are prompted to select the first plane.

2. Select the planar face of the Base; you are prompted to select the second plane. Select the bottom face of the Sliding Jaw, as shown in Figure 16-67.

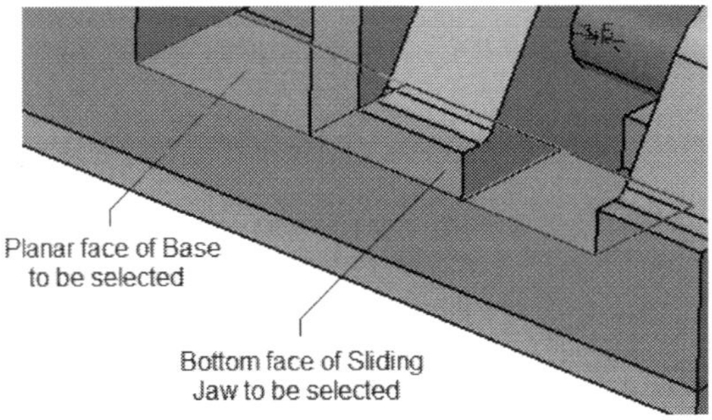

Figure 16-67 Faces to be selected for creating the Planar joint

3. Choose **OK** to create the joint and exit the dialog box.

Applying the Screw Joint to the Mechanism

As you rotate the Screw Bar in the Bench Vice assembly, the Jaw Screw will also rotate about its central axis and slide inside the threaded hole of Base as per the pitch value. If you create a Screw Joint between the Jaw Screw and the Base, it will fulfil the above mentioned requirements.

1. Choose the **Screw Joint** tool from **DMU Kinematics > Kinematics Joints** sub-toolbar; the **Joint Creation: Screw** dialog box is displayed and you are prompted to select the first line.

2. Select the central axis of the Jaw Screw; you are prompted to select the second line. Select the central axis of the hole from the Base, as shown in Figure 16-68.

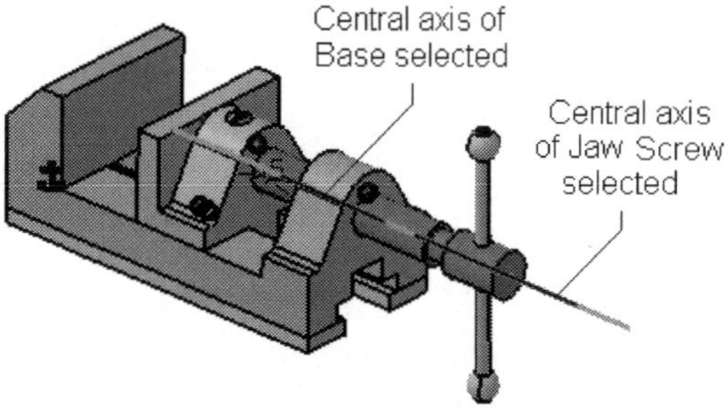

Figure 16-68 *Axes selected for creating the Screw joint*

3. Select the **Length driven** check box to specify that the movement of the Jaw Screw drives the mechanism in linear direction.

4. Enter **5** in the **Pitch** edit box.

5. Next, choose **OK** to create the Screw joint and exit the dialog box.

   After this joint gets created and the **Information** message box is displayed informing that you can now simulate the mechanism.

6. Choose **OK** from the **Information** message box to close it.

Simulating the Mechanism

After creating necessary joints for the mechanism, you can view the working of the mechanism by simulating it. However, if the mechanism does not perform according to the design requirements, you need to verify all joints.

1. Choose the **Simulation with Commands** tool from the **DMU Kinematics** toolbar; the **Kinematics Simulation** dialog box is displayed.

2. If the dialog box is not already expanded, choose the **More** button from the dialog box to expand it.

3. Select the **On request** radio button in the **Simulation** area of the **Kinematics Simulation** dialog box if not already selected.

4. Drag the slider in the **Command.1** area to extreme right.

5. Next, choose the **Play forward** button from the **Simulation** area of the **Kinematics Simulation** dialog box; the mechanism starts simulating.

You will notice that the Sliding Jaw has moved beyond the limit in the simulation. Therefore, you need to specify the limit.

6. Set the value to **0** in the **Command.1** spinner; the Sliding Jaw moves to its original position.

7. Again, choose the **Play forward** button in the **Simulation** area of the **Kinematics Simulation** dialog box; the Sliding Jaw comes to its original position while simulating the mechanism.

   Now you need to restrict the movement of the Sliding Jaw.

8. Choose the **Swatch** button from the **Command.1** area of the **Kinematics Simulation** dialog box; the **Slider: Command.1** dialog box is displayed.

9. In the **Slider : Command.1** dialog box, enter **-25** in the **Lowest Value** edit box to restrict the lowest limit of movement of the Sliding Jaw.

10. Enter **35** in the **Highest Value** edit box to restrict the highest limit of movement of the Sliding Jaw.

11. Choose **OK** to apply the limits and exit the dialog box.

12. Play the simulation again to observe the working of the mechanism.

13. Choose the **Close** button to close the **Kinematics Simulation** dialog box.

Saving and Closing the File

1. Choose the **Save** button from the **Standard** toolbar to save the file.

2. Close the file by choosing **File > Close** from the menu bar.

Tutorial 2

In this tutorial, you will create an assembly and its mechanism using the CV joint shown in Figure 16-69. The orthographic views and dimensions of the parts are shown in Figures 16-70 through 16-73. **(Expected time: 1.5 hr)**

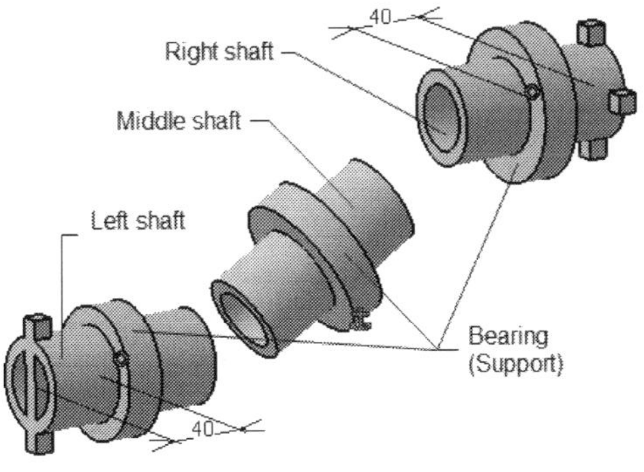

Figure 16-69 *Mechanism created using the CV joint*

The following steps are required to complete this tutorial:

a. Create all components of the mechanism as separate part files in the **Part Design** workbench.
b. Start a new file in the **Assembly Design** workbench.
c. Insert all parts in the **Assembly Design** workbench and place them appropriately using the Snap tool and Compass.
d. Invoke the **DMU Kinematics** workbench.
e. Fix a part for the mechanism.
f. Apply the CV Joint to the mechanism.
g. Apply the Revolute Joints to the mechanism.
h. Simulate the mechanism.
i. Save and close the file.

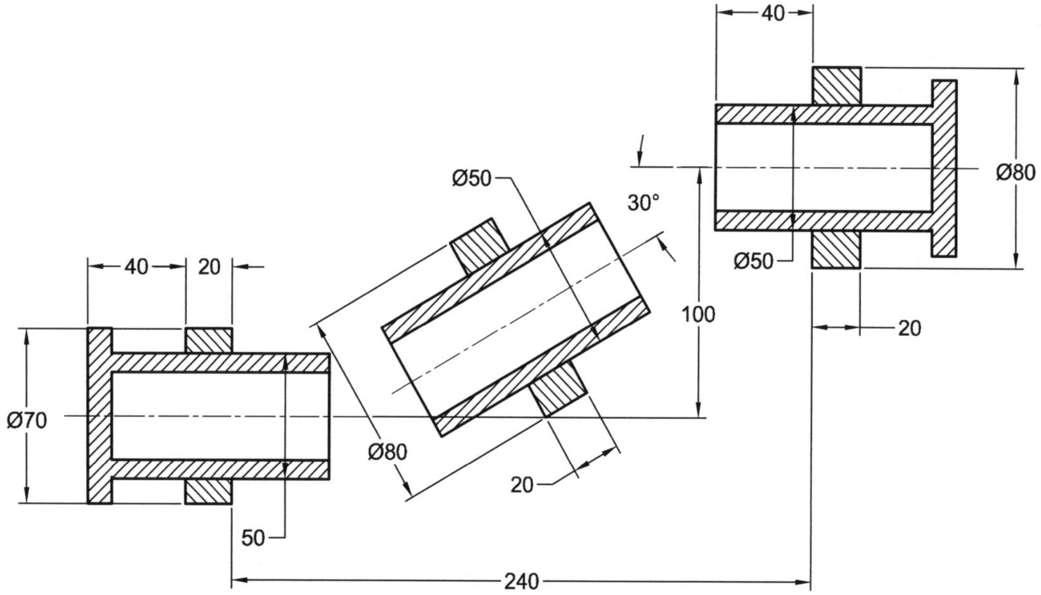

***Figure 16-70** Cross-sectional views and dimensions of the Bearing (Support)*

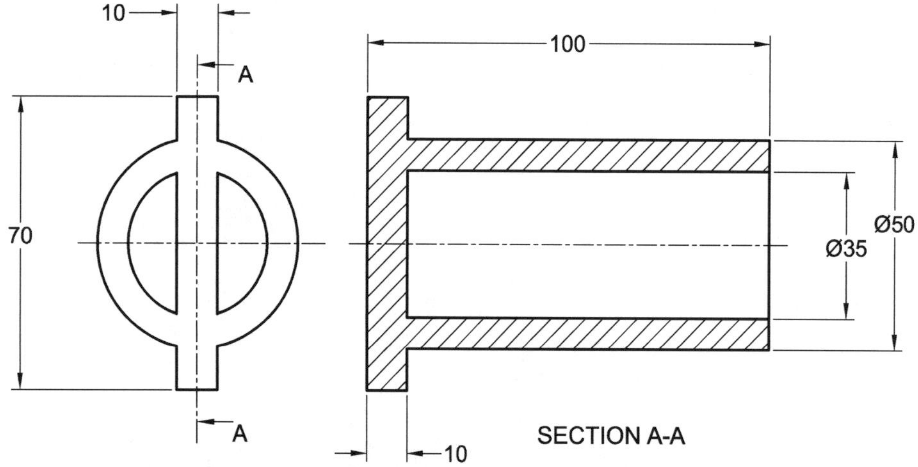

***Figure 16-71** Orthographic views and dimensions of the Left shaft*

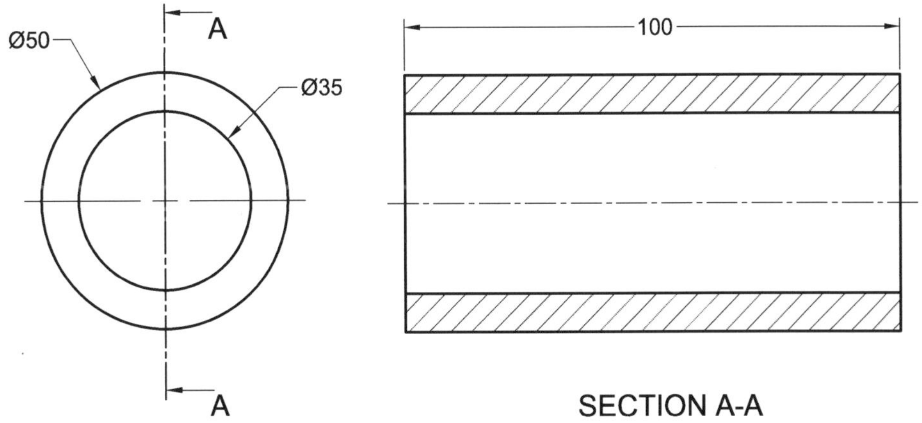

Figure 16-72 *Orthographic views and dimensions of the Middle shaft*

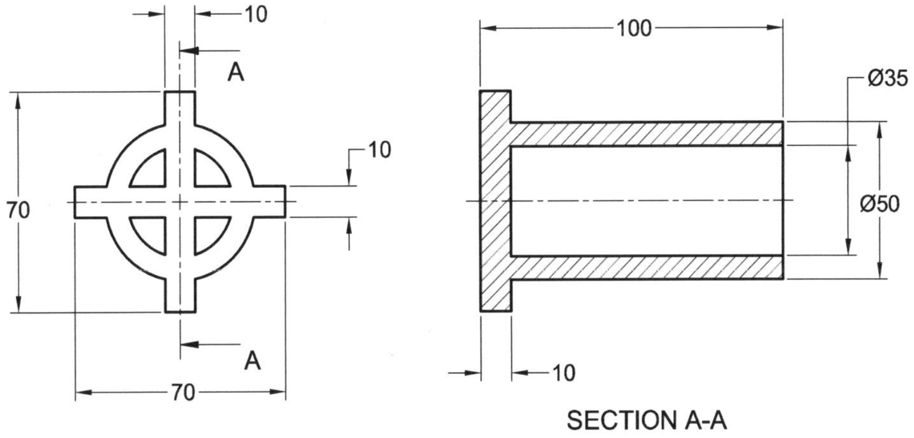

Figure 16-73 *Orthographic views and dimensions of the Right shaft*

Note

In this tutorial, for the simplification of the mechanism, the Bearing (Support) has been created as a single part, as shown in Figure 16-70. However, you can create this Bearing (Support) as three individual parts and assemble them in the ***Assembly Design*** *workbench. But then, the steps to create the mechanism will be different from the one discussed in this tutorial.*

Before creating the assembly parts for this tutorial, you will create the *CV Joint* folder at *CATIA\c16*. You need to save the parts of the CV Joint assembly in this folder. Note that you should change the number of every part before saving it. The process of changing the part number of a part has already been discussed in Chapter 12.

Creating the Components of the Mechanism

In this section, you need to create the CV Joint assembly using the bottom-up approach. In this approach, first parts are created as individual part files and then inserted in the assembly file.

1. Create all parts of the assembly and save them as separate part files in the *CV Joint* folder.

2. Close all part files if they are open.

Starting a New File in the Assembly Design Workbench

To create a CV Joint, you need to place all parts properly in the product file that you have created above. For placing them properly, you need to start a new file in the **Assembly Design** workbench.

1. Choose the **New** tool from the **Standard** toolbar; the **New** dialog box is displayed.

2. Select the **Product** option from the **List of Types** list box in the **New** dialog box.

3. Choose the **OK** button to start a new product file; a new product file is started in the **Assembly Design** workbench and **Product1** is displayed on top in the Specification tree. In case, the new product file starts in the **DMU Kinematics** workbench, choose **Start > Mechanical Design > Assembly Design** from the menu bar.

Inserting Parts and Placing them Properly

After the new product file is started, you need to insert the base component into the assembly. For this mechanism, the Bearing (Support) is the base component.

1. Choose the **Existing Component** tool from the **Product Structure Tools** toolbar.

2. Select **Product1** from the Specification tree; the **File Selection** dialog box is displayed. In this dialog box, browse to the location of the Bearing (Support) part file and open it.

3. Next, insert all remaining parts of the CV Joint mechanism in the **Assembly Design** workbench.

   You will notice that all parts inserted for the joint overlap each other. You can place them separately in 3D space by using the Compass.

4. Drag the red point of the Compass, place it over the Right shaft and then drag its axis. As the cursor moves, the selected part also moves along the selected axis.

5. Place the Middle shaft in 3D space, refer to Figure 16-74.

6. Choose the **Snap** tool from the **Move** toolbar and select the central axis of the Right shaft and its bearing, refer to Figure 16-74; both the selected axes are aligned.

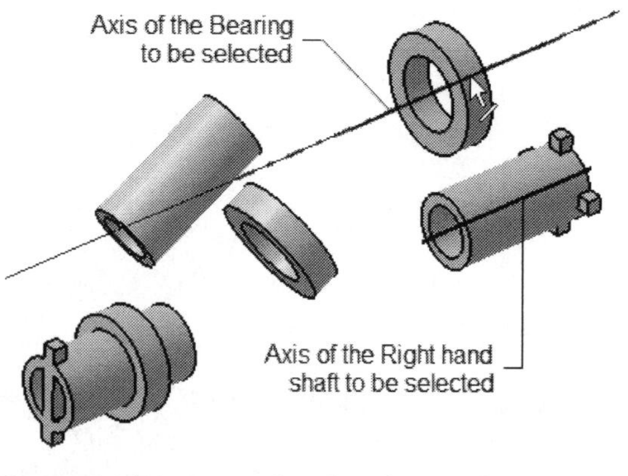

***Figure 16-74** Axes to be selected using the **Snap** tool*

7. Similarly, align the axis of the Middle shaft and its Bearing, refer to Figure 16-69.

8. Next, align the axis of the Left shaft and its Bearing.

   You can also use the central axis of the Compass to slide the shaft in its bearing for proper positioning, refer to Figure 16-69.

Invoking the DMU Kinematics Workbench

1. Choose **Start > Digital Mockup > DMU Kinematics** from the menu bar.

2. Choose **File > Save As** from the menu bar; the **Save As** dialog box is displayed.

3. Save the file at *CATIA/c16/CV Joint* with the name *mechanism_1*.

Fixing the Bearing(Support) in the DMU Kinematics Workbench

Now, you need to start a new mechanism.

1. Choose **Insert > New Mechanism** from the menu bar; the **Mechanisms** sub node is created under the **Applications** node in the Specification tree.

   Now, you need to fix the Bearing (Support).

2. Choose the **Fixed Part** tool from the **DMU Kinematics** toolbar; the **New Fixed Part** dialog box is displayed and you are prompted to select the fixed part.

3. Select the Bearing (Support) part from the Specification tree or from the graphics window; the selected part gets fixed in the mechanism.

Applying the CV Joint to the Mechanism

To apply the CV joint on three different parts, an axis from each part must lie on the same plane, and one of the axes must coincide with the other two axes. As you have already aligned all axes of shafts with their respective bearings, the above requirement gets fulfilled automatically. Now, you can apply the CV Joint to the mechanism.

1. Choose the **CV Joint** tool from **DMU Kinematics > Kinematics Joints** sub-toolbar; the **Joint Creation: CV Joint** dialog box is displayed and you are prompted to select the first line.

2. Select the axis of the Right shaft; you are prompted to select the second line.

3. Select the central axis of the Middle shaft as the second line; you are prompted to select the third line.

4. Select the central axis of the Left shaft, as shown in Figure 16-75.

5. Next, choose **OK** to create the joint and exit the dialog box.

Applying the Revolute Joint to the Mechanism

In the CV Joint mechanism when you drive one shaft, the other two shafts rotate automatically about their own axes. Therefore, to define all DOFs of the mechanism and create a command, you need to use two Revolute joints.

1. Choose the **Revolute Joint** tool from **DMU Kinematics > Kinematics Joints** sub-toolbar; the **Joint Creation: Revolute** dialog box is displayed and you are prompted to select the first line.

2. Select the central axis of the Bearing, as shown in Figure 16-76; you are prompted to select the second line.

3. Select the central axis of the Right shaft, refer to Figure 16-76; you are prompted to select the first plane.

4. Select the planar face of the Bearing, refer to Figure 16-76; you are prompted to select the second plane.

5. Select a planar face of the Right shaft, refer to Figure 16-76.

6. Next, select the **Offset** radio button in the **Joint Creation: Revolute** dialog box; the offset distance between the selected planes is displayed in the **Offset** spinner.

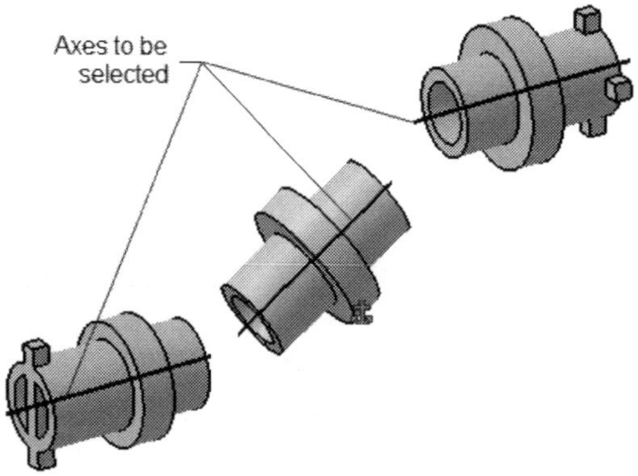

Figure 16-75 Axes to be selected for creating the CV Joint

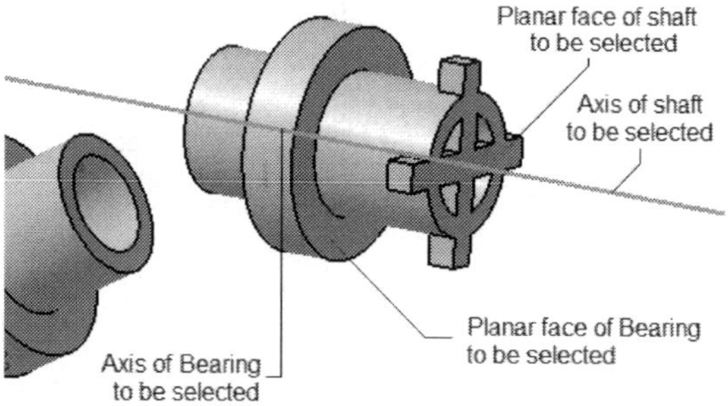

Figure 16-76 Axes and planes to be selected for creating the Revolute Joint

7. Enter **40** in the **Offset** edit box of the **Joint Creation: Revolute** dialog box and select the **Angle driven** check box.

8. Next, choose **OK** to create the joint and exit the dialog box.

   Note that **DOF=4** is displayed in the **Mechanism** node of the Specification tree. Therefore, you need to define these undefined DOFs and create a command to simulate the mechanism. To do so, create a Revolute Joint between the Left shaft and its Bearing.

9. Choose the **Revolute Joint** tool again; the **Joint Creation: Revolute** dialog box is displayed and you are prompted to select the first line.

10. Specify the axes and planar faces from the Left shaft and its Bearing, as shown in Figure 16-77.

11. Next, select the **Offset** radio button in the **Joint Creation: Revolute** dialog box; the offset distance between the selected planes is displayed in the **Offset** spinner.

12. Enter **40** in the **Offset** spinner of the **Joint Creation: Revolute** dialog box and choose **OK** to create the joint and exit the dialog box.

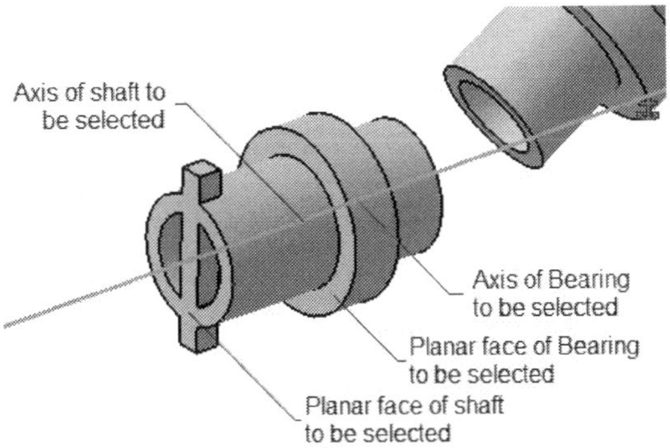

Figure 16-77 *Axes and planes to be selected for creating the Revolute Joint*

Note that after the joint is created and all DOFs are defined, the **Information** message box informing that the mechanism is ready for simulation is displayed.

13. Next, choose **OK** from the **Information** message box to close the message box.

Simulating the Mechanism

Simulate the mechanism as discussed in *Tutorial 1*.

Saving and Closing the File

1. Choose the **Save** button from the **Standard** toolbar to save the file.

2. Close the file by choosing **File > Close** from the menu bar.

Tutorial 3

In this tutorial, you will create the Cam and Follower mechanism shown in Figure 16-78. The orthographic views and dimensions of the parts are shown in Figures 16-79 through 16-81. You need to place a sketch point on the cam, as shown in Figure 16-80. This point will be used as a reference point while creating the mechanism. **(Expected time: 2 hr)**

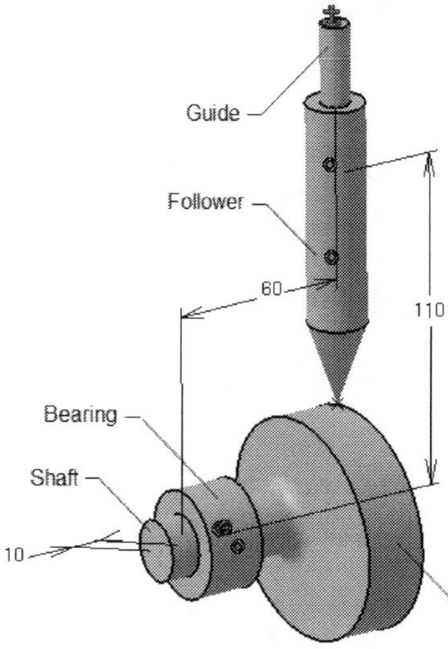

Figure 16-78 The Cam and Follower mechanism

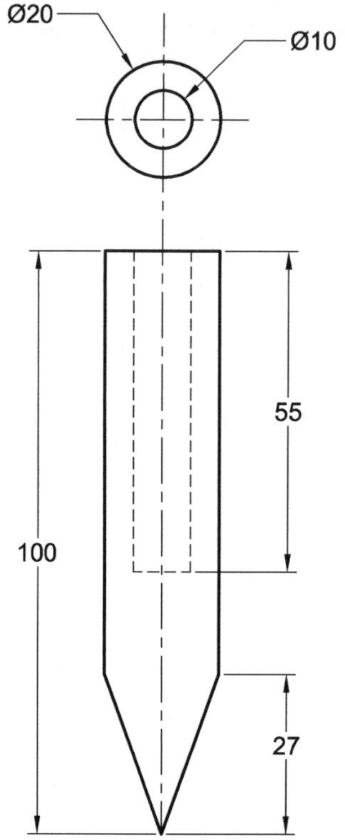

Figure 16-79 Views and dimensions of the Follower

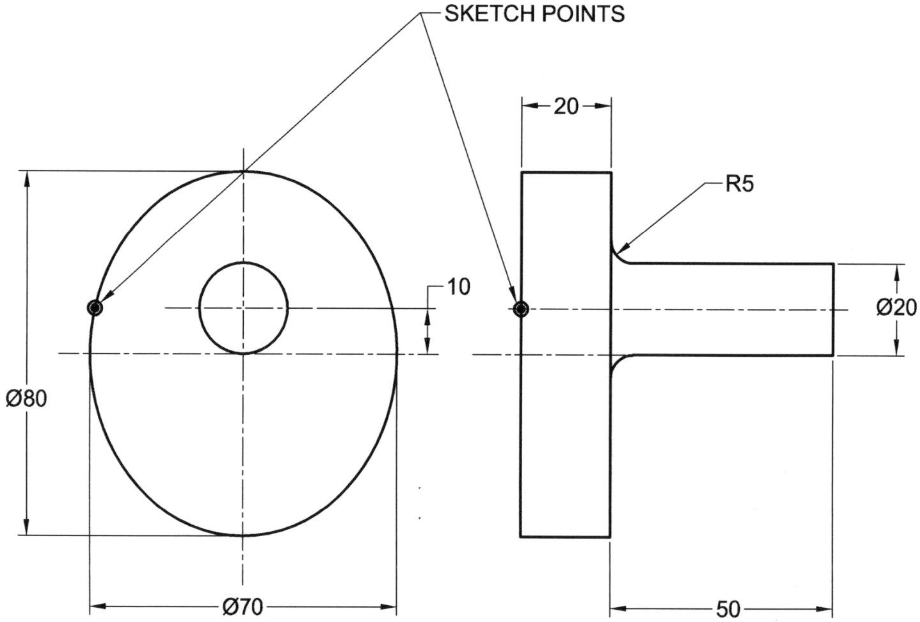

Figure 16-80 Views and dimensions of the Cam

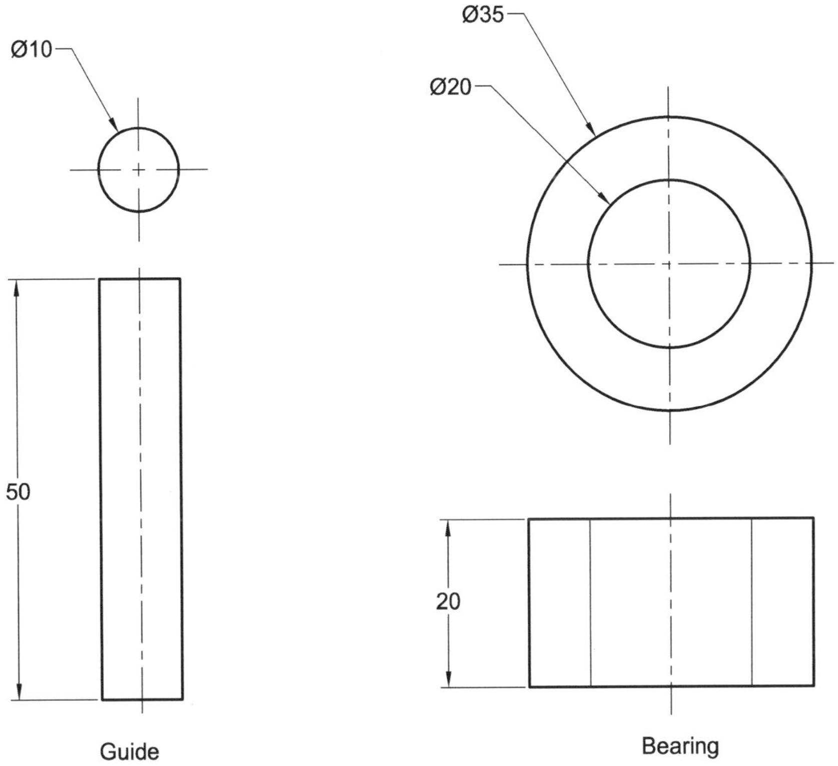

Figure 16-81 *Views and dimensions of the Guide and the Bearing*

The following steps are required to complete this tutorial:

a. Create all parts of the mechanism as separate part files in the **Part Design** workbench.
b. Start a new file in the **Assembly Design** workbench.
c. Insert the required parts in the **Assembly Design** workbench and assemble them using assembly constraints.
d. Invoke the **DMU Kinematics** workbench.
e. Fix a part for the mechanism.
f. Convert assembly constraints into joints.
g. Insert remaining parts.
h. Apply the Revolute joint to the mechanism.
i. Apply the Cylindrical joint to the mechanism.
j. Apply the Spherical joint to the mechanism and then delete it.
k. Apply the Point Curve joint to the mechanism.
l. Simulate the mechanism.
m. Save and close the file.

Before creating parts for this tutorial, you will create the *Cam Follower mechanism* folder at *CATIA\c16*. You need to save the parts of the Cam and Follower mechanism in this folder. Note that you should change the part number of every part before you save it.

Creating Parts of the Mechanism

In this section you need to create the Cam Follower assembly using the bottom-up approach. In this approach, all parts are created as individual part files and then inserted in the assembly file.

1. Create all parts of the assembly and then save them as separate part files at *CATIA/c16/Cam Follower mechanism*. (While creating a Cam, create a sketch point over the edge of the Cam, refer to Figure 16-80).

2. Close all part files if they are open.

Starting a New File in the Assembly Design Workbench

The parts that you have created need to be placed at proper position in the assembly environment to create joints for the mechanism. For this, you need to start a new file in the **Assembly Design** workbench.

1. Choose **Start > Mechanical Design > Assembly Design** from the menu bar.

Inserting Parts and Placing them Properly Using Assembly Constraints

After the new product file is started, you can insert the base part into the assembly environment. In this case, Guide is the base part. Next, you can apply assembly constraints and convert them into joints.

1. Choose the **Existing Component** tool from the **Product Structure Tools** toolbar.

2. Select **Product1** from the Specification tree; the **File Selection** dialog box is displayed. In this dialog box, browse to the location of the file of the Guide and open it.

3. Next, insert the Bearing in the **Assembly Design** workbench, as shown in Figure 16-82. Note that the edges have been hidden in this figure for better visualization.

   Since both parts overlap each other, you need to place them separately in 3D space by using the manipulation tool from the move toolbar or using the Compass, as discussed in Tutorial 2.

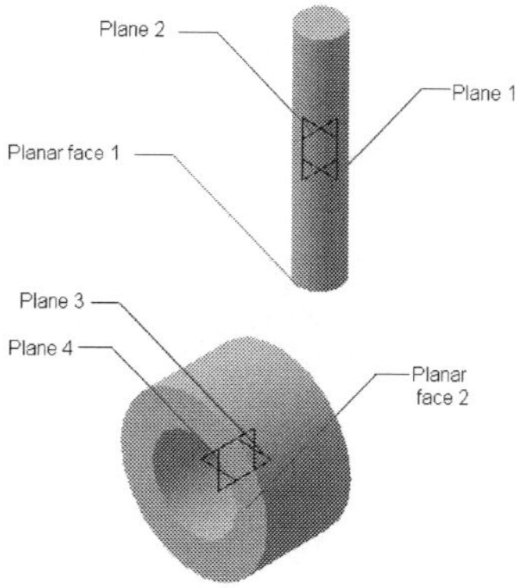

***Figure 16-82** Parts inserted in the **Assembly Design** workbench*

4. Choose the **Fix Component** tool from the **Constraints** toolbar and select Guide; the Guide in the assembly becomes fixed.

5. Now, apply the **Coincidence Constraint** between Plane 2 and Plane 4, refer to Figure 16-82.

6. Apply the **Offset Constraint** between Planar face 1 and Plane 3 with an offset distance of 110, refer to Figure 16-82.

7. Similarly, apply the Offset Constraint between Planar face 2 and Plane 1 with an offset distance of 60, refer to Figure 16-82 and Figure 16-83, and then update the assembly.

8. Choose the **Fix Together** tool from the **Constraints** toolbar and select the Guide and the Bearing. Then, choose the **OK** button; the relative position of the Guide and the Bearing gets fixed in the assembly.

   Now you need to remove the **Coincidence Constraint** and both **Offset Constraint** as these constraints were used only for positioning of Guide and Bearing.

9. Select the **Coincidence Constraint** and both **Offset Constraint** and then choose **Delete** from the menu displayed on right clicking.

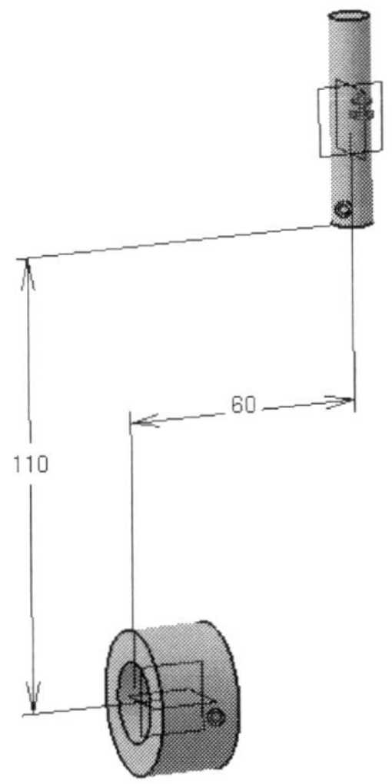

Figure 16-83 *Assembly of the Guide and Bearing*

Invoking the DMU Kinematics Workbench

1. Choose **Start > Digital Mockup > DMU Kinematics** from the menu bar.

2. Choose **File > Save As** from the menu bar; the **Save As** dialog box is displayed.

3. Save the file at *CATIA\c16\Cam Follower mechanism* with the name *mechanism_1*.

Converting Assembly Constraints into Joints

1. Choose the **Assembly Constraints Conversion** tool from the **DMU Kinematics** toolbar; the **Assembly Constraints Conversion** dialog box is displayed.

2. Choose the **New Mechanism** button from the **Assembly Constraints Conversion** dialog box; the **Mechanism Creation** dialog box is displayed with the default name **Mechanism.1** in the **Mechanism name** edit box.

3. Accept the default name and choose **OK** from the **Mechanism Creation** dialog box.

4. Next, choose the **More** button from the **Assembly Constraints Conversion** dialog box, if this dialog box is not expanded.

5. Select the **Fix Together.1** constraint from the **Constraints list** list box which is in the left pane of the **Assembly Constraints Conversion** dialog box; **Rigid** is displayed in the **Resulting type** display box.

6. Choose the **Create Joint** button below the **Resulting type** display box; **Rigid.1** is displayed in the **Joints list** list box which is in the right pane of the **Assembly Constraints Conversion** dialog box.

   You can fix any one of the parts for the mechanism. In this tutorial, we will fix the guide that was fixed using the **Fixed Component** tool in the assembly environment.

7. Choose the **Create Fixed Part** button from the **Assembly Constraints Conversion** dialog box; **Guide.1** is displayed in the **Current fixed part** display box.

8. Choose **OK** from the **Assembly Constraints Conversion** dialog box; the Rigid joint is created and the Guide is fixed for the mechanism.

Inserting Remaining Parts

1. Insert the Cam and the Follower in the **DMU Kinematics** workbench for the mechanism and then place them in 3D space using the Compass.

Applying the Revolute Joint to the Mechanism

To drive the shaft of the Cam for simulating the mechanism, you need to create the Revolute joint.

1. Choose the **Revolute Joint** tool from **DMU Kinematics > Kinematics Joints** sub-toolbar; the **Joint Creation: Revolute** dialog box is displayed and you are prompted to select the first line.

2. Select the central axis of the Bearing, as shown in Figure 16-84; you are prompted to select the second line.

3. Select the central axis of the shaft of the Cam, refer to Figure 16-84; you are prompted to select the first plane.

4. Select the planar face of the Bearing, refer to Figure 16-84; you are prompted to select the second face.

5. Select the planar face of the shaft, refer to Figure 16-84.

6. Next, select the **Offset** radio button in the **Joint Creation: Revolute** dialog box and set the value of the **Offset** spinner to **10**. Note that orientation should be the same as shown in Figure 16-84.

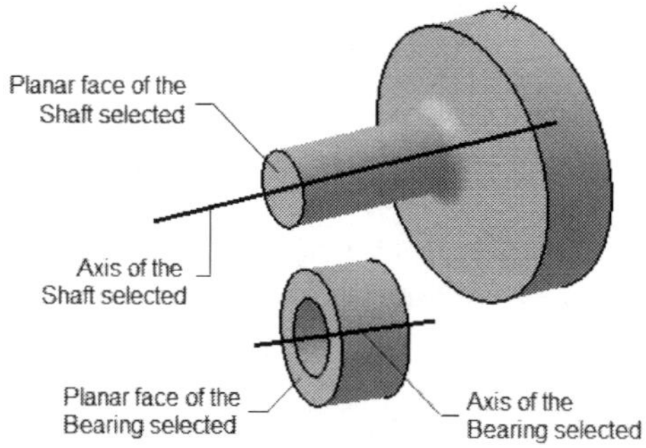

***Figure 16-84** Planes and axes selected for applying the Revolute joint*

7. Select the **Angle driven** check box in the **Joint Creation: Revolute** dialog box to create a command and then choose **OK**; the **Information** message box informing that the mechanism can be simulated is displayed.

8. Choose **OK** from the **Information** message box to close it.

Applying the Cylindrical Joint to the Mechanism

In a Cam and Follower mechanism, the movement of the Follower is controlled by the Guide. Therefore, in such cases, you need to create a cylindrical joint between them.

1. Choose the **Cylindrical Joint** tool from **DMU Kinematics > Kinematics Joints** sub-toolbar; the **Joint Creation: Cylindrical** dialog box is displayed and you are prompted to select the first line.

2. Select the central axis of the Guide, as shown in Figure 16-85; you are prompted to select the second line.

3. Select the central axis of the Follower, refer to Figure 16-85. Note that orientation should be same as shown in the figure.

In a Cam and Follower mechanism, the Pointer of the Follower slides over the profile of the Cam such that the vertical movement of the Follower is controlled by the revolution of the Cam. Therefore, it is recommended that instead of selecting the **Length driven** check box, you should select the **Angle driven** check box from the **Joint Creation: Cylindrical** dialog box to define the rotational movement of the Follower.

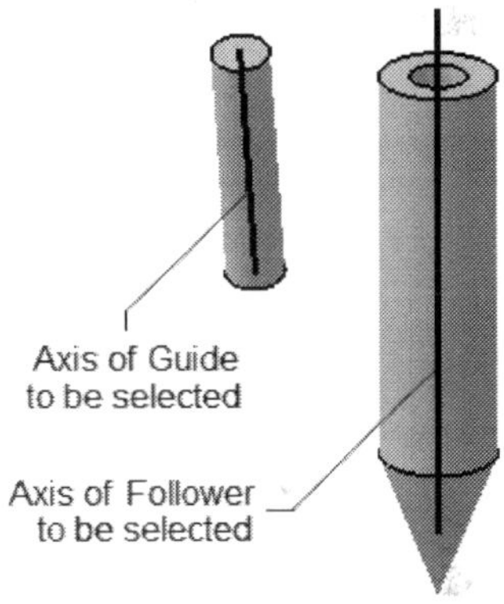

Figure 16-85 *Axes to be selected for creating the Cylindrical joint*

4. Select the **Angle driven** check box to define the angular movement.

5. Choose **OK** to create a joint and exit the dialog box.

Applying the Spherical Joint to the Mechanism

You can create the Point Curve joint between the Cam and Follower only when the pointer of the Follower is at the curve of the Cam. You can bring the pointer of the Follower and the curve of the Cam together by using the Spherical joint.

1. Choose the **Spherical Joint** tool from **DMU Kinematics > Kinematics Joints** sub-toolbar; the **Joint Creation: Spherical** dialog box is displayed and you are prompted to select the first point.

2. Select the pointer of the Follower; you are prompted to select the second point.

3. Select the sketch point created on the Cam, as shown in Figure 16-86, and choose **OK**; the **Information** message box is displayed, informing that the mechanism has too many commands. To resolve this problem, you need to create one more joint and delete unwanted constraints of the mechanism.

4. Choose **OK** from the **Information** message box to close it.

5. Choose the **Update Positions** tool from the **Kinematics Update** toolbar; the **Update Mechanism** dialog box is displayed with the default name *Mechanism.1*.

6. Choose **OK** from the **Update Mechanism** dialog box; the mechanism is updated and both the selected points coincide with each other.

   As the pointer of the Follower is resting on the Cam, you can apply the Point Curve joint between the Cam and the Follower. Also, now you do not need the Spherical joint, therefore, you can remove it.

7. Right-click on **Spherical.1** in the **Joints** node of the Specification tree; a contextual menu is displayed.

8. Choose the **Delete** option from the contextual menu displayed; the **Delete** dialog box is displayed.

9. Make sure the **Delete all children** check box is selected in the **Delete** dialog box and then choose **OK** to delete the joint and exit the dialog box.

Applying the Point Curve Joint to the Mechanism

As the pointer of the Follower rests over the curve of the Cam, you can apply the Point Curve joint to the mechanism.

1. Choose the **Point Curve Joint** tool from **DMU Kinematics > Kinematics Joints** sub-toolbar; the **Joint Creation: Point Curve** dialog box is displayed and you are prompted to select the first curve. 

2. Select the edge of the Cam, as shown in Figure 16-87; you are prompted to select the first point.

3. Select the Pointer of the Follower.

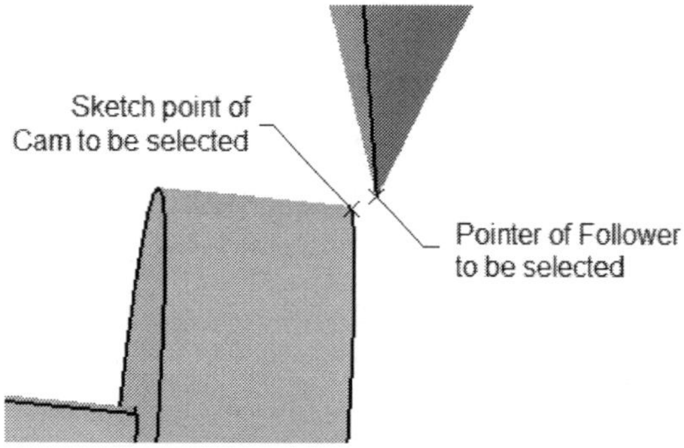

Figure 16-86 Points to be selected for applying the Spherical joint

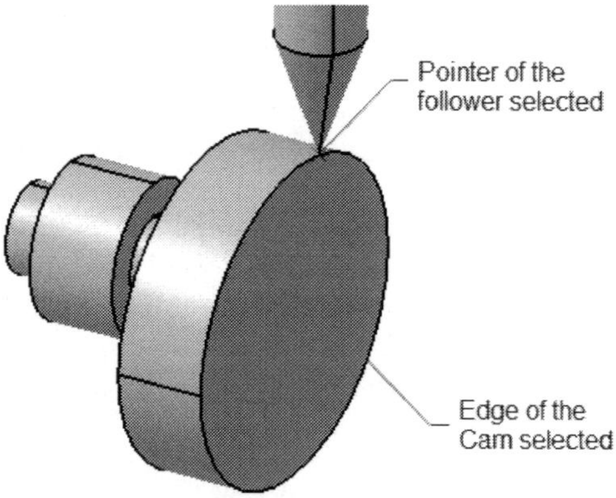

Figure 16-87 Edge and Pointer selected for creating the Point Curve joint

4. Next, choose **OK** to create the joint and exit the dialog box. Also, the **Information** message box is displayed again informing that the mechanism can be simulated.

5. Choose **OK** from the **Information** message box to close the message box.

Simulating the Mechanism

Update the mechanism and simulate it as discussed earlier.

Saving and Closing the File

1. Choose the **Save** button from the **Standard** toolbar to save the file.

2. Close the file by choosing **File** > **Close** from the menu bar.

Tutorial 4

In this tutorial, you will create the Bearing mechanism shown in Figure 16-88, using the Roll Curve joint. The views and dimensions of the parts are shown in Figures 16-89 through 16-91. Draw a sketch curve and place a sketch point on the model as specified in the orthographic views. The curve and the point will be used as reference while creating the mechanism.

(Expected time: 1 hr)

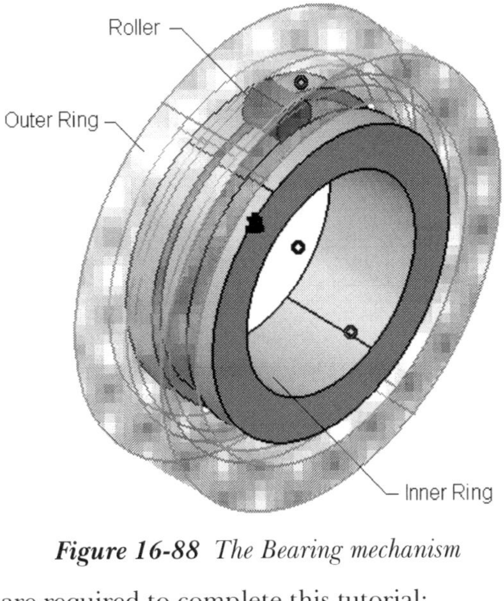

Figure 16-88 *The Bearing mechanism*

The following steps are required to complete this tutorial:

a. Create all parts of the mechanism as separate part files in the **Part Design** workbench.
b. Start a new file in the **Assembly Design** workbench.
c. Insert all parts in the **Assembly Design** workbench and assemble them.
d. Invoke the **DMU Kinematics** workbench.
e. Fix a part for the mechanism.
f. Apply the Roll Curve joint to the mechanism.

g. Apply the Revolute joint to the mechanism.
h. Simulate the mechanism.
i. Save and close the file.

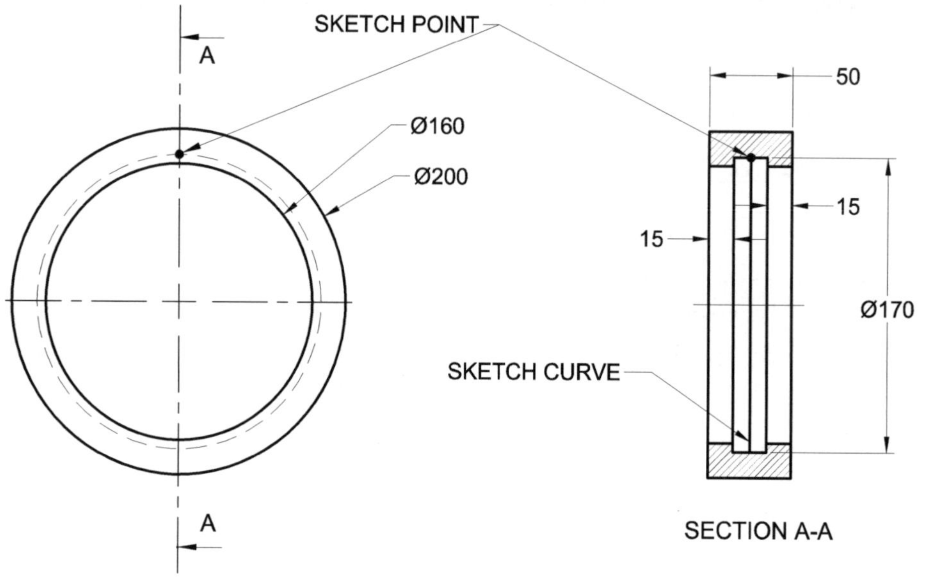

Figure 16-89 Orthographic views and dimensions of Outer Ring

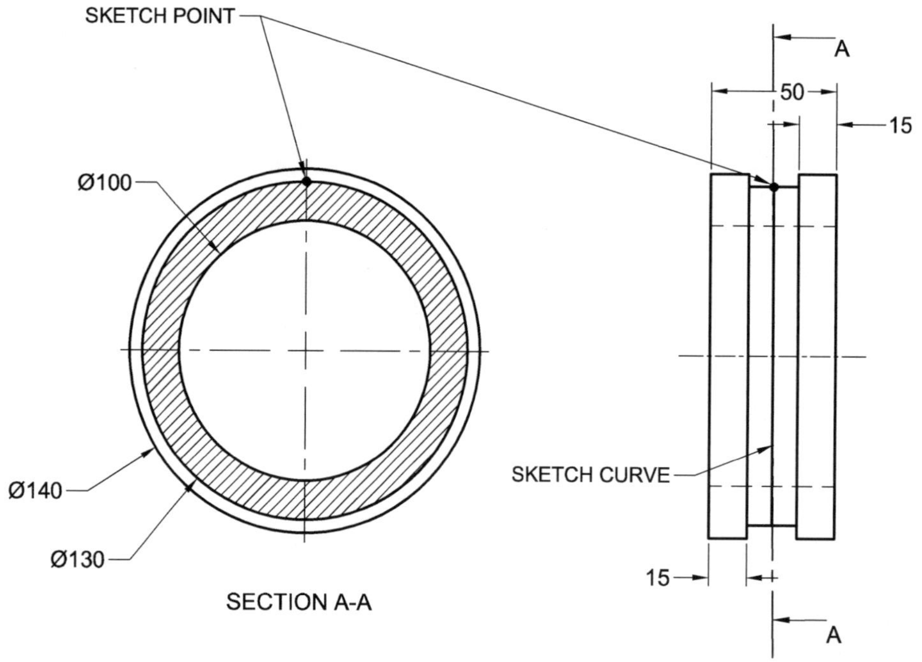

Figure 16-90 Orthographic views and dimensions of Inner Ring

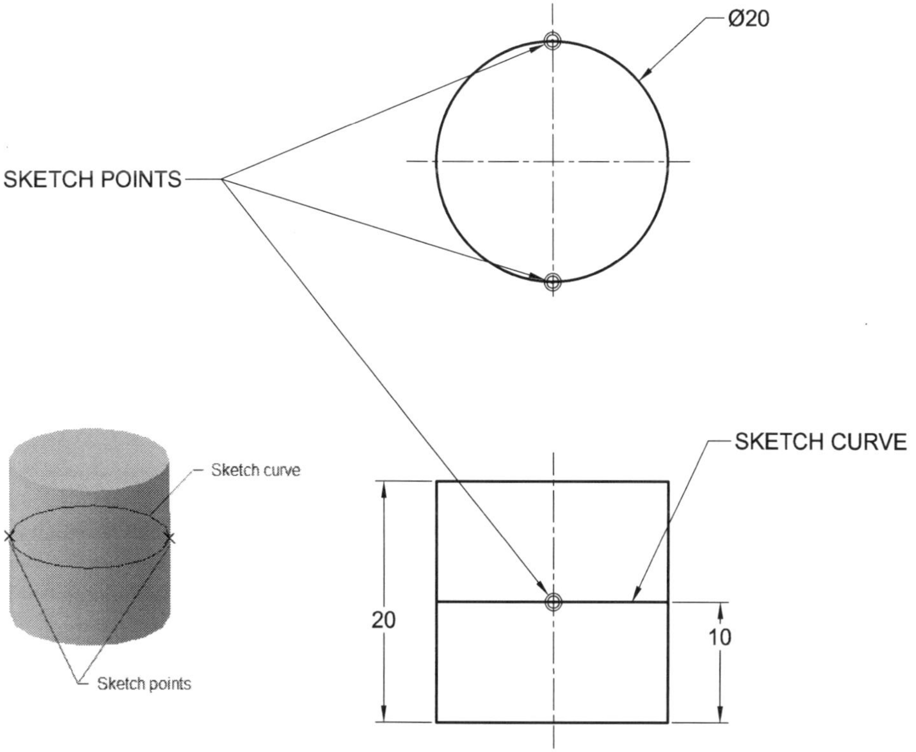

Figure 16-91 *Orthographic views and dimensions of the Roller*

Before creating parts for this tutorial, you will create the *Bearing* folder at *\CATIA\c16*. You will save the parts of the Bearing mechanism in this folder.

Creating the Components of the Mechanism

You will create the assembly of the Bearing mechanism using the bottom-up approach. In this approach, you will first create all parts as individual part files and then insert them in the assembly file.

While creating parts, create sketch curves and sketch points over parts, as shown in Figures 16-89 through 16-91. These curves and points will be used to assemble the parts and to create the mechanism.

1. Create all parts of the assembly and then save them as separate part files at *CATIA\c16\Bearing*.
2. Close all part files if they are open.

Starting a New File in the Assembly Design Workbench

All the components that you have created so far are required to be placed appropriately to create joints for the mechanism. Therefore, you need to start a new file in the **Assembly Design** workbench to place the parts properly.

1. Choose the **New** tool from the **Standard** toolbar; the **New** dialog box is displayed.

2. Select the **Product** option from the **List of Types** list box in the **New** dialog box.

3. Next, choose the **OK** button from the **New** dialog box; a new file is started in the **Assembly Design** workbench and **Product1** is displayed on the top of the Specification tree. In case the new file starts in the **DMU Kinematics** workbench, choose **Start > Mechanical Design > Assembly Design** from the menu bar to invoke the **Assembly Design** workbench.

Inserting Parts and Assembling Them

After starting the new Product file, you need to insert the base component into the assembly. In this case, the Inner Ring is the base component.

1. Insert all parts of the Roller Curve Joint mechanism in the **Assembly Design** workbench and place them in 3D space using the Compass, as shown in Figure 16-92. In this figure, the edges of parts are hidden for better visualization.

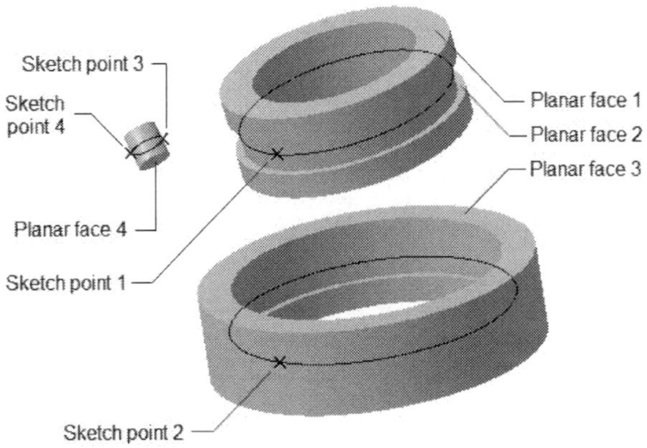

Figure 16-92 *Parts inserted in the* ***Assembly Design*** *workbench*

2. Make the axes of the Inner Ring and Outer Ring coaxial by using the **Coincidence Constraint** tool.

3. Make the Planar face 1 and the Planar face 3 coplanar by using the **Coincidence Constraint** tool, refer to Figure 16-92.

4. Apply Coincidence Constraint between the Sketch point 1 and the Sketch point 3.

5. Apply Coincidence Constraint between the Sketch point 2 and the Sketch point 4.

6. Finally, apply Coincidence Constraint between Planar face 2 and Planar face 4, refer to figure 16-92.

   Update the assembly. The final assembly of the Roller Curve mechanism looks similar to the one shown in Figure 16-93. In this figure, the transparency of the Outer Ring is changed for better visualization.

Invoking the DMU Kinematics Workbench

1. Choose **Start > Digital Mockup > DMU Kinematics** from the menu bar.

2. Choose **File > Save As** from the menu bar; the **Save As** dialog box is displayed.

3. Save the file in the *Bearing* folder of the *c16* folder with the name *mechanism_1*.

Fixing a Part in the DMU Kinematics Workbench

Now, you need to start a new mechanism.

1. Choose **Insert > New mechanism** from the menu bar; the **Mechanisms** sub-node is created under the **Applications** node in the Specification tree under the **Application > Mechanism** node.

   Now, you need to fix the base part.

2. Choose the **Fixed Part** tool from the **DMU Kinematics** toolbar; the **New Fixed Part** dialog box is displayed and you are prompted to select the fixed part.

3. Select Inner Ring from the Specification tree or from the graphics window; the selected part is fixed in the mechanism.

Applying the Roll Curve Joint to the Mechanism

In this mechanism, the Roller revolves about its own axis as well as about the axis of the Inner Ring. Therefore, you need to apply the Roll Curve joint between the curves only when they are tangentially connected to each other.

1. Choose the **Roll Curve Joint** tool from **DMU Kinematics > Kinematics Joints** sub-toolbar; the **Joint Creation: Roll Curve** dialog box is displayed and you are prompted to select the first curve.

2. Select the sketch curve of the Inner Ring, as shown in Figure 16-93; you are prompted to select the second curve.

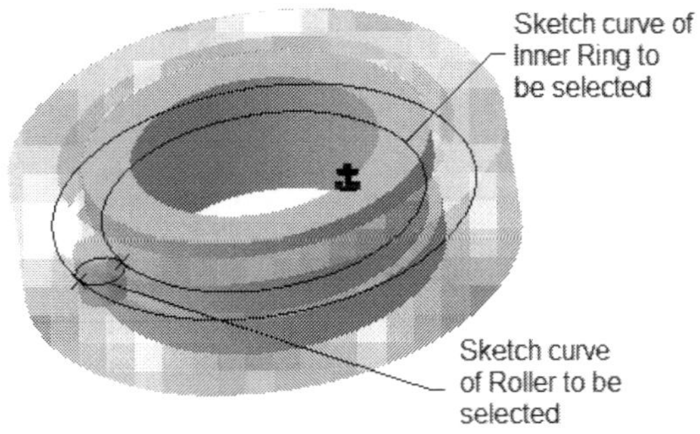

Figure 16-93 *Sketch curves selected for the* ***Roll Curve Joint***

3. Select the sketch curve of the Roller, as shown in Figure 16-93.

4. Choose **OK** to create the joint and exit the dialog box.

5. Similarly, apply the Roll Curve joint between the sketch curve of the Roller and the sketch curve of the Outer Ring.

Applying the Revolute Joint to the Mechanism

In this mechanism, the Outer Ring revolves about a common axis of the Outer Ring and the Inner Ring at a common plane. Therefore, you need to create the Revolute joint for the mechanism.

1. Choose the **Revolute Joint** tool from **DMU Kinematics > Kinematics Joints** sub-toolbar; the **Joint Creation: Revolute** dialog box is displayed and you are prompted to select the first line. 

2. Select the central axis of the Inner Ring, as shown in Figure 16-94; you are prompted to select the second line.

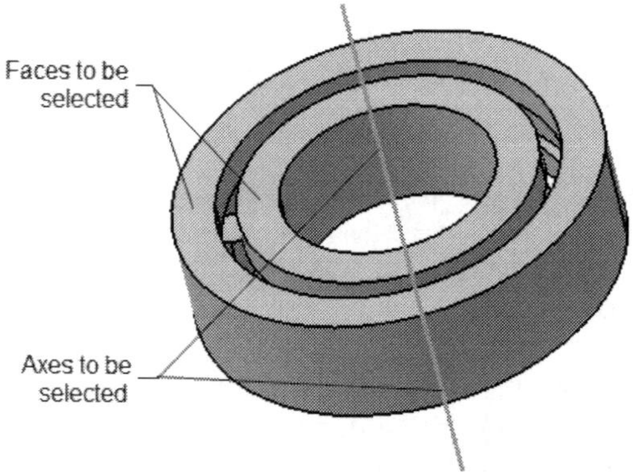

Figure 16-94 *Axes and Faces to be selected for applying the Revolute joint*

3. Now, select the central axis of the Outer Ring, refer to Figure 16-94; you are prompted to select the first plane.

4. Select the front face of the Inner Ring, refer to Figure 16-94; you are prompted to select the second plane.

5. Next, select the front face of the Outer Ring, refer to Figure 16-94.

6. Select the **Angle driven** check box to create a command, and then choose **OK**; the **Information** message box is displayed informing that the mechanism can be simulated.

7. Next, choose **OK** from the **Information** message box to close it.

Simulating the Mechanism

Update the mechanism and simulate it as discussed earlier.

Saving and Closing the File

1. Choose the **Save** button from the **Standard** toolbar to save the file.

2. Close the file by choosing **File > Close** from the menu bar.

Tutorial 5

In this tutorial, you will create the assembly for the Slide Curve joint mechanism, as shown in Figure 16-95. The orthographic views and dimensions of the parts are shown in Figures 16-96 through 16-98. Place the sketch points on the model as shown in orthographic views. These points will be used as references while creating the mechanism. (**Expected time: 1.5 hr**)

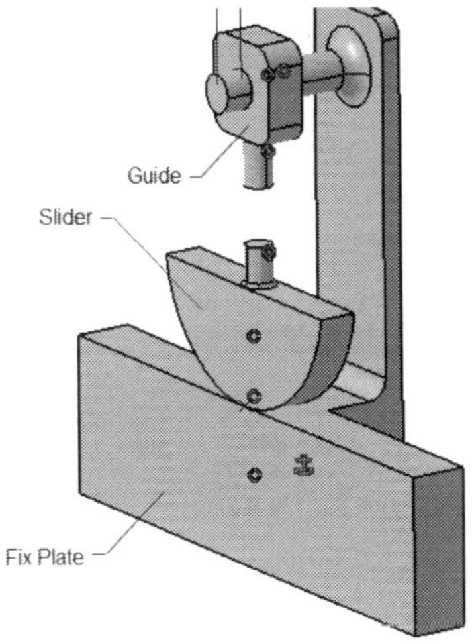

Figure 16-95 The Slide Curve joint mechanism

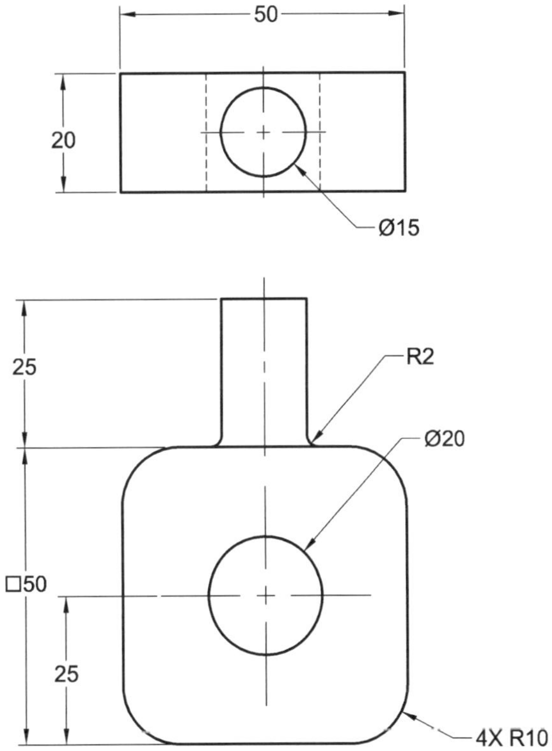

Figure 16-96 Orthographic views and dimensions of the Guide

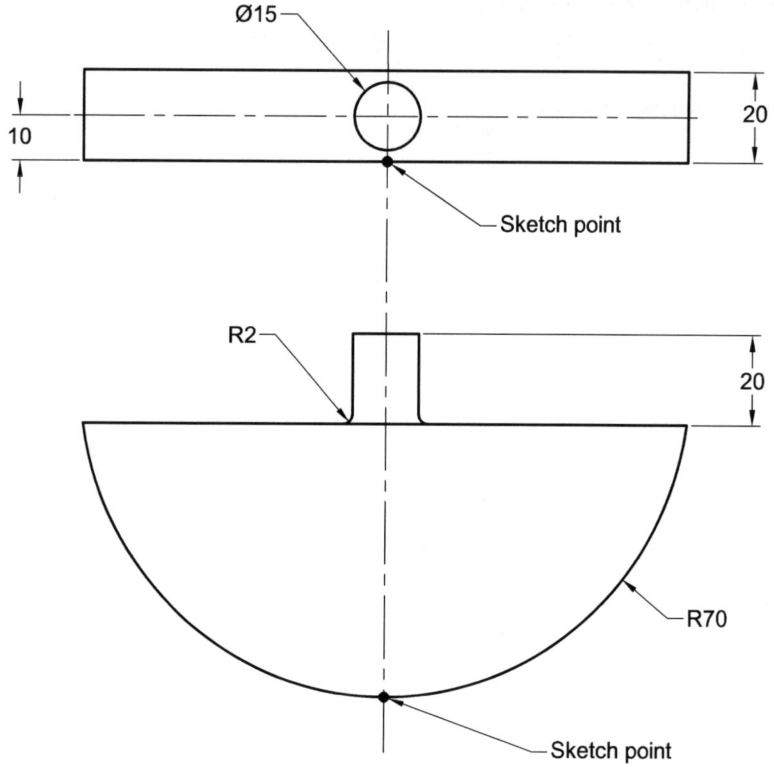

Figure 16-97 Orthographic views and dimensions of the Slider

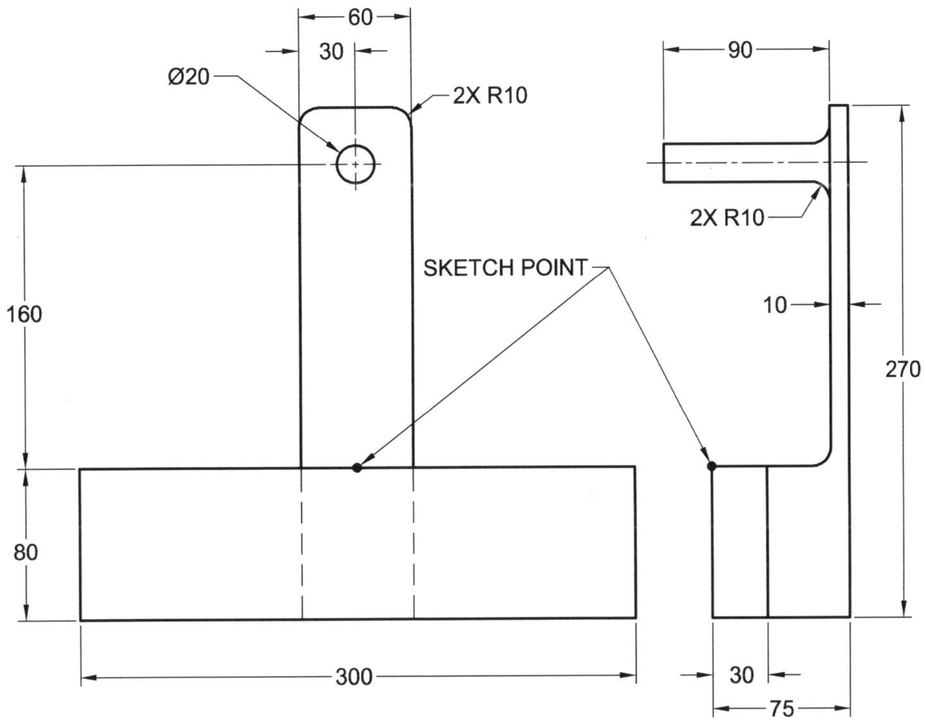

***Figure 16-98** Orthographic views and dimensions of the Fix Plate*

The following steps are required to complete this tutorial:

a. Create all parts of the mechanism as separate part files in the **Part Design** workbench.
b. Start a new file in the **Assembly Design** workbench.
c. Insert all parts in the **Assembly Design** workbench and assemble them.
d. Invoke the **DMU Kinematics** workbench.
e. Fix a part for the mechanism.
f. Apply the Slide Curve joint to the mechanism.
g. Apply the Prismatic Curve joint to the mechanism.
h. Apply the Revolute joint to the mechanism.
i. Simulate the mechanism.
j. Save and close the file.

Before creating components for this tutorial, you will create the *Slide Curve* folder at *CATIA\c16*. You need to save the parts of the Slide Curve mechanism in this folder.

Creating the Components of the Mechanism

In this section, you will create an assembly of Roll Curve joint mechanism using the bottom-up approach. In this approach, you need to create all parts as individual part files and then insert them in the assembly file.

While creating parts, create sketch lines and sketch points over them, refer to Figures 16-97 and 16-98. These curves and points will be used to assemble parts as well as create mechanism.

1. Create all parts of the assembly and save them as separate part files at *CATIA\c16\Slide Curve* folder.

2. Close all part files.

Starting a New File in the Assembly Design Workbench

All parts that you have created so far are required to be placed properly to create the mechanism. You need to start a new file in the **Assembly Design** workbench to place the parts properly at appropriate positions.

1. Choose the **New** tool from the **Standard** toolbar; the **New** dialog box is displayed.

2. In this dialog box, select the **Product** option from the **List of Types** list box.

3. Next, choose the **OK** button; a new file is started in the **Assembly Design** workbench, and **Product1** is displayed on top of the Specification tree. In case, the new file starts in the **DMU Kinematics** workbench, choose **Start > Mechanical Design > Assembly Design** from the menu bar to invoke the **Assembly Design** workbench.

Inserting Parts and Assembling Them

After starting the new Product file, you need to insert the base component into the assembly. In this case, the Fix Plate is the base component.

1. Insert all parts of the Slide Curve mechanism in the **Assembly Design** workbench and place them in 3D space using the Compass, as shown in Figure 16-99.

2. Make the Plane 1 and Plane 2 coplanar by using the **Coincidence Constraint** tool.

3. Make the hole of the Guide and the shaft of the Fix Plate coaxial by using the **Coincidence Constraint** tool.

4. Now, apply Coincidence Constraint between the Sketch point 1 and the Sketch point 2.

5. Similarly, apply Coincidence Constraint between the shaft of the Guide and the shaft of the Slider.

6. Finally, apply Coincidence Constraint between the Planar face 1 and the Planar face 2, and update the assembly.

   The final assembly of the Slide Curve mechanism is shown in Figure 16-100. In this figure, constraints are hidden for better visualization.

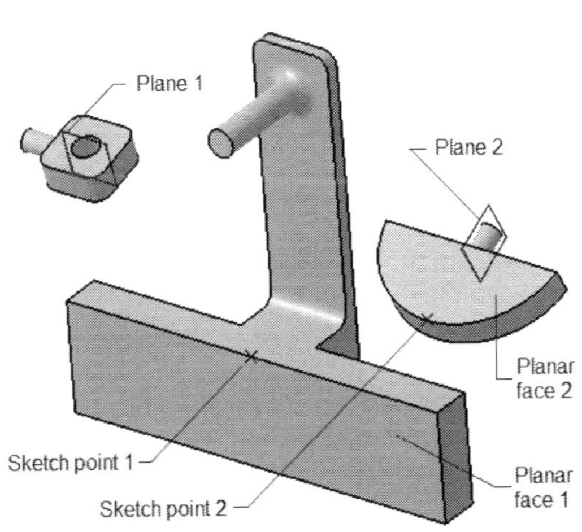

Figure 16-99 *Parts inserted in the* ***Assembly Design*** *workbench*

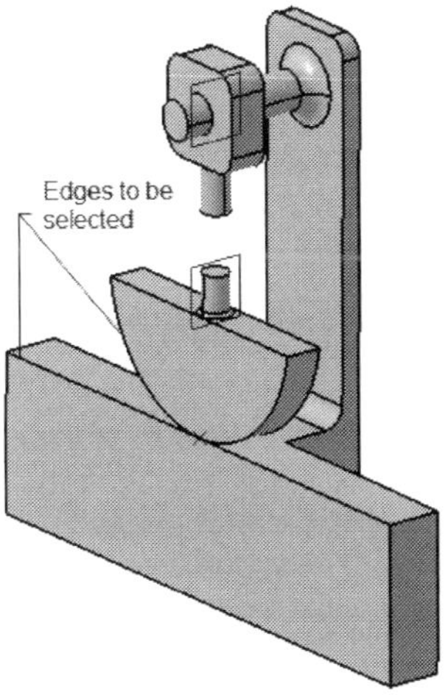

Figure 16-100 *Edges to be selected for creating the Slide Curve joint*

Invoking the DMU Kinematics Workbench

1. Choose **Start > Digital Mockup > DMU Kinematics** from the menu bar.

2. Choose **File > Save As** from the menu bar; the **Save As** dialog box is displayed.

3. Save the file with the name *mechanism_1* in the *Slide Curve* folder of the *c16* folder.

Fixing a Part in the DMU Kinematics Workbench

1. Choose the **Fixed Part** tool from the **DMU Kinematics** toolbar; the **New Fixed Part** dialog box is displayed.

2. Choose the **New Mechanism** button from the **New Fixed Part** dialog box; the **Mechanism Creation** dialog box is displayed. By default, **Mechanism.1** is displayed in the **Mechanism name** edit box.

3. Accept the default name and choose **OK** from the **Mechanism Creation** dialog box.

4. Select the Fix Plate in the Specification tree or in the graphics window; the selected part becomes the fixed part in the mechanism.

Applying the Slide Curve Joint to the Mechanism

As you already know that you can apply the Slide Curve joint between two curves only when they are tangentially connected with each other. Therefore, in this section, you can apply this joint between the edges of the Fix plate and the Slider

1. Choose the **Slide Curve Joint** tool from **DMU Kinematics > Kinematics Joints** sub-toolbar; the **Joint Creation: Slide Curve** dialog box is displayed and you are prompted to select the first curve.

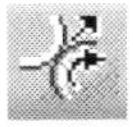

2. Select the edge of the Fix Plate, refer to Figure 16-100; you are prompted to select the second curve.

3. Select the outer edge of the Slider, refer to Figure 16-100.

4. Choose **OK** to create the joint and exit the dialog box.

Applying the Prismatic Joint to the Mechanism

Next, you need to apply the Prismatic joint between the Guide undefined DOFs.

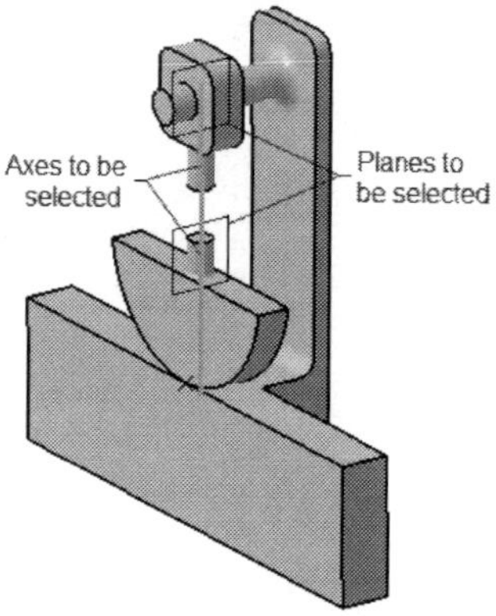

Figure 16-101 Axes and planes to be selected for creating the Prismatic joint

1. Choose the **Prismatic Joint** tool from **DMU Kinematics > Kinematics Joints** sub-toolbar; the **Joint Creation: Prismatic** dialog box is displayed and you are prompted to select the first line.

2. Select the central axis of the shaft of the Guide, as shown in Figure 16-101; you are prompted to select the second line.

3. Select the central axis of the shaft of the Slider, refer to Figure 16-101; you are prompted to select the first plane.

4. Select the datum plane of the Guide, refer to Figure 16-101; you are prompted to select the second plane.

5. Select the plane of the Slider, refer to Figure 16-101.

6. Choose **OK** to create the joint and exit the dialog box.

Applying the Revolute Joint to the Mechanism

After creating the Prismatic joint, **DOF=4** is displayed beside the **Mechanism.1** node under the **Applications > Mechanism** node of the Specification tree indicating that there are four undefined DOFs in the mechanism. Out of these, two are rotational DOFs and the other two are translational DOFs. To simulate the mechanism, you need to define the undefined DOFs and create a command. You can define these DOFs by applying the Revolute joint between the Guide and the Fix Plate.

1. Choose the **Revolute Joint** tool from **DMU Kinematics > Kinematics Joints** sub-toolbar; the **Joint Creation: Revolute** dialog box is displayed and you are prompted to select the first line.

2. Select the axis of the hole in the Guide, as shown in Figure 16-102; you are prompted to select the second line. Note that, in this figure, the hidden edges of the parts are shown for better visualization.

3. Select the axis of the shaft of the Fix Plate, refer to Figure 16-102; you are prompted to select the first plane.

4. Select the planar face of the Guide, refer to Figure 16-102; you are prompted to select the second plane.

5. Select the planar face of the shaft, refer to Figure 16-102.

6. Select the **Offset** radio button and specify the offset distance as **15**.

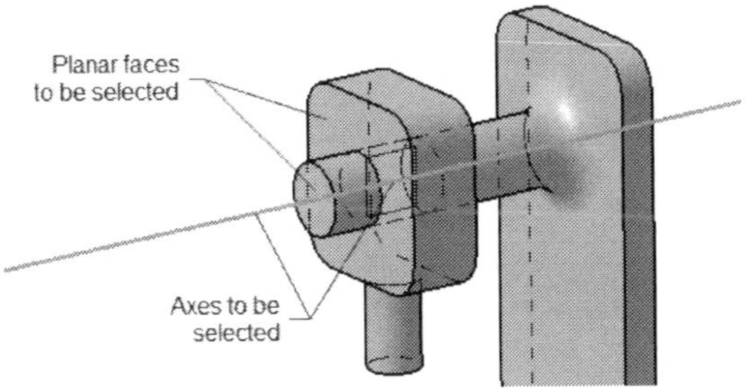

***Figure 16-102** Axes and planar faces to be selected for creating the Revolute joint*

7. Select the **Angle driven** check box from the **Joint Creation: Revolute** dialog box for creating a command and then choose **OK**; the joint is created.

   Since all DOFs have been defined, the **Information** message is displayed informing that now you can simulate the mechanism.

8. Choose **OK** from the **Information** message box to close the message box.

Simulating the Mechanism

Update the mechanism and simulate it as discussed earlier.

Saving and Closing the File

1. Choose the **Save** button from the **Standard** toolbar to save the file.

2. Close the file by choosing **File > Close** from the menu bar.

Self-Evaluation Test

Answer the following questions and then compare them to those given at the end of this chapter:

1. Initially each part of a mechanism has __________ degrees of freedom (DOFs).

2. You need one __________ joint and one __________ joint to create a Rack joint.

3. You need two __________ joints to create a Gear joint.

4. The __________ tool is used to create two universal joints among three parts.

5. You can generate various joints for a single mechanism by using the tools from the **DMU Kinematics** toolbar. (T/F)

6. The **Revolute Joint** tool is used to create a joint between two parts such that one part can rotate with respect to the fixed part about a common axis. (T/F)

7. You can simulate a mechanism without keeping a part fixed. (T/F)

8. The Prismatic joint allows rotation of a part with respect to the fixed part at a common plane. (T/F)

9. For defining a Screw joint between two parts, first you need to specify one axis from each part and then the pitch value. (T/F)

10. Choose Insert > New mechanism from the menu bar to start a new mechanism. (T/F)

Review questions

Answer the following questions:

1. Which of the following tools is used to simulate a mechanism?

   (a) **Simulation with Command** (b) **Fixed Part**
   (c) **Revolute Joint** (d) **Gear Joint**

2. Which of the following check boxes should be selected to create a command in the **Joint Creation: Roll Curve** dialog box?

   (a) **Length Driven** (b) **Angle driven**
   (c) **Offset** (d) None of these

3. Which of the following radio buttons of the **Joint Creation: Gear** dialog box is used to drive parts in the same direction?

   (a) **Offset** (b) **Opposite**
   (c) **Same** (d) **Null Offset**

4. Which of the following dialog boxes will be displayed if you choose the **Define** button from the **Joint Creation: Gear** dialog box?

   (a) **Gear Radius Definition** (b) **Rack Ratio Definition**
   (c) **Gear Ratio Definition** (d) **New Fixed Part**

5. The __________ tool is used to create a universal joint between two parts.

6. The __________ joint is used to maintain a contact between a surface and a point in the simulation.

7. The __________ tool is used to create a joint between two parts such that one point from each part always remains in touch with the other.

8. The __________ tool is used to create joints by using assembly constraints.

9. The __________ joint is used to fix two parts rigidly.

10. The __________ tools are used to bring two planar faces together.

EXERCISES

Exercise 1

In this exercise, you will create a mechanism for the V-Block assembly that was created in Exercise 1 of Chapter 13. **(Expected time: 30 min)**

Hint

1. Fix the V-Block for the mechanism.
2. Apply the Rigid joint between the V-Block and the U-Clamp.
3. Apply the Screw joint between the U-Clamp and the Fastener.

Exercise 2

In this exercise, you will create a mechanism such that the pointer of the follower can move anywhere over the Surface when the mechanism is simulated, refer to Figure 16-103. The orthographic views and dimensions of the parts are shown in Figures 16-104 through Figure 16-107. **(Expected time: 2 hr)**

Hint

1. Fix the Surface for the mechanism.
2. Apply the Prismatic joint between the Guide and the Surface.
3. Apply the Point Surface joint between the Pointer and the Surface.
4. Apply the Cylindrical joint between the Guide and the Pointer.
5. Apply the Prismatic joint between the Sketch line and the Arm Guide.

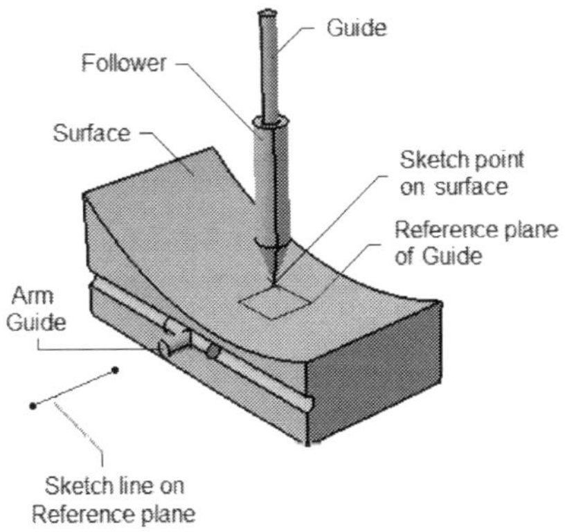

Figure 16-103 *Mechanism created between the Pointer and the Surface*

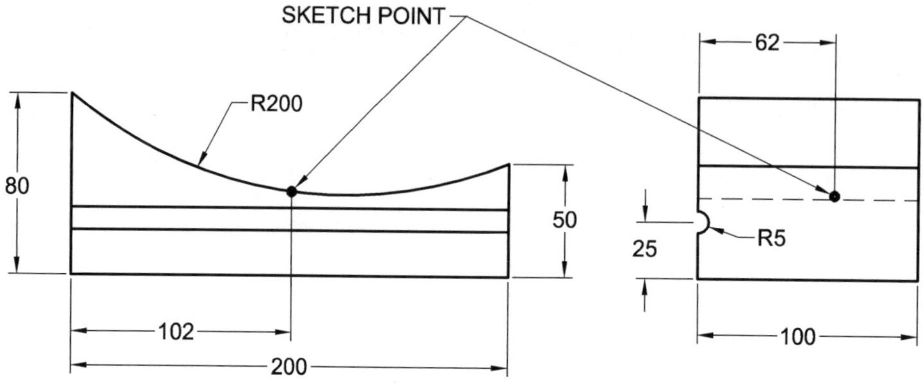

Figure 16-104 Orthographic views and dimensions of the Surface

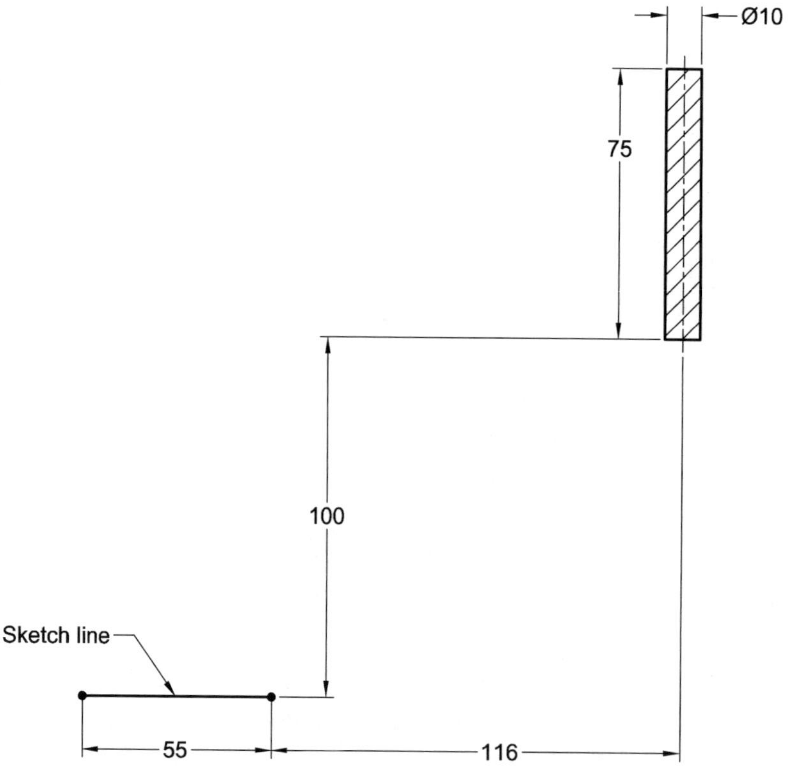

Figure 16-105 Cross-sectional view and dimensions of the Guide and its Sketch line

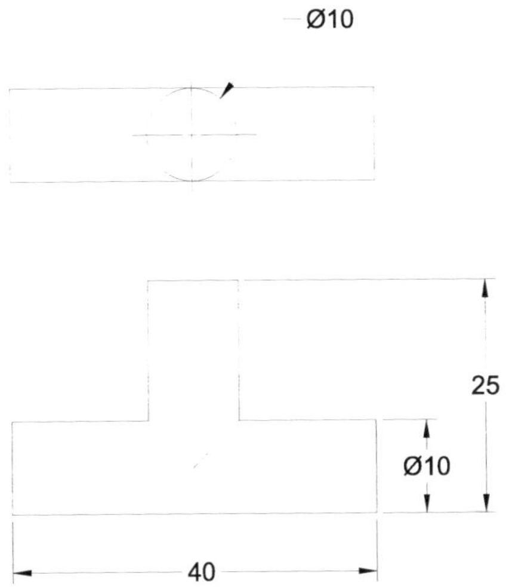

Figure 16-106 *Orthographic views and dimensions of the Arm Guide*

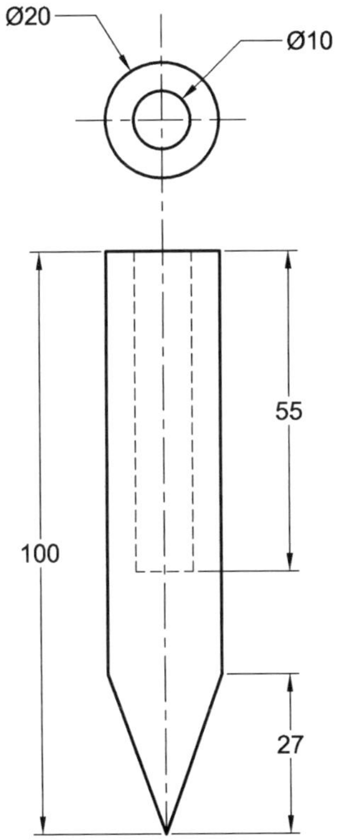

Figure 16-107 *Orthographic views and dimensions of the Follower*

Answers to Self-Evaluation Test

1. six, **2.** Prismatic, Revolute, **3.** Revolute, **4. CV Joint**, **5.** T, **6.** T, **7.** F, **8.** F, **9.** T, **10.** T

Index

Symbols

A

B

C

D

E

F

G

H

I

J

S

T

U

V

W

Z

Other Publications by CADCIM Technologies

The following is the list of some of the publications by CADCIM Technologies. Please visit *www.cadcim.com* for complete listing.

AutoCAD Textbooks

- AutoCAD 2023: A Problem-Solving Approach, Basic and Intermediate, 29th Edition
- AutoCAD 2022: A Problem-Solving Approach, Basic and Intermediate, 28th Edition
- AutoCAD 2021: A Problem-Solving Approach, Basic and Intermediate, 27th Edition
- AutoCAD 2020: A Problem-Solving Approach, Basic and Intermediate, 26th Edition
- AutoCAD 2019: A Problem-Solving Approach, Basic and Intermediate, 25th Edition
- AutoCAD 2018: A Problem-Solving Approach, Basic and Intermediate, 24rd Edition
- Advanced AutoCAD 2018: A Problem-Solving Approach, 3D and Advanced, 24th Edition

Autodesk Inventor Textbooks

- Autodesk Inventor Professional 2023 for Designers, 23rd Edition
- Autodesk Inventor Professional 2022 for Designers, 22nd Edition
- Autodesk Inventor Professional 2021 for Designers, 21st Edition
- Autodesk Inventor Professional 2020 for Designers, 20th Edition
- Autodesk Inventor Professional 2019 for Designers, 19th Edition
- Autodesk Inventor Professional 2018 for Designers, 18th Edition

AutoCAD MEP Textbooks

- AutoCAD MEP 2022 for Designers, 6th Edition
- AutoCAD MEP 2020 for Designers, 5th Edition
- AutoCAD MEP 2018 for Designers, 4th Edition
- AutoCAD MEP 2016 for Designers, 3rd Edition

Solid Edge Textbooks

- Solid Edge 2023 for Designers, 20th Edition
- Solid Edge 2022 for Designers, 19th Edition
- Solid Edge 2021 for Designers, 18th Edition
- Solid Edge 2020 for Designers, 17th Edition
- Solid Edge 2019 for Designers, 16th Edition
- Solid Edge ST10 for Designers, 15th Edition
- Solid Edge ST9 for Designers, 14th Edition

NX Textbooks

- Siemens NX 2022 for Designers, 15th Edition
- Siemens NX 2021 for Designers, 14th Edition
- Siemens NX 2020 for Designers, 13th Edition
- Siemens NX 2019 for Designers, 12th Edition
- Siemens NX 12.0 for Designers, 11th Edition
- NX 11.0 for Designers, 10th Edition

NX Nastran Textbook
- NX Nastran 9.0 for Designers

NX Mold Textbook
- Mold Design Using NX 11.0: A Tutorial Approach

SOLIDWORKS Textbooks
- SOLIDWORKS 2023 for Designers, 21st Edition
- SOLIDWORKS 2022 for Designers, 20th Edition
- SOLIDWORKS 2021 for Designers, 19th Edition
- SOLIDWORKS 2020 for Designers, 18th Edition
- SOLIDWORKS 2019 for Designers, 17th Edition
- SOLIDWORKS 2018 for Designers, 16th Edition
- SOLIDWORKS 2017 for Designers, 15th Edition
- SOLIDWORKS 2022: A Tutorial Approach
- Learning SOLIDWORKS 2022: A Project Based Approach

CATIA Textbooks
- CATIA V5-6R2021 for Designers, 19th Edition
- CATIA V5-6R2020 for Designers, 18th Edition
- CATIA V5-6R2019 for Designers, 17th Edition
- CATIA V5-6R2018 for Designers, 16th Edition
- CATIA V5-6R2017 for Designers, 15th Edition
- CATIA V5-6R2016 for Designers, 14th Edition
- CATIA V5-6R2015 for Designers, 13th Edition

Creo Parametric and Pro/ENGINEER Textbooks
- Creo Parametric 9.0 for Designers, 8th Edition
- Creo Parametric 8.0 for Designers, 7th Edition
- Creo Parametric 7.0 for Designers, 7th Edition
- Creo Parametric 6.0 for Designers, 6th Edition
- Creo Parametric 5.0 for Designers, 5th Edition
- Creo Parametric 4.0 for Designers, 4th Edition

ANSYS Textbooks
- ANSYS Workbench 2022 R1: A Tutorial Approach, 5th Edition
- ANSYS Workbench 2021 R1: A Tutorial Approach
- ANSYS Workbench 2019 R2: A Tutorial Approach

Creo Direct Textbook
- Creo Direct 2.0 and Beyond for Designers

Autodesk Alias Textbooks
- Learning Autodesk Alias Design 2016, 5th Edition
- Learning Autodesk Alias Design 2015, 4th Edition
- Learning Autodesk Alias Design 2012
- Learning Autodesk Alias Design 2010
- AliasStudio 2009 for Designers

AutoCAD LT Textbooks
- AutoCAD LT 2022 for Designers, 14th Edition
- AutoCAD LT 2020 for Designers, 13th Edition
- AutoCAD LT 2017 for Designers, 12th Edition
- AutoCAD LT 2016 for Designers, 11th Edition
- AutoCAD LT 2015 for Designers, 10th Edition
- AutoCAD LT 2014 for Designers
- AutoCAD LT 2013 for Designers

EdgeCAM Textbooks
- EdgeCAM 11.0 for Manufacturers
- EdgeCAM 10.0 for Manufacturers

AutoCAD Electrical Textbooks
- AutoCAD Electrical 2023 for Electrical Control Designers, 14th Edition
- AutoCAD Electrical 2022 for Electrical Control Designers, 13th Edition
- AutoCAD Electrical 2021 for Electrical Control Designers, 12th Edition
- AutoCAD Electrical 2020 for Electrical Control Designers, 11th Edition
- AutoCAD Electrical 2019 for Electrical Control Designers, 10th Edition
- AutoCAD Electrical 2018 for Electrical Control Designers, 9th Edition
- AutoCAD Electrical 2017 for Electrical Control Designers, 8th Edition
- AutoCAD Electrical 2016 for Electrical Control Designers, 7th Edition

Autodesk Revit Architecture Textbooks
- Exploring Autodesk Revit 2023 for Architecture, 19th Edition
- Exploring Autodesk Revit 2022 for Architecture, 18th Edition
- Exploring Autodesk Revit 2021 for Architecture, 17th Edition
- Exploring Autodesk Revit 2020 for Architecture, 16th Edition
- Exploring Autodesk Revit 2019 for Architecture, 15th Edition
- Exploring Autodesk Revit 2018 for Architecture, 14th Edition
- Exploring Autodesk Revit 2017 for Architecture, 13th Edition

Autodesk Revit Structure Textbooks
- Exploring Autodesk Revit 2023 for Structure, 13th Edition
- Exploring Autodesk Revit 2022 for Structure, 12th Edition
- Exploring Autodesk Revit 2021 for Structure, 11th Edition
- Exploring Autodesk Revit 2020 for Structure, 10th Edition
- Exploring Autodesk Revit 2019 for Structure, 9th Edition
- Exploring Autodesk Revit 2018 for Structure, 8th Edition
- Exploring Autodesk Revit 2017 for Structure, 7th Edition
- Exploring Autodesk Revit Structure 2016, 6th Edition

AutoCAD Civil 3D Textbooks
- Exploring AutoCAD Civil 3D 2023, 13th Edition
- Exploring AutoCAD Civil 3D 2022, 12th Edition
- Exploring AutoCAD Civil 3D 2021, 11th Edition
- Exploring AutoCAD Civil 3D 2020, 10th Edition

- Exploring AutoCAD Civil 3D 2019, 9th Edition
- Exploring AutoCAD Civil 3D 2018, 8th Edition
- Exploring AutoCAD Civil 3D 2017, 7th Edition

AutoCAD Map 3D Textbooks
- Exploring AutoCAD Map 3D 2022, 9th Edition
- Exploring AutoCAD Map 3D 2018, 8th Edition
- Exploring AutoCAD Map 3D 2017, 7th Edition
- Exploring AutoCAD Map 3D 2016, 6th Edition
- Exploring AutoCAD Map 3D 2015, 5th Edition
- Exploring AutoCAD Map 3D 2014
- Exploring AutoCAD Map 3D 2013

Autodesk 3ds Max Design Textbooks
- Autodesk 3ds Max 2017 for Beginners: A Tutorial Approach, 17th Edition
- Autodesk 3ds Max 2017: A Comprehensive Guide, 17th Edition
- Autodesk 3ds Max 2016: A Comprehensive Guide, 16th Edition
- Autodesk 3ds Max 2016 for Beginners: A Tutorial Approach, 16th Edition
- Autodesk 3ds Max Design 2015: A Tutorial Approach, 15th Edition
- Autodesk 3ds Max Design 2014: A Tutorial Approach
- Autodesk 3ds Max Design 2013: A Tutorial Approach
- Autodesk 3ds Max Design 2012: A Tutorial Approach
- Autodesk 3ds Max Design 2011: A Tutorial Approach

Autodesk 3ds Max Textbooks
- Autodesk 3ds Max 2023: A Comprehensive Guide, 23rd Edition
- Autodesk 3ds Max 2022: A Comprehensive Guide, 22nd Edition
- Autodesk 3ds Max 2021: A Comprehensive Guide, 21st Edition
- Autodesk 3ds Max 2020: A Comprehensive Guide, 20th Edition
- Autodesk 3ds Max 2019: A Comprehensive Guide, 19th Edition

Autodesk Maya Textbooks
- Autodesk Maya 2023: A Comprehensive Guide, 14th Edition
- Autodesk Maya 2022: A Comprehensive Guide, 13th Edition
- Autodesk Maya 2020: A Comprehensive Guide, 12th Edition
- Autodesk Maya 2019: A Comprehensive Guide, 11th Edition
- Autodesk Maya 2018: A Comprehensive Guide, 10th Edition
- Autodesk Maya 2017: A Comprehensive Guide, 9th Edition

Pixologic ZBrush Textbooks
- Pixologic ZBrush 2022: A Comprehensive Guide, 8th Edition
- Pixologic ZBrush 2020: A Comprehensive Guide, 6th Edition
- Pixologic ZBrush 2018: A Comprehensive Guide, 5th Edition
- Pixologic ZBrush 4R8: A Comprehensive Guide, 4th Edition

Fusion Textbooks

- Blackmagic Design Fusion 7 Studio: A Tutorial Approach
- The eyeon Fusion 6.3: A Tutorial Approach

Flash Textbooks

- Adobe Flash Professional CC2015: A Tutorial Approach, 3rd Edition
- Adobe Flash Professional CC: A Tutorial Approach
- Adobe Flash Professional CS6: A Tutorial Approach

Computer Programming Textbooks

- Introduction to C^{++} programming, 2nd Edition
- Introduction to C^{++} programming
- Learning Oracle 12c- APL/SQL Approach, 2nd Edition
- Learning Oracle 12C- APL/SQL Approach
- Learning ASP.NET AJAX
- Introduction to Java Programming, 2nd Edition
- Introduction to Java Programming
- Learning Visual Basic.NET 2008

AutoCAD Textbooks Authored by Prof. Sham Tickoo and Published by Autodesk Press

- AutoCAD: A Problem-Solving Approach: 2013 and Beyond
- AutoCAD 2012: A Problem-Solving Approach
- AutoCAD 2011: A Problem-Solving Approach
- AutoCAD 2010: A Problem-Solving Approach
- Customizing AutoCAD 2010
- AutoCAD 2009: A Problem-Solving Approach

Textbooks Authored by CADCIM Technologies and Published by Other Publishers

3D Studio MAX and VIZ Textbooks

- Learning 3DS Max: A Tutorial Approach, Release 4
  Goodheart-Wilcox Publishers (USA)
- Learning 3D Studio VIZ: A Tutorial Approach
  Goodheart-Wilcox Publishers (USA)

CADCIM Technologies Textbooks Translated in Other Languages

SolidWorks Textbooks

- SolidWorks 2008 for Designers (Serbian Edition)
  Mikro Knjiga Publishing Company, Serbia
- SolidWorks 2006 for Designers (Russian Edition)
  Piter Publishing Press, Russia

- SolidWorks 2006 for Designers (Serbian Edition)
  Mikro Knjiga Publishing Company, Serbia

NX Textbooks

- NX 6 for Designers (Korean Edition)
  Onsolutions, South Korea
- NX 5 for Designers (Korean Edition)
  Onsolutions, South Korea

Pro/ENGINEER Textbooks

- Pro/ENGINEER Wildfire 4.0 for Designers (Korean Edition)
  HongReung Science Publishing Company, South Korea
- Pro/ENGINEER Wildfire 3.0 for Designers (Korean Edition)
  HongReung Science Publishing Company, South Korea

Autodesk 3ds Max Textbook

- 3ds Max 2008: A Comprehensive Guide (Serbian Edition)
  Mikro Knjiga Publishing Company, Serbia

AutoCAD Textbooks

- AutoCAD 2006 (Russian Edition)
  Piter Publishing Press, Russia
- AutoCAD 2005 (Russian Edition)
  Piter Publishing Press, Russia
- AutoCAD 2000 Fondamenti (Italian Edition)

Coming Soon from CADCIM Technologies

- Flow Simulation Using SOLIDWORKS 2021
- Finite Element Analysis Using ANSYS Workbench 2021

Made in United States
Orlando, FL
04 November 2026

90218265R00446